6th International Workshop DICE2012 Spacetime - Matter – Quantum Mechanics: From the Planck Scale to Emergent Phenomena

Journal of Physics: Conference Series Volume 442

Castiglioncello, Tuscany, Italy
17-21 September 2012

ISBN: 978-1-62748-674-3
ISSN: 1742-6588

Printed from e-media with permission by:

Curran Associates, Inc.
57 Morehouse Lane
Red Hook, NY 12571

Some format issues inherent in the e-media version may also appear in this print version.

Printed by Curran Associates, Inc. (2013)

Institute of Physics
Dirac House, Temple Back
Bristol BS1 6BE UK

Phone: 44 1 17 929 7481
Fax: 44 1 17 920 0979

techtracking@iop.org

Additional copies of this publication are available from:

Curran Associates, Inc.
57 Morehouse Lane
Red Hook, NY 12571 USA
Phone: 845-758-0400
Fax: 845-758-2634
Email: curran@proceedings.com
Web: www.proceedings.com

TABLE OF CONTENTS

PAPERS

INVITED LECTURES

012001 Gravity-related Wave Function Collapse: Mass Density Resolution 1
L Diósi

012002 On the Completeness of Quantum Mechanics and the Interpretation of the State Vector 8
G C Ghirardi, R Romano

012003 Open Quantum Walks: A Short Introduction 18
I Sinayskiy, F Petruccione

012004 Entang-bling: Observing Quantum Correlations in Room-temperature Solids 28
I A Walmsley, K C Lee, M Sprague, B Sussman, J Nunn, N Langford, X-M Jin, T Champion, P Michelberger, K Reim, D UK, D Jaksch

012005 in Search of Superluminal Quantum Communications: Recent Experiments and Possible Improvements 34
B Cocciaro, S Faetti, L Fronzoni

012006 Hydrogen Atom As a Quantum-classical Hybrid System 50
F Zhan, B Wu

012007 Quantum-classical Hybrid Dynamics – A Summary 58
H-T Elze

012008 Entropy of Conditional Tomographic Probability Distributions for Classical and Quantum Systems 65
M A Man'ko, V I Man'ko

012009 Cosmology with Quantum Matter and a Classical Gravitational Field: the Approach of Configuration-space Ensembles 75
M Reginatto

012010 Pathways Toward Understanding Macroscopic Quantum Phenomena 90
B L Hu, Y Subasi

012011 Measurement Problem: from De Broglie to Theory of Classical Random Fields Interacting with Threshold Detectors 101
A Khrennikov

012012 "Systemic Nonlocality" from Changing Constraints on Sub-quantum Kinematics 114
G Grössing, S Fussy, J M Pascasio, H Schwabl

012013 The Phenomenon of Spontaneous Replica Symmetry Breaking in Complex Statistical Mechanics Systems 129
F Guerra

012014 Does a Quantum Particle Know Its Own Energy? 136
R D Sorkin

012015 Noncommutative Spectral Geometry: A Short Review 145
M Sakellariadou

012016 Noncommutative Spectral Geometry and the Deformed Hopf Algebra Structure of Quantum Field Theory 156
M Sakellariadou, A Stabile, G Vitiello

012017 Topology, Cosmic Strings and Quantum Dynamics – A Case Study with Graphene 167
B Chakraborty, K S Gupta, S Sen

012018 Amplitudes for Spacetime Regions and the Quantum Zeno Effect: Pitfalls of Standard Path Integral Constructions 181
J J Halliwell, J M Yearsley

012019 Fractional Field Equations for Highly Improbable Events 197
H Kleinert

012020 CPT-violating Leptogenesis Induced by Gravitational Backgrounds 206
N E Mavromatos, S Sarkar

012021 Status of the OPERA Experiment 223
E Medinaceli

012022 Beyond Classical Limits by Exploiting Quantum Light: A Short Review of INRIM Results 230
G Brida, I P Degiovanni, M Genovese, E D Lopaeva, A Meda, I Ruo Berchera

012023 From Single Constituents to Correlation: Phase Transitions in Few Atom Systems 237
E Prati

012024 Entanglement Generation in Relativistic Cavity Motion 242
D E Bruschi, J Louko, D Faccio
012025 The Semiclassical Approximation to Quantum Gravity and Its Observational Consequences 250
C Kiefer

CONTRIBUTED PAPERS

012026 Detecting the Relative Localisation of Quantum Particles 257
P A Knott, J Sindt, J A Dunningham
012027 Hybrid Dynamics As a Constrained Quantum System 264
N Buric, I Mendaš, D B Popovic, M Radonjic, S Prvanovic
012028 Coherent Structures in Liquid Water Close to Hydrophilic Surfaces 272
E D Giudice, A Tedeschi, G Vitiello, V Voeikov
012029 Non-hermitian Approach to Decaying Ultracold Bosonic Systems 277
S Wimberger, C A Parra-Murillo, G Kordas
012030 Density-functional Theory, Finite-temperature Classical Maps, and Their Implications for Foundational Studies of Quantum Systems 282
M W C Dharma-Wardana
012031 The Supramolecular Structure of Liquid Water and Quantum Coherent Processes in Biology 290
A D Ninno, A C Castellano, E D Giudice
012032 Telegraph Signals As a Solution of the Time Dependent Schrödinger Equation Let Standard Copenhagen Quantum Mechanics Emerge 299
G Doyen, D Drakova
012033 Emergence from Irreversibility 311
P F de Córdoba, J M Isidro, M H Perea
012034 Topological Ordered Spaces As a Foundation for a Quantum Spacetime Theory 319
E Minguzzi
012035 Toward a Real Synthesis of Quantum and Relativity Theories: Experimental Evidence for Absolute Simultaneity 323
S A Emelyanov
012036 Gravitational Contraction and Curvature Singularities in f(R) Gravity 334
L Reverberi
012037 Quantization of the Schwarzschild Geometry 344
E Melas
012038 Quantum Gravity Contributions to the Cosmological Constant 352
S Zonetti
012039 Short Review on Noncommutative Spacetimes 359
A Pachol
012040 Superconductivity, Superfluidity and Holography 367
A Salvio
012041 De Donder-Weyl Hamiltonian Formulation and Precanonical Quantization of Vielbein Gravity 377
I V Kanatchikov
012042 An Example of Numerical Simulation in Causal Set Dynamics 388
A L Krugly, I A Tserkovnikov
012043 Realist Model Approach to Quantum Mechanics 398
P Hájícek
012044 The Coevent Formulation of Quantum Theory 406
P Wallden
012045 Is the Consistent Histories Approach to Quantum Mechanics Consistent? 425
E Okon
012046 Beyond Bell's Theorem: Admissible Hidden-variable Models for the Spin-singlet 433
A Di Lorenzo
012047 Modeling "Quantum" Field Theory Through "Classical" Cellular Automaton 443
R Sverdlov
012048 Intrinsic Periodicity: The Forgotten Lesson of Quantum Mechanics 453
D Dolce
012049 Low Temperature Motion of Hydrogen on Metal Surfaces Signals Breakdown of Quantum Mechanics in 3+1 Dimensions 463
D Drakova, G Doyen
012050 Quantum Mechanics and Permutation Invariants of Finite Groups 476
V V Kornyak

012051 Classical and Quantum-mechanical Axioms with the Higher Time Derivative Formalism 487
T Kamalov
012052 Noncommutative CP^N and CH^N and Their Physics 493
A Sako, T Suzuki, H Umetsu
012053 R^2 Theory of Gravity 503
F Payandeh, M Fathi
012054 Granular Universe: Statistical Origin of Special and Doubly Special Relativity 513
P Jizba, F Scardigli
012055 An Introduction to the Quantum Backflow Effect 522
J M Yearsley, J J Halliwell
012056 Graphene: QFT in Curved Spacetimes Close to Experiments 532
A Iorio
012057 The PVLAS Experiment: Detecting Vacuum Magnetic Birefringence 544
G Zavattini, F D Valle, U Gastaldi, G Messineo, E Milotti, R Pengo, L Piemontese, G Ruoso
012058 Generalized Creation and Annihilation Operators Via Complex Nonlinear Riccati Equations 558
D Schuch, O Castaños, O Rosas-Ortiz
012059 Solving the Nonlocality Riddle by Conformal Quantum Geometrodynamics 568
E Santamato, F De Martini
012060 Is the De Broglie-bohm Interpretation of Quantum Mechanics Really Plausible? 576
K Jung
012061 Entanglement and Complexity of Measurement 584
C V Usenko
012062 Quantum Driving of a Two Level System: Quantum Speed Limit and Superadiabatic Protocols – An Experimental Investigation 594
N Malossi, M G Bason, M Viteau, E Arimondo, D Ciampini, R Mannella, O Morsch
012063 Resonant Driving of a Nonlinear Hamiltonian System 603
C Palmisano, G Gervino, M Balma, D Devona, S Wimberger
012064 Microscopic Theory of Warm Ionized Gases: Equation of State and Kinetic Schottky Anomaly 608
A Capolupo, S M Giampaolo, F Illuminati
012065 Power-law Quantum Distributions in Protoneutron Stars 616
G Gervino, A Lavagno, D Pigato
012066 Primordial Planck Mass Black Holes (PPMBHs) As Candidates for Dark Matter? 621
G M Obermair
012067 Acceleration Factor for Propagation of a Stationary Wave in Its Wave Medium: Movement of Energy in the 3-D Space 625
S Nagao
012068 Exciting Peculiarities of the Planck Scale Physics 635
P Di Sia
012069 Geometric Invariants As Detector of Hawking and Unruh Effects and Quantum Field Theory in Curved Space 640
A Capolupo, G Vitiello
012070 Entanglement in Quantum Field Theory: Particle Mixing and Oscillations 647
M Blasone, F Dell'Anno, S D Siena, F Illuminati
012071 The SSC Formalism and the Collapse Hypothesis for Inflationary Origin of the Seeds of Cosmic Structure 656
D Sudarsky
012072 Environment Induced Time Arrow and the Closed Time Path Method 671
J Polonyi
Author Index

PREFACE

Presented in this volume are the Invited Lectures and the Contributed Papers of the Sixth International Workshop on Decoherence, Information, Complexity and Entropy – DICE 2012, held at Castello Pasquini, Castiglioncello (Tuscany), 17–21 September 2012.

These proceedings may document to the interested public and to the wider scientific community the stimulating exchange of ideas at the meeting. The number of participants has been steadily growing over the years, reflecting an increasing attraction, if not need, of such conference. Our very intention has always been to bring together leading researchers, advanced students, and renowned scholars from various areas, in order to stimulate new ideas and their exchange across the borders of specialization.

In this way, the series of meetings successfully continued from the beginning with DICE 2002[1], followed by DICE 2004[2], DICE 2006[3], DICE 2008[4], and DICE 2010[5], Most recently, DICE 2012 brought together more than 120 participants representing more than 30 countries worldwide.

It has been a great honour and inspiration to have Professor Yakir Aharonov (Tel Aviv) with us, who presented the opening Keynote Lecture 'The two-vector quantum formalism'.

With the overarching theme 'Spacetime – Matter – Quantum Mechanics – from the Planck scale to emergent phenomena', the conference took place in the very pleasant and inspiring atmosphere of Castello Pasquini – in beautiful surroundings, overlooking a piece of Tuscany's coast. The 5-day program covered these major topics:

- Quantum Mechanics, Foundations and Quantum-Classical Border
- Quantum-Classical Hybrids and Many-Body Systems
- Spectral Geometry, Path Integrals and Experiments
- Quantum –/– Gravity –/– Spacetime
- Quantum Mechanics on all Scales?

A Roundtable Discussion under the theme 'Nuovi orizzonti nella ricerca scientifica. Ci troviamo di fronte ad una rivoluzione scientifica?' formed an integral part of the program. With participation of E Del Giudice (INFN & Università di Milano), F Guerra (Università 'La Sapienza', Roma) and G Vitiello (Università di Salerno), this event traditionally dedicated to the public drew a large audience involved in lively discussions until late.

The workshop was organized by L Diósi (Budapest), H-T Elze (Pisa, chair), L Fronzoni (Pisa), J J Halliwell (London), E Prati (Milano) and G Vitiello (Salerno), with most essential help from our conference secretaries L Fratino, N Lampo, I Pozzana, and A Sonnellini, all students from Pisa, and from our former secretaries M Pesce-Rollins and L Baldini.

Several institutions and sponsors supported the workshop and their representatives and, in particular, the citizens of Rosignano/Castiglioncello are deeply thanked for the generous help and kind hospitality:

- Comune di Rosignano – A Franchi (Sindaco di Rosignano), S Scarpellini (Segreteria sindaco), L Benini (Assessore ai lavori pubblici), M Pia (Assessore all' urbanistica)
- REA Rosignano Energia Ambiente s.p.a. – F Ghelardini (Presidente della REA), E Salvadori and C Peccianti (Segreteria)
- Associazione Armunia – A Nanni (Direttore), G Mannari (Programmazione), C Perna, F Bellini, M Nannerini, P Bruni and L Meucci (Tecnici). Special thanks go to G Mannari and her collaborators for advice and great help in all the practical matters that had to be dealt with, in order to run the meeting at Castello Pasquini smoothly

Funds made available by Università di Pisa, Domus Galilaeana (Pisa), Centro Interdisciplinare per lo Studio dei Sistemi Complessi – CISSC (Pisa), Dipartimento di Ingegneria Industriale (Università di Salerno), Istituto Italiano per gli Studi Filosofici – IISF (Napoli), Solvay Italia SA (Rosignano), Institute of Physics Publishing – IOP (Bristol), Springer Verlag (Heidelberg), and Hungarian Scientific Research Fund OTKA are gratefully acknowledged.

Last, but not least, special thanks are due to Laura Pesce (Vitrium Galleria, San Vincenzo) for the exposition of her artwork 'arte e scienza' at Castello Pasquini during the conference.

The papers submitted in the wake of the conference have been edited by L Diósi, H-T Elze, L Fronzoni, J J Halliwell, E Prati, G Vitiello and J Yearsley. The proceedings follow essentially the order of presentation during the conference, separating, however, invited lectures and contributed papers[6]. In the name of all participants, we would like to thank S Toms with her collaborators at IOP Publishing (Bristol) for friendly advice and most valuable immediate help during the editing process and, especially, for their continuing efforts to make the *Journal of Physics: Conference Series* available to all.

Budapest, Pisa, London, Milano, Salerno, Cambridge, April 2013
Lajos Diósi, Hans-Thomas Elze, Leone Fronzoni, Jonathan Halliwell, Enrico Prati, Giuseppe Vitiello and James Yearsley

[1] *Decoherence and Entropy in Complex Systems* ed H-T Elze *Lecture Notes in Physics* **633** (Berlin: Springer, 2004)
[2] *Proceedings of the Second International Workshop on Decoherence, Information, Complexity and Entropy – DICE 2004* ed H-T Elze *Braz. Journ. Phys.* **35** A & 2B (2005) pp 205–529; free access at: www.sbfisica.org.br/bjp

[3] *Proceedings of the Third International Workshop on Decoherence, Information, Complexity and Entropy – DICE 2006* eds H-T Elze, L Diósi and G Vitiello *Journal of Physics: Conference Series* **67** (2007); free access at: www.iop.org/EJ/toc/1742-6596/67/1

[4] *Proceedings of the Fourth International Workshop on Decoherence, Information, Complexity and Entropy – DICE 2008*> eds H-T Elze, L Diósi, L Fronzoni, J J Halliwell and G Vitiello *Journal of Physics: Conference Series* **174** (2009); free access at: http://www.iop.org/EJ/toc/1742-6596/174/1

[5] *Proceedings of the Fifth International Workshop on Decoherence, Information, Complexity and Entropy – DICE 2010* eds H-T Elze, L Diósi, L Fronzoni, J J Halliwell, E Prati, G Vitiello and J Yearsley *Journal of Physics: Conference Series* **306** (2011); free access at: http://iopscience.iop.org/1742-6596/306/1

[6] We regret that invited lectures by Y Aharonov, J Barbour, G Casati and X-G Wen could not be reproduced here, partly for copyright reasons

Peer review statement

All papers published in this volume of *Journal of Physics: Conference Series* have been peer reviewed through processes administered by the proceedings Editors. Reviews were conducted by expert referees to the professional and scientific standards expected of a proceedings journal published by IOP Publishing.

Gravity-related wave function collapse: mass density resolution

Lajos Diósi

Wigner Research Center for Physics, H-1525 Budapest 114, P.O.Box 49, Hungary

E-mail: diosi.lajos@wigner.mta.hu

Abstract. Selected issues of the concept of spontaneous collapse are discussed, with the emphasis on the gravity-related model. We point out that without spontaneous collapses the Schrödinger cat states would macroscopically violate the standard conservation laws even in the presence of environmental noise. We prove that the collapse time of condensed matter c.o.m. superpositions is not sensitive to the natural uncertainty of the nuclear locations whereas we formulate the conjecture that superfluid He may show an anomalous low rate of spontaneous collapse compared to common condensed matter.

1. Introduction

If we apply quantum mechanics for macroscopic variables, e.g., to the c.o.m. motion of a given massive object, we face paradoxical as well as concrete physical issues. The hypothesis of spontaneous collapse (SC) has been thought a remedy. According to it, all macroscopic superpositions, i.e.: the Scrödinger cat (Cat) states, are quickly eliminated as if a quantum measurement were performed on them. This time, however, the measurement happens spontaneously and universally, without the presence of any measurement device. SC models contain a further innovation compared to standard wave function collapse. The latter is a sudden one-shot event whereas SCs take time, they have a certain stochastic dynamics. Formally, they are time-resolved standard one-shot collapses. The SC models take out the concept of standard one-shot collapse.

We discuss some old problems within the gravity-related model, proposed independently by Penrose and the author [1, 2, 3, 4, 5, 6, 7], of the yet hypothetical SC. Some problems are acute because of related experimental considerations and experiments [8, 9, 11, 12, 10, 13, 14]. Some are common for almost all SC models [15], some are specific for ours. For the sake of a minimum comparison, we introduce the elements of the gravity-related SC model together with the so-called continuous spontaneous localization (CSL) [16] model, using the notion of 'catness' (Sec. 2). Sec. 3 presents two novel results for the gravity-related model. First, we show that the equilibrium collapse rate of the Cat c.o.m. is a classical expression, independent of $\hbar$. Second, we prove that the calculated SC rate does not depend on the spread of nuclear locations even if the mass density spatial resolution assumes the nuclear size. Many question the physical relevance of SCs of the Cat states because their unavoidable entanglement with the environment prevents Cat states from being created. We point out in Sec. 4, that the immediate environmental decoherence may not help Cat states avoid their macroscopic violation of conservation laws.

2. Spontaneous collapse models

Most SC models are similar to each other. Let us introduce the notion of 'catness' for the superposition of two different macroscopic configurations, say, two mass densities $f(\mathbf{r})$ and $f'(\mathbf{r})$. Consider a typical Cat state:

$$|\mathrm{Cat}\rangle = |f\rangle + |f'\rangle\,, \tag{1}$$

where f and f' are 'macroscopically' different. Macroscopicity, i.e.: catness, is measured by a distance $\ell(f,f')$. Let, for concreteness, ℓ^2 be of dimension of energy. SC models assume that Nature makes $|\mathrm{Cat}\rangle$ decay (collapse) stochastically at mean life-time

$$\tau = \frac{\hbar}{\ell^2(f,f')}. \tag{2}$$

Each collapse model has its own choice of the measure ℓ of catness. The choice of the distance $\ell(f,f')$ must be careful: there should be practically no decay (extreme large life-time τ) for atomic cats, and immediate decay (very short life-time τ) for 'macroscopic' Cats. When specifying catness, the spatial mass density resolution is a crucial issue. The models set up a finite cutoff σ of spatial resolution.

The gravity-related (G-related, or DP) model defines

$$\ell_G^2(f,f') = G\int\int[f(\mathbf{r})-f'(\mathbf{r})][f(\mathbf{s})-f'(\mathbf{s})]\frac{d\mathbf{r}d\mathbf{s}}{|\mathbf{r}-\mathbf{s}|}, \tag{3}$$

where f, f' are coarse-grained with the spatial resolution σ. The model is further discussed in Sec. 3. The closest alternative to the DP model is the mass-proportional CSL model [16]. The CSL choice of catness reads

$$\ell_{CSL}^2(f,f') = \frac{\hbar\lambda\sigma^3}{m_0^2}\int[f(\mathbf{r})-f'(\mathbf{r})]^2 d\mathbf{r}, \tag{4}$$

where f, f' are Gaussian coarse-grained with the spatial resolution $\sigma \sim 10^{-5}$cm. The CSL model uses the nucleon mass m_0 as a reference mass and the further fenomenological parameter $\lambda{\sim}10^{-17}$/s.

The predictions of the DP and CSL models may differ by orders of magnitudes, but they fulfill the main request. Practically infinite life-times are preserved for superpositions of microscopic systems whereas the Cat states (those apparently missing from Nature) will decay immediately. According to both DP and CSL models, the competition of SCs with the standard unitary dynamics may reach a regime of equilibrium. The DP model is unique in that the equilibrium collapse rate is classical, $\hbar$ cancels from it (Sec. 3.1).

3. G-related spontaneous collapse

We can re-express the (squared) catness (3) as the following combination of three Newton interaction energies:

$$\ell_G^2(f,f') = 2U(f,f') - U(f,f) - U(f',f'), \tag{5}$$

where

$$U(f,f') = -G\int\int\frac{f(\mathbf{r})f'(\mathbf{s})}{|\mathbf{r}-\mathbf{s}|}d\mathbf{r}d\mathbf{s}. \tag{6}$$

Suppose our system is a single rigid homogeneous ball of mass M and radius R and we consider its c.o.m. wave function. The popular 'mechanical' Cat state [1, 2, 3, 4, 5, 6, 7, 9] is the superposition of two macroscopically different c.o.m. locations $\mathbf{x}$ and $\mathbf{x}'$:

$$|\mathrm{Cat}\rangle = |\mathbf{x}\rangle + |\mathbf{x}'\rangle\,. \tag{7}$$

The corresponding two mass densities of the Cat state (1) read

$$\begin{aligned} f(\mathbf{r}) &= \rho\theta(|\mathbf{r}-\mathbf{x}|\leq R), \\ f'(\mathbf{r}) &= \rho\theta(|\mathbf{r}-\mathbf{x}'|\leq R), \end{aligned} \tag{8}$$

where $\rho = M/(4\pi R^3/3)$ is the constant mass density, θ is the step-function. The central quantity is the mean collapse time τ_G, cf. (2), of the c.o.m. wave function. It becomes the function of the c.o.m. displacement $\Delta\mathbf{x} = \mathbf{x} - \mathbf{x}'$; we calculate the mean collapse rate in the first non-vanishing order [1, 2, 3]:

$$\frac{1}{\tau_G} = \frac{1}{2}\frac{M\omega_G^2}{\hbar}(\Delta\mathbf{x})^2, \tag{9}$$

valid if $|\Delta\mathbf{x}| \ll R$. The R-dependence has been absorbed into the parameter

$$\omega_G = \sqrt{4\pi G\rho/3} \tag{10}$$

which we call the frequency of the Newton oscillator (cf. Fig. 1). This frequency is purely classical.

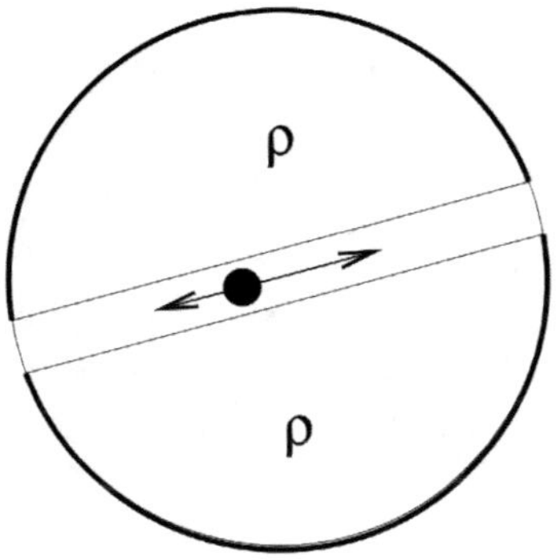

Figure 1. Schematic view of the Newton Oscillator: a homogeneous ball with an infinite narrow diagonal hole where the probe is oscillating under the linear directional force of the Newton field of the ball. The frequency of the oscillator $\omega_G = \sqrt{4\pi G\rho/3}$ does not depend on the size but on the density ρ of the ball.

3.1. Equilibrium collapse rate

The existence of equilibrium is a generic feature of the SC models. We are going to use the heuristic arguments of [2] which anticipated the exact derivations [3].

Let us start from the collapse rate (9) of a mechanical Cat, previously determined in the leading order of the coherent displacement $\Delta\mathbf{x} = \mathbf{x} - \mathbf{x}'$. We can heuristically extend this expression for a c.o.m. wave packet of coherent dispersion $\Delta\mathrm{x}$: the SCs at about the rate (9) tend to shrink the width $\Delta\mathrm{x}$. Now we take the unitary evolution of the wave packet into the account: it tends to spread $\Delta\mathrm{x}$ at the rate about $\hbar/M(\Delta\mathrm{x})^2$. The two contrary tendencies get balanced when the two rates coincide:

$$\frac{\hbar}{M(\Delta\mathrm{x})^2} \sim \frac{M\omega_G^2(\Delta\mathrm{x})^2}{\hbar}. \tag{11}$$

This is the condition of equilibrium between SC and unitary dynamics, resulting in a sort of natural quantum state of the massive object. The corresponding natural coherent size of the c.o.m. wave packet yields

$$\Delta\mathrm{x}^{\mathrm{eq}} \sim \sqrt{\hbar/M\omega_G}, \tag{12}$$

which is a very tiny size for massive objects due to the smallness of $\hbar$. Similar strong localization of the macro-objects follows from the CSL model, too. Nevertheless, we find a unique feature of the DP model if we calculate the equilibrium collapse rate. Let's just take the geometric mean of both sides of the balance condition above, it yields the equilibrium collapse rate directly:

$$\frac{1}{\tau_G^{\mathrm{eq}}} \sim \omega_G. \tag{13}$$

The equilibrium collapse rate coincides approximately with the frequency of the Newton oscillator, does not depend on the mass M and size R, moreover, it is fully classical, independent of $\hbar$.

3.2. Mass resolution issue, He superfluid

Catness (3) and collapse rates would diverge for point-like objects, that's why we need the finite spatial resolution σ of $f(\mathbf{r})$. It's crucial to decide that we resolve the microscopic structure or we don't. Adopting the resolution $\sigma{\sim}10^{-5}$cm of the CSL model, the mass density in condensed matter becomes the smooth macroscopic density ρ equal to a few times $1\mathrm{g/cm}^3$. We can estimate the parameter ω_G which governs the collapse rate (9) and yields the equilibrium collapse rate (13):

$$\frac{1}{\tau_G^{\mathrm{eq}}} = \omega_G^{\mathrm{eq}} \sim \frac{1}{\mathrm{h}}. \tag{14}$$

This extreme low rate questions the significance of DP collapses in common objects. The microscopic structure of the mass density may be a loophole.

We can take a much finer resolution with the 'nuclear' cutoff $\sigma{\sim}10^{-12}$cm. Then $f(\mathbf{r})$ shows the granular structure of localized nuclear contributions. For simplicity, we consider that the mass M is carried by the nuclei which are homogeneous balls of radius σ and density $\rho^{\mathrm{nucl}} = (R/\sigma)^3\rho$. For small displacements $|\Delta\mathbf{x}| \ll \sigma$, each nucleus will contribute separately to the (squared) catness, like a homogeneous ball of radius σ and density ρ^{nucl}. At the end of the day we obtain the collapse rate similar to (9):

$$\frac{1}{\tau_G} = \frac{1}{2} \times \frac{M(\omega_G^{\mathrm{nucl}})^2}{\hbar}(\Delta\mathbf{x})^2, \tag{15}$$

just with ω_G^{nucl} instead of ω_G [18]. The expression is valid if $|\Delta\mathbf{x}| \ll \sigma$. Here

$$\omega_G^{\mathrm{nucl}} = \sqrt{4\pi G\rho^{\mathrm{nucl}}/3} \tag{16}$$

is the frequency of the Newton oscillator in nuclear matter. Its order of magnitude is $\omega_G^{\mathrm{nucl}}{\sim}10^3/s$.

With the 'nuclear' resolution, the Cat life-time τ_G becomes cca 10^{12} times shorter! Without this amplification, first shown in [18], the experimental significance of the SCs may remain illusory for another while, as recognized in [11, 12, 13, 14]. It seems therefore a sort of necessity that we define the DP model with the 'nuclear' cutoff to take granular structure of the mass density $f(\mathbf{r})$ into the account.

We must note that the exact interpretation of the nuclear mass distribution does not exist. We don't know the principles either. In the best case we believe that high energy gravitons would copy how high energy photons explore the electronic charge density. But the available non-relativistic interpretation of nuclear mass density is fenomenological, we have to rely upon the simple fenomenology.

Finally, we ought to discuss whether or not the thermal and quantum blurredness of the nuclear locations matters when we calculate the DP collapse rate of the Cat state. In solids, the blurredness of nuclear location is about one tenth of the interatomic distance. If this broadening of mass distribution did matter then the above said factor 10^{12} of amplification would be lost and replaced by some 10^3. Fortunately, this is not the case: the DP collapse of the c.o.m. superposition is always independent of the quantum spread of microscopic constituents. The general proof is trivial in the density matrix formalism. Let us introduce the coordinates $\mathbf{q}_1, \mathbf{q}_2, \dots, \mathbf{q}_N$ of the N nuclei w.r.t. the c.o.m. $\mathbf{x}$. We choose the following set of independent coordinates: $\mathbf{x}, \mathbf{q}_1, \mathbf{q}_2, \dots, \mathbf{q}_{N-1}$. The corresponding density matrix reads

$$\rho(\mathbf{x}, \mathbf{q}_1, \mathbf{q}_2, \dots, \mathbf{q}_{N-1}|\mathbf{x}', \mathbf{q}'_1, \mathbf{q}'_2, \dots, \mathbf{q}'_{N-1}). \tag{17}$$

The full DP collapse rate of this state of the Cat is described by the following term [1, 2, 3]:

$$\frac{\ell_G^2(f,f')}{\hbar}\rho(\mathbf{x}, \mathbf{q}_1, \mathbf{q}_2, \dots, \mathbf{q}_{N-1}|\mathbf{x}', \mathbf{q}'_1, \mathbf{q}'_2, \dots, \mathbf{q}'_{N-1}), \tag{18}$$

where f is the (σ-coarse-grained) mass density of the configuration $(\mathbf{x}, \mathbf{q}_1, \mathbf{q}_2, \dots, \mathbf{q}_{N-1})$ and f' is the (σ-coarse-grained) mass density of the configuration $(\mathbf{x}', \mathbf{q}'_1, \mathbf{q}'_2, \dots, \mathbf{q}'_{N-1})$. We are not interested in the collapse rate of the integral degrees of freedom, therefore we trace over them, yielding

$$\int \frac{\ell_G^2(f,f')}{\hbar}\rho(\mathbf{x}, \mathbf{q}_1, \mathbf{q}_2, \dots, \mathbf{q}_{N-1}|\mathbf{x}', \mathbf{q}_1, \mathbf{q}_2, \dots, \mathbf{q}_{N-1}) d\mathbf{q}_1 d\mathbf{q}_2 \dots d\mathbf{q}_{N-1}, \tag{19}$$

where f is the (σ-coarse-grained) mass density of the configuration $(\mathbf{x}, \mathbf{q}_1, \mathbf{q}_2, \dots, \mathbf{q}_{N-1})$ and f' is the (σ-coarse-grained) mass density of the configuration $(\mathbf{x}', \mathbf{q}_1, \mathbf{q}_2, \dots, \mathbf{q}_{N-1})$. The expression (19) is the exact c.o.m. collapse rate. This shows that the coherence of the spread of the nuclear locations is completely irrelevant. And there is an even more important result. We must not blur the nuclear locations first. Rather we must calculate the c.o.m. squared catness and the corresponding collapse rate with the sharp locations $\mathbf{q}_1, \mathbf{q}_2, \dots$ of the nuclei. Second, the effect of the spread of nuclear locations will be taken into the account by blurring the above squared catness (i.e., the collapse rate) over the spread of the nuclear locations. In our special case of very small displacements $D\mathbf{x} \ll \sigma$, the squared catness $\ell_G^2(f,f')$ is, as we said before, well approximated by the sum of separate contributions of the nuclei. The contribution of a single nucleus, say the 'first' one, depends on the displacement between the two superposed locations $\mathbf{x} + \mathbf{q}_1$ and $\mathbf{x}' + \mathbf{q}_1$, which is just $|\Delta\mathbf{x}| = |\mathbf{x} - \mathbf{x}'|$. Neither the precise nuclear locations nor their spreads will influence the c.o.m. SC of the Cat.

Liquid He is an exception. In superfluid He the quantum dispersion of the nuclei is of the order of the interatomic distance, the nuclei have no identity and the above proof does not apply. Distinction between internal and c.o.m. collapses, respectively, is more involved. In liquid He the c.o.m. collapse rate may be much smaller than the value (15) valid in common condensed matter. Superfluid He may have an anomalous low rate of SCs in it. Of particular interest is the temperature dependence of the collapse rate. Too high temperatures would push back the coherent dispersion of the nuclei below the interatomic scale, resulting in increased c.o.m. collapse rate. It is a further question yet to be answered whether cold liquid He is sufficient for the anomalous low collapse rates or superfluidity is essential for it.

4. Environment does not mask spontaneous collapse

We know that typical environmental noise and entanglement used to decohere any Cat state much earlier than SCs would kill it. Environmental noise in Nature used to be much stronger than SCs. The experimental test of SCs appears very hard. Since SCs are masked by environmental noise, one might think that SCs are irrelevant, the corresponding models are superfluous. This conclusion is not correct.

Let us concentrate on the mechanical Cat. In the naive Cat state (7) we considered the c.o.m. but ignored all internal and environmental degrees of freedom. Now we refine the naive Cat state and include the environmental degrees of freedom:

$$|\mathrm{Cat}\rangle = |\mathbf{x}\rangle \otimes |E\rangle + |\mathbf{x}'\rangle \otimes |E'\rangle , \tag{20}$$

where $|E\rangle$ and $|E'\rangle$ stand for two different states of the environment. To distill the central issue, we assume that the only massive object is the Cat itself (a rigid homogeneous ball with c.o.m. $\mathbf{x}, \mathbf{x}'$) and the environment is massless compared to the Cat. The above state is a macroscopic superposition of different mass configurations (8), so it is a genuine Cat. The two states $|\mathbf{x}\rangle$ and $|\mathbf{x}'\rangle$ are entangled with the environmental states $|E\rangle$ and $|E'\rangle$, respectively. Since in practice the environmental states $|E\rangle$ and $|E'\rangle$ are orthogonal, the coherence of the macroscopic superposition is not observable locally, our Cat is not observable locally. To see the Cat, we need to observe the remote parts of the entangled environment that we are not able to do. How can the Cats keep annoying us? Why should we deploy SCs or any mechanism against them at all?

Suppose the Cat state in the following form:

$$|\mathrm{Cat}\rangle = |\mathbf{x}\rangle \otimes |E\rangle \otimes |1\rangle + |\mathbf{x}'\rangle \otimes |E'\rangle \otimes |2\rangle , \tag{21}$$

where $|1\rangle, |2\rangle$ are orthogonal single photon states. Say, the photon (like billions of other environmental photons) bounced back from the Cat some time ago and went to the states $|1\rangle, |2\rangle$ entangled with the Cat c.o.m. states $|\mathbf{x}\rangle, |\mathbf{x}'\rangle$, respectively. Suppose that we measure the photon so that, according to the standard quantum measurement theory, the Cat collapses into $|\mathbf{x}\rangle$ or into $|\mathbf{x}'\rangle$ at random. The c.o.m. of the Cat (and of the whole system) is $(\mathbf{x}+\mathbf{x}')/2$ before the measurement. After the measurement it becomes either $\mathbf{x}$ or $\mathbf{x}'$. We got a macroscopic shift of the c.o.m. of a closed macroscopic system whose only interaction with the outer world was that we measured a single photon that belonged to the system. This is not merely paradoxical. Actually all (non-superselected) conservation laws are macroscopically violated by the mere existence of certain Cats whether they are naive isolated ones or real ones entangled with the environment. Violation of energy-momentum conservation in standard collapse has been discussed earlier [17]. Note that our present concern is the possible macroscopicity of the violation, not the violation itself. A microscopic defect of conservation can always be attributed to (and balanced by) the measuring device. But macroscopic defects are annoying.

The SC models, like DP and CSL, offer a simple treatment of the Cat problem. They collapse the Cats (20), kill them before they can come up, much before we could measure them and provoque the macroscopic violation of conservation laws. The collapse models themselves impose violation of the conservation laws at a tinier scale, but they prevent macroscopic violations at least.

5. Closing remarks

We have added some new facts, thoughts and conjectures to the gravity-related SC model. On one end, sound analysis has clarified how the thermal and quantum motion of the nuclei constituting a massive body should enter the calculation of the c.o.m. collapse rate. On the other extreme, we have conjectured that in superfluid He the c.o.m. SC rate may be much lower than in common condensed matter. Ref. [19] by the present author discusses similar topics and a related possible extension of the DP model.

Acknowledgments

Support by the Hungarian Scientific Research Fund under Grant No. 75129 and support by the EU COST Action MP100 are acknowledged.

References

[1] Diósi L 1986 *A quantum-stochastic gravitation model and the reduction of the wavefunction* (Thesis in Hungarian)
http://www.rmki.kfki.hu/~diosi/thesis1986.pdf
[2] Diósi L 1987 A universal master equation for the gravitational violation of the quantum mechanics *Phys. Lett.* A **120** 377
[3] Diósi L 1989 Models for universal reduction of macroscopic quantum fluctuations *Phys. Rev.* A **40** 1165
[4] Penrose R 1994 *Shadows of the Mind* (Oxford: Oxford University Press)
[5] Penrose R 1996 On gravity's role in quantum state reduction it Gen. Rel. Gra. **28** 581
[6] Penrose R 1998 Quantum computation, entanglement and state reduction *Phil. Trans. R. Soc. Lond.* A **356** 1927
[7] Penrose R 2004 *The Road to Reality* (London: Jonathan Cape Publishers)
[8] Marshall W, Simon C, Penrose R and Bouwmeester D 2003 Towards quantum superpositions of a mirror *Phys. Rev. Lett.* **91** 130401
[9] Adler SL 2007 Comments on proposed gravitational modifications of Schrödinger dynamics and their experimental implications *J. Phys.* A **40** 755
[10] Vanner MR, Pikovski I, Cole GD, Kim MS, Brukner Č, Hammerer K, Milburn GJ and Aspelmeyer M 2011 Pulsed quantum optomechanics *PNAS* **108** 16182
[11] Romero-Isart O 2011 Quantum superposition of massive objects and collapse models *Phys. Rev.* A **84** 052121
[12] Li T, Kheifets S and Raizen MG 2011 Millikelvin cooling of an optically trapped microsphere in vacuum *Nat. Phys.* **7** 527
[13] Pepper B, Ghobadi R, Jeffrey E, Simon C and Bouwmeester D 2012 Optomechanical superpositions via nested interferometry *Phys. Rev. Lett.* **109** 023601
[14] Yang H, Miao H, Lee D, Helou B and Chen Y 2012 Macroscopic quantum mechanics in a classical spacetime *Preprint* arXiv:1210.0457
[15] Bassi A and Ghirardi GC 2003 Dynamical reduction models *Phys. Rep.* **379** 257
[16] Ghirardi GC, Grassi R and Benatti F 1995 Describing the macroscopic world: Closing the circle within the dynamical reduction program *Found. Phys.* **25** 5
[17] Pearl P 2000 Wavefunction collapse and conservation laws *Found. Phys.* **30** 1145
[18] Diósi L 2007 Notes on certain Newton gravity mechanisms of wave function localisation and decoherence *J. Phys.* A **40** 2989
[19] Diósi L 2013 Gravity-related wave function collapse: Is superfluid He exceptional? *Preprint* arXiv:1302.5364

On the completeness of quantum mechanics and the interpretation of the state vector

GianCarlo Ghirardi[1] and Raffaele Romano[2]

[1] Department of Physics, University of Trieste, and the Abdus Salam ICTP, Trieste (Italy)
[2] Department of Mathematics, Iowa State University, Ames, IA (USA)

E-mail: ghirardi@ictp.it, rromano@iastate.edu

Abstract. Recently, it has been argued that quantum mechanics is complete, and that quantum states vectors are necessarily in one-to-one correspondence with the elements of reality, under the assumptions that quantum theory is correct and that measurement settings can be freely chosen. In this work, we argue that the adopted form of the free choice assumption is stronger than needed. In our perspective, there are hidden assumptions underlying these results, which limit their range of validity. We support our argument by a model for the bipartite two-level system, reproducing quantum mechanics, in which the free will assumption is respected, and different quantum states can be connected to the same state of reality.

1. Introduction

Quantum mechanics is one of the most successful theories ever developed. It has been elaborated to account for the phenomena of the microscopic world and it has been experimentally verified to an impressive degree of accuracy. It has required a radical change in the scientific conception of nature particularly in connection with its indeterminism and nonlocality. However, from a conceptual point of view, the theory is not immune of drawbacks. The crucial point stays in the fact that it appears more as a set of operational prescriptions to predict outcomes of prospective measurement procedures than a coherent description of reality. Its range of validity is not clear, it relies on two different kinds of evolution depending on the rather vague notion of measurement. As a consequence of these problematic aspects it is not clear what is the conceptual status of the quantum state vector ψ, the mathematical entity associated to the state of any physical system. After the famous incompleteness argument by Einstein, Podolski and Rosen [1], the question of whether ψ represents a state of reality or rather a state of knowledge, as suggested by its updating following a measurement procedure, remains unanswered.

In order to address some of these questions, several *ontological models* of quantum mechanics have been introduced. These are theories predictively equivalent to quantum mechanics, but providing a possibly richer description of microscopic reality in terms of the so-called *ontic state*, i.e., the most accurate specification of the physical state of the system [2]. In these theories, the state vector ψ might embody only partial information on the ontic state. In particular, it is associated to a distribution $\rho_\psi(\lambda)$ on the space of the ontic variables, with $\rho_\psi(\lambda) \geqslant 0$ and

$$\int \rho_\psi(\lambda) d\lambda = 1 \qquad \text{for all } \psi. \tag{1}$$

The ontic state λ contains all the information about the elements of reality of the underlying theory, and, accordingly, it provides the complete description of the state of the system [1]. In principle, λ can be not fully accessible. Ignorance of its precise value is deemed to be the origin of various of the aforementioned problematic aspects of quantum mechanics.

In the past literature, ontological models were usually referred to as *hidden variable theories*, meaning that additional variables, absent in the standard quantum formalism (and for this reason named "hidden"), were expected to supplement the information given by ψ, providing in this way a complete description of the physical situation. From the beginning, this program had supporters as well as opponents. J. von Neumann proved that deterministic hidden variable models were unviable [3], but later D. Bohm provided an explicit, fully deterministic, hidden variable model predictively equivalent to quantum mechanics [4, 5]. Later J.S. Bell appropriately stressed that von Neumann's impossibility theorem was circular, since it was based on logically unnecessarily strong assumptions [6]. He also proved that any theory predictively equivalent to quantum mechanics must violate some inequalities, named *Bell's inequalities* after him, which are satisfied by any local theory [7].

Recently, a new argument supporting the completeness of quantum mechanics has been presented in [8], based only on two assumptions: that quantum mechanics is correct (QM), and that a precise mathematical condition (FR) - which the authors believe to express the request that measurement settings can be freely chosen - be satisfied. According to [8], any completion of quantum mechanics which consistently reproduces its outcomes, necessarily comes at the expenses of the free will of the observers in choosing the measurement settings. Here, we argue that, similarly to the case of von Neumann argument against hidden variables theories, also this result is inconclusive due to the unappropriate formalization of the assumption (FW) of free will, which we will make precise in what follows. Specifically, we provide evidence that the physically meaningful request of FW is not correctly expressed in mathematical terms in [8], since the constraint FR, which stays at the basis of the authors' argument, is unnecessarily stronger than FW. For related comments see [9, 10, 11].

Our criticism has relevant consequences concerning the meaning of the quantum state vector ψ in ontological models. First of all, let us recall that the distributions $\rho_\psi(\lambda)$ might have overlapping supports for different ψ, a property which characterizes the so-called *ψ-epistemic models.* Alternatively, when the supports are always disjoint we are dealing with *ψ-ontic models* [2]. According to this classification the state vector ψ acquires a strongly different conceptual status. In fact, in a ψ-ontic theory different state vectors necessarily correspond to different ontic states, whereas in a ψ-epistemic theory they could correspond to the same ontic state. As a consequence, one can associate to ψ well defined elements of reality only in the first case.

As it should be obvious, the distinction between ψ-epistemic and ψ-ontic models implies completely different interpretations of the quantum state vector. For this reason, these classes of models have recently been the subject of several investigations. A first argument arguing that ψ-epistemic models necessarily contrast with quantum mechanics was developed in [12], but its conclusions are severely limited by the assumption that factorized quantum states correspond

[1] To clarify the problem of attributing elements of physical reality to a physical system it is useful to make reference, once more, to the EPR paper. In the case in which ψ is an eigenstate of an observable Ω pertaining to the eigenvalue ω_k, we know that the standard quantum theory predicts that a measurement of Ω gives with certainty the outcome ω_k, and, on the basis of this fact, EPR claim that the system possesses $\Omega = \omega_k$ as an element of physical reality. In such a case, with reference to the ontic state λ, we must impose that the value $\Omega(\lambda)$ of the observable Ω must be defined and must coincide with ω_k. Obviously, it may very well happen (and to which extent this actually occurs depends on the ontic theory one has in mind) that for the given ψ and λ, the knowledge of λ allows the specification of further elements of reality besides those based on the EPR criterion. Typically, ψ might be an eigenstate of the z spin component of a particle, but knowledge of λ might allow one to associate elements of physical reality also to the x and/or y spin components.

to factorized states of the underlying theory. In fact, an explicit ψ-epistemic model has been built by relaxing this assumption [13]. Another no-go theorem for ψ-epistemic models has been developed in [14], based on the completeness result presented in [8]. Our arguments against the assumption FR make also this argument inconclusive in determining the ontological status of the quantum state vector, since a weaker and more significant condition of free will is still compatible with ψ-epistemic models. We substantiate our argument by presenting an explicit ψ-epistemic model for a pair of two-level systems, in which the experimenters can freely choose their local settings, so that FW holds, even though condition FR is violated.

Before proceeding, a relevant remark is at order. Here, and anywhere in this work, when using the term *reality* we do not mean *what the world is*, but rather *what the theory is about*, exactly in the spirit of Bell's beables [15]:

> *I use the term beable rather than some more committed term like being or beer to recall the essentially tentative nature of any physical theory.*

We further assume that a complete description can be obtained by confining our attention to a specific set of physical systems, the rest of the universe not contributing at all, as appropriately required by Shimony, Horne and Clauser [16]:

> *Unless we proceed under the assumption that hidden conspiracies [...] do not occur, we have abandoned in advance the whole enterprise of discovering the laws of nature by experimentation.*

Accordingly, we do not consider superdeterministic models in our analysis.

In Section 2 we review the completeness argument of [8], and its implications on the interpretation of the state vector [14], by slightly modifying the original notations for sake of clarity. We only describe the steps which are relevant to our analysis, and refer the interested reader to the original papers for further details. In Section 3 we criticize the assumption that the condition FR introduced in [8] correctly embodies the request FW of free will. Actually, we shall prove that FR implies more than the free will of the experimenters, and this will make clear how our analysis limits the results of [8, 14]. In Section 4 we substantiate our arguments by providing an explicit model in which the quantum state vector is not in one-to-one correspondence with the elements of reality. Finally, in Section 5, we criticize a more recent argument supporting the completeness of quantum mechanics, and we present our final remarks.

2. The completeness argument of Colbeck and Renner

We consider the standard EPR-like scenario: two space-like separated observers perform local measurements on the two parties of an entangled state ψ. The measurement settings at the two sides are given by vectors A and B, and the measurement outcomes by X and Y, whose values are in $\{-1, 1\}$. We assume that additional information on the ontic state λ can be accessed through a measurement procedure with setting C and output Z. All these quantities are assumed to be space-time random variables, whose values can be associated to well-defined space-time points (t, r_1, r_2, r_3), making precise the statement that two measurement procedures and the associated outcomes are space-like with respect to one another. All these quantities are described by joint conditional probabilities of the form $P_{XYZ|ABC}$, and similar ones. Generally, these probabilities are conditioned by ψ, but usually this dependence will be left implicit for notational convenience.

The argument presented in [8] relies essentially on two assumptions, which we now review. The first one, denoted by FR, is that the observers can independently choose the measurements they perform. As explicitly stated in [8]:

> *Assumption FR is that the input, A, of a measurement process can be chosen such that it is uncorrelated with certain other space-time random variables, namely all those whose coordinates lie outside the future light-cone of the coordinates of A.*

This assumption holds true for B and C as well, and it is mathematically expressed (and this is actually what the author's mean by the request FR) through the following requirements on the conditional probabilities:

$$P_{A|BCYZ} = P_A, \qquad P_{B|ACXZ} = P_B, \qquad P_{C|ABXY} = P_C. \tag{2}$$

The second assumption, denoted by QM, is that quantum mechanics is correct [8]:

> *Measurement outcomes obey quantum statistics and [...] all processes within quantum theory can be considered as unitary evolutions, if one takes into account the environment. [Moreover] the second part of the assumption need only hold for microscopic processes on short timescales.*

Now, by relying only on these two assumptions, the authors of [8] prove that any knowledge on the ontic state besides the one given by ψ is useless, since it cannot modify the outcome distributions predicted by quantum mechanics: $P_{XY|ABCZ} = P_{XY|AB}$, $P_{X|ABCZ} = P_{X|A}$ and $P_{Y|ABCZ} = P_{Y|B}$. They conclude that [8]

> *Any attempt to better explain the outcomes of quantum measurements is destined to fail. Not only is the universe not deterministic, but quantum theory provides the ultimate bound on how unpredictable it is. [...] under the assumption that measurement settings can be chosen freely, quantum theory really is complete.*

The technical details of their proof are not relevant here, since our criticism focuses on the adoption of the assumption FR to formalize the request FW of the possibility of freely choosing the apparatus settings. We point out that all the three constraints in (2) are needed to derive the result and that these constraints imply the non-signalling conditions:

$$P_{YZ|ABC} = P_{YZ|BC}, \qquad P_{XZ|ABC} = P_{XZ|AC}, \qquad P_{XY|ABC} = P_{XY|AB}, \tag{3}$$

so that no experimenter can extract, form his outcomes, any information about the measurement settings which have been chosen by the other experimenter. This fact is physically relevant since it avoids superluminal communication between them.

By adapting their argument, in [14] the same authors provide a no-go theorem for ψ-epistemic models, ultimately based on the same assumptions FR and QM. They show that $P_{X|\lambda A} = P_{X|\psi A}$, and then conclude that there should be a one-to-one correspondence between λ and ψ. As before, the technical machinery leading to this conclusion is irrelevant here, but we stress once more that also this result is a direct consequence only of the two assumptions: FR and QM. In the following, we assume that QM is fulfilled, and we criticize the identification of FR with the assumption FW of the free choice of measurement settings, arguing that the latter can be satisfied even though FR is violated.

3. Criticism of the FR assumption and its consequences

As noticed by the authors of [8], their result has, among others, relevant 1 consequences for existing hidden variables models, notably, for Bohmian mechanics. Since this is a deterministic completion of quantum mechanics, following the analysis of J.P. Jarrett, its non-local features necessarily arise from a violation of the condition of parameter independence (PI), which is expressed as

$$P_{X|AB\lambda} = P_{X|A\lambda}, \qquad P_{Y|AB\lambda} = P_{Y|B\lambda}. \tag{4}$$

In the context of [8], PI reduces to $P_{X|ABZ} = P_{X|AZ}$ and $P_{Y|ABZ} = P_{Y|BZ}$, since it is assumed that the information represented by λ is fully accessible (and obtained by knowing Z). This is by itself a constraint which limits the applicability of the results presented in [8], since it ignores the important distinction between controllable and uncontrollable hidden variables, which has been strongly emphasized by A. Shimony [17]. In fact, it is generally assumed that the hidden variables are not completely accessible; that is, there can be some physical principles which limit their knowledge and their manipulation, ensuring, first of all, the impossibility of superluminal signalling. Maintaining the authors' perspective on the accessibility of λ, we observe that the non-signalling conditions (3) automatically imply PI, and therefore the assumption FR trivially forbids all the deterministic completions of quantum mechanics. In [8], this fact is taken as a proof that these models violate the assumption of free choice. Nonetheless, as we have already mentioned and as we shall shortly prove, the FR assumption cannot be considered as the appropriate formalization of the free will assumption FW, since it embodies more than this assumption. Therefore, the violation of FR cannot be automatically related to lack of free will.

To start with, we make precise the condition FW in a way, which, in our opinion, provides a more suitable expression for the assumption of free choice . We observe that, when dealing with correlation experiments of this type, one is merely interested in the freedom of choice of the measurement settings A and B, located at the two wings of the experimental set-up. This condition, when expressed in terms of conditional probabilities, reads

$$P_{A|B\lambda} = P_A, \qquad P_{B|A\lambda} = P_B, \tag{5}$$

and it fully accounts for the fact that the two experimenters can freely and independently choose which observable to measure, since it implies the factorization $P_{AB\lambda} = P_A P_B P_\lambda$. Notice that the assumption FW does not involve a third party, and its measurement setting C. Actually, the authors of [8] remark that the additional information supplementing ψ

> *[...] must be static, that is, its behavior cannot depend on where or when it is observed [...] so, we can consider the case where its observation is also space-like separated from the measurements specified by A and B.*

This statement intends to stress that FR is the appropriate way to express the free will request, since it is consistent with the condition $P_{C|ABXY} = P_C$. Nonetheless, for future reference, we take it as an independent assumption, denoted as ST, and expressed through the condition

$$P_{CZ|ABXY} = P_{CZ}. \tag{6}$$

Notice that in [8] ST does not appear as an assumption, nor the mathematical expression (6) appears somewhere: it is introduced by us make more clear our critical remarks on the assumption FR.

There is another striking difference between the conditions FR and FW: the former involves the random variables X and Y, the latter does not. We believe that it is important to avoid this dependence, since the extra variables X and Y could bring spurious correlations, completely independent from the free will assumption. To clarify the idea, let us assume that we are interested in the free choice of A, and let us raise the following question: if $P_{A|BCYZ} \neq P_A$, can we conclude that A cannot be freely chosen? We do not think that this is the case. For instance, we might suppose that the two Stern-Gerlach apparatuses located at the opposite wings of the experiment could superluminally communicate, at a suitable finite speed. In the rest frame of the two experimenters, the communication happens after the free choice of A and B, and before the generation of the outputs X and Y . In this way, correlation between A and Y might be produced. Even though it is highly implausible that a physical process as the one just mentioned has any physical meaning, its consideration serves the purpose of making clear that the request $P_{A|BCYZ} \neq P_A$ does not forbid that A and B can be freely and independently chosen. In other words, we believe that a good mathematical formulation of the free measurements choice should not automatically reject situations where free will and superluminal communication coexist.

Unfortunately, this is not the case when the FR assumption is considered, since it implies the non-signalling constraints (3), as stated in Section 2.

In accordance with the previous analysis, we find convenient to write non-signalling conditions (which are independent from the additional information on the ontic state, and from the setting of its measuring device) as follows:

$$P_{X|AB} = P_{X|A}, \qquad P_{Y|AB} = P_{Y|B}. \tag{7}$$

They are weaker than relations (3), still they fully express the impossibility of superluminal communication in the standard EPR scenario. In the following, we refer to Eq. (7) as the NS assumption. As proven in [11], the relation between FR, FW, ST and NS is given by

$$FW \wedge NS \wedge ST \Rightarrow FR, \tag{8}$$

and, if the ontic state λ is fully accessible, also the inverse implication holds,

$$FR \Rightarrow FW \wedge NS \wedge ST. \tag{9}$$

Therefore, in the case considered in [8], the FR assumption is equivalent to the logical conjunction of our assumptions FW of free will, staticity ST of the ontic state, and impossibility of superluminal communication, NS. This result proves our statement that FR is more than the free choice assumption, and shows that, at least when the ontic state is (partially or totally) unaccessible, negation of FR does not necessarily imply absence of free will. It might depend on a violation of free will, on the fact that the additional information on λ is not static, or on a violation of the impossibility of superluminal communication. Of those, the second condition appears the easiest to digest: why the extra information on λ, which is complementing the information provided by the state vector ψ, should be static, when ψ itself, and the measurement procedure involved in its preparation, are not space-like separated with respect to A and B?

Therefore, the completeness argument presented in [8] has not the claimed generality, although it is formally correct. When considering Bohmian mechanics, the statement that measurement settings cannot be freely chosen is not justified. If we assume that λ is fully accessible (that is, that all positions are known), it is a well known fact that the theory allows superluminal communication. On the other hand, the additional information on the ontic state is not static (in fact, it should be a partial information on the positions, which are distributed according to ψ, which in turn is certainly a non-static quantity). In both cases, violation of FR does not require lack of free will.

The no-go theorem for ψ-epistemic theories derived in [14] is based on the assumptions QM and FR. Therefore, the same conclusions apply to this case as well, and the result is not general at all. Both the questions of whether quantum mechanics is a complete theory, and the interpretation of the quantum state as a state of knowledge or an element of reality, remain open. In the following section we present a ψ-epistemic model for a pair of two-level systems, fully consistent with quantum mechanics, in which the measurement settings A and B can be freely chosen.

4. A ψ-epistemic model for the bipartite system

The model is a trivial application of the results of [13], where, however, it has not been realized that the family of ψ-epistemic models therein presented are valid beyond the single system case. Following our reasoning in the previous section, we can claim that superluminal influences of measurement choices upon ontic variables, in the multipartite case, are not necessarily expected. Despite the fact that in [13] a system of arbitrary dimension is considered, here, for simplicity, we limit our attention to a pair of two-level systems.

The model is inspired by a generalization, to the case of two subsystems, of a Bell model for a single two-level system. We start by describing this generalization. If we denote by $\mathcal{H}$ the Hilbert space of the single system, an arbitrary quantum state vector is given by $\varphi \in \mathcal{H} = \mathcal{H}_A \otimes \mathcal{H}_B$. The hidden variable is a real parameter $\tau \in T = [0,1]$, and the ontic state is given by the ordered pair $\lambda = (\varphi, \tau) \in \mathcal{H} \times T$. Its distribution corresponding to the quantum state ψ is given by

$$\rho_\psi(\lambda) = \rho_\psi(\varphi, \tau) = \delta(\varphi - \psi), \tag{10}$$

that is, τ is uniformly distributed over T for all states ψ. We consider now the local observables $\hat{A} = A \cdot \sigma_A$ and $\hat{B} = B \cdot \sigma_B$, where the settings A and B are unit real vectors in $\mathbb{R}^3$ and σ_A, σ_B are the vectors of Pauli matrices acting on $\mathcal{H}_A$ and $\mathcal{H}_B$ respectively. We denote the common eigenvectors of the commuting observables $\hat{A}$, $\hat{B}$ and $\hat{A} \otimes \hat{B}$ as ϕ_j, and we order them according to

$$|\langle \phi_j | 0 \rangle|^2 \geqslant |\langle \phi_{j+1} | 0 \rangle|^2 \qquad \text{with } j = 0, \dots, 3, \tag{11}$$

where $|0\rangle \in \mathcal{H}$ is an arbitrary reference state. The states ϕ_j are factorized, and they are associated with well defined outcomes (X, Y) for the observables $\hat{A}$ and $\hat{B}$, which are in the set $\{(-1,-1), (-1,1), (1,-1), (1,1)\}$. The outcome of $\hat{A} \otimes \hat{B}$ is given by the product XY. The specific correspondence between j and the pairs of possible outcomes for X and Y depends on the local observables taken into account, in the way we describe now. In terms of the ontic state $\lambda = (\varphi, \tau)$ the values of the projectors $\hat{\Phi}_j = |\phi_j\rangle\langle\phi_j|$ are defined as

$$\Phi_j(\lambda) = 1, \qquad \text{if } \sum_{k=0}^{j-1} |\langle \phi_k | \varphi \rangle|^2 \leqslant \tau < \sum_{k=0}^{j} |\langle \phi_k | \varphi \rangle|^2, \tag{12}$$

and $\Phi_j(\lambda) = 0$ otherwise; when j = 0, the lower bound for τ is defined to be 0 (in this case the former expression is meaningless). Therefore, with any λ there is associated a value of j. Since the spectral decomposition of the local operators $\hat{A}$ and $\hat{B}$ (as well as that of $\hat{A} \otimes \hat{B}$) can be written through the projectors $\hat{\Phi}_j$, the corresponding outcomes $X = X(\lambda)$ and $Y = Y(\lambda)$ (and, of course, their product) are unambiguously determined. They are distributed according to the quantum mechanical rules [13], since

$$\int d\lambda \rho_\psi(\lambda) \Phi_j(\lambda) = \int d\tau \Phi_j(\psi, \tau) = |\langle \phi_j | \psi \rangle|^2. \tag{13}$$

As the original Bell model for a single two-level system, this generalized model is ψ-ontic. In fact, according to (10), the distributions corresponding to different state vectors never overlap. However, this property is not needed at all in order to produce a model predictively equivalent to quantum mechanics; for this it is sufficient that the measure of the support of the function $\Phi_j(\lambda)$ over $\{(\varphi, \tau), \tau \in T\}$ equals the quantum probability $|\langle \phi_j | \psi \rangle|^2$. Following [13], by suitably redistributing this support, it is possible to turn the model into a ψ-epistemic one. After noticing that $|\langle \phi_0 | 0 \rangle|^2 \geqslant 1/4$, the authors introduce

$$z(\varphi) = \inf_{\phi : \langle \phi | 0 \rangle|^2 \geqslant 1/4} |\langle \phi | \varphi \rangle|^2, \tag{14}$$

which is used to define a subset of the ontic space,

$$\mathcal{E}_0 = \{(\varphi, \tau) : |\langle \varphi | 0 \rangle|^2 > 3/4 \text{ and } 0 \leqslant \tau < z(\varphi)\}. \tag{15}$$

This set has the property that for all $\lambda \in \mathcal{E}_0$ the states φ are in the support of $\Phi_0 = |\phi_0\rangle\langle\phi_0|$. Therefore, these ontic states have all the same outcome, which is the one associated with ϕ_0. If

Journal of Physics: Conference Series **442** (2013) 012002 doi:10.1088/1742-6596/442/1/012002

we redistribute over the whole of $\mathcal{E}_0$ the probabilities which the distribution $\rho_\psi(\lambda)$ in (10) assigns to ontic states within $\mathcal{E}_0$ itself, we achieve the desired target. For example, by denoting by $\rho_{\mathcal{E}_0}(\lambda)$ the uniform distribution over $\mathcal{E}_0$, and by $\Theta(x)$ the unit-step function, such that $\Theta(x) = 1$ for $x \geqslant 0$ and $\Theta(x) = 0$ for $x < 0$, it is possible to define

$$\tilde{\rho}_\psi(\lambda) = \tilde{\rho}_\psi(\varphi, \tau) = \begin{cases} \delta(\varphi - \psi), & \text{for } |\langle\psi|0\rangle|^2 \leqslant 3/4 \\ \delta(\varphi - \psi)\Theta(x - z(\psi)) + z(\psi)\rho_{\mathcal{E}_0}(\lambda), & \text{for } |\langle\psi|0\rangle|^2 > 3/4 \end{cases} \tag{16}$$

which, by construction, satisfies

$$\int d\lambda \tilde{\rho}_\psi(\lambda)\Phi_j(\lambda) = |\langle\phi_j|\psi\rangle|^2, \tag{17}$$

but in general produces overlapping supports for distinct state vectors. Therefore, if we replace $\rho_\psi(\lambda)$ by $\tilde{\rho}_\psi(\lambda)$, we obtain a ψ-epistemic model for a pair of two-level systems. There is not reason why the conditions FW and NS introduced in the previous section should be violated, therefore we conclude that it is possible to build ψ-epistemic models which do not imply violations of the free will assumption, or superluminal communication.

In this section we have adopted a weak notion of ψ-epistemic model: there must be at least one pair of quantum state vectors whose corresponding distributions have overlapping supports. A stronger definition requires that all non-orthogonal quantum states have overlapping distributions. While a detailed treatment of this case is out of the scopes of this paper, we point out that our remarks on the expression of the free will assumption holds true also in this case.

5. Final remarks and conclusions

Recently, the authors of [8, 14] have provided a different argument leading to the same conclusion of completeness of quantum mechanics [18]. This work is presented as an expanded and more pedagogical version of [8, 14], nonetheless it appears as a distinct derivation. For instance, the result that quantum mechanics is maximally informative (that is, complete), makes use of different resources in the cases of maximally and non-maximally entangled states, unlike [8, 14]. Here we are not interested in a general analysis of this work, but in investigating the assumption of free choice which appears in it, which is the most significative requirement to draw the conclusions expressed in the paper. Following the authors [18],

> *[...] a parameter of the theory, say A, is considered free if it is possible to choose A such that it is uncorrelated with all other values (described by the theory) except those that lie in the causal future of A.*

They consistently express this idea through a factorization of the probabilities $P_{A\Gamma_A} = P_A P_{\Gamma_A}$, where Γ_A is the set of all random variables which are not in the causal future of A. This condition must hold also for B, that is $P_{B\Gamma_B} = P_B P_{\Gamma_B}$, where Γ_B is the set of all random variables which are not in the causal future of B.

Differently from the former assumption FR, this condition does not involve the random variables C and Z, and therefore it does not require the assumption ST of static variables. Nonetheless, as FR did, also this condition automatically satisfies NS, and therefore it cannot coexist, even in principle, with a theory allowing superluminal communication. This can be derived by expressing P_{ABX} in different ways:

$$P_{ABX} = P_{X|AB}P_{AB} = P_{X|AB}P_A P_B, \tag{18}$$

where, in the second step, we have considered that $B \in \Gamma_A$; but also

$$P_{ABX} = P_{AX}P_B = P_{X|A}P_A P_B, \tag{19}$$

where the first step follows from $A \in \Gamma_B$ and $X \in \Gamma_B$. By comparing (18) and (19), we derive the non-signalling condition $P_{X|AB} = P_{X|A}$; a similar argument proves that $P_{Y|AB} = P_{Y|B}$, and finally NS is derived.

Therefore, according to our former discussion, we believe that this is not a legitimate expression of the free will, and also the results of this work are not as general as stated. Notice that, to motivate their mathematical expression of the free will, the authors of [18] mention how J.S. Bell characterized the concept of free variables [19]:

> *for me this means that the values of such variables have implications only in their future light cones.*

However, "having no implications" is not the same as "being uncorrelated". The fact that I decide to go out with my umbrella is correlated with the weather forecasts, but it has not implications on them. Therefore, in our opinion, the expression of the free will adopted in [18] does not expresses the Bell's point of view, and is not adequate.

For maximally entangled states, there is independent evidence that no theory consistent with quantum mechanics can produce probabilities differing from those predicted by the quantum theory [20, 21, 22, 23]. Nonetheless, this result is in general false for non-maximally entangled states [10]. Notice that, in [18], the predictive equivalence between the quantum theory and the underlying model, in the case of non-maximally entangled states, is obtained by reducing this case to that of maximally entangled states. This step relies on the extra resource of *embezzling states*, a special set of states whose precise characterization requires an infinite-dimensional Hilbert space [24]. By using these states, the two experimenters can transform an arbitrary state of some Hilbert space into a maximally entangled state of the same Hilbert space, via local operations. We believe that the use of this resource is not legitimate in this context, since it does not allow a self-contained description of any finite-dimensional Hilbert space.

We conclude that the issue of completeness of quantum mechanics, and the problem of the interpretation of the quantum state vector, are still open.

Acknowledgments
We thank R. Colbeck for a critical reading of the paper and for his comments. This research is supported by the ARO MURI grant W911NF-11-1-0268.

References

[1] Einstein A Podolsky B and Rosen N 1935 *Phys. Rev.* **47** 777
[2] Harrigan N and Spekkens R W 2010 *Found. Phys.* **40** 125
[3] von Neumann J 1932 *Mathematische Grundlagen der Quantenmechanik* (Berlin: Springer-Verlag)
[4] Bohm D 1952 *Phys. Rev.* **85** 166
[5] Bohm D 1952 *Phys. Rev.* **85** 180
[6] Bell J S 1966 *Rev. Mod. Phys.* **38** 447
[7] Bell J S 1964 *Physics* **1** 195
[8] Colbeck R and Renner R 2011 *Nature Comm.* **2** 411
[9] Ghirardi G C and Romano R 2013 About possible extensions of quantum theory *Preprint* arXiv:1301.5040
[10] Ghirardi G C and Romano R 2013 Onthological models predictively inequivalent to quantum theory *Preprint* arXiv:1301.2695
[11] Ghirardi G C and Romano R 2013 Comment on "Is a system's wave function in one-to-one correspondence with its elements of reality?" *Preprint* arXiv:1302.1635
[12] Pusey M F Barrett J and Rudolph T 2012 *Nature Phys.* **8** 476
[13] Lewis P G Jennings D Barrett J and Rudolph T 2012 *Phys. Rev. Lett.* **109** 150404
[14] Colbeck R and Renner R 2012 *Phys. Rev. Lett.* **108** 150402
[15] Bell J S 1984 Beables for quantum field theory *Preprint* CERN-TH 4035/84
[16] Shimony A Horne M A and Clauser J F 1976 Comment on "The Theory of Local Beables", in *Epistemological Letters Hidden Variables and quantum Uncertainty* **9** ed F Bonsack
[17] Shimony A 1983 *Proc. Int. Symp. Foundations of Quantum Mechanics* (Tokyo) 225

[18] Colbeck R and Renner R 2012 The completeness of quantum theory for predicting measurement outcomes *Preprint* arXiv:1208.4123
[19] Bell J S 1987 Free variables and local causality , in *Speakable and unspeakable in quantum mechanics* chapter 12 p 100 (Cambridge: Cambridge University Press)
[20] Colbeck R and Renner R 2008 *Phys. Rev. Lett.* **101** 050403
[21] Branciard C Ling A Gisin N Kurtsiefer C Lamas-Linares A and Scarani V 2007 *Phys. Rev. Lett.* **99** 210407
[22] Ghirardi G C and Romano R 2012 *Phys. Rev.* A **86** 022107
[23] di Lorenzo A 2012 *Phys. Rev.* A **86** 042119
[24] van Dam W and Hayden P 2003 *Phys. Rev.* A **67** 060302(R)

Journal of Physics: Conference Series **442** (2013) 012003 doi:10.1088/1742-6596/442/1/012003

Open Quantum Walks: a short introduction

Ilya Sinayskiy and Francesco Petruccione

National Institute for Theoretical Physics and Quantum Research Group, School of Chemistry and Physics, University of KwaZulu-Natal, Durban, South Africa

E-mail: sinayskiy@ukzn.ac.za, petruccione@ukzn.ac.za

Abstract. The concept of open quantum walks (OQW), quantum walks exclusively driven by the interaction with the external environment, is reviewed. OQWs are formulated as discrete completely positive maps on graphs. The basic properties of OQWs are summarised and new examples of OQWs on $\mathbb{Z}$ and their simulation by means of quantum trajectories are presented.

1. Introduction

The mathematical concept of classical random walks (CRW) [1] has found wide application in various fields of modern science, i.e. physics [2], computer science [3], economy [4] and biology [5]. In this case, the trajectory of the walker is fully determined by the transition matrix of the underlying graph. A quantum extension of this concept was also suggested [6, 7]. In this case the trajectory of the walker is defined not only by the transition matrix but also by some inner state of the quantum walker, e.g., spin or polarisation. Unitary quantum walks are one of the ways to implement quantum computing algorithms [8, 9]. Although the practical implementation of any complex quantum system is very difficult due to unavoidable interaction with an environment [10], successful realisations of the unitary quantum walks have been reported. A more adequate description of the experimental realisations of the unitary quantum walks can be obtained by including the effects of decoherence and dissipation. Usually, in these approaches, the effects of decoherence and dissipation are small and the influence the dynamics of the quantum walker needs to be minimised or eliminated.

Recently, the formalism of discrete-time open quantum walks (OQW) was introduced [11, 12]. In an OQWs the quantum walker is exclusively driven by the interaction with an environment. The formalism of OQWs rests upon the implementation of completely positive maps. OQWs can be used to perform efficient dissipative quantum computation and state engineering [11, 13]. The properties of the asymptotic probability distribution were analysed as well [14, 15, 16].

The paper has the following structure. In Section 2 we introduce the OQW formalism. In Section 3 we present a simple example of an OQWs on a two-node graph and of an OQWs on $\mathbb{Z}$. In Section 4 we demonstrate the simulation of OQWs by the means of quantum trajectories. In Section 5 the *realisation procedure* to obtain OQWs from unitary dynamics is presented. In Section 6 we show the connection of OQWs to unitary quantum walks and classical random walks. In Section 7 results of the limit theorems are presented. In Section 8 the application of OQW to dissipative quantum computing is described and in Section 9 we conclude.

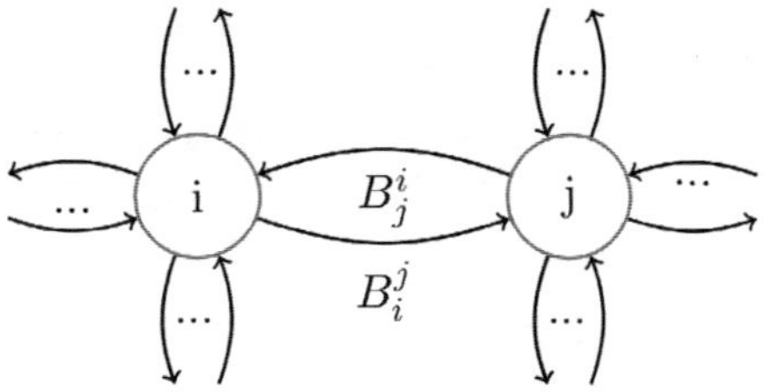

Figure 1. A schematic representation of an open quantum walk on a graph with nodes i and j. The operators B^j_i represent the transition operators of the walker.

2. Open Quantum Walks

OQWs are introduced on a set of vertices $\mathcal{V}$ with oriented edges $\{(i,j)\,;\; i,j \in \mathcal{V}\}$. The number of nodes is considered to be finite or countable infinite. The space of states corresponding to the dynamics on the graph specified by the set of nodes $\mathcal{V}$ will be denoted by $\mathcal{K} = \mathbb{C}^{\mathcal{V}}$. If $\mathcal{V}$ is an infinite countable set, the space of states $\mathcal{K}$ will be any separable Hilbert space with an orthonormal basis $(|i\rangle)_{i\in\mathcal{V}}$ indexed by $\mathcal{V}$. The internal degrees of freedom of the quantum walker, e.g. the spin, polarisation, angular momenta or n-energy levels, will be described by a separable Hilbert space $\mathcal{H}$ attached to each vertex of the graph. Any state of the quantum walker will be described on the direct product of the Hilbert spaces $\mathcal{H} \otimes \mathcal{K}$.

To describe the dynamics of the internal degree of freedom of the quantum walker for each edge (i,j) we introduce a bounded transition operator $B^i_j \in \mathcal{H}$. This operator describes the change in the internal degree of freedom of the walker due to the "quantum jump" from node j to node i (see Fig. 1). By imposing for each j that,

$$\sum_i {B^i_j}^\dagger B^i_j = I, \tag{1}$$

we insure, that for each node of the graph $j \in \mathcal{V}$ there is a corresponding trace preserving completely positive map on the operators of $\mathcal{H}$:

$$\mathcal{M}_j(\tau) = \sum_i B^i_j \tau {B^i_j}^\dagger. \tag{2}$$

The transition operators B^i_j act only on the "quantum coin" Hilbert space $\mathcal{H}$ and do not perform transitions from node j to node i, they can be easy extended to operators $M^i_j \in \mathcal{H} \otimes \mathcal{K}$ acting on total Hilbert space in the following way

$$M^i_j = B^i_j \otimes |i\rangle\langle j|\,. \tag{3}$$

It is clear that, if the condition expressed in Eq. (1) is satisfied, then $\sum_{i,j} {M^i_j}^\dagger M^i_j = 1$. This condition defines a trace preserving completely positive map for density matrices on $\mathcal{H}\otimes\mathcal{K}$, i.e.,

$$\mathcal{M}(\rho) = \sum_i \sum_j M^i_j \,\rho\, {M^i_j}^\dagger. \tag{4}$$

This map defines the discrete time *Open Quantum Walk* (OQW) [11, 12]. It is clear that for an arbitrary initial state the density matrix $\sum_{i,j} \rho_{i,j} \otimes |i\rangle\langle j|$ will take a diagonal form after just one step of the open quantum random walk Eq. (4). One can see that,

$$\begin{aligned}\mathcal{M}\left(\sum_{k,m} \rho_{k,m} \otimes |k\rangle\langle m|\right) &= \sum_{i,j,k,m} B^i_j \otimes |i\rangle\langle j| \left(\rho_{k,m} \otimes |k\rangle\langle m|\right) B^{i\dagger}_j \otimes |j\rangle\langle i| \\ &= \sum_{i,j,k,m} B^i_j \rho_{k,m} B^{i\dagger}_j \otimes |i\rangle\langle i| \delta_{j,k}\delta_{j,m} \\ &= \sum_i \left(\sum_j B^i_j \rho_{j,j} B^{i\dagger}_j\right) \otimes |i\rangle\langle i|. \end{aligned} \tag{5}$$

Hence, in the following, we will assume that the initial state of the system has the form

$$\rho = \sum_i \rho_i \otimes |i\rangle\langle i|, \tag{6}$$

with

$$\sum_i \mathrm{Tr}(\rho_i) = 1. \tag{7}$$

It is easy to give an explicit iteration formula for the OQW from step n to step $n+1$

$$\rho^{[n+1]} = \mathcal{M}(\rho^{[n]}) = \sum_i \rho_i^{[n+1]} \otimes |i\rangle\langle i|, \tag{8}$$

where

$$\rho_i^{[n+1]} = \sum_j B^i_j \rho_j^{[n]} B^{i\dagger}_j. \tag{9}$$

The above iteration formula gives a clear physical meaning to the mapping that we introduced: the state of the system on site i is determined by the conditional shift from all connected sites j, which are defined by the explicit form of the open quantum coin operators B^i_j. Also, one can see that $\mathrm{Tr}[\rho^{[n+1]}] = \sum_i \mathrm{Tr}[\rho_i^{[n+1]}] = 1$.

3. Examples of OQW

3.1. OQW on a 2-node graph

As an example of open quantum walk we consider the simplest case of a walk on a 2-node graph (see Fig. 2a). In this case the transition operators B^j_i $(i, j = 1, 2)$ satisfy:

$$B^{1\dagger}_1 B^1_1 + B^{2\dagger}_1 B^2_1 = I, \quad B^{2\dagger}_2 B^2_2 + B^{1\dagger}_2 B^1_2 = I. \tag{10}$$

The state of the walker $\rho^{[n]}$ after n steps is given by,

$$\rho^{[n]} = \rho_1^{[n]} \otimes |1\rangle\langle 1| + \rho_2^{[n]} \otimes |2\rangle\langle 2|, \tag{11}$$

where the particular form of the $\rho_i^{[n]}$ $(i = 1, 2)$ is found by recursion,

$$\begin{aligned}\rho_1^{[n]} &= B^1_1 \rho_1^{[n-1]} B^{1\dagger}_1 + B^1_2 \rho_2^{[n-1]} B^{1\dagger}_2, \\ \rho_2^{[n]} &= B^2_1 \rho_1^{[n-1]} B^{2\dagger}_1 + B^2_2 \rho_2^{[n-1]} B^{2\dagger}_2.\end{aligned} \tag{12}$$

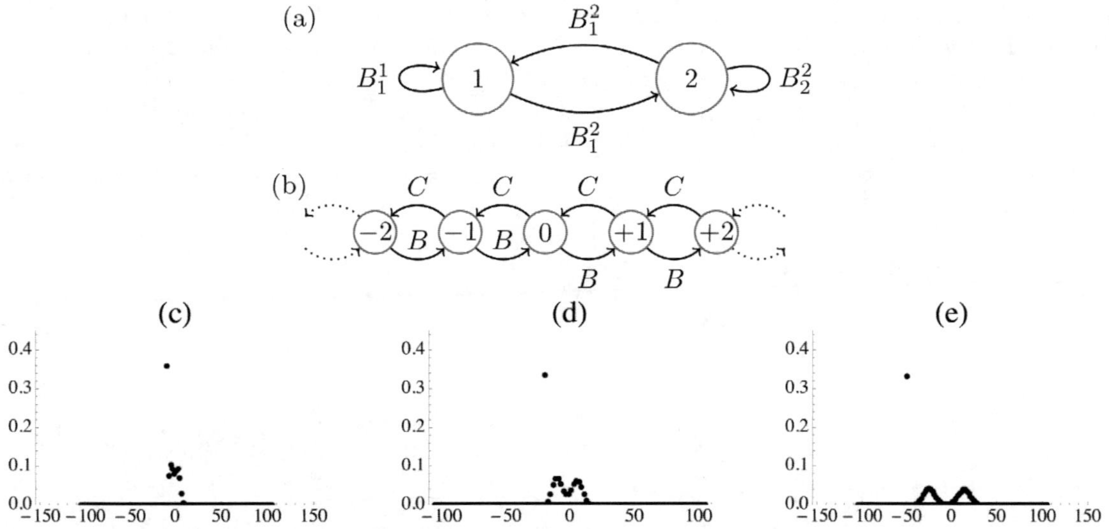

Figure 2. (a) A schematic representation of an open quantum walk on a 2-node graph. The operators B_i^j $(i, j = 1, 2)$ represent the transition operators of the walk. Open quantum walk on Z. (b) A schematic representation of the OQW on $\mathbb{Z}$: all transitions to the right are induced by the operator B, while all transitions i to the left are induced by the operator C; Figures (c)-(e) show the occupation probability distribution for the walker with the initial state $\rho^{[0]} = \frac{1}{3}I_3 \otimes |0\rangle\langle 0|$ and transition operators given by Eq. (18) after 10, 20 and 50 steps, respectively.

3.2. OQW on $\mathbb{Z}$

In this subsection we consider the case of OQWs on $\mathbb{Z}$ with transition between neighbouring nodes (see Fig. 2b). In this case, the general expression for the open quantum walk on the graph reads,

$$\rho^{[n+1]} = \mathcal{M}(\rho^{[n]}) = \sum_i \rho_i^{[n+1]} \otimes |i\rangle\langle i|, \tag{13}$$

where

$$\rho_i^{[n+1]} = B_{i+1}^i \rho_{i+1}^{[n]} B_{i+1}^{i\dagger} + B_{i-1}^i \rho_{i-1}^{[n]} B_{i-1}^{i\dagger}. \tag{14}$$

Here we consider a homogenous open quantum walk (which implies that $\forall i, B_i^{i+1} \equiv B$ and $B_i^{i-1} \equiv C$). For an initially localised walker in the node 0, namely

$$\rho^{[0]} = \rho_0^{[0]} \otimes |0\rangle\langle 0|, \tag{15}$$

the density matrix after one $\rho^{[1]}$ and two steps $\rho^{[2]}$ reads,

$$\rho^{[1]} = B\rho_0^{[0]}B^\dagger \otimes |1\rangle\langle 1| + C\rho_0^{[0]}C^\dagger \otimes |-1\rangle\langle -1|, \tag{16}$$

and

$$\rho^{[2]} = B\rho_1^{[1]}B^\dagger \otimes |2\rangle\langle 2| + (C\rho_1^{[1]}C^\dagger + B\rho_{-1}^{[1]}B^\dagger) \otimes |0\rangle\langle 0| + C\rho_{-1}^{[1]}C^\dagger \otimes |-2\rangle\langle -2|. \tag{17}$$

In order to demonstrate the dynamics of the probability distribution, we choose B and C in the following way,

$$B = \begin{pmatrix} 1 & 0 & 0 \\ 0 & \sqrt{3}/2 & 0 \\ 0 & 0 & 3/5 \end{pmatrix}, \quad C = \begin{pmatrix} 0 & 0 & 0 \\ 0 & 1/2 & 0 \\ 0 & 0 & 4/5 \end{pmatrix}. \tag{18}$$

If we choose the initial state of the walker to be localised in the node 0 with unpolarised state of an internal degree of freedom, i.e., $\rho^{[0]} = \frac{1}{3}I_3 \otimes |0\rangle\langle 0|$, then by iteration we obtain the probability distribution of the walker after an arbitrary number of steps. In Fig. 2(c)-(e) we show the probability to find a walker on a particular lattice site for different numbers of steps. Already, after 20 steps (Fig. 2(d)) one can see the formation of a "soliton"- like distribution and two gaussian packets moving in different directions. After 50 steps (Fig. 2(e)) this behaviour is even more certain.

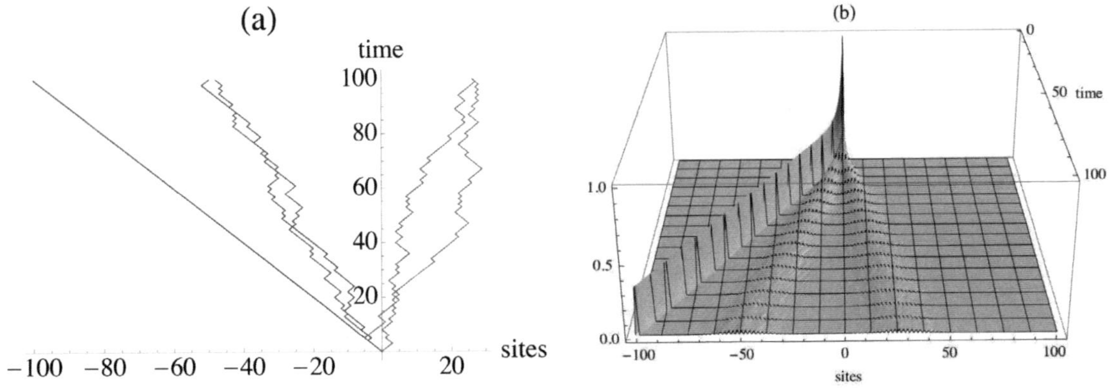

Figure 3. Simulation of the OQW in term of quantum trajectories. Examples of trajectories are shown in (a); (b) shows the average over 10^5 trajectories

4. Unravelling of OQW

It is interesting to mention that OQW can be simulated by the means of *Quantum Trajectories* [12]. To introduce this formalism we start by considering a particular case of initial state of the system, namely, a walker that is localized at a single node,

$$\rho^{[0]} = \rho \otimes |i\rangle\langle i|. \tag{19}$$

After one step of OQW the state of the walker is given by

$$\rho^{[1]} = \sum_j (B_i^j \rho B_i^{j\dagger}) \otimes |j\rangle\langle j|\,. \tag{20}$$

The probability to find the walker at the site j is given by $p_j = \mathrm{tr}(B_i^j \rho B_i^{j\dagger})$. If one performs measurements of the position of the walker at node j the state of the walker reads

$$\frac{1}{p_j}(B_i^j \rho B_i^{j\dagger}) \otimes |j\rangle\langle j|\,. \tag{21}$$

Repetition of this procedure gives rise to a classical Markov chain, valued in the set of states of the form $\rho \otimes |i\rangle\langle i|$. One can see that this procedure on average will simulate a master equation driven by $\mathcal{M}$:

$$\mathcal{E}[\rho_{n+1}] = \sum_j p_j \frac{1}{p_j}(B_i^j \rho_n B_i^{j\dagger}) \otimes |j\rangle\langle j| = \sum_j (B_i^j \rho_n B_i^{j\dagger}) \otimes |j\rangle\langle j| = \mathcal{M}(\rho_n)\,. \tag{22}$$

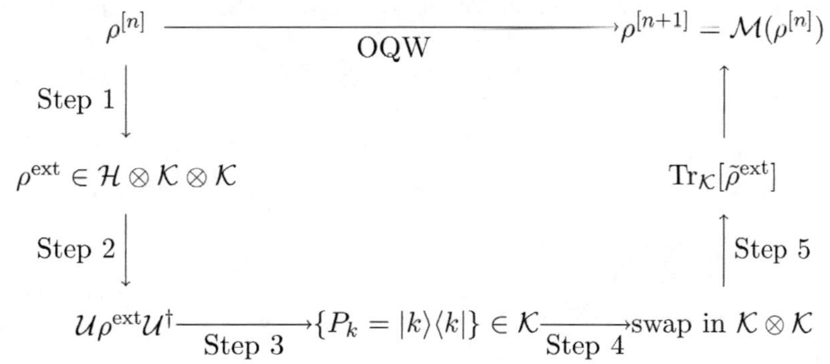

Figure 4. Scheme of the various steps required for the *realisation procedure* of the Open Quantum Walk.

If one choose the initial state of the system in the pure state $\rho = |\phi\rangle\langle\phi| \otimes |i\rangle\langle i|$, then the realization remains in a pure state. It is easy to see that, an arbitrary initial pure state $|\phi\rangle \otimes |i\rangle$ will jump randomly to one of the states

$$\frac{1}{\sqrt{p_i^j}} B_i^j |\phi\rangle \otimes |j\rangle \tag{23}$$

with probability

$$p_i^j = ||B_i^j|\phi\rangle||^2 \,. \tag{24}$$

This procedure leads to a classical Markov chain valued in the space of wave functions of the form $|\phi\rangle \otimes |i\rangle$. On average, this random walk simulates the OQW driven by $\mathcal{M}$.

As an example of this unravelling we consider an OQW on a line. The transition matrices B and C are given by the Eq. (18) and the initial state of the walker is

$$|\psi_0\rangle = \frac{1}{\sqrt{3}}(|1\rangle + |2\rangle + |3\rangle) \otimes |0\rangle.$$

The results of the simulations are presented in Fig. 3. In Fig. 3(a) we show five different quantum trajectories, one can clearly see the different qualitative character of some of them. The average over 10^5 trajectories is shown in Fig. 3(b).

5. Realisation procedure

It is natural to ask how to obtain an OQW by reduction of the unitary dynamics of an appropriately enlarged system. The situation is depicted in Fig. 4 and will be described step by step [12]. To achieve this goal we need to extend the Hilbert space of the system by an ancilla-graph identical to the graph described by the space $\mathcal{K}$ and define on this extended Hilbert space $\mathcal{H} \otimes \mathcal{K} \otimes \mathcal{K}$ a specially chosen unitary operator $\mathcal{U}$. It is always possible for any given k to define a unitary operator $U(k) \in \mathcal{H} \otimes \mathcal{K}$, such that the first column of the operator is given by B_k^i and by filling the rest with corresponding elements to fulfil the unitary condition

$$U_1^i(k) = B_k^i, \qquad \sum_j U_j^{i'}(k)^\dagger U_j^i(k) = \delta_{i,i'} I. \tag{25}$$

The unitary operator $\mathcal{U}$ is a diagonal sum of the operators $U(k)$ with corresponding projector representation

$$\mathcal{U} = \sum_k U(k) \otimes |k\rangle\langle k|. \tag{26}$$

Again, we assume the initial state $\rho = \sum_i \rho_i \otimes |i\rangle\langle i| \in \mathcal{H} \otimes \mathcal{K}$.

The first step is to extend the state into a larger Hilbert space in the following way

$$\rho = \sum_i \rho_i \otimes |i\rangle\langle i| \to \rho^{\text{ext}} = \sum_i \rho_i \otimes |1\rangle\langle 1| \otimes |i\rangle\langle i|, \tag{27}$$

so that the extended density matrix $\rho^{\text{ext}} \in \mathcal{H} \otimes \mathcal{K} \otimes \mathcal{K}$.

Step 2 is the application of the unitary operator $\mathcal{U}$ to the extended density matrix ρ^{ext}, i.e.,

$$\mathcal{U}\rho^{\text{ext}}\mathcal{U}^\dagger = \sum_{i,j,p} B_i^j \rho_i B_i^{p\dagger} \otimes |j\rangle\langle p| \otimes |i\rangle\langle i|. \tag{28}$$

In *Step 3* we perform a full set of measurements $P_k = I_{\mathcal{H}} \otimes |k\rangle\langle k| \otimes I_{\mathcal{K}}$ on the "extra" position space or we subject the system to decoherence in this "extra" position subspace. As result of Step 3 we obtain the following density matrix

$$\sum_{i,j} B_i^j \rho_i B_i^{j\dagger} \otimes |j\rangle\langle j| \otimes |i\rangle\langle i|. \tag{29}$$

In Step 4 we swap position subspaces. The result of this operation is the following density matrix

$$\sum_{i,j} B_i^j \rho_i B_i^{j\dagger} \otimes |i\rangle\langle i| \otimes |j\rangle\langle j|. \tag{30}$$

The last step (Step 5) is tracing out the "extra" position subspace and one obtains the OQW as promised

$$\sum_{i,j} B_i^j \rho_i B_i^{j\dagger} \otimes |j\rangle\langle j|. \tag{31}$$

6. Connection to Unitary Quantum Walk and Classical Random Walk

6.1. Connection to unitary random walk

The *unitary quantum walk* (UQW) can be recovered from the OQW realisation procedure by a special choice of the transition operators B_i^j and by excluding from the realisation procedure illustrated in Fig. 4 Step 3 (measurements or decoherence in position space). This is easily demonstrated for a UQW on $\mathbb{Z}$. In this case of a UQW on the line an additional condition needs to be imposed on the transition operators B and C: next to the normalisation condition $B^\dagger B + C^\dagger C = I$ we need to request that $C^\dagger B = 0$. In this case the sum of B and C is a unitary operator. Explicitly, in order to recover the Hadamard UQW on the line we may choose

$$B = \begin{pmatrix} \alpha & \beta \\ 0 & 0 \end{pmatrix}, \quad C = \pm \begin{pmatrix} 0 & 0 \\ -\beta^* & \alpha^* \end{pmatrix}, \tag{32}$$

where $|\alpha|^2 + |\beta|^2 = 1$ and $\alpha, \beta \in \mathbb{C}$. Starting from the state $|\psi_0\rangle \otimes |0\rangle$ and following the realization procedure (Fig. 4) while skipping Step 3, the state of the system after one step is given by $B|\psi_0\rangle \otimes |+1\rangle + C|\psi_0\rangle \otimes |-1\rangle$. This is exactly a Hadamard UQW on $\mathbb{Z}$. The general condition for obtaining UQWs from OQWs can be found in [12].

Journal of Physics: Conference Series **442** (2013) 012003 doi:10.1088/1742-6596/442/1/012003

6.2. The classical random walk

It is interesting to note that the OQWs contains, as a special case, the classical random walk as well. To see this, we consider the case $\mathcal{H} = \mathcal{K} = \mathbb{C}^{\mathcal{V}}$, and we introduce a matrix $P = \{P_{i,j}\}$ of classical transition probabilities on the graph $\mathcal{V}$ with standard normalisation condition $\sum_i P_{j,i} = 1$ and add an arbitrary family of unitary operators, $U_i^j \in \mathbb{C}^{\mathcal{V}}$. In this case, if we choose the transition operators B_i^j to have the form $B_i^j = \sqrt{P_{i,j}} U_i^j$. Then, for the initial state $\rho = \sum_k \rho_k \otimes |k\rangle\langle k|$ the probability to find a "walker" after one step on the site i reads,

$$p^{[1]}(i) = \mathrm{Tr}(\mathcal{M}(\rho)|i\rangle\langle i|) = \sum_k P_{k,i}\mathrm{Tr}(\rho_k). \tag{33}$$

After two steps the probability to find walker on the site i will be given by

$$p^{[2]}(i) = \mathrm{Tr}(\mathcal{M}(\mathcal{M}(\rho))|i\rangle\langle i|) = \sum_{k,m} p_{m,k}p_{k,i}\mathrm{Tr}(\rho_m). \tag{34}$$

Thus, the probability of transition does not depend on the internal degrees of freedom of the "walker". It only depends on the classical transition probability matrix P, as expected for a classical random walk.

7. Limit theorems for OQWs

Recently Attal *et al* [14] and Konno and Yoo [15] have formulated central limit theorems for OQWs. In particular, Attal *et al* have shown that if we have an OQW on $\mathbb{Z}^d$ and ρ_∞ is a unique invariant state of the OQW, then the normalised quantum trajectory of the walk will converge to the Gaussian distribution $\mathcal{N}(0, C)$ in $\mathbb{R}^d$ with covariance matrix C. Similar results were obtained by Konno and Yoo for OQW on $\mathbb{Z}$. Sinayskiy and Petruccione [16] have analysed the limit distribution of the OQW on $\mathbb{Z}$ for the case of simultaneously diagonalisable transition operators. They found that the number of Gaussian distributions in the limiting case is bounded by the dimension of the transition operator and defined by its spectrum.

The asymptotic behaviour of the example of OQW on $\mathbb{Z}$ considered in this proceedings fits into the case studied by Sinayskiy and Petruccione in Ref. [16]. The transition operators B and C Eq. (18) are simultaneously diagonalizable. The dimension of the tradition operators is 3 which implies maximally three different asymptotic distributions (see Fig.2(e)). The condition for a "soliton" like distribution is the presence of zeros in spectrum of one of the transition operators. In this case transition operator C has a zero eigenvalue. Two other non-zero eigenvalues are describing the velocity of drift and the width of two Gaussian distributions.

8. Quantum Computing application

Recently, Verstraete *et al.* [17] proposed a dissipative model of quantum computing (DQC), capable of performing universal quantum computation. The dissipative quantum computing setup consists of a linear chain of time registers. Initially, the system is in a time register labeled by 0. The result of the computation is measured in the last time register labeled by T. Neighboring time registers are coupled to local baths. The result of the computation can be read out from the time-register T. In particular, for a quantum circuit given by the set of unitary operators $\{U_t\}_{t=1}^T$ the final state of the system is given by $|\psi_T\rangle = U_T U_{T-1} \ldots U_2 U_1 |\psi_0\rangle$. It is shown that the unique steady state of the system in this case will be

$$\rho = \frac{1}{T+1}\sum_t |\psi_t\rangle\langle\psi_t| \otimes |t\rangle\langle t|. \tag{35}$$

Recently, it was shown that, using the formalism of OQWs one can perform dissipative quantum computations with higher efficiency than the original scheme of DQC. The explicit open quantum walk implementation of the Toffoli gate and the Quantum Fourier Transform has been reported [13]. If one needs to implement a circuit with a set of unitaries so that initial and final states are connected as $|\psi_T\rangle = U_T U_{T-1} \dots U_2 U_1 |\psi_0\rangle$, then the corresponding OQW implementation will be given by a linear chain of $T+1$ nodes with corresponding iteration formula:

$$\rho_j^{[n]} = \omega U_j \rho_{j-1}^{[n-1]} U_j^\dagger + \lambda U_{j+1}^\dagger \rho_{j+1}^{[n-1]} U_{j+1}, j = 1 \dots (T-1), \tag{36}$$

$$\rho_0^{[n]} = \lambda \rho_0^{[n-1]} + \lambda U_1^\dagger \rho_1^{[n-1]} U_1, \tag{37}$$

$$\rho_T^{[n]} = \omega \rho_T^{[n-1]} + \omega U_{T-1} \rho_{T-1}^{[n-1]} U_{T-1}^\dagger, \tag{38}$$

where the constants ω and λ are positive constants such that $\omega + \lambda = 1$.

In this case the unique steady state of the OQW will have the following form,

$$\rho_{SS} = \sum_{i=0}^{T} p_i |\psi_i\rangle\langle\psi_i|, \tag{39}$$

where p_i denote the steady state probabilities of detecting the walker at the node i. In the case of $\omega = \lambda$ all probabilities of detection will be the same, namely $p_i = 1/(T+1)$. In the case $\omega > \lambda$, the probability of detection of the walker at site T will be bounded between $1/(T+1) < p_T < 1$.

As an example of application of OQW implementation of dissipative quantum computing model we consider a phase estimation algorithm with an unknown unitary operator acting on 4-qubit Hilbert Space [18]. The quantum phase estimation algorithm is implemented through a sequence of Hadamard operations (1 unitary), a conditional unknown unitary gate on each ancilla qubit (4 unitaries) and a 4-qubit inverse Quantum Fourier Transform (15 unitaries). In summary, the 4-qubit phase estimation algorithm is implemented with help of 21 unitary operation. This means that an equivalent OQW scheme will contains 22 nodes (from node 0 for initial conditions to node 21 for the results of computation). Fig. 5(a) shows the dependence of the probability of successful performance of the phase estimation as a function of the number of steps of the walk. Curves (1)-(4) in Fig. 5(a) correspond to different values of the parameter $\omega = 0.5, 0.6, 0.8, 0.9$, respectively. Curve (1) corresponds to the case $\omega = 0.5$ which is the conventional dissipative quantum computing model. In Fig. 5(b) we analyse the necessary number of steps to reach the steady state and the success probability of measurement as a function of the parameter ω. We observe that with increasing ω the number of steps to reach the steady state is decreasing and the probability of successful detection is increasing.

9. Conclusion

In this paper we review the concept of Open Quantum Walks. We briefly present the formalism, give an example of the OQW on $\mathbb{Z}$ and apply unravelling procedure to this example. We simulate a particular case of an open quantum walk by means of quantum trajectories, give an example of quantum trajectories and perform an average over 10^5 realisations. We summaries *realisation procedure* and demonstrate a connection of OQWs to unitary quantum walks and classical random walks. We briefly mention the results of central limit theorems and show the OQW implementation of the dissipative quantum computing. As a particular example we present OQW implementation of the 4-qubit phase estimation algorithm.

This work is based upon research supported by the South African Research Chair Initiative of the Department of Science and Technology and National Research Foundation.

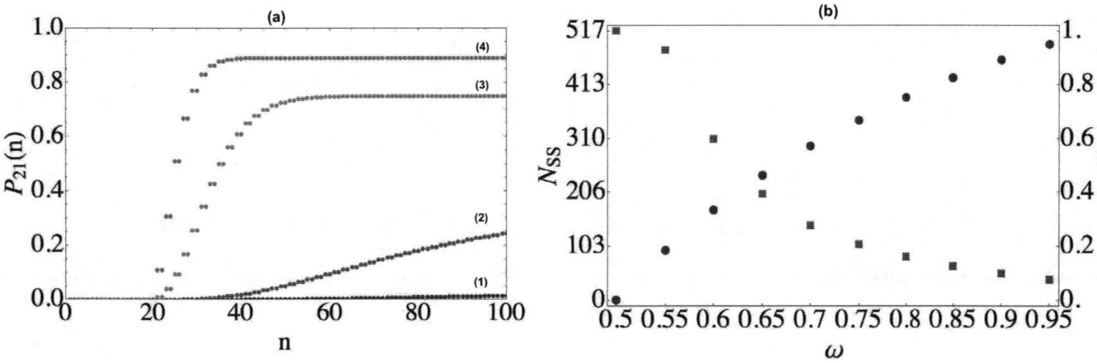

Figure 5. Simulation of the efficiency of the OQW implementation of 4-qubit phase estimation algorithm; (a) shows the dynamics of the detection probability in the final node 21 as function of the number of steps of the OQW. Curves (a1) to (a4) correspond to different values of the parameter $\omega = 0.5, 0.6, 0.8, 0.9$, respectively; (b) shows the number of steps needed to reach the steady state (squares) and the probability of detection of the successful implementation of the quantum algorithm (circles) as function of the parameter ω. The number of steps to reach a steady states is simulated with 10^{-6} accuracy.

References

[1] Feller W 1968 *An Introduction to Probability Theory and its Applications*, Vol. 1 (New York: Wiley)
[2] Barber M and Ninham B W 1970 *Random and Restricted Walks: Theory and Applications* (New York: Gordon and Breach)
[3] Motwani R and Raghavan P 1995 *Randomized Algorithms* (Cambridge: Cambridge University Press)
[4] Malkiel B 1973 *A Random Walk Down Wall Street* (New York: W. W. Norton)
[5] Berg H C 1993 *Random Walks in Biology* (Princeton: Princeton University Press)
[6] Aharonov Y, Davidovich L and Zagury N 1993 *Phys. Rev.* A **48** 1687
[7] Kempe J 2003 *Contemp. Phys.* **44** 307
[8] Childs AM 2009 *Phys. Rev. Lett.* **102** 180501
[9] Lovett NB, Cooper S, Everitt M, Trevers M and Kendon V 2010 *Phys. Rev.* A **81** 042330
[10] Breuer H-P and Petruccione F 2002 *The Theory of Open Quantum Systems* (Oxford: Oxford University Press)
[11] Attal S, Petruccione F and Sinayskiy I 2012 *Phys. Lett.* A **376** 1545
[12] Attal S, Petruccione F, Sabot C and Sinayskiy I 2012 *J. Stat. Phys.* **147** 832
[13] Sinayskiy I and Petruccione F 2012 *Quant. Inf. Proc.* **11** 1301-1301
[14] Attal S, Guillotin - Plantard N and Sabot C 2012 Central Limit Theorems for Open Quantum random Walks *Preprint* math.PR/1206.1472
[15] Konno N and Yoo HJ 2013 *J. Stat. Phys.* **150** 299
[16] Sinayskiy I and Petruccione F 2012 *Phys. Scr.* **T151** 014077
[17] Verstraete F, Wolf MM and Cirac JI 2009 *Nature Phys.* **5** 633
[18] Nielsen MA and Chuang IL 2000 *Quantum Computation and Quantum Information* (Cambridge: Cambridge University Press)

Entang-bling: Observing quantum correlations in room-temperature solids.

I. A. Walmsley, K. C. Lee, M. Sprague, B. Sussman, J. Nunn, N. Langford, X.-M. Jin, T. Champion, P. Michelberger, K. Reim, D. England and D. Jaksch.

University of Oxford, Department of Physics, Clarendon Laboratory, Parks Rd. Oxford, OX3 0BU

walmsley@physics.ox.ac.uk

Abstract. Quantum entanglement in the motion of macroscopic solid bodies has implications both for quantum technologies and foundational studies of the boundary between the quantum and classical worlds. Entanglement is usually fragile in room-temperature solids, owing to strong interactions both internally and with the noisy environment. We generated motional entanglement between vibrational states of two spatially separated, millimeter-sized diamonds at room temperature. By measuring strong nonclassical correlations between Raman-scattered photons, we showed that the quantum state of the diamonds has positive concurrence with 98% probability. Our results show that entanglement can persist in the classical context of moving macroscopic solids in ambient conditions.

1. Macroscopic quantum correlations

Our intuition about the nature of the physical world is strongly conditioned by the experience that macroscopic bodies move according to the rules of classical mechanics. Quantum theory, however, asserts that superpositions and entanglement are possible even for large objects. Therefore, exploration of the persistence of quantum correlations in the traditionally classical realm is important for both fundamental science and technology, because of the implications for physics beyond conventional quantum mechanics and for quantum information processing, which requires sustained coherence across many particles.

A long-standing challenge is to reveal quantum features in "ordinary" materials under ambient conditions, without a particularly exotic preparation or engineering. The requirements for this are that any experiment must satisfy two conditions. First that it take place on a timescale much shorter than the decoherence time τ of the degree of freedom in which we expect to observe the correlations. Second that it should take place on a much shorter timescale $\tau' = \tau/n$ than it takes for the degree of freedom to begin to thermalize with its environment, where n is the mean number of excitations in the bath (say a set of bosonic modes) that are coupled to the degree of freedom of interest and cause the decoherence.

Since τ is small in many solid materials – typical electronic and inter-atomic excitations decays occur on femto- to pico-second timescales – we require a particularly high frequency degree of freedom so that any bath will have a very low thermal occupancy, and thus $\tau' = \tau$. Optical phonons provide such a degree of freedom. A number of very tightly bonded crystals have appropriate phonon frequencies and decay times that are accessible using modern ultrafast laser techniques. This allows us both to prepare and to measure single quanta of vibrational excitation.

2. Preparation of nonclassical vibrational states in diamond

Diamond is a particularly promising material for the study of room temperature quantum vibrational states since it has an optical phonon branch with phonon energies up to 1332 cm^{-1}, with lifetimes of the order of 10 ps. This combination means that the phonons can be non-impulsively excited with optical pulses of duration great than about 10 fs.

The principle behind the experiment is to use intense, brief, laser pulses to excite the diamond and set it ringing by means of Raman scattering. Raman scattering occurs when the light beam incident on a material is scattered into a different frequency, shifted from that of the incident light by a characteristic frequency of the material. This is typically a vibrational frequency in molecules or crystals. The effect arises from the coupling of the electronic and vibrational degrees of freedom of the material, leading to the dependence of the shift of the electronic resonance frequency of the optical transitions in the material on the amplitude of vibrations. Often Raman scattering is initiated by the thermal population of the material. Crystals, for example, usually have significant population in low frequency vibrational modes at room temperature. However, even at zero degrees Kelvin, when there are no material excitations, Raman scattering can still occur, due to vacuum fluctuations of the electromagnetic field. In that case, a photon from the incident light beam has its energy divided into a single down-shifted photon - the Stokes photon - and a single quantum of excitation of the material. Thus the number of Stokes photons and the number of material quanta are tightly correlated.

We used a modelocked Ti:Sapphire laser, producing 80 fs duration pulses at 805 nm mean wavelength, to excite optical phonons near the center of the Brillouin zone of a diamond crystal. Diamond has a particularly large Raman scattering cross-section, and an electronic band gap of several eV, making it well suited to optical manipulation. During scattering, the diamond absorbs some energy from the laser pulse and begins to vibrate. The photon emerging from the diamond, called the Stokes photon, is red-shifted, with less energy than an incident laser pulse photon. The detection of a Stokes photon signals that an optical phonon has been created in the diamond.

Similarly, the reverse process, whereby a photon from the incident light beam gains energy from the material to scatter into a higher-energy, shorter-wavelength photon - the anti-Stokes photon - may also occur, once the material has some excitation. The ratio of Stokes to anti-Stokes intensity provides therefore a means to measure the degree of excitation of the material. We next send in another laser pulse, and we look for an anti-Stokes photon — that is, a blue-shifted Raman scattered photon. Seeing this second photon confirms that we initially created a single phonon with the first pulse, since it could not be produced without a phonon in the diamond. In fact, there is a strict limit to how often two colors can be detected together, because, according to classical physics waves cannot be perfectly correlated, unless the Stokes and anti-Stokes are themselves highly auto-correlated. The quantum prediction is that for every blue photon, there is a red photon: 100% correlation, with zero auto-correlation.

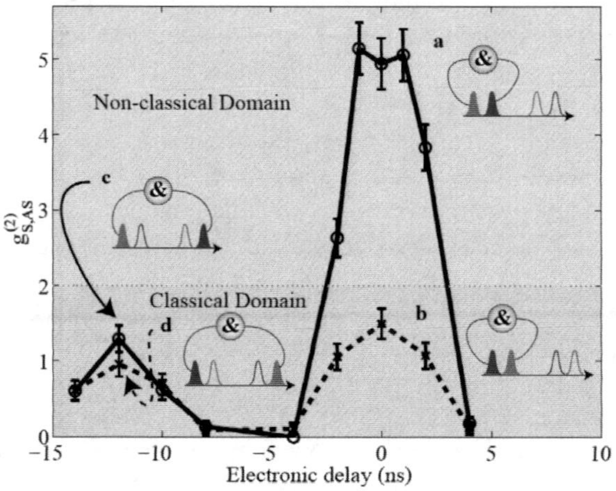

Fig 1. Nonclassical correlations between Stokes and anti-Stokes photons scattered from a diamond crystal. The peak near zero delay shows that it is more likely than allowed by classical physics that detection of a Stokes photon is succeeded by the detection of an anti-Stokes photon, indicating the generation of a single vibrational quantum in the diamond during the time between these pulses.

Of course, in practice, the correlations

are never perfect – detector noise, poor mode matching, background light all contribute to the signal. However, it was possible to show that the probability with which anti-Stokes photons were detected once a Stokes photon had been detected was much greater than any classical model would allow.

The outcome of a measurement of the Stokes/anti-Stokes correlations are shown in Fig. 1 [1]. The strength of the correlations is specified by the ratio $g^{(2)}_{S,aS}$ of the joint probability $P_{S,aS}$ for detecting both Stokes and anti-Stokes photons, to the product of the marginal probabilities for detection of either Stokes photons (P_S) or anti-Stokes photons (P_{aS}): $g^{(2)}_{S,aS} = \frac{P_{S,aS}}{P_S P_{aS}}$.

Classically, the magnitude of the correlation is bounded by the Cauchy-Schwartz inequality $g^{(2)}_{S,aS} \le \sqrt{g^{(2)}_{S,S} g^{(2)}_{aS,aS}}$, where the autocorrelations of the Stokes and the anti-Stokes fields appear under the radical. These fields are approximately thermal, with $g^{(2)}_{S,S} = g^{(2)}_{aS,aS} = 2$, so that measured values of $g^{(2)}_{S,aS}$ exceeding 2 are indicative of nonclassicality. This allows us to infer that the vibrational state generated by the scattering of a Stokes photon is itself nonclassical. Indeed, the results are consistent with it being a single vibrational quantum, as shown in Fig. 1.

3. Entangling the vibrational states of two diamonds

This approach can be extended to prepare a single phonon distributed across two diamonds. This "sharing" of a single quantum of excitation between the two spatially separated crystals is described by an entangled state.

This may be understood by an optical analog, and shown in Fig. 2(a). A single photon incident on a beam splitter may exit by either port. Since there is only a single photon, and the mode transformation is unitary, it ends up in a superposition: it has a probability amplitude for being found in either mode, with a definite phase relation between the two amplitudes. The combination of properties of a single excitation of the optical field (just one photon) distributed between the two paths (in a coherent superposition state of the two modes) is a sufficient condition to prove entanglement [2].

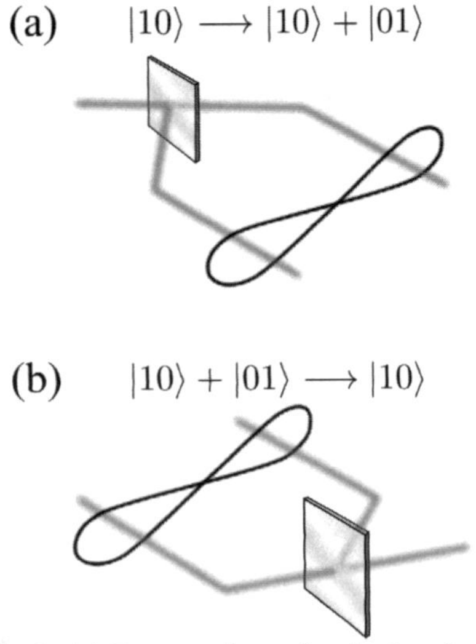

Fig 2. (a) Preparation of a single-photon entangled state by means of a beam splitter. (b) Analysis of a single-photon entangled state. If, on multiple repetitions of the process using identical states, the photon always exits through the same port of the beamsplitter, this proves that it was prepared in an entangled state.

This may be confirmed experimentally by reversing the process, illustrated in Fig 2.(b). If a single photon in a path superposition state is incident on the beam splitter, then the photon will always exit a particular port. Which port is determined by the phase difference between the two paths. This can be adjusted by inserting a phase shifter in one of the paths, so that the photon always exits a specific port. A photon that occupied either path at random (described by a mixed state) would exit through the two output ports at random, in an uncontrollable fashion.

Of course revealing path interference by adjusting phases between the two paths is not in itself a specifically quantum interference – two coherent states occupying the paths would exhibit the same phenomenon. But, when combined with a measurement of $g^{(2)}_{aS,aS}$ of the conditionally prepared anti-Stokes beam, this is enough to show entanglement, provided the photon statistics are sub-Poissonian. The density operator of the state in the photon number basis then shows ideally nothing in the zero and two photon sectors ($p_{00} = p_{22} = 0$), but full quantum coherence in the one photon sector ($p_{10}, p_{01} \neq 0$). This means that the concurrence of the state, defined as $C = 2\max\left(0, \left|V\left(p_{10} + p_{01}\right)/2\right| - \sqrt{p_{00}p_{22}}\right)$ is positive.

The same measurement strategy may be used to demonstrate the entanglement of two diamonds by means of a shared phonon. The arrangement is shown schematically in Fig. 3. First, a single phonon is generated in a pair of diamonds by illuminating them with two laser pulses generated by splitting a single input pulse on a beamsplitter. The spontaneous scattering of a single Stokes photon from the pair of incident laser pulses heralds the excitation of a single vibrational quantum. Erasing the which-path information of the Stokes photon by means of a beam splitter leaves that the diamond in which the phonon resides unknowable, even in principle. Thus the quantum state describing the diamonds is a superposition state in which the phonon is in one diamond and the other at the same time. The Stokes photon may be detected at either output port, and this varies randomly from shot to shot.

The state of the phonon in the two diamonds can be determined by "reading it out" using anti-Stokes Raman scattering. The single phonon is converted into a single photon by this means. If the anti-Stokes photon always exits the beamsplitter by the same port as the Stokes photon did, then this is sufficient proof that the diamonds were entangled, because it shows that the anti-Stokes photon was in an entangled state.

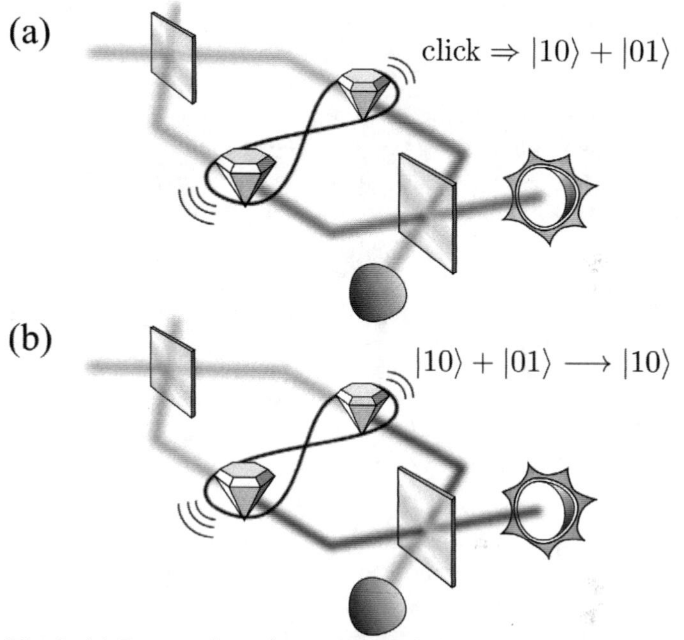

Fig 3. (a) Preparation of an entangled phonon state. Detection of a single Stokes photon heralds the preparation of a single vibrational quantum distributed between the two diamonds. (b) Measurement of the diamond phonon entanglement by means of anti-Stokes scattering. The registration of single anti-Stokes photons at the same output port as the Stokes photons were detected indicates the coherence of the two paths and the coherence of the vibrational state of the diamonds. This is sufficient to determine that the two are entangled.

Of course, experimental imperfections mean that this is never the case, but a measurement of the concurrence of the state of the anti-Stokes beam by the methods described above is possible, and shows that the anti-Stokes photon is entangled. Fig. 4 shows the experimentally measured density matrix of the anti-Stokes beam, conditioned on the detection of a single Stokes photon. The diagonal elements are direct measurements of the zero-, one- and two-photon probabilities (shown in blue, red and green, respectively). The one-photon sector off-diagonal elements are obtained from the visibility of the interference fringes obtained by adjusting the relative phase of the two anti-Stokes paths by means of a waveplate. The grey elements were unmeasured components of the density matrix. These are difficult to access for small photon number states (they require, for instance, an operation that can rotate a two-photon state into a one-photon state, for instance). However, for our purposes we can set them to zero and still obtain a lower bound on the degree to which the anti-Stokes beam is in an entangled state. If they are non-zero, the entanglement only increases [2].

Since we have already shown that the anti-Stokes photons are tightly correlated with the existence of a phonon in the diamonds, we can infer that the diamonds are entangled [3].

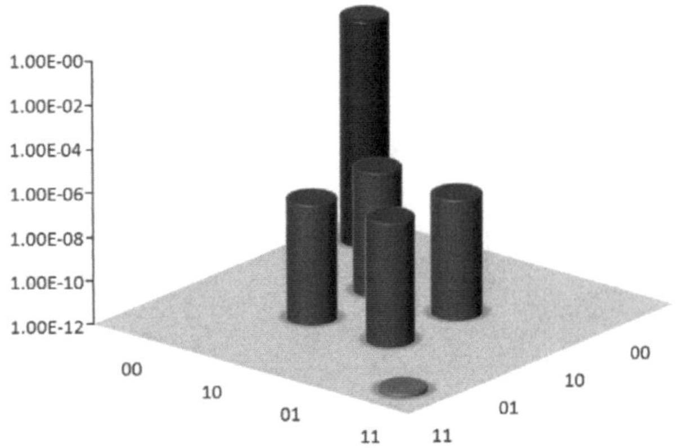

Fig. 4. Experimental measurements of the density matrix of the anti-Stokes beam. The zero- (blue), one- (red) and two- (green) photon probabilities show that the anti-Stokes photon statistics are sub-Poissonian. The off-diagonal elements of the one-photon sector indicate path coherence. The grey areas represent unmeasured components of the density matrix where the values are set to zero. A lower bound of the concurrence of the anti-Stokes beam, and thus the diamond state, can be calculated from this density matrix, and is positive. This demonstrates the entanglement of the two diamonds.

3.1. Collapse vs retrodiction

One feature of our experiment that can be puzzling is the fact that the time for the Stokes and anti-Stokes photons to propagate to the detector and thus to be registered is much longer than the lifetime of the phonons in the diamonds, which have decohered after less than 10 picoseconds.

In quantum physics, dynamics and measurements are generally decomposed into a set of temporally instantaneous operations (unitaries and projectors) acting globally on Hilbert space without explicit reference to space-time. Such analysis can lead to the seemingly straightforward notion that a measurement on part of a distributed system (say, the registration of a photon emitted from one of the diamonds at point D in Fig. 5) causes an instantaneous collapse of the global quantum state across all spatial positions at the same moment (point C in Fig. 5). However the idea that the registration of a Stokes photon "prepares" the single phonon state in the diamond by "collapsing" the wavefunction is problematic [3]. A causal effect would "generate" the entanglement at a later time within the light cone of point A – say point F in Fig. 5.

The answer to this enigma is well understood, but is nonetheless worth rehearsing since the untenable position of action at a distance implied by the collapse hypothesis is rather starkly illustrated by our apparatus. The collapse interpretation leads to some uncomfortable conclusions that are strongly at odds with a standard notion of causality. Specifically, in other inertial reference frames, such spatially separated "collapse" events would occur at different times. Indeed, it is always possible to find reference frames in which the quantum state at other locations collapses before the detection event itself takes place. In fact, measurements in quantum

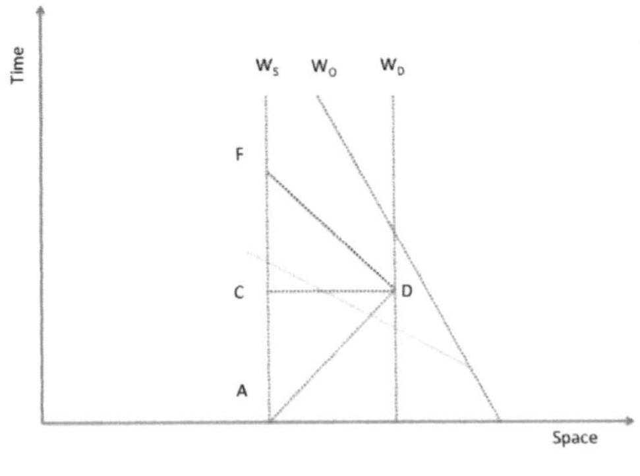

Fig. 5. World lines for the experiment, showing that the "collapse" hypothesis leads to the conclusion that in some reference frames in motion with respect to the laboratory frame, there are observers (green line) who see the generation of entanglement before the detection of any photons. Therefore it cannot be said that the detection "causes" the entanglement.

mechanics are not related to causality; projection operations update the state to enforce consistency with classical data. In general, a measurement projection can always be back-propagated to any arbitrary time during the system evolution [4]. In the case of photodetection, the physically significant time is the moment at which the photons were created, and the purpose of such a measurement is to provide information about the system at this instant. In our experiment, it is therefore natural to treat the propagation of signals by evolving the Stokes measurement operator backwards in time from the moment of detection, τ_0, (point D) to the time τ_S (point A) at which the Stokes photon was scattered. The resulting projection leaves the diamonds in an entangled state at τ_S, and it is this entanglement that is probed by anti-Stokes scattering at time $\tau_A = \tau_S + T$, where T is less than the decoherence time of the phonon.

4. Conclusions

We have experimentally demonstrated that two room-temperature solids can be prepared in an entangled state. In order to show this we combined ultrafast interferometric pump-probe scheme with photon-counting techniques to generate and prepare short-lived quantum correlations in a strongly decohering environment. This approach lays the foundation for future studies of quantum phenomena in many-body, strongly interacting systems in ambient conditions, which may prove to be helpful in understanding the emergence of the classical world from the quantum.

Acknowledgements

This work was undertaken at the University of Oxford with support from EPSRC, the Royal Society and the European Union. B. J. Sussman and D. England are currently at the National Research Council of Canada in Ottawa, K. C. Lee is at TeraView, Cambridge, and N. Langford is at Royal Holloway College, London.

References

[1] Lee K C, Sussman B J, Sprague M R, Michelberger P, Reim K F, Nunn J, Langford N K, Bustard P J, Jaksch D and Walmsley I A 2012 Macroscopic non-classical states and terahertz quantum processing in room-temperature diamond *Nat. Phot.* **6** 41 (2012)

[2] van Enk S J, Lutkenhaus N and Kimble H J 2007 On Experimental Procedures for Entanglement Verification *Phys. Rev.* A **75** 052318

[3] Lee K C, Sprague M R, Sussman B J, Nunn J, Langford N K, Jin X -M, Champion T, Michelberger P, Reim K F, England D, Jaksch D and Walmsley I A 2011 Entangling Macroscopic Diamonds at Room Temperature *Science* **334** 1253

[4] Barnett S M, Pegg D T , Jeffers Jand Jedrkiewicz O 2001 Master equation for retrodiction of quantum communication signals,*Phys. Rev. Lett.* **86** 2455

Journal of Physics: Conference Series **442** (2013) 012005 doi:10.1088/1742-6596/442/1/012005

In search of superluminal quantum communications: recent experiments and possible improvements.

B Cocciaro[1], **S Faetti**[2] **and L Fronzoni**[2]

[1] High School XXV Aprile, Via Milano 2, I-56025 Pontedera (Pisa), Italy
[2] Department of Physics Enrico Fermi, Largo Pontecorvo 3, I-56127 Pisa, Italy

E-mail: b.cocciaro@comeg.it, faetti@df.unipi.it, fronzoni@df.unipi.it

Abstract. As shown in the famous *EPR* paper (Einstein, Podolsky e Rosen, 1935), Quantum Mechanics is non-local. The Bell theorem and the experiments by Aspect and many others, ruled out the possibility of explaining quantum correlations between entangled particles using local hidden variables models (except for implausible combinations of loopholes). Some authors (Bell, Eberhard, Bohm and Hiley) suggested that quantum correlations could be due to superluminal communications (tachyons) that propagate isotropically with velocity $v_t > c$ in a preferred reference frame. For finite values of v_t, Quantum Mechanics and superluminal models lead to different predictions. Some years ago a Geneva group and our group did experiments on entangled photons to evidence possible discrepancies between experimental results and quantum predictions. Since no discrepancy was found, these experiments established only lower bounds for the possible tachyon velocities v_t. Here we propose an improved experiment that should lead us to explore a much larger range of possible tachyon velocities v_t for any possible direction of velocity $\vec{V}$ of the tachyons preferred frame.

1. Introduction

The non local character of Quantum Mechanics (QM) has been object of a great debate starting from the famous Einstein-Podolsky-Rosen (EPR) paper [1]. Consider, for instance, a quantum system made by two photons a and b that are in the polarization entangled state

$$|\psi>= \frac{1}{\sqrt{2}}\left(|H,H> + e^{i\phi}|V,V>\right) \tag{1}$$

where H and V stand for horizontal and vertical polarization, respectively, and ϕ is a constant phase coefficient. The two entangled photons are created at point O, propagate in space far away one from the other (see Fig.1) and reach at the same time points A and B that are equidistant from O as schematically drawn in Fig.1.

Suppose the polarization of the two photons is measured at the same time at points A and B. According to QM, a measurement of horizontal polarization of photon a (or b) leads to the collapse of the entangled state to $|H,H>$, then, also photon b (or a) must collapse to the horizontal polarization. This behaviour suggests the existence of a sort of "action at a distance" qualitatively similar to that introduced in the past to describe interactions between either electric charges or masses. However, according to the Maxwell electromagnetic theory and to the Einstein General Relativity theory, it is now commonly accepted that interactions between electric charges or masses are not instantaneous but occur through exchange of signals

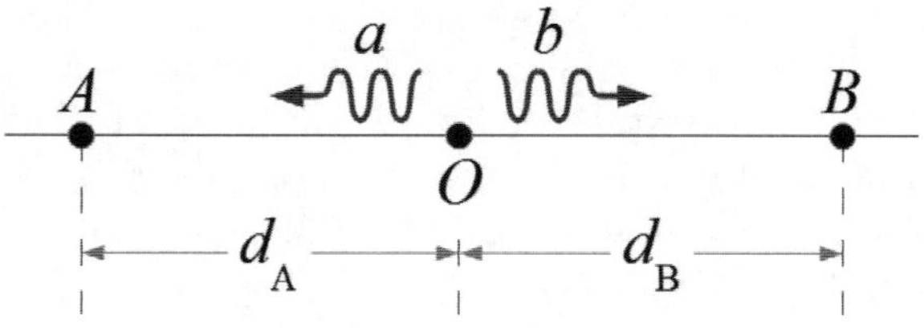

Figure 1: O: point where a couple of entangled photons (a and b) are created; A and B: points equidistant from O ($d_A = d_B$) where the polarization of the entangled photons is measured.

(photons or gravitons). This means that classical physical phenomena are described by local models. Many physicists are unsatisfied of the non local character of QM and alternative local models have been proposed to explain quantum correlations. The simplest way to explain the non local aspects of QM as well as its probabilistic behaviour is to assume a lack of complete information about the actual state of the system in analogy with what happens for thermodynamic systems. Then, one could explain the probabilistic behaviour predicted by QM as due to a not complete knowledge of all influences affecting the system (*hidden variables*). In particular, the polarizations of the entangled photons a and b would be already well defined (by their hidden variables) when they are created at point O. However, as shown by Bell [2] and other authors [3, 4], the correlations between entangled particles predicted by any theory based on local hidden variables must satisfy some inequalities that are not satisfied by QM. The existence of these inequalities permits to perform experiments to decide unambiguously between hidden variables theories and QM. Many experiments of this kind have been performed before and after the famous Aspect experiments [5–14]. All the experiments (except a few old ones [15, 16], see cap. 11 of [17] for a detailed bibliography) demonstrated that the Bell inequalities are violated. Although the locality loophole and the detection loophole have not yet completely closed using a single experimental apparatus [18], experiments have separately closed both the locality loophole [6–10] and the detection loophole [12–14]. Then, it is reasonable to think that hidden variables alone cannot justify the experimentally observed correlations (except if implausible combinations of loopholes are supposed to exist). For many classical systems, correlations between two events are often explained as the consequence of communications and, thus, also quantum correlations between entangled particles could be due to some communication. However, the Aspect experiments and many other *EPR* experiments were performed in space-like conditions and, thus, if Quantum correlations would be due to communications, the communications velocity should exceed the light velocity c. For this reason, after the Aspect experiment, Bell said "*in these EPR experiments there is the suggestion that behind the scenes something is going faster than light*" [19]. Successively, well defined models for QM based on the presence of superluminal communications have been proposed [20, 21]. The possibility of the existence of particles going faster than light (*tachyons*) has been proposed some years ago by some authors [22–24]. Tachyons are known to lead to causal paradoxes [25] (the present occurring in a given point can be affected by the future occurring in the same point). Consider, for instance, a first person that exits from his home and is wetted by rain. He could send a tachyon to inform a second person that send a replay tachyon that is received by the first person before he exits from the home. Then, he could decide to get an umbrella to be not wetted by the rain. Obviously such a behaviour is unrealistic. However, no causal paradox arises if tachyons are supposed to propagate in a preferred frame where the tachyon velocity $v_t = \beta_t c$ ($\beta_t > 1$) is the same in all directions (see, for instance, [26–30]). Note that also in this case the Special Relativity predicts that a tachyon that is emitted at point A at the local time $t = 0$ can get point B at time $t < 0$. However, this does not represent a paradoxical result since the time ordering between separated points of the space has no direct physical meaning. Indeed, the time ordering between events occurring in different space points is conventional and depends on the conventional procedure used to synchronize far clocks. We assume here the conventionalist thesis on the synchronization of distant clocks [31]. It is true that we cannot yet consider closed the debate on this topic [32], but we believe that the conventionalist thesis is the correct one.

Consider, now, two photons that are in the entangled state of Eq.(1) and suppose that photon a passes through a polarizing filter with horizontal polarization axis. According to the superluminal models of QM, when photon a passes through the polarizing filter, it collapses to the horizontally polarized state, then a tachyon is sent to the entangled photon b that collapses to the horizontally oriented state only after this communication has been received. Therefore, the QM correlations between entangled photons can be recovered only if it has been sufficient time to communicate between the two entangled photons. Consider, for instance, an ideal experiment performed in the tachyon preferred frame S' where two polarizing filters lie at points A and B at the same optical distances $d'_A = d'_B$ from source O of the entangled photons as shown in Fig.1. In these conditions photons a and b get both polarizers at the same time (in the PF) and, thus, if the tachyon velocity in the PF has a finite value, no communication is possible and correlations between entangled particles should differ appreciably from the predictions of QM. Of course, there is always communication if the tachyon velocity is $v_t \to \infty$, then no experiment satisfying QM can invalidate the superluminal models provided $v_t \to \infty$. In such a case superluminal models are completely equivalent to QM as occurs for the Bohm model [33, 34]. It has been recently shown [35, 36] that, if QM correlations are due to superluminal signals with finite velocity v_t, then superluminal signalling becomes possible, that is communications at faster than light velocity must be possible at a macroscopic level and they cannot be confined to "hidden" variables.

From the experimental point of view, equality $d'_A = d'_B$ can be only approximatively verified within a given uncertainty $\Delta d'$. Consequently, photons a and b could get the polarisers at two different times ($\Delta t' = {}^{\Delta d'}/_{c}$) and, thus, they could communicate only if the tachyon velocity exceeds a lower bound $v_{t,min} = {}^{d'_{AB}}/_{\Delta t'}$ where d'_{AB} is the distance between points A an B in the PF. The possible results of such an experiment are: *i)* a lack of quantum correlations is observed; *ii)* quantum correlations are always satisfied. In the first case (*i)*) one can conclude that orthodox QM is not correct and that quantum correlations are due to exchange of superluminal messages. By suitably changing distances d'_A and d'_B one could obtain a measure of the tachyon velocity v_t. In the second case (*ii)*), due to the experimental uncertainty, one cannot invalidate the tachyon model of QM but can only establish a lower bound $v_{t,min}$ for the tachyon velocity. So far we assumed that the experiment is carried out in the preferred frame, but velocity vector $\vec{V}$ of PF is unknown and, thus, the experiment cannot be performed in this frame. It will be shown in Section 2 that this drawback can be bypassed performing the experiment on the Earth with the A-B axis aligned along the West-East direction. Such an experiment could provide both the tachyon velocity v_t and the the velocity vector $\vec{V}$ of the PF.

A long-distance (10.6 km) EPR experiment to detect possible effects of superluminal quantum communications has been performed by Scarani et al. [37] using energy-time entangled photons. No deviation from the predictions of QM was observed and, thus, the authors obtained only a lower bound for the tachyon velocities in the PF. The experimental results were analyzed under the assumption that the preferred frame is the frame of cosmic microwave background radiation. With this assumption, the authors obtained a lower bound $v_{t,min} = 1.5 \times 10^4\, c$. Successively, similar long-distance measurements have been performed by Salart et al. [38] improving some features of the previous experiment and using detectors aligned close to West-East direction (at angle $\alpha = 5.8°$). The authors found a lower bound for the tachyon velocity for many different possible directions of velocity $\vec{V}$ of the PF. More recently [39] we performed EPR measurements on polarization entangled photons in a laboratory experiment with small distances $d_A \approx d_B \approx 1\,\text{m}$ and with the AB axis in Fig.1 precisely aligned along the West-East direction ($|\alpha| < 0.2°$). In our experiment, too, no deviation from the predictions of QM was found and, thus, we obtained only a lower bound for the tachyon velocity. The choice of aligning the measurement points A and B just along the West-East direction allowed us to obtain a lower bound for the tachyon velocity for any possible orientation of the velocity of the preferred frame. The experiment in [38] and in [39] are somewhat complementary. Indeed, the Salart et al. experiment used very large distances ($d_A \approx d_B \approx 10\,\text{km}$) but somewhat large acquisition times ($\delta t \approx 360\,\text{s}$), while our experiment used shorter distances ($d_A \approx d_B \approx 1\,\text{m}$) but much shorter acquisition times ($\delta t \approx 4\,\text{s}$). With these features, the Salart et al. experiment was much more sensitive to tachyons propagating in PF that have velocities much smaller than the light velocity whilst our experiment was more sensitive to PF travelling at higher velocities (see Fig.4). As will be shown in Section 2, an EPR experiment similar to our previous experiment but with entangled photons that propagate in air at much larger distances (of order of 1 km) and with much smaller acquisition times (of order of 0.1 s) could increase greatly the range of detectable tachyon velocities. To get this goal, we plan to perform such an experiment inside the long galleries of the EGO structure (European Gravitational Observatory) [40]. In this paper, we will analyse the main features of the proposed experiment. The possible results of this experiment are either the detection of possible discrepancies between experiment and QM due to a finite velocity superluminal communication or the increase by about two orders of magnitude of the actual lower bounds for the tachyon velocities.

The main features of the experiment are discussed in Section 2. The critical points and the main sources of experimental uncertainty are discussed in Section 3. Finally, the conclusions are given in Section 4.

2. The main features of the experimental method.

Consider the geometry of Fig.1. We start considering the ideal case where the experiment is performed in the preferred frame S' with tachyons that propagate with the same velocity $v_t = \beta_t c$ ($\beta_t > 1$) along any direction [38]. For simplicity, in the following, we will call "tachyon velocity" the reduced velocity $\beta_t = \frac{v_t}{c}$. Two polarizing filters lie at points A and B aligned along a x'-axis and at optical distances d'_A and d'_B from the source O of the entangled photons (the apostrophe denotes the parameters measured in the PF). The two entangled photons will get both the polarisers at the same times $t'_A = t'_B$ if $d'_A = d'_B$ and, thus, no superluminal communication can be possible if the tachyon velocity has a finite value. Due to the experimental uncertainty $\Delta d'$, distances d'_A and d'_B can never be exactly equalized and $|\Delta t'| = |t'_A - t'_B| = {}^{\Delta d'}/_{c} \neq 0$. In these conditions, a superluminal communication between points A and B will be not possible only if the time that a tachyon spend to go from A to B (or from B to A) is greater than $|\Delta t'|$, that is if the tachyon velocity β_t is smaller than

$$\beta_{t,min} = \frac{d'_{AB}}{|\Delta ct'|}, \tag{2}$$

where d'_{AB} is the distance between points A and B. QM correlations will be always established if the tachyon velocity β_t is higher than $\beta_{t,min}$ of Eq.(2) but appreciable differences between QM correlations and experimentally observed correlations is expected if $\beta_t < \beta_{t,min}$. In this latter case, the experimental correlations should satisfy the Bell inequality. Therefore, to detect possible discrepancies between experiment and QM due to the finite velocity of tachyons, one has to increase the lower limit in Eq.(2) as much as possible to satisfy condition $\beta_t < \beta_{t,min}$. This goal could be achieved either reducing $\Delta d' = |\Delta ct'|$ or increasing distance d'_{AB} between the polarisers. If discrepancies between QM and experiment would be found for given values of $\Delta d'$ and d'_{AB}, one could conclude that the tachyon velocity is lower than the lower bound in Eq.(2). In such a case, one could measure the tachyon velocity β_t changing distances d'_A, d'_B and d'_{AB} to reduce $\beta_{t,min}$ of Eq.(2) until condition $\beta_{t,min} < \beta_t$ is verified and QM correlations are re-established. The tachyon velocity β_t would correspond to this critical value of $\beta_{t,min}$.

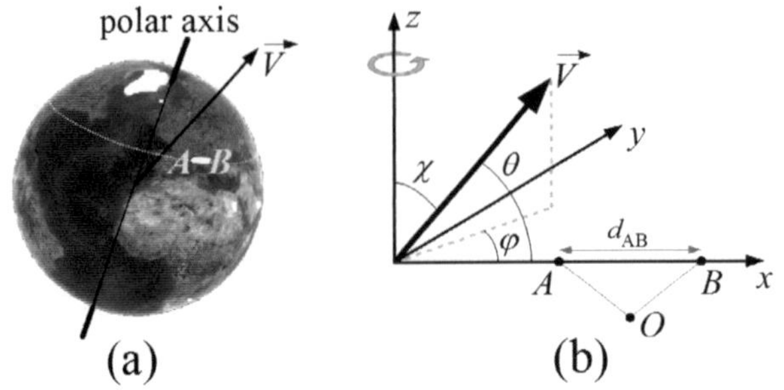

Figure 2: (a) geometry of the experiment with segment AB oriented along the West-East direction on the Earth; (b) detail of the geometric parameters that characterize the experiment. z is the North-South axis and x is the West-East axis. $\vec{V}$ is the velocity vector of the PF with respect to the laboratory, θ denotes the angle of $\vec{V}$ with the x-axis, φ is the azimuthal angle and χ is the polar angle. The polarizing filters that collect the entangled photons lie at points A and B aligned along the x-axis at the same optical distances from the source O of the entangled photons.

The results above were obtained assuming that the experiment is performed in the tachyons PF. Of course, this is not possible because we do not know what is the direction and the magnitude of the PF velocity $\vec{V} = \vec{\beta}c$. However, this apparent difficulty can be overcome if one takes advantage of the rotation motion of the Earth around its axis and aligns points A and B along the West-East x-axis on the Earth as shown in Fig.2. Suppose that points A and B are precisely aligned along the x-axis and that optical distances d_A and d_B of A and B from the entangled photon source O have the same values on the Earth frame. In this condition, the two entangled photons get the two polarizing filters at the same times t_A and t_B in the Earth frame but the arrival times t'_A and t'_B can be different in the tachyon PF. Due to the rotation of the Earth with the angular velocity ω, angle θ between the the velocity $\vec{V}$ of the PF and the x-axis (West-East direction) changes periodically with time t according to the simple law:

$$\theta(t) = \arccos\left[\sin\chi \cos\omega\left(t - t_0\right)\right], \tag{3}$$

where t_0 is the unknown time which gives $\varphi(t) = 0$ in Fig.2. Angle θ oscillates periodically between a minimum value $\pi/2 - \chi$ ($\varphi(t)$=0 in Fig.2) and a maximum value $\pi/2 + \chi$ ($\varphi(t)$=π in Fig.2). Then, whatever is the orientation of the velocity vector $\vec{V}$ of the PF, there are two times t_1 and t_2 during each sidereal day where $\vec{V}$ becomes perpendicular to the West-East x-axis. At these two times, according to the Special Relativity, the distances of A and B from O are equal also in the PF frame ($d'_A = d'_B$) and, thus, the arrival of the entangled photons at points A and B is simultaneous also in the PF. Then, deviations of correlations from the predictions of the QM should be observed at the special times t_1 and t_2 if the optical paths of the entangled photons would be equal. Of course, due to the experimental uncertainty Δd on the equalization of the optical paths, deviations from the predictions of QM could be only observed if the tachyon velocity β_t is lower than a lower bound $\beta_{t,min}$. Furthermore, the need of performing the experiment in the Earth frame introduces also another source of uncertainty because vector $\vec{V}$ becomes orthogonal to the West-East axis only at two well defined times t_1 and t_2. However, to detect a statistically significant number of coincidences of entangled photons, a sufficiently long acquisition time δt is needed. Also if this acquisition time is centred around the special times t_1 and t_2, the velocity vector $\vec{V}$ does not remain exactly perpendicular to the x-axis during the whole acquisition time. Therefore, the finite acquisition time leads to an adjunctive uncertainty on the equality of the optical paths with a consequent decreasing of the value of $\beta_{t,min}$. In conclusion, two main parameters determine the lower limit of $\beta_{t,min}$ obtainable with

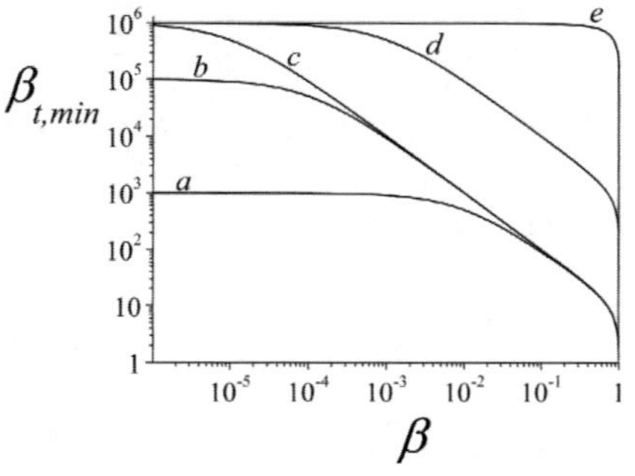

Figure 3: The typical plot of function $\beta_{t,min}$ versus β is shown for the unfavourable case $\chi = {}^{\pi}/_{2}$ and for some values of the experimental parameters $\bar{\rho}$ and δt. Curves a, b and c correspond to the fixed acquisition time $\delta t = 10^{-1} \times \frac{T}{\pi}$ and to decreasing values of $\bar{\rho}$ ($a : \bar{\rho} = 10^{-3}$, $b : \bar{\rho} = 10^{-5}$, $c : \bar{\rho} = 10^{-6}$). Curves c, d and e correspond to a fixed value $\bar{\rho} = 10^{-6}$ and to decreasing values of δt ($c : \delta t = 10^{-1} \times \frac{T}{\pi}$, $d : \delta t = 10^{-3} \times \frac{T}{\pi}$, $e : \delta t = 10^{-7} \times \frac{T}{\pi}$). Note that curve e satisfies condition (7) and, thus, it depends on β only for PF moving at improbable relativistic velocities.

the measurements on the Earth frame: 1) the accuracy Δd on the equalization of the optical paths; 2) the finite acquisition time δt. The lower bound $\beta_{t,min}$, defined by Eq.(2), can be rewritten in terms of physical parameters measured in the Earth frame and becomes [38, 39]:

$$\beta_{t,min} = \sqrt{1 + \frac{(1-\beta^2)\,[1-\bar{\rho}^2]}{\left[\bar{\rho} + \beta \sin\chi \sin\frac{\pi\delta t}{T}\right]^2}}, \tag{4}$$

where $\bar{\rho} = \frac{\Delta d}{d_{AB}}$, T is the sidereal day, δt is the acquisition time, χ is the polar angle between the North-South axis of the Earth and velocity $\vec{V}$ of the PF (see Fig.2) and β is the reduced velocity ($\beta = {}^{V}/_{c}$) of the PF. A quick way to obtain (4) is shown in the Appendix. In typical experimental conditions [38, 39], the acquisition time δt is much smaller than the sidereal day T and $\beta_{t,min}$ is a decreasing function of both $\bar{\rho}$ and δt and reaches a minimum value if $\chi = {}^{\pi}/_{2}$. Our following considerations and figures will be restricted to $\delta t \ll T$ and to the unfavourable condition $\chi = {}^{\pi}/_{2}$. $\beta_{t,min}$ is also a decreasing function of β that assumes its maximum value $\beta_{t,min} = \frac{1}{\bar{\rho}}$ for $\beta = 0$ and approaches the minimum value $\beta_{t,min} = 1$ for $\beta \to 1$. The typical plot of function $\beta_{t,min}$ versus the reduced velocity β of the PF for $\chi = {}^{\pi}/_{2}$ is drawn in Fig.3 for some values of the experimental parameters $\bar{\rho}$ and δt. $\beta_{t,min}$ keeps an almost constant value $\beta_{t,min} \approx \frac{1}{\bar{\rho}}$ for $\beta \ll \beta_0 = {}^{\bar{\rho}T}/_{(\pi\delta t)}$, then it decreases rapidly as β increases above β_0. Eq.(4) and Fig.3 evidence what is the optimal strategy to increase $\beta_{t,min}$ as much as possible. First of all we must reduce $\bar{\rho}$ minimizing the uncertainty Δd and increasing distance d_{AB} as far as possible. Then, we must use an acquisition time δt sufficiently small to render negligible the δt-contribution in Eq.(4). This contribution is always negligible if:

$$\delta t \ll \bar{\rho}\frac{T}{\pi}. \tag{5}$$

If condition (5) is satisfied, the lower limit $\beta_{t,min}$ becomes insensitive to the acquisition time δt.

In the Geneva experiment [38] the very small value $\bar{\rho} = 5.4 \times 10^{-6}$ was obtained using Telecom optical fibres connecting two Telecom stations at a distance of about 20 km and aligned approximatively along the West-East direction. The experiment was performed using energy-time entangled photons. The two Telecom stations (A and B) were not exactly aligned along the West-East axis but did the angle $\gamma = 5.8°$ with this axis and, thus, the experiment was poorly sensitive to tachyons PF travelling with velocity $\vec{V}$ lying in a cone of aperture $\approx 6°$ around the North south axis. Indeed, due to the Earth rotation, the angle between the AB axis and the velocity vector $\vec{V}$ oscillates with time between the minimum value ${}^{\pi}/_{2} - \chi - \gamma$ and the maximum value ${}^{\pi}/_{2} + \chi - \gamma$. Then, the AB axis can become orthogonal to $\vec{V}$ only if $\chi \geq \gamma$. In this long path experiment the main effect that limited a further reduction of $\bar{\rho}$ was due to the appreciable optical dispersion of the photon wave-packet in the long Telecom fibres. On the other hand, due to losses in the optical fibres, the photons count rate was relatively

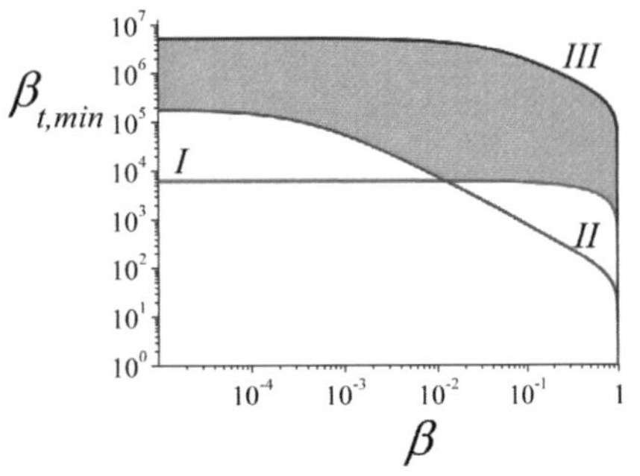

Figure 4: The three curves *I* (red), *II* (green) and *III* (blue) correspond to the experimental results for the lower bound $\beta_{t,min}$ versus β in different experiments. *I*: our previous results ($\bar{\rho} = 1.6\times10^{-4}$, $\delta t = 4$s) ; *II*: the Geneva group results ($\bar{\rho} = 5.4\times10^{-6}$, $\delta t = 360$s) ; *III*: the predicted results for the experiment proposed here ($\bar{\rho} = 1.9\times10^{-7}$, $\delta t = 0.1$s). The gray region represents the new region of tachyons velocities β_t that would become accessible with the new experiment.

small and a somewhat long measurement time ($\delta t \approx 360\,$s) was needed to obtain a statistically significant number of coincidences. In these conditions the experiment was greatly sensitive to tachyons *PF* travelling at speeds much smaller than that the light velocity and much less sensitive to relativistic *PF* (see curve *II* in Fig.4). The Pisa experiment [39] was performed with entangled photons that propagate in air over much shorter distances (about 2 m) and the polarization correlations were measured instead of energy-time correlations. In these conditions photon losses were minimized and the typical acquisition time was $\delta t \approx 4\,$s. An interferometric method was used to minimize the uncertainty on the equalization of the photons optical paths and the main source of uncertainty Δd was due to the 220 μm thickness of the absorbing layer of the polarizing filters. The obtained value of parameter $\bar{\rho}$ was $\bar{\rho} = 1.6\times10^{-4}$ that is about 30 times higher than the value characterizing the experiment of [38]. Then, our experiment is much less sensitive than that in [38] to tachyons propagating in a low speed *PF* but is more sensitive to tachyons propagating in high speed *PF* (see curve *I* in Fig.4). In this sense, our experiment was somewhat complementary to that of the Geneva Group. In the present paper we propose a new experiment exploiting the propagation of polarized entangled photons in air at distances about 850 times longer than in our previous experiment. The experiment should be performed inside the long tunnels of the European Gravitational Observatory (*EGO*) that host the *VIRGO* experiment on the detection of gravitational waves [41]. Using an interferometric method and feedback procedures we will hold the uncertainty Δd on the equality of the optical paths of the entangled photons well below the basic uncertainty due to the thickness of the absorbing layer of the polarizing filters (220 μm). The small uncertainty and the long optical paths will allow us to decrease our previous value $\bar{\rho} = 1.6\times10^{-4}$ up to the much smaller value $\bar{\rho} = 1.9\times10^{-7}$. As shown in the discussion above, the lower bound for the tachyon velocities is also determined by the acquisition time δt that must satisfy the condition that the number of detected coincidences must be statistically relevant. δt can be reduced increasing the brightness of the source of the entangled photons and minimizing the losses. In some papers, the Kwiat group [42–44] developed a very efficient and simple method to obtain highly bright sources of entangled photons with a high degree of entanglement. Using this technique and a suitable optical configuration we plan to increase by more than 1000 times the number of measured coincidences and to reduce the acquisition time to less than $\delta t = 0.1\,$s. The lower bounds obtained in the previous experiments together with the new lower bound that should be reached with the experiment proposed here are shown in Fig.4. Curve *I* (red) corresponds to our previous experimental results, curve *II* (green) to the results of the Geneva group and curve *III* (blue) to the results expected with the proposed experiment. The grey region corresponds to the new region of tachyon velocities that could become accessible.

In the next section, the critical points of the proposed experiment and the main sources of experimental uncertainty will be discussed.

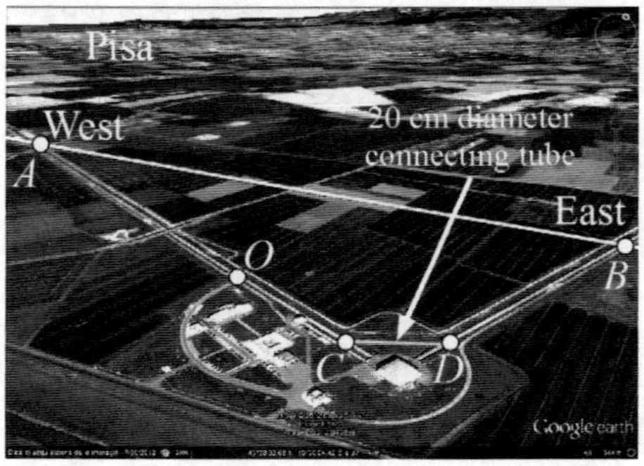

Figure 5: A Google earth view of the *EGO* structure with the two orthogonal tunnels is shown together with the West-East line. Point O represents the position of the source of entangled photons, while A and B denote the positions of the two polarizing filters in the two *EGO* arms. The full line represents the West-East direction. The entangled photon that travel at the right of point O will be deviated from the tunnel containing source O to the other tunnel passing through a suitable CD tube (20 cm diameter) that will be built to connect the two arms of the *EGO* structure.

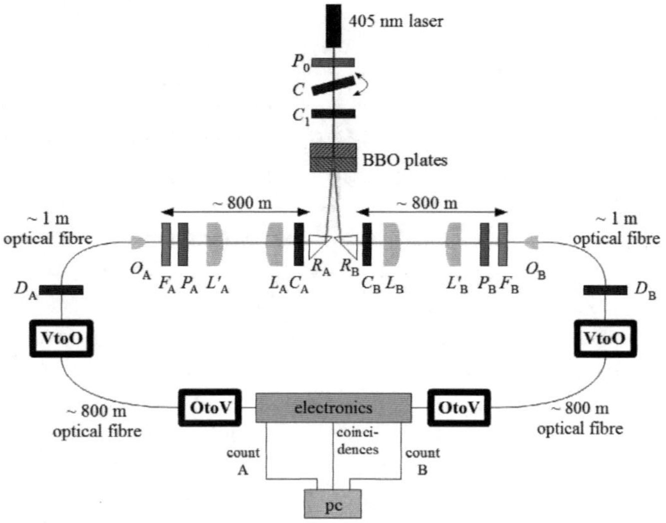

Figure 6: Schematic view of the experimental apparatus. To simplify the drawing, the optical path followed by the left entangled photon has been represented by a straight line and the optical prisms that deviate the beam from a tunnel to the other (see Fig.5) are not shown. A 405 nm laser beam impinges on two adjacent *BBO* plates with orthogonal optical axes and produces two 810 nm entangled beams preferentially emitted at the symmetric angles $\alpha_A = -\alpha_B \approx 2°$. C, C_1, C_A and C_B are optical compensators, R_A and R_B are two right angle prisms; L_A, L'_A, L_B, L'_B are large diameter (18 cm) planoconvex lenses, P_A and P_B are polarizing filters that are at the same distance from source O, F_A and F_B are optical interferometric filters with 10 nm line-width, O_A and O_B are aspheric lenses and D_A and D_B are single photons counters. $V\text{to}O$ and $O\text{to}V$ are voltage pulses to optical pulses converters and vice-versa.

3. The experiment: critical points and main sources of experimental uncertainty

As stated above, the polarizing filters should be aligned along the West-East direction, but the *EGO* tunnels are oriented at 19° and 109° with respect to this direction, respectively. Therefore, the polarizing filters must be placed in two different tunnels and one of the entangled photons must be deviated from a *EGO* tunnel to the other through the small CD tube (20 cm diameter, 100 m length) built to connect them. The source of the entangled photons (point O) will be placed in the 19° tunnel together with polariser P_A at distance $\approx$ 800 m while polariser P_B will be placed at point B in the other *EGO* tunnel at the same distance from O as shown schematically in Fig.5 ($\overline{OA} = \overline{OC} + \overline{CD} + \overline{DB}$).

The experimental apparatus is schematically shown in Fig.6 where, for simplicity of drawing, the path $OCDB$ shown in Fig.5 has been represented by a straight line and the prisms needed to deviate the photons from one *EGO* arm to the other are not represented. A blue diode laser beam ($\lambda = 405\,\text{nm}$) is polarized at 45° with respect to the vertical axis by polariser P_0, passes through a tilting plate optical compensator (C) with vertical extraordinary axis and the quartz compensator (C_1) with horizontal extraordinary axis and, finally, impinges at normal incidence on two thin (0.5 mm) adjacent non-linear optical crystals (*BBO*) cut for type-I phase matching [45]. The optic axes of the adjacent *BBO* crystals are tilted at the angle 29.2° and lie in planes perpendicular to each other with the first plane that is horizontal. The pump beam induces down conversion at $\lambda = 810\,\text{nm}$ in each crystal [45] with a maximum of emission at the two symmetric angles $\alpha_A = -\alpha_B \approx 2°$ with respect to the pump laser beam. The down converted photons are created in the maximally entangled state $\left(|H,H> + e^{i\phi}|V,V>\right)/\sqrt{2}$, where phase ϕ can be adjusted tilting the optical compensator C. R_A and R_B are two right angle prisms, C_A and C_B are optical *BBO* plates with tilt angle 29.2°, P_A and P_B are thin near infrared polarizing films (LPNIR, Thorlabs), F_A and F_B are interference filters ($\lambda = 810\,\text{nm} \pm 5\,\text{nm}$) and D_A and D_B are single photons counters (Perkin Elmer SPCM-AQ4C). L_A, L_B, L'_A and L'_B are large plano-convex optical lenses (18 cm-diameter) that ensure that all the entangled photons emitted within a cone of $\approx 0.35°$ aperture angle (around the two maximum emission directions $\alpha_A = -\alpha_B \approx 2°$) can be collected by aspheric objectives O_A and O_B and sent to detectors D_A and D_B. The centres of polarisers P_A and P_B are aligned along a x-axis in the West-East direction within $\approx 0.1°$. The role of P_0, C, P_A, P_B, F_A and F_B is shown in [45]; the role of C_1, C_A and C_B is explained in [42–44] and is resumed in Section 3.1. The outputs electric pulses of the detectors (20 ns width) are transformed into optical pulses, sent via optical fibres to the central region close to the photon source, transformed again to electric pulses and sent to electronic counters and to a coincidence circuit connected to a *PC*. A labview program controls any experimental feature.

The results shown in curve *III* of Fig.4 were obtained assuming that the uncertainty Δd can be maintained much lower than the thickness of the polarizing layers ($\approx 220\,\mu\text{m}$) and the acquisition time can be small enough ($\delta t \approx 0.1\,\text{s}$). Here below we will describe how both these conditions can be satisfied.

3.1. Minimization of the acquisition time δt.

The minimum acquisition time δt is determined by the condition that the number of detected entangled photons must be statistically significant during time δt (> 1000 measured coincidences). Then, low values of δt can be obtained only if: *a*) the number of entangled photons collected by lenses L_A and L_B is large enough, *b*) the losses of the entangled photons in the path from the source to the detectors are as smaller as possible. The momentum conservation would require that the entangled photons are emitted at two well defined angles ($\alpha_A = -\alpha_B \approx 2°$), but, due to the finite thickness of the *BBO* plates (0.5 mm), a $\approx \pm 0.35°$ spreading around these directions occurs. Then, a large fraction of the entangled photons can be collected only if the aperture angle of lenses L_A and L_B with respect to the photon source is $\approx 0.35°$. However, due to the great optical anisotropy of the *BBO* plates, the relative phase between entangled photons (ϕ in Eq.(1)) depends greatly on the emission direction. This means that the entanglement of the photons is satisfactory (close to 100% fidelity) only if the aperture angle is $\ll 0.05°$. Furthermore, the number of entangled photons greatly increases with the power of the pump laser diode and, thus, a high power laser diode should be used. However, high power laser diodes (> 50 mW) are characterized by a small coherence length which is typically ≈ 0.2 mm. In these conditions, the V-polarized down converted photons that are produced in the first *BBO* plate (0.5 mm thick) are essentially uncorrelated with respect to the H-polarized photons produced in the second adjacent plate and, thus, the entanglement is very poor. This latter drawback could be avoided using high coherence length laser diodes but, in this case, the laser power would be smaller than 50 mW. The Kwiat group [42–44] showed that the drawbacks described above can be bypassed using suitable optical compensators (C_1, C_A and C_B in Fig.6). C_1 is a quartz plate that introduces a retardation between the V and the H components of the pump beam virtually equal to the difference between the emission times of down converted photons from the two BBO plates. In this way, the losses of entanglement due to the low coherence of the laser beam are virtually eliminated. C_A and C_B are two anisotropic plates (BBO) with a suitable thickness to compensate the phase differences between entangled photons propagating along different directions. Using these compensation methods with a 280 mW laser beam and collecting entangled photons with a 0.35° aperture angle, the Kwiat group obtained an ultra bright and high fidelity (99 %) source of entangled photons with 1.02×10^6 detected coincidences/s. In our experiment we require also that all the entangled photons that are emitted within the 0.35° aperture angle get photodetectors D_A and D_B at a great distance ($\approx$ 800 m) from the source. Furthermore, it is convenient to maintain the diameter of the entangled beam sufficiently smaller when passes through the 20 cm-diameter connecting tube. Both these conditions are satisfied using large diameter (18 cm) plano-convex lenses with long focal length $f = 10$ m and slightly focusing the pump beam on the BBO plates to have a source of entangled photons with a small diameter D ($D \lesssim 0.5$ mm). Lenses L_A and L_B are put at a distance from the source slightly higher than f to produce real images of the source approximatively at distance $d_i \approx 400$ m from L_A and L_B. In this condition, the image produced by lens L_B occurs in the centre of the connecting tube. The image diameter is $D_i \approx D\, {}^{d_i}\!/_f = 2$ cm that is much smaller than the 20 cm-diameter of the connecting tube. Furthermore, due to the large diameter of lenses, the enlargement produced by diffraction is negligible. In these conditions all the entangled photons emitted in the cones of aperture angle 0.35° are collected by the large diameter lenses L_A, L'_A and L_B, L'_B and real images of the source with diameter $D'_i \approx D = 0.5$ mm are generated on the surfaces of polarisers P_A and P_B at distance slightly higher than $f = 10$ m from lenses L'_A and L'_B. The optical rays exiting from these images are collected by the aspheric objectives O_A and O_B that focus the entangled photons on two 60 μm-diameter optical fibres connected to detectors D_A and D_B. Using the Zemax optical design program we have verified that this geometric arrangement ensures that all the entangled photons emitted by the source within the 0.35° aperture angle are collected by detectors.

The previous analysis was performed disregarding any effect due to the presence of air. Density fluctuations of air produce wander of the photons beams and beam size variations that could reduce our collection efficiency. To estimate the relevance of these effects we consider the experimental results obtained by Resch et al. [46] that performed measurements with entangled photons propagating between two buildings on the Vienna sky at a distance of 7.8 km. Due to air turbulences, the authors observed large beam size variations (displacement of the centre of the beam and changes of diameter) of about 25 cm at distance 7.8 km from the source. Making a linear extrapolation of the Resch et al. results to our experimental lengths (≈ 0.8 km $\ll 7.8$ km), we expect that beam size variations should be reduced to less than 2.6 cm in our experiment. The diameter of lenses L'_A and L'_B is 18 cm whilst the diameter of entangled beams impinging on lenses L'_A and L'_B is expected to be < 14 cm, then, the 2.6 cm displacements should not affect appreciably the collection efficiency.

Another effect that could reduce the collection efficiency is the air absorption at the wavelengths of the entangled photons ($\lambda = 810\,\text{nm} \pm 5\,\text{nm}$). However, air is known to exhibit a transmission window at these frequencies (see Fig.8 in [47]) and, thus, losses due to absorption are expected to be not dramatic. Precise in loco measurements would be needed to evaluate losses due to absorption. However, one can obtain a very rough upper estimate of the losses due to absorption yet using the results by Resch et al. obtained with 810 nm entangled photons propagating on the Vienna sky. In their experiment with the 7.8 km distance they observed that only a 1.4% fraction of the photons emitted by the source actually reached the detectors at the 7.8 km distance. Note that the average beam displacements in their experiment were very large

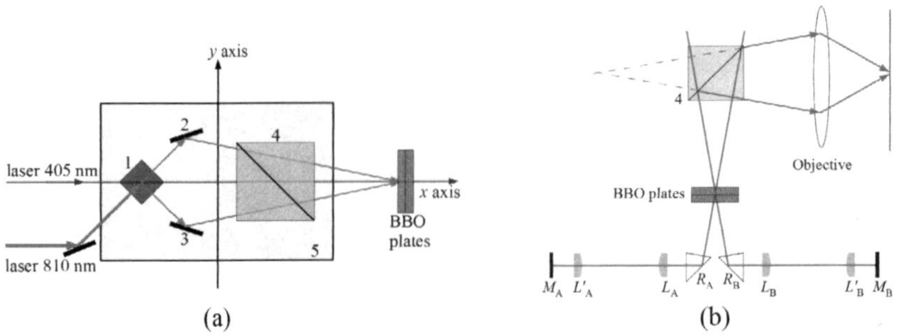

Figure 7: (a): system used to obtain two laser beams impinging at $\pm 2°$. 1 and 4 are beam splitters, 2,3 are mirrors and 5 is a rigid plate that is mounted on a x-y carriage that allows its insertion and removal. Beam splitter 4 needs to collect the beams reflected by mirrors M_A and M_B as shown in Fig.(b). (b): A schematic view of the interferometer. The laser beams are reflected by mirrors M_A and M_B, go back and produce a interference pattern in the surface corresponding to the central plane of the *BBO* plate. Then, beams are reflected by beam splitter 4 and a magnified image of the interference pattern is produced by the objective.

(25 cm) and greater than the diameter of their collecting telescopes. Then, beam displacements are expected to appreciably reduce the photon counts in their experiment. However, also assuming that losses would be entirely due to absorption, one would obtain the absorption coefficient $\alpha = 5.5 \times 10^{-4}\,\mathrm{m}^{-1}$ that would correspond to a 65% collection efficiency for a beam that propagates at a 800 m distance. Tacking into account for the entangled photon intensities reported in [46], we can conclude that, using a 120 mW laser diode and using suitable compensation procedures and a proper optical design, the number of measured coincidences should be $\gg 10^5$ coincidences/s that is more than 10^4 coincidences during the acquisition time $\delta t = 0.1\,\mathrm{s}$.

3.2. Equalization of the optical paths.

The optical apparatus in Fig.6 provides the collection of a large number of entangled photons (99%-fidelity) emitted within the 0.35° aperture angle during the short acquisition time $\delta t = 0.1\,\mathrm{s}$. However, a high value of the lower bound $\beta_{t,min}$ can be obtained only if also parameter $\bar{\rho}$ has a very small value, that is if the uncertainty on the equalization of the optical paths OA and OB is small enough. In our experiment, the thickness of the sensitive layer of the polarizing filters is 220 μm and this leads to an intrinsic uncertainty $\Delta d \lesssim 220\,\mu\mathrm{m}$, that defines the minimum obtainable value of $\bar{\rho}$ in Eq.(4). Then, it is sufficient to require that the difference between the optical paths of the entangled photons is much smaller than $\Delta d \lesssim 220\,\mu\mathrm{m}$. This goal will be reached using the interferometric method described below. Initially, polarisers P_A and P_B will be positioned approximatively at the same distance (within one centimeter) using *GPS* and, successively, the distances will be equalized with the interferometric method with an estimated accuracy better that 10 μm.

To obtain the paths equalization, we will use a 810 nm laser diode with low coherence length (0.1 – 0.2 mm). Using the simple optical system in Fig.7(a), we will obtain two beams that impinge with equal phases on the central point of the *BBO* plates with incidence angles $\pm 2°$. In such a way the outgoing beams follow the same paths of the entangled beams. Then, the *BBO* plates will be removed using a translator and polarisers P_A and P_B will be replaced by two mirrors M_A and M_B. The laser beams reflected by mirrors M_A and M_B will go back and will again meet the position of the central plane of the *BBO* plates (now removed) where they form an interference pattern with a $\approx 10\,\mu\mathrm{m}$ interline. The transmitted beams will be collected by beam splitter 4 in Fig.7(b) and a magnified image of fringes will be produced by

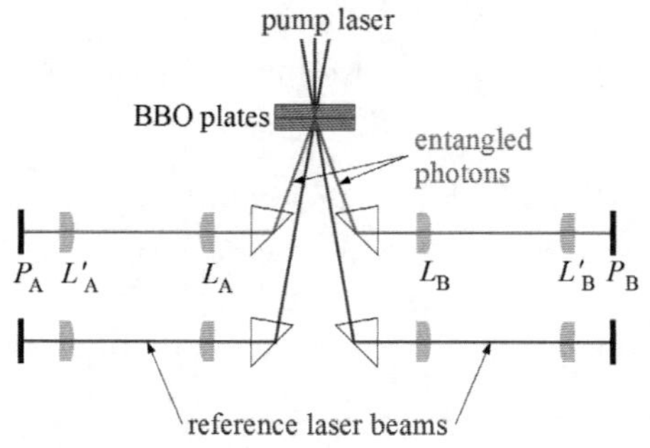

Figure 8: Schematic view of the paths of the entangled photons and of the reference laser beams.

the objective. Fringes will be visible only if the difference of optical paths is lower than the coherence length of the laser beam ($0.1 - 0.2\,\text{mm}$) with a maximum visibility when the optical paths are equalized. Translating one of the mirrors with a motorized micrometer carriage and doing real-time measurements of the fringes visibility, the optical paths can be equalized within $\pm 10\,\mu\text{m}$ (for a detailed analysis see [39]). Satisfactory preliminary tests of the proposed optical scheme have been carried out in our laboratory over distances of about 10 m. Once the optical paths are equalized, mirrors M_A and M_B should be replaced by polarizes P_A and P_B, using the procedures described in [39] that ensure that the surfaces of the polarizing layers just coincide with those of the mirrors within $\pm 10\,\mu\text{m}$.

The above procedure to equalize the optical paths leads to a satisfactory equalization (within $\pm 10\,\mu\text{m}$) at a given time. However, due to the long optical paths characterizing our experiment, temperature variations of the air refractive index can produce the optical paths variation:

$$\Delta L_{ott} = \left(\frac{\partial n^*}{\partial T}\right) L_0 \Delta T, \tag{6}$$

where L_0 is the initial path length, n^* is the group refractive index of air at the wavelength $\lambda = 810\,\text{nm}$, $\partial n^*/\partial T = 9.47 \times 10^{-7}\,\text{K}^{-1}$ [48] and ΔT is the difference between the mean temperatures along the paths of the two entangled photons. Substituting the path length value $L_0 = 800\,\text{m}$ in Eq.(6), we get $\Delta L_{ott} = 0.76\,\text{mm}$ for $\Delta T = 1°\,\text{C}$. Other contributions are expected from changes of air pressure and humidity, from thermal expansion and from Earth tides. The resulting expected variations of the optical paths are much greater than the accuracy required on the equalization of the optical paths. Then, a suitable feedback procedure to keep constant the paths difference is needed. To reach this goal, polariser P_B will be placed on a carriage driven by an electric motor and a proper feedback voltage will be sent to the electric motor. To obtain the feedback signal we want to build an interferometer with reference laser beams ($\lambda \approx 780\,\text{nm}$) moving along the arms on paths parallel to those of the entangled photons but spatially separated at an average distance of $20 - 30\,\text{cm}$ (see Fig.8). Since the paths are sufficiently close, we think that the optical paths variations due to the above effects are similar for the entangled photons and for the reference beams. If this assumption is correct, one can measure the paths variations of the reference beams to obtain a feedback signal to translate the P_B polariser. Of course, the validity of this assumption has to be firstly checked. If the assumption would not be verified, it would be needed to use more complex feedback schemes as, for example, those used in [49] where the reference beams follow the same paths of the entangled photons. Finally, due to the long optical paths, a special care has to be devoted to take under control the laser beam pointing and possible drifts of the entangled beams trajectories.

4. Conclusions

In this paper, we propose a long distance *EPR* experiment performed exploiting the long paths that characterize the *EGO* structures to test the superluminal models of *QM*. Such an experiment should increase by about two orders of magnitude the actual lower bounds for the tachyon velocities. This goal can be reached using special compensation methods to obtain high intensity sources of entangled photons and interferometric methods to minimize the uncertainty on the optical paths. An important feature of this experiment is that the measuring points A and B are exactly aligned along the West-East axis. This alignment ensures that the experimental apparatus is sensitive to any possible orientation of the velocity vector of the preferred frame of tachyons. Therefore, if the Quantum Mechanic correlations between entangled photons would be entirely or partially due to superluminal communications and if the tachyon velocity in the *PF* would be lower than the $\beta_{t,min}$ value shown in the curve *III* in Fig.4, the measured correlations should exhibit appreciable deviations from the predictions of *QM*. In such a latter case, also the actual velocity of tachyons and the actual velocity of the *PF* could be obtained from further experimental measurements. If no deviation from the predictions of *QM* would be found, our experiment would provide a lower bound for the tachyon velocities two orders of magnitude higher than the actual ones. In our previous laboratory experiment [39] we tested only the simplest model of superluminal quantum communications where all correlations between entangled photons are only due to superluminal communications and no correlation exists if there is no communication. In such a special case, the model can be tested orienting the polarizing axes of polarisers P_A and P_B both at a $\pi/4$ angle with respect to the horizontal direction [H in Eq.(1)] and measuring the coincidences between photons passing through both the polarisers. However, according to Eberhard [20], this is only one of the possible superluminal models but more complex models are also possible. In particular, some correlation between entangled particles could be already present at the beginning (hidden variables) and only a part of quantum correlations could be due to superluminal communications. In such a case, also if particles have not sufficient time to communicate, some correlations between entangled particles remain. According to the Bell theorem, correlations due to hidden variables alone cannot reproduce entirely *QM* correlations and, in particular, they have to satisfy the Bell inequalities. Then, a general test of any kind of possible superluminal model can be made only measuring a Bell-like inequality. The Bell inequalities and other equivalent inequalities proposed in the literature, need measurements of coincidences between entangled photons passing through polarisers P_A and P_B for at least four different orientations of their axes. In our experiment, one channel polarisers (polarizing filters) are used and special B.C.H.S.H. inequalities must be considered [50]. In particular, for any hidden variables model, it has been shown that the quantity:

$$M = \frac{N(a,b) - N(a,b') + N(a',b) + N(a',b') - N(a',\infty) - N(\infty,b)}{N(\infty,\infty)} \tag{7}$$

must satisfy the inequality

$$-1 \le M \le 0. \tag{8}$$

N represents the number of measured coincidences between photons passing through polarisers P_A and P_B for some different orientations of the polarisers. a and a' are two different orientations of polariser P_A, b and b' are two different orientations of polariser P_B and symbol ∞ corresponds to a removed polariser. The maximum deviation of *QM* correlations from condition (8) occurs if the polarization directions a, a', b and b' are represented by angles $\theta_a = 0°, \theta_{a'} = 45°, \theta_b = 22.5°$ and $\theta_b = 67.5°$ with respect to the vertical axis. With this choice, a positive value of M is predicted by *QM* in disagreement with inequality (8).

From Eq.(7) we see that 7 different measurements of coincidences have to be performed to verify inequality (8). The measurements will be performed in this way: polarisers P_A and P_B

will be oriented along the first combination (a,b) occurring in Eq.(7) and coincidences will be detected during an entire sidereal day, then the polarisers orientations will be changed to the second combination (a,b') and measurements will be repeated until all the seven combinations have been considered. Finally, the value of quantity M at the different sidereal times t will be obtained substituting in Eq.(7) the coincidences measured at the same sidereal times of different sidereal days. In such a way, our experiment will provide a complete test of any possible model of QM superluminal communications.

Appendix

The Alice (A) and Bob (B) polarization measurements will be represented by the space-time events $(\vec{x}_A, ct_A)$, $(\vec{x}_B, ct_B)$ in the laboratory frame R and by $(\vec{x}'_A, ct'_A)$, $(\vec{x}'_B, ct'_B)$ in the tachyons preferred frame R', respectively. The invariance of the space-time intervals writes

$$(d_{AB})^2 - (\Delta ct)^2 = (d'_{AB})^2 - (\Delta ct')^2 \tag{A.1}$$

where $d_{AB} = |\Delta\vec{x}|$, with $\Delta\vec{x} = \vec{x}_B - \vec{x}_A$, $\Delta ct = ct_B - ct_A$, $d'_{AB} = |\vec{x}'_B - \vec{x}'_A|$ and $\Delta ct' = ct'_B - ct'_A$. Dividing both sides of Eq.(A.1) by $(\Delta ct')^2$ we get

$$\left(\frac{d'_{AB}}{\Delta ct'}\right)^2 = 1 + \frac{d^2_{AB} - (\Delta ct)^2}{(\Delta ct')^2}, \tag{A.2}$$

where $\Delta ct'$ satisfies the Lorentz equation:

$$\Delta ct' = \gamma\left(\Delta ct - \vec{\beta}\cdot\Delta\vec{x}\right), \tag{A.3}$$

with $\vec{\beta} = \vec{V}/c$ ($\vec{V}$ is the velocity of R' with respect to R) and $\gamma = 1/\sqrt{1-\beta^2}$. Substituting $\Delta ct'$ of Eq.(A.3) in the right-hand side of Eq.(A.2) we obtain

$$\left(\frac{d'_{AB}}{\Delta ct'}\right)^2 = 1 + \frac{(1-\beta^2)\left[1-\left(\frac{\Delta ct}{d_{AB}}\right)^2\right]}{\left(\frac{\Delta ct - \vec{\beta}\cdot\Delta\vec{x}}{d_{AB}}\right)^2}. \tag{A.4}$$

If $v_t = \beta_t c$ is the tachyons velocity in R', the quantum communication cannot occur (the Bell inequalities are verified) only if

$$\beta_t < \frac{d'_{AB}}{|\Delta ct'|}. \tag{A.5}$$

Inequality (A.5) would be always verified for any finite β_t provided the two conditions

$$\Delta ct = 0 \quad \wedge \quad \vec{\beta}\cdot\Delta\vec{x} = 0 \tag{A.6}$$

would be satisfied that lead to $\frac{d'_{AB}}{|\Delta ct'|} \to \infty$ in Eq.(A.4). However, due to the experimental features, conditions (A.6) can be only approximatively satisfied and, thus, discrepancies with the QM will occur only if

$$\beta_t < \beta_{t,min}, \tag{A.7}$$

where $\beta_{t,min}$ is the minimum value of $\frac{d'_{AB}}{|\Delta ct'|}$ in Eq.(A.4) that can be effectively reached in the experiment. The basic experimental features are the uncertainty Δd on the equality of the optical paths, that leads to the condition

$$-\Delta d \le \Delta ct \le \Delta d, \tag{A.8}$$

and the acquisition time δt that leads to the condition

$$-\beta\left|\Delta\vec{x}\right|\sin\chi\sin\frac{\pi\delta t}{T}\leq\vec{\beta}\cdot\Delta\vec{x}\leq\beta\left|\Delta\vec{x}\right|\sin\chi\sin\frac{\pi\delta t}{T},\qquad\text{(A.9)}$$

where χ is the angle between vector $\vec{V}$ and the polar axis while T is the sidereal day. Eq.(A.9) is obtained (without lack of generality) using the expressions $\Delta\vec{x}=\left|\Delta\vec{x}\right|(1,0,0)$ and $\vec{\beta}=\beta\left(\sin\chi\cos\frac{2\pi t}{T},\sin\chi\sin\frac{2\pi t}{T},\cos\chi\right)$ that lead to

$$\vec{\beta}\cdot\Delta\vec{x}=\beta\left|\Delta\vec{x}\right|\sin\chi\cos\frac{2\pi t}{T}.\qquad\text{(A.10)}$$

From Eq.(A.10) we infer that $\vec{\beta}\cdot\Delta\vec{x}$ is minimized if the acquisition time interval δt is centred on times $t_1=\frac{T}{4}$ or $t_2=\frac{3}{4}T$ that is if $t_{1,2}-\frac{\delta t}{2}\leq t\leq t_{1,2}+\frac{\delta t}{2}$. Then, in both cases $\vec{\beta}\cdot\Delta\vec{x}$ in Eq.(A.10) will satisfy conditions (A.9). The minimum value of $\left|\frac{d'_{AB}}{\Delta ct'}\right|$ in Eq.(A.4) ($\beta_{t,min}$ in Eq.(A.7)) occurs if ratio $1-\left(\frac{\Delta ct}{d_{AB}}\right)^2/\left(\frac{c\Delta t-\vec{\beta}\cdot\Delta\vec{x}}{d_{AB}}\right)^2$ is minimized with Δct and $\vec{\beta}\cdot\Delta\vec{x}$ satisfying conditions (A.8) and (A.9). It can be easily seen that this occurs if $(\Delta ct)^2=(\Delta d)^2$ and $\left(c\Delta t-\vec{\beta}\cdot\Delta\vec{x}\right)^2=\left(\Delta d+\beta\left|\Delta\vec{x}\right|\sin\chi\sin\frac{\pi\delta t}{T}\right)^2$. Then, we find

$$\beta_{t,min}=\sqrt{1+\frac{(1-\beta^2)(1-\bar{\rho}^2)}{\left(\bar{\rho}+\beta\sin\chi\sin\frac{\pi\delta t}{T}\right)^2}},\qquad\text{(A.11)}$$

where $\bar{\rho}\equiv\frac{\Delta d}{d_{AB}}$. Eq.(A.11) just corresponds to Eq.(4).

References

[1] Einstein A, Podolsky B and Rosen N 1935 *Phys. Rev.* **47**(10) 777–780
[2] Bell J S 1964 *Physics* **1** 195–200
[3] Clauser J F, Horne M A, Shimony A and Holt R A 1969 *Phys. Rev. Lett.* **23**(15) 880–884
[4] Clauser J F and Horne M A 1974 *Phys. Rev.* D **10**(2) 526–535
[5] Freedman S J and Clauser J F 1972 *Phys. Rev. Lett.* **28**(14) 938–941
[6] Aspect A, Dalibard J and Roger G 1982 *Phys. Rev. Lett.* **49**(25) 1804–1807
[7] Zeilinger A 1986 *Phys. Lett.* A **118** 1–2
[8] Tittel W, Brendel J, Zbinden H and Gisin N 1998 *Phys. Rev. Lett.* **81**(17) 3563–3566
[9] Weihs G, Jennewein T, Simon C, Weinfurter H and Zeilinger A 1998 *Phys. Rev. Lett.* **81**(23) 5039–5043
[10] Aspect A 1999 *Nature* **398** 189–190
[11] Pan J, Bouwmeester D, Daniell M, Weinfurter H and Zeilinger A 2000 *Nature* **403** 515–519
[12] Grangier P 2001 *Nature* **409** 774–775
[13] Rowe M A, Kielpinski D, Meyer V, Sackett C A, Itano W M, Monroe C and Wineland D J 2001 *Nature* **409** 791–794
[14] Matsukevich D N, Maunz P, Moehring D L, Olmschenk S and Monroe C 2008 *Phys. Rev. Lett.* **100**(15) 150404
[15] Faraci G, Gutkowski D, Notarrigo S and Pennisi A 1974 *Lett. al Nuovo Cim.* **9**(15) 607–611
[16] Clauser J 1976 *Nuovo Cim.* B **33**(2) 740–746
[17] Selleri F 2003 *La fisica nel Novecento: per un bilancio critico* 2nd ed (Bari: Progedit)
[18] Genovese M 2005 *Phys. Rep.* **413** 319–396

[19] Davies P C W and Brown J R 1993 *The ghost in the atom: a discussion of the mysteries of quantum physics* canto ed. ed (Cambridge; New York: Cambridge University Press)

[20] Eberhard P H 1989 A realistic model for quantum theory with a locality property *Quantum theory and pictures of reality: foundations, interpretations, and new aspects* ed Schommers W (Berlin; New York: Springer-Verlag)

[21] Bohm D and Hiley B J 1991 *The undivided universe: an ontological interpretation of quantum mechanics* (Routledge)

[22] Bilaniuk O M P, Deshpande V K and Sudarshan E C G 1962 *Am. J. Phys.* **30** 718–723

[23] Feinberg G 1967 *Phys. Rev.* **159** 1089–1105

[24] Bilaniuk O M P and Sudarshan E C G 1969 *Phys. Today* **22** 43–51

[25] Møller C 1955 *The theory of relativity* (Oxford: Clarendon)

[26] Kowalczyński J 1984 *Int. J. Theor. Phys.* **23**(1) 27–60

[27] Reuse F 1984 *Ann. Phys.* **154** 161–210

[28] Caban P and Rembieliński J 1999 *Phys. Rev.* A **59**(6) 4187–4196

[29] Maudlin T 2001 *Quantum non-locality & relativity* (Oxford: Blackwell)

[30] Cocciaro B 2012 (*Preprint* `1209.3685`)

[31] Anderson R, Vetharaniam I and Stedman G 1998 *Phys. Rep.* **295** 93 – 180

[32] Jammer M 2006 *Concepts of simultaneity from antiquity to Einstein and beyond* (Baltimore, Md.: Johns Hopkins University Press)

[33] Bohm D 1952 *Phys. Rev.* **85**(2) 166–179

[34] Bohm D 1952 *Phys. Rev.* **85**(2) 180–193

[35] Bancal J, Pironio S, Acín A, Liang Y, Scarani V and Gisin N 2012 *Nat. Phys.* **8** 867–870

[36] Gisin N 2012 (*Preprint* `1210.7308`)

[37] Scarani V, Tittel W, Zbinden H and Gisin N 2000 *Phys. Lett.* A **276** 1–7

[38] Salart D, Baas A, Branciard C, Gisin N and Zbinden H 2008 *Nature* **454** 861–864

[39] Cocciaro B, Faetti S and Fronzoni L 2011 *Phys. Lett.* A **375** 379–384

[40] European Gravitational Observatory URL `http://www.ego-gw.it/`

[41] The Virgo collaboration 1997 *Final Design Report* (Virgo Technical Note VIR-TRE-DIR-1000-13)

[42] Altepeter J, Jeffrey E and Kwiat P 2005 *Opt. Express* **13** 8951–8959

[43] Akselrod G M, Altepeter J B, Jeffrey E R and Kwiat P G 2007 *Opt. Express* **15** 5260–5261

[44] Rangarajan R, Goggin M and Kwiat P 2009 *Opt. Express* **17** 18920–18933

[45] Kwiat P G, Waks E, White A G, Appelbaum I and Eberhard P H 1999 *Phys. Rev. A* **60**(2) R773–R776

[46] Resch K *et al.* 2005 *Opt. Express* **13** 202–209

[47] Gisin N, Ribordy G, Tittel W and Zbinden H 2002 *Rev. Mod. Phys.* **74**(1) 145–195

[48] Ciddor Equation URL `http://emtoolbox.nist.gov/wavelength/ciddor.asp`

[49] Peng C Z *et al.* 2005 *Phys. Rev. Lett.* **94** 150501

[50] Aspect A 2002 Bell's theorem: The naive view of an experimentalist *Quantum (Un)speakables: From Bell to Quantum Information* ed Bertlmann R A and Zeilinger A (Springer)

Hydrogen atom as a quantum-classical hybrid system

Fei Zhan[1,2] **and Biao Wu**[1]

[1] International Center for Quantum Materials, Peking University, 100871, Beijing, China
[2] Centre for Engineered Quantum Systems, School of Mathematics and Physics, The University of Queensland, St Lucia QLD 4072, Australia

E-mail: wubiao@pku.edu.cn

Abstract. Hydrogen atom is studied as a quantum-classical hybrid system, where the proton is treated as a classical object while the electron is regarded as a quantum object. We use a well known mean-field approach to describe this hybrid hydrogen atom; the resulting dynamics for the electron and the proton is compared to their full quantum dynamics. The electron dynamics in the hybrid description is found to be only marginally different from its full quantum counterpart. The situation is very different for the proton: in the hybrid description, the proton behaves like a free particle; in the fully quantum description, the wave packet center of the proton orbits around the center of mass. Furthermore, we find that the failure to describe the proton dynamics properly can be regarded as a manifestation of the fact that there is no conservation of momentum in the mean-field hybrid approach. We expect that such a failure is a common feature for all existing approaches for quantum-classical hybrid systems of Born-Oppenheimer type.

1. Introduction

Though quantum mechanics and its ensuing developments have been verified by all the experiments to be the correct description of the physical universe, classical physics is still a useful, convenient, and sometimes irreplaceable tool to describe many systems. For example, it is very convenient and also accurate to describe the motion of the Earth around the Sun with Newton's equations of motion instead of the Schrödinger equation. There are even various situations, where one finds it convenient and necessary to separate one system into two subsystems: one subsystem is described quantum mechanically and the other is described classically. This gives rise to quantum-classical hybrid systems.

Quantum-classical hybrid systems come in three main categories: (*i*) In unifying gravity with the other fundamental interactions, one simply gives up quantizing gravity and proposes a unification theory where gravity is treated as a classical field and the other three forces as quantum fields [1, 2, 3]. (*ii*) In the standard Copenhagen interpretation of quantum mechanics, one always deals with a hybrid system, where a quantum system interacts with a classical measuring apparatus. In such a case, the classical apparatus can cause the collapse of the wave function of the quantum system, which is still beyond mathematical description [4]. (*iii*) The hybrid system in this category is typified by systems in the field of solid state physics and chemistry [5, 6, 8]. In the Born-Oppenheimer approximation [9], the nuclei are treated classically because their motion is much slower due to their large masses while the electrons are treated as quantum objects. This kind of systems are now also arising in nano-science [10, 11], where a classical detector of tens of nanometers in size interacts with a quantum object. For these systems, the interaction from the classical subsystem does not cause dramatic changes in the quantum subsystem such as the collapse of the wave function. For systems in this category, we call them Born-Oppenheimer systems.

We are only interested in the Born-Oppenheimer hybrid systems. There have been many different approaches proposed to describe these hybrid systems[5, 6, 7, 8, 12, 13, 14, 15, 16, 17, 18, 19, 20, 21, 22]; some of these approaches are shown to be equivalent to each other [23]. There are discussions on what conditions need to be satisfied by a self-consistent theory for a hybrid system[24, 25, 26, 27, 28]. In this work, we use hydrogen atom as a concrete example to explore the possible inconsistency or inadequacy of a theory for hybrid systems. As is well known, the full quantum mechanical description of hydrogen atom, the simplest Born-Oppenheimer system in nature, can be found analytically. This full quantum description can be used to benchmark the hybrid description.

We use the mean-field approach that was proposed in Ref. [11] to describe hydrogen atom. Its theoretical structure was fully analyzed in Ref. [27]. With this approach, we find that the electron dynamics is very similar to the corresponding full quantum description. However, we find that the proton in the hybrid approach behaves like a free particle, very different from its full quantum description where the wave packet center of the proton orbits around the center of mass. This failure to describe the proton motion properly is rooted in the fact that there is no conservation of momentum in the hybrid approach. Our further analysis shows that this failure or inadequacy in the hybrid approach is intrinsic: it is caused by the loss of entanglement of the electron and proton dynamics in the mean-field hybrid approach. As this loss of entanglement exists in all known hybrid approaches, we expect that all the hybrid approaches fail to describe properly the dynamics of the classical subsystem.

2. Full quantum solution for hydrogen atom

Before we study the hydrogen atom as a hybrid system, we briefly review its general full quantum solution and then apply it to a special case, where the electron is described by a wave packet that moves and evolves around a circle [29].

The Schrödinger equation for a hydrogen atom can be written as

$$i\hbar\frac{\partial}{\partial t}\psi(\boldsymbol{r}_e,\boldsymbol{r}_p,t)=\left[-\frac{\hbar^2}{2m_e}\nabla^2_{\boldsymbol{r}_e}-\frac{\hbar^2}{2m_p}\nabla^2_{\boldsymbol{r}_p}+V(|\boldsymbol{r}_e-\boldsymbol{r}_p|)\right]\psi(\boldsymbol{r}_e,\boldsymbol{r}_p,t), \tag{1}$$

where m_p ($\boldsymbol{r}_p$) is the mass (coordinate) of the proton, m_e ($\boldsymbol{r}_e$) is the mass (coordinate) of the electron, and $V(|\boldsymbol{r}_e-\boldsymbol{r}_p|)$ is the Coulomb potential that depends only on the distance between the proton and the electron. As done in every textbook, we introduce the relative coordinate $\boldsymbol{r}$ and the coordinate of the center of mass $\boldsymbol{R}$,

$$\boldsymbol{r}=\boldsymbol{r}_e-\boldsymbol{r}_p,\quad \boldsymbol{R}=\frac{m_e\boldsymbol{r}_e+m_p\boldsymbol{r}_p}{M}, \tag{2}$$

where $M=m_e+m_p$ is the total mass of the hydrogen atom. In these new coordinates, the Schrödinger equation (1) becomes

$$i\hbar\frac{\partial}{\partial t}\psi(\boldsymbol{R},\boldsymbol{r},t)=\left[-\frac{\hbar^2}{2M}\nabla^2_{\boldsymbol{R}}-\frac{\hbar^2}{2\mu}\nabla^2_{\boldsymbol{r}}+V(|\boldsymbol{r}|)\right]\psi(\boldsymbol{R},\boldsymbol{r},t), \tag{3}$$

where $\mu=m_em_p/(m_e+m_p)$ is the reduced mass. So, the wave function of a hydrogen atom can be written as

$$\psi(\boldsymbol{R},\boldsymbol{r},t)=\psi_r(\boldsymbol{r},t)\psi_c(\boldsymbol{R},t)\,, \tag{4}$$

where $\psi_r(\boldsymbol{r},t)$ and $\psi_c(\boldsymbol{R},t)$ satisfy, respectively,

$$i\hbar\frac{\partial}{\partial t}\psi_r(\boldsymbol{r},t) = -\frac{\hbar^2}{2\mu}\nabla^2_{\boldsymbol{r}}\psi_r(\boldsymbol{r},t) + V(|\boldsymbol{r}|)\psi_r(\boldsymbol{r},t)\,, \tag{5}$$

$$i\hbar\frac{\partial}{\partial t}\psi_c(\boldsymbol{R},t) = -\frac{\hbar^2}{2M}\nabla^2_{\boldsymbol{R}}\psi_c(\boldsymbol{R},t)\,. \tag{6}$$

This shows that the motion of a hydrogen atom can be separated into two independent parts: $\psi_r(\boldsymbol{r},t)$ describes the motion of a particle in a Coulomb potential $V(|\boldsymbol{r}|)$ while $\psi_c(\boldsymbol{R},t)$ describes the motion of a free particle. However, this by no means implies that the motions of the electron and the proton can be described by two independent wave functions, which will become much clearer in the later analysis.

We go back to the coordinate system of $\boldsymbol{r}_e$ and $\boldsymbol{r}_p$. The density distribution of the electron is

$$\begin{aligned}
|\psi_e(\boldsymbol{r}_e,t)|^2 &= \int |\psi_r(\boldsymbol{r},t)|^2|\psi_c(\boldsymbol{R},t)|^2 d\boldsymbol{r}_p \\
&= \int |\psi_r(\boldsymbol{r}_e-\boldsymbol{r}_p,t)|^2|\psi_c(\frac{m_e\boldsymbol{r}_e+m_p\boldsymbol{r}_p}{M},t)|^2 d\boldsymbol{r}_p \\
&= \int |\psi_r(\boldsymbol{r}_e-\boldsymbol{r}_p,t)|^2|\psi_c(\boldsymbol{r}_e-\frac{m_p}{M}(\boldsymbol{r}_e-\boldsymbol{r}_p),t)|^2 d(\boldsymbol{r}_p-\boldsymbol{r}_e) \\
&= \int |\psi_r(\boldsymbol{x},t)|^2|\psi_c(\boldsymbol{r}_e-\frac{m_p}{M}\boldsymbol{x},t)|^2 d\boldsymbol{x}\,,
\end{aligned} \tag{7}$$

where we have plugged in Eq. (2). Similarly, the proton density is given by

$$|\psi_e(\boldsymbol{r}_p,t)|^2 = \int |\psi_r(\boldsymbol{x},t)|^2|\psi_c(\boldsymbol{r}_p-\frac{m_e}{M}\boldsymbol{x},t)|^2 d\boldsymbol{x}\,. \tag{8}$$

The above results are very illuminating. As the center of mass motion is free particle-like, we assume that $\psi_c(\boldsymbol{R},t)$ is a Gaussian wave packet with width σ. So, the density of electron is just the density of the relative motion $|\psi_r(\boldsymbol{r},t)|^2$ coarse-grained with a Gaussian function of width $M\sigma/m_p \approx \sigma$. In contrast, the density of proton is the density of the relative motion $|\psi_r(\boldsymbol{r},t)|^2$ coarse-grained with a Gaussian function of a much larger width $M\sigma/m_e \approx 1837\sigma$. In other words, the electron density and the proton density are very similar to each other but the proton density looks about 1837 times fuzzier.

In the reference frame where the hydrogen atom is motionless, the center of the wave packet of electron is

$$\langle \boldsymbol{r}_e\rangle = \frac{m_p}{M}\int |\psi_r(\boldsymbol{r},t)|^2\boldsymbol{r}d\boldsymbol{r} = \frac{m_p}{M}\langle \boldsymbol{r}\rangle\,, \tag{9}$$

where $\langle \boldsymbol{r}\rangle$ is the center of the wave packet for the relative motion. The center of the wave packet of proton is

$$\langle \boldsymbol{r}_p\rangle = -\frac{m_e}{M}\langle \boldsymbol{r}\rangle\,. \tag{10}$$

We now consider a special case. In this case, the initial state of the center of mass of the hydrogen atom is described by a Gaussian function

$$\psi_c(\boldsymbol{R},t=0) = \frac{1}{(2\pi\sigma^2)^{1/4}}\exp\left[-\frac{|\boldsymbol{R}|^2}{4\sigma^2}\right] \tag{11}$$

with σ being the width; the relative motion at the beginning is depicted by the following wave function [29],

$$\psi_r(\boldsymbol{r}, t=0) = \frac{1}{(2\pi\sigma_{\bar{n}}^2)^{1/4}} \sum_{n=1}^{\infty} \exp\left[-\frac{(n-\bar{n})^2}{4\sigma_{\bar{n}}^2}\right] u_{n(n-1)(n-1)}(\boldsymbol{r}), \tag{12}$$

where $\bar{n}$ and $\sigma_{\bar{n}}$ are the mean and the width of the Gaussian distribution, respectively, and u_{nlm} is the standard energy-eigenstate for the hydrogen atom[30]

$$u_{nlm} = \sqrt{\left(\frac{2}{na_B}\right)^3 \frac{(n-l-1)!}{2n[(n+1)!]^3}} e^{-r/na_B} \left(\frac{2r}{na_B}\right)^l L_{n-l-1}^{2l+1}\left(\frac{2r}{na_B}\right) Y_l^m(\theta, \phi) \tag{13}$$

with L being associated Laguerre polynomial, Y the spherical harmonics, and a_B the Bohr radius.

The ensuing dynamics of the circular wave packet in Eq.(12) under the influence of the Coulomb potential $V(|\boldsymbol{r}|)$ has been studied in detail in Ref.[29]. For the sake of self-containment, we summarize their results here. The wave packet is localized on a circle with radius $\sim \bar{n}^2 a_B$. This wave packet remains localized and moves on the circle for several T_{Kepler}, the period of the corresponding classical motion on the same circle. At time $T_{\text{spread}} \sim 10 T_{\text{Kepler}}$, the spreading of the wave packet becomes so severe that it distributes rather uniformly on the circle. This is characterized by $\langle \boldsymbol{r} \rangle = 0$. The wave packet can recover its localized form and revive at $T_{\text{rev}} = (\bar{n}/3) T_{\text{Kepler}}$, and repeats its previous dynamics afterwards. With this in mind, it is straightforward to picture the quantum dynamic motion for both the electron and the proton in this special case.

For the electron, its wave packet is the wave packet in Eq.(12) coarse-grained with the Gaussian wave packet for the center of mass motion (see Eq.(7)). As long as the width σ is not too large, its wave packet dynamics should be very similar: it is localized and orbits on a circle of radius $\sim \bar{n}^2 a_B$ before T_{spread}. During this period, the wave packet center $\langle \boldsymbol{r}_e \rangle$ oscillates periodically with its amplitude decreasing. After T_{spread} and before T_{rev}, the wave packet spreads over the circle, which is characterized by $\langle \boldsymbol{r}_e \rangle = 0$. This dynamics repeats itself after T_{rev}.

For the proton, its wave packet center has a similar motion as the electron's according to Eqs.(9,10). The difference is that the proton moves in the opposite direction and $\langle \boldsymbol{r}_p \rangle$ varies with time on a circle of much smaller radius. However, the dynamics of the proton wave packet is very different. The wave packet of the proton is the result of coarse-graining with a Gaussian function of large width $M\sigma/m_e \approx 1837\sigma$ (see Eq.(8)). This large-size coarse-graining makes the wave packet much less localized. At the same time, the proton moves on a much smaller circular orbit (about 1837 times smaller). With these two factors combined, it is clear that the wave packet of the proton is always spread and smeared out over the entire circular orbit. No distinct peak and other structure can be seen. The three different time scales of the wave packet dynamics, T_{Kepler}, T_{spread}, and T_{rev}, which are used to characterize the distinct features of the wave packet at different times, become rather meaningless for the proton.

3. Hybrid dynamics in hydrogen atom

We now treat the hydrogen atom as a hybrid system, where the proton is regarded as a classical object while the electron is treated quantum mechanically. With the approach in Ref.[11], the hybrid Hamiltonian for the hydrogen atom is

$$H = \langle \varphi_e(\boldsymbol{r}_e, t)| - \frac{\hbar^2}{2m_e} \nabla_{\boldsymbol{r}_e}^2 + V(|\boldsymbol{r}_e - \boldsymbol{r}_p|) |\varphi_e(\boldsymbol{r}_e, t)\rangle + \frac{\boldsymbol{p}_p^2}{2m_p}. \tag{14}$$

We expand the wave function $|\varphi_e\rangle$ in a set of complete orthonormal basis,

$$|\varphi_e(\boldsymbol{r}_e,t)\rangle = \sum_j \varphi_j(t)|j\rangle. \tag{15}$$

With the classical canonical Hamiltonian structure introduced by Heslot[31], we have the following Poisson brackets,

$$\{\varphi_j,\varphi_k^*\} = i\delta_{jk}/\hbar,\quad \{r_{pj},p_{pk}\} = \delta_{jk}, \tag{16}$$

$$\{\varphi_j,\varphi_k\} = \{r_{pj},r_{pk}\} = \{p_{pj},p_{pk}\} = 0, \tag{17}$$

where $r_{pj}(p_{pj})$ is the jth component of coordinate (momentum) vector of the proton. With these Poisson brackets, we can obtain the hybrid equations of motion,

$$i\hbar\frac{d}{dt}|\varphi_e(\boldsymbol{r}_e,t)\rangle = \left[-\frac{\hbar^2}{2m_e}\nabla^2_{\boldsymbol{r}_e} + V(\boldsymbol{r}_e-\boldsymbol{r}_p)\right]|\varphi_e(\boldsymbol{r}_e,t)\rangle\,, \tag{18}$$

$$\dot{\boldsymbol{r}}_p = \frac{\partial H}{\partial \boldsymbol{p}_p} = \frac{\boldsymbol{p}_p}{m_p}\,, \tag{19}$$

$$\dot{\boldsymbol{p}}_p = -\frac{\partial H}{\partial \boldsymbol{r}_p} = -\nabla_{\boldsymbol{r}_p}\langle\varphi_e(\boldsymbol{r}_e,t)| - \frac{\hbar^2}{2m_e}\nabla^2_{\boldsymbol{r}_e} + V(\boldsymbol{r}_e-\boldsymbol{r}_p)|\varphi_e(\boldsymbol{r}_e,t)\rangle\,. \tag{20}$$

Note that $\boldsymbol{p}_p$ and $\boldsymbol{r}_p$ are independent dynamical variables in the above equations of motion while $\boldsymbol{r}_e$ is just some external parameter. The dynamical variables for the electrons are φ_j's in Eq.(15).

Similar to the full quantum treatment, we focus on the case where the initial state for the electron is given by the circular wave packet in Eq.(12). In terms of the electron coordinate $\boldsymbol{r}_e$, the circular wave packet has the following form,

$$\varphi_e(\boldsymbol{r}_e,t=0) = \frac{1}{(2\pi\sigma^2_{\bar{n}})^{1/4}}\sum_{n=1}^{\infty}\exp\left[-\frac{(n-\bar{n})^2}{4\sigma^2_{\bar{n}}}\right]u_{n(n-1)(n-1)}(\boldsymbol{r}_e-\boldsymbol{r}_{p0})\,, \tag{21}$$

where $\boldsymbol{r}_{p0}$ is the initial position of the proton. Since the proton is much more massive than the electron, its motion is rather slow and will not cause quantum transition between different quantum states of the electron. In other words, the weight before each eigenstate $u_{n(n-1)(n-1)}$ will not be changed by the proton motion. This is just the famous quantum adiabatic theorem[32] or the essence of the Born-Oppenheimer approximation. As a result, the dynamics of the electron wave packet is just the same as the electron dynamics in the full quantum description. The only difference is that the wave function of the electron in this hybrid approach is not coarse-grained.

In contrast, the proton is very different: the classical motion of a proton in the hybrid treatment does not agree with the motion of the wave packet center $\langle\boldsymbol{r}_p\rangle$ for any meaningful period of time. Let us examine the right hand side of Eq.(20). $E_e = \langle\varphi_e| - \frac{\hbar^2}{2m_e}\nabla^2_{\boldsymbol{r}_e} + V(\boldsymbol{r}_e - \boldsymbol{r}_p)|\varphi_e\rangle$ is the electron energy. It is clear that the eigen-energy E_{nlm} for each eigenstate u_{nlm} is independent of the proton position $\boldsymbol{r}_p$. As mentioned above, the proton motion is so slow that it will not cause quantum transition between different electronic states u_{nlm}. This means that E_e is independent of $\boldsymbol{r}_p$. As a result, the right hand side of Eq.(20) is zero and the momentum of the proton does not change with time. So, the proton in the hybrid dynamics is like a free-particle, motionless or moves along a straight line. In contrast, the wave packet center $\langle\boldsymbol{r}_p\rangle$ makes circular motion around the center of mass before time T_{spread}. This shows that the hybrid approach fails to describe the proton dynamics properly.

This failure to describe the proton dynamics can be illustrated from a different angle. Our analysis shows that this failure essentially has the root in the fact that there is no conservation of momentum in the hybrid dynamics of hydrogen atom. We define the total momentum for this system as

$$\boldsymbol{P} = \boldsymbol{p}_p + \langle \varphi_e | \hat{\boldsymbol{p}}_e | \varphi_e \rangle \,. \tag{22}$$

In the adiabatic limit, we know that $\boldsymbol{p}_p$ is a constant while $\langle \varphi_e | \hat{\boldsymbol{p}}_e | \varphi_e \rangle$ changes with time significantly at least for the first several T_{Kepler}. This means that the total momentum $\boldsymbol{P}$ changes with time. The easiest way to appreciate this result to set the initial momentum $\boldsymbol{p}_{p0}$ of the proton be zero. Then according to Eqs.(18,19,20), the proton remains motionless while the electron wave packet moves on the circle and its change of momentum $d\langle \varphi_e | \hat{\boldsymbol{p}}_e | \varphi_e \rangle / dt$ has a finite value for the first several T_{Kepler}.

4. Self-consistency of hybrid approaches

The above discussion leads to immediate questions, such as, "How general is the conclusion that the hybrid approach is inadequate in describing the dynamics of the classical subsystem?", "Is it possible that hybrid approaches are intrinsically flawed?" There is already some attempt to answer these questions [28]. Again we do not do general analysis here; we instead use hydrogen atom as an example and hope that our analysis with hydrogen atom may shed some light on the general questions.

First, the analysis on hydrogen atom can be generalized to multi-nucleus systems. The electronic states clearly do not depend on the position of the center of mass of all the nuclei. When the nuclear center of mass moves slowly as in the usual case, it will not cause quantum transition between different electronic states. As a result, the center of mass of all the nuclei feels no force and there is no conservation of the total momentum. It seems that this shortcoming can be remedied by moving the derivative in Eq.(20) inside the bra-ket, that is, replacing Eq.(20) with

$$\dot{\boldsymbol{p}}_p = -\frac{\partial H}{\partial \boldsymbol{r}_p} = -\langle \varphi_e(\boldsymbol{r}_e, t) | \nabla_{\boldsymbol{r}_p} \left[-\frac{\hbar^2}{2m_e} \nabla^2_{\boldsymbol{r}_e} + V(\boldsymbol{r}_e - \boldsymbol{r}_p) \right] | \varphi_e(\boldsymbol{r}_e, t) \rangle \,. \tag{23}$$

This is nothing but the well-known Ehrenfest equation[33, 34]. It is easy to show that the total momentum is conserved if this Ehrenfest equation is used instead of Eq.(20). As the Ehrenfest equation is derived from Eq.(20) with an argument that appears right but is fundamentally flawed upon close examination[34], we now have a very intriguing situation: the total momentum is not conserved for the "correct" Eq.(20) while the total momentum is conserved for the "flawed" Eq.(23). This may be regarded as an indication that there exists intrinsic inconsistency in a hybrid theory.

Secondly, there is loss of quantum entanglement in any hybrid theory. In the full classical treatment, a pair of independent dynamical variables $\{\boldsymbol{r}_e(t),\ \boldsymbol{r}_p(t)\}$, one for the electron and the other for the proton, are enough to specify completely the dynamics of the system. In the hybrid approach, there also exists such a dynamical pair $\{\boldsymbol{r}_p(t), \varphi_e(\boldsymbol{r}_e, t)\}$, for the proton and the electron, respectively, which can completely describe the whole dynamics.

In contrast, in the full quantum mechanical approach, there exists no such a dynamical pair, one for the electron and the other for the proton, with which the full quantum dynamics is completely determined. It is true according to Eq.(4) that the full quantum dynamics can be specified by two independent wave functions. However, one wave function is for the relative motion and the other is for the center of mass, instead of for the electron and the proton, respectively. The motion of the electron and the proton is always entangled together. This entanglement between the electron and the proton is manifested at two levels. For the first level, it is the entanglement understood in the usual sense that one can not have the total wave function as the product of the electron wave function and the proton wave function,

$\psi(t) = \psi_e(\boldsymbol{r}_e, t)\psi_p(\boldsymbol{r}_p, t)$. Maybe temporarily at a given moment, one can have this kind of direct product but not for any meaningful period of time. For the second level, which is much weaker, the total wave function can not be the functional of two independent dynamical functions, one for the electron $f_e(\boldsymbol{r}_e, t)$ and the other for the proton $f_p(\boldsymbol{r}_p, t)$. In other words, we can never write the total wave function as $\psi(t) = \psi[f_e(\boldsymbol{r}_e, t), f_p(\boldsymbol{r}_p, t)]$. To see this clearly, we introduce the reduced wave functions for the electron and the proton. Similar to the densities, they are also coarse-grained from the wave function for the relative motion,

$$\widetilde{\psi}_e(\boldsymbol{r}_e, t) = \int \psi_r(\boldsymbol{r}, t)\psi_c(\boldsymbol{R}, t)d\boldsymbol{r}_p = \int \psi_r(\boldsymbol{x}, t)\psi_c(\boldsymbol{r}_e - \frac{m_p}{M}\boldsymbol{x}, t)d\boldsymbol{x}\,, \tag{24}$$

$$\widetilde{\psi}_p(\boldsymbol{r}_p, t) = \int \psi_r(\boldsymbol{r}, t)\psi_c(\boldsymbol{R}, t)d\boldsymbol{r}_e = \int \psi_r(\boldsymbol{x}, t)\psi_c(\boldsymbol{r}_p - \frac{m_e}{M}\boldsymbol{x}, t)d\boldsymbol{x}\,. \tag{25}$$

It is clear that these two functions contain all the dynamical information that we can have for the electron and the proton. However, as they are obtained from the coarse-graining, these two reduced wave functions alone can not specify the total wave function $\psi(t)$. The entanglement between the electron and the proton at the second level is immediately destroyed in the hybrid approach, where the dynamics of the whole system can be determined completely by the proton dynamics $\boldsymbol{r}_p$ and the electron dynamics $\varphi_e(\boldsymbol{r}_e, t)$. Although our analysis is done for hydrogen atom, it can be generalized. In any hybrid approach, there is no entanglement between the quantum subsystem and the classical subsystem at the second level. This loss of entanglement is general and intrinsic.

5. Conclusion

In summary, we have studied the dynamics in a hydrogen atom with two different approaches. One is the full quantum theory that can be found in the standard textbook and the other is the hybrid mean-field approach, where the proton is regarded as a classical object and the electron is a quantum object. We have found that there is only marginal difference for the dynamics of the electron between these two approaches. However, the dynamics for the proton is very different between the two: the proton in the hybrid approach behaves like a free particle, not moving or moving along a straight line; the wave packet center for the proton in the full quantum approach can make circular motion for a limited time. These differences are summarized in Table 1.

	full quantum approach	hybrid approach
electron dynamics	coarse-grained $\psi_r(\boldsymbol{r}, t)$	$\psi_r(\boldsymbol{r}, t)$
proton dynamics	$\langle \boldsymbol{r}_p \rangle$ moves around center of mass	behaves as a free particle
	conservation of total momentum	no conservation of total momentum

Table 1. Comparison between the full quantum dynamics and the hybrid dynamics of hydrogen atom

The failure of proper description of the proton dynamics is the manifestation that there is no conservation of total momentum in the hybrid approach. We acknowledge that this failure may just be to the interest of theorists as there is no experimental way to probe the proton dynamics. For systems with more than one nuclei, the non-physical artifacts caused by the hybrid treatment may also be not essential to experiments. The reason is that the hybrid approach may only fail to capture the proper motion for the center of mass of the nuclei. For the relative motions between the nuclei, the hybrid approach may be adequate. All the vibrational and rotational frequencies for the relative nuclear motion are computed by treating the nuclei as classical objects. If this approximation can not capture physics that can be measured experimentally, it would be found and known for a long time.

Acknowledgments
We thank Yinhan Zhang and Shaoqi Zhu for useful discussions. B.W. is supported by the National Basic Research Program of MOST (NBRP) of China (2012CB921300, 2013CB921900), the National Natural Science Foundation (NSF) of China (11274024), the Research Fund for the Doctoral Program of Higher Education (RFDP) of China (20110001110091).

References
[1] Anderson A 1995 *Phys. Rev. Lett.* **74** 621
[2] Kuo C I and Ford L H 1993 *Phys. Rev.* D **47** 4510
[3] Calzetta E and Hu B L 1994 *Phys. Rev.* D **49** 6636
[4] Zurek W H 2003 *Rev. Mod. Phys.* **75** 715
[5] Tully J C and Preston R K 1971 *J. Chem. Phys.* **55** 562
[6] Tully J C 1990 *J. Chem. Phys.* **93** 1061
[7] Jones K R W 1994 *Phys. Rev.* A **50** 1062
[8] Prezhdo O V and Rossky P J 1997 *J. Chem. Phys.* **107** 825
[9] Born M and Oppenheimer R 1927 *Ann. Phys.* **84** 30
[10] Rugar D, Budakian R, Mamin H J and Chui B W 2004 *Nature* **430** 329
[11] Zhang Q and Wu B 2006 *Phys. Rev. Lett.* **97** 190401
[12] Aleksandrov I V 1981 *Z. Naturforsch., A: Phys. Sci.* **36** 902
[13] Gindensperger E C M and Beswick J A 2000 *J. Chem. Phys.* **113** 9369
[14] Delos J B and Knudson S K 1972 *Phys. Rev.* A **6** 709
[15] Delos J B and Thorson W R 1972 *Phys. Rev.* A **6** 720
[16] Halliwell J J 1998 *Phys. Rev.* D **57** 2337
[17] Jones K R W 1992 *Phys. Rev.* D **45** R2590
[18] Kapral R and Ciccotti G 1999 *J. Chem. Phys.* **110** 8919
[19] Nielsen S R K and Ciccotti G 2001 *J. Chem. Phys.* **115** 5805
[20] Prezhdo O V and Brooksby C 2001 *Phys. Rev. Lett.* **86** 3215
[21] Bittner E R and Rossky P J 1995 *J. Chem. Phys.* **103** 8130
[22] Alonso J L, Castro A, Clemente-Gallardo J, Cuchí J C, Echenique P and Falceto F 2011 *J. Phys. A: Math. Theor.* **44** 396004
[23] Zhan F, Lin Y and Wu B 2008 *J. Chem. Phys.* **128** 204104
[24] Caro J and Salcedo L L 1999 *Phys. Rev.*A **60** 842
[25] Diósi L, Gisin N and Strunz W T 2000 *Phys. Rev.* A **61** 022108
[26] Elze H T, Gambarotta G and Vallone F 2011 *J. Phys.: Conf. Ser.* **306** 012010
[27] Elze H T 2012 *Phys. Rev.* A **85** 052109
[28] Salcedo L L 2012 *Phys. Rev.* A **85** 022127
[29] Gaeta Z D and Stroud Jr C R 1990 *Phys. Rev.* A **42** 6308
[30] Schiff L I 1955 *Quantum Mechanics* Second Edition (New York: McGraw-Hill)
[31] Heslot A 1985 *Phys. Rev.* D **31** 1341
[32] Born M and Fock V 1928 *Z. Phys.* **51** 165
[33] Alonso J L, Clemente-Gallardo J, Cuchí J C, Echenique P and Falceto F 2012 *J. Chem. Phys.* **137** 054106
[34] Bornemann F A, Nettesheim P and Schütte C 1996 *J. Chem. Phys.* **105** 1074

Quantum-classical hybrid dynamics – a summary

Hans-Thomas Elze

Dipartimento di Fisica "Enrico Fermi", Università di Pisa, Largo Pontecorvo 3, I-56127 Pisa, Italia

E-mail: elze@df.unipi.it

Abstract. A summary of a recently proposed description of quantum-classical hybrids is presented, which concerns quantum and classical degrees of freedom of a composite object that interact directly with each other. This is based on notions of classical Hamiltonian mechanics suitably extended to quantum mechanics.

1. Introduction

We briefly review an attempt to construct a theory that describes *quantum-classical hybrids*, consisting of quantum mechanical and classical objects that interact directly with each other [1, 2, 3]. Hybrids might exist as a fundamentally different species of composite objects "out there", with consequences for the range of applicability of quantum mechanics, or they may serve in an approximate description of certain complex quantum systems.

The concept of quantum-classical hybrids has recently been employed, in order to explore a hypothetical direct coupling of classical time machines to quantum objects [4], while the application of hybrid dynamics to the experiment proposed by Penrose *et al.* [5], which is intended to test the existence of "Schrödinger cat" states of spatially displaced macroscopic objects, and to the manipulation of entanglement among two q-bits is presently under study [6].

2. Quantum-classical hybrids

The direct coupling of quantum mechanical (QM) and classical (CL) degrees of freedom – *"hybrid dynamics"* – departs from quantum mechanics. Details of our approach can be found in Refs. [1, 2, 3], where also additional references and discussion of related works are presented.

Hybrid dynamics has been researched extensively for various reasons. – For example, the Copenhagen interpretation of quantum mechanics entails the measurement problem which, together with the fact that quantum mechanics needs interpretation, in order to be operationally well defined, may indicate that it needs amendments. Such as a theory of the *dynamical* coexistence of QM and CL objects. This should have impact on the measurement problem [7] as well as on the description of the interaction between quantum matter and (possibly) classical spacetime [8].

Furthermore, it is of great practical interest to better understand QM-CL hybrids appearing in QM approximation schemes addressing many-body systems or interacting fields, which are naturally separable into QM and CL subsystems; for example, representing fast and slow degrees of freedom, mean fields and fluctuations, *etc.* (See also the article by Hu and Subasi in this volume for an informative discussion of some related issues [9].)

Concerning the hypothetical emergence of quantum mechanics from some coarse-grained deterministic dynamics (see Refs. [10, 11, 12] with numerous references to related work), the quantum-classical backreaction problem might appear in new form, namely regarding the interplay of fluctuations among underlying deterministic and emergent QM degrees of freedom. Which can be rephrased succinctly as: *"Can quantum mechanics be seeded?"*

2.1. Consistency requirements

Thus, there is ample motivation for the numerous attempts to formulate a satisfactory hybrid dynamics. Generally, however, they are deficient in one or another respect. Which has led to various no-go theorems, in view of the lengthy list of desirable properties or consistency requirements that "*the*" hybrid theory should fulfil, see, for example, Refs. [13, 14]:

- Conservation of energy.
- Conservation and positivity of probability.
- Separability of QM and CL subsystems in the absence of their interaction, recovering the correct QM and CL equations of motion, respectively.
- Consistent definitions of states and observables; existence of a Lie bracket structure on the algebra of observables that suitably generalizes Poisson and commutator brackets.
- Existence of canonical transformations generated by the observables; invariance of the classical sector under canonical transformations performed on the quantum sector only and *vice versa.*
- Existence of generalized Ehrenfest relations (*i.e.* the correspondence limit) which, for bilinearly coupled CL and QM oscillators, are to assume the form of the CL equations of motion ("Peres-Terno benchmark" test [15]).
- 'Free Will' [16].
- Locality.
- No-signalling.
- QM / CL symmetries and ensuing separability carry over to hybrids.

These issues have also been discussed for the hybrid ensemble theory of Hall and Reginatto, which does conform with the first six points listed [17, 18] but is in conflict with the last two [1, 19, 20].

We have proposed an alternative theory of hybrid dynamics based on notions of phase space [1]. This extends work by Heslot, demonstrating that quantum mechanics can entirely be rephrased in the language and formalism of classical analytical mechanics [21]. Introducing unified notions of states on phase space, observables, canonical transformations, and a generalized quantum-classical Poisson bracket, this has led to the intrinsically linear hybrid theory to be summarized in the following, which allows to fulfil *all* of the above consistency requirements.

Recently Burić and collaborators have shown that the dynamical aspects of our proposal can indeed be derived for an all-quantum mechanical composite system by imposing constraints on fluctuations in one subsystem, followed by suitable coarse-graining [22, 23, 24].

3. A representation of quantum mechanics in classical terms

We recall that evolution of a *classical* object can be described in relation to its $2n$-dimensional phase space, its *state space.* A real-valued regular function on this space defines an *observable,* *i.e.*, a differentiable function on this smooth manifold.

There always exist (local) systems of *canonical coordinates*, commonly denoted by (x_k, p_k), $k = 1, \dots, n$, such that the *Poisson bracket* of any pair of observables f, g assumes the standard form (Darboux's theorem):

$$\{f,g\} \;=\; \sum_k \Big(\frac{\partial f}{\partial_{x_k}}\frac{\partial g}{\partial_{p_k}} - \frac{\partial f}{\partial_{p_k}}\frac{\partial g}{\partial_{x_k}}\Big) \;. \tag{1}$$

This is consistent with $\{x_k, p_l\} = \delta_{kl}$, $\{x_k, x_l\} = \{p_k, p_l\} = 0$, $k, l = 1, \dots, n$, and has the properties defining a Lie bracket operation, mapping a pair of observables to an observable.

General transformations $\mathcal{G}$ of the state space are restricted by compatibility with the Poisson bracket structure to so-called *canonical transformations*, which do not change physical properties of an object. They form a Lie group and it is sufficient to consider infinitesimal transformations. An *infinitesimal transformation* $\mathcal{G}$ is *canonical*, if and only if for any observable f the map $f \rightarrow \mathcal{G}(f)$ is given by $f \rightarrow f' = f + \{f, g\}\delta\alpha$, with some observable g, the so-called *generator* of $\mathcal{G}$, and $\delta\alpha$ an infinitesimal real parameter. – Thus, for example, the canonical coordinates transform as follows:

$$x_k \;\rightarrow\; x'_k = x_k + \frac{\partial g}{\partial p_k}\delta\alpha \;\;,\;\;\; p_k \;\rightarrow\; p'_k = p_k - \frac{\partial g}{\partial x_k}\delta\alpha \;\;. \tag{2}$$

This illustrates the fundamental relation between observables and generators of infinitesimal canonical transformations in classical Hamiltonian mechanics.

Following Heslot's work, we learn that the previous analysis can be generalized and applied to quantum mechanics; this concerns the dynamical aspects as well as the notions of states, canonical transformations, observables, and measurements [21].

The following derives from the fact that the *Schrödinger equation* and its adjoint are *Hamiltonian equations* following from an action principle [1]. We must add the *normalization condition*, $\mathcal{C} := \langle\Psi(t)|\Psi(t)\rangle \stackrel{!}{=} 1$, for all state vectors $|\Psi\rangle$, which is essential for the probability interpretation of amplitudes; state vectors that differ by an unphysical constant phase are to be identified. Thus, the *QM state space* is formed by the rays of the underlying Hilbert space.

3.1. Oscillator representation

A unitary transformation describes QM evolution, $|\Psi(t)\rangle = \hat{U}(t-t_0)|\Psi(t_0)\rangle$, with $U(t-t_0) = \exp[-i\hat{H}(t-t_0)/\hbar]$, solving the Schrödinger equation. Thus, a stationary state, characterized by $\hat{H}|\phi_i\rangle = E_i|\phi_i\rangle$, with real energy eigenvalue E_i, performs a harmonic motion, *i.e.*, $|\psi_i(t)\rangle = \exp[-iE_i(t-t_0)/\hbar]|\psi_i(t_0)\rangle \equiv \exp[-iE_i(t-t_0)/\hbar]|\phi_i\rangle$. We assume a denumerable set of such states. Following these observations, it is quite natural to introduce what we may call the *oscillator representation.*

We expand state vectors with respect to a complete orthonormal basis, $\{|\Phi_i\rangle\}$:

$$|\Psi\rangle = \sum_i |\Phi_i\rangle(X_i + iP_i)/\sqrt{2\hbar} \;\;, \tag{3}$$

where the time dependent coefficients are separated into real and imaginary parts, X_i, P_i. This expansion allows to evaluate the *Hamiltonian function* defined by $\mathcal{H} := \langle\Psi|\hat{H}|\Psi\rangle$:

$$\mathcal{H} = \frac{1}{2\hbar}\sum_{i,j}\langle\Phi_i|\hat{H}|\Phi_j\rangle(X_i - iP_i)(X_j + iP_j) =: \mathcal{H}(X_i, P_i) \;\;. \tag{4}$$

Choosing the set of energy eigenstates, $\{|\phi_i\rangle\}$, as basis of the expansion, we obtain:

$$\mathcal{H}(X_i, P_i) = \sum_i \frac{E_i}{2\hbar}({P_i}^2 + {X_i}^2) \ , \qquad (5)$$

hence the name *oscillator representation.* – Evaluating $|\dot\Psi\rangle = \sum_i |\Phi_i\rangle(\dot X_i + i\dot P_i)/\sqrt{2\hbar}$ according to Hamilton's equations with $\mathcal{H}$ of Eq. (4) or (5), gives back the Schrödinger equation. – Furthermore, the *normalization condition* becomes:

$$\mathcal{C}(X_i, P_i) = \frac{1}{2\hbar}\sum_i ({X_i}^2 + {P_i}^2) \stackrel{!}{=} 1 \ . \qquad (6)$$

Thus, the vector with components given by (X_i, P_i), $i = 1, \ldots, N$, is confined to the surface of a $2N$-dimensional sphere with radius $\sqrt{2\hbar}$, which presents a major difference to CL Hamiltonian mechanics.

The (X_i, P_i) may be considered as *canonical coordinates* for the state space of a QM object. Correspondingly, we introduce a *Poisson bracket*, cf. Eq.(1), for any two *observables* on the *spherically compactified state space*, *i.e.* real-valued regular functions F, G of the coordinates (X_i, P_i):

$$\{F, G\} \ = \ \sum_i \Big(\frac{\partial F}{\partial_{X_i}}\frac{\partial G}{\partial_{P_i}} - \frac{\partial F}{\partial_{P_i}}\frac{\partial G}{\partial_{X_i}}\Big) \ . \qquad (7)$$

As usual, time evolution of an observable O is generated by the Hamiltonian: $\mathrm{d}O/\mathrm{d}t = \partial_t O + \{O, \mathcal{H}\}$. In particular, we find that the constraint of Eq. (6) is conserved: $\mathrm{d}\mathcal{C}/\mathrm{d}t = \{\mathcal{C}, \mathcal{H}\} = 0$.

3.2. Canonical transformations and quantum observables

In the following, we recall briefly the compatibility of the notion of observable introduced in passing above – as in classical mechanics – with the usual QM one. This can be demonstrated rigourously by the implementation of canonical transformations and analysis of the role of observables as their generators. For details, see Refs. [1, 2, 3, 21].

The Hamiltonian function has been defined as observable in Eq. (4), which relates it directly to the corresponding QM observable, namely the expectation of the self-adjoint Hamilton operator. This is indicative of the general structure with the following most important features.

3.2.1. Compatibility of unitary transformations and Poisson structure. – Classical canonical transformations are automorphisms of the state space which are compatible with the Poisson bracket. Automorphisms of the QM Hilbert space are implemented by unitary transformations. This implies a transformation of the canonical coordinates (X_i, P_i) here. From this, one derives the invariance of the Poisson bracket defined in Eq. (7) under unitary transformations. Consequently, the *unitary transformations on Hilbert space are canonical transformations on the* (X, P) *state space.*

3.2.2. Self-adjoint operators as observables. – Any infinitesimal unitary transformation $\hat U$ can be generated by a self-adjoint operator $\hat G$, such that: $\hat U = 1 - (i/\hbar)\hat G\delta\alpha$, which leads to the QM relation between an observable and a self-adjoint operator. By a simple calculation, one obtains:

$$X_i \ \to \ X_i' = X_i + \frac{\partial\langle\Psi|\hat G|\Psi\rangle}{\partial P_i}\delta\alpha \ , \quad P_i \ \to \ P_i' = P_i - \frac{\partial\langle\Psi|\hat G|\Psi\rangle}{\partial X_i}\delta\alpha \ . \qquad (8)$$

From these equations, the relation between an observable G, defined in analogy to classical mechanics (as above), and a self-adjoint operator $\hat G$ can be inferred:

$$G(X_i, P_i) = \langle\Psi|\hat G|\Psi\rangle \ , \qquad (9)$$

i.e., by comparison with the classical result. Hence, a *real-valued regular function G of the state is an observable, if and only if there exists a self-adjoint operator $\hat{G}$ such that Eq. (9) holds.* This implies that *all QM observables are quadratic forms* in the X_i's and P_i's, which are essentially fewer than in the corresponding CL case. Interacting QM-CL hybrids require additional discussion, cf. below.

3.2.3. Commutators as Poisson brackets. – From the relation (9) between observables and self-adjoint operators and the Poisson bracket (7) one derives:

$$\{F, G\} = \langle\Psi|\frac{1}{i\hbar}[\hat{F}, \hat{G}]|\Psi\rangle \ , \tag{10}$$

with both sides of the equality considered as functions of the variables X_i, P_i and with the commutator defined as usual. Hence, the *QM commutator is a Poisson bracket with respect to the (X, P) state space* and relates the algebra of its observables to the algebra of self-adjoint operators.

In conclusion, quantum mechanics shares with classical mechanics an even dimensional state space, a Poisson structure, and a related algebra of observables. It differs essentially by a restricted set of observables and the requirements of phase invariance and normalization, which compactify the underlying Hilbert space to the complex projective space formed by its rays.

4. Quantum-classical Poisson bracket, hybrid states and their evolution

The far-reaching parallel of classical and quantum mechanics, as we have seen, suggests to introduce a *generalized Poisson bracket* for QM-CL hybrids:

$$\{A, B\}_\times \ := \ \{A, B\}_{\mathrm{CL}} + \{A, B\}_{\mathrm{QM}} \tag{11}$$

$$:= \ \sum_k \Big(\frac{\partial A}{\partial x_k}\frac{\partial B}{\partial p_k} - \frac{\partial A}{\partial p_k}\frac{\partial B}{\partial x_k}\Big) + \sum_i \Big(\frac{\partial A}{\partial X_i}\frac{\partial B}{\partial P_i} - \frac{\partial A}{\partial P_i}\frac{\partial B}{\partial X_i}\Big) \ , \tag{12}$$

of any two observables A, B defined on the Cartesian product of CL *and* QM state spaces. It shares the usual properties of a Poisson bracket. – Note that due to the convention introduced by Heslot [21], to which we adhered in Sect. 3, the QM variables X_i, P_i have dimensions of $(\text{action})^{1/2}$ and, consequently, no $\hbar$ appears in Eqs. (11)–(12). At the expense of introducing appropriate rescalings, these variables could be made to have their usual dimensions and $\hbar$ to appear explicitly here. – For the remainder of this article, instead, we choose units such that $\hbar \equiv 1$.

Let an *observable "belong" to the CL (QM) sector, if it is constant with respect to the canonical coordinates of the QM (CL) sector.* Then, the $\{\ ,\ \}_\times$-bracket has the important properties:

- It reduces to the Poisson brackets introduced in Eqs. (1) and (7), respectively, for pairs of observables that belong *either* to the CL *or* the QM sector.
- It reduces to the appropriate one of the former brackets, if one of the observables belongs only to either one of the two sectors.
- It reflects the *separability* of CL and QM sectors, since $\{A, B\}_\times = 0$, if A and B belong to different sectors.

Hence, *if a canonical tranformation is performed on the QM (CL) sector only, then observables that belong to the CL (QM) sector remain invariant.*

4.1. The hybrid phase space density

The hybrid density ρ for a self-adjoint density operator $\hat{\rho}$ in a given state $|\Psi\rangle$ is defined by [1] :

$$\rho(x_k, p_k; X_i, P_i) := \langle\Psi|\hat{\rho}(x_k, p_k)|\Psi\rangle = \frac{1}{2}\sum_{i,j} \rho_{ij}(x_k, p_k)(X_i - iP_i)(X_j + iP_j) \ , \tag{13}$$

using Eq. (3) and $\rho_{ij}(x_k, p_k) := \langle\Phi_i|\hat{\rho}(x_k, p_k)|\Phi_j\rangle = \rho^*_{ji}(x_k, p_k)$. It describes a *QM-CL hybrid ensemble* by a real-valued, positive semi-definite, normalized, and possibly time dependent regular function on the Cartesian product state space canonically coordinated by $2(n+N)$-tuples $(x_k, p_k; X_i, P_i)$; the variables $x_k, p_k,\ k = 1, \ldots, n$ and $X_i, P_i,\ i = 1, \ldots, N$ are reserved for CL and QM sectors, respectively.

Note that the relation between an observable in (X, P)-space and a self-adjoint operator, Eq. (9), can be written as: $G(X_i, P_i) = \mathrm{Tr}(|\Psi\rangle\langle\Psi|\hat{G})$, with the QM pure state acting as one-dimensional projector here. Concerning the hybrid density ρ, we may use the representation of $\hat{\rho}$ in terms of its eigenstates, $\hat{\rho} = \sum_j w_j|j\rangle\langle j|$, to obtain:

$$\rho(x_k, p_k; X_i, P_i) = \sum_j w_j(x_k, p_k)\mathrm{Tr}(|\Psi\rangle\langle\Psi|j\rangle\langle j|) = \sum_j w_j(x_k, p_k)|\langle j|\Psi\rangle|^2 \ , \tag{14}$$

with $0 \leq w_j \leq 1$ and $\sum_j \int \Pi_l(\mathrm{d}x_l\mathrm{d}p_l)w_j(x_k, p_k) = 1$. – This shows that $\rho(x_k, p_k; X_i, P_i)$ is the *probability density to find in the hybrid ensemble the QM pure state* $|\Psi\rangle$, parametrized by X_i, P_i through Eq. (3), *together with the CL state* given by a point in phase space, specified by coordinates x_k, p_k.

Further remarks about superposition, pure/mixed, or separable/entangled QM states that may enter the hybrid density can be found in Ref. [2, 23]. Generally, the simplest observable-like form of ρ – a bilinear function of QM "phase space" variables X_i, P_i, as above – has to be replaced for interacting hybrids, allowing for the form of an *almost-classical observable*; see Sect. 5.4 in [1] and a more detailed study discussing measurements in [3].

In particular, the density ρ evolves according to the hybrid Liouville equation (see the following subsection). This will lead to a solution $\rho(x_k, p_k; X_i, P_i; t)$ that is not necessarily bilinear in the QM variables X_i, P_i. However, we can (re)construct the self-adjoint density operator for the QM subsystem of the hybrid by:

$$\hat{\rho}(x_k, p_k; t) = \Gamma_N^{-1} \int_{\delta S_{2N}(\sqrt{2})} \Pi_i(\mathrm{d}X_i\mathrm{d}P_i)\ \rho(x_k, p_k; X_i, P_i; t)|\Psi(X_i, P_i)\rangle\langle\Psi(X_i, P_i)| \ , \tag{15}$$

i.e., conditioned on the state of the CL subsystem parametrized by (x_k, p_k) and consistent with Eq. (13); the measure of integration here has been evaluated in detail in Ref. [1].

4.2. Phase space evolution

The *Liouville equation* for the evolution of hybrid ensembles [1] must be based on the generalized Poisson bracket defined in Eqs. (11)–(12) and Liouville's theorem. This leads us to:

$$-\partial_t\rho = \{\rho, \mathcal{H}_\Sigma\}_\times \ , \tag{16}$$

with $\mathcal{H}_\Sigma \equiv \mathcal{H}_\Sigma(x_k, p_k; X_i, P_i)$ and:

$$\mathcal{H}_\Sigma := \mathcal{H}_{\mathrm{CL}}(x_k, p_k) + \mathcal{H}_{\mathrm{QM}}(X_i, P_i) + \mathcal{I}(x_k, p_k; X_i, P_i) \ , \tag{17}$$

defining the relevant Hamiltonian function, including a hybrid interaction; $\mathcal{H}_\Sigma$ is required to be an *observable*, in order to have a meaningful notion of energy. – Note that *energy conservation* follows from $\{\mathcal{H}_\Sigma, \mathcal{H}_\Sigma\}_\times = 0$.

Finally, the Liouville equation describes a Hamiltonian flow, which implies that:

- The normalization and positivity of the probability density ρ are conserved in presence of a hybrid interaction; hence, its interpretation remains valid under hybrid evolution.

5. Conclusion

Besides constructing the QM-CL hybrid formalism, as outlined here, and showing how it conforms with the strong consistency requirements presented in Sec. 2.1, we earlier discussed the possibility to have classical-environment induced decoherence, quantum-classical backreaction, a deviation from the Hall-Reginatto proposal [17, 18] in presence of translation symmetry, and closure of the algebra of hybrid observables [1, 3, 22]. Questions of locality, symmetry vs. separability, incorporation of superposition, separable, and entangled QM states, and 'Free Will' were considered in Ref. [2, 16, 23], while an exotic application to (a class of) time machines coupled to quantum mechanical degrees of freedom has been shown in Ref. [4] and further studies of the Penrose *et al.* experiment, designed to study reduction (or not) of macroscopic Schrödinger cat states [5], and of entanglement control are underway [6].

Acknowledgments

It is a pleasure to thank N. Burić and T. Skinner for discussions and correspondence.

References

[1] Elze H-T 2012 Linear dynamics of quantum-classical hybrids *Phys. Rev.* A **85** 052109
[2] Elze H-T 2012 Four questions for quantum-classical hybrid theory *J. Phys.: Conf. Ser.* **361** 012004
[3] Elze H-T 2012 Proliferation of observables and measurement in quantum-classical hybrids *Int. J. Qu. Inf. (IJQI)* **10** No. 8, 1241012
[4] Elze H-T 2013 Discrete mechanics, "time machines" and hybrid systems *Europ. Phys. Journ. : Web of Sci.* to appear
[5] Marshall W, Simon C, Penrose R and Bouwmeester D 2003 Towards quantum superpositions of a mirror *Phys. Rev. Lett.* **91** 130401
[6] Elze H-T, Fratino L and Lampo A 2013 *in preparation*
[7] Sherry T N and Sudarshan E C G 1978 Interaction between classical and quantum systems: A new approach to quantum measurement. I. *Phys. Rev.* D **18** 4580; do. 1979 II. *Phys. Rev.* D **20** 857
[8] Boucher W and Traschen J 1988 Semiclassical physics and quantum fluctuations *Phys. Rev.* D **37** 3522
[9] Hu B-L and Subasi Y 3013 Pathways toward understanding macroscopic quantum phenomena *J. Phys.: Conf. Ser.* in this volume
[10] 't Hooft G 2010 Classical cellular automata and quantum field theory *Int. J. Mod. Phys.* A **25** 4385; *do.* 2009 *Preprint* arXiv:0908.3408
[11] Elze H-T 2009 Does quantum mechanics tell an atomistic spacetime? *J. Phys.: Conf. Ser.* **174** 012009; *do.* 2006 Deterministic models of quantum fields *J. Phys.: Conf. Ser.* **33** 399
[12] Adler S L 2005 *Quantum Mechanics as an Emergent Phenomenon* (Cambridge, UK: Cambridge U. Press)
[13] Caro J and Salcedo L L 1999 Impediments to mixing classical and quantum dynamics *Phys. Rev.* A **60** 842
[14] Diósi L, Gisin N and Strunz W T 2000 Quantum approach to coupling classical and quantum dynamics *Phys. Rev.* A **61** 022108
[15] Peres A and Terno D R 2001 Hybrid classical-quantum dynamics *Phys. Rev.* A **63** 022101
[16] Diósi L 2012 Classical-quantum coexistence: a 'Free Will' test *J. Phys.: Conf. Ser.* **361** 012028
[17] Hall M J W and Reginatto M 2005 Interacting classical and quantum ensembles *Phys. Rev.* A **72** 062109
[18] Hall M J W 2008 Consistent classical and quantum mixed dynamics *Phys. Rev.* A **78** 042104
[19] Chua A J K, Hall M J W and Savage C M 2011 Interacting classical and quantum particles *Phys. Rev.* A **85** 022110
[20] Hall M J W, Reginatto M and Savage C M 2012 Nonlocal signaling in the configuration space model of quantum-classical interactions *Phys. Rev.* A **86** 054101
[21] Heslot A 1985 Quantum mechanics as a classical theory *Phys. Rev.* D **31** 1341
[22] Radonjić M, Prvanović S and Burić N 2012 Hybrid quantum-classical models as constrained quantum systems *Phys. Rev.* A **85** 064101
[23] Burić N, Mendas I, Popović D B, Radonjić M and Prvanović S 2012 Statistical ensembles in Hamiltonian formulation of hybrid quantum-classical systems *Phys. Rev.* A **86** 034104
[24] Burić N *et al.* 2013 Hybrid dynamics as a constrained quantum system *J. Phys.: Conf. Ser.* in this volume

Entropy of conditional tomographic probability distributions for classical and quantum systems

Margarita A Man'ko and Vladimir I Man'ko
P. N. Lebedev Physical Institute, Leninskii Prospect 53, Moscow 119991, Russia
E-mail: mmanko@sci.lebedev.ru

Abstract. The possibility to describe hybrid systems containing classical and quantum subsystems by means of conditional tomographic probability distributions (tomograms) is discussed. Tomographic Shannon and Rényi entropies associated with the tomograms are studied, and new tomographic uncertainty relations are obtained.

1. Introduction

The possibility to describe the hybrid systems containing both classical and quantum subsystems was considered, for example, in [1–5]. The problems presented in the process of considering combined systems are related to a difference in the description of classical and quantum states. The classical states are associated with the probability density on the phase space. The quantum states are associated either with the wave function or with the density matrix [6, 7].

Recently the probability representation of quantum states was introduced using the tomographic approach [8] (see also the review [9]). Within the framework of this approach, the states in quantum mechanics are identified with the probability density of the particle position (called the tomogram) measured in a rotated and scaled reference frame in the phase space. It is important that the states in classical mechanics can also be identified with tomograms [10, 11].

In view of the mentioned above, the difference in the concept of classical and quantum states in the tomographic approach is reduced to the difference of the sets where the classical and quantum tomograms are located. This circumstance was used to make an attempt to describe a hybrid system consisted of the classical and quantum subsystems by a joint tomographic probability distribution depending on the positions of the classical and quantum particles. Also it was introduced that the tomograms can be interpreted as conditional probability distributions [12, 13]. For example, the optical tomogram $w(X,\theta)$, being dependent on a random homodyne quadrature X and the local oscillator phase θ [14, 15], which is a measurable parameter (see, e.g., [16, 17]), can be interpreted as the conditional probability density $w(X,\theta) = w(X \mid \theta)$.

In view of Bayesian formula, the conditional probability distribution is connected with a joint probability distribution $P(X,\theta)$ of the both random variables.

The aim of our work is to study the possibility to associate the states of hybrid systems consisted of classical and quantum subsystems with the conditional probability densities and joint probability distributions related to the densities by the Bias formula. Using the tomographic entropies associated with the probability distributions, we consider following [18] the entropic inequalities for hybrid systems.

This paper is organized as follows.

In section 2, we review the tomographic approach and present optical and symplectic tomograms of classical and quantum systems. In section 3, we consider the tomograms of hybrid systems. In section 4, we study the tomographic entropies. In section 5, we discuss the properties of possible kinetic equations for the tomograms of the hybrid systems. We give our conclusions and prospectives in section 6.

2. Tomograms of classical and quantum systems

Following [13] we review in this section the properties of tomograms.

By definition, the optical tomogram $w(\vec{X} \mid \vec{\theta})$ of a classical system with N degrees of freedom, i.e., $\vec{X} = (X_1, X_2, \dots, X_N)$ and $\vec{\theta} = (\theta_1, \theta_2, \dots, \theta_N)$, where $-\infty < X_j \leq \infty$ and $0 \leq \theta_j \leq 2\pi$, is the joint conditional probability density of random positions X_j, having in mind that the measurements of the positions are made in the set of reference frames in the system phase space with rotated initial axes $q_j \to X_j = q_j \cos\theta_j + p_j \sin\theta_j$ and $p_j \to \mathcal{P}_j = -q_j \sin\theta_j + p_j \cos\theta_j$. The set of N angles θ_j determines a new set of reference frames rotated in the phase space.

The nonnegative symplectic tomogram $M(\vec{X} \mid \vec{\mu}, \vec{\nu})$ of a classical system is the joint conditional probability density of random positions X_j measured in the canonically transformed reference frame in the system phase space, having in mind that new scales in the space and time for each degree of freedom are introduced, i.e., $q_j \to \mu_j q_j$, $t \to \nu_j^{-1} t$, and the measured positions are $X_j = \mu_j q_j + \nu_j p_j$. The scaling parameters satisfy the condition $-\infty < \mu_j, \nu_j \leq \infty$.

For the case $\mu_j = \cos\theta_j$ and $\nu_j = \sin\theta_j$, the symplectec tomogram provides the optical tomogram, i.e., $M(\vec{X} \mid \cos\vec{\theta}, \sin\vec{\theta}) = w(\vec{X} \mid \vec{\theta})$, where $\sin\vec{\theta} = (\sin\theta_1, \sin\theta_2, \dots \sin\theta_N)$ and $\cos\vec{\theta} = (\cos\theta_1, \cos\theta_2, \dots \cos\theta_N)$.

The joint probability density $f(\vec{q}, \vec{p}) \geq 0$ of N positions and N momenta is determined by the symplectic tomogram due to the Fourier transform

$$f(\vec{q}, \vec{p}) = \int M(\vec{X} \mid \vec{\mu}, \vec{\nu}) \left\{ \prod_{k=1}^{N} \exp[i(X_k - \mu_k q_k - \nu_k p_k)] \frac{dq_k dp_k}{4\pi^2} \right\}. \tag{1}$$

Since $f(\vec{q}, \vec{p}) \geq 0$, only such symplectic tomograms $M(\vec{X} \mid \vec{\mu}, \vec{\nu})$ are admissible for describing the classical system states for which Fourier integral (1) yields a nonnegative function.

The symplectic tomogram can be found in view of the inverse Radon transform

$$M(\vec{X} \mid \vec{\mu}, \vec{\nu}) = \int f(\vec{q}, \vec{p}) \left\{ \prod_{k=1}^{N} \delta(X_k - \mu_k q_k - \nu_k p_k)\, dq_k dp_k \right\}. \tag{2}$$

Both optical and symplectic tomograms are normalized functions

$$\int M(\vec{X} \mid \vec{\mu}, \vec{\nu})\, d\vec{X} = \int w(\vec{X} \mid \vec{\theta})\, d\vec{X} = 1. \tag{3}$$

The conditional probability densities can be associated with joint probability densities via Bayesian formula. For example, one has the optical tomogram

$$w(\vec{X} \mid \vec{\theta}) = \widetilde{w}(\vec{X}, \vec{\theta}) \left\{ \int \widetilde{w}(\vec{X}, \vec{\theta})\, d\vec{X} \right\}^{-1} \tag{4}$$

in terms of a joint probability density $\widetilde{w}(\vec{X}, \vec{\theta})$ of random positions X_j and rotations θ_j.

An analogous relation takes place for symplectic tomograms

$$M(\vec{X} \mid \vec{\mu}, \vec{\nu}) = \widetilde{M}(\vec{X}, \vec{\mu}, \vec{\nu}) \left\{ \int \widetilde{M}(\vec{X}, \vec{\mu}, \vec{\nu})\, d\vec{X} \right\}^{-1}, \tag{5}$$

where $\widetilde{M}(\vec{X}, \vec{\mu}, \vec{\nu})$ is a joint probability density of random positions X_j and scaling parameters μ_j and ν_j.

For a given joint probability density, there exists the conditional probability density determined without an ambiguity, but for a given conditional probability density there exist many joint probability densities. For example, by choosing an arbitrary marginal probability distribution in the denominator of (4) as

$$\mathcal{P}(\vec{\theta}) = \int \widetilde{w}(\vec{X}, \vec{\theta})\, d\vec{X}, \tag{6}$$

we have

$$\widetilde{w}(\vec{X}, \vec{\theta}) = w(\vec{X} \mid \vec{\theta})\mathcal{P}(\vec{\theta}). \tag{7}$$

Analogously, for symplectic tomogram we have

$$\widetilde{w}(\vec{X}, \vec{\mu}, \vec{\nu}) = M(\vec{X} \mid \vec{\mu}, \vec{\nu})\Pi(\vec{\mu}, \vec{\nu}), \tag{8}$$

where $\Pi(\vec{\mu}, \vec{\nu})$ is an arbitrary joint probability density of random variables μ_j and ν_j.

There exists the other tomographic construction called the center-of-mass tomography (see, e.g., [9]). The center-of-mass tomogram $C(X \mid \vec{\mu}, \vec{\nu})$ depends only on one random position X interpreted as an analog of the system center-of-mass coordinate. Thus, this tomogram is the probability density of the coordinate X which is considered in a reference frame in the phase space with rescaled time and positions for all degrees of freedom. Then the conventional probability density $f(\vec{q}, \vec{p})$ is given by the Fourier integral

$$f(\vec{q}, \vec{p}) = (2\pi)^{-2N} \int C(X \mid \vec{\mu}, \vec{\nu}) \exp[i(X - \vec{\mu}\vec{q} - \vec{\nu}\vec{p})]\, dX\, d\vec{\mu}\, d\vec{\nu}. \tag{9}$$

The inverse Radon transform determines the center-of-mass tomogram

$$C(X \mid \vec{\mu}, \vec{\nu}) = \int f(\vec{q}, \vec{p})\delta(X - \vec{\mu}\vec{q} - \vec{\nu}\vec{p})\, d\vec{q}\, d\vec{p}. \tag{10}$$

The tomogram is nonnegative and normalized conditional probability density

$$\int C(X \mid \vec{\mu}, \vec{\nu})\, dX = 1. \tag{11}$$

The joint probability density $\widetilde{C}(X,\vec{\mu},\vec{\nu})$ can be reconstructed

$$\widetilde{C}(X,\vec{\mu},\vec{\nu}) = C(X \mid \vec{\mu},\vec{\nu})\Pi(\vec{\mu},\vec{\nu}), \tag{12}$$

where $\Pi(\vec{\mu},\vec{\nu})$ is an arbitrary joint probability density of random variables μ_j and ν_j.

For classical systems, the probability densities $C(X \mid \vec{\mu},\vec{\nu})$ are admissible as the state tomograms only if the Fourier integral given by (9) provides a nonnegative function.

For quantum states, one has the same tomograms given in terms of the density operators $\widehat{\rho}$. For example, the optical tomogram reads

$$w(\vec{X} \mid \vec{\theta}) = \mathrm{Tr}\,\widehat{\rho}\prod_{j=1}^{N}\delta(X_j\widehat{1} - \widehat{q}_j\cos\theta_j - \widehat{p}_j\sin\theta_j). \tag{13}$$

The symplectic tomogram is given by an analogous formula

$$M(\vec{X} \mid \vec{\mu},\vec{\nu}) = \mathrm{Tr}\,\widehat{\rho}\prod_{j=1}^{N}\delta(X_j\widehat{1} - \mu_j\widehat{q}_j - \nu_j\widehat{p}_j). \tag{14}$$

The center-of-mass tomogram is

$$C(X \mid \vec{\mu},\vec{\nu}) = \mathrm{Tr}\,\widehat{\rho}\;\delta(X\widehat{1} - \vec{\mu}\widehat{\vec{q}} - \vec{\nu}\widehat{\vec{p}}). \tag{15}$$

The density operator $\widehat{\rho}$ can be found by means of quantum Radon transform. For example, the symplectic tomogram provides the reconstruction of the density operator as follows:

$$\widehat{\rho} = (2\pi)^{-2N}\int M(\vec{X} \mid \vec{\mu},\vec{\nu})\left\{\prod_{k=1}^{N}\exp[i(X_k\widehat{1} - \mu_k\widehat{q}_k - \nu_k\widehat{p}_k)]\,dX_k\,d\mu_k\,d\nu_k\right\}. \tag{16}$$

For the center-of-mass tomography, one has the density operator in the form

$$\widehat{\rho} = (2\pi)^{-2N}\int C(X \mid \vec{\mu},\vec{\nu})\exp[i(X\widehat{1} - \vec{\mu}\widehat{\vec{q}} - \vec{\nu}\widehat{\vec{p}})]\,dX\,d\vec{\mu}\,d\vec{\nu}. \tag{17}$$

Quantum states are associated with nonnegative density operators. Thus, the probability densities $M(\vec{X} \mid \vec{\mu},\vec{\nu})$ and $C(X \mid \vec{\mu},\vec{\nu})$ are admissible as tomograms of quantum states only if integrals (16) and (17) yield the Hermitian operators with nonnegative eigenvalues. In the quantum case, the joint tomographic probability distributions can be constructed in view of the same relations (8) and (12) used for classical systems. It is worth pointing out that the nonnegativity conditions for the Fourier integrals of the tomograms of classical and quantum systems are different. There can exist the probability distributions satisfying the both conditions, either only classical or only quantum condition, as well as neither classical nor quantum one.

In spite of the fact that, in the tomographic picture of classical and quantum mechanics, the states (both classical and quantum) are associated with the same tomographic probability densities like optical, symplectic, and center-of-mass tomograms, sets of the tomograms are different. The sets are determined by the nonnegativity conditions of the corresponding integrals reconstructing the probability density in the classical domain and the density operator in the quantum domain. This is just the essence of the difference in the classical and quantum cases in the context of tomographic description of the states.

3. Hybrid systems

In [19, 20], it was suggested to describe hybrid systems consisted of classical and quantum subsystems by tomographic probability distributions. Below we develop this proposal on an example of the center-of-mass tomogram.

We consider a classical subsystem with N_1 degrees of freedom and a quantum subsystem with N_2 degrees of freedom. With the hybrid-system state, we associate the cluster center-of-mass tomogram $C(X, Y \mid \vec{\mu}, \vec{\nu}, \vec{m}, \vec{n})$, where $\vec{\mu} = (\mu_1, \mu_2, \ldots, \mu_{N_1})$, $\vec{\nu} = (\nu_1, \nu_2, \ldots, \nu_{N_1})$, $\vec{m} = (m_1, m_2, \ldots, m_{N_2})$, and $\vec{n} = (n_1, n_2, \ldots, n_{N_2})$. The cluster center-of-mass tomogram $C(X, Y \mid \vec{\mu}, \vec{\nu}, \vec{m}, \vec{n})$ is nonnegative normalized distribution function

$$\int C(X, Y \mid \vec{\mu}, \vec{\nu}, \vec{m}, \vec{n})\, dX\, dY = 1. \tag{18}$$

The state of the classical subsystem is associated with the marginal center-of-mass tomogram

$$C_{\mathrm{cl}}(X \mid \vec{\mu}, \vec{\nu}) = \int C(X, Y \mid \vec{\mu}, \vec{\nu}, \vec{m}, \vec{n})\, dY, \tag{19}$$

and the state of the quantum subsystem is associated with the following marginal center-of-mass tomogram

$$C_{\mathrm{q}}(Y \mid \vec{m}, \vec{n}) = \int C(X, Y \mid \vec{\mu}, \vec{\nu}, \vec{m}, \vec{n})\, dX. \tag{20}$$

The tomogram of the hybrid-system state must satisfy the conditions

$$f(\vec{q}, \vec{p}) = (2\pi)^{-2N_1} \int C(X, Y \mid \vec{\mu}, \vec{\nu}, \vec{m}, \vec{n}) \exp[i(X - \vec{\mu}\vec{q} - \vec{\nu}\vec{p})]\, dX\, dY\, d\vec{\mu}\, d\vec{\nu} \geq 0, \tag{21}$$

$$\widehat{\rho} = (2\pi)^{-2N_2} \int C(X, Y \mid \vec{\mu}, \vec{\nu}, \vec{m}, \vec{n}) \exp[i(Y - \vec{m}\widehat{\vec{q}} - \vec{n}\widehat{\vec{p}})]\, dX\, dY\, d\vec{m}\, d\vec{n} \geq 0, \tag{22}$$

where $\vec{\mu}\,\vec{q} = \sum_{k=1}^{N_1} \mu_k q_k$, $\vec{\nu}\,\vec{p} = \sum_{k=1}^{N_1} \nu_k p_k$, $\vec{m}\,\widehat{\vec{q}} = \sum_{j=1}^{N_2} m_j \widehat{q}_j$, and $\vec{n}\,\widehat{\vec{p}} = \sum_{j=1}^{N_2} n_j \widehat{p}_j$, with $\widehat{q}_j$ and $\widehat{p}_j$ being the position and momenta operators of quantum degrees of freedom. Analogous conditions can be written for optical and symplectic tomograms [20]. Conditions (21) and (22) correspond to the nonnegativity of the probability density $f(\vec{q}, \vec{p})$ on the phase space of the classical subsystem state and the nonnegativity of the density operator $\widehat{\rho}$ of the quantum subsystem state.

The simplest case of the hybrid-system-state tomogram is the tomographic probability density without correlations, i.e.,

$$C(X, Y \mid \vec{\mu}, \vec{\nu}, \vec{m}, \vec{n}) = C_{\mathrm{cl}}(X \mid \vec{\mu}, \vec{\nu}) C_{\mathrm{q}}(Y \mid \vec{m}, \vec{n}). \tag{23}$$

The case of correlated classical quantum states can be described by tomograms which are convex sums of uncorrelated tomograms, e.g.,

$$C(X, Y \mid \vec{\mu}, \vec{\nu}, \vec{m}, \vec{n}) = \sum_k P(k) C_{\mathrm{cl}}^{(k)}(X \mid \vec{\mu}, \vec{\nu}) C_{\mathrm{q}}^{(k)}(Y \mid \vec{m}, \vec{n}), \tag{24}$$

where $P(k) \geq 0$ is the probability distribution satisfying the normalization condition $\sum_k P(k) = 1$, and the index k can be either discrete or continuous one. The form of tomogram (24) corresponds to separable probability distributions.

Also other kinds of the probability distributions can exist, which cannot be presented in the form (24), but can be presented in the form of entangled probability distributions like

$$\begin{aligned} C(X,Y \mid \vec{\mu},\vec{\nu},\vec{m},\vec{n}) &= (1+\delta)\sum_k P(k) C_{\rm cl}^{(k)}(X \mid \vec{\mu},\vec{\nu}) C_{\rm q}^{(k)}(Y \mid \vec{m},\vec{n}) \\ &\quad -\delta \sum_j \mathcal{P}(j) C_{\rm cl}^{(j)}(X \mid \vec{\mu},\vec{\nu}) C_{\rm q}^{(j)}(Y \mid \vec{m},\vec{n}), \end{aligned} \tag{25}$$

where $\delta > 0$, and $P(k)$ and $\mathcal{P}(j)$ are normalized probability distributions of the corresponding indices k and j.

Analogous arguments provide the possibility to write a structure of the optical tomogram $w(\vec{X},\vec{Y},\vec{\theta},\vec{\varphi})$ of a hybrid-system state with N_1 classical degrees of freedom $\vec{X}$ and N_2 quantum degrees of freedom $\vec{Y}$, i.e.,

$$\begin{aligned} w(\vec{X},\vec{Y} \mid \vec{\theta},\vec{\varphi}) &= (1+\delta)\sum_k P(k) w_{\rm cl}^{(k)}(\vec{X} \mid \vec{\theta}) w_{\rm q}^{(k)}(\vec{Y} \mid \vec{\varphi}) \\ &\quad -\delta \sum_j \mathcal{P}(j) w_{\rm cl}^{(j)}(\vec{X} \mid \vec{\theta}) w_{\rm q}^{(j)}(\vec{Y} \mid \vec{\varphi}). \end{aligned} \tag{26}$$

In (25) and (26), at $\delta = 0$, one has a separable form of the tomographic conditional probability distributions of the hybrid-system states.

For the symplectic tomogram, the corresponding formula reads

$$\begin{aligned} M(\vec{X},\vec{Y} \mid \vec{\mu},\vec{\nu},\vec{m},\vec{n}) &= (1+\delta)\sum_k P(k) M_{\rm cl}^{(k)}(\vec{X} \mid \vec{\mu},\vec{\nu}) M_{\rm q}^{(k)}(\vec{Y} \mid \vec{m},\vec{n}) \\ &\quad -\delta \sum_j \mathcal{P}(j) M_{\rm cl}^{(j)}(\vec{X} \mid \vec{\mu},\vec{\nu}) M_{\rm q}^{(j)}(\vec{Y} \mid \vec{m},\vec{n}). \end{aligned} \tag{27}$$

4. Admissible dynamics of the hybrid-system tomograms

Now we discuss a possible dynamics of tomograms of the hybrid-system states starting from expression (24) for the center-of-mass tomogram. Since the tomogram is the probability distribution, its dynamics can be described as a map $\Phi(t)$ of the probability distribution onto a probability distribution of the same kind. Thus, we determine some admissible maps given by the formula

$$\begin{aligned} \Phi(t) C(X,Y \mid \vec{\mu},\vec{\nu},\vec{m},\vec{n}) &= C(X,Y,t \mid \vec{\mu},\vec{\nu},\vec{m},\vec{n}) \\ &= \sum_k P(k,t) C_{\rm cl}^{(k)}(X,t \mid \vec{\mu},\vec{\nu}) C_{\rm q}^{(k)}(Y,t \mid \vec{m},\vec{n}). \end{aligned} \tag{28}$$

The above dynamics is coded by the dynamics of the probability distribution $P(k)$ as well as by the time dependence of the classical subsystem-state tomograms $C_{\rm cl}^{(k)}(X,t \mid \vec{\mu},\vec{\nu})$ and quantum subsystem-state tomograms $C_{\rm q}^{(k)}(Y,t \mid \vec{m},\vec{n})$. If the map $\Phi(t)$ is factorized in the form of direct product of the classical and quantum maps of the tomographic probability distributions $\Phi(t) = \Phi_{\rm cl}(t) \otimes \Phi_{\rm q}(t)$ and the probability distributions $P(k)$ and $\mathcal{P}(j)$ do not vary with time, the kinetic equation for the tomogram can be considered as a combination of independent kinetic equations for classical and quantum subsystems. In this case, for the factorized initial states the classical and quantum subsystems evolve according to the classical and quantum evolution equations. Thus, the zero initial correlations of classical and quantum degrees of freedom remain to be zero at any time moment. The map $\Phi_{\rm cl}(t)$ acts only on the classical degrees of freedom of the hybrid system while the map $\Phi_{\rm q}(t)$ acts only on the quantum degrees of freedom of the hybrid system. This means that the maps preserve the conditions of tomogram admissibility for any time moment.

The map $\Phi(t)$ can be constructed as a convex sum of the direct product maps

$$\Phi(t) = \sum_k \Pi(k,t)\Phi_{\rm cl}^{(k)}(t) \otimes \Phi_{\rm q}^{(k)}(t), \qquad (29)$$

where $\Pi(k,t) \geq 0$ and $\sum_k \Pi(k,t) = 1$. In this case, due to the dynamics, the initial uncorrelated state of the hybrid system starts to have nonzero correlations of classical and quantum observables.

The set of all admissible maps of the hybrid-system states has a structure of the semigroup. The tomograms of the hybrid-system states belong to the domain in the simplex of all probability distributions. This domain is determined for the center-of-mass tomograms by conditions (21) and (22). The admissible dynamics of the hybrid-system states is determined by the invariance semigroup of the domain in the simplex.

We consider a simple example of the dynamics of a system with one classical X and one quantum Y degrees of freedom. Let the initial state to be described by a factorized symplectic tomogram

$$M(X,Y \mid \mu,\nu,m,n) = [\pi(m^2+n^2)]^{-1/2}\delta(X-\mu q_0-\nu p_0)\exp[-Y^2/(m^2+n^2)]. \qquad (30)$$

The initial state tomogram (30) corresponds to the Gaussian state of the quantum degree of freedom with equal dispersions of position and momentum $\sigma_{qq} = \sigma_{pp} = 1/2$ and the classical system with position q_0 and momentum p_0.

Let the dynamics to be described by the map

$$\Phi(t) = \alpha 1 \otimes \Phi_{\rm q}(t) + (1-\alpha)\Phi_{\rm cl}(t) \times 1, \qquad 0 \leq \alpha \leq 1, \qquad (31)$$

where $\Phi_{\rm q}(t)$ provides the free-motion map of the quantum degree of freedom and $\Phi_{\rm cl}(t)$ provides the free-motion map of the classical system. After the action of dynamical map (31), the tomogram of the hybrid-system state (30) becomes

$$\begin{aligned} M(X,Y,t \mid \mu,\nu,m,n) &= \Phi(t)M(X,Y,\mid \mu,\nu,m,n) \\ &= \alpha\delta(X-\mu q_0-\nu p_0)[2\pi\sigma_{YY}(t)]^{-1/2}\exp[-Y^2/2\sigma_{YY}(t)] \\ &+(1-\alpha)\delta(X-\mu[q_0+p_0t]-\nu p_0)[\pi(m^2+n^2)]^{-1/2}\exp[-Y^2/(m^2+n^2)]. \end{aligned} \qquad (32)$$

The dispersion parameter $\sigma_{YY}(t)$ reads

$$\sigma_{YY}(t) = [(m^2+n^2)/2] + mnt + m^2t^2/2. \qquad (33)$$

The constant parameter α can be replaced by the time-dependent probability $\alpha \to \cos^2 \Omega t$. In this case, the classical–quantum correlations are modeled by the map $\Phi(t)$, which provides the variation in time of the probability distribution of uncorrelated products in the convex sum of such products.

Other examples of admissible dynamics of hybrid-system tomograms can be constructed. In this way, the classical subsystem can follow the dynamics given by kinetic equations with dissipation, and the quantum subsystem can follow the dynamics provided by Gorini–Kossakowski–Sudarshan–Lindblad equations [21, 22]. The time dependence of the probability distribution $P_k(t)$ can be modeled by an arbitrary stochastic map (both Markovian or non-Markovian). The physical meaning and possible applications of the discussed dynamical models of the hybrid-system states need extra study.

5. Tomographic entropy of hybrid systems

The optical, symplectic, and center-of-mass tomograms under discussion are standard probability distributions. For classical and quantum systems, the Shannon and Rényi tomographic entropies and their specific properties were reviewed, e.g., in [18]. Some new inequalities for the tomographic entropies were checked in experiments [17] with homodyne detection of photon states where the optical tomograms are measured.

The tomographic entropy can also be introduced for hybrid systems. For example, the Rényi entropy associated with the center-of-mass tomogram of the hybrid-system state is defined as follows:

$$R_C(q) = \frac{1}{1-q}\ln\left\{\int [C(X, Y \mid \vec{\mu}, \vec{\nu}, \vec{m}, \vec{n})]^q dX\, dY\right\}. \tag{34}$$

The limit $q \to 1$ provides the Shannon tomographic entropy for the hybrid-system state $H_C = R_C(1)$.

An analogous tomographic Rényi entropy can be associated with the optical tomogram of the hybrid-system state

$$R_w(q) = \frac{1}{1-q}\ln\left\{\int [w(\vec{X}, \vec{Y} \mid \vec{\theta}, \vec{\varphi})]^q d\vec{X}\, d\vec{Y}\right\}. \tag{35}$$

The Shannon tomographic entropy associated with the optical tomogram of the hybrid-system state $H_w = R_w(1)$.

The symplectic tomographic Rényi entropy is associated with the symplectic tomogram of the hybrid-system state $R_M(q)$

$$R_M(q) = \frac{1}{1-q}\ln\left\{\int [M(\vec{X}, \vec{Y} \mid \vec{\mu}, \vec{\nu}, \vec{m}, \vec{n})]^q d\vec{X}\, d\vec{Y}\right\}. \tag{36}$$

The corresponding Shannon entropy reads $H_M = R_M(1)$.

All three entropies depend on the parameters $\vec{\mu}$, $\vec{\nu}$, $\vec{m}$, $\vec{n}$ and $\vec{\theta}$, $\vec{\varphi}$, respectively.

It is obvious that the introduced tomographic entropies for the hybrid-system states must satisfy some inequalities. We present here a new inequality for the optical tomographic entropy of the hybrid-system state; it reads

$$\begin{gathered}
-\int\left[\int w(\vec{X}, \vec{Y} \mid \vec{\theta}, \vec{\varphi})\, d\vec{X}\right]\ln\left[\int w(\vec{X}, \vec{Y} \mid \vec{\theta}, \vec{\varphi})\, d\vec{X}\right] d\vec{Y} \\
-\int\left[\int w(\vec{X}, \vec{Y} \mid \vec{\theta}, \vec{\varphi}')\, d\vec{X}\right]\ln\left[\int w(\vec{X}, \vec{Y} \mid \vec{\theta}, \vec{\varphi}')\, d\vec{Y}\right] d\vec{X} \geq N_2 \ln \pi e, \\
\vec{\varphi}' = (\varphi_1 + \pi/2, \varphi_2 + \pi/2, \ldots, \varphi_{N_2} + \pi/2).
\end{gathered} \tag{37}$$

The bound $N_2 \ln \pi e$ in this inequality corresponds to quantum fluctuations of the quantum degrees of freedom. It is worth pointing out that there is no such a bound for fluctuations of the classical subsystem.

6. Conclusions

To conclude, we point out the main results of our work.

We showed the possibility to describe the states of hybrid systems consisted of classical and quantum subsystems by tomographic probability distributions. We discussed three equivalent variants of the tomographic probability distributions: optical tomograms, symplectic tomograms, and center-of-mass tomograms. We demonstrated that these tomograms are admissible for describing the states of hybrid systems if they satisfy specific constrains. These constrains have the form of inequalities for the marginal tomographic probability densities. These inequalities determine a domain in the simplex where the hybrid-system-state tomograms are located. We modeled the dynamics of the hybrid-system states by stochastic maps which connect the initial tomogram with the tomogram at any future time moment. We considered the example of such dynamics related to the free motion of classical and quantum subsystems in an explicit form. We introduced the tomographic entropies associated with the hybrid-system-state tomograms. For the optical tomogram of the hybrid-system state, we obtained the new inequality extending the entropic uncertainty relations known for quantum systems. We will study the other examples of dynamical maps of the tomographic probability distributions and their connections with other models of the hybrid systems and their evolution suggested by Elze et al. [2, 3, 23–25] in the future publication.

Acknowledgments

This study was supported by the Russian Foundation for Basic Research under Projects Nos. 11-02-00456 and 10-02-00312. The authors are grateful to the Organizers of the Sixth International Workshop DICE 2012 Spacetime–Matter–Quantum Mechanics from the Planck Scale to Emergent Phenomena (Castello Pasquini/Castiglioncello, Tuscany, Italy, September 17–21, 2012) and especially Prof. Thomas Elze for invitation and kind hospitality.

References

[1] Sherry T N and Sudarshan E C G 1978 *Phys. Rev.* D **18** 4580; *do.* 1979 *Phys. Rev.* D **20** 857
[2] Elze H-T 2012 *Phys. Rev.* A **85** 052109
[3] Elze H-T, Gambarotta G and Vallone F 2011 *J. Phys.: Conf. Ser.* **306** 012010
[4] Peres A and Temo D 2001 *Phys. Rev.* A **64** 022101
[5] Chruściński D, Kossakowski A, Marmo G and Sudarshan E C G 2011 *Open Syst. Inform. Dynamics* **18** 399
[6] Schrödinger E 1926 Quantisierung als Eigenwertproblem [tr. Quantization as an Eigenvalue Problem] *Annalen d. Physik* **79** 361-U8
[7] Landau L D 1927 *Z. Phys.* **45** 430
The damping problem in wave mechanics, in: Ter Haar D (ed.) *Collected Papers of L. D. Landau* 1965 pp. 8–18 (New York: Gordon & Breach)
J. von Neumann, 1927 *Nach. Ges. Wiss. Göttingen* **11** 245
von Neumann J 1932 *Mathematische Grundlagen der Quantummechanik* (Berlin: Springer)
[8] Mancini S, Man'ko V I and Tombesi P 1996 *Phys. Lett.* A **213** 1
[9] Ibort A, Man'ko V I, Marmo G, Simoni A and Ventriglia F 2009 *Phys. Scr.* **79** 065013
[10] Man'ko O V and Man'ko V I 1997 *J. Russ. Laser Res.* **18** 407
[11] Man'ko V I and Vilela Mendes R 2000 *Physica* D **145** 330
[12] Man'ko M A and Man'ko V I 2012 "Tomographic entropic inequalities in the probability representation of quantum mechanics" *AIP Conf. Proc.* **1488** 110 (*Preprint* arXiv:1208.5695 [quant-ph])
[13] Man'ko M A 2013 "Joint probability distributions and conditional probabilities in the tomographic representation of quantum states" *Phys. Scr.* T **153**
[14] Bertrand J and Bertrand P 1989 *Found. Phys.* **17** 397
[15] Vogel K and Risken H 1989 *Phys. Rev.* A **40** 2847
[16] Smithey D T, Beck M, Raymer M G and Faridani A 1993 *Phys. Rev. Lett.* **70** 1244

[17] Bellini M, Filippov S N, Man'ko V I, Coelho A S and Zavatta A 2012 *Phys. Rev.* A **85** 052129
[18] Man'ko M A and Man'ko V I 2011 *Found. Phys.* **41** 330
[19] Man'ko O V and Man'ko V I 2004 *J. Russ. Laser Res.* **25** 477
[20] Chernega V N and Man'ko V I 2012 "System with classical and quantum subsystems in tomographic probability representation" *AIP Conf. Proc.* **1424** 33
[21] Gorini V, Kossakowski A, and Sudarshan E C G 1976 *J. Math. Phys.* **17** 821
[22] Lindblad G 1976 *Comm. Math. Phys.* **48** 119
[23] Elze H-T, Gambarotta G and Vallone F 2011 *Int. J. Quantum Inform.* **9** Suppl. 1, 203
[24] Elze H-T 2009 *J. Phys.: Conf. Ser.* **174** 012009
[25] Elze H-T 2012 *J. Phys.: Conf. Ser.* **361** 012004 (*Preprint* arXiv:1202.3448)

Cosmology with quantum matter and a classical gravitational field: the approach of configuration-space ensembles

Marcel Reginatto

Physikalisch-Technische Bundesanstalt, Bundesallee 100, 38116 Braunschweig, Germany

E-mail: marcel.reginatto@ptb.de

Abstract. I consider the formulation of hybrid cosmological models that consists of a classical gravitational field interacting with a quantized massive scalar field in the formalism of ensembles on configuration space. This is a viable approach that provides an alternative to semiclassical gravity. I discuss a particular, highly nonclassical solution in two approximations, minisuperspace and spherically-symmetric midisuperspace. In both cases, the coupling of the quantum scalar field and classical gravitational field leads to a cosmological model which has a quantized radius of the universe.

1. Introduction

There are a few good reasons to consider hybrid systems in which the gravitational field remains classical while matter is assumed to consist of quantum fields. A full theory of quantum gravity is not yet available, and an approximation in which spacetime remains classical while matter is described in terms of quantum fields is often physically and computationally appropriate. Furthermore, since the quantization of gravity does not appear to follow from consistency arguments alone [1], it is of interest to investigate to what extent a hybrid system may provide a consistent, satisfactory description of matter and gravitation. The study of such systems can provide valuable clues that may help in the search for a full quantum theory of gravity. Finally, one must also consider the possibility that the gravitational field may not be quantum in nature [2, 3, 4]. For example, Butterfield and Isham, while putting forward the point of view that some type of theory of quantum gravity should be sought, have concluded that there is arguably no definitive proof that general relativity has to be quantized [5]. Dyson has argued that it might be impossible in principle to observe the existence of individual gravitons, and this has lead him to the conjecture that "the gravitational field described by Einstein's theory of general relativity is a purely classical field without any quantum behaviour" [6]. His observations regarding the impossibility of detecting gravitons have been confirmed by detailed calculations [7, 8]. If Dyson's conjecture is true, hybrid models become unavoidable.

In the standard approach used for coupling quantum fields to a classical gravitational field (i.e., semiclassical gravity), the energy momentum tensor that serves as the source in the Einstein equations is replaced by the expectation value of the energy momentum operator $\widehat{T}_{\mu\nu}$ with respect to some quantum state Ψ:

$$ {}^4R_{\mu\nu} - \frac{1}{2} g_{\mu\nu}\, {}^4R + \lambda g_{\mu\nu} = \frac{\kappa}{2} \langle \Psi | \widehat{T}_{\mu\nu} | \Psi \rangle \tag{1} $$

where ${}^4R_{\mu\nu}$ is the curvature tensor, ${}^4R$ the curvature scalar and $g_{\mu\nu}$ the metric tensor in spacetime, λ is the cosmological constant and $\kappa = 16\pi G$ (in units where $c = 1$ and G is the gravitational constant). This approach, however, presents a number of well known difficulties. An alternative to semiclassical gravity is provided by the formalism of ensembles on configuration space [9]. The formalism can be applied to both quantum and classical systems and allows a general and consistent description of interactions between them. When applied to hybrid systems, it can be shown that the approach overcomes difficulties arising in previous attempts; in particular, the correct equations of motion for the quantum and classical sectors are recovered in the limit of no interaction, conservation of probability and energy are satisfied, uncertainty relations hold for conjugate quantum variables, and the formalism allows a back reaction of the quantum system on the classical system [10, 11]. The approach can be generalized to describe the coupling of a quantized field to a classical gravitational field; it is therefore an appropriate formalism to use if one wants to investigate hybrid cosmological models.

In the next section, I describe the coupling of a quantized scalar field to a classical gravitational field using ensembles on configuration space. I then consider a particular, highly nonclassical solution in two approximations, minisuperspace and spherically-symmetric midisuperspace. In the last section, I summarize the results and give conclusions. Technical details are provided in the three Appendices.

2. Equations for a quantized scalar field interacting with a classical metric field

In the approach followed here, the interaction of quantum matter with a classical gravitational field will be described using the formalism of ensembles on configuration space. A description of physical systems by means of ensembles on configuration space may be introduced at a very fundamental level, requiring only the notions of probability and an action principle. Consider the situation in which the configuration of a physical system is subject to uncertainty. Under these circumstances, the system must be described by an ensemble of configurations with probability P. Introduce an action principle to specify the dynamics. In the Hamiltonian formulation, this implies the existence a function, call it S, that is conjugate to P, a symplectic structure with its corresponding Poisson bracket algebra, and an ensemble Hamiltonian that is the generator of time evolution. The state of a given system is therefore completely described by the pair of functions P and S and time evolution is determined by the choice of ensemble Hamiltonian.

Since the formalism of ensembles on configuration space is not widely known, a detailed, self-contained introduction is given in Appendix A, where the approach is illustrated using ensembles that describe classical, quantum and hybrid classical-quantum systems of particles [9, 10, 11]. In this section, the focus will be on the application of the formalism to the coupling of a quantized scalar field to a classical gravitational field. I first illustrate the approach for the case of vacuum gravity [12, 13, 14] and then discuss the formulation for hybrid systems.

The most direct way of introducing a classical configuration space ensemble for gravity is to start from the Einstein-Hamilton-Jacobi equation, which in the metric representation takes the form [15]

$$ \mathcal{H}_h = \kappa G_{ijkl} \frac{\delta S}{\delta h_{ij}} \frac{\delta S}{\delta h_{kl}} - \frac{1}{\kappa} \sqrt{h}\,(R - 2\lambda) = 0, \tag{2} $$

where R is the curvature scalar and h_{kl} the metric tensor on a three-dimensional spatial hypersurface, and $G_{ijkl} = (2h)^{-1/2} (h_{ik}h_{jl} + h_{il}h_{jk} - h_{ij}h_{kl})$ is the DeWitt supermetric. The functional S is assumed to be invariant under the gauge group of spatial coordinate transformations, which is equivalent to satisfying the momentum constraints of the canonical formulation of general relativity. Eq. (2) corresponds to an infinity of constraints, one at each point. It is possible to introduce an alternative viewpoint [16, 17] in which Eq. (2) is regarded as an equation to be integrated with respect to a "test function" in which case we are dealing with one equation for each choice of lapse function N,

$$\int d^3x \; N\mathcal{H}_h = 0; \tag{3}$$

i.e. for each choice of foliation. Such an alternative viewpoint is extremely useful: although it may be impossible to find the general solution (which requires solving the Einstein-Hamilton-Jacobi equation for all choices of lapse functions), it may be possible to find particular solutions for specific choices; for example, the choice $S \sim \int d^3x \sqrt{h}$ is a particular solution that describes de Sitter spacetime in a flat foliation [17].

An appropriate ensemble Hamiltonian for vacuum gravity is given by

$$\tilde{H}_h = \int d^3x \int Dh \, P \; N\mathcal{H}_h \tag{4}$$

(technical issues are discussed in more detail in Appendix B and in ref. [14]). The functional P is also assumed, like S, to be invariant under the gauge group of spatial coordinate transformations. The corresponding equations have the form

$$\frac{\partial P}{\partial t} = \frac{\delta \tilde{H}_h}{\delta S}, \quad \frac{\partial S}{\partial t} = -\frac{\delta \tilde{H}_h}{\delta P} \tag{5}$$

where $\delta/\delta F$ denotes the variational derivative with respect to a functional F [12]. Assuming the constraints $\frac{\partial S}{\partial t} = \frac{\partial P}{\partial t} = 0$, these equations lead to Eq. (3), as required, and to a continuity equation,

$$\int d^3x \, N \frac{\delta}{\delta h_{ij}} \left(P \, G_{ijkl} \frac{\delta S}{\delta h_{kl}} \right) = 0. \tag{6}$$

These two equations define an ensemble on configuration space for the case of vacuum gravity.

A hybrid system where a quantum scalar field ϕ couples to the classical metric h_{kl} requires a generalization of Eq. (4) in which [9]

$$\tilde{H}_{\phi h} = \int d^3x \int Dh \, P \, N \left[\mathcal{H}_{\phi h} + F_\phi \right], \tag{7}$$

where

$$\mathcal{H}_{\phi h} = \mathcal{H}_h + \frac{1}{2\sqrt{h}} \left(\frac{\delta S}{\delta \phi} \right)^2 + \sqrt{h} \left[\frac{1}{2} h^{ij} \frac{\partial \phi}{\partial x^i} \frac{\partial \phi}{\partial x^j} + V(\phi) \right] \tag{8}$$

is the ensemble Hamiltonian for gravity with a classical scalar field and

$$F_\phi = \frac{\hbar^2}{4} \frac{1}{2\sqrt{h}} \left(\frac{\delta \log P}{\delta \phi} \right)^2 \tag{9}$$

is an additional, non-classical kinetic energy term. Assuming again the constraints $\frac{\partial S}{\partial t} = \frac{\partial P}{\partial t} = 0$, the coresponding equations are given by

$$\int d^3x \, N \left[\mathcal{H}_{\phi h} - \frac{\hbar^2}{2\sqrt{h}} \left(\frac{1}{A} \frac{\delta^2 A}{\delta \phi^2} \right) \right] = 0, \tag{10}$$

where $A \equiv \sqrt{P}$, and a continuity equation that is identical to Eq. (6).

3. Hybrid cosmological model in minisuperspace

Before introducing the cosmological model in spherical gravity that is discussed in the next section, it will be instructive to look at the corresponding minisuperspace model [9].

Consider a closed Robertson-Walker universe with a massive scalar field. The line element can be written in the form

$$ds^2 = -N^2(t)\,dt^2 + a^2(t)\,d\Omega_3^2 \tag{11}$$

where a is the scale factor and $d\Omega_3^2$ is the standard line element on S^3. Symmetry reduction leads to a minisuperspace formulation in a finite dimensional configuration space with coordinates a and ϕ [18]. Following Ref. [9], I restrict to a potential term that is quadratic in ϕ. Then, the classical Hamilton-Jacobi equation takes the form $\mathcal{H}_{\phi a} = 0$, with

$$\mathcal{H}_{\phi a} = -\frac{1}{a}\left(\frac{\partial S}{\partial a}\right)^2 + \frac{1}{a^3}\left(\frac{\partial S}{\partial \phi}\right)^2 - a + \frac{\lambda a^3}{3} + m^2 a^3 \phi^2, \tag{12}$$

where m is the mass of the field and the units of this section have been chosen so that $2G/3\pi = 1$ (to agree with those of Ref. [9]).

The corresponding ensemble Hamiltonian for a quantized field interacting with the classical metric is given by [9]

$$\tilde{H}_{\phi a} = \int da d\phi P \left[\mathcal{H}_{\phi a} + \frac{\hbar^2}{4}\frac{1}{a^3}\left(\frac{\partial \log P}{\partial \phi}\right)^2\right]. \tag{13}$$

Assuming the constraints $\frac{\partial S}{\partial t} = \frac{\partial P}{\partial t} = 0$, the equations of motion take the form

$$\mathcal{H}_{\phi a} - \frac{\hbar^2}{a^3}\frac{1}{\sqrt{P}}\frac{\partial^2 \sqrt{P}}{\partial \phi^2} = 0 \tag{14}$$

and

$$-\frac{\partial}{\partial a}\left(\frac{P}{a}\frac{\partial S}{\partial a}\right) + \frac{\partial}{\partial \phi}\left(\frac{P}{a^3}\frac{\partial S}{\partial \phi}\right) = 0. \tag{15}$$

An *exact solution* can be derived for the case $S = 0$. This is a highly non-classical solution: Eq. (12), the classical Hamilton-Jacobi equation, does not admit any solutions with this ansatz. When $S = 0$, Eq. (14) reduces to

$$-\frac{\hbar^2}{a^3}\frac{1}{\sqrt{P}}\frac{\partial^2 \sqrt{P}}{\partial \phi^2} - a + \frac{\lambda a^3}{3} + a^3 m^2 \phi^2 = 0 \tag{16}$$

while Eq. (15) is automatically satisfied. The non-negative, normalizable solutions take the form

$$P_n(\phi, a) = \delta(a - a_n)\frac{\alpha_n}{\sqrt{\pi}2^n n!}\exp\left(-\alpha_n^2\phi^2\right)\left[H_n(\alpha_n\phi)\right]^2 \tag{17}$$

where the H_n are Hermite polynomials, $\alpha_n^2 = a_n^3 m/\hbar$ and the a_n satisfy the condition

$$a_n - \frac{\lambda a_n^3}{3} = 2\hbar m\left(n + \frac{1}{2}\right) \tag{18}$$

for $n = \{0, 1, 2, ...\}$. If the term proportional to the cosmological constant λ can be neglected, the quantization condition takes the simple form $a_n = 2\hbar m\left(n + \frac{1}{2}\right)$.

While the transformation $\psi = \sqrt{P}$ leads, via Eq. (16), to a Schrödinger equation for ψ, it is not possible to introduce solutions that are linear superpositions of the ψ_n because the potential term in the equation ends up being a function of a_n.

This solution that has been derived in this section has some remarkable features. One can see that the coupling of the quantum field to a purely classical metric leads to a *quantization condition* for the scale factor a. Furthermore, the classical singularity at $a = 0$ is *excluded* from these solutions. Finally, notice that there is a natural ordering of the solutions $\{P_n\}$ in terms of n and one may argue that this ordering leads to a thermodynamic arrow of time. This follows from the observation that the amount of structure associated with a solution P_n (as determined, for example, by counting the number of nodes in ψ_n or by evaluating the entropy expression $-\int d\phi\, P_n \log P_n$ for different values of n) increases with increasing n. Note that this thermodynamic arrow of time coincides with the arrow of time as determined by an expanding universe whenever the non-linear term proportional to λ can be neglected.

4. Hybrid cosmological model in spherically symmetric gravity

Consider now a midisuperspace hybrid cosmological model in spherically symmetric gravity. In the case of spherical symmetry, the line element may be written in the form

$$g_{\mu\nu}dx^\mu dx^\nu = -N^2dt^2 + \Lambda^2\left(dr + N^r dt\right)^2 + R^2 d\Omega^2. \tag{19}$$

The lapse function N and the shift function N^r are functions of the radial coordinate r and the time coordinate t. The configuration space for the gravitational field consists of two fields, R and Λ. Under transformations of r, R behaves as a scalar and Λ as a scalar density. Spherically symmetric gravity is discussed in detail in a number of papers, mostly in reference to the canonical quantization of black hole spacetimes. For discussions using the metric representation, see for example [19, 20, 21]. For discussions of the Einstein-Hamilton-Jacobi equation in the context of the WKB approximation of quantized spherically symmetric gravity, see for example [22, 23].

I now set $\hbar = c = G = 1$. The Einstein-Hamilton-Jacobi equation for the case of a vacuum gravity takes the form

$$\mathcal{H}_{\Lambda R} = -\frac{1}{R}\frac{\delta S}{\delta R}\frac{\delta S}{\delta \Lambda} + \frac{1\Lambda}{2R^2}\left(\frac{\delta S}{\delta\Lambda}\right)^2 + \lambda\frac{\Lambda R^2}{2} + V = 0 \tag{20}$$

where S is assumed to be invariant under diffeomorphisms. V is related to the curvature scalar by ${}^4R = -4\,\Lambda R^2\, V$, and is given by

$$V = \frac{RR''}{\Lambda} - \frac{RR'\Lambda'}{\Lambda^2} + \frac{R'^2}{2\Lambda} - \frac{\Lambda}{2} \tag{21}$$

where primes indicate derivatives with respect to r.

The ensemble Hamiltonian of a hybrid system where the matter is in the form of a minimally coupled quantized radially symmetric scalar field of mass m is given by

$$\tilde{H}_{\phi\Lambda R} = \int dr \int Dh\, P\, N\left[\mathcal{H}_{\phi\Lambda R} + F_\phi\right], \tag{22}$$

where

$$\mathcal{H}_{\phi\Lambda R} = \mathcal{H}_{\Lambda R} + \frac{1}{2\Lambda R^2}\left(\frac{\delta S}{\delta\phi}\right)^2 + \frac{R^2}{2\Lambda}\phi'^2 + \frac{\Lambda R^2 m^2}{2}\phi^2, \tag{23}$$

and

$$F_\phi = \frac{1}{8\Lambda R^2}\left(\frac{\delta \log P}{\delta\phi}\right)^2. \tag{24}$$

Eq. (22) is the analogous of Eq. (7) for the case of spherically symmetric gravity. Assuming again the constraints $\frac{\partial S}{\partial t} = \frac{\partial P}{\partial t} = 0$, the corresponding equations are

$$\int dr\, N \left[\mathcal{H}_{\phi \Lambda R} - \frac{1}{2\Lambda R^2} \left(\frac{1}{A} \frac{\delta^2 A}{\delta \phi^2} \right) \right] = 0, \tag{25}$$

where $A \equiv P^{1/2}$, and the continuity equation

$$\int dr\, N \left[\frac{\delta}{\delta R} \left(P \frac{1}{R} \frac{\delta S}{\delta \Lambda} \right) + \frac{\delta}{\delta \Lambda} \left(P \frac{1}{R} \frac{\delta S}{\delta R} - P \frac{\Lambda}{R^2} \frac{\delta S}{\delta \Lambda} \right) - \frac{\delta}{\delta \phi} \left(P \frac{1}{\Lambda R^2} \frac{\delta S}{\delta \phi} \right) \right] = 0. \tag{26}$$

I now want to consider a class of solutions that is analogous to the class of minisuperspace solutions described in the previous section. To do this, I will look for a solution that satisfies the following requirements:

(i) it corresponds to choosing a foliation of spaces of constant positive curvature and a lapse function N that is constant, and

(ii) it satisfies $S = 0$.

In this case, Eqs. (25) and (26) reduce to a single Schrödinger functional equation for A; i.e., of the type

$$-\frac{1}{2\Lambda R^2} \left(\frac{\delta^2 A}{\delta \phi^2} \right) + \left[\lambda \frac{\Lambda R^2}{2} + V + \frac{R^2}{2\Lambda} \phi'^2 + \frac{\Lambda R^2 m^2}{2} \phi^2 \right] A = 0. \tag{27}$$

This equation can be solved using standard techniques developed for the Schrödinger functional representation of quantum field theory [18, 24, 25, 26].

To get an explicit solution that corresponds to the lowest state of the minisuperspace model that I considered in the previous section, I will assume that A has the form of a ground state Gaussian functional; i.e.,

$$A_{(0)} \sim \exp \left\{ -\frac{1}{2} \int \int dy\, dz\, \Lambda_y\, \Lambda_z\, R_y^2\, R_z^2\, \phi_y\, K_{yz}\, \phi_z \right\}. \tag{28}$$

Instead of $A_{(0)}$, one may also consider the excited states which solve the functional Schrödinger equation; I discuss the consequences of making this alternative choice at the end of this section.

As shown in Appendix C, the equation that determines the functional $A_{(0)}$ can be mapped to a functional Schrödinger equation in a space of constant curvature and the kernel K_{xy} can be expressed in the simple form

$$K_{xy} = \frac{1}{2a_0^4} \sum_n \sqrt{\gamma_n}\, \psi_x^{(n)} \psi_y^{(n)}, \tag{29}$$

where the $\psi_r^{(n)}$ are solutions of a Schrödinger-type equation in a space of constant curvature,

$$-\frac{1}{\sin^2 r} \frac{\partial}{\partial r} \left(\sin^2 r \frac{\partial \psi_r^{(n)}}{\partial r} \right) + m^2 a_0^2\, \psi_r^{(n)} = \gamma_n \psi_r^{(n)}. \tag{30}$$

The eigenvalues γ_n are given by

$$\gamma_n = n^2 - 1 + m^2 a_0^2, \qquad n = 1, 2, 3... \ . \tag{31}$$

As described in Appendix C, in principle it is possible to get an expression for a_0 that depends on the cosmological constant λ and the energy $E_{(0)}$ of the quantized scalar field, which is given by [26]

$$E_{(0)} \sim a_0^3 \int dr \, \sin^2 r \, K_{rr}. \tag{32}$$

However, $\int dr \, \sin^2 r \, K_{rr} \sim \sum_n \gamma_n$, which diverges. This is a consequence of the infinite zero-point energy of the quantum field. It is necessary to use a renormalization procedure to extract a finite result.

In this particular case, the equation for $A_{(0)}$ is similar in form to a functional Schrödinger equation for a quantum scalar field in an Einstein universe, so it is possible to use previous results from the literature where such renormalization procedures have been carried out (for a thorough analysis, see reference [27]). Note that this is not the generic case: since the equations of hybrid cosmology are not linear, in general they will not map to a Schrödinger-type functional equation. The simplification that is achieved here is a direct consequence of choosing $S = 0$. With a different ansatz, the equations can not be solved in this way. A discussion of renormalization procedures for this solution is outside of the scope of this paper and will be the subject of a future publication.

A similar analysis may be carried out where the ground state functional $A_{(0)}$ is replaced by an *excited* state. Consider a first excited state $A_{(1)}$ specified by the eigenfunction $\psi_r^{(n)}$. This state will differ in energy from the (divergent) ground state energy by a finite amount $\Delta E(n)$ which depends on the value of γ_n, with $\Delta E(n) = \sqrt{\gamma_n}$ [26]. $\Delta E(n)$ can only take discrete values because γ_n is quantized, and this means that a_0 will also be restricted to *discrete values.* Therefore, the coupling of the quantum scalar field and classical gravitational field leads to the quantization of the radius of the universe, not only for the minisuperspace model but also for the midisuperspace model.

5. Concluding remarks

The main result of this paper is a formulation of hybrid cosmological models that makes use of the formalism of ensembles on configuration space and thus provides an alternative to other approaches (e.g., semiclassical gravity) which couple quantum matter to classical gravitational fields. In particular, I have examined the case in which a classical gravitational field interacts with a quantized massive scalar field. I have derived a particular, highly nonclassical solution using two approximations, minisuperspace and spherically-symmetric midisuperspace.

The hybrid cosmological model examined here was previously considered in reference [9] but only within the minisuperspace approximation. One of the aims of the present work was to reexamine this model using the next level of approximation; i.e., in the context of a midisuperspace formulation like spherically symmetric gravity. It appears that the space of solutions in midisuperspace may be very rich by comparison. This is a consequence of the renormalization procedure that needs to be carried out to achieve a finite result. This complexity is absent from the minisuperspace solution.

In both minisuperspace and midisuperspace, the coupling of the quantum scalar field to the classical gravitational field leads to the quantization of the radius of the universe, a remarkable result. It would be interesting to know whether this is a generic feature of hybrid closed cosmological models or whether this is a feature that is particular to the family of solutions considered here (i.e., those with $S = 0$). In the minisuperspace solution, it is clear that a singular solution with $a = 0$ is excluded; the situation is more complicated in the case of the midisuperspace solution due to the need to introduce renormalization.

In the Einstein universe of classical relativity, the cosmological constant acts as a repulsive force which balances the gravitational attraction of matter; it is essential to include it in order to achieve a static solution. In the hybrid solutions considered here, it is possible to achieve static solutions with a vanishing cosmological constant and it would appear that quantum fluctuations are sufficient to balance the gravitational attraction. It would be interesting to extend these results to consider perturbative solutions without the restriction $S = 0$, which should describe solutions that are not static.

Acknowledgments
I thank M. J. W. Hall and Stephen Boughn for valuable discussions.

Appendix A. Ensembles on configuration space

This appendix provides a brief summary of the formalism of ensembles on configuration space. Derivations of the results stated in this appendix can be found in Refs. [9, 11, 10].

Appendix A.1. Ensembles on configuration space describing classical, quantum and mixed classical-quantum systems

Start with the assumption that, as seems to be implied by quantum mechanics, the configuration of a physical system is an inherently statistical concept. The system is therefore described by an ensemble of configurations with probability density P, where $P \geq 0$ and $\int dx\, P(x,t) = 1$. To derive equations of motion, introduce an *ensemble Hamiltonian* $\tilde{H}[P,S]$, where S is an auxiliary field that is canonically conjugate to P. The equations of motion take the form

$$\frac{\partial P}{\partial t} = \left\{P, \tilde{H}\right\}_{PB} = \frac{\delta \tilde{H}}{\delta S}, \qquad \frac{\partial S}{\partial t} = \left\{S, \tilde{H}\right\}_{PB} = -\frac{\delta \tilde{H}}{\delta P}, \tag{A.1}$$

where $\{A, B\}_{PB}$ is the Poisson bracket of the fields A and B.

The following ensemble Hamiltonians lead to equations that describe the evolution of quantum and classical non-relativistic systems,

$$\begin{aligned} \tilde{H}_C[P,S] &= \int dx\, P\left[\frac{|\nabla S|^2}{2m} + V(x)\right], \\ \tilde{H}_Q[P,S] &= \tilde{H}_C[P,S] + \frac{\hbar^2}{4}\int dx\; P\frac{|\nabla \log P|^2}{2m}. \end{aligned} \tag{A.2}$$

For example, the equations of motion derived from $\tilde{H}_Q[P,S]$ are given by

$$\frac{\partial P}{\partial t} + \nabla.\left(P\frac{\nabla S}{m}\right) = 0, \quad \frac{\partial S}{\partial t} + \frac{|\nabla S|^2}{2m} + V + \frac{\hbar^2}{2m}\frac{\nabla^2 P^{1/2}}{P^{1/2}} = 0 \tag{A.3}$$

while the equations of motion derived from $\tilde{H}_C[P,S]$ are the same as Eq. (A.3) but with $\hbar = 0$. The first equation in Eq. (A.3) is a continuity equation, the second equation is the classical Hamilton-Jacobi equation when $\hbar = 0$ and a modified Hamilton-Jacobi equation when $\hbar \neq 0$. Defining $\psi = \sqrt{P}\, e^{iS/\hbar}$, Eq. (A.3) takes the form

$$i\hbar\frac{\partial \psi}{\partial t} = \frac{-\hbar^2}{2m}\nabla^2\psi + V\psi, \tag{A.4}$$

which is the usual form of the Schrödinger equation. Therefore, in this formalism, quantum and classical particles are treated on an equal footing, with differences being primarily due to the different forms of the respective ensemble Hamiltonians.

It is possible to extend the formalism in a natural way to allow for mixed quantum-classical systems. A mixed quantum-classical ensemble Hamiltonian on a configuration space with coordinates q, y is given by

$$\begin{aligned}\tilde{H}_{QC}[P,S] &= \int dq\, dy\, P\left[\frac{|\nabla_y S|^2}{2M} + \frac{|\nabla_q S|^2}{2m}\right] \\ &+ \int dq\, dy\, P\left[\frac{\hbar^2}{4}\frac{|\nabla_q \log P|^2}{2m} + V(q,y,t)\right]. \end{aligned} \tag{A.5}$$

Here q denotes the configuration space coordinate of a quantum particle of mass m and y that of a classical particle of mass M, and $V(q,y,t)$ is a potential energy function describing the quantum-classical interaction. The equations of motion for P and S derived from $\tilde{H}_{QC}$ are

$$\begin{aligned}\frac{\partial P}{\partial t} &= -\nabla_q \cdot \left(P\frac{\nabla_q S}{m}\right) - \nabla_y \cdot \left(P\frac{\nabla_y S}{M}\right), \\ \frac{\partial S}{\partial t} &= -\frac{|\nabla_q S|^2}{2m} - \frac{|\nabla_y S|^2}{2M} - V + \frac{\hbar^2}{2m}\frac{\nabla_q^2 P^{1/2}}{P^{1/2}}. \end{aligned} \tag{A.6}$$

The state of a system is described by specifying the two fields P and S. While the interpretation of P is straightforward, there are some subtle issues concerning the physical interpretation of S. For the cases discussed here (i.e., classical, quantum and mixed systems), it is possible to define local energy and momentum densities in terms of S. If $\tilde{H}[\lambda P, S] = \lambda \tilde{H}[P,S]$ (which holds for the ensemble Hamiltonians that we considered above), then

$$\tilde{H} = \int dx\, P\frac{\delta \tilde{H}}{\delta P} = -\int dx\, P\frac{\partial S}{\partial t} = -\langle \partial S/\partial t\rangle, \tag{A.7}$$

which shows that $-P\partial S/\partial t$ is a local energy density. Furthermore, $\int dx\, P\nabla S$ is the canonical infinitesimal generator of translations, since

$$\begin{aligned}\delta P(x) &= \delta\mathbf{x}\cdot\left\{P, \int dx\, P\nabla S\right\}_{PB} = -\delta\mathbf{x}\cdot\nabla P, \\ \delta S(x) &= \delta\mathbf{x}\cdot\left\{S, \int dx\, P\nabla S\right\}_{PB} = -\delta\mathbf{x}\cdot\nabla S, \end{aligned} \tag{A.8}$$

under action of the generator, and therefore $P\nabla S$ can be considered a local momentum density. To maintain full generality, S should not be regarded as a "momentum potential". In particular, for an ensemble of classical particles with uncertainty described by the probability P, it will not be assumed that the momentum of a member of the ensemble is a well-defined quantity proportional to the gradient of S, as it is done in the usual deterministic interpretation of the Hamilton–Jacobi equation. This avoids forcing a similar deterministic interpretation in the quantum and quantum-classical cases. A deterministic picture can be recovered for classical ensembles precisely in those cases in which trajectories are operationally defined.

Appendix A.2. Observables and local densities

Observables are functionals of P and S. Given two functionals $A[P,S]$ and $B[P,S]$, define their Poisson bracket in the standard way,

$$\{A,B\}_{PB} = \int dx\left(\frac{\delta A}{\delta P}\frac{\delta B}{\delta S} - \frac{\delta A}{\delta S}\frac{\delta B}{\delta P}\right), \tag{A.9}$$

which gives an algebra of obervables.

Arbitrary functionals $A[P,S]$ are not necessarily observables because observables have to satisfy certain mild requirements. For example, the infinitesimal canonical transformation generated by any observable A must preserve the normalization and positivity of P. This implies the two conditions

$$A[P,S+c] = A[P,S], \qquad \delta A/\delta S = 0 \text{ if } P(x) = 0. \tag{A.10}$$

Note that the first equation implies gauge invariance of the theory under $S \to S + c$.

A more general condition that may be imposed on observables is that they be homogenous of degree one in P; i.e., $A[\lambda P, S] = \lambda A[P,S]$. This is a condition that is satisfied by both classical and quantum systems, and it will be assumed here that it is also valid for mixed classical-quantum systems. Then, it follows that

$$A[P,S] = \int dx\, P\, (\delta A/\delta P) = \langle \delta A/\delta P \rangle. \tag{A.11}$$

That is, one can associate with each observable A a local density on the configuration space, and the value of A can be calculated by integrating over this local density.

As a simple example, consider position and momentum observables. In all three cases (classical, quantum, and mixed classical-quantum), they are given by $X[P,S] = \int dx Px$ and $\Pi[P,S] = \int dx P\nabla S$, with corresponding local densities Px and $P\nabla S$. Such an equivalence does not hold of course for more complicated observables. For example, different functionals are needed to describe the kinetic energy of classical, quantum, and mixed classical-quantum systems, as is apparent from their corresponding ensemble Hamiltonians.

Appendix A.3. An example of a mixed classical/quantum system: Measurement of the position of a quantum particle carried out by a classical apparatus

To illustrate the application of the formalism, I consider a simple model: a classical apparatus which measures the position of a quantum particle [9].

To model a measurement of position, introduce the ensemble Hamiltonian

$$\tilde{H}_{\text{position}} = \tilde{H}_{QC} + \kappa(t) \int dq\, dx\, P\, q.\nabla_x S. \tag{A.12}$$

For an interaction over a short time period $[0,T]$ such that $\tilde{H}_{QC}$ can be ignored during the interaction,

$$\frac{\partial P}{\partial t} = -\kappa(t)\, q.\nabla_x P, \qquad \frac{\partial S}{\partial t} = -\kappa(t)\, q.\nabla_x S, \tag{A.13}$$

which integrates to ($K = \int_0^T dt\, \kappa(t)$)

$$P(q,x,T) = P(q, x - Kq, 0), \qquad S(q,x,T) = S(q, x - Kq, 0). \tag{A.14}$$

Consider the case where the initial position of the pointer is sharply defined, $P(q,x,0) = \delta(x - x_0)P_Q(q)$. Then, after the measurement, $\Delta x \neq 0$, with probability $P_Q(q)$ of finding $x = x_0 - Kq$. In other words, *uncertainty is transferred* from the quantum ensemble to the classical ensemble. Transfer of uncertainty is the norm whenever two systems interact.

To interpret the outcome it will be helpful to introduce some additional concepts. Consider the conditional probability $P(q|x) = P(q,x)/P(x)$. The conditional wave function is defined by

$$\psi(q|x) = \sqrt{P(q|x)}\, e^{iS(q,x)/\hbar}, \qquad |\psi_x\rangle = \int dq\, \psi(q|x)\, |q\rangle. \tag{A.15}$$

Introduce a conditional density operator,

$$\rho_{Q|C} = \int dx\, P(x)\, |\psi_x\rangle\langle\psi_x|. \tag{A.16}$$

It is important to keep in mind that $\psi(q|x)$ and $\rho_{Q|C}$ do *not* satisfy linear Schrödinger and Liouville equations, nor unitary invariance properties. These quantities only contain *partial* information.

The conditional density for the quantum component is diagonal in the position basis,

$$\rho_{Q|C} = \int dq\, P_Q(q)\, |q\rangle\langle q| \tag{A.17}$$

and thus "decoheres" with respect to position.

The main features of this simple measurement model are: (i) the measuring apparatus is described classically; (ii) information about quantum ensembles is obtained via an appropriate interaction with an ensemble of classical measuring apparatuses, which correlates the classical configuration with a corresponding quantum property, and (iii) there is a conditional decoherence of the quantum ensemble relative to the classical ensemble, which depends upon the nature of the quantum-classical interaction.

Appendix A.4. Entanglement

Entangled states in quantum mechanics are those states of a multi-particle state that can not be expressed as a product of single-particle states. A non-entangled state is therefore one that has a wave function ψ that can be written in the form

$$\psi(t, x_1, x_2, ..., x_n) = \prod_{k=1}^{n} \psi_k(t, x_k) \tag{A.18}$$

where the ψ_k are single-particle wave functions. In other words, a non-entangled state is one in which P and S satisfy the conditions

$$\begin{aligned} P(t, x_1, x_2, ..., x_n) &= \prod_{k=1}^{n} P_k(t, x_k), \\ S(t, x_1, x_2, ..., x_n) &= \sum_{k=1}^{n} S_k(t, x_k) \end{aligned} \tag{A.19}$$

There is no difficulty in extending this distinction between entangled and non-entagled states to physical states described in the ECS formalism; in particular, classical and mixed classical-quantum states can be entangled since they need not satisfy the requirements of Eq. (A.19). Therefore, entanglement in this sense is a general feature of ensembles in configuration space and it is not restricted to quantum mechanical systems only.

Appendix B. Classical ensembles of gravitational fields

A Hamilton-Jacobi formulation for the gravitational field can be defined in terms of the functional equations

$$\begin{aligned} \mathcal{H} &= \kappa G_{ijkl} \frac{\delta S}{\delta h_{ij}} \frac{\delta S}{\delta h_{kl}} - \frac{1}{\kappa}\sqrt{h}\,(R - 2\lambda) = 0, \\ \mathcal{H}_i &= D_j \left(h_{ik} \frac{\delta S}{\delta h_{kj}} \right) = 0, \end{aligned} \tag{B.1}$$

where R is the curvature scalar and h_{kl} the metric tensor on a three-dimensional spatial hypersurface, λ is the cosmological constant, $\kappa = 16\pi G$ (in units where $c = 1$ and G is the gravitational constant), and $G_{ijkl} = (2h)^{-1/2}(h_{ik}h_{jl} + h_{il}h_{jk} - h_{ij}h_{kl})$ is the DeWitt supermetric [15].

As a consequence of the Hamiltonian constraint $\mathcal{H} = 0$ and the momentum constraints $\mathcal{H}_i = 0$, S must satisfy an infinite set of constraints, numbering four per three-dimensional point. S also satisfies the condition $\frac{\partial S}{\partial t} = 0$ [28]. The momentum constraints are equivalent to requiring the invariance of the Hamilton-Jacobi functional S under spatial coordinate transformations. One may therefore formulate the theory in an equivalent way by keeping the Hamiltonian constraint, ignoring the momentum constraints, and requiring instead that S be invariant under the gauge group of spatial coordinate transformations.

To define classical ensembles for gravitational fields, it is necessary to introduce some additional mathematical structure: a measure Dh over the space of metrics h_{kl} and a probability functional $P[h_{kl}]$. A standard way of defining the measure [29, 30] is to introduce an invariant norm for metric fluctuations that depends on a parameter ω,

$$\|\delta h\|^2 = \int d^n x\, [h(x)]^{\omega/2} H^{ijkl}[h(x);\omega]\, \delta h_{ij}\delta h_{kl} \tag{B.2}$$

where n is the number of dimensions and

$$H^{ijkl} = \frac{1}{2}[h(x)]^{(1-\omega)/2}\left[h^{ik}h^{jl} + h^{il}h^{jk} + \lambda h^{ij}h^{kl}\right] \tag{B.3}$$

is a generalization of the inverse of the DeWitt supermetric (in [29] the particular case $\omega = 0$ was considered). This norm induces a local measure for the functional integration given by

$$\int d\mu[h] = \int \prod_x [\det H(h(x))]^{1/2} \prod_{i\geq j} dh_{ij}(x) \tag{B.4}$$

where

$$\det H(h(x)) \propto \left(1 + \frac{1}{2}\lambda n\right)[h(x)]^{\sigma}, \tag{B.5}$$

and $\sigma = (n+1)[(1-\omega)n - 4]/4$ (one needs to impose the condition $\lambda \neq -2/n$, otherwise the measure vanishes). Therefore, up to an irrelevant multiplicative constant, the measure takes the form

$$\int d\mu[h] = \int \prod_x \left[\sqrt{h(x)}\right]^{\sigma} \prod_{i\geq j} dh_{ij}(x). \tag{B.6}$$

Without loss of generality, one may set Dh equal to $d\mu[h]$ with $\sigma = 0$, since a term of the form $\left[\sqrt{h(x)}\right]^{\sigma}$ may be absorbed into the definition of $P[h_{kl}]$.

It is natural to require also that $\int DhP$ be invariant under the gauge group of spatial coordinate transformations. Since the family of measures defined by eq. (B.6) is invariant under spatial coordinate transformations [30, 31], the invariance of $\int DhP$ leads to a condition on P that is similar to the one required of S. To show this, consider an infinitesimal change of coordinates $x'^k = x^k + \epsilon^k(x)$ and the corresponding transformation of the metric, $h_{kl} \to h_{kl} - (D_k\epsilon_l + D_l\epsilon_k)$. The variation of $\int DhP$ can be expressed as

$$\delta_\epsilon \int DhP = \int Dh\left[D_k\left(\frac{\delta P}{\delta h_{kl}}\right)\epsilon_l + D_l\left(\frac{\delta P}{\delta h_{kl}}\right)\epsilon_k\right]. \tag{B.7}$$

Therefore, $\delta_\epsilon \int DhP = 0$ requires

$$D_k\left(\frac{\delta P}{\delta h_{kl}}\right) = 0. \tag{B.8}$$

or the gauge invariance of P. In addition to eq. (B.8), it will be assumed that $\frac{\partial P}{\partial t} = 0$ also holds.

Finally, it should be pointed out that one must factor out the infinite diffeomorphism gauge group volume out of the measure to calculate finite averages using the measure Dh and probability functional P. This can be achieved using the geometric approach described in [32]. This issue will not be discussed further here, since it does not affect the derivation of the equations of motion.

An appropriate ensemble Hamiltonian for the gravitational field is given by

$$\tilde{H}_c = \sum_x \int DhP\mathcal{H} \sim \int d^3x \int DhP\mathcal{H}. \tag{B.9}$$

The equations of motion derived from eq. (B.9) are of the form

$$\frac{\partial P}{\partial t} = \frac{\Delta \tilde{H}_c}{\Delta S}, \quad \frac{\partial S}{\partial t} = -\frac{\Delta \tilde{H}_c}{\Delta P} \tag{B.10}$$

where $\Delta/\Delta F$ denotes the variational derivative with respect to the functional F. With $\frac{\partial S}{\partial t} = \frac{\partial P}{\partial t} = 0$, the equations of motion take the form

$$\mathcal{H} = 0, \tag{B.11}$$

and

$$\int d^3x \frac{\delta}{\delta h_{ij}} \left(PG_{ijkl} \frac{\delta S}{\delta h_{kl}} \right) = 0. \tag{B.12}$$

Eq. (B.11) is the Hamiltonian constraint, and eq. (B.12) may be interpreted as a continuity equation.

It is of interest that the interpretation of eq. (B.12) as a continuity equation leads to a rate equation that relates $\frac{\partial h_{kl}}{\partial t}$ and $\frac{\delta S}{\delta h_{kl}}$. This follows from the observation that such an interpretation is only possible if the "field velocity" $\frac{\partial h_{kl}}{\partial t}$ is related to $G_{ijkl}\frac{\delta S}{\delta h_{kl}}$ in a linear fashion. Indeed, the most general rate equation for the metric h_{ij} that is consistent with the interpretation of eq. (B.12) as a continuity equation is of the form

$$\delta h_{ij} = \left(\alpha G_{ijkl} \frac{\delta S}{\delta h_{kl}} + \delta_\epsilon h_{ij} \right) \delta t \tag{B.13}$$

where α is an arbitrary function of x (I have include a term $\delta_\epsilon h_{kl} = -(D_k \epsilon_l + D_l \epsilon_k)$ which allows for gauge transformations of h_{kl}, which is permitted because the gauge transformations are assumed to leave $\int DhP$ invariant, as discussed before). Therefore, one may write the rate equation for h_{kl} in the standard form

$$\frac{\partial h_{ij}}{\partial t} = NG_{ijkl} \frac{\delta S}{\delta h_{kl}} + D_i N_j + D_j N_i. \tag{B.14}$$

Eq. (B.14) agrees with the equations derived from the ADM canonical formalism, provided N is identified with the lapse function and N_k with the shift vector [33].

It is remarkable that the rate equations for the metric, eq. (B.14), can be shown to be a direct consequence of applying the theory of ensembles on configuration space to classical general relativity. By contrast, the derivation of the rate equations from the Hamilton-Jacobi formalism alone [34] requires a much more lengthy derivation.

Appendix C. Ground state Gaussian functional solution

Consider the Schrödinger functional equation

$$-\frac{1}{2\Lambda R^2}\left(\frac{\delta^2 A}{\delta\phi^2}\right)+\left[\lambda\frac{\Lambda R^2}{2}+V+\frac{R^2}{2\Lambda}\phi'^2+\frac{\Lambda R^2 m^2}{2}\phi^2\right]A=0. \tag{C.1}$$

with the ansatz

$$A\sim\exp\left\{-\frac{1}{2}\int\int dy\,dz\,\Lambda_y\,\Lambda_z\,R_y^2\,R_z^2\,\phi_y\,K_{yz}\,\phi_z\right\}. \tag{C.2}$$

If I calculate $\frac{\delta^2 A}{\delta\phi^2}$ and collect terms that have the same powers of ϕ, I get the following two equations for the kernel K_{xy},

$$\int dr\,N\left[\lambda\frac{\Lambda R^2}{2}+V(R,\Lambda)+\frac{\Lambda_r R_r^2}{2}K_{rr}\right]=0 \tag{C.3}$$

and

$$\int dr\,\Lambda_r R_r^2\,N\left[\frac{\phi_r'^2}{2\Lambda_r^2}+\frac{m^2}{2}\phi_r^2-\frac{1}{2}\int\int dydz\,\Lambda_y R_y^2\Lambda_z R_z^2\,\phi_y K_{yr}K_{rz}\phi_z\right]=0. \tag{C.4}$$

After an integration by parts in Eq. (C.4), the equations for K_{xy} are of a type that is standard in the context of the Schrödinger functional representation of quantum field theory in curved spacetimes [18, 24, 25, 26].

I want a solution that is valid for a foliation of spaces of constant positive curvature and a lapse function N that is constant. Note that the Λ and R that appear in the line element of Eq. (19) satisfy $\sqrt{h}=\Lambda R^2$ and $h^{rr}=\Lambda^{-2}$, where h^{kl} is the inverse metric tensor on the three-dimensional spatial hypersurface of constant curvature. Then, Eq. (C.4) can be written in the form

$$\int dr\sqrt{h_r}\left[\frac{1}{2}h^{rr}\frac{\partial\phi_r}{\partial r}\frac{\partial\phi_r}{\partial r}+\frac{m^2}{2}\phi_r^2-\frac{1}{2}\int\int dydz\sqrt{h_y}\sqrt{h_z}\,\phi_y K_{yr}K_{rz}\phi_z\right]=0, \tag{C.5}$$

and one can show that K_{xy} satisfies

$$\int dr\sqrt{h_r}K_{yr}K_{rz}=\left[-\frac{1}{\sqrt{h_y}}\frac{\partial}{\partial y}\left(h^{yy}\sqrt{h_y}\frac{\partial}{\partial y}\right)+m^2\right]\delta(y,z) \tag{C.6}$$

where $\delta(y,z)=\frac{1}{\sqrt{h_y}}\delta(y-z)$ is the delta function on the hypersurface.

To get an explicit expression for K_{xy} that solves Eq. (C.6), introduce a fixed, particular set of coordinates for the line element of Eq. (19). Let

$$N=1,\qquad N^r=0,\qquad \Lambda=a_0,\qquad R=a_0\sin r, \tag{C.7}$$

where $r\in[0,2\pi)$ and a_0 can be interpreted as the scale factor of a closed Robertson-Walker universe. Then the solution of Eq. (C.6) is given by

$$K_{xy}=\frac{1}{2a_0^4}\sum_n\sqrt{\gamma_n}\,\psi_x^{(n)}\psi_y^{(n)}, \tag{C.8}$$

where the basis functions $\psi_r^{(n)}$ are solutions of a Schrödinger-type equation in a space of constant curvature,

$$-\frac{1}{\sin^2 r}\frac{\partial}{\partial r}\left(\sin^2 r\,\frac{\partial\psi_r^{(n)}}{\partial r}\right)+m^2a_0^2\,\psi_r^{(n)}=\gamma_n\psi_r^{(n)}. \tag{C.9}$$

The $\psi_n(r)$ satisfy orthonormality and completeness relations. The eigenvalues γ_n are given by

$$\gamma_n = n^2 - 1 + m^2 a_0^2, \qquad n = 1, 2, 3... \quad . \tag{C.10}$$

Given the solution of Eq. (C.4), one can use Eq. (C.3) to express a_0 in terms of the cosmological constant and the energy E of the quantized scalar field, since [26]

$$E \sim a_0^3 \int dr \, \sin^2 r \, K_{rr}. \tag{C.11}$$

However, $\int dr \, \sin^2 r \, K_{rr} \sim \sum_n \gamma_n$, which diverges. This is a consequence of the infinite zero-point energy of the quantum field. Therefore, to extract a finite result for a_0 it becomes necessary to introduce renormalization procedures.

References

[1] Albers M, Kiefer C and Reginatto M 2008 *Phys. Rev.* D **78** 064051
[2] Rosenfeld L 1963 *Nucl. Phys.* **40**
[3] Carlip S 2008 *Class. Quantum Grav.* **25** 154010
[4] Boughn S 2009 *Found. Phys.* **39** 331
[5] Butterfield J and Isham C 2001 *Physics meets philosophy at the Planck scale* ed C. Callender and N. Huggett (Cambridge: Cambridge University Press)
[6] Dyson F 2004 *The New York Review of Books* **LI:8** 16
[7] Rothman T and Boughn S 2006 *Found. Phys.* **36** 1801
[8] Boughn S and Rothman T 2006 *Class. Quantum Grav.* **23** 5839
[9] Hall M J W and Reginatto M 2005 *Phys. Rev.* A **72** 062109
[10] Hall M J W 2008 *Phys. Rev.* A **78** 042104
[11] Reginatto M and Hall M J W 2009 *J. Phys.: Conf. Ser.* **174** 012038
[12] Hall M J W, Kumar K and Reginatto M 2003 *J. Phys.* A **36** 9779
[13] Hall M J W 2005 *Gen. Rel. Grav.* **37** 1505
[14] Reginatto M 2005 *Braz. J. Phys.* **35** 476
[15] Misner C W, Thorne K S and Wheeler J A 1973 *Gravitation* (San Francisco: Freeman).
[16] Kiefer C 1993 *From Newton's bucket to quantum gravity* ed J. B. Barbour and H. Pfister (Boston: Birkhuser)
[17] Kiefer C 1993 The semiclassical approximation to quantum gravity *Preprint* gr-qc/9312015
[18] Kiefer C 2004 *Quantum Gravity* (Oxford: Clarendon Press)
[19] Romano J D 1995 Spherically Symmetric Scalar Field Collapse: An Example of the Spacetime Problem of Time *Preprint* gr-qc/9501015 v1
[20] Kuchař K 1994 *Phys. Rev.* D **50**3961.
[21] Lau S R 1996 On the Canonical Reduction of Spherically Symmetric Gravity *Preprint* gr-qc/9508028 v3
[22] Fischler W, Morgan D and Polchinski J 1990 *Phys. Rev.* D **42** 4042
[23] Brotz T and Kiefer C 1997 *Phys. Rev.* D **55** 2186
[24] Jackiw R 1995 *Diverse Topics in Theoretical and Mathematical Physics* (Singapore: World Scientific)
[25] Hatfield B 1992 *Quantum field theory of point particles and strings* (Cambridge: Perseus Books)
[26] Long D V and Shore G M 1998 *Nucl. Phys.* B **530** 247
[27] Herdeiro C A R, Ribeiro R H and Sampaio M 2008 *Class. Quantum Grav.* **25** 165010
[28] Bergmann P G 1966 *Phys. Rev.* **144** 1078
[29] DeWitt B 1979 *General Relativity: An Einstein Centenary Survey* ed S. W. Hawking and W. Israel (Cambridge: Cambridge University Press)
[30] Hamber H W and Williams R M 1999 *Phys. Rev.* D **59** 064014
[31] Fadeev L and Popov V 1974 *Sov. Phys. Usp.* **16** 777
[32] Mottola E 1995 *Jour. Math. Phys.* **36** 2470
[33] Wald R 1984 *General Relativity* (Chicago: University of Chicago Press)
[34] Gerlach U H 1969 *Phys. Rev.* **177** 1929

Journal of Physics: Conference Series **442** (2013) 012010 doi:10.1088/1742-6596/442/1/012010

Pathways toward understanding Macroscopic Quantum Phenomena[1]

B L Hu and Y Subaşı

Maryland Center for Fundamental Physics and Joint Quantum Institute
University of Maryland, College Park, Maryland 20742, USA

E-mail: blhu@umd.edu,ysubasi@umd.edu

Abstract. Macroscopic quantum phenomena refer to quantum features in objects of 'large' sizes, systems with many components or degrees of freedom, organized in some ways where they can be identified as macroscopic objects. This emerging field is ushered in by several categories of definitive experiments in superconductivity, electromechanical systems, Bose-Einstein condensates and others. Yet this new field which is rich in open issues at the foundation of quantum and statistical physics remains little explored theoretically (with the important exception of the work of A J Leggett [1], while touched upon or implied by several groups of authors represented in this conference. Our attitude differs in that we believe in the full validity of quantum mechanics stretching from the testable micro to meso scales, with no need for the introduction of new laws of physics.) This talk summarizes our thoughts in attempting a systematic investigation into some key foundational issues of quantum macroscopic phenomena, with the goal of ultimately revealing or building a viable theoretical framework. Three major themes discussed in three intended essays are the large N expansion [2], the correlation hierarchy [3] and quantum entanglement [4]. We give a sketch of the first two themes and then discuss several key issues in the consideration of macro and quantum, namely, a) recognition that there exist many *levels of structure* in a composite body and only by judicious choice of an appropriate set of *collective variables* can one give the best description of the dynamics of a specific level of structure. Capturing the quantum features of a macroscopic object is greatly facilitated by the existence and functioning of these collective variables; b) *quantum entanglement*, an exclusively quantum feature [5], is known to persist to high temperatures [6] and large scales [7] under certain conditions, and may actually decrease with increased connectivity in a quantum network [8]. We use entanglement as a measure of quantumness here and pick out these somewhat counter-intuitive examples to show that there are blind spots worthy of our attention and issues which we need to analyze closer. Our purpose is to try to remove the stigma that quantum only pertains to micro, in order to make way for deeper probes into the conditions whereby quantum features of macroscopic systems manifest.

Macroscopic quantum phenomena (MQP) manifest in a number of systems, superconductivity is probably the oldest often referred to, Bose-Einstein condensate (BEC) [9] and electro- and opto-mechanical devices [10, 11] are amongst the recent excitements. It is a relatively new research venue, with exciting ongoing experiments and bright prospects, yet with surprisingly little theoretical activity. From the traditional view that macroscopic objects are classical and quantum describes the microscopic realm, MQP appears like a transgression. This of course is

[1] Invited talk presented by BLH

what makes it interesting intellectually. This simplistic and hitherto rarely challenged view needs to be scrutinized anew, perhaps eventually with much of the conventional wisdoms repealed. In a series of papers in preparation we attempt to explore systematically into some key foundational issues of MQP, with the hope of finding a viable theoretical framework for this new endeavour. The three major themes discussed in these three essays are the large N expansion, the correlation hierarchy and quantum entanglement for systems of "large" sizes, with many components or degrees of freedom. Before delving into the subject proper, to orient ourselves, we excerpt the beginning of [2] which charts the interloping and intersecting domains of quantum /classical and micro/ Macro.

1. Quantum / classical, micro / macro

There are many ways to deal with the issue of quantum-classical correspondence [12]. In the most common and traditional view the classical limit corresponds to $\hbar \to 0$, or, invoking the Bohr correspondence principle, the principal quantum number of a system $n \to \infty$, or regarding the coherent state as the "most classical" quantum state, or the Wigner function as the "closest to classical" distribution. Less precise criteria also abound, such as the loose concept that a system at high temperature behaves classically, or viewing the thermodynamic / hydrodynamic limits (of a quantum system) as classical. (For a description of the various criteria, see, e.g., [13]). There are holes in almost all of the above common beliefs. A more sophisticated viewpoint invokes decoherence, the process whereby a quantum system loses its coherence (measured by its quantum phase information) through interaction with its environment [14]. In this work we examine an alternative perspective, as the folklore goes, that quantum pertains to the small (mass, scale) while classical to the large (size, multiplicity). This common belief now requires a much closer scrutiny in the face of new challenges from macroscopic quantum phenomena (MQP), viz, quantum features may show up even at macroscopic scales. A common example is superconductivity where the Cooper pairs can extend to very large scales compared to interatomic distances and Bose-Einstein condensate (BEC) where a large number N of atoms occupy the same quantum state, the N-body ground state. Other examples include nanoelectromechanical devices [15] where the center of mass of a macroscopic classical object, the cantilever, obeys a quantum mechanical equation of motion. Experiments to demonstrate the quantum features such as the existence of interference between two macroscopic objects have been carried out, e.g., for C^{60} molecules passing through two slits [16] or proposed mirror superposition experiments [17, 18].

A most direct account of the difference between the microscopic and the macroscopic behaviours of a quantum system is by examining N, the number of physically relevant (e.g. for atomic systems, forgetting about the tighter-bound substructures) quantum particles or components in a macroscopic object. One may ask: At what number of N will it be suitable to describe the object as mesoscopic with qualitatively distinct features from microscopic and macroscopic? For classical systems significant advances in the recent decade have been made in providing a molecular dynamics basis to the foundations of thermodynamics [19], relating the macroscopic thermodynamic behaviour of a gas to the chaotic dynamics of its molecular constituents. One could even calculate the range in the number of molecules where a microscopic system begins to acquire macroscopic behaviour and hence identify the approximate boundaries of mesoscopia [20]. For quantum systems one needs to deal with additional concerns of quantum coherence and entanglement which are critically important issues in quantum information processing (QIP) [21]. A fundamental issue in QIP is how the performance of a quantum information processor alters as one scales the system up. This dependence on N is known as the "scaling" problem [22].

There are many important and interesting issues of MQP, one subset of special interest to us is how quantum expresses itself in the macroscopic domain since usually macro conjures classicality. Thus even the simplest yet far from naive questions need to be reconsidered properly. For example, why is it that an ostensibly macroscopic object such as a cantilever should follow a quantum equation of motion. This "center-of-mass axiom" is implicitly assumed in many descriptions of MQP but rarely justified. The conditions upon which this can be justified are explored in [23] with the derivation of a master equation for N coupled harmonic oscillators (NHO) in a finite temperature harmonic oscillator bath, in the form of the HPZ master equation [24], extending earlier work for 2HO [25]. (A mathematically more vigorous and complete treatment of NHO system is given in [26].) Presently we are continuing to explore the conditions where one could infer macroscopic quantum behavior, specifically in terms of the existence and degree of quantum entanglement in this coupled NHO model. One aspect is in terms of entanglement at finite temperature [27, 28, 29, 30] and large distance [7, 31], the other in terms of entanglement between different levels of structure (micro to meso to macro) [32, 33] and the crucial role in a judicious choice of the appropriate collective variables [34, 4]. This is discussed in Section 3. In Section 4 we use the results of a recent paper on complex quantum network [8] to illustrate the somewhat counterintuitive finding that entanglement does not necessarily increase with connectivity but varies with the strength of coupling and the type of connectivity. (See also [35].) Our understanding of this aspect will be described in a future work [36].

2. Pathways toward understanding macroscopic quantum phenomena

In what follows we present two pathways as explored in two recent essays [2, 3], the first concerns what macroscopic means – large size? number? What about the degree of complexity of its constituents? What if the constituents are non-interacting versus interacting? Weakly interacting versus strongly interacting? The second pathway explores how quantum correlations and fluctuations impact on MQP using the n-particle-irreducible (nPI) representation.

2.1. Pathway 1: From the large N perspective

In this paper we use different theories in a variety of contexts to examine the conditions or criteria whereby a macroscopic quantum system may take on classical attributes, and, more interestingly, that it keeps some of its quantum features. The theories we consider here are, the $O(N)$ quantum mechanical model, semiclassical stochastic gravity and gauge / string theories; the contexts include that of a "quantum roll" in inflationary cosmology, entropy generation in quantum Vlasov equation for plasmas, the leading order and next-to-leading order large N behaviour, and hydrodynamic / thermodynamic limits. The criteria for classicality in our consideration include the use of uncertainty relations, the correlation between classical canonical variables, randomization of quantum phase, environment-induced decoherence, decoherent histories of hydrodynamic variables, etc. All this exercise is to ask only one simple question: Is it really so surprising that quantum features can appear in macroscopic objects? By examining different representative systems where detailed theoretical analysis has been carried out, we find that there is no a priori good reason why quantum phenomena in macroscopic objects cannot exist.

2.2. Pathway 2: From the correlation, coupling and criticality perspectives

In this sequel paper we explore how macroscopic quantum phenomena can be measured or understood from the behavior of quantum correlations which exist in a quantum system of many particles or components and how the interaction strengths change with energy or scale, under ordinary situations and when the system is near its critical point. We use the nPI (master) effective action related to the Boltzmann-BBGKY / Schwinger-Dyson hierarchy of equations as a tool for systemizing the contributions of higher order correlation functions to the dynamics of lower order correlation functions. Together with the large N expansion discussed in our first paper [2] we explore 1) the conditions whereby an H-theorem is obtained, which can be viewed as a signifier of the emergence of macroscopic behavior in the system. We give two more examples from past work: 2) the nonequilibrium dynamics of N atoms in an optical lattice under the large $\mathcal{N}$ (field components), $2PI$ and second order perturbative expansions, illustrating how N and $\mathcal{N}$ enter in these three aspects of quantum correlations, coherence and coupling strength. 3) the behavior of an interacting quantum system near its critical point, the effects of quantum and thermal fluctuations and the conditions under which the system manifests infrared dimensional reduction. We also discuss how the effective field theory concept bears on macroscopic quantum phenomena: the running of the coupling parameters with energy or scale imparts a dynamical-dependent and an interaction-sensitive definition of "macroscopia".

3. Levels of structure consideration in the micro-macro divide

Before we describe the third pathway toward understanding MQP, namely, using quantum entanglement in a system as a measure of its "quantumness", it is useful to ponder on a related issue we explored a bit along the first pathway, namely, what exactly is "largeness" which usually is connected with macroscopic? Do all the basic constituents of a large object contribute equally towards its quantum feature? (This point is highlighted in footnote 2 of [23].) We actually know how these basic constituents are and how they are organized. A C^{60} molecule is made of carbon atoms, each atom is made of nuclei and electrons, each nucleus contains a certain number of protons and neutrons, each of them in turn is made up of quarks and gluons. Are we to simply count the number of quarks /gluons or protons /neutrons when we say an object is macroscopic? Obviously the tight binding of them to form a nucleus enters into our consideration when we treat the nucleus as a unit which maintains its own more or less distinct features and dynamics. Thus when one talks about the mesoscopic or macroscopic behavior of an object one needs to specify which level of structure is of special interest and how important each level contributes to these characteristics. The coupling strength between constituents at each level of structure (e.g., inter-atomic) compared to that structure's coupling with the adjacent and remaining levels (which can be treated as an environment to this specific level of structure in an effective theory description, and its influence on it represented as noise [37]) will determine the relative weight of each level of structure's partaking of the macroscopic object's overall quantum behavior. The best description of the behavior and dynamics of a particular level of structure is given by an effective theory for the judiciously chosen "collective variables". Nuclear forces in terms of QCD is an illustrative example.

3.1. Choose the right collective variables before considering their quantum behavior

Same consideration should enter when one looks for the "quantumness" of an object, be it of meso or macro scale. One can quantize any linear system of whatever size, even macroscopic objects, such as sound waves from their vibrations. Giving it a name which ends with an "on" such as phonon and crowning it into a quantum variable is almost frivolous compared to the task of identifying the correct level of structure and finding the underlying constituents, the atoms in a lattice in this example, and their interactions. Constructing the relevant collective variables which best capture the salient physics of interest should come before one considers their quantum features. Thus, viewed in this perspective in terms of collective variables, we see that quantum features need not be restricted to microscopic objects. In fact 'micro' is ultimately also a relative concept as new "elementary" particles are discovered which make up the once regarded 'micro' objects.

We illustrate this idea with two examples below, one on the relevance of the center of mass variable in capturing the quantum features of a macroscopic object, the other on the description of entanglement between two macroscopic objects.

3.1.1. The quantum and macroscopic significance of center of mass variable. We can ask the question: what are the conditions upon which the mechanical and statistical mechanical properties of a macroscopic object can be described adequately in terms mainly of its center-of-mass (COM) variable kinematics and dynamics as captured by a master equation (for the reduced density matrix with the environmental variables integrated out). This is an implicit assumption made in many MQP investigations, namely, that the quantum mechanical behavior of a macroscopic object, like the nanoeletromechanical oscillator [10, 15], mirror [17], or a C^{60} molecule [16], placed in interaction with an environment – behavior such as quantum decoherence, fluctuations, dissipation and entanglement– can be captured adequately by its COM behavior. For convenience we refer to this as the "COM axiom". This assertion is intuitively reasonable, as one might expect it to be true from normal- mode decompositions familiar in classical mechanics, but when particles (modeled by NHO) interact with each other (such as in a quantum bound state problem) in addition to interacting with their common environment, all expressed in terms of the reduced density matrix, it is not such a clear-cut result. At least we have not seen a proof of it.

With the aim of assessing the validity of the COM axiom the authors of [23] considered a system modeled by N harmonic oscillators interacting with an environment consisting of n harmonic oscillators and derived an exact non-Markovian master equation such a system in a bath with arbitrary spectral density and temperature. The authors outlined a procedure to find a canonical transformation to transform from the individual coordinates (x_i, p_i) to the collective coordinates $(\tilde{X}_i, \tilde{P}_i), i = 1, ..., N$ where $\tilde{X}_1, \tilde{P}_1$ are the center-of-mass coordinate and momentum respectively. In fact they considered a more general type of coupling between the system and the environment in the form $f(x_i)q_j$ (instead of the ordinarily assumed x_iq_j) and examined if the COM variable dynamics separates from the reduced variable dynamics. They noted that if the function $f(x)$ has the property $\sum_{i=1}^{N} f(x_i) = \tilde{f}(\tilde{X}_1) + g(\tilde{X}_2, ..., \tilde{X}_N)$, for example $f(x) = x$ or $f(x) = x^2$, one can split the coupling between the system and environment into couplings containing the COM coordinate and the relative coordinates. Tracing out the environmental degrees of freedom q_i, one can easily get the influence action which characterizes the effect of the environment on the system.

However, as the authors of [23] emphasized, the coarse-graining made by tracing out the environmental variables q_i does not necessarily lead to the separation of the COM and the relative variables in the effective action. When they are mixed up and can no longer be written as the sum of these two contributions, the form of the master equation will be radically altered as it would contain both the relative variable and the center-of-mass variable dynamics.

With these findings they conclude that for the N harmonic oscillators quantum Brownian model, the coupling between the system and the environment need be bi-linear, in the form x_iq_j, for this axiom to hold. They also proved that the potential $V_{ij}(x_i - x_j)$ is independent of the center-of-mass coordinate. In that case, one can say that the quantum evolution of a macroscopic object in a general environment is completely described by the dynamics of the center-of-mass canonical variables $(\tilde{X}_1, \tilde{P}_1)$ obeying a master equation of the HPZ [24] type.

What is the relevance of this finding to MQP? Within the limitation of the N harmonic oscillator model it conveys at least two points: 1) For certain types of coupling the center of mass (COM) variable of an object composed of a large number of constituents does play a role in capturing the collective behavior of this object 2) Otherwise, more generally, the environment-induced quantum statistical properties of the system such as decoherence and entanglement could be more complicated. (For a similar conclusion considering the cross level (of structure) coarse-graining, see [32].)

We now look at the quantum entanglement between two macroscopic objects, comparing the entanglement between the micro-variables of their constituents in two types of couplings, one to one and one to many. The very different nature between these two types serves to illustrate the relevance of how the micro-constituents organize into a macro object and how entanglement between collective variables reveals the quantum features of a macroscopic entity. This issue is raised by Martins [34] in the consideration of the entanglement between two wavepackets each containing sub-levels.

3.2. M-M, m-m and m-M entanglement

We now apply the methods developed in [23] to the study of the entanglement between the COM variables of two macroscopic objects. Each macroscopic object is modeled by N identical harmonic oscillators (NHO). However, unlike [23], we do not include an environment in our discussion because our focus is on the entanglement between the two objects induced by various types of direct interactions between their microscopic constituents. We denote the coordinates and the momenta of the microscopic constituents of the first and second macroscopic object by $\left\{x_{[1]i}, p_{[1]i}\right\}$ and $\left\{x_{[2]i}, p_{[2]i}\right\}$ respectively. The interactions between the microscopic constituents of one macroscopic object are assumed to be functions of the difference of variables only and we restrict ourselves to bilinear couplings between the microscopic constituents of the two macroscopic objects. The total Hamiltonian is then given by:

$$H_1 = \sum_{i=1}^{N}\left(\frac{p_{[1]i}^2}{2M} + \frac{1}{2}M\Omega^2 x_{[1]i}^2\right) + \sum_{i\neq j}^{N} V_{ij}\left(x_{[1]i} - x_{[1]j}\right), \quad (3.1)$$

$$H_2 = \sum_{i=1}^{N}\left(\frac{p_{[2]i}^2}{2M} + \frac{1}{2}M\Omega^2 x_{[2]i}^2\right) + \sum_{i\neq j}^{N} V_{ij}\left(x_{[2]i} - x_{[2]j}\right), \quad (3.2)$$

$$H_I = \sum_{i,j}^{N} G_{ij} x_{[1]i} x_{[2]j}. \quad (3.3)$$

The canonical transformation described in the Appendix A. of [23] can be applied to each object separately to yield a new set of phase space variables $\left\{\tilde{X}_{[1]i}, \tilde{P}_{[1]i}\right\}$ and $\left\{\tilde{X}_{[2]i}, \tilde{P}_{[2]i}\right\}$ and the associated masses $\tilde{M}_i$. Here $\tilde{X}_{[1]1} = \frac{1}{N}\sum_{n=1}^{N} x_{[1]i}$ and $\tilde{X}_{[2]1} = \frac{1}{N}\sum_{n=1}^{N} x_{[2]i}$ are the COM variables. The Hamiltonians of the macroscopic objects can be written in terms of these variables as:

$$H_1 = \sum_{i=1}^{N}\left(\frac{\tilde{P}_{[1]i}^2}{2\tilde{M}_i} + \frac{1}{2}\tilde{M}_i\Omega^2\tilde{X}_{[1]i}^2\right) + \tilde{V}\left(\tilde{X}_{[1]2}, \cdots, \tilde{X}_{[1]N}\right) = H_{1COM} + H_{1REL}, \quad (3.4)$$

$$H_2 = \sum_{i=1}^{N}\left(\frac{\tilde{P}_{[2]i}^2}{2\tilde{M}_i} + \frac{1}{2}\tilde{M}_i\Omega^2\tilde{X}_{[2]i}^2\right) + \tilde{V}\left(\tilde{X}_{[2]2}, \cdots, \tilde{X}_{[2]N}\right) = H_{2COM} + H_{2REL}. \quad (3.5)$$

It has been proven in [23] that the potential $\tilde{V}$ is not a function of the COM variable. This is a consequence of the form assumed for the potential energy. For a general bilinear coupling characterized by G_{ij} the interaction Hamiltonian H_I can take a complicated form, possibly mixing the COM variable with the relative variables. Inspired by Martins[34], in what follows we will focus on two particular choices of G_{ij}. The use of the new set of canonical variables which include the COM will help interpret the behaviour of macroscopic entanglement.

3.2.1. Pairwise interaction pattern. The pairwise interaction pattern is defined by $G_{ij} = G\delta_{ij}$. In other words one oscillator from object one couples to one oscillator from object two, and all such couplings have the same strength. Using the canonical transformation of [23] it can be shown that the interaction Hamiltonian takes the form:

$$H_I \;= \textstyle\sum_i^N \frac{G}{M}\tilde{M}_i\tilde{X}_{[1]i}\tilde{X}_{[2]i}. \tag{3.6}$$

Note that pairwise interactions among the original variables translate into pairwise interactions among the transformed variables. A very important difference is that whereas the pairwise interactions in the original variables were all equal strength, the strength of the interactions scale with the mass of the variables after the transformation. As a result the relative strength of interactions between variable pairs are the same for all the variables, including the COM. To see this explicitly let us consider the case with $\tilde{V} = 0$ for simplicity, namely the micro-constituents of each macroscopic object do not interact with each other. Then we rescale the coordinates by $\tilde{X}_{[1]i} \to \tilde{X}_{[1]i}\sqrt{M/\tilde{M}_i}$ and $\tilde{X}_{[2]i} \to \tilde{X}_{[2]i}\sqrt{M/\tilde{M}_i}$, after which the Hamiltonian takes the form:

$$H \;= \textstyle\sum_{i=1}^N \left(\frac{\tilde{P}^2_{[1]i}}{2M} + \frac{1}{2}M\Omega^2\tilde{X}^2_{[1]i} + \frac{\tilde{P}^2_{[2]i}}{2M} + \frac{1}{2}M\Omega^2\tilde{X}^2_{[2]i} + G\tilde{X}_{[1]i}\tilde{X}_{[2]i}\right). \tag{3.7}$$

In this form it is easy to see that the effective strength of interactions in the COM variable is the same as the effective strength of interactions in all the other variables. Hence the pairwise interaction pattern will induce the same amount of entanglement between pairs of transformed variables, without distinguishing the COM variable. Entanglement between non-COM variables would be effected if the interactions among the oscillators within the same object, i.e. V_{ij}, are not set to zero.

If we only focus on the effect of the pairwise interactions, it is fair to say that such interactions couple the pairwise transformed variables with equal effective strength independent of the size N of the macroscopic objects. As a consequence we expect the behavior of entanglement between the corresponding variables of the objects to be independent of the size of the macroscopic objects, even for the COM variable, as well. For instance, at a given temperature the amount of entanglement between the two corresponding variables of the objects will not depend on N. The critical temperature above which the entanglement ceases to exist also should not depend on N.

3.2.2. One-to-all interaction pattern. The one-to-all interaction pattern is characterised by $G_{ij} = G$. Then it is easy to see that the interaction Hamiltonian in the transformed variables takes the form:

$$H_I \;= N^2G\tilde{X}_{[1]1}\tilde{X}_{[2]1}. \tag{3.8}$$

Note that one-to-all interaction pattern corresponds to a coupling only between the COM variables of the macroscopic objects, the relative variable Hamiltonian is unaffected. Thus one-to-all pattern differs from the pairwise pattern in that it distinguishes the COM variable. Moreover if we perform the same rescaling of the previous section to determine the effective strength of this coupling we get:

$$H_{COM} = \frac{\tilde{P}^2_{[1]1}}{2M} + \frac{1}{2}M\Omega^2\tilde{X}^2_{[1]1} + \frac{\tilde{P}^2_{[2]1}}{2M} + \frac{1}{2}M\Omega^2\tilde{X}^2_{[2]1} + NG\tilde{X}_{[1]i}\tilde{X}_{[2]i}. \tag{3.9}$$

We see that the effective strength of the coupling increases with increasing N for the one-to-all pattern. Thus in this case we expect the entanglement between the COM variables to increase with increasing size of the macroscopic objects and survive at higher temperatures. The one-to-all interaction pattern is crucial for the scaling of the entanglement of COM variables with N. Hence it is important to investigate if this type of interaction pattern can occur in realistic situations and if so how common it is.

4. Macroscopic quantum phenomena from the entanglement perspective

Entanglement is considered as a uniquely quantum feature [5], and quantum is habitually viewed as a zero or low temperature phenomenon or pertaining only to small systems. Both of these conditions are now being reconsidered, the "small system" restriction facing new challenges from macroscopic quantum phenomena (MQP) we are considering here. The "low temperature" restriction is lifted by theoretical observations [6, 38] and experimental proposals [39] that entanglement can survive at high temperatures, some even speculate that it is witnessed in biological systems [40].

4.1. Quantum entanglement at high temperatures and long distances?

Theoretical analysis of this issue for such systems has been carried out for coupled oscillator chain (1D) [27] or lattices (2D or 3D), where bounds and phase diagrams showing entangled and separable states were obtained by Anders [28]. For a nice exposè of the general issues on this topic we refer readers to the Discussion and Conclusion Section of [29].

Thermal or hot entanglement can be studied by generalizing the (zero-temperature) quantum field mimicking a harmonic lattice to a thermal (finite temperature) field. In terms of model description quantum entanglement between two inertial harmonic oscillators interacting via a zero-temperature quantum field was studied earlier by Lin and Hu [41] who showed that in addition to the temporal evolution of their entanglement there is also a dependence on their spatial separation. This generalization is done recently in [30] wherein both the temporal and spatial dependence of quantum entanglement studied before are shown to be sensitive to temperature variation. These authors also considered thermal entanglement in a harmonic lattice but with strong coupling, extending the comprehensive study of [29].

Another aspect is how much quantum entanglement can survive at large distance. *Long range entanglement* in a coupled oscillator chain was claimed by Wolf et al [7]. Their setup of two harmonic oscillators interacting with a one dimensional harmonic lattice in a Gibbs state and their choice of parameters (continuum limit) map snugly to the thermal field model mentioned above. There are advantages in approaching the thermal entanglement issue for continuum systems from a quantum field theory perspective. Besides the technical ease to perform integrals over finite sums, the special properties of lower-dimensional systems (such as the Coleman-Mermin-Wegner theorem and the Berezinskii-Kosterlitz-Thouless phase transition) are well known, in some cases aided by elegant conformal field theory properties. This calculation is presently carried out in [31] where existence of zeros in the spectral density is found to be the cause of long range entanglement. How general is this tie has yet to be decided.

4.2. Quantum networks: more connected not always more entangled

Finally we mention the results from a recent paper to illustrate a point on the relation between connectivity in a quantum network and entanglement. The following are excerpted from [8].

A network is defined as a set of N nodes and E edges accounting for their pairwise interactions. The network is usually characterized by its adjacency matrix, A, with elements $A_{ij} = 1$ if an edge connects nodes i and j while $A_{ij} = 0$ otherwise. We restrict attention to the undirected network where $A_{ij} = A_{ji}$. The Laplacian is related to the adjacency matrix by $L_{ij} = k_i\delta_{ij} - A_{ij}$, where $k_i = \sum_j A_{ij}$ is the connectivity of node i, *i.e.*, the number of nodes connected to i.

We can represent the nodes of the network by identical quantum oscillators interacting in accordance to the network topology encoded in L. The Hamiltonian of the harmonic quantum network is given by:

$$H_{\text{network}} = \frac{1}{2}\left(\mathbf{p}^{\mathrm{T}}\mathbf{p} + \mathbf{x}^{\mathrm{T}}(\mathbb{I} + 2cL)\mathbf{x}\right), \tag{4.1}$$

here $\mathbb{I}$ is the $N \times N$ identity matrix, c is the coupling strength between connected oscillators while $\mathbf{p}^{\mathrm{T}} = (p_1, p_2, ..., p_N)$ and $\mathbf{x}^{\mathrm{T}} = (x_1, , x_2, ..., x_N)$ are the operators corresponding to the momenta and positions of nodes respectively, satisfying the usual commutation relations: $[\mathbf{x}, \mathbf{p}^{\mathrm{T}}] = i\hbar\,\mathbb{I}$.

The properties of the ground state of Hamiltonian (4.1) can be studied to quantify the amount of information each element of a network shares with the rest of the system via quantum fluctuations. Even at zero temperature the nodes are not at rest due to Heisenberg uncertainty principle. Their spatial fluctuations depends on the pattern of physical interactions, *i.e.*, the network structure. To show this, the authors of [8] consider the partition of the network into a node, say i, and its complement i^{c}, *i.e.* the rest of the network. The mutual information shared by the two parties is given by:

$$\mathcal{I}(i|i^{\mathrm{c}}) = S_i + S_{i^{\mathrm{c}}} - S_{\mathrm{tot}} \,. \tag{4.2}$$

Here S_i and $S_{i^{\mathrm{c}}}$ are marginal entropies and S_{tot} is the total entropy of the network. It is natural to choose the Von Neumann entropy to quantify the quantum information of the system, yielding $S_{\mathrm{tot}} = 0$ for the ground state (as it is a *pure state*). Since the total network is in its ground (and pure) state we have $S_i = S_{i^{\mathrm{c}}} = \mathcal{I}(i|i^{\mathrm{c}})/2$. Therefore, the information that a node shares with the network is intrinsically due to quantum correlations. Equivalently, the mutual information is, itself, a measure of the entanglement (quantified by S_i) between a single node and the rest of the system.

The authors then quantify the entanglement entropies of nodes embedded in different network topologies. They consider two homogeneous network substrates: *(i)* Random Regular Graphs (RRG), in which all the nodes have the same number of contacts ($k_i = \langle k \rangle, \forall i$), and *(ii)* Erdős-Rényi (ER) networks [42], for which the probability of finding a node with k neighbors, $P(k)$, follows a Poisson distribution so that most of the nodes have a degree k close to the average $\langle k \rangle$. They also analyze two networks having a scale-free (SF) pattern for the probability distribution, $P(k) \sim k^{-3}$, constructed by means of a configurational random model (SF-CONF) [43] and the Barabási-Albert model (SF-BA) [44]. Their results are presented in plots of the average entanglement entropy of a node with connectivity k, i.e $\langle S_k \rangle$, vs k for the three network models: ER, SF-CONF and SF-BA.

Interesting features can be gleaned from the figures in [8]: Fig.2 plots $\langle S_k \rangle$ for fixed average connectivity $\langle k \rangle$ and 4 different values of coupling strength c. It shows that the average entanglement of a node with given connectivity k increases with increasing coupling strength. As a check the case $c = 0$ corresponds to non-interacting oscillators which in their ground state are not entangled. It is expected that as the interactions get stronger the ground state becomes more and more entangled. Fig.3 plots $\langle S_k \rangle$ vs k for fixed c and different values of $\langle k \rangle$. It can be seen that for fixed k the entanglement $\langle S_k \rangle$ increases for decreasing $\langle k \rangle$ for all graphs.

Here we offer some tentative explanations on such qualitative behaviors. We can understand this using the idea of monogamy of entanglement, which says that a system which is fully entangled to another system cannot be entangled to a third system. Keeping k fixed while decreasing $\langle k \rangle$ amounts to reducing the connections the neighbors of the node of interest has. Thus its neighbors have less neighbors to get entangled with. As a result they can be entangled more with the node of interest.

Another observation we can make from both Fig.2 and Fig.3, which is less intuitive, is the fact that $\langle S_k \rangle$ flattens out for ER for large $\langle k \rangle$ and first rises and then falls for SF. This indicates that for ER the nodes with large connectivity have all the same amount of entanglement with the rest of the network. On the other hand for SF there is an optimal number for the connectivity such that those nodes with the optimal number of connections have the highest amount of entanglement with the rest of the network.

How can we make sense of this? Naively one expects the entanglement to increase with increasing number of connections, because more connections means more correlations. However entanglement is not just correlations. There may be a competition between correlations and monogamy of entanglement (or some argument using properties of quantum mutual information) that causes the rise and fall of entanglement in SF and the saturation in ER. Our own investigation into the entanglement behavior of quantum oscillator networks is under way [36].

5. Conclusion

In this talk we presented several pathways toward understanding MQP, identified some key issues we need to address or be concerned with and provided some examples to illustrate possibly counter-intuitive behavior. For the present issue of quantum entanglement, specifically, using its extent and behavior to measure the quantumness of a system, we pointed out the necessity to recognize the levels of structure and the usefulness of collective variables in describing a macroscopic composite object when we try to identify its quantum features. One needs to be aware of the qualitative differences between the entanglement amongst the micro-constituents and that between collective variables which reveal MQP. We mentioned entanglement at finite temperature and at long ranges, and used quantum coupled oscillator networks to illustrate the varying degrees of entanglement with different types of connectivity. We hope with these sampling of ideas, approaches and illustrative examples we can stimulate greater interest in how to think about the quantum nature of macroscopic objects, and, perhaps along the way, gain a deeper understanding of quantum physics itself.

Acknowledgments
BLH wishes to thank Professor Thomas Elze for his invitation to this rather unique series of meetings where issues at the foundations of gravitation and quantum and statistical mechanics are explored. He is also grateful to Professor M. C. Chu for his gentle hospitality at the Chinese University of Hong Kong where this paper was finished.

[1] Leggett A J 2002 *Journal of Physics: Condensed Matter* **14** R415
[2] Chou C, Hu B and Subaşı Y 2011 Macroscopic quantum phenomena from the large n perspective *Journal of Physics: Conference Series* vol 306 (IOP Publishing) p 012002
[3] Chou C, Hu B and Subaşı Y 2011 Macroscopic quantum phenomena from the correlation, coupling and criticality perspectives *Journal of Physics: Conference Series* vol 330 (IOP Publishing) p 012003
[4] Chou C, Hu B and Subaşı Y Macroscopic quantum phenomena from the quantum entanglement perspective (in preparation for Physica A)
[5] Schrödinger E 1935 *Mathematical Proceedings of the Cambridge Philosophical Society* **31**(04) 555–563 ISSN 1469-8064
[6] Amico L, Fazio R, Osterloh A and Vedral V 2008 *Reviews of Modern Physics* **80** 517
[7] Wolf A, De Chiara G, Kajari E, Lutz E and Morigi G 2011 *EPL (Europhysics Letters)* **95** 60008
[8] Cardillo A, Galve F, Zueco D and Gómez-Gardeñes J 2012 *arXiv preprint arXiv:1211.2580*
[9] He Q Y, Drummond P D and Reid M D 2012 Einstein-podolsky-rosen entanglement in two-well bec systems: ground states (*Preprint* `arXiv:1202.5752`)
[10] Gröblacher S, Hammerer K, Vanner M R and Aspelmeyer M 2009 *Nature* **460** 724–727
[11] Chen Y 2013 *arXiv preprint arXiv:1302.1924*
[12] Feng D H and Hu B L (eds) 1997 *Quantum Classical Correspondence: Proceedings of the 4th Drexel Symposium on Quantum Nonintegrability, Drexel University, Philadelphia, USA, September* (International Press of Boston) ISBN 1571460993
[13] Hu B L and Zhang Y 1995 *International Journal of Modern Physics A* **10** 4537–4561
[14] Zurek W H 2003 *Rev. Mod. Phys.* **75**(3) 715–775
[15] Armour A, Blencowe M and Schwab K 2002 *Physical Review Letters* **88** 148301
[16] Arndt M, Nairz O, Vos-Andreae J, Keller C, Van der Zouw G, Zeilinger A *et al.* 1999 *Nature* **401** 680–682
[17] Marshall W, Simon C, Penrose R and Bouwmeester D 2003 *Physical Review Letters* **91** 130401
[18] Müller-Ebhardt H, Rehbein H, Schnabel R, Danzmann K and Chen Y 2008 *Phys. Rev. Lett.* **100**(1) 013601
[19] Dorfman J R 1999 *An introduction to chaos in nonequilibrium statistical mechanics* vol 14 (Cambridge University Press)

[20] Gaspard P 1998 *Chaos, Scattering and Statistical Mechanics, by Pierre Gaspard, pp. 495. ISBN 0521395119. Cambridge, UK: Cambridge University Press, June 1998.* **1**
[21] Hu B L and Yu T (guest eds) 2009 *Quantum Information Processing* vol 8, Special Issue on Quantum Decoherence and Entanglement (Springer)
[22] DiVincenzo D P *et al.* 2000 *arXiv preprint quant-ph/0002077*
[23] Chou C H, Hu B and Yu T 2008 *Physica A: Statistical Mechanics and its Applications* **387** 432–444
[24] Hu B, Paz J P, Zhang Y *et al.* 1992 *Physical Review D* **45** 2843–2861
[25] Chou C H, Yu T and Hu B L 2008 *Physical Review E* **77** 011112
[26] Fleming C, Roura A and Hu B 2011 *arXiv preprint arXiv:1106.5752*
[27] Audenaert K, Eisert J, Plenio M B and Werner R F 2002 *Phys. Rev. A* **66**(4) 042327
[28] Anders J 2008 *Physical Review A* **77** 062102
[29] Anders J and Winter A 2008 *Quantum Information & Computation* **8** 245–262
[30] Zhou R, Hsiang J T and Hu B L Thermal entanglement and macroscopic quantumness in coupled harmonic oscillator system (in preparation)
[31] Subasi Y, Zhou Z and Hu B L Long-range quantum entanglement and zeros of spectral density function (in preperation)
[32] Raeisi S, Sekatski P and Simon C 2011 *Physical Review Letters* **107** 250401
[33] Subasi Y, Zhou Z and Hu B L Quantum thermodynamics examined from entanglement in mesoscopic systems (in preperation)
[34] Martins A M 2012 Macroscopic entanglement between wave-packets at finite temperature (*Preprint* `arXiv:1210.2828`)
[35] Schopfer F, Mallet F, Mailly D, Texier C, Montambaux G, Bäuerle C and Saminadayar L 2007 *Physical Review Letters* **98** 26807
[36] Subasi Y and Hu B L Entanglement in a quantum network from a Brownian motion analysis (in preperation)
[37] Calzetta E and Hu B L 1997 *Phys. Rev. D* **55**(6) 3536–3551
[38] Wieśniak M, Vedral V and Brukner Č 2008 *Physical Review B* **78** 064108
[39] Galve F, Pachón L A and Zueco D 2010 *Physical Review Letters* **105** 180501
[40] Whaley K B, Sarovar M and Ishizaki A 2011 *Procedia Chemistry* **3** 152–164
[41] Lin S Y and Hu B L 2009 *Phys. Rev. D* **79**(8) 085020
[42] Erdős P and Rényi A 1960 *Magyar Tud. Akad. Mat. Kutató Int. Közl* **5** 17–61
[43] Molloy M and Reed B 1995 *Random Structures & Algorithms* **6** 161–180
[44] Barabási A L and Albert R 1999 *Science* **286** 509–512

Measurement problem: from De Broglie to theory of classical random fields interacting with threshold detectors

Andrei Khrennikov
International Center for Mathematical Modeling
in Physics, Engineering, Economics, and Cognitive Science
Linnaeus University, Växjö, S-35195, Sweden

E-mail: Andrei.Khrennikov@lnu.se

Abstract. The quantum measurement problem as was formulated by von Neumann in 1933 can be solved by going beyond the operational quantum formalism. In our "prequantum model" quantum systems are symbolic representations of classical random fields. The Schrödinger's dynamics is a special form of the linear dynamics of classical fields. Measurements are described as interactions of classical fields with detectors. Discontinuity, the "collapse of the wave function", has the trivial origin: usage of threshold type detectors. The von Neumann projection postulate can be interpreted as the formal mathematical encoding of the absence of coincidence detection in measurement on a single quantum system, e.g., photon's polarization measurement. Our model, prequantum classical statistical field theory (PCSFT), in combination with measurements by threshold detectors satisfies the quantum restriction on coincidence detections: the second order coherence is less than one (opposite to all known semiclassical and classical feld models). The basic rule of quantum probability, the Born's rule, is derived from properties of prequantum random felds interacting with threshold type detectors. Comparison with De Broglie's views to quantum mechanics as theory of physical waves with singularities is presented.

1. Introduction

The measurement problem in quantum mechanics [1] is the unresolved problem of how (or if) wave function collapse occurs. The inability to observe this process directly has given rise to different interpretations of quantum mechanics and poses a key set of questions that each interpretation must answer. (See [2]-[7] for hot debates on the "right interpretation".) The wave function in quantum mechanics evolves according to the Schrödinger equation which preserves linear superposition of different states, but actual measurements always find the physical system in a definite state. Any future evolution is based on the state the system was discovered to be in when the measurement was made, meaning that the measurement "did something" to the process under examination. Whatever that "something" may be does not appear to be explained by the basic conventional quantum theory.

During the last 12 years quantum foundations were discussed at the series of conferences which took place in Växjö (South-East Sweden), see, e.g., [2]-[7]. And the most exciting spectacle started each time when the question of interpretations of the wave function attracted the attention. Finally, it became clear that the number of different interpretations is in the best case equal to the number of participants. If you meet two people who say that they are advocates of, e.g., the Copenhagen interpretation of QM, ask them about the details. You will see immediately that their views on what is the Copenhagen interpretation can differ very much. The same is true for other interpretations. If two scientists tell that they are followers of Albert Einstein's ensemble interpretation, ask them about the details... At one of the round tables (after two hours of debates with opinions for and against completeness of QM) we had decided to vote on this problem. Incompleteness advocates have won, but only because a few advocates of completeness voted for incompleteness. The situation is really disappointing: the basic notion of QM has not yet been properly interpreted (after 100 years of exciting, but not very productive debates).

We can also mention the Växjö interpretation of QM. This interpretation evolved essentially from the Växjö interpretation-2000 [8] which was based on "naive Einsteinian realism" (the values of physical observables can be assigned to a quantum system before measurement) to the Växjö interpretation 2009 in which measurement context played a fundamental role [9] (in particular, the values of physical observables cannot be assigned to a quantum system before measurement). The latter combines the views of Einstein on incompleteness of QM with the views of Bohr on contextual structure of quantum measurements. By the Växjö interpretation subquantum reality exists. However, for a moment we cannot approach it by means of available observables. The presently used observables, "quantum observables", are not "elements of subquantum reality": for example, the components of the electric and magnetic fields of photon or the density of electron's charge, see W. Hofer [10] for the latter. In our model the presently used "quantum observables" are contextual: their values cannot be assigned to subquantum systems in advance, before measurement.[1] The results of quantum measurements are determined by measurement contexts. This is really surprising, because typically contextuality was considered as a quantum feature. The basic classical physical model, classical statistical mechanics, is not contextual, the values of physical observables are considered as the values of an object. However, as the reader will see in the present paper, already classical wave models combined with measurement theory based on usage of threshold type detectors can exhibit contextual features.

This is well known (starting with the work of S. Gudder [12], see also K. Svozil [13] and the author's book [9]) that contextual models can violate Bell's inequality [14]. Therefore this is not surprising that PCSFT in combination with measurements of the threshold type can peacefully coexist with Bell's no-go theorem. (This problem will not be considered in the present paper; see, however, [15] for the detailed presentation.)[2]

There is a plenty of approaches to solve the measurement problem by using the formalism of quantum mechanics, e.g., [19], see also [20] for the detailed review. However, we speculate that it seems to be impossible to find the "real solution of this problem" in the standard quantum framework, since the quantum formalism (as was pointed by N. Bohr on many occasions, e.g., [21], [22]) is a formal operational formalism describing measurements for microsystems and not physical processes in the microworld [22]: "Strictly speaking, the mathematical formalism of quantum mechanics and electrodynamics merely offers rules of calculation for the deduction of expectations pertaining to observations obtained under well-defined experimental conditions specified by classical physical concepts", cf. [23], [24], [25]. By using the operator representation

[1] This is a good place to point to a general scientific methodology which was advertised during many years by Atmanspacher and Primas [11]. Any scientific theory is based on two levels of description of reality: ontic (reality as it is) and epistemic (the image of reality obtained with the aid of a special class of observables). The QM-formalism is an example of an epistemic model. However, existence of an epistemic model does not prevent scientists to go beyond it to approach the ontic level.

[2] See [16], [17], [18] for recently constructed prequantum models which do not contradict to the Bell's argument. We do not comment these models in the present paper.

Journal of Physics: Conference Series **442** (2013) 012011 doi:10.1088/1742-6596/442/1/012011

of observables we escape a detailed description of the process of measurement and such a description could not recovered in the quantum formalism. To solve the measurement problem of quantum theory, one has to go beyond the operational quantum formalism, cf. [19], [20], [26].

In this paper, we "solve the quantum measurement problem"[3] by interpreting quantum systems as formal operational representations of classical prequantum fields (waves) and by describing the measurement process as the process of interaction of a classical wave with a threshold type detector. In section 2 we analyze the process of measurement of a classical wave by a detector which selects one of eigenfunctions from the coherent superposition representing the wave, see (3). After such a measurement this superposition is "collapsed". This consideration is motivated by chapter 5 of L. De Broglie's book [27], pp. 51-54; see also [28]. The "collapse" of a sound wave (emitted e.g., by a vibrating string) which interacts with the turning fork having one of the basic string's frequencies can be operationally described in the same way as the collapse of the wave function of a photon (emitted in the state of superposition of a few frequencies) which interacts with a photo-detector. However, opposite to the quantum wave function, the classical wave is collapsed not to a single eigenfunction of the corresponding stationary equation, but to the decoherent mixture of such eigenfunctions. Thus the formal mathematical description is similar to the description of quantum measurement without selection of the fixed value of the observable. In the later case we also obtain not a single eigenvector, but a mixture, described by the density operator.

Roughly speaking in our approach the main difference between classical and quantum collapses is degeneration of the former, the impossibility to select a single output component of the signal ("quantum particle"). In quantum theory this problem is known as the problem of coincidence detection. It was intensively studied experimentally to reject (semi)classical models pretending to reproduce quantum predictions, see Grangier et al. [42], [41], also [43]. The quantum prediction that coincidence detections should occur relatively rarely was supported by experiments. It is commonly accepted that these experiments demonstrated that classical or semiclassical models without coincidence detection do not exist.

However, the field model elaborated in [29]-[36], prequantum classical statistical field theory (PCSFT), in combination with usage of the threshold type and properly calibrated detectors solves the coincidence detection problem, opposite to known (semi)-classical models. In [37] we found that quantum statistics can be obtained in a simple way: by combination of stationary random field describing spatial or internal degrees of freedom with the Brownian motion (Wiener process) describing temporal fluctuations, cf. with [15], where a more complicated stochastic processes were in use. In this paper we essentially improved the rigorousness of the presentation of the threshold detection scheme for random waves by coupling this problem with the well studied problem of probability theory, namely, the first hitting time problem. Unfortunately, probabilistic studies on hitting times were restricted to real valued stochastic processes. (These studies were essentially stimulated by applications to finances and here the real valued processes are in usage.) Development of theory of hitting times for complex valued stochastic processes is a complicated mathematical problem and we hope that our paper will stimulate research in this direction. For a moment, we can apply only results on hitting times for real processes and therefore we restrict our consideration to the case of real Hilbert space. In this paper we reproduce quantum statistics for density matrices with real elements. This is merely a question of the mathematical justification; in [37] we proceeded in the general case of complex Hilbert space, but not in the completely rigorous mathematical framework.

We remark that the aforementioned classical-quantum analogy between measurements of classical and quantum waves is valid only for disturbative classical measurements. It is commonly postulated that in classical physics it is possible to perform measurements with an arbitrary

[3] Of course, one staying at the Copenhagen position would not consider our quantum solution of the measurement problem as the real solution of this problem.

precision; in particular, it is possible to determine the form of the wave without to destroy coherent superposition. In this paper, we do not criticize this postulate[4]. We just remark that quantum observables form only a subclass of observables for prequantum fields. These are coarse and disturbabtive measurements. PCSFT's class of observables is essentially larger and it contains measurements of fields components, e.g., the electric and magnetic fields components of a photon. The modern experimental technology is still far from realization of such measurements, cf., however, with recent experimental results presented by W. Hofer (Conference "Quantum Mechanics as Emergent Phenomenon", Vienna, November 2011) on a possibility to violate Heisenberg's uncertainty relation.

At the very end of section 2, we compare the De Broglie's Double Solution theory [28], [27] with PCSFT (combined with threshold-type measurements) and with Bohmian mechanics. There is a rather common opinion that the De Broglie's Double Solution theory is simply an early version of a more advanced theory, namely, Bohmian mechanics. However, the careful study of works of De Broglie shows that this viewpoint to the inter-relation between the *De Broglie's Double Solution theory* and *Bohmian mechanics* is rather primitive, see section 2.

2. Classical fields: superposition, linear dynamics, measurement

In this section we consider classical waves with linear dynamics (in vacuum or some media). The process of measurement induces effects[5] which formally can be described by the von Neumann projection postulate: linear superposition "collapses". We claim that the essence of the "collapse" of superposition is the transition from one linear dynamics in the absence of measurement to another dynamics (nonlinear or even linear) of interaction with a measurement device.

A possibility to form linear superposition of waves (fields) and to split a wave into superposing summands is the basic feature of the classical wave theory. A crucial point is the existence of dynamics which preserve linear superpositions in the process of evolution. We can mention the dynamics of the string or the classical electromagnetic field (Maxwell's equations)[6].

Suppose that in the absence of interaction with measurement devices the field dynamics is described by the differential equation:

$$\gamma \frac{d\phi(t,x)}{dt} = L\phi(t,x), \tag{1}$$

where L is a linear partial differential operator and γ is a constant. Since, at each instant of time t, the field's energy is finite, $\int_{\mathbf{R}^3} |\phi(t,x)|^2 dx < \infty$, the equation (1) can be considered as a linear (ordinary) differential equation in the L_2- space. This space has the Hilbert space structure. Denote it by the symbol H. For all basic physical processes, we can assume that the operator L is self-adjoint in H.

Sometimes it is convenient to proceed with complex vector valued fields, e.g., for the electromagnetic field, we can consider the Riemann-Silberstein representation, $\phi(t,x) = E(t,x) + iB(t,x)$, where E and B are the electric and magnetic components, respectively. We remark that the system of the Maxwell equations can be written in such a form with $\gamma = i$ [30], i.e., in the same form as the Schrödinger equation [38], [30]. (We also mention the work of Strocchi [39] who demonstrated that the Schrödinger equation can be written as the system of linear

[4] One of the main objections is the presence of noise; for prequantum classical fields their coupling with the background field, vacuum fluctuations, is irreducible [29]–[36]. For a moment, we ignore this problem.

[5] The interaction between a measurement device and the input wave can be linear; so nonlinearity is not crucial for collapse.

[6] We remark that all these linear dynamics are approximate. Although these are very good approximations, one should not overestimate the role of linear dynamics. At the fundamental level the majority of processes are nonlinear, cf. De Broglie [28].

Hamiltonian equations, see also [30].) In general, we work with real valued vector fields. The wave equation and Klein-Gordon equation cannot be written in the complex form, here fields are real and $\gamma = 1$. Solution of a linear dynamical equation with self-adjoint operator L can be represented in the form of superposition of solutions $\phi_k = \phi_k(x)$ of the stationary equation:

$$L\phi_k(t,x) = \omega_k \phi_k(t,x); \tag{2}$$

here

$$\phi(t,x) = \sum_k c_k(t)\phi_k(t,x), c_k(t) = e^{\frac{\omega_k t}{\gamma}} c_{k0}, \tag{3}$$

where

$$\phi(0,x) = \sum_k c_{k0}\phi_k(x) \tag{4}$$

is the expansion of the initial wave. The main statement is that this form of expansion, (4), is preserved in the process of evolution. If, for some k, the term with $\phi_k(x)$ was present in (4) at $t = 0$, then it will never disappear from (3), for any instant of time $t > 0$. (If the operator L has continuous spectrum, then the sign of sum is changed to the sign of integral.)

As was pointed out by L. De Broglie [27], the expansion (3) is a formal mathematical representation. The field $\phi(t,x)$ cannot be imagined as "physical superposition" of fields $\phi_k(x)$. He stressed that the contributions of summands ϕ_k in (3) can be extracted from the field ϕ only through interaction with measurement devices, cf. Roychoudhuri [40].

In [27] the example of vibrating string was presented. The waves ϕ_k correspond to the basic frequencies ω_k of the vibrating string (for some type of boundary conditions). However, until we start measurement the waves ϕ_k do not present in the integral wave ϕ so to say physically. In this case measurement can be done with the aid of a turning fork. Put a turning fork nearby the vibrating string and by adjustment of turning fork[7] we can find one of the basic frequencies, say ω_{k_0} . The sound emitted by turning fork at the frequency ω_{k_0} corresponds to extracting from the integral wave ϕ its fixed component ϕ_{k_0}. Formally, this process can be described as the orthogonal projection P_{k_0} (in Hilbert space H) of the integral wave ϕ onto the one dimensional subspace corresponding to the wave ϕ_{k_0}. One can call this process the collapse of the classical wave ϕ or more precisely the collapse of the linear superposition (3). Of course, there is nothing mysterious in this collapse; in particular, it is completely clear that it is not instantaneous, the process of interaction with the turning fork has a finite duration; neither mystery is destruction of superposition (3): the dynamics of interaction with the turning fork is different from the dynamics (1) preceding interaction and hence the original superposition need not be preserved and it can be destroyed, collapsed. (The interaction dynamics need not be nonlinear. It can be linear as the original dynamics (1), but with another linear operator, say L_1, Dynamics with L and L_1 can be unified through linear dynamics with time dependent generator.)

In fact, the situation is more complicated. Here the string plays the role of a source of sound waves (cf. with a source of quantum systems). Hence, the turning fork interacts not directly with the string, but with the emitted sound wave. To proceed rigorously, we have to use a different notation for this (sound) wave, say $\varphi(t,x)$ and the corresponding solutions of the stationary equation $\varphi_k(x)$. Before measurement, the dynamics of the sound wave $\varphi(t,x)$ can be well described by the linear partial differential equation, the wave equation. Hence, as well as (3), the representation

$$\varphi(t,x) = \sum_k c_k(t)\varphi_k(x), \tag{5}$$

[7] We can use either a turning fork which frequency can be changed or what is may be even easier a collection of turning forks representing scale of frequencies.

can be used for mathematical calculations. Here we stress again that the physical (sound) wave $\varphi(t, x)$ is not composed of physical stationary (sound) waves $\varphi_k(x)$.

To simplify consideration, we shall proceed with original vibrations of the string (by having in mind that, in fact, the measurement device turning fork destroys the coherent superposition (5) in the sound wave). We state again that one has to be very careful with terminology. The expression "destroys coherent superposition" is related to the "mathematical wave" and not the physical one. (Hence, "collapse" takes place in mathematical space.) In the presented considerations we stressed the analogy between "collapses" of the classical wave and the quantum wave function. The similarity of operational descriptions of "collapses", i.e., by using projection operators in Hilbert space, is especially important. However, although classical superposition (3) collapsed as the result of interaction with the turning fork which is used for measurement and in the operational formalism the extraction of the component $\phi_{k_0}(x)$ can be described by the projection operator b P_{k_0}, the reader experienced in quantum theory would point out that there is a crucial difference between the collapse of the quantum wave function and the classical wave. In the classical case, although the component $\phi_{k_0}(x)$ is extracted from the signal, the signal is not completely reduced to this component. If we put two turning forks, we can select two basic frequencies, $\omega_j, j = 1, 2$. The selection of each component ϕ_j can be operationally represented by the corresponding projector P_j in H. However, opposite to the quantum case (we consider the case of nondegenerate spectrum, $\omega_i \neq \omega_j, i \neq j$) the output signal is not reduced to one of the selected components. And if we put many (and even infinitely many) turning forks, it is possible to select corresponding components of the wave ϕ.

By the conventional interpretation of quantum mechanics, in the process of interaction with a detector, a quantum system exhibits particle properties, hence, it could not be detected simultaneously by two different detectors. In fact, the starting point was the analysis of the two slit experiment by N. Bohr. This analysis played a fundamental role in elaboration of the complementarity principle. In the two slit experiment by placing detectors directly behind slits one does not observe coincidence detections, detectors never click simultaneously.

Of course, this is the idealization of the real physical situation. The first basic assumption is that the used source is really *a single-photon source.* At least in 1920th such sources did not exist; even nowadays one can only approximately assume that a source is of the one-photon type (with sufficiently high approximation). Another problem is noise. In any event, there are coincidence detections and their number is not negligibly small. Nevertheless, it has to be relatively small.

The corresponding experiments have been done. The first experiment was done by Grangier [41], [42], see also [43] for a review, in the framework of quantum optics: detection of the outputs of two channels of the polarization beam splitter (PBS). [8] By quantum mechanics a single photon passing PBS could not be split and coincidence detection for two detectors placed in the two outputs of PBS is impossible. Grangier demonstrated that the relative probability of coincidence detection, the *second order coherence*:

$$g^{(2)}(0) = \frac{P_{12}}{P_1 P_2} \tag{6}$$

is less than 1. At the same time all known (semi)classical field models predicted that this coefficient exceeds 1, see [43] for review.

In fact, the von Neumann projection postulate (for observables with nondegenerate spectrum) is the formal Hilbert space description of the absence of coincidence detections. As the result of quantum measurement only one detector can click. (We state again that we

[8] Up to author's knowledge the experiment on the coincidence detection in the real two slit experiment has never been done, at least with high precision.

consider only noncompound systems.) This physical process is represented in the operational quantum formalism by projection of the state ϕ onto one fixed state ϕ_{k_0}. The quantum collapse of superposition differs from the "classical collapse" of the waves superposition by its nondegeneration, the output is only a single state ϕ_{k_0}. Hence, in quantum theory only a single result of measurement, say ω_{k_0}, can be obtained. The state ψ of the quantum system is "collapsed" to the eigenstate ϕ_{k_0}. (Opposite to classical wave theory, all states are assumed to be normalized.) In the mathematical formalism this process is described by the von Neumann projection postulate:

$$\psi_{k0} = \frac{P_{k_0}\psi}{\|P_{k_0}\psi\|} \tag{7}$$

The probability to get the result ω_{k_0} is given by Born's rule

$$P(\omega_{k_0}) = |\langle\psi|\psi_{k_0}\rangle|^2. \tag{8}$$

We remark that in quantum theory this rule is a postulate, i.e., it is not derived from other natural physical assumptions. And there is no satisfactory derivation of this rule from other ("naturally justified principles"), in spite of numerous attempts to do this.[9]

In classical wave theory, in principle, it is possible to measure simultaneously all values ω_k. (We stress the role of simultaneous measurement. We shall come back to this crucial point later.) Measurements induce extractions of the components Φ_k which are proportional to eigenwaves ϕ_k from the coherent superposition ϕ. see (3):

$$\Phi_k = P_k\phi. \tag{9}$$

Opposite to the quantum case this components are not normalized; the quantity

$$\mathcal{E}_k = \|\Phi_k\|^2 = |\langle\phi|\Phi_k\rangle|^2 \tag{10}$$

is the energy of the kth output wave, corresponding to the result of measurement $\omega = \omega_k$.

The common between the quantum and classical cases is that in any event the *coherent superposition is destroyed.* In the quantum case, it is transformed into one of eigenstates, in the classical case into decoherent mixture of eigenwaves ϕ_k. We remark that in the quantum case if the result of the fixed measurement, $\omega = \omega_k$, is not selected, then we also get decoherent mixture of eigenstates:

$$\rho = \sum_k P(\omega_k)P_k. \tag{11}$$

We point out that in the classical framework the Born rule for quantum probabilities can be formally written by using relative energies of the classical output waves:

$$\frac{\mathcal{E}_k}{\mathcal{E}_\phi} \tag{12}$$

where $\mathcal{E}_\phi$ is the total energy of the field. We state again that in the classical wave ϕ is not normalized. If now we use the normalized (by its total energy) wave, i.e.,

$$\phi \to \psi = \frac{\phi}{\|\phi\|}, \tag{13}$$

[9] "The conclusion seems to be that no generally accepted derivation of the Born rule has been given to date, but this does not imply that such a derivation is impossible in principle". See [44]; cf. 't Hooft [45], [46] and Hofer [10].

we see that the formal mathematical expression for classical relative energy of the kth output channel coincides with (8), the Born's rule.[10] Consider the classical wave normalized by its total energy, i.e., the wave ψ. In this case the output wave is the mixture of normalized waves. It can be written in the same way as (11) by using relative energies, instead of probabilities:

$$\rho = \sum_k \frac{\mathcal{E}_k}{\mathcal{E}_\phi} P_k. \tag{14}$$

Heuristically it is clear that the number of detector's clicks in kth channel is proportional to the energy distributed to this channel. However, it is not a trivial task to find a class of classical waves which produce quantum statistics of clicks given by the rule:

$$P(\omega_k) = \frac{\mathcal{E}_k}{\mathcal{E}_\phi}. \tag{15}$$

It is evident that waves have to be random and detectors have to be (similar to quantum detectors) of the threshold type – to produce individual clicks and not continuous output signals. Another crucial constraint is that there should be no coincidence detections. Hence, the temporal structure of the random field has to be selected in such a way that, although the field is continuous and so to say it is present everywhere – in any detector, simultaneous clicks (corresponding to random energy spikes) would occur relatively rarely.

This is a good place to compare our model, PCSFT, with the De Broglie's *Theory of the Double Solution*. By the latter quantum mechanics can be considered as an operational formalism for a deeper theory, theory of classical physical waves containing singularities. The singularities in waves correspond to quantum particles. In PCSFT, particles do not present at all. Singularities appear only at the level of measurements, as clicks of the threshold type detectors. However, the common point is that PCSFT-fields are also singular; these are stochastic processes valued in the L_2-space. So, the prequantum fields are nonsmooth and discontinuous for almost all values of a random parameter.

This is also a good place to make a remark about nonlocality. The Theory of the Double Solution is closely related to Bohmian mechanics. The latter is definitely nonlocal, since the quantum potential is nonlocal. On the other hand, PCSFT is a local classical field theory, see [29]–[36]. In PCSFT a sort of classical nonlocality can be assigned to the background random field. This field is not considered in the present paper, it started to play a crucial role in classical wave modeling of compound systems which we do not study in this paper, see [33]–[35]. However, such a background field is a classical field, as in stochastic electrodynamics [48], and has nothing to do with nonlocality of the Bohmian type. There is a rather common opinion that creation of Bohmian mechanics was the improvement of the Theory of the Double Solution, so nonlocality was also firmly incorporated in De Broglie's views. However, the real inter-relation between views of De Broglie and Bohm is more complicated. In particular, by reading the De Broglie's book [27] I was not able to find any trace of nonlocality. In the description of a compound quantum system, De Broglie [27], p. 74-76, of course, used the nonlocal quantum potential induced by the wave function of a compound system. However, the wave function by itself and hence the potential corresponding to it were treated as related to the formal mathematical description. The propagations of real physical waves corresponding to a compound system are also coupled, but through random fluctuations of the field of random media, a kind of the background field. In this sense De Broglie's views are close to the views of the author of this

[10] We suspect that our consideration is similar to Born's motivation of the rule for probabilities (8); the motivation which he did not present in his paper [47]. This is a pity that Born did not present any motivations for the rule (8). This formal postulation of the basic probabilistic rule of quantum theory makes the impression of something completely new and even mysterious.

Journal of Physics: Conference Series **442** (2013) 012011 doi:10.1088/1742-6596/442/1/012011

paper. I also stress that both for De Broglie and for me the starting point of model's creation was Einstein's attempt to create a purely field model of physical reality, see, especially, Einstein and Infeld [49]; cf. De Broglie [27], p. 43: "My first attempts to interpret wave mechanics in terms of the Theory of the Double Solution in 1926-1927, were undoubtedly suggested to me by Einstein's work on general relativity. Einstein believed that the physical world should be described wholly by means of fields, well defined at every point of space-time and obeying well-defined equations of a non-random nature." As we shall see in section 3, the later statement is not valid for PCSFT, we shall operate with random fields. However, "late De Broglie" also rejected "naive Einsteinian determinism" and stressed the role of the random background field, see [27], p. viii.

3. Threshold detection of classical random fields

We consider a threshold type detector with the threshold $\mathcal{E}_d$. It interacts with a random field $\phi(s;\omega)$, where s is time and ω is a chance parameter describing randomness. For a moment, we consider the **C**-valued random field (complex stochastic process). Later we shall consider random fields valued in finite and infinite dimensional complex Hilbert spaces. The finite dimensional case corresponds to detection of internal degrees of freedom such as, e.g., polarization. We stress that the real physical situation corresponds to random fields with infinite-dimensional state space, e.g., $H = L_2(\mathbf{R}^3)$; the space of complex valued fields $\phi : \mathbf{R}^3 \to \mathbf{C}$ (or $\mathbf{C}^k$). The energy of the field is given by $\mathcal{E}(s;\omega) = |\phi(s;\omega)|^2$ (hence, the random field has the physical dimension $\sim \sqrt{\text{energy}}$). A threshold detector clicks at the first moment of time $\tau = \tau(\omega)$, when field's energy $\mathcal{E}$ exeeds the threshold:

$$\mathcal{E}(\tau(\omega), \omega) \geq \mathcal{E}_d. \tag{16}$$

In the mathematical model the detection moment is defined as the *first hitting time* [50]

$$\tau(\omega) = \inf\{s \geq 0 : \mathcal{E}(\tau(\omega), \omega) \geq \mathcal{E}_d\}. \tag{17}$$

We consider the following detection scheme. After arriving to a threshold type detector a classical random field behaves inside this detector as the Brownian motion in the space of (complex) fields. Thus $\phi(s,\omega)$ is the Wiener process: the Gaussian process having zero average at any moment of time $E\phi(s,\omega) = 0$; and the covariance function

$$E\phi(s_1,\omega)\overline{\phi(s_2,\omega)} = \min(s_1,s_2)\sigma^2; \tag{18}$$

in particular, we can find average of its energy

$$E\mathcal{E}(s,\omega) = \sigma^2 s. \tag{19}$$

From this equation, we see that the coefficient $\sigma^2 = \frac{E\mathcal{E}(s,\omega)}{S}$ has the physical dimension of *power*. We are interested in average of the moments of the $\mathcal{E}_d$-threshold detection for the energy of the Brownian motion. Since moments of detection are defined formally as hitting times, we can apply theory of hitting times for the Wiener process, see [50]:

$$\bar{\tau} \equiv E\tau = \frac{\mathcal{E}_d}{\sigma^2} \tag{20}$$

or

$$\frac{1}{\bar{\tau}} = \frac{\sigma^2}{\mathcal{E}_d}\,. \tag{21}$$

Hence, during a long period of time T such a detector clicks N_σ-times, where

$$N_\sigma \approx \frac{T}{\bar{\tau}} = \frac{\sigma^2 T}{\mathcal{E}_d} \,. \tag{22}$$

Consider now a random $\phi(s,\omega)$ valued in the m-dimensional complex Hilbert space H, where m can be equal to infinity. Let (e_j) be an orthonormal basis in H. The vector-valued $\phi(s,\omega)$ can be expanded with respect to this basis

$$\phi(s,\omega) = \sum_j \phi_j(s,\omega)e_j, \tag{23}$$

where $\phi_j(s,\omega) = \langle\phi(s,\omega)|e_j\rangle$. This mathematical operation is physically realized as splitting of the field $\phi(s,\omega)$ into components $\phi_j(s,\omega)$ These components can be processed through mutually disjoint channels, $j = 1, 2, ..., m$.[11] We now assume that there is a threshold detector in each channel, $D_1, ..., D_m$ We also assume that all detectors have the same threshold $\mathcal{E}_d > 0$.

Suppose now that $\phi(s,\omega)$ is the Wiener process valued in H. This process is determined by the covariance operator $B : H \to H$. Any covariance operator is Hermitian, positive, and trace-class and vice versa. The complex Wiener process is characterized by Hermitian covariance operator. We have, for $y \in H$, $E\langle y, \phi(s,\omega)\rangle = 0$, and, for $y_j \in H, j = 1, 2$, $E\langle y_1, \phi(s_1,\omega)\rangle\langle\phi(s_2,\omega), y_2\rangle = \min(s_1,s_2)\langle By_1, y_2\rangle$. The latter is the covariance function of the stochastic process; in the operator form: $B(s_1,s_2) = \min(s_1,s_2)B$. We note that the dispersion of the H-valued Wiener process (at the instant of time s) is given by $\Sigma^2(s) = E\|\phi(s,\omega)\|^2 = s\mathrm{TrB}$. The quantity $\mathcal{E}(s,\omega) = \|\phi(s,\omega)\|^2$ is the total energy of the Brownian motion signal at the instant of time s. Hence, the quantity

$$\Sigma^2 \equiv \frac{\Sigma^2(s)}{s} = \mathrm{TrB} \tag{24}$$

is the average power of this random signal. We stress that the average power is time-independent.

We also remark that by normalization of the covariance function for the fixed s by the dispersion we obtain the operator,

$$\rho = B(s,s)/\Sigma^2(s) = B/\mathrm{TrB}, \tag{25}$$

which formally has all properties of the *density operator* used in quantum theory to represent quantum states. Its matrix elements have the form $\rho_{ij} = b_{ij}/\Sigma^2$. These are dimensionless quantities. The relation (25) plays a fundamental role in our approach : *each classical random process generates a quantum state (in general mixed) which is given by the normalized covariance operator of the process.* One can proceed the other way around as well: each density operator determines a class of classical random processes.

Consider components $\phi_j(s,\omega)$ of the vector valued signal $\phi(s,\omega)$. Then

$$E\phi_i(s,\omega)\overline{\phi_j(s,\omega)} = \min(s_1,s_2)\langle Be_i, e_j\rangle = b_{ij}.$$

In particular, $\sigma_j^2(s) \equiv E\mathcal{E}_j(s,\omega) \equiv E|\phi_j(s,\omega)|^2 = sb_{jj}$. This is the average energy of the jth component at the instant of time s. We also consider the average powers of components

$$\sigma_j^2 \equiv \frac{\sigma_j^2(s)}{s} = b_{jj}. \tag{26}$$

[11] We consider separation of a signal into disjoint channels as a part of the measurement procedure. These channels correspond to separate detectors that are looking at the input state through some kind of beam splitter. And it is assume that an input state is a single mode photon state. (Hence, aforementioned channels do not correspond to different photon modes.)

We remark that the average power of the total signal is equal to the sum of the average powers of its components.

$$\Sigma^2 = \sum_j \sigma_j^2. \tag{27}$$

Consider now a run of experiment of the duration T. The average number of clicks for the jth detector can be approximately expressed as

$$N_j \equiv N_{\sigma_j} \approx \frac{\sigma_j^2 T}{\mathcal{E}_d}. \tag{28}$$

The total number of clicks is (again approximately) given by $N = \sum_j N_j \approx \frac{\Sigma_j^2 T}{\mathcal{E}_d}$. Hence, for the detector D_j, the probability of detection can be expressed as

$$P_j \approx \frac{N_j}{N} \approx \frac{\sigma_j^2}{\Sigma_j^2} = \rho_{jj}. \tag{29}$$

This is, in fact, *the Born's rule for the quantum state* ρ *and the projection operator* $\hat{C}_j = |e_j\rangle\langle e_j|$ *on the vector* e_j. For the detector D_j, the probability of detection can be expressed as

$$P_j = \mathrm{Tr}\rho\, \hat{\mathrm{C}}_{\mathrm{j}}. \tag{30}$$

4. Coincidence counts

The relative number of coincidence counts is given by second order coherence:

$$g^{(2)}(0) = \frac{P_{12}}{P_1 P_2} \tag{31}$$

Probabilities $P_j, j = 1, 2$, for singles were found in (30); they coincide with corresponding quantum probabilities. For a single photon source, the P_{12}, the probability of clicks in both channels after a beam splitter, equals to zero. But, of course, nobody has yet seen really single photon sources. Therefore in reality P_{12} differs from zero. Our threshold detection model definitely contradicts to quantum mechanics as a theory: the coincidence probability is positive (in any case for $P_1 = P_2 = 1/2$). However, we are not disappointed by this situation. We are sure that quantum mechanics is just an approximate model to describe probabilistic data. Moreover, this model cannot take into account experimental technicalities, such as e.g. the size of the discrimination threshold. Therefore its theoretical prediction $g^{(2)}(0) = 0$, for "single photon sources", is far from the real experimental data. Nevertheless, even with the aid of this rough prediction it was possible to discard the semiclassical model. The latter predicts that $g^{(2)}(0) \geq 1$. And by getting $g^{(2)}(0)$ sufficiently small, experiments claim that their data match with the predictions of quantum mechanics and mismatch with predictions of the semicalssical model. By taking into account the above discussion, we have to test our model contra experiment and not contra the theoretical quantum formalism.

We were not able to obtain a formula for P_{12} in the framework of classical Brownian model fro the prequantum field. We were able only to obtain an estimate from above, see [37]. This estimate shows that in some range of variation of the detection threshold $\mathcal{E}_d$ the second order coherence $g^{(2)}(0)$ decreases with the increase of $\mathcal{E}_d$. Hence, $g^{(2)}(0) < 1$ for sufficiently large $\mathcal{E}_d$. Thus our model does not say that there are no coincidence counts, neither that their number is relatively small irrelatively to magnitudes of experimental parameters. However, by our model these parameters can always be selected in such a way that detected data would be described (with a good approximation) by the quantum probabilistic formalism. Thus Devil is really in detectors and experimental technicalities.

Unfortunately, it seems that Grangier's type experiments with detailed monitoring of dependence of the coincidence probability on the value of the threshold have never been done.

5. Conclusion

By following L. De Broglie we stressed the formal analogy between the operational descriptions of the processes of measurement of classical waves and quantum systems, including the wave collapse. Our analysis showed that the main experimental difference between measurements of classical and quantum waves is the absence of coincidence detections in the latter. Then we presented the classical wave model which reproduces the quantum detection probabilities (described by the Born's rule) and at the same time the number of coincidence detections is relatively small (for sufficiently large detection threshold).

Acknowledgments

I would like to thank Borje Nilsson and Sergei Polyakov for numerous discussions and constructive criticism. This paper was written under support of the grant "Mathematical Modeling of Complex Hierarchic Systems" of the faculty of Natural Science and Engineering of Linnaeus University.

[1] von Neumann J 1955 *Mathematical Foundations of Quantum Mechanics* (Princeton: Princeton Univ. Press)
[2] Khrennikov A (ed) 2001 *Foundations of Probability and Physics. Quantum Probability and White Noise Analysis* vol 13 (Singapore: WSP)
[3] Khrennikov A (ed) 2002 *Quantum Theory: Reconsideration of Foundations.* Ser. Math. Modelling vol 2 (Växjö: Växjö Univ. Press)
[4] Khrennikov A (ed) 2003 *Quantum Theory: Reconsideration of Foundations-2.* Ser. Math. Modelling vol 10 (Växjö: Växjö Univ. Press)
[5] Khrennikov A (ed) 2005 *Foundations of Probability and Physics-3.* vol 750
[6] Jaeger G, Khrennikov A, Schlosshauer M, Weihs G (eds) 2011 *Advances in Quantum Theory* vol 1327 (Melville, NY: AIP)
[7] Khrennikov A (ed) 2010 *Quantum Theory: Reconsideration of Foundations-5* vol 1232 (Melville, NY: AIP)
[8] Khrennikov A 2002 On foundations of quantum theory. Quantum Theory: Reconsideration of Foundations. Ser. Math. Modelling vol 2 (Växjö: Växjö Univ. Press) pp 163-196
[9] Khrennikov A 2009 *Contextual Approach to Quantum Formalism* (Berlin-Heidelberg-New York: Springer)
[10] Hofer W 2011 *Foundations of Physics* **41** 754
[11] Atmanspacher H and Primas H 2005 Epistemic and ontic quantum realities *Foundations of Probability and Physics-3* vol 750 (Melville, NY: AIP) pp 49-62
[12] Gudder S 1970 *J. Math. Physics* **11** 431
[13] Svozil K 2005 On counterfactuals and contextuality. *Foundations of Probability and Physics-3* vol 750 (Melville, NY: AIP) pp 351-360
[14] Bell J 1987 *Speakable and Unspeakable in Quantum Mechanics* (Cambridge: Cambridge Univ. Press)
[15] Khrennikov A 2011 Quantum probabilities and violation of CHSH-inequality from classical random signals and threshold type properly calibrated detectors *Preprint* arxiv.org/abs/1111.1907
[16] Accardi L 2005 Some loopholes to save quantum nonlocality. *Foundations of Probability and Physics-3* vol 750 (Melville, NY: AIP) pp 1-20
[17] Adenier G. 2008 *Am. J. Phys.* **76** 147
[18] Hess K, Michielsen K and De Raedt H 2009 *EPL* **87** 60007
[19] Allahverdyan A E, Balian R, and Nieuwenhuizen Th M 2005 The quantum measurement process in an exactly solvable model *Foundations of Probability and Physics-3* vol 750 (Melville, NY: AIP) pp 16-24
[20] Allahverdyan A E, Balian R and Nieuwenhuizen Th M 2011 Understanding quantum measurement from the solution of dynamical models *Preprint* arxiv.org/abs/1107.2138
[21] Bohr N 1935 *Phys. Rev.* **48** 696
[22] Bohr N 1987 *The Philosophical Writings of Niels Bohr,* 3 vols (Woodbridge, Conn.: Ox Bow Press)
[23] De Muynck W M 2002 *Foundations of Quantum Mechanics, an Empiricists Approach* (Dordrecht: Kluwer)
[24] D'Ariano G M and Tosini A 2010 *Quant. Inf. Proc.* **9** 95
[25] D'Ariano G M 2010 Probabilistic theories: What is special about quantum mechanics? *Philosophy of Quantum Information and Entanglement* (Cambridge: Cambridge Univ. Press) pp 85-126
[26] Mehmani B and Nieuwenhuizen Th M 2010 An overview on single apparatus quantum measurements *Preprint* arxiv.org/abs/1007.3635
[27] de Broglie L and Andrade E Silva J (the author of a chapter) 1964 *The Current Interpretation of Wave Mechanics. A Critical Study* (Amstredam-London-New York: Elsevier)
[28] de Broglie L 1960 *Non-linear Wave Mechanics: A Causal Interpretation.* (Amsterdam: Elsevier)

[29] Khrennikov A 2005 *J. Phys.* A **38** 9051
[30] Khrennikov A 2006 *Phys. Lett.* A **357** 171
[31] Khrennikov A 2008 *Phys. Lett.* A **372** 6588
[32] Khrennikov A, Ohya M, and Watanabe N 2010 *J. Russian Laser Research* **31** 462
[33] Khrennikov A 2010 *J. Math. Phys.* **51** 082106
[34] Khrennikov A 2010 *EPL* **90** 40004
[35] Khrennikov A 2010 *Found. Phys.* **40** 1051
[36] Khrennikov A 2010 *Physica* E **42** 287
[37] Khrennikov A, Nilsson B, and Nordebo S 2011 Classical signal model reproducing quantum probabilities for single and coincidence detections *Preprint* arxiv.org/abs/1112.5591
[38] Bialynicki-Birula I 1994 *Acta Phys. Polonica* A **86** 97
[39] Strocchi F 1996 *Rev. Mod. Phys.* **38** 36
[40] Roychoudhuri Ch 2007 Shall we climb on the shoulders of the giants to extend the reality horizon of physics? *Quantum Theory: Reconsideration of Foundations-4* vol 962 (Melville, NY: AIP) 195-205
[41] Grangier p 1986 *Etude expérimentale de propriétés non-classiques de la lumière: interférence à un seul photon* (Centre D'Orsay: Université de Paris-Sud)
[42] Grangier P, Roger G and Aspect A 1986 *EPL* **1** 173
[43] Beck M (2007) *J. Opt. Soc. Am.* B **24** 2972
[44] Landsman N P 2008 Compendium of Quantum Physics. Weinert F, Hentschel K, Greenberger D and Falkenburg B (eds) (Berlin: Springer)
[45] 't Hooft G 2006 The mathematical basis for deterministic quantum mechanics (*Preprint* arXiv: quant-ph/0604008); *do.* 2007 *J. Phys.: Conf. Series* **67** 012015
[46] 't Hooft G 2011 *Herald of Russian Acad. Sc.* **81** 907
[47] Born M 1926 *Zeitschrift für Physik* **37** 863
[48] De la Pena L and Cetto A M 1996 *The Quantum Dice: An Introduction to Stochastic Electrodynamics* (Dordrecht: Kluwer)
[49] Einstein A and Infeld L 1961 *Evolution of Physics: The Growth of Ideas from Early Concepts to Relativity and Quanta* (New-York: Simon and Schuster)
[50] Mansuy R and Yor M 2006 *Random Times and Enlargements of Filtrations in a Brownian Setting* (Berlin: Springer)

"Systemic nonlocality" from changing constraints on sub-quantum kinematics

G Grössing[1], S Fussy[1], J Mesa Pascasio[1,2] and H Schwabl[1]

[1] Austrian Institute for Nonlinear Studies, Akademiehof, Friedrichstr. 10, 1010 Vienna, Austria
[2] Institute for Atomic and Subatomic Physics, Vienna University of Technology, Operng. 9, 1040 Vienna, Austria

E-mail: ains@chello.at

Abstract. In a new approach to explain double-slit interference "from the single particle perspective" via *"systemic nonlocality"*, we answer the question of how a particle going through one slit can "know" about the state of the other slit. We show that this comes about by changed constraints on assumed classical sub-quantum currents, which we have recently employed [1] to derive probability distributions and Bohm-type trajectories in standard double-slit interference on the basis of a modern, 21st century classical physics. Despite claims in the literature that this scenario is to be described by a *dynamical nonlocality* that could best be understood in the framework of the Heisenberg picture [2], we show that an explanation can be cast within the framework of the intuitively appealing Schrödinger picture as well. We refer neither to potentials nor to a "quantum force" or some other dynamics, but show that a *"systemic nonlocality"* may be obtained as a phenomenon that emerges from an assumed sub-quantum kinematics, which is manipulated only by changing its constraints as determined by the changes of the apparatus. Consequences are discussed with respect to the prohibition of superluminal signaling by standard relativity theory.

1. Introduction

In two seminal papers [3, 4], Aharonov *et al.* more than 40 years ago introduced an approach to explain double-slit interference "from the single particle perspective" via *dynamical nonlocality*, thereby answering the question of how a particle going through one slit can "know" about the state of the other slit (i.e., being open or closed, for example). Dynamical nonlocality may lead to (causality preserving) changes in probability distributions and is thus distinguished from the *kinematic nonlocality* implicit in many quantum correlations, which, however, does not cause any changes in probability distributions.

In 2009, Tollaksen *et al.* [2] proposed some gedanken experiments to test the validity of this approach, and in a recent paper by Spence and Parks first experimental evidence for dynamical nonlocality was presented with the aid of weak measurements [5]. Essentially, it turns out that there is a nonlocal effect of an open or a closed first slit on a particle going through the second slit, thereby shifting the so-called modular momentum of the particle whereas the expectation values of (moments of) its momentum are left unchanged.

In the papers introducing dynamical nonlocality, the basis for a mechanism to explain how a particle at the right slit can "know" about what goes on at the left slit (or *vice versa*) is given by the nonlocal Heisenberg equations of motion for modular variables like the above-mentioned modular momentum (to be discussed below). Tollaksen *et al.* [2] claim that one thus arrived at a fundamental difference between classical and quantum mechanics that was "often missed when the Schrödinger picture is taught and classical intuitions are applied to interference." In contrast, we want to show here not only that a corresponding phenomenon of a *"systemic nonlocality"* can be accommodated within the Schrödinger picture, too, but also that a completely "classical" approach is feasible if one understands "classical" in the sense of present-day, "21st century classical physics", i.e., including recent developments such as diffusion wave fields, superstatistics, or ballistic diffusion, for example. It is clear that both the Heisenberg and the Schrödinger pictures, respectively, have their advantages and disadvantages, for example w.r.t. applications to more complex scenarios, but they also complement each other, thereby highlighting different aspects with different useful insights. In this sense, our intention is not to criticize the use of the Heisenberg picture for the problem in question, but to show the usefulness of the Schrödinger picture as well, thus throwing light on "systemic nonlocality" from a perspective which is more in line with common intuitions.

In [2, 3, 4], the Heisenberg equations of motion are shown to be nonlocal because of the dependence on the potential at two distinct locations (i.e., of the slits). Modeled like the scalar Aharonov-Bohm effect, one finds also here that there are no forces acting on the particle. For example, when a particle reaches a slit and the other slit is suddenly closed, the authors claim that the only effect is a nonlocal change of the particle's modular momentum due to the associated change of the potential at the location of the other (now closed) slit.

In this paper, we substitute the description via potentials in double slit interference by considering only constraints, or boundary conditions, on an assumed sub-quantum kinematics. Thus, putting a hypothesized sub-quantum medium into a systemic context, we shall attempt to model "systemic nonlocality" not via changes of potential differences, but by changes in sub-quantum kinematics due to their changed constraints. This amounts to a conceptual switch from a formal consideration of the macroscopic potentials of the experimental setup to a more "material" consideration of microscopic, i.e., sub-quantum currents whose modifications upon changing boundary conditions will be shown to co-determine the "nonlocal effects". (The reason for using the terms such as "systemic nonlocality" or "nonlocal effects" only within quotation marks is given by the circumstance, to be discussed in Section 4, that these descriptions are supposed to only hold for the time resolutions of present-day experiments, but to eventually break down at very small time scales.) Just as in the above-quoted papers on dynamical nonlocality, we shall obtain the usual quantum mechanical results by introducing no potential-related forces whatsoever. Although our approach has many features in common with a Bohmian one, this is where an essential difference must be pointed out. Whereas in Bohmian theory, a nonlocal "quantum force" is made responsible for the genuine quantum effects, in our model no such force exists, since the equivalent of the Bohmian quantum potential is given by contributions from purely kinetic energy terms. As we employ no "quantum force", therefore, we consider *"systemic nonlocality" as a phenomenon that emerges from a sub-quantum kinematics*, which is manipulated only by changing its constraints as determined by the changes of the apparatus.

In fact, with our approach we have in a series of papers obtained essential elements of quantum theory [1, 6, 7, 8, 9, 10, 11]. They derive from the assumption that a particle of energy $E = \hbar\omega$ is actually an oscillator of angular frequency ω phase-locked with the zero-point oscillations of the surrounding environment, the latter of which containing both regular and fluctuating components and being constrained by the boundary conditions of the experimental setup via the buildup and maintenance of standing waves. The particle in this approach is an off-equilibrium steady-state maintained by the throughput of zero-point energy from its vacuum surroundings. This is in close analogy to the bouncing/walking droplets in the experiments of Couder's group [12, 13, 14, 15], which in many respects can serve as a classical prototype guiding our intuition. We have recently applied our model to the case of interference at a double slit [1], thereby obtaining the exact quantum mechanical probability distributions on a screen behind the double slit, the average particle trajectories (which because of the averaging are shown to be identical to the Bohmian ones), and the involved probability density currents. One aim of the present paper is to extend the applicability of our model to a scenario that is only apparently dynamical, i.e., to describe "systemic nonlocality" as introduced above.

This paper is structured as follows. In Section 2, we give a short introduction to our sub-quantum model, with explanations of Gaussian dispersion and interference at a double slit, among others. Section 3 presents a review of dynamical nonlocality, its explanations in the recent literature via the Heisenberg picture, and our own explanation via the Schrödinger picture. Finally, in Section 4 we show how a corresponding "systemic nonlocality" can be understood as emerging from a classical sub-quantum kinematics.

2. A short introduction to our sub-quantum approach to quantum mechanics

In older approaches to quantum theory by de Broglie or Bohm, for example, one basically focused on a role of the waves, although sometimes possibly "empty", as essentially "guiding" the particle through the experimental apparatus. However, the Couder experiments explicitly point at a more complex scenario by an observed partial decoupling of wave and particle propagation. While the particle (or bouncer/walker, respectively) still is guided through interfering and fluctuating waves, the latter constitute a landscape that is not only present in the vicinity of the particle, but throughout the *whole apparatus*. (In a rare illustration of a similar circumstance in quantum physics, Bohm and Hiley write: "...it is only through the existence of... pools of information which are not expressible solely in terms of relationships of actual particles that the notion of an objective whole can be given meaning." [16])

We transfer the corresponding insight from the Couder experiments into our modeling of quantum systems and assume that the waves are a space-filling phenomenon involving the whole experimental setup. Thus, one can imagine a partial decoupling of the physics of waves and particles in that the latter still may be "guided" through said landscape, but the former may influence other regions of the landscape by providing specific phase information independently of the propagation of the particle. This is why a remote change in the experimental setup, when mediated to the particle via de- and/or re-construction of standing waves, can potentially amount to a nonlocal effect on a particle via the thus modified guiding landscape.

In [8] we presented a model for the classical explanation of the quantum mechanical dispersion of a free Gaussian wave packet. In accordance with the classical model, we shall now relate it more directly to a "double solution" analogy gleaned from Couder and Fort [15]. For, as is shown, e.g., in [17], one can construct various forms of classical analogies to the quantum mechanical Gaussian dispersion. Originally, the expression of a "double solution" refers to an early idea of de Broglie [18] to model quantum behavior by a two-fold process, i.e., by the movement of a hypothetical point-like singularity solution of the Schrödinger equation, and by the evolution of the usual wavefunction that would provide the empirically confirmed statistical predictions. Recently, Couder and Fort [15] used this ansatz to describe the behaviors of their bouncer- (or walker-) droplets: on an individual level, one observes particles surrounded by circular waves they emit through the phase-coupling with an oscillating bath, which provides, on a statistical level, the emergent outcome in close analogy to quantum mechanical behavior (like, e.g., diffraction or double-slit interference).

In the context of the double solution idea, which is related to correlations on a statistical level between individual uncorrelated particle positions x and momenta p, respectively, we consider the free Liouville equation. It provides a phase-space distribution $f(x,p,t)$ that shows the emergence of correlations between x and p from an initially uncorrelated product function of non-spreading (classical) Gaussian position distributions as well as momentum distributions. The motivation for their introduction comes exactly from what one observes in the Couder experiments. In an idealized scenario, we assume that at each point x an unbiased emission of momentum fluctuations π_0 in all possible directions takes place, thus mimicking (in a two-dimensional scenario) the circular waves emitted from the "particle as bouncer". If we compare the typical frequency of the bouncers in the Couder experiments (i.e., roughly 10^2 Hz) with that of an electron, for example (i.e., roughly 10^{20} Hz), we see that a continuum ansatz is practically plausible, particularly if we are interested in statistical averages over a long series of experimental runs.

Thus, one can construct said phase-space distribution, with σ_0 being the initial x–space standard deviation, i.e., $\sigma_0 = \sigma(t=0)$, and $\pi_0 := mu_0$ the momentum standard deviation, such that

$$f\left(x,p,t\right) = \frac{1}{2\pi\sigma_0 m u_0} \exp\left\{-\frac{\left(x-pt/m\right)^2}{2\sigma_0^2}\right\} \exp\left\{-\frac{p^2}{2m^2u_0^2}\right\}. \tag{1}$$

Now, the above-mentioned correlations between x and p emerge when one considers the probability density in x–space. It turns out that

$$P\left(x,t\right) = \int f \,\mathrm{d}p = \frac{1}{\sqrt{2\pi}\sigma} \exp\left\{-\frac{x^2}{2\sigma^2}\right\}, \tag{2}$$

with the standard deviation at time t given by

$$\sigma^2 = \sigma_0^2 + u_0^2 t^2. \tag{3}$$

In other words, the distribution (2) with (3) describing a spreading Gaussian is obtained from a continuous set of classical, i.e., non-spreading, Gaussian position distributions of particles whose associated momentum fluctuations also have non-spreading Gaussian distributions. One thus obtains the exact quantum mechanical dispersion formula for a Gaussian, as we have obtained also previously from our classical ansatz. For confirmation with respect to our diffusion model [8, 9], we note that with the usual definition of the "osmotic" velocity field $u = -D\frac{\nabla P}{P}$, one obtains with (2), and with bars denoting averages,

$$\overline{u^2} = D^2 \overline{\left(\frac{\nabla P}{P}\right)^2} = \frac{D^2}{\sigma^2}, \qquad \text{and thus also} \qquad u_0 = \frac{D}{\sigma_0}, \tag{4}$$

so that one can rewrite Eq. (3) in the more familiar form

$$\sigma^2 = \sigma_0^2\left(1 + \frac{D^2t^2}{\sigma_0^4}\right). \tag{5}$$

Note also that by using the Einstein relation $D = \hbar/(2m) = h/(4\pi m)$ the norm in (1) thus becomes the invariant expression (reflecting the "exact uncertainty relation" [19])

$$\frac{1}{2\pi\sigma_0 m u_0} = \frac{1}{2\pi m D} = \frac{2}{h}. \tag{6}$$

Following from (5), in references [1, 8] we obtained for smoothed-out trajectories (i.e., averaged over a very large number of Brownian motions) a sum over an Ehrenfest-type and a fluctuations term, respectively, for the motion in the x–direction

$$x_{\text{tot}}(t) = vt + x(t) = vt + x(0)\frac{\sigma}{\sigma_0} = vt + x(0)\sqrt{1 + \frac{D^2t^2}{\sigma_0^4}}. \tag{7}$$

Thus one obtains the *average velocity field* of a Gaussian wave packet as

$$v_{\rm tot}(t) = v(t) + \frac{{\rm d}x(t)}{{\rm d}t} = v(t) + [x_{\rm tot}(t) - vt]\,\frac{u_0^2 t}{\sigma^2}. \tag{8}$$

Note that Eqs. (7) and (8) are derived solely from statistical physics. Still, they are in full accordance with quantum theory, and in particular with Bohmian trajectories [17]. Note also that one can rewrite Eq. (5) such that it appears like a linear-in-time formula for Brownian motion,

$$\overline{x^2} = \overline{x^2(0)} + D_t\, t, \tag{9}$$

where a time dependent diffusivity

$$D_{\rm t} = u_0^2\, t = \frac{\hbar^2}{4m^2\sigma_0^2}\, t \tag{10}$$

characterizes Eq. (9) as *ballistic diffusion.* This makes it possible to simulate the dispersion of a Gaussian wave packet on a computer by simply employing coupled map lattices for classical diffusion, with the diffusivity given by Eq. (10). (For detailed discussions, see refs. [8] and [9].)

Moreover, one can easily extend this scheme to more than one slit, like, for example, to explain interference effects at the double slit[1, 11]. For this, we chose similar initial situations as in [17], i.e., electrons (represented by plane waves in the forward y–direction) from a source passing through soft-edged slits 1 and 2 in a barrier (located along the x–axis) and recorded at a screen. In our model, we therefore note two Gaussians representing the totality of the effectively "heated-up" so-called *path excitation field* (to be detailed below), one for slit 1 and one for slit 2, whose centers have the distances $+X$ and $-X$ from the plane spanned by the source and the center of the barrier along the y–axis, respectively.

With the total amplitude R of two coherent waves with (suitably normalized) amplitudes $R_i = \sqrt{P_i}$, and the local phases φ_i, $i = 1$ or 2, one has as usual that

$$R = R_1 \cos(\omega t + \varphi_1) + R_2 \cos(\omega t + \varphi_2)\,. \tag{11}$$

Introducing an arbitrarily chosen unit vector $\hat{\mathbf{n}}$, one may also define $\cos(\omega t + \varphi_i(\mathbf{x}))= \hat{\mathbf{n}} \cdot \hat{\mathbf{k}}_{\mathbf{i}}(\mathbf{x}, {\rm t})$, such that along with the system's evolution, the emergent outcome of the time evolution of (11) can be written as

$$R(\mathbf{x}, {\rm t}) = \hat{\mathbf{n}} \cdot \left(R_1 \hat{\mathbf{k}}_1(\mathbf{x}, {\rm t}) + R_2 \hat{\mathbf{k}}_2(\mathbf{x}, {\rm t})\right), \tag{12}$$

which we shall use later on. According to classical textbook wisdom, the *averaged total intensity* becomes

$$P_{\rm tot} := R^2 = R_1^2 + R_2^2 + 2R_1R_2\cos\varphi_{12} = P_1 + P_2 + 2\sqrt{P_1P_2}\cos\varphi_{12}, \tag{13}$$

where φ_{12} is the relative phase $\varphi_{12} = \varphi_1 - \varphi_2 = (\mathbf{k}_1 - \mathbf{k}_2)\cdot\mathbf{r}$. Note that φ_{12} enters Eq. (13) only via the cosine function, such that, e.g., even if the total wave numbers (and thus also the total momenta) $\mathbf{k}_i$ were of vastly different size, the cosine effectively makes Eq. (13) independent of said sizes, but dependent only on an angle modulo 2π. This will turn out as essential for our discussion further below.

The x–components of the centroids' motions from the two alternative slits 1 and 2, respectively, are given by the particle velocity components

$$v_x = \pm\frac{\hbar}{m}\, k_x, \tag{14}$$

respectively, such that the relative group velocity of the Gaussians spreading into each other is given by $\Delta v_x = 2v_x$. However, in order to calculate the phase difference φ_{12} descriptive of the interference term of the intensity distribution (13), one must take into account the total momenta involved, i.e., one must also include the wave packet dispersion as described in the previous Section. Thus, one obtains with the displacement $\pm x\,(t) = \mp\,(X + v_x t)$ in Eq. (8) the total relative velocity of the two Gaussians as

$$\Delta v_{\mathrm{tot},x} = 2\left[v_x - (X + v_x t)\frac{u_0^2 t}{\sigma^2}\right]. \tag{15}$$

Therefore, the total phase difference between the two possible paths 1 and 2 (i.e., through either slit) becomes

$$\varphi_{12} = \frac{1}{\hbar}(m\Delta v_{\mathrm{tot},x}\,x) = 2mv_x\frac{x}{\hbar} - (X + v_x t)x\frac{1}{D}\frac{u_0^2 t}{\sigma^2}. \tag{16}$$

In our earlier papers [6, 7, 8], we have shown that, apart from the ordinary particle current $\mathbf{J}(\mathbf{x},t) = P(\mathbf{x},t)\mathbf{v}$, we are now dealing with two additional, yet opposing, currents $\mathbf{J}_u = P(\mathbf{x},t)\mathbf{u}$, which are on average orthogonal to $\mathbf{J}$ [6, 7, 8, 10], and which are the emergent outcome from the presence of numerous corresponding velocities

$$\mathbf{u}_{\pm} = \mp\frac{\hbar}{2m}\frac{\nabla P}{P}. \tag{17}$$

We denote with $\mathbf{u}_+$ and $\mathbf{u}_-$, respectively, the two opposing tendencies of the diffusion process. Moreover, when we take the averages, we obtain a smoothed-out *average velocity field*

$$\overline{\mathbf{u}(\mathbf{x},t)} = \int P\mathbf{u}(\mathbf{x},t)\,\mathrm{d}^n x, \tag{18}$$

which is all that we need for our further considerations. Similarly, based on the fact that we have an initial Gaussian distribution of velocity vectors $\mathbf{v}(\mathbf{x},t)$, we define an average velocity field $\overline{\mathbf{v}}$ of the wave propagation as

$$\overline{\mathbf{v}(\mathbf{x},t)} = \int P\mathbf{v}(\mathbf{x},t)\,\mathrm{d}^n x, \tag{19}$$

and for the free particle make use of an average orthogonality between the two velocity fields, $\mathbf{u}$ and $\mathbf{v}$, [6, 7, 8, 9],

$$\overline{\mathbf{v}\cdot\mathbf{u}} = \int P\mathbf{v}\cdot\mathbf{u}\,\mathrm{d}^n x = 0. \tag{20}$$

In effect, then, the combined presence of both velocity fields $\mathbf{u}$ and $\mathbf{v}$ can be denoted as a *path excitation field*: via diffusion, the bouncer in its interaction with already existing wave-like excitations of the environment creates an "agitated", or "heated-up", thermal landscape, which can also be pictured by interacting wave configurations all along between source and detector of an experimental setup. Recall that our prototype of a walking bouncer, i.e., from the experiments of Couder's group, is always driven by its interactions with a superposition of waves emitted at the points it visited in the past. Couder *et al.* denote this superposition of in-phase waves the "path memory" of the bouncer [20]. This implies, however, that the bouncers at the points visited in the present necessarily create new wave configurations which will form the basis of a path memory in the future. In other words, the wave configurations of the past determine the bouncer's path in the present, whereas its bounces in the present co-determine the wave configurations at any of the possible locations it will visit in the future. Therefore, we call the latter configurations the *path excitation field*, which may also be described as heated-up thermal field. Ideally, as in the coupling of an oscillator with classical diffusion, non-relativistic diffusion wave fields arise with instantaneous field propagation [7, 21], one has elements of the whole setup which may be nonlocally oscillating ("breathing") in phase. This means that the Gaussian of Eq. (2) can be said to represent an idealized nonlocal path excitation field in that it is a physically existing and effective entity responsible for where the bouncing particle can possibly go. (We shall discuss a less idealized approach in the last Section of this paper.)

Let us now consider a single, classical particle (bouncer) following the propagation of a set of waves of equal amplitude R_i, each representing one of i possible alternatives according to our principle of path excitation, and focus on the specific role of the velocity fields. To describe the required details, each path i be occupied by a Gaussian wave packet with a forward momentum $\mathbf{p}_i = \hbar\mathbf{k}_i = m\mathbf{v}_i$. Moreover, due to the stochastic process of path excitation, the latter has to be represented also by a large number N of consecutive Brownian shifts, $\mathbf{p}_{u,\alpha} = m\mathbf{u}_\alpha$. Defining (with indices $i = 1$ or 2 referring to the two slits, and with + and − referring to the right and the left from the average direction of $\mathbf{v}_i$, respectively)

$$\overline{\mathbf{v}}_{\text{tot},i} := \overline{\mathbf{v}}_i + \overline{\mathbf{u}}_{i+} + \overline{\mathbf{u}}_{i-}, \tag{21}$$

and with two Gaussian distributions $P_1 = R_1^2$ and $P_2 = R_2^2$, one has with (12)

$$R_{\text{tot}}^2 = \left(R_1 \hat{\overline{\mathbf{v}}}_{\text{tot},1} + R_2 \hat{\overline{\mathbf{v}}}_{\text{tot},2}\right)^2 , \tag{22}$$

which after a few calculational steps provides (similarly to [1], but now with slightly different labellings which apply more generally) the total average current

$$\overline{\mathbf{J}}_{\text{tot}} = P_1\overline{\mathbf{v}}_1 + P_2\overline{\mathbf{v}}_2 + \sqrt{P_1P_2}\,(\overline{\mathbf{v}}_1 + \overline{\mathbf{v}}_2)\cos\varphi_{12} + \sqrt{P_1P_2}\,(\overline{\mathbf{u}}_1 - \overline{\mathbf{u}}_2)\sin\varphi_{12}. \tag{23}$$

A more detailed account of Eq. (23) and its extension to n slits is in preparation [Fussy *et al.* (2013)]. Note that Eq. (23), upon the identification of $\overline{\mathbf{u}}_i = -\frac{\hbar}{m}\frac{\nabla R_i}{R_i}$ from Eq. (17) and with $P_i = R_i^2$, turns out to be in perfect agreement with a comparable "Bohmian" derivation [17, 22]. In fact, with $\overline{\mathbf{v}}_i = \frac{\nabla S_i}{m}$, one can rewrite (23) as

$$\overline{\mathbf{J}}_{\text{tot}} = R_1^2\frac{\nabla S_1}{m} + R_2^2\frac{\nabla S_2}{m} + R_1R_2\left(\frac{\nabla S_1}{m} + \frac{\nabla S_2}{m}\right)\cos\varphi_{12} + \frac{\hbar}{m}\left(R_1\nabla R_2 - R_2\nabla R_1\right)\sin\varphi_{12}. \tag{24}$$

The formula for the averaged particle trajectories, then, simply results from

$$\overline{\mathbf{v}}_{\text{tot}} = \frac{\overline{\mathbf{J}}_{\text{tot}}}{P_{\text{tot}}}. \tag{25}$$

Although we have obtained the usual quantum mechanical results, we have so far not used the quantum mechanical formalism in any way. However, upon employment of the Madelung transformation for each path j ($j = 1$ or 2),

$$\Psi_j = R e^{\mathrm{i}S/\hbar}, \tag{26}$$

and thus $P_j = R_j^2 = |\Psi_j|^2 = \Psi_j^*\Psi_j$, with the definitions (17) and $v_j := \nabla S_j/m$, $\varphi_{12} = (S_1 - S_2)/\hbar$, and recalling the usual trigonometric identities such as $\cos\varphi = \frac{1}{2}\left(e^{\mathrm{i}\varphi} + e^{-\mathrm{i}\varphi}\right)$, etc., one can rewrite the total average current (23) immediately as

$$\begin{aligned}\overline{\mathbf{J}}_{\text{tot}} &= P_{\text{tot}}\overline{\mathbf{v}}_{\text{tot}} \\ &= (\Psi_1+\Psi_2)^*(\Psi_1+\Psi_2)\frac{1}{2}\left[\frac{1}{m}\left(-\mathrm{i}\hbar\frac{\nabla(\Psi_1+\Psi_2)}{(\Psi_1+\Psi_2)}\right) + \frac{1}{m}\left(\mathrm{i}\hbar\frac{\nabla(\Psi_1+\Psi_2)^*}{(\Psi_1+\Psi_2)^*}\right)\right] \\ &= -\frac{\mathrm{i}\hbar}{2m}\left[\Psi^*\nabla\Psi - \Psi\nabla\Psi^*\right] = \frac{1}{m}\mathrm{Re}\left\{\Psi^*(-\mathrm{i}\hbar\nabla)\Psi\right\},\end{aligned} \tag{27}$$

where $P_{\text{tot}} = |\Psi_1 + \Psi_2|^2 =: |\Psi|^2$. The last two expressions of (27) are the exact well-known formulations of the quantum mechanical probability current, here obtained without any quantum mechanics, but just by a re-formulation of (23). In fact, it is a simple exercise to insert the wave functions (26) into (27) to re-obtain (23).

3. "Systemic" versus dynamical nonlocality in double slit interference

In reference [2], Tollaksen *et al.* discuss interference at a double slit "from a single particle perspective" and ask the following question: If a particle goes through one slit, how does it "know" whether the second slit is open or closed? We shall here first recapitulate the arguments providing these authors' answer and later provide our own arguments and answer. Of course, the question is about the phase information and how it affects the particle. We know from quantum mechanics that phases cannot be observed on a local basis and that a common overall phase has no observational meaning. Assuming that two initially non-overlapping Gaussian wave functions, Ψ_1 and Ψ_2, describe the probability amplitudes for particles emerging from slits 1 or 2, respectively, which are separated by a distance D, the total wave function for the particle exiting the double slit may be written as

$$\Psi = \mathrm{e}^{\mathrm{i}\alpha_1}\Psi_1 + \mathrm{e}^{\mathrm{i}\alpha_2}\Psi_2, \tag{28}$$

but since a common overall phase is insignificant, one writes the total wave function as

$$\Psi_\varphi = \Psi_1 + \mathrm{e}^{\mathrm{i}\varphi}\Psi_2, \tag{29}$$

where $\varphi := \alpha_2 - \alpha_1$ is the physically significant *relative phase*. Tollaksen *et al.* now ask where the relative phase appears in the form of (deterministic) observables that describe interference. When looking at the expectation value (in the one-dimensional case, for simplicity)

$$\begin{aligned} \overline{x} = \left[\int \Psi_\varphi^* \Psi_\varphi \,\mathrm{d}x\right]^{-1} \int \Psi_\varphi^* x \Psi_\varphi \,\mathrm{d}x &\equiv \int \left(\Psi_1^* + \mathrm{e}^{-\mathrm{i}\varphi}\Psi_2^*\right) x \left(\Psi_1 + \mathrm{e}^{\mathrm{i}\varphi}\Psi_2\right) \mathrm{d}x \\ &= \int \left(\Psi_1^* x \Psi_1 + \Psi_2^* x \Psi_2\right) \mathrm{d}x + \int \Psi_1^* x \mathrm{e}^{\mathrm{i}\varphi}\Psi_2 \,\mathrm{d}x + \text{c.c.}, \end{aligned} \tag{30}$$

one sees that it is independent of φ because of the vanishing of the last term due to the assumed non-overlapping of Ψ_1 and Ψ_2. Similarly, this also holds for all moments $\overline{x^n}$, and for all moments $\overline{p^n}$ as well. In particular, one has for the expectation value of the momentum

$$\begin{aligned} \overline{p} &= \left[\int \Psi_\varphi^* \Psi_\varphi \,\mathrm{d}x\right]^{-1} \mathrm{Re} \int \Psi_\varphi^* \mathrm{i}\hbar \frac{\partial}{\partial x} \Psi_\varphi \,\mathrm{d}x \\ &\equiv \mathrm{Re} \int \left(\Psi_1^* \mathrm{i}\hbar \frac{\partial}{\partial x}\Psi_1 + \Psi_2^* \mathrm{i}\hbar \frac{\partial}{\partial x}\Psi_2\right) \mathrm{d}x + \mathrm{Re} \int \Psi_1^* \mathrm{i}\hbar \frac{\partial}{\partial x} \mathrm{e}^{\mathrm{i}\varphi}\Psi_2 \,\mathrm{d}x + \text{c.c.}, \end{aligned} \tag{31}$$

where the φ–dependent term vanishes identically, because $\frac{\partial}{\partial x}\Psi_2 = 0$ for $\Psi_2 = 0$. So, again, where does the relative phase appear? The answer of Tollaksen *et al.* is given by a "shift operator" that shifts the location of, say Ψ_1, over the distance D to its new location coinciding with that of Ψ_2. The expectation value of the shift operator is thus given by

$$\overline{\mathrm{e}^{-\mathrm{i}\frac{pD}{\hbar}}} \equiv \int \left(\Psi_1^* \mathrm{e}^{-\mathrm{i}\frac{pD}{\hbar}}\Psi_1 + \Psi_2^* \mathrm{e}^{-\mathrm{i}\frac{pD}{\hbar}}\Psi_2\right) \mathrm{d}x + \int \Psi_1^* \mathrm{e}^{-\mathrm{i}\frac{pD}{\hbar}}\Psi_2 \mathrm{e}^{\mathrm{i}\varphi} \,\mathrm{d}x + \int \Psi_2^* \mathrm{e}^{-\mathrm{i}\varphi} \mathrm{e}^{-\mathrm{i}\frac{pD}{\hbar}}\Psi_1 \,\mathrm{d}x, \tag{32}$$

where all but the last term vanish identically, thus providing (with the correct normalization)

$$\overline{\mathrm{e}^{-\mathrm{i}\frac{pD}{\hbar}}} = \mathrm{e}^{-\mathrm{i}\varphi}/2 \quad \text{and} \quad \overline{\mathrm{e}^{-\mathrm{i}\frac{pD}{\hbar}}} + \overline{\mathrm{e}^{\mathrm{i}\frac{pD}{\hbar}}} = \cos\varphi. \tag{33}$$

In order to make sense of the shift operator, the authors now shift to the Heisenberg picture, thereby providing with a Hamiltonian $H = \frac{p^2}{2m} + V(x)$ the time evolution of the operator as

$$\frac{\mathrm{d}}{\mathrm{d}t}\,\mathrm{e}^{-\mathrm{i}\frac{pD}{\hbar}} = \frac{\mathrm{i}}{\hbar}\left[\mathrm{e}^{-\mathrm{i}\frac{pD}{\hbar}}, \frac{p^2}{2m} + V(x)\right] = \frac{-\mathrm{i}D}{\hbar}\mathrm{e}^{-\mathrm{i}\frac{pD}{\hbar}}\left\{\frac{V(x) - V(x+D)}{D}\right\}. \tag{34}$$

With its dependence on the distance D between the two slits, Eq. (34) is claimed to be a description of *dynamical nonlocality* showing how a particle can "know" about the presence of the other slit. Tollaksen *et al.* maintain that it is possible to understand this dynamical nonlocality only by employing the Heisenberg picture. However, we shall now show that an equivalent understanding is possible also within the Schrödinger picture, and even more intuitively accessible, too. For, there is one assumption in the foregoing analysis that is not guaranteed to hold in general, i.e., the non-overlapping of the wave functions Ψ_1 and Ψ_2. On the contrary, we now shall assume that the two Gaussians representing the probability amplitudes for the particle immediately after passing one of the two slits do not have any artificial cut-off, but actually extend across the whole slit system, even with only very small (and practically often negligible) amplitudes in the regions further away from the slit proper. (We shall give arguments for this assumption further below.) In other words, we now ask: what if Ψ_1 and Ψ_2 do overlap, even if only by a very small amount? To answer this question, we consider the expectation value of the momentum, or the total current $\mathbf{J}_{\mathrm{tot}} = P_{\mathrm{tot}}\frac{\bar{\mathbf{P}}}{m}$, respectively, and we obtain from Eq. (31) that the terms $\mathbf{p}_\varphi$ involving the relative phase φ stem from the interference terms of $\mathbf{J}_{\mathrm{tot}}$, i.e., they provide with $\Psi_j = R_j e^{\mathrm{i}\varphi_j}$ and $\varphi_j = \frac{S_j}{\hbar}$

$$P_{\mathrm{tot}}\frac{\overline{\mathbf{p}_\varphi}}{m} = R_1 R_2\left(\frac{\nabla S_1}{m} + \frac{\nabla S_2}{m}\right)\cos\varphi + \frac{\hbar}{m}\left(R_1\nabla R_2 - R_2\nabla R_1\right)\sin\varphi. \tag{35}$$

First of all one notes upon comparison of Eq. (35) with Eqs. (23)–(25) the exact correspondence of (35) with our classically obtained expression for the interference terms of the emerging current, or of the expression for $\overline{\mathbf{v}}_{\mathrm{tot}}$, respectively. Moreover, although the product $R_1 R_2$ is in fact negligibly small for regions where only a long tail of one Gaussian overlaps with another Gaussian (i.e., such that the non-overlapping assumption would be largely justified), nevertheless the second term in (35) can be very large despite the smallness of R_1 or R_2. It is this latter part which is responsible for the genuinely quantum-like nature of the average momentum, i.e., for its nonlocal nature. This is similar in the Bohmian picture, but here given a more direct physical meaning in that this last term refers to a difference in diffusive currents as explicitly formulated in the last term of Eq. (23). Because of the mixing of diffusion currents from both channels, we call this decisive term in $\mathbf{J}_{\mathrm{tot}}$ the "entangling current", $\mathbf{J}_{\mathrm{e}}$ [23].

For illustration, Figs. 1–6 show our classical computer simulations of interference and the role of the entangling current $\mathbf{J}_{\mathrm{e}}$ in different situations. Figs. 1 and 2 show the emerging interference pattern and the average trajectories without, and Figs. 3 and 4 with, an applied extra phase shift at one slit. To bring out the shifting of the interference pattern more clearly, in Figs. 5 and 6 we apply – mainly for didactic reasons – the phase shift at much later times than in Figs. 3 and 4. Thereby, also a decoupling of wave and particle behaviors becomes visible.

Finally, there are substantial arguments against the non-overlapping scenario in Tollaksen *et al.* Firstly, experiments by Rauch *et al.* have shown that in interferometry interference does not only happen when the main bulks of the Gaussians overlap, but also when a Gaussian interferes with the off-bulk plane-wave components of the other wave function as well [24]. On a theoretical side, we have repeatedly stressed that the diffusion processes employed in our model can ideally be described by nonlocal diffusion wave fields [21, 25] which thus require small but non-zero amplitudes across the whole experimental setup. Our scenario

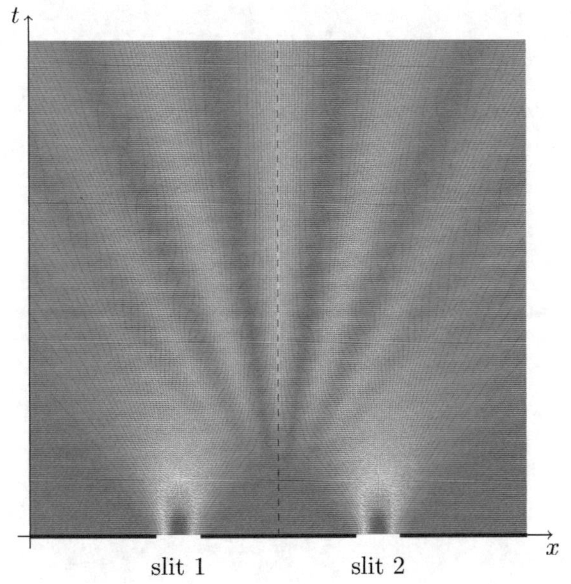

Figure 1. Classical computer simulation of the interference pattern with equal slit widths: intensity distribution with increasing intensity from white through yellow and orange, with averaged trajectories (red) for two Gaussian slits ($v_{x,1} = v_{x,2} = 0$).

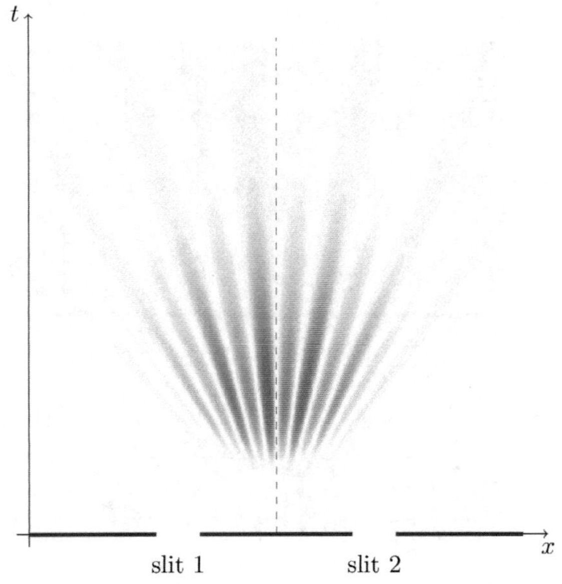

Figure 2. The corresponding entangling current density $\mathbf{J}_{\mathrm{e}}$, i.e., the last term on the r.h.s. of Eq. (23). As this term is characterized by the difference of the diffusive velocities u_i, the entangling current is responsible for the "systemically nonlocal" nature of the process forming the interference pattern.

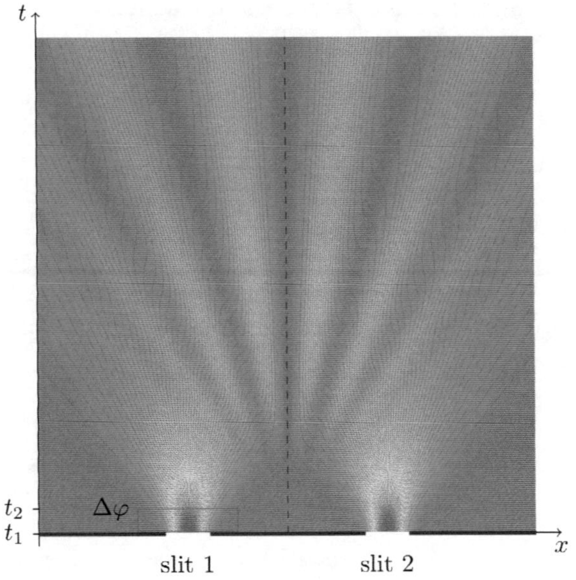

Figure 3. Classical computer simulation as in Fig. 1, but with additional phase shift $\Delta\varphi = 3\pi$ gradually accumulated during the time interval between t_1 and t_2 at slit 1.

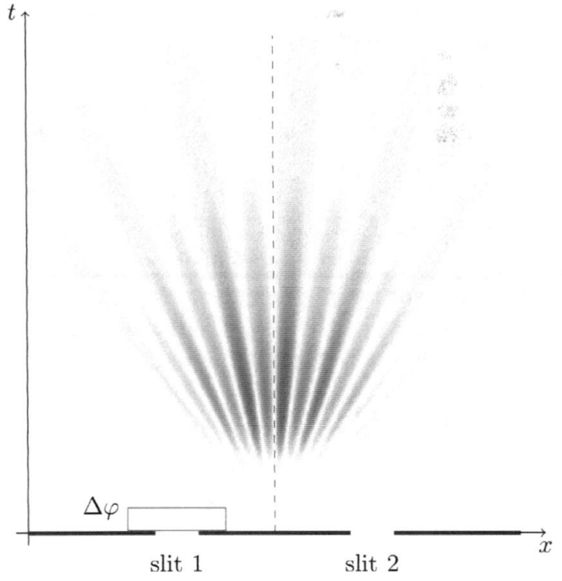

Figure 4. The corresponding entangling current density $\mathbf{J}_{\mathrm{e}}$. Note the shift of maxima and minima in the emerging pattern, as compared to Fig. 2.

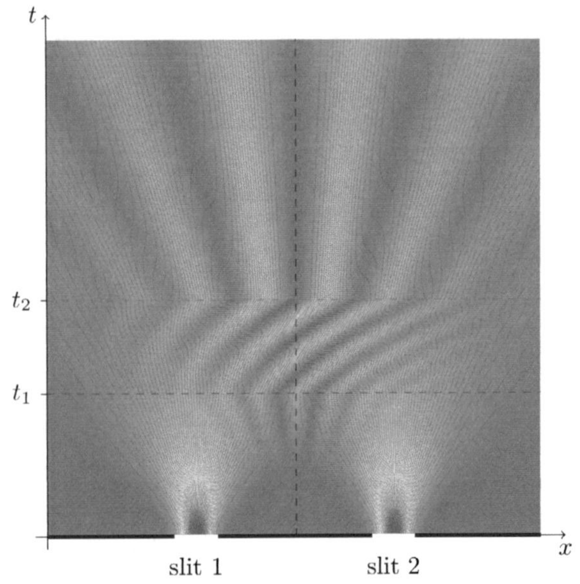

Figure 5. Classical computer simulation as in Fig. 3, but with different times t_i and with gradually accumulated additional phase $\Delta\varphi = 5\pi$. This results in the same distributions of P and $\mathbf{J}_e$ for times $t > t_2$ and shows the effect of the shifting of the interference fringes more clearly than Fig. 3. Note the radically different behaviors of the probability density related to wave-like interference on one hand, and that of the average particle trajectories on the other.

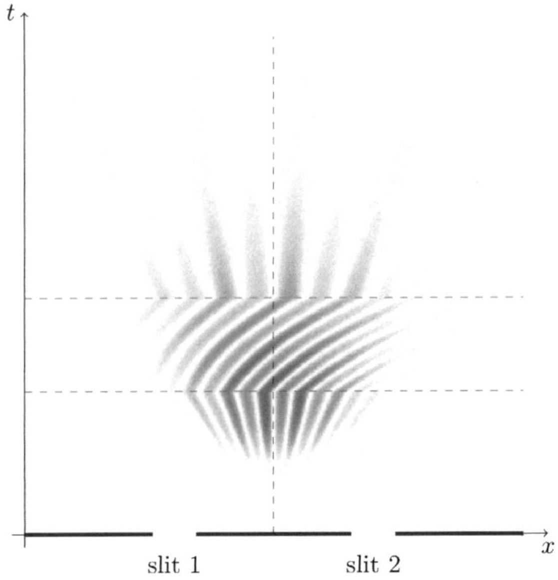

Figure 6. Although the currents $\mathbf{J}_e$ dramatically cross the central symmetry line separating the areas of the two slits, the average particle trajectories (Fig. 5) strictly obey the *no crossing rule* familiar from, but not restricted to, the deBroglie-Bohm interpretation. This is a clear demonstration of the partial decoupling of wave and particle behavior as envisaged in our model.

may also relate to the recently developed models of superstatistics, where a macroscopically observed emergent behavior is the result of a superposition of two or more statistical systems operating at vastly different space-time scales. ([26], and Jizba and Scardigli, this volume) Moreover, and more specifically, we have shown [8, 9] that quantum propagation can be identified with sub-quantum anomalous (i.e., "ballistic") diffusion which is characterized by infinite mean displacements $\overline{x} = \infty$ despite the finite drifts $\overline{x^2} = u^2t^2$ and $u < c$. In sum, these arguments speak in favor of using small, but non-zero amplitudes from the Gaussian of the other slit which can interfere with the Gaussian at the particle's location.

4. Emergence of "systemic nonlocality" from sub-quantum kinematics

Let us now see how we can understand "systemic nonlocality" within our sub-quantum approach. For example, we can ask the following question in our context: what if we start with one slit only, and when the particle might pass it we open the second slit? Let us assume for the time being, and without restriction of generality, that the x–component of the velocity v_x is zero. Then, according to (16), upon opening the second slit (with the same σ), we obtain a term proportional to the distance $2X$ between the two slits. Thus, there will be a shift in momentum on the particle passing the first slit given by

$$\frac{\Delta p}{\hbar} = \pm\frac{1}{2}\nabla\varphi_{12} = \pm\frac{1}{2\hbar}\nabla\left(S_1 - S_2\right) = \frac{\Delta p_{\mathrm{mod}}}{\hbar}, \tag{36}$$

where one effectively uses the "modular momentum" $p_{\rm mod} = p \bmod \frac{\hbar}{2X} = p - 2n\pi\frac{\hbar}{2X}$ because an added or subtracted phase difference $\varphi_{12} = 2n\pi$ does not change anything. In other words, by splitting X into a component X_n providing $\varphi_{12} = 2n\pi$ on the one hand, and the modular remainder ΔX on the other hand providing $\varphi_{12} = p_{\rm mod}x/\hbar$, one rewrites (16) with $X := X_n + \Delta X$ (and with $v_x = 0$ for simplicity) as

$$\varphi_{12} = -(X_n + \Delta X)x\frac{1}{D}\frac{u^2 t}{\sigma_0^2} =: -2n\pi - \Delta X x\frac{\hbar}{2m}\frac{t}{\sigma^2\sigma_0^2}\,. \tag{37}$$

Therefore, one can further on substitute Δp by $\Delta p_{\rm mod}$, and one obtains a momentum shift (with the sign depending on whether the right or left of the two slits is opened as the second one)

$$\Delta p_{\rm mod} = \pm\frac{\hbar}{2}\nabla\varphi_{12} = \pm\frac{\Delta X}{2}\frac{\hbar^2}{2m}\frac{t}{\sigma^2\sigma_0^2} = \pm m\Delta X\frac{D^2 t}{\sigma^2\sigma_0^2} = \pm m\Delta X\frac{\dot\sigma}{\sigma}\,. \tag{38}$$

For large times, one has $\dot\sigma \simeq \sigma/t$ such that

$$\Delta p_{\rm mod} \simeq \pm m\frac{\Delta X}{t}\,. \tag{39}$$

In a separate paper [23] we show that, with $\xi(t) = x - vt$ describing the location of a particle in a Gaussian field, the action is given by

$$S = \int m v_{\rm tot}(t)\,{\rm d}x - \int E\,{\rm d}t = S = mvx + \frac{mu_0^2}{2}\left[\frac{\xi(t)}{\sigma(t)}\right]^2 t - Et = mvx + \frac{mu_0^2}{2}\left[\frac{\xi(0)}{\sigma_0}\right]^2 t - Et, \tag{40}$$

with E being the system's total energy. Thus, the expression $\nabla\varphi_{12}$ of (38) essentially refers to the gradient of the thermal fluctuation field, with the kinetic temperature $kT = mu_0^2\left[\frac{\xi(t)}{\sigma(t)}\right]^2$ of what we have termed the path excitation field.

It follows that, due to the vacuum pressure stemming from the opened second slit, there is an *emergent* "nonlocal force" which does not derive from a potential but from the impinging of the second slit's sub-quantum diffusive momenta on the particle at the first slit. Although there is a corresponding shift in the latter's velocity distribution, i.e.,

$$\frac{\partial}{\partial t}\Delta p_{\rm mod} = \pm m\Delta X\frac{1}{\sigma^2}\left(\sigma\ddot\sigma - \dot\sigma^2\right) = \pm m\Delta X\left(\frac{\ddot\sigma}{\sigma} - \left[\frac{\dot\sigma}{\sigma}\right]^2\right), \tag{41}$$

which reduces for large times to a gradually diminishing deceleration, or acceleration, respectively,

$$\frac{\partial}{\partial t}\Delta p_{\rm mod} \simeq \mp m\Delta X\left[\frac{\dot\sigma}{\sigma}\right]^2 \simeq \mp m\frac{\Delta X}{t^2}, \tag{42}$$

in the vicinity of the slits which interests us here primarily, the expression (41) practically vanishes identically, i.e., there is no additional force on the particle near the slit.

Now that we have found a possible "nonlocal momentum transfer", the question immediately arises whether this would imply superluminal signaling. To answer this question, we recall that contained in the phase difference (37) is the osmotic velocity u, or the diffusion-related momentum fluctuation $\delta p = mu$, respectively, for which it holds that it must be unbiased [27], i.e.,

$$\int P\delta p\,{\rm d}x = 0. \tag{43}$$

In other words, both with regard to the directions and with regard to the sizes, the average over all momentum fluctuations and positions of them must vanish identically, i.e., there is a "complete uncertainty" of δp, just as Tollaksen *et al.* refer to a "complete uncertainty" of the relative phase [2]. This means that, since an experimenter can neither know where exactly within the Gaussian distribution the particle is located nor what quantity $\delta p = mu$ would actually be transferred in any single run of the experiment upon opening the second slit, one can only obtain information about the average momentum transfer $\overline{\Delta p_{\rm mod}}$ in a sufficiently large statistical sample. As the momentum-shifted particle does not allow to extract a "signal" from its experiencing that shift, i.e., as there is no way to distinguish between an unambiguous state before and an unambiguous state after the shift, one can have a "nonlocal momentum transfer" which could at best imply a "hidden" superluminal signaling.

Although we are therefore dealing with an epistemic indeterminism in this case, we nevertheless have to face the problem that on an ontic level our approach is a deterministic one (i.e., based on an assumed stochastic, and thus essentially deterministic, sub-quantum thermodynamics). As nonlocality can "peacefully co-exist" with relativity only if one also has an underlying indeterminism, we shall in the remainder of this paper discuss the possible consequences for and of our model.

Firstly, the foregoing analysis was restricted to the one-dimensional momentum transfer, i.e., in the x–direction normal to the forward particle propagation. However, with the spatial components of Gaussians being independent of each other, the corresponding dispersion and momentum transfer processes must also hold in the other directions, e.g., in the y–direction of particle propagation when considering the two-dimensional case.

Generally, the question arises of how the calculated momentum transfer can be nonlocal, while discarding the distances X_n which just provide the ineffective phase differences $\varphi_{12} = 2n\pi$. Here, the analogy to the Couder experiments provides another clue. For there, it is clear that the modes of the standing waves emerging between the walls of the bath container depend on the distance between the walls. In other words, changing that distance means selecting another set of modes. Now, it is possible to transfer this reference to the boundary conditions also to the quantum systems we experiment with in the laboratory. For, the whereabouts of the quanta are always restricted to the space between sources and detectors. Changing the experimental setup, e.g., by introducing a phase shifter, by closing a slit, or the like, therefore always amounts to changing boundary conditions in such a way that old modes of standing waves are substituted by new ones to whose phases the particle/bouncer now locks in.

In principle, this implies essentially only two options with respect to the question of superluminal signaling (with the addition of a third option briefly mentioned at the end of this paper): i) either, one has a strict nonlocality, i.e., with instantaneous changes of boundary conditions across spacelike distances, or ii) the changes take a small, but non-zero time to become effective, thus also implying possible superluminal effects. Option i) is an idealized one that is argued for by using the physics of nonrelativistic oscillator-driven diffusion wave fields as introduced above. For example, in [28] it was shown that for the one-dimensional problem of a pulsating particle in a box, the heat distribution turns out classically to be of the same sinusoidal type as the quantum mechanical probability distribution. Then, a shifting of a wall results in an instantaneous shifting of all the nodes of the distribution just like in the quantum case. Therefore, one has in this idealized classical scenario a nonlocality identical with the quantum one, but originating from the interplay of oscillator-driven diffusion wave fields with macroscopic boundaries of the experimental setup. As mentioned, ontologically this would lead to superluminal signaling, even if it was only a "hidden" one for the observers because of the epistemic indeterminism due to the unknowability of the exact (and, on the sub-quantum level: existing!) initial conditions.

This leaves us with option ii) as the possibly more realistic scenario of an emergent "nonlocality" that we describe as "systemic". In other words, with the bouncer always oscillating harmonically with the standing waves of the zero-point field, and with the latter being co-determined by the experimental setup, a sudden change in the corresponding standing wave pattern will "practically instantaneously" (i.e., during times of the order $t \sim \frac{1}{\omega}$) affect the particle. In the case of the particle at one slit being affected by the opening of the other slit, therefore, it does not matter how far apart the two slits are: As long as the emerging standing wave configurations result in a "systemic nonlocality" (which is a basic assumption of our model with regard to the physical meaning of the zero-point field), the modular remainder ΔX will make itself visible in a shift of the particle's velocity distribution. Although framed in a "systemically nonlocal" setup, therefore, the latter will just reflect the "local" shift ΔX in the vicinity of the particle. As this is accompanied by a complete uncertainty of the relative phase, no signal can in practice be extracted from this process. This constitutes again an explanation of the *momentum transfer which excludes superluminal signaling in practice, but not in principle.* However, in this scenario the simultaneity (i.e., instantaneousness) of nonlocal effects as described by quantum theory is lost. Still, this is very likely a problem that can be pinned down to the problem of too idealistic applications of concepts of relativity. In other words, the emergent quantum mechanics we envisage may very well have to be accompanied by an *emergent relativity* as it is discussed, for example, within the paradigm of superstatistics mentioned above. Particularly with regard to a "double special relativity" [26, and Jizba and Scardigli, this volume], classical relativity may turn out as a limiting case of a more general approach, just as ordinary quantum theory may be a limiting case of a "deeper level theory" as envisaged by *emergent quantum mechanics.*

To conclude, we can now compare the metaphysical assumptions underlying standard quantum mechanics with those of our emergent quantum mechanics [29, 30]. Roughly, quantum theory may be classified via two types of approaches: i) indeterministic and ii) deterministic ones. The first type of approaches is consistent with standard relativity, while the second type is not. More specifically, the first type, which one can call *standard quantum mechanics* (i.e., including interpretations of its formalism like the Copenhagen or the deBroglie-Bohm versions), is characterized by *nonlocal indeterminism,* i.e., by a principal indeterminism that guarantees that nonlocality cannot be used for superluminal signaling. The second, deterministic type is characteristic for various forms of *emergent quantum mechanics.* These may either be characterized by *local determinism* with only apparent nonlocality (see, e.g., the approach of "stochastic electrodynamics" as in [31, 32]) or by a *determinism with "systemic nonlocality"*, as it characterizes one variant of our model presented here, for example.

Within our model, we discussed in the present paper two options for possible further elaboration. One option involves the idealized version of nonlocal diffusion wave fields. The latter describe a classical analogue representing nonlocality, which combined with an underlying ontic determinism implies "hidden" superluminal signaling (i.e., despite its epistemic impossibility due to the epistemic indeterminism). The other option is systemically nonlocal-like with only "practical instantaneousness" and thus superluminal signaling, which in combination with emergent relativity still may be constructed as free from causal paradoxes. Finally, as mentioned, a third option can be thought of. It would constitute a hybrid of models via the mixing of deterministic and indeterministic elements, which is suggested as a possibility by the assumed partial decoupling of wave and particle physics in our model. Thus, it is at least conceivable that the deterministic "systemic nonlocality" described in one of our discussed options refers only to the proposed wave-like physics, including that involving the boundary conditions, whereas the particle-like bouncer/walker has indeterministic degrees of freedom of its own. In this case, one would have a "nonlocal momentum transfer" *which excludes superluminal signaling both in practice and in principle.* Hence, the whole theory would effectively be indeterministic and comply with standard relativity.

References

[1] Grössing G, Fussy S, Mesa Pascasio J and Schwabl H 2012 *Ann. Phys.* **327**(2) 421–437 (*Preprint* quant-ph/1106.5994v3)
[2] Tollaksen J, Aharonov Y, Casher A, Kaufherr T and Nussinov S 2010 *New J. Phys.* **12**(1) 013023 (*Preprint* quant-ph/0910.4227v1)
[3] Aharonov Y, Pendleton H and Petersen A 1969 *Int. J. Theor. Phys.* **2**(3) 213–230
[4] Aharonov Y, Pendleton H and Petersen A 1970 *Int. J. Theor. Phys.* **3**(6) 443–448
[5] Spence S and Parks A 2012 *Found. Phys.* **42**(6) 803–815 (*Preprint* quant-ph/1010.3289v1)
[6] Grössing G 2008 *Phys. Lett.* A **372**(25) 4556–4563 (*Preprint* quant-ph/0711.4954v2)
[7] Grössing G 2009 *Physica* A **388** 811–823 (*Preprint* quant-ph/0808.3539v1)
[8] Grössing G, Fussy S, Mesa Pascasio J and Schwabl H 2010 *Physica* A **389**(21) 4473–4484 (*Preprint* quant-ph/1004.4596v1)
[9] Grössing G, Fussy S, Mesa Pascasio J and Schwabl H 2011 *J. Phys.: Conf. Ser.* **306** 012046 (*Preprint* physics.gen-ph/1005.1058v2)
[10] Grössing G, Mesa Pascasio J and Schwabl H 2011 *Found. Phys.* **41**(9) 1437–1453 (*Preprint* quant-ph/0812.3561v4)
[11] Grössing G, Fussy S, Mesa Pascasio J and Schwabl H 2012 *J. Phys.: Conf. Ser.* **361** 012008 (*Preprint* quant-ph/1205.3393v1)
[12] Couder Y, Protière S, Fort E and Boudaoud A 2005 *Nature* **437** 208–208
[13] Couder Y and Fort E 2006 *Phys. Rev. Lett.* **97** 154101
[14] Couder Y, Boudaoud A, Protière S and Fort E 2010 *Europhys. News* **41**(1) 5
[15] Couder Y and Fort E 2012 *J. Phys.: Conf. Ser.* **361**(1) 012001
[16] Bohm D and Hiley B J 1993 *The undivided universe: An ontological interpretation of quantum theory* repr. ed (London: Routledge) ISBN 0415065887
[17] Holland P R 1993 *The Quantum Theory of Motion: An account of the de Broglie-Bohm causal interpretation of quantum mechanics* (Cambridge: Cambridge University Press) ISBN 0-521-35404-8
[18] de Broglie L V P R 1960 *Non-Linear Wave Mechanics: A Causal Interpretation.* (Amsterdam: Elsevier)
[19] Hall M J W and Reginatto M 2002 *J. Phys. A: Math. Gen.* **35**(14) 3289–3303
[20] Fort E, Eddi A, Boudaoud A, Moukhtar J and Couder Y 2010 *PNAS* **107**(41) 17515–17520
[21] Mandelis A, Nicolaides L and Chen Y 2001 *Phys. Rev. Lett.* **87**(2) 020801
[22] Sanz A S and Miret-Artés S 2008 *J. Phys. A: Math. Gen.* **41**(43) 435303 (*Preprint* quant-ph/0806.2105)
[23] Mesa Pascasio J, Fussy S, Schwabl H and Grössing G 2013 to be published in *Physica* A
[24] Rauch H 1993 *Phys. Lett.* A **173**(3) 240–242
[25] Mandelis A 2001 *Diffusion-wave fields: Mathematical methods and Green functions* (New York, NY: Springer) ISBN 0387951490
[26] Jizba P and Scardigli F 2012 *Phys. Rev.* D **86**(2) 025029 (*Preprint* hep-th/1105.3930)
[27] Grössing G 2004 *Found. Phys. Lett.* **17**(4) 343–362 (*Preprint* quant-ph/0311109v2)
[28] Grössing G 2010 *Entropy* **12**(9) 1975–2044
[29] Walleczek J 2013 to be published in *Found. Phys.*
[30] Walleczek J and Grössing G 2013 to be published in *Found. Phys.*
[31] de la Peña L, Valdés-Hernández A, Cetto A and França H 2011 *Phys. Lett.* A **375**(16) 1720–1723
[32] Cetto A M and de la Peña L 2012 *J. Phys.: Conf. Ser.* **361**(1) 012013

Journal of Physics: Conference Series **442** (2013) 012013 doi:10.1088/1742-6596/442/1/012013

The phenomenon of spontaneous replica symmetry breaking in complex statistical mechanics systems

Francesco Guerra

Dipartimento di Fisica, Università di Roma "La Sapienza" and Istituto Nazionale di Fisica Nucleare, Sezione di Roma, Piazzale Aldo Moro 5, 00185 Roma (Italy)

E-mail: `francesco.guerra@roma1.infn.it`

Abstract. We analyze the main aspects of the phenomenon of spontaneous replica symmetry breaking, introduced by Giorgio Parisi. We work in the frame of real replicas, by taking into account the simple case of the random energy model. In particular, we study the phase space diagram for systems of coupled replicas, and the connected phase transitions. Our considerations can be generalized to the more complicated models of mean field spin glasses and neural networks. We report also about a letter of Ettore Majorana, written in December 1937 to his uncle Dante, very interesting for its methodological content.

The phenomenon of "spontaneous replica symmetry breaking", for some mean field models for spin glasses, as the Sherrington-Kirkpatrick model [1], was discovered by Giorgio Parisi, in the frame of his celebrated "replica trick" (see for example [2]). Quite recently a rigorous treatment has been possible by using interpolation methods [3].

"Spontaneous replica symmetry breaking" can be extended to other cases of interest. Typical cases are:

– Spin glasses with coupled replicas. Here a partial answer has been given by Michel Talagrand [4], in the frame of his treatment of the replica symmetric region and the rigorous establishment of the Parisi solution for the free energy. However, the general structure of spontaneous replica symmetry breaking for coupled replicas has not been achieved yet. This would be very important, also for the property of ultra-metricity of the states.

– Bipartite spin glasses. Here we find that the replica symmetric Ansatz, and the fully broken replica case, require a sort of mini-max principle for the free energy (see for example [5]).

– One can consider also neural networks of the Hopfield type. We modify the interaction, so that the learned words, which act as quenched noise, are not made by a string of ± 1 variables, but by a string of unit Gaussian random variables. This helps in the treatment of interpolation, which would be more cumbersome in the standard case. We see here how the mini-max principle emerges already at the level of the replica symmetric Ansatz. Replica symmetry breaking is introduced in the natural way, and still implies a mini-max structure [6, 7].

– Of course the mini-max structure adds difficulty to the rigorous treatment of these models. The nice interpolation inequalities for the mean field spin glass no longer hold. This gives a real challenge for future research.

Coming from spontaneous replica symmetry breaking, the main features of the Parisi variational principle for the free energy are the following:

– the order parameter is a functional order parameter (while for example in the elementary mean field ferromagnetic model, the Curie-Weiss model, the order parameter is simply a constant, with the meaning of the magnetization of the system). The functional order parameter gives all physical properties of the model: thermodynamic quantities (as free energy per site, internal energy, entropy), the states, the fluctuations.

– the variational principle is expressed through a *sup* for the free energy, and NOT through an *inf*, as it happens for example in the case of the entropy principle of equilibrium statistical mechanics. In the Curie-Weiss model the variational principle is given through an *inf*, so it is a genuine entropy principle.

Therefore, the Parisi variational principle is a completely new structure.

We have studied some of its properties, by finding the Legendre form of the variational principle and its Legendre dual. These results will appear in a paper now in preparation. In this frame, let us only recall that the Legendre dual of the Parisi functional order parameter is connected with the general shape of the interaction.

As we have already remarked, the problem of extending the broken replica structure to more complicated cases has not found a general complete solution yet.

In this paper, we will stress that spontaneous replica symmetry breaking is a real physical phenomenon, with its physical consequences, and with the meaning literally expressed.

This is particularly evident if we work with real replicas, as it is possible to do, and not inside a frame where the number of replicas is allowed to go to zero.

In order to simplify the treatment, we work in a completely elementary frame, that of the so called random energy model, introduced by Derrida some years ago [8], in order to mimic some of the features of spin glasses. Our aims are pedagogical.

Let me recall also that I was told by Giorgio Parisi that a kind of random energy model was considered in the 70-ies by Nicola Cabibbo, who did not publish any result because it was considered too simple. This simplicity will give us an elementary frame in order to discuss the phenomenon of spontaneous replica symmetry breaking.

The configuration space is given by an Ising system on N sites

$$\sigma : (1, 2, \ldots, N) \ni i \to \sigma_i = \pm 1.$$

Let us introduce a system of (quenched) random variables J_σ, indexed by the configurations σ, which are independent centered unitary Gaussian, so that we have the averages $\mathbb{E}(J_\sigma) = 0$, and $\mathbb{E}(J_\sigma J_{\sigma'}) = \delta_{\sigma\sigma'}$. The symbol $\delta_{\sigma\sigma'}$ (overlap between the two configurations) means 1 if the two configurations are equal, and zero otherwise. Now the partition random function is given by

$$Z_N(\beta, J) \equiv \sum_\sigma \exp(\beta \sqrt{\frac{N}{2}} J_\sigma),$$

where $\beta \geq 0$ is a parameter with the physical meaning of inverse temperature. Of course, we have the random free energy $F_N(\beta, J)$ related to Z in the known form

$$Z_N(\beta, J) = \exp(-\beta F_N(\beta, J)).$$

Call $\mathbb{E}$ the average with respect to the quenched $J_{..}$ variables. Quenched means that they do not participate to thermal equilibrium, but act as an external random environment.

Of course this is a very simple and completely solvable model. Let us see some results. Firstly, a thermodynamically corrected annealed inequality is easily established.

For any $0 < m \leq 1$, we have for the quenched free energy per site the inequality

$$N^{-1} \mathbb{E} \log \sum_\sigma \exp(\beta \sqrt{\frac{N}{2}} J_\sigma) \leq m^{-1} N^{-1} \mathbb{E} \log \sum_\sigma \exp(m\beta \sqrt{\frac{N}{2}} J_\sigma),$$

which holds for any spin system for purely thermodynamic reasons (positivity of the entropy).

Now we apply the annealed inequality, coming from convexity, $\mathbb{E}\log .. \leq \log\mathbb{E}...$ The $\mathbb{E}$ of the *Boltzmannfaktor* is immediately calculated.

$$\mathbb{E}\exp(\beta\sqrt{\frac{N}{2}}J_\sigma) = \exp(\frac{\beta^2}{4}N).$$

By taking into account that $\sum_\sigma = 2^N$, we end with the inequality

$$N^{-1}\mathbb{E}\log\sum_\sigma \exp(\beta\sqrt{\frac{N}{2}}J_\sigma) \leq \frac{\log 2}{m} + \frac{\beta^2}{4}m.$$

The inequality is made the best possible by taking the minimum with respect to the parameter m, for each value of β. The result is then the following. Define the critical temperature through $\beta_c = 2\sqrt{\log 2}$. Then, for $\beta \leq \beta_c$, the minimum is at $m(\beta) = 1$, and its value is $\log 2 + \beta^2/4$. While for $\beta \geq \beta_c$, the minimum is at $m(\beta) = \beta_c/\beta$, and its value is $\beta\sqrt{\log 2}$. Therefore, the optimal $m(\beta) = \max(1, \beta_c/\beta)$ is continuous in β, but its first derivative is discontinuous at the critical point.

With some little additional effort it can be shown that these bounds, given by the stated variational principle, are in fact the true values, in the thermodynamic limit $N \to \infty$.

Therefore we have

$$\lim_{N\to\infty} N^{-1}\mathbb{E}\log\sum_\sigma \exp(\beta\sqrt{\frac{N}{2}}J_\sigma) \equiv A(\beta),$$

where $A(\beta) = \log 2 + \beta^2/4$, for $\beta \leq \beta_c$, and $A(\beta) = \beta\sqrt{\log 2}$, for $\beta \geq \beta_c$.

Fluctuations are easily controlled, so that the above limit holds also without the $\mathbb{E}$ average, J-almost surely.

Let us show the meaning of the optimal order parameter $m(\beta)$. Introduce, as usual, the random Boltzmann-Gibbs average ω for observables F depending on σ:

$$\omega(F) = Z^{-1}\sum_\sigma F\exp..,$$

where exp .. is the *Boltzmannfaktor*. By taking s, $s = 1, 2, \ldots,$ replicas of the system, indexed by the variables σ_i^a, $a = 1, \ldots, s$, $i = 1, \ldots, N$, we can introduce the product state Ω on observables F explicitly as follows

$$\Omega(F) = Z^{-s}\sum_{\sigma^1\ldots\sigma^s} F\exp(\beta\sqrt{\frac{N}{2}}(J_{\sigma^1} + \cdots + J_{\sigma^s}),$$

where the *Boltzmannfaktor* is the product for each replica.

We introduce also the quenched averages $\langle F\rangle = \mathbb{E}\Omega(F)$. The thermodynamic limits are well defined.

Then, an easy calculation, involving integration by parts on the J variables, shows:

$$\partial_\beta A(\beta) = \frac{\beta}{2}(1 - \langle\delta_{12}\rangle).$$

Here, two replicas are involved σ^1 and σ^2, and δ_{12} is their $0-1$ overlap.

From the explicit expression given above we have also $\langle\delta_{12}\rangle = 0$, below the critical value for β, and $\langle\delta_{12}\rangle = 1 - \beta_c/\beta = 1 - m(\beta)$, above the critical value for β. Therefore, by taking into account that δ_{12} takes the values $(0,1)$, we see that the physical meaning of $m(\beta)$ is simply the probability that δ_{12} takes the value 0 under the global average $\langle..\rangle$.

Analogous considerations can be developed for the calculation of the $\langle..\rangle$ averages for all kinds of overlap products among different replicas. For example we have

$$\langle\delta_{12}\delta_{13}\rangle = \frac{1}{2}\langle\delta_{12}\rangle + \frac{1}{2}\langle\delta_{12}\rangle^2 = 1 - \frac{3}{2}m(\beta) + \frac{1}{2}m(\beta)^2,$$

$$\langle\delta_{12}\delta_{34}\rangle = \frac{1}{3}\langle\delta_{12}\rangle + \frac{2}{3}\langle\delta_{12}\rangle^2 = 1 - \frac{5}{3}m(\beta) + \frac{2}{3}m(\beta)^2,$$

as an elementary instance of the so called Ghirlanda-Guerra identities [9].

Of course, we have complete symmetry among all replicas, so that for example $\langle\delta_{12}\rangle = \langle\delta_{13}\rangle$, $\langle\delta_{12}\delta_{23}\rangle = \langle\delta_{34}\delta_{35}\rangle$, $\langle\delta_{12}\delta_{34}\rangle = \langle\delta_{25}\delta_{36}\rangle$, and so on.

Now we show that the region $\beta > \beta_c$ is really characterized by spontaneous replica symmetry breaking, in the literal sense.

Since here we have a kind of mean field system, replica symmetry breaking should be achieved along these steps: firstly we break explicitly the symmetry according to some small parameter λ, then we go to the limit $N \to \infty$, then we bring $\lambda \to 0$, and see what happens. Therefore, we start from a system made by the σ variables for two replicas, for example σ^1 and σ^2, and introduce, for some $\lambda \geq 0$,

$$\tilde{A}_N(\beta,\lambda) = (2N)^{-1}\mathbb{E}\log\sum_{\sigma^1\sigma^2}\exp(\beta\sqrt{\frac{N}{2}}(J_{\sigma^1} + J_{\sigma^2}))\exp(\frac{1}{2}\lambda N\delta_{\sigma^1\sigma^2}).$$

There is an explicit coupling between the σ^1 and the σ^2. Due to the N factor the coupling does have thermodynamic effects. The replicated systems will contain the couples of replicas $(12), (34), (56), \ldots$. If $\lambda > 0$ replica symmetry is explicitly broken. The surviving symmetries are $(12), (34)$ and so on. But there is no (13) symmetry for example.

However, the model with coupled replicas is also explicitly solvable. In order to understand this, let us write

$$\exp(\frac{1}{2}\lambda N\delta_{\sigma^1\sigma^2}) = (\exp(\frac{1}{2}\lambda N) - 1)\delta_{\sigma^1\sigma^2} + 1 \simeq \exp(\frac{1}{2}\lambda N)\delta_{\sigma^1\sigma^2} + 1,$$

since the -1 term has no thermodynamic effect.

Therefore, by substitution in the previous

$$\tilde{A}_N(\beta,\lambda) \simeq (2N)^{-1}\mathbb{E}\log(Z_N(2\beta)\exp(\frac{1}{2}\lambda N) + Z_N^2(\beta)).$$

Now consider the dominant terms in the thermodynamic limit

$$Z_N(2\beta)\exp(\frac{1}{2}\lambda N) \simeq \exp((A(2\beta) + \frac{1}{2}\lambda)N),$$

$$Z_N^2(\beta) \simeq \exp(2A(\beta)N).$$

As $N \to \infty$, between the two competitors in the sum, the highest will prevail, so that we have

$$\lim_{N\to\infty}\tilde{A}_N(\beta,\lambda) = \tilde{A}(\beta,\lambda) = \max(\frac{1}{2}A(2\beta) + \frac{1}{4}\lambda, A(\beta)).$$

Since $A(2\beta)$ and $A(\beta)$ are explicitly known, as we have shown before, we can easily calculate $\tilde{A}(\beta,\lambda)$. The final result is the following.

There is a critical line for λ, defined by two interlaced parabolas in β, as follows: $\lambda_c(\beta) = 2\log 2 - \beta^2$ for $0 \leq \beta \leq \sqrt{\log 2} = \beta_c/2$, $\lambda_c(\beta) = (\beta - \beta_c)^2$ for $\beta_c/2 \leq \beta \leq \beta_c$, $\lambda_c(\beta) = 0$ for $\beta \geq \beta_c$.

For $\lambda \leq \lambda_c, \beta \leq \beta_c$ we have $\tilde{A}(\beta, \lambda) = A(\beta)$, i.e. the λ coupling has absolutely no effect.

On the other hand, if $\lambda \geq \lambda_c$, then we have:

$$\tilde{A}(\beta, \lambda) = \frac{1}{2}\log 2 + \frac{1}{2}\beta^2 + \frac{1}{4}\lambda,$$

for $0 \leq \beta \leq \beta_c/2$, and

$$\tilde{A}(\beta, \lambda) = \beta\sqrt{\log 2} + \frac{1}{4}\lambda,$$

for $\beta_c/2 \leq \beta \leq \beta_c$. Therefore for $\lambda > \lambda_c$ the coupling is effective and $\tilde{A}(\beta, \lambda)$ does depend on λ. Notice that in any case

$$\lim_{\lambda \to 0} \tilde{A}(\beta, \lambda) = A(\beta).$$

This procedure can be easily generalized to the case where we couple any number of s replicas through parameters $\lambda_{12}, \ldots \lambda_{1s}, \ldots \lambda_{s-1s}$, and let them go to zero according to different patterns, after the thermodynamic limit.

Now comes the interesting surprise. By a direct calculation we have

$$\partial_\lambda \tilde{A}(\beta, \lambda) = \frac{1}{4}\langle \delta_{12} \rangle_{\beta\lambda},$$

$$\partial_\beta \tilde{A}(\beta, \lambda) = \frac{\beta}{2}(1 + \langle \delta_{12} \rangle_{\beta\lambda} - 2\langle \delta_{13} \rangle_{\beta\lambda}),$$

where integration by parts on J is exploited again (but now replicas $(12), (34), \ldots$ are coupled). In the notation $\langle .. \rangle_{\beta\lambda}$ we have emphasized the $\beta\lambda$ dependence of the averages.

Since the explicit expression of $\tilde{A}(\beta, \lambda)$ is know, by making the derivatives, through simple calculations, we get $\langle \delta_{12} \rangle_{\beta\lambda}$ and $\langle \delta_{13} \rangle_{\beta\lambda}$ in the following explicit form:

$$\langle \delta_{12} \rangle_{\beta\lambda} = 0, \langle \delta_{13} \rangle_{\beta\lambda} = 0,$$

for $\lambda < \lambda_c, \beta \leq \beta_c$, as expected. Moreover

$$\langle \delta_{12} \rangle_{\beta\lambda} = 1, \langle \delta_{13} \rangle_{\beta\lambda} = 0,$$

for $\lambda > \lambda_c, 0 \leq \beta \leq \beta_c/2$. Crossing the critical line $\langle \delta_{12} \rangle_{\beta\lambda}$ gets a bang-bang transition, from the minimal value 0 to the maximal value 1. On the other hand, the interaction is not strong enough to displace $\langle \delta_{13} \rangle_{\beta\lambda}$ from its 0 value. However, if $\lambda > \lambda_c, \beta_c/2 \leq \beta \leq \beta_c$, then we have

$$\langle \delta_{12} \rangle_{\beta\lambda} = 1, \quad \langle \delta_{13} \rangle_{\beta\lambda} = 1 - \frac{\beta_c}{2\beta}.$$

Notice the factor $1/2$. From these explicit expressions, we have the spontaneous replica symmetry breaking in the most direct form. In fact for $\beta > \beta_c$ we have

$$\langle \delta_{12} \rangle_{\beta 0+} = \lim_{\lambda \to 0} \langle \delta_{12} \rangle_{\beta\lambda} = 1 > \langle \delta_{12} \rangle_\beta,$$

$$\langle \delta_{13} \rangle_{\beta 0+} = \lim_{\lambda \to 0} \langle \delta_{13} \rangle_{\beta\lambda} = 1 - \frac{1}{2}\frac{\beta_c}{\beta} > \langle \delta_{12} \rangle_\beta = 1 - \frac{\beta_c}{\beta},$$

where we have indicated with $\langle..\rangle_\beta$ the averages for uncoupled replicas ($\lambda = 0$ from the beginning) at inverse temperature β. Clearly $\langle\delta_{12}\rangle_{\beta 0^+}$ is different from $\langle\delta_{13}\rangle_{\beta 0^+}$. Replica symmetry is spontaneously broken! The chain $\lambda > 0, N \to \infty, \lambda \to 0$ gives different results from those where $\lambda = 0$ from the beginning.

Of course, we have also to understand how the broken quantities are connected with the original ones, so to recognize the pattern of the spontaneously breaking of the symmetry. Fortunately in this case the problem is very simple. We have already seen the validity of

$$\lim_{\lambda\to 0} \tilde{A}(\beta, \lambda) = A(\beta, \lambda),$$

telling us that the free energy is continuous in λ. But we have more. In fact, we can easily check that we have also for the derivatives

$$\lim_{\lambda\to 0} \partial_\beta \tilde{A}(\beta, \lambda) = \partial_\beta A(\beta).$$

Therefore, we have continuity also for the internal energy and the entropy. Spontaneous replica symmetry breaking is able to change the state at $\lambda = 0^+$, with respect to $\lambda = 0$, but preserves the free energy, the internal energy and the entropy.

We leave to the interested reader the enquire about the fate of the Ghirlanda-Guerra identities for systems with coupled replicas, and for generic spontaneously broken states.

Analogous considerations can be developed for more complicated cases, as the Sherrington-Kirkpatrick model, and neural networks. Here Parisi functional order parameter plays a central role.

Acknowledgments
This work was supported in part by MIUR (Italian Minister of Instruction, University and Research), University of Rome "La Sapienza", and INFN (Italian National Institute for Nuclear Physics).

Appendix
As a kind of introduction to the topics of complex systems, in a very peculiar historical perspective, we would like to report here about an important letter of Ettore Majorana to his uncle Dante, dated December 27, 1937, relevant for its methodological considerations.

Ettore Majorana (1906-1939), the famous Theoretical Physicists, is worldwide known for his remarkable achievements in the fields of atomic and molecular physics, nuclear physics and elementary particles [10]. His ideas are well up to date, as for example the ongoing search for the Majorana neutrino shows. Moreover, his personal history attracted the interest of many people, because he disappeared suddenly, under still mysterious circumstances, in March 1938, only few months after his direct appointment as Full Professor of Theoretical Physics at the Royal University of Naples.

The letter is written to the uncle Dante, brother of the father Fabio, in December 1937, only one month after the appointment in Naples. A copy can be found at the Majorana Fund of the Regional University Library in Catania, the city in Sicily where the Majorana Family took origin. This is a gigantic Fund, amounting to more than one hundred thousand documents, not fully classified in a complete catalog yet. It is an invaluable source of information about the Majorana Family, starting from the Patriarch Salvatore Majorana Calatabiano (1825-1897), *lawyer, member of the Parliament, minister, senator, professor of the University of Catania, economist, jurist, orator, writer*, and his sons Giuseppe (1863-1940), Angelo (1865-1910), Quirino (1871-1957), Dante (1874-1955), Fabio (1875-1934), the father of Ettore, and the daughters Elvira, Emilia. There are many important documents in this Fund, not all fully explored yet, related to Ettore, and his disappearance, as seen from the Catania observatory of the uncle Giuseppe.

Here is the text of the letter, copied by hand of a member of the Majorana Family in Catania, in our literal English translation.

Roma 27 - 12 - 1937 XVI

Dear uncle Dante,

I thank you very warmly for your greetings and wishes, which are cordially reciprocated to you and your relatives. Thank you also for your comments on the method. Allow me to add an impression of mine. I believe in the unity of science, but just because I believe seriously in it, I think that, until different sciences will practically exist with different objects, no error would be so pernicious as the confusion of methods. In particular, the mathematical method can not be of any substantial utility in sciences that are presently extraneous to physics. In other words, if some day the mathematical face of the simplest facts of life and conscience will be discovered, most certainly this will not happen for a natural evolution of biology and psycology, but only because some additional radical renewal of the general principles of physics will allow to extend its domain in fields, that are still extraneous to it. The most significant example is given by chemistry, that, after a long and very glorious life as an independent science, it has been completely absorbed by physics in the last years. This has been made possible by the rise of quantum mechanics, while no useful connection could be established between chemistry and classical mechanics. While waiting that physics will perform new miracles, it would be right to recommend to the scholars of other disciplines to rely on the methods proper to each discipline, without searching for models or suggestions in today physics, even less in yesterday's. Because the kind of physics, that someday will tell the definite truth on biological or moral facts, is something completely outside of our understanding.

Affectionate greetings

Ettore

These deep and precise methodological considerations are very impressive. In the present times, some typical biological problems are considered in the frame of methods typical of Physics, as for example neural networks, immune systems, animal behavior. This is perhaps an indication that the development, foresighted by Majorana, has taken place, at least in part. The modern Physics of complex systems (see for example [11] for an inspired overview) is perhaps the bridge toward the beginning of the assimilation of Biology into Physics.

References

[1] Sherrington D and Kirkpatrick S 1975 Solvable Model of a Spin-Glass *Phys. Rev. Lett.* **35** 1792
[2] Mézard M, Parisi G and Virasoro M A 1987 *Spin Glass Theory and Beyond* (Singapore: World Scientific)
[3] Guerra F 2003 Broken Replica Symmetry Bounds in the Mean Field Spin Glass Model *Comm. Math. Phys.* **233** 1
[4] Talagrand M 2003 *Spin Glasses: a Challenge for Mathematicians. Cavity and Mean Field Models* (Berlin: Springer)
[5] Barra A, Genovese G and Guerra F 2011 Equilibrium statistical mechanics of bipartite spin systems *J. Phys. A: Math. Theor.* **44** 245002
[6] Barra A, Genovese G and Guerra F 2010 The replica symmetric behavior of the analogical neural network *J. Stat. Phys.* **140** 784
[7] Barra A, Genovese G, Guerra F and Tantari D 2012 How glassy are neural networks? *J. Stat. Mech.* P07009
[8] Derrida B 1981 Random-Energy Model: an Exactly Solvable Model of Disordered Systems *Phys. Rev.* B **24** 2613
[9] Ghirlanda S and Guerra F 1998 General Properties of Overlap Probability Distributions in Disordered Spin Systems. Towards Parisi Ultrametricity *J. Phys. A: Math. Gen.* **31** 9149
[10] Guerra F and Robotti N 2008 *Ettore Majorana. Aspects of his scientific and academic activity* (Pisa: Edizioni della Scuola Normale Superiore)
[11] Parisi G 1999 Complex Systems: a Physicist's Viewpoint *Physica* A **263** 557

Journal of Physics: Conference Series **442** (2013) 012014 doi:10.1088/1742-6596/442/1/012014

Does a quantum particle know its own energy?

Rafael D Sorkin

Perimeter Institute, 31 Caroline Street North, Waterloo ON, N2L 2Y5 Canada
Department of Physics, Syracuse University, Syracuse, NY 13244-1130, U.S.A.

E-mail: `rsorkin@perimeterinstitute.ca`

Abstract. If a wave function does not describe microscopic reality then what does? Reformulating quantum mechanics in path-integral terms leads to a notion of "precluded event" and thence to the proposal that quantal reality differs from classical reality in the same way as a set of worldlines differs from a single worldline. One can then ask, for example, which sets of electron trajectories correspond to a Hydrogen atom in its ground state and how they differ from those of an excited state. We address the analogous questions for simple model that replaces the electron by a particle hopping (in discrete time) on a circular lattice.

Should we try to form for ourselves an image of the quantum world? Or must our theories find their meaning solely in assertions about laboratory instruments and their readings? In other words, is it permissible to ask the question, *To what 'reality' does the quantum formalism refer?* I believe that this question is not only a legitimate one, but it is one that we must ask. If we try to avoid it, we fall into a vicious circle because instruments are made of atoms and not vice versa. More importantly, quantum gravity and, especially, cosmology need to deal with parts of nature where one finds neither observers nor instruments, and to which an "operational point of view" therefore seems unsuited.

Beyond the more or less familiar reasons just adduced to support the claim that the problems of quantum gravity and "quantum foundations" are intertwined, there's another connection that could be important via the concept of relativistic causality. If the condition that "physical influences propagate causally" could be given an intrinsic formulation, free of references to external observers, and if the resulting criterion were formulated in terms of histories (as Bell's "local causality" is, for example), then one should be able to decide whether or not quantum mechanics and quantum field theory satisfied this condition. If they did, then it would make sense to require it also of quantum gravity, and this in turn could be the key to constructing a viable quantum dynamics for causal sets.

Perhaps the challenge of "quantum foundations" is not so urgent for quantum computing, which is concerned more with the manipulation of information than with microscopic reality as such. But even there, it seems possible that a more definite picture of the micro-world could someday lead us to widen our conceptions of measurement and of computation.

What I hope to illustrate in this paper is a possible answer to the italicized question posed above, an answer that arose from thinking of quantum mechanics in the language of histories, i.e. the language of the path integral. According to this answer, the micro-world is described by something called a *coevent*, but instead of attempting to define that concept in the general case,

I will consider a very simple model of a particle hopping on a lattice, and in that setting, will present a short calculation that I hope will indicate more concretely how the proposal is meant to go. In the particular scheme I will present, reality will be something like a trajectory or worldline of the hopping particle, but instead of being a single trajectory, as in classical physics, it will be a set of trajectories. This particular choice is not necessarily the best one, but it is the simplest and therefore appropriate to illustrate the main idea.

Notice that in a coevent formulation, reality is *not* represented by a wave function ψ. Therefore, although a concept like position will have a straightforward meaning, a concept like momentum or energy will not. But then we have to ask how an electron in a hydrogen atom "knows" for example, whether the atom is in its ground state or in an excited state. Is there something about the trajectories that carries this information? Hence the title of this paper.

1. A simple unitary model — the n-site hopper

A simple model allowing one to refer to trajectories and ground/excited states is the "n-site hopper", by which I mean a particle residing on an n-site periodic lattice, and at each of a discrete succession of moments either staying where it is or jumping to some other site, the respective amplitudes being those given by the simple "transfer matrix" reproduced below [1, 2]. In this "toy world", nothing exists beyond the hopper itself, and since both space and time are discrete, the possible realities or *coevents* can be computed with minimal difficulty.

In order to describe the hopping amplitudes, let us identify the nodes of the lattice with the elements of $\mathbf{Z}_n$, the integers modulo n. Further let $x \in \mathbf{Z}_n$ be the location of the particle at some moment and let x' be its location at the next moment, and write for brevity $\exp(2\pi i z) \equiv \mathbf{1}^z$. The amplitude to go from x to x' in a single step is then

$$\frac{1}{\sqrt{n}}\,\mathbf{1}^{(x-x')^2/n} \tag{1a}$$

for n odd, and

$$\frac{1}{\sqrt{n}}\,\mathbf{1}^{(x-x')^2/2n} \tag{1b}$$

for n even. For example, for $n = 6$ and with $q = \mathbf{1}^{1/12}$, the (un-normalized) amplitudes to hop by 0, 1, 2 or 3 sites respectively are $q^0 = 1$, $q^1 = q$, q^4, and $q^9 = -i$.

It is not difficult to verify that these amplitudes (more precisely the matrix they comprise) are unitary. Interestingly, they take precisely the form of the propagator of a non-relativistic free particle in one-dimension, suggesting that in a suitable $n \to \infty$ limit, this hopper model could provide a fully self-consistent regularization of the path-integral for such a particle.

For the 2- and 3-site hoppers, the amplitudes are particularly simple, yielding for $n = 3$ the matrix

$$\frac{1}{\sqrt{3}}\begin{pmatrix} 1 & \omega & \omega \\ \omega & 1 & \omega \\ \omega & \omega & 1 \end{pmatrix} \qquad (\omega = \mathbf{1}^{1/3})$$

and for $n = 2$ the matrix

$$\frac{1}{\sqrt{2}}\begin{pmatrix} 1 & i \\ i & 1 \end{pmatrix}.$$

2. Review of anhomomorphic coevents and the "Multiplicative Scheme"

In the formulation I am advocating, "nature" is represented in terms of *histories*, which for present purposes means trajectories of the hopping particle [3, 4, 5]. Physical reality — a "possible world" — is then described by a *coevent* which specifies which events happen and which don't, where an *event* is by definition a set of histories.[1] More formally, a coevent will be a function ϕ that assigns either 0 or 1 to each event, according as the event doesn't or does happen in the world described by ϕ.

From this perspective, our description of the physical world would be complete if we were able to specify fully the actual coevent ϕ. The role of "dynamics", then, is to help us toward a fuller such specification by placing conditions on ϕ that narrow down the range of possibilities that need to be considered. I will assume that the input to this dynamics takes the form of a path integral (or in the case of the hopper, a path sum). Wave functions ψ, insofar as they play a role at all, will provide provisional initial amplitudes that go into the computation of the path-sum (as an approximate summary of past).

But is it possible to base a dynamics on a path-sum alone? If we wish to do so then, plainly, we must construe the latter as something more than a technical device to compute transition amplitudes between some initial wave-function and some final one. Instead we will interpret it as providing for any event A, the *quantal measure* $\mu(A)$ of that event.[2] Once again, I will omit the formal definition [6, 7, 8], since we will see very soon how concretely to compute $\mu(A)$ in the case at hand.

2.1. The preclusion principle

It is not hard to convince oneself that in one special case $\mu(A)$ has the meaning of an ordinary probability, namely when the event A can be described as a possible outcome ("instrument reading") of a given laboratory experiment. In such a case, the analyses of numerous gedankenexperiments over the years have made it plausible (albeit people don't always express it this way) that $\mu(A)$ coincides with the Born-rule probability of the outcome A. But if this is accepted, then it follows immediately that a zero value of $\mu(A)$ implies that the corresponding instrument-event almost surely does not occur — it is *precluded*.[3] Extending this conclusion to the case of an arbitrary — macroscopic or microscopic — event, we arrive at a dynamical principle of general applicability: *If* $\mu(A) = 0$ *then the event* A *cannot happen.*

2.2. Preclusion and primitivity

The preclusion principle requires of any dynamically viable coevent ϕ that it deny every event whose quantal measure vanishes: $\phi(A) = 0$ if $\mu(A) = 0$. We have seen that this principle flows naturally from the path-integral, but by itself it is still rather weak, in the sense that a vast number of coevents can satisfy it even when preclusions abound. In our hopper example, for instance, we will have 27 histories, and consequently $2^{2^{27}} = 2^{134217728}$ coevents in toto. The number of precluded events is also large (2017807), but even so, there remain $2^{132199921}$ coevents which are *preclusive* in the sense that they satisfy our condition. On the other hand, $2^{132199921}$ is only a tiny fraction of $2^{134217728}$, so one might feel on the contrary that the preclusion principle is rather strong.

Be that as it may, I think the weightiest reason why the preclusion principle cannot stand alone is that it seems incapable of yielding the classical conception of reality-as-a-single-history when the measure μ is classical (and possibly "deterministic"). Thus, preclusion alone would not exclude coevents for which an experiment had more than one macroscopic outcome, or for which no definite outcome at all happened. In other words it would not resolve the "measurement problem".

[1] not to be confused with the word "event" used to denote a point of spacetime. The definition here follows the usage in probability theory, which in turn is more in tune with the everyday meaning of the word.

[2] $\mu(A)$ is also known as the diagonal element of the decoherence functional $D(A, A)$.

[3] To this extent, $\mu(A)$ is a sort of "propensity" for the event A to happen. One cannot go further and interpret it as a genuine probability, because, thanks to quantal interference, it is not additive on disjoint events.

To complete the dynamical story, then, we will supplement preclusion with a further principle of "minimality" or *primitivity* designed to remedy the deficiencies just cited. How properly to frame such a supplementary condition is a question not yet settled, but the simplest proposal is that of the so-called *multiplicative scheme*, and this is the one I will adopt for the present analysis.

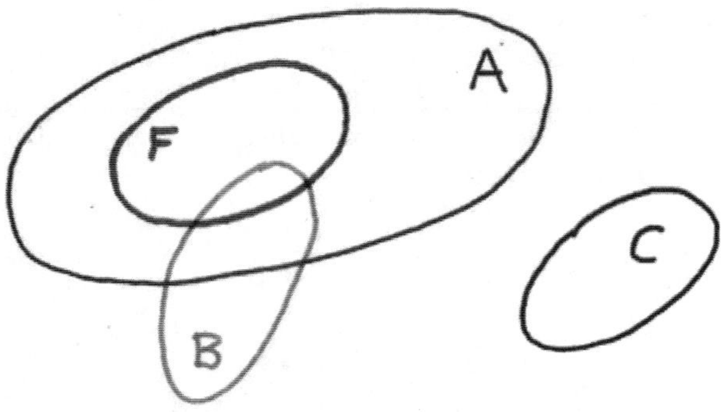

Figure 1. Three events and a multiplicative coevent. The three events are three sets of histories, A, B, C, while the coevent ϕ corresponds to a further set of histories F called its *support*. In the "reality" described by ϕ, A happens, while B and C do not happen. In formulas, $\phi = F^*$, $\phi(A) = 1$, and $\phi(B) = \phi(C) = 0$.

2.3. The multiplicative scheme

Classical physics identified reality with a single history, but that no longer seems possible quantum mechanically because the characteristic phenomenon of *interference* produces non-classical patterns of preclusion which seem to demand a modified conception of reality.

In place of a single history, the multiplicative scheme describes the physical world by a coevent of the form $\phi = F^*$, where F is now a *set* of histories that reduces to a singleton set only in very special (effectively classical) circumstances. As illustrated in Figure 1, the coevent F^* assigns 1 ('true') to an event A if and only if A is a superset of F. Thus in the diagram, if $\phi = F^*$ then $\phi(A) = 1$ while $\phi(B) = \phi(C) = 0$. When $\phi = F^*$ I will refer to F as the *support* of ϕ. Within the multiplicative scheme, a coevent is thus fully determined by its support.[4]

Now when is a coevent $\phi = F^*$ "dynamically viable" within the multiplicative scheme? By assumption it must be preclusive, and this means precisely that its support F must not fall wholly within any precluded event. Beyond this, we require further that F be as small as possible consistent with the preclusivity condition just stated. A coevent (or its support) that fulfills all these conditions I will call *primitive preclusive*, or for short just *primitive*. A primitive coevent thus describes a "possible world", where "possible" is to be understood relative to the given set of preclusions.

One sees immediately that in the absence of any preclusions (other than the empty event itself) a primitive preclusive support will consist solely of a single history, and the same holds whenever the pattern of preclusions is of the type that occurs in either classical deterministic, or classically stochastic theories. Much more than this could be said about the multiplicative

[4] A little thought should convince you that the rule that $\phi(A) = 1$ iff $F \subseteq A$ also holds classically, with F being the set that contains the "actual history" as its sole element.

scheme and its consequences [9, 10, 11], but now I want to focus on a very concrete example and on our specific question: Does the particle know its own energy?

3. Histories and amplitudes for the 3-site hopper

In simple cases, preclusion is decided by whether the sum of the amplitudes vanishes. This will be true for our example, and it will make it easy to find the primitive supports.

To simplify as much as possible, let the hopper take three steps and then stop. And let there be nothing else in the world beside this hopper. Our space of histories then comprises exactly 81 trajectories, depending on where the hopper starts from and where it lands at each of the three subsequent moments. In fact, however, we only need to consider the 27 histories shown in Figure 2 , because one can prove, as a general feature of the multiplicative scheme, that any primitive support must correspond to a sharp final position. That is, one of the three events, "the hopper terminates at 0", "the hopper terminates at 1", or "the hopper terminates at 2", must happen. Without loss of generality we can suppose it is the first of these events, and this is what the figure illustrates.

With the final position fixed at $x = 0$, it is also very easy to decide whether a given set of histories is precluded: this occurs iff the amplitudes of the constituent histories sum to zero. To work out the pattern of preclusions, it thus suffices to know all the amplitudes.

Now given a history γ, its net amplitude is the amplitude it inherits from its starting location, multiplied by the amplitudes of the individual hops, the former being given by what one might call the "initial wave-function" $\psi_{initial}$. For $\psi_{initial}$ let us consider two possible choices, ψ_0 and ψ_+, which we may call by way of analogy "ground state" and "traveling wave". The amplitudes for these are respectively $(\psi(0), \psi(1), \psi(2)) = (1, 1, 1)$ and $(\psi(0), \psi(1), \psi(2)) = (1, \omega, \omega^2)$, where $\omega = \mathbf{1}^{1/3}$ as before. Clearly both ψ_0 and ψ_+ are eigenvectors of the "transfer matrix" defined earlier. (The overall normalization of $\psi_{initial}$ is immaterial since it has no effect on preclusion.)

It is now straightforward to compute the amplitudes of our 27 histories in each case. The figure exhibits them for the traveling wave, and those for the ground-state are similar (but even simpler to compute since the dependence on the starting position is absent.) Curiously, the amplitudes per se (though not of course the way they are distributed among the histories) turn out to be exactly the same for both cases. Counting them up, one obtains the following multiset:

$$\overline{\omega}^{\{12\}} \qquad 1^{\{9\}} \qquad \omega^{\{6\}} ,$$

meaning 12 histories with amplitude $\omega^2 = \overline{\omega}$, 9 with amplitude 1, and 6 with amplitude ω.

4. Primitive coevents for the ground state and the traveling wave

Which combinations of the histories illustrated in Figure 2 yield primitive coevents? If we think in terms of the amplitudes abstractly, it is easy to answer this question, and the same answer will then apply unchanged to both cases of traveling wave and ground state. An event E will correspond to a set of amplitudes (strictly speaking a multiset), and E will be precluded precisely when the amplitudes sum to zero. Because the only real-linear relation among the complex numbers 1, ω, $\overline{\omega}$ is $1 + \omega + \overline{\omega} = 0$, this will occur when, and only when, the three amplitudes occur in equal numbers within E. It is then easy to see what are the maximal precluded multisets of amplitudes. They are those consisting of 6 copies each of 1, ω, and $\overline{\omega}$.

Now let F be the support of a preclusive coevent. In order to be preclusive, F must not fall wholly within any precluded event, and this means that it must not be possible to adjoin further amplitudes to those of F such that the resulting multiset sums to zero. Plainly F will be protected in this way iff it contains at least 7 copies of 1 or 7 copies of $\overline{\omega}$. On the other hand, we also want F to be primitive, meaning minimal among the preclusive supports. Again it is easy to see what this means: it must comprise *precisely* 7 copies of 1 or 7 copies of $\overline{\omega}$.

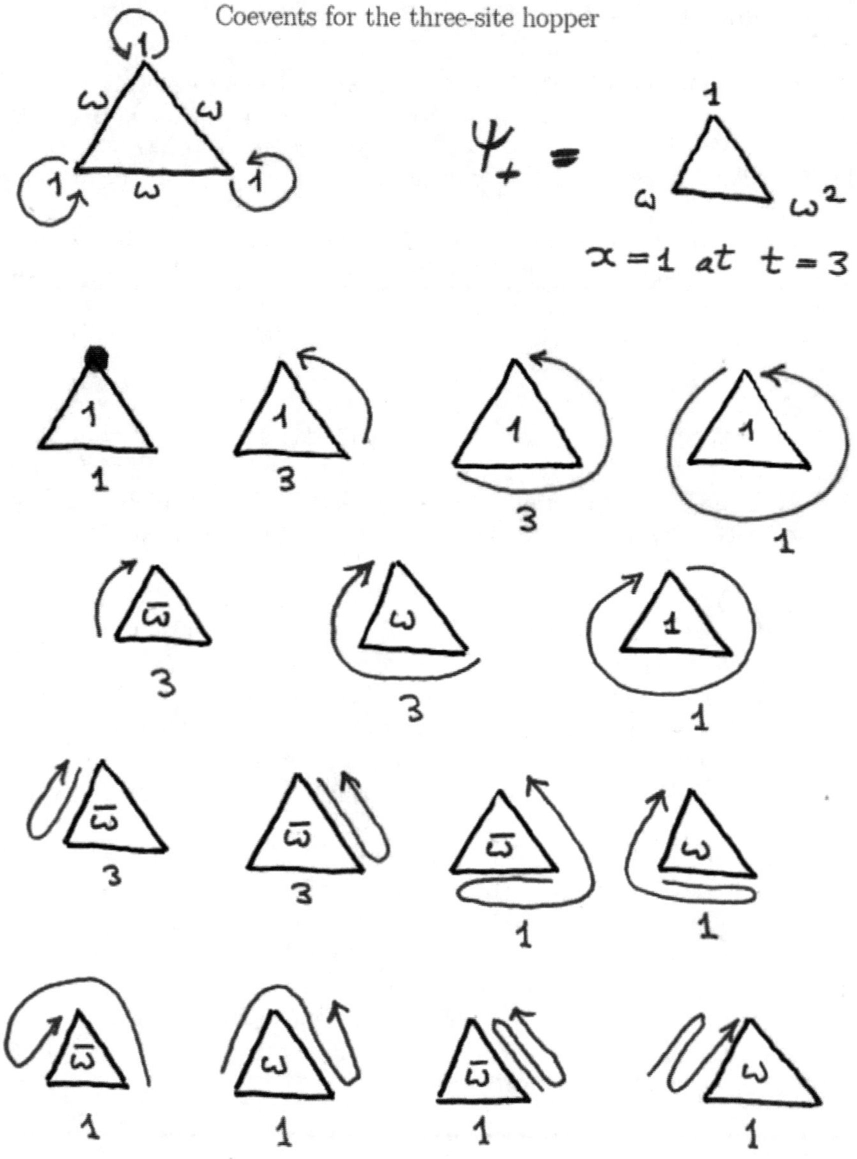

Figure 2. The 27 histories, and their amplitudes for the case of the "traveling wave". Each path indicated by an arrow represents from 1 to 3 possible histories differing from each other by the moments at which the hopper chooses to rest. The resulting multiplicity is shown under the triangle, while the number inside the triangle is the amplitude itself.

In this way, we find a total of $\binom{12}{7} + \binom{9}{7} = 828$ primitive coevents of the multiplicative form $\phi = F^*$, each made up of a total of 7 histories.

In the case of the traveling wave, for example, one such set of histories corresponds to the last three patterns in the first row of Figure 2 . Interestingly, all seven of these trajectories move in the positive (counterclockwise) direction and none of them in the contrary direction. Thus the event, $P =$ "The particle circulates exclusively in the positive sense", happens in the reality described by this coevent.

Happily enough, this sense of circulation corresponds perfectly with the "phase velocity" of ψ_+, but we cannot assert that all of the 828 traveling-wave coevents also affirm this same event P. In order to quantify the tendency toward counterclockwise motion, then, let us associate a "net circulation" with each coevent, as the total number of "forward" hops less the total number of "backward" ones. (So for example the above coevent has a net circulation of $3 \times 1 + 3 \times 2 + 1 \times 3 = 12$.) Averaging this quantity over all 828 primitive coevents yields an average net circulation of 7/23. A tendency toward counterclockwise motion is therefore present but not extremely pronounced – just as one might expect since a lattice of only $n = 3$ positions can lend no more than a very rudimentary meaning to a derivative like "d(phase)/d(angle)".

The primitive preclusive coevents for the ground-state are again supported on sets of seven histories, as we have already remarked. By symmetry we cannot expect a favoured sense of circulation in this case, but it is also interesting to ask how "restless" the particle proves to be. Here, it is natural to compare with "Bohmian" or "pilot wave" conceptions of reality, since they also give meaning to the notion of particle-trajectory. As far as I know, the Bohmian "guidance equation" is limited to continuum space and time, where the nearest analog to our hopper ground-state might be the ground-state of a particle in a box. Since the phase of the Schrödinger wave-function is independent of position in that case, the Bohmian particle does not move at all. Rather than explore its surroundings, it just stays put, wherever it happens to find itself.[5].

It turns out that the coevents of the multiplicative scheme paint a very different picture. The event, "The hopper never moves" is of course denied by all 828 coevents, while the contrary event that "The hopper never rests" is affirmed by eight of them. (For such a coevent, all seven of the histories in its support are in constant motion.) Of the other 820 primitive coevents, 28 of them affirm that the hopper either never rests or never moves (6 histories vs. 1), while the remaining 792 primitive supports consist entirely of histories such that the hopper rests once and hops twice. Moreover, for none of the coevents does the event "the hopper avoids some lattice-site" happen.[6] All in all, very peripatetic indeed.

5. Does the hopper know its energy?

If the multiplicative scheme (or any other coevent scheme) is a kind of "equation of motion for coevents", then a primitive preclusive coevent is a kind of "solution of the equations of motion". In celestial mechanics such a solution would be the orbit of a planet. One can deduce the binding energy of the planet from a knowledge of its orbit, but the converse is impossible since the energy is only one among a number of orbital parameters. Now consider some microscopic counterpart of this problem, like a hydrogen atom. By analogy one should not expect to deduce a unique

[5] The first excited state, and indeed every standing wave of pure frequency, also leads to static trajectories. In this sense a Bohmian particle in a box with reflecting walls would be about as far as possible from knowing its own energy. However, the strictly stationary wave-functions are very special, and their superpositions might generically lead to different typical trajectories [12]

[6] Which of course doesn't mean that the complementary event, "the hopper visits all three lattice sites" ever happens either. That's why the coevents are "anhomomorphic": with $\phi = F^*$ defined as above, it can easily happen that two events A and B are complementary subsets of the space of histories, but both $\phi(A)$ and $\phi(B)$ vanish.

Journal of Physics: Conference Series **442** (2013) 012014 doi:10.1088/1742-6596/442/1/012014

coevent ϕ from a knowledge of the total energy, but one might wonder whether, conversely, the coevent determines the energy unambiguously.

Quantum mechanically, energies are determinate only for eigenfunctions of the Hamiltonian operator, and the nearest analogs for these in our hopper-world are the initial-amplitude sets, ψ_0 and ψ_+ (plus, of course, the parity-reversed set ψ_-, together with linear combinations like the "standing wave", $\psi_+ + \psi_-$). We are thus led to ask whether the primitive coevents pertaining to ψ_0 overlap with those pertaining to ψ_+. To the extent that the answer is negative one can indeed deduce the hopper's energy from a knowledge of the coevent that describes its motion. Based on the above analysis of the primitive coevents, the required calculation is straightforward, and the result is that the degree of overlap is zero. No coevent is common to both ψ_0 and ψ_+.

Moreover it is possible to distinguish ψ_0 from ψ_+ in terms of relatively elementary consequences. For example if the hopper never rests (meaning $\phi(E) = 1$, where E is the event consisting of the fourth, seventh, and tenth through fifteenth histories shown in Figure 2), then ϕ pertains to $\psi = \psi_0$. Similarly, if the hopper moves only counterclockwise, or if it rests exactly once, then $\psi = \psi_+$.

In this sense, we can say that the hopper does know its own energy. We can also say that it "knows its own angular momentum", since a similar comparison reveals that the primitive coevents pertaining to ψ_+ are disjoint from those pertaining to ψ_-.

Of course the example we've studied is exceptionally simple, both with respect to the hopping amplitudes and the amplitudes of the "initial states" $\psi_{initial}$ which we have considered. It would be good to analyze also the case of the standing wave, and more generally to extend the analysis to include longer times, and larger lattice sizes n. (The case of shorter times is also instructive. For only two time-steps, one finds that there do exist primitive coevents common to ψ_+ and ψ_0. This strengthens the impression that with the passage of time the hopper would "know" more and more about $\psi_{initial}$.)

Our hopper-world is exceptionally simple in another way too, that relates to the radical inseparability (or "interconnectedness") that seems to show up in the quantum world. Were we to include a second system or process in our idealized world, say for example a four-site hopper, the coevents would change. Obviously the global coevents would change, but even those induced for the three-site hopper would in general be different. This at least is a feature of the multiplicative scheme, and it is likely true more generally. It is therefore important to study extensions of our model of this type.

Extensions of our model in any of the directions just mentioned would of course be interesting. But even without them, I hope the examples studied in this paper suffice to illustrate how one can start to think about the quantum world without invoking the ideas of either evolving wave-functions or external observers.

If these examples are not misleading us, then we can already draw some conclusions of more general validity, concerning first of all the relation between wave-functions and descriptions of reality. A histories-based or "path-integral" formulation of the sort we have been working with has no use for the Schrödinger equation at a fundamental level. The basic concept of precluded event has a "spacetime character" and refers directly to the histories and their amplitudes, not to any evolving wave-function. On the other hand, something like a wave-function does enter into the "initial conditions" needed in setting up the path-integral. In principle one should probably replace these initial amplitudes with cosmological boundary conditions imposed directly on the histories, but in our hopper-world there is no moment earlier than $t = 0$, and we have assumed instead that in a more complete model, we would be able to condense the effect of the true boundary conditions into a set of three initial amplitudes for the hopper at $t = 0$. This is the wave-function $\psi_{initial}$ in its role as "effective summary of the past".

What our hopper model teaches us about $\psi_{initial}$ is that it is far from furnishing a detailed description of physical reality. Rather, the coevent which by definition does furnish such a description is determined by $\psi_{initial}$ only to a very limited extent. Reality possesses far more "internal structure" than is reflected in a wave-function.

This observation in turn resolves an old paradox that has recently been emphasized in a cosmological setting by Daniel Sudarsky [13]. In its terrestrial form the paradox asks how the spherically symmetric wave-function resulting from the decay of a spinless nucleus can be compatible with the fact that the daughter nuclei will be found to be localized in angle if one sets up detectors. How does the symmetry break? But in terms of coevents, there's no problem. Just because the ψ-function is symmetric, that doesn't imply the same of the individual coevents. Indeed, we see exactly such a phenomenon in our hopper model, where $\psi_{initial}$ exhibits perfect rotational symmetry while the individual coevents are completely asymmetric. No single coevent shares the symmetry of $\psi_{initial}$, but only the ensemble of all 3×828 of them taken together.

Acknowledgments

For discussions of these questions over a long period, I would like to thank Fay Dowker, Cohl Furey, Yousef Ghazi-Tabatabai, Stan Gudder, Joe Henson, David Rideout, Sumati Surya, and Petros Wallden. I also thank the Foundational Questions Institute (FQXI) for a grant, and the maintainers of Steel Belt Common Lisp (SBCL), whose software I used at several points in the analysis reported above. Finally, I thank the members of the Raman Research Institute (RRI) for their warm hospitality while parts of this paper were being written.

This research was supported in part by NSERC through grant RGPIN-418709-2012. Research at Perimeter Institute for Theoretical Physics is supported in part by the Government of Canada through NSERC and by the Province of Ontario through MRI.

References

[1] Sorkin R D 2012 Toward a 'fundamental theorem of quantal measure theory' *Mathematical Structures in Computer Science* **22(5)** 816-852 (*Preprint* arXiv:1104.0997) http://www.pitp.ca/personal/rsorkin/some.papers/141.fthqmt.pdf
[2] Gudder S and Sorkin R D 2011 Two-site quantum random walk *General Relativity and Gravitation* **43** 3451-3475 (*Preprint* arXiv:1105.0705) http://www.pitp.ca/personal/rsorkin/some.papers/
[3] Sorkin R D 2007 Quantum dynamics without the wave function *J. Phys. A: Math. Theor.* **40** 3207-3221 (http://stacks.iop.org/1751-8121/40/3207) (*Preprint* arXiv:quant-ph/0610204) http://www.pitp.ca/personal/rsorkin/some.papers/
[4] Wallden P 2013 The coevent formulation of quantum theory *Preprint* arXiv:1301.5704
[5] Sorkin R D 2012 Logic is to the quantum as geometry is to gravity, in *Foundations of Space and Time: Reflections on Quantum Gravity ed* G F R Ellis, J Murugan and A Weltman (Cabridge, UK: Cambridge Univ. Press) (*Preprint* arXiv:1004.1226 [quant-ph]) http://www.pitp.ca/personal/rsorkin/some.papers/
[6] Sorkin R D 1994 Quantum Mechanics as Quantum Measure Theory *Mod. Phys. Lett.* A **9** (No. 33) 3119-3127 (*Preprint* arXiv:gr-qc/9401003) http://www.pitp.ca/personal/rsorkin/some.papers/80.qmqmt.pdf
[7] Salgado R B 2002 Some Identities for the Quantum Measure and its Generalizations *Mod. Phys. Lett.* A **17** 711-728 (*Preprint* arXiv:gr-qc/9903015)
[8] Surya S and Wallden P 2010 Quantum Covers in Quantum Measure Theory *Foundations of Physics* **40** 585-606 (*Preprint* arXiv:0809.1951)
[9] Ghazi-Tabatabai Y and Wallden P 2009 Dynamics & Predictions in the Co-Event Interpretation *J. Phys. A: Math. Theor.* **42** 235303 (*Preprint* arXiv:0901.3675)
[10] Ghazi-Tabatabai Y and Wallden P 2009 The emergence of probabilities in anhomomorphic logic *Journal of Physics: Conf. Ser.* **174** 012054 (*Preprint* arXiv:0907.0754 (quant-ph))
[11] Dowker F and Ghazi-Tabatabai Y 2008 The Kochen-Specker Theorem Revisited in Quantum Measure Theory *J. Phys.* A **41** 105301 (*Preprint* arXiv:0711.0894 (quant-ph))
[12] Lucien Hardy, remark at a conference
[13] Sudarsky D 2011 Shortcomings in the Understanding of Why Cosmological Perturbations Look Classical *Int. J. Mod. Phys.* D **20** 509-552 (*Preprint* arXiv:0906.0315 [gr-qc])

DICE2012 IOP Publishing
Journal of Physics: Conference Series **442** (2013) 012015 doi:10.1088/1742-6596/442/1/012015

Noncommutative spectral geometry: a short review

Mairi Sakellariadou

Department of Physics, King's College, University of London, Strand WC2R 2LS, London, U.K.

E-mail: `mairi.sakellariadou@kcl.ac.uk`

Abstract. We review the noncommutative spectral geometry, a gravitational model that combines noncommutative geometry with the spectral action principle, in an attempt to unify General Relativity and the Standard Model of electroweak and strong interactions. Despite the phenomenological successes of the model, the discrepancy between the predicted Higgs mass and the current experimental data indicate that one may have to go beyond the simple model considered at first. We review the current status of the phenomenological consequences and their implications. Since this model lives by construction at high energy scales, namely at the Grand Unified Theories scale, it provides a natural framework to investigate early universe cosmology. We briefly review some of its cosmological consequences.

1. Introduction

Given the plethora of precise cosmological and astrophysical data and the measurements obtained by particle physics experiments reaching constantly higher energy scales, we are presently in a position to falsify early universe cosmological models. In return, by comparing the theoretical predictions against current data, we are able to constrain the fundamental theories upon which our cosmological models were based. A fruitful such example is the cosmological model based on Noncommutatitive Geometry [1, 2] and the Spectral Action principle, leading to the Noncommutative Spectral Geometry (NCSG), a theoretical framework that can provide [3] a purely geometric explanation for the Standard Model (SM) of strong and electroweak interactions. This model lives by construction in high energy scales, the Grand Unified Theories (GUTs) scale, offering a natural framework to study early universe cosmology [4]-[11]. Hence, instead of postulating a Lagrangian upon which we will build our cosmological model, we will adopt the one dictated by NCSG, with the constraints imposed by NCSG itself, and within the framework of this gravitational theory we will address some cosmological issues. Clearly, this is a solid approach, since the cosmological model is inspired and controlled from a fundamental theory. In return, by comparing the predictions of the model against high energy physics measurements and cosmological data, we will be able to constrain some of the free parameters of NCSG and/or its basic element, namely the choice of the algebra.

In this presentation, we will first briefly review [12] some elements of NCSG and then discuss some cosmological consequences of the model. Our purpose is not to give a full and detailed analysis, but rather highlight some results and open questions and guide the reader through this powerful interplay between mathematics, high energy physics, cosmology and astrophysics [1].

[1] We refer the reader to the contribution by Sakellariadou, Stabile and Vitiello [13], presented in the same meeting, for a discussion on firstly, the physical meaning of the choice of the geometry and its relation to quantisation, and

Let us clarify that the approach we will follow here is *a priori* distinctive from the noncommutative approach based upon $[x^i, x^j] = i\theta^{ij}$, where θ^{ij} is an anti-symmetric real $d \times d$ matrix, used to implement fuzziness of space-time (with d the space-time dimensionality). One finds in the literature that noncommutative space is often of Moyal type, involving noncommutative tori or Moyal planes. Note however that the Euclidean version of Moyal noncommutative field theory is compatible with the spectral triples formulation of noncommutative geometry. This statement holds for compact noncommutative spaces, while it has been argued [15] that the compactness restriction is merely a technical one and noncompact noncommutative spin geometry can be built, implying that the classical background of noncommutative field theory can be recast in the spectral triple approach developed by Connes to describe noncommutative spaces.

2. Elements of noncommutative spectral geometry

It is reasonable to argue that the notion of geometry, as we are familiar with, loses its meaning at very high energy scales, namely near and above Planck scale. The simple classical picture and the notion of a continuous space should cease to be valid as quantum gravity effects turn on. Thus, according to one such school of thought, one may argue that at sufficiently high energy scales, spacetime becomes discrete and coordinates no longer commute. The noncommutative spacetime can be thus seen as a quantum effect of gravity, an approach that may shed some light on the regularisation of quantum field theory. One can, to a first approximation, consider the simplest class of noncommutative spaces (almost commutative), which are not incompatible with low energy physics (namely, today's physics) and study their consistency with experimental and observational data within the realm of high energy physics and cosmology, respectively. If such a programme passes successfully this nontrivial test, as a second step, one should attempt to construct less trivial noncommutative spaces whose limit is this simple but successful case studied first. All current studies remain at present within the first step of this promising programme.

The choice of the noncommutative space, followed by Alain Connes and his collaborators, was such that at low energy scales one recovers the SM action. There is an intrinsic difference between this approach and other ones that attempt to capture the effects of quantum gravity, which can no longer be switched off once we reach Planck energy scale. More precisely, in the NCSG approach one does not postulate the physics at very high energy scales, but instead one is guided by low energy physics. Hence, within NCSG the SM is considered as a phenomenological model which dictates the geometry of space-time so that the Maxwell-Dirac action functional leads to the SM action. To be more specific, in the framework of NCSG we are following here, gravity and the SM fields are put together into matter and geometry on a noncommutative space made from the product of a four-dimensional commutative manifold by a noncommutative internal space. Combining noncommutative geometry with the spectral action principle, and choosing the smallest finite dimensional algebra that can account for the SM particles, Connes and his collaborators have obtained a purely geometric explanation for the SM Lagrangian coupled to gravity. Thus, the model we will briefly present here, has been tailored in order to give the SM of particle physics. This implies that if this model leads to small discrepancies with respect to the data, one could then first examine a larger algebra (that could accommodate particles beyond the SM sector), before considering large deviations from commutative spaces.

Noncommutative spectral geometry is composed by a two-sheeted space, made from the product of a smooth four-dimensional manifold $\mathcal{M}$ (with a fixed spin structure), by a discrete noncommutative space $\mathcal{F}$ composed by only two points. The internal space $\mathcal{F}$ has dimension 6 to allow fermions to be simultaneously Weyl and chiral, while it is discrete to avoid the infinite tower of massive particles that are produced in string theory. The noncommutative nature of $\mathcal{F}$ is given

secondly, the relation of NCSG to the gauge structure of the theory and to dissipation, summarising the results of Ref. [14].

by a spectral triple, introduced by Connes as an extension of the notion of Riemannian manifold to noncommutative geometry. The real spectral triple is given by $(\mathcal{A}_\mathcal{F}, \mathcal{H}_\mathcal{F}, D_\mathcal{F})$. The algebra $\mathcal{A}_\mathcal{F} = C^\infty(\mathcal{M})$ of smooth functions on $\mathcal{M}$ is an involution of operators on the finite-dimensional Hilbert space $\mathcal{H}_\mathcal{F}$ of Euclidean fermions; it is essentially the algebra of coordinates. The operator $D_\mathcal{F}$ is the Dirac operator $\partial\!\!\!/_\mathcal{M} = \sqrt{-1}\gamma^\mu\nabla^s_\mu$ on the spin Riemannian manifold $\mathcal{M}$. It is such that $J_\mathcal{F}\mathcal{D}_\mathcal{F} = \epsilon'\mathcal{D}_\mathcal{F}J_\mathcal{F}$, where $J_\mathcal{F}$ is an anti-linear isometry of the finite dimensional Hilbert space, with the properties $J^2_\mathcal{F} = \epsilon$, $J_\mathcal{F}\gamma_\mathcal{F} = \epsilon''\gamma_\mathcal{F}J_\mathcal{F}$, with $\gamma_\mathcal{F}$ the chirality operator and $\epsilon, \epsilon', \epsilon'' \in \{\pm 1\}$. The operator $D_\mathcal{F}$ is a linear self-adjoint unbounded operator and corresponds to the inverse of the Euclidean propagator of fermions; it is given by the Yukawa coupling matrix which encodes the masses of the elementary fermions and the Kobayashi–Maskawa mixing parameters.

The spectral geometry is then given by the product rules:

$$\mathcal{A} = C^\infty(\mathcal{M}) \oplus \mathcal{A}_\mathcal{F} \ , \ \ \mathcal{H} = L^2(\mathcal{M}, S) \oplus \mathcal{H}_\mathcal{F} \ , \ \ \mathcal{D} = \mathcal{D}_\mathcal{M} \oplus 1 + \gamma_5 \oplus \mathcal{D}_\mathcal{F} \ ,$$

where $L^2(\mathcal{M}, S)$ is the Hilbert space of L^2 spinors and $\mathcal{D}_\mathcal{M}$ is the Dirac operator of the Levi-Civita spin connection on the four-dimensional manifold $\mathcal{M}$. Note that the chirality operator is $\gamma = \gamma_5 \oplus \gamma_\mathcal{F}$ and the anti-unitary operator on the complex Hilbert space is $J = J_\mathcal{M} \oplus J_\mathcal{F}$, with $J_\mathcal{M}$ the charge conjugation.

Since in the following we only consider the noncommutative discrete space $\mathcal{F}$, we omit the subscript $_\mathcal{F}$ to keep the notation lighter.

Assuming the algebra $\mathcal{A}$ to be symplectic-unitary, it reads [17]

$$\mathcal{A} = M_a(\mathbb{H}) \oplus M_k(\mathbb{C}) \ , \tag{1}$$

with $k = 2a$; $\mathbb{H}$ is the algebra of quaternions, which encodes the noncommutativity of the manifold. The first possible value for k is 2, corresponding to a Hilbert space of four fermions. This choice is however ruled out from the existence of quarks. The next possible value is $k = 4$ leading to the correct number of $k^2 = 16$ fermions in each of the three generations; the number of generations is a physical input in the theory. The model developed by Connes and his collaborators is the minimal one ($k = 4$) that can account for the Standard Model.

To obtain the NCSG action we apply the spectral action principle to the product geometry $\mathcal{M} \times \mathcal{F}$. Thus, the bare bosonic Euclidean action is simply

$$\mathrm{Tr}(f(D_\mathcal{A}/\Lambda)) \ , \tag{2}$$

where $D_\mathcal{A} = D + \mathcal{A} + \epsilon' J\mathcal{A}J^{-1}$ are uni-modular inner fluctuations, f is a cutoff function and Λ fixes the energy scale. This action can be seen *à la* Wilson as the bare action at the mass scale Λ. The fermionic term can be included in the action functional by adding $(1/2)\langle J\psi, D\psi\rangle$, where J is the real structure on the spectral triple and ψ is a spinor in the Hilbert space $\mathcal{H}$ of the quarks and leptons.

Using heat kernel methods, the trace $\mathrm{Tr}(f(\mathcal{D}_A/\Lambda))$ can be written in terms of the geometrical Seeley-de Witt coefficients a_n, which are known for any second order elliptic differential operator, as $\sum_{n=0}^\infty F_{4-n}\Lambda^{4-n}a_n$, where the function F is defined such that $F(\mathcal{D}^2_A) = f(\mathcal{D}_A)$. To be more precise, the bosonic part of the spectral action can be expanded in powers of Λ in the form [18, 19]

$$\mathrm{Tr}\left(f\left(\frac{\mathcal{D}_A}{\Lambda}\right)\right) \sim \sum_{k\in \mathrm{DimSp}} f_k\Lambda^k {\int\!\!\!\!\!\!-} |\mathcal{D}_A|^{-k} + f(0)\zeta_{\mathcal{D}_A(0)} + \mathcal{O}(1) \ , \tag{3}$$

where f_k are the momenta of the smooth even test (cutoff) function which decays fast at infinity:

$$\begin{aligned} f_0 &\equiv f(0) \ , \\ f_k &\equiv \int_0^\infty f(u)u^{k-1}\mathrm{d}u \ , \quad \text{for} \ \ k > 0 \ , \\ \mathrm{f}_{-2k} &= (-1)^k\frac{k!}{(2k)!}f^{(2k)}(0) \ . \end{aligned}$$

The noncommutative integration is defined in terms of residues of zeta functions $\zeta_{\mathcal{D}_A}(s) = \mathrm{Tr}(|\mathcal{D}_A|^{-s})$ at poles of the zeta function and the sum is over points in the dimension spectrum of the spectral triple. For a four-dimensional Riemannian geometry, the trace $\mathrm{Tr}(f(\mathcal{D}_A/\Lambda))$ can be expressed perturbatively as [20]-[21]

$$\mathrm{Tr}(f(\mathcal{D}_A/\Lambda)) \sim 2\Lambda^4 f_4 a_0 + 2\Lambda^2 f_2 a_2 + f_0 a_4 + \cdots + \Lambda^{-2k} f_{-2k} a_{4+2k} + \cdots \ . \tag{4}$$

Since the Taylor expansion of the cutoff function vanishes at zero, the asymptotic expansion of Eq. (4) reduces to

$$\mathrm{Tr}(f(D/\Lambda)) \sim 2\Lambda^4 f_4 a_0 + 2\Lambda^2 f_2 a_2 + f_0 a_4 \ . \tag{5}$$

Hence, the cutoff function f plays a rôle only through its three momenta f_0, f_2, f_4, which are three real parameters, related to the coupling constants at unification, the gravitational constant, and the cosmological constant, respectively. More precisely, the first term in Eq. (5) which is in in Λ^4 gives a cosmological term, the second one which is in Λ^2 gives the Einstein-Hilbert action functional, and the third one which is Λ-independent term yields the Yang-Mills action for the gauge fields corresponding to the internal degrees of freedom of the metric.

As it has been shown in Ref. [3], the NCSG summarised above, offers a purely geometric approach to the SM of particle physics, where the fermions provide the Hilbert space of a spectral triple for the algebra and the bosons are obtained through inner fluctuations of the Dirac operator of the product $\mathcal{M} \times \mathcal{F}$ geometry. More precisely, the computation of the asymptotic expression for the spectral action functional results to the full Lagrangian for the Standard Model minimally coupled to gravity, with neutrino mixing and Majorana mass terms; supersymmetric extensions have been also considered [22].

3. Phenomenological consequences

We will briefly review the phenomenological consequences of the NCSG model. We assume that the function f is well approximated by the cutoff function and ignore higher order terms. Normalisation of the kinetic terms implies

$$\frac{g_3^2 f_0}{2\pi^2} = \frac{1}{4} \quad \text{and} \quad g_3^2 = g_2^2 = \frac{5}{3} g_1^2 \ ,$$

which coincide with the relations obtained in GUTs, while

$$\sin^2\theta_{\mathrm{W}} = \frac{3}{8} \ , \tag{6}$$

which is also found for the SU(5) and SO(10) groups. Assuming the validity of the big desert hypothesis, the running of the couplings $\alpha_i = g_i^2/(4\pi)$ with $i = 1, 2, 3$ is obtained via the RGE.

Considering only one-loop corrections, hence the β-functions are $\beta_{g_i} = (4\pi)^{-2} b_i g_i^3$ with $i = 1, 2, 3$ and $b = (41/6, -19/6, -7)$, it was shown [3] that the gauge couplings and the Newton constant do not meet at a point, the error being within just few percent. The lack of a unification scale implies that the big desert hypothesis is only approximately valid and new physics are expected between the unification and today's energy scales. Phrasing it differently, the lack of a unique unification scale implies that even though the function f can be approximated by the cutoff function, there exist small deviations. On the positive side, the NCSG model leads to the correct representations of the fermions with respect to the gauge group of the SM, the gauge bosons appear as inner fluctuations along the continuous directions while the Higgs doublet appears as part of the inner fluctuations of the metric. Spontaneous Symmetry Breaking mechanism for the electroweak symmetry arises naturally with the negative mass term without any tuning. The see-saw mechanism is obtained and the 16 fundamental fermions (the number of states on the Hilbert space) are recovered. At unification scale $\Lambda \sim 1.1 \times 10^{17}$ GeV, with $g \sim 0.517$, the Renormalisation Group Equations (RGE) lead to a top quark mass of ~ 179 GeV.

However, in zeroth order, the model predicts a heavy Higgs mass of ~ 170 GeV, which is ruled out by current experimental data. One has though to keep in mind that the Higgs mass is sensitive to the value of unification scale, as well as to deviations of the spectral function from the cutoff function we have considered. Hence, the Higgs mass should be determined by considering higher order corrections and incorporating them to the appropriate RGE. One may argue that the reason for which the top quark mass is consistent with experimental data while the predicted Higgs mass is ruled out, is simply because the top quark mass is less sensitive to the ambiguities of the unification scale than the Higgs mass is. This may indeed be the case since the bosonic part of the action is given by an infinite expansion assuming convergence of higher order terms.

In a more drastic modification of the NCSG model at hand, one may consider a bigger algebra than the one considered so far, which was chosen so that it leads to the Standard Model particles. Thus, one may argue that the discrepancy between the predicted Higgs mass and the experimental constraints may be resolved by considering models beyond the SM. One may construct [23] such a model based on a minimal spectral triple which contains the SM particles, but it has also new vector-like fermions and a new U(1) gauge subgroup. In the model presented in Ref. [23] it also appears a new complex scalar field that couples to the right-handed neutrino, the new fermions and the standard Higgs field. For the case of a nonzero vacuum expectation value the new scalar and the Higgs field mix and the mass eigenstates may consist of a light scalar particle with $m_{H_1} \sim 120$ GeV and a heavy particle with $m_{H_2} \leq 170$ GeV.

More recently, Chamseddine and Connes have argued [24] that including a real scalar singlet, strongly coupled to the Higgs doublet, they can accommodate a Higgs mass of order 125 GeV. This singlet field is associated with the Majorana mass of the right-handed neutrino. Even though this singlet is responsible for the breakdown of the symmetry of the discrete space, it has been neglected in the original calculations [3] of the phenomenological consequences of the spectral action. Note that as we have shown in Ref. [7] this scalar singlet cannot play the rôle of the inflaton field and, at the same time, provide the seeds of temperature anisotropies.

Nevertheless, it is fair to say that even the simplest version of NCSG predicts a Higgs mass of the correct order of magnitude, which is certainly a nontrivial result. Finally, let me note that this approach to unification does not provide any explanation of the number of generations, nor leads to constraints on the values of the Yukawa couplings.

4. Cosmological consequences

Since the NCSG gravitational model lives by construction are high energy scales, it provides a natural framework to investigate early universe cosmology [4]-[11]. We will review some of these consequences in what follows.

Within NCSG it is natural to obtain the bosonic action in Euclidean signature. It reads [3]

$$\mathcal{S}^{\rm E} = \int \Big(\frac{1}{2\kappa_0^2} R + \alpha_0 C_{\mu\nu\rho\sigma} C^{\mu\nu\rho\sigma} + \gamma_0 + \tau_0 R^\star R^\star + \frac{1}{4} G^i_{\mu\nu} G^{\mu\nu i} + \frac{1}{4} F^\alpha_{\mu\nu} F^{\mu\nu\alpha} + \frac{1}{4} B^{\mu\nu} B_{\mu\nu} + \frac{1}{2} |D_\mu \mathbf{H}|^2 - \mu_0^2 |\mathbf{H}|^2 - \xi_0 R |\mathbf{H}|^2 + \lambda_0 |\mathbf{H}|^4 \Big) \sqrt{g}\, d^4 x \ , \tag{7}$$

where

$$\kappa_0^2 = \frac{12\pi^2}{96 f_2\Lambda^2 - f_0\mathfrak{c}} \quad , \quad \alpha_0 = -\frac{3f_0}{10\pi^2} \quad ,$$
$$\gamma_0 = \frac{1}{\pi^2}\left(48 f_4\Lambda^4 - f_2\Lambda^2\mathfrak{c} + \frac{f_0}{4}\mathfrak{d}\right) \quad , \quad \tau_0 = \frac{11 f_0}{60\pi^2} \quad ,$$
$$\mu_0^2 = 2\Lambda^2\frac{f_2}{f_0} - \frac{\mathfrak{e}}{\mathfrak{a}} \quad , \quad \xi_0 = \frac{1}{12} \quad , \quad \lambda_0 = \frac{\pi^2\mathfrak{b}}{2f_0\mathfrak{a}^2} \; ; \tag{8}$$

$\mathbf{H}$ is a rescaling $\mathbf{H} = (\sqrt{af_0}/\pi)\phi$ of the Higgs field ϕ to normalise the kinetic energy. The geometric parameters $\mathfrak{a}, \mathfrak{b}, \mathfrak{c}, \mathfrak{d}, \mathfrak{e}$ correspond to the Yukawa parameters (which run with the RGE) of the particle physics model and the Majorana terms for the right-handed neutrinos. Extrapolations to lower energy scales are possible through RGE analysis, however at low (today's) energy scales nonperturbative effects can no longer be neglected. Thus, any results based on the asymptotic expansion and on RGE analysis can only be valid for early universe cosmology. Note also that the relations in Eq. (8) are tied to the scale of the asymptotic expansion; there is no reason for these constraints (boundary conditions) to hold at scales below the unification scale Λ.

To apply the NCSG action in cosmology we must express it in Lorentzian signature. Assuming that one can perform a Wick rotation in imaginary time, the Lorentzian version of the gravitational part of the asymptotic expression for the bosonic sector of the NCSG action reads [3]

$$\mathcal{S}_{\rm grav}^{\rm L} = \int \left(\frac{1}{2\kappa_0^2}R + \alpha_0 C_{\mu\nu\rho\sigma}C^{\mu\nu\rho\sigma} + \tau_0 R^\star R^\star \xi_0 R|\mathbf{H}|^2\right)\sqrt{-g}\; d^4x \; , \tag{9}$$

leading to the equations of motion [4]:

$$R^{\mu\nu} - \frac{1}{2}g^{\mu\nu}R + \frac{1}{B^2}\delta_{\rm cc}\left[2C^{\mu\lambda\nu\kappa}_{\;;\lambda;\kappa} + C^{\mu\lambda\nu\kappa}R_{\lambda\kappa}\right] = \kappa_0^2\delta_{\rm cc}T^{\mu\nu}_{\rm matter} \; ,$$

where $B^2 \equiv -(4\kappa_0^2\alpha_0)^{-1}$. The nonminimal coupling between the Higgs field and the Ricci curvature scalar is captured by the parameter $\delta_{\rm cc}$, defined by $\delta_{\rm cc} \equiv [1 - 2\kappa_0^2\xi_0\mathbf{H}^2]^{-1}$.

In the low energy weak curvature regime, the nonminimal coupling between the background geometry and the Higgs field can be neglected, leading to $\delta_{\rm cc} = 1$. In this regime, noncommutative corrections do not occur at the level of a Friedmann-Lemaître-Robertson-Walker (FRLW) background, since in this case the modified Friedmann equation reduces to its standard form [4]. Any modifications to the background equation will be apparent at leading order only for anisotropic and inhomogeneous models. For instance, consider the Bianchi type-V model, for which the space-time metric, in Cartesian coordinates, reads

$$g_{\mu\nu} = {\rm diag}\left[-1, \{a_1(t)\}^2 e^{-2nz}, \{a_2(t)\}^2 e^{-2nz}, \{a_3(t)\}^2\right] \; ; \tag{10}$$

$a_i(t)$ with $i = 1, 2, 3$ are arbitrary functions, denoting the scale factors in the three spatial coordinates and n is an integer. Defining $A_i(t) = {\rm ln} a_i(t)$ with $i = 1, 2, 3$, the modified Friedmann equation reads [4]:

$$\begin{aligned}
\kappa_0^2 T_{00} = \\
-\dot{A}_3\left(\dot{A}_1+\dot{A}_2\right)-n^2e^{-2A_3}\left(\dot{A}_1\dot{A}_2-3\right) \\
+\frac{8\alpha_0\kappa_0^2n^2}{3}e^{-2A_3}\left[5\left(\dot{A}_1\right)^2+5\left(\dot{A}_2\right)^2-\left(\dot{A}_3\right)^2-\dot{A}_1\dot{A}_2-\dot{A}_2\dot{A}_3-\dot{A}_3\dot{A}_1-\ddot{A}_1-\ddot{A}_2-\ddot{A}_3+3\right] \\
-\frac{4\alpha_0\kappa_0^2}{3}\sum_i\Bigg\{\dot{A}_1\dot{A}_2\dot{A}_3\dot{A}_i+\dot{A}_i\dot{A}_{i+1}\left(\left(\dot{A}_i-\dot{A}_{i+1}\right)^2-\dot{A}_i\dot{A}_{i+1}\right) \\
+\left(\ddot{A}_i+\left(\dot{A}_i\right)^2\right)\left[-\ddot{A}_i-\left(\dot{A}_i\right)^2+\frac{1}{2}\left(\ddot{A}_{i+1}+\ddot{A}_{i+2}\right)+\frac{1}{2}\left(\left(\dot{A}_{i+1}\right)^2+\left(\dot{A}_{i+2}\right)^2\right)\right] \\
+\left[\dddot{A}_i+3\dot{A}_i\ddot{A}_i-\left(\ddot{A}_i+\left(\dot{A}_i\right)^2\right)\left(\dot{A}_i-\dot{A}_{i+1}-\dot{A}_{i+2}\right)\right]\left[2\dot{A}_i-\dot{A}_{i+1}-\dot{A}_{i+2}\right]\Bigg\} \qquad (11)
\end{aligned}$$

with $i = 1, 2, 3$; the t-dependence of the terms has been omitted for simplicity. Any term containing α_0 in Eq. (11) encodes a modification from the conventional case. The correction terms can be divided into two types. The first one contains the terms in braces in Eq. (11), which are fourth order in time derivatives and hence, for the slowly varying functions that are usually considered in cosmology, these corrections can be neglected. The second type, which appears in the third line in Eq. (11), occurs at the same order as the standard Einstein-Hilbert terms. However, since this correction term is proportional to n^2, it vanishes for homogeneous versions of Bianchi type-V. In conclusion, the corrections to Einstein's equations can only be important for inhomogeneous and anisotropic space-times [4].

The coupling between the Higgs field and the background geometry can no longer be neglected once we reach energies of the Higgs scale. In this case, the nonminimal coupling of Higgs field to curvature leads to corrections to Einstein's equations even for homogeneous and isotropic cosmological models. To keep the analysis simpler, let us neglect the conformal term in Eq. (10), so that the equations of motion are [4]

$$R^{\mu\nu} - \frac{1}{2}g^{\mu\nu}R = \kappa_0^2\left[\frac{1}{1-\kappa_0^2|\mathbf{H}|^2/6}\right]T^{\mu\nu}_{\rm matter}\ . \qquad (12)$$

Thus, $|\mathbf{H}|$ plays the rôle of an effective gravitational constant [4]. Alternatively, the nonminimal coupling of the Higgs field to the curvature can be seen, for static geometries, as leading to an increase of the Higgs mass [4]. In the presence of this nonminimal coupling, the cosmological model built upon the NCSG action share some similarities with chameleon gravity and dilatonic cosmology [4].

The nonminimal coupling between the Higgs field and the Ricci curvature may turn out to be crucial in early universe cosmology [5, 7]. Such a coupling has been introduced *ad hoc* in the literature, in an attempt to drive inflation through the Higgs field. However, the value of this coupling should be dictated by an underlying theory and cannot be tuned by hand for a purely phenomenological convenience. Actually, even if classically the coupling between the Higgs field and the Ricci curvature could be set equal to zero, a nonminimal coupling will be induced once quantum corrections in the classical field theory are taken into account. A large coupling between the Higgs field and the background geometry is plagued by pathologies [25, 26], but this is not the case for a small coupling, as in the NCSG case. It is therefore worth investigating whether the Higgs field could play the rôle of the inflaton within the NCSG context.

In a FLRW metric, the Gravity-Higgs sector of the asymptotic expansion of the spectral action, in Lorentzian signature, reads

$$S^{\rm L}_{\rm GH} = \int \Big[\frac{1-2\kappa_0^2\xi_0 H^2}{2\kappa_0^2}R - \frac{1}{2}(\nabla H)^2 - V(H)\Big]\sqrt{-g}\, d^4x \ , \tag{13}$$

where

$$V(H) = \lambda_0 H^4 - \mu_0^2 H^2 \ , \tag{14}$$

with μ_0 and λ_0 subject to radiative corrections as functions of energy. For large enough values of the Higgs field, the renormalised value of μ_0 and λ_0 must be calculated. At high energies the mass term in Eq. (14) is sub-dominant and can be neglected. It has been shown [7] that for each value of the top quark mass there is a value of the Higgs mass where the effective potential is about to develop a metastable minimum at large values of the Higgs field and the Higgs potential is locally flattened. Calculating [7] the renormalisation of the Higgs self-coupling up to two-loops, we have constructed an effective potential which fits the renormalisation group improved potential around the flat region. The analytic fit to the Higgs potential around the minimum of the potential is [7]:

$$V^{\rm eff} = \lambda_0^{\rm eff}(H)H^4 = [a\ln^2(b\kappa H) + c]H^4 \ , \tag{15}$$

where the parameters a, b are related to the low energy values of top quark mass $m_{\rm t}$ as [7]

$$\begin{aligned} a(m_{\rm t}) &= 4.04704\times 10^{-3} - 4.41909\times 10^{-5}\left(\frac{m_{\rm t}}{\rm GeV}\right) + 1.24732\times 10^{-7}\left(\frac{m_{\rm t}}{\rm GeV}\right)^2 \ , \\ b(m_{\rm t}) &= \exp\Big[-0.979261\left(\frac{m_{\rm t}}{\rm GeV} - 172.051\right)\Big] \ . \end{aligned} \tag{16}$$

The third parameter, c, encodes the appearance of an extremum and depends on the values for top quark mass and Higgs mass. The region where the potential is flat is narrow, thus to achieve a long enough period of quasi-exponential expansion, the slow-roll parameters, ϵ and η, must be slow enough to allow sufficient number of e-folds. In addition, the amplitude of density perturbations $\Delta^2_{\mathcal{R}}$ in the Cosmic Microwave Background (CMB) must be within the window allowed from the most recent experimental data. More precisely, inflation predicts that at horizon crossing (denoted by stars), the amplitude of density perturbations is related to the inflaton potential through $(V_*/\epsilon_*)^{1/4} = 2\sqrt{3\pi}\ m_{\rm Pl}\ \Delta^{1/2}_{\mathcal{R}}$, where $\epsilon_* \leq 1$. Its value, as measured by WMAP7 [27], requires $(V_*/\epsilon_*)^{1/4} = (2.75\pm 0.30)\times 10^{-2}\ m_{\rm Pl}$, where $m_{\rm Pl}$ is the Planck mass.

Performing a systematic search in the parameter space using a Monte-Carlo chain, we have shown [7] that even though slow-roll inflation can be realised, the ratio of perturbation amplitudes is too large for any experimentally allowed values for the masses of the top quark and the Higgs boson. Note that running of the gravitational constant and corrections by considering the more appropriate de Sitter space-time, instead of the Minkowski geometry employed here, do not improve substantially the realisation of a successful slow-roll inflationary era [7].

Let us now proceed with the study of linear perturbations around a Minkowski background metric, which will allow us to constrain one of the three free parameters of the theory, namely the moment f_0 which is related to the coupling constants at unification. Note that we have to go beyond an FLRW space-time, since for a homogeneous and isotropic geometry the Weyl tensor vanishes, implying that the NCSG corrections to the Einstein equation vanish [4], rending difficult to restrict B via cosmology or solar-system tests. It is worth noting that imposing a lower limit on B would imply an upper limit to the moment f_0, and thus restrict particle physics at unification. To impose an upper limit to the moment f_0, we will study the energy lost to gravitational radiation by orbiting binaries [8, 9]. Considering linear perturbations around a Minkowski background metric, the equations of motion read [9]

$$\left(\Box - B^2\right)\Box h^{\mu\nu} = B^2\frac{16\pi G}{c^4}T^{\mu\nu}_{\rm matter} \ , \tag{17}$$

where $T^{\mu\nu}_{\rm matter}$ is taken to lowest order in $h^{\mu\nu}$. Since B plays the rôle of a mass, it must be real and positive, thus α_0 must be negative for Minkowski space to be a stable vacuum of the theory.

Consider the energy lost to gravitational radiation by orbiting binaries. In the far field limit, $|\mathbf{r}| \approx |\mathbf{r} - \mathbf{r}'|$ ($\mathbf{r}$ and $\mathbf{r}'$ denote the locations of observer and emitter, respectively), the spatial components of the general first order solution for a perturbation against a Minkowski background read [9]

$$h^{ik}(\mathbf{r},t) \approx \frac{2GB}{3c^4}\int_{-\infty}^{t-\frac{1}{c}|\mathbf{r}|} \frac{dt'}{\sqrt{c^2(t-t')^2 - |\mathbf{r}|^2}} \mathcal{J}_1\left(B\sqrt{c^2(t-t')^2 - |\mathbf{r}|^2}\right)\ddot{D}^{ik}(t')\ ; \tag{18}$$

D^{ik} is the quadrupole moment, defined as $D^{ik}(t) \equiv \frac{3}{c^2}\int x^i x^k T^{00}(\mathbf{r},t)\ d\mathbf{r}$, and $\mathcal{J}_1$ a Bessel function of the first kind. In the $B \to \infty$ limit, the NCSG gravitational theory reduces to the standard General Relativity (GR), while for finite B gravity waves radiation contains massive and massless modes, which are both sourced from the quadrupole moment of the system.

Considering a binary pair of masses m_1, m_2 in circular (for simplicity) orbit in the (xy)-plane, the rate of energy loss is

$$-\frac{\mathrm{d}\mathcal{E}}{\mathrm{d}t} \approx \frac{c^2}{20G}|\mathbf{r}|^2 \dot{h}_{ij}\dot{h}^{ij}\ , \tag{19}$$

with [9]

$$\dot{h}^{ij}\dot{h}_{ij} = \frac{128\mu^2|\rho|^4\omega^6G^2B^2}{c^8} \times \left[f_{\rm c}^2\left(B|\mathbf{r}|, \frac{2\omega}{Bc}\right) + f_{\rm s}^2\left(B|\mathbf{r}|, \frac{2\omega}{Bc}\right)\right]\ , \tag{20}$$

$$f_{\rm s}(x,z) \equiv \int_0^\infty \frac{ds}{\sqrt{s^2+x^2}}\mathcal{J}_1(s)\sin\left(z\sqrt{s^2+x^2}\right)\ , \tag{21}$$

$$f_{\rm c}(x,z) \equiv \int_0^\infty \frac{ds}{\sqrt{s^2+x^2}}\mathcal{J}_1(s)\cos\left(z\sqrt{s^2+x^2}\right)\ . \tag{22}$$

The orbital frequency ω, defined in terms of the magnitude $|\rho|$ of the separation vector between the two bodies, is constant and equal to $\omega = |\rho|^{-3/2}\sqrt{G(m_1+m_2)}$.

The integrals in Eqs. (21), (22), exhibit a strong resonance behavior at $z = 1$, which corresponds to the critical frequency [9]

$$2\omega_{\rm c} = Bc\ , \tag{23}$$

around which strong deviations from the GR results are expected. This maximum frequency results from the natural length scale, given by B^{-1}, at which NCSG effects become dominant.

One can find in the literature several binary pulsars for which the rate of change of the orbital frequency is well-known and the predictions of GR agree with the data to a high accuracy. Since the magnitude of the NCSG deviations from GR must be less than the allowed uncertainty in the data, we are able to constrain [8] B, namely

$$B > 7.55 \times 10^{-13}\ \mathrm{m}^{-1}\ . \tag{24}$$

This constraint is not too strong but since it is obtained from systems with high orbital frequencies, future observations of rapidly orbiting binaries, relatively close to the Earth, could improve it by many orders of magnitude.

5. Conclusions

We have briefly described the prescription of Connes and collaborators in order to recover the action of the SM of particle physics from purely gravitational considerations. In this approach one recovers the Einstein plus Yang-Mills and Weyl actions including the spin-1 bosons and the part induced by the spin-0 Higgs fields. Thus, gravity and the electroweak and strong forces are described as purely gravitational forces on a unified noncommutative spacetime. The trick consists of employing spectral triples, consisting of an algebra (that is equivalent of a topological space), a Dirac operator (that corresponds to the metric on the topological space) and a Hilbert space of Dirac 4-spinors, on which the algebra is represented and on which the Dirac operator acts. In his approach, Connes combined spacetime with an internal space, composed by only 2 points, a construction that can be seen as a discrete Kaluza-Klein space where the product manifold of spacetime with extra spatial dimensions is replaced by the product of spacetime with discrete spaces represented by matrices.

The phenomenological consequences of this gravitational theory are in a very good agreement with current data, however the predicted Higgs mass seems to be ruled out. The reason for this discrepancy may be due to the ambiguities of the unification scale, the approximation of the spectral function by a cutoff one, or the assumption of convergence of the higher order terms in the infinite expansion of the bosonic part of the NCSG action. A more drastic approach is to consider larger algebras which predict particles beyond the SM sector. Even though these directions deserve further investigations, it is clear that the NCSG approach offers a beautiful explanation of the SM, the most successful particle physics model we have at hand, providing in addition a geometric explanation for the Higgs field which is otherwise introduced by hand.

The NCSG action offers a natural framework to build a cosmological model, which may allow us to explain some open questions in cosmology, as for instance through investigations on the rôle of scalar fields which arise naturally in NCGS and they do not have to be introduced by hand. Moreover, comparison of NCSG predictions against astrophysical data may allow us to constrain the free parameters of the theory.

Even though, as we have shown, the Higgs field can play the rôle of the inflaton field, in terms of providing a slow-roll period of fast expansion, it cannot provide the seed of perturbations leading to the observed CMB anisotropies and the large-scale structure. If one wants to decouple the inflaton from the curvaton field, then one has to provide such a field with the appropriate potential. In this sense, one may investigate the rôle of other scalar fields within the NCSG model. It is particularly encouraging that by studying gravity waves propagation we were able to constrain one of the free parameters of the theory, namely the one related to the coupling constants at unification.

In conclusion, NCSG offers a fruitful interplay between mathematics, gravitational theories, particle physics and cosmology, tracing another approach to the goal of unification.

References

[1] Connes A 1994 *Noncommutative Geometry* (New York: Academic Press)
[2] Connes A and Marcolli M 2008 *Noncommutative Geometry, Quantum Fields and Motives* (India: Hindustan Book Agency)
[3] Chamseddine A H, Connes A and Marcolli M 2007 *Adv. Theor. Math. Phys.* **11** 991
[4] Nelson W and Sakellariadou M 2010 *Phys. Rev.* D **81** 085038
[5] Nelson W and Sakellariadou M 2009 *Phys. Lett.* B **680** 263
[6] Marcolli M and Pierpaoli E *Preprint* arXiv:hep-th/0908.3683
[7] Buck M, Fairbairn M and Sakellariadou M 2010 *Phys. Rev.* D **82** 043509
[8] Nelson W, Ochoa J and Sakellariadou M 2010 *Phys. Rev. Lett.* **105** 101602
[9] Nelson W, Ochoa J and Sakellariadou M 2010 *Phys. Rev.* D **82** 085021
[10] Sakellariadou M 2011 *Int. J. Mod. Phys.* D **20** 785
[11] Sakellariadou M 2010 *PoS CNCFG* **2010** 028 (*Preprint* arXiv:1101.2174 [hep-th])
[12] Sakellariadou M 2011 *PoS CORFU* **2011** 053 (*Preprint* [arXiv:1204.5772 [hep-th])

[13] Sakellariadou M, Stabile A and Vitiello G *Preprint* arXiv:1301.2563 [hep-th]
[14] Sakellariadou M, Stabile A and Vitiello G 2011 *Phys. Rev.* D **84** 045026
[15] Gayral V, Gracia-Bondia J M, Iochum B, Schucker T and Varilly J C 2004 *Commun. Math. Phys.* **246** 569 (*Preprint* arXiv:hep-th/0307241)
[16] Connes A 1995 *Lett. Math. Phys.* **34** 203.
[17] Chamseddine A H and Connes A 2007 *Phys. Rev. Lett.* **99** 191601
[18] Chamseddine A H and Connes A 1996 *Phys. Rev. Lett.* **77** 4868
[19] Chamseddine A H and Connes A 1977 *Comm. Math. Phys.* **186** 731
[20] Chamseddine A H and Connes A 2006 *J. Math. Phys.* **47** 063504
[21] Chamseddine A H and Connes A 2010 *Comm. Math. Phys.* **293** 867
[22] van den Broek T and van Suijlekom W 2011 *Commun. Math. Phys.* **303** 149 (*Preprint* arXiv:1003.3788 [hep-th])
[23] Stephan C A 2009 *Phys. Rev. D* **80** 065007 (*Preprint* arXiv:0905.0997 [hep-ph])
[24] Chamseddine A H and Connes A 2012 *JHEP* **1209** 104 (*Preprint* arXiv:208.1030 [hep-ph])
[25] Burgess C P, Lee H M, Trott M 2009 *JHEP* **0909** 103 (*Preprint* arXiv:0902.4465 [hep-ph])
[26] Barbon J L F and Espinosa J R 2009 *Phys.Rev.D* **79** 081302 (*Preprint* arXiv:0903.0355 [hep-ph])
[27] Bennett C L *et al* 2011 *Astrophys. J. Suppl* **192** 17 (*Preprint* arXiv:1001.4758 [astro-ph.CO])

Noncommutative spectral geometry and the deformed Hopf algebra structure of quantum field theory

Mairi Sakellariadou
Department of Physics, King's College, University of London, Strand WC2R 2LS, London, U.K.
E-mail: mairi.sakellariadou@kcl.ac.uk

Antonio Stabile
Dipartimento di Fisica "E.R. Caianiello" and Gruppo Collegato I.N.F.N., Università di Salerno, I-84100 Salerno, Italy
E-mail: anstabile@gmail.com

Giuseppe Vitiello
Dipartimento di Fisica "E.R. Caianiello" and Gruppo Collegato I.N.F.N., Università di Salerno, I-84100 Salerno, Italy
E-mail: vitiello@sa.infn.it

Abstract. We report the results obtained in the study of Alain Connes noncommutative spectral geometry construction focusing on its essential ingredient of the algebra doubling. We show that such a two-sheeted structure is related with the gauge structure of the theory, its dissipative character and carries in itself the seeds of quantization. From the algebraic point of view, the algebra doubling process has the same structure of the deformed Hops algebra structure which characterizes quantum field theory.

1. Noncommutative spectral geometry and the standard model

One may assume that near the Planck energy scale, the geometry of space-time ceases to have the simple continuous form we are familiar with. At high enough energy scales, quantum gravity effects turn on and they alter space-time. One can thus assume that at high energy scales, space-time becomes discrete and the coordinates do not longer commute. Such an approach could a *priori* be tested by its phenomenological and cosmological consequences.

Combining noncommutative geometry [1, 2] with the spectral action principle, led to Noncommutative Spectral Geometry (NCSG), used by Connes and collaborators [3] in an attempt to provide a purely geometric explanation for the Standard Model (SM) of electroweak and strong interactions. In their approach, the SM is considered as a phenomenological model, which dictates the geometry of space-time so that the Maxwell-Dirac action functional leads to the SM action. The model is constructed to hold at high energy scales, namely at unification scale; to get its low energy consequences which will then be tested against current data, one uses standard renormalization techniques. Since the model lives at high energy scales, it can be used to investigate early universe cosmology [4]-[11] [1]. The purpose of this contribution is twofold: firstly, to investigate the physical meaning of the choice of the *almost* commutative geometry and its relation to quantization [12], and secondly to explore the relation of NCSG with the gauge structure of the theory and with dissipation [12]. We will show that Connes construction is intimately related with the deformed Hopf algebra characterizing quantum field theory (QFT) [13] and therefore the seeds of quantization are built in such a NCSG construction.

We start by summarizing the main ingredients of NCSG, composed by a two-sheeted space, made from the product of a four-dimensional smooth compact Riemannian manifold $\mathcal{M}$ with a fixed spin structure, by a discrete noncommutative space $\mathcal{F}$ composed by only two points. Thus, geometry is specified by the product of a continuous manifold for space-time times an internal geometry for the SM. The noncommutative nature of the discrete space $\mathcal{F}$ is denoted by a spectral triple $(\mathcal{A}, \mathcal{H}, D)$. The algebra $\mathcal{A} = C^\infty(\mathcal{M})$ of smooth functions on $\mathcal{M}$ is an involution of operators on the finite-dimensional Hilbert space $\mathcal{H}$ of Euclidean fermions; it acts on $\mathcal{H}$ by multiplication operators. The operator D is the Dirac operator $\partial\!\!\!/_{\mathcal{M}} = \sqrt{-1}\gamma^\mu\nabla^s_\mu$ on the spin Riemannian manifold $\mathcal{M}$; D is a self-adjoint unbounded operator in $\mathcal{H}$. The space $\mathcal{H}$ is the Hilbert space $L^2(\mathcal{M}, S)$ of square integrable spinors S on $\mathcal{M}$. Thus one obtains a model of pure gravity with an action that depends on the spectrum of the Dirac operator, which provides the (familiar) notion of a metric.

The most important ingredient in this NCSG model is the choice of the algebra constructed within the geometry of the two-sheeted space $\mathcal{M} \times \mathcal{F}$; it captures all information about space. The product geometry is specified by:

$$\mathcal{A} = \mathcal{A}_1 \otimes \mathcal{A}_2 \ , \qquad \mathcal{H} = \mathcal{H}_1 \otimes \mathcal{H}_2 \ , \tag{1}$$

as a consequence of the noncommutative nature of the discrete space $\mathcal{F}$ composed by only two points. In other words, Eq. (1) expresses, at the algebraic level and at the level of the space of the states, the two-sheeted nature of the geometry space $\mathcal{M} \times \mathcal{F}$. Assuming $\mathcal{A}$ is symplectic-unitary, it can be written as [14]

$$\mathcal{A} = M_a(\mathbf{H}) \oplus M_k(\mathbf{C}) \ , \tag{2}$$

with $k = 2a$ and $\mathbf{H}$ being the algebra of quaternions, which encodes the noncommutativity of the manifold. The first possible value for the even number k is 2, corresponding to a Hilbert space of four fermions, but this choice is ruled out from the existence of quarks. The next possible value is $k = 4$ leading to the correct number of $k^2 = 16$ fermions in each of the three generations. Thus, we will consider the minimal choice that can accommodate the physics of the SM; certainly other choices leading to larger algebras which could accommodate particle beyond the SM sector could be possible.

The second basic ingredient of the NCSG is the spectral action principle stating that, within the context of the product $\mathcal{M} \times \mathcal{F}$, the bare bosonic Euclidean action is given by the trace of the heat kernel associated with the square of the noncommutative Dirac operator and is simply

$$\mathrm{Tr}(f(D/\Lambda)) \ , \tag{3}$$

where f is a cut-off function andΛ fixes the energy scale. This action can be seen à *la* Wilson as the bare action at the mass scale Λ. The fermionic term can be included in the action functional by adding $(1/2)\langle J\psi, D\psi\rangle$, where J is the real structure on the spectral triple and ψ is a spinor in the Hilbert space $\mathcal{H}$ of the quarks and leptons.

[1] See the contribution of M. Sakellariadou in the same conference.

Since we are considering a four-dimensional Riemannian geometry, the trace $\mathrm{Tr}(f(D/\Lambda))$ can be expressed perturbatively as [15]-[18]

$$\mathrm{Tr}(f(D/\Lambda)) \sim 2\Lambda^4 f_4 a_0 + 2\Lambda^2 f_2 a_2 + f_0 a_4 + \cdots + \Lambda^{-2k} f_{-2k} a_{4+2k} + \cdots \ , \tag{4}$$

in terms of the geometrical Seeley-deWitt coefficients a_n, known for any second order elliptic differential operator. It is important to note that the smooth even test function f, which decays fast at infinity, appears through its momenta f_k:

$$\begin{aligned} f_0 &\equiv f(0) \ , \\ f_k &\equiv \int_0^\infty f(u) u^{k-1} \mathrm{d}u \ , \quad \text{for} \ \ k > 0 \ , \\ \mathrm{f}_{-2k} &= (-1)^k \frac{k!}{(2k)!} f^{(2k)}(0) \ . \end{aligned}$$

Since the Taylor expansion of the cut-off function vanishes at zero, the asymptotic expansion of Eq. (4) reduces to

$$\mathrm{Tr}(f(D/\Lambda)) \sim 2\Lambda^4 f_4 a_0 + 2\Lambda^2 f_2 a_2 + f_0 a_4 \ . \tag{5}$$

Hence, the cut-off function f plays a rôle only through its three momenta f_0, f_2, f_4, which are three real parameters, related to the coupling constants at unification, the gravitational constant, and the cosmological constant, respectively. More precisely, the first term in Eq. (5) which is in in Λ^4 gives a cosmological term, the second one which is in Λ^2 gives the Einstein-Hilbert action functional, and the third one which is Λ-independent term yields the Yang-Mills action for the gauge fields corresponding to the internal degrees of freedom of the metric.

The NCSG offers a purely geometric approach to the SM of particle physics, where the fermions provide the Hilbert space of a spectral triple for the algebra and the bosons are obtained through inner fluctuations of the Dirac operator of the product $\mathcal{M} \times \mathcal{F}$ geometry. The computation of the asymptotic expression for the spectral action functional results to the full Lagrangian for the Standard Model minimally coupled to gravity, with neutrino mixing and Majorana mass terms. Supersymmetric extensions have been also considered.

In this report we closely follow the presentation of Ref. [12] (see also [19]). The relation between the algebra doubling and the deformed Hopf algebra structure of QFT are discussed in Section 2; dissipation, the gauge structure and quantization in Section 3. Section 4 is devoted to conclusions, where we also mention about the dissipative interference phase arising in the noncommutative plane in the presence of algebra doubling.

2. Noncommutative spectral geometry and quantum field theory

Our first observation is that the doubling of the algebra, $\mathcal{A} \rightarrow \mathcal{A}_1 \otimes \mathcal{A}_2$ acting on the "doubled" space $\mathcal{H} = \mathcal{H}_1 \otimes \mathcal{H}_2$, which expresses the two sheeted nature of the NCSG (cf. Eq. (1)), is a key feature of quantum theories. As observed also by Alain Connes in Ref. [1], already in the early years of quantum mechanics (QM), in establishing the "matrix mechanics" Heisenberg has shown that noncommutative algebras governing physical quantities are at the origin of spectroscopic experiments and are linked to the discretization of the energy of the atomic levels and angular momentum. One can convince himself that this is the case by observing that in the density matrix formalism of QM the coordinate $x(t)$ of a quantum particle is split into two coordinates $x_+(t)$ (going forward in time) and $x_-(t)$ (going backward in time). The forward in time motion and the backward in time motion of the density matrix $W(x_+, x_-, t) \equiv \langle x_+|\rho(t)|x_-\rangle = \psi^*(x_-, t)\psi(x_+, t)$, where $x_\pm = x \pm y/2$, is described indeed by "two copies" of the Schrödinger equation, respectively:

$$i\hbar \frac{\partial \psi(x_+, t)}{\partial t} = H_+ \psi(x_+, t), \qquad -i\hbar \frac{\partial \psi^*(x_-, t)}{\partial t} = H_- \psi^*(x_-, t), \tag{6}$$

which can be written as

$$i\hbar\frac{\partial\langle x_+|\rho(t)|x_-\rangle}{\partial t} = \hat{H}\ \langle x_+|\rho(t)|x_-\rangle, \tag{7}$$

where H is given in terms of the two Hamiltonian operators $H_\pm$ as

$$\hat{H} = H_+ - H_-\ . \tag{8}$$

The introduction of a doubled set of coordinates, $(x_\pm, p_\pm)$ (or (x, p_x) and (y, p_y)) and the use of the two copies of the Hamiltonian $H_\pm$ operating on the outer product of two Hilbert spaces $\mathcal{H}_+ \otimes \mathcal{H}_-$ thus show that the eigenvalues of $\hat{H}$ are directly the Bohr transition frequencies $h\nu_{nm} = E_n - E_m$, which are at the basis of the explanation of spectroscopic structure. We have observed elsewhere that the doubling of the algebra is implicit also in the theory of the Brownian motion of a quantum particle [20] (see also [12] and the references there quoted) and the doubled degrees of freedom are known [13] to allow quantum noise effect.

It is thus evident the connection with the two sheeted NCSG. Moreover, it has been shown [12] that as a consequence of the algebra doubling Eq. (1) the NCSG construction has an intrinsic gauge structure and is a thermal dissipative field theory. As we will discuss below, this suggests that Connes construction carries in itself the seeds of quantization, namely it is more than just a classical theory construction.

Let us start by discussing the simple case of the massless fermion and the $U(1)$ local gauge transformation group. We will see how in this case the doubling of the algebra is related to the gauge structure of the theory. Extension to the massive fermion case, the boson case and non-Abelian gauge transformation groups is possible [21, 22]. The system Lagrangian is

$$\hat{L} = L - \tilde{L} = -\bar{\psi}\gamma^\mu\partial_\mu\psi + \bar{\tilde{\psi}}\gamma^\mu\partial_\mu\tilde{\psi}. \tag{9}$$

The fermion tilde-field $\tilde{\psi}(x)$, which satisfies the usual fermionic anticommutation relations and anticommutes with the field $\psi(x)$, is a "copy" (with the same spectrum and couplings) of the ψ-system. We thus "double" the field algebra by introducing such a tilde-field $\tilde{\psi}(x)$.

For simplicity, no coupling term of the field $\psi(x)$ with $\tilde{\psi}(x)$ is assumed in $\hat{L}$. In the quantized theory, let $a^\dagger_{\mathbf{k}}$ and $\tilde{a}^\dagger_{\mathbf{k}}$ denote the creation operators associated to the quantum fields ψ and $\tilde{\psi}$, respectively (all quantum number indices are suppressed except momentum). The vacuum $|0(\theta)\rangle$ of the theory is:

$$|0(\theta)\rangle = \prod_k \left[\cos\theta_k + \sin\theta_k a^\dagger_{\mathbf{k}}\tilde{a}^\dagger_{\mathbf{k}}\right]|0\rangle, \tag{10}$$

namely a condensate of couples of $a^\dagger_{\mathbf{k}}$ and $\tilde{a}^\dagger_{\mathbf{k}}$ modes. Here $|0\rangle$ denotes the vacuum $|0,0\rangle \equiv |0\rangle \otimes |\tilde{0}\rangle$, with $|0\rangle$ and $|\tilde{0}\rangle$ the vacua annihilated by the annihilation operators $a_{\mathbf{k}}$ and $\tilde{a}_{\mathbf{k}}$, respectively. On the other hand, $|0(\theta)\rangle$ is the vacuum with respect to the fields $\psi(\theta; x)$ and $\tilde{\psi}(\theta; x)$ which are obtained by means of the Bogoliubov transformation:

$$\psi(\theta; x) = B^{-1}(\theta)\psi(x)B(\theta), \tag{11a}$$

$$\tilde{\psi}(\theta; x) = B^{-1}(\theta)\tilde{\psi}(x)B(\theta)\ , \tag{11b}$$

where $B(\theta) \equiv e^{-i\mathcal{G}}$, with the generator $\mathcal{G} = -i\sum_{\mathbf{k}}\theta_k(a^\dagger_{\mathbf{k}}\tilde{a}^\dagger_{\mathbf{k}} - a_{\mathbf{k}}\tilde{a}_{\mathbf{k}})$. For simplicity, θ is assumed to be independent of space-time. Extension to space-time dependent Bogoliubov transformations can be done [22]. $|0(\theta)\rangle$ is a $SU(2)$ generalized coherent state [23].

The Hamiltonian for the $\{\psi(x), \tilde{\psi}(x)\}$ system is $\hat{H} = H - \tilde{H}$ (to be compared with Eq. (8)), and is given by $\hat{H} = \sum_{\mathbf{k}}\hbar\,\omega_{\mathbf{k}}(a^\dagger_{\mathbf{k}}a_{\mathbf{k}} - \tilde{a}^\dagger_{\mathbf{k}}\tilde{a}_{\mathbf{k}})$. The θ-vacuum $|0(\theta)\rangle$ is the the zero eigenvalue eigenstate of $\hat{H}$. The relation $[a^\dagger_{\mathbf{k}} a_{\mathbf{k}} - \tilde{a}^\dagger_{\mathbf{k}} \tilde{a}_{\mathbf{k}}]|0(\theta)\rangle = 0$, for any $\mathbf{k}$ (and any θ) characterizes the θ-vacuum structure and it is called the θ-vacuum condition.

The space of states $\hat{\mathcal{H}} = \mathcal{H} \otimes \tilde{\mathcal{H}}$ is constructed by repeated applications of creation operators of $\psi(\theta; x)$ and $\tilde{\psi}(\theta; x)$ on $|0(\theta)\rangle$ and is called the θ-representation $\{|0(\theta)\rangle\}$. θ-representations corresponding to different values of the θ parameter are unitarily inequivalent representations of the canonical anti-commutation relations in QFT [13]. The state $|0(\theta)\rangle$ is known to be a finite temperature state [13, 21, 22, 24, 25], namely one finds that θ is a temperature-dependent parameter. This tells us that the algebra doubling leads to a thermal field theory. The Bogoliubov transformations then induce transition through system phases at different temperatures.

We now consider the subspace $\mathcal{H}_{\theta c} \subset \{|0(\theta)\rangle\}$ made of all the states $|a\rangle_{\theta c}$, including $|0(\theta)\rangle$, such that the *θ-state condition*

$$[a^\dagger_{\mathbf{k}} a_{\mathbf{k}} - \tilde{a}^\dagger_{\mathbf{k}} \tilde{a}_{\mathbf{k}}]|a\rangle_{\theta c} = 0 \;, \qquad \text{for any} \quad \mathbf{k}, \tag{12}$$

holds in $\mathcal{H}_{\theta c}$ (note that Eq. (12) is similar to the Gupta-Bleuler condition defining the physical states in quantum electrodynamics (QED)). Let $\langle ... \rangle_{\theta c}$ denote matrix elements in $\mathcal{H}_{\theta c}$; we have

$$\langle j_\mu(x) \rangle_{\theta c} = \langle \tilde{j}_\mu(x) \rangle_{\theta c}, \tag{13}$$

where $j_\mu(x) = \bar{\psi}\gamma^\mu\psi$ and $\tilde{j}_\mu(x) = \bar{\tilde{\psi}}\gamma^\mu\tilde{\psi}$. Equalities between matrix elements in $\mathcal{H}_{\theta c}$, say $\langle A \rangle_{\theta c} = \langle B \rangle_{\theta c}$, are denoted by $A \cong B$ and we call them θ-w-equalities (θ-weak equalities). They are classical equalities since they are equalities among c-numbers. $\mathcal{H}_{\theta c}$ is invariant under the dynamics described by $\hat{H}$ (even in the general case in which interaction terms are present in $\hat{H}$ provided that the charge is conserved).

The key point is that, due to Eq. (13), the matrix elements in $\mathcal{H}_{\theta c}$ of the Lagrangian Eq. (9) are invariant under the simultaneous local gauge transformations of ψ and $\tilde{\psi}$ fields given by

$$\psi(x) \to \exp\left[ig\alpha(x)\right]\psi(x), \qquad \tilde{\psi}(x) \to \exp\left[ig\alpha(x)\right]\tilde{\psi}(x) \;, \tag{14}$$

i.e.,

$$\langle \hat{L} \rangle_{\theta c} \to \langle \hat{L}' \rangle_{\theta c} = \langle \hat{L} \rangle_{\theta c} \;, \quad \text{in} \;\; \mathcal{H}_{\theta c}, \tag{15}$$

under the gauge transformations (14). The tilde term $\bar{\tilde{\psi}}\gamma^\mu\partial_\mu\tilde{\psi}$ thus plays a crucial rôle in the θ-w-gauge invariance of $\hat{L}$ under Eq. (14). Indeed it transforms in such a way to compensate the local gauge transformation of the ψ kinematical term, i.e.,

$$\bar{\tilde{\psi}}(x)\gamma^\mu\partial_\mu\tilde{\psi}(x) \to \bar{\tilde{\psi}}(x)\gamma^\mu\partial_\mu\tilde{\psi}(x) + g\partial^\mu\alpha(x)\tilde{j}_\mu(x). \tag{16}$$

This suggests to us to introduce the vector field A'_μ by

$$gj^{\bar{\mu}}(x)A'_{\bar{\mu}}(x) \cong \bar{\tilde{\psi}}(x)\gamma^{\bar{\mu}}\partial_{\bar{\mu}}\tilde{\psi}(x) \;, \quad \bar{\mu} = 0, 1, 2, 3. \tag{17}$$

Here and in the following, the bar over μ means no summation on repeated indices. Thus, the vector field A'_μ transforms as

$$A'_\mu(x) \to A'_\mu(x) + \partial_\mu\alpha(x), \tag{18}$$

when the transformations of Eq. (14) are implemented. In $\mathcal{H}_{\theta c}$, A'_μ can be then identified with the conventional U(1) gauge vector field and can be introduced it in the original Lagrangian through the usual coupling term $ig\bar{\psi}\gamma^\mu\psi A'_\mu$.

The position (17) does not change the θ-vacuum structure. Therefore, provided that one restricts himself/herself to matrix elements in $\mathcal{H}_{\theta c}$, matrix elements of physical observables, which are solely functions of the $\psi(x)$ field, are not changed by the position (17). Our identification of A'_μ with the U(1) gauge vector field is also justified by the fact that observables turn out to be invariant under gauge transformations and the conservation laws derivable from $\hat{L}$, namely in the simple case of Eq. (9) the current conservation laws, $\partial^\mu j_\mu(x) = 0$ and $\partial^\mu \tilde{j}_\mu(x) = 0$, are also preserved as θ-w-equalities when Eq. (17) is adopted. Indeed, one obtains [21, 22] $\partial^\mu j_\mu(x) \cong 0$ and $\partial^\mu \tilde{j}_\mu(x) \cong 0$. One may also show that

$$\partial^\nu F'_{\mu\nu}(x) \cong -g j_\mu(x), \qquad \partial^\nu F'_{\mu\nu}(x) \cong -g \tilde{j}_\mu(x), \tag{19}$$

in $\mathcal{H}_{\theta c}$. In the the Lorentz gauge from Eq. (19) we also obtain the θ-w-relations $\partial^\mu A'_\mu(x) \cong 0$ and $\partial^2 A'_\mu(x) \cong g j_\mu(x)$.

In conclusion, the "doubled algebra" Lagrangian (9) for the field ψ and its "double" $\tilde{\psi}$ can be substituted in $\mathcal{H}_{\theta c}$ by:

$$\hat{L}_{\rm g} \cong -1/4 F'^{\mu\nu} F'_{\mu\nu} - \overline{\psi}\gamma^\mu \partial_\mu \psi + ig\overline{\psi}\gamma^\mu \psi A'_\mu \ , \quad \text{in } \mathcal{H}_{\theta c} \ , \tag{20}$$

where, remarkably, the tilde-kinematical term $\overline{\tilde{\psi}}\gamma^\mu \partial_\mu \tilde{\psi}$ is replaced, in a θ-w-sense, by the coupling term $ig\overline{\psi}\gamma^\mu \psi A'_\mu$ between the gauge field A'_μ and the matter field current $\overline{\psi}\gamma^\mu \psi$.

Finally, in the case an interaction term is present in the Lagrangian (9), $\hat{L}_{\rm tot} = \hat{L} + \hat{L}_I$, $\hat{L}_I = L_{\rm I} - \tilde{L}_{\rm I}$, the above conclusions still hold provided $\mathcal{H}_{\theta c}$ is an invariant subspace under the dynamics described by $\hat{L}_{\rm tot}$.

Our discussion has thus shown that the "doubling" of the field algebra introduces a gauge structure in the theory: Connes two sheeted geometric construction has intrinsic gauge properties.

The algebraic structure underlying the above discussion is recognized to be the one of the noncommutative q-deformed Hopf algebra [24]. We remark that the Hopf coproduct map $\mathcal{A} \to \mathcal{A} \otimes \mathbf{1} + \mathbf{1} \otimes \mathcal{A} \equiv \mathcal{A}_1 \otimes \mathcal{A}_2$ is nothing but the map presented in Eq. (1) which duplicates the algebra. On the other hand, it can be shown [24] that the Bogoliubov transformation of "angle" θ relating the fields $\psi(\theta; x)$ and $\tilde{\psi}(\theta; x)$ to $\psi(x)$ and $\tilde{\psi}(x)$, Eqs. (11), are obtained by convenient combinations of the *q-deformed* Hopf coproduct $\Delta a_q^\dagger = a_q^\dagger \otimes q^{1/2} + q^{-1/2} \otimes a_q^\dagger$, with $q \equiv q(\theta)$ the deformation parameters and $a_q^\dagger$ the creation operators in the q-deformed Hopf algebra [24]. These deformed coproduct maps are noncommutative. All of this signals a deep physical meaning of noncommutativity in the Connes construction since the deformation parameter is related to the condensate content of $|0(\theta)\rangle$ under the constrain imposed by the θ-state condition Eq. (12). Actually, such a state condition is a characterizing condition for the system physical states. The crucial point is that a characteristic feature of quantum field theory [13, 24] is that the deformation parameter *labels* the θ-representations $\{|0(\theta)\rangle\}$ and, as already mentioned, for $\theta \neq \theta'$, $\{|0(\theta)\rangle\}$ and $\{|0(\theta')\rangle\}$ are unitarily inequivalent representations of the canonical (anti-)commutation rules [13, 25]. In turn, the physical meaning of this is that an order parameter exists, which assumes different θ-dependent values in each of the representations. Thus, the q-deformed Hopf algebra structure of QFT induces the *foliation* of the whole Hilbert space into physically inequivalent subspaces. From our discussion we conclude that this is also the scenario which NCSG presents to us.

One more remark in this connection is that in the NCSG construction the derivative in the discrete direction is a finite difference quotient [1, 2, 12] and it is then suggestive that the q-derivative is also a finite difference derivative. This point deserves further formal analysis which is in our plans to do.

We thus conclude that Connes NCSG construction is built on the same noncommutative deformed Hopf algebra structure of QFT. In the next Section we show that it is also related with dissipation and carries in it the seeds of quantization.

3. Dissipation, gauge field and quantization

In the second equation in (19) the current $\tilde{j}_\mu$ act as the source of the variations of the gauge field tensor $F'_{\mu\nu}$. We express this by saying that the tilde field plays the rôle of a "reservoir". Such a reservoir interpretation may be extended also to the gauge field A'_μ, which is known to act, indeed, in a way to "compensate" the changes in the matter field configurations due to the local gauge freedom.

When we consider variations in the θ parameter (namely in the q-deformation parameter), induced by the Bogoliubov transformation generator, we have (time-)evolution over the manifold of the θ-labeled (i.e. q-labeled) spaces and we have dissipative fluxes between the doubled sets of fields, or, in other words, according to the above picture, between the system and the reservoir. We talk of dissipation and open systems when considering the Connes conctruction and the Standard Model in the same sense in a system of electromagnetically interacting matter field, neither the energy-momentum tensor of the matter field, nor that of the gauge field, are conserved. However, one verifies in a standard fashion [26] that $\partial_\mu T^{\mu\nu}_{\rm matter} = eF^{\mu\nu} j_\mu = -\partial_\mu T^{\mu\nu}_{\rm gauge\ field}$, so that what it is conserved is the *total* $T^{\mu\nu}_{\rm total} = T^{\mu\nu}_{\rm matter} + T^{\mu\nu}_{\rm gauge\ field}$, namely the energy-momentum tensor of the *closed* system {matter field, electromagnetic field}. As remarked in Ref. [12], each element of the couple is *open* (dissipating) on the other one, although the *closeness* of the total system is ensured. Thus the closeness of the SM is not spoiled in our discussion.

In order to further clarify how the gauge structure is related to the algebra doubling and to dissipation we consider the prototype of a dissipative system, namely the classical one-dimensional damped harmonic oscillator

$$m\ddot{x} + \gamma\dot{x} + kx = 0 \ , \tag{21}$$

and its time-reversed ($\gamma \to -\gamma$) (doubled) image

$$m\ddot{y} - \gamma\dot{y} + ky = 0 \ , \tag{22}$$

with time independent m, γ and k, needed [27] in order to set up the canonical formalism for open systems. The system of Eq. (21) and Eq. (22) is then a closed system described by the Lagrangian density

$$L(\dot{x}, \dot{y}, x, y) = m\dot{x}\dot{y} + \frac{\gamma}{2}(x\dot{y} - y\dot{x}) + kx\, y \ . \tag{23}$$

It is convenient to use the coordinates $x_1(t)$ and $x_2(t)$ defined by

$$x_1(t) = \frac{x(t) + y(t)}{\sqrt{2}} \ , \qquad x_2(t) = \frac{x(t) - y(t)}{\sqrt{2}} \ . \tag{24}$$

The motion equations are rewritten then as

$$m\ddot{x}_1 + \gamma\dot{x}_2 + kx_1 = 0 \ , \tag{25a}$$

$$m\ddot{x}_2 + \gamma\dot{x}_1 + kx_2 = 0 \ . \tag{25b}$$

The canonical momenta are: $p_1 = m\dot{x}_1 + (1/2)\gamma x_2$; $p_2 = -m\dot{x}_2 - (1/2)\gamma x_1$; the Hamiltonian is

$$\hat{H} = H_1 - H_2 = \frac{1}{2m}(p_1 - \frac{\gamma}{2}x_2)^2 + \frac{k}{2}x_1^2 - \frac{1}{2m}(p_2 + \frac{\gamma}{2}x_1)^2 - \frac{k}{2}x_2^2 \ . \tag{26}$$

We then recognize [28, 29, 21, 22] that

$$A_i = \frac{B}{2}\epsilon_{ij}x_j \ , \quad (i,j = 1,2) \ , \qquad \epsilon_{ii} = 0 \ , \ \ \epsilon_{12} = -\epsilon_{21} = 1 \ , \tag{27}$$

with $B \equiv c\gamma/e$, acts as the vector potential and obtain that the system of oscillators Eq. (21) and Eq. (22) is equivalent to the system of two particles with opposite charges $e_1 = -e_2 = e$ in the (oscillator) potential $\Phi \equiv (k/2/e)({x_1}^2 - {x_2}^2) \equiv \Phi_1 - \Phi_2$ with $\Phi_i \equiv (k/2/e){x_i}^2$ and in the constant magnetic field $\boldsymbol{B}$ defined as $\boldsymbol{B} = \boldsymbol{\nabla} \times \boldsymbol{A} = -B\hat{\mathbf{3}}$. The Hamiltonian is indeed

$$\hat{H} = H_1 - H_2 = \frac{1}{2m}(p_1 - \frac{e_1}{c}A_1)^2 + e_1\Phi_1 - \frac{1}{2m}(p_2 + \frac{e_2}{c}A_2)^2 + e_2\Phi_2 \, . \tag{28}$$

and the Lagrangian of the system can be written in the familiar form

$$\begin{aligned} L &= \frac{1}{2m}(m\dot{x}_1 + \frac{e_1}{c}A_1)^2 - \frac{1}{2m}(m\dot{x}_2 + \frac{e_2}{c}A_2)^2 - \frac{e^2}{2mc^2}({A_1}^2 + {A_2}^2) - e\Phi \\ &= \frac{m}{2}(\dot{x}_1^2 - \dot{x}_2^2) + \frac{e}{c}(\dot{x}_1 A_1 + \dot{x}_2 A_2) - e\Phi \, . \end{aligned} \tag{29}$$

Note the "minus" sign of the Lorentzian-like (pseudoeuclidean) metric in Eq. (29) (cf. also Eqs. (8), (9) and(28)), not imposed by hand, but derived through the doubling of the degrees of freedom and crucial in our description (and in the NCSG construction).

The doubled coordinate x_2 thus acts as the gauge field component A_1 to which the x_1 coordinate is coupled, and *vice versa*. The energy dissipated by one of the two systems is gained by the other one and viceversa, in analogy to what happens in standard electrodynamics as observed above. The picture is recovered of the gauge field as the bath or reservoir in which the system is embedded [21, 22].

Our toy system of harmonic oscillators Eq. (21) and Eq. (22) offers also an useful playground to show how dissipation is implicitly related to quantization. 't Hooft indeed has conjectured that classical, deterministic systems with loss of information might lead to a quantum evolution [30, 31, 32] provided some specific energy conditions are met and some constraints are imposed. Let us verify how such a conjecture is confirmed for our system described by Eq. (21) and Eq. (22). We will thus show that quantization is achieved as a consequence of dissipation. We rewrite the Hamiltonian Eq. (26) as $\hat{H} = H_{\rm I} - H_{\rm II}$, with

$$H_{\rm I} = \frac{1}{2\Omega\mathcal{C}}(2\Omega\mathcal{C} - \Gamma J_2)^2 \, , \quad H_{\rm II} = \frac{\Gamma^2}{2\Omega\mathcal{C}}J_2^2 \tag{30}$$

where C is the Casimir operator and J_2 is the (second) $SU(1,1)$ generator [29]:

$$\mathcal{C} = \frac{1}{4\Omega m}\left[\left(p_1^2 - p_2^2\right) + m^2\Omega^2\left(x_1^2 - x_2^2\right)\right] \, , \qquad J_2 = \frac{m}{2}\left[(\dot{x}_1 x_2 - \dot{x}_2 x_1) - \Gamma r^2\right] \, . \tag{31}$$

$\mathcal{C}$ is taken to be positive and

$$\Gamma = \frac{\gamma}{2m} \, , \; \Omega = \sqrt{\frac{1}{m}(k - \frac{\gamma^2}{4m})} \, , \text{ with } k > \frac{\gamma^2}{4m} \, .$$

This $\hat{H}$ belongs to the class of Hamiltonians considered by 't Hooft and is of the form [33, 34]

$$\hat{H} \;=\; \sum_{i=1}^{2} p_i \, f_i(q) \, , \tag{32}$$

with $p_1 = \mathcal{C}$, $p_2 = J_2$, $f_1(q) = 2\Omega$, $f_2(q) = -2\Gamma$. Note that $\{q_i, p_i\} = 1$ and the other Poisson brackets are vanishing. The $f_i(q)$ are nonsingular functions of q_i and the equations for the q's, $\dot{q}_i = \{q_i, H\} = f_i(q)$, are decoupled from the conjugate momenta p_i. A complete set of observables, called *beables*, then exists, which Poisson commute at all times. Thus the system admits a deterministic description even when expressed in terms of operators acting on some functional space of states $|\psi\rangle$, such as the Hilbert space [31]. Note that this description in terms of operators and Hilbert space, does not imply *per se* quantization of the system. Physical states are defined to be those satisfying the condition $J_2|\psi\rangle = 0$. This guaranties that $\hat{H}$ is bounded from below and $\hat{H}|\psi\rangle = H_{\rm I}|\psi\rangle = 2\Omega\mathcal{C}|\psi\rangle$. $H_{\rm I}$ thus reduces to the Hamiltonian for the two-dimensional "isotropic" (or "radial") harmonic oscillator $\ddot{r} + \Omega^2 r = 0$. Indeed, putting $K \equiv m\Omega^2$, we obtain

$$\hat{H}|\psi\rangle = H_{\rm I}|\psi\rangle = \left(\frac{1}{2m}p_r^2 + \frac{K}{2}r^2\right)|\psi\rangle\,. \tag{33}$$

The physical states are invariant under time-reversal ($|\psi(t)\rangle = |\psi(-t)\rangle$) and periodical with period $\tau = 2\pi/\Omega$. $H_{\rm I} = 2\Omega\mathcal{C}$ has the spectrum $\mathcal{H}^n_{\rm I} = \hbar\Omega n$, $n = 0, \pm1, \pm2, ...$; since $\mathcal{C}$ has been chosen to be positive, only positive values of n are considered. Then one obtains

$$\frac{\langle\psi(\tau)|\hat{H}|\psi(\tau)\rangle}{\hbar}\tau - \phi = 2\pi n \quad , \quad n = 0, 1, 2, \ldots\,.$$

Using $\tau = 2\pi/\Omega$ and $\phi = \alpha\pi$, where α is a real constant, leads to

$$\mathcal{H}^n_{\rm I,eff} \equiv \langle\psi_n(\tau)|\hat{H}|\psi_n(\tau)\rangle = \hbar\Omega\left(n + \frac{\alpha}{2}\right)\,. \tag{34}$$

$\mathcal{H}^n_{\rm I,eff}$ gives the effective nth energy level of the system, namely the energy given by $\mathcal{H}^n_{\rm I}$ corrected by its interaction with the environment. We conclude that the dissipation term J_2 of the Hamiltonian is responsible for the zero point ($n = 0$) energy: $E_0 = (\hbar/2)\Omega\alpha$. In QM the zero point energy *is* the "signature" of quantization since it is formally due to the nonzero commutator of the canonically conjugate q and p operators. Our conclusion is thus that the (zero point) "quantum contribution" E_0 to the spectrum of physical states is due and signals the underlying dissipative dynamics: dissipation manifests itself as "quantization".

We remark that the "full Hamiltonian" Eq. (32) plays the rôle of the free energy $\mathcal{F}$, and $2\Gamma J_2$ represents the heat contribution in $\hat{H}$ (or $\mathcal{F}$). Indeed, using $S \equiv (2J_2/\hbar)$ and $U \equiv 2\Omega\mathcal{C}$ and the defining relation for temperature in thermodynamics: $\partial S/\partial U = 1/T$, with the Boltzmann constant $k_{\rm B} = 1$, from Eq. (32) we obtain $T = \hbar\Gamma$, i.e., provided that S is identified with the entropy, $\hbar\Gamma$ can be regarded as the temperature. Note that it is not surprising that $2J_2/\hbar$ behaves as the entropy since it controls the dissipative part of the dynamics (thus the irreversible loss of information). It is interesting to observe that this thermodynamical picture is also cofirmed by the results on the canonical quantization of open systems in quantum field theory [27].

On the basis of these results (confirming 't Hooft's conjecture) we have proved that the NCSG classical construction, due to its essential ingredient of the doubling of the algebra, is intrinsically related with dissipation and thus carries in itself the seeds of quantization [12].

4. Conclusions

In Ref. [35] it has been shown that a "dissipative interference phase" also appears in the evolution of dissipative systems. Such a feature exhibits one more consequence of the algebra doubling which plays so an important rôle in the Connes NCSG construction. In this report we have discussed how such a doubling process implies the gauge structure of the theory, its thermal characterization, its built in dissipation features and the consequent possibility to exhibit quantum evolution behaviors. Here we only mention that the doubling of the coordinates also provides a relation between dissipation and noncommutative geometry in the plane. We refer to Ref.[12] for details with reference to Connes construction (see also [19]). We only recall that in the (x_+, x_-) plane, $x_\pm = x \pm y/2$ (cf. Section 2), the components of forward and backward in time velocity $v_\pm = \dot{x}_\pm$, given by

$$v_\pm = \frac{\partial H}{\partial p_\pm} = \pm\frac{1}{m}\left(p_\pm \mp \frac{\gamma}{2}x_\mp\right) , \qquad (35)$$

do not commute: $[v_+, v_-] = i\hbar\,\gamma/m^2$. Thus it is impossible to fix these velocities v_+ and v_- as being identical [35]. Moreover, one may always introduce a set of canonical conjugate position coordinates (ξ_+, ξ_-) defined by $\xi_\pm = \mp L^2 K_\mp$, with $\hbar K_\pm = m v_\pm$, so that $[\xi_+, \xi_-] = iL^2$, which characterizes the noncommutative geometry in the plane (x_+, x_-). L denotes the geometric length scale in the plane.

One can show [35] that an Aharonov–Bohm-type phase interference can always be associated with the noncommutative (X, Y) plane where

$$[X, Y] = iL^2 \, . \qquad (36)$$

Eq. (36) implies the uncertainty relation $(\Delta X)(\Delta Y) \geq L^2/2$, due to the zero point fluctuations in the coordinates. Consider now a particle moving in the plane along two paths starting and finishing at the same point, in a forward and in a backward direction, respectively, forming a closed loop. The phase interference ϑ may be be shown [35] to be given by $\vartheta = \mathcal{A}/L^2$, where $\mathcal{A}$ is the area enclosed in the closed loop formed by the paths and Eq. (36) in the noncommutative plane can be written as

$$[X, P_X] = i\hbar \quad \text{where} \quad P_X = \left(\frac{\hbar Y}{L^2}\right) . \qquad (37)$$

The quantum phase interference between two alternative paths in the plane is thus determined by the length scale L and the area $\mathcal{A}$. In the dissipative case, it is $L^2 = \hbar/\gamma$, and the quantum dissipative phase interference $\vartheta = \mathcal{A}/L^2 = \mathcal{A}\gamma/\hbar$ is associated with the two paths in the noncommutative plane, provided $x_+ \neq x_-$. When doubling is absent, i.e., $x_+ = x_-$, the quantum effect disappear. It is indeed possible to show [20] that in such a classical limit case the doubled degree of freedom is associated with "unlikely processes" and it may thus be dropped in higher order terms in the perturbative expansion, as indeed it happens in Connes construction. At the Grand Unified Theories scale, when inflation took place, the effect of gauge fields is fairly shielded. However, since these higher order terms are the ones responsible for quantum corrections, the second sheet cannot be neglected if one does not want to preclude quantization effects, as it happens once the universe entered the radiation dominated era.

References

[1] Connes A 1994 *Noncommutative Geometry* (New York: Academic Press)
[2] Connes A and Marcolli M 2008 *Noncommutative Geometry, Quantum Fields and Motives* (India: Hindustan Book Agency)
[3] Chamseddine A H, Connes A and Marcolli M 2007 *Adv. Theor. Math. Phys.* **11** 991
[4] Nelson W and Sakellariadou M 2010 *Phys. Rev.* D **81** 085038
[5] Nelson W and Sakellariadou M 2009 *Phys. Lett.* B **680** 263
[6] Marcolli M and Pierpaoli E *Preprint* hep-th/0908.3683
[7] Buck M, Fairbairn M and Sakellariadou M 2010 *Phys. Rev.* D **82** 043509
[8] Nelson W, Ochoa J and Sakellariadou M 2010 *Phys. Rev. Lett.* **105** 101602
[9] Nelson W, Ochoa J and Sakellariadou M 2010 *Phys. Rev.* D **82** 085021
[10] Sakellariadou M 2011 *Int. J. Mod. Phys.* D **20** 785
[11] Sakellariadou M 2010 PoS CNCFG **2010** 028 [arXiv:1101.2174 [hep-th]]
[12] Sakellariadou M, Stabile A and Vitiello G 2011 *Phys. Rev.* D **84** 045026
[13] Blasone M, Jizba P and Vitiello G 2011 *Quantum Field Theory and its macroscopic manifestations* (London: Imperial College Press)

[14] Chamseddine A H and Connes A 2007 *Phys. Rev. Lett.* **99** 191601
[15] Chamseddine A H and Connes A 2006 *J. Math. Phys.* **47** 063504
[16] Chamseddine A H and Connes A 1996 *Phys. Rev. Lett.* **77** 4868
[17] Chamseddine A H and Connes A 1977 *Comm. Math. Phys.* **186** 731
[18] Chamseddine A H and Connes A 2010 *Comm. Math. Phys.* **293** 867
[19] Sakellariadou M, Stabile A and Vitiello G 2012 *J. of Phys.: Conf. Series* **361** 012025
[20] Blasone M, Srivastava Y N, Vitiello G and Widom A 1998 *Annals Phys.* **267** 61
[21] Celeghini E, Graziano E, Nakamura K and Vitiello G 1992 *Phys. Lett.* **B285** 98
[22] Celeghini E, Graziano E, Nakamura K and Vitiello G 1993 *Phys. Lett.* **B304** 121
[23] Perelomov A M 1986 *Generalized coherent states and their applications* (Berlin: Springer-Verlag)
[24] Celeghini E, De Martino S, De Siena S, Iorio A, Rasetti M and Vitiello G 1998 *Phys. Lett.* **A244** 455
[25] Umezawa H, Matsumoto H and Tachiki M 1982 *Thermo Field Dynamics and condensed states* (Amsterdam: North-Holland)
[26] Landau L D and Lifshitz E M 1980 *The Classical Theory of Fields* (Oxford: Butterworth-Heinemann)
[27] Celeghini E, Rasetti M and Vitiello G 1992 *Annals Phys.* **215** 156
[28] Tsue Y, Kuriyama A and Yamamura M 1994 *Prog. Theor. Phys.* **91** 469
[29] Blasone M, Graziano E, Pashaev O K and Vitiello G 1996 *Annals Phys.* **252** 115
[30] 't Hooft G 1999 *Class. Quant. Grav.* **16** 3263
[31] 't Hooft G 1999 *Basics and Highlights of Fundamental Physics (Erice)* [arXiv:hep-th/0003005].
[32] 't Hooft G 2007 *J. Phys.: Conf. Series* **67** 012015
[33] Blasone M, Jizba P and Vitiello 2001 *Phys. Lett.* A **287** 205
[34] Blasone M, Celeghini E, Jizba P and Vitiello G 2003 *Phys. Lett.* A **310** 393
[35] Sivasubramanian S, Srivastava Y, Vitiello G and Widom A 2003 *Phys. Lett.* **A311** 97

Topology, cosmic strings and quantum dynamics - a case study with graphene

Baishali Chakraborty[1], **Kumar S Gupta**[1] **and Siddhartha Sen**[2]
[1] Theory Division, Saha Institute of Nuclear Physics, 1/AF Bidhannagar, Calcutta 700064, India
[2] CRANN, Trinity College Dublin, Dublin 2, Ireland
E-mail: baishali.chakraborty@saha.ac.in, kumars.gupta@saha.ac.in, siddhartha.sen@tcd.ie

Abstract. We explore the possibility to study the quantum dynamics of Dirac fermions in presence of a cosmic string by introducing a conical topological defect in gapped graphene in the presence of a Coulomb charge. When the Coulomb charge exceeds a certain critical strength, quantum instability sets in. Below the critical regime and for certain values of the system parameters, the allowed boundary conditions in gapped graphene cone can be classified in terms of a single real quantity. Observables such as local density of states, scattering phase shifts and the bound state spectra are dependent on the value of this real parameter, which has to be determined empirically. For a supercritical Coulomb charge, we analyze the system with a regularized potential as well as with a zigzag boundary condition and find the effect of the sample topology on the observable features of the system.

1. Introduction

The study of the topological properties of lower dimensional quantum systems show remarkable properties in various physical systems [1, 2, 3]. In the presence of a topological defect such as a cosmic string, whose exterior space-time has a topology, the dynamics of Dirac fermions exhibit novel nonperturbative quantum features such as inequivalent quantizations [2]. Though these features are theoretically interesting, it is difficult to devise an experiment with cosmic strings which will exhibit such quantum effects. In 2004 the experimental fabrication of monolayer graphene [4, 5, 6] offered a chance to observe the effects of topological defects [7, 8, 9, 10, 11, 12, 13, 14, 15, 16, 17, 18, 19, 20, 21, 22, 23, 24, 25, 26] in the laboratory. The low energy excitations in pristine graphene obey a 2D massless Dirac equation[27, 28, 29, 30, 31, 32, 33]. However violation of sublattice symmetry in graphene due to various impurities or short distance effects can lead to a mass gap in the system [34, 35, 36, 37, 38, 39, 40, 41, 42, 43, 44, 45, 46, 47, 48, 49, 50, 51]. When a conical topological defect is introduced in such a system, some nontrivial holonomies [9, 10, 16] arise. The boundary conditions associated with the holonomies can be realized by introducing a suitable flux tube, analogous to a cosmic string, passing through the origin [2, 52, 53, 54, 55, 56, 57]. In our analysis, such a flux tube shall be used to model the conical topological defect on the 2D graphene sheet. Thus the gapped graphene cone in presence of an external Coulomb impurity offers a possibility to study the inequivalent quantization of massive Dirac fermions in the presence of cosmic strings.

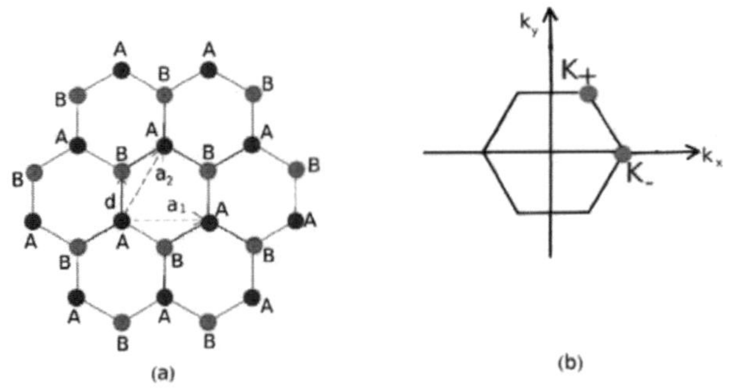

Figure 1. Hexagonal lattice structure of graphene and the Brillouin zone

The external Coulomb charge in a gapped graphene system can either be sub or supercritical. Here the critical charge refers to that value of Coulomb potential strength when the system dives into the negative energy continuum [58, 59, 60] indicating the quantum instability. The two different regimes of Coulomb potential are characterized by markedly different behaviour of the local density of states (LDOS) [58].

For a subcritical Coulomb charge impurity in the presence of the flux tube, we shall show that the quantization of the gapped graphene system is not unique and an additional parameter is required to fully characterize the boundary conditions at the origin, analogous to what was obtained for Dirac fermions in 2+1 dimensional gravity with a topological defect [2, 52].We shall also show that the experimental observables such as the LDOS, phase shifts and the bound state energies depend explicitly on that new parameter.

In the supercritical region the external Coulomb charge in gapped graphene cone can produce strong nonperturbative electric field effect [37, 58, 59, 60, 61, 62, 63, 64, 65, 66, 67, 68, 69, 70, 71, 72], at a relatively small value $Ze \sim 1$, due to the Fermi velocity $v_F \approx 10^6 m/s$ in graphene, which is 300 times smaller than the velocity of light. For supercritical charge, we study the system with a regularized Coulomb potential[58, 59] and also with a zigzag edge boundary condition. We shall show that the critical charge in gapped graphene cone is renormalized to a value higher than that of the gapless case and the value depends on the gap, the cut off parameter, the topology of the system and also on the boundary conditions used to obtain the quasibound state spectra in the supercritical region. It will be shown that with the increase in gap or cut off parameter the critical charge in presence of zigzag edge boundary condition increases more rapidly than in presence of a regularized Coulomb potential.

2. Low energy excitations in graphene cone with a Coulomb charge

Graphene has a hexagonal honeycomb lattice structure which is formed by two inter penetrating triangular sublattices [5, 6, 27, 29, 30] A and B. The low energy gapped excitations of planer graphene have minimum energy eigenvalues at the six vertices of the hexagonal first Brillouin zone of graphene and these vertices are known as the Dirac points. Among the six Dirac points, two are inequivalent [5, 6, 27, 29, 30]. We consider them to be situated at the opposite corners of the Brillouin zone and we denote their wave vectors by $\mathbf{K}_+$ and $\mathbf{K}_-$. The low-energy properties of the quasiparticle states in graphene near the Dirac point having valley index K_+, can be described by the Dirac equation

$$H\Psi = [-i(\sigma_1\partial_x + \sigma_2\partial_y) + m\sigma_3]\,\Psi = E\Psi, \tag{1}$$

where m denotes the Dirac mass generated due to sublattice symmetry breaking, E is the energy eigenvalue and we have set $\hbar = v_F = 1$. The Hamiltonian acts on the array of the slowly varying envelope functions $\Psi = \begin{pmatrix} \Psi_{K_+,A} \\ \Psi_{K_+,B} \end{pmatrix}$ and the Pauli matrices $\sigma_{1,2,3}$ act on the pseudospin indices A, B.

To study the effect of topology on this system, we consider to introduce local defects in the hexagonal lattice structure of graphene[9, 10] to form a cone.

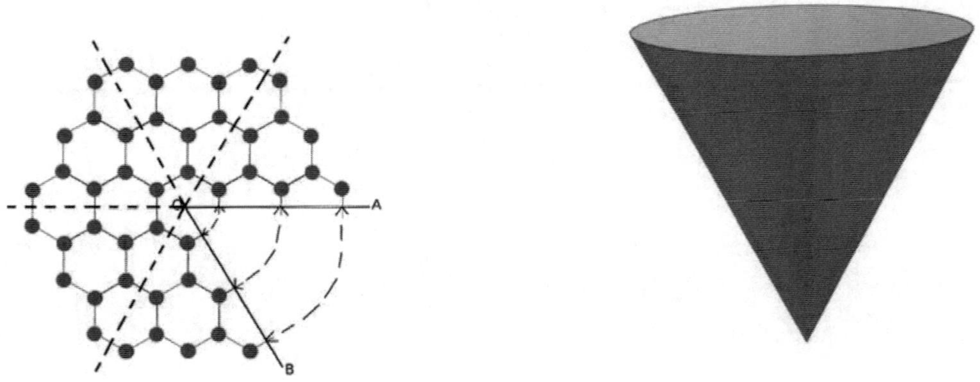

Figure 2. Formation of a cone from plane graphene sheet by cut and paste procedure.

The conical topology gives rise to nontrivial holonomies for the pseudoparticle wavefunctions in graphene. When a cone with angle of deficit $\frac{2n\pi}{6}$ is formed, where n can take only discrete values $1, 2, 3, 4, 5$, the angular boundary condition obeyed by the Dirac spinor as it goes around a closed path is given by

$$\Psi(r, \theta = 2\pi) = e^{i2\pi(1-\frac{n}{6})\frac{\sigma_3}{2}}\Psi(r, \theta = 0). \tag{2}$$

Here (r, θ) denotes the polar coordinate of the lattice points.

When the cone is formed by removing odd number of wedges of angle $\frac{2\pi}{6}$ an additional phase shift appears affecting the valley indices of the wave function in the boundary condition[9, 10, 11, 18]. The states with valley index K_- will be affected in the same manner as the states with valley index K_+ but there will be a relative phase difference of $180°$ between them (for details see [73]). Therefore this boundary condition can be described by involving a τ_2 matrix in it where the matrix τ_2 operate on the valley indices[9, 10, 74]. When n is even, this off diagonal matrix does not play any role and the exponential factor appearing in the boundary condition just gives ± 1 depending on the value of n. We diagonalize the matrix τ_2 for all allowed odd values of n. As a result the valley indices of the electronic states become mixtures of K_+ and K_-. Then the angular boundary condition satisfied for all values of n, by a branch of electronic states having a fixed Fermi index, is given by [9, 10]

$$\Psi(r, \theta = 2\pi) = e^{i2\pi[\pm\frac{n}{4}\sigma_0 + (1-\frac{n}{6})\frac{\sigma_3}{2}]}\Psi(r, \theta = 0). \tag{3}$$

Here σ_0 is an identity matrix which acts on the pseudospin indices A, B. $\Psi = \begin{pmatrix} \Psi_{A,K'} \\ \Psi_{B,K'} \end{pmatrix}$ where K' is a mixture of K_+ and K_-.

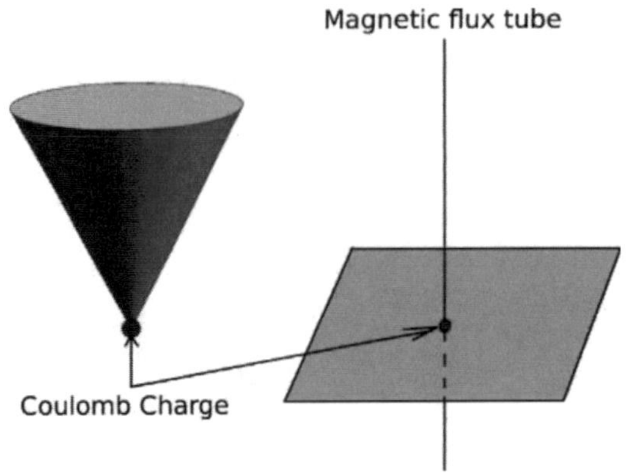

Figure 3. Graphene cone = planar graphene + flux tube

The effect of these holonomies can be modelled by introducing a fictitious magnetic flux tube [13] passing through the apex of the cone. The magnetic vector potential modifies the boundary condition on a Dirac spinor as

$$\Psi(r, \theta = 2\pi) = e^{ie \oint \vec{A}\cdot\vec{ds}}\,\Psi(r, \theta = 0). \tag{4}$$

Here $\vec{ds}$ is a line element on the circumference of the cone at a distance r from the apex, i.e.

$$\vec{ds} = \hat{e}_\theta \; r(1 - \frac{n}{6})d\theta. \tag{5}$$

Substituting (5) in (4) and assuming that the component A_θ of the magnetic vector potential is independent of the angle θ, we have from Equation(3)

$$A_\theta = \frac{1}{er}[\pm\frac{\frac{n}{4}\sigma_0}{(1-\frac{n}{6})} + \frac{\sigma_3}{2}]. \tag{6}$$

Thus an external Coulomb charge localized at the apex of the gapped graphene cone can be equivalently described by a suitable combination of electric charge and magnetic flux tube [74]. Replacing the ordinary derivatives in the Hamiltonian by the corresponding covariant derivatives, the Dirac equation for the low energy excitations of gapped graphene cone in presence of a Coulomb charge at its apex is given by

$$H\Psi(r,\theta) = \begin{pmatrix} m - \frac{\alpha}{r} & \partial_r - \frac{i}{r(1-\frac{n}{6})}\partial_\theta \pm \frac{\frac{n}{4}}{r(1-\frac{n}{6})} + \frac{1}{2r} \\ -\partial_r - \frac{i}{r(1-\frac{n}{6})}\partial_\theta \pm \frac{\frac{n}{4}}{r(1-\frac{n}{6})} - \frac{1}{2r} & -m - \frac{\alpha}{r} \end{pmatrix} \begin{pmatrix} \Psi_A(r,\theta) \\ \Psi_B(r,\theta) \end{pmatrix}$$

$$= E\begin{pmatrix} \Psi_A(r,\theta) \\ \Psi_B(r,\theta) \end{pmatrix}. \tag{7}$$

Let

$$\Psi(r,\theta) = \sum_j \begin{pmatrix} \Psi_A^{(j)}(r) \\ \Psi_B^{(j)}(r) \end{pmatrix} e^{ij\theta}, \tag{8}$$

where j is half-integer. Substituting (8) in (2), we obtain that the leading short distance behavior of the wavefunction is given by

$$\Psi^{(j)}_{A,B}(r) \sim r^{\beta-\frac{1}{2}} \quad \text{where} \quad \beta = \sqrt{\gamma^2-\alpha^2} \quad \text{and} \quad \gamma = \frac{(j \pm \frac{n}{4})}{(1-\frac{n}{6})}. \tag{9}$$

From (9) we can see that when $|\alpha|$ exceeds $|\gamma|$, β becomes imaginary. Therefore, the eigenstates $\Psi^{(j)}_A(r)$ and $\Psi^{(j)}_B(r)$ becomes wildly oscillatory and have no well defined limit as $r \to 0$. For massive excitations the critical coupling α_c corresponds to that value of α for which $E = -m$. When $m = 0$, the value of α_c is equal to the minimum allowed value of γ and depending on the magnitude of Dirac mass and boundary conditions α_c increases gradually from γ. It will be shown that the critical coupling for the gapped graphene cone explicitly depends on the angle of the cone and also on the product of gap and cutoff parameter.

Depending on the strength of the external Coulomb charge compared to that of the critical charge of a gapped graphene cone with a particular opening angle, the effect of the charge impurity on the cone can be analyzed in two separate regions: subcritical and supercritical.

3. Inequivalent quantizations for graphene cone in the presence of a subcritical Coulomb charge

In this Section we discuss the bound and scattering state solutions of the Dirac fermions in a gapped graphene cone in the presence of a subcritical Coulomb charge impurity. To get the solutions following [37], we consider the ansatz

$$\Psi^j_A(\rho) = \sqrt{m+E}e^{-\frac{\rho}{2}}\rho^{\beta-\frac{1}{2}}[F(\rho)+G(\rho)] \tag{10}$$

and

$$\Psi^j_B(\rho) = \sqrt{m-E}e^{-\frac{\rho}{2}}\rho^{\beta-\frac{1}{2}}[F(\rho)-G(\rho)], \tag{11}$$

where $\rho = 2\kappa r$, $\kappa = \sqrt{m^2-E^2}$, $\beta = \sqrt{\gamma^2-\alpha^2}$, $\gamma = \frac{(j\pm\frac{n}{4})}{(1-\frac{n}{6})}$ and total angular momentum j takes all half integer values. Bound states occur when the wavefunctions reduce to polynomials i.e. when

$$\beta - \frac{\alpha E}{\kappa} = -N, \tag{12}$$

where

$$N = \begin{cases} 0,1,2,..., & \text{when } \gamma > 0, \\ 1,2,3...., & \text{when } \gamma < 0. \end{cases} \tag{13}$$

The corresponding bound state spectra is obtained as

$$E_p = \frac{m \ \mathrm{sgn}(\alpha)}{\sqrt{1+\frac{\alpha^2}{(p+\beta)^2}}}. \tag{14}$$

Here the energy should be of the same sign (positive or negative) as α because otherwise the value of N will become negative and in our range of interest, it is not allowed (for details see [73]).

In the scattering state sector the parameter $\kappa = \sqrt{m^2-E^2}$ is purely imaginary, i.e. $\kappa = -ik$ [37], where k is defined as $k = \sqrt{E^2-m^2}$. Consequently, the variable ρ also becomes purely imaginary, $\rho = -2ikr$. Using the $r \to \infty$ limit of the scattering states the scattering matrix is obtained as

$$S(k) = (2ik)^{\frac{2i\alpha E}{k}} \frac{(\gamma + i\frac{m\alpha}{k})\,\Gamma\left(1+\beta-i\frac{\alpha E}{k}\right)}{\left(\beta - i\frac{E\alpha}{k}\right)\Gamma\left(1+\beta+i\frac{E\alpha}{k}\right)} e^{i\pi\left(\beta+i\frac{\alpha E}{k}\right)}. \tag{15}$$

From Equation(15) it can be seen that the poles of the S matrix determined by $\left(1+\beta-i\frac{\alpha E}{k}\right)=1-N$, where N is a nonzero positive integer and $\left(\beta-i\frac{E\alpha}{k}\right)=0$ when $\gamma>0$, gives back the corresponding bound states as expected.

The Dirac equation discussed in the previous section is valid only for low energy excitations. Therefore the short range interactions produced by the conical defect as well as the Coulomb charge impurity, cannot be incorporated as dynamical terms in the Dirac equation. However, a suitable choice of boundary conditions can take into account the combined effect of those short range interactions. In systems with unitary time evolution those allowed boundary conditions can be determined using the well defined prescription due to von Neumann[75, 76, 77, 78]. Below we recall the salient features of this idea.

A symmetric operator T with domain $D(T)\subset\mathcal{H}$ is self-adjoint if and only if

$$T=T^{*} \text{ and } D(T)=D(T^{*}).$$

where T^* denote the operator adjoint to T and $\mathcal{H}$ is the Hilbert space. To check whether T is self-adjoint or not we consider the equations

$$T^{*}\phi_{+}=+i\phi_{+}$$

$$T^{*}\phi_{-}=-i\phi_{-}.$$

Let $n_\pm$ denote the number of linearly independent square integrable solutions of the above two equations respectively. The pair (n_+,n_-) are called the deficiency indices for the operator T. The operator T can be classified in terms of the deficiency indices as follows:

1. T is essentially self-adjoint iff $(n_+,n_-)=(0,0)$.
2. T is not self-adjoint but has self-adjoint extensions iff $n_+=n_-=n\neq 0$.
3. If $n_+\neq n_-$,then T has no self-adjoint extensions.

If T admits self-adjoint extension, von Neumann's prescription tells us that its domain of self adjointness is given by

$$\begin{aligned} D_U(T) &= \left\{\phi+\phi_{+}+U\phi_{-}\middle|\phi\in D(T)\right.\\ &\qquad \left.\text{and } U \text{ is a } n\times n \text{ unitary matrix}\right\}. \end{aligned} \tag{16}$$

We can see from Equation(2) that the angular part of Dirac operator H operates on a domain $Y(\theta)$ which is spanned by the antiperiodic functions $e^{ij\theta}$ where j is a half integer and the corresponding boundary condition is kept unchanged. The radial Dirac operator H_ρ, given by

$$H_\rho=\begin{pmatrix} \rho\frac{d}{d\rho}+\left(\beta-\frac{\alpha E}{\kappa}\right) & -\left(\gamma+\frac{m\alpha}{\kappa}\right)\\ \left(-\gamma+\frac{m\alpha}{\kappa}\right) & \rho\frac{d}{d\rho}+\left(\beta-\rho+\frac{\alpha E}{\kappa}\right)\end{pmatrix},$$

is symmetric in the domain $\mathcal{D}_0=C_0^\infty(R^+)$ which consists of infinitely differentiable functions of compact support in the real half line R^+ and its adjoint operator $H_\rho^\dagger$ has the same expression as H_ρ but its domain can be different. Now to determine the domain of self-adjointness of the Dirac operator H, consider the equations

$$H^{\dagger}\Psi_{\pm}=\pm\frac{i}{l}\Psi_{\pm}, \tag{17}$$

where l has the dimension of length. The total number of square integrable, linearly independent solutions of Equation(17) gives the deficiency indices for H and they are denoted by $n_{\pm}$.

To find the deficiency indices $n_{\pm}$ let us first consider the following.

$$\Psi^{j}_{A\pm}(\rho) = \sqrt{m \pm \frac{i}{l}} e^{-\frac{\rho}{2}} \rho^{\beta - \frac{1}{2}} [F_{\pm}(\rho) + G_{\pm}(\rho)] \tag{18}$$

and

$$\Psi^{j}_{B\pm}(\rho) = \sqrt{m \mp \frac{i}{l}} e^{-\frac{\rho}{2}} \rho^{\beta - \frac{1}{2}} [F_{\pm}(\rho) - G_{\pm}(\rho)], \tag{19}$$

where $\rho = 2\kappa_1 r$, $\kappa_1 = \sqrt{m^2 + \frac{1}{l^2}}$, $\beta = \sqrt{\gamma^2 - \alpha^2}$ and $\gamma = \frac{(j \pm \frac{n}{4})}{(1 - \frac{n}{6})}$.

Using these expressions and proceeding as prescribed by von Neumann it can be shown that $\psi_{A\pm}$ and $\psi_{B\pm}$ are square integrable everywhere provided $0 < \beta < \frac{1}{2}$ and for this parameter range $n_{\pm} = 1$ (for details see [73]).

Thus we have a one parameter family of self-adjoint extensions, labeled by a real parameter $z \in R \bmod (2\pi)$ which enters the low energy dynamics of the Dirac fermions of gapped graphene cone through boundary conditions to ensure the unitary time evolution of the system. All the physically interesting quantities such as bound state energies, scattering phase shifts and LDOS depend on the choice of this self-adjoint extension parameter z. The parameter z cannot be determined theoretically, but instead must be obtained empirically. It would be interesting to see if there is any corresponding effect in the full tight-binding description of graphene. In terms of the system parameters and the self-adjoint extension parameter z the spectrum is determined by the equation

$$f(E) \equiv \left(\frac{\kappa^2}{1+m^2}\right)^{\beta} \frac{\left(1 - \beta - \frac{\alpha E}{\kappa}\right) \Gamma\left(1 + \beta - \frac{\alpha E}{\kappa}\right) \Gamma\left(1 - 2\beta\right)}{\left(1 + \beta - \frac{\alpha E}{\kappa}\right) \Gamma\left(1 - \beta - \frac{\alpha E}{\kappa}\right) \Gamma\left(1 + 2\beta\right)}$$

$$= \frac{\xi_1 \cos\left(\phi_1 + \frac{z}{2}\right)}{\xi_2 \cos\left(\phi_2 + \frac{z}{2}\right)}.$$

where

$$\sqrt{m + \frac{i}{l}}(P_+ + R_+) = \xi_1 e^{i\phi_1}$$

and

$$\sqrt{m + \frac{i}{l}}(Q_+ + S_+) = \xi_2 e^{i\phi_2}.$$

and P_+, Q_+, R_+ and S_+ are system parameter dependent constants (for detailed calculation see [73]). Though the equation cannot be solved analytically, from a typical plot of $f(E)$ it can be obtained numerically.

We have used the following expression given in Equation(20) for LDOS during the plotting.

$$\mu(E, r) = \frac{4}{\pi \hbar v_F} \sum_j |\Psi^{(j)}(k, r)|^2. \tag{20}$$

We have plotted the energy dependence of LDOS at a distance close to the charge impurity ($r = 1.2$). From these Figures we can observe that LDOS depend on the values of self-adjoint extension parameter z and also on the topology of the system. Therefore measurement of LDOS is a major experimental activity in graphene physics which should yield information about the self-adjoint extension parameter and the topology of the system.

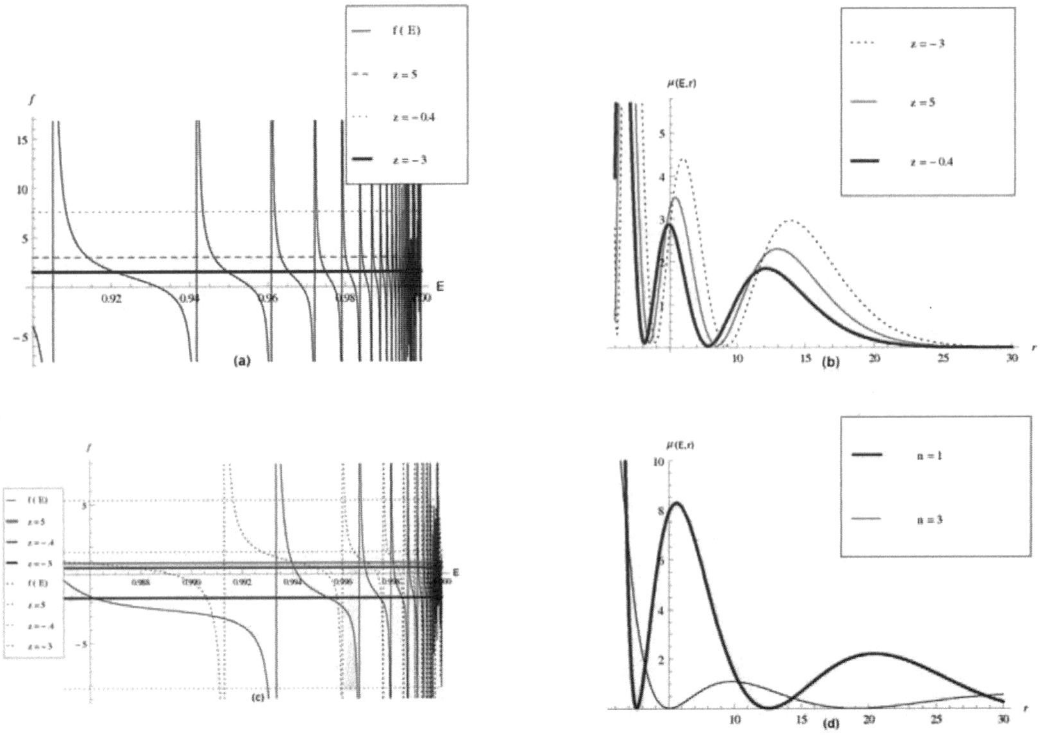

Figure 4. (a)Plot of $f(E)$ is shown for system parameters $j = \frac{3}{2}$, $n = 1$, $\alpha = 1.49$ and $m = 1$. The three horizontal line corresponds to the three different values of the self-adjoint extension parameter $z = 5, -0.4, -3$. (b) Dependence of LDOS in the bound state sector of the gapped graphene cone on the distance r from the external charge impurity is shown for three different values of bound state energy corresponding to three different values of self-adjoint extension parameter. Here the contribution from angular momentum channel $j = \frac{3}{2}$ is shown and the system parameters are $n = 1$, $\alpha = 1.49$ and $m = 1$. We have assumed $l = 1$. (c) Plot of $f(E)$ is shown for two different values of n (1 and 3) with system parameters $j = \frac{1}{2}$, $\alpha = 0.28$ and $m = 1$. The three horizontal line corresponds to the three different values of the self-adjoint extension parameter $z = 5, -0.4, -3$. The solid lines correspond to $n = 3$ and the dotted lines correspond to $n = 1$. (d) Dependence of LDOS on the distance r from the external charge impurity is shown for two different values of bound state energy corresponding to two different values of n with the self-adjoint extension parameter $z = -0.4$. Here the contribution from angular momentum channel $j = \frac{1}{2}$ is shown and the system parameters are $\alpha = 0.28$ and $m = 1$.

The scattering matrix $\mathbf{S}(k)$ for gapped graphene cone for the parameter range $0 < \beta < \frac{1}{2}$ is given as

$$\mathbf{S}(k) = (2ik)^{2i\frac{\alpha E}{k}} \frac{-\frac{\xi_1 \cos(\phi_1 + \frac{z}{2})}{\xi_2 \cos(\phi_2 + \frac{z}{2})}(2\kappa_1)^{2\beta}(2\kappa)^{-2\beta}\frac{1+f_2}{1+f_1} f_1 \frac{\Gamma(1+2\beta)}{\Gamma(1+\beta+i\frac{\alpha E}{k})} + f_2 \frac{\Gamma(1-2\beta)}{\Gamma(1-\beta+i\frac{\alpha E}{k})}}{-\frac{\xi_1 \cos(\phi_1 + \frac{z}{2})}{\xi_2 \cos(\phi_2 + \frac{z}{2})}(2\kappa_1)^{2\beta}(2\kappa)^{-2\beta}\frac{1+f_2}{1+f_1}\frac{\Gamma(1+2\beta)}{\Gamma(1+\beta-i\frac{\alpha E}{k})} e^{-i\pi(\beta+i\frac{\alpha E}{k})} + \frac{\Gamma(1-2\beta)}{\Gamma(1-\beta-i\frac{\alpha E}{k})} e^{-i\pi(-\beta+i\frac{\alpha E}{k})}},$$

where

$$f_1 \equiv \frac{\beta + \frac{i\alpha E}{k}}{\gamma - \frac{im\alpha}{k}}, \quad f_2 \equiv \frac{-\beta + \frac{i\alpha E}{k}}{\gamma - \frac{im\alpha}{k}}. \tag{21}$$

Thus we can see scattering phase shifts also depend explicitly on the self-adjoint extension

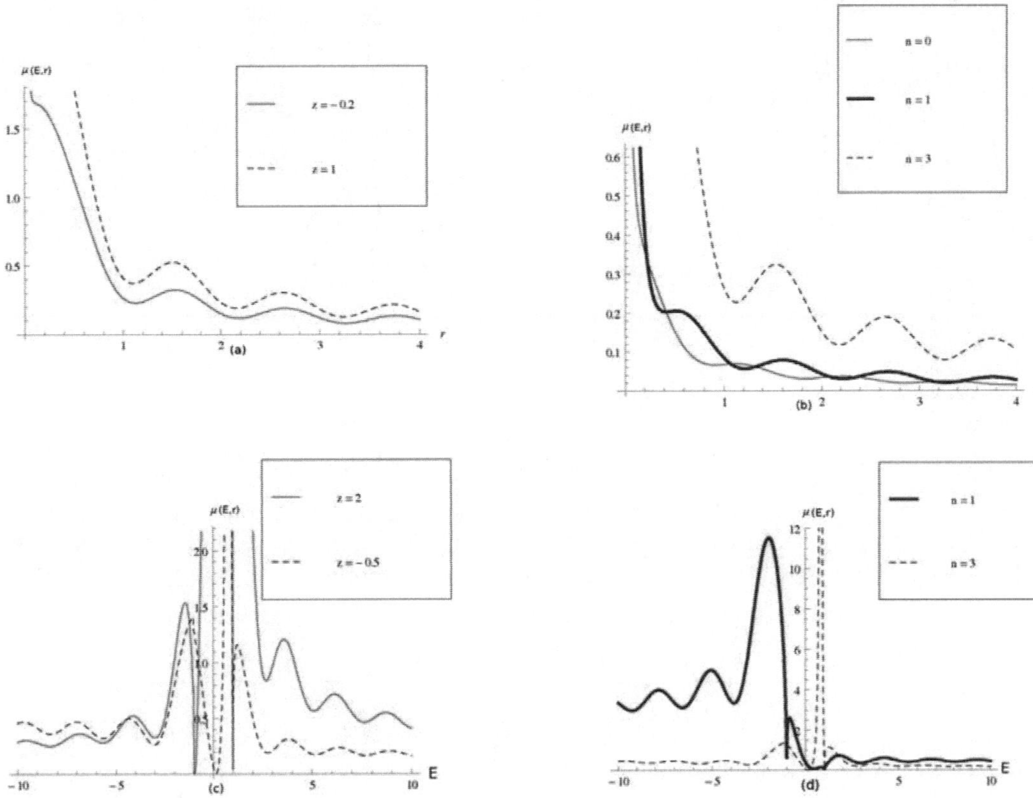

Figure 5. (a)Dependence of LDOS on the distance r from the Coulomb impurity is shown for two different values of self-adjoint extension parameter $z = -0.2, 1$ and a particular value of $E = 3$ and with $j = \frac{1}{2}$, $n = 3$, $\alpha = 0.28$ and $m = 1$. (b) Effect of topology on r dependence of LDOS is shown for self-adjoint extension parameter $z = -0.2$, $E = 3$ and system parameters $\alpha = 0.28$, $n = 0, 1, 3$ and $m = 1$ considering contribution coming from the angular momentum channel $j = \frac{1}{2}$. (c)Energy dependence of LDOS is shown for two different values of self-adjoint extension parameter $z = 5, -0.8$ at a distance $r = 1.2$ from the external Coulomb impurity. The system parameters used for the plot are $\alpha = 0.28$ and $m = 1$ and contribution coming from the angular momentum channel $j = \frac{1}{2}$ is considered. (d) Effect of topology on the energy dependence of LDOS is shown for self-adjoint extension parameter $z = -0.8$, angular momentum channel $j = \frac{1}{2}$ and system parameters $\alpha = -0.28$, $n = 1, 3$ and $m = 1$

parameter z and the topology of the system. For each value of z (mod 2π), we have an inequivalent set of the scattering data.

4. Effects of topological defects in the supercritical regime of gapped graphene

In the supercritical region β is always imaginary as the Coulomb potential strength α exceeds the value of γ. We denote $\beta = i\eta = \sqrt{\alpha^2 - \gamma^2}$. This region is marked by the quantum instability in the system. Therefore we need to regularize the Coulomb potential to study the effect of an external supercritical Coulomb charge on the gapped graphene cone. We shall also consider the zigzag edge boundary condition in our analysis.

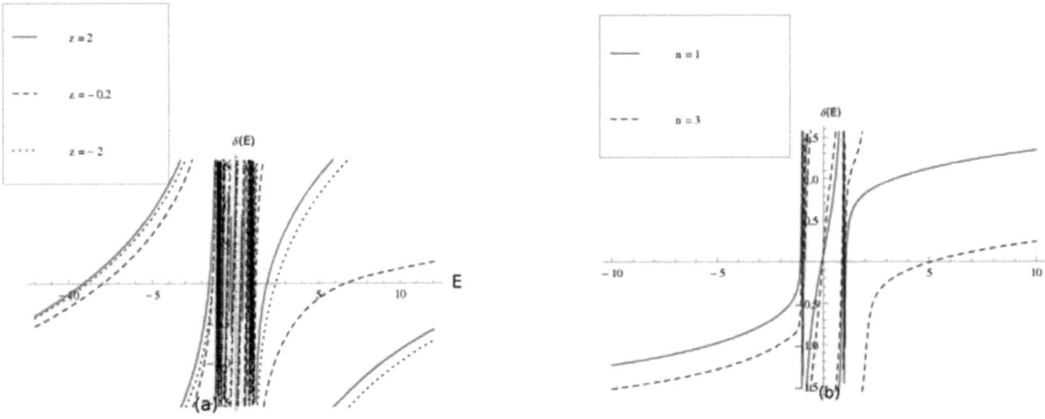

Figure 6. (a)Phase shifts in the gapped graphene cone is shown for three different values of the self-adjoint extension parameter $z = 2, -0.2, -2$ where the system parameters are $n = 1, j = \frac{3}{2}, \alpha = 1.46$, and $m = 1$. (b) Scattering phase shifts are shown for different angles of the gapped graphene cone with the self-adjoint extension parameter $z = -0.2$ and system parameters $j = \frac{1}{2}, \alpha = 0.28$ and $m = 1$.

4.1. Regularized Coulomb potential

In presence of a regularized Coulomb potential the bound states can be extended until the negative continuum $E = -m$ is reached[58, 59]. The regularization of the Coulomb potential is given by

$$V(r) = \begin{cases} -\alpha/r, & r > a, \\ -\alpha/a, & r \leq a. \end{cases} \tag{22}$$

Here a is the minimum distance of the Dirac electron from the apex where the Coulomb charge is placed and it is of the order of the lattice parameter. The Dirac equation for gapped graphene cone is solved for the two different regions $r > a$ and $r \leq a$ separately. Then we use the continuity condition of the wave function at $r = a$ to determine the quasibound state energy spectrum which is given by

$$f(E) \equiv f_r(E) = \text{Arg}[\Gamma(1 + 2i\eta)] \tag{23}$$

where

$$f_r(E) = \text{Arg}\left[\Gamma\left(1 + i\eta - \frac{E\alpha}{\kappa}\right)\right] + \text{Arg}\left[\gamma - \frac{\alpha}{\kappa}(m - E) + i\eta\right] + \eta \ln(2\kappa a)$$
$$+ \text{Arg}\left[\gamma - \alpha \frac{J_{|\gamma + \frac{1}{2}|}(\alpha)}{J_{|\gamma - \frac{1}{2}|}(\alpha)} - i\eta\right] + p\pi$$

and p is a positive integer (for details see [73]).

As the regularized potential allows the bound states to dive into negative energies and the critical charge of a gapped graphene cone refers to that value of a Coulomb potential for which $E = -m$, we have

$$\alpha_c = \gamma + \frac{\pi^2}{2\gamma \log^2[2m\gamma Ca]} \tag{24}$$

where

$$C = \exp\left[-2\psi(1) - \frac{J_{|\gamma-\frac{1}{2}|}(\gamma)}{\gamma(J_{|\gamma-\frac{1}{2}|}(\gamma) - J_{|\gamma+\frac{1}{2}|}(\gamma))}\right] \tag{25}$$

for the region near the critical potential. From Equation (24) we can see when $ma \to 0, \alpha_c \approx \gamma$ which agrees with the result obtained for massless case [74]. Also for $n = 0, \gamma = j$ and for $j = \frac{1}{2}$, the dependence of the critical charge on the gap and the cutoff parameter agrees with that obtained in [59].

4.2. Zigzag edge boundary condition

To find out the energies of the quasistationary states formed in the supercritical region now we use the zigzag edge boundary condition $\Psi_B^j(a) = 0$, where a is a distance from the apex, of the order of the lattice scale in graphene. The quasibound state energy spectrum is given by

$$f(E) \equiv f_z(E) = \text{Arg}[\Gamma(1 + 2i\eta)] \tag{26}$$

where

$$f_z(E) = \text{Arg}\left[\Gamma\left(i\eta - \frac{E\alpha}{\kappa}\right)\right] + \text{Arg}\left[\gamma + \frac{\alpha}{\kappa}(m + E) - i\eta\right] + \eta\ln(2\kappa a) + p\pi$$

and p is a positive integer (for detailed calculation see [73]). This quasibound state energy

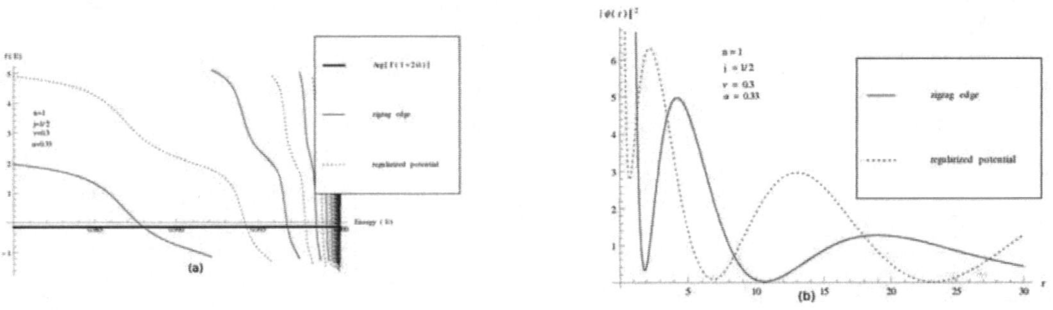

Figure 7. (a)Quasibound state energy spectrum with zigzag edge boundary condition and regularized Coulomb potential. Here the blue line represents $\text{Arg}[\Gamma(1 + 2i\eta)]$ and the dashed and the solid line represents $f_z(E)$ and $f_r(E)$ respectively. (b)Dependence of $|\Psi(r)|^2$ on the distance r from the charge impurity placed at the apex of the gapped graphene cone is shown for both the zigzag edge boundary condition and regularized Coulomb potential. The values of energy are obtained from the quasibound state energy spectrum.

spectrum for zigzag edge boundary can be compared with the spectrum obtained for the regularized potential and the effect of the two different boundary conditions on the spectrum can be observed from Fig.(7).

Proceeding in the same manner as for regularized potential, we can see with zigzag edge boundary the critical charge of the system is given by

$$\alpha_c = \gamma + \frac{\pi^2}{2\gamma\log^2[2m\gamma a\exp(-2\psi(1))]}. \tag{27}$$

Here also from Equation (27) we can see when $ma \to 0, \alpha_c \approx \gamma$ which agrees with the result obtained for massless case [74]. The critical charge has been shown to depend on the nonzero mass, cutoff parameter and the different opening angles of the gapped graphene cone. The dependence of the critical charge on ma is compared for two different boundary conditions.

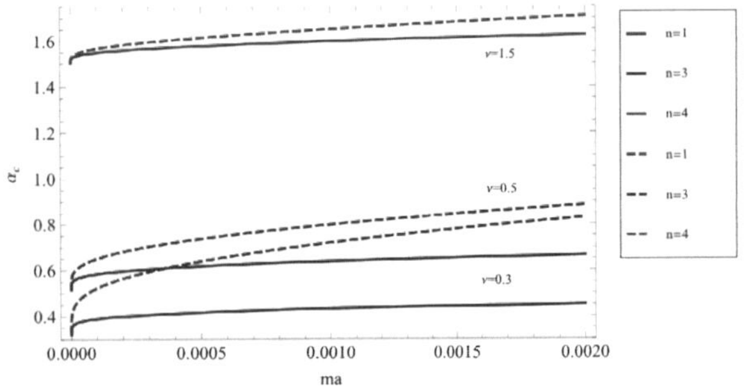

Figure 8. Dependence of critical charge on the nonzero mass and cutoff parameter are shown for both zigzag edge boundary condition and regularized Coulomb potential for different opening angles of the gapped graphene cone. The dotted lines show the dependence for zigzag edge boundary condition and the solid lines show the dependence for regularized Coulomb potential.

5. Conclusion

Inequivalent quantization arising from self-adjoint extensions is one of the earliest predictions of quantum mechanics. To experimentally realize that prediction, here in our analysis we have considered the opportunity provided by the massive Dirac fermions in gapped graphene in the presence of a conical defect and an external Coulomb impurity. The combination of this conical defect as well as the Coulomb impurity results in short distance interactions in the system. For a subcritical charge impurity, the effect of these short range interactions can be modelled through appropriate choice of boundary conditions which are labelled by a single real parameter and satisfies the requirement of a unitary time evolution. We have shown that the observables such as LDOS, scattering phase shifts and bound state energies depend explicitly on this parameter. This feature of gapped graphene cone agrees well with the prediction made for Dirac fermions in a plane in the presence of a cosmic string and thus the study of the gapped graphene cone considered here have established its importance for being more amenable to empirical analysis.

The supercritical regime of the Coulomb impurity have been analyzed with a regularized Coulomb potential as well as with a zigzag boundary condition. Quantum instability characterizes this regime and we have shown that the quasibound state spectra and the probability amplitude depend explicitly on the number of sectors removed from a planar graphene to form the cone. In addition, the dependence of the critical charge on the product of the Dirac mass m and the cutoff parameter a has been obtained. Though the nature of the dependence is similar for both the regularized Coulomb potential and the zigzag boundary condition but in the latter case the critical charge increases more rapidly with ma than the former case.

Thus the quantum dynamics of massive Dirac fermions of a graphene sample have been studied in presence of a conical topology and an external Coulomb charge and the study has offered a possibility to experimentally observe the inequivalent quantization of Dirac fermions in the presence of a string defect and many other interesting supercritical phenomena. More details of the analysis presented here can be found in [73].

References

[1] Jackiw R and Rebbi C 1976 *Phys. Rev.* D **13** 3398
[2] de Sousa Gerbert P 1989 *Phys. Rev.* D **40** 1346-49
[3] Hindmarsh M B and Kibble T W B 1995 *Rep. Prog. Phys* **58** 477
[4] Novoselov K S, Geim A K, Morozov S V, Jiang D, Katsnelson M I, Grigorieva I V and Firsov A A 2004 *Science* **306** 666
[5] Novoselov K S, Geim A K, Morozov S V, Jiang D, Katsnelson M I, Grigorieva I V, Dubonos S V and Firsov A A 2005 *Nature* **438** 197
[6] Zhang Y, Tan Y-W, Stormer H L and Kim P 2005 *Nature* **438** 201
[7] Gonzalez J, Guinea F and Vozmediano M A H 1992 *Phys. Rev. Lett.* **69** 172
[8] Gonzalez J, Guinea F and Vozmediano M A H 1993 *Nucl. Phys.* B **406** 771

[9] Lammert P E, Crespi V H 2000 *Phys. Rev. Lett.* **85** 5190
[10] Lammert P E, Crespi V H 2004 *Phys. Rev.* B **69** 035406
[11] Kolesnikov D V and Osipov V A 2006 *Eur. Phys. J.* B **49** 465
[12] Sitenko Y A and Vlasii N D 2007 *Nucl. Phys.* B **787** 241
[13] Cortijo A and Vozmediano M A H 2007 *Nucl. Phys.* B **763** 293
[14] de Juan F, Cortijo A and Vozmediano M A H 2007 *Phys. Rev.* B **76** 165409
[15] Hou C Y, Chamon C and Mudry C 2007 *Phys. Rev. Lett.* **98** 186809
[16] Pachos J K, Stone M and Temme K 2008 *Phys. Rev. Lett.* **100** 156806
[17] Furtado C, Moraes F, Carvalho A M de M 2008 *Phys. Lett.* A **372** 5368
[18] Roy A and Stone M 2010 *J. Phys.* A **43** 015203
[19] Vozmediano M A H, Katsnelson M I and Guinea F 2010 *Phys. Reports* **496** 109-148
[20] Gonzalez J and Herrero J 2010 *Nucl. Phys.* B **825** 426-443
[21] Yazyev O V and Louie S G 2010 *Phys. Rev.* B **81** 195420
[22] Fonseca J M, Moura-Melo W A and Pereira A R 2010 *Phys. Lett.* A **374** 4359
[23] de Juan F, Cortijo A, Vozmediano M A H and Cano A 2011 *Nature Physics* **7** 810
[24] Abedpour N, Asgari R and Guinea F 2011 *Phys. Rev.* B **84** 115437
[25] Bakke K, Petrov A Y and Furtado C 2011 Kaluza-Klein description of geometric phases in graphene *Preprint* arXiv:1111.0836 [cond-mat.mes-hall]
[26] Cortijo A, Guinea F and Vozmediano M A H 2012 *J. Phys.* A **45** 383001
[27] Wallace P R 1947 The band theory of graphite. *Phys. Rev.* **71** 622
[28] DiVincenzo D P and Mele E J 1984 *Phys. Rev.* B **29** 1685
[29] Semenoff G W 1984 *Phys. Rev. Lett.* **53** 2449
[30] Geim A K and Novoselov K S 2007 *Nature Materials* **6** 183
[31] Castro Neto A H, Guinea F, Peres N M R, Novoselov K S and Geim A K 2009 *Rev. Mod. Phys.* **81** 109
[32] Peres N M R 2010 *Rev. Mod. Phys.* **82** 2673
[33] Das Sarma S, Adam S, Hwang E H and Rossi E 2011 *Rev. Mod. Phys.* **83** 407
[34] Gorbar E V, Gusynin V P, Miransky V A and Shovkovy I A 2002 *Phys. Rev.* B **66** 045108
[35] Khveshchenko D V and Leal H 2004 *Nucl. Phys.* B **687** 323
[36] Kane C L and Mele E J 2005 *Phys. Rev. Lett.* **95** 226801
[37] Novikov D S 2007 *Phys. Rev.* B **76** 245435
[38] Gusynin V P, Miransky V A, Sharapov S G and Shovkovy I A 2008 *Phys. Rev.* B **77** 205409
[39] Khveshchenko D V 2009 *J. Phys.: Condens. Matter* **21** 075303
[40] Gamayun O V, Gorbar E V and Gusynin V P 2010 *Phys. Rev.* B **81** 075429
[41] Li W, Liu G-Z 2010 *Phys. Lett.* A **374** 2957
[42] Araki Y and Hatsuda T 2010 *Phys. Rev.* B **82** 121403 (R)
[43] Araki Y 2011 *J. Phys.: Conf. Series* **302** 012022
[44] Zhu J L, Sun S and Yang N 2012 *Phys. Rev.* B **85** 035429
[45] Han M Y, Ozyilmaz B, Zhang Y and Kim P 2007 *Phys. Rev. Lett.* **98** 206805
[46] Zhou S Y *et al* 2007 *Nat. Mater.* **6** 770
[47] Zhou S Y *et al* 2008 *Nat. Mater.* **7** 259
[48] Haberer D *et al* 2010 *Nano Letters*, **10** 3360
[49] Cooper D R *et al* 2012 *ISRN Condensed Matter Physics* Vol. **2012** 501686
[50] Andrei E Y *et al* 2012 *Rep. Prog. Phys.* **75** 056501
[51] Ki D-K, Morpurgo A F 2012 *Phys. Rev. Lett.* **108** 266601
[52] de Sousa Gerbert P and Jackiw R 1989 *Commun. Math. Phys.* **124** 229-260
[53] Yamagishi H 1983 *Phys. Rev.* D **27** 2383
[54] Jackiw R and Pi S Y 2007, *Phys. Rev. Lett.* **98** 266402
[55] Chamon C, Hou C Y, Jackiw R, Mudry C, Pi S Y and Semenoff G 2008 *Phys.Rev.* B **77** 235431
[56] Jackiw R and Pi S Y 2008 *Phys. Rev.* B **78** 132104
[57] Jackiw R, Milstein A I, Pi S Y and Terekhov I S 2009 *Phys. Rev.* B **80** 033413
[58] Pereira V M, Kotov V N and Castro Neto A H 2008 *Phys. Rev.* B **78** 085101
[59] Gamayun O V, Gorbar E V and Gusynin V P 2009 *Phys. Rev.* B **80** 165429
[60] Gamayun O V, Gorbar E V and Gusynin V P 2011 *Phys. Rev.* B **83** 235104
[61] Reinhardt J and Greiner W 1977 *Rep. Prog. Phys.* **40** 219
[62] Greiner W, Müller B and Rafelski J 1985 *Quantum Electrodynamics of Strong Fields* (Berlin: Springer-Verlag)
[63] Katsnelson M I, Novoselov K S and Geim A K 2006 *Nature Physics* **2** 620
[64] Katsnelson M I and Novoselov K S 2007 *Solid State Communications* **143** 3
[65] Pereira V M, Nilsson J and Castro Neto A H 2007 *Phys. Rev. Lett.* **99** 166802
[66] Shytov A V, Katsnelson M I and Levitov L S 2007 *Phys. Rev. Lett.* **99** 236801

[67] Shytov A V, Katsnelson M I and Levitov L S 2007 *Phys. Rev. Lett.* **99** 246802
[68] Shytov A, Rudner M, Gu N, Katsnelson M and Levitov L 2009 *Solid State Comm.* **149** 1087
[69] Gupta K S and Sen S 2009 *Mod. Phys. Lett.* A **24** 99
[70] Milstein A I, Terekhov I S 2010 *Phys. Rev.* B **81** 125419
[71] Wang J, Fertig H A and Murthy G 2010 *Phys. Rev. Lett.* **104** 186401
[72] Kotov V N, Uchoa B, Pereira V M, Guinea F and Castro Neto A H 2012 *Rev. Mod. Phys.* **84** 1067
[73] Chakraborty B, Gupta K S and Sen S 2012 *Preprint* arXiv:1207.5705v3 [hep-th], to appear in J. Phys. **A**.
[74] Chakraborty B, Gupta K S and Sen S 2011 *Phys. Rev.* B **83** 115412
[75] Reed M and Simon B 1972 *Methods of Modern Mathematical Physics* Vol. **2** (New York: Academic Press)
[76] Falomir H and Pisani P A G 2001 *J. Phys. A : Math. Gen.* **34** 4143
[77] Gupta K S and Sen S 2008 *Phys. Rev.* B **78** 205429
[78] Gupta K S, Samsarov A and Sen S 2010 *Eur. Phys. J.* B **73** 389

Journal of Physics: Conference Series **442** (2013) 012018 doi:10.1088/1742-6596/442/1/012018

Amplitudes for spacetime regions and the quantum Zeno effect: pitfalls of standard path integral constructions

J J Halliwell
Blackett Laboratory
Imperial College
London SW7 2BZ
UK

E-mail: j.halliwell@imperial.ac.uk

J M Yearsley
Centre for Quantum Information and Foundations, DAMTP, Centre for Mathematical Sciences, University of Cambridge, Wilberforce Road, Cambridge CB3 0WA, UK

E-mail: jmy27@cam.ac.uk

Abstract. Path integrals appear to offer natural and intuitively appealing methods for defining quantum-mechanical amplitudes for questions involving spacetime regions. For example, the amplitude for entering a spatial region during a given time interval is typically defined by summing over all paths between given initial and final points but restricting them to pass through the region at any time. We argue that there is, however, under very general conditions, a significant complication in such constructions. This is the fact that the concrete implementation of the restrictions on paths over an interval of time corresponds, in an operator language, to sharp monitoring at every moment of time in the given time interval. Such processes suffer from the quantum Zeno effect – the continual monitoring of a quantum system in a Hilbert subspace prevents its state from leaving that subspace. As a consequence, path integral amplitudes defined in this seemingly obvious way have physically and intuitively unreasonable properties and in particular, no sensible classical limit. In this paper we describe this frequently-occurring but little-appreciated phenomenon in some detail, showing clearly the connection with the quantum Zeno effect. We then show that it may be avoided by implementing the restriction on paths in the path integral in a "softer" way. The resulting amplitudes then involve a new coarse graining parameter, which may be taken to be a timescale ϵ, describing the softening of the restrictions on the paths. We argue that the complications arising from the Zeno effect are then negligible as long as $\epsilon \gg 1/E$, where E is the energy scale of the incoming state. Our criticisms of path integral constructions largely apply to approaches to quantum theory such as the decoherent histories approach or quantum measure theory, which do not specifically involve measurements. We address some criticisms of our approach by Sokolovksi, concerning the relevance of our results to measurement-based models.

1. Introduction

Consider the following question in non-relativistic quantum mechanics for a point particle in d dimensions: What is the amplitude $g_{\Delta}(\mathbf{x}_f, t_f|\mathbf{x}_0, t_0)$ for the particle to start at a spacetime point $(\mathbf{x}_0, t_0)$, pass through a spatial region Δ and end at a spacetime point $(\mathbf{x}_f, t_f)$? The seemingly-obvious answer to this question is surely the path integral expression,

$$g_\Delta(\mathbf{x}_f, t_f|\mathbf{x}_0, t_0) = \int_\Delta \mathcal{D}\mathbf{x}(t)\ \exp\left(i\int_{t_0}^{t_f} dt \left[\frac{1}{2}m\dot{\mathbf{x}}^2 - U(\mathbf{x})\right]\right) \tag{1}$$

(We choose units in which $\hbar = 1$). In this expression, the paths $\mathbf{x}(t)$ summed over satisfy the initial condition $\mathbf{x}(t_0) = \mathbf{x}_0$, the final condition $\mathbf{x}(t_f) = \mathbf{x}_f$ and pass, at any intermediate time, through the region Δ [1, 2, 3], as depicted in Fig.1. The point of this paper is to argue that

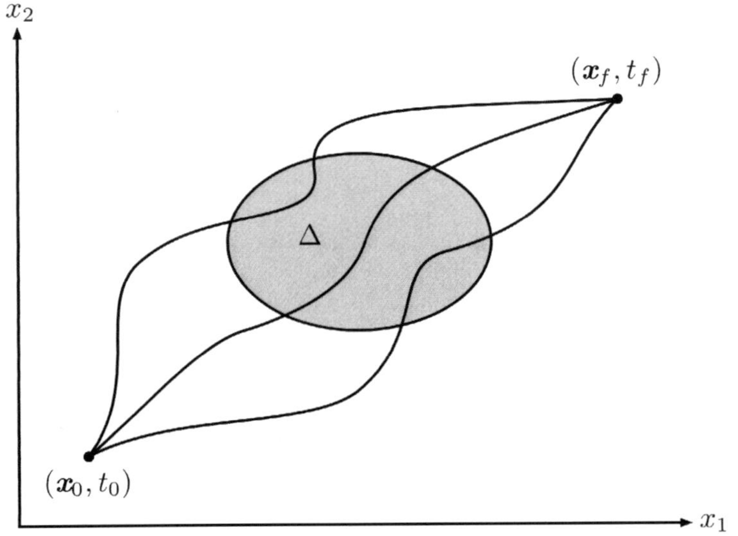

Figure 1. The amplitude Eq.(1) is obtained by summing over paths which enter the spatial region Δ at any time between t_0 and t_f.

there are potential problems with this seemingly-obvious answer.

We first develop these ideas further. We might similarly assert that the amplitude $g_r(\mathbf{x}_f, t_f|\mathbf{x}_0, t_0)$ for the particle never entering the region Δ is a given by a path integral expression of the form

$$g_r(\mathbf{x}_f, t_f|\mathbf{x}_0, t_0) = \int_r \mathcal{D}\mathbf{x}(t)\ \exp\left(i\int_{t_0}^{t_f} dt \left[\frac{1}{2}m\dot{\mathbf{x}}^2 - U(\mathbf{x})\right]\right) \tag{2}$$

This object, the restricted propagator, is given by a sum over paths restricted to lie always outside Δ. We then clearly have

$$g(\mathbf{x}_f, t_f|\mathbf{x}_0, t_0) = g_\Delta(\mathbf{x}_f, t_f|\mathbf{x}_0, t_0) + g_r(\mathbf{x}_f, t_f|\mathbf{x}_0, t_0) \tag{3}$$

where g is the usual propagator obtained by summing over all paths from initial to final point.

Objects such as Eq.(2) have to be given proper mathematical definition. For the moment, we have in mind a definition involving the usual time-slicing procedure, in which the time interval is divided up into n equal intervals of size ϵ so that $t_f - t_0 = n\epsilon$ and we consider propagation between slices labelled by times $t_k = t_0 + k\epsilon$, where $k = 0, 1 \cdots n$. Eq.(2) is then defined by a limit of the form

$$g_r(\mathbf{x}_f, t_f|\mathbf{x}_0, t_0) = \lim_{\epsilon\to 0, n\to\infty} \int_r d^d x_1 \cdots \int_r d^d x_{n-1} \prod_{k=1}^{n} g(\mathbf{x}_k, t_k|\mathbf{x}_{k-1}, t_{k-1}) \tag{4}$$

where $\mathbf{x}_n = \mathbf{x}_f$ and the integrals are over the region outside Δ. The propagator is approximated by

$$g(\mathbf{x}_k, t_k|\mathbf{x}_{k-1}, t_{k-1}) = \left(\frac{m}{2\pi i\epsilon}\right)^{d/2} \exp\left(iS(\mathbf{x}_k, t_k|\mathbf{x}_{k-1}, t_{k-1})\right) \tag{5}$$

for small times, where the exponent is the action between the indicated initial and final points and the limit is taken in such a way that $n\epsilon$ is fixed.

Writing the above amplitudes in operator form, $\hat{g}_\Delta$, $\hat{g}_r$, where, for example,

$$g_r(\mathbf{x}_f, t_f|\mathbf{x}_0, t_0) = \langle \mathbf{x}_f|\hat{g}_r(t_f, t_0)|\mathbf{x}_0\rangle \tag{6}$$

it seems reasonable to suppose that the probabilities for a particle in initial state $|\psi\rangle$ entering or not entering the region Δ during the time interval $[t_0, t_f]$ are given by

$$p_\Delta = \langle\psi|\hat{g}_\Delta(t_f, t_0)^\dagger \hat{g}_\Delta(t_f, t_0)|\psi\rangle \tag{7}$$

$$p_r = \langle\psi|\hat{g}_r(t_f, t_0)^\dagger \hat{g}_r(t_f, t_0)|\psi\rangle \tag{8}$$

To be sensible probabilities these expressions should obey the simple sum rule

$$p_\Delta + p_r = 1 \tag{9}$$

but using $\hat{g} = \hat{g}_\Delta + \hat{g}_r$, it is easy to see that this is generally not the case unless the interference between the two types of paths vanishes, which means that

$$\mathrm{Re}\ \langle\psi|\hat{g}_r(t_f, t_0)^\dagger \hat{g}_\Delta(t_f, t_0)|\psi\rangle = 0 \tag{10}$$

This condition can hold, perhaps approximately, for certain types of initial states, although this may be non-trivial to prove and there is no guarantee that those states are physically interesting ones.

Path integral expressions of the above general type have been postulated and used extensively in a wide variety of circumstances in which time is involved in a non-trivial way. Indeed, Feynman's original paper on path integrals was entitled, "Space-time approach to non-relativistic quantum mechanics", clearly suggesting that the spacetime features of the path integral construction should be made use of [1]. Path integral constructions have for many years played an important role in quantum cosmology and quantum gravity [4, 5, 6, 7, 8, 9, 10, 11, 12, 13, 14, 15, 16, 17, 18, 19, 20, 21, 22]. They have also been particularly useful in addressing issues concerning time in non-relativistic quantum mechanics, for example in studies of the arrival time [23, 24, 25, 26, 27, 28, 29, 30, 31, 32, 33, 34], dwell time and tunneling time [35, 36, 37, 38, 39]. Path integrals are also useful ways of formulating continuous quantum measurement theory [40, 41]

Many of these applications of path integrals are in the specific framework of the decoherent histories approach to quantum theory [43, 44, 45, 46, 47, 48, 49, 50, 51, 52, 53] (in which the relations Eqs.(9), (10) arise) and quantum measure theory [54, 55]. Also related are the various attempts to derive the Hilbert space formulation of quantum theory from path integral constructions [56]. Here we wish to study path integral expressions as entities in their own right without being tied to a specific approach to quantum theory.

However, as indicated, there is a problem with the innocent-looking and informally appealing path integral expressions Eqs.(1), (2), with the consequence that their properties can be very different to intuitive expectations. This difficulty goes beyond the problem of satisfying the no-interference condition, Eq.(10), although is related to it. To see it, let us focus on the amplitude for not entering the region Δ, Eq.(2). The key issue is that the innocent-looking restriction on paths effectively means that an initial state propagated by Eq.(2) is required to have zero support in the region Δ at *every moment of time* between t_1 and t_2. Evolution with this propagator therefore suffers from the quantum Zeno effect – the fact that continual monitoring of a quantum system in a Hilbert subspace prevents it from leaving that subspace [57, 58, 59, 60, 61, 62, 63, 64, 65]. The consequence is that the amplitude Eq.(2), or equivalently $\hat{g}_r(t_2, t_1)$, actually describes *unitary* propagation on the Hilbert space of states with support only in the region outside Δ and therefore gives probability $p_r = 1$ for any incoming state. This then means that either the sum rule Eq.(9) is not satisfied in which case the probabilities are not meaningful, or that it is satisfied but $p_\Delta = 0$, which means that any incoming state aimed at Δ has probability zero for entering that region, a physically nonsensical result.

Differently put, the innocent-looking restriction on paths in the restricted propagator Eq.(2) with the usual implementation Eq.(4) effectively sets *reflecting* boundary conditions on the boundary of Δ, which means that any incoming state is totally reflected under propagation by $\hat{g}_r$. To obtain the intuitively sensible result, we would need a propagator analogous to Eq.(2) in which the incoming state is *absorbed*.

One could perhaps look for another way out, which is to suppose that $\hat{g}_\Delta$ is related to some sort of measurement scheme, in which case there is no obligation to satisfy the probability sum rule and that Eq.(7) may then give a reasonable formula for the probability of entering. To this end it is useful to give a more detailed formula for g_Δ using the path decomposition expansion (PDX) [66, 67, 68]. The paths summed over in Eq.(1) may be partitioned according to the time t and location $\mathbf{y}$ at which they cross the boundary Σ of Δ for the first time. (See Fig.2) As a

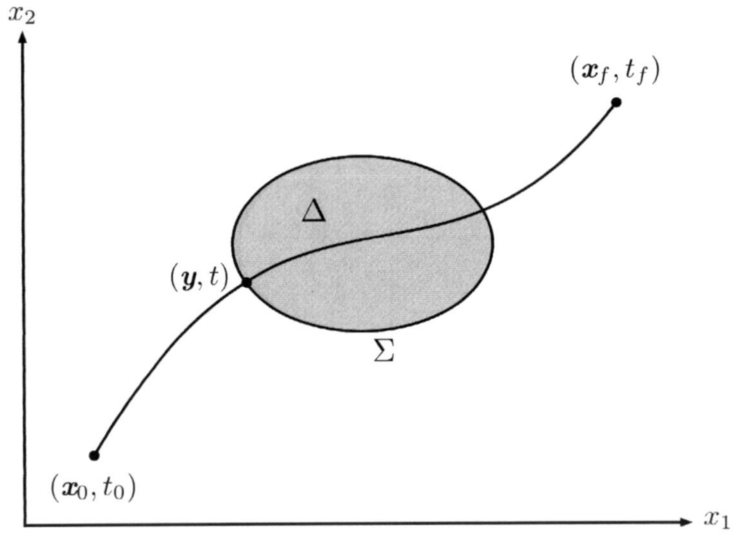

Figure 2. The path decomposition expansion Eq.(11). Each path from $(\mathbf{x_0}, \mathbf{t_0})$ to $(\mathbf{x}_f, t_f)$ passing through Δ may be labeled by its time t and location $\mathbf{y}$ of first crossing of the boundary Σ.

consequence, it is possible to derive the formula

$$g_\Delta(\mathbf{x}_f, t_f|\mathbf{x}_0, t_0) = \frac{i}{2m}\int_{t_0}^{t_f} dt \int_\Sigma d^{d-1}y \; g(\mathbf{x}_f, t_f|\mathbf{y}, t) \; \mathbf{n}\cdot\nabla_{\mathbf{x}} g_r(\mathbf{x}, t|\mathbf{x}_0, t_0)\big|_{\mathbf{x}=\mathbf{y}} \tag{11}$$

where $\mathbf{n}$ is the outward pointing normal to Σ. The restricted propagator $g_r(\mathbf{x}, t|\mathbf{x_1}, t_0)$ vanishes when either end is on Σ but its normal derivative does not. In fact, the normal derivative of the restricted propagator in Eq.(11) represents the sum over paths which are restricted to remain outside Δ but end on its boundary [66, 67, 68].

Although on the face of it, Eq.(11) is a plausible formula for the crossing amplitude, the problem with this expression, which is fully equivalent to Eq.(3), concerns the treatment of paths that are outside Δ before their first crossing at t. The fact that these paths are represented using the restricted propagator means that any such paths arriving at Σ before t are reflected, rather than simply dropped from the sum. This means an incoming wave packet propagated using Eq.(11) will be partly reflected and we anticipate that this formula could fail to give intuitively sensible results.

One way or another, we find that the simple and seemingly obvious notion of "restricting paths" leads to the quantum Zeno effect and hence to unphysical results. This is the sense in which we would say that path integral constructions may suffer from pitfalls.

The purpose of this paper is to give a concise statement of the general problem outlined above with naive path integral expressions of the form Eqs.(1), (2), and to point out how to construct practically useful modified path integrals which give a meaningful answer to the question of assigning amplitudes to spacetime regions with a sensible classical limit.

Many papers using path integral constructions of the above type appear to be oblivious to this effect and its consequences. Although the results of such papers are not necessarily wrong, they often do not have sensible classical limits and in measurement-based models, they may, unknowingly, only be valid in the regime of strong measurement. These problems with the Zeno effect in path integral constructions can be seen in a number of early attempts to define probabilities for spacetime regions in the context of the decoherent histories approach [23, 24, 25, 26, 27, 28, 32] and in some papers on the decoherent histories approach to quantum cosmology [10, 11, 12, 13, 14]. A possible resolution was given in Refs.[29, 15]. There are undoubtedly many other places in which this difficulty has been encountered.

Here, our aim is to give a general account of the problem and solution, independent of specific approaches to quantum theory and of specific applications. We give a detailed formulation of the problem in Section 2 and present the proposed solution in Section 3. Some examples are discussed in Section 4. In Section 5, we give a detailed discussion of the connections with other work. In particular, we briefly address a recent criticism by Sokolovski [69] of an earlier account of our work [70]. We summarize and conclude in Section 6.

2. Detailed formulation of the problem

We first discuss the generality of the problem with path integrals outlined above. The example in Eqs.(1), (2) is in non-relativistic quantum mechanics in any number of dimensions. The region Δ can consist of a number of disconnected pieces and then there could be many different ways of partitioning the paths according to how many different regions they enter or not. We will also assume that Δ is reasonably large and that its boundary is reasonably smooth, in comparison to any quantum-mechanical lengthscales set by the incoming state. Furthermore, in the situation depicted in Fig.1, the spatial region is constant in time, but there is no obstruction to allowing it to vary with time. For example, in relativistic quantum mechanics, it may be natural to look at regions whose boundaries are null surfaces [15]. One could also contemplate path integrals in curved spacetimes which may have unusual properties, such as closed timelike curves [10], but the basic ideas of path integration are still applicable. An important area of application of these ideas is to quantum cosmology, where there is no explicit physical time coordinate, but the basic object Eq.(1) is still the appropriate starting point for the construction of class operators in the decoherent histories analysis of quantum cosmology [10, 14, 15]. This is clearly not an exhaustive list of possibilities, but the arguments presented in what follows will apply to all of these cases, even though they are presented in the context of the example Eqs.(1), (2).

We now set out in more mathematical detail the argument outlined in the Introduction. We focus on the path integral representation of the restricted propagator Eq.(2) and its time-slicing representation Eq.(4). We will give an equivalent operator form for Eq.(4). We introduce the projector onto Δ,

$$P = \int_{\Delta} d^d x \, |\mathbf{x}\rangle\langle\mathbf{x}| \tag{12}$$

and its negation, $\bar{P} = 1 - P$, the projector onto the region $\bar{\Delta}$, the region of $\mathbb{R}^d$ outside of Δ. Again we divide the time interval up into n equal intervals of size ϵ so that $t_f - t_0 = n\epsilon$. It is then easy to see that, in terms of the operator $\hat{g}_r$ defined in Eq.(6), the time-slicing expression Eq.(4) is equivalent to

$$\hat{g}_r(t_f, t_0) = \lim_{\epsilon \to 0, n \to \infty} \bar{P} e^{-iH\epsilon} \bar{P} \cdots e^{-iH\epsilon} \bar{P} \tag{13}$$

where there are $n+1$ projectors and n unitary evolution factors $e^{-iH\epsilon}$ and the limit is taken in such a way that $n\epsilon = t_f - t_0$ is fixed. That is, inserted in Eq.(6), Eq.(13) gives the time-slicing definition of the path integral expression, Eq.(4). (Note that there are $n+1$ projectors $\bar{P}$ in Eq.(13) but only $n-1$ corresponding integrals in Eq.(4). The two extra projectors are redundant, and hence the two expressions completely equivalent, if we take $\mathbf{x}_0$ and $\mathbf{x}_f$ to be outside Δ).

One can clearly see from the operator form Eq.(13) that it involves "monitoring" of the particle to check if it is in $\bar{\Delta}$ at each instant of time. The limit in Eq.(13) may be computed explicitly [63] and leads to the explicit form

$$\hat{g}_r(t_f, t_0) = \bar{P} \exp\left(-i\bar{P}H\bar{P}(t_f - t_0)\right) \tag{14}$$

This propagator is, as claimed, unitary on the Hilbert space of states with support in $\bar{\Delta}$ [63]. It therefore describes the situation in which an incoming state never actually leaves $\bar{\Delta}$ due to monitoring becoming infinitely frequent, which is clearly the quantum Zeno effect. In simple examples of the restricted propagator, one can easily see that an incoming state is totally reflected off the boundary of the region Δ. Eq.(14) and its properties explain why the naive path integral expressions Eqs.(1), (1.2) have counter-intuitive properties which lead to unphysical results if not used sufficiently carefully.

It is also reasonable to consider other possible methods for defining the path integral Eq.(2). Another natural method is to consider the imaginary time version of Eq.(2) and then define the path integral in terms of the limit of a stochastic process involving random walks on a spacetime lattice (in the case $U = 0$, for simplicity) [26, 71]. Restricting the paths to lie outside Δ means finding a suitable boundary condition on the random walks at the boundary of Δ. Most studies of this problem have imposed reflecting boundary conditions, which appear to be the easiest ones to impose in practice. As a consequence, this method of defining the path integral once again leads to the restricted propagator of the form Eq.(14). However, it is not clear that one is compelled to use reflecting boundary conditions in such implementations of the path integral. We will discuss this further below.

3. Proposed solution

Since the Zeno effect is the root of the problem, the solution is clearly to limit or soften the monitoring of the system in some way in Eq.(13) so that reflection is minimized. The first obvious way to do this is to decline to take the limit in Eq.(13) and keep the time-spacing ϵ finite. The second way is to replace the exact projectors by POVMs. These solutions have been explored in the specific context of the arrival time problem in one dimension [29, 30], but the purpose here is to present these solutions in a more generally applicable way.

In the first approach, we therefore define a modified propagator for not entering Δ by

$$\hat{g}_r^{\epsilon}(t_f, t_0) = \bar{P} e^{-iH\epsilon} \bar{P} \cdots e^{-iH\epsilon} \bar{P} \tag{15}$$

where as before there are $n+1$ projectors and n unitary time operators and $n\epsilon = t_f - t_0$. This object can also be represented by a path integral expression of the form Eq.(2), except that the paths are required to be outside Δ only at the $n+1$ times $t_0 + k\epsilon$, where $k = 0 \cdots n$, but between these times the paths are unrestricted. This situation is depicted in Fig.3 for a one-dimensional example in which the region Δ is the interval $[a, b]$.

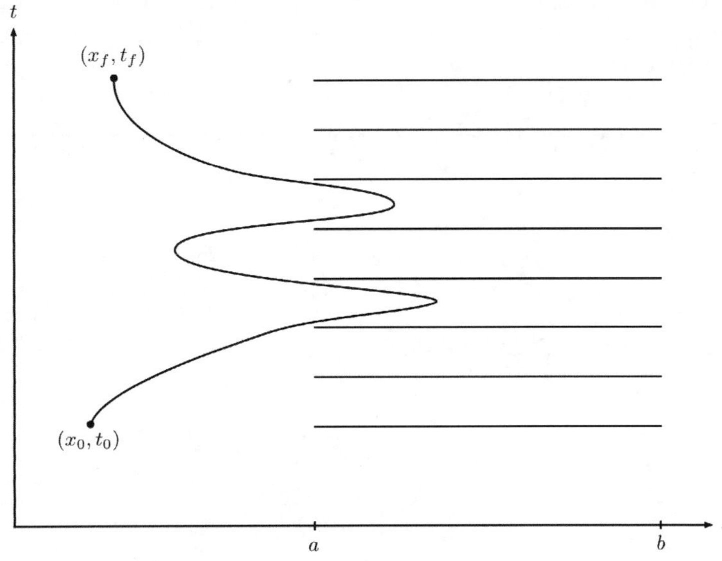

Figure 3. The paths summed over in the path integral representation of the modified propagator Eq.(15). The paths may enter the region $[a, b]$ but not at the times at which the projectors act. Very wiggly paths, which enter the region frequently, will generally have small contribution to the path integral in a semiclassical approximation, which means that there is an effective suppression of paths entering the region.

This modified propagator involves a *new coarse graining parameter* ϵ, describing the precision to within which the paths are monitored. The original propagator Eq.(2) is obtained in the limit of infinite precision $\epsilon \to 0$. The physically interesting case, however, is that in which ϵ is small enough to monitor the paths well, but sufficiently large that an incoming state is not significantly affected by reflection.

There is a very useful approximate alternative to Eq.(15), which is also helpful in terms of calculating the timescale required to define what "small" and "large" mean for ϵ. For the special case of projectors onto the positive x-axis in one dimension, $P = \theta(\hat{x})$, it has been shown [72, 73] that the string of projectors Eq.(15) is to a good approximation equivalent to evolution in the presence of a complex potential consisting of a window function on the region Δ, that is,

$$\bar{P}e^{-iH\epsilon}\bar{P} \cdots e^{-iH\epsilon}\bar{P} \approx \exp\left(-(iH + V_0 P)(t_f - t_0)\right) \tag{16}$$

Here, the real parameter V_0 depends on ϵ and it was shown numerically [73] that the best match is obtained with the choice

$$\epsilon V_0 \approx 4/3 \tag{17}$$

The approximate equivalence Eq.(16) is expected to hold when acting on states with energy width ΔH for which $\epsilon \ll 1/\Delta H$ (a timescale often called the Zeno time). Moreover, the general arguments given in Refs.[72, 73] for the equivalence Eq.(16) are not obviously tied to the one-dimensional case with $P = \theta(\hat{x})$, so we expect it to hold very generally.

This approximate equivalence means that a second natural candidate for the modified propagator is

$$\hat{g}_r^V(t_f, t_0) = \exp\left(-(iH + V_0 P)(t_f - t_0)\right) \tag{18}$$

As stated, this is approximately equal to Eq.(15), at least in simple one-dimensional models. However, it may be taken as an independently-postulated alternative propagator which is essentially equivalent to using the form Eq.(13) but replacing the exact projector $\bar{P}$ with POVMs of the form $\exp(-\epsilon V_0 P)$, which is our second obvious way of softening the monitoring so as to avoid the Zeno effect. Moreover, this expression too has a path integral form,

$$g_r^V(\mathbf{x}_f, t_f|\mathbf{x}_0, t_0) = \int \mathcal{D}\mathbf{x}(t)\ \exp\left(i\int_{t_0}^{t_f} dt\ \left[\frac{1}{2}m\dot{\mathbf{x}}^2 - U(\mathbf{x}) + iV_0 f_\Delta(\mathbf{x})\right]\right) \tag{19}$$

where $f_\Delta(\mathbf{x})$ is a window function on Δ. Here, the paths are unrestricted, except at their end points, but paths entering Δ are suppressed by a complex potential. This situation is depicted in Fig.4 for the above one-dimensional example, in which the region Δ is the interval $[a, b]$.

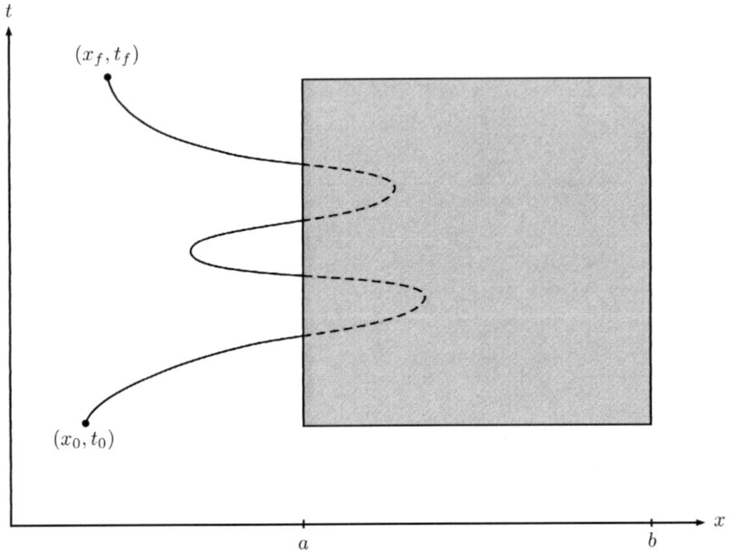

Figure 4. The paths summed over in the modified propagator Eq.(19). The paths may enter the region $[a, b]$ but their contributions to the sum over paths is exponentially supressed.

Complex potentials of this general type arise in a variety of contexts and have been extensively studied [74, 75, 76, 77]. They will in general still involve reflection, but they behave in the classically expected way, i.e. they are absorbing, for sufficiently small V_0, which is what is required for the propagator to have the intuitively correct properties. Generalizations of the above scheme are possible in which a real potential of step function form is included. This can help with minimizing reflection and indeed it is possible, for given classes of initial states, to find potentials which are almost perfect absorbers [77].

Given the relationship Eq.(16), we can now compute the scales associated with reflection, since at least in simple examples, scattering off a complex potential is easy to compute. In the simplest case of a free particle in one dimension scattering off a simple complex step potential $-iV_0\theta(x)$, reflection is negligible for $V_0 \ll E$, where E is the energy scale of the incoming state [76, 73]. Because V_0 is connected to ϵ by Eq.(17), this means that the Zeno effect in the string of projectors Eq.(15) is negligible as long as

$$\epsilon \gg \frac{1}{E} \tag{20}$$

Again on general grounds, we expect this to be true in a wide class of models. For more elaborate models in which there are length scales describing the size of the region, the requirement of negligible reflection may impose further conditions on E, in addition to Eq.(20). However, we assert that for sufficiently large regions Δ, Eq.(20) is the most important condition.

We thus see that there are two natural and simple ways of defining a restricted propagator outside Δ, in such a way that reflection is minimized but which remain as true as possible to the notion of paths not entering Δ. These definitions involve a new coarse graining parameter ϵ which must in general be chosen to be sufficiently large to avoid unphysical results. The earlier, problematic definitions of the propagators Eqs.(1), (2) are obtained in the limit $\epsilon \to 0$.

Given the modified propagator Eq.(18) describing restricted propagation outside Δ, one can also derive a corresponding propagator for entering Δ, analogous to Eq.(3) and Eq.(11). We define the modified propagator for entering by

$$\hat{g}^V_\Delta(t_f, t_0) = \exp\left(-iH(t_f - t_0)\right) - \exp\left(-(iH + V_0 P)(t_f - t_0)\right) \tag{21}$$

Some elementary calcuation [29, 15] leads to the equivalent form

$$\hat{g}^V_\Delta(t_f, t_0) = \int_{t_0}^{t_f} dt \ \exp\left(-iH(t_f - t)\right) \ V_0 P \ \exp\left(-(iH + V_0 P)(t - t_0)\right) \tag{22}$$

This may be rewritten

$$g^V_\Delta(\mathbf{x}_f, t_f | \mathbf{x}_0, t_0) = V_0 \int_{t_0}^{t_f} dt \int_\Delta d^d y \ g(\mathbf{x}_f, t_f | \mathbf{y}, t) g^V_r(\mathbf{y}, t | \mathbf{x}_0, t_0) \tag{23}$$

This is the generalization of Eq.(11) and tends to it as $V_0 \to \infty$. It is different in that firstly, the restricted propagation suppresses paths that enter Δ but does not completely exclude them, so the restriction is "softer". Secondly, the intermediate integral is over all of the region Δ, not just the boundary. With some straightforward calculation in essence the same as a similar case in quantum cosmology considered in Ref.[15], Eq.(22) may be simplified for small V_0 and has the form of an ingoing current operator on the boundary Σ, the anticipated semiclassical form, and gives intuitively sensible results.

4. Some simple examples

To illustrate the above ideas, we now give some simple one-dimensional examples. We consider a free particle in one dimension consisting of mainly positive momenta and initially perfectly localized in $x < 0$, so that $P|\psi\rangle = |\psi\rangle$, where $P = \theta(\hat{x})$. We take the spatial region Δ to be $x > 0$ and consider the following question: What is the probability that the particle either enters or never enters Δ during the time interval $[0, \tau]$ and ends in $x_f < 0$ at time τ? This question is closely related to the arrival time problem, addressed in many places [78, 79], but here we use it as a simple example of spacetime coarse graining.

The amplitude for not entering $x > 0$ is given by the restricted propagator, which in this case is given by the usual method of images expression

$$g_r(x, \tau | x_0, 0) = \theta(-x)\theta(-x_0) \left[g(x, \tau | x_0, 0) - g(x, \tau | -x_0, 0) \right] \tag{24}$$

and g is the usual free particle propagator

$$g(x,\tau|x_0,0) = \left(\frac{m}{2\pi i\tau}\right)^{1/2} \exp\left(\frac{im(x-x_0)^2}{2\tau}\right) \tag{25}$$

Note the restricted propagator Eq.(24) consists of direct and reflected pieces and hence describes reflection off the origin, as discussed earlier. The restricted propagator is also conveniently written in the operator form

$$\hat{g}_r(\tau,0) = \bar{P}(1-R)e^{-iH\tau}\bar{P} \tag{26}$$

where $\bar{P} = \theta(-\hat{x})$ and R is the reflection operator,

$$R = \int dx|x\rangle\langle -x| \tag{27}$$

Since, as stated in Section 2, $\hat{g}_r$ is unitary on states with support only in $x<0$ and since the initial state is perfectly localized in $x<0$, the probability for not entering is $p_r = 1$.

However, we may still explore the properties of the crossing amplitude to see what sort of result it gives, ignoring the fact that the sum rules will not be satisfied, as we discussed in the more general case Eq.(11). We therefore consider the propagator $\hat{g}_\Delta$ for entering $x>0$ during the given time interval, defined by summing over paths which start at $x_0<0$ at $t=0$, cross the origin at least once and end $x<0$ at τ. It is most simply expressed in the operator form

$$\hat{g}_\Delta(\tau,0) = \bar{P}\left(\hat{g}(\tau,0) - \hat{g}_r(\tau,0)\right)\bar{P} \tag{28}$$

where $\hat{g} = \exp(-iH\tau)$ is the free particle propagator. Using Eq.(26), this is equivalently given by

$$\hat{g}_\Delta(\tau,0) = \bar{P}Re^{-iH\tau}\bar{P} \tag{29}$$

(This is in fact equivalent to the PDX form of the crossing propagator, Eq.(11)). The probability for entering Δ is then given by

$$\begin{aligned} p_\Delta(0,\tau) &= \langle\psi|\hat{g}_\Delta(\tau,0)^\dagger\hat{g}_\Delta(\tau,0)|\psi\rangle \\ &= \langle\psi|\bar{P}P(\tau)\bar{P}|\psi\rangle \end{aligned} \tag{30}$$

because $R\bar{P}R = P = \theta(\hat{x})$. Now note that we may write

$$P(\tau) = P + \int_0^\tau dt\hat{J}(t) \tag{31}$$

where $\hat{J} = \frac{1}{2m}(\hat{p}\delta(\hat{x}) + \delta(\hat{x})\hat{p})$ is the current operator, so we finally have

$$p_\Delta(0,\tau) = \int_0^\tau dt\ \langle\psi|\hat{J}(t)|\psi\rangle \tag{32}$$

This is, on the face of it, a familiar semiclassical answer for the probability for entering $x>0$ during $[0,\tau]$ – the flux across the origin [78, 79]. It gives a probability close to 1 for incoming states which substantially cross during the time interval $[0,\tau]$ and probability close to zero for states which substantially miss the interval.

This result is problematic for two reasons. Firstly, because, as stated earlier, the sum rules are generally not obeyed, so that $p_r + p_c \neq 1$ in general. For example, for a state which substantially crosses during the time interval we get $p_r + p_\Delta \approx 2$, a rather striking violation of the sum rules!

The sum rules are respected only for states which substantially miss the interval, for which $p_\Delta = 0$.

Secondly, even aside from the failure of the sum rules, the result is misleading. Recall that the histories are required to end in $x < 0$ at the final time after entering $x > 0$. Semiclassically, such histories would have probability approximately zero for a free particle. Hence Eq.(32) is semiclassically incorrect for this model.

The point is emphasized by considering a slightly more complicated version of the above problem. We take the same initial state in $x < 0$ and ask a modified question: What is the probability that the particle either enters or does not enter $x > 0$ during the time interval $[0, \tau]$ and ends at any point x_f at time $t_f > \tau$? The situation is depicted in Fig.5. The propagator

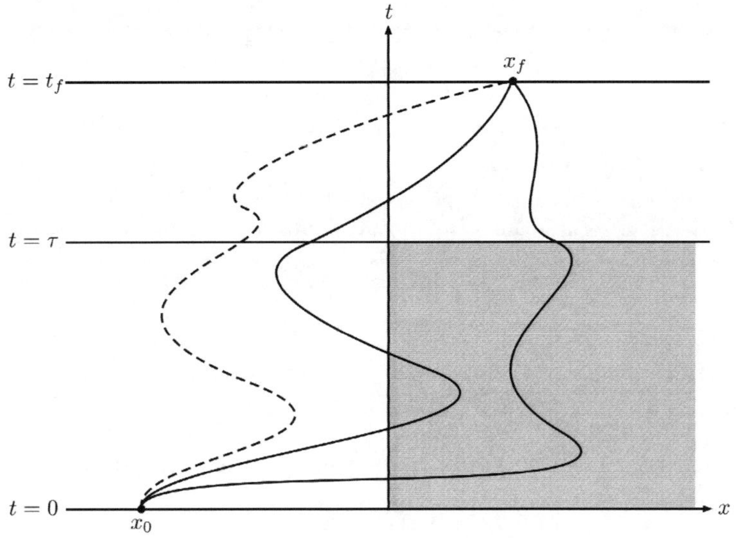

Figure 5. The amplitude for not entering or entering the region $x > 0$ during $[0, \tau]$ and ending at any final x_f at t_f is obtained by summing over paths which respectively, do not enter (dotted line) or enter (bold lines) $x > 0$ during $[0, \tau]$.

for not entering is very similar,

$$\hat{g}_r(t_f, 0) = e^{-iH(t_f - \tau)} \bar{P}(1 - R) e^{-iH\tau} \bar{P} \tag{33}$$

so we still have $p_r = 1$ and again the sum rules are not obeyed. The propagator for entering, however, acquires a new type of term,

$$\begin{aligned} \hat{g}_\Delta(t_f, 0) &= (\hat{g}(t_f, 0) - \hat{g}_r(t_f, 0)) \bar{P} \\ &= e^{-iH(t_f - \tau)} \left(\bar{P} R e^{-iH\tau} \bar{P} + P e^{-iH\tau} \bar{P} \right) \end{aligned} \tag{34}$$

because there is now the new possibility that the particle can be in $x > 0$ at time τ. It is easily seen that these two terms make identical contributions to the probability and we therefore have

$$p_\Delta(t_f, 0) = 2 \int_0^\tau dt \ \langle \psi | \hat{J}(t) | \psi \rangle \tag{35}$$

This is twice the expected semiclassical answer.

The underlying problem is, in essence, the reflection produced by the restricted propagator which persists into the crossing propagator as one can see in Eq.(34). As described in Section

3, a solution to this difficulty is obtained by softening the coarse graining by using a complex potential to characterize the restrictions on paths or by using projections not acting at every time. This is described in Refs.[29, 30] and we briefly summarize the key ideas here for the complex potential case applied to the second model considered above. The amplitude for not entering the region $x > 0$ during the time interval $[0, \tau]$ is then given by an expression of the form Eq.(18), where $P = \theta(\hat{x})$. For $V_0 \ll E$, where E is the energy scale of the incoming state, there is negligible reflection, so the part of the state crossing the origin during $[0, \tau]$ will be absorbed and the remainder of the state is unchanged. This effectively means that under propagator with Eq.(18),

$$\hat{g}_r^V(0,\tau)|\psi\rangle \approx \bar{P}(\tau)|\psi\rangle \tag{36}$$

This is the key property that makes the amplitudes defined with a complex potential give sensible physical results. For states consisting of single wave packets reasonably well peaked in position and momentum, the sum rules are satisfied approximately and the crossing probability is approximately Eq.(32) (with small modifications depending on V_0). The term involving reflection in Eq.(34) is essentially suppressed which is why the 2 becomes a 1 in Eq.(35).

A final related example is that of Yamada and Takagi [23], who took a general initial state and asked for the probability of either crossing or not crossing the origin in either direction during the time interval $[0, \tau]$. The non-crossing propagator is similar to that above, Eq.(26) and is given by

$$\hat{g}_r(\tau,0) = \bar{P}(1-R)e^{-iH\tau}\bar{P} + P(1-R)e^{-iH\tau}P \tag{37}$$

They found that the sum rules are satisfied only for states antisymmetric about the origin and that the crossing probability is exactly zero, once more a physical counter-intuitive result.

The analysis of this situation with one of the above modified propagators essentially follows from the work described in Ref.[31]. For superpositions of incoming wave packets (in either direction), the sum rules are approximately satisfied if the energy scale of the wave packet E satisfies $E \gg V_0$ and the probabilities are the expected semiclassical ones, so intuitive properties are restored.

Note that this model actually has *two* different regimes in which the sum rules are satisfied: approximately for small V_0 and exactly, in the Zeno limit of $V_0 \to \infty$ for antisymmetric initial states, but the resulting probabilities are very different in each case.

Note also that this simple model suggests that the considerations of this paper may have implications for quantum measure theory, in which it is asserted that histories with probability zero do not occur [54, 55]. Intuitively, one would expect a non-zero crossing probability in this model, but the crossing probability is zero in the Zeno limit case. Hence the predictions of quantum measure theory may conflict with the intuitively expected result in this limit.

5. Connections to other work

The potential difficulties with path integral expressions highlighted in this paper first arose in applications of the decoherent histories approach to spacetime coarse grainings in non-relativistic quantum mechanics [23, 24, 25, 26, 27, 28, 10]. The problems with the Zeno effect did not seem to be appreciated except that Hartle noted that some coarse grainings are "too strong for decoherence" [26]. The problems also persisted in applications of the decoherent histories approach to quantum cosmology and reparametrization invariant theories [10, 11, 12, 13, 14]. These issue also have consequences for the continuous tensor product structure in the decoherent histories approach, discussed by Isham and collaborators [52, 53].

The specific problem with the Zeno effect was noticed in Ref.[14]. It was also observed that it can in fact be avoided if the spacetime regions involved in the coarse graining are carefully chosen so that the initial state has zero current across the boundary. This resolution has been pursued [19].

In this paper we have presented a generally applicable resolution to the Zeno problem involving a softening of the coarse-graining. This was first proposed in the specific context of the arrival time problem in Refs.[29, 30, 80] and in a quantum cosmological model in Ref.[15].

Numerous studies of the dwell and tunneling time problem involve path integral expression of the type considered here (for example, Refs.[35, 37, 38, 39]). Some of these applications are typically connected to specific models of measurements for which the problems described in this paper may not apply. Indeed, specific models for the measurement of time frequently lead to path integral constructions in which the softening of the coarse graining of the type described in Section 3 is already implemented. (A detailed analysis of the dwell time problem will be given in another publication [81]).

We also stress that partitioning of the paths of the type involved in Eqs.(1) and (2) is often used as part of a given calculation (for example, when there is a potential of window function form concentrated in a given region), and the issues raised here are not problematic in such cases. The issues raised in the present paper concern the possible *interpretation* of amplitudes obtained by restricted sums over paths.

However, it remains an interesting question to investigate specific measurement models to assess whether the problems with path integrals described here arise. Of particular note in this regard is a recent paper by Sokolovski criticizing the present work [69], with particular reference to the definition of quantum traversal time [39]. He argues that when the spacetime coarse grainings of the type considered here are regarded as *measurements*, some of their unusual properties, and in particular the Zeno effect we discussed here, are not surprising since in the regime of strong measurement the probabilities are more a reflection of the action of the measuring device than a reflection of some underlying measurement-independent property. Indeed, in the limit of infinitely strongly measurement, the measuring device completely reflects an incoming state, as one would physically expect and also consistent with our results. On the other hand, measurement-based approaches tend to agree with a decoherent histories analysis in the regime of weak measurement or soft coarse graining, as has been seen for example, in the analysis of the arrival time problem [79]. These observation stress that, as suggested already, our criticisms are perhaps most relevant to the interpretation of approaches not directly based on measurements and may be less relevant to measurement-based models, but this is a matter for further investigation.

In terms of the proposed solutions to the problems with path integrals in Section 3, our considerations relating to modified coarse grainings have some connection to continuous quantum measurement theory [82, 40, 41, 42]. We note also that there could be other possible ways of softening the coarse graining to improve the properties of amplitudes constructed by path integrals. For example, Marchewka and Schuss have considered path integrals with absorbing boundary conditions [83]. We also note the interesting and perhaps relevant observations concerning the rigorous definition of path integrals by Sorkin [84], Geroch [85] and Klauder [86].

6. Summary and conclusions

We have argued that amplitudes constructed by path integrals for questions involving time in a non-trivial way can, if implemented in the simplest and most obvious way, lead to problems due to the Zeno effect. This has the consequence that they do not have a sensible classical limit and have properties very different to those expected from the underlying intuitive picture. When path integrals are used in the decoherent histories approach, the Zeno effect can have the consequence that the sum rules are not satisfied except for very trivial initial states. These problems have been observed in numerous examples and applications but here we have argued that the issue is a very general one to do with the use of path integrals in a wide variety of applications.

We outlined a successful solution to the Zeno problem, through a softening of the coarse graining. Again, this solution has been put forward in specific examples and applications, but here we stress that such a solution will offer a very general solution to the problem. We also note that the softening of the coarse graining introduces, perhaps unexpectedly, one or more new coarse graining parameters, in the simplest case a timescale ϵ, describing the precision to within which the paths are monitored in time. The Zeno problem is then avoided as long as $\epsilon \gg 1/E$, where E is the energy scale of the incoming state.

These observations about path integral amplitudes apply most strongly to any approach to quantum theory which involves constructing amplitudes for spacetime coarse grainings which do not refer to a particular measurement scheme, such as the decoherent histories approach and quantum measure theory. Many approaches to quantum gravity are of this type. However, these observations may be less relevant to measurement-based models of spacetime coarse grainings and indeed the possible consequences of our observations for such models have been criticized by Sokolovski [69].

We certainly do not claim that any of the papers in this area are wrong. Indeed many authors have noted the unphysical nature of their results. The present paper is, if anything, a cautionary note on the use of path integrals in space time coarse grainings: some types of "obvious" coarse-grainings are quite simply unphysical. However, amplitudes with sensible intuitive properties *may* be successfully constructed with proper attention to the implementation of the coarse graining and to the associated time scale.

7. Acknowledgements

JMY was supported by the Templeton Foundation. We are grateful to Charis Anastopoulos, Fay Dowker and Larry Schulman for useful comments on an earlier version of the manuscript.

References

[1] Feynman R P 1948 *Rev. Mod. Phys.* **20** 267
[2] Feynman R P and Hibbs A R 1965 *Quantum Mechanics and Path Integrals* (New York: McGraw Hill)
[3] Schulman L S 1981 *Techniques and Applications of Path Integration* (New York: Wiley and Sons)
[4] A useful list of references to the older work on path integals in quantum gravity may be found in Hartle J B 1993 *Vistas Astron.* **37** 569 (*Preprint* gr-qc/9210004)
[5] Hartle J B and Hawking S W 1983 *Phys. Rev.* D **28** 2960
[6] Hawking S W 1984 *Nucl. Phys.* B **239** 257
[7] Halliwell J J 1991 *Quantum Cosmology and Baby Universes*, ed S Coleman *et al* (Singapore: World Scientific) (*Preprint* gr-qc/0909.2566)
[8] Kiefer C and Sandhoefer B 2008 Quantum Cosmology *Preprint* quant-ph/0804.0672
Wiltshire D L 2003 An introduction to quantum cosmology *Preprint* gr-qc/0101003
[9] Hartle J B, Hawking S W and Hertog T 2008 *Phys. Rev.* D **77** 123537
Hartle J B, Hawking S W and Hertog T 2008 *Phys. Rev. Lett.* **100** 201301
[10] Hartle J B 1995 *Proc. of School on Gravity and its Quantizations (1992 Les Houches)*, ed B Julia and J Zinn-Justin (Amsterdam: Elsevier)
[11] Hartle J B and Marolf D 1997 *Phys. Rev.* D **56** 6247
[12] Halliwell J J and Thorwart J 2001 *Phys. Rev.* D **64** 124018
[13] Halliwell J J and Thorwart J 2002 *Phys. Rev.* D **65** 104009
[14] Halliwell J J and Wallden P 2006 *Phys. Rev.* D **73** 024011
[15] Halliwell J J 2009 *Phys. Rev.* D **80** 124032
[16] Craig D A and Hartle J B 2004 *Phys. Rev.* D **69** 123525
[17] Craig D A and Singh P 2010 *Phys. Rev.* D **82** 123526
[18] Whelan J 1994 *Phys. Rev.* D **50** 6344
[19] Christodoulakis T and Wallden P 2011 *J.Phys.: Conf. Series* **283** 012041
[20] Anastopoulos C and Savvidou K 2000 *Class. Quant. Grav.* **17** 2463
[21] Anastopoulos C and Savvidou K 2005 *Class. Quant. Grav* **22** 1841
[22] Schroeren D P *Decoherent histories of spin networks* (in preparation)
[23] Yamada N and Takagi S 1991 *Prog. Theor. Phys.* **85** 985

Yamada N and Takagi S 1991 *Prog. Theor. Phys.* **86** 599
Yamada N and Takagi S 1992 *Prog. Theor. Phys.* **87** 77
[24] Yamada N 1996 *Phys. Rev* A **54** 182
[25] Yamada N 1992 *Sci. Rep. Tôhoku Uni., Series 8* **12** 177
[26] Hartle J B 1991 *Phys. Rev.* D **44** 3173
[27] Micanek R J and Hartle J B 1996 *Phys. Rev.* A **54** 3795
[28] Halliwell J J and Zafiris E 1998 *Phys. Rev.* D **57** 3351
[29] Halliwell J J and Yearsley J M 2009 *Phys. Rev.* A **79** 062101
[30] Halliwell J J and Yearsley J M 2009 *Phys. Lett.* A **374** 154
[31] Yearsley J M 2011 *Aspects of time in Quantum Theory* (Imperial College PhD thesis) (*Preprint* quant-ph/1110.5790)
[32] Anastopoulos C and Savvidou K 2006 *J. Math. Phys.* **47** 122106
[33] Anastopoulos C and Savvidou K 2007 *J. Math. Phys.* **48** 032106
[34] Anastopoulos C and Savvidou K 2012 Time-of-arrival probabilities for general particle detectors *Preprint* quant-ph/1205.2781
[35] Fertig H A 1990 *Phys. Rev. Lett.* **65** 2321
Fertig H A 1993 *Phys. Rev.* B **47** 1346
[36] Yamada N 1999 *Phys. Rev. Lett.* **83** 3350
[37] Yamada N 2004 *Phys. Rev. Lett.* **93** 170401
[38] Garcia-Calderon G, Villavicencio J and Yamada N 2003 *Phys.Rev.* A **67** 052106
[39] Sokolovski D 1995 *Phys. Rev.* A **52** R5
Sokolovski D and Connor J N L 1993 *Phys. Rev.* A **47** 4677
Sokolovski, Brouard S and Connor J N L 1994 *Phys. Rev.* A **50** 1240
Sokolovski D and Connor J N L 1994 *Solid State Commun.* **89** 475
Sokolovski D and Connor J N L 1991 *Phys. Rev.* A **44** 1500
Sokolovski D and Baskin L M 1987 *Phys. Rev.* A **36** 4604
Sokolovski D 1997 *Phys. Rev. Lett.* **79** 4946
[40] Caves C M and Milburn G J 1987 *Phys. Rev.* A **36** 5543
[41] Mensky M B 1993 *Continuous Quantum Measurements and Path Integrals* (Bristol and Philadelphia: IOP Publishing)
Mensky M B, Onofrio R and Presilla C 1991 *Phys. Lett.* A **161** 236
Onofrio R, Presilla C and Tambini U 1993 *Phys. Lett.* A **183** 135
Mensky M B 1998 *Phys. Usp.* **41** 923
Mensky M B 1998 *Intern. J. Theor. Phys.* **37** 273
Mensky M B and von Borzeszkowski H 1995 *Phys. Lett.* A **208** 269
Mensky M B 1997 *Phys. Lett.* A **231** 1
[42] Mensky M B 2000 *Quantum Measurements and Decoherence. Models and Phenomenology* (Dordrecht: Kluwer Academic Publishers)
[43] Gell-Mann M and Hartle J B 1990 *Complexity, Entropy and the Physics of Information, SFI Studies in the Sciences of Complexity* vol VIII, ed W Zurek (Reading: Addison Wesley)
Gell-Mann M and Hartle J B 1990 *Proceedings of the Third International Symposium on the Foundations of Quantum Mechanics in the Light of New Technology*, ed S Kobayashi *et al* (Tokyo: Physical Society of Japan)
[44] Gell-Mann M and Hartle J B 1993 *Phys. Rev.* D **47** 3345
[45] Griffiths R B 1984 *J. Stat. Phys.* **36** 219
Griffiths R B 1993 *Phys. Rev. Lett.* **70** 2201
Griffiths R B 1987 *Am. J. Phys.* **55** 11
[46] Omnès R 1988 *J. Stat. Phys.* **53** 893
Omnès R 1988 *J. Stat. Phys.* **53** 933
Omnès R 1988 *J. Stat. Phys.* **53** 957
Omnès R 1989 *J. Stat. Phys.* **57** 357
Omnès R 1991 *J. Stat. Phys.* **62** 841
Omnès R 1990 *Ann. Phys.* **201** 354
Omnès R 1992 *Rev.Mod.Phys.* **64** 339
Omnès R 1994 *The Interpretation of Quantum Mechanics* (Princeton: Princeton University Press)
[47] Halliwell J J 1994 *Fundamental Problems in Quantum Theory* vol 775, ed D Greenberger and A Zeilinger (New York: Annals of the New York Academy of Sciences) p 726
[48] Dowker H F and Halliwell J J 1992, *Phys. Rev.* D **46** 1580
[49] Halliwell J J 1999 *Phys. Rev.* D **60** 105031

[50] Isham C J 1994 *J. Math. Phys.* **35** 2157
[51] Isham C J 1993 *J. Math. Phys.* **23** 2157
[52] Isham C J and Linden N 1994 *J. Math. Phys.* **35** 5452
Isham C and Linden N 1995 *J. Math. Phys.* **36** 5392
[53] Isham C J, Linden N, Savvidou K and Schreckenberg S 1998 *J. Math. Phys.* **39** 1818
[54] Sorkin R D 1994 *Mod. Phys. Lett.* A **9** 3119
Sorkin R D 1997 *Quantum Classical Correspondence: Proceedings of the 4th Drexel Symposium on Quantum Nonintegrability (September 8-11, 1994, Philadelphia)*, ed D H Feng and B-L Hu (Cambridge Mass.: International Press) pp 229-251 (*Preprint* gr-qc/9507057)
[55] Salgado R B 2002 *Mod. Phys. Lett.* A **17** 711
[56] Dowker H F, Johnston S and Sorkin R D 2010 *J. Phys.* A **43** 275302
[57] Khalfin L A 1957 *Zh. Eksp. Teor. Fiz.* **33** 1371
Khalfin L A 1958 *Sov. Phys. JETP* **6** 1053
Khalfin L A 1968 *Pis'ma Zh. Eksp. Teor. Fiz.* **8** 106
Khalfin L A 1968 *JETP Lett.* **8** 65
[58] Misra B and Sudarshan E C G 1977 *J. Math. Phys.* **18** 756
[59] Chiu C B, Sudarhsan E C G and Misra B 1977 *Phys. Rev.* D **16** 520
[60] Peres A 1980 *Am. J. Phys.* **48** 931
[61] Kraus K 1981 *Found. Phys.* **11** 547
[62] Sudbery A 1984 *Ann. Phys.* **157** 512
[63] Facchi P, Pascazio S, Scardicchio A and Schulman L S 2002 *Phys. Rev.* A **65** 012108
[64] Wallden P 2008 *Int. J. Theor. Phys.* **47** 1512
[65] Wallden P 2007 *J. Phys.: Conf. Series* **67** 012043
[66] Auerbach A and Kivelson S 1985 *Nucl. Phys.* B **257** 799
van Baal P 1993 *Lectures on Path Integration*, ed H A Cerdeira *et al* (Singapore: World Scientific) (also available from the website http://www.lorentz.leidenuniv.nl/vanbaal/HOME/publ.html)
[67] Halliwell J J and Ortiz M E 1993 *Phys. Rev.* D **48** 748
[68] Halliwell J J 1995 *Phys. Lett.* A **207** 23
[69] Sokolovski D (2013) *Preprint* quant-ph/1301.1244
[70] Halliwell JJ and Yearsley J M 2012 *Phys. Rev.* D **86** 024016
[71] Jaffe A and Glimm J 1981 *Quantum Physics* (New York: Springer)
Simon B 1979 *Functional Integration and Quantum Physics* (New York: Academic)
[72] Echanobe J, del Campo A and Muga J G 2008 *Phys. Rev.* A **77** 032112
[73] Halliwell J J and Yearsley J M 2010 *J. Phys. A. Math. Theor.* **43** 445303
[74] Blanchard P and Jadczyk A 1996 *Helv. Phys. Acta* **69** 613
Palao J P, Muga J G, Brouard S and Jadczyk A 1997 *Phys. Lett.* A **233** 227
Ruschhaupt A 2002 *J. Phys.* A **35** 10429
Muga J G, Palao J P, Navarro N and Egusquiza I L 2004 *Phys. Rep.* **395**
Halliwell J J 2008 *Phys. Rev.* A **77** 062103
[75] Halliwell J J 1999 *Prog. Th. Phys.* **102** 707
[76] Allcock G R 1969 *Ann. Phys* **53** 253
Allcock G R 1969 *Ann. Phys* **53** 286
Allcock G R 1969 *Ann. Phys* **53** 311
[77] Muga J G, Brouard S and Macias D 1995 *Ann. Phys.* **240** 351
Palao J P, Muga J G and Sala R 1998 *Phys. Rev. Lett.* **80** 5469
[78] *Time in Quantum Mechanics* 2002, ed J G Muga *et al* (Berlin: Springer)
Muga J G and Leavens C R 2000 *Phys. Rep.* **338** 353
[79] Yearsley J M, Downs D A, Halliwell J J and Hashagen A K 2011 *Phys. Rev.* A **84** 022109
[80] Yearsley J M 2010 *Phys. Rev.* A **82** 012116
[81] Halliwell J J and Yearsley J M *Decoherent Histories Analysis of the Dwell Time Problem* (in preparation)
[82] Jacobs K and Steck D A 2006 *Contemp. Phys.* **47** 279
[83] Marchewka A and Schuss Z 1998 *Phys. Lett.* A **240** 177 (*Preprint* quant-ph/9906003, quant-ph/9903076, quant-ph/9906078)
[84] Sorkin R D *Notes on path integrals and the histories framework*, available at http://www.perimeterinstitute.ca/personal/rsorkin/lecture.notes
[85] Geroch R *Path Integrals*, available at Sorkin's website, Ref.[84].
[86] Klauder J R 2011 *A Modern Approach to Functional Integration* (Boston: Birkhauser) (*Preprint* quant-ph/0303034)

Fractional field equations for highly improbable events

H Kleinert

Institut für Theoretische Physik, Freie Universität Berlin, Arnimallee 14, D14195 Berlin, Germany
ICRANeT, Piazzale della Republica 1, 10 -65122, Pescara, Italy

E-mail: h.k@fu-berlin.de

Abstract. Free and weakly interacting particles perform approximately Gaussian random walks with collisions. They follow a second-quantized nonlinear Schrödinger equation, or relativistic versions of it. By contrast, the fields of strongly interacting particles extremize more involved effective actions obeying fractional wave equations with anomalous dimensions. Their particle orbits perform universal Lévy walks with heavy tails, in which rare events are much more frequent than in Gaussian random walks. Such rare events are observed in exceptionally strong windgusts, monster or rogue waves, earthquakes, and financial crashes. While earthquakes may destroy entire cities, the latter have the potential of devastating entire economies.

1. Introduction

Since the bestselling book "The Black Swan" by N.N. Taleb [1], the "*disproportionate role of high-profile, hard-to-predict, and rare events that are beyond the realm of normal expectations in history, science, finance, and technology*", has moved into public awareness, thereby contrasting previous besetesellers focusing on Gaussian distributions [3]. Since the last financial crash and the recent "flash crash" [4] that has appeared in 2010 [4] with the approach of nanosecond-trading in stockmarkets, the dangers of such events for the world economy have become so worrisome that also politicians begin to get worried.

2. Second-quantized field theory

The purpose of this lecture is to incorporate them into our present description of particle physics. The quantum-mechanical phenomena of fundamental particles is explained with high accuracy by Schrödinger theory. The wave equation for many particles can convenienty be reformulated as a second-quantized *field theory*, with an action that is the sum of quadratic and an interacting term

$$\mathcal{A} = \mathcal{A}_2 + \mathcal{A}_{\rm int}, \tag{1}$$

where the term $\mathcal{A}_2$ has typically the form

$$\mathcal{A}_2 = \int d^D x dt \psi^*(\mathbf{x}, t)[i\partial_t + \hbar^2\nabla^2/2m - V(\mathbf{x})]\psi(\mathbf{x}, t), \tag{2}$$

with D being the space timension, m the mass, and $V(\mathbf{x})$ some external potential. The interaction term $\mathcal{A}_{\rm int}$ may be approximated in molecular systems by a fourth-order term in the field

$$\mathcal{A}_{\rm int} = \frac{1}{2}\int d^D x d^D x' dt \psi^*(\mathbf{x}', t)\psi^*(\mathbf{x}, t)V_{12}(\mathbf{x}, \mathbf{x}')\psi(\mathbf{x}, t)\psi(\mathbf{x}', t), \tag{3}$$

where $V_{12}(\mathbf{x}, \mathbf{x}')$ is some two-body potential.

If relativistic velocities are present, the field is generalized to a scalar Klein-Gordon field, or a quantized Dirac field. In molecular physics, the fourth-order term is due to the exchange of a minimally coupled quantized photon field and is proportional to e^2, where e is the electric charge. The field equations may be studied with any standard method of quantum field theory, and corrections can be derived using perturbation theory in powers of $\alpha \equiv e^2/\hbar \approx 1/137$. Since α is very small, this appeoch is quite successful.

If time is continued analytically to imaginary values $t = i\tau$, one is faced with the so-called Euclidean version of quantum field theory. Then perturbation theory may be understood as developing a theory of particle physics from an expansion around Gaussian random walks. Indeed, the relativistic scalar free-particle propagator of mass m in $D+1$-dimensional euclidean energy-momentum space $p^\mu = (\mathbf{p}, p_4)$, has the form

$$G(p) = \frac{1}{\mathbf{p}^2 + p_4^2 + m^2} = \int_0^\infty ds\, e^{-sm^2} e^{-s(\mathbf{p}^2+p_4^2)}, \tag{4}$$

where the energy has been continued analytically to $p_4 = -iE$. The Fourier transform of $e^{-s(\mathbf{p}^2+p_4^2)}$ is the distribution of Gaussian random walks of length s in $D+1$ euclidean dimensions

$$P(\mathbf{x}, x_4) = (4\pi s)^{-(D+1)/2} e^{-(|\mathbf{x}|^2+x_4^2)/4s}, \tag{5}$$

which makes the propagator (4) a superposition of such walks with lenghts distributed like e^{-sm^2} [5, 6, 7]. This propagator is the relativistic version of the free-field propagator of the action (2). The second-quantized field theory described by (1) accounts for grand-canical ensembles of orbits with their two-body interactions [8].

Gaussian random walks are a natural and rather universal starting point for many stochastic processes. For instance, they form the basis of the most important tool in the theory of financial markets, the Black-Scholes option price theory [9] (Nobel Prize 1997), by which a portfolio of assets is hoped to remain steadily growing through hedging. In fact, the famous *central-limit theorem* permits us to prove that many independent random movements of finite variance always pile up to display a Gaussian distribution [10].

However, since the last stock market crash and the still ongoing financial crisis it has become clear that realistic distributions belong to a more general universality class, the so-called Lévy stable distribution. They are the univarsal results of a pile up of random movements of infinite variance [11]. They account for the fact that rare events, which initiate crashes, are much more frequent than in Gaussian distributions. These are events in the so-called Lévy tails $\propto 1/|x|^{1+\lambda}$ of the distributions, whose description requires a Hamiltonian [7]

$$H = \text{const}\,(p^2)^{\lambda/2}. \tag{6}$$

Such tail-events are present in the self-similar distribution of matter in the universe [12, 13, 14], in velocity distributions of many body sytems with long-range forces [15], and in the distributions of windgusts [16] and earthquakes [17], with often catastrophic consequences. They are a consequence of rather general maximal entropy assumptions [18]. In the limit $\lambda \to 2$, the Lévy distributions reduce to Gaussian distributions.

3. Strong-coupling quantum field theory

At this point we observe that such distributions occur quite naturally also in many-particle systems, provided the interactions are very strong [19]. They have been observed in numerous experiments at second-order phase transitions. The most accurate measurement of this type was done in a satellite (the so called Infrared Atronomical Satellite IRAS) by studying the singularity of the specific heat of superfluid ^{4}He near the critical temperature [20]. The observation agreed extremely well with the theoretical strong-coupling prediction [21].

The field of a strongly interacting N-body system is usually a multivalued function. Singularities perforate the space via vortex lines (for instance in type II superconductors or in superfluid ^{4}He), or via line-like defects in the displacement field of a world-crystal formulation of Einstein(-Cartan) gravity [?]. If the positions of two particles are exchanged, one obtains a factor $+1$ for bosons or -1 for electrons. In two dimensions, one may even obtain a general phase $e^{i\phi}$ (anyons) [26].

A strongly interacting field system has a conformally invariant Green function [23, 24, 26]

$$G(\mathbf{p}, p_4) = [p_4^{1-\gamma}\phi(\mathbf{p}^2/p_4^z)]^{-1}. \tag{7}$$

If the dimension D differs only by a very small amount ϵ from the critical dimension D_c, where the theory is scale-invariant, i.e., $D = D_c + \epsilon$, then γ is of order ϵ and z differs from unity by a similar amount. Such a power behavior is assured near D_c if the Gell-Mann-Low function [27] has an infrared-stable fixed point in the renormalization flow of the coupling constant. Very close to the critical dimension, a lowest approximation to $G(\mathbf{p}, p_4)$ is

$$G(\mathbf{p}, p_4) = \{p_4^{1-\gamma}[1 + D_\lambda(\mathbf{p}^2/p_4^z)^{\lambda/2}]\}^{-1}, \tag{8}$$

where λ is close to 2, and D_λ is a generalization of the diffusion constant in the Fokker-Planck equation.

Time-independent propagators involve the limit $p_4 \to 0$, where the correlation function behaves like

$$G(\mathbf{p}, 0) \propto |\mathbf{p}|^{-2+\eta}. \tag{9}$$

The index η is the *anomalous dimension* of the field, which is also of order ϵ. The existence of this limit in (8) fixes the scaling relation

$$\lambda = (2 - \gamma)/z = 2 - \eta. \tag{10}$$

See Appendix for the calculation of the exponents to order ϵ. The Green function (8) determines the probability distribution of particle after a time t via the *double fractional Fokker-Planck equation*

$$[\hat{p}_4^{1-\gamma} + D_\lambda(\hat{\mathbf{p}}^2)^{\lambda/2}]P(\mathbf{x}, t) = \delta(t)\delta^{(D)}(\mathbf{x}), \tag{11}$$

where $\hat{p}_4 \equiv \partial_t$, $\hat{\mathbf{p}} \equiv i\partial_{\mathbf{x}} \equiv i\boldsymbol{\nabla}$. A convenient definition of the fractional derivatives uses the same formula as in the dimensional continuation of Feynman diagrams $(-\boldsymbol{\nabla}^2)^{\lambda/2} = \Gamma[\lambda/2]^{-1}\int d\sigma\sigma^{-\lambda/2-1}e^{\sigma\boldsymbol{\nabla}^2/2}$ [28, 29]. The solution of (11) is given in the literature [30] and reads

$$\frac{t^{-\gamma}}{\pi^{D/2}|\mathbf{x}|^{D/2}} H_{2,3}^{2,1}\left(\frac{|\mathbf{x}|^\lambda}{2^\lambda D_\lambda t^{1-\gamma}}\,\middle|\,\begin{matrix}(1,1);(1-\gamma,1-\gamma)\\(1,1),(D/2,\lambda/2);(1,\lambda/2)\end{matrix}\right), \tag{12}$$

where $H^{2,1}_{2,3}$ is a *Fox H-function* [31]. In the limits $\gamma = 0$ and $\lambda = 2$, this reduces to the standard quantum mechanical Gaussian expression $(4\pi D_\lambda t)^{-D/2}e^{-|\mathbf{x}|^2/4D_\lambda t}$. For $\gamma = 0, \lambda = 1$, the result is

$$P(\mathbf{x},t) = \frac{D_\lambda t}{\pi^{(D+1)/2}|\mathbf{x}|^{D+1}} H^{1,1}_{1,1}\left(\frac{D_\lambda^2 t^2}{|\mathbf{x}|^2}\Big|^{(1/2-D/2,1)}_{(0,1)}\right), \tag{13}$$

which is simply the Cauchy-Lorentz distribution function

$$[\Gamma(D/2+1/2)/\pi^{(D+1)/2}]D_\lambda t/[(D_\lambda t)^2 + |\mathbf{x}|^2]^{D/2+1/2}.$$

The probability (11) may be calculated from the *doubly fractional canonical path integral* over fluctuating orbits $t(s), \mathbf{x}(s)$ $p_4(s), \mathbf{p}(s)$ viewed as functions of some pseudotime s [32]:

$$\{\mathbf{x}_b t_b s_b|\mathbf{x}_a t_s s_a\} = \int \mathcal{D}\mathbf{x}\mathcal{D}t\mathcal{D}\mathbf{p}\mathcal{D}p_4 e^{-\mathcal{A}}, \tag{14}$$

where $\mathcal{A}$ is the euclidean action of the paths $t(s), \mathbf{x}(s)$:

$$\mathcal{A} = \int ds[i(\mathbf{p}\mathbf{x}' - ip_4 t') - \mathcal{H}(\mathbf{p}, p_4)]. \tag{15}$$

Here $t'(s) \equiv dt(s)/ds$, $\mathbf{x}'(s) \equiv d\mathbf{x}(s)/ds$, and $\mathcal{H}(\mathbf{p}, p_4) = p_4^{1-\gamma} + D_\lambda(\hat{\mathbf{p}}^2)^{\lambda/2}$. At each s, the integrals over the components of $\mathbf{p}(s)$ and $p_4(s)$ run from $-\infty$ to ∞, whereas those over $p_4(s)$ run from $-i\infty$ to $i\infty$. At the end we obtain $P(\mathbf{x},t)$ from the integral $\int_0^\infty ds\{\mathbf{x}\,t\,s|\mathbf{0}\,0\,0\}$.

If $\gamma = 0$, the path integral over $p_4(s)$ yields the functional $\delta[t'(s) - 1]$, which brings (14) to the canonical path integral

$$(\mathbf{x}_b t_b|\mathbf{x}_a t_a) = \int \mathcal{D}\mathbf{x}\mathcal{D}\mathbf{p} e^{-\mathcal{A}'}, \tag{16}$$

with

$$\mathcal{A}' = \int dt[i\mathbf{p}\dot{\mathbf{x}} - D_\lambda(\hat{\mathbf{p}}^2)^{\lambda/2}]. \tag{17}$$

Now $P(\mathbf{x},t) = (\mathbf{x}t|\mathbf{0}\,0)$ satisfies the ordinary fractional Fokker-Planck equation

$$[\hat{p}_4 + D_\lambda(\hat{\mathbf{p}}^2)^{\lambda/2}]P(t,\mathbf{x}) = \delta(t)\delta^{(D)}(\mathbf{x}). \tag{18}$$

This has been discussed at length in recent literature [33].

At this place it is worth mentioning that the probability can be written as a superposition $\int_0^\infty (d\sigma/\sigma) f_\lambda(\sigma t^{-2/\lambda}) P_{\rm G}(\sigma,\mathbf{x})$ of Gaussian distributions $P_{\rm G}(\sigma,\mathbf{x}) = (4\pi\sigma)^{-D/2}e^{-\mathbf{x}^2/4\sigma}$ with weight

$$f_\lambda(\sigma) = S_D \sum_{n=1}^{\infty} \frac{(-1)^n \sigma^{-n\lambda/2}}{(n+1)!\Gamma(D-1-n\lambda/2)} D_\lambda^{n/\lambda}, \tag{19}$$

where $S_D = 2\pi^{D/2}/\Gamma(D/2)$ is the surface of a sphere in D dimensions.

4. Stochastic Duru-Kleinert transformation

If $\gamma \neq 0$, the above functional δ-function is softened, and the relation between the pseudotime s and the physical time becomes stochastic. It is governed by the probability distribution that solves the path integral the

$$\{t_b s_b | t_a s_a\} = \int \mathcal{D}t \mathcal{D}p_4 \exp\left\{\left[\int ds\, [p_4 t' - p_4^{1-\gamma})]\right\}. \tag{20}$$

For imaginary $p_4 = -iE$, we define a *noise Hamiltonian* $\tilde{H}(\eta)$ which has the property that [32, 34]

$$e^{-p_4^{1-\gamma}} = \int_{-\infty}^{\infty} d\eta e^{-p_4\eta - \tilde{H}(\eta)}. \tag{21}$$

The inverse of the Fourier integral yields the *noise probability* $P(\eta) = \int_{-i\infty}^{i\infty} dp_4 e^{p_4\eta - p_4^{1-\gamma}}$, and a probability functional [35]:

$$P[\eta] \equiv e^{-\int ds \tilde{H}(\eta)} = \int \mathcal{D}p_4 \exp\left[\left[\int ds\, (p_4\eta - p_4^{1-\gamma})\right]\right]. \tag{22}$$

Using this we may solve the stochastic differential equation of the Langevin type

$$t'(s) = \eta(s), \tag{23}$$

in which the noise $\eta(s)$ has a zero expectation value for each s, and the correlation functions for $n = 2, 4, 6, \ldots$:

$$\langle \eta(s_1) \ldots \eta(s_{2n}) \rangle \equiv \int \mathcal{D}\eta\, \eta(s_1) \ldots \eta(s_{2n}) P[\eta]. \tag{24}$$

If $\gamma = 0$, the solution of (22) is $P[\eta] = \delta[\eta(s) - 1]$, implying that $\eta(s)$ ceases to fluctuate, and (23) becomes $t'(s) \equiv 1$, so that $t \equiv s$.

In the past, many nontrivial Schrödinger equations (for instance that of the $1/r$-potential) have been solved with path integral methods by re-formulating them on the pseudotime axis s, that is related to the time t via a *space-dependent differential equation* $t'(s) = f(x(t))$. This method, invented by Duru and Kleinert [36] to solve the path integral of the hydrogen atom, has recently been applied successfully to various Fokker-Planck equations [37, 38]. The stochastic differential equation (23) may be seen as a stochastic version of the Duru-Kleinert transformation that promises to be a useful tool to study non-Markovian systems.

Certainly, the solutions of Eq. (18) can also be obtained from a stochastic differential equation

$$\dot{\mathbf{x}} = \boldsymbol{\eta}, \tag{25}$$

whose noise is distributed with a fractional probability

$$P[\boldsymbol{\eta}] = \int \mathcal{D}^D x e^{\int dt (i\mathbf{p}\cdot \; -D_\lambda (\mathbf{p}^2)^{\lambda/2})}. \tag{26}$$

5. Experimental consequences

Experimentally, a system with in the strong-coupling limit can be produced by forming a Bose-Einstein condensate (BEC) in a magnetic field whose strength is tuned to a Feshbach resonance [39] of the two-particle interaction. In a BEC, the four-field term in the interaction (3) is local and parametrized by $V_{12}(\mathbf{x}, \mathbf{x}) \propto g\delta(\mathbf{x} - \mathbf{x}')$. At the Feshbach resonance, the bare coupling strength g goes to infinity [40], and the renormalized coupling g_R, multiplied by $6\mu^{-\epsilon}/(4\pi)^2$, converges to a fixed point $g^* \approx 0.503$ [see Fig. 17.1 in Ref. [24]), where μ is some mass scale.

The theoretical tool to describe the physics in this regime is the effective action $\Gamma[\Psi, \Psi^*]$. This a functional of the classical expectation values of the quantum fields $\Psi(t, \mathbf{x}) \equiv \langle \psi(t, \mathbf{x}) \rangle$, and contains all information of the full quantum theory [24, 41]. It is the Legendre transform of the generating functional $Z[\eta, \eta^*] = \int \mathcal{D}\psi \mathcal{D}\psi^* e^{-\mathcal{A} - \eta^*\psi - \eta\psi^*}$ of the full quantum theory, and is extremal on the physical field expections. All its vertex functions can be found from the functional derivatives of $\Gamma[\Psi, \Psi^*]$. In the strong-coupling limit, the effective interaction changes the interaction (3) to an anomalous power law $\Gamma^{\text{int}}[\Psi, \Psi^*] = (g_c/2) \int dt d^D x \, |\Psi(t, \mathbf{x})|^{\delta+1}$. The power δ is a critical exponent that is measured experimentally by the relation $B = |\Psi|^\delta$. Its value is determined by η via the so-called hyperscaling relation [42] $\delta = (D{+}2{-}\eta)/(D{-}2{+}\eta)$. The value of g_c is related to the critical value $g^* \approx 0.503$ by $g_c \mu^{-\eta D/(D-2+\eta)} = (2g^*)^{(\delta-1)/2}(4\pi)^2/24 \approx 6.7$. As a possible application we may study the behavior of a triangular lattice of vortices which form in a rotating Bose-Einstein condensate [43], and letting the magnetic field approach a Feshbach resonance.

The results may then be compared with a calculation based on a new field equation that generalizes the famous Gross-Pitaevskii equation [44]

$$\left[\hat{E} - \frac{1}{2m}\hat{\mathbf{p}}^2 - g|\Psi(t, \mathbf{x})|^2\right] \Psi(t, \mathbf{x}) = 0. \tag{27}$$

The new equation is obtained by extremizing the effective action $\Gamma[\Psi, \Psi^*] = \Gamma_0[\Psi, \Psi^*] + \Gamma^{\text{int}}[\Psi, \Psi^*]$, where

$$\Gamma_0 \equiv \int dt d^D x \, \Psi^\dagger(t, \mathbf{x})[\hat{E}^{1-\gamma} - D_\lambda(\hat{\mathbf{p}}^2)^{\lambda/2}]\Psi(t, \mathbf{x}). \tag{28}$$

By forming $\delta\mathcal{A}^{\text{eff}}/\delta\Psi^\dagger(t, \mathbf{x})$, we obtain what may be called the *fractional Gross-Pitaevskii equation*:

$$\left[\hat{E}^{1-\gamma} - D_\lambda(\hat{\mathbf{p}}^2)^{1-\eta/2} - \frac{\delta+1}{4\mu^\eta} g_c |\Psi(t, \mathbf{x})|^{\delta-1}\right]\Psi(t, \mathbf{x}) = 0. \tag{29}$$

The fractional Schrödinger equation has many problems, such as the nonvalidity of the quantum superposition law, the violation of unitarity of the time evolution, and the violation of probability conservation which can produce nonsensical probabilities > 1 [33]. However, these problems exist only if we restrict ourselves to the free effective action (28), but this is meaningless, since the entire theory is only defined by the effective action in the strong-coupling limit — and this contains necessarily additional nonquadratic terms. Hence it does not possess free quasiparticles as in the time-honored Landau theory of Fermi liquids [45]. There is always an interaction that invalidates the standard discussion of Schrödinger equations. In fact, the theory of high-T_c superconductivity must probably be built as a true strong-coupling theory of this type with electrons being non-Fermi liquids [45].

The relativistic version of the entire discussion is simpler since it is based on the euclidean Green function (9) in which $\mathbf{p}$ denotes the $D - 1$-dimensional vectors $(\mathbf{p}, p_4)$. The Fourier transform is the distribution fulfilling the Fokker-Planck equation

$$[\partial_s + (\hat{\mathbf{p}}^2)^{1-\eta/2}]P(s, \hat{\mathbf{x}}) = \delta(s)\delta^{(D+1)}(\mathbf{x}). \tag{30}$$

and possessing the path integral representation

$$P(s, \hat{\mathbf{x}}) = \int \mathcal{D}\mathbf{x}\mathcal{D}\mathbf{p} e^{\int ds[i\mathbf{p}\dot{\mathbf{x}} - (\hat{\mathbf{p}}^2)^{1-\eta/2}]}. \tag{31}$$

The ϵ-expansion is now around $D_c = 4$ in powers of $\epsilon = -(D - D_c)$. The critical exponent η is small of order ϵ^2: $\eta = \epsilon^2/50 + \cdots \approx 0.04$. It can be ignored for $\epsilon = 1$. The power δ in the interaction is $3 + \epsilon + 23\epsilon^2/50 + \cdots \approx 4.76$ [46].

The time-independent *fractional Gross-Pitaevskii equation* reads now

$$\left[(\hat{\mathbf{p}}^2)^{1-\eta/2} + \frac{\delta+1}{4\mu^\eta} g_c |\Psi(\mathbf{x})|^{\delta-1}\right] \Psi(\mathbf{x}) = 0, \tag{32}$$

with $g_c \approx 27$. For a $d = D - 1$ -dimensional vortex in $D = 3$ dimensions, it is solved by $\tilde{\Psi}(\mathbf{x}) = a|\mathbf{x}_\perp|^{-A}$ with $A = (2-\eta)/(\delta-1) = D/2 - 1 + \eta/2 \approx 1/2$ and for $\mu = 1$: $[(\delta+1)a^{\delta-1}/4]g_c = -{}^d c_{\lambda+A-d}\, {}^d c^{-1}_{A-d} \approx 0.2$, $\lambda = 2 - \eta$ [29].

Let us study the appearance of a reduced mass $\hat{m}^2 \propto (1 - r^2)$ in the trap. In the effective action, it will appear in (32) in the form $\mu^{2-\eta}(\hat{m}^2)^{\nu(2-\eta)} \times f(|\Psi|^2/(\hat{m}^2)^{2\beta})$ with a Taylor series of $f(x)$ (note that $\nu(2-\eta) = 1 + \frac{\epsilon}{5} + \cdots \approx 1.3$). For small $\hat{m}^2$, this may be resummed to a Widom type expression $[(\delta+1)/4\mu^\eta]g_c|\Psi|^{\delta-1}w(\hat{m}^2/|\Psi|^{1/\beta})$ [24]. This explains the earlier-stated steeper falloff $|\Psi|^2 \propto (\hat{m}^2)^{2\beta}$ of the density profiles in Fig. 1. The Widom function $w(\hat{m}^2/|\Psi|^{1/\beta})$ can be expanded as 1 plus a power series in $(\hat{m}^2)^{\omega/2\nu} \propto \xi^{-\omega}$ which contains the Wegner critical exponent $\omega \approx 0.8$ that governs the *approach to scaling* [25]. Thereby the interaction term $|\Psi|^{\delta-1}$ is modified to $|\Psi|^{\delta-1}(1 + \text{const} \times \xi^{-\omega}|\Psi|^{-\omega\nu/\beta})$. Similarly, the kinetic term $(\hat{\mathbf{p}}^2)^{1-\eta/2}$ in (27), (32) will receive ω-dependent correction terms and become $(\hat{\mathbf{p}}^2)^{1-\eta/2}[1 + \xi^{-\omega/\eta^*_m}(\hat{\mathbf{p}}^2)^{\omega/2} + \ldots]$ [42], with $\eta^*_m = 1 - 1/2\nu \approx 1/4$.

To compare our theory with experimental data, we must study the BEC in the scale-invariant strong-coupling limit. This is reached either by going to the temperature T_c of the second-order phase transition, or by raising the magnetic field B towards the field strength B_c of a Feshbach resonance. Then the coherence length ξ grows like $\xi \propto |t|^{-\nu}$ where $\nu \approx 2/3$ [24, 50], and $t \equiv 1 - T/T_c$ or $t \equiv 1 - B/B_c$ ([39]). If the BEC is enclosed in a weak harmonic trap, this adds in the brackets of (27) a term $\propto |\mathbf{x}|^2 = R^2$. This is normally observed by the condensate density going to zero linearly like $1 - r^2 \equiv 1 - R^2/R_\text{b}^2$ near the border R_b (in the Thomas-Fermi approximation) [47]. For B near B_c (or T near T_c), however, the anomalous power δ will lead to the steeper approach to zero $(1 - r^2)^{2\beta}$ where $2\beta \equiv \nu(D - 2 + \eta) = 1 - 3\epsilon/10 + \cdots \approx 0.7$, plotted in Fig. 1, as will be shown immediately. In addition, the central region is depleted.

Figure 1. Condensate density from Gross-Pitaevskii equation (27) (GP,dashed) and its fractional version (29 (FGP), both in Thomas-Fermi approximation where the gradients are ignored. The FGP-curve shows a marked depletion of the condensate. On the right hand, a vortex is included. The zeros at $r \approx 1$ will be smoothened by the gradient terms in (32), as shown on the left-hand plot.

6. Summary

Summarizing we have seen that a many-body system with strong couplings between the constituents satisfies a more general form of the Schrödinger equation, in which the momentum and the energy appear with a power different from $\lambda = 2$ and $\gamma = 0$, respectively. The associated Green function can be written as a path integral over fluctuating time and space orbits that are functions of some pseudotime s. This is a Markovian object, but non-Markovian in the physical time t that is related to s by a stochastic differential equation of the Langevin type. The particle distributions can also be obtained by solving a Langevin type of equation in which the noise has correlation functions whose probability distribution is specified.

Appendix: The lowest-order critical exponents can be extracted directly from the one-loop-corrected inverse Green function $G^{-1}(E, \mathbf{p})$ in $D = 2+\epsilon$ dimensions after a minimal subtraction of the $1/\epsilon$ -pole at [51]:

$$E-\mathbf{p}^2+a\left(\tfrac{1}{3}\mathbf{p}^2-E\right)^{D-1}. \tag{33}$$

For $\mathbf{p} = 0$, this has a power $-(-E)^{1-a\epsilon}$, so that $\gamma = a\epsilon$. For $E = 0$, on the other hand, we obtain $(-\mathbf{p}^2)^{1-a\epsilon/3}$, so that $(1-\gamma)/z - 1 \approx \gamma/3$.

Acknowledgements

I am grateful to H.-T. Elze, Luca Di Fiore, P. Jizba, N. Laskin, M. Lewenstein, A.Pelster, F. Scardigli, and M. Zwierlein for useful comments.

References

[1] See `http://en.wikipedia.org/wiki/Black_swan_theory`
[2] See `http://www.nytimes.com/2007/04/22/books/chapters/0422-1st-tale.html? pagewanted=all`.
[3] Contrasting the 1994 book by R.J. Herrnstein and C. Murray entitled "The Bell Curve" `http://de.wikipedia.org/wiki/The_Bell_Curve`.
[4] See `http://en.wikipedia.org/wiki/2010_Flash_Crash`.
[5] Fock V 1937 Physik. Zeits. Sowjetunion **12** 404
[6] Feynman R P 1950 Phys. Rev. **80** 440; Feynman R P and Hibbs A R 1965 *Quantum Mechanics and Path Integrals* (New York: McGraw-Hill)
[7] Kleinert H 2006 *Path Integrals in Quantum Mechanics, Statistics, Polymer Physics, and Financial Markets* (Singapore: World Scientific)
[8] Kleinert H 1989 *Gauge Fields in Condensed Matter* Vol. I (Singapore: World Scientific Publishing Co.,)
[9] Black F and Scholes M 1973 J. Pol. Economy **81** 637
[10] Feller W 1991 *An Introduction to Probability Theory and Its Applications* vol. 2 (New York: Wiley); Bouchaud J P and Potters M 2000 *Theory of Financial Risks, From Statistical Physics to Risk Management* (Cambridge: Cambridge U. Press). See also Ch. 20 in [7]
[11] A travelling pedestrian salesman is a Gaussian random walker, as a jetsetter he becomes a Lévy random walker
[12] Einasto J 2011 *Preprint* arXiv:1109.5580
[13] Angulo R E, Springel V, White S D M, Jenkins A, Baugh C M and Frenk C S 2012 *Preprint* arXiv:1203.3216
[14] Jiulin Du 2004 *Europhys. Lett.* **67** 893; *do.* 2004 *Phys. Lett.* A **329** 262
[15] Nardini C, Gupta S, Ruffo S, Dauxois T, and Bouchet F 2012 *J. Stat. Mech.* L01002 (*Preprint* arXiv:1111.6833)
[16] Boettcher F, Renner C, Waldl H P and Peinke J *Preprint* arXiv:physics/0112063
[17] Bhattacharyya P, Chatterjee A and Chakrabarti B K 2007 *Physica* A **381** 377 (*Preprint* arXiv:physics/0510038)
[18] Umarov S, Tsallis C, Gell-Mann M and Steinberg S *Preprint* arXiv:cond-mat/0606040
[19] Kleinert H 1998 *Phys. Rev.* D **57** 2264 *Preprint* arXiv:cond-mat/9801167)
[20] Lipa J A, Nissen J A, Stricker D A, Swanson D R and Chui T C P 2003 *Phys. Rev.* B **68** 174518; Barmatz M, Hahn I, Lipa J A and Duncan R V 2007 Rev. Mod. Phys. **79** 1. Also see picture on the titlepage of the textbook [24]

[21] Kleinert H 2000 *Phys. Lett.* A **277** 205 (cond-mat/9906107); *do.* 1999 *Phys. Rev.* D **60** 085001 *Preprint* arXiv:hep-th/9812197)
[22] Kleinert H 2008 *Multivalued Fields in Condensed Matter, Electromagnetism, and Gravitation* (Singapore: World Scientific) pp. 1–497 (`http://klnrt.de/b11`)
[23] Gribov V N and Migdal A A 1969 *Soviet Phys. JETP* **28** 7841; see also Abarbanel H D, Bronzan J D, Sugar R L and White A R 1975 *Phys. Reports* **21** 119 [in particular Eq. (4.27)]
[24] Kleinert H and Schulte-Frohlinde V 2001 *Critical Properties of ϕ^4-Theories* (Singapore: World Scientific) (`klnrt.de/b8`)
[25] See Eq. (10.191) in Ref. [24] and expand $f(t/M^{1/\beta}) \sim \tilde{f}(\xi\Phi^{2/(D-2+\eta)})$ like $\tilde{f}(x) = 1 + cx^{-\omega} + \dots$.
[26] In two dimensions, a sequence of critical exponents have been tabulated in Belavin A A, Polyakov A M, and Zamolodchikov A B 1984 *Nucl. Phys.* B **241** 333
[27] Gell-Mann M and Low F E 1954 *Phys. Rev.* **95** 1300
[28] For the so-called Riesz fractional derivative see Metzler R, Barkai E and Klafter J 1999 *Phys. Rev. Lett.* **82** 3564; West B J, Grigolini P, Metzler R and Nonnenmacher T F 1997 *Phys.Rev.* E **55** 99. For the so-called Weyl derivative, see Raina R K and Koul C L 1979 *Proc. Am. Math. Soc.* **73** 188: $\hat{p}_4^{1-\gamma} f(t) \equiv \Gamma^{-1}[1-\gamma] \int_t^\infty dt'(t-t'+i\epsilon)^{-2+\gamma} f(t')$.
[29] The relevant functional matrix is $\langle\mathbf{x}|(-\boldsymbol{\nabla}^2)^{\lambda/2}|\mathbf{x}'\rangle = \Gamma[-\lambda/2]^{-1}\int d\sigma\,\sigma^{-\lambda/2-1}(4\pi\sigma)^{-D/2}e^{R^2/4\sigma} = {}^Dc_\lambda R^{-\lambda-D}$, where ${}^Dc_\lambda = 2^\lambda\Gamma((D+\lambda)/2)/\pi^{D/2}\Gamma(-\lambda/2)$, and $R \equiv |\mathbf{x}-\mathbf{x}'|$. If λ is close to an even integer, it needs a small positive shift $\lambda \to \lambda_+ \equiv \lambda + \epsilon$ and we can replace $\epsilon R^{\epsilon-1}/2$ by $\delta(R) = S_D R^{D-1}\delta^{(D)}(\mathbf{R})$. For $A > 0$ we have $|\mathbf{x}'|^{-A} = {}^Dc_{\lambda_A}^{-1}\langle\mathbf{x}'|(-\boldsymbol{\nabla}^2)^{\lambda_A/2}|\mathbf{0}\rangle$ with $\lambda_A \equiv A - D$, so that we find $\int d^Dx'\langle\mathbf{x}|(-\boldsymbol{\nabla}^2)^{\lambda/2}|\mathbf{x}'\rangle|\mathbf{x}'|^{-A} = {}^Dc_{\lambda_A}^{-1}\langle\mathbf{x}|(-\boldsymbol{\nabla}^2)^{(\lambda+A-D)/2}|\mathbf{0}\rangle = {}^Dc_{\lambda+A-D}\,{}^Dc_{\lambda_A}^{-1}|\mathbf{x}|^{-A-\lambda}$.
[30] Jun-Sheng Duan 2005 *J. Math. Phys.* **46** 013504
[31] Fox C 1961 *Trans. Amer. Math. Soc.* **98** 395
[32] This technique is explained in Chapters 12 and 19 of Ref. [7]. The pseudotime s resembles the so-called Schwinger proper time used in relativistic physics.
[33] Laskin N 2010 *Preprint* arXiv:1009.5533; *do.* 2000 *Phys.Lett.* A **268** 298; *do.* 2000 *Phys. Rev.* E **62** 3135; *do.* 2002 *ibid.* E **66** 056108; *do.* 2000 *Chaos* **10** 780; *do.* 2007 *Communications in Nonlinear Science and Numerical Simulation* **12** 2
[34] There should be no danger of confusing the fluctuating noise variable η in this equation with the constant critical exponent η in (9).
[35] See Eq. (29.165) in Ref. [7].
[36] Duru I H and Kleinert H 1979 *Phys. Lett.* B **84** 30 (`klnrt.de/65/65.pdf`); *do.* 1982 *Fortschr. Phys.* **30** 401 (`klnrt.de/83/83.pdf`). See also Chaps. 13 and 14 in [7].
[37] Young A and DeWitt-Morette C 1984 *Ann. Phys. (N.Y.)* **169** 140; Kleinert H and Pelster A 1997 *Phys. Rev. Lett.* **78** 565
[38] Liang L Z J, Lemmens D and Tempere J 2011 *Preprint* arXiv:1101.3713
[39] For a pedagogical discussion see: Zwierlein M 2006 MIT PhD Thesis (klnrt.de/papers/ZW.pdf)
[40] We ignore the problem that near the resonance it is hard to confine the particles to the trap. See: Stenger J, Inouye S, Andrews M R, Miesner H J, Stamper-Kurn D M and Ketterle W 1999 *Phys. Rev. Lett.* **82** 2422
[41] For the universal use of this functional, see: Kleinert H 1982 *Fortschr. Phys.* 30 187 (klnrt.de/82)
[42] See Eq. (1.35) in Ref. [24].
[43] Bretin V, Stock V S, Seurin Y, Chevy F and Dalibard J 2004 *Phys. Rev. Lett.* **92** 050403
[44] Pitaevskii L P and Stringari S 2003 *Bose-Einstein Con densation* (Oxford: Clarendon Press); Yukalov V I 2009 *Laser Phys.* **19** 1 (*Preprint* arXiv:0901.0636)
[45] Schulz H J 1995 Fermi liquids and non–Fermi liquids, in *Proceedings of Les Houches Summer School LXI* Akkermans E, Montambaux G, Pichard J et Zinn-Justin J ed p.533 (Amsterdam: Elsevier) (*Preprint* arXiv:cond-mat/9503150)
[46] The decimal numbers are from seven-loop calculation in $D = 3$ dimensions in Table 20.2 of Ref. [24].
[47] Dalfovo F, Giorgini S, Pitaevskii P and Stringari S 1999 *Rev. Mod. Phys.* bf 71 463512 (Fig. 3); Hau L V, Busch B D, Liu C, Dutton Z, Burns M M and Golovchenko J A 1998 *Phys. Rev.* A **58** R54 (Fig. 2)
[48] Pollack S E, Dries D, Hulet R G, Magalhaes K M F, Henn E A L, Ramos E R F, Caracanhas M A and Bagnato V S 2010 *Phys. Rev. A* **82** 020701(R)
[49] See Section 10.8 in Ref. [24].
[50] Donner T, Ritter S, Bourdel T, Öttl A, Köhl M, and Esslinger T 2007 *Science* **315** 1556
[51] Abarbanel H D and Bronzan J D 1974 *Phys. Rev.* D **9** 3304; see their Eq. (40).

CPT-violating leptogenesis induced by gravitational backgrounds

Nick E Mavromatos[1] **and Sarben Sarkar**

Theoretical Particle Physics and Cosmology Group, Department of Physics, King's College London, Strand, London WC2R 2LS, UK.

E-mail: Nikolaos.Mavromatos@kcl.ac.uk; Sarben.Sarkar@kcl.ac.uk

Abstract. We explore leptogenesis induced by the propagation of neutrinos in non-trivial gravitational backgrounds that may characterise the Early Universe epochs of various theories, including string theory. The key point in all these models is that the background induces different populations of fermions as compared to antifermions, and hence CPT Violation (CPTV), already in thermal equilibrium. Depending on the model, then, such populations may freeze out at various conditions leading to leptogenesis and baryogenesis. Among the considered scenarios, is a stringy one, in which the CPTV is associated with a cosmological background with torsion provided by the Kalb-Ramond antisymmetric tensor field (axion) of the string gravitational multiplet. We also discuss CPTV models that go beyond the *local* effective lagrangian framework, such as a stochastic (Lorentz Violating) Finsler metric and D-particle foam, where the CPTV is due to populations of stochastically fluctuating point-like space-time defects that can be encountered in string/brane theory (D0 branes).

1. Introduction

One of the most important issues of fundamental physics, relates to an understanding of the magnitude of the observed baryon asymmetry $n_B - n_{\overline{B}}$ (where B denotes baryon, $\overline{B}$ denotes antibaryon, n_B is the number density of baryons and $n_{\overline{B}}$ the number density of antibaryons in the universe). The universe is overwhelmingly made up of matter rather than anti-matter. According to the standard Big Bang theory, matter and antimatter have been created in equal amounts in the early universe. However, the observed charge-parity (CP) violation in particle physics [1], prompted A. Sakharov [2] to conjecture that for Baryon Asymmetry in the universe (BAU) we need:

- Baryon number violation to allow for states with $\Delta B \neq 0$ starting from states with $\Delta B = 0$ where ΔB is the change in baryon number.
- If C or CP conjugate processes to a scattering process were allowed with the same amplitude then baryon asymmetry would disappear. Hence C and CP need to be broken.
- Chemical equilibrium does not permit asymmetries. Hence Sakharov required that chemical equilibrium does not hold during an epoch in the early universe.

Hence non-equilibrium physics in the early universe together with baryon number (B), charge (C) and charge-parity (CP) violating interactions/decays of anti-particles in the early universe,

[1] Invited lecture presented by N.E.M.

may result in the observed BAU. In fact there are two types of non-equilibrium processes in the early universe that can produce this asymmetry: the first type concerns processes generating asymmetries between leptons and antileptons (*leptogenesis*), while the second produces asymmetries between baryons and antibaryons (*baryogenesis*). The near complete observed asymmetry today, is estimated in the Big-Bang theory [3] to imply:

$$\Delta n(T \sim 1\text{ GeV}) = \frac{n_B - n_{\overline{B}}}{n_B + n_{\overline{B}}} \sim \frac{n_B - n_{\overline{B}}}{s} = (8.4 - 8.9) \times 10^{-11} \tag{1}$$

at the early stages of the expansion, e.g. for times $t < 10^{-6}$ s and temperatures $T > 1$ GeV. In the above formula s denotes the entropy density. Unfortunately, the observed CP violation within the Standard Model (SM) of particle physics (found to be of order $\epsilon = O(10^{-3})$ in the neutral Kaon experiments [1]) induces an asymmetry much less than that in (1) [4]. There are several ideas that go beyond the SM (*e.g.* grand unified theories, supersymmetry, extra dimensional models *etc.*) which involve the decays of right handed sterile neutrinos. For relevant important works on this see [5, 6, 7, 8, 9, 10, 11]. These ideas lead to extra sources for CP violation that could generate the observed BAU. Some degree of fine tuning and somewhat *ad hoc* assumptions are involved in such scenarios and the quest for an understanding of the observed BAU still needs further investigation. An example of fine tuning is provided by the choice of the hierarchy of the right-handed Majorana neutrino masses. For instance, enhanced CP violation, necessary for BAU, can be achieved in models with three Majorana neutrinos, by assuming two of these neutrinos are nearly degenerate in mass.

The requirement of non-equilibrium is on less firm ground [12] than the other two requirements of Sakharov, *e.g.* if the non-equilibrium epoch occurred prior to inflation then its effects would be hugely diluted by inflation. A basic assumption in the scenario of Sakharov is that *CPT symmetry* [13] (where T denotes time reversal operation) holds in the very early universe which leads to the production of matter and antimatter in equal amounts. Such *CPT invariance* is a cornerstone of all known *local* effective *relativistic* field theories without gravity, which current particle-physics phenomenology is based upon. It should be noted that the necessity of non-equilibrium processes in CPT invariant theories can be dropped if the requirement of CPT is relaxed [14]. This violation of CPT (denoted by CPTV) is the result of a breakdown of Lorentz symmetry (which might happen at ultrahigh energies [15]). For many models with CPTV, in the time line of the expanding universe, CPTV generates first lepton asymmetries (*leptogenesis*); subsequently through sphaleron processes [16] or Baryon-Lepton (B-L) number conserving processes in Grand Unified Theories (GUT), the lepton asymmetry can be communicated to the baryon sector to produce the observed BAU.

Thus, CPTV in the early universe may also obviate the need for including extra sources of CP violation, such as sterile neutrinos and/or supersymmetry, in order to obtain the observed BAU. In this talk, which is based on joint work appeared in [17], we will consider CPTV leptogenesis from a non-Riemannian point of view, inspired by a stringy model of gravitational defects and backgrounds interacting with neutral fermions. The explicit non-Riemannian structure that we investigate is Finsler geometry [18] where momentum as well as position are explicitly involved in its metric. The defects that we will consider are point-like solitonic structures in some string theories; they are known as D0-branes (or *D-particles*) [19]. Our generic model involves effectively three-space-dimensional brane universes, obtained from compactification of higher-dimensional branes, which are embedded in a bulk space punctured by D-particles (see e.g. [20]). One of these brane universes constitutes our observable world, which moves in the bulk. As a consequence of this motion, D-particles cross the brane. Open strings for electrically neutral particles on the brane can attach an end to the D-particle; subsequently this string can detach [21, 22]. Scattering off a *population* of D-particles (D-foam) affects the kinematics of the stringy matter and leaves an imprint on the background geometry [23] on the brane world. This geometry is similar to Finsler metrics but with stochastic parameters [24, 25, 26]. We first investigate the consequences for gravitational leptogenesis of this metric structure in the general setting of Finsler geometry without reference explicitly to D-foam. For the underlying stringy model however, since it is microscopic, we can consider in addition the kinematics of D particle scattering. This kinematical aspect leads naturally to the asymmetry between the particle and anti-particle abundances having the right sign. The kinematical argument, which involves recoil kinetic energy, does not fit into an effective local field theory approach and represents a new approach that is relevant to leptogenesis.

The structure of the talk will be as follows: in the next section 2 we shall briefly review some relevant existing models for fermionic asymmetry, which entail CPTV-induced differences in the dispersion relations between particles and antiparticles propagating in curved gravitational backgrounds in the early universe. In section 3, we review a new model for gravitational leptogenesis, proposed in [17], which follows broadly an earlier framework [27, 28], but differs crucially in that the full gravitational multiplet [29] that arises in string theory is used. The leptogenesis in this model is due to CPTV dispersion relations between fermions/antifermions, induced by the (constant) *torsion* associated with the antisymmetric Kalb-Ramond tensor field in the gravitational multiplet of the string. In section 4, we consider another scenario for gravitational leptogenesis that involves a non-Riemannian Finsler metric, with stochastically fluctuating parameters, a variant of which appears in our string/brane ("*D-foam*") model in section 5. In section 5 we present our D-foam model for the stringy universe and the induced CPTV and leptogenesis/baryogenesis. The model is characterised by CPTV and matter dominance over antimatter *naturally*, without the need for an adjustment of the sign of the lepton/antilepton asymmetry. This is an interesting feature of our model which differentiates it from earlier proposals on gravitational leptogenesis/baryogenesis [30, 31, 32, 33, 28, 27, 34, 35, 36], where the sign of the asymmetry is implicitly chosen. In common with earlier discussions of CPTV in baryogenesis [37], this class of models involves violation of Lorentz symmetry, but only because of a non-zero *variance* of the stochastic parameter.

2. Models of CPT violation (CPTV) in the early universe

In this section we shall review some existing models of CPTV induced asymmetry between matter and antimatter in the early universe, which can be contrasted with our approach in this article. We shall be brief in our exposition, referring the interested reader to the relevant literature for more details.

2.1. CPTV models with particle-antiparticle mass difference

The simplest possibility [38] for inducing CPTV in the early universe is through particle-antiparticle mass differences $m \neq \overline{m}$. These would affect the particle phase-space distribution function $f(E,\mu)$, $f(E,\mu) = [\exp(E-\mu)/T) \pm 1]^{-1}$, $E^2 = \vec{p}^2 + m^2$, and antiparticle phase-space distribution function $f(\overline{E},\bar{\mu}) = [\exp(\bar{E}-\bar{\mu})/T) \pm 1]^{-1}$, $\bar{E}^2 = \vec{p}^2 + \bar{m}^2$, with $\vec{p}$ being the 3−momentum. (Our convention will be that an overline over a quantity will refer to an antiparticle, + will denote a fermionic (anti-)particle and − will denote a bosonic (anti-)particle). (Anti)particles will have, by definition, (negative) positive energies. Mass differences between particles and antiparticles, $\overline{m} - m \neq 0$, generate a matter-antimatter asymmetry in the relevant number densities n and $\overline{n}$ $n - \overline{n} = g_{d.o.f.} \int \frac{d^3p}{(2\pi)^3}[f(E,\mu) - f(\overline{E},\bar{\mu})]$, where $g_{d.o.f.}$ denotes the number of degrees of freedom of the particle species under study. In the case of spontaneous Lorentz violation [37] there is a vector field A_μ with a non-zero time-like expectation value which couples to a global current J^μ such as baryon number through an interaction lagrangian density

$$\mathcal{L} = \lambda A_\mu J^\mu. \tag{2}$$

This leads to $m \neq \bar{m}$ and $\mu \neq \bar{\mu}$. Alternatively, following [38] we can make the assumption that the dominant contributions to baryon asymmetry come from quark-antiquark mass differences, and that their masses "run" with the temperature i.e. $m \sim gT$ (with g the QCD coupling constant). One can provide estimates for the induced baryon asymmetry on noting that the maximum quark-antiquark mass difference is bounded by the current experimental bound on the proton-antiproton mass difference, $\delta m_p (= |m_p - \overline{m}_p|)$, known to be less than $2 \cdot 10^{-9}$ GeV. Taking $n_\gamma \sim 0.24 T^3$ (the photon equilibrium density at temperature T) we have [38]:

$$\beta_T = \frac{n_B}{n_\gamma} = 8.4 \times 10^{-3} \frac{m_u \, \delta m_u + 15 m_d \, \delta m_d}{T^2} \quad , \quad \delta m_q = |m_q - \overline{m}_q| \; . \tag{3}$$

β_T is too small compared to the observed one. To reproduce the observed $\beta_{T=0} \sim 6 \cdot 10^{-10}$ one would need $\delta m_q (T = 100 \text{ GeV}) \sim 10^{-5} - 10^{-6}$ GeV $\gg \delta m_p$, which is somewhat unnatural.

However, active (*light*) neutrino-antineutrino mass differences alone may reproduce BAU; some phenomenological models in this direction have been discussed in [39], considering, for instance, particle-antiparticle mass differences for active neutrinos compatible with current oscillation data. This leads to the result

$$n_B = n_\nu - n_{\overline{\nu}} \simeq \frac{\mu_\nu T^2}{6} \tag{4}$$

yielding $n_B/s \sim \frac{\mu_\nu}{T} \sim 10^{-11}$ at $T \sim 100$ GeV, in agreement with the observed BAU. (Here s, n_ν, and μ_ν are the entropy density, neutrino density and chemical potential respectively.)

2.2. CPTV decoherence models

Particle-antiparticle mass differences however may not be the only way by which CPT is violated. As discussed in [40, 41], quantum gravity fluctuations in the structure of space-time, may be strong in the early universe; the fluctuations may act as an *environment* inducing decoherence for the (anti-)neutrinos. However the couplings between the particles and the environment are different for the neutrino and antineutrino sectors. Once there is decoherence for an observer with a low energy (compared to the Planck scale $M_P \sim 10^{19}$ GeV), the effective CPT symmetry generator may be *ill-defined* as a quantum mechanical operator, according to a theorem by R. Wald [42], leading to an intrinsic violation of CPT symmetry. This type of violation may characterise models of quantum gravity with stochastic space-time fluctuations due, for instance, to gravitational space-time defects, as is the case of certain brane models [22, 20, 25]. In such a case, a slight mismatch in the strength of the stochastic space-time fluctuations between particle and antiparticle sectors, can lead to different decoherence parameters to describe the interaction of the gravitational environment with matter.

In [40, 41], simple models of Lindblad decoherence [43], conjectured to characterise quantum-gravity-induced CPTV decoherent situations [44, 45], have been considered for neutrinos [46]. It was assumed on phenomenological grounds, that non-trivial decoherence parameters were *only* present in the antiparticle sector: this is consistent with the lack of any experimental evidence to date [47, 48, 49] for vacuum decoherence in the particle sector. The antineutrino decoherence parameters (with dimension of energy) had a mixed energy dependence. The model of [40, 41] assumes a diagonal $\mathcal{L}_{\mu\nu}$. A diagonal Lindblad decoherence matrix for three-generation neutrinos requires eight coefficients $\overline{\gamma}_i$. Some of the eight coefficients were assumed for simplicity in [40, 41] to be proportional to the antineutrino energy $\overline{\gamma}_i = \frac{T}{M_P} E$, $i = 1, 2, 4, 5$, while the remaining (subdominant) ones were inversely proportional to it $\overline{\gamma}_j = \frac{10^{-24}\,(\text{GeV})^2}{E}$, $j = 3, 6, 7, 8$. The model was proposed without any microscopic justification; its choice was originally motivated by fitting the LSND "anomalous data" in the antineutrino sector [50] with the rest of the neutrino data.

and this required T to be $T/M_P \sim 10^{-18}$,i.e. in the temperature range of electroweak symmetry breaking. One can derive [40, 41] an active (light) $\nu - \overline{\nu}$ asymmetry of order

$$\mathcal{A} = \frac{|n_\nu - n_{\overline{\nu}}|}{n_\nu + n_{\overline{\nu}}} = \frac{\overline{\gamma}_1}{\sqrt{\Delta m^2}} = \frac{T}{M_P} \cdot \frac{E}{\sqrt{\Delta m^2}} \,, \tag{5}$$

where Δm^2 denotes the (atmospheric) neutrino mass squared difference, which plays the role of a characteristic low mass scale in the problem. This lepton number violation is communicated to the baryon sector by means of baryon number (B) plus lepton number (L) conserving sphaleron processes (in fact, in this case one needs an antilepton excess in order to produce a baryon excess). These processes lead to an estimate [40] for the current value of B to be

$$B = \frac{n_\nu - n_{\overline{\nu}}}{s} \sim \mathcal{A}\, \frac{n_\nu}{g^\star\, n_\gamma} \tag{6}$$

with n_γ the photon number density, $g^\star$ the effective number of degrees of freedom (at the temperature where the asymmetry developed, i.e. the electroweak symmetry breaking temperature in the model of [40]). $g^\star$ depends on the matter content of the model (with a typical range $g^\star \in [10^2 - 10^3]$). For such parameter values $\mathcal{A} \sim 10^{-6}$ and so the observed BAU may be reproduced in this case without the need for extra sources of CP violation e.g. sterile neutrinos. Such models, however, do not provide an underlying microscopic understanding. In particular there is missing an understanding of the preferential role of the neutrino compared to other particles of the Standard Model in the CPT violating decoherence process. Within some microscopic models of space-time foam, involving populations of point-like brane defects (D-particles) puncturing three(or higher)-spatial dimension brane worlds [22, 20, 25], such a preferred role may be justified as we shall discuss in section 5.

2.3. CPTV-induced by curvature effects in background geometry

Although the role of gravity was alluded to in the last subsection, associated features of space-time were not discussed. In the literature the role of gravity has been explicitly considered within a local effective action framework which is essentially that of (2) A coupling to scalar curvature R [31, 27, 34, 32] through a CP violating interaction Lagrangian $\mathcal{L}$:

$$\mathcal{L} = \frac{1}{M_*^2} \int d^4x \sqrt{-g}\, (\partial_\mu R)\, J^\mu \tag{7}$$

where M_* is a cut-off in the effective field theory and J^μ could be the current associated with baryon (B) number. There is an implicit choice of sign in front of the interaction (7), which has been fixed so as to ensure matter dominance.

It has been shown that [31]

$$\frac{n_{B-L}}{s} = \frac{\dot{R}}{M_*^2 T_d}, \tag{8}$$

T_d being the freeze-out temperature for $B - L$ interactions. The idea then is that this asymmetry can be converted to baryon number asymmetry provided the $B + L$ electroweak sphaleron interaction has not frozen out. To leading order in M_*^{-2} we have $R = 8\pi G(1 - 3w)\rho$ where ρ is the energy density of matter and the equation of state is $p = w\rho$ where p is pressure. For radiation $w = 1/3$ and so in the radiation dominated era of the Friedmann-Robertson-Walker cosmology $R = 0$. However w is precisely $1/3$ when $T^\mu_\mu = 0$. In general $T^\mu_\mu \propto \beta(g) F^{\mu\nu} F_{\mu\nu}$ where $\beta(g)$ is the beta function of the running gauge coupling g in a $SU(N_c$ gauge theory with N_c colours. This allows $w \neq 1/3$. Further issues in this approach can be found in [31, 27, 34, 32].

Another approach involves an axial vector current [28, 33, 35, 36] instead of J_μ. The scenario is based on the well known fact that fermions in curved space-times exhibit a coupling of their spin to the curvature of the background space-time.The Dirac Lagrangian density of a fermion can be re-written as:

$$\mathcal{L} = \sqrt{-g}\,\overline{\psi}\left(i\gamma^a\partial_a - m + \gamma^a\gamma^5 B_a\right)\psi\ , \quad B^d = \epsilon^{abcd}e_{b\lambda}\left(\partial_a e^\lambda_{\ c} + \Gamma^\lambda_{\nu\mu}\,e^\nu_{\ c}\,e^\mu_{\ a}\right)\ , \tag{9}$$

in a standard notation, where $e^\mu_{\ a}$ are the vielbeins, $\Gamma^\mu_{\alpha\beta}$ is the Christoffel connection and Latin (Greek) letters denote tangent space (curved space-time) indices. The space-time curvature background has, therefore, the effect of inducing an "axial" background field B_a which can be non-trivial in certain anisotropic space-time geometries, such as Bianchi-type cosmologies [28, 33, 35, 36]. For an application to particle-antiparticle asymmetry it is necessary for this axial field B_a to be a constant in some local frame. The existence of such a frame has not been demonstrated. As before if it can be arranged that $B_a \neq 0$ for $a = 0$ then for constant B_0 CPT is broken: the dispersion relation of neutrinos in such backgrounds differs from that of antineutrinos. Explicitly we have

$$E = \sqrt{(\vec{p} - \vec{B})^2 + m^2} + B_0\ , \quad \overline{E} = \sqrt{(\vec{p} + \vec{B})^2 + m^2} - B_0\ . \tag{10}$$

The relevant neutrino asymmetry emerges on following the same steps used when there was an explicit particle-antiparticle mass difference, As a consequence the following neutrino-antineutrino density difference is found in Bianchi II Cosmologies [28, 33, 35, 36]:

$$\Delta n_\nu \equiv n_\nu - n_{\overline{\nu}} \sim g^\star\,T^3\left(\frac{B_0}{T}\right) \tag{11}$$

with $g^\star$ the number of degrees of freedom for the (relativistic) neutrino. An excess of particles over antiparticles is predicted only when $B_0 > 0$, which had to be assumed in the analysis of [28, 33, 35, 36]; we should note, however, that the sign of B_0 and its constancy have not been justified in this phenomenological approach [2].

At temperatures $T < T_d$, with T_d the decoupling temperature of the lepton-number violating processes, the ratio of the net Lepton number ΔL (neutrino asymmetry) to entropy density (which scales as T^3) remains constant,

$$\Delta L(T < T_d) = \frac{\Delta n_\nu}{s} \sim \frac{B_0}{T_d} \tag{12}$$

which, for $T_d \sim 10^{15}$ GeV and $B_0 \sim 10^5$ GeV, implies a lepton asymmetry (leptogenesis) of order $\Delta L \sim 10^{-10}$, in agreement with observations. The latter can then be communicated to the baryon sector to produce the observed BAU (baryogenesis) by a B-L conserving symmetry in the context of either Grand Unified Theories (GUT) [28], or sphaleron processes in the standard model. (Note that in the scenario of decoherence-induced CPT Violating case of [40, 41], mentioned previously, $B + L$ was assumed to be conserved in the corresponding sphaleron processes; in such a case one needs an antilepton excess to produce baryogenesis.)

[2] The above considerations concern the dispersion relations for any fermion, not only neutrinos. However, when one considers matter excitations from the vacuum, as relevant for leptogenesis, we need chiral fermions to get non trivial CPTV asymmetries in *populations* of particle and antiparticles, because $<\psi^\dagger\gamma^5\psi> = - <\psi_L^\dagger\gamma^5\psi_L> + <\psi_R^\dagger\gamma^5\psi_R>$.

3. CPTV-induced in (string-inspired) background geometry with torsion

In this section we will discuss the case of a constant B^0 "axial" field that appears due to the interaction of the fermion spin with a string-theory background geometry with *torsion*. This is a novel observation, which (as far as we are aware) was discussed for first time in [17]. In the presence of torsion the Christoffel symbol contains a part that is antisymmetric in its lower indices: ${\Gamma^{\lambda}}_{\mu\nu} \neq {\Gamma^{\lambda}}_{\nu\mu}$. Hence the last term of the right-hand side of the Eqn.(9) is *not* zero. Since the torsion term is of gravitational origin it couples universally to all fermion species. The effect of the coupling to neutrinos will be clarified below.

The massless gravitational multiplet in string theory contains the dilaton (spin 0, scalar), Φ, the graviton (spin 2, symmetric tensor), $g_{\mu\nu}$, and the spin 1 antisymmetric tensor $B_{\mu\nu}$. The (Kalb-Ramond) field B appears in the string effective action only through its totally antisymmetric field strength, $H_{\mu\nu\rho} = \partial_{[\mu} B_{\nu\rho]}$, where $[\ldots]$ denotes antisymmetrization of the indices within the brackets. The calculation of string amplitudes [51] shows that $H_{\mu\nu\rho}$ plays the role of *torsion* in a generalised connection $\overline{\Gamma}$:

$$ {\overline{\Gamma}^{\lambda}}_{\mu\nu} = {\Gamma^{\lambda}}_{\mu\nu} + e^{-2\Phi} H^{\lambda}_{\mu\nu} \equiv {\Gamma^{\lambda}}_{\mu\nu} + {T^{\lambda}}_{\mu\nu} \ . \tag{13} $$

${\Gamma^{\lambda}}_{\mu\nu} = {\Gamma^{\lambda}}_{\nu\mu}$ is the torsion-free Einstein-metric connection, and ${T^{\lambda}}_{\mu\nu} = -{T^{\lambda}}_{\nu\mu}$ is the *torsion*.

In ref. [52] exact solutions to the conformal invariance conditions (to all orders in α') of the low energy effective action of strings have been presented. In four "large" (uncompactified) dimensions of the string, the antisymmetric tensor field strength can be written uniquely as

$$ H_{\mu\nu\rho} = e^{2\Phi} \epsilon_{\mu\nu\rho\sigma} \partial^{\sigma} b(x) \tag{14} $$

with $\epsilon_{0123} = \sqrt{g}$ and $\epsilon^{\mu\nu\rho\sigma} = |g|^{-1}\epsilon_{\mu\nu\rho\sigma}$, with g the metric determinant. The field $b(x)$ is a "pseudoscalar " *axion*-like field. The dilaton Φ and axion b fields are fields that appear as Goldstone bosons of spontaneously broken scale symmetries of the string vacua, and so are exactly massless classically. In the effective string action such fields appear only through their derivatives.The exact solution of [52] in the *string frame* requires that both dilaton and axion fields are linear in the target time X^0, $\Phi(X^0) \sim X^0$, $b(X^0) \sim X^0$. This solution will shift the minima of all fields in the effective action which couple to the dilaton and axion by a space-time independent amount.

In the "physical" *Einstein frame* , relevant for cosmological observations, the temporal component of the metric is normalised to $g_{00} = +1$ by an appropriate change of the time coordinate. In this setting, the solution of [52] leads to a Friedmann-Robertson-Walker (FRW) metric, with scale factor $a(t) \sim t$, where t is the FRW cosmic time. Moreover, the dilaton field Φ behaves as $-\ln t + \phi_0$, with ϕ_0 a constant, and the axion field $b(x)$ is linear in t. There is an underlying world-sheet conformal field theory with central charge $c = 4 - 12Q^2 - \frac{6}{n+2} + c_I$ where $Q^2 (> 0)$ is the central-charge deficit and c_I is the central charge associated with the world-sheet conformal field theory of the compact "internal" dimensions of the string model [52]. The condition of cancellation of the world-sheet ghosts that appear because of the fixing of reparametrisation invariance of world-sheet co-ordinates requires that $c = 26$. The solution for the axion field is

$$ b(x) = \sqrt{2} e^{-\phi_0} \sqrt{Q^2} \frac{M_s}{\sqrt{n}} t \ , \tag{15} $$

where M_s is the string mass scale and n is a positive integer, associated with the level of the Kac-Moody algebra of the underlying world-sheet conformal field theory. For non-zero Q^2 there is an additional dark energy term in the effective target-space time action of the string [52] of the form $\int d^4x \sqrt{-g} e^{2\Phi}(-4Q^2)/\alpha'$. The linear axion field (15) *remains* a non-trivial solution *even* in the *static* space-time limit with a constant dilaton field [52]. In such a case space time is an Einstein universe with positive cosmological constant and constant positive curvature proportional to $6/(n+2)$.

For the solutions of [52], the covariant torsion tensor $e^{-2\Phi}H_{\mu\nu\rho}$ is *constant.* (This follows from (13) and (14) since the exponential dilaton factors cancel out in the relevant expressions.) Only the spatial components of the torsion are nonzero in this case,

$$T_{ijk} \sim \epsilon_{ijk}\dot{b} = \epsilon_{ijk}\sqrt{2Q^2}e^{-\phi_0}\frac{M_s}{\sqrt{n}}\ , \tag{16}$$

where the overdot denotes derivative with respect to t.

As discussed in [17], in the framework of the target-space effective theory, the relevant Lagrangian terms for fermions (to lowest order in α') will be of the form (9), with the vector B^0 being associated with the spatial components of the constant torsion part $B^0 \sim \epsilon^{ijk}T_{ijk}$, where From (13), (14) and (9), we also observe that only the temporal component B^0 of the B^d vector is nonzero. Note that the torsion-free gravitational part of the connection (for the FRW or flat case) yields a vanishing contribution to B^0. From (9) and (16) then we obtain a constant B^0 of order

$$B^0 \sim \sqrt{2Q^2}e^{-\phi_0}\frac{M_s}{\sqrt{n}}\ \text{GeV} > 0. \tag{17}$$

We follow the conventions of string theory for the sign of B^0 . From phenomenological considerations M_s and $g_s^2/4\pi$ are taken to be larger than $\mathrm{O}(10^4)$ GeV and about 1/20 respectively. A stringent constraint on the Kac-Moody level is not imposed from the requirement of CPTV-induced leptogenesis. We observe from (17) that the central charge deficit Q^2 of the underlying conformal field theory determines the order of particle-antiparticle asymmetries in this model.

The particle-antiparticle asymmetry occurs already in thermal equilibrium, due to the background-induced [28] difference in the dispersion relations between particles and antiparticles. Since the coupling of fermions to torsion is *universal*, the axion background would also couple to quarks and charged leptons.For CPT-Violating leptogenesis at the GUT scales [28] $B^0 \sim 10^5$ GeV. Hence a constant torsion-induced B^0 could lead directly to *baryogenesis* at the quark decoupling temperature $T_q \sim 100$ MeV, provided B^0 is much lower than $B^0 \sim 10^5$ GeV.In such a scenario, the observed BAU could be realised directly through the quark-H-torsion interactions; unfortunately, the corresponding neutrino asymmetry would be too high, given that the neutrino decoupling temperatures are at 1 MeV scale. Note that standard model B-L conserving sphaleron processes freeze-out much earlier (at temperatures of order 100 GeV), but such interactions cannot produce sufficient baryon asymmetry in the universe without extra sources for CP violation [4].

A phenomenologically viable alternative scenario is one in which B^0 is not constant with time. In such a scenario the universe undergoes a phase transition at GUT scales ($\sim 10^{15}$ GeV). Above the GUT temperature heavy right-handed Majorana neutrinos could decouple or lepton number violating (but B-L conserving) processes could occur. At the phase transition the value of the H-torsion background could change from a large to a smaller value. This approach is based on the possibility that conformal field theories with different central charges characterise different epochs of the early universe [52]. Transitions between such conformal (fixed) points in moduli space correspond to non-conformal (Liouville) time evolution [53] and represent phase transitions in the early universe [3]. In this scenario, leptogenesis occurs at GUT scales (just after

[3] An enlightening example of such a scenario is provided by the quantum Hall system in condensed matter where, as the temperature varies, there are transitions between plateaux with fixed values of the conductivity. The plateaux correspond to underlying conformal field theories of a "stringy" nature [54, 55], with different values of the central charge. Transitions between the plateaux are not described by conformal field theories. In the approach of [53] the cosmic time evolution in the universe is associated with a world-sheet renormalization group flow, whose description lies beyond the scope of the talk

inflation). Following the idea of Ref.[28], this leptogenesis can be communicated to the baryon sector by means of B-L conserving sphaleron processes. The difference between the fermion and antifermion populations is due to the interaction with a CPTV background torsion and occurs already in thermal equilibrium. Provided the background torsion is constant in time over the appropriate epoch of the evolution of the universe (until the freeze out of the pertinent processes characterising the epoch), such differences cannot be eroded during the expansion of the universe. No enhanced CP violation in the lepton sector is essential for leptogenesis in this approach, in contrast to the case of conventional leptogenesis [56, 57, 58, 5, 59].(This observation does not completely remove the need for right-handed sterile neutrinos.The latter may be essential, for instance, for explaining the smallness of the active neutrino masses, through see-saw mechanisms, or may play the role of dark matter [5, 59].)

4. CPTV in stochastic Finsler geometries

Although all the models displaying CPTV that we have considered so far are based on local effective field theories, there is no compelling reason for a restriction to such a framework. In fact a microscopic model involving space-time defects based on string-brane theory suggests the use of a non-Riemannian metric background similar to that which occurs in Finsler geometry [18, 60].(This model will be discussed in a subsequent section.) Independently there has been much interest in Finsler geometry [61, 62, 63, 64, 65, 66, 67, 68, 69, 70] for characterising the Early universe [71, 72, 73, 74, 75] and for descriptions of modified dispersion relations for particle probes [76, 72, 60, 77]. Finsler geometry has a metric which, in addition to space-time coordinates, depends also on "*velocities*". Lorentz symmetry is broken through some fixed vectors in the metric. We explore the consequences of making such vectors having components which are stochastic with possibly zero mean. This is a feature that arises in the defect model of D-foam [76, 20, 22] that we will consider in the next section. The defects stochastically fluctuate, due to both statistical and quantum stringy effects in large populations of such D-particles that can populate eras of the early universe. As we shall discuss below, the result of the interaction of neutrinos with these defects, leads to stochastically fluctuating Finsler-like metrics. However we wish to consider the consequences for CPTV and matter-antimatter asymmetry of this stochasticity in a general context. This underlying model provides our main motivation to study this class of space-times in this section and to contrast our findings on the induced CPTV for such cases with the corresponding ones for the D-foam model.

Below we shall be brief in our description of the stochastic CPTV backgrounds, concentrating only on the main features leading to modified dispersion relations that are different between particles and antiparticles. For the interested reader details can be found in [17].

We focus on on-shell neutrinos (which are now known to have small masses). Again, this is motivated by our desire to discuss leptogenesis in such geometries. Moreover, for reasons that will become clear in section 5, it is neutrinos that play a preferential rôle in interacting non-trivially with the D-particle foam background, which induces stochastically fluctuating Finsler-like space times. As our main motivation is to compare the generic Finsler-like case with the D-foam model, as far as CPTV is concerned, we restrict our attention here on the effects of stochastically fluctuating Finsler geometries on dispersion relations of neutrinos and antineutrinos. We shall consider a particular type of Finsler metric on a manifold M which is known as the Randers metric [18] [4]. It was noted in [80] that the geodesics of this metric coincided with the minimum

[4] We mention, for completeness, that the other popular class of Finsler geometries, that appears in the General Relativistic version [63, 66, 64, 65] of the so-called Very Special Relativety Model [78], and cosmological extensions thereof [71, 73] are *not* characterised by CPTV in the dispersion relations, nor of the spin-curvature type discussed in section 2.3. In fact such VSR-related models have been prop[osed in the past as candidates for the generation of Lepton-number conserving neutrino masses [79], and hence our Lepton-number violating considerations in this work do not apply.

time trajectories of a particle moving on a Riemannian manifold in the presence of a time independent drift given by a vector field. This is similar to Fermat's principle for propagation in refractive media. We mention in passing that similarities of D-particle foam to a refracting medium have been discussed in the literature [76, 81, 82]. If we were to assume that the result on minimum time trajectories was true for a pseudo-Riemannian situation and the drift was given by collisions due to D-particle scattering, then at a heuristic level a stochastic drift could be a reasonable generic phenomenological model of the back-reaction of low dimensional recoiling branes on matter.

The metric in this models is given by [17] (below, x^μ denotes space-time coordinates, while the variables y^ν play the rôle of "velocities" in the Finsler framework)

$$g_{\mu\nu}(x,y) = r_{\mu\nu}(x) + \phi^2(x)\, l_\mu l_\nu + \left(\frac{r_{\mu\nu}(x)}{\alpha(x,y)} - \frac{r_{\mu\varrho}(x)\, r_{\nu\sigma}(x)\, y^\varrho y^\sigma}{\alpha(x,y)^3}\right)\phi(x)\, l_c y^c$$
$$+ \frac{1}{\alpha(x,y)}\left(r_{\mu\varrho}(x)\, y^\varrho \phi(x)\, l_\nu + r_{\nu\varrho}(x)\, y^\varrho \phi(x)\, l_\mu\right) \;,\; \alpha(x,y) = \sqrt{r_{\mu\nu}(x)\, y^\mu\, y^\nu} \;,$$

where l_μ is a constant vector. In our model $\phi(x)$ will be a gaussian stochastic variable. On average the metric will be like a Riemannian metric if the mean of $\langle\phi\rangle$ vanishes.

We shall consider a situation with $r_{\mu\nu}(x) = \eta_{\mu\nu}$ where $\eta_{\mu\nu}$ is the diagonal Minkowski matrix with entries $(1,-1,-1,-1)$. (The summation convention of repeated indices will be always understood unless explicitly stated otherwise.) Within the framework of a Robertson-Walker metric we shall ignore effects on the time-scale of the inverse expansion We have assumed a homogeneous ϕ with ϕ being x independent. The mass shell conditions of generalised plane-wave solutions for particles in this background $h^{\mu\nu}\omega_\mu\omega_\nu = m^2$, where $\omega_\mu = g_{\mu\nu}(x,y)y^\nu$ is a "phase-space" variable, dual to y^μ, and $h^{\mu\nu}(x,\omega)$ is the inverse of the Finsler metric in phase-space [17], lead to the generalised dispersion relations. In the model it is possible to choose l_μ. Not all choices will lead to asymmetric population distributions between particles and anti-particles.The space-like choice $l_0 = 0$ gives a degenerate spectrum for particle and anti-particle and hence no CPTV in dispersion relations. Therefore this case cannot be used for Leptogenesis in our framework.

More generally the dispersion relation is [17]

$$\omega_0 = \pm\frac{\phi}{m} l_0 \left(2\vec{\omega}^2 + m^2\right) + \mathfrak{K}(\phi,\omega,m) \tag{18}$$

where $\mathfrak{K}(\phi,\omega,m) = \left(\vec{\omega}^2 + m^2 + \frac{2\phi}{m}\left(2\vec{\omega}^2+m^2\right)(l_1\omega_1 + l_2\omega_2 + l_3\omega_3)\right)^{1/2}$. The +sign is for the particle and the −sign is for the antiparticle. For the "time-like" case $l_0 = 1, l_1 = l_2 = l_3 = 0$ the dispersion relation reduces to e $\omega_0 = \sqrt{\vec{\omega}^2 + m^2} \pm \frac{\phi}{m}\left(2\vec{\omega}^2 + m^2\right)$. The sign of l_0 can be reabsorbed in ϕ.

For the "null" case $l_0 = l_1 = 1$ and $l_2 = l_3 = 0$ the dispersion relation reduces to $\omega_0 = \sqrt{\vec{\omega}^2+m^2}\left(1 + \frac{\phi}{m}\left(2 - \frac{m^2}{\vec{\omega}^2+m^2}\right)\omega_1\right) \pm \frac{\phi}{m}\left(2\vec{\omega}^2+m^2\right)$. The parameters $\vec{\omega}$ play the rôle of momenta $\vec{p}$ in our case of neutrinos of mass $m = m_\nu$ propagating in these space-times, and the Finsler metric may be seen as sort of back reaction on the space-time of such a propagation (to better appreciate this, the reader is invited to the discussion in the next section 5, where a particular model of D(efect)-foam is considered as a medium for neutrino propagation in the early universe, leading to Finsler-like metric distortions as a consequence of medum/particle interactions).

Corresponding to such models involving D-foam, the parameter ϕ is modelled as a stochastic gaussian process with a mean $\mathfrak{a}$ and standard deviation σ. The fermion number distribution n from equilibrium statistical mechanics is given by $n = g_{d.o.f.}\int \frac{d^3p}{(2\pi)^3}\frac{1}{\exp(\beta(\omega_0-\mu))+1}$. where we have ignored degeneracy factors. First of all, it immediately follows from the corresponding dispersion relation that for the "time-like" case, when $\mathfrak{a} = 0$, there is no particle/antiparticle asymmetry. This is to be expected, given that $\mathfrak{a} \neq 0$ corresponds in a sense to an averaged Lorentz violation in this stochastic geometry, and hence one of the basic assumptions for CPT Invariance of the effective theory of neutrinos in this "medium" is relaxed.

For the "time-like" case, when $\mathfrak{a} \neq 0$ and β small, we obtain to leading order in $T/m \gg 1$ [17]:

$$\ll \Delta n \gg \sim -\frac{2}{\pi^2}\mathfrak{a}\, g_{d.o.f.} T^3 \left(\frac{T}{m}\right) \int_0^\infty \frac{dx\, x^4\, e^x}{(1+e^x)^2} = -\mathfrak{a}\, g_{d.o.f.} T^3 \frac{7\pi^2}{15}\left(\frac{T}{m}\right) \tag{19}$$

We require $\mathfrak{a} < 0$ in order to have a particle-antiparticle asymmetry where the particle distribution dominates the antiparticle distribution. This yields the following Lepton (neutrino) asymmetry, assumed to freeze at the neutrino decoupling temperature T_d

$$\Delta L(T \sim T_d) = \frac{\Delta n_\nu}{s} \sim -10\, \mathfrak{a}\, \frac{T_d}{m_\nu} \tag{20}$$

where, as usual, s denotes the entropy density, which for relativistic species is assumed to be $s \sim g_{d.o.f.} \frac{2\pi^2}{45} T^3$. Thus, from (20) we see that there is no CPTV asymmetry in the Lorentz invariant (on average) case $\mathfrak{a} = 0$.

To obtain the phenomenologically correct value of $\Delta L(T \sim T_d) \sim 10^{-10}$, which is then communicated to the baryon sector via B-L conserving sphaleron processes, or B-L conserving grand unified models (assumed appropriately embedded in such space-time geometries), one needs to fix appropriately the value of $\mathfrak{a}$, consider appropriate conditions for freeze-out, that depend on the underlying microscopic model, and take into account that, according to current data, the masses of the active neutrinos that are assumed to participate in (20) must be smaller than $m_\nu < 0.2$ eV. For example, in GUT-scale lepton-number violation models, as we discussed before, $T_d \sim 10^{15}$ GeV, which implies that one needs only an extremely small in magnitude violation of Lorentz symmetry on average in this stochastic Finsler space time, $\mathfrak{a} \sim -10^{-36}$, in order to reproduce the observed Baryon Asymmetry in the universe. The assumption of fixing the sign of $\mathfrak{a}$ is considered as fine tuning, and is a feature that is common in the models of gravitational leptogenesis/baryogenesis that exist in the current literature, as discussed briefly above [30, 31, 28, 33, 35, 36, 27, 34, 32]. One can calculate the asymmetry for the null case with results similar to the time-like case considered above [17].

5. Stringy-defect (D-)foam-induced CPTV and leptogenesis

In this section we shall consider a population of D0-branes (or lower dimensional compactified D-branes which are effectively point-like from from the point of view of a brane world observer [5]) interacting with neutral fermions such as the neutrino and anti-neutrino. This interaction leads to different dispersion relations for neutrinos and anti-neutrinos which in turn leads to an excess of the population of neutrinos over anti-neutrinos. The freeze-out of neutrinos at the decoupling temperature of neutrinos leads to leptogenesis given by the standard cosmological considerations. The latter results, through standard Baryon (B) and Lepton (L) number violating sphaleron processes or B-L conserving interactions in grand unified models describing matter excitations on the brane, to the observed Baryon Asymmetry in the Unverse, with complete dominance of matter over antimatter, in a rather natural way, as we shall discuss below. Moreover, as we shall explain below, in this model of D(efect)-foam, the prevalence of matter over antimatter, *i.e.* the *positive sign* of the asymmetry $\Delta n > 0$, follows naturally, as a consequence of loss of

[5] It should be remarked that for the effective compactified D-"particles" the interactions with the charged matter excitations are suppressed relative to the neutral ones [82]. Hence, even in this case, it is the electrically neutral excitations which interact primarily with the D-foam.

energy of neutrinos during their interactions with the space-time defects, due to recoil of the latter. Thus, the sign of the induced asymmetry need not be fixed by hand, unlike the cases of gravitational leptogenesis discussed in previous sections. For instructive purposes, we first discuss the properties of the foam model, in the next subsection, before moving onto issues of CPTV and leptogenesis.

D-foam models [22, 20, 25] are stringy models of space-time foamy geometries, which involve brane universes, propagating in higher-dimensional bulk geometries. The bulk contains point-like D-brane defects ("D-particles" or D0 branes) whose population density is constrained by the amount of CPTV that we observe. In many string theories (such as bosonic and type IIA string theories) they are stable zero-dimensional defects. However for our purposes we will consider them to be present in string theories of phenomenological interest [81] since, even when elementary D-particles cannot exist consistently, as is the case of type IIB string models, there can be effective D-particles formed by the compactification of higher dimensional D-branes [82] (*e.g.* three-branes wrapped around three-cycles, with relatively small radii).

The preferential rôle of neutrinos in feeling the full effects of D-foam, and hence the CPTV, is attributed to electric charge conservation: the representation of SM particles as open strings, with their ends attached to the brane worlds, prevents capture and splitting of open strings carrying electric fluxes by the D-particles. (We should recall that in string theory the electric charge is at the end point of an open string.) D-particles are electrically neutral and thus electric charge would not have been conserved if such processes had taken place. This is also consistent with the effective D-particles which may have formed as a result of nucleation [30]. Hence, the D-particle foam is transparent to charged excitations of the SM, leaving neutral particles, in particular neutrinos, susceptible to the foam effects. This different behaviour of neutrinos from charged leptons implies a background-induced breaking of the SU(2) gauge symmetry of the standard model. In type IIB string theories, effective D-particles [82] can interact, but in a suppressed manner, with the entire SU(2) lepton doublet, so the SU(2) symmetry is not broken. However, in such scenarios these interactions are suppressed compared to photons, as discussed in [82]. The D-foam interactions with sterile right-handed Majorana neutrinos, though, remain unsuppressed, since the sterile neutrinos do not have any standard model charges. Such heavy states can participate in lepton-number violating processes freezing out at GUT scales.

For our purposes in this work we may consider that statistically significant populations of D-particles existed in the early eras of the brane universe. As the time elapses, the brane universe, which propagates in the higher-dimensional bulk, enters regions characterised by D-particle depletion, in such as way that the late eras cosmology of the universe is not affected. Nevertheless, as we shall discuss below, the early D-particle populations may still have important effects in generating neutrino-antineutrino populations differences (asymmetries), which are then communicated to the baryon sector via the standard sphaleron processes [16] or B-L conserving GUT symmetries in unified particle physics models.

To this end, we need to consider the *effective dispersion relation* of a (anti)neutrino field in a brane space-time punctured with statistically significant populations of D-particles. The latter is a dynamical population, consisting of defects crossing the brane all the time, thereby appearing to a brane observer as flashing "on" and "off" space-time "foamy" structures. The (anti)neutrino excitations are represented as matter open strings with their ends attached on the brane.The number density of (anti) neutrinos on the brane world is limited by the requirement that they do not overclose the universe.

In [17] we estimated the *modification of the dispersion relations* of neutrinos in such a "media" of D-particles in the early universe. The interaction of a string with a D-particle implies that at least one of the ends of the string is attached to the D-particle defect. Furthermore, the simultaneous creation of virtual strings stretched between the defect and the brane, describes the recoil of the D-particle. During the interaction time, the D-particle undergoes motion characterized by non-trivial "recoil" velocities, $u_{\|} = \frac{g_s}{M_s}\Delta p_i = \frac{g_s}{M_s} r_i \, p_i$ along the brane longitudinal dimensions, where r_i denotes the proportion of the incident neutrino momentum that corresponds to the momentum transfer Δp_i during the scattering, and $v_\perp$ in directions transverse to the brane world [24] .

As discussed in [76, 25, 83] the non-trivial capture and splitting of the open string during its interaction with the D-particle, and the recoil of the latter, result in a *local* effective metric distortions, which acquire a Finsler-like form

$$ds^2 = g_{\mu\nu}dx^\mu dx^\nu = (\eta_{\mu\nu} + h_{\mu\nu})dx^\mu dx^\nu \ , \qquad h_{0i} = (u^a_{i\,\|}\sigma_a) \ , \tag{21}$$

where $u_{i\,\|}$ is the recoil velocity of the D-particle *on* the D-brane world, with $i = 1, 2, 3$ a spatial space-time index, σ_a, with $a = 1, 2, 3$, are the 2×2 Pauli flavour matrices, whose presence is necessitated by the assumption that neutrinos can undergo flavour oscillations during their capture by D-particles (we assume here two-flavour oscillations for definiteness and simplicity).

However, the effects of D-foam go beyond those encoded in the induced Finsler like metric. The fine tuning that is required in stochastic Finsler metrics to get the correct sign for the particle-antiparticle asymmetry is a feature, although commonplace in other approaches, that is not entirely satisfactory. However because we have a microscopic model we can consider the kinematics of D-particle scattering. On considering string theory scattering amplitudes we find that the four momentum is conserved in the scattering of D-particles and strings. Indeed, upon averaging $\langle\langle \ldots \rangle\rangle$ over a statistically significant number of events, due to multiple scatterings in a D-foam background, we may use the following stochastic hypothesis [25]

$$\ll u_{i\,\|} \gg = 0 \ , \qquad \ll u_{i\,\|} u_{j\,\|} \gg = \sigma^2 \delta_{ij} \ . \tag{22}$$

implying that Lorentz invariance holds only as an average symmetry over large populations of D-particles in the foam. At a microscopic level, (22) translates to momentum conservation on average.

The dispersion relation then of a neutrino of mass m propagating on such a deformed isotropic space-time, then, reads:

$$p^\mu p^\nu g_{\mu\nu} = p^\mu p^\nu (\eta_{\mu\nu} + h_{\mu\nu}) = -m^2 \Rightarrow E^2 - 2E\vec{p} \cdot u_{\|} - \vec{p}^2 - m^2 = 0 \ , \tag{23}$$

which upon averaging over D-particle populations, can be solved to yield the following dispersion relations for particles and antiparticles

$$\ll E_\nu \gg = \sqrt{p^2 + m_\nu^2}\left(1 + \frac{1}{2}\sigma^2\right) - \frac{1}{2}\frac{M_s}{g_s}\sigma^2 \ , \ \ll E_{\overline{\nu}} \gg = \sqrt{p^2 + m_\nu^2}\left(1 + \frac{1}{2}\sigma^2\right) + \frac{1}{2}\frac{M_s}{g_s}\sigma^2 \tag{24}$$

with $E_{\overline{\nu}} > 0$ representing the positive energy of a physical antiparticle. In our analysis above we have made the symmetric assumption that the recoil-velocities fluctuation strengths are the same between particle and antiparticle sectors. (Scenarios for which this symmetry was not asssumed have also been considered in an early work [25].) There can thus be *local* CPTV in the sense that the effective dispersion relation between neutrinos and antineutrinos are different. This is a consequence of the local violation of Lorentz symmetry (LV), as a result of the non-trivial recoil velocities of the D-particle, leading to the LV space-time distortions (21).

The discussion of CPTV in such foamy universes now follows the line of argument adopted by others: the difference in the dispersion relations between particles and antiparticles will imply differences in the relevant populations of neutrinos (n) and antineutrinos ($\overline{n}$), (*cf.* the dispersion (24)). This difference between neutrino and antineutrino phase-space distribution functions in D-foam backgrounds generates a matter-antimatter lepton asymmetry in the relevant densities

$$\ll n - \overline{n} \gg = g_{d.o.f.} \int \frac{d^3p}{(2\pi)^3} \ll [f(E) - f(\overline{E})] \gg \ , \tag{25}$$

where $g_{d.o.f.}$ denotes the number of degrees of freedom of relativistic neutrinos, and $\ll \cdots \gg$ denotes an average over suitable populations of stochastically fluctuating D-particles (22).

The result for the D-foam-induced lepton asymmetry can be estimated from (25), using (24). Ignoring neutrino mass terms and $(1 + \frac{\sigma^2}{2})$ square-root prefactors in (24), as yielding subleading contributions for $\sigma^2 \ll 1$ we are considering here, setting the (anti)neutrino chemical potential to zero (which is a sufficient approximation for relativistic light neutrino matter) and performing a change of variables $|\vec{p}|/T \to \tilde{u}$ we obtain from (25) the result [17]:

$$\Delta n_\nu = \frac{g_{d.o.f.}}{2\pi^2} T^3 \int_0^\infty d\tilde{u}\, \tilde{u}^2 \left[\frac{1}{1 + e^{\tilde{u} - \frac{M_s \sigma^2}{2 g_s T}}} - \frac{1}{1 + e^{\tilde{u} + \frac{M_s \sigma^2}{2\, g_s T}}}\right] \simeq \frac{g_{d.o.f.}}{\pi^2} T^3 \left(\frac{M_s \sigma^2}{g_s T}\right) > 0, \qquad (26)$$

to leading order in σ^2. We thus observe that the CPTV term $-\frac{1}{2}\frac{M_s}{g_s}\sigma^2$ in the dispersion relation (24) for the neutrino, which corresponds to the energy 'loss' due to the D-particle recoil kinetic energies, comes with the right sign ('loss') so as to guarantee an excess of particles over antiparticles. Unlike the model of [28, 33, 35, 36], then, where the sign of the B_0 parameter had to be assumed, in our D-foam case there is no such freedom, and the positive Δn_ν is derived from first principles. We consider this a nice feature of our model.

As in standard scenarios of Leptogenesis, the Lepton asymmetry (26) decreases with decreasing temperature up to a freeze-out point, which occurs at temperatures T_d at which the Lepton-number violating processes decouple. For reasons already stated in section 3 the phenomenologically realistic scenario in string theory frameworks, like the present one, seems to be the one in which one has GUT scale lepton-number violating processes, which freeze out the lepton-antilepton number at temperatures $T_d \sim 10^{15}$ GeV. In the context of D-particle brane Universe one may assume that at such early epochs one has a sufficiently dense D-particle gas in the bulk and on the brane, so that phenomenologically realistic Leptogenesis takes place. The latter then can be communicated via B-L conserving GUT processes to the baryon sector. After the GUT scale freeze out the bulk population of D-particles can be depleted, in the sense that the brane world representing our Universe passes through an area with scarce D-particle populations, so that late eras of the Universe are characterised by a more-or-less conventional Cosmology.

The above estimate ignores an important fact, namely the dependence of the stochastic variable σ^2 on the neutrino energy. Indeed, as discussed in detail in [24, 25], one may parameterise the momentum transfer by the fraction parameter of the incident momentum r, which is in turn assumed stochastic, that is

$$u_i = \frac{g_s}{M_s}\Delta p_i \to g_s\, r_i \frac{p_i}{M_s}\ \text{, no sum over i}\ , \quad \ll r_i \gg = 0\ , \quad \ll r_i r_j \gg = \Delta^2 \delta_{ij}\ .$$

In this case, the dispersion relations (24) are modified by the replacement of $\sigma^2 \to \frac{g_s^2}{M_s^2}\Delta^2 p^2$, which is now momentum dependent: A detailed analysis in [17] leads in this case to the following result for the lepton asymmetry (assuming a freeze-out temperature T_d of the Lepton-Violating processes):

$$\Delta L(T < T_d) = \frac{\Delta n_\nu}{s} = \frac{2\Delta^2 g_s T_d}{M_s}. \qquad (27)$$

From (27), we observe that for a freeze-out temperature $T_d \sim 10^{15}$ GeV, which characterises GUT scale Lepton-Violating processes (*e.g* involving heavy Majorana neutrinos $N_I \to H\,\nu,\ \overline{H}\,\overline{\nu}$), the phenomenological value $\Delta L \sim 10^{-10}$ is attained for $\frac{M_s}{g_s} \sim 10^{25}\,\Delta^2$ GeV . For $\Delta^2 \sim 10^{-6}$ a Planck size D-particle mass $M_s/g_s \sim 10^{19}$ GeV is required so that the D-foam provides the physically observed Lepton and, thus, Baryon Asymmetry. For the unnaturally small $\Delta^2 < 10^{-21}$ one arrives at $M_s/g_s \sim 10$ TeV. Unfortunately, for $\Delta^2 \sim \mathcal{O}(1)$ transplanckian D-particle masses are required.

Our approach to leptogenesis is distinguished from others in that a local effective field theoretical description is not adopted. Because of D-particle recoil when scattering off matter strings the background of D-particles can be modelled as a stochastic medium [24, 26, 25]. The underlying string theoretic description provides the rigorous description of the scattering of D-particles. The D-particles backreact (as seen from infra-red divergences in perturbation theory) and change the metric which influences the space in which matter is moving. Furthermore as discussed at length in [26], and mentioned briefly above, the D-particle foam model does *not* lead to overclosing the universe. Hence despite having statistically significant populations of D-particles in the early universe, which provide the CPTV background on which neutrinos propagate, the assumption of a subcritical energy density for the universe can still hold.

Acknowledgments
N.E.M. thanks the organisers of DICE2012 for the invitation to give a plenary talk. The work of N.E.M. was supported in part by the London Centre for Terauniverse Studies (LCTS), using funding from the European Research Council via the Advanced Investigator Grant 267352, and by STFC UK under the research grant ST/J002798/1.

References
[1] Christenson J, Cronin J, Fitch V and Turlay R 1964 *Phys.Rev.Lett.* **13** 138–140
[2] Sakharov A 1967 *Pisma Zh.Eksp.Teor.Fiz.* **5** 32–35 reprinted in *Kolb, E.W. (ed.), Turner, M.S. (ed.): The early universe* 371-373, and in *Lindley, D. (ed.) et al.: Cosmology and particle physics* 106-109, and in Sov. Phys. Usp. 34 (1991) 392-393 [Usp. Fiz. Nauk 161 (1991) No. 5 61-64]
[3] Gamow G 1946 *Phys.Rev.* **70** 572–573
[4] Kuzmin V, Rubakov V and Shaposhnikov M 1985 *Phys.Lett.* **B155** 36
[5] Shaposhnikov M 2009 *Prog.Theor.Phys.* **122** 185–203
[6] Shaposhnikov M and Tkachev I 2006 *Phys.Lett.* **B639** 414–417 (*Preprint* hep-ph/0604236)
[7] Lindner M, Merle A and Niro V 2011 *JCAP* **1101** 034 (*Preprint* 1011.4950)
[8] Kusenko A, Takahashi F and Yanagida T T 2010 *Phys.Lett.* **B693** 144–148 (*Preprint* 1006.1731)
[9] Randall L and Sundrum R 1999 *Phys.Rev.Lett.* **83** 3370–3373 (*Preprint* Randall:1999ee)
[10] Merle A and Niro V 2011 *JCAP* **1107** 023 (*Preprint* 1105.5136)
[11] Barry J, Rodejohann W and Zhang H 2011 *JHEP* **1107** 091 (*Preprint* 1105.3911)
[12] Carmona J M, Cortes J L, Das A K, Gamboa J and Mendez F 2006 *Mod.Phys.Lett.* **A21** 883–892 (*Preprint* hep-th/0410143)
[13] Streater R and Wightman A 1989 *PCT, spin and statistics, and all that* (Princeton University Press)
[14] Bertolami O, Colladay D, Kostelecky V A and Potting R 1997 *Phys.Lett.* **B395** 178–183 (*Preprint* hep-ph/9612437)
[15] Mavromatos N E 2010 and references therein *Int.J.Mod.Phys.* **A25** 5409–5485 (*Preprint* 1010.5354)
[16] Arnold P and McLerran L 1987 *Phys. Rev. D* **36**(2) 581–595 URL http://link.aps.org/doi/10.1103/PhysRevD.36.581
[17] Mavromatos N E and Sarkar S 2012 (*Preprint* 1211.0968)
[18] Bao D, Chern S and Shen Z 2000 *An Introduction to Riemann-Finsler Geometry* Graduate Texts in Mathematics (Springer) ISBN 9780387989488 URL http://books.google.fr/books?id=m11kEYr_V_AC
[19] Johnson C 2002 *D-Branes* Cambridge Monographs on Mathematical Physics (Cambridge University Press) ISBN 9780521809122 URL http://books.google.co.uk/books?id=LeJCOY7VGDOC
[20] Ellis J R, Mavromatos N E and Westmuckett M 2004 *Phys.Rev.* **D70** 044036 (*Preprint* gr-qc/0405066)
[21] Douglas M R, Kabat D N, Pouliot P and Shenker S H 1997 *Nucl.Phys.* **B485** 85–127 (*Preprint* hep-th/9608024)
[22] Ellis J R, Mavromatos N E and Westmuckett M 2005 *Phys.Rev.* **D71** 106006 (*Preprint* gr-qc/0501060)
[23] Mavromatos N E 2004 (*Preprint* hep-th/0407026)
[24] Mavromatos N and Sarkar S 2005 *Phys.Rev.* **D72** 065016 (*Preprint* hep-th/0506242)
[25] Bernabeu J, Mavromatos N E and Sarkar S 2006 *Phys.Rev.* **D74** 045014 (*Preprint* hep-th/0606137)
[26] Mavromatos N E, Mitsou V A, Sarkar S and Vergou A 2012 *Eur.Phys.J.* **C72** 1956 (*Preprint* 1012.4094)
[27] Lambiase G and Mohanty S 2007 *JCAP* **0712** 008 (*Preprint* astro-ph/0611905)

[28] Debnath U, Mukhopadhyay B and Dadhich N 2006 *Mod.Phys.Lett.* **A21** 399–408 (*Preprint* hep-ph/0510351)
[29] Zwiebach B 2004 *A First Course in String Theory* (Cambridge: Cambridge Univ. Press)
[30] Dvali G and Gabadadze G 1999 *Phys.Lett.* **B460** 47–57 (*Preprint* hep-ph/9904221)
[31] Davoudiasl H, Kitano R, Kribs G D, Murayama H and Steinhardt P J 2004 *Phys.Rev.Lett.* **93** 201301 (*Preprint* hep-ph/0403019)
[32] Li H, Li M z and Zhang X m 2004 *Phys.Rev.* **D70** 047302 (*Preprint* hep-ph/0403281)
[33] Mukhopadhyay B 2005 *Mod.Phys.Lett.* **A20** 2145–2156 (*Preprint* astro-ph/0505460)
[34] Lambiase G and Mohanty S 2011 *Phys.Rev.* **D84** 023509 (*Preprint* 1107.1213)
[35] Mukhopadhyay B 2007 *Class.Quant.Grav.* **24** 1433–1442 (*Preprint* gr-qc/0702062)
[36] Sinha M and Mukhopadhyay B 2008 *Phys.Rev.* **D77** 025003 (*Preprint* 0704.2593)
[37] Carroll S M and Shu J 2006 *Phys.Rev.* **D73** 103515 (*Preprint* hep-ph/0510081)
[38] Dolgov A 2010 *Phys.Atom.Nucl.* **73** 588–592 (*Preprint* 0903.4318)
[39] Barenboim G, Borissov L, Lykken J D and Smirnov A 2002 *JHEP* **0210** 001 (*Preprint* hep-ph/0108199)
[40] Barenboim G and Mavromatos N E 2004 *Phys.Rev.* **D70** 093015 (*Preprint* hep-ph/0406035)
[41] Barenboim G and Mavromatos N E 2005 *JHEP* **0501** 034 (*Preprint* hep-ph/0404014)
[42] Wald R M 1980 *Phys.Rev.* **D21** 2742–2755
[43] Lindblad G 1976 *Commun.Math.Phys.* **48** 119
[44] Ellis J R, Hagelin J, Nanopoulos D V and Srednicki M 1984 *Nucl.Phys.* **B241** 381
[45] Ellis J R, Lopez J L, Mavromatos N and Nanopoulos D V 1996 *Phys.Rev.* **D53** 3846–3870 (*Preprint* hep-ph/9505340)
[46] Benatti F and Floreanini R 2001 *Phys.Rev.* **D64** 085015 (*Preprint* hep-ph/0105303)
[47] Fogli G, Lisi E, Marrone A, Montanino D and Palazzo A 2007 *Phys.Rev.* **D76** 033006 (*Preprint* 0704.2568)
[48] Lisi E, Marrone A and Montanino D 2000 *Phys.Rev.Lett.* **85** 1166–1169 (*Preprint* hep-ph/0002053)
[49] Barenboim G, Mavromatos N E, Sarkar S and Waldron-Lauda A 2006 *Nucl.Phys.* **B758** 90–111 (*Preprint* hep-ph/0603028)
[50] Aguilar A *et al.* (LSND Collaboration) 2001 *Phys.Rev.* **D64** 112007 (*Preprint* hep-ex/0104049)
[51] Gross D J and Sloan J H 1987 *Nucl.Phys.* **B291** 41
[52] Antoniadis I, Bachas C, Ellis J R and Nanopoulos D V 1989 *Nucl.Phys.* **B328** 117–139
[53] Ellis J R, Mavromatos N, Nanopoulos D V and Westmuckett M 2006 *Int.J.Mod.Phys.* **A21** 1379–1444 (*Preprint* gr-qc/0508105)
[54] Lutken C and Ross G 2007 *Phys.Lett.* **B653** 363–365
[55] Lutken C and Ross G 2011 *Nucl.Phys.* **B850** 321–338
[56] Pilaftsis A and Underwood T E 2004 *Nucl.Phys.* **B692** 303–345 (*Preprint* hep-ph/0309342)
[57] Pilaftsis A 2005 *Phys.Rev.Lett.* **95** 081602 (*Preprint* hep-ph/0408103)
[58] Pilaftsis A and Underwood T E 2005 *Phys.Rev.* **D72** 113001 (*Preprint* hep-ph/0506107)
[59] Boyarsky A, Ruchayskiy O and Shaposhnikov M 2009 *Ann.Rev.Nucl.Part.Sci.* **59** 191–214 (*Preprint* 0901.0011)
[60] Girelli F, Liberati S and Sindoni L 2007 *Phys.Rev.* **D75** 064015 (*Preprint* gr-qc/0611024)
[61] Vacaru S I 2002 (*Preprint* hep-th/0211068)
[62] Vacaru S I 2003 (*Preprint* hep-th/0310132)
[63] Bogoslovsky G Y and Goenner H 1998 *Phys.Lett.* **A244** 222–228 (*Preprint* gr-qc/9804082)
[64] Bogoslovsky G Y 2007 (*Preprint* 0706.2621)
[65] Bogoslovsky G 2007 (*Preprint* 0712.1718)
[66] Gibbons G, Gomis J and Pope C 2007 *Phys.Rev.* **D76** 081701 (*Preprint* 0707.2174)
[67] Sindoni L 2008 *Phys.Rev.* **D77** 124009 (*Preprint* 0712.3518)
[68] Anastasiei M and Vacaru S I 2009 *J.Math.Phys.* **50** 013510 (*Preprint* 0710.3079)
[69] Vacaru S I 2012 (*Preprint* 1206.4012)
[70] Vacaru S I 2011 *Class.Quant.Grav.* **28** 215001 (*Preprint* 1008.4912)
[71] Kouretsis A, Stathakopoulos M and Stavrinos P 2009 *Phys.Rev.* **D79** 104011 (*Preprint* 0810.3267)
[72] Magueijo J and Smolin L 2004 *Class.Quant.Grav.* **21** 1725–1736 (*Preprint* gr-qc/0305055)
[73] Kouretsis A, Stathakopoulos M and Stavrinos P 2010 *Phys.Rev.* **D82** 064035 (*Preprint* 1003.5640)
[74] Stavrinos P C and Vacaru S I 2012 (*Preprint* 1206.3998)
[75] Vacaru S I 2012 *Int.J.Mod.Phys.* **D21** 1250072 (*Preprint* 1004.3007)
[76] Ellis J R, Mavromatos N and Nanopoulos D V 2000 *Gen.Rel.Grav.* **32** 127–144 (*Preprint* gr-qc/9904068)
[77] Skakala J and Visser M 2009 *J.Phys.Conf.Ser.* **189** 012037 (*Preprint* 0810.4376)
[78] Cohen A G and Glashow S L 2006 *Phys.Rev.Lett.* **97** 021601 (*Preprint* hep-ph/0601236)
[79] Cohen A G and Glashow S L 2006 (*Preprint* hep-ph/0605036)
[80] Gibbons G, Herdeiro C, Warnick C and Werner M 2009 *Phys.Rev.* **D79** 044022 (*Preprint* 0811.2877)
[81] Ellis J R, Mavromatos N and Nanopoulos D 2008 *Phys.Lett.* **B665** 412–417 (*Preprint* 0804.3566)

[82] Li T, Mavromatos N E, Nanopoulos D V and Xie D 2009 *Phys.Lett.* **B679** 407–413 (*Preprint* 0903.1303)
[83] Mavromatos N E, Sarkar S and Tarantino W 2009 *Phys.Rev.* **D80** 084046 (*Preprint* 0907.5122)

Status of the OPERA experiment

E Medinaceli for the OPERA Collaboration
INFN and University of Padova
E-mail: medinaceli@pd.infn.it

Abstract. The OPERA neutrino detector at the Gran Sasso underground laboratory (LNGS) has been designed to perform the first detection of neutrino oscillations in direct appearance mode, through the study of the $\nu_\mu \rightarrow \nu_\tau$ channel. Tau leptons produced in the charged current interactions are identified by their decay topologies, using nuclear emulsions. After a brief description of the experimental setup, the first two tau candidates found in the runs 2008 to 2010 are reported. The OPERA experiment has measured the velocity of neutrinos sent from the CERN CNGS beam over a baseline of about 730 km. The measurement is based on data taken from 2009 to 2011. An arrival time of CNGS muon neutrinos with respect to the one computed, assuming the speed of light in vacuum, of $(6.5 \pm 7.4(stat)^{+3.4}_{-3.3}(sys))$ ns was measured.

1. The CNGS beam and the OPERA detector

The CNGS beam at CERN [1], is a high energy beam ($< E_{\nu_\mu} >\sim 17GeV$) optimized for the study of the appearance of ν_τ starting from a pure beam of ν_μ. At the CNGS energies, the average τ decay length is submillimetric, therefore nuclear emulsions are used as high precision tracking devices. In terms of interactions, the $\bar{\nu}_\mu$ contamination is 2.1%, the ν_e and $\bar{\nu}_e$ contaminations together are smaller than 1%, while the prompt number of ν_τ is negligible.

OPERA's detector basic unit is the so-called "brick", which is composed of 56 lead layers, 1mm thick, interleaved with 57 nuclear emulsion layers ($0\mu m$ accuracy). The submicrometer spatial resolution of the nuclear emulsion allows a precise 3D reconstruction of the neutrino interaction point, and of the decay vertex associated with short-lived particle produced, like the τ lepton. The overall target is composed of about 150000 bricks for a total mass of 1.25 kton. Each brick is a stand-alone detector used to estimate several kinematical variables of each neutrino interaction, *e.g.* particle momentum through their multiple Coulomb scattering in the lead plates [2], and the energy of electromagnetic showers. OPERA is a hybrid detector [3] made of a veto plane followed by two identical Super Modules (SM). Each one includes a target section made of arrays of bricks and a scintillator Target Tracker detector (TT) [4] to trigger the read-out and localize neutrino interactions within the target. Each target section is followed by a muon spectrometer [6], which is a dipolar magnet instrumented with resistive plate chambers (RPCs) and drift tubes detectors, to measure the muon charge and momentum. See Fig. 1.

2. Event selection and analysis

All electronic detector triggers recorded on-time with the CNGS, are classified as internal or external events by an online algorithm *i.e.* as interactions inside or outside the OPERA target (*e.g* in material along the beam-line preceding the OPERA target), respectively. Only internal events are used for oscillation studies. The algorithm further classify the events as charged current (CC) or neutral current (NC) interactions, through the spectrometers muon tracks identification or the amount of traversed material [6].

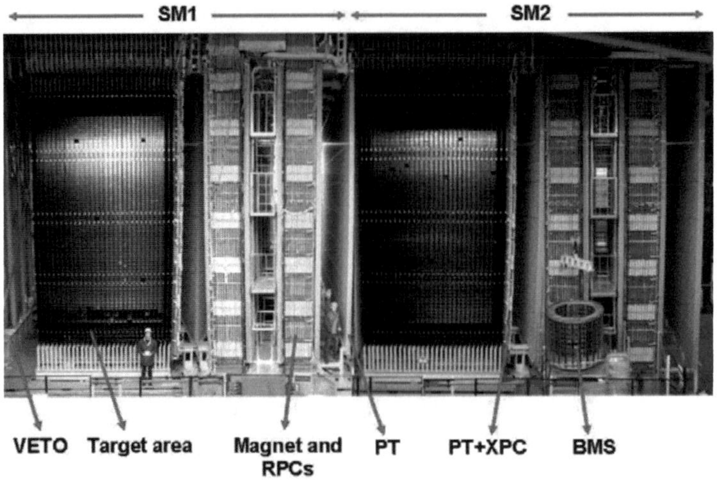

Figure 1. The OPERA detector at the Gran Sasso laboratory. The picture shows both supermodules (SM1 and SM2), and the target and muon spectrometers in each SM.

The neutrino vertex position is reconstructed, and the brick with the highest probability of containing it is extracted from the target. After a possitive scan result of two "trigger" emulsions (a detachable set of two emulsion films, called Changeable Sheets, or CS doublet) where tracks compatible with TT data are found, the brick is developed and dispatched to a scanning laboratory in Europe or in Japan. Tracks found in the CS doublet are extrapolated to the most downstream film of the brick, then followed upstream using its prediction from the scanning, until they reach their stopping point *i.e.* when the track is not found in the next three consecutive films. The vertex confirmation is done by scanning a large volume of $\sim 2cm^3$ around the stopping point.

A procedure called decay search is applied afterwards, in order to search for: (1) charged or neutral decay topologies, (2) secondary interactions or (3) gamma-ray conversions. If a secondary vertex is found, a full kinematical analysis is performed combining the measurements in the nuclear emulsion with data from the electronic detectors. The momentum of charged particles can be measured in emulsions up to 6 GeV/c. It can be measured up to 12 GeV/c with a resolution better than 33% using position deviations. For muons crossing the spectrometers, the momentum is measured with a resolution better than 22% up to 30 GeV/c, the muon charge is also determined [6]. The hint of a decay topology is the observation of an impact parameter greater than 10 μm, defined as the minimum distance between the track and the reconstructed vertex, excluding low momentum tracks.

3. Oscillation results

By applying the procedure, described in section [2], to 4190 events acquired between 2008-2011, two events pass the selection criteria defined for the τ lepton decay in two hadronic channels [3, 5].

3.1. First tau candidate

The first ν_τ candidate event was observed in the 2008-2009 dataset, details can be found in [7]. The event, shown in Fig. 2, has seven prongs at the primary vertex; four of them originate from hadrons, three have a probability lower than 0.1% of being caused by a muon, and none of them are left by an electron. The parent track (labeled 4 and displayed in red in Fig. 2) exhibits a kink topology, and the daughter track (shown also in Fig. 2) is identified as produced by a hadron through its interaction. The daughter impact parameter (IP) with respect to the primary vertex

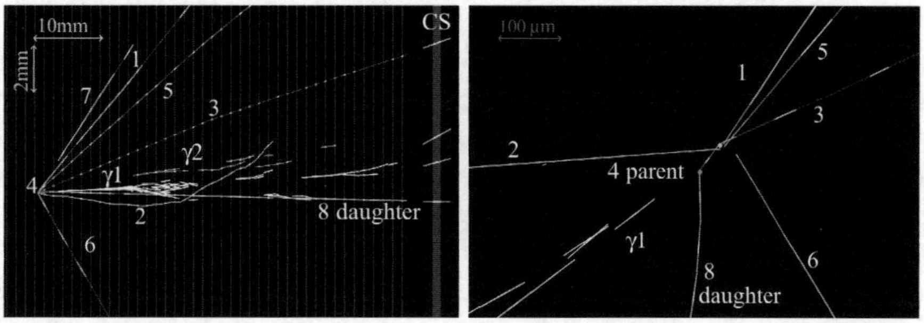

Figure 2. *Left:* Sketch of the fisrt ν_τ candidate, showing seven tracks attached to the primary vertex. Red track number 4, performing a kink, is the parent of a single prong decay. The daughter and both γ-rays (γ_1 and γ_2) are attached to the secondary vertex. *Right:* Zoom-in of the event, showing both primary and secondary vertices, displayed with dots. The kink of the parent can be clearly seen.

is 55±4 μm, while the other tracks have and IP<7 μm. Two γ-rays (labeled $\gamma 1$ and $\gamma 2$ in the plot) point to the secondary vertex with probabilities greather than 99% and 90% respectively. The invariant mass of the two γ-rays is $(120 \pm 20(\text{stat}) \pm 35(\text{sys}))$ MeV/c^2 compatible with the π^0 mass. Assuming the secondary hadron is a π^-, and adding the two γ-rays, the invariant parent's mass is compatible with the mass of a $\rho(770)$. The decays mode $\tau \to \rho\nu_\tau$ branching ratio is about 25%. Different kinematical variables used for the event classification are listed in table 1, as well as the selection criteria applied for the single hadron decay channel.

Variable	Value	Selection Criteria
kink [mrad]	41±2	>20
decay length [μm]	1335 ±35	≤2600
P daughter [GeV/c]	12^{+8}_{-3}	>2
P_t [MeV/c]	470^{+230}_{-120}	>300 (γ attached)
missing P_t [MeV/c]	570^{+320}_{-170}	<1000
ϕ [deg]	173±2	>90

Table 1. Kinematical variables (and selection criteria) for the first ν_τ candidate in the $\tau \to$1h decay channel.

3.2. Second tau candidate

The second ν_τ candidate event was found in the 2011 dataset. The event, displayed in Fig. 3 is a two prongs interaction with production of a short track (by a possible τ lepton) with a flight length of 1.54 mm, and a track identified as a hadron. One nuclear fragment was associated to the primary vertex. The short track decay vertex (shown in Fig. 3) is located in the plastic base in bewteen the two sentitive layers of an emulsion film. The parent lepton decays in three hadrons (shown in Fig. 3), one that reinteracts producing two charged tracks and four back-scattered nuclear fragments. The charged hadron momentum at the primary vertex is $2.8^{+2.1}_{-2.5}$ GeV/c and the hypothesis of a muon on the basis of consistency between momentum and range is very unlikely. All daughter particles coming from the decay vertex were also identified as hadrons. The kinematical analysis of the event satisfies the specified criteria for the $\tau \to 3h$ decay channel. Kinematical variables used in the event classification are listed in table 2.

3.3. Charm background

The main background source comes from ν_μ interactions with production of a charmed particle. The study of decay vertices of charm particles, that have lifetimes and decay topologies close to

Figure 3. Event display of the second ν_τ candidate, showing a two prong primary vertex, The red track is the decay vertex parent, with 3 prongs attached.

Variable	Value	Selection
kink [mrad]	87.4±1.5	<500
P at 2^{ry} vertex [GeV/c]	8.4±1.7	>3.0
P_t at 1^{ry} vertex [GeV/c]	0.31±0.11	<1.0
Min invariant mass [GeV/c^2]	0.96±0.13	0.5<m<2.0
Invariant mass [GeV/c^2]	0.8±0.12	0.5<m<2.0
ϕ(τ-hadron) [deg]	167.8±1.1	>90

Table 2. Kinematical variables of the second ν_τ candidate satisfying specified criteria for the $\tau \rightarrow$3h decay channel.

those of the τ lepton, offer an opportunity to check τ efficiency. The number of charm decays found so far in the data taken from 2008 to 2011 is 49. This number is in good agreement with the Monte Carlo expectation of 51±7.5. The statistical significance of the two ν_τ candidate event observations is still under evaluation, along with the efficiencies for signal and background, taking into account different sample selections. The preliminary number of expected ν_τ in the analysed data sample is 2.1, with background of 0.2 events.

4. Non oscillation results, neutrino velocity

For this analysis the neutrino velocity is defined as the ratio of the measured distance from CERN to OPERA (*i.e.* the baseline), to neutrinos' time of flight, ToF_ν, traveling through the Earth's crust. High statistics data taken by OPERA between 2009 and 2011 is used for the analysis. The systematic uncertainties were reduced to the level of the statistical error, by dedicated upgrates of the timing systems, for time tagging and synchronisation of the CNGS beam at CERN and of the OPERA detector.

4.1. Neutrino time of flight measurement

The neutrino starting time is defined by time tagging the extractions of the SPS beam. The proton time structure is accurately measured by a fast Beam Current Transformer (BCT) detector (400 MHz bandwidth) read out by a 1 GS/s Wave Form Digitizer (WFD) with a 250 MHz bandwidth. The waveforms recorded for each extraction by the WFD are UTC (Universal Time Coordinates) time-stamped with a Symmetricon GPS receiver and stored in the CNGS database. The ToF_ν is given by the difference between the arrival time in OPERA and the start time at CERN. At LNGS, every millisecond a pulse (PPmS) is derived from the one second periodic pulse (1PPS) of a ESAT2000 GPS system and is transmitted from the surface laboratory to the OPERA Master Clock in the underground laboratory through 8.3 km of optical fibre. The time base of the Master Clock is transmitted to OPERA detector sensors, by which neutrino interaction time is defined. The required ns accuracy for relative time tagging between CERN and the OPERA detector is achieved by adopting two identical systems, installed in both sites, composed of a GPS receiver for time-transfer applications operating in common-view mode and a Cs atomic clock. Both systems were calibrated in 2008 by the Swiss Federal Metrology Institute (METAS) and established a permanent time link at the level of 2 ns between t_{CERN} and t_{OPERA} as reference points. This time link was independently verified in 2011 by the German Federal Metrology Institute (PTB) by taking data at CERN and LNGS with a portable time-transfer device. The difference between the time base of the both GPS receivers was measured as (2.3±1.7) ns. A detailed description of these procedures can be found in Ref. [8].

4.2. Neutrino baseline measurement

The neutrino baseline is defined as the distance between the point where the proton time-structure is measured (by the BCT at CERN) and the origin of the OPERA reference frame ORF. The travel path of protons from the BCT to the focal point of the CNGS target has been measured with millimetric accuracy, $L_{BCT-target} = (743.391 \pm 0.002)$ m. When these positions are transformed into the global geodesy reference frame (ETRF2000) by relating them to the external GPS benchmarks at CERN, they are known whitin 2 cm accuracy. The coordinates of the ORF origin were measured in 2010 by establishing GPS benchmarks at the both sides of the 10 km long Gran Sasso highway tunnel, and by transporting their positions with a terrestrial traverse down to OPERA. A common analysis in the ETRF2000 for 3D coordinates of OPERA origin and of the target focal point allowed their distance determination, $L_{target-OPERA} = (730534.61 \pm 0.20)$ m. The 20 cm uncertainty is dominated by a long underground link between the outdoor GPS and the OPERA detector benchmarks. Then the baseline considered for measuring ToF_ν is the sum of the distance between CNGS target focal point and the origin of the OPERA detector reference frame, $L_{target-OPERA}$, and the distance between BCT and the focal point, $L_{BCT-target}$, *i.e.* (731278.0±0.2) m.

4.3. Data analysis and results

Neutrinos ToF_ν from the standard CNGS beam can not be treated at the single interaction level, due to the 10.5 μs pulse duration of the SPS extraction. Any extracted proton may produce a neutrino detected by OPERA. Therefore for each neutrino interaction measured in OPERA its corresponding proton extraction waveform is considered. Those waveforms were individually normalized to unity and summed up in order to build a PDF $w(t)$ of neutrinos' emission time. The neutrino interaction time is defined as the time of the earliest hit in the TT. ToF_ν is obtained by comparing the distribution of protons crossing the BCT with neutrino events detected by OPERA.

Assuming a propagation at the speed of light of protons over the baseline ToF_c, the deviation of ToF_ν with respect to ToF_c ($\delta t = ToF_c - ToF_\nu$), can be obtained by a maximum likelihood analysis of the time tags of OPERA events, with respect to the PDF as a function of δt. For the analysis 15223 events (7235 NC and CC events contained inside the detector, and 7988 external CC events) were used, corresponding to $\sim 10^{20}$ p.o.t. The arrival time of neutrinos with respect to the time evaluated considering the speed of light for the propagation is:

$$\delta t = ToF_c - ToF_\nu = 6.5 \pm 7.4(stat)^{+8.3}_{-8.0}(sys) \quad [ns] \tag{1}$$

In order to exclude possible biases affecting the statistical analysis based on the neutrino emision PDF, the neutrino velocity was measured using a dedicated CNGS beam generated by an special proton beam set-up, from October 22 to November 6, 2011. The modified SPS super-cycle consisted of a single extraction including bunches $\sim$ 3 ns long (FWHM) separated by 524 ns, yielding a total of 1.1×10^{12} protons per cycle. Given the short bunch length and the relatively long inter-bunch distance one could unambiguously associate each neutrino event to its corresponding proton bunch. The average time difference for 20 events (6 contained plus 14

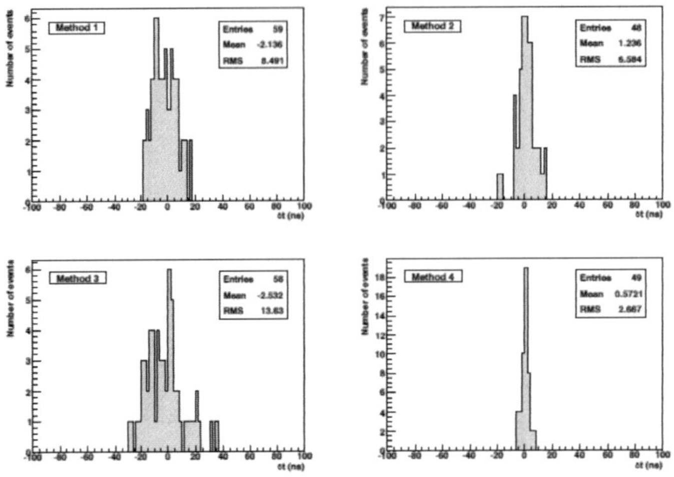

Figure 4. Distribution of the time difference δ_t between the light and the neutrino times of flight. Neutrino time distribution. First row: selected events for the TT analysis using the first hit (Method 1), and using muon hits only (Method 2). Second row: events selected for the RPC analysis using muon hits (Method 3), and events analyzed using a dedicated timing board (Method 4).

external) measured with the TT is $\delta t = -1.9 \pm 3.7(stat)^{+8.3}_{-8.0}(sys)$ ns, in agreement with previous analysis. At first order, systematic uncertainties related to the bunched beam operation are equal or smaller than those affecting the result with the nominal CNGS beam. Further details can be found in Ref. [8].

CERN provided in 2012 a new narrow-bunch short-spacing beam between 10th and 24th of May. In each 13.2 s CNGS cycle, a single extraction delivers 16 proton bunches. Each 2.8 ns long bunch contained $\sim 10^{11}$ protons, giving an intensity a factor 6 higher than the 2011 bunched beam run. During this two week period 1.8×10^{17} p.o.t. were delivered. In OPERA 104 on-line events were detected using TT (67 events) and RPC (62 events). An independent anlysis was performed for each detector type [9].

For the analysis using TT data, two methods were used to compute neutrinos' interaction time. "Method 1" relies on the information of the earliest hit. While "Method 2 " exploits the information of all hits of the reconstructed 3D muon track only. In the second method NC-like events are therefore not considered. On the top row of Fig. 4 are plotted the distributions of δt for Method 1 (left) yielding $\delta t = -2.1 \pm 1.1(stat) \pm 4.4(sys)$ ns, and for Method 2 (right) yielding $\delta t = 1.2 \pm 1.0(sta) \pm 3.3(sys)$ ns. The analysis using RPC data was performed using the standard OPERA DAQ "Method 3". In this case only neutrino interactions producing a clear muon track in the RPC's are used. The corresponding δt distribution is shown in Fig. 4 (bottom left), the final result is $\delta t = -2.5 \pm 1.8(stat) \pm 5.3(sys)$. "Method 4" uses RPC data taken with a dedicated timing board TB which gives complementary information to events acquired using the standard OPERA DAQ. Details about this timing system can be found in Ref. [9]. The δt distribution for this method is shown in Fig. 4 (bottom right), the final result is $\delta t = 0.6 \pm 0.4(stat) \pm 3.6(sys)$. Methods 1 and 2 are statistically correlated since they use the same DAQ system and sub-detector. For the same reason they are also correlated with Method 3. Method 4 is almost uncorrelated with the other methods because of the different timing system used. Residual correlation arises from the systematic error of the common part of the timing chain. In order to obtain a single combined result, Methods 2 and 4 were used since they have the smallest statistical and systematical error, and because they are almost completely uncorrelated thus providing maximum information. Values obtained with Method 1 were used when Method 2 could not be used because of the absence of a muon track. Summing in quadrature systematic errors, separately for ν and $\bar{\nu}$ contributions, leads to $\delta t_\nu = (0.6 \pm 0.4(stat) \pm 3.0(syst))$ ns and $\delta t_{\bar{\nu}} = (1.7 \pm 1.4(stat) \pm 3.1(syst))$ ns.

5. Conclusions

The OPERA experiment started collecting data in 2008, with the aim to make the first direct observation of neutrino oscillations in appearance mode. In the $\nu_\mu \rightarrow \nu_\tau$ channel two events were observed in the data from 2008 to 2010 and part of 2011. The significance of this observation is still under evaluation; the preliminary expectancy is 2.1 ν_τ events with 0.2 background events.

The OPERA experiment measured the neutrino velocity over the CNGS baseline (~730 Km). After correcting and updating its timing system [10, 8], it was able to confirm the results obtained from standard CNGS data from 2009 to 2011 with data from two dedicated bunched beam runs preformed in 2011 Ref. [8] and 2012 Ref. [9]. A limit on the speed of light deviation was derived $-1.6 \times 10^{-6} < (v_\nu - c)/c < 2.3 \times 10^{-6}$ showing no significant difference between neutrino and antineutrino velocities and speed of light.

References

[1] CNGS project, http://proj-cngs.web.cern.ch/proj-cngs/
[2] Agafanova N *et al.* (OPERA Collab.) 2012 *New J. Phys.* **14** 013026
[3] Guler M *et al.* (OPERA Collaboration) 2000 *CERN-SPSC-2000-028, LNGS P025/2000*
[4] Adam T *et al.* 2007 *Nucl. Instrum. Meth.* A **577** 523-539
[5] Guler M *et al.* (OPERA Collab.) 2001 *CERN-SPSC-2001-025, LNGS-EXP 30/2001 Add. 1/01*
[6] Agafanova N *et al.* (OPERA Collab.) 2011 *New J. Phys.* **13** 053051
[7] Agafanova N *et al.* (OPERA Collab.) 2010 *Phys. Lett.* B **691** 138
[8] Adam T *et al.* (OPERA Collab.) 2012 *JHEP* **1210** 093
[9] Adam T *et al.* (OPERA Collab.) 2012 *Preprint* arXiv:1212.1276 [hep-ex]
[10] Agafanova N *et al.* (OPERA and LVD Collab.) 2012 *Eur. Phys. J. Plus* **127** 71

Beyond classical limits by exploiting quantum light : a short review of INRIM results

G Brida, I P Degiovanni, M Genovese, E D Lopaeva, A Meda and I Ruo Berchera
INRIM, strada delle Cacce 91, 10135 Torino, Italy

E-mail: m.genovese@inrim.it

Abstract. Quantum properties of light and, in particular, quantum correlations have recently disclosed the possibility of realising protocols addressed to overcome limits of classical imaging, a field collectively christened quantum imaging. In particular quantum correlations between twin beams represent a fundamental resource for these studies. Here we present three experimental applications of these properties.

1. Introduction

Quantum properties of the optical field represent a resource of the utmost relevance for the development of quantum technologies, allowing unprecedented results in disciplines ranging from quantum information and metrology [1,2,3] to quantum imaging [4,5].

A very interesting example is offered by the possibility of overcoming measurement limits pertaining classical states. In this proceedings we review three achievements that could have relevant practical application one concerning the imaging, the other the detection of weak objects by exploiting the quantum correlations of parametric down conversion (PDC) emission and split thermal light.

2. Sub shot noise quantum imaging

The principle of the first technique (see Fig.1), dubbed sub shot noise quantum imaging, is to take advantage of the correlation in the noise of two conjugated branches of PDC emission for imaging a weak absorbing object, otherwise lost in the noise. The image of the object, eventually previously hidden in the noise, could be restored [5,6,7,8] by subtracting the noise measured on one branch from the image of a weak object obtained in the other branch

When operating under shot noise this method allows in principle a full reconstruction of the absorption pattern of an object with a sensitivity superior to that available with classical techniques at the same illumination level. In order to reach this goal, in view of important practical application, one should obtain a very high level of spatial quantum correlation, that requires a precise characterization/control of speckles structures and high quantum detection efficiency.

The degree of correlation is quantified by the figure of merit $\sigma = \langle \delta^2 (N_i - N_s) \rangle / \langle N_i + N_s \rangle$, N_i and N_s being the photon number measured in correlated pixels, whose theoretical value, in term of the transmission of the optical channel η (including the quantum efficiency of the detector) for PDC it is $1-\eta$. Thus, in principle, in an ideal situation, it can approach zero, while in classical case it is larger than 1. Thus, for $\sigma<1$, one is allowed beating the performances of the analogous differential classical scheme, while when $\sigma<0.5$, one is allowed beating every classical scheme (the best classical scheme is still the direct illumination with a shot noise limited source).

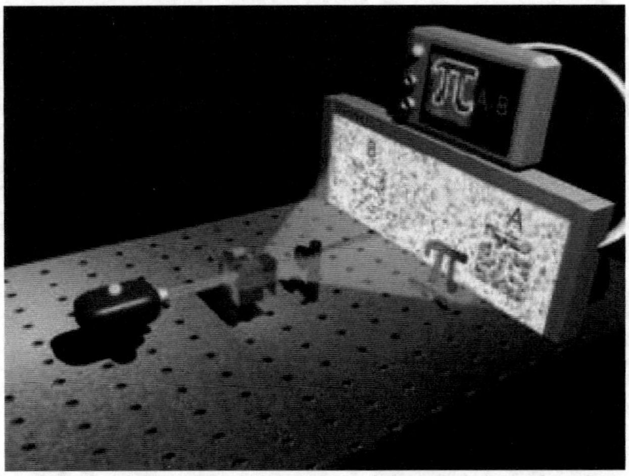

Figure 1. Simplified schematic of the set-up for imaging under shot noise level.

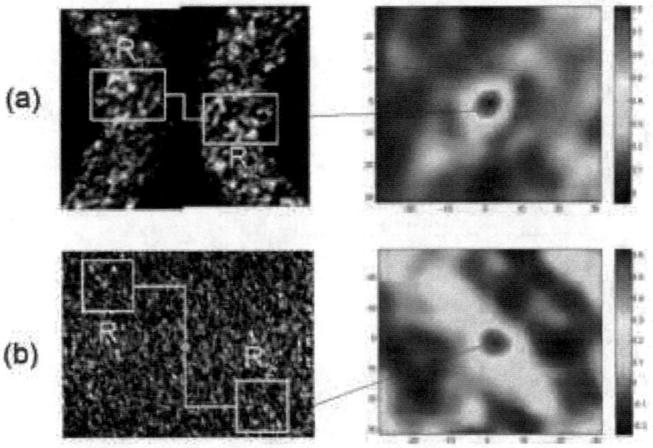

Figure 2. Example of correlations in noise pattern in twin beams, with (a) and without (b) an interference filter. In the second raw we report the cross correlation function.

Our experimental set up consisted of a 355 nm laser beam that, after a spatial filtering, pumped a type II BBO crystal producing PDC. After eliminating the UV beam, one correlated beam crossed a weak absorbing object and it was addressed to a CCD array. The other beam (reference) was directly addressed to another area of CCD camera.

After a study on the speckle structure of twin beams and their control [9,10] (Fig.2), in a first phase we have evaluated the capability to reach a sub shot noise level (σ <0.5). Our results clearly demonstrated this achievement, see Fig.3.

Since this preliminary step showed that we had achieved the regime needed for obtaining a higher sensitivity respect to every classical imaging, we have then inserted a weak absorbing object on the optical path of the signal beam, a titanium deposition on glass with a uniform absorption 5%.

In order to verify the theoretical predictions we collected a large number of frames (about 1200) and grouped the frames in classes with respect their value of σ. Fig.4 compares the performances of SSNQI with classical schemes for a single shot picture.

A quantitative statistical analysis of the advantages of the SSNQI is shown in Fig.5 , where we report the ratio R of the quantum to the classical Signal to Noise Ratio (SNR) plotted in terms of the average correlation degree σ of the data set: the SSNQI has, as expected, a clear advantage when σ<0.5 with respect to the direct classical imaging, and is definitely better than differential classical imaging for σ < 1. For the best achieved values of the correlation degree in the quantum regime, we obtain an improvement of the SNR larger than 30% compared with the best classical imaging scheme and more than 70% better than the differential classical scheme.

These results clearly demonstrate the potentialities of quantum systems for overcoming the classical limits of measurements, paving the way to further future developments in this sense.

Furthermore, they prompt for further developments of this scheme for reaching levels interesting for commercial applications, in measurements concerning photosensible samples (as, for instance, biological or thin films samples).

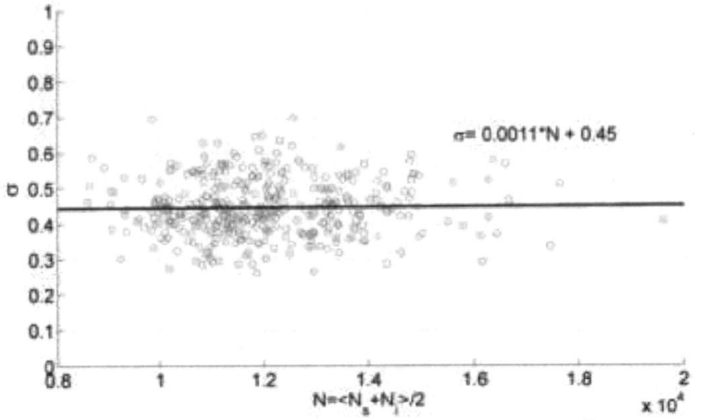

Figure 3. The degree of correlation. The degree of correlation $\sigma = \langle \delta^2 (N_i - N_s) \rangle / \langle N_i + N_s \rangle$ is plotted in function of the average number of photons of the single beam. One can observe how it is always largely under the classical limit 1, and on average under 0.5: this represents the condition for realising an advantageous quantum imaging.

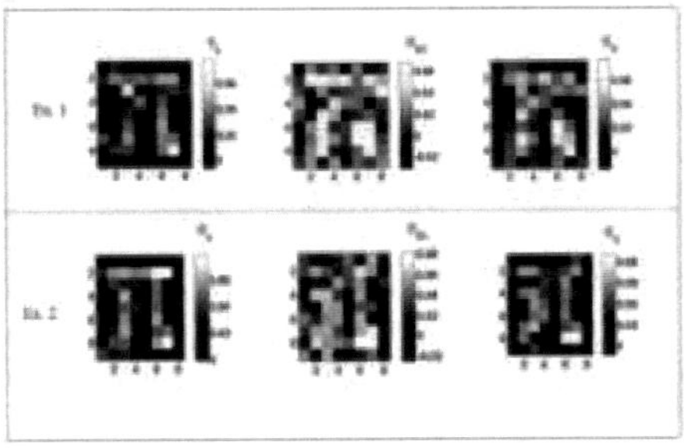

Figure 4. Example of the imaging of a weak absorbing object, a π, with quantum imaging (left) compared with two classical schemes (direct and differential).

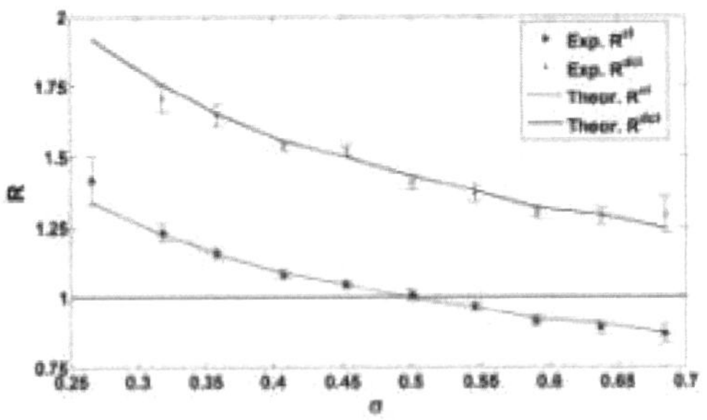

Figure 5. Ratio of signal to noise ratios in SSNQI and differential (black curve) and direct (red curve) classical protocols.

3. Toward quantum illumination

The principle of the second technique, dubbed quantum illumination, is to take advantage of the correlation in the noise of two conjugated branches of PDC emission for detecting the presence of a weak absorbing object, otherwise lost in a preponderant thermal background.

In particular in [3] Lloyd demonstrated, expanding some previous theoretical ideas [11,12,13], that the use of entanglement can in principle improve the possibility of detecting an object hidden in a background. This idea was then further elaborated in [2], where it was shown that quantum illumination may offer a significant performance gain respect to optimum reception coherent-state systems, and in [14], where it was demonstrated that the quantum protocol advantages are significant in multi- photon regime. Finally, application to secure quantum key distribution was proposed in [15]. However, these works did not specify a feasible model for the detection apparatus. An attempt in this sense was then presented in [16] where two detectors were proposed: one based on an optical parametric amplifier and an ideal photon-counter, the other based on the use of phase conjugation followed by balanced dual detection. Nevertheless, both the schemes are of very difficult experimental realisation, limiting their applicability.

Here, we explore the capability of a quantum illumination scheme and realize it in a more realistic characterized by a dominant and unknown background, severe losses, and when the receivers are photo-counting detector available nowadays. We present preliminary data demonstrating that the use of simple second order correlation measurements of the joint photon numbers distribution already suffices in guaranteeing strong advantages to the quantum protocol, then we realise quantum target detection both with twin beams and thermal light pointing out unequivocally the experimental advantage of the quantum illumination protocol in mesoscopic regime, with respect to its classical counterpart. This achievement, beyond paving the way of future practical application of quantum illumination, also provides a significant example of ancilla assisted quantum protocol besides the few previous realisations [15].

In the first configuration of our set up, type II Parametric Down Conversion light (with correlated photon pairs of orthogonal polarisations) is generated by means of a BBO non-linear crystal pumped with the third harmonic (355 nm) of a Q-switched Nd-Yag laser, with a repetition rate of 10 Hz and 5 ns of pulse width, after a spatial filtering. The correlated emissions are then addressed to a high quantum efficiency (about 80% at 710 nm) CCD camera. The exposure time of CCD and the synchronization with the pump are chosen such that each acquired frame corresponds to the emission generated by a single laser shot. Thus, the number of temporal modes collected is of some thousands, estimating the coherence time of PDC process around one picosecond. On one of the two paths it is posed a beam splitter that represents the target object and combines the PDC light with a thermal background produced by addressing a laser beam to an Arecchi's rotating ground glass.

In a second configuration the twin beam is substituted by beam split thermal light: this configuration allows the comparison with classical case.

The evident advantage of the quantum configuration is demonstrated in Figure 6.

3. Ghost Imaging towards medical diagnostic

Ghost Imaging [1,17] represents the first techniques where quantum correlations were exploited for imaging.

In this technique a light beam crosses (or it is reflected) by an object, that one wants to image.

However, the beam that crossed the object is detected by a detector without any spatial resolution (bucket detector). The image of the object is retrieved when the bucket detector signal is correlated with the signal of a spatial resolving detector measuring a light beam whose spatial noise is correlated to the previous beam. The first demonstration of this technique was achieved with quantum optical states of light, known as twin beams, produced by Parametric Down Conversion (PDC). Then it was shown, both theoretically and experimentally, that this result can be achieved also with beam-split thermal light [18,19,20,21,22,23], even if with a smaller visibility. Nevertheless, non-classicality of these states can be quantified by discord showing its relation with visibility [24].

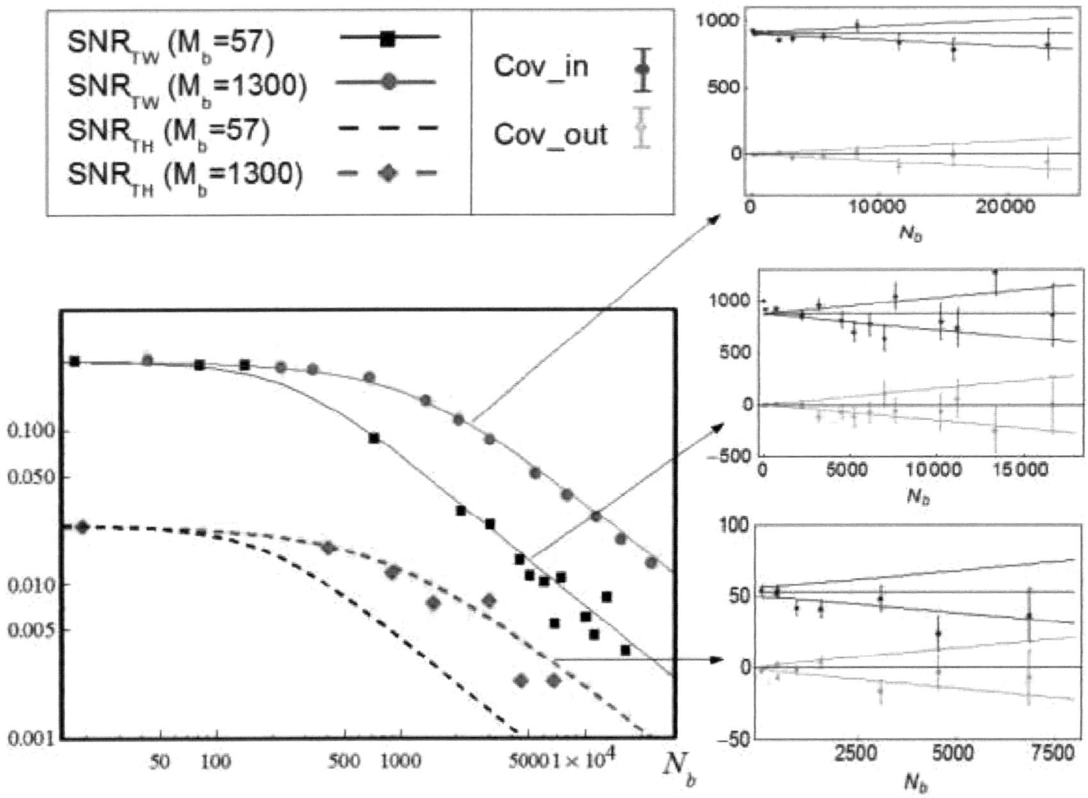

Figure 6. Signal to noise ratio (SNR) versus the total number of photons of the thermal bath normalized by the number of repetitions. The red (black) markers are the data for 1300 and 57 modes o thermal bath respectively,. The solid (dashed) theoretical curve corresponds to quantum (classical) illuminating beams. The lower curve of the classical protocol has not been compared with the experimental data because the SNR is so low that a very large number of images (out of the possibility of the actual setup) is required to have reliable points. The inset small graphs represent the value of the covariance for each measurement in presence and absence of the Target.

The interesting point now is realising configuration suited for practical applications. An idea is to retrieve the image of an object in a scattering media for medical diagnosis application, in particular for the diagnosis of breast cancer. In fact, multiple scattering influences the quality of such images, with degradation in resolution and contrast. Moreover, the most widespread method for medical and clinical diagnosis is X ray radiation imaging, with the drawback of ionizing and unsafe radiation. Hence, the use of optical photons is becoming an interesting method even if an accurate image deteriorated by multiple scattering of the blood and tissues can be retrieved only using second order correlation techniques such as Ghost Imaging.

In our work, we are investigating the use of a compact thermal light ghost imaging set-up for realistic application in breast cancer detection.

In our setup, a thermal beam is produced by addressing 1 ns laser pulses at 532 nm on an Arecchi disk. The thermal beam is divided by a 50-50 beam splitter; one part is sent through the object (a test bar of 1 mm x 5 m) in the scattering media. Both beams are imaged by means of a lens of focal length of 15 cm on two portions of a CCD array, synchronized with the laser; the portion with the direct image of the object is summed to obtain the bucket detector.

Ghost image is reconstructed by means of second order correlation coefficient. Preliminary results show that the image of the object can be retrieved, with a visibility:

$$V = \frac{\left| c^{(2)}{}_{in} - {}^{(2)}{}_{out} \right|}{c^{(2)}{}_{in} + {}^{(2)}{}_{out}},$$

of 50% where the indexes $c^{(2)}{}_{in}$ and $c^{(2)}{}_{out}$ indicates the mean second order correlation coefficients calculated respectively in a region with the object and outside it.

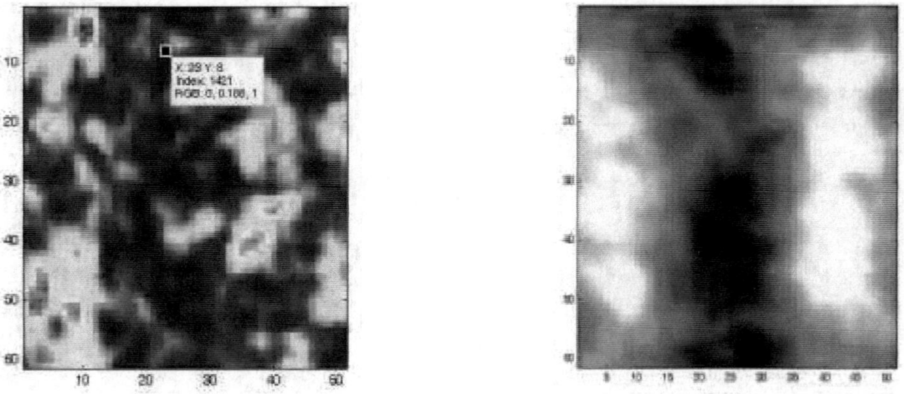

Figure 7. The left part shows the image of the object, deteriorated by the scattering of the light in the medium. In the right part we observe the reconstructed ghost image by means of second order correlation coefficient. The object is recognizable even if the limit of this reconstructed image is the current resolution of the imaging system. In fact, resolution is limited by the size of the speackles of the thermal beam that are of 200 μm of diameter. These are preliminary results and we are now working to reduce the size of the speackles.

4. Conclusion

We have reported the experiments performed in INRIM to overcome the limits imposed by classical physics by exploiting quantum properties of light. In particular, we described three approaches that exploit quantum correlation to improve imaging technique and the detection of weak objects.

References

[1] Pittman T, Shih Y H, Strekalov D V and Sergienko A V 1995 *Phys. Rev. A* **52** R3429-R3432
[2] Tan S, Erkmen B I, Giovannetti V, Guha S, Lloyd S, Maccone L, Pirandola S and Shapiro J H *Phys. Rev. Lett.* **101** 253601
[3] Lloyd S 2008 *Science* **321** 1463
[4] Meyers R, Deacon K and Shih Y 2008 *Phys. Rev. A* **77** 041801
[5] Brambilla E, Caspani L, Jedrkiewicz O, Lugiato L A and Gatti A 2008 *Phys. Rev. A* **77** 053807
[6] Brida G, Caspani L, Gatti A, Genovese M, Meda A and Ruo Berchera I 2009 *Phys. Rev. Lett.* **102** 213602

[7] Brida G, Genovese M and Ruo Berchera I 2010 *Nature Photonics* **4** 227 – 230
[8] Brida G, Genovese M, Meda A and Ruo Berchera I 2011 *Phys. Rev. A* **83** 033811
[9] Brida G, Genovese M, Meda A, Predazzi E and Ruo Berchera I 2009 *Int. Journ. Quant. Inf.* **7** 139
[10] Brida G, Genovese M, Meda A, Predazzi E and Ruo Berchera I 2009 *Journ. of Mod. Opt.* **56** 201
[11] Sacchi M F 2005 *Phys. Rev. A* **71** 062340
[12] Sacchi M F 2005 *Phys. Rev. A* **72** 014305
[13] D'Ariano G and Lo Presti P 2001 *Phys. Rev. Lett.* **86** 4195-4198
[14] Shapiro J H and Lloyd S 2009 *New Journ. of Phys*. **11**, 063045
[15] Guha S and Erkmen B I 2009 *Phys. Rev.* A **80** 052310
[16] Shapiro J 2009 *Phys. Rev. A* **80**, **022320**
[17] Belinskii A and Klyshko D 1994 *Sov. Phys. JETP* **78** 259
[18] Lugiato L A, Gatti A and Brambilla E 2002 *J. Opt. B* **4** S176-S183
[19] Boto A, Kok P, Abrams D S, Braunstein L, Williams C P and Dowling J P 2000 *Phys. Rev. Lett.* **85** 2733-2736
[20] Brida G, Chekhova M V, Fornaro G A, Genovese M, Lopaeva E D and Ruo Berchera I 2011 *Phys. Rev. A* **83** 063807
[21] Chen X, Liu Q, Luo K, Wu L. 2009 *Optics Letters* **34** 695
[22] Valencia A, Scarcelli G, D'Angelo M. Shih Y 2005 *Phys. Rev. Lett*. **94** 063601
[23] Meyers R, Deacon k, Shih Y 2008 *Phys. Rev*. A **77** 041801
[24] Raggi S and Adesso G 2012 *Scient. Rep.* **2** 651

From single constituents to correlation: phase transitions in few atom systems

Enrico Prati

Laboratorio Materiali e Dispositivi per la Microelettronica, Consiglio Nazionale delle Ricerche - IMM, Via Olivetti 2, I-20864 Agrate Brianza, Italy

E-mail: enrico.prati@cnr.it

Abstract. The advancements of nanofabrication, together with the control of implantation of individual atoms at nanometric precision, have opened the experimental study of phase transitions of electronic systems constituted by six electrons and less. Here I review the recent advancements made in the field of phase transitions at the few atoms scale, including the discretized version of the Anderson-Mott quantum phase transition and the melting of electrons arranged in a Wigner-like phase into a Fermi glass realized in a $Si : X$ quantum device where X is a donor element. The rise of collective phenomena is directly probed by controlling the distance between few donor atoms. Four atoms are sufficient to observe emergent phenomena such as Hubbard bands instead of single particle behaviour.

1. Introduction

The study of the Anderson-Mott transition [1, 2] have raised a considerable experimental and theoretical interest for decades.[3, 4] One of its manifestations, given by the transition from delocalized to localized electrons in semiconductors (a metal-insulator transition) triggered by the doping concentration by a different donor element, has represented a hard problem to solve theoretically and experimentally, even in the case of the most simple and well known case of doped silicon (Si:X where $X = P, As$).[4] The possibilities opened by nanofabrication of semiconductor devices [5, 6] have been exploited by engineering the ability to control the position of impurity atoms at nanometric precision by a one-by-one method of implantation called single ion implantation (SII). Taking advantage of this method, developed in the last decade by three groups in the world [7, 8, 9], it has been possible to create chains of individually implanted ions in silicon [10], and to investigate the effects caused by their proximity and the residual disorder affecting their position and ground state energy distribution [11]. Therefore, the transition from a single particle behaviour to collective phenomena is directly observed. In the following, I recall the basic ingredients of the models involved in the Anderson-Mott transition, i. e. the Anderson transition and the Mott transition. Next, I describe the physics observed in single atom transistors, the band formation in arrays of donors, and finally the phase transitions observed in such systems. To conclude, only four elements are sufficient to exhibit the emergence [12] of a collective many-body behaviour.

2. Short review on metal insulator transitions

With metal-insulator transitions (MIT) are generally intended transitions of the electron states from delocalized to localized electron wavefunctions. MIT phase transitions are generated by means of two possible mechanisms: a structural change due to a change in the ionic lattice, or purely electronic. In this second case, two limiting cases are classified: Mott phase transitions, triggered by electron correlation and Anderson phase transitions, triggered by disorder. Transitions in which the competition of the two latter mechanisms concur, the transition is referred as Anderson-Mott transition. In the following we consider only the purely electronic phase transitions. The Mott transition is captured by the Hubbard Hamiltonian:

$$H = t \sum_{\langle i,j \rangle} \left(a_{i\uparrow}^{+} a_{j\uparrow} + a_{i\downarrow}^{+} a_{j\downarrow} \right) + U \sum_{i} n_{i\uparrow} n_{i\downarrow} \tag{1}$$

where

$$n_{i\sigma} = a_{i\sigma}^{+} a_{j\sigma}, \tag{2}$$

t is the hopping matrix element and U is the on-site repulsion energy ($U > 0$). The Anderson transition is governed by the Hamiltonian

$$H = t \sum_{\langle i,j \rangle} a_i^{+} a_j + \sum_i \epsilon_i a_i^{+} a_i \tag{3}$$

which is a non-interacting electrons model, so spin here only provides a trivial factor of two which is omitted. The energies ϵ_i are distributed within a bandwidth W: Anderson predicted for W/t large but finite a transition from delocalized states to insulating regime. Delocalization is achieved by either decreasing the density of scatterers, or, for soft scattering potential, increasing energy of scattered particles. Pure Anderson transition is not expected to be realized in nature, as Coulomb interaction between electrons is always present. To describe the effects of the Anderson transition in semiconductors such as silicon in which disorder is introduced by the addition of donor atoms, such as P and As, the Mott transition, which would be inadequate if considered alone, has to be taken into account. Conclusion that has been reached over the years is that neither Andersons nor Motts picture itself is sufficient to understand the observed MIT in semiconductors[4]. The resulting quantum phase transition, which carries aspects of both types of transitions, is called an Anderson-Mott transition.

3. Single atom transistors

Single electron phenomena in defects have been originally observed in silicon/silicon oxide interfaces from random telegraph signal in small devices [13, 14, 15, 16, 17, 18]. Later, single electron effects have been observed in donor at the silicon/silicon oxide interface.[19] Individual atoms centered between the source and the drain of a small device provide the energy levels for sequential tunneling. The first observation of single impurities in silicon were obtained by using a small field effect transistor [19, 20] or a small flash memory [21] with a silicon channel, in which an impurity was randomly diffused from the doping of one of the contacts. The presence of the individual atom is observed by looking at the quantum transport of the device at cryogenic temperature [22, 23, 24]. The condition necessary to observe the quantum transport is that the overlap integral between the electron wavefunction and the conduction electron wavefunction exponentially decaying out of the doped region is sufficiently large so that the probability of creating sequential tunneling from the contact to the donor site and from the donor site to the second contact is measurable. Generally the host device, which is equipped with a top gate, at 4 K behaves as an electrostatic quantum dot (QD) in Coulomb blockade regime. The energy scale governing the Coulomb blockade oscillations in a QD is set by the charging energy

$U = E_C + (E_{n+1} - E_n) = e^2/2C$, where C is the total capacitance of the QD island relative to the surrounding conductors, E_n is the energy of the nth level and $-e$ is the electron charge. If the single donor is positioned along the channel, it provides a second set of localized eigenstates of electrons corresponding to additional permitted energy levels below the conduction band edge. In Ref. [11] transistors fabricated on 100 silicon-on-insulator (SOI) substrates with 125-nm-thick buried oxide (BOX), which acts as a back-gate oxide, are considered. The nominal channel length is 200 nm, the width and thickness of the channel are 100 nm and 90 nm, respectively. Here only the Coulomb blockade of the atoms is observed, while no quantum dots are formed. The channels are in contact with highly doped n-type source and drain regions and are controlled by a gate bias (V_g) from the substrate through the buried oxide. The device exhibits accumulation-mode n-type transistor operation. The arsenic and phosphorous ions implanted two by two at 60 keV into the channel through the surface oxide with 10nm thick are expected to distribute around the depth of 53 nm from the interface on the back side. Transistors with only two atoms implanted along the channel, for which their average distance is of the order of 20-40 nm, are trated as isolated particles. This is experimentally confirmed by looking at the quantum trasport below the threshold voltage V_G which probes the silicon band gap. There, Coulomb blockade is observed at low temperature (4-20 K). Coulomb blockade reveals that one electron occupies the ground state of each donor (first peak) and a second may occupy the second available state, as the charging energy is sufficiently low so that a second electron can stay bound to the donor (the D^- state, which has a 5 times larger extension). To resume, in this regime the single particle behaviour is observed through the Coulomb blockade caused by two isolated donors which create two pairs of discrete energy levels in the band gap of the silicon.

4. Hubbard band formation in arrays of atoms in silicon

If one moves from the extremely dilute impurity concentration of previous paragraph, which is conventionally defined as $n < 1 \cdot 10^{16}$ cm^{-3} in bulk silicon, such that r_C >46.4 nm where $r_C = n^{-1/3}$.[25], to higher concentrations, one first achieves a linear density which corresponds to a cubic lattice density n between $1 \cdot 10^{16}$ cm^{-3} and 10^{17} cm^{-3} (r_C = 21.5 nm), called semidilute regime, characterised in bulk silicon by the formation of pairs and complexes because of the randomness of the donor distribution. At higher linear density, which corresponds to approximately 10^{17} cm^{-3}, Hubbard bands are formed (intermediate regime) with activation energies of ϵ_2 and ϵ_3 for the upper and lower bands, respectively, triggered by temperature. Such a classification has been created for bulk semiconductor. The novel single ion implanted quantum devices have shown that the Hubbard impurity bands are observed also for arrays of few atoms. This is observed experimentally as flat bands below the conduction band (the threshold voltage of the transistor) in I-Vg characteristics (Figure 1, top-right). The number of atoms needed to create such bands is very small. Four and six atoms are sufficient to observe band formation. As each donor provides a double-spin-degenerate ground state that is also six-fold valley-degenerate, a system of 4 atoms has 48 available $1s$ states to create the band. Consequently the discretize version of the Anderson-Mott transition is observed by doubling the linear average density of the atoms in the device. In Figure 1 the formation of the Hubbard bands in a transistor of effective length of about 150 nm doped with an array of 4 atoms is reported as first derivative of the I-V_G curves at many temperatures. The gate voltage V_G scans the upper part of the silicon bandgap of the channel of the device. The conduction band edge is represented by the threshold voltage V_T. On the left, at low temperature, the contrast shows the Coulomb blockade transport through the localized energy levels provided by the four As donors. By increasing the temperature, the combination of the correlation and of the thermal activation determines the rise of the two partially overlapped Hubbard bands. As the frozen electron system is created at 4 K, we may call it a Wigner-like phase (each electron occupies a site of the potential landscape created by the 4 donors). By raising the temperature, the

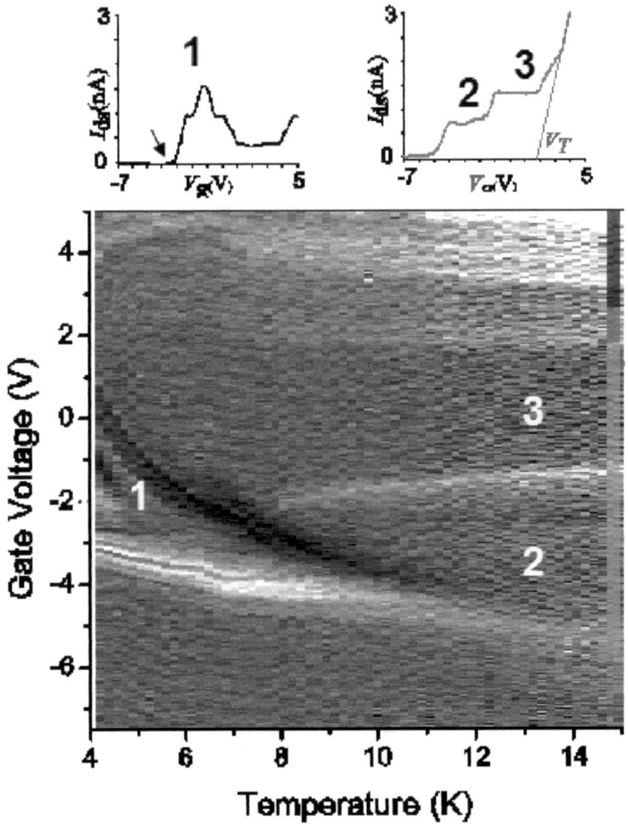

Figure 1. Band formation in an array of 4 atoms below the threshold voltage V_G. From Ref. [11]. In region 1, the sequential tunneling through the array of 4 atoms energy levels dominates. By raising the temperature, the lower Hubbard band (2) and the upper Hubbard band (3) are formed.

electron system becomes better described in terms of Fermi glass.[2]

5. Conclusion

To conclude, recent experiments have shown that the emergence of phenomena typical of many-particle physics is possible with only four electrons confined thanks to an array of donors contained in a silicon quantum devices operated at cryoenic temperature. This kind of device opens the exploration of emergence at the nanoscale, when only few elementary constituents are employed and controlled.

References

[1] Mott N F and Twose W D 1961 The theory of impurity conduction *Adv. in Phys.* **10** 38 107-163

[2] Anderson P W 1958 Absence of diffusion in certain random lattices *Phys. Rev.* **109** 1492-1505

[3] Hubbard J 1963 Electron Correlations in Narrow Energy Bands *Proc. R. Soc. Lond.* A **276** 1365 238-257

[4] Belitz D and Kirkpatrick T R 1994 The Anderson-Mott transition *Rev. Mod. Phys.* **66** 261-380

[5] Prati E et al. 2012 Few Electron Limit of n-type Metal Oxide Semiconductor Single Electron Transistors *Nanotechnology* **23** 215204

[6] De Michielis M, Prati E, Fanciulli M, Fiori G and Iannaccone G 2012 Geometrical Effects on Valley-Orbital Filling Patterns in Silicon Quantum Dots for Robust Qubit Implementation *Appl. Phys. Expr.* **5** 1

[7] Shinada T, Okamoto S, Kobayashi T and Ohdomari I 2005 Enhancing semiconductor device performance using ordered dopant arrays *Nature* **437** 1128-1131

[8] Tan K Y et al. 2010 Transport spectroscopy of single phosphorus donors in a silicon nanoscale transistor *Nano Lett.* **10** 11-15

[9] Schenkel T, Persaud A, Park SJ, Nilsson J, Bokor J, Liddle JA, Keller R, Schneider DH, Cheng DW and Humphries DE 2003 Solid state quantum computer development in silicon with single ion implantation *J. Appl. Phys.* **94** 7017-7024

[10] Shinada T, Hori M, Guagliardo F, Ferrari G, Komatubara A, Kumagai K, Tanii T, Endo T, Ono Y and Prati E 2011 Quantum transport in deterministically implanted single-donors in Si FETs. 2011 IEEE International Electron Devices Meeting (IEDM) 5-7 Dec. 2011 30.4.1-30.4.4

[11] Prati E, Hori M, Guagliardo F, Ferrari G and Shinada T 2012 Anderson-Mott transition in arrays of a few dopant atoms in a silicon transistor *Nature Nano.* **7** 443-447

[12] Licata I and Sakaji A 2008 *Physics of Emergence and Organization* (Singapore: World Scientific Publ.)

[13] Prati E, Fanciulli M, Calderoni A, Ferrari G and Sampietro M 2007 Microwave irradiation effects on random telegraph signal in a MOSFET *Phys. Lett.* A **370** 491-493

[14] Prati E and Fanciulli M 2008 Manipulation of localized charge states in n-MOSFETs with microwave irradiation *Phys. Lett.* A **372** 3102-3104

[15] Prati E, Fanciulli M, Ferrari G and Sampietro M 2008 Giant Random Telegraph Signal Generated by Single Charge Trapping in sub-micron n-MOSFETs *J. Appl. Phys.* **103** 123707

[16] Prati E, Fanciulli M, Calderoni A, Ferrari G and Sampietro M 2008 Effects of microwave irradiation on the emission and capture dynamics in silicon metal oxide semiconductor field effect transistors *J. Appl. Phys.* **103** 104503

[17] Prati E et al. 2010 Measuring the Temperature of a Mesoscopic Electron System by means of Single Electron Statistics *Applied Physics Letters* **96** 113109

[18] Prati E 2010 The finite quantum grand canonical ensemble and temperature from single-electron statistics for a mesoscopic device *J. Stat. Mech.* **1** P01030

[19] Sellier H et al. 2006 Transport spectroscopy of a single dopant in a gated silicon nanowire *Phys. Rev. Lett.* **97** 206805

[20] Pierre M, Wacquez R, Jehl X, Sanquer M, Vinet M and Cueto O 2010 Single-donor ionization energies in a nanoscale CMOS channel *Nature Nanotechnology* **5** 133-137

[21] Prati E, Latempa R and Fanciulli M 2009 Microwave Assisted Transport in a Single Donor Silicon Quantum Dot, *Phys. Rev.* B **80** 165331

[22] Prati E et al. 2011 Adiabatic charge control in Silicon single donor quantum dots *Appl. Phys. Lett.* **98** 053109

[23] Mazzeo G et al. 2012 Charge dynamics of a single donor coupled to a few electrons quantum dot in silicon *Appl. Phys. Lett.* **100** 213107

[24] Leti G, Prati E, Belli M, Petretto G, Fanciulli M, Wacquez R, Vinet M and Sanquer M 2011 Switching quantum transport in a three donors silicon Fin-field effect transistor *Appl. Phys. Lett.* **99** 242102

[25] Thomas GA, Capizzi M, DeRosa F, Bhatt RN and Rice MT 1981 Optical study of interacting donors in semiconductors *Phys. Rev.* B **23** 10 5471-4394

Entanglement generation in relativistic cavity motion

David Edward Bruschi[1,2], **Jorma Louko**[1] **and Daniele Faccio**[3]

[1] School of Mathematical Sciences, University of Nottingham, Nottingham NG7 2RD, United Kingdom

[2] School of Electronic and Electrical Engineering, University of Leeds, Leeds LS2 9JT, United Kingdom

[3] School of Engineering and Physical Sciences, David Brewster Building, Heriot-Watt University, SUPA, Edinburgh EH14 4AS, United Kingdom

Abstract. We analyse particle creation and mode mixing for a quantum field in an accelerated cavity, assuming small accelerations but allowing arbitrary velocities, travel times and travel distances, and in particular including the regime of relativistic velocities. As an application, we identify a desktop experimental scenario where the mode mixing resonance frequency in linear sinusoidal motion or in uniform circular motion is significantly below the particle creation resonance frequencies of the Dynamical Casimir Effect, and arguably at the threshold of current technology. The mode mixing acts as a beamsplitter quantum gate, experimentally detectable not only via fluxes or particle numbers but also via entanglement generation.

1. Introduction and overview

As anyone caught by a police speed camera can testify, spacetime kinematics affects photons that are present in spacetime. For a relativistic quantum field, spacetime kinematics can also create particles where initially there were none, and the very notion of a "particle" depends on the motion of an observer. In flat spacetime, celebrated examples are the thermality seen in Minkowski vacuum by uniformly accelerated observers, known as the Unruh effect [1, 2], and the creation of particles by moving boundaries, known as the Dynamical (or non-stationary) Casimir Effect (DCE) [3, 4]. In curved spacetime, a celebrated example is the Hawking radiation emitted by black holes [5].

There has been significant recent interest in harnessing consequences of spacetime kinematics to serve quantum information tasks. The broad issue is how entanglement and the fidelity of quantum communication are affected by observer motion and by spacetime curvature [6, 7, 8, 9, 10, 11, 12, 13]. A specific question is how entanglement may be created and used to construct quantum gates [14, 15, 16, 17]. A recent survey can be found in [18].

In this contribution we consider a quantum field in an accelerated, perfectly confining cavity in Minkowski spacetime. As the cavity walls shield the field from acceleration horizons, the cavity field cannot be expected to exhibit thermality via the Unruh effect when the acceleration is linear and uniform [19]. While the absence of the conventional Unruh effect can hence be regarded as a disadvantage of a cavity field, the field will nevertheless respond to the cavity's acceleration in a nontrivial fashion, and the confinement has technical advantages in that a discrete spectrum makes a number of entanglement quantifiers mathematically well defined [20] and conceptual advantages related to localisation issues in relativistic quantum measurement theory [21, 22].

Finally, a practical advantage of a cavity field is that the cavity can be given a variety of travel scenarios of interest, including periodic shakes that appear in the DCE literature [3, 4].

In Sections 2 and 3 we review a recently-developed formalism [23] for a relativistic field in a moving cavity that is assumed mechanically rigid, in the sense that the cavity's shape in its instantaneous rest frame remains unchanged even though the velocity may change in time. The acceleration is assumed to be small compared with c^2 divided by the cavity's size but may otherwise have arbitrary time dependence, both in magnitude and in direction. The velocities, travel times and travel distances are unrestricted, so that the treatment remains valid even when the velocities become relativistic. To linear order in the acceleration, the Bogoliubov coefficients between inertial motions at early and late times can then be expressed in terms of the Fourier transform of the cavity's acceleration.

In Section 4 we apply the formalism to two periodic motions, a linear sinusoidal motion and a uniform circular motion [23]. We identify a configuration that brings the mode mixing resonance frequency in these motions into an apparently experimentally feasible regime, significantly below the experimentally prohibitive particle creation resonance frequencies known from the DCE literature [3, 4]. A geometrically optimised desktop experimental scenario at optical wavelengths is seen to produce significant mode mixing within a millisecond of shaking at megahertz frequencies. From the quantum information viewpoint the mode mixing acts as a beamsplitter quantum gate. The experimental verification opportunities can hence utilise entanglement phenomena, in addition to conventional observations of fluxes or particle numbers.

We conclude in Section 5 by summarising the results and discussing the prospects to bridge the gap to an actual experiment that would create entanglement by mechanical motion on the desktop.

We set from now on $c = \hbar = 1$, reinstating units by dimensional analysis where appropriate.

2. Rigid cavity at small accelerations: (1+1) Minkowski

2.1. Setup

We start with a cavity in $(1 + 1)$-dimensional Minkowski spacetime. The cavity is assumed mechanically rigid, in the sense that it maintains constant length L in its instantaneous rest frame. We denote by τ the proper time at the centre of the cavity and by $a(\tau)$ the proper acceleration at the centre of the cavity. To maintain rigidity, the acceleration must be bounded by $|a|L < 2$ [11].

The cavity contains a real scalar field ϕ of mass $\mu_0 \geq 0$, with Dirichlet boundary conditions at the walls.

We assume a to vanish at early and late times but to be nonvanishing at some intermediate times. An example where a is piecewise constant with exactly two points of discontinuity is shown in Figure 1.

The inertial segments of the cavity worldtube at early and late times are static: the cavity is dragged along a Minkowski time translation Killing vector. These timelike Killing vectors provide a distinguished definition of positive and negative frequencies tailored to the cavity's motion at early and late times, resulting to field modes with the angular frequencies $\omega_n = \sqrt{\mu_0^2 + (\pi n/L)^2}$, $n = 1, 2, \ldots$, and quantisation leads to the familiar Fock space of an inertial cavity. However, because a is nonvanishing at intermediate times, the early time and late time vacua do not need to coincide, nor do the early time and late time notions of particles. The task is to find the Bogoliubov transformation that relates the early time and late time Fock spaces.

2.2. Solution

When the accelerations are so small that $|a|L \ll 1$, the problem has a solution that builds on the following three key observations [11, 23].

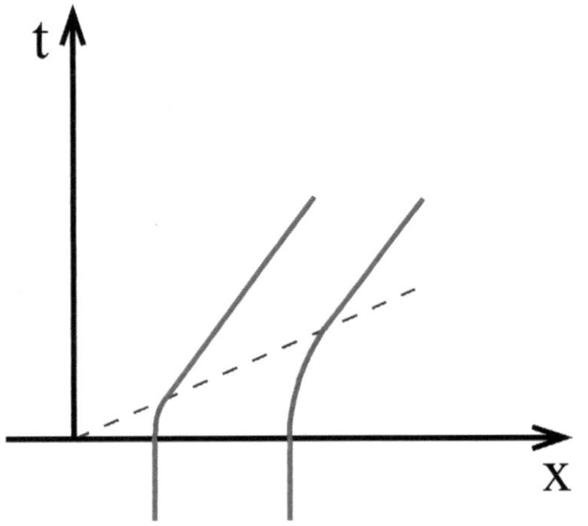

Figure 1. A $(1+1)$ cavity worldtube in which the initial and final inertial segments are joined by one segment of uniform acceleration. During the accelerated segment the cavity is static with respect to the boost Killing vector $x\partial_t + t\partial_x$, whereas during each of the inertial segments the cavity is static with respect to a Minkowski time translation Killing vector.

First, while a cavity with constant nonvanishing a is not inertial, it is nevertheless static: the cavity is dragged along a boost Killing vector, as shown in Figure 1. The cavity can be regarded as uniformly accelerated, even though the value of the acceleration varies between spatial points within the cavity, attaining the values $2a/(2 \pm aL)$ at the two walls and being equal to a only at the centre. The boost Killing vector is timelike and provides now a distinguished definition of positive and negative frequencies tailored to the cavity's motion. For $\mu_0 > 0$, the field modes can be written down in terms of modified Bessel functions and the angular frequencies are determined implicitly by a transcendental equation. For $\mu_0 = 0$, the field modes and their angular frequencies have elementary expressions [11].

Second, the Bogoliubov transformation between an inertial segment and a uniformly accelerated segment has a Maclaurin expansion in the dimensionless parameter aL, involving at each order only elementary expressions, and for $|a|L \ll 1$ the linear order terms provide already a good approximation [11]. Cavity world tubes for which a is piecewise constant with a finite number of discontinuities can hence be treated by composing this inertial-to-uniform transformation, its inverse, and the free evolution of the field modes during the uniformly-accelerated segments [11, 15]. The prototype cavity worldtube in which the initial and final inertial segments are joined by a single segment of uniform acceleration is shown in Figure 1.

Third, acceleration that is not necessarily piecewise constant can be handled by passing to the limit [23]. The Bogoliubov coefficient matrices α and β from the initial inertial segment to the final inertial segment have the expansions

$$\alpha = e^{i\boldsymbol{\omega}(\tau_1-\tau_0)}(1 + \hat{A} + O(h^2)), \tag{1}$$

$$\beta = e^{i\boldsymbol{\omega}(\tau_1-\tau_0)}\hat{B} + O(h^2), \tag{2}$$

where $h(\tau) := La(\tau)$, $\boldsymbol{\omega} = \mathrm{diag}(\omega_1, \omega_2, \cdots)$,

$$\hat{A}_{nn} = 0\ , \tag{3}$$

$$\hat{A}_{mn} = -i\frac{\pi^2 mn(1-(-1)^{m+n})}{L^4\,(\omega_m-\omega_n)^2\,\sqrt{\omega_m\omega_n}} \int_{\tau_0}^{\tau_1} e^{-i(\omega_m-\omega_n)(\tau-\tau_0)}\,h(\tau)\,d\tau \quad \text{for } m \neq n, \tag{4}$$

$$\hat{B}_{mn} = i\frac{\pi^2 mn(1-(-1)^{m+n})}{L^4\,(\omega_m+\omega_n)^2\,\sqrt{\omega_m\omega_n}} \int_{\tau_0}^{\tau_1} e^{-i(\omega_m+\omega_n)(\tau-\tau_0)}\,h(\tau)\,d\tau\ , \tag{5}$$

the initial inertial segment ends at $\tau = \tau_0$ and the final inertial segment begins at $\tau = \tau_1$. To linear order in h, the Bogoliubov coefficients are hence obtained by just Fourier transforming the acceleration.

2.3. Outcomes

While the perturbative solution (1)–(5) for the Bogoliubov coefficients assumes the acceleration to be so small that $|h| \ll 1$ over the cavity's worldtube, the velocities, travel times and travel distances remain unrestricted. The solution remains in particular valid even when the velocities grow relativistic. In this sense our perturbative treatment is complementary to the small distance approximations often considered in the DCE literature [3, 4], while of course overlapping in the common domain of validity.

$\hat{A}_{mn}$ and $\hat{B}_{mn}$ scale linearly in h, but their magnitudes depend also crucially on whether h changes slowly or rapidly compared with the oscillating integral kernels in (4) and (5). It is useful to consider the two extreme limits individually.

One extreme is the limit of slowly-varying h. In this limit $\hat{A}_{mn}$ and $\hat{B}_{mn}$ vanish, as is seen from (4) and (5) by the Riemann-Lebesgue lemma, and as can also be argued on more general adiabaticy grounds [19, 24]. The Bogoliubov coefficients indeed evolve by pure phases over any segment in which h is constant, not just to linear order in h as seen from (4) and (5) but also nonperturbatively in h [11, 23]. Particles in the cavity are hence created by *changes* in the acceleration, rather than by acceleration itself. The cavity is in this respect similar to a single accelerating mirror, which radiates only when its acceleration changes in time [25].

The other extreme is the case of piecewise constant h. In this case the contributions to $\hat{A}_{mn}$ and $\hat{B}_{mn}$ come entirely from the discontinuous jumps in h. While this extreme may not be experimentally realisable by a material cavity, it can be simulated by a cavity whose walls are static dc-SQUIDs undergoing electric modulation that simulates mechanical motion [26, 27].

When h is sinusoidal and its angular frequency ω_r equals the angular frequency of the integral kernel in (4) or (5), the corresponding Bogoliubov coefficient will grow linearly in the duration of the sinusoidal motion. These resonance conditions read

$$\hat{A}_{mn} : \ \omega_r = |\omega_m - \omega_n| \ , \tag{6}$$

$$\hat{B}_{mn} : \ \omega_r = \omega_m + \omega_n \ , \tag{7}$$

where in each case $m - n$ needs to be odd in order for the coefficient to be nonvanishing. The particle creation resonance (7) is well known in the DCE literature [3, 4, 28, 29, 30, 31, 32, 33, 34, 35, 36, 37, 38]. The mode mixing resonance (6) is also known [29, 30, 31, 32, 33, 34, 35] but seems to have received significant attention only in situations where it happens to coincide with a particle creation resonance.

3. Rigid cavity at small accelerations: (3+1) Minkowski

The above $(1 + 1)$-dimensional analysis generalises immediately to a rectangular cavity of edge lengths (L_x, L_y, L_z) in $(3+1)$-dimensional Minkowski spacetime provided the acceleration keeps pointing in one of the cavity's three principal directions. The quantum numbers in the transverse directions remain inert, just contributing to the effective $(1+1)$-dimensional mass μ_0. Note that μ_0 is hence positive even when the $(3 + 1)$-dimensional field in massless.

It is further immediate to compose segments in which the acceleration points in one of the cavity's principal directions, as seen in the cavity's instantaneous rest frame, even if this direction differs from segment to segment.

Acceleration of unrestricted magnitude and direction would require new input regarding how the shape of the cavity responds to such acceleration. To *linear* order in the acceleration, however, boosts in different spatial directions commute, and we can treat acceleration as a vector superposition of accelerations in the three principal directions in the cavity's instantaneous rest

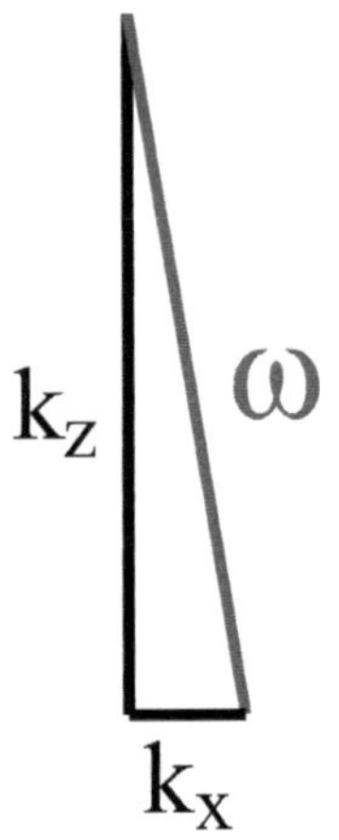

Figure 2. A momentum space configuration in which the mode mixing resonance angular frequency (8) is significantly lower than the angular frequencies of the individual cavity modes. The momenta of the quanta are aligned in the z-direction to a high degree of accuracy, so that $k_z \gg (k_x^2 + k_y^2)^{1/2}$. When k_x and k_y change by the respective small amounts Δk_x and Δk_y, the changes in the angular frequency ω are much smaller than Δk_x and Δk_y.

frame [23]. For sinusoidal acceleration with angular frequency ω_r, the resonance conditions (6) and (7) generalise to

$$\hat{A}_{mnp,m'n'p'} : \ \omega_r = |\omega_{mnp} - \omega_{m'n'p'}| \ , \tag{8}$$

$$\hat{B}_{mnp,m'n'p'} : \ \omega_r = \omega_{mnp} + \omega_{m'n'p'} \ , \tag{9}$$

where $\omega_{mnp} = \sqrt{\mu^2 + (\pi m/L_x)^2 + (\pi n/L_y)^2 + (\pi p/L_z)^2}$, μ is the mass of the field, the quantum numbers (m, n, p) are positive integers, and the difference in the quantum number in the direction of oscillation needs to be odd.

4. Desktop mode mixing experiment with periodic motion

The particle creation resonance angular frequency (9) is comparable in magnitude to the angular frequencies of the individual cavity modes and appears prohibitively difficult to realise in experiments with mechanical oscillations [3, 4]. The mode mixing resonance angular frequency (8) can however be significantly lower when the cavity modes are highly transverse to the direction of the oscillation, as illustrated in Figure 2. We now outline a scenario that optimises this lowering [23].

4.1. Scenario

We consider a massless field in $(3 + 1)$ dimensions and focus on quanta whose wavelength λ is much smaller than any of the cavity's edge lengths. The crucial input is to align the momenta close to the z-direction, as shown in Figure 2, so that $\omega_{mnp} \approx 2\pi/\lambda + \frac{1}{4}\pi\lambda[(m/L_x)^2 + (n/L_y)^2]$. For small changes in m and n, the changes in ω_{mnp} are then much smaller than the changes in the horizontal wave numbers $\pi m/L_x$ and $\pi n/L_y$.

We let the cavity undergo linear or circular harmonic oscillation orthogonal to the z-direction, with amplitude r_x (r_y) in the x-direction (y-direction). For motion in x-direction, the mode mixing resonance angular frequency (8) between modes m and m', with $m - m'$ odd, is

$$\omega_r \approx \tfrac{1}{4}\pi\lambda L_x^{-2} \left|m^2 - (m')^2\right| , \tag{10}$$

and the growth rate of the mode mixing Bogoliubov coefficient is

$$\tfrac{d}{d\tau}|\hat{A}_{\text{res}}| \approx \tfrac{1}{2}\pi m m' r_x \lambda L_x^{-3} . \tag{11}$$

The lowest resonance occurs for $m = 1$ and $m' = 2$. Similar formulas ensue for the y-resonance, and for circular motion both resonances are present.

The experimental scenario starts by first trapping one or more quanta in the cavity, in modes whose momenta are aligned close to the z-direction. The cavity is then made to oscillate perpendicularly to the z-direction, linearly or circularly. Finally, a measurement on the quantum state of the cavity is performed. The resonance mode mixing that takes place during the acceleration is assumed to dominate any effects due to the initial trapping and the final releasing of the quanta.

4.2. Desktop numbers

For concrete numbers, we take $\lambda = 600\,\mathrm{nm}$ and $L_x = L_y = 1\,\mathrm{cm}$ (the value of L_z does not need to be specified provided $L_z \gg \lambda$). The lowest resonance angular frequency is $\omega_r \approx 4.2 \times 10^6\,\mathrm{s}^{-1}$, corresponding to an oscillation frequency 0.7 MHz. With amplitude $1\,\mu\mathrm{m}$, (11) and its y-counterpart show that the mode mixing Bogoliubov coefficient grows to order unity within a millisecond. Following the growth further would require theory beyond our perturbative treatment.

For linear motion, oscillation of micron amplitude at megahertz frequency may be achievable by using ultrasound to accelerate the cavity. Storing the quantum in the cavity for a millisecond could be challenging although recent advances indicate that it may be feasible [39].

For circular motion, the threshold angular velocity $\omega_r \approx 4.2 \times 10^6\,\mathrm{s}^{-1} \approx 4 \times 10^7$ rpm exceeds the angular velocity 1.5×10^5 rpm achieved by medical ultracentrifuges [40], although only by about two orders of magnitude, and it is conceivable that this gap could be bridged by a specifically designed system.

5. Conclusions and outlook

We have quantised a scalar field in a rectangular cavity that is accelerated arbitrarily in $(3+1)$-dimensional Minkowski spacetime, in the limit of small accelerations but arbitrary velocities and travel times. The Bogoliubov coefficients between inertial motion at early and late times were expressed in terms of the Fourier transform of the acceleration. For linear or circular periodic motions, we identified a configuration in which the mode mixing resonance frequency is significantly below the frequencies of the cavity modes.

It can be verified that our scalar field analysis adapts in a straightforward way to a Maxwell field with perfect conductor boundary conditions: the two polarisation modes reduce to $(1+1)$ scalar fields with respectively Dirichlet and Neumann boundary conditions, and the Neumann condition Bogoliubov coefficients obey estimates that are qualitatively similar to those given above [41]. The mode mixing effects appear hence to be within the reach of a desktop experiment with photons, achievable with current technology in its mechanical aspects, if perhaps not yet in the storage capabilities required of a mechanically oscillating optical cavity.

Assuming the cavity to be rectangular allowed us to treat the cavity as spatially rigid in the small acceleration limit even when the velocities may be relativistic and the direction of the acceleration may vary in time. We anticipate that the particle creation and mode mixing effects are not qualitatively sensitive to the detailed shape of the cavity, and this freedom could be utilised in the development of a concrete laboratory implementation.

The experimental prospects could be further improved by filling the cavity with a medium that slows light down [42]. As the medium breaks Lorentz-invariance, our analysis would not be directly applicable at relativistic velocities, but at nonrelativistic velocities it may suffice to just include appropriate slowing-down numerical factors in the angular frequencies.

Our experimental scenario involves significant mode mixing but no significant particle creation. One might therefore be tempted to dismiss the scenario as a glorified speed camera effect and question the interest of a laboratory observation. However, mode mixing without particle creation is known in quantum optics as a passive transformation [43], normally implemented by passive optical elements such as beam splitters and phase plates, and these transformations have a well understood capability to create and degrade entanglement [14, 16, 44, 45]. For example, the mode mixing generates entanglement from an initial Gaussian state only if this state is squeezed [16, 44, 45]. The oscillating cavity can hence be tuned to act as a beam splitter quantum gate, creating or degrading entanglement in situations where particles are initially present. This means that the mode mixing could be verified experimentally by observations of entanglement, and the entangling power of the system could be potentially harnessed to quantum information tasks. We anticipate that observations of entanglement will generally provide opportunities for experimental verification of both particle creation and mode mixing effects that are complementary to observations of fluxes or particle numbers.

Acknowledgments

We thank Ivette Fuentes for the collaboration [23] on which this contribution is based. J. L. thanks the organisers of DICE2012 for the invitation to present this work. J. L. acknowledges financial support from STFC [Theory Consolidated Grant ST/J000388/1]. D. F. acknowledges financial support from EPSRC project EP/J00443X/1.

References

[1] Unruh W G 1976 *Phys. Rev.* D **14** 870
[2] Crispino L C B, Higuchi A and Matsas G E A 2008 *Rev. Mod. Phys.* **80** 787 (*Preprint* 0710.5373 [gr-qc])
[3] Dodonov V V 2001 *Adv. Chem. Phys.* **119** 309 (*Preprint* quant-ph/0106081)
[4] Dodonov V V 2010 *Phys. Scripta* **82** 038105
[5] Hawking S W 1975 *Commun. Math. Phys.* **43** 199 [Erratum 1976 *Commun. Math. Phys.* **46** 206 (1976)]
[6] Alsing P M and Milburn G J 2003 *Phys. Rev. Lett.* **91** 180404 (*Preprint* quant-ph/0302179)
[7] Fuentes-Schuller I and Mann R B 2005 *Phys. Rev. Lett.* **95** 120404 (*Preprint* quant-ph/0410172)
[8] Bruschi D E, Louko J, Martín-Martínez E, Dragan A and Fuentes I 2010 *Phys. Rev.* A **82** 042332 (*Preprint* 1007.4670 [quant-ph])
[9] Downes T G, Fuentes I and Ralph T C 2011 *Phys. Rev. Lett.* **106** 210502 (*Preprint* 1007.4035 [quant-ph])
[10] Bruschi D E, Dragan A, Fuentes I and Louko J 2012 *Phys. Rev.* D **86** 025026 (*Preprint* 1205.5296 [quant-ph])
[11] Bruschi D E, Fuentes I and Louko J 2012 *Phys. Rev.* D **85** 061701 (*Preprint* 1105.1875 [quant-ph])
[12] Friis N, Lee A R, Bruschi D E and Louko J 2012 *Phys. Rev.* D **85** 025012 (*Preprint* 1110.6756 [quant-ph])
[13] Downes T G, Ralph T C and Walk N 2012 *Preprint* 1203.2716 [quant-ph]
[14] Friis N, Bruschi D E, Louko J and Fuentes I 2012 *Phys. Rev.* D **85** 081701 (*Preprint* 1201.0549 [quant-ph])
[15] Bruschi D E, Dragan A, Lee A R, Fuentes I and Louko J 2012 *Preprint* 1201.0663 [quant-ph]
[16] Friis N and Fuentes I 2012 *J. Mod. Opt.* online (*Preprint* 1204.0617 [quant-ph])
[17] Friis N, Huber M, Fuentes I and Bruschi D E 2012 *Phys. Rev.* D **86** 105003 (*Preprint* 1207.1827 [quant-ph])
[18] Rideout D *et al* 2012 *Class. Quantum Grav.* **29** 224011 (*Preprint* 1206.4949 [quant-ph])
[19] Schützhold R and Unruh W G 2005 *Preprint* quant-ph/0506028
[20] Plenio M B and Virmani S 2007 *Quant. Inf. Comput.* **7** 1 (*Preprint* quant-ph/0504163)
[21] Sorkin R D 1993 *Directions in General Relativity: Proceedings of the 1993 International Symposium, College Park,Maryland, 1993* vol 2 ed B L Hu and T A Jacobson (Cambridge, England: Cambridge University Press) p 293 (*Preprint* gr-qc/9302018)
[22] Benincasa D M T, Borsten L, Buck M and Dowker F 2012 *Preprint* 1206.5205 [quant-ph]
[23] Bruschi D E, Louko J, Faccio D and Fuentes I 2012 *Preprint* 1210.6772 [quant-ph]
[24] Unruh W G 2012 private communication
[25] Davies P C W and Fulling S A 1976 *Proc. Roy. Soc. Lond.* A **348** 393
[26] Wilson C M, Johansson G, Pourkabirian A, Simoen M, Johansson J R, Duty T, Nori F and Delsing P 2011 *Nature* **479** 376 (*Preprint* 1105.4714 [quant-ph])
[27] Friis N, Lee A R, Truong K, Sabín C, Solano E, Johansson G and Fuentes I 2012 *Preprint* 1211.5563 [quant-ph]
[28] Lambrecht A, Jaekel M-T and Reynaud S 1996 *Phys. Rev. Lett.* **77** 615
[29] Mundarain D F and Maia Neto P A 1998 *Phys. Rev.* A **57** 1379 (*Preprint* quant-ph/9808064)
[30] Crocce M, Dalvit D A R and Mazzitelli F D 2001 *Phys. Rev.* A **64** 013808
[31] Crocce M, Dalvit D A R and Mazzitelli F D 2002 *Phys. Rev.* A **66** 033811 (*Preprint* quant-ph/0205104)
[32] Ruser M 2006 *Phys. Rev.* A **73** 043811 (*Preprint* quant-ph/0509030)

[33] Yuce C and Ozcakmakli Z 2008 *J. Phys.* A **41** 265401 (*Preprint* 0802.3765 [hep-th])
[34] Dodonov A V and Dodonov V V 2001 *Phys. Lett.* A **289** 291 (*Preprint* quant-ph/0109019)
[35] Yuce C and Ozcakmakli Z 2009 *J. Phys.* A **42** 035403 (*Preprint* 0811.0665)
[36] Schützhold R, Plunien G and Soff G 1998 *Phys. Rev.* A **57** 2311 (*Preprint* quant-ph/9709008)
[37] Dodonov V V and Andreata M A 1999 *J. Phys.* A **32** 6711 (*Preprint* quant-ph/9908038)
[38] Dodonov A V, Dodonov V V and Mizrahi S S 2005 *J. Phys.* A **38** 683
[39] Kuhr S *et al* 2007 *Appl. Phys. Lett.* **90** 164101 (*Preprint* quant-ph/0612138)
[40] Website `http://www.thermoscientific.com`, visited 29 September 2012
[41] Louko J 2013 (in preparation)
[42] Lauprêtre T, Proux C, Ghosh R, Schwartz S, Goldfarb F and Bretenaker F 2011 *Opt. Lett.* **36** 1551 (*Preprint* 1104.0158 [physics.optics])
[43] Simon R, Mukunda N and Dutta B 1994 *Phys. Rev.* A **49** 1567
[44] Wolf M M, Eisert J and Plenio M B 2003 *Phys. Rev. Lett.* **90** 047904 (*Preprint* quant-ph/0206171)
[45] Alsing P M and Fuentes I 2012 *Class. Quantum Grav.* **29** 224001 (*Preprint* 1210.2223 [quant-ph])

The semiclassical approximation to quantum gravity and its observational consequences

Claus Kiefer
Institut für Theoretische Physik, Universität zu Köln, Zülpicher Straße 77, 50937 Köln, Germany
Copernicus Center for Interdisciplinary Studies, ul. Sławkowska 17, 31-016 Kraków, Poland
E-mail: kiefer@thp.uni-koeln.de

Abstract. We study possible observational effects from the Wheeler–DeWitt equation of quantum geometrodynamics. For this purpose, we perform a semiclassical expansion and derive quantum-gravitational correction terms that are inversely proportional to the Planck mass squared. We apply these results to cosmology and calculate the resulting modification of the CMB power spectrum. Although the correction terms are too small to be currently observable, they could provide the key for future tests of quantum gravity.

1. Introduction

The construction of a consistent theory of quantum gravity is among the major open problems of fundamental physics. In addition to conceptual and mathematical difficulties, the main obstacle is the lack of empirical tests so far. It is thus not surprising that many different approaches to quantum gravity exist, which all possess their strengths and weaknesses [1]. A decision between these approaches can eventually be made only on the basis of experiments or observations. For this reason, it is important to calculate as many concrete predictions as possible, even if they are currently too small to be tested.

One example for a concrete prediction is the quantum-gravitational correction to the Newtonian potential between two masses m_1 and m_2. This was calculated at the one-loop level of quantum gravity as defined by the path integral [2]. The result for the potential is

$$V(r) = -\frac{Gm_1m_2}{r}\left(1 + 3\frac{G(m_1+m_2)}{rc^2} + \frac{41}{10\pi}\frac{G\hbar}{r^2c^3}\right).$$

The first correction term is a well-known correction from classical general relativity, while the second term is proportional to $\hbar$ and is thus a genuine quantum gravity correction term. One can, of course, not measure this correction in the laboratory, because this would require that the two masses approach each other down to the Planck length. The main merit of this result (and others) is that a concrete prediction can be made.

Predictions can also be made from other approaches to quantum gravity. Here, I will discuss the case of canonical quantum gravity in geometrodynamical variables, on which I have worked myself. The central equation of this approach is the Wheeler–DeWitt equation, which is of the form [1]

$$\hat{H}\Psi = 0, \tag{1}$$

where $\hat{H}$ denotes the full Hamiltonian of gravity and non-gravitational fields, and Ψ depends on the three-dimensional metric and matter variables.

In the following, I shall first briefly review the semiclassical approximation, on which the derivation of the quantum-gravitational effects is based. I then discuss the quantum-gravitational correction terms that lead to the modification of the power spectrum for the anisotropies of the cosmic microwave background (CMB). So far, the observations only lead to upper limits for these terms and I quote some of these limits using the most recent data from the PLANCK satellite mission. These results are based on the research articles [3] and [4], to which I refer the reader for more details.

2. Semiclassical approximation

In the semiclassical approximation to the Wheeler–DeWitt equation, one starts with the ansatz

$$|\Psi[h_{ab}]\rangle = C[h_{ab}]\mathrm{e}^{\mathrm{i}m_{\mathrm{P}}^2 S[h_{ab}]}|\psi[h_{ab}]\rangle \tag{2}$$

and performs an expansion with respect to the inverse Planck mass squared m_{P}^{-2}. This is inserted into (1), and consecutive orders in this expansion are considered [5, 6]. This is close to the Born–Oppenheimer (BO) approximation scheme in molecular physics. One may also employ the direct BO scheme, which leads to analogous, but not equivalent results [7]. We shall here restrict ourselves to the m_{P}^2-expansion scheme. In (2), h_{ab} denotes the three-metric, and the Dirac bra-ket notation refers to the non-gravitational fields, for which the usual Hilbert-space structure is assumed [6].

The highest orders of the BO scheme lead to the following picture. One evaluates $|\psi[h_{ab}]\rangle$ along a solution of the classical Einstein equations, $h_{ab}(\mathbf{x}, t)$, which corresponds to a solution $S[h_{ab}]$ of the Hamilton–Jacobi equations. One can then define

$$\dot{h}_{ab} = NG_{abcd}\frac{\delta S}{\delta h_{cd}} + 2D_{(a}N_{b)},$$

where N is the lapse function and N^a is the shift vector. In this way, one can recover a classical spacetime as an approximation, a spacetime that satisfies Einstein's equations in this limit. The time derivative of the matter wave function is then defined by

$$\frac{\partial}{\partial t}|\psi(t)\rangle := \int \mathrm{d}^3x\, \dot{h}_{ab}(\mathbf{x}, t)\, \frac{\delta}{\delta h_{ab}(\mathbf{x})}|\psi[h_{ab}]\rangle.$$

This leads to a functional Schrödinger equation for quantized matter fields in the chosen external classical gravitational field,

$$\mathrm{i}\hbar\frac{\partial}{\partial t}|\psi(t)\rangle = \hat{H}^{\mathrm{m}}|\psi(t)\rangle, \tag{3}$$

$$\hat{H}^{\mathrm{m}} := \int \mathrm{d}^3x \left\{N(\mathbf{x})\hat{\mathcal{H}}_{\perp}^{\mathrm{m}}(\mathbf{x}) + N^a(\mathbf{x})\hat{\mathcal{H}}_a^{\mathrm{m}}(\mathbf{x})\right\}, \tag{4}$$

where $\hat{H}^{\mathrm{m}}$ denotes the matter-field Hamiltonian in the Schrödinger picture, which depends parametrically on the (generally non-static) metric coefficients of the curved space–time background recovered from $S[h_{ab}]$. The 'WKB time' t controls the dynamics in this approximation.

3. Quantum-gravitational corrections

The next order in the m_{P}^{-2}-expansion yields quantum-gravitational correction terms [5, 6]. In this way, one finds a 'corrected' Schrödinger equation (3) in which the matter Hamiltonian is replaced according to

$$\hat{H}^{\mathrm{m}} \to \hat{H}^{\mathrm{m}} + \frac{1}{m_{\mathrm{P}}^2} \times (\text{various terms})\,.$$

The general correction terms can be found in explicit form in [6]. We shall specify them below for the case of our cosmological model. As an example, I mention the quantum-gravitational corrections to the trace anomaly in de Sitter space as calculated in [8],

$$\delta\epsilon \approx -\frac{2G\hbar^2 H_{\rm dS}^6}{3(1440)^2\pi^3 c^8},$$

where $H_{\rm dS}$ denotes the constant Hubble parameter of de Sitter space. Again, one recognizes that concrete effects can be calculated, even if they are too tiny to be currently observable.

We shall now review the calculations that lead to the quantum-gravitational corrections for the CMB anisotropy power spectrum [3, 4]. We consider the Wheeler–DeWitt equation for small fluctuations in a flat Friedmann–Lemaître universe with scale factor $a \equiv \exp(\alpha)$ and inflaton field ϕ. In fact, we assume the presence of an inflationary regime in the semiclassical limit.

For the inflaton, we choose the simplest potential given by

$$\mathcal{V}(\phi) = \frac{1}{2}m^2\phi^2\ ;$$

however, any other potential obeying at the classical level the slow-roll condition $\dot{\phi}^2 \ll |\mathcal{V}(\phi)|$ should lead to similar results.

The Wheeler–DeWitt equation for the Friedmann–Lemaître ('minisuperspace') background reads

$$\mathcal{H}_0\Psi_0(\alpha,\phi) \equiv \frac{\mathrm{e}^{-3\alpha}}{2}\left[\frac{1}{m_{\rm P}^2}\frac{\partial^2}{\partial\alpha^2} - \frac{\partial^2}{\partial\phi^2} + \mathrm{e}^{6\alpha}m^2\phi^2\right]\Psi_0(\alpha,\phi) = 0. \tag{5}$$

In the following, we choose units with $\hbar = c = 1$ and a modified Planck mass defined by $m_{\rm P} := \sqrt{3\pi/2G} \approx 2.65\times 10^{19}$ GeV. The scalar field is redefined for convenience by $\phi \to \phi/\sqrt{2\pi}$.

To implement the slow-roll condition at the quantum level, we assume in the following that $\partial^2\Psi_0/\partial\phi^2 \ll \mathrm{e}^{6\alpha}m^2\phi^2\Psi_0$, and we substitute $m\phi$ by $m_{\rm P}H$, where H is the quasistatic Hubble parameter of inflation (this holds in the Born–Oppenheimer approximation). Many investigations in quantum cosmology make use of this quantum slow-roll condition [1].

We now introduce inhomogeneities for the scalar field according to

$$\phi \to \phi(t) + \delta\phi(\mathbf{x},t)$$

and perform a decomposition into Fourier modes with wave vector $\mathbf{k}$, $k \equiv |\mathbf{k}|$,

$$\delta\phi(\mathbf{x},t) = \sum_k f_k(t)\,\mathrm{e}^{\mathrm{i}\mathbf{k}\cdot\mathbf{x}}\,.$$

A full treatment of the problem should make use of the gauge-invariant variables of cosmological perturbation theory, but this is not required for our purpose.

The Wheeler–DeWitt equation including the fluctuation modes then reads [9]

$$\left[\mathcal{H}_0 + \sum_{k=1}^{\infty}\mathcal{H}_k\right]\Psi(\alpha,\phi,\{f_k\}_{k=1}^{\infty}) = 0, \tag{6}$$

with

$$\mathcal{H}_k = \frac{1}{2}\mathrm{e}^{-3\alpha}\left[-\frac{\partial^2}{\partial f_k^2} + \left(k^2\,\mathrm{e}^{4\alpha} + m^2\,\mathrm{e}^{6\alpha}\right)f_k^2\right]. \tag{7}$$

Since the fluctuations are small, they do not interact among each other, but only with the background. One can thus make the ansatz

$$\Psi(\alpha, \phi, \{f_k\}_{k=1}^{\infty}) = \Psi_0(\alpha, \phi) \prod_{k=1}^{\infty} \widetilde{\Psi}_k(\alpha, \phi, f_k).$$

The components $\Psi_k(\alpha, \phi, f_k) := \Psi_0(\alpha, \phi)\widetilde{\Psi}_k(\alpha, \phi, f_k)$ then obey the equation

$$\frac{1}{2}\,\mathrm{e}^{-3\alpha}\left[\frac{1}{m_{\mathrm{P}}^2}\frac{\partial^2}{\partial\alpha^2} + \mathrm{e}^{6\alpha}\, m_{\mathrm{P}}^2\, H^2 - \frac{\partial^2}{\partial f_k^2} + W_k(\alpha) f_k^2\right]\Psi_k(\alpha, \phi, f_k) = 0 \tag{8}$$

with

$$W_k(\alpha) := k^2\,\mathrm{e}^{4\alpha} + m^2\,\mathrm{e}^{6\alpha}.$$

Following the general scheme outlined above, we make the ansatz

$$\Psi_k(\alpha, f_k) = \mathrm{e}^{\mathrm{i}\,S(\alpha, f_k)}$$

and expand $S(\alpha, f_k)$ in terms of powers of m_{P}^2,

$$S(\alpha, f_k) = m_{\mathrm{P}}^2\, S_0 + m_{\mathrm{P}}^0\, S_1 + m_{\mathrm{P}}^{-2}\, S_2 + \ldots .$$

We insert this ansatz into the full Wheeler–DeWitt equation (6) and compare consecutive orders of m_{P}^2. The highest order is $\mathcal{O}(m_{\mathrm{P}}^4)$ and leads to S_0 being independent of the f_k. The order $\mathcal{O}(m_{\mathrm{P}}^2)$ then demands that S_0 must obey the Hamilton–Jacobi equation

$$\left[\frac{\partial S_0}{\partial\alpha}\right]^2 - V(\alpha) = 0\,, \quad V(\alpha) := \mathrm{e}^{6\alpha} H^2,$$

which is easily solved by $S_0(\alpha) = \pm\mathrm{e}^{3\alpha} H/3$. At order $\mathcal{O}(m_{\mathrm{P}}^0)$, we write $\psi_k^{(0)}(\alpha, f_k) \equiv \gamma(\alpha)\,\mathrm{e}^{\mathrm{i}\,S_1(\alpha, f_k)}$ and impose a condition on $\gamma(\alpha)$ that makes it equal to the standard WKB prefactor. After introducing the 'WKB time' according to

$$\frac{\partial}{\partial t} := -\,\mathrm{e}^{-3\alpha}\frac{\partial S_0}{\partial\alpha}\frac{\partial}{\partial\alpha}\,, \tag{9}$$

one finds that each $\psi_k^{(0)}$ obeys a Schrödinger equation,

$$\mathrm{i}\,\frac{\partial}{\partial t}\,\psi_k^{(0)} = \mathcal{H}_k \psi_k^{(0)}\,. \tag{10}$$

This corresponds to the general case (3) above as specialized to our model.

The next order $\mathcal{O}(m_{\mathrm{P}}^{-2})$ then leads to the quantum-gravitational correction terms. We decompose for this purpose $S_2(\alpha, f_k)$ as

$$S_2(\alpha, f_k) \equiv \varsigma(\alpha) + \eta(\alpha, f_k)$$

and demand that $\varsigma(\alpha)$ be the standard second-order WKB correction. The wave functions

$$\psi_k^{(1)}(\alpha, f_k) := \psi_k^{(0)}(\alpha, f_k)\,\mathrm{e}^{\mathrm{i}\,m_{\mathrm{P}}^{-2}\,\eta(\alpha, f_k)}$$

then obey the quantum-gravitationally corrected Schrödinger equation

$$\begin{aligned} \mathrm{i}\,\frac{\partial}{\partial t}\,\psi_k^{(1)} &= \mathcal{H}_k \psi_k^{(1)} - \\ &\frac{\mathrm{e}^{3\alpha}}{2 m_{\mathrm{P}}^2 \psi_k^{(0)}}\left[\frac{(\mathcal{H}_k)^2}{V}\psi_k^{(0)} + \mathrm{i}\,\frac{\partial}{\partial t}\left(\frac{\mathcal{H}_k}{V}\right)\psi_k^{(0)}\right]\psi_k^{(1)}. \end{aligned} \tag{11}$$

The second correction term leads to a possible unitarity violation, which either may be neglected because it is small or may be absorbed into a redefinition of the wave function [4]. We can now calculate from (3) the modifications to the CMB power spectrum.

4. Modification of the CMB power spectrum

We look for a solution of the uncorrected Schrödinger equation of the Gaussian form

$$\psi_k^{(0)}(t, f_k) = \mathcal{N}_k^{(0)}(t)\, \mathrm{e}^{-\frac{1}{2}\Omega_k^{(0)}(t)\, f_k^2}, \tag{12}$$

which corresponds to the assumption that the fluctuations f_k are initially in their ground state. From this, one finds the usual power spectrum which is approximately given by

$$\mathcal{P}^{(0)}(k) := \frac{k^3}{2\pi^2}\left|\delta_k(t_{\mathrm{enter}})\right|^2 \propto \left.\frac{H^4}{|\dot{\phi}(t)|^2}\right|_{t_{\mathrm{exit}}}, \tag{13}$$

where 'exit' and 'enter' refer to the times when a particular modes leaves and re-enters the Hubble scale, respectively. One thus recovers also in this framework the (approximate) scale-invariant power spectrum following from inflation.

For the solution of the corrected Schrödinger equation (11), we make again a Gaussian ansatz and consider corrections in the exponent and in the normalization. The technical details are somewhat lengthy and can be found in [3, 4]. As a boundary condition, we assume that the corrections to (12) vanish for late times, corresponding to the idea that quantum-gravitational effects become negligible for the present universe.

When solving the corrected Schrödinger equation with this boundary condition, one encounters a subtlety [4]. There exist two solutions with the same asymptotics, one approaching the value zero for the corrections continuously, the other making a jump on the imaginary axis when approaching zero. The formal origin of this difference is the choice in the definition of the exponential integral that occurs in the solution of this equation. We shall give below the consequences for the power spectrum from both solutions.

The quantum-gravitational corrections to the power spectrum can be described by the ansatz

$$\mathcal{P}^{(1)}(k) = \mathcal{P}^{(0)}(k)\, C_k^2. \tag{14}$$

Following [10], we can write

$$C_k^2 = 1 + \delta^{\pm}_{\mathrm{WDW}}(k) + \frac{1}{k^6}\,\mathcal{O}\!\left(\left(\frac{H}{m_{\mathrm{P}}}\right)^4\right), \tag{15}$$

where $\delta^{\pm}_{\mathrm{WDW}}(k)$ either takes the form

$$\delta^{+}_{\mathrm{WDW}}(k) = \frac{179.09}{k^3}\left(\frac{H}{m_{\mathrm{P}}}\right)^2, \tag{16}$$

which follows from the continuous solution, or the form

$$\delta^{-}_{\mathrm{WDW}}(k) = -\frac{247.68}{k^3}\left(\frac{H}{m_{\mathrm{P}}}\right)^2, \tag{17}$$

which follows from the discontinuous solution. One thus recognizes that the continuous solution leads to an *enhancement* of the power at large scales (small k), whereas the discontinuous solution leads to a *suppression* of power at large scales. One also recognizes that the power spectrum is no longer scale-invariant, but exhibits a characteristic k^{-3}-dependence.

Let us now consider the spectral index and its running. These parameters are defined by

$$n_s - 1 := \frac{\mathrm{d}\log\mathcal{P}}{\mathrm{d}\log k} \approx 2\eta - 4\epsilon - 3\delta^{\pm}_{\mathrm{WDW}} \tag{18}$$

and

$$\alpha_s := \frac{\mathrm{d}n_s}{\mathrm{d}\log k} \approx 2(5\epsilon\eta - 4\epsilon^2 - \Xi^2) + 9\delta^{\pm}_{\mathrm{WDW}}, \tag{19}$$

where we have introduced the slow-roll parameters

$$\epsilon = -\frac{\dot{H}}{H^2} = \frac{4\pi G\,|\dot{\phi}(t)|^2_{t_{\mathrm{exit}}}}{H^2} \tag{20}$$

and

$$\eta := -\frac{\ddot{\phi}}{H\dot{\phi}}, \qquad \Xi^2 := \frac{1}{H^2}\frac{\mathrm{d}}{\mathrm{d}t}\frac{\ddot{\phi}}{\dot{\phi}}. \tag{21}$$

So far, we have treated the scale k as dimensionless. For the comparison with observations, we introduce a reference wavenumber, which can either correspond to the largest observable scale, $k_{\mathrm{min}} \sim 1.4 \times 10^{-4}\,\mathrm{Mpc}^{-1}$, or to the pivot scale used in the WMAP9 analysis, $k_0 = 0.002\,\mathrm{Mpc}^{-1}$ [11]. We now write k/k_{min} or k/k_0 instead of k. Since H/m_{P} has to be smaller than about 3.5×10^{-6} because of the observational bound on the tensor-to-scalar ratio $r < 0.11$ for $k_0 = 0.002\,\mathrm{Mpc}^{-1}$ from the Planck 2013 results [12], we find that for $k \to k/k_0$ the absolute value of the quantum-gravitational corrections is bounded by [4]

$$\left|\delta^{+}_{\mathrm{WDW}}(k_0)\right| \lesssim 2.2 \times 10^{-9}, \quad \left|\delta^{-}_{\mathrm{WDW}}(k_0)\right| \lesssim 3.0 \times 10^{-9}, \tag{22}$$

while with $k \to k/k_{\mathrm{min}}$ this limit becomes even more stringent,

$$\left|\delta^{+}_{\mathrm{WDW}}(k_0)\right| \lesssim 7.5 \times 10^{-13}, \quad \left|\delta^{-}_{\mathrm{WDW}}(k_0)\right| \lesssim 1.0 \times 10^{-12}. \tag{23}$$

These values are too tiny to be currently observable.

5. Outlook

In my contribution, I have tried to emphasize that one must derive concrete predictions from approaches to quantum gravity and compare them with observations. I have limited myself here to one approach: quantum geometrodynamics with the Wheeler–DeWitt equation as its central equation. From this, one can derive quantum-gravitational corrections to the CMB anisotropy power spectrum. Assuming a scenario with an inflationary regime in the early universe, these corrections depend on the ratio $(H/M_{\mathrm{P}})^2$, where H is the quasi-static Hubble parameter of inflation.

Given the known limits on this ratio, the corrections turn out to be too small to be observable. The situation will also not change with the release of further data from the PLANCK mission, since the main source of uncertainty for large scales comes from cosmic variance, which is a fundamental limit. One might, of course, speculate that the quantum-gravitational correction terms can be seen in other situations, such as the galaxy-galaxy correlation functions.

If one did not have the limit on $(H/M_{\mathrm{P}})^2$, the situation would be different. From the non-observation of the correction terms, one would then get a constraint that H/M_{P} must be smaller than a value of the order of 10^{17} GeV. It is evident that the quantum-gravitational effects would be stronger if the inflationary scale were closer to the Planck scale. If one attributed the power suppression at large scales, as reported in [12], to the quantum-gravitational corrections, one would predict for H/M_{P} a value of this order.

Other approaches to quantum gravity have also been applied to this situation. Investigations in loop quantum cosmology, for example, lead to a prediction that the power at large scales is enhanced [10, 13]. The observation of small effects may thus eventually provide the key for the correct quantum theory of gravity [14].

Acknowledgments
I thank the organizers of the DICE12 meeting, and in particular Thomas Elze, for inviting me to this most interesting and inspiring meeting. I am grateful to Manuel Krämer, Donato Bini, Giampiero Esposito, and Francesco Pessina for our collaboration on this subject. I also thank the Copernicus Center for Interdisciplinary Studies, Cracow, and in particular Michael Heller, for kind hospitality while part of this contribution was written.

[1] Kiefer C 2012 *Quantum Gravity* third edition (Oxford: Oxford University Press)
[2] Bjerrum-Bohr N E J, Donoghue J F and Holstein BR 2003 *Phys. Rev.* D **67** 084033
[3] Kiefer C and Krämer M 2012 *Phys. Rev. Lett.* **108** 021301
[4] Bini D, Esposito G, Kiefer C, Krämer M and Pessina F 2013 *Preprint* arXiv:1303.0531 [gr-qc]
[5] Kiefer C and Singh T P 1991 *Phys. Rev.* D **44** 1067
[6] Barvinsky A O and Kiefer C 1998 *Nucl. Phys.* B **526** 509
[7] Bertoni C, Finelli F and Venturi G 1996 *Class. Quantum Grav.* **13** 2375
[8] Kiefer C 1996 *New frontiers in gravitation* ed G A Sardanashvily (Palm Harbor: Hadronic Press) pp 203–214
[9] Halliwell J J and Hawking S W 1985 *Phys. Rev.* D **31** 1777
[10] Calcagni G 2012 *Preprint* arXiv:1209.0473 [gr-qc]
[11] Hinshaw G *et al.* 2012 *Preprint* arXiv:1212.5226v2 [astro-ph.CO]
[12] Planck Collaboration 2013 *Preprint* arXiv:1303.5076v1 [astro-ph.CO]
[13] Bojowald M, Calcagni G and Tsujikawa S 2011 *Phys. Rev. Lett.* **107** 211302
[14] Kiefer C and Krämer M 2012 *Int. J. Mod. Phys.* D **21** 1241001

Detecting the relative localisation of quantum particles

P A Knott, J Sindt and J A Dunningham

School of Physics and Astronomy, University of Leeds, Leeds LS2 9JT, United Kingdom

E-mail: j.a.dunningham@leeds.ac.uk

Abstract. One interpretation of how the classical world emerges from quantum physics involves the build-up of certain robust entangled states between particles due to scattering events [1]. This is intriguing because it links classical behaviour with the uniquely quantum effect of entanglement and differs from other interpretations that say classicality arises when quantum correlations are lost or neglected in measurements. However, the problem with this new interpretation has been finding an experimental way of verifying it. Here we outline a straightforward scheme that enables just that and should, in principle, allow experiments to confirm the theory to any desired degree of accuracy.

1. Introduction

A lot of work has been done to understand why apparently different physical theories should apply to microscopic and macroscopic systems and how one theory crosses over to the other in the mesoscopic regime. The prevailing view is that the emergence of classical physics from a more fundamental quantum reality can be interpreted in terms of decoherence [2, 3, 4, 5]. Simply put, this says that quantum systems tend to interact with their environments and become entangled with them. The total system including the environment is therefore properly treated with quantum physics. However if we are interested only in the quantum subsystem we neglect the information about which environmental states are correlated with which subsystem states. We then find that the subsystem appears to behave more and more classically the more it has interacted with the environment. In effect, by throwing away information about the quantum correlations we are left with a system that behaves classically.

Another interpretation that extends this idea was put forward a few years ago [1]. It showed the emergence of classicality without having to throw away all the quantum correlations. In fact it showed that even classical objects are entangled with one another but with a special type of robust entanglement sometimes called "fluffy-bunny" entanglement [6, 7]. This is an appealing view since it gives one consistent theory that describes both quantum and classical systems. It is also intriguing that in this theory classicality is associated with entanglement, which is usually thought of as a purely quantum feature. Furthermore, this interpretation gives a clear basis to the idea that we should only think in terms of relative (rather than absolute) positions in physics.

The problem with this formalism up until now is that it has not been at all clear how it could be tested experimentally. In this paper we resolve this issue by providing a simple,

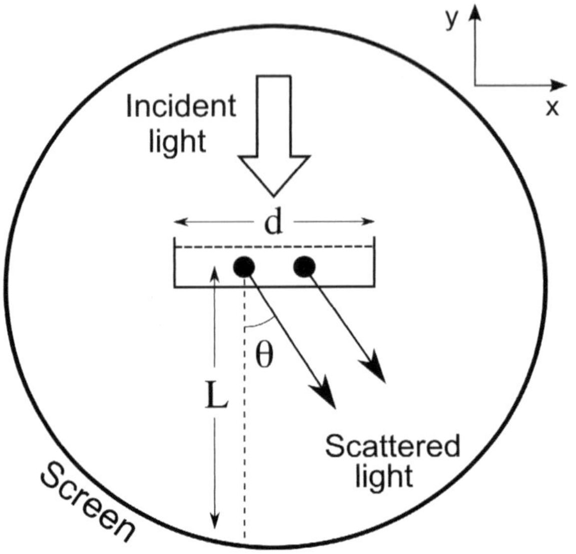

Figure 1. A schematic showing the setup. Two massive particles are delocalised over some region d and are illuminated by plane wave incident light. The scattered light is detected at an angle θ on a screen located at a distance L away from the particles. For clarity, the diagram is not to scale: we consider the case where $L \gg d$.

experimentally accessible scheme that could clearly demonstrate this process. In principle, it should allow experiments to confirm to any level of certainty that this process has occurred.

2. The scheme

We begin by reviewing the scheme [1] for measurement-induced relative-position localisation through entanglement. The setup is shown in Fig. 1. Two distinguishable massive particles are delocalised over some region d in the x-direction in the sense that their de Broglie wavelengths are comparable to d in this dimension. To simplify the following analysis we will take these particles to be tightly confined in the y and z directions, although this assumption is not necessary. These two particles will form the sub-systems of the system we are interested in. They are illuminated with plane-wave light with wavelength λ incident along the y-axis, which scatters from them and is detected at an angle θ on a screen located at a distance L away. We will consider the far-field limit where $L \gg d$.

The initial wave function of the particles in the relative position representation can be written as

$$|\psi\rangle = \int_{-d}^{d} c(x)|x\rangle\, dx, \tag{1}$$

where $|x\rangle$ represents the *relative* position of the two particles in the x-direction, $c(x)$ is the probability amplitude for x and $\int_{-d}^{d} |c(x)|^2 dx = 1$. Since the particles are delocalised over d, the position of one particle relative to the other lies in the range $[-d, d]$. Strictly, to give a full spatial description of the system we should specify the centre-of-mass position as well as the relative position. However the centre-of-mass remains unentangled from the relative position coordinate throughout the scattering process and so can be conveniently neglected. When a photon of wavelength λ scatters off a particle into direction θ, its momentum in the x-direction changes by $\Delta p = h \sin\theta/\lambda$, where h is Planck's constant. The particle from which it scatters therefore gets an equal and opposite kick. In relative momentum space the particles therefore receive a kick of $\pm h \sin\theta/\lambda$ depending on which particle the photon scatters from and, since we do not know, we get a superposition of both possibilities.

This allows us to write the overall state of the system after a photon has scattered as

$$|\psi\rangle = \int_{-d}^{d} dx\, c(x)|x\rangle \otimes \left[\frac{1}{2\sqrt{2\pi}} \int_0^{2\pi} d\theta \left(e^{i2\pi x \sin\theta/\lambda} + e^{-i2\pi x \sin\theta/\lambda}\right)|\theta\rangle + A(x)|0\rangle\right]. \quad (2)$$

The first part of the tensor product is the state of the two particles in relative position space and the second part (in square brackets) is the state of the scattered photon where $|\theta\rangle$ represents a photon scattered at angle θ. The term proportional to $A(x)$, defined as

$$A(x) = \left[\frac{1}{2\pi}\int_0^{2\pi} \sin^2(2\pi x \sin\theta'/\lambda)\, d\theta'\right]^{1/2}, \quad (3)$$

represents a nonscattering event that leaves the photon in the undeflected state $|0\rangle$. This term is necessary because the total rate of scattering (integrated over all angles) depends on the separation of the particles, x. Odd as it seems at first sight, this means that detecting a photon that is not scattered gives us information about the relative position of the particles. The $A(x)$ term is required to properly account for this.

We can now understand the localisation process. The probability density for detecting a scattered photon at angle $\theta \neq 0$ is

$$P_S(\theta) = \langle\psi|a_\theta^\dagger a_\theta|\psi\rangle = \frac{1}{2\pi}\int_{-d}^{d} |c(x)|^2 \cos^2(2\pi x \sin\theta/\lambda)\, dx,$$

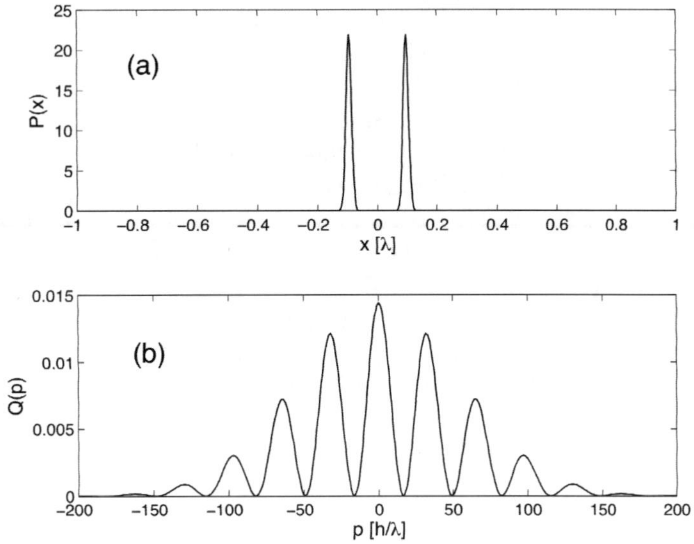

Figure 2. The case of light scattering causing relative localisation. (a) Probability density, $P(x)$, for the relative position of the particles after the scattering and detection of 150 photons. The position is given in units of the wavelength, λ, of the scattered light. (b) Probability density, $Q(p)$ for the corresponding relative momentum of the particles.

where $|\psi\rangle$ is given by Eq. (2) and a_θ is the annihilation operator for the photonic mode $|\theta\rangle$. The probability of detecting a nonscattered photon, i.e. $\theta = 0$, is $P_{NS} = \int_{-d}^{d} |c(x)|^2 A^2\, dx = 1 - \int_0^{2\pi} P_s(\theta)\, d\theta$.

To simulate the localisation process, we draw a random number from this distribution to see whether the photon is scattered and, if so, at what angle. If it is not scattered the (unnormalised) new state is $a_0|\psi\rangle = \int_{-d}^{d} dx\, c(x)A(x)|x\rangle$ and if it is scattered at an angle θ_1, the (unnormalised) new state is $a_{\theta_1}|\psi\rangle = \int_{-d}^{d} dx\, c(x)\cos(2\pi x \sin\theta_1/\lambda)|x\rangle$. We then normalise the state and repeat for the next photon.

3. Results

We choose to start our simulations with a flat distribution, $c(x) = 1/\sqrt{2d}$ because we want to choose the 'hardest' case and show that relative localisation builds up even when there is none to begin with. This choice does not restrict the generality of the results and qualitatively similar outcomes are obtained for different choices. The probability distribution, $P(x)$, for the relative position of the two particles is shown in Fig. 2(a) for a typical run after 150 photons have been detected and for $d = \lambda$. We assume that the 150 photons are all incident on the particles in a sufficiently short time period that we do not need to consider the dynamics of the particles between detection events. Initially the distribution is completely flat and we see that the measurement process has induced localisation. The distribution shown in Fig. 2(a) after 150 photons have been detected has two peaks that are symmetric about the origin. This is what we would expect since the detection process cannot determine which particle is on the left and which is on the right and so the relative position must have a positive and negative component. If the two particles had a well-defined relative position to begin with, then there would be only one peak in this distribution[1] as shown, for example, in Fig. 3(a).

The relative-position localisation process is analogous to the build-up of relative phase between two number state Bose-Einstein condensates when interference patterns are detected between them [8, 9, 10]. Just like in that case we cannot distinguish from the detected particles whether the position (or phase of the BECs) was well-defined to begin with or created by the measurements. We need a way of doing this in order to experimentally verify that the localisation process takes place. Although the distinction between Fig. 2(a) and Fig. 3(a) is clear, an experimentalist would not have direct access to this since the detected photons cannot tell these distributions apart. One possible solution is to look in the conjugate space – in this case relative momentum. A similar idea has been applied to BECs [11, 12, 13].

The relative momentum distributions corresponding to the relative position distributions in Figs. 2(a) and 3(a) are shown as the solid lines in Figs. 2(b) and 3(b) respectively. For ease of comparison, the result from Fig. 2(b) is superimposed on Fig. 3(b) as a dashed line. We see that the two distributions have the same envelope, but the case where localisation is induced has interference fringes. For particles that are *a priori* perfectly localised, the distribution in Fig. 3(a) would be a delta function and the momentum distribution would be completely flat. We have chosen the relative position distribution shown because it is an upper limit to the width possible based on the photons detected. In other words, it is the 'hardest' case to distinguish from that shown in Fig. 2(a). We want to demonstrate that our technique works even in this worst-case scenario.

The measurement scheme proceeds as follows. After scattering the photons from the particles,

[1] Strictly, in the case where the particles are each initially localized, we have a mixture of the two different relative positions. This mixture reflects the classical uncertainty in our knowledge of the relative position of the particles based on detecting the scattered photons. This does not change the shape of Q_2. The case where the particles are initially delocalized is quite different and gives a coherent superposition of the two relative positions, which leads to the distribution Q_1.

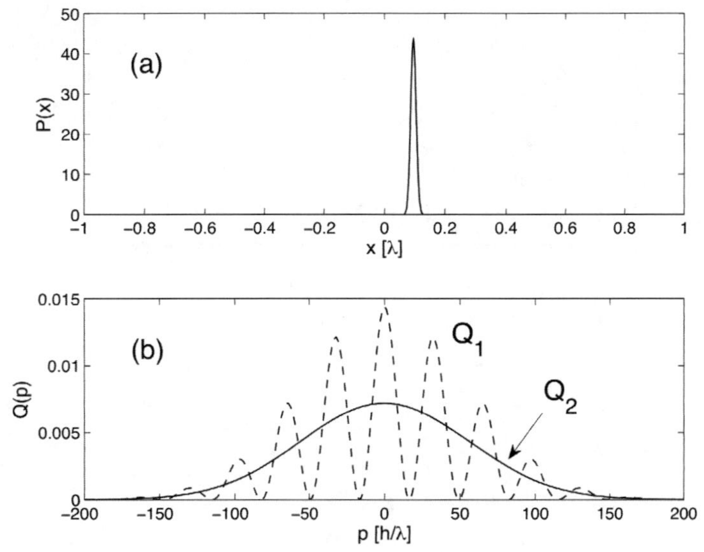

Figure 3. The same as in Figure 2 but with the particles initially localised before the photons are scattered. (a) There is now only one peak in the relative position or, strictly, an equally weighted mixture of the two peaks, both of which give the same relative momentum distribution. (b) The corresponding relative momentum probability density is shown as a solid line (labelled Q_2) and is compared to the result in Figure 2 shown as a dashed line (labelled Q_1).

we want to distinguish the two relative momentum distributions shown in Fig. 3(b). To do this, we switch off any trapping potential and allow the particles to move freely. By detecting their positions in the x-direction after some time of flight, we can infer the x-components of their momenta and hence the relative momentum of the particles in that direction. By repeating the whole process from the beginning many times, we should be able to build up a probability distribution and so distinguish the two cases. However the stochastic nature of the process means that the particles localise to a different relative position on each run and so the relative momentum fringes are different each time. If we were to just naïvely add the results from each run, the fringes would wash out. Instead we can use Bayesian analysis to precisely distinguish the two scenarios.

Suppose on a particular run we detected scattered photons on the screen that meant the relative momentum distribution was either $Q_1(p)$ or $Q_2(p)$ depending on whether or not the scattering process induced relative localisation (see Fig. 3(b)). To begin with, we do not know whether the particles are localised or not so we take our prior probability of them initially being localised, P_l to be the same as the prior probability of them not initially being localised, P_{nl}, i.e. $P_l = P_{nl} = 0.5$. Now suppose, upon releasing the particles, we measure their relative momentum to be p_1. This gives us some information about which scenario is more likely. In particular, Bayes theorem tells us that the updated probabilities are $P_{nl} \propto Q_1(p_1) \times 0.5$ and $P_l \propto Q_2(p_1) \times 0.5$. Normalising, we get

$$P_{nl} = \frac{Q_1(p_1)}{Q_1(p_1) + Q_2(p_1)} \times 0.5 \tag{4}$$

and $P_l = 1 - P_{nl}$. We can then iterate this process by using these updated probabilities as the prior probabilities in the next step. By repeating many times we increasingly refine our

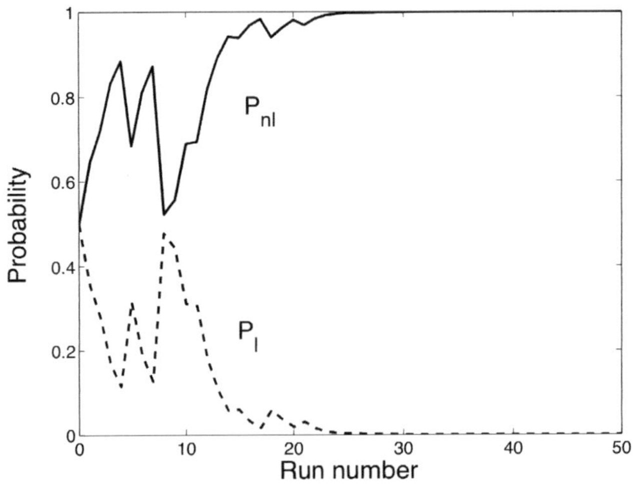

Figure 4. A simulated experiment showing the Bayesian analysis of the probability P_{nl}, that the photon scatterings caused relative localisation of the particles (solid line) and the probability, P_l, that they were localised to begin with (dashed line). In our simulation, we have taken the particles to start off with no well-defined relative position.

knowledge of which process is occurring.

A sample simulation is shown in Fig. 4 for the case that the particles do not initially have a well-defined relative position. We see that initially $P_{nl} = P_l = 0.5$ and that as more and more runs are performed our knowledge of what process is occurring is refined. The probabilities initially jump around for a while before settling down after about 25 runs. The information in this figure is what would be directly accessible to experimentalists and so, in this case, they would be quite certain after about 25 runs that they had observed measurement-induced relative-position localisation. The experimentalists can stop the experiment as soon as the desired level is reached.

4. Conclusion

We have presented a conceptually simple scheme that should enable experiments to detect relative position localisation for quantum particles. This detection scheme has been the missing element in an interpretation for how scattering events could lead to the emergence of classical-like behaviour in quantum systems. This idea could have important consequences for our understanding of the boundary between quantum and classical physics and the role of relative observables in nature.

Acknowledgments

This work was partly supported by DSTL

References

[1] Rau A V, Dunningham J A and Burnett K 2003 *Science* **301** 1081
[2] Joos E and Zeh H D 1985 *Z. Phys.* B **59** 223
[3] Ghirardi G C, Rimini A and Weber T 1986 *Phys. Rev.* D **34** 470
[4] Zurek W H 1991 *Phys. Today* **44** 36
[5] Zurek W H 2003 *Rev. Mod. Phys.* **75** 715

[6] Wiseman H M, Bartlett S D and Vaccaro J A 2004 in Laser Spectroscopy (New Jersey: World Scientific) pp. 307-31.
[7] Dunningham J A, Rau A V and Burnett K 2005 *Science* **307** 872
[8] Javanainen J and Yoo S M 1996 *Phys. Rev. Lett.* **76** 161
[9] Castin Y and Dalibard J 1997 *Phys. Rev.* A **55** 4330
[10] Dunningham J A and Burnett K 1999 *Phys. Rev. Lett.* **82** 3729
[11] Dunningham J A, Burnett K, Roth R and Phillips W D 2006 *New J. Phys.* **8** 182
[12] Mullin W J and Laloë F 2010 *Phys. Rev. Lett.* **104**, 150401
[13] Mullin W J and Laloë F 2010 *Phys. Rev.* A **82** 013618

Hybrid dynamics as a constrained quantum system

N Burić, I Mendaš, D B Popović, M Radonjić and S Prvanović

Institute of Physics, University of Belgrade, P.O.Box 68, 11000 Belgrade, Serbia

E-mail: buric@ipb.ac.rs

Abstract. Hamiltonian formulation of quantum dynamics and nonlinear constraints are used to derive dynamical equations of a hybrid classical-quantum system. Starting with a compound quantum system in the Hamiltonian formulation, conditions for classical behavior are imposed on one of its subsystems and the corresponding hybrid dynamical equations are derived. The dynamical equations for hybrid systems in pure and in mixed states indicate that the hybrid systems have properties that are not exhausted by those of quantum and classical systems.

1. Introduction

Classical and quantum theories have developed different formalisms to successfully describe interactions between systems belonging to their respective domains. Correlations between quantum objects are kinematically captured by direct product structure of Hilbert spaces adopted to the linear evolution. On the other hand compound classical systems are described on the Cartesian product of the component's phase spaces. In this work the framework of Hamiltonian dynamical systems is used to treat the hybrid quantum-classical systems and to develop a description of the interactions within such systems which is consistent with the main physically justified requirements. The Hamiltonian framework was used in [1] to develop a description of the hybrid classical-quantum systems by treating both, quantum and classical, formally as Hamiltonian systems described in the Hamiltonian language. The coupling between the systems is introduced somewhat *ad hoc* as if both systems were classical, just because they are both described in the framework of the Hamiltonian dynamical systems. This assumption about the treatment of compound systems is not trivially obvious. For example, such treatment of coupling between two quantum systems, both separately described in the Hamiltonian framework, would be incorrect. In this paper we start with the total compound quantum system in the geometric Hamiltonian framework [2]. The next step is to consider a classical limit of one of the component systems. To this purpose we utilize our recently developed theory of general quantum constraints within the Hamiltonian approach [3], and the corresponding description of the classical limit [4, 5]. The Hamiltonian form of the derived evolution equations of the hybrid system turns out to be the same as the one postulated in Ref. [1] and therefore satisfies a list of standard requirements collected and tested in [1]. We also study the evolution of general ensembles of hybrid system, and demonstrate that, consistently with the nonlinear pure states evolution, the ensembles evolve in such a way that initially equivalent mixed states of the quantum degrees of freedom become nonequivalent. Non-unitary evolution of the quantum degrees of freedom in a hybrid system suggests that the hybrid systems, if existent, must be considered as conceptually independent class and not as such whose properties are exhausted by the properties of quantum and of classical systems.

2. Quantum dynamics as a Hamiltonian dynamical system

Schrödinger dynamical equation on $\mathcal{H}$ generates a Hamiltonian dynamical system on an appropriate symplectic manifold [6, 7, 8, 9]. In fact the Hilbert space $\mathcal{H}$ is viewed as a real manifold $\mathcal{M}$. In general $\mathcal{M}$ is an infinite dimensional Euclidean manifold. The point from $\mathcal{M}$ corresponding to the vector $|\psi\rangle \in \mathcal{H}$ is denoted by X_ψ. The real manifold $\mathcal{M}$ has Riemannian and symplectic structure. The scalar product on $\mathcal{H}$ is decomposed into it's real and imaginary parts

$$\langle\psi_1|\psi_2\rangle = G(\psi_1,\psi_2) + i\Omega(\psi_1,\psi_2). \tag{1}$$

G is Riemannian metric on $\mathcal{M}$ and Ω is symplectic form on $\mathcal{M}$.

$\mathcal{M}$ associated with the Hilbert space $\mathcal{H}$ can be viewed as a phase space of a Hamiltonian dynamical system.

Quantum observables $\hat{A}$ are represented by functions of the form $A(\psi) = \langle\psi|\hat{A}|\psi\rangle$. Only functions of this form, i.e. bilinear, have physical interpretation of quantum observables.

The Poisson bracket of two functions relates to the commutator between corresponding observables

$$\{A_1, A_2\} = \frac{1}{i\hbar}\langle[\hat{A}_1, \hat{A}_2]\rangle. \tag{2}$$

The Schrödinger evolution generated by a Hamiltonian $\hat{H}$

$$i\hbar|\dot{\psi}\rangle = \hat{H}|\psi\rangle, \tag{3}$$

is equivalent to the Hamilton's equations on $\mathcal{M}$

$$\dot{X}^a_\psi = \Omega^{ab}\nabla_b H(X_\psi). \tag{4}$$

2.1. *Constrained quantum systems*

The Hamiltonian framework for quantum systems opens up a possibility to treat nonlinear constraints [3, 10]. In particular, the formalism of nonlinear constraints provides a natural framework to study the classical limit of a quantum system [4, 5].

Constraints are generally given by a set of k independent functional equations

$$f_l(X) = 0,\ l = 1, 2, \ldots, k. \tag{5}$$

which define a submanifold Γ of $\mathcal{M}$

Equations of motion of the constrained system are obtained using the method of Lagrange multipliers. In the Hamiltonian form:

$$\dot{X} = \Omega(\nabla X, \nabla H_{tot}), \qquad H_{tot} = H + \sum_{l=1}^{k}\lambda_l f_l, \tag{6}$$

that should be solved together with the equations of the constraints (5).

The Lagrange multipliers λ_l are to be determined from the following conditions

$$0 = \dot{f}_l = \Omega(\nabla f_l, \nabla H_{tot}) \tag{7}$$

$$= \Omega(\nabla f_l, \nabla H) + \sum_{m=1}^{k}\lambda_m\Omega(\nabla f_l, \nabla f_m). \tag{8}$$

If one does not know the structure of Γ than one can follow Dirac's procedure. For this, the constraints have to be regular. The case of interest here involves precisely the irregular constraints that cannot be easily replaced, in the general case, by an equivalent set of regular constraints. However, one often knows that the manifold determined by the constraints Γ is a symplectic submanifold of $\mathcal{M}$. In this case one knows that: $H_{tot}|_\Gamma = H|_\Gamma$, i.e. the constrained system is a Hamiltonian system on Γ and it's Hamiltonian is simply $H|_\Gamma$.

2.2. *Example: System of oscillators*

The dynamical algebra is $\otimes_i h_{4,i}$ represented in the Hilbert space $\mathcal{H} = L_2(R^n)$. The fundamental observables are represented by $2n$ operators $(\hat{Q}_i, \hat{P}_i)$, $i = 1, 2, \ldots n$, satisfying $[\hat{Q}_i, \hat{P}_j] = i\delta_{i,j}$ and the Hamiltonian is

$$\begin{aligned} \hat{H} &= \sum_{i=1}^{n} \frac{1}{2m_i}\hat{P}_i^2 + V(\hat{Q}_1, \hat{Q}_2, \ldots, \hat{Q}_n) \\ &= \sum_{i=1}^{n} \frac{1}{2m_i}\hat{P}_i^2 + \frac{m_i\omega_i^2}{2}\hat{Q}_i^2 + \ldots, \end{aligned} \tag{9}$$

The symplectic phase space $\mathcal{M}$ of the Hamiltonian formulation of the quantum oscillators system is the space of fields on $\mathbf{R}^n$ with the symplectic coordinates given by $\phi(x_1, \ldots, x_n), \pi(x_1, \ldots, x_n)$ $(x_i \in \mathbf{R})$ where $\langle x|\psi\rangle = \phi(x) + i\pi(x) \in \mathcal{H}$. By X_ψ we denote an element from $\mathcal{M}$.

For notational simplicity we continue with only one nonlinear oscillator. General case is commented later.

The constraints are given by the following equation

$$f(X) = f_q(X) + f_p(X) = 0 \tag{10}$$

where

$$f_q(X) = (\Delta\hat{Q})^2 - \frac{\hbar}{2m\omega} = 0, \tag{11}$$

$$f_p(X) = (\Delta\hat{P})^2 - \frac{m\omega\hbar}{2} = 0, \tag{12}$$

The constraint implies an equivalence relation on $\mathcal{M}$ (or on $\mathcal{H}$).

$$X_1 \sim X_2 \Leftrightarrow q(X_1) = q(X_2) \wedge p(X_1) = p(X_2) \tag{13}$$

where $q(X_\psi) = \langle\psi|\hat{Q}|\psi\rangle \ldots$.

The manifold determined by the constraint is the symplectic submanifold $\Gamma \subset \mathcal{M}$ if fact $\Gamma = \mathcal{M} / \sim$, where $\sim$ is the equivalence relation (13). Each equivalence class contains a single coherent state $|p, q\rangle$. The manifold is parameterized by $\{p, q\}$.

The constrained system defined by the Hamiltonian (9) and the constraints (10) preserve the equivalence classes and preserve the dispersions of the fundamental quantum observables $\hat{P}, \hat{Q}$. The constrained system on Γ represents the coarse-grained description of the (quantum) nonlinear oscillator.

Γ is symplectic and therefore $H_{tot}|_\Gamma = H|_\Gamma$ i.e. the Hamiltonian of the constrained system on Γ is simply $H(p, q) = \langle p, q|\hat{H}|p, q\rangle$.

H_{tot} preserves constant and minimal quantum fluctuations of fundamental observables, while the evolution with H can in general make them quite large.

We turn to the macro-limit. The constrained system satisfies

$$\langle V(\hat{Q})\rangle_\alpha = V(q) + \sum_{k=1}^{\infty} \frac{(\Delta\hat{Q})_\alpha^{2k}}{2^k k!} V^{(2k)}(q), \tag{14}$$

where $q = \langle\hat{Q}\rangle_\alpha$ and $(\Delta\hat{Q})_\alpha = \sqrt{\hbar/(2m\omega)}$. Thus, the total Hamiltonian in a point $\alpha \equiv (q, p)$ on the constrained manifold is

$$\begin{aligned} H_{tot} &= \frac{p^2}{2m} + V(q) + \sum_{k=1}^{\infty} \frac{1}{2^k k!} \frac{\hbar^k V^{(2k)}(q)}{(2m\omega)^k} \\ &\equiv h_{cl} + \sum_{k=1}^{\infty} \frac{1}{2^k k!} \frac{\hbar^k V^{(2k)}(q)}{(2m\omega)^k}. \end{aligned} \tag{15}$$

For the system with more than one oscillators, that might be nonlinear and interacting, the condition that $\Delta\hat{Q}_i$ and $\Delta\hat{P}_i$ are simultaneously minimal implies that each of the oscillators is always in some pure H_4 coherent state $|\alpha_i(t)\rangle$. Thus, the total state $|\psi(t)\rangle$ is always given by the tensor product of the single oscillator's pure coherent states $|\psi(t)\rangle = \otimes_i |\alpha_i(t)\rangle$.

To summarize: We see that the classical system emerges because of:
a) the coarse-grained description of the quantum system and then
b) the macroscopic limit.
It is important to note that the two factors, i.e. the coarse-graining and the macro-limit, are independent and both are necessary.

3. Hybrid systems

3.1. Derivation of pure states evolution equations

The Hilbert space of a bipartite quantum-quantum system is defined as $\mathcal{H}_1 \otimes \mathcal{H}_2 = \mathcal{H}$. The corresponding phase space is obtained as $\mathcal{M} \neq \mathcal{M}_1 \times \mathcal{M}_2$; The corresponding point from $\mathcal{M}$ has complex canonical coordinates $(\psi(x), \psi^*(x))$ where $\psi(x) \in \mathcal{H}$.

The Poisson bracket of two functions on $\mathcal{M}$ is given by

$$\{f_1, f_2\}_{\mathcal{M}} = \frac{1}{i\hbar}\int dx \left(\frac{\delta f_1}{\delta\psi(x)}\frac{\delta f_2}{\delta\psi^*(x)} - \frac{\delta f_2}{\delta\psi(x)}\frac{\delta f_2}{\delta\psi^*(x)}\right). \tag{16}$$

For simplicity we consider the first system to be given by the $k-$fold product of the Heisenberg algebras, that is by the basic operators $(\hat{Q}_1, \ldots, \hat{Q}_k, \hat{P}_1, \ldots, \hat{P}_k) \equiv (\hat{Q}, \hat{P})$.

In a hybrid classical-quantum system the total quantum fluctuations of the first subsystem, that is the sum of dispersions of the basic observables $(\hat{Q}, \hat{P})$ must be preserved minimal during the evolution:

$$F(X_\psi) = \sum_{i=1}^{k}((\Delta\hat{Q}_i)^2_\psi + (\Delta\hat{P}_i)^2_\psi) - min = 0. \tag{17}$$

The evolution of the fully quantum composite system must be modified in such a way that the constraint is respected.

The manifold $\bar{\Gamma}$ of the constraint is a nonlinear symplectic submanifold of $\mathcal{M}$ locally isomorphic with the Cartesian product $\Gamma_1 \times \mathcal{M}_2$. Γ_1 is the manifold of the standard Heisenberg algebra minimal uncertainty coherent states of the first subsystem, denoted by $|\alpha\rangle$ or $|q,p\rangle$, and $\mathcal{M}_2 \sim \mathcal{H}_2$ is the quantum phase space of the second subsystem.

At each point $|\mathcal{C}\rangle\rangle$ of $\bar{\Gamma}$ given by

$$|\mathcal{C}\rangle\rangle = |\alpha\rangle|\omega_2\rangle \equiv |q,p\rangle|\omega_2\rangle, \tag{18}$$

there are local symplectic coordinates $(q, p, \omega_2(x_2), \omega_2^*(x_2))$ expressed in terms of $|\mathcal{C}\rangle\rangle$ as $q = \langle\langle\mathcal{C}|\hat{Q}|\mathcal{C}\rangle\rangle$, $p = \langle\langle\mathcal{C}|\hat{P}|\mathcal{C}\rangle\rangle$ and $\omega_2(x_2) = \langle x_2|\langle q,p|\mathcal{C}\rangle\rangle$.

The constrained manifold $\bar{\Gamma}$ is symplectic. Therefore, the constrained system is Hamiltonian with the Hamilton's function given by the original Hamilton's function $\langle\langle\psi|\hat{H}|\psi\rangle\rangle$ evaluated on the constrained manifold. Therefore, the dynamics is generated by the Poisson bracket on $\mathcal{M}$ and the Hamiltonian

$$\begin{aligned} H_t &= \langle\langle\mathcal{C}(\psi)|\hat{H}|\mathcal{C}(\psi)\rangle\rangle = \langle\langle\psi|q,p\rangle\langle q,p|\hat{H}|q,p\rangle\langle q,p|\psi\rangle\rangle \\ &\equiv \langle\langle\psi|\hat{H}_\alpha(q,p)|\psi\rangle\rangle, \end{aligned} \tag{19}$$

where $\hat{H}_\alpha(q,p) \equiv |q,p\rangle\langle q,p| \otimes \langle q,p|\hat{H}|q,p\rangle$. In fact the constrained evolution of an arbitrary function-observable $A(\psi) = \langle\langle\psi|\hat{A}|\psi\rangle\rangle$ on the constrained manifold is obtained by reducing the following equation

$$\begin{aligned}\dot{A}(\psi) &= \{A(\psi), H_t\}_{\mathcal{M}} \\ &= \frac{1}{i\hbar}\int dx \left(\frac{\delta A(\psi)}{\delta\psi(x)}\frac{\delta H_t}{\delta\psi^*(x)} - \frac{\delta H_t}{\delta\psi(x)}\frac{\delta A(\psi)}{\delta\psi^*(x)}\right)\end{aligned} \tag{20}$$

on the constrained manifold $\bar{\Gamma}$.

For example, before reduction on $\bar{\Gamma}$ the dynamical equation for $q = \langle\langle\psi|\hat{Q}|\psi\rangle\rangle$ and $p = \langle\langle\psi|\hat{P}|\psi\rangle\rangle$ are given by

$$\dot{q} = \frac{1}{i\hbar}\langle\langle\psi|[\hat{Q},\hat{H}_\alpha]|\psi\rangle\rangle + \frac{\partial H_t}{\partial p}, \tag{21}$$

$$\dot{p} = \frac{1}{i\hbar}\langle\langle\psi|[\hat{P},\hat{H}_\alpha]|\psi\rangle\rangle - \frac{\partial H_t}{\partial q}. \tag{22}$$

In fact, for an arbitrary operator $\hat{A}_1$ acting only in $\mathcal{H}_1$ one has $\langle\langle\psi|[\hat{A}_1,\hat{H}_\alpha]|\psi\rangle\rangle|_{\bar{\Gamma}} = 0$. Therefore, the dynamical equations for the first system's coordinates and momenta are

$$\dot{q} = \frac{\partial H_t}{\partial p}, \quad \dot{p} = -\frac{\partial H_t}{\partial q}, \tag{23}$$

where $H_t = \langle\langle\mathcal{C}(\psi)|\hat{H}|\mathcal{C}(\psi)\rangle\rangle$.

Dynamical equations for functions of the form

$$\omega_2(x_2) \equiv \langle x_2|\omega_2(\psi)\rangle = \langle x_2|\langle q,p|\psi\rangle\rangle \tag{24}$$

are obtained as follows.

Starting again with the equation

$$\dot{\omega}_2(x_2) = \frac{1}{i\hbar}\int dx \left(\frac{\delta\omega_2}{\delta\psi(x)}\frac{\delta H_t}{\delta\psi^*(x)} - \frac{\delta H_t}{\delta\psi(x)}\frac{\delta\omega_2}{\delta\psi^*(x)}\right) \tag{25}$$

and after somewhat lengthy calculation one obtains before the reduction on $\bar{\Gamma}$

$$\begin{aligned} i\hbar\,\dot{\omega}_2(x_2) &= \langle x_2|\langle q,p|\hat{H}|q,p\rangle|\omega_2\rangle \\ &+ \left(\frac{q}{2}\frac{\partial H_t}{\partial q} + \frac{p}{2}\frac{\partial H_t}{\partial p}\right)\omega_2(x_2) \\ &+ \frac{i}{\hbar}\langle x_2|\langle q,p|(\hat{p}-p/2)|\psi\rangle\rangle\langle\langle\psi|[\hat{q},\hat{H}_\alpha]|\psi\rangle\rangle \\ &- \frac{i}{\hbar}\langle x_2|\langle q,p|(\hat{q}-q/2)|\psi\rangle\rangle\langle\langle\psi|[\hat{p},\hat{H}_\alpha]|\psi\rangle\rangle. \end{aligned} \tag{26}$$

Upon reduction on the constrained manifold $\bar{\Gamma}$ the last two terms are annulled and after dropping the pure phase term the relevant dynamical equations can be written in the form

$$i\hbar\,\dot{\omega}_2(x_2;\psi) = \langle x_2|\langle\alpha(\psi)|\hat{H}|\alpha(\psi)\rangle|\omega_2(\psi)\rangle. \tag{27}$$

The equation has the form of a Schröedinger equation for the state vector $\omega_2(x_2;\psi) = \langle x_2|\langle q,p|\psi\rangle\rangle \in \mathcal{H}_2$, with the Hamiltonian operator $\langle\alpha(\psi)|\hat{H}|\alpha(\psi)\rangle$ acting on $\mathcal{H}_2$ and depending on $q = \langle\langle\psi|\hat{Q}|\psi\rangle\rangle$ and $p = \langle\langle\psi|\hat{P}|\psi\rangle\rangle$.

The Poisson bracket on $\bar{\Gamma}$ for arbitrary functions on $\bar{\Gamma}$ represented in the local coordinates $(q, p, \omega_2, \omega_2^*)$ is

$$\{f_1, f_2\}_{\bar{\Gamma}} = \sum_{i=1}^{k} \left(\frac{\partial f_1}{\partial q_i} \frac{\partial f_2}{\partial p_i} - \frac{\partial f_2}{\partial q_i} \frac{\partial f_1}{\partial p_i} \right) + \frac{1}{i\hbar} \int dx_2 \left(\frac{\delta f_1}{\delta \omega_2(x_2)} \frac{\delta f_2}{\delta \omega_2^*(x_2)} - \frac{\delta f_2}{\delta \omega_2(x_2)} \frac{\delta f_1}{\delta \omega_2^*(x_2)} \right). \quad (28)$$

The Hamiltonian form of the hybrid dynamics on the constrained phase space manifold $\bar{\Gamma}$ reads

$$\dot{q} = \{q, H_t\}_{\bar{\Gamma}}, \quad \dot{p} = \{p, H_t\}_{\bar{\Gamma}}, \quad (29)$$

$$\dot{\omega}_2 = \{\omega_2, H_t\}_{\bar{\Gamma}}, \quad \dot{\omega}_2^* = \{\omega_2^*, H_t\}_{\bar{\Gamma}}, \quad (30)$$

where the Hamilton's function $H_t(q, p, \omega_2(x_2), \omega_2^*(x_2))$ is given by $H_t = \langle\langle \mathcal{C}(\psi) | \hat{H} | \mathcal{C}(\psi) \rangle\rangle$ and $|\mathcal{C}\rangle\rangle = |q, p\rangle |\omega_2\rangle$.

The constrained dynamics which preserves minimal value of the quantum fluctuations of one of the subsystems is only the first step. The second step is the relevant macro-limit so that the minimal quantum fluctuations, still present in the corresponding coherent states, can be neglected when compared with actual values of the dynamical variables.

Therefore in the macro-limit the replacement

$$\langle\langle \mathcal{C}(\psi) | \hat{F}(\hat{Q}, \hat{P}) | \mathcal{C}(\psi) \rangle\rangle \to F(q, p) \quad (31)$$

should be applied in the equations relevant for the first subsystem.

3.2. Ensembles of hybrid systems

For convenience, we denote the canonical coordinates in $\mathcal{M}_q$ of $\mathcal{M}_c \times \mathcal{M}_q$ by (x, y). Quantities defined exclusively as functions on $\mathcal{M}_q$ ($\mathcal{M}_c$) are referred on as quantum degrees of freedom or QDF.

Consider a general probability density $\rho(p, q, x, y)$ on the total hybrid phase space $\mathcal{M} = \mathcal{M}_c \times \mathcal{M}_q$[11].

In general, following the Hamiltonian formulation of the hybrid system dynamics, the evolution of $\rho(p, q, x, y; t)$ considered as a statistical ensemble on $\mathcal{M}$ is given by the Liouville equation

$$\frac{\partial}{\partial t} \rho(p, q, x, y; t) = \{H_t(p, q, x, y), \rho(p, q, x, y; t)\}_{\mathcal{M}}. \quad (32)$$

We shall argue that the most general statistical ensembles of hybrid systems need to be represented by general probability densities $\rho(p, q, x, y; t)$.

The density $\rho(p, q, x, y; t)$ generates a unique positive operator valued function (POVF):

$$\hat{\rho}(p, q; t) = \int_{\mathcal{M}_q} \rho(p, q, x, y; t) \hat{\Pi}(x, y) dM_q, \quad (33)$$

where $\hat{\Pi}(x, y)$ is the projection onto the vector form $\mathcal{H}_2$ corresponding to the point (x, y). $\hat{\rho}(p, q; t)$ can be called the hybrid statistical operator. It contains less information about the hybrid system state then the density $\rho(p, q, x, y; t)$.

The unconditional mixed state of the quantum subsystem of the hybrid in the state $\rho(p, q, x, y; t)$ is also uniquely obtained as

$$\hat{\rho}(t) = \int_{\mathcal{M}} \rho(p, q, x, y; t) \hat{\Pi}(x, y) dM. \quad (34)$$

Many hybrid ensembles, represented by different $\rho(p,q,x,y;t_0)$, have the quantum subsystem in the same conditional or unconditional mixed state. Each different $\rho(p,q,x,y;t_0)$ describes physically different ensembles of hybrid systems with the quantum subsystem in the same mixed state. The differences are manifested in the evolution of $\hat{\rho}(x,y;t)$ and $\hat{\rho}(t)$.

In the purely quantum case all different $\rho(x,y;t_0)$ with the same first moment correspond to the physically equivalent quantum mixture $\hat{\rho}(t_0)$, and generate unique von Neumann evolution of $\hat{\rho}(t)$ which is obtained from the Liouville evolution of any such $\rho(x,y;t)$. Therefore, all such $\rho(x,y;t)$ are equivalent in the purely quantum case. In the hybrid case, different $\rho(p,q,x,y;t_0)$ which give the same $\hat{\rho}(p,q;t_0)$ (or $\hat{\rho}(t_0)$) generate different evolution of $\hat{\rho}(p,q;t)$ (or $\hat{\rho}(t)$) and thus must be considered as physically different.

The evolution equation for $\hat{\rho}(p,q)$ is

$$\begin{aligned}\frac{\partial\hat{\rho}(p,q;t)}{\partial t} &= \frac{1}{i\hbar}[\hat{H}_q+\hat{V}_{int}(p,q),\hat{\rho}(p,q;t)] + \{H_c(p,q),\hat{\rho}(p,q;t)\}_{p,q} \\ &+ \int_{\mathcal{M}_q}\{V_{int}(p,q,x,y),\rho(p,q,x,y;t)\}_{p,q}\hat{\Pi}(x,y)dM_q. \end{aligned} \tag{35}$$

The solution remains a well defined statistical operator on $\mathcal{H}$ for all t, which is a desirable property not shared by some other hybrid system theories.

The dynamical equation for $\hat{\rho}(t)$ is

$$\begin{aligned}\frac{d\hat{\rho}(t)}{dt} &= \frac{1}{i\hbar}[\hat{H}_q,\hat{\rho}(t)] + \frac{1}{i\hbar}\int_{\mathcal{M}_c}[\hat{V}_{int}(p,q),\hat{\rho}(p,q;t)]dM_c \\ &+ \int_{\mathcal{M}_c}\{H_c(p,q),\hat{\rho}(p,q;t)\}_{p,q}dM_c \\ &+ \int_{\mathcal{M}}\{V_{int}(p,q,x,y),\rho(p,q,x,y;t)\}_{p,q}\hat{\Pi}(x,y)dM \\ &= \frac{1}{i\hbar}[\hat{H}_q,\hat{\rho}(t)] + \frac{1}{i\hbar}\int_{\mathcal{M}_c}[\hat{V}_{int}(p,q),\hat{\rho}(p,q;t)]dM_c. \end{aligned} \tag{36}$$

The first term on the right side of (36) generates the unitary part of the evolution and the second term does not preserve the norm of $\hat{\rho}$ and is responsible for non-unitary effects. Notice that the evolution of $\hat{\rho}(p,q;t)$ ($\hat{\rho}(t)$) cannot be expressed only in terms of $\hat{\rho}(p,q;t)$ ($\hat{\rho}(t)$), but irreducibly involves the probability density $\rho(p,q,x,y;t)$.

4. Conclusion

The evolution of QDF of a hybrid system is fundamentally different from the linear evolution of a quantum subsystem of a quantum system. The characteristic main features of the QDF evolution are expressed by the nonlinearity of the pure state evolution, or by the dynamically induced differences of $\hat{\rho}(t)$ with different initial convex mixture representations. It is well known that the linearity of the Schrödinger equation and the equivalence of different convex mixtures are both necessary in order to prevent superluminal communication in ordinary quantum mechanics of bipartite systems. If either of the two properties is violated, without further modification of the quantum formalism, superluminal communication between entangled parts of a bipartite system is possible. The nonlinear pure state evolution and the evolution dependence on the initially equivalent different ensembles appears quite naturally in the Hamiltonian description of hybrid systems, and in the same time the QDF of the hybrid might be in an entangled state [12]. Therefore, superluminal communication can be avoided only by some further modification of the scheme. It has been argued that the direct product might not be the natural type of coupling between systems with nonlinear evolution [13], and that nonlinear evolution might suggest non-standard computation of correlations [14]. Alternatively, one might consider the model of hybrid

Journal of Physics: Conference Series **442** (2013) 012027 doi:10.1088/1742-6596/442/1/012027

systems presented here as insufficient to describe fully the true features of coupled real quantum and real classical systems. One might try to incorporate back-reaction of the classical system on quantum fluctuations, like for example in [15], or one might explore the possibilities opened up by replacing a simple classical system by truly complex classical systems with many degrees of freedom [1, 12].

Acknowledgments

This work was supported by the Ministry of Science and Education of the Republic of Serbia, contracts No. 171017, 171028, 171038, 171006 and 45016. and by COST (Action MP1006).

References

[1] Elze H-T 2012 *Phys. Rev.* A **85** 052109
[2] Radonjić M, Prvanović S and Burić N 2012 *Phys. Rev.* A **85** 064101
[3] Burić N 2008 *Ann. Phys. (NY)* **233** 17
[4] Radonjić M, Prvanović S and Burić N 2011 *Phys. Rev.* A **84** 022103
[5] Radonjić M, Prvanović S and Burić N 2012 *Phys. Rev.* A **85** 022117
[6] Kibble T W B 1978 *Commun. Math. Phys.* **64** 73
[7] Heslot A 1985 *Phys. Rev.* D **31** 1341
[8] Weinberg S 1989 *Phys. Rev. Lett.* **62** 485
[9] Brody D C and Hughston L P 2001 *J. Geom. Phys.* **38** 19
[10] Brody D C, Gustavsson A C T and Hughston L P 2008 *J. Phys.* A **41** 475301
[11] Burić N, Mendaš I, Popović D B, Radonjić M and Prvanović S 2012 *Phys. Rev.* A **86** 034104
[12] Burić N, Popović D B, Radonjić M and Prvanović S 2013 Entanglement dynamics in a hybrid quantum-classical system, *Preprint*
[13] Mielnik B 2001*Phys. Lett.* A **289** 1
[14] Doebner H D 2002 *Phys. Lett.* A **301** 139
[15] Diósi L, Gisin N and Strunz W T 2000*Phys. Rev.* A **61** 022108

Journal of Physics: Conference Series **442** (2013) 012028 doi:10.1088/1742-6596/442/1/012028

Coherent structures in liquid water close to hydrophilic surfaces

Emilio Del Giudice
Centro Studi Eva Reich,Via Orti 5, I-20122 Milano, Italy
E-mail: emilio.delgiudice@mi.infn.it

Alberto Tedeschi
WHITE Holographic Bioresonance, Via Francesco Petrarca 16, I-20123, Milano, Italy
E-mail: gowhite@usa.net

Giuseppe Vitiello
Dipartimento di Fisica "E.R. Caianiello" and Gruppo Collegato I.N.F.N., Università di Salerno, I-84100 Salerno, Italy
E-mail: vitiello@sa.infn.it

Vladimir Voeikov
Lomonosov Moscow State University, Faculty of Biology, Vorobevy Gory, Ru-119995 Moscow, Russia
E-mail: v.l.voeikov@gmail.com

Abstract. Quantum Electrodynamics (QED) predicts the occurrence of a number of coherent dynamical phenomena in liquid water. In the present paper we focus our attention on the joint oscillation of the almost free electrons produced by the coherent oscillation of the electron clouds of water molecules, which has been described in previous publications, and of the negative electric charges lying on the solid surfaces wet by water. This joint coherent oscillation gives rise to a number of phenomenological consequences which are found to exist in the physical reality and coincide with the layers of Exclusion Zone (EZ) water experimentally observed close to hydrophilic surfaces.

1. Introduction

In the last decades the structure of liquid water has been investigated in the framework of Quantum Electrodynamics (QED) [1]-[4]. It has been shown that an ensemble of water molecules in a gaseous state, above a density threshold and below a critical temperature, transits to a lower energy coherent state where the electron clouds oscillate between their individual ground configuration and an excited configuration where one electron per molecule is almost free (the binding energy is about 0.4 eV). The coherent state is a superposition of the ground configuration, with a weight of almost 90%, and of the excited configuration with a weight of about 10%. The molecule coherent oscillation is in tune with an electromagnetic field (emf), trapped within the molecule ensemble and therefore unable to be irradiated outwards. The molecules and the trapped emf are confined within a region (termed Coherence Domain (CD)), whose size is the wavelength of the emf. In the case of liquid water this size is 0.1 microns. Each CD is a pool of quasi-free electrons, which, as in a semi-conductor, can be coherently excited giving rise to a CD spectrum of coherent excited levels produced by the coherent ("cold") vortices of quasi-free electrons, whose magnetic dipoles are aligned with the external (e.g. the terrestrial) magnetic field. Because of the coherence the vortices have no internal friction and therefore have a very long lifespan, so that it is possible to pile up within the CD a large number of excitations which sum up to a unique vortex whose rotational frequency is just the sum of the frequencies of the component vortices. In this way, as it has been shown in [5], the CD becomes a device able to transform ambient noise, namely an ensemble of a large number of low frequency excitations, into a unique high frequency excitation. This property implements the intuition of Albert Szent-Gyorgyi [6], who traced back the presence in living organisms of excitations at molecular scale to the capability of biological water to sum up the small perturbations contained in ambient noise.The molecule condensation produced by coherence is counteracted at any non vanishing temperature T by thermal collisions which push some molecules out of tune. Consequently at each T there is a coherent fraction $F_c(T)$ of molecules and a non-coherent fraction $F_{nc}(T)$ whose sum is 1. Molecules cross over continuously between the two fractions leaving constant the total number of coherent and non-coherent molecules. Besides the above coherence which involves electron clouds, another coherent dynamics has been shown [7] to involve the molecule rotational levels producing a coherent oscillation between the molecule ground state and the low lying states at about 20 wave-numbers. This coherence spans a length of slightly less than 500 microns and gives rise to an extended oscillating polarization field since it is able to correlate together the individual molecule electric dipoles. In the absence of an external electric field, the time average of the above polarization field is zero. However in the presence of an electric field E, the polarization field acquires a permanent component along E. In aqueous systems macromolecules suspended in water or solid surfaces on the boundary could be sources of electric fields so that according to QED layers of polarized water as deep as some hundreds of microns can be expected. We will see in the following that this prediction is corroborated by laboratory observations. Furthermore an additional coherent dynamics can be found in aqueous systems. As shown in [8] plasmas, which are ensembles of charged particles (either negative like electrons or positive like protons) can give rise to coherent oscillations when suitable conditions are met. We examine the dynamics in aqueous systems of the ensemble of quasi-free electrons of water CDs coupled with the ensemble of negative charges lying on the solid surface of the walls. In the present paper we will investigate the condition for the emergence of coherence in the above plasmas and compare the predictions with the observed facts.

2. Plasma coherent oscillations

Stable matter is electrically neutral, made up, however, of constituents electrically charged. Plasma dynamics starts with the small oscillations of charges with a pulsation ω_R around their equilibrium positions. Charges are kept correlated by the emf. As shown in Chapter 5 of Ref. [8], the plasma driven by the coupled emf could "run away" from the initial perturbative regime and transit to a coherent state where all charges oscillate in unison, in tune with a coherent resonating emf at a common pulsation

$$\omega_r = |1 - \dot{\phi}|\, \omega_R < \omega_R\,, \tag{1}$$

where ϕ is the change of the phase of the emf induced by the onset of coherence and the symbol $|...|$ denotes, as usual, the absolute value.

Furthermore, a relevant parameter is the plasma oscillation amplitude α, connected to the emf amplitude A by the equation

$$A^2(1-\dot{\phi})+\alpha^2=0\,. \tag{2}$$

When it is assumed that each set of charges moves within a homogeneous ensemble of the opposite charges we have the simplest case of the "ideal plasma", whose pulsation is

$$\omega_P=\frac{e}{\sqrt{m}}\sqrt{\frac{N}{V}}\,, \tag{3}$$

where the natural unit system ($\hbar = 1 = c$) is used, e and m are the charge and the mass of plasma constituents, respectively, and N/V their density. An ideal plasma can be shown to meet always the condition for coherence, namely the existence of the runaway. However, in ideal plasmas the oscillation amplitude is not bounded from above so that, as discussed in Ref. [8], there is no energy minimum; ideal plasmas therefore cannot exist as physical systems. Real plasmas, which are real physical systems, should consequently exhibit dis-homogeneities; the oscillating charges should stay confined within finite volumes, for instance the matrices of opposite charges.

In this case the analysis of the Chapter 5 of Ref. [8] shows that the oscillation amplitude is bounded from above, so that the system has an energy minimum. The condition for coherence is:

$$\frac{\omega_R}{\omega_P}<1.88\,, \tag{4}$$

where ω_P is the pulsation of the ideal plasma having the same average density.

If condition (4) is not met, there is no runaway so that the field amplitude does not increase; however, the field, although weak, becomes coherent and "weak coherence" is produced.

When coherence emerges, the energy of the system decreases and consequently the new coherent configuration is protected against instabilities by the energy gap produced by the emergence of coherence. The physical requirement of minimization of energy implies that a physical system could always reach a coherent configuration provided that spontaneous fluctuations allow the suitable change of its parameters necessary to meet the requirement for the onset of coherence.

By using the above concepts, let us examine what happens when liquid water, made up of Coherence Domains containing a pool of almost free electrons and correspondingly an ensemble of almost free protons (the widows of the quasi free electrons), becomes close to a solid surface (a wall, or a suspended microsphere surface, or a molecular backbone) on which electric charges able to oscillate are distributed. We denote by r the size of the charge spot on the surface, δ the depth of the charge layer, z the total number of the charges contained in the spot, λ a parameter measuring the dis-homogeneity of the charge distribution. The pulsation ω_R of the surface plasma is

$$\omega_R=\frac{e\,\lambda}{\sqrt{m}}\sqrt{\frac{z}{4\pi\,r^2\delta}}\,. \tag{5}$$

According to Ref. [8] the energy gain of the energy spot is

$$\Delta E=-15\pi\sqrt{\frac{2\pi}{3}}\,\omega_R. \tag{6}$$

The above coherent oscillation of the solid surface charges occurs close to a corresponding coherent oscillation of the quasi free charges of the water CDs. The parameters (pulsation and amplitude) of this last oscillation depend on the excitation levels where the CD finds itself due to its own dynamics. As a matter of fact, as said in the Introduction, vortices can be induced in the ensemble of quasi free electrons of CDs by the uptake of energy from the environment. Such vortices control the maximum displacement of the oscillations of quasi free electrons in the CD and therefore the maximum amplitude and the pulsation ω_r of the coherent oscillation (see Eqs. (1) and (2)). It is then possible for the system to self-adjust these parameters in order to produce a resonance between the two charge ensembles and consequently the onset of a compound coherence of the two plasmas on a common frequency, producing the advantage of a larger energy gap due to the larger number of correlated oscillators. The larger energy gap would weaken the binding of the charges oscillating on the surface to their matrix, making possible a charge separation. On the other hand, the increased energy gap enhances the protection of the water CDs against the assaults of thermal collisions, responsible for the formation of the water non-coherent fraction. Water close to surfaces should therefore exhibit a much larger coherent fraction than bulk water. Its coherent inner structure should remain stable in time, allowing, contrary to bulk water where a continuous cross-over between the two fractions occurs, a direct observation of the consequences of the presence of coherence. The depth of the coherent layer close to a surface is governed, according to [7], by the strength of the electric field emitted by the surface, which correlates coherently the CD electric dipoles. The depth of the layer could therefore reach values as high as hundreds of microns, much larger than the depths of a few molecular layers predicted by conventional theories based on H-bonding [9].

Finally, the charge dynamics on the interface water-solid surface should exhibit a feature totally unexpected in the usual framework, namely the attraction among like charges produced by the coherent dynamics which overcomes the repulsion expected according to the Coulomb law of electrostatics. In the next section, we will compare the above predictions of the QED analysis with the results of observations.

3. Phenomenology of interfacial water

In the last years Gerald Pollack and his coworkers [10] - [12] have reported that water adjacent to hydrophilic surfaces (interfacial water) is dynamically different from bulk water. Solutes and microspheres are excluded from this kind of interfacial water which has been termed for this reason exclusion zone (EZ) water. The depth of the layer of EZ water on the solid surface is in the order of tens and hundreds of microns, up to 500 microns. With respect to bulk water, EZ water has higer (up to tenfold) viscosity, diminished infrared emissivity, shorter spin-lattice T_1 relaxation time, retarded spin-spin T_2 relaxation time, as probed by NMR and smaller self-diffusion coefficient [10, 11]. All these features are compatible with the presence of coherent water as described in [1] - [4]. Actually, in a coherent system one could not touch a single molecule without touching also the other correlated molecules, so that viscosity becomes higher as well as the correlations detected by NMR, whereas self-diffusion which depends on the mutual independence of molecules is also depressed. The decrease of infrared emissivity is justified by the presence of the infrared oscillations of CDs which absorb energy in this frequency range.

EZ water is electrically charged with respect to bulk water, which is well known to be neutral. If EZ water is close to a surface bearing a net negative charge, it acquires a negative charge, too, whereas it acquires a positive charge near a surface bearing a positive charge [12]. It is interesting to observe that when EZ water becomes negatively charged near a negative surface, positive charges (protons) appear beyond EZ water on the side of the EZ layer facing bulk water. On the contrary, negative charges appear on the side of the layer facing the bulk water when EZ water is positively charged. It is also interesting to realize that no electric opposite charges have been detected in the interstice between the solid surface and the aqueous surface, suggesting that the attraction water-solid surface underlying the hydrophily of the surface is not produced by electrostatics but by a dynamic collective attraction such as the one produced by QED coherence. **Like likes like**, contrary to the Coulomb law of electrostatics.

Moreover, the expulsion of non-resonating particles from coherence domains oscillating on a definite frequency accounts for the phenomenon of the exclusion of solutes and also of opposite electric charges since

$$\frac{\omega_{neg}}{\omega_{pos}} \propto \sqrt{\frac{m_{proton}}{m_{electron}}} \simeq 45\,. \qquad (7)$$

The charge separation occurring at the interface is understood in the QED framework by the weakening of the binding of the oscillating charges to their matrices as a consequence of the energy gap produced by the combined coherence of the quasi free charges of the water CDs and of the solid surfaces.

Additional evidence for the existence of charge separation on the interface between water and solid surface comes from a recent report [13] that almost all the water droplets constituting the aerosols present in the waterfalls in Austrian Alps were electrically charged (85 % negatively, 15 % positively). It is conceivable that the charge separation occurred on the interface between the flowing water of the river and its rocky banks and the turbulence of the river stripped the charged droplets from the banks.

4. Conclusions and outlook

The predictions of the QED analysis are in striking agreement with the findings of experimental investigations on interfacial water. The so far mysterious features of EZ water have found an explanation which on the contrary cannot be found in the framework of models of water based purely on electrostatics. The attraction among like charges finds a rationale in the context of a long range many-body dynamics emerging above a critical density threshold and below a critical temperature. This result is particularly useful when we try to understand the dynamics existing in living organisms. Since no point of them is more distant than a fraction of a micron from a surface or a molecular backbone, we can safely presume that the whole of biological water is interfacial water. Therefore, all the conclusions of the present article apply to biological water. In particular, we can understand why cells, whose surface is negatively charged can attract each other giving rise to tissues.

Finally, another application of the present analysis could explain the stability of clouds constituted by negatively charged water droplets.

References

[1] Arani R, Bono I, Del Giudice E and Preparata G 1995 *Int. J. Mod. Phys.* B **9** 1813-1841
[2] Del Giudice E, Preparata G 1998 In: Sassaroli E, Srivastava Y, Swain J, Widom A (eds) *Macroscopic Quantum Coherence* (Singapore: World Scientific) pp. 108-129
[3] Del Giudice E, Spinetti P R, Tedeschi A 2010 *Water* **2** 566-586
[4] Bono I, Del Giudice E, Gamberale L and Henry M 2012 *Water* **4** 510-532
[5] Del Giudice E, Pulselli R M, Tiezzi E 2009 *Ecological Modelling* **220** 1874-1879
[6] Szent-Gyorgyi A 1957 *Bioenergetics* (New York: Academic Press)
[7] Del Giudice E, Preparata G, Vitiello G 1988 *Physical Review Letters* **61** 1085-1088
[8] Preparata G 1995 *QED, Coherence in Matter* (Singapore:World Scientific)
[9] Franks F 1972-1982 *Water, a Comprehensive Treatise* (New York: Plenum)
[10] Zheng J M and Pollack G H 2003 *Phys. Rev.* E **68** 031408.10.1103
[11] Zheng J M , Chin W C , Khijniak E, Khijniak E Jr and Pollack G H 2006 *Adv. Colloid Interface Sci.* **23** 1927
[12] Zheng J M , Wexler A and Pollack G H 2009 *J Colloid Interface Sci.* **332** 511-514
[13] Madl P, Kolarz P, Del Giudice E, Tedeschi A, Hartl A, Gaisberger M, Hofmann W 2013 The formation of coherence domains for aerosolized water molecules at alpine waterfalls, submitted to *Water*, online Journal

DICE2012 IOP Publishing
Journal of Physics: Conference Series **442** (2013) 012029 doi:10.1088/1742-6596/442/1/012029

Non-hermitian approach to decaying ultracold bosonic systems

Sandro Wimberger, Carlos A Parra-Murillo and Georgios Kordas
Institut für Theoretische Physik and Center for Quantum Dynamics, Universität Heidelberg, D–69120 Heidelberg

E-mail: s.wimberger@thphys.uni-heidelberg.de

Abstract. A paradigm model of modern atom optics is studied, strongly interacting ultracold bosons in an optical lattice. This many-body system can be artificially opened in a controlled manner by modern experimental techniques. We present results based on a non-hermitian effective Hamiltonian whose quantum spectrum is analyzed. The direct access to the spectrum of the metastable many-body system allows us to easily identify relatively stable quantum states, corresponding to previously predicted solitonic many-body structures.

1. Introduction

Open quantum systems are ubiquitous in microscopic physics. Any system can only be decoupled from its environment in an abstract sense. There exist many successful approaches to describe open quantum systems. Let us just mention the theory of scattering [1], techniques based on Master equations [2], heavily used e.g. in quantum optics [3], or effective Hamiltonians which contain absorbing boundary conditions [4]. The latter approach has the advantage that it allows one to access a generalized quantum spectrum, thus it extends spectral methods, normally applied to closed systems, to open ones. Keeping in mind that the method of effective non-hermitian Hamiltonians is based on perturbative arguments, such as a weak decay or coupling to the environment, it nevertheless gives the possibility to predict the evolution of the system for short times at least. The qualitative understanding based on the analysis of the spectrum is at least as useful as more accurate methods which are based on numerical propagation in most cases.

In this paper we concentrate on systems of ultracold atoms loaded into periodic lattice structures. Such systems are naturally open when one, for instance, starts to tilt or to accelerate the lattice [6] since this induces decay or quantum tunneling. Alternatively, the system may be opened artificially to observe interesting effects induced by dissipation and/or quantum noise. Figure 1 sketches the mentioned setups. While we studied the case (a) of tilted lattices for strongly correlated bosons in previous works [5–8], in the following we investigate the case (b) in more detail with an effective non-hermitian Hamiltonian for the many-body system. Experimentally, the loss of particles as shown in figure 1(b) can be realized easily in a controlled manner. Single particle decay channels are provided by shining in appropriate laser beams [9] or even electronic beams [10, 11]. The latter ionize single atoms in the condensate, which, together with the produced electrons, are afterward accelerated by electric fields, hence they leave the system extremely fast without strong backaction to the remaining bosons in the lattice environment.

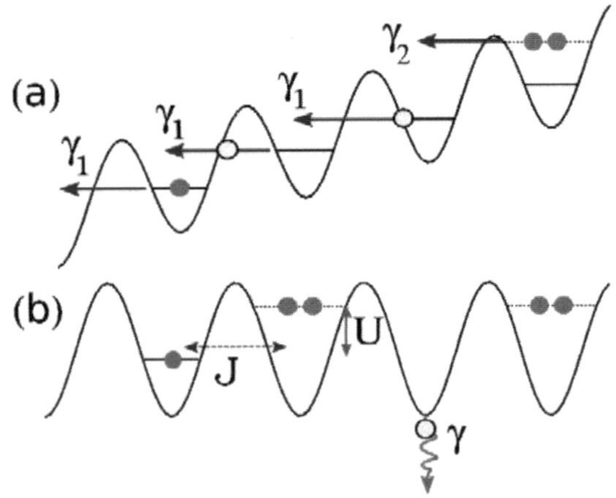

Figure 1. Sketch of two typical realization of strongly correlated many-body open quantum systems: (a) Particles tunneling from a tilted optical lattice. The local decay rates depend on the initial particle numbers in the wells represented by the indices, see reference [5] for details. (b) engineered particle loss with decay rate γ in one of the potential wells. The terms J and U denote the hopping and the two-body interaction energy respectively. Interplay of the various processes can lead to interesting dynamics in the open many-body systems, as discussed in section 2 for case (b).

2. A dissipative Bose-Hubbard system

2.1. Motivation

Ultracold bosons at zero temperature can be described by the following Bose-Hubbard Hamiltonian [12]

$$\hat{H} = -\frac{J}{2}\sum\nolimits_j \left(\hat{a}_{j+1}^\dagger \hat{a}_j + \hat{a}_j^\dagger \hat{a}_{j+1}\right) + \frac{U}{2}\sum\nolimits_j \hat{a}_j^\dagger \hat{a}_j^\dagger \hat{a}_j \hat{a}_j \,, \tag{1}$$

provided the optical lattice holding them is sufficiently deep (such that the approximation by discrete modes denoted by the index j is good). Here $\hat{a}_j$ and $\hat{a}_j^\dagger$ are the bosonic annihilation and creation operators in the jth well. $\hbar$ is set to one measuring all energies in frequency units. This Hamiltonian can be made dimensionless by rescaling with one of the two energy scales J or U respectively. Hence, without loss of generality, we set $J/2 = 1$ in the following, measuring all energies in units of $J/2$.

As briefly mentioned in the introduction and motivated in a series of previous papers on dissipative Bose condensates [14–17], a tunable source of single-particle dissipation can be added to the system. With the spatially local dissipation rates γ_j, the dynamics is then best described by a quantum Master equation [2, 18] for the many-body density $\hat{\rho}$:

$$\dot{\hat{\rho}} = -i[\hat{H}, \hat{\rho}] - \frac{1}{2}\sum\nolimits_j \gamma_j \left(\hat{a}_j^\dagger \hat{a}_j \hat{\rho} + \hat{\rho}\hat{a}_j^\dagger \hat{a}_j - 2\hat{a}_j \hat{\rho} \hat{a}_j^\dagger\right) \,. \tag{2}$$

This Master equation may be "unraveled", i.e. solved exactly, by a quantum jump approach [2]. Alternatively, it can be solved approximately by methods which take into account effects beyond the mean-field expansion, see, e.g. [15–17], where results of both methods are shown and compared.

Here we will follow a different approach based on an effective Hamiltonian in which the dissipative terms of equation (2) are included as complex absorbing potentials. In contrast to the above mentioned methods for solving the Master equation, this gives us access to the quantum spectrum of the non-hermitian Hamiltonian studied in the next subsection. Within this approximation, the real parts of the spectrum correspond to the energy levels – just as for any closed quantum system – and the imaginary parts describe the decay widths of the metastable eigenstates, i.e. their stability in the presence of the loss [4].

2.2. A spectral approach

As a study case, we now look at a particular situation of three potential wells (also called sites) with relatively large filling factors, i.e. particles per site, of the order ten or larger. For this setup, reference [15] predicts the *dynamical* formation of very stable collective states of many bosons within a single well, despite strong atom losses. More precisely, we choose $\gamma_1 = \gamma_3 = \gamma$ at the two boundary wells and $\gamma_2 = 0$ in the middle site. For weak atomic interactions, i.e. for the limit $U \to 0$, all particles will be lost steadily as time evolves. This is just the situation for independent (almost) non-interacting particles. For strong interactions, i.e. $U > U_{\text{crit}}$, however, the particles at the leaky sites one and three are quickly lost, while the well in the middle is protected against the decay. This may be expected by a simple mean-field argument, saying that the bosons cannot get rid of their high energy in the middle well by just gaining the energy $J = 2$ by hopping to the leaky sites. Then, $U_{\text{crit}}\bar{n} \gtrsim J = 2$ in our dimensionless units, which is the so-called self-trapping criterion where mean-field tunneling is suppressed by the mentioned energetic argument [19]. $\bar{n}$ is the filling factor of the corresponding well. The big surprise reported in [15, 16] is yet that the surviving part of the initially prepared condensate is very coherent, implying that non-trivial many-body dynamics occurred during the formation of this very stable, so-called breather or solitonic state. This strongly correlated state is well known for its stability with respect to decay [13], and this property we are going to use in order to identify them using our spectral analysis. Thus dissipation can be used to prepare almost pure condensed many-body states in the middle well [15, 16].

Those breather states can be easily detected (without any time-propagation as done in [15, 16]), by studying the spectrum of the following effective Hamiltonian

$$\hat{H}_{\text{eff}} = -\sum_{j=1}^{3}\left(\hat{a}_{j+1}^{\dagger}\hat{a}_j + \hat{a}_j^{\dagger}\hat{a}_{j+1} + \frac{U}{2}\hat{a}_j^{\dagger}\hat{a}_j^{\dagger}\hat{a}_j\hat{a}_j - i\frac{\gamma}{2}\hat{n}_j\right), \tag{3}$$

with $\hat{n}_j = \hat{a}_j^{\dagger}\hat{a}_j$ being the number operator. In the framework of the quantum jump method, this effective Hamiltonian gives the continuous evolution between two quantum jumps. Nevertheless, one can use it to identify and study the breather state, which is predicted as an attractively stable state of the evolution of the full open system. Here, for simplicity, we use periodic boundary conditions for $\hat{H}_{\text{eff}}$, identifying the well four by the well one. The results presented below qualitatively remain unchanged when using open (also called hard-wall) boundary conditions, since the decay affects both ends of the lattice in exactly the same way.

Figure 2 reports in panel (a) the eigenspectrum $\{E_k\}_k$ of the effective Hamiltonian $\hat{H}_{\text{eff}}$ for typical parameters. There, the decay rates, i.e. $\Gamma_k = -2\text{Im}E_k$, are plotted versus the real parts of the spectrum $\epsilon_k = \text{Re}E_k$. One observes immediately one state which is far separated from the rest of the spectrum. This is the very stable breather state discussed above. Its stability is given by the rate Γ_1, where the index starts to count from the most stable state. As the effective interaction $U\bar{n}$ increases, this state moves away from the rest of the spectral band. A criterion for the stability is therefore the distance of the imaginary parts from the next stable state (which actually is degenerate because of the spatial symmetry of the problem, and both are denoted by the index two in panel (a)). This distance, or spectral gap $\Delta_{21} = \Gamma_2 - \Gamma_1$, is plotted in panel (b) as a function of the effective interaction UN, where N denotes the total particle number in the system (which is fixed to a specific value). We can indeed confirm that the breather state forms for a sufficiently large effective interaction strength. The critical value may be estimated by the maxima of the curves shown in panel (b). These maxima come about because all the lower lying states in the spectrum have decreasing decay rates Γ_k as a function of UN, yet the one which decreases fastest is exactly the breather state, the one with minimal Γ_1 in panel (c).

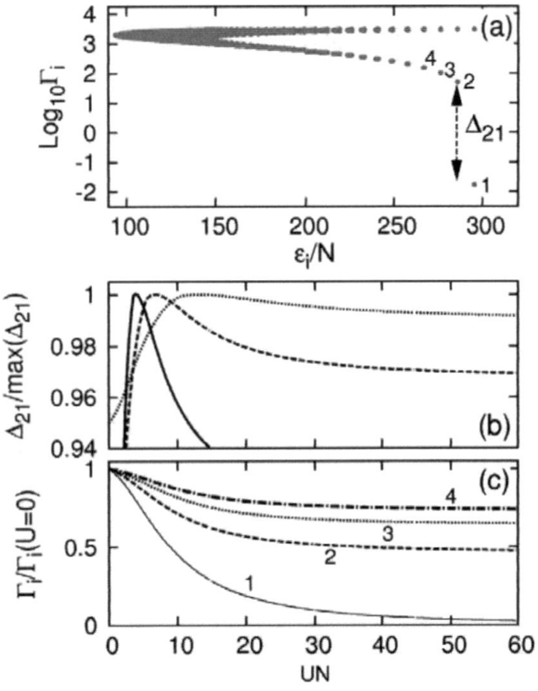

Figure 2. (a) Spectrum of the effective Hamiltonian (3) for a loss rate $\gamma = 20$, interaction strength $U = 2$ and fixed number of particles $N = 60$. Shown are $\Gamma_i = -2\mathrm{Im}E_i$ vs $\epsilon_i/N = \mathrm{Re}E_i/N$. (b) The spectral gap Δ_{21} from (a) as a function of UN for $\gamma = 5$ (solid line), $\gamma = 10$ (dashed line), and $\gamma = 20$ (dotted line). The values on the y-axis are normalized with respect to their maximum in order to allow a better comparison. (c) The decay rates of the four most stable states vs UN for $\gamma = 20$. In (b,c) $N = 60$ is always kept fixed and U was varied. Doing the same for a different value of $N \gg 1$ produces similar results, hence the results are robust as long as $N \gg 1$ (which is necessary for the scaling with UN to be approximately valid).

The qualitative picture remains unchanged for a large range of single-particle loss rates γ in the effective Hamiltonian of equation (3), as shown in figure 2(b). Increasing γ tends to widen the peak in the spectral gap $\Delta_{21} = \Gamma_2 - \Gamma_1$ and to shift the peak position toward larger values of UN (since more interaction energy is necessary to balance the increasing loss at the outer sites, i.e. to avoid decay into them). An interesting effect can be observed for the very large dissipation rate $\gamma = 20$. As one can see in figure 2(b), the normalized spectral gap is almost one even for zero interaction in this case. This means that the decay rate is almost independent of the interaction strength. This happens because strong dissipation blocks the tunneling to the two leaky sites, an effect discussed in detail already in [15, 16].

The most stable eigenstate of $\hat{H}_{\text{eff}}$ with decay rate Γ_1 corresponds to almost one hundred percent to the Fock state with all particles in the middle site, represented by $|0, N, 0\rangle$. This is expected since this state has the largest real part, given essentially by the on-site effective interaction UN, and hence it is most stable by energetic constraints which forbid the hopping of single particles to the leaky boundary wells.

As a final remark, one may find analogies to the phenomenon of resonance trapping in open systems often described by effective non-hermitian Hamiltonians [4, 20]. This phenomenon describes the formation of a gap in the imaginary parts of the spectrum when the opening of the system is increased. The consequence is that very few very stable states exist as compared with a large number of fast decaying ones. This effect appears as the opening, described by γ in our case, increases, whilst in our problem presented above, the stable breather state rather forms when increasing the interparticle interaction strength (and remaining qualitative unchanged over a wide range of γ). Nevertheless it would certainly be interesting to further investigate this analogy for many-body quantum systems.

3. Conclusions

We have shown that the method based on effective Hamiltonians, as the one of equation (3), is a useful tool for the quick identification of states with certain properties, in our case very stable many-body modes. A direct diagonalisation is always possible for small systems, i.e. not too many bosons and lattice sites. For larger many-body systems, a diagonalisation based on the Lanczos method may be used to find interesting states locally in the energy spectrum [8]. For this, the scaling of eigenstates with the system size for smaller systems can guide the search for the position in the complex energy plane. In the future our results will be extended to larger systems, comparing with the approximate evolution techniques used, for instance, in references [15–17] and extending results on one-band systems [5] such as sketched in figure 1(a) to several energy bands.

Acknowledgments

S W and G K thank with pleasure the organiser Thomas Elze for his kind invitation to DICE2012. We are very grateful as well to Dirk Witthaut for our longstanding collaboration. Our work is supported by the the DFG (FOR760), the Helmholtz Alliance Program EMMI (HA-216), and the HGSFP (GSC 129/1).

References

[1] Taylor J R 1972 *Scattering theory: the quantum theory on nonrelativistic collisions* (New York: Wiley)
[2] Breuer HP and Petruccione F 2007 *The theory of open quantum systems* (Oxford: Oxford Univ. Press)
[3] Walls D F and Milburn G J 1995 *Quantum optics* (Heidelberg: Springer)
[4] Rotter I 2009 *J. Phys.* A **42** 153001 (Topical Review)
[5] Tomadin A, Mannella R and Wimberger S 2007 *Phys. Rev. Lett.* **98** 130402; 2008 *Phys. Rev. A* **77** 013606
[6] Arimondo E and Wimberger S 2011 Tunneling of ultracold atoms in time-independent potentials, in *Dynamical Tunneling* ed Keshavamurthy S and Schlagheck P (Boca Raton: Taylor & Francis – CRC Press)
[7] Plötz P, Madroñero J and Wimberger S 2010 *J. Phys.* B **43** 08001(FTC); Plötz P, Schlagheck P and Wimberger S 2011 *Eur. Phys. J.* D **63** 47
[8] Parra-Murillo C A, Madroñero J and Wimberger S 2012 *Preprint* arXiv:1207.4699
[9] Sherson J F, Weitenberg C, Endres M, Cheneau M, Bloch I and Kuhr S 2010 *Nature* **467** 68
[10] Gericke T, Würtz P, Reitz D, Langen T and Ott H 2008 *Nat. Phys.* **4** 949
[11] Würtz P, Langen T, Gericke T, Koglbauer A and Ott H 2009 *Phys. Rev. Lett.* **103** 080404
[12] Bloch I, Dalibard J and Zwerger W 2008 *Rev. Mod. Phys.* **80** 885
[13] Flach S and Gorbach A V 2008 *Phys. Rep.* **467** 1
[14] Witthaut D, Trimborn F and Wimberger S 2008 *Phys. Rev. Lett.* **101** 200402; *do.* 2009 *Phys. Rev.* A **79** 033621
[15] Witthaut D, Trimborn F, Hennig H, Kordas G, Geisel T and Wimberger S 2011 *Phys. Rev.* A **83** 063608
[16] Trimborn F, Witthaut D, Hennig H, Kordas G, Geisel T and Wimberger S 2011 *Eur. Phys. J.* D **63** 63
[17] Kordas G, Wimberger S and Witthaut D 2012 *EPL* **100** 30007
[18] Anglin J 1997 *Phys. Rev. Lett.* **79** 6
[19] Leggett A J 2001 *Rev. Mod. Phys.* **73** 307
[20] Persson E, Rotter I, Stöckmann H J and Barth M 2000 *Phys. Rev. Lett.* **85** 2478

Journal of Physics: Conference Series **442** (2013) 012030 doi:10.1088/1742-6596/442/1/012030

Density-functional theory, finite-temperature classical maps, and their implications for foundational studies of quantum systems

M W C Dharma-wardana

National Research Council of Canada, Ottawa, Canada, K1A 0R6

E-mail: chandre.dharma-wardan@nrc-cnrc.gc.ca

Abstract. The advent of the Hohenberg-Kohn theorem in 1964, its extension to finite-T, Kohn-Sham theory, and relativistic extensions provide the well-established formalism of density-functional theory (DFT). This theory enables the calculation of all static properties of quantum systems *without* the need for an n-body wavefunction ψ. DFT uses the one-body density distribution instead of ψ. The more recent time-dependent formulations of DFT attempt to describe the time evolution of quantum systems without using the time-dependent wavefunction. Although DFT has become the standard tool of condensed-matter computational quantum mechanics, its foundational implications have remained largely unexplored. While all systems require quantum mechanics (QM) at T=0, the pair-distribution functions (PDFs) of such quantum systems have been accurately mapped into classical models at effective finite-T, and using suitable non-local quantum potentials (e.g., to mimic Pauli exclusion effects). These approaches shed light on the quantum → hybrid → classical models, and provide a new way of looking at the existence of non- local correlations without appealing to Bell's theorem. They also provide insights regarding Bohmian mechanics. Furthermore, macroscopic systems even at 1 Kelvin have de Broglie wavelengths in the micro-femtometer range, thereby eliminating macroscopic cat states, and avoiding the need for *ad hoc* decoherence models.

1. Introduction.

In 1964, Hohenberg and Kohn [1] proved a theorem asserting that the ground- state properties of a stationary, non-relativistic system can be calculated from a variational principle involving only the one-body density $n(\vec{r})$ of the system, without recourse to the Schrödinger equation and its wavefunction. The theorem was soon generalized to finite-T [2], and to relativistic systems [3]. Thus it became clear that all thermodynamic properties of n-particle quantum systems, entangled, interacting, or not, could be calculated without recourse to the n-body wavefunction. The method, re-written in the form of the Kohn-Sham theory has now become the preferred method in computational quantum mechanics (QM). Furthermore, since linear transport properties depend only on equilibrium correlation functions (c.f., the Kubo relations), a large class of quantum phenomena depends only on the density $n(r)$, a measurable quantity having a familiar 'ontology' in quantum as well as classical mechanics. Unlike the many-electron wavefunction of conventional QM, the one-body density is an observable. For instance, the X-ray scattering from a specimen of matter gives direct information regarding $n(r)$. The development of 'time-dependent' density functional theory [4], if substantiated, implies that density functional theory may enable a very different discussion of quantum phenomena than has been the custom in foundational studies. In this article, for the sake of brevity and simplicity, we limit ourselves mainly to non-relativistic equilibrium systems. In the following presentation we look at DFT and compare it with relevant aspects of Bohmian mechanics, treatment of hybrid systems within DFT, as well as classical representations inspired by DFT ideas. Furthermore, finite-T physics leads us to conclude that there is no need for *ad hoc* decoherence theories to eliminate macroscopic cat states from QM.

2. Density functionals and Bohmian mechanics

To be specific, we consider a system of electrons subject to an external potential $v_{ext}(\vec{r})$, and interacting with one another via the Coulomb potential. In many cases this external potential is produced by a fixed set of nuclei, as in molecular physics or quantum chemistry calculations at $T = 0$. However, the ions cannot be treated as fixed in dealing with liquid metals and hot plasmas. Hence, in discussing true hybrid systems, we allow the nuclei (or ions) to be a second interacting subsystem, interacting with the electron subsystem, and the finite-T discussion becomes not only appropriate, but also necessary. However, in simplified models, e.g., the jellium model, the ions provide a fixed uniform positive distribution of charge that exactly neutralizes the negative charge of the electron distribution. The $q \to 0$ singularities in the Coulomb potential $4\pi/q^2$ for electron-electron or ion-ion interactions are exactly canceled by the corresponding term in the ion-electron interactions.

However, while keeping such specific models in mind, we can simply work with an unspecified external potential v_{ext} and a corresponding one-body electron distribution $n(r)$. Then according to Hohenberg and Kohn [1], the exact ground state density $n(r)$ and the ground-state energy $E_v[n]$ corresponding to that external potential v_{ext} are obtained by the minimization of the following energy functional.

$$E_v[n] = T[n] + U_{cou}[n] + \int d\vec{r} v_{ext}(\vec{r})n(\vec{r}) \tag{1}$$

The kinetic energy contribution, viz., $T[n]$, and the Coulomb electrostatic energy $U_{coul}[n]$ are said to be 'universal functionals' of $n(r)$, in the sense that they do not depend on the external potential $v_{ext}(\vec{r})$ specific to the system. The original Hohenberg-Kohn theorem was established for non-degenerate ground states at $T = 0$. This has since been restated within the language of functional analysis [5], while degenerate ground states, spin-polarized systems, finite-T systems, relativistic systems, multi-component hybrid systems, superconductors, liquids and plasmas etc., have been studied within the first three decades of DFT [7].

It is clear that a direct minimization of $E_v[n]$ via $\delta E_v/\delta n(\vec{(r})$ would require a knowledge of the kinetic energy functional $T[n]$, and the electrostatic potential $U_{cou}[n]$. The latter is a known functional of $n(r)$, but $T[n]$ is unknown. Although this is a universal functional of $n(\vec{r})$, it turns out to be highly non-local. Hence gradient expansions merely lead to Thomas-Fermi like theories. These, sometimes known as 'orbital-free DFT' [6], presently lack the full accuracy of methods that use a wavefunction. Since DFT is presented via a one-body density, the Kohn-Sham formulation of DFT uses a one-body 'Kohn-Sham wavefunction' to construct a non-local kinetic energy, where a single Kohn-Sham electron moves in the electrostatic potential of the other electrons and an additional exchange-correlation potential $V_{xc}[n]$ which is a functional of the one-body density. Thus, the many-electron problem is replaced by an effective one electron problem in Kohn-Sham theory. Here the Kohn-Sham electron moves in a non-local Kohn-Sham potential that is made up of the electrostatic potential $U_{cou}[n]$, and the non-local $V_{xc}[n]$. However, unlike $T[n]$, approximate models of $V_{xc}[n]$ have proved to be quite successful, when used in the Kohn-Sham equation where $T[n]$ is constructed for each case, instead of relying on a knowledge of the universal functional $T[n]$.

It is of interest at this stage to examine Bohmian mechanics, where the Schrödinger many-particle wavefunction $\psi(\vec{x}_1, \cdots \vec{x}_n)$ is used to construct a non-linear equation of motion containing the external potential as well as a quantum potential $Q(\vec{x}_1, \cdots \vec{x}_n)$. Let us limit ourselves to a single particle problem in 1-D for the sake of brevity. Bohm [8] writes the wavefunction in the form:

$$\phi(x,t) = R(x,t)\exp\{iS(x,t)/\hbar\} \quad (2)$$

$$R(x,t) = \sqrt{n}(x,t);\ n(x,t) = |\phi(x,t)|^2 \quad (3)$$

so that R (positive), and S are real functions. Substituting this form of the wavefunction into the Schrödinger equation, one obtains Bohm's form of the equations as:

$$\frac{\partial n}{\partial t} + \nabla.\left[n\frac{\nabla S}{m}\right] = 0 \quad (4)$$

$$\frac{\partial S}{\partial t} + v_{ext} + \frac{(\nabla S)^2}{2m} + Q(x,t) = 0 \quad (5)$$

$$Q(x,t) = \frac{-\hbar^2}{2m}\frac{\nabla^2 R}{R} \quad (6)$$

Eq. 4 is simply the equation of continuity. The term $(\nabla S)^2/2m$ in Eq. 5 is interpreted as the kinetic energy of a point-like particles with momentum $p = \nabla S$. In effect, Eq. 5 has the form:

$$\partial S/\partial t + H(p,x) + Q(x,t) = 0 \quad (7)$$

This is exactly like a classical Hamilton-Jacobi equation with the extra potential $Q(x,t)$, called the quantum potential, or the Bohm potential. Since we are interested in the simplest form of DFT (systems in equilibrium) in this presentation, let us consider an electron confined within an infinite potential well of width a. The eigenfunctions are of the form $\psi_n(x) = \sqrt{(2/a)}\sin(k_n x)$; $n = 1, 2, \cdots$. It is easily shown that the particle current $j(x,t)$ is zero, and the Bohmian particle is *at rest* in some unknown location x in the well, with the probability $|\psi_n(x)|^2$ of *being* there. Einstein felt that the particle being at rest in the well was not the correct physical picture. A more detailed discussion of the Bohmian particle in a well, and a possible resolution to Einstein's objections [9] are given in Chapter 6 of Ref. [10].

Does the Bohmian particle in the well have no kinetic energy? It is easy to show that the kinetic energy is completely resident in Bohm's quantum potential while the particle current is zero. In fact, the kinetic energy functional $T[n(x)]$ per electron confined in the quantum well, given by Hohenberg-Kohn theory (or conventional QM) is essentially the quantum potential of Bohm. Although some advocates of Bohmian mechanics have moved away from the quantum potential (mainly due to philosophical reasons), we find it to be of a sort similar to the effective potentials based on density-functionals used in condensed-matter many-body physics.

For instance, one of the objections adduced against the quantum potential is that it has the 'unusual effect' of making the particle density n influence the dynamics of individual particles. However, that is precisely the message of DFT. It is typical of many-particle physics that the individual dynamics is influence by the over-all particle distribution. For instance, a low-order (mean-field) many-body theory like the Hartree theory of many electrons requires that an electron moves in the many-body field of the other electrons produced by the density distribution $n(x)$. Hartree-Fock theory involves a highly non-local Hartree-Fock potential dependent on the density $n(r)$.

Interestingly, Dürr, Goldstein and Zhangi [11] summarize their attitude to the quantum potential with the statement, 'We believe that the most serious flaw in the quantum-potential formulation of Bohmian mechanics is that it gives a completely wrong impression of the lengths to which we must go in order to convert orthodox quantum theory into something more rational. The quantum potential suggests ... we must incorporate into the theory a quantum potential of a *grossly* non-local character'. We emphasized the word 'grossly' in the above to high-light where the attitudinal squeamishness resides. In contrast, most physicists who use the highly non-local Hartree-Fock potential as a basic first step in their calculations have given up any such squeamishness. Regarding 'the lengths' needed to reach the classical Hamilton-Jacobi form, as seen from above, we need only a few lines.

The insights from DFT enable us to consider the Bohm potential as a functional of the density, without invoking a wavefunction. Furthermore, Eq. 6 may be used as a differential equation for the direct determination of Q given in terms of $R = \sqrt{n}$, subject to satisfying the equation of continuity, and minimizing the energy. Once R is determined by iteratively solving Eq. 6, S can be obtained from the continuity equation. In effect, a time-dependent system is described by the four vector consisting of the current $j(x,t)$ and the density $n(x,t)$. These are precisely the quantities treated in DFT, and in Bohmian mechanics.

The above considerations suggest that Bohmian mechanics can be formulated entirely in a language free of the wavefunction, using only density functionals, exactly as in DFT.

3. Pair-distributions and classical maps

The Hohenberg-Kohn theorem reduces the n-electron problem to that of a functional of an effective one-body problem and an exchange-correlation potential $V_{xc}[n]$. Given that the basic electron-electron interactions (that lead to quantum correlations including those of the Bell type) are pair-interactions and Pauli exclusion effects, the reduction to a one-body problem achieved in DFT appears a bit mysterious, and one would expect that fundamental two-particle correlations are embedded in the V_{xc}. In fact, the latter is most transparently evaluated as a coupling constant integration over the pair-distribution function (PDF), i.e., $g(\vec{r}_1, \vec{r}_2)$ of the system. This is the probability of finding an electron at r_2, given an electron at r_1.

The PDF is well-defined both classically and quantum mechanically. In QM, the PDF is calculated from the many-particle wavefunction by integrating over all except two space variables $\vec{r}_1, \vec{r}_2$ of two electrons. In a classical system, we can consider a uniform electron fluid (as in jellium), with a mean density n and Wigner-Seitz radius $r_s = \{3/(4\pi n)\}^{1/3}$. We position the origin on the first electron, so that $\vec{r}_1 = 0$. Then, invoking radial symmetry and uniformity, we can use $g(r)$ rather than $g(\vec{r})$ for the PDF, where r is the radial distance of the (second) electron from the origin. When classical mechanics holds (i.e., for $T > 0$ and for very small de Broglie wavelengths), this $g(r)$ for an interacting Coulomb fluid can be calculated using the modified-hyper-netted-chain (MHNC) equation [21], or via molecular dynamics (MD).

In QM, even the non-interacting electron system is of interest if all the particles have the same spin. Then the Pauli exclusion principle controls the entanglement of the n-electron system. This non-interacting set of n-entangled electrons is described by a Slater determinantal wavefunction. The fully spin-polarized system (parallel spins σ), and the fully unpolarized system (anti-parallel spins $\sigma_1 \neq \sigma_2$) are of interest. They can be analytically treated at $T = 0$ as in Ref. [12], and extended to finite-T as in Ref. [13]. Furthermore, the Pauli exchange effect (which leads to entanglement since the wavefunction is a Slater determinant) can be re-written as a non-local quantum potential which is a universal function of r/r_s. This 'Pauli-exclusion' potential $P(r/r_s)$ can be constructed such that it reproduces the quantum $g(r)$ when used in the classical MHNC equation. This provides an interesting and useful classical map of the quantum problem that has been very fruitful in applications to interacting quantum problems ranging from 2-D electron layers [14], hydrogen plasmas, electrons in graphene, and quantum dots [15]. That is, correlated, interacting quantum problems can be replaced by equivalent classical problems where pair-potentials involve additional potentials containing quantum effects which are usually non-local. Bohmian mechanics achieves this directly from the $R = \sqrt{(n)}$ part of the wavefunction. DFT

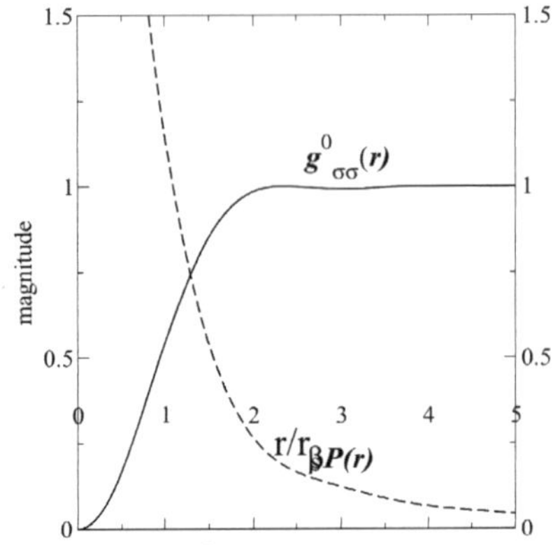

Figure 1. The pair-distribution function $g^0_{\sigma\sigma}(r)$ gives the probability of finding a σ-spin electron at r if there is already a σ-spin electron at the origin (Coulomb repulsion neglected). We plot r/r_s on the x axis, and the result is valid for any density, r_s being the sphere-radius containing an average of one electron. Thus the 'entanglement' due to Pauli exclusion holds even if r_s is a lunar distance. The Pauli exclusion is exactly mimicked by a scale-independent repulsive potential $\beta P(r)$ when used in a classical MHNC equation.

achieves this via functionals of the one-body density $n(r)$. Naturally, non-local contributions are usually quite important in both DFT and Bohmian mechanics.

The PDF of non-interacting fully spin-polarized entangled electrons, denoted by $g^0_{\sigma,\sigma}(r)$ is shown in Fig. 1. It has the form:

$$g^0_{\sigma\sigma'}(r) = 1; \quad \sigma \neq \sigma' \tag{8}$$
$$g^0_{\sigma\sigma}(r) = 1 - \{3j_1(rk_F)/(rk_F)\}^2 \tag{9}$$

(The large-r behavior is normalized to unity, whereas if $\zeta = 0$, it is natural to normalize the spin-resolved functions to 1/2). Here $j_1(x)$ is the spherical Bessel function, while its argument x contains $k_F = 1/(\alpha r_s), \alpha = 0.52106$, where k_F is the Fermi momentum. The probability of finding an electron at r, given that there is an electron at the origin of coordinates is affected by their mutual Coulomb interaction. This is not included in the above calculation, and hence the superfix 'zero' in g^0. A treatment including the Coulomb potential is given in Ref. [13], where it is shown that remarkably good agreement with Monte-Carlo quantum simulations is obtained for these classical maps.

Since the PDF scales as r/r_s, the correlations remain undiminished, whatever be the mean-separation $R = 2r_s$ that is imposed on any two electrons in the entangled system. That is, *local realism* as required by the Einstein-Podolsky-Rosen gedanken experiment does not hold. Thus the correlations unearthed by the Bell inequalities are very transparently and simply revealed by an examination of the PDF of correlated electron systems. This topic is discussed in greater detail in Chapters 7 and 8 of Ref. [10].

4. Finite temperatures, hybrid systems and decoherence

Formal discussions of hybrid systems (i.e., systems containing quantum particles as well as classical particles) [16, 17] rarely bother to discuss the role of temperature, partly because some of these discussions deal with model problems dealing with just one or two particles. In reality, all particles in any system obey quantum mechanics. However, if some particles have de Broglie wavelengths λ smaller than some assigned length scales associated with the system, then such particles among themselves can be treated within classical mechanics. The interactions between the particle deemed to be classical, and those deemed to be quantum occur via a Hamiltonian H_{int} whose characteristic length scales and energy scales have to be investigated separately. In the following we look at systems at finite temperature T.

If a system is coupled to a heat bath and held at some finite temperature T, the average kinetic energy of the particles is a function of T. In the classical limit the average kinetic energy of an atom is $3T/2$, using energy units for T where the Boltzmann constant becomes unity. At $T = 0$, for our model system of fermions (jellium), the kinetic energy becomes $3E_F/5$, where E_F is the Fermi energy. The latter can be thought of as a temperature $T = 2E_f/5$. Hence we see that we can ascribe a de Broglie wavelength λ and a thermal momentum mv to any particle in a system at an effective temperature T, with v and λ given by:

$$v = \sqrt{3T/m}, \quad \lambda = h/p = h/\sqrt{3mT} \tag{10}$$

It is clear that λ is infinite for systems at $T = 0$, or for sufficiently dilute quantum systems ($E_F \to 0$ as $n \to 0$). Thus all systems become quantum systems as $T \to 0$ since their de Broglie wavelengths become infinite. However, if at some finite (even if low) temperature, e.g., 1 Kelvin, if the mass of a particle is sufficiently large, then if $\lambda \ll 1$ atomic unit, we may assume that superposition and entanglement become unimportant. Such heavy particles are classical particles. Hence, since all systems are quantum systems at $T = 0$, and heavy particles become classical at some finite-T, any quantum theory of hybrid systems [16] has to ultimately incorporate a quantum treatment of thermal fields.

The conventional quantum approach to finite-T problems is difficult, and involves a typical doubling of the Hilbert space as in Umezawa's thermofield dynamics [18], or in the Martin-Schwinger-Kedysh contour technique [19]. These methods are useful for tackling a class of non-equilibrium physics. However, even static finite-T quantum calculations (e.g., via the Matsubara method [12]) are restricted to perturbation theory.

On the other hand, finite-T density functional theory [2] presents a computationally and conceptually much simpler approach to finite-T problems, as well as two-temperature quasi-equilibria. Here the finite T one-body density functionals $n(\vec{x}, T)$ determine all the thermodynamics and linear transport properties of the system *without* the need for a wavefunction.

The explicit construction of a DFT and the calculation of such finite-T hybrid systems for a system of protons (classical) and electrons (quantum) at finite-T has been given in Ref. [20]. The method is applicable to any electron-ion system, where the ions are treated classically. The usual Born-Oppenheimer approximation is not needed. If the ion distribution is denoted by $\rho(\vec{r})$, and the electron distribution is $n(\vec{r})$, the total free energy $F([\rho],[n])$ is a functional of both distributions. Hence we have two coupled Hohenberg-Kohn-Mermin type equations to determine the finite-T distributions of the hybrid system. That is, the functional derivatives with respect to variations $\delta n, \delta\rho$ lead to two coupled Euler-Lagrange equations.

$$\frac{\delta F[\rho, n]}{\delta n} = 0, \quad \frac{\delta F[\rho, n]}{\delta \rho} = 0. \tag{11}$$

The first of these equation can be reduced to a Kohn-Sham equation for electrons at finite-T, or simply retained as a Hohenberg-Kohn variational form. The second equation leads to a classical Hohenberg-Kohn equation for the ion distribution that is identified as a form of the MHNC equation. In effect, it is a Boltzmann distribution for the density ρ in terms of the 'potential of mean force' (PMF) used in the statistical mechanics of liquids [21]. The PMF is the appropriate Kohn-Sham potential for this classical system. The full practical implementation of this approach to a typical hybrid system (an interacting gas of electrons and protons at finite T inclusive of the formation of hydrogenic bound states) is given in Ref. [20].

4.1. Decoherence and macroscopic cat states

The equation 10 is of great importance in showing us that there is no need for *ad hoc* theories of decoherence to eliminate macroscopic cat states from QM. Let us consider a 1 kg cat at room temperature (300 K, i.e., 9.500×10^{-4} Hartees). The atoms in the cat have internal energies with approximately $3T/2$ arising from translational motion. The center of mass will also make random oscillatory motions about an equilibrium position consistent with its temperature and any motion of the cat. The corresponding center-of-mass de Broglie wavelength λ_c turns out to be

$$\lambda_c \simeq 9.45 \times 10^{-23} \text{ meters} \tag{12}$$

The 'radius' of a proton is about 0.88 femtometers (1 fm=10^{-15} meters). The de Borglie wavelength of the cat is about one hundred-millionth of the size of a proton! Even when cooled to one Kelvin, the de Broglie wavelength remains totally negligible. Even a free electron close to a cat's body would not get entangled with the cat's quantum energy states. Matter has to be squashed to densities where nuclear reactions begin before superpositions are possible at such small values of λ_c. Clearly, Schrödinger's cat states of macroscopic bodies (entanglements or superpositions) do not exist in nature except at T=0, or possibly in white dwarfs.

One may argue that the mass M of the center of mass is not the proper quantity to use. One may claim that QM should be applied to some molecules of the cat that are specific to it being dead or alive in Schrödinger's cat paradox. In that case we are no longer applying QM to a macroscopic system and then there is no difficulty in having superpositions or entangled states. Chemical reactions (e.g., cyanide reacting with the cat) are simply processes where reactive atomic groups enter into entanglements. Such processes pose no difficulty and are not 'paradoxical'.

One may also question the use of a momentum calculated from $3T/2$. This is the classical kinetic energy of an ideal gas. We know from Dulong and Petit's law (or from the more modern Einstein-Debye theory of solids) that classically, a system held together by chemical bonds, e.g., a cat, has three degrees of vibration per atom, and three degrees of translation, each having an energy $T/2$. For solids in the low-temperature limit, we need the Fermi energy of the system. Such detailed treatments merely introduce numerical factors of little consequence. The basic conclusion that λ_c is very very small remains firm.

These considerations imply that explicit decoherence theories like those of Penrose ('objective reduction' due to quantum gravity) [22], or the 'spontaneous localization' model of Ghirardi, Rimini and Weber [23] are unnecessary additions to quantum mechanics, if we grant that the relevant de Broglie lengths are much smaller than even nuclear radii.

5. Conclusion

The tradition of studying just one or two quantum particles, usually in a two-state model has a long history in foundational studies. However, if the many-body problem is examined, QM can be approached using one-body density functionals, without recourse to wavefunctions, both at $T = 0$ and at finite-T. An examination of Bohmian mechanics within the perspective of DFT is conceptually rewarding and clarifies the nature of the Bohmian quantum potential. Finite-T quantum mechanics leads us to consider the thermal de Broglie wavelength which, being negligible for macroscopic systems, ensures that there are no macroscopic Schrödinger cat states.

References

[1] Hohenberg P and Kohn W 1964 *Phys. Rev.* B **136** 864

[2] Mermin N D 1965 *Phys. Rev.* A **137** 1441
[3] Rajagopal A K and Callaway J 1973 B *Phys. Rev* . **7** 1912;
MacDonald A H and Vosko S H 1979 *J. Phys.* C **12** 2977
[4] Runge E and Gross E K U 1984 *Phys. Rev. Lett.* **52** 997
[5] Lieb E 1985 *Density functional methods in Physics* NATO Series **B123** ed R M Dreizler and J Providencia (New York: Plenum)
[6] Karasiev V V and Trickey S B 2012 *Preprint* arXiv:1109.6602
[7] Gross E K U and Dreizler R M 1993 *Density Functional Theory* NATO ASI Series **337** (New York: Plenum)
[8] Bohm D 1952 *Phys. Rev.* **85** 166; *do.* 1953 *Phys. Rev.* **89** 1040
[9] Einstein A 1953 *Scientific papers presented to Max Born* (*Festschrift*) (New York: Hafner) p 33 and Bohm D in the same volume, p 13-19
[10] Dharma-wardana M W C 2013 *A physicist's view of Matter and Mind* (World Scientific Singapore)
[11] Dürr D Goldstein S and Zhangi N 1995 *Bohmian Mechanics and Quantum Theory: An Appraisal* ed J T Cushing, A Fine and S Goldstein (New York: Kluwer Academic Press) (*Preprint* quant-ph/9511016)
[12] Mahan G D 1981 *Many-Particle Physics* (New York: Plenum)
[13] Dharma-wardana M W C and Perrot F 2000 *Phys. Rev. Lett.* **84** 959
[14] Perrot F and Dharma-wardana M W C 2001 *Phys. Rev. Lett.* **87** 206404
[15] Dharma-wardana M W C 2012 *Int. J. Quant. Chem.* **112** 53
[16] Heslot A 1985 *Phys. Rev.* D **31** 1341
[17] Elze H-T 2012 *Phys. Rev.* A **85** 052109
[18] Umezawa H 1993 *Advanced field theory: Micro, Macro and Thermal Physics* (New York: AIP)
[19] Rammer J and Smith H 1986 *Rev. Mod. Phys.* **58** 323
[20] Dharma-wardana M W C and Perrot F 1982 *Phys. Rev.* A **26** 2096
[21] Hansen J-P and MacDonald I R 2006 *Theory of Simple Liquids* (London: Academic)
[22] Penrose R 1996 *Gen. Relativ. Grav.* **28** 581
[23] Ghirardi C, Rimini A and Weber T 1986 *Phys. Rev.* D **34** 470

The supramolecular structure of liquid water and quantum coherent processes in biology

A De Ninno[1], A Congiu Castellano[2] and E Del Giudice[3]

[1] Centro Ricerche ENEA Frascati, via Enrico Fermi, 27 00044 - Roma, Italy
[2] Dept. of Physics, Uni. of Rome Sapienza, p. Aldo Moro, 00100 Roma, Italy
[3] via Friuli 21, 20135 Milano, Italy

E-mail: antonella.deninno@enea.it

Abstract. Vibrational spectroscopy provides a powerful tool to understand the molecular structures. When applied to the liquid water, this technique reveals so many details which can also shed a light on the supramolecular arrangement of the most ubiquitous of the substances. In particular, the two fluid model of water, proposed several decades ago, founds experimental evidence. Moreover, some fundamental parameters calculated in the realm of the theory of Quantum ElectroDynamics applied to liquid water can be actually measured showing an excellent agreement with the theory. This allows to add a dynamical origin to the mixed cluster model of water well known by the biologists for fifty years and opens the way to the dawn of a real quantum biology.

1. Introduction

The study and the comprehension of the dynamics of water have been often thought to be at the basis of most outstanding problems in modern biology, such as protein folding and the fast communication pathways between biological molecules. Several models of liquid water have been proposed so far, starting from the suggestive two-phase picture proposed by Roentgen in 1892. Some decades later, Linus Pauling introduced the concept of the hydrogen bond (HB), and the properties of the continuum were explained by a network formed of HBs and distorted by the temperature. For many decades, water has been described as a collection of independent molecules whose cohesion was only secured by short-range forces. Long-range cohesion was explained by introducing the concept of a statistically interconnected network that consisted of random HBs that were subject to continuous restructuring as a result of thermal collisions. The widespread diffusion of the computer modelling and the very smart calculation capability of contemporary informatics have been the basis of a considerable number of Molecular Dynamic (MD) simulations. The one most commonly applied are derived from non-polarisable model of water which consists of three or four interaction sites located on the atomic centres of oxygen and hydrogen, the fourth being located along the H-O-H bisector. Using this conceptual framework it is possible to calculate the component of the electric field exerted on the proton by all the surrounding molecules and projected onto the OH covalent bond. The emerging picture is that water is made up of a single population of absorbers whose energy varies continuously. However, if the MD is not dynamic in nature then it cannot provide a dynamic scenario. Moreover by fitting a number of different sets of experimental data, it has been shown that all the anomalies of liquid water may be explained in terms of domains of different density. Such a model assumes the existence of two states that have different specific volumes at a given pressure p and temperature T. In this paper we explore the features of the IR spectra of liquid water with the aim to show that the experimental results obtained by several experimentalists can hardly be discussed and strongly suggest the existence of two kinds of H_2O molecules in liquid water. Moreover these two kinds of molecules can be attributed to the two quantum levels system described by Giuliano Preparata and coworkers in their theory formulated in last two decades.

2. The vibrational spectrum of water

The IR spectrum of liquid water is composed by two bands, one centred around 1650 cm^{-1} having a a Gaussian line shape and another much wider ranging from 3000 to 3800 cm^{-1} (see fig. 1). The first one has been attributed to the bending motion of the H_2O molecule, while the latter to the stretching vibrations. For many decades this spectrum has been used both to support and to confute the two-phase nature of liquid water and its opposite, the continuum picture. Theories of the two-phase nature were first proposed in the 19th century by Roentgen but they were discarded according to the statement of Bernal and Fowler in 1933 that molecules having the same thermodynamic properties cannot show any differences in the behaviour. More recently computer simulations have been extensively used to check models calibrated against sets of experimental data (e.g. radial distribution function, heat of vaporization). However, none of the existing models is able to reproduce the dynamic properties of the liquid such as the rate of formation of H-bonds and even the vibrational spectrum is obtained by Molecular Dinamics (MD) only with the use of "ad hoc" hypothesis. The MD modelling of the shapes of the H_2O IR and Raman spectra and their simple temperature dependencies could seem to support the continuum model of water [1, 2, 3]. According to the viewpoint of liquid-state statistical mechanics the continuum picture emerges quite naturally in case the system is at a sufficiently high temperature and low density in order to assume the mutual independence of the component molecules so that the ensemble of molecules should be described by a density matrix. In case of "isolated" chromophores, the transition frequency of a vibrational mode will fluctuate on some characteristic time scale. In case this time rate is sufficiently slow the line shape will be simply the distribution of the frequencies, at the opposite, when the rate is much higher, the effect of fluctuation dynamics will appear into the spectrum and the line shape will be narrower than the distribution of the frequencies. However, in case of liquid water the situation is quite more complicated because of the intermolecular and intramolecular coupling. One recently proposed solution calculates the line shape as a distribution of time averaged frequencies where the time of averaging is a parameter obtained by fitting the experimental line shape [4]. However, a careful inspection of the line shape does not show a unique Gaussian line, actually it is well known that the broad absorption peak from 2800 to 3800 cm^{-1}, attributed to the OH stretching mode, can be deconvoluted into line shapes which are usually assigned to different bond ordering parameters [5] . The deconvolution in sub-bands could be done mathematically in infinitely many ways; this is a trivial consequence of the central limit theorem which states that independent random variables (in the present case the number of H-bonds) are normally distributed, i.e. tend to be distributed around a mean value. However every hypothesis should be physically reasonable and experimentally proved. Moreover, the inspection of the second derivative of the IR spectrum reveals three minima located respectively to 3200, 3400 and 3600 cm^{-1} and the mean squared error χ^2 of the deconvolution in three sub-bands is much smaller than in case of a single Gaussian fit. Thus it can be safely assumed that the best deconvolution of the stretching peak is done by using three (Gaussian) substructures respectively located at approximately 3200, 3400 and 3600 cm^{-1}. It is quite interesting to address the problem of the dependence of the peaks on the co-ordination of the involved molecules by comparing the spectroscopic features of gaseous, liquid and solid water. It is apparent that in vapour, where molecules can be assumed safely to be independent, the peaks at 3200 and 3400 cm^{-1}, observed in condensed water are absent,

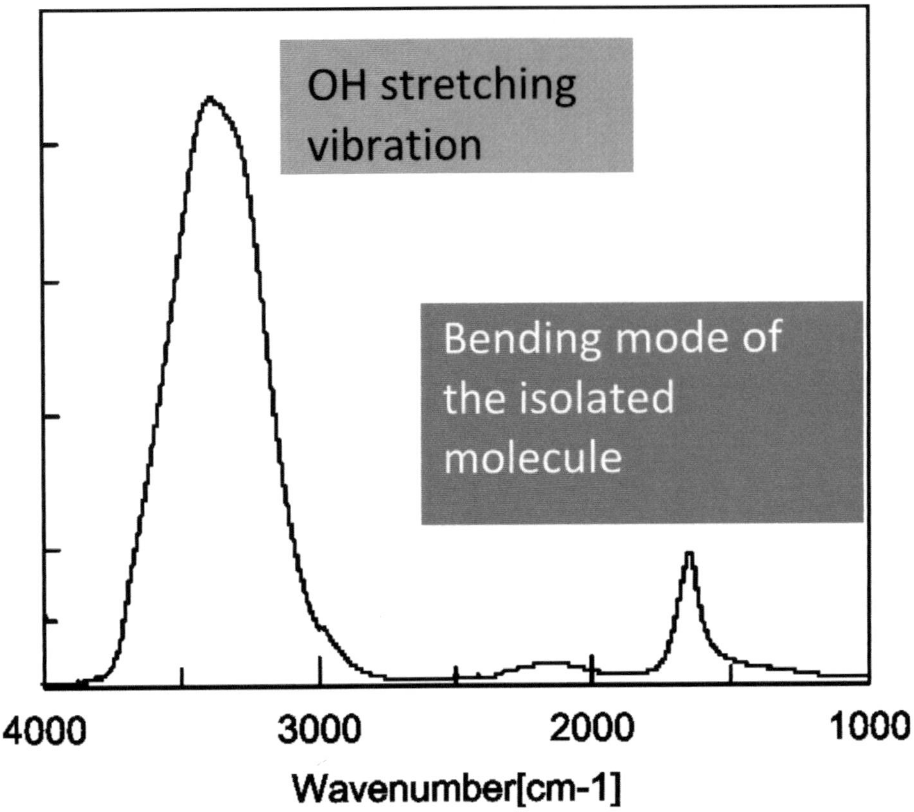

Figure 1. IR spectrum of liquid water at room temperature

whereas we can observe a number of peaks whose position is slightly higher than 3600 cm^{-1}. On the contrary in ice the dominant feature is the peak at 3200 cm^{-1} while the other features fade away by decreasing temperature. Since we know that component molecules in ice are mutually correlated we can assume that the peak at 3200 cm^{-1} is produced by the sub ensemble of molecules which are mutually correlated. In liquid, where we can observe three peaks, we should assume the simultaneous presence of a fraction of correlated molecules, a fraction of uncorrelated molecules and an intermediate fraction (responsible of the peak at ~ 3400 cm^{-1}) made up of water molecules that are connected with other molecules, but are unable to develop the full connections of the previous network.

3. The two-phase model of liquid water

On the basis of the above experimental observations, let's assume that the spectral intensity $I(\omega, T)$ can be decoupled in two components: correlated water (1) and non-correlated water (2), the latter being the sum of what we call intermediate and free molecules, whose fractions has been observed to depend on the temperature T [5, 7]:

$$I(\omega, T) = I_1(\omega, T) + I_2(\omega, T) \tag{1}$$

In a two components system (in which the two components are in chemical equilibrium) an isosbestic point exists, which is a point of equal absorption, invariant with respect to the relative concentrations of the two components. This is a very basic property of the system in equilibrium and it has been discussed widely in literature[8, 9, 10]. A van 't Hoff plot can be

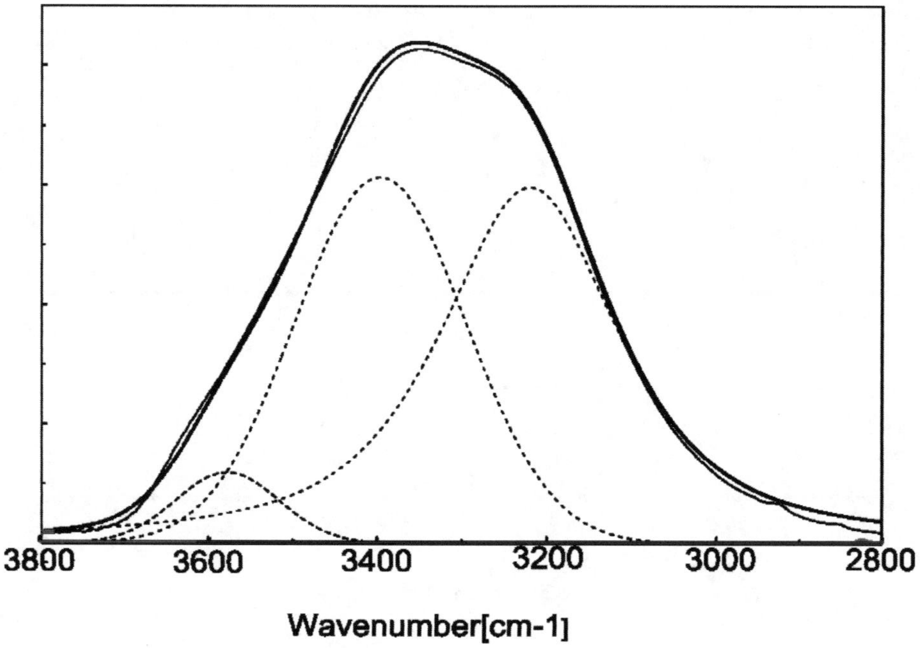

Figure 2. Deconvolution in sub-bands of the stretching peak in fig. 1.

done by bisecting the spectrum at the isosbestic point and plotting the natural logarithm of the ratio of the areas above and below that point versus 1/T. The slope of that curve gives the energy difference between the two vibrational states 1 and 2:

$$ln\frac{I_1(\omega,T)}{I_2(\omega,T)} = \frac{E_2 - E_1}{RT} + c \tag{2}$$

We have measured the IR spectrum of water in a range 25-60 C^o. Fig. 3 shows the Van't Hoff plot we obtained for the model of the two populations, namely correlated and non correlated (including the intermediate population). The picture shows a perfect linearity thus confirming that the difference of energy of the two phases does not depend on temperature in the range under investigation. The slope of the curve $\triangle E/R$ (R is the ideal gas constant) gives an enthalpy difference between the two states

$$\triangle E = 2.9 \pm 0.6 Kcal/mol \tag{3}$$

which is in reasonable agreement with various estimates in the literature $\triangle E = 2.5 \pm 0.6$ Kcal/mol [9] and 2.8 ± 0.4 Kcal/mol [17]; this value is usually associated to the enthalpy change in the hydrogen-bond rupture for water. However, this value is quite lower than 4.5 Kcal/mol the value for the rupture of the H-bond estimated by Linus Pauling from the enthalpy of sublimation of ice. The intercept of the curve is equal to molar entropy change $\triangle S/R$ between the two populations, from our data we found

$$T\triangle S = 2.7 \pm 0.2 Kcal/mol \tag{4}$$

per particle at T=300K, which almost coincides with the above estimate for $\triangle E$ showing that the decrease of energy of the coherent fraction with respect to the non coherent comes from

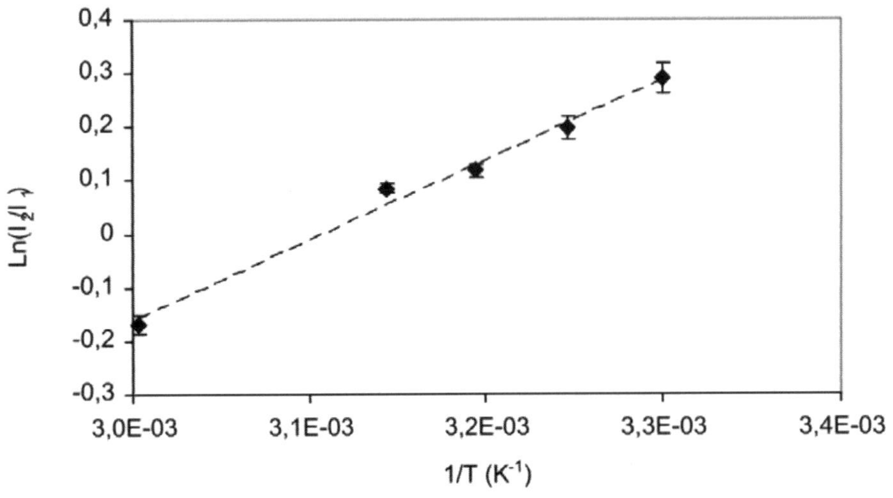

Figure 3. Van't Hoff plot

the vanishing of the entropic component[1]

Since the beginning of the last century, it has been widely accepted that the attractive forces acting between two neutral atoms or molecules separated by a distance large in comparison with their dimensions (Van der Waals Forces) are electromagnetic in origin. A theory of long-range attractive forces in condensed matter was first proposed in 1961 by Dzyaloshinskii, Lifshitz and Pitaevskii, using Quantum Field Theory [11]. They proposed that "the interaction between bodies is considered to take place through a fluctuating electromagnetic field". However this interaction does not induce a phase correlation among components.

The existence of water in two different populations made up by molecules having different degree of mutual correlation has been demonstrated in the framework of Quantum Electrodynamics (QED). According to general principles of Quantum Field Theory, an ensemble of electrically charged atoms or molecules (nuclei, electrons) is coupled with the quantum electromagnetic fluctuation of the vacuum. It can be proved that when the temperature T is lower than a certain critical temperature $T \leq T_{crit}$ and the density is grater than a critical density $N \geq N_{crit}$ the ensemble is subjected to a collective coherent oscillation between a couple of internal levels of its components, in tune with the electromagnetic field [12, 13, 14, 15]. These fields induce long-range forces which account for the actual existence of the condensed state. The short-range static attractions come into play only after the molecules have been brought sufficiently close together by the attraction of the long-range force. Liquid water is peculiar because of its electronic configuration. In the ground state all electrons are tightly bound, whereas in the excited configuration E = 12.06 eV electrons are only 0.54 eV below the ionization threshold E_{exc}= 12.6 eV, thus they can be considered as "quasi-free" or very loosely

[1] Several authors have interpreted the existence of the isosbestic point as the result of changes in a continuous distribution of hydrogen bonding geometries and of the microscopic nature of the fluctuations of the (single state) thermal bath. They also stated that a single specie coupled to a Gaussian bath satisfies the Van't Hoff behaviour equation. However the demonstration always implies the existence of two species artificially dividing the spectrum at a cutoff frequency. In order to guarantee that the energy difference in eq. (2) depends weakly on temperature, it must be required that the fluctuations are dominated by a few bath modes. In such a way the origin of the effect is moved from the two-state picture of the liquid structure to the two-state fluctuations of the thermal bath whose energy are below and above the average.

bound. The coherent state is a superposition of the ground state and the excited state with weights 0.87 and 0.13 respectively. The size of the region where molecules are phase locked is the wavelength of the e.m. mode

$$\lambda = \frac{hc}{E_{excit}} = 0.1\mu \tag{5}$$

This region has been termed Coherence Domain (CD). At the border of the CD the thermal collisions extract molecules from the coherent phase thus the fractions of the coherent and non-coherent states is a function of T:

$$F_{coh}(T) + F_{non-coh}(T) = 1 \tag{6}$$

Calculations show [16] that the energy gap of the coherent water at T = 0K is

$$\triangle E = -0.17 \pm 0.05eV \tag{7}$$

This value is ˜ 6.8 KT (T = 300K), this means that the gap is well protected against thermal collisions at room temperature. Accordingly, the lifetime of the coherent structure, i.e. the time each single molecule spend in the coherent phase is ˜ $2.3x10^{-14}$ sec. The calculated gap must be compared to the energy gap measured by the Van't Hoff plot between correlated and non correlated fraction of dipoles in the FT-IR spectra, $\triangle E = 0.12 \pm 0.02$eV per molecule (2.9Kcal/mol). [2] The fair agreement with the value calculated for the energy gap by the Preparata and Del Giudice theory allows us to identify the correlated population with the coherent fraction of water and the non-correlated population with the non-coherent fraction. In such a picture, even the so called intermediate population could find a rationale. The measured spectrum emerges from a dipole-dipole transition between two specific quantum states. We have seen that the first and third peak describe the transitions occurring within the coherent and non coherent fraction respectively. It is very natural to assign the second (intermediate) peak to the transitions where the initial state is in the coherent fraction and the final state is in the non-coherent fraction and vice versa.

On the basis of such assignment we can also evaluate the energy content of a solution. In fact we have previously shown [7] that when a solute is added to pure water the percentage of the two components is changed. In Fig. 4 we show the relative abundance of the correlated and non-correlated water upon the addition of NaCl to the solution. While the non-correlated plus intermediate fraction is almost unchanged, the correlated fraction is reduced .

In Fig. 5 the ratio between the non-coherent and coherent populations is shown. The addition of NaCl into water increases this ratio and affects the energy content of the system, in other words the energy content of a solution increases when the solute content is increased. This observations may be used qualitatively to control the lowering of the freezing point.

4. The mixed cluster model in biology

The structure of liquid water and of aqueous solutions of non polar and polar molecules it is argument of debate among the biologists because it influences the structure and the reactivity of protein molecules. Actually, in order to explain long range cohesion of liquid water we are compelled to consider clusters of structured H-bonded molecules, nevertheless, biologists describes hydration processes as interactions of independent water molecules with ions or non polar molecules. Therefore the explanation of two different properties of aqueous systems demands two different internal structures of liquid water. A model describing water as a rapidly

[2] The value in (7) has been calculated at T=0 K where the cross over (of the boundaries of CDs) is not allowed. At T≠0 the energy gap is decreased by the thermal collisions.

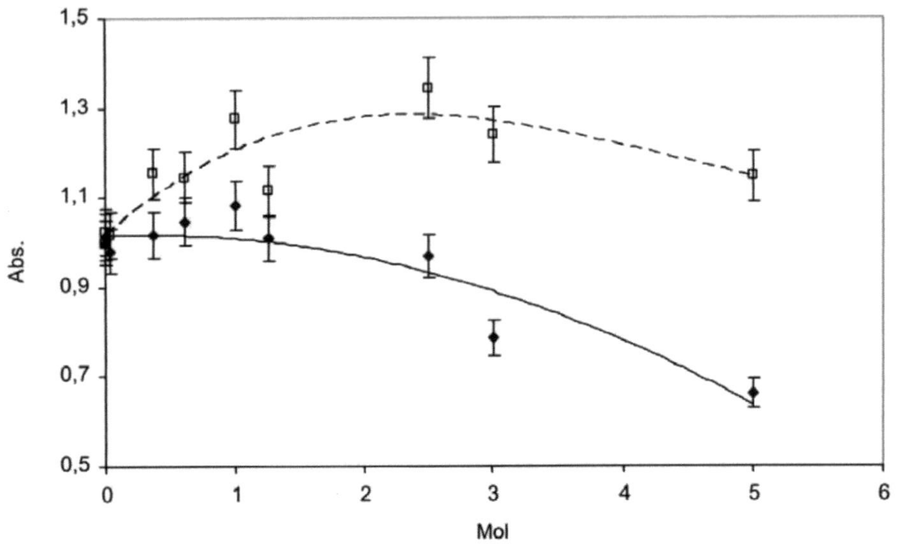

Figure 4. Relative abundance of the two populations of water in NaCl solution vs molarity. Open squares: non-correlated plus intermediate fractions; black diamonds: correlated fraction.

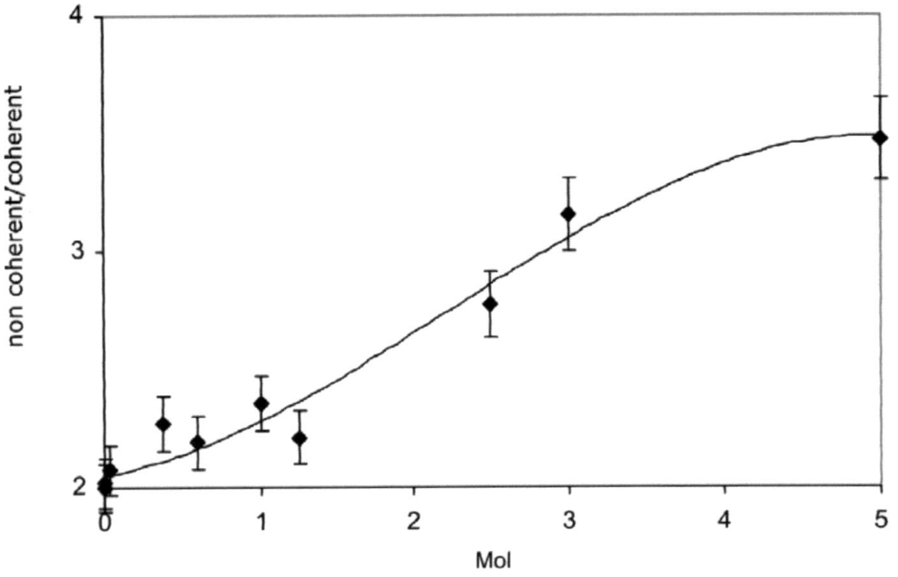

Figure 5. Ratio between the non-coherent to coherent fractions of water in NaCl solution vs molarity. This ratio is directly related to the energy content of the solution (see text).

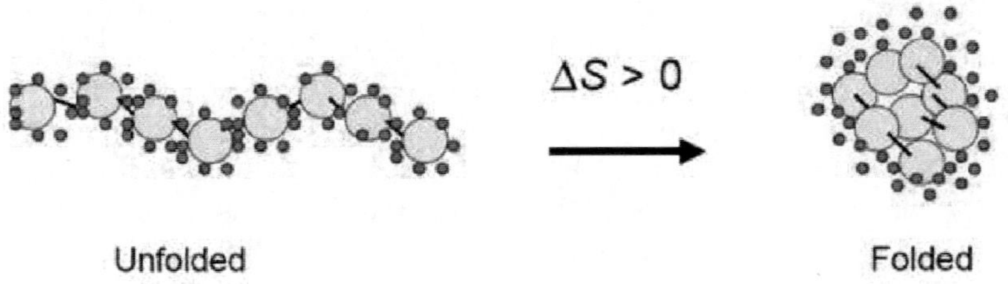

Figure 6. Formation of protein hydrophobic core in water.

fluctuating mixture of variously hydrogen bonded aggregates of molecules is known as "flickering cluster" model [18, 19]. It was assumed that liquid water consists of a mixture of monomers and clusters uniform in size. The size of the cluster, the relative numbers of cluster and monomers and their thermodynamic properties are assumed to be a function of the temperature. The "flickering clusters" models deals with this necessary double nature of liquid water. However, in order to calculate the partition function it uses several adjustable parameters. The Preparata-Del Giudice theory provides for the first time a theory for the water structure without adjustable parameters. In this picture the CDs and are responsible for the long range structure whereas isolated molecules belong to the non-coherent fraction, in fact, as explained above, solutes dissolve exclusively into the non-coherent fraction. Another key issue in the understanding of the living systems is the existence of what is called "vicinal water" [20]. It is well known that a water layer exists near the surface of an organic macro-molecule having peculiar properties: 1) the effective dielectric constant decreases dramatically to 10 (from 80 of bulk water); 2) the entropy content is lower than in bulk phase. Electrostatic interactions may justify the existence of a structured layer having a thickness of 0.1- 0.2 nm. However, in order for biological processes take place, the changes in the structure of water must extend over a zone large enough to accommodate the reaction species in a layer larger than the one obtained only via short-range forces. The coherent phase of water is compatible with all the peculiar properties observed by biologists at surfaces: the dielectric constant is decreased by the polarizability for deformation of the CDs[3] and the entropy is lowered by the ordering. Also the folding process of biomolecules can be understood with a two phase system: in the formation of macromolecules the surface available for the interaction with water may decrease and water molecules are expelled (the simplest case is the formation of the peptidic bond). From a thermodynamic point of view this means a decrease of the entropy of the macromolecule, which becomes more structured and a decrease of the thickness of the surface water layer with a major increase of the entropy of water. As a result, the entropy of the whole system increases, therefore, these kind of processes may occur spontaneously and are called *entropy driven* (see fig. 6):

$$\triangle G = \triangle H_{system} - T\triangle S_{system} \tag{8}$$

We are able to estimate from (4) the contribution of water to the entropy change ˜11.3 KJ/mol, however, it is not easy to evaluate the amount of water molecules expelled by the surface because of the side-chain rotations or movements of fragments of polypeptide chains, so that the quantitative evaluation of the ΔG of protein folding remains still hard to do.

[3] A detailed description is out of the scope of the present paper but extensive descriptions can be found in the quoted literature, in particular in Ref.12.

5. Conclusions

The vibrational spectra of liquid water give a strong support to the picture of water as a two phase liquid. The value of the energy gap of the coherent water calculated by Preparata and co-workers is in good agreement with the energy gap measured by the van 't Hoff plot between correlated and non correlated fraction of dipoles in the FT-IR spectra. The two phase picture also allows to explain several features of the biologic systems such as the protein folding and the structured layer of water around cells and membranes. Details of the formation of such a layer require a detailed analysis of the electronic structure of the Coherence Domains. This appear to be a very stimulating challenge and will be the core of our research in the next future.

References

[1] Geissler P L 2005 *J. Am. Chem. Soc.* B **127** 14930
[2] Smith J D, Cappa C D, Wilson K R, Cohen R C, Geissler P L and Saykally R J 2005 *Proc. Natl. Acad. Sci. U.S.A.*. **102** 14171
[3] Soper A K; Teixeira J and Head-Gordon T 2010 *Proc. Natl. Acad. Sci. U.S.A.* **107** E44
[4] Auer B M, Skinner J L 2007 *J Chem. Phys.* **127** 104105
[5] Brubach J B, Mermet A, Filabozzi A, Gersche A, Lairez D, Krafft M and Roy P 2001 *J. Chem. Phys.* **105** 430
[6] Brubach J B, Mermet A, Filabozzi A, Gerschel A and Roy P 2005 *J. Chem. Phys.* **122**184509
[7] De Ninno A, Congiu Castellano A 2011 *J. Mol. Struct.* **1006** 434
[8] Marechal Y 1991 *J. Chem. Phys.* **95** 5565
[9] Walrafen G E 2004 *J. Chem. Phys.* **120** 4868
[10] Walrafen G, Fisher M, Hokmabadi M and Yang W H 1986 *J. Chem. Phys.* **85** 6970
[11] Dzyaloshinskii I, Lifshitz E and Pitaevskii L P 1961 *Physics-Uspekhi* **4** 153
[12] Preparata G 1995 *QED Coherence in Matter* (Singapore: World Scientific)
[13] Del Giudice E, Preparata G and Vitiello G 1988 *Phys. Rev. Lett.* **61** 1085
[14] Arani R, Bono I, Del Giudice E and Preparata G 1995 *Int. J. Mod. Phys.* B **9** 1813
[15] Del Giudice E and Vitiello G 2006 *Phys. Rev.* A **74** 022105
[16] Bono I, Del Giudice E, Gamberale L and Henry M 2011 *Water* **2** 1-x
[17] Walrafen G, Yang W H, Chu Y and Hokmabadi M 1996 *J. Phys. Chem.* **100** 1381
[18] Nmethy G and Scheraga H A 1962 *J. Chem. Phys.* **36** 3382
[19] Hagler A T, Scheraga H A and Nemethy G 1972 *J. Chem. Phys.* **76** 3229
[20] Wiggins P 2008 *PLos ONE* **3** e1406

DICE2012 IOP Publishing
Journal of Physics: Conference Series **442** (2013) 012032 doi:10.1088/1742-6596/442/1/012032

Telegraph signals as a solution of the time dependent Schrödinger equation let standard Copenhagen quantum mechanics emerge

G Doyen and D Drakova
Ludwig-Maximilians Universität, München, Germany
Faculty of Chemistry, University of Sofia, Bulgaria
E-mail: nhdd@chem.uni-sofia.bg

Abstract. A particle switching between two sites of a symmetric system in weak interaction with an environmental continuum of high density of states exhibits a telegraph-signal-like time development without the need of the Born-Bohr principle of reduction on eigenstates of the measuring equipment. The origin of the telegraph signal is a very weak local coupling of the particle to the continuum of gravitons which is connected with an enormous slow down of the particle motion. The proposed mechanism envisages entanglement to gravitons in high dimensional spacetime, which, in accord with Gauss law, gives rise to a local and much stronger gravitational law compared to classical gravitation. The physics behind the quantum jumps and the telegraph-signal-like time development is elucidated. The model might be useful for studying environmental decoherence effects on adsorbate localization and quantum diffusion on solid surface.

1. Introduction

Standard Copenhagen quantum mechanics needs in addition to the unitary time development according to Schrödinger's equation a strange dynamics called collapse or wave function reduction. This dynamics is not predictable by any theoretical method but has statistical character. It leads to telegraph signals in the recorded experimental data. The Copenhagen formalism is believed to describe any known quantum phenomenon. "It is often said that quantum theory is eminently successful, that its predictions are verified in countless cases and that no phenomenon has been found which contradicts it" [1]. And further "A fundamental fact of orthodox quantum theory is that the evolution of the state of the physical system between the collapse events is mathematically very different from the evolution of this state associated with the collapses: the former are "unitary" and "local" whereas the latter are neither" [2].

The message of the present contribution will be that there is no random statistical element in the time development resulting from the time dependent Schrödinger equation. The time development of a local quantum system entangled to quantum fields in its environment of high dimensions and high density of states resembles telegraph-signal-like "quantum jumps".

Telegraph-signal-like quantum jumps have been observed in experiments of very different nature. They are displayed as sudden changes of a measured signal between two or more values with time, appearing in a random, statistical manner (cf. for instance [3]-[11]). The time development of two-level Rydberg atoms entangled with a finite small number of electromagnetic quanta in optical cavities shows also statistical quantum jumps between the two states, rather than Rabi oscillations [12]-[16].

To our knowledge a telegraph-signal-like behaviour has not been reported theoretically within a unitary coherent quantum description. We present a coherent dynamics of an entangled adsorbate-graviton system in the framework of quantum field theory leading to telegraph-signal-like changes of state of an adsorbate on a solid surface. This theory is completely different from the treatment within the Jaynes-Cummings quantum mechanical theory in earliar approaches to collapse and revival (cf. refs. [17]-[23] and the theory presented in refs. [24, 25]).

In decoherence theory the environmental coupling leads to vanishing interference between the two quantum states, because the corresponding pointer states in the environment become orthogonal [26]. However, there exist abundant experiments which cannot be explained in the spirit of the decoherence interpretation. These are experiments in the low temperature scanning tunnelling microscopy (STM) concerning adsorbates. For instance Eigler et al. [27] find that an isolated adsorbed Xe atom on a solid Ni(110) surface remains localized on the same adsorption site for days on end. A Xe atom interacts weakly with metal surfaces and the potential hardly varies parallel to the surface. Applying Schrödinger's equation to follow the time dependence of an initially localized Xe wave packet one finds that already after 10^{-10} s the Xe wave packet will be delocalized over several hundred lattice sites [28]. Localization of atoms and molecules, which according to Schrödinger's equation should be delocalized, is explained by decoherence theory as due to a mechanism of continuous "measurement" by environmental particles [29]. However, the time needed to localize a Xe atom via interaction with the rest gas in the best laboratory vacuum in the low temperature STM experiments amounts to ≈ 10 days. During this time, Schrödinger's law would have delocalized the Xe atom over the whole surface, the Xe atom would have completely vanished for the STM. Other quasi particles in the nickel surface, available for localization of the Xe atom, might be the phonons, the tomonagons, excitons, etc. However, at the low temperature in the STM experiments (below 5 K) according to the dispersion relations of the excitations of the solid only quasi particles of long wave length are available and they cannot localize adsorbed atoms on one of several closely situated adsorption sites.

We will demonstrate that gravitons can localize adsorbates on solid surfaces. As it has been suggested by string theory, gravitons and only gravitons exist in hidden dimensions of high-dimensional spacetime. If there are additional spatial dimensions then the gravitational force law will change in the higher dimensional world as compared to the 3+1 dimensional one. The motivation for our approach comes from the fact that the missing diffusion of Xe [27] or the very slow surface diffusion of hydrogen on a copper surface [30] in the low temperature STM cannot be understood by conventional environmental decoherence theory. The hidden dimensions of string theory provide a possibility to explain the experiment within the theoretical framework of quantum field theory.

The question is: can a coherent quantum mechanical description of a local system in interaction with the environment lead to the telegraph signal. In the present paper we give a positive answer to this question. We will demonstrate that it is possible for a system, comprising a local part, entangled to the continua of the environmental excitations, to change state suddenly in an apparently statistical fashion, which comes out of the solution of the time dependent Schrödinger equation. It will be shown that the conditions for a telegraph-signal-like change of state of the system is weak and local coupling to a continuum of gravitons in the environment. The time development is coherent and deterministic in the phase space of the total system including the environmental excitations. Focusing on the time development of the local system alone, which is assumed to be accessible to measurement in experiment, it appears as if it changes state in a random way.

In the next section we define the model and the Hamiltonian. The method of solution is outlined, describing the entanglement of the adsorbate motion to the continuum of gravitons in 11 dimensional spacetime. The solution of a model with two adsorption sites and two local states, where the interaction with gravitons is effective, reproduces the extremely slow diffusion of a hydrogen atom on the Cu(001) surface in the quantum diffusion regime as measured by Lauhon and Ho below 60 K [30]. As the results show, the adsorption site changes occur in a quantum-jump-like manner and not as the Rabi oscillations predicted by quantum mechanics. The physical background and the conditions for the quantum-jump-like time development of the entangled system are elucidated in a next section.

2. Theory of telegraph-signal-like time development of an adsorate entangled with gravitons in hidden dimensions

The model describes an adsorbate which can jump between two adsorption sites on a solid surface, site α and site β. We investigate the dynamics of site change. Gravitons in high spatial dimensions, including hidden spatial dimensions are described in ref. [31].

2.1. The Hamiltonian

As the modified gravitational force law is of very short range [32] it is only effective if the adsorbate is near a substrate atom. This is modelled by two localized adsorbate field excitations in the regions w_α and w_β in fig. 1, respectively. Only when the adsorbate is in one of these localized regions it can entangle to the gravitons. We assume a symmetric model with the same number and kind of localized regions for the adsorbate hopping between two adsorption sites (cf. fig. 1):

(i) One local region g_α or g_β on each adsorption site, respectively. From these local regions the adsorbate can propagate to the places w_α on the same site or w_β on the other site.

(ii) One localized place w_α or w_β. At these places the adsorbate can either entangle to the graviton continua $\{\kappa_\alpha\}$ or $\{\kappa_\beta\}$ on the same site or it can develop amplitude in the local regions g_α and g_β on the same adsorption site or on the opposite site.

(iii) Continua of N gravitons $\{\kappa_\alpha\}$ or $\{\kappa_\beta\}$. After being entangled with gravitons near the places w_α and w_β the adsorbate can develop amplitude in the local regions g_α and/or g_β. Due to the large number of spatial dimensions there is a high density of graviton modes with the required wave length of approximately 1 Å [33].

The explicit form of the Hamiltonian is:

$$\begin{aligned}
H &= H_\alpha + H_\beta + H_{\alpha-\beta} \\
H_\alpha &= E_g n_{g\alpha} + E_w n_{w\alpha} + \sum_\kappa \varepsilon_\kappa n_{\kappa\alpha} + V(c^+_{g\alpha} c_{w\alpha} + h.c.) \\
&+ W \sum_\kappa (c^+_{\kappa\alpha} c_{w\alpha} + h.c.) \\
H_\beta &= E_g n_{g\beta} + E_w n_{w\beta} + \sum_\kappa \varepsilon_\kappa n_{\kappa\beta} + V(c^+_{g\beta} c_{w\beta} + h.c.) \\
&+ W \sum_\kappa (c^+_{\kappa\beta} c_{w\beta} + h.c.) \\
H_{\alpha-\beta} &= (V - \Delta V)(c^+_{g\alpha} c_{w\beta} + c^+_{g\beta} c_{w\alpha} + h.c.)
\end{aligned} \tag{1}$$

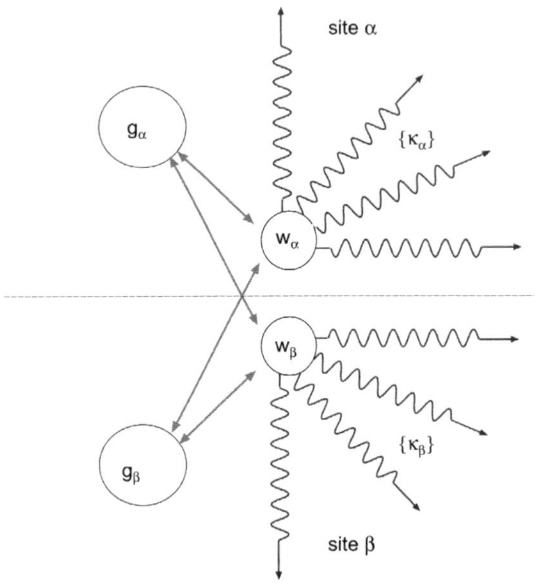

Figure 1. Model of two adsorption sites α and β where entanglement to gravitons $\{\kappa_\alpha\}$ and $\{\kappa_\beta\}$ occurs in two regions w_α and w_β, respectively.

where E_g and E_w are the energies of the adsorbate field excitations created by $c^+_{g\alpha}$, $c^+_{g\beta}$ and $c^+_{w\alpha}$, $c^+_{w\beta}$, respectively. ε_κ are the excitation energies for the gravitons. $n_{g\alpha}$ and $n_{w\alpha}$ are the number of excited field quanta in the regions g_α, w_α; $c^+_{g\alpha}$, $c_{g\alpha}$, $c^+_{w\alpha}$, $c_{w\alpha}$: creation and destruction operators for the adsorbate field quanta in the respective regions; $c^+_{\kappa\alpha}$ and $c_{\kappa\alpha}$: creation and destruction operators for the gravitons; V and $V - \Delta V$: interaction matrix elements between the various local regions; W: interaction matrix elements for coupling to the gravitons. $\{\kappa_\alpha\}$, $\{\kappa_\beta\}$ constitute a quasi-continuum which is modified due to the presence of the adsorbate (cf. ref. [31]). The coupling of the adsorbate field to the environmental degrees of freedom, the gravitons, is extremely weak and extremely localized in the local regions w_α and w_β. More details can be found in the description in ref. [31].

2.2. Solution method: coherent time evolution of an adsorbate entangled with a continuum of gravitons

The time evolution of the system is according to Schrödinger's time-dependent equation:

$$i\hbar \frac{\partial \Psi(t)}{\partial t} = H\Psi(t) \tag{2}$$

The eigenfunctions Ψ_I, determined by an expansion in the basis states, are used to derive the time dependence according to the unitary time evolution:

$$\Psi_I(t) = \Psi_I(t_0)e^{-iE_I t/\hbar} \tag{3}$$

A basis for the many particle wave function is given by $|\, g_\alpha 0\rangle$, $|\, g_\beta 0\rangle$, $|\, w_\alpha 0\rangle$, $|\, w_\beta 0\rangle$, $|\, w_\alpha \kappa_\alpha\rangle$, $|\, w_\beta \kappa_\beta\rangle$, where the respective adsorbate field excitations and graviton excitations are indicated.

Assume that the system starts in $\mid g_\alpha 0\rangle$. This state, expanded in the eigenfunctions $\{\mid I\rangle\}$ of the total Hamiltonian is:

$$\mid g_\alpha 0\rangle = c_{1,g_\alpha} \mid 1\rangle + c_{2,g_\alpha} \mid 2\rangle + ... \tag{4}$$

where c_{I,g_α} is the expansion coefficient of the basis state $\mid g_\alpha 0\rangle$ in the I-th eigenfunction of the Hamiltonian. Starting from the many particle state $\mid g_\alpha 0\rangle$ at $t = 0$, the further time development of this state is denoted by $\mid \Psi(t)\rangle$ and is obtained as:

$$\mid \Psi(t)\rangle = c_{1,g_\alpha} e^{-iE_1 t/\hbar} \mid 1\rangle + c_{2,g_\alpha} e^{-iE_2 t/\hbar} \mid 2\rangle + ... \tag{5}$$

The density operator at time t is:

$$\rho(t) = \mid \Psi(t)\rangle\langle\Psi(t) \mid \tag{6}$$

The occupation of site α as displayed in fig. 2 (red curves) is then defined by:

$$\text{occupation of site } \alpha = \langle g_\alpha 0 \mid \rho(t) \mid g_\alpha 0\rangle + \langle w_\alpha 0 \mid \rho(t) \mid w_\alpha 0\rangle + \sum_{\kappa_\alpha} \langle w_\alpha \kappa_\alpha \mid \rho(t) \mid w_\alpha \kappa_\alpha\rangle \tag{7}$$

Repeating this procedure for each next time step provides the coherent time evolution of the density operator (in the chosen basis) which can be used to illustrate the state changes of the adsorbate entangled with the continuum of gravitons.

3. Results

The model, solved neglecting the entanglement to gravitons, leads to the time development displayed in the left panel of fig. 2. The quantity plotted is the expression eq. (7) over site α (red curve) and site β (blue curve) as a function of time. The interaction parameters are chosen so that it takes approximately 0.1 seconds for a hydrogen atom to change site: $V = 0.05$ peV, $V - \Delta V = 0.032$ peV and $W = 0$ peV [34]. The solution of the Schrödinger equation eq. (2) has the sinusoidal behaviour with time, representing Rabi-oscillation, as it is expected. These are just coherent Rabi oscillations between the two sites α and β, mediated by coupling to the local regions w_α and w_β. The frequency of site transition is determined by the energy difference $\Delta\varepsilon$ between the eigenstates of the system involving no gravitons.

Only with entanglement to the gravitons included, does a telegraph-signal-like site switching arise, as it is seen in the right panel of fig. 2 (the value of $W = 0.007$ peV is used). Apparently statistical quantum jumps of the adsorbate between the two adsorption sites occur, reminiscent of collapses, however, resulting from the solution of Schrödinger's equation. Furthermore, a significant slow down of the hopping rate between the adsorption sites results, compared to the period of the Rabi oscillation, neglecting the entanglement to the gravitons. The residence time on one adsorption site has increased by roughly a factor of 10^4. In the regime leading to the telegraph-signal-like adsorbate hopping the average frequency of the telegraph signal is determined by a much smaller energy difference $\Delta\varepsilon = 2.2 \times 10^{-6}$ peV equal to the energy separation of the gravitons which entangle to the adsorbate movement. This behaviour resembles the experimental signal obtained in the quantum jumps of adsorbates in low temperature STM measurements ([7]-[11]).

4. Analysis of the theoretical telegraph-signal-like time development

The experimental observation of quantum jumps is interpreted as collapse, which allegedly cannot be described by unitary solutions of Schrödinger's equation. To the contrary, the analysis in the present section shows that, when certain limiting conditions concerning the density of

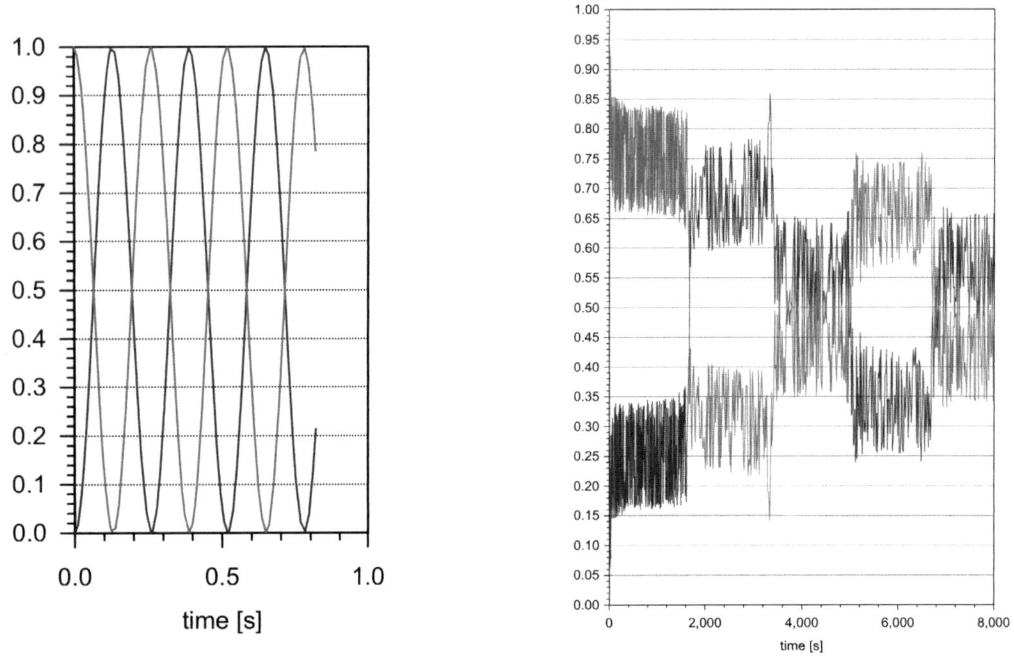

Figure 2. Left panel: coherent Rabi-like oscillations of the adsorbate between two sites without entanglement with the environment. Right panel: adsorbate telegraph-signal-like hopping between two sites including the entanglement with the gravitons. The amplitude squared of the total wave function is summed over site α (red curves) and site β (blue curves).

states and the coupling strength with the environmental continua are fulfilled, the unitary time development of a local quantum system entangled with the environmental excitations can display quantum jumps resembling collapse. More details of the derivation are avilable in ref. [34]. First we discuss the entanglement of the adsorbate movement to a degenerate graviton continuum, which, however does not yield the quantum jumps. The weak coupling to a non-degenerate graviton continuum results in graviton scattering which underlies the telegraph-signal-like quantum diffusion of the adsorbate.

4.1. A degenerate continuum of gravitons has no effect on the adsorbate dynamics

The symmetry of our geometric model permits two possible solutions: either symmetric or antisymmetric with respect to the interchange of site α and site β. The Hamiltonian can be rewritten in terms of the symmetric c^+_{g+}, c^+_{w+}, $c^+_{\kappa+}$, c_{g+}, c_{w+}, $c_{\kappa+}$ and antisymmetric combinations c^+_{g-}, c^+_{w-}, $c^+_{\kappa-}$, c_{g-}, c_{w-}, $c_{\kappa-}$ of the creation and destruction operators as:

$$\begin{aligned}
H &= H_+ + H_- \\
H_+ &= E_g n_{g+} + E_w n_{w+} + \sum_\kappa \varepsilon_\kappa n_{\kappa+} \\
&+ (2V - \Delta V)(c^+_{g+}c_{w+} + c^+_{w+}c_{g+}) + W\sum_\kappa (c^+_{\kappa+}c_{w+} + c^+_{w+}c_{\kappa+}) \\
H_- &= E_g n_{g-} + E_w n_{w-} + \sum_\kappa \varepsilon_\kappa n_{\kappa-} \\
&+ \Delta V(c^+_{g-}c_{w-} + c^+_{w-}c_{g-}) + W\sum_\kappa c^+_{\kappa-}c_{w-} + c^+_{w-}c_{\kappa-})
\end{aligned} \quad (8)$$

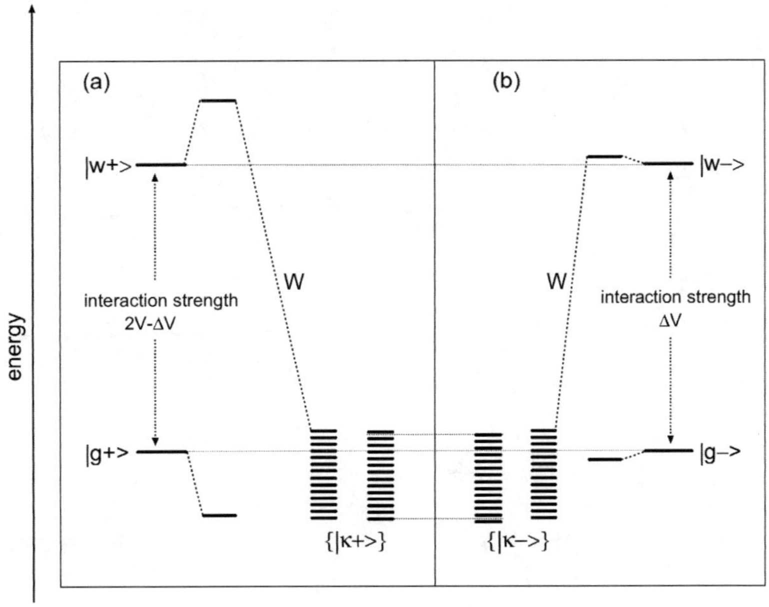

Figure 3. Schematic energy diagram resulting from the coupling within the symmetric (a) and antisymmetric (b) basis states. The degeneracy between the transformed graviton continua $\{| \ \kappa+\rangle\}$ and $\{| \ \kappa-\rangle\}$ is lifted due to coupling to $| \ w+\rangle$ (a) and to $| \ w-\rangle$ (b), respectively. The adsorbate in the $| \ g-\rangle$-component alone stays approximately on-shell with the initial state because of the weaker interaction of the order of ΔV with $| \ w-\rangle$.

The symmetry adapted form of the Hamiltonian is split into two independent parts. For each part a schematic level diagram is displayed in fig. 3. After interaction with the local states the antisymmetric combination of the adsorbate states is energetically much closer to the energy shell than the symmetric combination. Since the coupling to the gravitons is very weak, the antisymmetric solution has to dominate the dynamics. To surmount the problem with the enormous number of nearly degenerate gravitons as a first attempt we consider all graviton states exactly degenerate and construct a single graviton state which alone couples to the antisymmetric adsorbate states:

$$| \ w_\kappa-\rangle = \frac{1}{\sqrt{N}} \sum_{\kappa-} | \ \kappa-\rangle \quad \text{with} \tag{9}$$

$$| \ \kappa-\rangle = \frac{1}{\sqrt{2}}(| \ w_\alpha \kappa_\alpha\rangle - | \ w_\beta \kappa_\beta\rangle). \tag{10}$$

Defining

$$| \ w-\rangle = \frac{1}{\sqrt{2}}(| \ w_\alpha 0\rangle - | \ w_\beta 0\rangle) \quad \text{and} \tag{11}$$

$$| \ g-\rangle = \frac{1}{\sqrt{2}}(| \ g_\alpha 0\rangle - | \ g_\beta 0\rangle) \tag{12}$$

we end up with a 3×3 matrix which has to be diagonalized:

	$\| \ w-\rangle$	$\| \ w_\kappa-\rangle$	$\| \ g-\rangle$
$\langle w- \ \|$	E_w	$\sqrt{N}W$	ΔV
$\langle w_\kappa- \ \|$	$\sqrt{N}W$	0	0
$\langle g- \ \|$	ΔV	0	0

By forming a suitable linear combination out of the graviton state and the antisymmetric adsorbate state:

$$\mid w_{\kappa g}-\rangle = \frac{\sqrt{N}W \mid w_{\kappa}-\rangle + \Delta V \mid g-\rangle}{\sqrt{NW^2 + (\Delta V)^2}} \tag{13}$$

and orthogonalizing the adsorbate state to this linear combination:

$$\mid g_{\perp}-\rangle = \sqrt{\frac{NW^2 + (\Delta V)^2}{NW^2}} \left(\mid g-\rangle - \langle w_{\kappa g}- \mid g-\rangle \mid w_{\kappa g}-\rangle\right), \tag{14}$$

we see that the orthogonalized adsorbate state $\mid g_{\perp}-\rangle$ exactly decouples from the gravitons:

	$\mid w-\rangle$	$\mid w_{\kappa g}-\rangle$	$\mid g_{\perp}-\rangle$
$\langle w- \mid$	E_w	$\sqrt{NW^2 + (\Delta V)^2}$	0
$\langle w_{\kappa g}- \mid$	$\sqrt{NW^2 + (\Delta V)^2}$	0	0
$\langle g_{\perp}- \mid$	0	0	0

This kind of Hamiltonian signifies that the adsorbate state $\mid g_{\perp}-\rangle$ eq. (14) would never couple to gravitational degrees of freedom if the graviton continuum is degenerate. The error in this analysis arises, because the gravitons have been assumed exactly degenerate. Environment of this kind cannot have significant influence on the adsorbate dynamics.

4.2. Induced scattering in the graviton continuum leads to telegraph signals

We apply Dyson's equation to study the influence of lifting the degeneracy and accounting for the scattering of graviton states.

In an expansion over the graviton eigenstates (exact or approximate) $\{\mid \tilde{\kappa}_{\perp}-\rangle\}$ the coefficients $\langle g_{\perp}- \mid \tilde{\kappa}_{\perp}-\rangle$ can be evaluated using Lippmann-Schwinger equation [35]:

$$\begin{aligned}\langle g_{\perp}- \mid \tilde{\kappa}_{\perp}-\rangle &= \langle g_{\perp}- \mid \kappa_{\perp}-\rangle \\ &+ \sum_{\lambda} \langle g_{\perp}- \mid G(E = E_{\tilde{\kappa}-}) \mid \lambda_{\perp}-\rangle\langle \lambda_{\perp}- \mid H_{-} - H_o \mid \kappa_{\perp}-\rangle\end{aligned} \tag{15}$$

where H_o is the last but one line of eq. (8) and the summation over λ is over all states in the antisymmetric on-shell continuum. $G(E) = (E - H_{-})^{-1}$ is the Green operator. Using Dyson's equation and expanding the Green function in a Born series in the environmental states $\{\mid \kappa_{\perp}-\rangle\}$:

$$\mid \kappa_{\perp}-\rangle = \frac{1}{\sqrt{1- \mid \langle \kappa- \mid w_{\kappa g}-\rangle \mid^2}} \left(\mid \kappa-\rangle - \frac{W}{\sqrt{NW^2 + (\Delta V)^2}} \mid w_{\kappa g}-\rangle\right) \tag{16}$$

yields:

$$\langle\lambda_\perp - \mid G \mid \mu_\perp -\rangle = \langle\lambda_\perp - \mid G_o \mid \lambda_\perp -\rangle\langle\lambda_\perp - \mid H_- - H_o \mid \mu_\perp -\rangle\langle\mu_\perp - \mid G_o \mid \mu_\perp -\rangle$$
$$+\langle\lambda_\perp - \mid G_o \mid \lambda_\perp -\rangle\sum_\nu\langle\lambda_\perp - \mid H_- - H_o \mid \nu_\perp -\rangle\langle\nu_\perp - \mid G_o \mid \nu_\perp -\rangle\langle\nu_\perp - \mid H_- - H_o \mid \mu_\perp -\rangle\langle\mu_\perp - \mid G_o \mid \mu_\perp -\rangle$$
$$+\langle\lambda_\perp - \mid G_o \mid \lambda_\perp -\rangle\sum_\nu\langle\lambda_\perp - \mid H_- - H_o \mid \nu_\perp -\rangle\langle\nu_\perp - \mid G_o \mid \nu_\perp -\rangle\sum_\iota\langle\nu_\perp - \mid H_- - H_o \mid \iota_\perp -\rangle\langle\iota_\perp - \mid G_o \mid \iota_\perp -\rangle$$
$$\times\langle\iota_\perp - \mid H_- - H_o \mid \mu_\perp -\rangle\langle\mu_\perp - \mid G_o \mid \mu_\perp -\rangle + \cdots$$

Terminating the Born series at any finite order does not yield the telegraph-signal-like behaviour. Instead we assume short range scattering:

$$\langle\iota_\perp - \mid H_- - H_o \mid \mu_\perp -\rangle = U \tag{17}$$

which allows to sum up the infinite series:

$$\langle\lambda_\perp - \mid G \mid \mu_\perp -\rangle = \langle\lambda_\perp - \mid G_o \mid \lambda_\perp -\rangle\langle\mu_\perp - \mid G_o \mid \mu_\perp -\rangle\frac{U}{1 - U\sum_\iota\langle\iota_\perp - \mid G_o \mid \iota_\perp -\rangle} \tag{18}$$

It has to be emphasized that this assumption implies, of course, *a very local interaction with the gravitons.* The expansion coefficient eq. (15) then becomes

$$\langle g_\perp - \mid \tilde\kappa_\perp -\rangle \approx i\frac{\Delta\varepsilon}{\pi}\frac{\Delta V}{W}\frac{1}{E_{\kappa-}} \tag{19}$$

and has the right dependence on the energy $E_{\kappa-}$ to yield the telegraph signals (cf. fig. 4). $\Delta\varepsilon$ is the energy separation between two neighbouring states in the $\{\mid \kappa-\rangle\}$ continuum. Defining the on-shell projected state from the graviton continuum $\{\mid \tilde\kappa_\perp -\rangle\}$:

$$\mid A\kappa-\rangle = \frac{1}{\sqrt{N}}\sum_{\kappa-}\mid \tilde\kappa_\perp -\rangle \tag{20}$$

and projecting the time dependent wave packet on $\mid A\kappa-\rangle$, one obtains:

$$\begin{aligned}\langle A\kappa- \mid \Psi(t)\rangle &= i\frac{\Delta\varepsilon}{\pi}\frac{\Delta V}{W}\sum_{\tilde\kappa-}\frac{1}{E_{\kappa-}}\sqrt{\frac{NW^2}{NW^2+(\Delta V)^2}}\frac{1}{\sqrt{2}}\frac{1}{\sqrt{N}}e^{-iE_{\tilde\kappa-}t/\hbar}\\ &= \frac{1}{\pi}\frac{\Delta V}{W}\frac{1}{\sqrt{2N}}\sqrt{\frac{NW^2}{NW^2+(\Delta V)^2}}\\ &\times\left(\sin(\Delta\varepsilon t/\hbar)+\frac{\sin(3\Delta\varepsilon t/\hbar)}{3}+\frac{\sin(5\Delta\varepsilon t/\hbar)}{5}+\cdots\right).\end{aligned} \tag{21}$$

The Fourier series in the last line yields a telegraph-signal-like behaviour.

Neglecting all other components, $\mid \Psi(t)\rangle$ can be approximated as:

$$\mid \Psi(t)\rangle \approx \langle A\kappa- \mid \Psi(t)\rangle \mid A\kappa-\rangle + \langle A\kappa+ \mid \Psi(t)\rangle \mid A\kappa+\rangle \tag{22}$$

with

$$\mid A\kappa+\rangle = \frac{1}{\sqrt{N}}\sum_{\kappa+}\mid \tilde\kappa_\perp +\rangle \tag{23}$$

Projecting on the states $\mid g_\alpha 0\rangle$, $\mid w_\alpha 0\rangle$, $\mid w_\alpha\kappa_\alpha\rangle$ as required by eq. (7), one obtains e.g.:

$$\langle w_\alpha\kappa_\alpha \mid \Psi(t)\rangle \approx \langle A\kappa- \mid \Psi(t)\rangle\langle w_\alpha\kappa_\alpha \mid A\kappa-\rangle + \langle A\kappa+ \mid \Psi(t)\rangle\langle w_\alpha\kappa_\alpha \mid A\kappa+\rangle \tag{24}$$

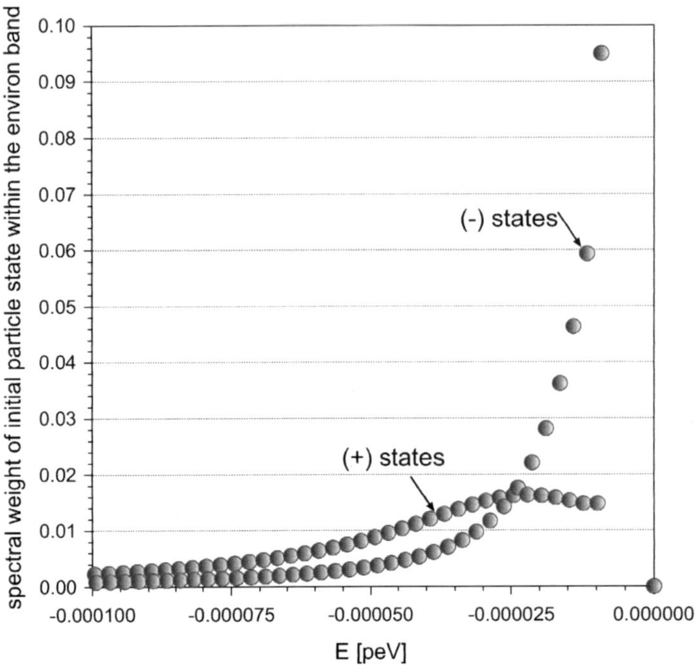

Figure 4. Spectral distribution of the coupling to the environmental gravitons. The coupling in the subspace of the antisymmetric states varies with the inverse energy of the gravitons $\frac{1}{E_{\kappa-}}$.

The contribution to the occupation of site α (cf. eq. 7) is:

$$\begin{aligned} &\langle w_\alpha \kappa_\alpha \mid \rho(t) \mid w_\alpha \kappa_\alpha \rangle = \mid \langle w_\alpha \kappa_\alpha \mid \Psi(t) \rangle \mid^2 \\ = &\mid \langle A\kappa- \mid \Psi(t) \rangle \mid^2 \mid \langle w_\alpha \kappa_\alpha \mid A\kappa- \rangle \mid^2 + \mid \langle A\kappa+ \mid \Psi(t) \rangle \mid^2 \mid \langle w_\alpha \kappa_\alpha \mid A\kappa+ \rangle \mid^2 \\ + &2\mathcal{R}e \langle \Psi(t) \mid A\kappa+ \rangle \langle A\kappa- \mid \Psi(t) \rangle \langle A\kappa+ \mid w_\alpha \kappa_\alpha \rangle \langle w_\alpha \kappa_\alpha \mid A\kappa- \rangle \end{aligned} \tag{25}$$

The first term on the r.h.s. in the last but one line is approximately time independent, because the square of the infinite Fourier series in eq. (21) is exactly constant. The second term on the r.h.s. in the last but one line is approximately time independent, because $\langle A\kappa+ \mid \Psi(t)\rangle$ is nearly time independent with respect to long time variations (only high frequency noise-like variations). The last term in eq. (25) has telegraph character, because $\langle A\kappa- \mid \Psi(t)\rangle$ has telegraph character. Similar time dependent terms with telegraph character are found in the expressions for $\langle g_\alpha 0 \mid \rho(t) \mid g_\alpha 0\rangle$ and $\langle w_\alpha 0 \mid \rho(t) \mid w_\alpha 0\rangle$. These are the time dependent contributions to the quantity plotted in fig. 2, implying that the occupation of site α will jump between two values.

5. Conclusion

Concluding, we showed that 3+1 dimensional quantum theory is incomplete in the case of adsorbate localization and surface diffusion. Introducing the entanglement of the adsorbate motion to the field of gravitons in high spatial dimensions (for instance 11 spacetime dimensions) results in the slow down of adsorbate quantum diffusion between different adsorption sites and in quantum-jump-like adsorption site changes, rather than Rabi oscillations, with time. The necessary conditions for the telegraph-signal-like variation of adsorption site with time can be summarized:

- On-shell (**degeneracy coupling**) between the initial many particle configuration and the graviton modes is effective.
- **Weak coupling** between the many particle initial state and the graviton modes.
- The weak coupling leads to a time dependent many particle wave function which is an entanglement between the adsorbate movement state and the continuum of gravitons. The frequency of the soft local graviton states determines the lifetime of the entangled state on a given site in the local region and the diffusion rate.

Schrödinger's equation is local and determines a unitary time development. The effect of the entanglement of the adsorbate motion to the gravitons is a telegraph-signal-like change of adsoption site, resembling a quantum jump. However, a quantum jump is a collapse on a different state. This is usually considered to indicate non-locality as it is implied within the Bohr-Born concept in quantum mechanics: collapse is a nonlocal effect.

We emphasize that in our theory nonlocality and quantum jumps of an adsorbate are the outcome of Schrödinger's quantum mechanics in the limit, when the following conditions are fulfilled:

- extremely *localized coupling* of the adsorbate to a continuum of environmental excitations;
- extremely *high density of states* of the continuum of environmental excitations as described in ref. [33];
- extremely *weak coupling* to the continuum of environmental excitations.

In this limiting case near perfect telegraph-signal-like site changes of an adsorbate between different adsorption sites are obtained. If the above requirements are relaxed, deviations from the zero-time quantum jumps will result. As suggested here environmental excitations which closely approach the requirements for telegraph-signal-like change of state of the quantum system are the gravitons.

References

[1] Haag R 1997 *Quantum future - from Volta and Como to the present and beyond* eds Ph Blanchard and A Jadczyk (Berlin: Springer) p 60

[2] Stapp H 1997 *Quantum future - from Volta and Como to the present and beyond* eds Ph Blanchard and A Jadczyk (Berlin: Springer) p 173

[3] Nirmal M, Dabbousi B O, Bawendi M G, Macklin J J, Trautman J K, Harris T D and Brus L E 1996 *Nature* **383** 802

[4] Dorozhkin S I, Pfeiffer L, West K, von Klitzing K and Smet J H 2011 *Nature Physics* **7** 336

[5] Prati E, Belli M, Fanciulli M and Ferrari G 2010 *Appl. Phys. Lett.* **96** 113109

[6] Mazzeo G, Prati E, Belli M, Leti G, Cocco S, Fanciulli M, Guagliardo F and Ferrari G 2012 *Appl. Phys. Lett.* **100** 213107

[7] Liljeroth P, Repp J and Meyer G 2007 *Science* **317** 1203

[8] Heinrich A J, Lutz C P, Gupta J A and Eigler D M 2002 *Science* **298** 1381

[9] Jewell A D, Tierney H L, Baber A E, Iski E V, Laha M M and Sykes E C H 2010 *J. Phys. Cond. Matter* **22** 264006

[10] Tierney H L, Baber A E, Jewell A D, Iski E V, Boucher M B and Sykes E C H 2009 *Chem. - Eur. J.* **15** 9678

[11] Nacci Ch, Folsch S, Zenichowski K, Dokic J, Klamroth T and Saalfrank P 2009 *Nano Lett.* **9** 2996

[12] Walther H 1997 *Quantum future - from Volta and Como to the present and beyond* eds Ph Blanchard and A Jadczyk (Berlin: Springer) p 220

[13] Haroche S, Dotsenko I, Gleyzes S, Brune M and Raimond J M 2010 in *Perspective of mesoscopic physics* eds A Aharony and O Entin-Wohlman (Singapore: World Sci.)chapter 6 pp 95-114

[14] Gleyzes S, Kuhr S, Guerlin C, Bernu J, Deléglise S, Hoff U B, Brune M, Raimond J M and Haroche S 2007 *Nature* **446** 297

[15] Reick S, Mølmer K, Alt W, Eckstein M, Kampschulte T, Kong L, Reimann R, Thobe A, Widera A and Meschede D 2010 *J. Opt. Soc. Am.* B **27** A152

[16] Berkeland D J, Raymondson D A and Tassin V M 2004 *Phys. Rev.* A **69** 059103

[17] Jaynes E T and Cummings F W 1963 *Proc. IEEE* **51** 89
[18] Cummings F W 1965 *Phys. Rev.* **140** A1051
[19] von Foerster Th 1975 *J. Phys.* A **8** 95
[20] Knight P L and Radmore P M 1982 *Phys. Lett.* A **90** 342
[21] Eberly J H, Narozhny N B and Sanchez-Mondragon J J 1980 *Phys. Rev. Lett.* **44** 1323
[22] Narozhny N B, Sanchez-Mondragon J J and Eberly J H 1981 *Phys. Rev.* A **23** 236
[23] Yoo H I and Eberly J H 1985 *Phys. Rep.* **118** 239
[24] Metz J, Trupke M and Beige A 2006 *Phys. Rev. Lett.* **97** 040503
[25] Metz J and Beige A 2007 *Phys. Rev. Lett.* **76** 022331
[26] Zurek W H 2003 *Rev. Mod. Phys.* **75** 715
[27] Eigler D M and Schweizer E K 1990 *Nature* **344** 524
[28] Doyen G and Drakova D 2009 *Surface Sci.* **603** 1579
[29] Joos E and Zeh H D 1985 *Z. Phys.* B **59** 223
[30] Lauhon L J and Ho W 2000 *Phys. Rev. Lett.* **85** 4566; Erratum: Lauhon L J and Ho W 2002 *Phys. Rev. Lett.* **89** 079901
[31] Drakova D and Doyen G 2013 *present volume*
[32] Zee A 2003 *Quantum field theory in a nutshell* (Princeton and Oxford: Princeton Univ. Press)
[33] Doyen G and Drakova D 2011 *J. Phys: Conf. Series* **306** 012033
[34] Drakova D and Doyen G *Preprint* arXiv:1204.5606v1 [quant-ph] 25 Apr 2012
[35] Lippmann B A and Schwinger J 1950 *Phys. Rev.* **79** 469

Journal of Physics: Conference Series **442** (2013) 012033 doi:10.1088/1742-6596/442/1/012033

Emergence from irreversibility

P Fernández de Córdoba[1,a], J M Isidro[1,b] and Milton H Perea[1,2,c]

[1] Instituto Universitario de Matemática Pura y Aplicada, Universidad Politécnica de Valencia, Valencia 46022, Spain

[2] Departamento de Matemáticas y Física, Universidad Tecnológica del Chocó, Colombia

E-mail: [a]`pfernandez@mat.upv.es`, [b]`joissan@mat.upv.es`, [c]`milpecr@posgrado.upv.es`

Abstract. The emergent nature of quantum mechanics is shown to follow from a precise correspondence with the classical theory of irreversible thermodynamics. Specifically, the linear (or Gaussian) regime of the latter can be put in a 1-to-1 map with the semiclassical approximation to quantum mechanics. The very possibility of reinterpreting quantum mechanics as a thermodynamics proves that the former is an emergent phenomenon. That is, quantum mechanics is a coarse–grained description of some underlying degrees of freedom.

1. Introduction

The aim of this work is to establish a correspondence between quantum mechanics, on the one hand, and the classical thermodynamics of irreversible processes, on the other. This we do in order to provide an independent proof of the statement that *quantum mechanics is an emergent phenomenon.*

Under the notion of an *emergent phenomenon* one usually understands the result of coarse–graining, or averaging over some deeper–level structures. These finer degrees of freedom, sometimes also called *microscopic* in order to distinguish them from the *emergent* degrees of freedom, usually carry information in such a way that they can be grouped together into higher structures, no longer microscopic, that give rise to the emergent phenomena observed in macroscopic experiments. In coarse graining microscopic information into higher structures, one is renouncing a detailed knowledge at a deeper level, in favour of some less–detailed averages, which are usually more tractable data. The canonical example of an emergent theory is classical thermodynamics, because it makes absolutely no assumption at all regarding the microscopics of the models it applies to. The corresponding microscopic theory is statistical mechanics, out of which thermodynamics emerges.

That quantum mechanics qualifies as an emergent theory has been suspected for some time [18, 19, 20, 21, 22]. In fact this topic has been the subject of a vast literature in recent years; a very incomplete list of refs. would include [9, 10, 11, 12, 13, 16, 17, 25, 26, 31, 32, 33]. The importance of the link between quantisation and dissipation has been explained in [4, 5, 6, 7, 8]. Some of the authors' previous work on this subject is collected in refs. [1, 2, 3].

In this work we draw on the fundamental link between quantum and dissipative phenomena. By exhibiting a 1–to–1 map between quantum mechanics and the *classical* thermodynamics of *irreversible* processes, the emergent nature of quantum mechanics is laid bare. Especially relevant here are the *classical* and the *irreversible* properties of the thermodynamics involved.

"Classical" means absence of $\hbar$; "irreversible" implies entropy production. As will be seen below, these two facts have a clear counterpart in the quantum–mechanical description.

In section 2 we review some background information on the classical thermodynamics of irreversible processes. This is necessary to establish the announced correspondence, which is presented in section 3. The implications of this correspondence for the quantum theory are examined in section 4. The discussion in section 5 rounds up our presentation with some suggestions for future work.

2. Basics in irreversible thermodynamics

We first summarise, for later use, some basic elements of the classical thermodynamics of irreversible processes in the linear regime [28].

Let a thermodynamical system be given, deviating only slightly from equilibrium. Assume that its entropy S depends on N extensive variables $y^1, \ldots, y^N$, so we can write $S = S(y^1, \ldots, y^N)$. The tendency of the system to seek equilibrium is measured by the *thermodynamic forces* Y_k, defined to be the components of the gradient of the entropy:

$$Y_k := \frac{\partial S}{\partial y^k}. \tag{1}$$

Now our system is away from equilibrium, but not too far away, so we can assume linearity between the fluxes $\dot{y}^k$ and the forces Y_j:

$$\dot{y}^i := \frac{\mathrm{d}y^i}{\mathrm{d}\tau} = \sum_{j=1}^{N} L^{ij} Y_j, \qquad Y_i = \sum_{j=1}^{N} R_{ij} \dot{y}^j, \qquad R_{ij} = (L^{ij})^{-1}. \tag{2}$$

We use τ to denote thermodynamical time, and we suppose the above relation between forces and fluxes to be invertible. A well–known result is Onsager's reciprocity theorem: the matrix L is symmetric,

$$L^{ij} = L^{ji}. \tag{3}$$

By (2), the rate of entropy production can be written either as a quadratic form in the fluxes, or as a quadratic form in the forces:

$$\dot{S} = \sum_{j=1}^{N} \frac{\partial S}{\partial y^j} \dot{y}^j = \sum_{j=1}^{N} Y_j \dot{y}^j = \sum_{i,j=1}^{N} R_{ij} \dot{y}^i \dot{y}^j = \sum_{i,j=1}^{N} L^{ij} Y_i Y_j. \tag{4}$$

We can Taylor expand the entropy S around equilibrium and truncate the series at second order, to find

$$S = S_0 - \frac{1}{2} \sum_{i,j=1}^{N} s_{ij} y^i y^j + \ldots, \tag{5}$$

where the matrix $s_{ij} = -\partial^2 S / \partial y^i \partial y^j |_0$ (the negative Hessian evaluated at equilibrium) is positive definite. This truncation has the consequence that fluctuations around equilibrium are Gaussian. Indeed, by Boltzmann's principle, the probability $P(y^1, \ldots, y^N)$ of finding the values $y^1, \ldots, y^N$ of the extensive variables is given by

$$P(y^1, \ldots, y^N) = Z^{-1} \exp\left(\frac{S}{k_B}\right) = Z^{-1} \exp\left(-\frac{1}{2k_B} \sum_{i,j=1}^{N} s_{ij} y^i y^j\right), \tag{6}$$

where Z is a normalisation factor.

For simplicity we set $N = 1$ in all that follows. Our aim is to calculate the probability of any path $y = y(\tau)$ in the thermodynamical configuration space. A *cumulative* distribution function $F_n\begin{pmatrix} y_1 \cdots y_n \\ \tau_1 \cdots \tau_n \end{pmatrix}$ is defined such that it yields the probability that the thermodynamical path $y(\tau)$ lie below the barriers $y_1, \ldots, y_n$ at times $\tau_1 < \tau_2 < \ldots < \tau_n$:

$$F_n\begin{pmatrix} y_1 \cdots y_n \\ \tau_1 \ldots \tau_n \end{pmatrix} := P\left(y(\tau_k) \le y_k,\ k = 1, \ldots, n\right). \tag{7}$$

A stationary process is defined to be one such that F_n is invariant under time shifts $\delta\tau$:

$$F_n\begin{pmatrix} y_1 \cdots y_n \\ \tau_1 \ldots \tau_n \end{pmatrix} = F_n\begin{pmatrix} y_1 & \cdots & y_n \\ \tau_1 + \delta\tau & \ldots & \tau_n + \delta\tau \end{pmatrix}. \tag{8}$$

In other words, the system that has been left alone long enough that any initial conditions have been forgotten. An *unconditional* probability density function $f_n\begin{pmatrix} y_1 \cdots y_n \\ \tau_1 \cdots \tau_n \end{pmatrix}$ is defined, such that the product

$$f_n\begin{pmatrix} y_1 \cdots y_n \\ \tau_1 \ldots \tau_n \end{pmatrix} \mathrm{d}y_1 \cdots \mathrm{d}y_n \tag{9}$$

measures the probability that a thermodynamical path $y = y(\tau)$ pass through a gate of width $\mathrm{d}y_k$ at instant τ_k, for all $k = 1, \ldots n$. Similarly, the *conditional* probability density function $f_1\left(\begin{smallmatrix} y_k \\ \tau_k \end{smallmatrix}\middle|\begin{smallmatrix} y_{k-1} \\ \tau_{k-1} \end{smallmatrix}\right)$ is such that the product

$$f_1\left(\begin{matrix} y_k \\ \tau_k \end{matrix}\middle|\begin{matrix} y_{k-1} \\ \tau_{k-1} \end{matrix}\right) \mathrm{d}y_k\, \mathrm{d}y_{k-1} \tag{10}$$

gives the probability that $y = y(\tau)$ pass through $\mathrm{d}y_k$ at τ_k, given that it passed through $\mathrm{d}y_{k-1}$ at τ_{k-1}. Finally a Markov process is defined to be *one that has a short memory* or, more precisely, one such that its cumulative, conditional probability function satisfies

$$F_1\left(\begin{matrix} y_{n+1} \\ \tau_{n+1} \end{matrix}\middle|\begin{matrix} y_1 \cdots y_n \\ \tau_1 \cdots \tau_n \end{matrix}\right) = F_1\left(\begin{matrix} y_{n+1} \\ \tau_{n+1} \end{matrix}\middle|\begin{matrix} y_n \\ \tau_n \end{matrix}\right). \tag{11}$$

One can prove that, for a Markov process, the following factorisation theorem holds [28]:

$$f_n\begin{pmatrix} y_1 \cdots y_n \\ \tau_1 \ldots \tau_n \end{pmatrix} = f_1\left(\begin{matrix} y_n \\ \tau_n \end{matrix}\middle|\begin{matrix} y_{n-1} \\ \tau_{n-1} \end{matrix}\right) \cdots f_1\left(\begin{matrix} y_2 \\ \tau_2 \end{matrix}\middle|\begin{matrix} y_1 \\ \tau_1 \end{matrix}\right) f_1\begin{pmatrix} y_1 \\ \tau_1 \end{pmatrix}. \tag{12}$$

Interesting about this factorisation theorem is the fact that $f_1\left(\begin{smallmatrix} y_1 \\ \tau_1 \end{smallmatrix}\right)$ is known from Boltzmann's principle. Therefore, by stationarity, all we need to know is

$$f_1\left(\begin{matrix} y_2 \\ \tau + \delta\tau \end{matrix}\middle|\begin{matrix} y_1 \\ \tau \end{matrix}\right), \tag{13}$$

and solving the n–gate problem $f_n\left(\begin{smallmatrix} y_1 \cdots y_n \\ \tau_1 \cdots \tau_n \end{smallmatrix}\right)$ nicely reduces to solving the 2–gate problem $f_1\left(\begin{smallmatrix} y_2 \\ \tau+\delta\tau \end{smallmatrix}\middle|\begin{smallmatrix} y_1 \\ \tau \end{smallmatrix}\right)$.

Now, under the assumption that our irreversible thermodynamical processes is stationary, Markov and Gaussian, the conditional probability density (13) has been computed in [28], with the result

$$f_1\left(\begin{matrix} y_2 \\ \tau + \delta\tau \end{matrix}\middle|\begin{matrix} y_1 \\ \tau \end{matrix}\right) = \frac{1}{\sqrt{2\pi}} \frac{s/k_B}{\sqrt{1 - \mathrm{e}^{-2\gamma\delta\tau}}} \exp\left[-\frac{s}{2k_B} \frac{\left(y_2 - \mathrm{e}^{-\gamma\delta\tau} y_1\right)^2}{1 - \mathrm{e}^{-2\gamma\delta\tau}}\right]. \tag{14}$$

Here we have defined the thermodynamical frequency γ,

$$\gamma := \frac{s}{R}, \tag{15}$$

with R given as in (2) and $s = -\mathrm{d}^2 S/\mathrm{d}y^2|_0$. Furthermore, one can reexpress the probability density (14) in terms of path integrals over thermodynamical configuration space: up to normalisation factors one finds [28]

$$f_1 \left(\begin{matrix} y_2 \\ \tau_2 \end{matrix} \middle| \begin{matrix} y_1 \\ \tau_1 \end{matrix} \right) = \int_{y(\tau_1)=y_1}^{y(\tau_2)=y_2} \mathrm{D}y(\tau) \exp \left\{ -\frac{1}{2k_B} \int_{\tau_1}^{\tau_2} \mathrm{d}\tau \, \mathcal{L}\left[\dot{y}(\tau), y(\tau)\right] \right\}. \tag{16}$$

Above we have defined the thermodynamical Lagrangian function $\mathcal{L}$

$$\mathcal{L}\left[\dot{y}(\tau), y(\tau)\right] := \frac{R}{2}\left[\dot{y}^2(\tau) + \gamma^2 y^2(\tau)\right], \tag{17}$$

whose actual dimensions are entropy per unit time.

3. Irreversible thermodynamics *vs.* quantum theory

We can now establish a precise map between quantum mechanics and classical, irreversible thermodynamics. Let t denote mechanical time, m the mass of the quantum particle under consideration, and ω the frequency of a harmonic potential experienced by the particle.

In the first place, the thermodynamical time variable τ must be analytically continued into $\mathrm{i}t$:

$$\tau \leftrightarrow \mathrm{i}t. \tag{18}$$

Second, the thermodynamical frequency γ becomes the mechanical frequency ω of the harmonic oscillator:

$$\gamma \leftrightarrow \omega. \tag{19}$$

Next we map the thermodynamical variable y onto the mechanical variable x:

$$y \leftrightarrow x. \tag{20}$$

As a rule, x will be a position coordinate. Hence there might be some dimensional conversion factor between x and y above, that we will ignore for simplicity. Bearing this in mind, we will finally make the identification

$$\frac{s}{2k_B} \leftrightarrow \frac{m\omega}{\hbar} \tag{21}$$

between thermodynamical and mechanical quantities. We have expressed all the above replacements with a double arrow $\leftrightarrow$ in order to indicate the bijective property of our map between quantum mechanics and classical, irreversible thermodynamics.

On general grounds, applying the replacements (18), (19), (20) and (21), one expects thermodynamical conditional probabilities to map onto mechanical conditional probabilities[1],

$$f_1 \left(\begin{matrix} y_2 \\ \tau_2 \end{matrix} \middle| \begin{matrix} y_1 \\ \tau_1 \end{matrix} \right) \leftrightarrow K(x_2, t_2 | x_1, t_1), \tag{22}$$

while thermodynamical unconditional probabilities are expected to map onto mechanical unconditional probabilities:

$$f_1 \begin{pmatrix} y \\ \tau \end{pmatrix} \leftrightarrow |\psi(x,t)|^2. \tag{23}$$

[1] While f_1 is a probability density, K is a probability density amplitude; see ref. [2] for a discussion of this issue.

Here $K(x_2, t_2|x_1, t_1)$ denotes the quantum–mechanical propagator, and $\psi(x, t)$ is the wavefunction. As in (20) above, one must allow for possible numerical factors between probabilities on the thermodynamical and on the mechanical sides; otherwise bijectivity is perfectly preserved.

Our expectations (22), (23) are borne out by experiment—experiment in our case being explicit computation. Indeed one finds the following. For $\gamma \to 0$, the irreversible thermodynamics corresponds to the free quantum–mechanical particle:

$$K^{(\text{free})}(x_2, t|x_1, 0) = \sqrt{\frac{k_B}{s}}\, f_1 \left(\begin{matrix} x_2 \\ \text{i}t \end{matrix} \middle| \begin{matrix} x_1 \\ 0 \end{matrix} \right)_{\gamma\to 0}, \tag{24}$$

while, for $\gamma \neq 0$, the irreversible thermodynamics corresponds to the quantum mechanics of a harmonic oscillator:

$$f_1 \left(\begin{matrix} x_2 \\ \text{i}t \end{matrix} \middle| \begin{matrix} x_1 \\ 0 \end{matrix} \right) = \exp\left(\frac{\text{i}\omega t}{2} - \frac{\Delta V}{\hbar\omega} \right) \sqrt{\frac{2m\omega}{\hbar}}\, K^{(\text{harmonic})}(x_2, t|x_1, 0) . \tag{25}$$

Above, $V = m\omega^2 x^2/2$ is the harmonic potential, and $\Delta V = V(x_2) - V(x_1)$. Moreover, if $\psi_0(x) = \exp\left(-m\omega x^2/2\hbar\right)$ is the harmonic oscillator groundstate, then it holds that, up to normalisation,

$$f_1 \left(\begin{matrix} x \\ \text{i}t \end{matrix} \right) = \exp\left(-\frac{m\omega}{\hbar} x^2 \right) = |\psi_0^{(\text{harmonic})}(x)|^2, \tag{26}$$

as expected.

Finally the path–integral representation of quantum–mechanical propagators,

$$K\left(x_2, t_2|x_1, t_1\right) = \int_{x(t_1)=x_1}^{x(t_2)=x_2} \text{D}x(t)\, \exp\left\{ \frac{\text{i}}{\hbar} \int_{t_1}^{t_2} \text{d}t\, L\left[x(t), \dot{x}(t)\right] \right\}, \tag{27}$$

has a nice reexpression in terms of classical, irreversible thermodynamics. Indeed, applying our dictionary (18), (19), (20) and (21) to the mechanical path integral (27), the latter becomes the thermodynamical path integral already seen in (16). This leads us to the following relation between the action integral I of the mechanical system and the entropy S of its thermodynamical counterpart:

$$\frac{\text{i}}{\hbar} I \leftrightarrow \frac{1}{k_B} S. \tag{28}$$

It should be remarked that both I and S independently satisfy an extremum principle. In the Gaussian approximation considered here, the respective fluctuations (measured with respect to the corresponding mean values of I and S as given by their extremals) are obtained upon taking the exponentials. We thus obtain the quantum–mechanical wavefunction and the Boltzmann distribution function:

$$\psi = \sqrt{\rho}\, \exp\left(\frac{\text{i}}{\hbar} I \right), \qquad \rho_B = Z^{-1} \exp\left(\frac{1}{k_B} S \right). \tag{29}$$

As usual, Z denotes some normalisation factor. Since, by the Born rule, we must have $\rho_B = |\psi|^2$, this provides us with an elegant expression combining thermodynamics and quantum mechanics into a single equation:

$$\psi = Z^{-1/2} \exp\left(\frac{1}{2k_B} S \right) \exp\left(\frac{\text{i}}{\hbar} I \right). \tag{30}$$

Eqs. (28) and (30) are very inspiring, as they reveal a fundamental complementarity between the mechanical action integral (on the mechanical side) and the entropy (on the thermodynamical side). We will later on return to the complementarity between these two descriptions, a feature already foreseen by Prigogine [30]. For the moment let us simply remark the following consequence of this complementarity, namely, the symmetrical role played by Planck's constant $\hbar$ and Boltzmann's constant k_B. This latter property, and the ensuing entropy quantisation, have been discussed at length in refs. [1, 2].

4. Emergence from irreversibility

It has been claimed that *quantisation is dissipation* [3, 4, 5, 6, 7, 8, 10, 18, 19, 31]—this claim is central to the emergence approach to quantum mechanics. In more precise terms, the previous statement implies that quantum behaviour can be expected from certain deterministic systems exhibiting information loss. One could compare this state of affairs to the relation between (equilibrium) thermodynamics and (classical) statistical mechanics. Namely, information loss in a microscopic theory (statistical mechanics) arises as the result of averaging out over many degrees of freedom; the emergent theory (thermodynamics) contains less information than its microscopic predecessor.

Thanks to the map established in section 3, the picture presented here features quantumness as an intrinsic property of dissipative systems. Conversely, by the same map, any quantum system features dissipation. In our picture, irreversibility and quantumness arise as the two sides of the same coin, thus becoming *complementary* descriptions of a given system (*complementarity* being understood here in Bohr's sense of the word). As opposed to the emergence property discussed above, the two theories (quantum mechanics and irreversible thermodynamics) contain exactly the same amount of information. It is interesting to observe that closely related views regarding the complementarity between mechanics and thermodynamics were defended long ago by Prigogine [30].

Now it has been (rightly) pointed out that correspondence and emergence are not quite the same concept [23]. This notwithstanding, we can still argue that quantum mechanics continues to arise as an *emergent phenomenon* in our picture. This is so because Boltzmann's dictum applies: *If something heats up, it has microstructure*. In other words, every thermodynamics is the coarse graining of some underlying statistical mechanics. Thus the mere possibility of recasting a given theory in thermodynamical language proves that the given theory is the coarse–grained version of some finer, microscopic theory.

5. Discussion

As a technical remark, we should point out that we have worked throughout in the Gaussian approximation. On the thermodynamical side of our map this corresponds to the linear response theory; on the mechanical side this refers to the harmonic approximation. Within the regime of applicability of this assumption we can safely claim to have provided a rigorous proof of the statement that *quantum mechanics is an emergent phenomenon, at least in the Gaussian approximation*.

Using the fact that any potential can be transformed into the free potential or into the harmonic potential by means of a suitable coordinate transformation (as in Hamilton–Jacobi theory [14, 15]), one would naively state that the Gaussian approximation is good enough to "prove" that quantum mechanics is an emergent phenomenon also beyond the Gaussian regime. However, this "proof" overlooks the fact that quantisation and coordinate changes do not generally commute. Therefore the previous reasoning invoking Hamilton–Jacobi can only be seen as a plausibility argument to support the statement that quantum mechanics must remain an emergent phenomenon *also beyond* the Gaussian approximation. There is, however, abundant literature dealing with the emergent nature of quantum mechanics, *regardless of* the Gaussian approximation, using techniques that are very different from those presented here, and with a spectrum of applicability that ranges from the smallest [11, 24] to the largest [27, 29]. We will therefore content ourselves with the proof of emergence presented here, the expectation being that some suitable extension thereof (possibly using perturbative techniques) will also apply beyond the Gaussian approximation.

Acknowledgments

J.M.I. would like to thank the organisers of the *Sixth International Workshop DICE 2012: Spacetime–Matter–Quantum Mechanics: from the Planck scale to emergent phenomena* (Castiglioncello, Italy, September 2012) for the invitation to present this work, for stimulating a congenial atmosphere of scientific exchange, and for the many interesting discussions that followed.
O muse, o alto ingegno, or m'aiutate;
o mente che scrivesti ciò ch'io vidi,
qui si parrà la tua nobilitate.—Dante Alighieri.

References

[1] Acosta D, Fernández de Córdoba P, Isidro J M and Santander J L G 2012 An entropic picture of emergent quantum mechanics *Int. J. Geom. Meth. Mod. Phys.* **9** 1250048 (*Preprint* 1107.1898 [hep-th])
[2] Acosta D, Fernández de Córdoba P, Isidro J M and Santander J L G 2012 A holographic map of action onto entropy *J. Phys.: Conf. Ser.* **361** 012027 (*Preprint* 1112.4353 [hep-th])
[3] Acosta D, Fernández de Córdoba P, Isidro J M and Santander J L G 2012 Emergent quantum mechanics as a classical, irreversible thermodynamics *Preprint* 1206.4941 [math-ph]
[4] Blasone M, Jizba P and Vitiello G 2001 Dissipation and quantization *Phys. Lett.* A **287** 205 (*Preprint* hep-th/0007138)
[5] Blasone M, Jizba P and Vitiello G 2004 Dissipation, emergent quantization, and quantum fluctuations, *Decoherence and Entropy in Complex Systems, Selected Lectures from DICE 2002*, ed H-T Elze *Lecture Notes in Physics* **633** (Berlin: Springer)
[6] Blasone M, Jizba P and Kleinert H 2005 Quantum behavior of deterministic systems with information loss: path integral approach *Ann. Phys.* **320** 468 (*Preprint* quant-ph/0504200)
[7] Blasone M, Jizba P and Scardigli F 2009 Can quantum mechanics be an emergent phenomenon? *J. Phys.: Conf. Ser.* **174** 012034 (*Preprint* 0901.3907 [quant-ph])
[8] Blasone M, Jizba P, Scardigli F and Vitiello G 2009 Dissipation and quantisation for composite systems *Phys. Lett.* A **373** 4106 (*Preprint* 0905.4078 [quant-ph])
[9] Carroll R 2010 *On the Emergence Theme of Physics* (Singapore: World Scientific)
[10] Elze H-T 2009 The attractor and the quantum states *Int. J. Qu. Info.* **7** 83 (*Preprint* 0806.3408 [quant-ph])
[11] Elze H-T 2009 Does quantum mechanics tell an atomistic spacetime? *J. Phys.: Conf. Ser.* **174** 012009 (*Preprint* 0906.1101 [quant-ph])
[12] Elze H-T 2012 Linear dynamics of quantum–classical hybrids *Phys. Rev.* A **85** 052109 (*Preprint* 1111.2276 [quant-ph])
[13] Elze H-T 2012 Four questions for quantum–classical hybrid theory *J. Phys.: Conf. Ser.* **361** 012004 (*Preprint* 1202.3448 [quant-ph])
[14] Faraggi A and Matone M 1998 Equivalence principle, Planck length and quantum Hamilton–Jacobi equation *Phys. Lett.* B **445** 77 (*Preprint* hep-th/9809125)
[15] Faraggi A and Matone M 2009 The equivalence postulate of quantum mechanics: main theorems *Preprint* arXiv:0912.1225 [hep-th]
[16] Groessing G, Fussy S, Mesa Pascasio J and Schwabl H 2012 The quantum as an emergent system *J. Phys.: Conf. Ser.* **361** 012008 (*Preprint* arXiv:1205.3393 [quant-ph])
[17] Groessing G, Fussy S, Mesa Pascasio J and Schwabl H 2012 A classical framework for nonlocality and entanglement *Preprint* arXiv:1210.4406 [quant-ph]
[18] 't Hooft G 1999 Quantum gravity as a dissipative deterministic system *Class. Quant. Grav.* **16** 3263 (*Preprint* arXiv:gr-qc/9903084)
[19] 't Hooft G 2007 Emergent quantum mechanics and emergent symmetries *AIP Conf. Proc.* **957** 154 (*Preprint* arXiv:0707.4568 [hep-th])
[20] 't Hooft G 2010 The emergence of quantum mechanics *AIP Conf. Proc.* **1446** 341
[21] 't Hooft G 2012 Quantum mechanics from classical logic *J. Phys.: Conf. Ser.* **361** 012024
[22] 't Hooft G 2012 Relating the quantum mechanics of discrete systems to standard canonical quantum mechanics *Preprint* arXiv:1204.4926 [quant-ph]

[23] Hu B 2011 Gravity and nonequilibrium thermodynamics of classical matter *Int. J. Mod. Phys.* D **20** 697 (*Preprint* arXiv:1010.5837 [gr-qc])

[24] Hu B 2009 Emergent/quantum gravity: macro/microstructures of spacetime *J. Phys.: Conf. Ser.* **174** 012015 (*Preprint* arXiv:0903.0878 [gr-qc])

[25] Khrennikov A 2012 "Einstein's dream" – quantum mechanics as theory of classical random fields *Preprint* arXiv:1204.5172 [quant-ph]

[26] Khrennikov A 2012 Växjö Interpretation of Wave Function: 2012 *Preprint* arXiv:1210.2390 [quant-ph]

[27] Kiefer C 2010 Can quantum theory be applied to the universe as a whole? *Found. Phys.* **40** 1410

[28] Onsager L and Machlup S 1953 Fluctuations and irreversible Processes *Phys. Rev.* **91** 1505

[29] Penrose R 2009 Black holes, quantum theory and cosmology *J. Phys.: Conf. Ser.* **174** 012001

[30] Prigogine I 1977 Time, structure and fluctuations, Nobel Prize Lecture

[31] Sakellariadou M, Stabile A and Vitiello G 2012 Noncommutative spectral geometry, dissipation and the origin of quantization *J. Phys. Conf. Ser.* **361** 012025

[32] Smolin L 2006 Could quantum mechanics be an approximation to another theory? *Preprint* arXiv:quant-ph/0609109

[33] Wetterich C 2009 Emergence of quantum mechanics from classical statistics *J. Phys.: Conf. Ser.* **174** 012008 (*Preprint* arXiv:0811.0927 [quant-ph])

Topological ordered spaces as a foundation for a quantum spacetime theory

E Minguzzi
Dipartimento di Matematica e Informatica "U. Dini", Università degli Studi di Firenze, Via S. Marta 3, I-50139 Firenze, Italy
E-mail: ettore.minguzzi@unifi.it

Abstract. I outline some recent progress on the theory of topological preordered spaces, and comment on the role of these spaces for an approach to a quantum theory of spacetime.

1. Ideas for a quantum spacetime theory

Physical theories rely in a hierarchy of specialized mathematical structures. Our current model for spacetime describes it as a 4-dimensional time-oriented Lorentzian manifold, usually denoted (M, g). Here we can recognize the following progressively less restrictive structures: (a) that of a Lorentzian manifold with its induced causal order determined by the distribution of light cones; (b) the deeper differentiable structure, typical of manifolds, which serves to give a meaning to differentiability; and the even deeper (c) topological structure, which gives a meaning to the notions of closeness and continuity.

The history of Science suggests that some of the deepest and more general structures can actually turn out to be dynamical. For example, the old rigid Euclidean space has been replaced by a dynamical Lorentzian manifold, and many other old rigid concepts such as distant simultaneity have lost they absolute status. We expect this process to continue and hence we are led to ask which concepts are more likely to survive when we remove mathematical specialization to our present model of spacetime.

Trying to answer this question leads us to consider quantum physics, and with it the idea that spacetime may lose its smooth character, which would be, therefore, just an apparent approximate feature for low energy processes. If we accept that smoothness should be lost at a quantum regime then we must also ready to accept that the concept of a Lorentzian metric should not make sense in that limit. We would like, however, to retain the causal relation as this seems to be a fundamental concept without which the mathematics would become too poor to lead to any interesting physical consequence, as even the interpretation of the theory would become too obscure. We are therefore lead to the idea that the following ingredients should be present in a fundamental spacetime theory

- Topology: gives a notion of continuity to work with.
- Order: to be interpreted in a causal fashion; essential for the interpretation of the theory.
- Measure: most of this element is already contained in the topology which determines the σ-algebra.

These ingredients are indeed those selected in other approaches to quantum gravity such as *Causal Set Theory* [1, 2]. However, in our approach we are not going to infer topology from order, as in that theory. Rather, we are going to impose compatibility conditions between topology and order. The approach by Causal Set Theory is, therefore, more restrictive as the known topologies that one can construct from the order are quite few in number, e.g. Alexandrov, Scott, etc.. We wish to show that our approach is indeed more natural since topology and order might be considered as two faces of the same entity: the *quasi-uniformity*. The theory that studies this connection is that of *topological preordered spaces*, which is still unknown to physicists and in fact also to most mathematicians. In this contribution I will give an introduction to this theory. First, however, let me mention how I came to study this theory and its relevance for relativity.

2. Some motivation from mathematical relativity

In mathematical relativity two events $p, q \in M$ are said to be causally related, in formula $p \le q$, if there is a future directed causal curve from p to q or $p = q$. The causal relation is a subset of $M \times M$ made by all the causally related pairs and is denoted J^+. It turns out that J^+ is not necessarily closed (e.g. consider Minkowski spacetime minus a point), and that closing it to get $\overline{J^+}$ might spoil transitivity. Early relativists looked for a better causal relation and Seifert [3] came up with the idea of considering $J_S^+ := \bigcap_{g'>g} J_{g'}^+$, where $g' > g$ if the timelike cones of g' contain the lightlike cones of g. The Seifert relation is indeed closed and transitive and it is a partial order if and only if the spacetime is stably causal, namely if there is $g' > g$ such that (M, g') is causal, i.e. opening the light cones does not spoil causality. Under stable causality J_S^+ is also the smallest closed and transitive relation which contains J^+, see [4].

Hawking proved that every stably causal spacetime admits a time function, namely a continuous function which increases over every causal curve [5]. His proof is technically quite complicated and uses in an essential way the Lorentzianity of the spacetime manifold. I realized that the existence of a time function had instead more to do with the existence of the closed partial order J_S^+. After some research I found [6] the existence of the following result in the theory of topological preordered spaces (I removed the Hausdorff condition [7])

Theorem 2.1 (Levin [8, 9]). *Every second countable locally compact space endowed with a closed order (e.g. a stably causal spacetime endowed with the Seifert relation) admits a strictly increasing continuous function.*

We are therefore able to infer the existence of time functions without using the Lorentzianity of the manifold or even the smoothness of the manifold. This is one of those results which would make perfect sense in a quantum theory of spacetime.

3. The mathematical theory

A topological preordered space is a triple $(E, \mathscr{T}, \le)$ where $(E, \mathscr{T})$ is a topological space and $\le$ is a reflexive and transitive relation over E (preorder), namely its graph $G(\le) := \{(x, y) : x \le y\}$ satisfies $G \circ G \subset G$, and $\Delta \subset G$. Here $\circ$ is the composition of relations, and $\Delta := \{(x, y) : y = x\}$ is the diagonal of $E \times E$, which plays the role of the identity for the composition. If $G \cap G^{-1} = \Delta$ we say that $\le$ is an order (or partial order).

Given $x \in E$ the set $i(x) := \{y : x \le y\}$ is the increasing hull of x and $d(x) := \{y : y \le x\}$ is the decreasing hull. A set $S \subset E$ for which $i(S) = S$ is said *increasing*, and dually, a set for which $d(S) = S$ is said *decreasing*.

Given a topological preordered space one has the upper topology $\mathscr{U}$, whose open sets are the open increasing sets, and the lower topology $\mathscr{L}$ which is defined dually. Sometimes from the triple $(E, \mathscr{U}, \mathscr{L})$ one can recover $(E, \mathscr{T}, \le)$, thus there is a connection between the theory of *bitopological spaces* and that of *topological preordered spaces*. A topological preordered space is said to be *convex* if $\mathscr{T} = \sup(\mathscr{U}, \mathscr{L})$, see [10].

Apart from convexity, one can place several inequivalent conditions between topology and order. We speak of *closed preordered space* if G is closed as a subset of $E \times E$ with the product topology. Let us observe that the usual topology theory can be understood as equivalent to this theory where the order is the discrete one, namely $G = \Delta$. In topology it is known that the Hausdorff condition is equivalent to the closure of Δ, thus any Hausdorff topological space is a *closed preordered space* for which the relation is the trivial one. It is also useful to observe that if $\le$ is a closed order then as $G \cap G^{-1} = \Delta$ is closed, that is, $\mathscr{T}$ is Hausdorff.

We speak of *normally preordered space* if G is closed and if any disjoint pair made of a closed increasing set B and a closed decreasing set A, can be separated by a disjoint pair made of an open increasing set $V \supset B$ and an open decreasing set $U \supset A$. In this case, for A and B as above, there is always a non-decreasing (isotone) continuous function $f : E \to [0, 1]$ such that $f(A) = 0$ and $f(B) = 1$. This is Nachbin's generalization of Urysohn's lemma [11, 12, 13].

Let $\mathcal{F}$ be the family of isotone continuous functions in $[0, 1]$. Another compatibility condition is that defining a *Tychonoff (or completely regularly) preordered space.* This is a closed preordered space for which (a) $\mathcal{F}$ determines the topology (the coarsest topology which makes all the functions in $\mathcal{F}$ continuous is $\mathscr{T}$), and (b) $\mathcal{F}$ determines the order, that is, $x \le y$ iff for every $f \in \mathcal{F}$, $f(x) \le f(y)$. These spaces are important because they can be compactified in a precise way [14, 15].

One of the reasons for the success of topology lies in the possibility of improving the separability properties of the space using countability or compactness conditions. Consider for instance the two results

Theorem 3.1. *Every locally compact Hausdorff space is completely regular (Tychonoff).*

and

Theorem 3.2 (Urysohn)**.** *Every second countable regular space is metrizable.*

Under local compactness and second countability one can therefore prove that the Hausdorff condition is improved to metrizability which implies normality.

In the preordered case these theorems do not hold and in any case, rather puzzlingly, a completely regularly preordered space need not be regularly preordered [16, Example 1]. One could prove that compact preordered spaces are normally preordered [11, 17] but this result is rather limited. For these difficulties the theory could not be applied to practical problems, as the manifolds which are usually encountered in applications are quite general, and certainly not compact. Therefore, I was led to follow a different path proving [17]

Theorem 3.3. *Every locally compact σ-compact space equipped with a closed preorder is normally preordered.*

Since manifolds satisfy these assumptions, this theorem proved that topological preordered spaces appearing in applications share good separability properties. These applications are in fact not restricted to relativity, as Thermodynamics, Microeconomics and Computer Science present situations in which topology and order are the basic elements of the theory.

Convex normally preordered spaces are completely regularly preordered spaces, thus one would like to prove convexity. I proved that a locally compact σ-compact closed ordered space for which the order is compactly generated is indeed convex and hence completely regularly preordered [18]. Here an order is compactly generated if it can be recovered from its specification over compact sets, that is if, so to speak, the information required to determine the order is local in nature. Cones structures over topological manifolds determine orders which are of this type.

In order to understand the relation between topology and preorder let us mention the notion of *uniformity* as it was introduced by Weyl. A uniformity is a kind of topology for which two far away open sets can be compared. Metric spaces induce a uniformity.

More precisely, a uniformity $\mathcal{U}$ is a filter of neighborhoods V of the diagonal $\Delta \subset E \times E$, such that (a) if $V \in \mathcal{U}$ then $V^{-1} := \{(x, y) : (y, x) \in V\} \in \mathcal{U}$; and (b) if $V \in \mathcal{U}$ then there is $U \in \mathcal{U}$ such that $U \circ U \subset V$. Every uniformity determines a topology over E, whose neighborhoods of x admits the base $U(x) := \{y : (x, y) \in U\}$ with $U \in \mathcal{U}$. The topology is Hausdorff if $\bigcap \mathcal{U} = \Delta$.

Nachbin [11] realized that one can actually drop axiom (a) getting a very interesting concept: the *quasi-uniformity*. Quasi-uniformities allow us to naturally define not only a topology, as the topology $\mathscr{T}$ of the symmetric uniformity $\sup(\mathcal{U}, \mathcal{U}^{-1})$, but also a preorder through $G = \bigcap \mathcal{U}$. It is easy to verify that this is indeed a preorder, which is an order if and only if the topology is Hausdorff. Thus topology and preorder are two aspects of the same entity, the quasi-uniformity. Also it turns out that the quasi-uniformizable topological preordered spaces are the completely regularly preordered spaces introduced above [11].

One can also introduce the concept of quasi-pseudo-metric $p(x, y)$ which satisfies: (i) $p(x, x) = 0$ and (ii) $p(x, z) \le p(x, y) + p(y, z)$. The question is whether a topological preordered space $(E, \mathscr{T}, \le)$ admits a quasi-pseudo-metric, in the sense that the topology $\mathscr{T}$ comes from the symmetric pseudo-metric $d(x, y) = p(x, y) + p(y, x)$, and the preorder satisfies $x \le y$ if and only if $p(x, y) = 0$. I proved that the answer is affirmative if $(E, \mathscr{T}, \le)$ is a quasi-uniformizable space with a second countable topology [19].

4. Conclusion

I have given a short account of the theory of topological preordered spaces. A study of this theory reveals that topology and order are two aspects of the same mathematical object and, therefore, should be studied jointly. This result is particularly interesting because these are exactly the mathematical ingredients which are believed to make sense in a quantum spacetime theory, namely at a limit in which the differentiability of the manifold is lost but the concepts of causal order and continuity still make sense.

Acknowledgments

This work has been partially supported by GNFM of INDAM.

References

[1] Bombelli L, Lee J H, Meyer D and Sorkin R D 1987 *Phys. Rev. Lett.* **59** 521–524
[2] Sorkin R D 1991 *Proceedings of the ninth Italian conference in General Relativity and Gravitational Physics* (Singapore: World Scientific) pp 68–90
[3] Seifert H 1971 *Gen. Relativ. Gravit.* **1** 247–259
[4] Minguzzi E 2009 *Commun. Math. Phys.* **290** 239–248
[5] Hawking S W and Ellis G F R 1973 *The Large Scale Structure of Space-Time* (Cambridge: Cambridge University Press)
[6] Minguzzi E 2010 *Commun. Math. Phys.* **298** 855–868
[7] Minguzzi E 2012 *Appl. Gen. Topol.* **13** 81–89 (*Preprint* arXiv:1108.5123)
[8] Levin V L 1983 *Soviet Math. Dokl.* **28** 715–718
[9] Levin V L 1983 *Soviet Math. Dokl.* **27** 639–643
[10] Nachbin M L 1948 *Comptes Rendus Hebdomadaires des Séances de l'Académie des Sciences* **226** 381–382
[11] Nachbin L 1965 *Topology and order* (Princeton: D. Van Nostrand Company, Inc.)
[12] Bonsall F F 1960 *Proc. London Math. Soc.* **10** 122–140
[13] Bonsall F F 1961 *Proc. Int. Symposium on Linear Spaces, Jerusalem* (Oxford: Pergamon Press) pp 101–114
[14] Fletcher P and Lindgren W 1982 *Quasi-uniform spaces* (*Lect. Notes in Pure and Appl. Math.* vol 77) (New York: Marcel Dekker, Inc.)
[15] Minguzzi E 2012 *Appl. Gen. Topol.* **13** 207–223 (*Preprint* arXiv:1209.1839)
[16] Künzi H P A and Watson S 1994 *Comment. Math. Univ. Carolinae* **35** 773–778
[17] Minguzzi E 2011 Normally preordered spaces and utilities Order, published online 06 Sept. 2011, DOI: 10.1007/s11083-011-9230-4 (*Preprint* arXiv:1106.4457v2)
[18] Minguzzi E 2012 Convexity and quasi-uniformizability of closed preordered spaces (*Preprint* arXiv:1212.3776)
[19] Minguzzi E 2012 *Topol. Appl.* **159** 2888–2898

Journal of Physics: Conference Series **442** (2013) 012035 doi:10.1088/1742-6596/442/1/012035

Toward a real synthesis of quantum and relativity theories: experimental evidence for absolute simultaneity

S. A. Emelyanov

Division of Solid State Electronics, Ioffe Institute, 194021 St. Petersburg, Russia

E-mail: sergey.emelyanov@mail.ioffe.ru

Abstract. We have demonstrated spatially-discontinuous quantum jumps of electrons at a distance as long as about 1cm. The effect occurs in a modified integer quantum Hall system consisted of a great number of extended Laughlin-Halperin-type states. Our observations directly contradict the no-aether Einstein's interpretation of special relativity together with the Minkowski's model of spacetime. However they are consistent with the aether-related Lorentz-Poincare's interpretation that allows absolute simultaneity. We thus strongly challenge the fundamental status of Lorentz invariance and hence break the basic argument against de Broglie-Bohm realistic quantum theory. We argue that both de Broglie-Bohm and Lorentz-Poincare theories are capable of providing a real synthesis of quantum and relativity theories. This synthesis is of such kind that quantum theory appears the most fundamental physical theory for which relativity is only a limiting case. In accordance with this hierarchy, quantum theory naturally resolves the problem of aether in Lorentz-Poincare's relativity. The role of aether could be played by a deeper Bohm-type undivided quantum pre-space, the relevance of which at any lengthscale directly follows from our observations.

1. Introduction

It may be that a real synthesis of quantum and relativity theories requires not just technical developments but radical conceptual renewal. J. S. Bell (1987)

Today it is generally believed that there is no reasonable alternative to Einstein's interpretation of special relativity as well as to the Minkowski's model of spacetime, which prohibit absolute simultaneity even in principle. The Lorentz-Poincare's aether-based interpretation of relativity is rare in literature despite the fact that it does not contradict any experiments known up to now. This situation is partly due to the method of Einstein's original presentation of the special relativity in his seminal work of 1905 [1]. The method looks like an elegant, almost mathematical proof where macroscopic rod and clock are regarded as fundamental entities which need no any further clarification. However, the inconsistence of such method was stressed even by the author. In 1949, he wrote: *"The theory.... introduces two kinds of physical things, i.e., (1) measuring rods and clocks, (2) all other things, e.g., the electro-magnetic field, the material point, etc. This, in a certain sense, is inconsistent; strictly speaking measuring rods and clocks would have to be represented as solutions of the basic equations…, not, as it were, as theoretically self-sufficient entities…"* [2]. Nevertheless, this comment does not alter general belief in the ultimate truth of Einstein's theory.

However, the problem is that the interpretation of the special relativity is not "a matter of taste" but leads to significant physical consequences. One of them is the status of Lorentz transformations. Following Einstein's theory, Lorentz transformations reflect a fundamental symmetry, the so-called Lorentz invariance, which is relevant to the other physical theories, for instance, to quantum mechanics (QM). At the same time, even the founders of QM were aware of the fact that their theory is hardly consistent with the relativistic model of spacetime. At the Solvay Conference of 1927, Niels Bohr directly pointed out that "… *The whole foundation for causal spacetime description is taken away by quantum theory, for it is based on [the] assumption of observations without interference…*" [3]. An even more formidable challenge is to reconcile relativity with the notion of instantaneous quantum jumping, which is an essential part of Schrödinger's wave mechanics. Schrödinger himself was so shocked by this notion that he was willing to give up the whole quantum theory. In this context, he addressed Bohr the following phrase: *"If all this damned quantum jumping were really here to stay, I should be sorry I ever got involved with quantum theory…"* Bohr's response is also well-known:*"… remember the Einstein derivation of Planck's [blackbody] radiation law. This derivation demands that the energy of the atom should assume discrete values and change discontinuously from time to time... You can't seriously be trying to cast doubt on the whole basis of quantum theory!"* [4]. So, on the one hand, the unprecedented success of QM in prediction of experimental observations left no doubt regarding its ultimate relevance but, on the other hand, it was hard to reconcile it with Einstein's relativity.

An extraordinary solution of the problem was proposed at the Solvay Conference by Bohr, Heisenberg and some other participants. It is known as the Copenhagen interpretation of QM, which actually is the standard QM known from textbooks. The key point of the solution is a radical denial of objectivity of the quantum world which appears to be real only insofar as some events, the so-called "measurements", take place in the macroscopic world. Bohr himself clarified this idea as follows: "…*There is no quantum world. There is only an abstract quantum mechanical description…"* [5]. Heisenberg spoke along the same line:"… *Reality is in the observation, not in the electron …"* [6]. In fact, it was an unprecedented case when realism, as a philosophical basis, was consciously rejected by a physical theory. In a sense, Bohr's solution was triggered by the fact that, at the time of Bohr, QM had a well-defined domain of applicability, that is, the microcosm. So, it seemed possible to separate subjective quantum world from objective classical world using some lengthscale criteria. At the same time, the advantage of Bohr's solution seems quite evident: henceforth any violation of Lorentz invariance in the quantum world cannot challenge a fundamental nature of this symmetry simply because this world is a subjective thing by definition.

However, the Copenhagen interpretation inevitably leads to serious conceptual problems. The most famous one is the so-called "measurement problem". In essence, the problem is that the formerly clear term "measurement" now acquires a truly mystical sound. Through the "measurement" some "measurer" literally creates, just like God, "something out of nothing" where something is a quantum object. Of course, a number of questions immediately arise in the context of the above "act of creation". One of them is who is the creator which could play the role of "measurer"? John Bell, for example, not without irony, tried to guess: *"…Was the world wave function waiting to jump for thousands of millions of years until a single-celled living creature appeared? Or did it have to wait a little longer for some more highly qualified measurer – with a Ph.D.?"* [7]. But that is not all.

According to the well-known Bohr's complementarity principle, the "measurer" not only creates some object out of nothing but, simply by selecting the measuring devices, he (or she) literally "causes" the object to be either a corpuscle or a wave. As a brief comment to this principle, Bell put in: *"The justly immense prestige of Bohr has led to the mention of complementarity in most text books of quantum theory. One is tempted to suspect that the authors do not understand the Bohr philosophy sufficiently to find it helpful..."* [7].

Certainly, the above questions are not the only ones that appear in the context of the Copenhagen interpretation. However *"... the pioneers of quantum mechanics were not unaware of these questions, but quite rightly did not wait for agreed answers before developing the theory ..."* [7]. But the point is the "agreed answers" haven't been found even up to now, i.e. almost hundred years later. As a result, a part of scientific community has come to the conclusion that we need no "agreed answers" at all simply because both mysticism and indeterminism are indeed the fundamental properties of microcosm. The other part of the community intuitively tries to avoid any mysticism reasonably believing that it strongly challenges one of the basic pillars of science – the principle of knowability. So, consciously or not, they follow a doctrine that could really appear only in the context of Bohr's QM. The doctrine sounds as "shut-up-and-calculate". This means that the formalism of Bohr's QM is indeed extremely successful but, at the same time, it is meaningless to think of what really lies behind it. Perhaps only a few persons refused to join any of the two camps. And one of them was Albert Einstein. On the one hand, he was convinced that "... *He [God] does not throw dice..."* [8]. But, on the other hand, he was no less convinced that "... *independence created by philosophical insight is ... the mark of distinction between a mere artisan or specialist and a real seeker after truth ..."* [9]. Unfortunately, the current trend is that the Einstein's decisive authority extends only over the theory of relativity but it breaks immediately when we are dealing with the quantum theory where his views are regarded as a stubborn conservatism.

The reason for the disliking of pilot-wave theory is actually well-known. While the violation of Lorentz invariance in Bohr's theory is regarded to be more or less remissible, the same thing in the pilot-wave theory is regarded to be absolutely inadmissible. And the key point here is the status of wavefunction. In the pilot-wave theory, it is not an abstract description but rather a real physical entity. As it was especially stressed by Bell, *"... no one can understand this [Bohm] theory until he is willing to think of ψ as a real objective field rather than just a "probability amplitude"* [7]. From the perspective of Einstein's relativity, this point is truly unacceptable. Indeed, let us take the familiar EPR experiment with entangled photons. It is seen immediately that, in terms of pilot-wave theory, EPR nonlocality acquires the characteristic features of a real physical effect, that is, a real action-at-a-distance. In this case, the impossibility of nonlocal signalling seems a peculiarity of EPR situation rather than a fundamental physical low.

Accordingly, the results of actual EPR experiments are currently viewed as a strong evidence for the ultimate truth of both Bohr's QM and Einstein's relativity [13-15]. On the one hand, they are fully consistent with the predictions of Bohr's QM but, on the other hand, EPR nonlocal correlations appear essentially non-causal, i.e. they cannot be used for any signalling. Moreover, the last point allows a purely mathematical proof known today as Eberhard's theorem [16]. However, in accordance with the common belief that quantum entanglement is the only possible physical reason for nonlocality, the Eberhard's theorem is usually interpreted in a wider sense, that is, as a mathematic proof of the impossibility of nonlocal signalling at all. The theorem thus appears to be quite fundamental so that it is often called the "no-communication theorem". And this status is clearly consistent with the fundamental status of Lorentz invariance.

However, the current relations between quantum and relativity theories are clouded. The point is they only coexist but not more. Almost centennial efforts to unify them have failed and it becomes more and more clear that unification is even inaccessible at least until both theories are in their generally-recognized forms. Today this point is declared to be one of the most fundamental physical challenges. However, a promising way to reach such unification was almost intuitively felt by John Bell who, just like Einstein, was unable to accept quantum mysticism. The idea was expressed in one of his interviews of 1986: "*Behind the apparent Lorentz invariance of the phenomena, there is a deeper level which is not Lorentz invariant ... what is not sufficiently emphasized in textbooks, in my opinion, is that the pre-Einstein position of Lorentz and Poincare, Larmor and Fitzgerald was perfectly coherent, and is not inconsistent with relativity theory ... I want to say there is a real causal sequence which is defined in the aether...*" [17]. So the idea is that we should rehabilitate the realistic quantum theory through the return to the aether-based Lorentz-Poincare relativity where Lorentz transformations are an emergent phenomenon.

In fact, taking a realistic view of the world, this status of Lorentz transformations seems quite natural. Indeed, they clearly are relevant only in the case of a movement that could be characterized by a certain speed. However, say, in the EPR situation, nothing could be characterized in such a way. It is interesting to note that, even in 1982, Karl Popper, who was rather a philosopher than physicist, made the following comment to the EPR experiment: "*... We have to give up Einstein's interpretation of special relativity and return to Lorentz's interpretation and with it to ... absolute space and time... Whether or not an infinite velocity can be attained in the transmission of signals is irrelevant for this argument...*" [18]. However, his view is rather an exception. Most physicists still believe that any alternative to Einstein's relativity is impossible and this belief is so deep that even the most radical supporters of quantum realism still try to squeeze pilot-wave theory into the Procrustean bad of Lorentz invariance [19-21].

So the current situation is that the only thing that could really break a truly universal belief in Einstein's relativity is the observation of a physical phenomenon which would directly contradict this theory but, at the same time, would be consistent with the Lorentz-Poincare's theory. Actually, it would be only the observation of a purely quantum communication which currently is regarded to be fundamentally impossible. And in this work we report on such observation. The effect has become possible insofar as it is not related to quantum entanglement but rests on such a well-known thing as the spatially-discontinuous transport of quantum particles. So the novelty of our work is that we directly demonstrate that this transport may well be macroscopic and hence should be regarded as a peculiar quantum dynamics capable of providing absolute simultaneity.

2. Beyond the no-communication theorem: spatially-discontinuous macroscopic quantum transport jumps of quantum particles

One of the main "surprises" which is presented by quantum mechanics is a possibility for quantum particles to transit from one spatial region to another without staying in the intermediate regions. The simplest example of that kind is an electron transition between atomic orbits when not only energy changes discontinuously but also the spatial position of the electron. However in such form, in which this effect is known today, it seems an absurd to interpret it in terms of a peculiar quantum transport. The point is the characteristic lengthscale of these transitions is quite microscopic in all known cases and this fact is generally regarded as being of a fundamental character. At the same time, according to the ideas of standard QM, the microscopicity of spatially-discontinuous transitions ultimately implies that we are dealing not with objective processes but rather with a method of description. So, it seems quite natural that no one has considered these transitions in the context of their compatibility with relativity principles.

However, the common belief in microscopicity of spatially-discontinuous transitions is intuitive at best. The point is there are no fundamental limitations on their lengthscale in the formalism of standard QM. This fact allows us to propose the following *gedanken* experiment. Let us consider a one-dimensional electron orbit C which can be approximated by a circle with a macroscopic diameter d. Let there be also two local quantum levels A and B with lower energies, which are located as shown in Fig. 1. Initially, only level A is occupied by electron while the other states are vacant. We expose the region A to light capable of providing quantum transition $A \rightarrow C$ while in the region B there is a number of local scatterers capable of providing quantum transition $C \rightarrow B$ during a characteristic time τ. In terms of QM, this means that after the absorption of light quantum, the electron will appear on the level B during the time τ.

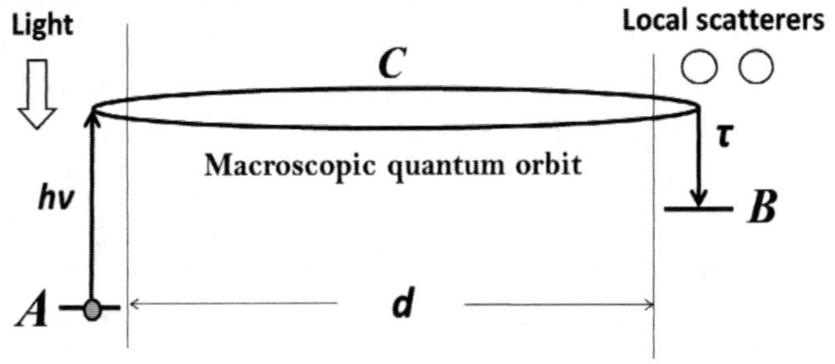

Figure 1. *Gedanken* experiment with a spatially-discontinuous quantum transport of electron between two local quantum levels (A and B) via a macroscopic-scale orbit-like state (C).

Actually, it is just the spatially-discontinuous transition from region A to region B avoiding any intermediate regions and we can talk about a real transport insofar as the distance d is macroscopic. But the most interesting point is that the characteristic time of the transport is independent of the distance d because it is determined by such local parameters as the concentration of scatterers in the region B and the efficiency of the scattering. Therefore, we potentially could fulfill the following relation: $d / \tau > c$. Thus, under the conditions of our *gedanken* experiment, the predictions of standard QM directly contradict Einstein's relativity as well as Minkowski's model of spacetime.

2.1 Realisability of the 'gedanken' experiment

Certainly, the most subtle point of our *gedanken* experiment is the existence of a macroscopic quantum orbit with an easy-to-control length. Indeed, at the Bohr-Einstein's times, such orbits were unknown so that our *gedanken* experiment could at best be regarded as a mind game. Today, however, it is not quite so. Recent development of nanotechnologies allows one to achieve such states of matter, in which the wavefunctions of quantum particles have an unusual shape and moreover their lengthscale may well be macroscopic. One of such states is the so-called integer quantum Hall (IQH) state [22-23].

As a rule, the IQH state appears in two-dimensional electronic systems at the temperatures below 10K and in a strong magnetic field higher than 1T. In the IQH state, there are two kinds of electron quantum states. Most electrons are in strongly localized microscopic orbits known as the cyclotron orbits. However, there is a relatively small number of quasi-one-dimensional orbits that are extended along the sample edges. This means that potentially the length of these orbits can be arbitrarily large.

In the generally-recognized Laughlin-Halperin's theory, such orbits are often called the extended current-carrying states and just these orbits are responsible for the quantum Hall effect that emerges in presence of an electric bias [24-25]. As is well-known, for the discovering of this effect Klaus von Klitzing received the Nobel Prize in 1985 [26].

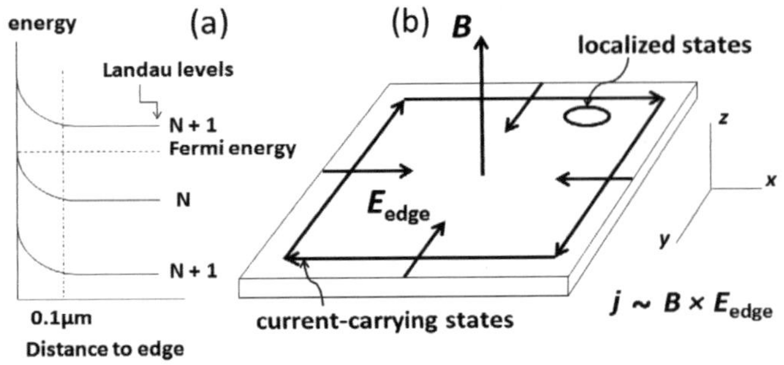

Figure 2. The origin of Laughlin-Halperin macroscopic-scale quasi-one-dimensional states. **(a)** Energy-band diagram of conventional IQH system close to sample edges. **(b)** Two types of quantum states in IQH system: localized states in the system interior and the Laughlin-Halperin states close to sample edges.

To clarify the origin of Laughlin-Halperin macroscopic edge states, let us address the energy diagram of an ideal IQH system (Fig. 2a). Here the electrons are in equidistant energy levels, the so-called Landau levels, which are fully occupied below the Fermi level while otherwise they are empty. On Landau levels, the ordinary states are just the above-mentioned microscopic cyclotron orbits and the number of these orbits per level is roughly determined by their close packing within the sample area (Fig. 2b). But the point is that near the sample edges there is a band bending that is an equivalent of the presence of a strong electric field pointed toward the sample center (see Fig. 2a). As a result, close to the sample edges the electrons are in crossed electric and magnetic fields. In terms of a semi-classical model, this leads to an electron drift along the sample edges so that the direction of the drift is determined by the direction of external magnetic field. However, in terms of quantum mechanics, we should rather talk about quasi-one-dimensional current-carrying states with a characteristic width of the order of cyclotron radius. These are just the Laughlin-Halperin macroscopic orbit-like states.

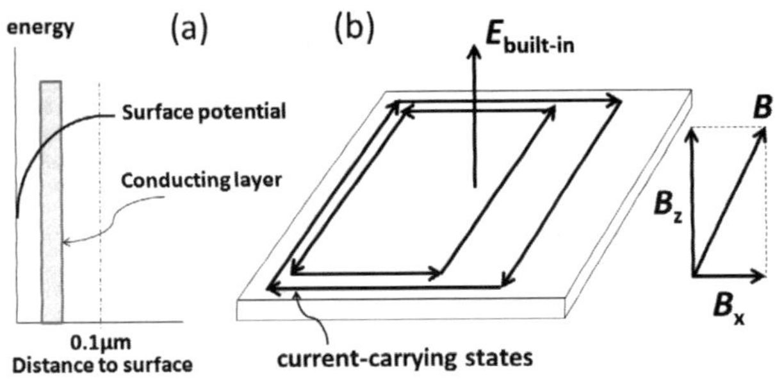

Figure 3. A method to provide Laughlin-Halperin-type states in the interior of an asymmetric IQH system. **(a)** Potential profile close to sample surface and typical position of a conducting two-dimensional layer. **(b)** Spatial distribution of Laughlin-Halperin-type states. Only two orbit-like states are shown with the opposite directions of spontaneous one-electron currents.

Generally speaking, the Laughlin-Halperin states seem suitable to implement our *gedanken* experiment. Actually, however, such implementation faces serious methodological difficulties. The point is that, in contrast to conventional electric measurements, in optical measurements we are always dealing with the effects related to the localized microscopic states. However, it turns out that the situation can be significantly improved if the external magnetic field has both quantizing and in-plane components while the entire two-dimensional layer is strongly asymmetric. The latter means that there is the so-called "built-in" transverse electric field within the layer. The field is due to the asymmetry of confining potential and may be as high as up to 10^5V/cm [27]. One of the reasons for the asymmetry is the penetration of surface potential into the layer which may be very close to the sample surface as it is shown in Fig. 3a [28]. In this case, the origin of the built-in field is similar to that of the edge field in a conventional IQH system. So we have got a system where not only edge electrons are in crossed electric and magnetic fields but the entire electron system (Fig. 3b).

The solution of Schrödinger equation for a very simplified infinite model allows one to expect that, instead of the strongly localized cyclotron orbits, now we have got spatially-separated spontaneous currents throughout the entire conducting layer [29]. The currents are expected to be perpendicular to the in-plane component of magnetic field. Their origin is similar to that of Laughlin-Halperin spontaneous edging currents so that their characteristic width is also of the order of electron cyclotron radius. They should strongly differ from each other and even could have the opposite signs. In the case the above currents really exist in a finite system, they should be a mixture with the true Laughlin-Halperin edging states. So, their expected spatial distribution is shown in Fig. 3b though this picture certainly should be evidenced experimentally.

From this perspective, it is crucial that the spatial distribution of spontaneous current loops should not be exactly the same at different Landau levels. As a result, even in the absence of an electric bias, the resonant optical transitions between the Landau levels should give rise to local ohmic currents which could be detected. Moreover, these light-induced currents are easy to identify through their characteristic feature: they should be a strong non-periodic function of their position with respect to the sample borders.

Our real experiments strongly support the picture shown in Fig. 3b and their detailed description can be found in [30-31]. So, we are truly dealing with a great number of spatially-separated quasi-one-dimensional macroscopic orbits each of which is characterized by its own value of a one-electron spontaneous current. In a sense, the system is thus reminiscent of a gigantic *single* atom where the crystal lattice together with the valence electrons plays the role of a positively-charged nucleus while free electrons are distributed among atomic orbitals. Thus, in contrast to conventional solid state systems, these free electrons appear to be spatially-separated and therefore to be distinguishable.

2.2 Demonstration of spatially-discontinuous macroscopic quantum transport

Certainly, the system in Fig. 3b is much more suitable than a conventional IQH system to implement our *gedanken* experiment. With this system, one would propose the following version of the experiment in Fig. 1. Let us expose to light not the whole sample but only a part of it, which thus plays the role of the region A (see Fig. 1). Then, the role of the region B could play a distant region where the emergence of photo-excited electrons could be detected through the detection of a local ohmic current initiated by them. According to our guess, after the absorption of light quanta in the region A, photo-excited electrons could emerge in the region B, if their orbits cross this region and if they are lucky enough to interact with a local scatterer from this region. So, if a local current will truly be detected in the region B, then we will only have to prove that it can only be associated with a spatially-discontinuous transport of photo-excited electrons.

Our real experiment in accordance with the above scheme is shown in Fig. 4. We use single quantum well structures of InAs/AlGaSb type. The details of their structure as well as the experimental evidence for their asymmetry can be found in [31]. Typically, the width of the InAs conducting layer is 15nm and the electron sheet concentration is about $10^{12}cm^{-2}$. The sample ($19\times12mm^2$) has two pairs of short (1mm each) contacts which are located as shown in Fig. 4. The sample is immersed in a superconducting magnet at the temperature as low as about 2K. We use the magnetic field of about 5T, which is tilted from the normal by the angle of 15° to provide the in-plane component. The sample is exposed to a spatially-uniform terahertz laser radiation. The source of radiation is an Место для формулы.optically-pumped pulsed ammonia laser with the intensity of incident radiation of up to 200W/cm^2 and with the pulse duration of about 40ns. The energy of light quanta is 13.7meV which roughly is the energy gap between the Landau levels under the experimental conditions so that resonant optical transitions to the lowest empty level have become possible. Electric signals from both contact pairs will be synchronously detected in a short-circuit regime when the resistance of load (50Ω) is about two order lower than that of conducting layer in the experiment.

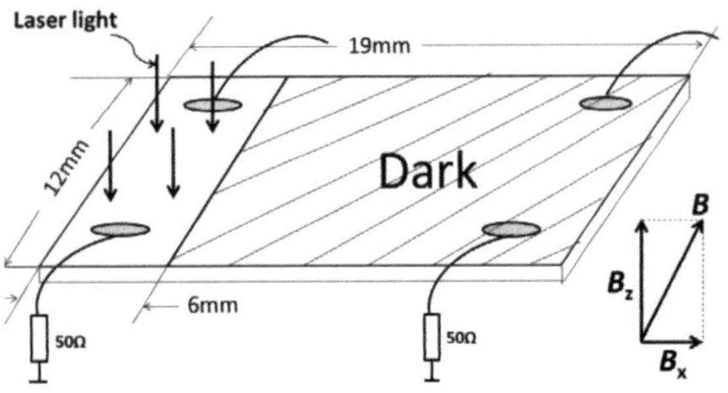

Figure 4. Realization of *gedanken* experiment in a modified IQH system with Laughlin-Halperin-type orbit-like states in the system interior. Only one third of the sample is exposed to light (region *A*). Signals are detected from two contact pairs: one pair is inside the laser spot while another one is about 1cm away from the spot so that the photo-excited electrons by no means can reach detectable area (region *B*) through continuous movement.

Following our scheme, we will expose not the whole sample but only a one third of it, as shown in Fig. 4. So, one contact pair is inside the laser spot while another one is about 1cm away from the excited region. This means that the distance from the region *A* to the region *B* is *five orders higher* than the characteristic electrons' mean free path determined by local scatterers. In other words, the concentration of local scatterers is so high that any remarkable diffusion of photo-excited electrons is impossible at least during their characteristic lifetime on a higher Landau level, which is expected to be less than 0.1ns [32].

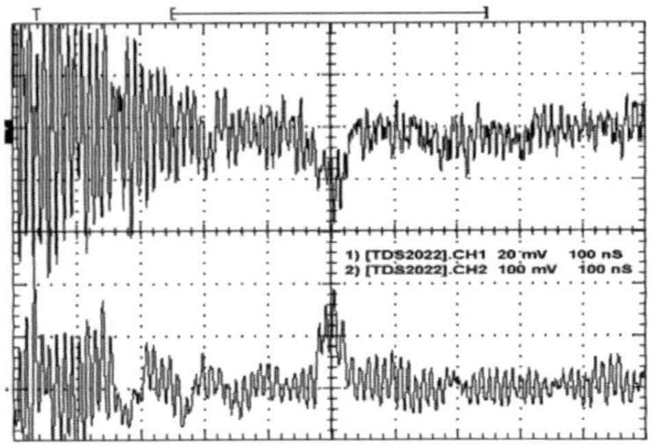

Figure 5. Typical tracks of synchronously-detection signals from two contact pairs in the experiment shown in Fig. 4. Upper track – signal from the pair in the light (20mV/div.). Lower track – signal from the pair in the dark (100mV/div.). Timescale is 100ns/div.

Typical tracks from both contact pairs are shown in Fig. 5 and, to be honest, the effect exceeds even our most optimistic expectations. Indeed, it is clearly seen that the signal in the dark is even *higher* than the signal in the light. This fact is consistent with the above ideas regarding a non-uniform distribution of currents throughout the sample and it clearly demonstrates that the current in the dark cannot be a "secondary effect" with respect to the current in the light. It is also clearly seen that the kinetics of both signals is so fast that any long-time diffusion or thermal effects are definitely irrelevant. Moreover, there is no delay between the signals at least with an accuracy of about 10ns. This means that to reach the region *B*, the photo-excited electrons should move toward this region with the speed of no less than 10^8cm/s *without any collisions* at a distance of about 1cm. This is absolutely impossible because, as we mentioned above, their mean free path is about 10^{-5}cm.

Finally, to avoid any loopholes, we repeat the experiment in Figure 4 but in a classic regime when the external magnetic field is fully in the well plane and therefore there is no Landau quantization. In this case, we also observe a signal from the pair in the light due to the so-called photo-voltaic effect [33]. However, despite the fact that the "classic" signal is two orders higher than the "quantum" signal from the pair in the light, no signal is observed from the pair in the dark.

Now it seems interesting to estimate the nominal "speed" of the transport from the region A to the region B as if we are truly dealing with a continuous movement of electrons. According to our ideas, the characteristic time of this transport is the so-called quantum relaxation time, that is, the time to destroy the coherence. In our structures it is about 0.3ps [27]. Thus, we immediately obtain the "speed" of about $3 \cdot 10^{12}$cm/s, i.e. two orders faster than the speed of light. Moreover, we can easily exceed this "speed", say, through the incorporation of additional scatterers into the region B.

3. Fundamental consequences: a new way to unify quantum and relativity theories

Our observations clearly have an applied aspect. But in this work we focus only on fundamental consequences which seem to have a greater importance.

One of these consequences is that the Einstein's relativity is experimentally-distinguishable from the Lorentz-Poincare's relativity and the experiment unambiguously supports the latter. However, this fact is crucial not only from the perspective of relativity but also from the perspective of quantum theory. The point is we break the basic argument against pilot-wave theory, that is, the lack of Lorentz invariance. We have thus got two quantum theories, standard QM and the pilot-wave theory, and both are equally successful in prediction of experimental observations.

However, their principle difference is that the former rests on mysticism and indeterminism while the latter rests on realism and determinism which is the basis of all current physical theories. As a result, the former encounters serious conceptual problems and can hardly be unified with relativity theory. Conversely, the latter naturally avoids conceptual problems and can easily be unified with relativity or, more precisely, with the Lorentz-Poincare's version of this theory. In this situation, the choice in favour of one of these theories seems self-evident though it clearly will take some courage to overcome the existing prejudices about quantum theory.

The important point is that the unification we mean is not a synthesis of equally fundamental theories, quantum mechanics and relativity, into a more general theory for which both ones are only approximations or limiting cases. Rather, we imply a hierarchy in the sense that quantum theory incorporates relativity as an approximation or a limiting case. Such a hierarchy allows one to overcome the natural difference between these theories on the level of their basic physical (but not philosophical!) concepts. This difference is related to the fact that relativity is inherently dealing with more or less autonomic objects. Otherwise, such notions as the "speed" or the "signalling" would appear meaningless. By contrast, quantum theory concerns essentially undivided systems for which any autonomy is a limiting case indeed.

Certainly, the indivisibility of quantum is the basis of both the standard QM and the pilot-wave theory. In a sense, Bohr's rejection of realism may be regarded as a sacrifice to reconcile quantum indivisibility with relativity. And this sacrifice seemed unavoidable at least at Bohr's times. This significant point was especially stressed by Bohm who wrote that "... *Bohr's rejection of hidden variables is ... based on a very radical revision of the notion of what a physical theory is supposed to mean, a revision that in turn follows from the fundamental role which he assigns to the indivisibility of the quantum* ..." [12]. Only now Bohr's sacrifice appears unnecessary. Moreover, instead of reconciliation, now we could rather talk about a harmony between quantum and relativity theories. The best manifestation of this harmony is the fact that quantum theory could help relativity to solve an important conceptual problem. It is the problem of physical meaning of aether.

However, to show the way of solving that problem we should first note the following thing. If the spatially-discontinuous jumps of quantum particles can indeed be macroscopic then we actually face a peculiar quantum dynamics. But any peculiar dynamics requires an adequate model of space which should be able to account for its characteristic features. As for the quantum dynamics, here the basic characteristic feature is the indivisibility of quantum. So, we need a model which would be adopted to account for this feature. But it turns out that such a model already exists and it is hardly surprising that the author of this model is David Bohm, i.e. one of the authors of the realistic quantum theory [11-12].

The basic idea of Bohm's model is that we give up such a seemingly eternal idea as the Cartesian order of space, which ultimately expresses the classical concept of spatially-separated interacting physical objects. Bohm himself characterized the "quantum" space (or pre-space) as follows: *"... pre-space would have a structure that was inherently conformable to the laws of the quantum theory. We would thus be free of the present incoherence of trying to force quantum laws into a framework of a Cartesian order that is really only suitable for classical mechanics..."* [12]. The "quantum" concept of spatial order was called the implicate order. Following Bohm's figurative comparison, this order differs from Cartesian order just like a holographic image differs from a mirror reflection. In the former case, each part of the image contains some information about the whole object and this information can potentially be unfolded. By contrast, in the latter case, each part of the reflection contains no information about the rest of the object. Leaving aside further details of Bohm's concept, we only would like to note that just this undivided "quantum" pre-space or undivided Universe could naturally play the role of aether or Bell's "deeper level" which is not Lorentz invariant.

Finally, it should be noted that Bohm did not actually suspect that the idea of quantum pre-space is so global. The point is he shared the common view, according to which quantum mechanics essentially concerns only microcosm. That is why he supposed that **"...** *processes that were associated with what we now call the Planck length would have to be described in what is an essentially quantum mechanical kind of space in which Cartesian notions of order would not even arise...*". Moreover, he supposed that *"... as long as the quantum theory is valid, there is no way to demonstrate ... non-Lorentz invariance experimentally..."* [12]. However, our demonstration of macroscopic-scale non-Lorentz invariance clearly shows that quantum mechanics is not only and even not so much the theory relevant to microcosm. Rather, it is the deepest physical theory which is relevant at any lengthscale and therefore gives rise to a new – non-mechanistic – insight into the world.

Acknowledgements
The author is grateful to H.-T. Elze as well as to the other participants of the 6th International Workshop DICE2012 (Spacetime-Matter-Quantum Mechanics) for the friendly atmosphere during the Conference.

References

[1] Einstein A 1905 *Ann. der Phys.* **17** 891
[2] Einstein A 1949 *Autobiographical notes*, in: *Albert Einstein: Philosopher-Scientist,* ed. Schilpp P A (Open Court, Illinois)

[3] Bacciagaluppi G and Valentini A 2009 *Quantum Theory at the Crossroads: Reconsidering the 1927 Solvay Conference* (Cambridge: Cambridge University Press)
[4] Heisenberg W 1971 *Physics and Beyond* (Harper and Row, Glasgow)
[5] Quote: Petersen A 1963 "The philosophy of Niels Bohr", in: The Bulletin of the Atomic Scientists, vol. 19
[6] Heisenberg W 1958 Physics and Philosophy: The Revolution in Modern Science in: Lectures at University of St. Andrews (University of St. Andrews, Scotland)
[7] Bell J S 1987 *Speakable and Unspeakable in Quantum Mechanics* (Cambridge University Press, Cambridge)
[8] Born M 1971 *The Born-Einstein Letters* (Walker and Company, New York)
[9] Einstein A *Letter to Robert A. Thornton*, 7 December 1944
[10] Bohm D 1952 *Phys. Rev.* **85** 166
[11] Bohm D 1980 *Wholeness and the Implicate Order* (Routledge, London)
[12] Bohm D and Hiley B J 1993 The Undivided Universe: An Ontological Interpretation of Quantum Theory (Routledge, London)
[13] Aspect A, Grangier P and Roger G 1981 *Phys. Rev. Lett.* **47** 460
[14] Groblacher S *et al* 2007 *Nature* **446** 871
[15] Zeilinger A 2005 *Nature* **438** 743
[16] Eberhard P H 1978 *Nuovo Cimento* **46B** 392
[17] Bell J S 1986 "interview", in: *The Ghost in the Atom*, eds. Davies P C W and Brown J R (Cambridge University Press, Cambridge)
[18] Popper K R 1982 *The Open Universe: An Argument for Indeterminism*, ed. Bartley W W (Hutchinson, London)
[19] Goldstein S 2012 Bohmian Mechanics in: *Stanford Encyclopedia of Philosophy*, ed. Zalta E N
[20] Dürr D, Goldstein S, Münch-Berndl K, and Zanghì N 1999 *Phys. Rev.* A **60** 2729
[21] Nikolic H 2005 *Foundations of Physics Letters* **18** 561
[22] Prange R and Girvin S 1986 *The Quantum Hall Effect* (Springer, Berlin)
[23] Yennie D R 1987 *Rev. Mod. Phys.* **59** 781
[24] Laughlin R B 1981 *Phys. Rev.* B **23** 5632
[25] Halperin B I 1982 *Phys. Rev.* B **25** 2185
[26] von Klitzing K 1986 *Rev. Mod. Phys.* **58** 519
[27] Brosig S *et al* 1999 *Phys. Rev.* B **60** R13989
[28] Altarelli M, Maan J S, Chang L L and Esaki L 1987 *Phys. Rev.* B **35** 9867
[29] Gorbatsevich A A, Kapaev V V and Kopaev Y V 1993 *JETP Lett.* **57** 580: 1994 *Ferroelectrics* **161** 303
[30] Emelyanov S A and Ivanov S V 2011 *AIP Conf. Proc.* **1399** 1049
[31] Emelyanov S A 2012 *Phys. Scr.* **T151** 014012
[32] Singh S K *et al* 1999 *Phys. Rev.* B **58** 7286
[33] Ganichev S D and Prettl W 2006 *Intense Terahertz Excitation of Semiconductors*, in: *Series on Semiconductor Science and Technology*, vol. **14** (Oxford University Press, Oxford)

Gravitational contraction and curvature singularities in $f(R)$ gravity

L Reverberi[1,2]
[1]Dipartimento di Fisica e Scienze della Terra, Università degli Studi di Ferrara
Polo Scientifico e Tecnologico - Edificio C, Via Saragat 1, 44122 Ferrara, Italy
[2]Istituto Nazionale di Fisica Nucleare (INFN), Sezione di Ferrara
Polo Scientifico e Tecnologico - Edificio C, Via Saragat 1, 44122 Ferrara, Italy

E-mail: reverberi@fe.infn.it

Abstract. The discovery of the accelerated expansion of the Universe has had a vast resonance on a number of physical disciplines. In recent years a few viable modified gravity models have been proposed, which naturally lead to a late-time de Sitter stage while reducing to General Relativity in the early Universe. We study two of these models during the contraction of a homogeneous cloud of pressureless dust. We show how the increasing energy/mass density may lead to a curvature singularity and derive the typical timescales for its development.

1. Introduction

The biggest mystery in modern physics is perhaps the nature of Dark Energy, which is responsible for the accelerated expansion of the Universe [1] and accounts for about 75% of its total content. This unknown form of energy seems to have an equation of state close to $w = -1$, which yields "anti-gravity" and makes it phenomenologically analogous to a cosmological Λ-term or to a vacuum energy component (for a review on Dark Energy, see e.g. [2] and references therein).

From a theoretical standpoint, both possibilities are rather natural, since a Λ-term is well within the framework of standard General Relativity (GR) and a zero-point energy is to be expected in the Standard Model as well as in any of its extensions. There remains, however, the terrible discrepancy between the expected value of the vacuum energy density, which can quite easily reach $\varrho_{vac} \sim m_{Pl}^4$, and its observed value, $\varrho_{vac} \sim 10^{-123} m_{Pl}^4$. This fine-tuning problem is maybe the best motivation to study alternative mechanisms to generate the observed accelerated cosmological expansion.

Modified $f(R)$ gravity theories were first introduced as ultraviolet corrections to GR, originated by loop quantum corrections to matter fields in curved spacetime, as discussed in the seminal paper [3] and in following works (e.g. [4]). These terms are particularly relevant in the early Universe, and may have driven the primordial inflationary stage.

The first infrared $f(R)$ models proposed relied on negative powers of R, which become dominant at small curvatures, in the gravitational action to generate the *late-time* de Sitter stage [5], but suffered from severe instabilities in the presence of matter [6, 7, 8]. In the last few years, the limits for the cosmological viability of $f(R)$ models have been widely studied [9], and recently at least three models evading all such tests have been proposed [10, 11, 12].

Studying these models in an astronomical/astrophysical situation has evidenced the possibility of non-trivial modifications of the standard theory. Investigating both the stability of static, spherically symmetric, dense structures [13, 14] and the contraction of large, rarefied collisionless dust clouds [15, 16], it was found that in these theories it is quite possible to develop a curvature singularity. The additional scalar, massive degree of freedom (dubbed *scalaron*) of $f(R)$ gravity theories moves in a time dependent potential, and the singular point corresponding to $|R| \to \infty$ can be reached with a finite energy. Due to the dependence of the potential on the external energy/mass density, the situation worsens as the contraction proceeds. The same mechanism is responsible for the past, finite-time singularity found in a cosmological scenario [17]. The structure of the singularity is peculiar, because although $|R| \to \infty$, the energy density and the metric remain finite; nevertheless, the appearance of such singularity remains hardly desirable in a valid theory. Furthermore, it seems likely that similar features may apply to an extended class of modified gravitational theories, not only $f(R)$ gravity [18].

In this paper, which extends the work of [15, 16], we will focus on the two models of [10, 11], and discuss their behaviour during the contraction of a spherically symmetric, nearly homogeneous cloud of pressureless dust. The possibility of forming a singularity is confirmed, and we propose an improved method to predict the behaviour of curvature while approaching the singularity. These results could be used to constrain models after a careful comparison with observational data, and also to be the starting point for an investigation of further effects of $f(R)$ gravity, e.g. gravitational particle production and the related possibility of detecting cosmic rays [19, 20], which may in principle provide additional signatures of physics beyond GR.

2. Curvature dynamics in $f(R)$ gravity

2.1. Action and field equations

We start with the general gravitational action[1]

$$A_{grav} = -\frac{m_{Pl}^2}{16\pi}\int d^4x\,\sqrt{-g}\,f(R) \equiv -\frac{m_{Pl}^2}{16\pi}\int d^4x\,\sqrt{-g}\,\left[R+F(R)\right]. \tag{1}$$

Varying (1) with respect to the metric $g_{\mu\nu}$ yields the modified Einstein equations

$$f'(R)R_{\mu\nu} - \frac{1}{2}f(R)\,g_{\mu\nu} + \left(g_{\mu\nu}\mathcal{D}^2 - \mathcal{D}_\mu\mathcal{D}_\nu\right)f'(R) = T_{\mu\nu}\,, \tag{2}$$

where $\mathcal{D}$ denotes covariant derivative, a prime denotes derivative with respect to R, and

$$T_{\mu\nu} \equiv \frac{8\pi}{m_{Pl}^2}\frac{2}{\sqrt{-g}}\frac{\delta A_m}{\delta g^{\mu\nu}}\,.$$

Note that the matter action A_m was not explicitly indicated in Eq. 1. The trace of (2) reads

$$3\mathcal{D}^2F' + RF' - 2F - (R+T) = 0\,. \tag{3}$$

Although some results will be rather general, we will specifically consider the two models

$$F_{HS}(R) = -\frac{\lambda R_c}{1+(R/R_c)^{-2n}}\,, \quad [10] \tag{4a}$$

$$F_S(R) = \lambda R_c\left[\left(1+\frac{R^2}{R_c^2}\right)^{-n} - 1\right]\,. \quad [11] \tag{4b}$$

[1] We use natural units $c = \hbar = k = 1$, and the Planck mass is defined as $m_{Pl}^2 = G_N^{-1}$. The metric has signature $(+\,-\,-\,-)$, and we use the conventions $\Gamma^\alpha_{\mu\nu} = \frac{1}{2}g^{\alpha\beta}(\partial_\mu g_{\beta\nu} + \cdots)$, $R^\alpha{}_{\mu\beta\nu} = \partial_\beta\Gamma^\alpha_{\mu\nu} + \cdots$, $R_{\mu\nu} = R^\alpha{}_{\mu\alpha\nu}$, $R = R^\mu_\mu$. With these conventions, $R < 0$ for matter- and vacuum-dominated Universes.

For $\lambda \sim 1$, in both models R_c is of the order of the present cosmological constant; since $f(0) = 0$, the cosmological constant "disappears" in vacuum (using the words of [11]). The late-time constant curvature solutions are roots of the algebraic equation

$$\bar{R} - \bar{R}\,F'(\bar{R}) + 2F(\bar{R}) = 0\,, \tag{5}$$

where the matching between theory and observational must be given by

$$\Lambda = \frac{8\pi\,\Omega_\Lambda\,\varrho_{vac}}{m_{Pl}^2} = -\frac{\bar{R}}{4}\,, \tag{6}$$

Ignoring further subtleties, for the sake of definiteness and simplicity, we will assume that λ is of order unity and that $R_c \sim \Lambda_0$; for mode information on the cosmological behaviour of the models, we refer the reader to the original papers [10, 11].

2.2. Contracting systems

Let us consider a spherically symmetric cloud of pressureless dust, hence having $T = 8\pi\,\varrho/m_{Pl}^2$. We make the additional assumption of homogeneity, in the sense that at each instant of time the time derivatives of T and R dominate over the space derivatives; this approach, albeit somewhat rudimentary, is intuitively clear and theoretically justifiable [15].
For definiteness, we assume that such cloud contracts on a typical timescale t_{contr} following:

$$T(t) = T_0\left(1 + \frac{t}{t_{contr}}\right) = \frac{8\pi\,\varrho_0}{m_{Pl}^2}\left(1 + \frac{t}{t_{contr}}\right)\,. \tag{7}$$

We must stress that this evolution is not supposed to be physically accurate and is only necessary to perform explicit calculations. Nonetheless, results obtained applying more sophisticated dynamical analyses should be qualitatively in agreement with our results, provided that the contraction law is not terribly different from (7) and in any case until $t \sim t_{contr}$. For a recent dynamical analysis of gravitational contraction in $f(R)$ gravity, see e.g. [21] and references therein.

We also consider the limit $|R/R_c| \gg 1$, which is usually appropriate since the typical densities in pre-stellar and pre-galactic clouds are much larger than the present critical energy density $\varrho_{vac} \approx 10^{-29}$ g cm^{-3}, in which $|F/R| \ll 1$ and $|F'| \ll 1$. Moreover, for typical values of energy density we are in low-gravity regime, that is (denoting with M the total mass of the object and with r its radius) $GM/r \ll 1$, so the spacetime is nearly Minkowski. This allows us to rewrite (3) as

$$\frac{\partial^2 F'}{\partial t^2} + \frac{1}{3}(R + T) \simeq 0\,. \tag{8}$$

2.3. Trace equation and scalaron potential

Defining

$$\xi \equiv -3F'\,, \tag{9}$$

we write (8) as an oscillator equation for the field ξ:

$$\ddot{\xi} + R + T = 0 \quad\Leftrightarrow\quad \ddot{\xi} + \frac{\partial U}{\partial \xi} = 0\,, \tag{10}$$

with U being the time-dependent potential

$$U(\xi, t) = T(t)\,\xi + \int^{\xi} R(\xi')\,d\xi'\,. \tag{11}$$

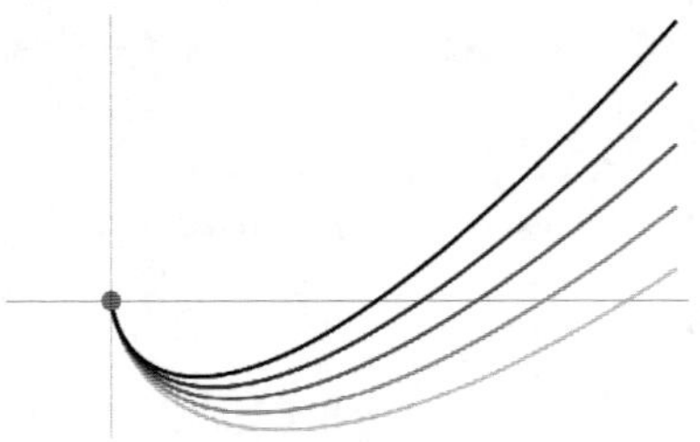

Figure 1. Qualitative shape of potentials for models (4), assuming $R/R_c \gg 1$. For both models $\xi = \xi_{sing} = 0$ (red dot) corresponds to the singular point $|R| \to \infty$. The typical time evolution is also shown, from light grey (initial times) to black (later times). Note that the bottom of the potential approaches ξ_{sing}.

The field ξ, sometimes dubbed *scalaron*, is the additional scalar, massive degree of freedom originated by the introduction of $F(R)$ in the action, which for both models considered (in the limit $|R/R_c| \gg 1$) is roughly equal to

$$\xi(R) \simeq 6n\lambda \left(\frac{R_c}{R}\right)^{2n+1} . \tag{12}$$

Evidently, the limit $|R| \to \infty$ corresponds to the finite value $\xi \to \xi_{sing} = 0$.
In GR this field vanishes identically, and R and T are related simply algebraically; in $f(R)$ gravity, instead, the relation between geometry and matter is differential, and in fact (10) describes oscillations around the GR solution

$$R + T = 0\,, \tag{13}$$

with a frequency given approximately by the square root of the second derivative of the potential with respect to the field, that is

$$\omega^2 \simeq \frac{\partial^2 U}{\partial \xi^2} = \frac{\partial R}{\partial \xi} \simeq \left.\frac{1}{\partial \xi/\partial R}\right|_{R+T=0} \simeq -\frac{R_c}{6n\lambda(2n+1)} \left(-\frac{T}{R_c}\right)^{2n+2} . \tag{14}$$

In this equation, we have assumed that Note that since $R_c < 0$, the square of the frequency is always positive; this prevents the instabilities of [7].

As we have mentioned, the potential U contains an explicit time-dependence, contained in the term proportional to $T(t)$, and in fact it will change with time or to be more precise with varying energy/mass density; beside the fact that its minimum moves following $R + T = 0$, the value of U in its minimum and the overall shape change, as well. Using the explicit expression (12), we find up to an additive constant, which can be set equal to zero without loss of generality:

$$U(\xi, t) \simeq T(t)\,\xi + 3\lambda(2n+1)R_c \left(\frac{\xi}{6n\lambda}\right)^{\frac{2n}{2n+1}} . \tag{15}$$

The qualitative shape of this potential is depicted in Fig. 1. Remarkably, we also notice that $U(\xi_{sing} = 0)$ does not vary with time.

We can define the "average" value of the field

$$\xi_a(t) \equiv \xi\,(R + T = 0)\,, \tag{16}$$

in order to quantify the displacement of solutions from the GR expectation. Note that ξ_a is not necessarily a solution of (10) with $R+T=0$, so it is not meant to be an actual physical solution, it simply corresponds to the position of the bottom of the potential. Since T is increasing in contracting systems, Eq. (12) indicates that ξ_a decreases with time, approaching the singular point $\xi_{sing}=0$. This will have important consequences, as we will see.

2.4. Energy conservation

From (10) we can also derive the exact "energy" conservation equation

$$\frac{1}{2}\dot{\xi}^2 + U(\xi,t) - \int^t dt'\,\frac{\partial T}{\partial t'}\,\xi(t') = \text{const}\,. \tag{17}$$

The last non-local, integral term on the l.h.s. is due to the explicit time-dependence of the external energy density; in the simple case (7), $\partial T/\partial t = T_0/t_{contr}$ and this term is proportional to the integral of ξ over time. In contracting systems, its overall contribution is to increase the "canonical" total energy, that is the sum of kinetic and potential energies[2].

We study the problem in two different regimes: the "adiabatic" regime and the "slow-roll" regime. Intuitively, the separation depends on how much the potential $U(\xi,t)$ varies during one oscillation of the scalaron, which occurs on a timescale

$$t_{osc} \simeq \frac{2\pi}{\omega}\,, \tag{18}$$

where ω is given by (14). On the other hand, the typical timescale for substantial variations in the potential is t_{contr}. Therefore, we expect the boundary between the two regimes to be somewhere near the equality

$$\frac{t_{contr}}{t_{osc}} = \frac{\omega\, t_{contr}}{2\pi} = 1\,. \tag{19}$$

3. Adiabatic regime

Let us start assuming that t_{contr} is very large compared to ω^{-1}, which means that U does not change noticeably over many scalaron oscillations. This also translates into the fact that the variation of the integral in (17) during a few oscillations is very small. Therefore, the oscillations of ξ are approximately adiabatic, in the sense that at each oscillation ξ moves between two values ξ_{min} and ξ_{max} having, ignoring the time dependence of U,

$$U(\xi_{min}) \simeq U(\xi_{max})\,. \tag{20}$$

In this situation, we expand ξ around ξ_a as

$$\xi(t) = \xi_a(t) + \xi_1(t) \tag{21a}$$
$$\equiv \xi_a(t) + \alpha(t)\sin\Phi(t)\,, \tag{21b}$$

where

$$\Phi(t) \simeq \int^t dt'\,\omega\,. \tag{22}$$

The function α, which is the amplitude of the scalaron oscillation around its average value, is also assumed to be relatively slowly-varying, that is

$$\frac{\dot{\alpha}}{\alpha} \ll \omega\,. \tag{23}$$

[2] Note that this is true for $\xi>0$, whereas for $\xi<0$ it would give the opposite behaviour. However, it has been shown that the condition $F'<0$, corresponding to $\xi>0$, is crucial for the correct behaviour of modified gravity models at (relatively) low curvatures [17]. Indeed, this is clearly the case in the considered models.

Journal of Physics: Conference Series **442** (2013) 012036 doi:10.1088/1742-6596/442/1/012036

The singularity point $\xi = \xi_{sing}$ is reached if at any moment

$$|\xi_a(t) - \xi_{sing}| \leq |\alpha| \,, \tag{24}$$

which in our case is equivalent, taking for simplicity and without loss of generality $\alpha > 0$, to

$$\xi_a(t) \leq \alpha(t) \,. \tag{25}$$

3.1. Linear approximation

At first order in ξ_1, Eq. (10) reads

$$\ddot{\xi}_1 + \omega^2 \xi_1 \simeq -\ddot{\xi}_a \,. \tag{26}$$

The term on r.h.s. only depends on the external energy density, and acts as a source term for the oscillations of the "perturbation" ξ_1. In first approximation, however, it can be neglected since it is approximately equal to ξ_0/t^2_{contr}, and we have assumed that $\omega\, t_{contr} \gg 1$. For analogous reasons, we can neglect $\ddot{\alpha}$. Using expansion (21b) then yields

$$\frac{\dot{\omega}}{\omega} \simeq -2\frac{\dot{\alpha}}{\alpha} \quad \Rightarrow \quad \alpha(t) \simeq \alpha_0 \sqrt{\frac{\omega_0}{\omega(t)}} \,. \tag{27}$$

This can be considered a rather general result, and the specific $F(R)$ model will determine the particular evolution of the oscillations. The value α_0 in Eq. (27) is related to the initial conditions we impose, that is to the initial displacement from the GR behaviour. We will set the initial values of R and $\dot{R}$, and from those derive the initial values of ξ and $\dot{\xi}$. Thus, α_0 can be calculated differentiating Eq. (21b), which gives

$$\dot{\xi}_0(R_0, \dot{R}_0) \simeq \dot{\xi}_{a,0} + \alpha_0 \omega_0 \;\Rightarrow\; \alpha_0 \simeq \frac{\dot{\xi}_0 - \dot{\xi}_{a,0}}{\omega_0} \,. \tag{28}$$

This leads to the explicit solution

$$\alpha(t) \simeq \left|\dot{\xi}_0 - \dot{\xi}_{a,0}\right| \left[\omega_0\, \omega(t)\right]^{-1/2} \,. \tag{29}$$

The modulus in the r.h.s. entails no loss of generality and ensures that $\alpha > 0$.

It is clear from Eq. (29) that if $\dot{\xi}_0 = \dot{\xi}_{\alpha,0}$ the amplitude of oscillations should vanish at all times. This can be seen to be completely wrong, for instance numerically. The disagreement with our estimate is due to the fact that we have neglected $\ddot{\xi}_a$, and in general terms proportional to $1/t^2_{contr}$, in Eq. (26). When α is initially very small, however, those terms should be kept and the approximations used are no longer valid. Therefore, Eq. (29) is reliable when $|\dot{\xi}_0 - \dot{\xi}_{\alpha,0}|$ is large enough, say of the order of $\dot{\xi}_{\alpha,0}$ or slightly less. In figure 2 we show a comparison of numerical solution of (10) with our analytical result.

We stress that the linear approach must not be extended to the analysis of $R(t)$, in fact we are studying the very process of $|R| \to \infty$. Moreover, as R evolves, $|\xi_1/\xi_a|$ increases so the linear approximation will eventually cease to hold. This happens when ξ approaches $\xi_{sing} = 0$, so it is indeed near the singularity that this behaviour will be less accurate. Still, as one can easily check numerically, Eq. (29) works reasonably well at all stages of the evolution of R.

3.2. Generation of the singularity

We solve the system giving the initial conditions

$$\begin{cases} R_0 = -T_0 \\ \dot{R}_0 = f\,\dot{T}_0 = -f\,T_0/t_{contr} \,, \end{cases} \tag{30}$$

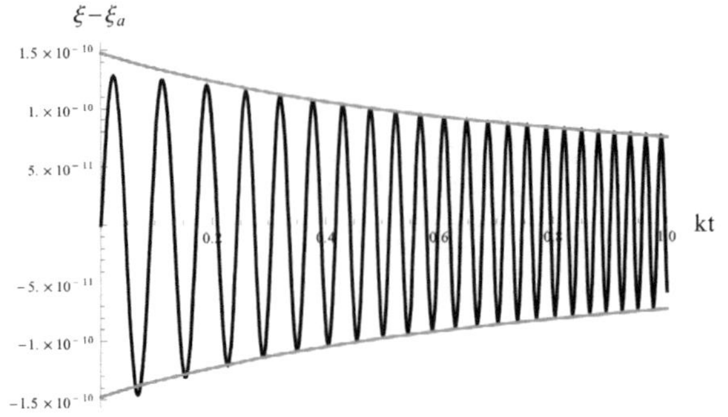

Figure 2. Comparison of the evolution of ξ with the predicted result, see equations (21) and (29). The values of parameters used are: $n = 1$, $\lambda = 1$, $R_{29} = 1$, $\varrho_{29} = 10^3$ and $t_{10} = 5 \cdot 10^{-4}$, $f = 0.5$.

where f is a free parameter quantifying the initial displacement from the GR behaviour $R + T = 0$; in particular, $f = 1$ corresponds to the situation in which R behaves initially exactly as if there were no $F(R)$ at all. Because of the considerations made at the end of the previous paragraph and noting that with these initial conditions

$$\dot{\xi}_0 = f\,\dot{\xi}_{a,0}\,, \tag{31}$$

our results will be particularly reliable for values of f not too close to unity.

Substituting ξ and ω from (12) and (14), respectively, equation (25) becomes

$$6n\lambda\left(-\frac{R_c}{T}\right)^{2n+1} \le \frac{[6n\lambda(2n+1)]^{3/2}\,|1-f|\,(-R_c)^{3n+3/2}}{T_0^{(5n+3)/2}\,T^{(n+1)/2}\,t_{contr}}\,. \tag{32}$$

For convenience, we define the following dimensionless quantities:

$$\begin{aligned} R_{29} &\equiv \frac{m_{Pl}^2}{8\pi}\frac{(-R_c)}{10^{-29}\text{ g cm}^{-3}}\,,\\ \varrho_{29} &\equiv \frac{\varrho_0}{10^{-29}\text{ g cm}^{-3}}\,,\\ t_{10} &\equiv \frac{t_{contr}}{10^{10}\text{ years}}\,. \end{aligned} \tag{33}$$

Substituting these expressions in (32) assuming (7), we find the value of the energy/mass density T_{sing} at which the curvature singularity forms:

$$\begin{aligned} \frac{T_{sing}}{T_0} &= \left(\frac{T_0^{n+1}\,t_{contr}}{\sqrt{6n\,|1-f|\,\lambda(2n+1)^3}\,R_c^{n+1/2}}\right)^{2/(3n+1)}\\ &\simeq \left(0.53\,\frac{\varrho_{29}^{n+1}\,t_{10}}{\sqrt{n\lambda\,|1-f|\,(2n+1)^3}\,R_{29}^{n+1/2}}\right)^{2/(3n+1)}\,. \end{aligned} \tag{34}$$

Typically we would have $\lambda, R_{29} \sim \mathcal{O}(1)$, so this result tells us that if the initial energy density is not too large and/or the contraction is fast enough, the overall contraction required to reach the singularity is quite realistic, with T_{sing}/T_0 possibly even only slightly larger than unity. Moreover, the larger n, the easier it is to develop the singularity with relatively small T_{sing}.

Once again, we should stress that (7) is not expected to be completely accurate, but our estimate (34) should be more or less correct even with different, more sophisticated contraction laws.

4. Slow-roll regime

Let us now relax the assumption of adiabaticity made above, which can be summarised in the statement

$$\frac{\dot{T}}{T}, \frac{\dot{\alpha}}{\alpha} \sim t_{contr}^{-1} \ll \omega\,. \tag{35}$$

We will instead focus on the opposite regime, that is (see Eq.19)

$$\frac{\omega\, t_{contr}}{2\pi} \lesssim 1\,, \tag{36}$$

showing that a singularity can form in this case, as well. We assume that ξ initially sits on the bottom of the potential (as in the initial conditions of Eq. 30), in $\xi = \xi_0$. Let us also denote ξ_{eq} the point corresponding to the equality

$$U(\xi_{eq}) = U(\xi_{sing})\,. \tag{37}$$

Clearly, ξ_{eq} can change with time, as the reader can see looking at figure 1. There, ξ_{eq} is the point other than ξ_{sing} at which the potential crosses the x axis; we can clearly see that ξ_{eq} is changing, and specifically decreasing, with time.

If there exists a time τ at which

$$\xi_{eq}(\tau) = \xi_0\,, \tag{38}$$

with

$$\frac{\omega\,\tau}{2\pi} \lesssim 1\,, \tag{39}$$

then $\xi(\tau)$ is, roughly speaking, still equal to $\xi_0 = \xi_{eq}(\tau)$. In the meantime the bottom of the potential has moved, so that now ξ finds itself on a slope dragging the field down towards $\xi = \xi_a$, from a height equal to the height of the singular point ξ_{sing}. In the next oscillation, therefore, ξ will most likely reach ξ_{sing} and hence the singularity.

The mechanism is the following: ξ starts at ξ_0, and the potential changes, due to the increasing energy density, more rapidly than ξ can oscillate, so that ξ_{eq} reaches ξ_0 before ξ has had the time to move away significantly from its initial position. Eventually, on a timescale roughly equal to ω^{-1}, ξ will start moving and reach ξ_{sing} on the other side of its oscillation; this will also be the typical timescale for the formation of the singularity. Evidently, the smaller the value of $\omega\,\tau$, the more accurate these assumptions will be.

For the models in exam, we have $U = 0$ in $\xi = \xi_{sing}$ and

$$\xi_{eq} \simeq 6n\lambda \left(-\frac{2n+1}{2n}\frac{R_c}{T}\right)^{2n+1}\,, \tag{40}$$

while initially the field is in

$$\xi_0 \simeq 6n\lambda \left(-\frac{R_c}{T_0}\right)^{2n+1}\,. \tag{41}$$

The request that $\xi_{eq} = \xi_0$ then becomes

$$\frac{T}{T_0} = 1 + \frac{1}{2n}\,, \tag{42}$$

which is always quite close to unity, and even more so for increasing n. This equality occurs at a time (see Eq. 7)

$$\tau = \frac{t_{contr}}{2n}\,. \tag{43}$$

Note that with this result equation (36) automatically implies (39). We must substitute this value in (39) to find the minimum contraction time, or on the other hand the maximum contraction rate, allowed; otherwise, a singularity would most probably develop within the first oscillation. This results in

$$t_{contr,\,min} \approx 2.4 \times 10^{11} \text{ years} \frac{n\sqrt{(2n+1)\lambda n}\, R_{29}^{n+1/2}}{\left(\dfrac{2n+1}{2n}\varrho_{29}\right)^{n+1}} . \tag{44}$$

Despite seeming at first sight very large, about twenty times the age of the Universe, this number is usually much shorter. In fact, for λ and R_{29} of order unity and $\varrho_{29} \gtrsim 10^3 - 10^5$, which are typical values for densities in (pre-)galactic clouds, the minimum contraction time already shortens by several orders of magnitude. However, if $t_{contr,\,min}$ becomes too short, basically every observed contracting system in the Universe would fulfil condition (44); in this case, we must use the analysis of section 3.

Nonetheless, for a reasonable portion of parameter space, in particular the region with small n and not-too-large ϱ_{29}, a contracting structure would develop a singularity within the first scalaron oscillation, thus on a timescale

$$t_{osc} \simeq \frac{2\pi}{\omega(\tau)} \approx 1.2 \times 10^{11} \text{ years} \frac{\sqrt{(2n+1)\lambda n}\, R_{29}^{n+1/2}}{\left(\dfrac{2n+1}{2n}\varrho_{29}\right)^{n+1}} . \tag{45}$$

For typical values of physical and model parameters (see above), this timescale could well be shorter than the age of the Universe, hence in principle observable.

5. Conclusions and discussion

The possibility of curvature singularities in DE $f(R)$ gravity models has been confirmed and studied in a rather simple fashion. The trace of the modified Einstein equations has been rewritten, under the simplifying assumptions of homogeneity and low-gravity, as an oscillator equation for the scalaron field ξ, which moves in a potential U depending on the external energy/mass density and thus on time. In the two models considered [10, 11], the potential is finite in the point corresponding to the curvature singularity $|R| \to \infty$, that is $\xi_{sing} = 0$; the energy conservation equation associated with ξ indicates that in contracting systems the probability of reaching this singularity increases with increasing energy/mass density.

The ratio between the typical contraction time and the inverse frequency of the scalaron determines two distinct regimes. In the adiabatic regime ξ oscillates very fast compared to relevant variations of U, and we can perform a linear analysis to estimate the scalaron amplitude and frequency analytically. The singularity is expected to be reached when the amplitude of the oscillations of ξ exceeds the separation between the "average" value ξ_a and the singular point. Results show that this can happen for reasonable values of physical and model parameters provided that the contraction lasts long enough. In the slow-roll regime, the field ξ is practically "frozen" in its initial value for a time longer than the time required by the potential to "lift" the field up to values which make ξ_{sing} accessible on the next oscillation. Basically, the energy of the field has increased enough for the singularity to be reached, which happens on a timescale comparable to the inverse frequency of the scalaron oscillations.

In principle, the results of this article could provide simple methods to constrain and possibly rule out models [10, 11], and most likely the same technique could be applied to other models already proposed. However, two effects could on one hand hinder the development of singularities, and on the other hand provide additional methods to constrain models: ultraviolet

gravity modifications and gravitational particle production.
Ultraviolet corrections to the gravitational action start dominating at large R, and should set a limit to its growth; in turn, R would never reach the singularity (for details see e.g. [15, 17]). Gravitational particle production, as is well known, is universal whenever curvature oscillates, and could in principle be a detectable source of high energy cosmic rays [20]. The back-reaction on curvature is a damping of its oscillations, so this damping may also prevent R from reaching infinity. This is particularly important in the adiabatic regime, where there can be very many oscillations before ξ reaches ξ_{sing} and therefore a large amount of energy could be released into SM particles. Fortunately, these cosmic rays would carry model-dependent signatures which could provide us valuable information to improve the constraints on the known models and maybe even suggest new gravitational theories.

Additional investigations involving, for instance, the High Energy Cosmic Rays "ankle" problem and the GZK cutoff, could stimulate further research on this very fascinating subject.

Acknowledgments

The author is grateful to Prof. A.D. Dolgov and an anonymous referee for valuable discussions and criticism.

References

[1] Riess A G *et al* 1998 *Astron. J.* **116** 1009 (*Preprint* astro-ph/9805201); Perlmutter S *et al* 1998 *Nature* **391** 51 (*Preprint* astro-ph/9712212); Schmidt B P *et al* 1998 *Astrophys. J.* **507** 46 (*Preprint* astro-ph/9805200); Perlmutter S *et al* 1999 *Astrophys. J.* **517** 565 (*Preprint* astro-ph/9812133); Riess A G *et al* 2004 *Astrophys. J.* **607** 665 (*Preprint* astro-ph/0402512)
[2] Amendola L and Tsujikawa S 2010 *Dark Energy: Theory and Observations* (Cambridge: Cambridge University Press)
[3] Starobinsky A A 1980 *Phys. Lett.* B **91** (1980) 99
[4] Whitt B 1984 *Phys. Lett.* B **145** 176; Maeda K 1989 *Phys. Rev.* D **39** 3159
[5] Capozziello S, Carloni S and Troisi A 2003 *Rec. Res. Dev. Astron. Astrophys.* **1** 625 (*Preprint* astro-ph/0303041); Capozziello S, Cardone V F, Carloni S and Troisi A 2003 *Int. J. Mod. Phys.* D **12** 1969 (*Preprint* astro-ph/0307018); Carroll S M, Duvvuri V, Trodden M and Turner M S 2004 *Phys. Rev.* D **70** 043528 (*Preprint* astro-ph/0306438)
[6] Chiba T 2003 *Phys. Lett.* B **575** 1 (*Preprint* astro-ph/0307338)
[7] Dolgov A D and Kawasaki M 2003 *Phys. Lett.* B **573** 1 (*Preprint* astro-ph/0307285)
[8] Faraoni V 2006 *Phys. Rev.* D **74** 104017 (*Preprint* astro-ph/0610734)
[9] Amendola L, Gannouji R, Polarski D and Tsujikawa S 2007 *Phys. Rev.* D **75** 083504 (*Preprint* gr-qc/0612180); Sawicki I an Hu W 2007 *Phys. Rev.* D **75** 127502 (*Preprint* astro-ph/0702278)
[10] Hu W and Sawicki I 2007 *Phys. Rev.* D **76** 064004 (*Preprint* arXiv:0705.1158)
[11] Starobinsky A A 2007 *JETP Lett.* **86** 157 (*Preprint* arXiv:0706.2041)
[12] Appleby S A and Battye R A 2007 *Phys. Lett.* B **654** 7 (*Preprint* arXiv:0705.3199)
[13] Frolov A V 2008 *Phys. Rev. Lett.* **101** (2008) 061103 (*Preprint* arXiv:0803.2500); T. Kobayashi and Maeda K 2008 *Phys. Rev.* D **78** 064019 (*Preprint* arXiv:0807.2503); Thingkool I, Sami M, Gannouji R G and Jhingan S 2009, *Phys. Rev.* D **80** 043523 (*Preprint* arXiv:0906.2460); Thingkool I, Sami M and Choudhury S 2009 *Phys. Rev.* D **80** 127501 (*Preprint* arXiv:0908.1693)
[14] Babichev E and Langlois D 2009 *Phys. Rev.* D **80** 121501 (*Preprint* arXiv:0904.1382); Babichev E and Langlois D 2010 *Phys. Rev.* D **81** 124051 (*Preprint* arXiv:0911.1297); Capozziello S, De Laurentis M, Odintsov S D and Stabile A 2011 *Phys. Rev.* D **83** 064004 (*Preprint* arXiv:1101.0219); Capozziello S, De Laurentis M, De Martino I, Formisano M and Odintsov S D 2012 *Phys. Rev.* D **85** 044022 (*Preprint* arXiv:1112.0761)
[15] Arbuzova E V and Dolgov A D 2011 *Phys. Lett. B* **700** 289 (*Preprint* arXiv:1012.1963)
[16] Bamba K, Nojiri S and Odintsov S D 2011 *Phys. Lett. B* **698** 451 (*Preprint* arXiv:1101.2820)
[17] Appleby S A, Battye R A and Starobinsky A A 2010 *JCAP* **1006** 005 (*Preprint* arXiv:0909.1737)
[18] Seifert M D 2007 *Phys. Rev.* D **76** 064002 (*Preprint* gr-qc/070360)
[19] Arbuzova E V, Dolgov A D and Reverberi L 2012 *JCAP* **02** 049 (*Preprint* arXiv:1112.4995)
[20] Arbuzova E V, Dolgov A D and Reverberi L 2012 *Eur. Phys. J. C* **72** (2012) 2247 (*Preprint* arXiv:1211.5011)
[21] Cembranos J A R, de la Cruz-Dombriz A, Montes Nunez B 2012 *JCAP* **1204** (2012) 021 (*Preprint* arXiv:1201.1289)

Quantization of the Schwarzschild geometry

Evangelos Melas

University of Athens, Department of Economics, Unit of Mathematics and Informatics, Pesmazoglou 8, 10559 Athens, Greece

E-mail: evangelosmelas@yahoo.co.uk

Abstract. The conditional symmetries of the reduced Einstein–Hilbert action emerging from a static, spherically symmetric geometry are used as supplementary conditions on the wave function. Based on their integrability conditions, only one of the three existing symmetries can be consistently imposed, while the unique Casimir invariant, being the product of the remaining two symmetries, is calculated as the only possible second condition on the wave function. This quadratic integral of motion is identified with the reparametrization generator, as an implication of the uniqueness of the dynamical evolution, by fixing a suitable parametrization of the r-lapse function. In this parametrization, the determinant of the supermetric plays the role of the mesure. The combined Wheeler - DeWitt and linear conditional symmetry equations are analytically solved. The solutions obtained depend on the product of the two "scale factors".

1. Introduction

A vast amount of work on the use of symmetries in the quantization of minisuperspace cosmological models fills the relevant literature of the last fifteen years. In fact, the application of the variational symmetries approach to classical Bianchi cosmologies has started in [1]. For the case of classical and quantum cosmology, in either Bianchi or higher derivative models, the first relevant works are [2], [3], [4], [5] and [6] (where conditional symmetries are used), while a revival of recent work on the subject can be found in [7], [8], [9] and [10]. Essentially, the method consists in the application of the standard theory of variational symmetries [11], [12] to the Lagrangian of some minisuperspace model. The steps of the usual procedure are (i) gauge fixing the lapse function to some convenient value, (ii) applying the first prolongation in the velocity phase space of a vector in the configuration space to this gauge fixed Lagrangian, and (iii) demanding its action to be zero. A slightly different approach exploiting the notions of the special projective and/or the homothetic group is adopted in [13], [14] and [15].

In the present work, we do not gauge fix the lapse function. This is justified by the fact that the full content of the corresponding reparametrization generator, i.e. the quadratic constraint, is kept intact by the presence of the lapse. This fact is important not only for the symmetries but also in the solution of the classical equations of motion, a task that is greatly facilitated by inserting the algebraic solution for N from the quadratic constraint into the other equations of motion [16], [17], [18] and [19]. In [20] a phase space point of view is adopted and a conditional symmetry is defined [21] as a conformal field of both the supermetric and the potential. Here, it is proven that this definition leads to the uncovering of all variational symmetries.

An application of the method to the case of static, spherically symmetric geometries is presented in this work. A minisuperspace Lagrangian for the aforesaid set of metrics has been given in [22], [23] where a $3+1$ decomposition is introduced along the radial coordinate r which is taken to be the dynamical variable. In [24], [25] the symmetries of the reduced Schwarzschild action have been used in the quantization procedure. Here, the quantum analogues of the linear integrals of motion are also adopted as supplementary conditions imposed on the wave function. A careful examination of their role and integrability conditions leads to the unique Casimir invariant of their algebra. Through a particular redefinition of the lapse function, the Casimir invariant is identified to the corresponding r - reparametrization generator.

The paper is organized as follows: In section 2 the reduced valid Lagrangian reproducing Einstein's field equations for static, spherically symmetric configurations is given, and passing to its Hamiltonian formulation [26], the three conditional symmetries are revealed. In section 3, the dynamical system is quantized by use of Dirac's canonical quantization procedure for constrained systems [27]. Finally, some concluding remarks are presented in the discussion.

2. Hamiltonian formulation of static, spherically symmetric geometries

Our starting point is the static, spherically symmetric line element

$$ds^2 = -a(r)^2 dt^2 + n(r)^2 dr^2 + b(r)^2 (d\theta^2 + \sin^2\theta\, d\phi^2). \tag{1}$$

In the usual ADM $3+1$ decomposition one foliates space-time in t-hypersurfaces and the coefficient of dt^2 is the lapse function. Here we adopt a $3+1$ foliation in the r coordinate and therefore the role of the lapse is attributed to the coefficient of dr^2 in the line element. Thus we consider $n(r)$ in (1) to be the r-lapse function while $a(r)$ and $b(r)$ are the dependent "dynamical" variables on the r-hypersurface. Previous authors use also a "shift" term $2B(r)\,dr\,dt$. However, this is not relevant for the following discussion, as $B(r)$ does not enter the Einstein tensor while it can be absorbed by a time redefinition of the form $t = \tilde{t} + \int \frac{B(r)}{a(r)^2} dr$.

The Einstein - Hilbert action $A_{E-H} = \int \sqrt{-g}\, R\, d^4x$ for the geometries (1) leads to the reduced action $\mathcal{A} = \int L dr$ with the following Lagrange function $L(a, b, a', b', n)$:

$$L = 2\,a\,n + \frac{4\,b\,a'\,b'}{n} + \frac{2\,a\,b'^2}{n}, \tag{2}$$

where $'$ denotes differentiation with respect to the spatial coordinate r. It is easy to verify that the Euler - Lagrange equations obtained from (2) are identical to Einstein's equations $\mathcal{G}_{\mu\nu} = 0$ for the line element (1).

In order to proceed with the Hamiltonian formalism we calculate the conjugate momenta,

$$\begin{aligned}
\pi_n &= \frac{\partial \mathcal{L}}{\partial n'} = 0, \\
\pi_a &= \frac{\partial \mathcal{L}}{\partial a'} = \frac{4b\,b'}{n}, \\
\pi_b &= \frac{\partial \mathcal{L}}{\partial b'} = \frac{4b\,a'}{n} + \frac{4\,a\,b'}{n}.
\end{aligned} \tag{3}$$

Obviously, π_n is a primary constraint. The Legendre transformation leads to the Hamiltonian

$$H = n\,\mathcal{H}_c,$$

where

$$\mathcal{H}_c = -2\,a - \frac{a\,\pi_a^2}{8\,b^2} + \frac{\pi_a\,\pi_b}{4\,b}. \tag{4}$$

The preservation of the primary constraint π_n in the $r-$evolution, i.e.

$$\pi_n' = \{\pi_n, H\} \approx 0,$$

leads to the secondary constraint

$$\mathcal{H}_c \approx 0. \tag{5}$$

The minisuperspace metric inferred from (4) is

$$G^{\alpha\beta} = \begin{pmatrix} -\frac{a}{4b^2} & \frac{1}{4b} \\ \frac{1}{4b} & 0 \end{pmatrix}. \tag{6}$$

According to our definition of a conditional symmetry [20], as a vector ξ on the configuration space the following relations must hold:

$$\mathcal{L}_\xi G^{\alpha\beta} = \phi(q) G^{\alpha\beta} \quad , \quad \mathcal{L}_\xi V(q) = \phi(q) V(q) \tag{7}$$

where q's are the configuration space variables. Thus we are led to the following three conformal Killing fields of both $G_{\alpha\beta}$ and the potential $V = -2a$:

$$\xi_1 = (-a, b), \quad \xi_2 = \left(\frac{1}{a\,b}, 0\right), \quad \xi_3 = \left(-\frac{a}{2\,b}, 1\right) \tag{8}$$

which, contracted with (π_a, π_b), provide us with the three integrals of motion:

$$Q_1 = -a\,\pi_a + b\,\pi_b, \quad Q_2 = \frac{\pi_a}{a\,b}, \quad Q_3 = -\frac{a\,\pi_a}{2\,b} + \pi_b. \tag{9}$$

We calculate the Poisson brackets of these conserved quantities with the canonical Hamiltonian H and the Poisson algebra that they satisfy:

$$\{Q_1, H\} = n\mathcal{H}_c, \quad \{Q_2, H\} = -\frac{n}{a^2\,b}\mathcal{H}_c, \quad \{Q_3, H\} = \frac{n}{2\,b}\mathcal{H}_c, \tag{10a}$$

$$\{Q_1, Q_3\} = Q_3, \quad \{Q_2, Q_1\} = Q_2, \quad \{Q_3, Q_2\} = 0. \tag{10b}$$

As expected, the Poisson brackets (10a) are weakly vanishing on the constraint surface $\mathcal{H}_c \approx 0$ and therefore the three Q_I's are constants of motion.

At this point it is interesting, and useful for what follows in the quantization, to adopt a new parametrization of the lapse $n(r) = \frac{\bar{n}(r)}{2a(r)}$ which makes the potential constant. The Lagrangian and the corresponding Hamiltonian are now given by

$$\bar{L} = \bar{n} + \frac{8\,a\,b\,a'\,b'}{\bar{n}} + \frac{4\,a^2\,b'^2}{\bar{n}} \quad , \quad \bar{H} = \bar{n}\bar{\mathcal{H}}_c = \bar{n}\frac{1}{2a}\mathcal{H}_c. \tag{11}$$

If the value of $\bar{n}$ specified by the constrained equation is substituted into the Euler - Lagrange equations for $a(r)$ and $b(r)$ the system can be solved for only one acceleration, say $a''(r)$, and the general solution of the entire system is:

$$\bar{n}(r) = 2\,c\,b'(r) \quad , \quad a(r) = c\sqrt{1 - \frac{2M}{b(r)}}, \tag{12}$$

where the constants of integration have been rearranged so that the ensuing line element

$$ds^2 = -c^2\left(1 - \frac{2M}{b(r)}\right)dt^2 + \left(1 - \frac{2M}{b(r)}\right)^{-1} b'(r)^2 dr^2 + b(r)^2(d\theta^2 + \sin^2\theta\, d\phi^2) \tag{13}$$

bears the closest possible resemblance to the standard form of the Schwarzschild metric, while the presence of the arbitrary function $b(r)$ reflects the r reparametrisation covariance of the system.

The new supermetric is now given by

$$\bar{G}^{\alpha\beta} = \begin{pmatrix} -\frac{1}{8b^2} & \frac{1}{8a\,b} \\ \frac{1}{8a\,b} & 0 \end{pmatrix}, \tag{14}$$

and the fields (8) are now turned into Killing fields of $\bar{G}^{\alpha\beta}$; they are also, trivially, symmetries of the constant potential $\bar{V} = 1$ as well. The algebra satisfied by the new quantities $\bar{Q}_I = \xi_I^\alpha \bar{\pi}_\alpha$ and $\bar{\mathcal{H}}_c$ can be easily seen to be

$$\{\bar{Q}_1, \bar{\mathcal{H}}_c\} = 0, \quad \{\bar{Q}_2, \bar{\mathcal{H}}_c\} = 0, \quad \{\bar{Q}_3, \bar{\mathcal{H}}_c\} = 0. \tag{15}$$

Therefore the quantities $\bar{Q}_I$ still remain constants of motion. It is in this conformal gauge that one can explicitly verify that the prolonged ξ_I's, X_I, are satisfying a form of the standard condition of variational symmetries $pr^{(1)}X_I(\bar{L}) = 0$, thus ensuring that they are generators of Noether symmetries. It is important to note that if one had inserted the X_I's into the full relation $pr^{(1)}X(L) + L\text{Div}\Xi = \text{Div}F$ with L given by (2) the result would be negative.

If we invert the definitions for $\bar{\pi}_\alpha$ and use (12) in $\bar{Q}_I$ we calculate the integrals' values on the solution space:

$$Q_1 = \bar{Q}_1 = 4\,c\,M \quad , \quad Q_2 = \bar{Q}_2 = \frac{4}{c} \quad , \quad Q_3 = \bar{Q}_3 = 2\,c \tag{16}$$

As it is known the Schwarzschild solution involves only one essential constant, the mass M. The second constant appearing in the solution (12), (13) can be seen to be absorbable by a scaling of the time coordinate $t \to \frac{t}{c}$, allowing to set $c = 1$ but not $c = 0$. So, on the solution space we can set $\bar{Q}_1 = 4M$, $\bar{Q}_2 = 4$, $\bar{Q}_3 = 2$. It is noteworthy that the values of the last two integrals of motion can be changed at will and that they do not involve the essential parameter M which characterizes the geometry. The above argument relies on the underlying geometry.

But what if we were deprived of the line element and we were just given the dynamical system (11)? How could we differentiate between the constants κ_1 and κ_2, κ_3? Interestingly enough, there is an argument that leads to a distinction between them: The crucial observation is that $\bar{Q}_1$ has a vanishing Poisson bracket with each kinetic term of $\bar{H}$, while $\bar{Q}_2$, $\bar{Q}_3$ need the entire kinetic term in order to produce vanishing Poisson brackets. This fact is reflected in the following property concerning the one-parameter family of canonical transformations generated by the charge Q_1 (see [29], [30] for the generalization of Noether symmetries for constrained systems and for Noether's theorem in phase space),

$$a \to e^\lambda a, \; \bar{\pi}_a \to e^{-\lambda}\bar{\pi}_a, \; b \to e^{-\lambda} b, \; \bar{\pi}_b \to e^\lambda \bar{\pi}_b \tag{17}$$

Under such transformations $\bar{Q}_1$ remains, of course, unchanged, while $\bar{Q}_2$, $\bar{Q}_3$ are scaled by $e^{-\lambda}$ and e^λ respectively. One can thus use the freedom of λ to arbitrarily change the values κ_2, κ_3, but not κ_1. Furthermore, the Hamiltonian $\bar{\mathcal{H}}_c$ remains, due to the particular scaling of $\bar{Q}_2$, $\bar{Q}_3$, unchanged.

If we return to the phase space, we can write

$$\bar{Q}_{23} = \bar{Q}_2\,\bar{Q}_3 = -\frac{1}{2\,b^2}\bar{\pi}_\alpha^2 + \frac{1}{a\,b}\,\bar{\pi}_\alpha\,\bar{\pi}_\beta \tag{18}$$

with the relevant Poisson bracket algebra now becoming

$$\{\bar{Q}_{23}, \bar{\mathcal{H}}_c\} = 0 \tag{19a}$$
$$\{\bar{Q}_I, \bar{Q}_{23}\} = 0, \qquad I = 1, 2, 3. \tag{19b}$$

As expected $\bar{Q}_{23}$ is a quadratic integral of motion. From a group theoretical view, $\bar{Q}_{23}$ is an element of the centre of the universal enveloping algebra (uea) generated by $\bar{Q}_I$'s, i.e. it is the Casimir invariant. The Hamiltonian $\bar{\mathcal{H}}_c$ belongs also to the centre of uea (it commutes with all $\bar{Q}_I$'s), thus it can only differ from $\bar{Q}_{23}$ by an additive and/or a multiplicative constant. It is an easy matter for one to check that indeed

$$\bar{Q}_{23} = 8\,(\bar{\mathcal{H}}_c + 1). \tag{20}$$

To sum up, we have constructed a gauge independent, quadratic in the momenta, integral of motion which commutes with the only integral of motion that entangles the sole essential constant of the Schwarzschild solution. In the next section, and in order to proceed with the quantization, we will rely on these two quantities.

3. Quantization

In order to quantize our system, we must turn into operators $\bar{\mathcal{H}}_c = \frac{1}{8}\left(\bar{Q}_{23} - 8\right)$, $\bar{Q}_1$, $\bar{Q}_2$ and $\bar{Q}_3$ (hereafter, for the shake of simplicity, we will omit the *bars* from the symbols of the corresponding operators). The corresponding quantum operators can be inferred from the general definitions

$$\widehat{\mathcal{H}}_c := -\frac{1}{2\mu}\partial_\alpha(\mu\, G^{\alpha\beta}\partial_\beta) + V(q) \tag{21}$$

$$\widehat{Q}_I := -\frac{\mathtt{i}}{2\mu}(\mu\,\xi_I^\alpha\,\partial_\alpha + \partial_\alpha\mu\,\xi_I^\alpha), \tag{22}$$

where μ is a suitable measure which transforms as a scalar density. Definition (22) supplies us with a Hermitian operator [28] for the previously mentioned conditional symmetries (8). It can be seen [20] that, the measure entering the quantum operators ought to be taken as $\mu(a,b) = \lambda\sqrt{det|\bar{G}_{\alpha\beta}|} \propto a\,b$. The constancy of λ is forced by the combined requirement that the Q_I's must be realized as Hermitian operators and at the same time retain their classical geometrical character by acting as derivatives. Thus the extra term $\xi_I^\alpha\partial_\alpha\lambda$ must vanish for all $I = 1,2,3$, which leads to a constant λ. Two further arguments in favor of this choice of measure are:

- The fact that the quantum analogue of the algebra (19b) is made isomorphic to the classical, i.e.
$$[\widehat{Q}_I, \widehat{Q}_{23}]F(a,b) = 0, \quad I = 1,2,3 \quad \text{for any } F(a,b) \tag{23}$$
a fact that is highly non trivial, since it depends on the choice of both the factor ordering and the measure.
- At the classical level, the only linear integral of motion involving the essential constant is Q_1. If we seek the functions on the configuration space which are invariant under the point transformations generated by Q_1 we find $\{Q_1, f(a,b)\} = 0 \Rightarrow f(a,b) = f(a\,b)$.

The above arguments lead to the following linear operators corresponding to the elements of the classical algebra, the Casimir invariant and the Hamiltonian:

$$\widehat{Q}_1 = -\mathtt{i}\,(b\,\partial_b - a\,\partial_a) \tag{24}$$

$$\widehat{Q}_2 = -\frac{\mathtt{i}}{a\,b}\partial_a \tag{25}$$

$$\widehat{Q}_3 = -\mathtt{i}\left(\partial_b - \frac{a}{2\,b}\,\partial_a\right) \tag{26}$$

$$\widehat{Q}_{23} = \frac{2}{b^2}\,\partial_a\partial_a - \frac{1}{a\,b}\,\partial_a\partial_b + \frac{1}{2a\,b^2}\,\partial_b \tag{27}$$

$$\widehat{\mathcal{H}}_c = \frac{1}{8}\left(\widehat{Q}_{23} - 8\right) \tag{28}$$

It is an easy task to check that these operators satisfy not only the relations

$$[\widehat{Q}_I, \widehat{Q}_J]F(a,b) = \mathtt{i}\, C^K{}_{IJ} \widehat{Q}_K F(a,b) \tag{29}$$

for any test function $F(a,b)$, but also (23) as well.

The integrability condition for the equations emerging from the enforcement of the $\widehat{Q_I}$'s upon the wave function

$$\widehat{Q}_I \Psi = \kappa_I \Psi \tag{30}$$

leads to the selection rule

$$C^M_{IJ} \kappa_M = 0. \tag{31}$$

Due to the constraint condition (31), applied to the specific structure constants inferred from (10b), we conclude that only the eigenvalue κ_1 is free and κ_2, κ_3 must necessarily be zero. However, the latter is impossible since on the classical solution space (16) hold. This means that not all three operators can be applied simultaneously. The latter results in the need to consider the two- and/or one- dimensional subalgebras together with the Wheeler - DeWitt equation $\widehat{\mathcal{H}}_c \Psi = 0$. The investigation of these cases can be easily carried out.

As far as the $2d$ subalgebras are concerned the results obtained are briefly the following:

(a) For the two non Abelian subgroups either κ_2 or κ_3 are forced to be zero, something that is inconsistent with their classical values.

(b) For the Abelian subgroup, the two linear equations lead to the solutions $\Psi(a,b) = A \exp\left(\frac{\mathtt{i}}{2}(\kappa_2\, a^2\, b + 2\kappa_3\, b)\right)$, where A is constant and the quadratic constraint enforces the restriction $\kappa_2\, \kappa_3 = 8$. It is of course doubtful if one can accept such a wave function to represent the geometry, knowing that it does not contain the essential constant M. However, one could interpret it as plane waves representing the limiting flat space-time $M = 0$.

The one-dimensional subalgebras spanned by $\widehat{Q}_2$, $\widehat{Q}_3$ give solutions which are special cases of the solution described in b), as expected since they commute.

Consequently, the only possibility is to adopt $\widehat{\mathcal{H}}_c$ and $\widehat{Q}_1$ as conditions on the wave function. This is indeed possible, since they commute with each other and therefore can be considered as physical quantities on the phase space that can be measured "simultaneously" (our dynamical parameter is the distance r). The ensuing eigenvalue equations are

$$\widehat{\mathcal{H}}_c \Psi = 0 \Rightarrow a\, \partial_a \partial_a\, \Psi - 2\, b\, \partial_a \partial_b\, \Psi + \partial_a\, \Psi - 16\, a\, b^2\, \Psi = 0 \tag{32a}$$

$$\widehat{Q}_1 \Psi = \kappa_1\, \Psi \Rightarrow \mathtt{i}\,(-a\, \partial_a\, \Psi + b\, \partial_b\, \Psi) = \kappa_1\, \Psi \tag{32b}$$

The solution of the linear partial differential equation (32b) is

$$\Psi(a,b) = a^{\mathtt{i}\kappa_1}\, S(a\, b). \tag{33}$$

If we insert the above solution into the Hamiltonian constraint (32a) we arrive at the following ordinary differential equation for $S(u)$ $(u = a\, b)$:

$$u^2\, S''(u) + u\, S'(u) + \left(\kappa_1^2 + 16\, u^2\right)\, S(u) = 0 \tag{34}$$

which has the general solution

$$S(u) = c_1\, J_{\mathtt{i}\kappa_1}(4\, u) + c_2\, Y_{\mathtt{i}\kappa_1}(4\, u), \tag{35}$$

in terms of the Bessel functions of imaginary order.

In order to gain some insight on the normalizability of the formal probability, instead of these Bessel functions and because of their imaginary order, we can use the functions $F_{\mathtt{i}\kappa_1}(4\,u)$ and $G_{\mathtt{i}\kappa_1}(4\,u)$ defined in [31] through the Hankel functions $H^{(1)}_\mu(u) = J_\mu(u) + \mathtt{i}\,Y_\mu(u)$ and $H^{(2)}_\mu(u) = J_\mu(u) - \mathtt{i}\,Y_\mu(u)$, $\mu \in \mathbb{C}$. Thus, the solution can be written as

$$S(u) = c_1\,F_{\mathtt{i}\kappa_1}(4\,u) + c_2\,G_{\mathtt{i}\kappa_1}(4\,u) \tag{36}$$

with

$$F_{\mathtt{i}\kappa_1}(4\,u) = \frac{1}{2}\left(e^{-\kappa_1\pi/2}\,H^{(1)}_{\mathtt{i}\kappa_1}(4u) + e^{\kappa_1\pi/2}\,H^{(2)}_{\mathtt{i}\kappa_1}(4u)\right) \tag{37}$$

$$G_{\mathtt{i}\kappa_1}(4\,u) = \frac{1}{2\mathtt{i}}\left(e^{-\kappa_1\pi/2}\,H^{(1)}_{\mathtt{i}\kappa_1}(4u) - e^{\kappa_1\pi/2}\,H^{(2)}_{\mathtt{i}\kappa_1}(4u)\right). \tag{38}$$

These functions are linearly independent solutions of (34) and have the following properties: a) when $u \in (0, +\infty)$ they are real, b) they are oscillatory with a phase difference of $\frac{\pi}{2}$ and c) when both u, and/or κ_1 tend to zero, $F_{\mathtt{i}\kappa_1}(4\,u)$ tends to 1, while $G_{\mathtt{i}\kappa_1}(4\,u)$ becomes infinite.

The final form of the wave function $\Psi(a,b)$ is

$$\Psi(a,b) = a^{\mathtt{i}\kappa_1}\,S(a\,b), \tag{39}$$

so we can define a probability density of the form

$$\mu(a\,b)\,\Psi^*(a,b)\,\Psi(a,b) \propto u\,S^*(u)\,S(u). \tag{40}$$

4. Discussion

In this paper we used a method for the quantizations of *minisuperspace* actions which can be summarized in the following steps:

(i) Go over to the Hamiltonian $\bar{H}$.

(ii) Calculate the Noether symmetries as Killing fields of the metric $\bar{G}^{\alpha\beta}(q)$ and, trivially, symmetries of the constant potential.

(iii) Identify the essential constants of the metric. Promote the allowed Q_I's to operators according to (22) with $\mu = |\bar{G}|^{1/2}$.

(iv) Promote the Hamiltonian constrain $\bar{\mathcal{H}}_c$ (which is a linear function of the Casimir invariant) via (21) to a Hermitian operator acting on the wave function.

We applied the above method for the case of static, spherically symmetric geometries. First, we begin from the Lagrangian (2) emanating from the line element (1). We find the simultaneous conformal Killing fields (8) of the supermetric and the potential, which define the three conserved charges (9). The unique Casimir invariant of their algebra is Q_{23}. In order to make it numerically proportional to the kinetic part of the Hamiltonian we are led to (11).

Our specific example (described in sections 2 and 3) has also been the subject of [24]. In that work, Vakili finds two of the Q_I's but then he uses a linear combination of them in order to reproduce the essential constant M of the Schwarzchild metric. He thus reaches to the unique acceptable linear quantum operator equivalent to our $\widehat{Q}_1$. The clever choice of the lapse function, his equation (4), along with the somewhat unorthodox choice of factor ordering for the operators (see below his equation (55)) leads essentially to a constant potential (his equation (53)) and to the Laplacian operator (his equation (56)). As a result, the solution spaces found both by us and Vakili essentially coincide. Of course, our general theory given in [20] constitutes a systematic explanation of the various choices of his work.

References

[1] Capozziello S, Marmo G, Rubano C and Scudellaro P 1997 *Int. J. Mod. Phys.* D **6** 4 491-503.
[2] Capozziello S and Lambiase G 2000 *Gen. Rel. Grav.* **32** 295-311
[3] Capozziello S and Lambiase G 2000 *Gen. Rel. Grav.* **32** 673-96
[4] Cotsakis S, Leach P G L and Pantazi H 1998 *Grav. Cosmol.* **4** 314-25
[5] Leach P G L, Cotsakis S and Miritzis J 2001 *Grav. Cosmol.* **7** 311-20
[6] Christodoulakis T, Gakis T and Papadopoulos G O 2002 *Class. Quantum Grav.* **19** 101325
[7] Vakili B and Khazaie F, 2012 *Class. Quantum Grav.* **29** 035015
[8] Falciano F T, Pereira R, Pinto-Neto N and Santini E S 2012 *Phys. Rev.* D **86** 063504
[9] Capozziello S, De Laurentis M and Odintsov S D 2012 *Preprint* gr-qc/1206.4842v1
[10] Sarkar K, Nayem Sk, Debnath S, and Sanyal A K 2012 *Preprint* astro-ph.CO/1207.3219v1
[11] Olver P J 2000, *Applications of Lie Groups to Differential Equations* 2nd ed. (Berlin, Heidelberg, New York: Springer - Verlag)
[12] Olver P J 1995 *Equivalence, Invariants and Symmetry* (Cambridge: Cambridge University Press) p 237
[13] Tsamparlis M and Paliathanasis A 2010 *Gen. Rel. Grav.* **42** 2957-80
[14] Tsamparlis M and Paliathanasis A 2011 *J. Phys.* A **A44** 175202
[15] Tsamparlis M and Paliathanasis A 2011 *Gen. Rel. Grav.* **43** 1861-81
[16] Christodoulakis T and Terzis P A 2006 *J. Math. Phys.* **47** 102502
[17] Christodoulakis T and Terzis P A 2007 *Class. Quantum Grav.* **24** 87587
[18] Terzis P A and Christodoulakis T 2009 *Gen. Rel. Grav.* **41** 46995
[19] Terzis P A and Christodoulakis T 2010 *Preprint* gr-qc/1007.1561
[20] Christodoulakis T, Dimakis N, Terzis P A, Doulis G, Grammenos Th, Melas E and Spanou A 2012 *Preprint* gr-qc/1208.0462
[21] Kŭchar K V 1982 *J. Math. Phys.* **23** 1647-61
[22] Cavaglià M, de Alfaro V, and Filippov A T 1995 *Int. J. Mod. Phys.* D **4** 661-72
[23] Cavaglià M, de Alfaro V and Filippov A T 1996 *Int. J. Mod. Phys.* D **5** 227-50
[24] Vakili B 2012 *Int. J. Theor. Phys.* **51** 1 133-45
[25] Jalalzadeh S, Vakili B 2012 *Int. J. Theor. Phys.* **51** 1 263-75
[26] Sundermeyer K 1982 *Constrained Dynamics* (Berlin, Heidelberg, New York: Springer - Verlag)
[27] Dirac P A M 1964 *Lectures on Quantum Mechanics* (New York: Yeshiva University, Academic Press)
[28] Christodoulakis T and Korfiatis E 1991 *Phys. Let.* B **256** 484-90
[29] Garcia J A and Pons J M 2000 *Int. J. Mod. Phys.* A **15** 4681
[30] Struckmeier J 2012 *Preprint* math-ph/1206.4449v1
[31] Dunster T M 1990 *SIAM J. Math. Anal.* **21** 4 995-1018

Journal of Physics: Conference Series **442** (2013) 012038 doi:10.1088/1742-6596/442/1/012038

Quantum gravity contributions to the cosmological constant

Simone Zonetti

Centre for Cosmology, Particle Physics and Phenomenology (CP3), Institut de Recherche en Mathématique et Physique, Université catholique de Louvain, Chemin du Cyclotron 2, B-1348, Louvain-la-Neuve, Belgium

E-mail: simone.zonetti@uclouvain.be

Abstract. In the standard QFT approach to the calculation of the cosmological constant contributions from the gravitational sector are usually neglected. Using a simple two-dimensional scenario, but in connection with higher dimensions, we discuss the possibility that quantum gravity can play an important role in the calculation of Λ, providing large quantum corrections and renormalizations. We reproduce the QFT result as a sub-case, while allowing a very large set of values.

1. Introduction

The cosmological constant problem [1, 2] and references therein), in its striking simplicity, is perhaps the strongest evidence that the present understanding of everything we generally refer to as "quantum gravity" is far from being satisfactory: while we know with a great accuracy that Λ is very small in Planck units ($\Lambda \sim 10^{-123}$) [3, 4], there is no theoretical approach that seems to be able to predict such a measurement in a convincing fashion. In the standard approach the cosmological constant is calculated in the semiclassical approximation as the vacuum expectation value of the stress energy tensor of the Standard Model of particle physics, but the results of this calculation turn out to be catastrophic, with a discrepancy between the theoretical and experimental value of Λ between 60 and 120 orders of magnitude. There is no debate that something has to be improved to overcome this situation.

A point of concern with the standard QFT analysis of the vacuum energy is represented by the semi-classical framework. Contributions to Λ are calculated on a flat background geometry, discarding completely any contribution which could be given by the gravitational sector. While it is true that the vacuum fluctuations of a quantum gravitational field are suppressed by the Planck scale with respect to (w.r.t.) the vacuum fluctuations of the matter fields, the dynamics of quantum gravity is completely ignored, precluding the possibility of some mechanism induced by quantum gravity that might explain the value of Λ that we measure today.

In this paper we discuss how the cosmological constant can be determined in a quantum theory of gravity, accounting for quantum dynamics of matter, gauge and the gravitational fields, including their vacuum fluctuations, possibly improving our understanding of what Λ is really made of. In order to put to practice our idea we will focus on the case of two-dimensional dilaton gravity, taking advantage of the simplified framework. Most of the technical results are contained in [5, 6, 7, 8].

This paper is organized as follows: in Section 2 we introduce two-dimensional dilaton gravity, its dual action in terms of Liouville fields, the inclusion of scalar matter. The BRST formulation of the model is briefly discussed. In Section 3 we deal with the quantization of the theory, determining the central extensions of the Virasoro algebra, while Section 4 presents the realization of the quantum constraints and describes the contributions to the cosmological constant. We summarize and discuss our results in Section 5.

2. 2d Dilaton Maxwell gravity coupled with scalar matter

We consider the most general action principle of two-dimensional dilaton gravity in the presence of a Maxwell field:

$$S_{DM} = \int dx^2\sqrt{-g}\left(XR - U(X)X_{,\mu}X^{,\mu} - 2V(X) - \frac{1}{4}G(X)F_{\mu\nu}F^{\mu\nu}\right) , \qquad (1)$$

where X is the dilaton, $U(X)$, $V(X)$ and $G(X)$ are arbitrary functions of X, and $F_{\mu\nu}$ is the usual field strength for the vector gauge field A_μ. Commas denote derivatives. With specific restrictions on the U and G potentials, there exists a dual formulation of (1) in terms of two decoupled Liouville fields on flat space-time [6]. In particular the dual action principle takes the form:

$$S_{dual} = \int d^2x\sqrt{-g_\flat}\left[\frac{1}{2}\left(Z_{,\mu}Z^{,\mu} - Y_{,\mu}Y^{,\mu}\right) - 2\Lambda e^{Z/\xi} - \frac{e^{-Y/\xi}}{16q\Lambda}F_{\mu\nu}F^{\mu\nu} + \xi\left(Z - Y\right)R_\flat\right] , \qquad (2)$$

where Z and Y are two real scalar fields, Λ is the cosmological constant and q and ξ are non-vanishing (coupling) constants. The line element reads:

$$dx^2 = -\lambda_0\lambda_1 dt^2 + (\lambda_0 - \lambda_1)dt\ ds + ds^2 , \qquad (3)$$

which is pure gauge. The subscript $\flat$ indicates gravitational quantities that are calculated w.r.t. this metric. In this formulation the cosmological constant can always be reabsorbed by a constant shift of Z, Y, as in $Z \to Z - \xi\ln\Lambda$, but it is left explicit in the following.
The inclusion of massless scalar matter directly in the dual formulation is trivial and it is simply done by adding an arbitrary number N of terms in the form:

$$\mathcal{S}_\phi = -\frac{1}{2}\int_{\mathcal{M}} d^2x\sqrt{-g_\flat}\phi_{,\mu}\phi^{,\mu} . \qquad (4)$$

The existence of this duality is of fundamental importance: a highly coupled non-linear system is cast into a much simpler decoupled one, opening the way for canonical non-perturbative quantization. This comes at the fair price of a restriction on the U, G potentials. In particular one requires:

$$U(X) = \tfrac{1}{2\xi^2} - \partial_X \ln(V(X)) , \qquad (5)$$

$$G(X) = \tfrac{\xi^2 e^{X/\xi^2}}{4qV(X)} . \qquad (6)$$

Furthermore we also ask $V(X) \neq 0$. Such a restriction allows still for enough freedom to cover a large classes of dilaton gravity models, for instance a subset of the so-called *ab*-family (including the Witten black hole and the CGHS models [9, 10]) and Liouville gravity [11].

We will focus on a model described by (2) with the addition of N massless real scalar fields (4). In order to take advantage of the formal equivalence of field modes and quantum harmonic oscillators, which will allow to express quantum field operators in terms of creation

and annihilation operators on a suitably defined Fock space of quantum states, we will restrict our attention to the case of a space-time with a cylindrical topology. In particular we will consider $\mathcal{M} = \mathbb{R} \otimes S^1$, where the time coordinate t takes values on the real line and the space coordinate s is replaced by an angular coordinate limited to the interval $[0, 2\pi)$. In this process we are implicitly introducing a length scale ℓ_c characteristic of the size of the compactification. In the following we will work in units which give $\ell_c = 1$[1], therefore we can simply replace the spatial coordinate $s \in \mathbb{R}$ with $s \in [0, 2\pi)$ with no ambiguities. All fields will be then required to be periodic in the space coordinate, so that for any field x one has $x(t, s = 0) = x(t, s = 2\pi)$.
To deal with the gauge invariances it is useful to look at the BRST formulation of the model [12]; after the introduction of ghost degree of freedom and the BRST charge, the BRST extended constraints are readily calculated and, as expected from the properties of the duality, it is easy to check that their algebra is first class. In particular the (smeared) Poisson brackets read:

$$\begin{aligned} \{L^\pm(f), L^\pm(g)\} &= \pm L^\pm(g_{,s}f - f_{,s}g) , \\ \{L^\pm(f), L^\emptyset(g)\} &= 0 , \\ \{L^\pm(f), L^\mp(g)\} &= -2q\left(e^{Y/\xi}P_{A_1}L^\emptyset\right)(fg) , \end{aligned} \tag{7}$$

where $\pm$ label the two Virasoro generators and $\emptyset$ indicates the constraint related to the $U(1)$ gauge invariance, while P_{A_1} is the momentum conjugate to the component A_1 of the Maxwell field. As expected these brackets are vanishing on the constraints surface. In the following we will choose to work in the conformal gauge for the gravitational sector, $\lambda_0 = \lambda_1 = 1$, and the Coulomb gauge in the $U(1)$ sector, $A_0 = A_{1,s} = 0$. Gauge fixing is implemented by a suitable choice of the gauge fixing fermion.
Per effect of the decoupling, the BRST extended Virasoro generators are a sum of terms from the different sectors of the model:

$$L^\pm = L^{\pm,Z} + L^{\pm,Y} + L^{\pm,g} + \sum_{n=1}^{N} L_n^{\pm,\phi} , \tag{8}$$

where the g superscript indicates the contributions of the BRST ghosts. This feature is of paramount importance in the quantization procedure, since it allows to quantize each sector separately. The different contributions take the form:

$$L^{\pm,Z} = -\frac{1}{4}\left(P_Z \mp Z_{,s}\right)^2 \mp \xi\left(P_Z \mp Z_{,s}\right)_s + \Lambda e^{Z/\xi} , \tag{9}$$

$$L^{\pm,Y} = \frac{1}{4}\left(P_Y \pm Y_{,s}\right)^2 \mp \xi\left(P_Y \pm Y_{,s}\right)_s + q\Lambda e^{Y/\xi}P_{A_1}^2 , \tag{10}$$

$$L^{\pm,\phi} = \frac{1}{4}\left(P_\phi \pm \phi_{,s}\right)^2 , \tag{11}$$

$$L^{\pm,g} = \frac{\mathrm{i}}{2\pi}\left(c^\pm b_{\pm,s} + 2c^\pm_{,s}b_\pm\right) \tag{12}$$

where P are the conjugate momenta of the fields and $c^\pm, b_\pm$ are the ghost fields associated with the Virasoro generators, satisfying:

$$\{c^\pm(s), b_\pm(s')\}^+ = -2\mathrm{i}\pi\delta_{2\pi}(s - s') , \tag{13}$$

with $\delta_{2\pi}(s - s')$ being the 2π-periodic Dirac δ distribution on the unit circle. Gauss' constraint on the other hand has no BRST extension and is simply given by $L^\emptyset = P_{A_1,s}$.

[1] We are entitled to this choice in the two-dimensional case: the *natural* units are usually chosen by fixing $c = \hbar = G = 1$. In two dimensions, however, Newton's constant is dimensionless, so that instead of fixing G we can fix the compactification scale ℓ_c.

3. The central charge of the Virasoro algebra

Quantization follows the usual canonical procedure, by choosing a polarization for phase space and a Hilbert space, by promoting observables to operators (using normal ordering when required, with the annihilation operators to the right of the creation ones) and by substituting Poisson brackets with commutators or anti-commutators, as the case may be, inclusive of the extra factor i multiplying the values of the corresponding classical brackets.

We will require the classical symmetries to be unbroken at the quantum level. The $L^{\emptyset}$ constraint has identically vanishing brackets with the two Virasoro generators and itself, hence the corresponding operator is quantum mechanically trivial. The Virasoro algebra, on the other hand will exhibit the typical central extension in each different sector x. Looking at the Fourier modes we will have:

$$\left[L_r^{\pm,x}, L_q^{\pm,x}\right] = (r-q)\hbar_* L_{r+q}^{\pm,x} + \hbar_*^2 c_x \delta_{r+q}\ , \tag{14}$$

where $\hbar_*$ is a place holder for $\hbar = 1$. The ghosts and scalar matter degrees of freedom reproduce the usual results known from string theory [13, 14]:

$$c_g = -\frac{13}{6}r^3 + \frac{1}{6}r\ , \qquad c_\phi = \frac{1}{12}r^3 + \frac{1}{6}r\ , \tag{15}$$

In the Liouville sector, in order to ensure the closure of the quantum algebra, the possibility of quantum corrections to the coupling constant ξ needs to be considered [15], in a manner dependent on the fields. As a matter of fact, only terms of (9) (10) involving the fields linearly need to be corrected, with the replacements $\xi \to \xi_Z = \xi + \delta_Z$ and $\xi \to \xi_Y = \xi + \delta_Y$ for the corresponding couplings, respectively. The factor ξ appearing in the exponential Liouville term contributions to (9) (10) remains unchanged. The commutators can be calculated explicitly by using expansions of the fields in terms of creation and annihilation operators and by fixing the value of the quantum corrections we can get a Virasoro algebra with a central extension. By choosing $\xi_Z = \xi_Y = \xi - \frac{\hbar_*}{8\pi\xi}$ we obtain the central charges:

$$c_Z = -\left(\tfrac{1}{12} + \tfrac{4\pi}{\hbar_*}\left(\xi - \tfrac{\hbar_*}{8\pi\xi}\right)^2\right) r^3 - \tfrac{1}{6}r\ , \tag{16}$$

$$c_Y = \left(\tfrac{1}{12} + \tfrac{4\pi}{\hbar_*}\left(\xi - \tfrac{\hbar_*}{8\pi\xi}\right)^2\right) r^3 + \tfrac{1}{6}r\ , \tag{17}$$

Note how the central charges in the Z and Y sector are identical in absolute value but appear with opposite signs. When summing up all contributions to the central charge the c_Z and c_Y cancel out and the only contributions are then given by the ghosts and the scalars:

$$c = \frac{N-26}{12}r^3 + \frac{N+1}{6}r\ . \tag{18}$$

The central charge c breaks the Virasoro algebra at the quantum level. However, it is possible to eliminate the r^3 term in c by tuning the number of scalars to $N = 26$, in the same way as the number of space-time dimensions is tuned in bosonic string theory. In this way the cubic term of the central charge is cancelled and the remaining linear term can be reabsorbed with a shift of the zero modes $L_0^{\pm} \to L_0^{\pm} - \hbar_* 9/4$ so that the quantum Virasoro algebra is closed.

Such a shift of the zero modes is nothing else than the contributions of the quantum vacuum fluctuations to the total energy of the system, which are finite in two dimensions.

4. Quantum constraints

Once the quantum Virasoro algebra is obtained, it is possible to find the quantum realization of the constraints on Hilbert space, following the usual Dirac prescription that physical states have

to be annihilated by the constraints. As a matter of fact the cosmological constant Λ and the coupling constants ξ and q are still free parameters: by requiring certain quantum states to be physical Λ can be constrained to take a specific value.
The presence of the Liouville potentials involving the Z and the Y fields prevents one from following the standard string theory approach, *i.e.*, extracting the modes $L_n^\pm$ of the quantum constraints with a discrete Fourier transform and looking at $L_n^\pm|\psi_{phys}\rangle = 0$ with $n = 0, 1, 2$. However, it is sufficient to look the weaker condition:

$$\langle\psi_{phys}|L^\pm(s)|\psi_{phys}\rangle = 0. \tag{19}$$

Under the requirement that a given quantum state of the Universe is physical, we can determine the value that the cosmological constant has to take for the quantum constraints to be satisfied. Considering linear combinations of the shifted Virasoro generators, the quantum constraints for an arbitrary quantum physical state will be:

$$\langle L^+ + L^-\rangle = 0\ , \qquad \langle L^+ - L^-\rangle = 0\ , \qquad \langle L^\emptyset\rangle = 0\ , \tag{20}$$

with:

$$L^\pm = L^{\pm,Z} + L^{\pm,Y} + \sum_{n=1}^{N} L^{\pm,\phi_n} - c_m/2\ , \tag{21}$$

inclusive of the quantum corrected coupling constants $\xi_Z = \xi_Y$ and the shift of the zero modes determined above. The ghost sector is omitted by taking advantage of the BRST invariance. As the cosmological constant enters the expressions for $L^\pm$ only through the Liouville potential terms (cf. (9) (10)), which are identical for the + and − cases, only $\langle L^+ + L^-\rangle$ will depend on Λ. We have:

$$\begin{aligned}\langle L^+ + L^-\rangle = &-\frac{1}{2}\langle P_Z^2 + Z_{,s}^2\rangle + \frac{1}{2}\langle P_Y^2 + Y_{,s}^2\rangle + \sum_{n=1}^{N}\frac{1}{2}\langle P_n^2 + \phi_{n,s}^2\rangle + \\ &+2\xi\left(\langle Z_{,ss}\rangle - \langle Y_{,ss}\rangle\right) - \frac{\hbar_*}{4\pi\xi}\left(\langle Z_{,ss}\rangle - \langle Y_{,ss}\rangle\right) - \\ &-c_m + \Lambda\left[2\langle e^{Z/\xi}\rangle + 2q\langle e^{Y/\xi}P_{A_1}^2\rangle\right],\end{aligned} \tag{22a}$$

$$\begin{aligned}\langle L^+ - L^-\rangle = &\langle P_Z Z_{,s}\rangle + \langle P_Y Y_{,s}\rangle + \sum_{n=1}^{N}\langle P_{\phi_n}\phi_{n,s}\rangle - \\ &-2\left(\xi - \frac{\hbar_*}{8\pi\xi}\right)\left(\langle P_{Z,s}\rangle - \langle P_{Y,s}\rangle\right)\ ,\end{aligned} \tag{22b}$$

$$\langle L^\emptyset\rangle = \langle P_{A_1,s}\rangle\ , \tag{22c}$$

where $c_m = \hbar_* 9/2$. A qualitative analysis of the quantum constraints reveals interesting features. By solving $22a$ for the cosmological constant we can see that contributions to Λ are of three kinds (up to the factor in square brackets): classical contributions (the first line in ($22a$)), quantum fluctuations of the vacuum (the c_m term) and additional quantum contributions determined by the quantum corrections to the coupling constant (the second term in the second line of ($22a$)). Schematically we can write the cosmological constant as:

$$\Lambda = -\left[\langle Kin_{grav}\rangle + \langle Kin_\phi\rangle + 2\left(\xi - \frac{\hbar_*}{8\pi\xi}\right)\langle\partial^2_{grav}\rangle - \delta_m\right]\langle Liouville\rangle^{-1} \tag{23}$$

The QFT approach is recovered if all fields are taken to be in the vacuum state, so that only vacuum fluctuations of ghosts and scalars (finite in two dimensions when the regularization is

removed) are left in the square brackets and the Liouville potential factor is unity.
It is more interesting, however, to look at the cosmological constant for a generic state away from the vacuum, when all terms can contribute to the value of Λ. We can expect the Liouville potentials, which are exponential in the fields and the inverse of the coupling constant ξ, to be very sensitive to excitations of the gravitational sector and/or small values of ξ itself. For instance in the case of small ξ we have $\langle Liouville \rangle \gg 0$, suppressing all terms in square brackets in (23). Let us choose $\xi = 0.0035$ and let us require to be physical a quantum state in which all fields but Z are in the vacuum and $\langle Z \rangle = 1$. Then (22c) and (22c) vanish identically and (22a) gives:

$$\Lambda \sim 1.85 \times 10^{-124} \ . \tag{24}$$

This quantum state is nothing else than the vacuum used in the QFT approach supplemented by an excitation of the one gravitational degree of freedom in (1) (see [6] for details). A physical interpretation of ξ can be given from the condition (5) for the potential $U(X)$. Going a step back, the dilaton field X in two-dimensional dilaton gravity can be seen, generally speaking, as the field that encodes the collective behaviour of the degrees of freedom that have been integrated out in the reduction process from higher dimensions: *e.g.* in spherically reduced gravity X is what is left of the angular degrees of freedom. On the other hand the single physical degree of freedom contained in the two-dimensional metric can always be taken to be a conformal mode, rescaling a Minkowski metric. From (5) we can see that ξ regulates the relative weight of the kinetic term for the dilaton in the dilaton gravity action (1). A larger value for ξ corresponds to a more "frozen" dynamics for X w.r.t. the dynamics of the two-dimensional metric, while for small ξ the kinetic term for X dominates the action. In the same way, from (6), we can see that a smaller ξ corresponds to a stronger coupling of the dilaton with the Maxwell sector. Explicitly, for $\xi \ll 1$, we can rescale $X \to \xi X$ and $q \to q\xi^2$ and the classical action (1) can be approximated as:

$$S_{\xi\ll 1} = \int_{\mathcal{M}} dx^2 \sqrt{-g} \left(-\frac{1}{2} X_{,\mu} X^{,\mu} - 2\Lambda - \frac{e^{X/\xi}}{16 q \Lambda} F_{\mu\nu} F^{\mu\nu} + O(\xi) \right) \ , \tag{25}$$

which at the zeroth order is a massless scalar non-linearly coupled to a Maxwell field on a conformally flat background. In our cosmological observations this seems a good approximation: at large scales the Universe is well described by a Minkowski metric and considering the short time scale during which observation are made we can approximate the conformal mode/scale factor as a constant. The remaining degrees of freedom then dominate the dynamics, justifying a choice of $\xi \ll 1$.
On the other hand for small ξ an important contribution to Λ can be given by the second of the $\langle \partial^2_{grav} \rangle$ terms, generated by the quantum corrections to the coupling constant. This suggests that quantum gravity contributions to the cosmological constant can play a fundamental role. If we take a closer look at the specific form of the contributions to Λ in (22a) we notice that interestingly the Z and Y field appear with opposite signs in all non-exponential terms, so that $\langle Kin_{grav} \rangle$, $\langle \partial^2_{grav} \rangle$ can have both positive or negative, allowing for cancellations of the c_m contributions to Λ. Let us also point out that the Maxwell field A_1 exhibits a continuous spectrum in this model.

If we take the leap of thinking the models treated here as effective descriptions of highly symmetrical four-dimensional theories we can consider the possibility similar features to appear, including quantum gravity contributions to Λ and excitations of the gravitational degrees of freedom that would renormalize the value of the cosmological constant in the fashion of $\langle Liouville \rangle \gg 0$.

5. Discussion

In this work we discussed the cosmological constant problem in quantum gravity, with a particular focus on the case of two-dimensional models, intended as an effective description of highly symmetric higher dimensional models.

By requiring the expectation values of the quantum constraints to vanish for generic quantum states we were able to see that the value of the cosmological constant is determined not only by the vacuum fluctuations of the quantum fields (both in the matter and gravitational sectors, even if cancellations occur), but also by dynamical contributions, including quantum corrections to effective coupling constants, and a renormalization factor of gravitational origin. In particular for small values of the effective coupling constant ξ, which we argue to be an approximate description of the regime in which we can measure the cosmological constant, the renormalization factor can be very small, resulting in a very small value for Λ in natural units. The usual result of QFT of a cosmological constant proportional to the vacuum energy is recovered if the state required to be physical is the vacuum of the theory.

While the validity of this result is restricted to the specific framework of the two-dimensional dilaton gravity models considered (even though they can be motivated from higher dimensional considerations), it is a non-trivial example that quantum gravity can in fact contribute in a fundamental way to the value of the cosmological constant. In particular we cannot exclude the possibility of non-perturbative contributions in higher dimensions, such as the quantum corrections to the coupling constants, that can deeply change the picture.

At this point we are able to determine Λ for any given quantum states. If we believe that this approach might in fact predict a meaningful value for the cosmological constant, we are now faced with the issue of determining which one is the quantum state the Universe is in.

Acknowledgments

SZ benefits from a PhD research grant of the Institut Interuniversitaire des Sciences Nucléaires (IISN, Belgium); his work is supported by the Belgian Federal Office for Scientific, Technical and Cultural Affairs through the Interuniversity Attraction Pole P6/11.

References

[1] Carroll S M 2001 *Living Rev.Rel.* **4** 1 (*Preprint* astro-ph/0004075)
[2] Bousso R 2012 *Preprint* 1203.0307
[3] Perlmutter S *et al.* (Supernova Cosmology Project) 1999 *Astrophys. J.* **517** 565–86 (*Preprint* astro-ph/9812133)
[4] Riess A G *et al.* (Supernova Search Team) 1998 *Astron. J.* **116** 1009–38 (*Preprint* astro-ph/9805201)
[5] Govaerts J and Zonetti S 2011 *Class. Quantum Grav.* **28** 185001 (*Preprint* 1102.4957v1)
[6] Zonetti S and Govaerts J 2012 *J. Phys. A: Math. Theor.* **45** (2012) 042001 (*Preprint* 1111.1612)
[7] Zonetti S 2012 *Two-dimensional quantum dilaton gravity and the quantized cosmological constant* Ph.D. thesis Université catholique de Louvain (*Preprint* 1209.4472)
[8] Govaerts J and Zonetti S 2012 *Preprint* 1212.4538
[9] Witten E 1991 *Phys. Rev.* D **44** 314–24
[10] Callan C, Giddings S, Harvey J and Strominger A 1992 *Phys. Rev.* D **45** R1005
[11] Nakayama Y 2004 *Int. J. Mod. Phys.* A **19** 2771–930 (*Preprint* hep-th/0402009)
[12] Govaerts J 1991 *Hamiltonian Quantisation and Constrained Dynamics* (*Leuven Notes in Mathematical and Theoretical Physics* vol 4) (Leuven: Leuven University Press)
[13] Polchinski J 1998 *String Theory Vol. 1: An Introduction to the Bosonic String* (Cambridge, UK: Cambridge Univ. Pr.)
[14] Green M B, Schwarz J H and Witten E 1987 *Superstring Theory Vol. 1: Introduction* Cambridge Monographs On Mathematical Physics (Cambridge, UK: Cambridge Univ. Pr.)
[15] Curtright T L and Thorn C B 1982 *Phys. Rev. Lett.* **48** 1309–13

Short review on noncommutative spacetimes

A Pachoł

Science Institute, University of Iceland, Dunhaga 3, 107 Reykjavik, Iceland

E-mail: pachol@hi.is

Abstract. This review presents noncommutative spacetimes as one of the approaches to Planck scale physics, with the main assumption that at this energy scale spacetime becomes quantized. Spacetime coordinates become noncommutative as observables in Quantum Mechanics. The basic elements of Drinfeld twist deformation theory are reminded. The Hopf algebra language provides natural framework for deformed relativistic symmetries which constitute Quantum Group of symmetry and noncommutative spacetime is in fact Hopf module algebra. The notion of realization for noncommutative coordinates in terms of differential operators is also presented.

1. Introduction

The so called "Planck scale" is the scale at which gravitational effects are equivalently strong as quantum ones and it relates to either a very big energy scale or equivalently to a tiny size scale. There are many theories trying to consistently describe that region. One of the possible assumptions is that below Planck scale spacetime has more general structure, a noncommutative one where (as with quantum mechanics phase-space) uncertainty relations naturally arise. A natural quantization of manifolds can be described in the language of Noncommutative Geometry. Motivation for this fact starts with Gelfand-Naimark theorem [1] which states that there is a one-to-one relation between certain commutative algebras and certain spaces. One can say that the idea of Noncommutative Geometry is to consider noncommutative algebras as noncommutative geometric spaces, i.e. to algebralize geometric notions and then generalize them to noncommutative algebras. Minkowski spacetime can be seen in that picture as well.

Geometrically Minkowski spacetime is a 4-dim affine space over vector space equipped with a nondegenerate, symmetric bi-linear form with Lorentzian signature $\eta_{\mu\nu}$. The position of an event in spacetime is given by point $p = (x^0, x^1, x^2, x^3)$.

However, algebraically Minkowski "spacetime" is an Abelian algebra $\mathcal{A} = \mathfrak{x}^4$ of coordinate functions $x^\mu(p)$ on 4-dim real vector space described by commutation relations: $[x^\mu, x^\nu] = 0$. Algebra $\mathcal{A} = \mathfrak{x}^4 \equiv Poly(x^\mu) \equiv \mathbb{C}[x^0, \ldots, x^3]$ of spacetime coordinates x^μ.

At the Planck scale classical Minkowski spacetime (as commutative algebra) becomes quantized and it is described by noncommutative algebra, i.e. $x_\mu \to \hat{x}_\mu$. We get in such a way noncommutative model of quantum space-time $[x_\mu\, x_\nu] = 0 \longrightarrow [\hat{x}_\mu\, \hat{x}_\nu] \neq 0$. For example one can get the so-called Moyal-Weyl (θ) spacetime $\mathcal{A}^\theta$ [2],[3] described by:

$$[\hat{x}^\mu, \hat{x}^\nu] = ih\theta^{\mu\nu} \tag{1}$$

with the deformation parameter h which is of length^2 dimension and it is usually related with Planck length L_P. Another option is to consider the so-called Lie-algebraic one

$$[\hat{x}^\mu, \hat{x}^\nu] = ih\theta^{\mu\nu}_\rho \hat{x}^\rho \tag{2}$$

deformation parameter h of mass dimension which could correspond, e.g. to Planck mass or Quantum Gravity scale. Its special example is the so-called κ-Minkowski spacetime $\mathcal{A}^{\kappa}$ [4],[5]:

$$[\hat{x}^0, \hat{x}^k] = ih\hat{x}^k; \qquad [\hat{x}^i, \hat{x}^j] = 0 \tag{3}$$

(where one usually takes $h = \frac{1}{\kappa}$, hence the name of this type of noncommutative spacetime). Those two types ($\mathcal{A}^{\theta}$ and $\mathcal{A}^{\kappa}$) of noncommutative spacetimes have been widely investigated in the literature, since via the deformation parameter, e.g $\kappa \sim M_P$, they can be naturally connected with the Planck scale regime.

In general such noncommutative algebras can be obtained via deformation procedure from commutative ones. Quantum deformations, which lead to noncommutative algebras (as noncommutative spacetimes), are connected with the Quantum Groups (Hopf algebras) formalism, which constitutes the description of deformed symmetries. Quantum Group as generalized symmetry is described as Hopf algebra $\mathcal{H}(m, \eta, \Delta, \epsilon, S)$. It is composed by an unital associative algebra $(\mathcal{H}, m, \eta)$ and counital coassociative coalgebra $(\mathcal{H}, \Delta, \epsilon)$. Δ and ϵ are algebra homomorphisms, m and η are coalgebra homomorphisms. The simplest example of Hopf algebra is provided by universal enveloping algebra for given Lie algebra $\mathfrak{g}$. The universal enveloping algebra $\mathcal{U}_{\mathfrak{g}}$ can be equipped with the primitive coproduct: $\Delta_0(u) = u \otimes 1 + 1 \otimes u$, counit: $\epsilon(u) = 0$, $\epsilon(1) = 1$ and antipode: $S_0(u) = -u$, $S_0(1) = 1$, for $u \in \mathfrak{g}$, and extending them by multiplicativity property to the entire $\mathcal{U}_{\mathfrak{g}}$. Recall that the universal enveloping algebra is a result of the factor construction $\mathcal{U}_{\mathfrak{g}} = T\mathfrak{g}/J_{\mathfrak{g}}$ where $T\mathfrak{g}$ denotes tensor (free) algebra of the vector space $\mathfrak{g}$ quotient out by the ideal $J_{\mathfrak{g}}$ generated by elements $\langle X \otimes Y - Y \otimes X - [X, Y]\rangle$: $X, Y \in \mathfrak{g}$. Noncommutative spacetime $\mathcal{A}$ in this framework is a Hopf module algebra over Hopf algebra $\mathcal{H}$ with the (left) module action $\triangleright : \mathcal{H} \otimes \mathcal{A} \to \mathcal{A}$; $L \triangleright f$ of $L \in \mathcal{H}$ on $f \in \mathcal{A}$ satisfying: $L \triangleright (f \cdot g) = (L_{(1)} \triangleright f) \cdot (L_{(2)} \triangleright g)$ (for more details see, e.g., [6, 7]). One usually says that such noncommutative spacetime stays in this way invariant under quantum group of transformations in analogy to the classical case.

Hopf module algebras and the Hopf algebras can be deformed e.g. by relevant twisting element which leads to noncommutative spacetime as covariant quantum space over deformed group of symmetry. For Hopf algebra $\mathcal{H}$, we can consider the twisting two-tensor $\mathcal{F}$ (the so-called Drinfeld twist), as an invertible element in $\mathcal{H} \otimes \mathcal{H}$ such that:

$$\mathcal{F} = \mathrm{f}^{\alpha} \otimes \mathrm{f}_{\alpha} \in \mathcal{H} \otimes \mathcal{H} \qquad \text{and} \qquad \mathcal{F}^{-1} = \bar{\mathrm{f}}^{\alpha} \otimes \bar{\mathrm{f}}_{\alpha} \in \mathcal{H} \otimes \mathcal{H} \tag{4}$$

As a result of deformation quantized Hopf algebra $\mathcal{H}^{\mathcal{F}}$ has non-deformed algebraic sector (commutators), while coproducts and antipodes are subject of the deformation:

$$\Delta^{\mathcal{F}}(\cdot) = \mathcal{F}\Delta(\cdot)\mathcal{F}^{-1}, \qquad S^{\mathcal{F}}(\cdot) = uS(\cdot)u^{-1}, \tag{5}$$

where $u = \mathrm{f}^{\alpha}S(\mathrm{f}_{\alpha})$. $\mathcal{F}$ satisfies the 2-cocycle and normalization conditions [8, 9]:

$$\mathcal{F}_{12}(\Delta \otimes \mathrm{id})\mathcal{F} = \mathcal{F}_{23}(\mathrm{id} \otimes \Delta)\mathcal{F}, \qquad (\epsilon \otimes \mathrm{id})\mathcal{F} = 1 = (\mathrm{id} \otimes \epsilon)\mathcal{F}, \tag{6}$$

which guarantee co-associativity of the deformed coproduct $\Delta^{\mathcal{F}}$ and associativity of the corresponding twisted star-product in the twisted module algebra $\mathcal{A}^{\mathcal{F}}$:

$$f \star g = m \circ \mathcal{F}^{-1} \triangleright (f \otimes g) = (\bar{\mathrm{f}}^{\alpha} \triangleright f) \cdot (\bar{\mathrm{f}}_{\alpha} \triangleright g) \tag{7}$$

for $f, g \in \mathcal{A}$ and the Hopf action $\triangleright$ remains unchanged. $\mathcal{A}^{\mathcal{F}}$ can be represented by deformed $\star$-commutation relations

$$[x^{\mu}, x^{\nu}]_{\star} \equiv x^{\mu} \star x^{\nu} - x^{\nu} \star x^{\mu} = ih\theta^{\mu\nu}(x) \equiv ih(\theta^{\mu\nu} + \theta^{\mu\nu}_{\lambda}x^{\lambda} + \ldots) \tag{8}$$

replacing the nondeformed (commutative) one: $[x^{\mu}, x^{\nu}] = 0$ where the coordinate functions (x^{μ}) play a role of generators for the corresponding algebras: deformed and nondeformed. Instead of $[x^{\mu}, x^{\nu}]_{\star}$ one sometimes writes $[\hat{x}^{\mu}, \hat{x}^{\nu}]$, as in e.g. (1) or (3).

Journal of Physics: Conference Series **442** (2013) 012039 doi:10.1088/1742-6596/442/1/012039

From the general framework [8], a twisted deformation of Lie algebra $\mathfrak{g}$ requires a topological extension of the corresponding enveloping algebra $\mathcal{U}_\mathfrak{g}$ into an algebra of formal power series $\mathcal{U}_\mathfrak{g}[[h]]$ in the formal parameter h, providing the so-called h-adic topology (see e.g., [6, 7, 9]), it is mainly due to the fact that twisting element has to be invertible. The Hopf module algebra $\mathcal{A}$ has to be extended by h-adic topology to $\mathcal{A}[[h]]$ as well and then deformed into $\mathcal{A}^{\mathcal{F}}[[h]]$. There is a correspondence between twisting element, which can be rewritten as a power series expansion

$$\mathcal{F} = 1 \otimes 1 + \sum_{\alpha=1}^{\infty} h^\alpha \mathrm{f}^\alpha \otimes \mathrm{f}_\alpha \qquad \text{and} \qquad \mathcal{F}^{-1} = 1 \otimes 1 + \sum_{\alpha=1}^{\infty} h^\alpha \bar{\mathrm{f}}^\alpha \otimes \bar{\mathrm{f}}_\alpha, \tag{9}$$

$\mathrm{f}^\alpha, \mathrm{f}_\alpha, \bar{\mathrm{f}}^\alpha, \bar{\mathrm{f}}_\alpha \in \mathcal{U}_\mathfrak{g}$; h is deformation parameter, and classical r-matrix $\mathfrak{r} \in \mathfrak{g} \wedge \mathfrak{g}$ satisfying classical Yang–Baxter equation and universal (quantum) r-matrix $\mathcal{R}$:

$$\mathcal{R} = \mathcal{F}^{21}\mathcal{F}^{-1} = 1 + h\mathfrak{r} \mod (h^2) \tag{10}$$

satisfying quantum Yang–Baxter equation. Classical r-matrices classify non-equivalent deformations. We distinguish between two types of the classical r-matrix:

1) Abelian one, if it has the form

$$r_A = \sum_{i=1}^{n} a_i \wedge b_i \tag{11}$$

where all elements a_i, b_i $(i = 1, ..., n)$ commute among themselves. The corresponding twist is given as follows

$$F_{r_A} = exp(\frac{1}{2} r_A). \tag{12}$$

2) Another situation appears when the classical r-matrix $r = r_{J_n}(h)$ has the form

$$r_{J_n} = h\Big(\sum_{v=0}^{n} b_v \wedge a_v\Big), \tag{13}$$

where the elements a_v, b_v $(v = 0, 1, \ldots, n)$ satisfy the relations

$$\begin{aligned} [a_0, b_0] &= b_0, \quad [a_0, a_i] = (1 - t_i)a_i, \quad [a_0, b_i] = t_i b_i, \\ [a_i, b_j] &= \delta_{ij} b_0, \quad [a_i, a_j] = [b_i, b_j] = 0, \quad [b_0, a_j] = [b_0, b_j] = 0, \end{aligned} \tag{14}$$

$(i, j = 1, \ldots, n)$, $(t_i \in \mathbb{C})$. Such an r-matrix is called of Jordanian type. The corresponding twist is given as follows (see e.g. [10])

$$F_{r_{J_n}} = \exp\Big(h \sum_{i=1}^{n} a_i \otimes b_i \, e^{-t_i \sigma}\Big) \exp(a_0 \otimes \sigma), \tag{15}$$

where $\sigma := \ln(1 + hb_0)$.

One can notice that the zero component $r_{J_0}(h) := hb_0 \wedge a_0$ in (13) is the classical Jordanian r-matrix and the corresponding Jordanian twist is given by the formula (15) for $n = 0$, i.e. $F_{r_{J_0}} = \exp(a_0 \otimes \sigma)$ (and this shorter form will be used later on).

2. Quantum symmetries

The Poincaré symmetry is the full symmetry of special relativity and it includes translations, rotations and boosts. Algebraically it is described by the Poincaré Lie algebra (Lorentz generators $M_{\mu\nu}$ and momenta P_μ), usually denoted as $\mathfrak{iso}(1,3)$, and defined by the following commutation relations

$$[M_{\mu\nu}, M_{\lambda\rho}] = i\left(M_{\mu\lambda}\eta_{\nu\rho} - M_{\mu\rho}\eta_{\nu\lambda} - M_{\nu\lambda}\eta_{\mu\rho} + M_{\nu\rho}\eta_{\mu\lambda}\right), \tag{16}$$

$$[M_{\mu\nu}, P_\lambda] = i\left(P_\nu\eta_{\mu\lambda} - P_\mu\eta_{\nu\lambda}\right), \tag{17}$$

$$[P_\mu, P_\nu] = 0, \tag{18}$$

where $\eta_{\mu\nu}$ is the metric tensor with Lorentzian signature. As it is known (and shown above) any Lie algebra provides an example of undeformed Hopf algebra. First one has to extend it to universal enveloping algebra and then equip it in comultiplication, counit and antipode maps. Therefore, the universal enveloping algebra of the Poincaré Lie algebra $\mathcal{U}_{\mathfrak{iso}(1,3)}$ together with

$$\Delta_0 M_{\mu\nu} = M_{\mu\nu}\otimes 1 + 1\otimes M_{\mu\nu} \quad \text{and} \quad \Delta_0 P_\mu = P_\mu\otimes 1 + 1\otimes P_\mu \tag{19}$$

$$S(M_{\mu\nu}) = -M_{\mu\nu}; \qquad S(P_\mu) = -P_\mu; \qquad \epsilon(M_{\mu\nu}) = \epsilon(P_\mu) = 0 \tag{20}$$

defined on the generators and then extended to the whole $\mathcal{U}_{\mathfrak{iso}(1,3)}$ constitutes the undeformed Poincaré Hopf algebra. Such Quantum Group can be further deformed.

First deformations of Poincaré symmetry appeared in the early 90's. The simplest is the so-called θ-deformation corresponding to the Moyal-Weyl noncommutative spacetime (1) and it can be obtained by twist deformation:

$$\mathcal{F} = \exp\left(-\frac{i}{2}h\theta^{\mu\nu}P_\mu\otimes P_\nu\right); \qquad \mathcal{F}^{-1} = \exp\left(\frac{i}{2}h\theta^{\mu\nu}P_\mu\otimes P_\nu\right) \tag{21}$$

where $\theta^{\mu\nu}$ is a constant antisymmetric matrix. We say that this twist has support in Poincare algebra, i.e. $\mathcal{F}\in\mathcal{U}_{\mathfrak{iso}(1,3)}\otimes\mathcal{U}_{\mathfrak{iso}(1,3)}$. After twisting the Hopf algebra structure becomes twisted Poincaré algebra:

$$\Delta_\theta\left(P_\mu\right) = \Delta_0\left(P_\mu\right); \qquad \Delta_\theta\left(M_{\mu\nu}\right) = \Delta_0(M_{\mu\nu}) - h(P\theta)_\mu\wedge P_\nu + h(P\theta)_\nu\wedge P_\mu \tag{22}$$

where $(P\theta)_\mu = P_\rho\theta^{\rho\lambda}\eta_{\lambda\mu}$. One can easily notice that for the deformation parameter $h\to 0$ the coalgebra becomes undeformed (19) as classical limit should provide. It has been shown (e.g. see [3]) that the θ-spacetime $\mathcal{A}^\theta$ (1) stays covariant under θ- deformed version of the Poincaré quantum group. The module algebra $\mathcal{A}$ is also deformed accordingly into $\mathcal{A}^{\mathcal{F}} = \mathcal{A}^\theta$ (the module action stays the same). In $\mathcal{A}^\theta$ we have new ($\star$) multiplication (7) which leads to the algebra described by (1).

The other type deformation (not coming from twist) is provided by the so-called κ-deformation of Poincaré algebra [11, 12, 13] which originally was obtained by contraction procedure from q-deformed $SO_q(3,2)$. In the deformed case the κ−Poincaré $\mathcal{U}_{\mathfrak{iso}(1,3)}{}^\kappa$ algebra consists of (16) - (18) as in undeformed case but coalgebra structure is no longer primitive and has the form:

$$\Delta_\kappa(M_i) = \Delta_0(M_i) = M_i\otimes 1 + 1\otimes M_i, \qquad \Delta_\kappa(N_i) = N_i\otimes 1 + \Pi_0^{-1}\otimes N_i - \frac{1}{\kappa}\epsilon_{ijm}P_j\Pi_0^{-1}\otimes M_m \tag{23}$$

$$\Delta_\kappa(P_i) = P_i\otimes\Pi_0 + 1\otimes P_i, \qquad \Delta_\kappa(P_0) = P_0\otimes\Pi_0 + \Pi_0^{-1}\otimes P_0 + \frac{1}{\kappa}P_m\Pi_0^{-1}\otimes P^m, \tag{24}$$

$$S_\kappa(M_i) = -M_i, \qquad S_\kappa(N_i) = -\Pi_0 N_i - \frac{1}{\kappa}\epsilon_{ijm}P_jM_m, \qquad S_\kappa(P_i) = -P_i\Pi_0^{-1} \tag{25}$$

The counit ϵ is undeformed, i.e., $\epsilon(A) = 0$ for $A\in(M_i, N_i, P_i)$. We have used the notation: $M_{0i} = iN_i$; $M_{ij} = \epsilon_{ijk}M_k$; $\Pi_0 = \left(\frac{P_0}{\kappa} + \sqrt{1-\frac{P_\mu P^\mu}{\kappa^2}}\right)$.

Such quantum group constitutes deformed symmetry for κ-Minkowski spacetime defined by the commutation relations (3). The above Hopf algebra is given in the so-called classical basis [14], but there exist other bases in which κ-Poincare algebra looks different and even the commutation relations are modified (are no longer of (16) - (18) form). For the so-called bicrossproduct basis see e.g. [13] or for the so-called standard basis see e.g. [11]. Depending on the purpose of application one can use different forms of this quantum group, for example in κ-deformed quantum field theories (see e.g. [15]), doubly special relativity (see e.g. [16]) or in the most recent relative-locality effect (see e.g. [17]) the bicrossproduct basis form was used the most extensively. κ -Minkowski spacetime defined by the relations (3) constitutes a Hopf module algebra over κ-Poincare algebra defined by (16) - (18), (23) - (25).

The κ -deformation of Poincaré algebra is characterised by the inhomogeneous classical YB equation, which implies that one should not expect to get $\kappa-$Minkowski space from a Poincaré twist. However, twists belonging to extensions of the Poincaré algebra are not excluded, e.g. one can consider the twists with the support in $\mathcal{U}_{\mathfrak{igl}}$ [18, 19] or in the Weyl-Poincaré algebra [20, 21], the minimal one-generator extension of the Poincaré algebra which includes dilatation generator. For those twists we also obtain κ-Minkowski spacetime (3) however no longer as κ-Poincare module algebras, i.e. with different Hopf algebra as quantum symmetry.

3. Realizations and representations

Another commonly used feature in the noncommutative spacetimes approach is the realization for the noncommutative coordinates. One can express noncommutative coordinates as operators acting on algebra of functions in analogy to Quantum Mechanics when observables of position and momenta become operators on a Hilbert space. The Heisenberg algebra (i.e. quantum mechanical phase space) H can be defined as a free algebra of n coordinate generators and n generators of momenta, satisfying the following relations:

$$\left[x_\mu, x_\nu\right] = 0; \qquad \left[P_\mu, x_\nu\right] = -i\eta_{\mu\nu}\cdot 1; \qquad \left[P_\mu, P_\nu\right] = 0; \tag{26}$$

where $\eta_{\mu\nu} = (-, +, +, +)$ is diagonal metric tensor with Lorentzian signature.

Now for example let's consider $\mathfrak{g} = \mathfrak{igl}(n, \mathbb{R}) = \mathfrak{gl}(n, \mathbb{R}) \ni \mathfrak{t}^n$ of the inhomogeneous general linear group as a semidirect product of $\mathfrak{gl}(n, \mathbb{R})$ with translations $\mathfrak{t}^n$. We choose a basis $L^\mu_\nu, P_\mu \in \mathfrak{igl}(n, \mathbb{R})$ with the following standard set of commutation relations:

$$[L^\mu_\nu, L^\rho_\lambda] = -i\delta^\rho_\nu L^\mu_\lambda + i\delta^\mu_\lambda L^\rho_\nu\,, \qquad [L^\mu_\nu, P_\lambda] = i\delta^\mu_\lambda P_\nu \tag{27}$$

$\mu, \nu, \ldots = 0, \ldots, n-1$ $\mathfrak{igl}(n, \mathbb{R})$ as Lie algebra acting on $\mathcal{A}$ (algebra of (complex-valued) functions on the spacetime manifold $\mathbb{M} = \mathbb{R}^n$) via first-order differential operators (derivations $\equiv$ vectors fields). It has the following realization: $L^\nu_\mu = x^\nu P_\nu$ [1]. The action of differential operators on functions induced by derivations (vector fields) remains the same in deformed and nondeformed cases.

Therefore by defining the map $L^\nu_\mu \to x^\nu P_\nu$ which is a Lie algebra isomorphism we obtain "Heisenberg algebra realization" of $\mathcal{U}_{\mathfrak{igl}(n)}$. Moreover the Heisenberg realization described above induces Heisenberg representation via $P_\mu = -\imath\partial_\mu = \imath\frac{\partial}{\partial x^\mu}$; $\quad L^\nu_\mu = -\imath x^\nu\partial_\mu$ acting in the vector space $C^\infty(\mathbb{R}^n)$ [2].

In noncommutative spacetimes approach the noncommutative coordinates can be introduced as operators acting on algebra of functions as well. It comes from the fact that the twisted star product

[1] We will use the same letter to denote an abstract element in $\mathfrak{igl}(n)$ and its (first-order) differential operator realization in $\mathcal{X}\mathbb{M}$ (i.e. assuming the embedding of the $U_{\mathfrak{igl}(n)} \hookrightarrow U_{\mathcal{X}\mathbb{M}}$).

[2] The best known representations are given on the space of (smooth) functions on $\mathbb{R}^n$ in terms of multiplication and differentiation operators. For this reason one can identify Heisenberg algebra with an algebra of linear differential operators on $\mathbb{R}^n$ with polynomial coefficients.

enables us to 'realize' the algebra $\mathcal{A}^{\mathcal{F}}$ in terms of (formal) differential operators on a manifold $\mathbb{M}$, i.e. in algebra of vector fields $\mathcal{X}\mathbb{M}$. Operator realizations are naturally defined by $\hat{x} \in \mathcal{U}_{\mathcal{X}\mathbb{M}}$:

$$\hat{x}^\mu = x^\mu + \sum_{m=1}^{\infty} h^m (\bar{\mathrm{f}}^\alpha \triangleright x^\mu) \cdot \bar{\mathrm{f}}_\alpha \tag{28}$$

which can be written also as $\hat{x}^\mu(g) = x^\mu \star g, \qquad g \in \mathcal{A}$.

Summarising, one can say that we deal with the realization of given algebra in space of linear operators over $\mathfrak{h}$, i.e. $\mathscr{L}(\mathfrak{h})$ (where $\mathfrak{h}$ is the Hilbert space) when we determine homomorphism from one algebra to another. Hence the name, e.g. "Heisenberg algebra realization". When linear space is determined we speak about representations (for more detailed analysis and examples, see e.g. [22]).

For example in θ-deformation (corresponding to θ-twist (21)) the (Heisenberg) realization for coordinates is the following:

$$\hat{x}^\mu = x^\mu - \frac{1}{2} h\theta^{\mu\nu} P_\nu \tag{29}$$

One can easily check that such coordinates x^μ satisfy commutation relations (1) by using only (26).

The noncommutative κ–Minkowski space (3) also can be realized in terms of generalized differential operators. In fact there exist a huge amount of such Heisenberg realizations of the κ -Minkowski algebra. Particularly important is the so-called non-covariant family of realizations [19, 23, 24]:

$$\hat{x}^i = x^i \phi(A), \qquad \hat{x}^0 = x^0 \psi(A) + \imath h x^k \partial_k \gamma(A) \tag{30}$$

where $A = ih\partial_0$. Functions ϕ, ψ, and γ are taken to be real analytic obeying initial conditions $\phi(0) = 1$ and $\psi(0) = 1$, and $\gamma(0)$ has to be finite in order to ensure a proper classical limit at $h \mapsto 0$. The κ–Minkowski commutation relations [cf. (3)] are equivalent to the property that functions ϕ, ψ, and γ do satisfy the equation [23]:

$$\gamma = 1 + (ln\phi)'\psi \tag{31}$$

where $\phi' \equiv \frac{d\phi}{dA}$ There are few examples of twists providing κ-Minkowski spacetime (3) with support in $\mathcal{F} \in \mathcal{U}_{\mathfrak{igl}} \otimes \mathcal{U}_{\mathfrak{igl}}$ for which the realization of noncommutative coordinates fall into the non-covariant class (see e.g. [18, 19, 20]).

The simplest twist is of Abelian type $(\mathcal{F}_s^A)^{-1} = \exp\left[-\imath h\,(sP_0 \otimes D - (1-s)\,D \otimes P_0)\right]$ (with s being a numerical parameter labelling different twisting tensors) which provides the following realizations for noncommutative coordinates:

$$\hat{x}_s^i = x^i e^{(1-s)hP_0}, \qquad \hat{x}_s^0 = x^0 - hsx^k P_k \tag{32}$$

where $D = x^k P_k$ in Heisenberg realization. All twists $\mathcal{F}_s^A$ (for any s) correspond to the same classical r-matrix: $\mathfrak{r}_A = D \wedge P_0$ and they have the same universal quantum r-matrix which is of exponential form: $\mathcal{R} = (\mathcal{F}_s^A)^{21}(\mathcal{F}_s^A)^{-1} = e^{\imath D \wedge P_0}$. The deformed symmetry algebra corresponding to this twist can be found in [18, 20, 22]. Some interesting applications for symmetrized version of this twist (for the choice of parameter $s = \frac{1}{2}$) were considered e.g. in [25].

Another example of κ-space realizations falling into non-covariant class (30) come from Jordanian twist: $(\mathcal{F}_r^J)^{-1} = \exp(-J_r \otimes \sigma_r)$, and have the form:

$$\hat{x}_r^i = x^i (1 - rhP_0)^{-\frac{1}{r}} \qquad \text{and} \qquad \hat{x}_r^0 = x^0(1 - rhP_0). \tag{33}$$

where $J_r = \imath(\frac{1}{r}x^k P_k - x^0 P_0)$ (in Heisenberg realization) with a numerical factor $r \neq 0$ labelling different twists and $\sigma_r = \ln(1 - hrP_0)$ (cf. the remark in point 2) in Introduction). Direct calculations show that, regardless of the value of r, twisted commutation relations (8) take the form of that for κ-Minkowski spacetime (3). The corresponding classical r-matrices are the following:

$$\mathfrak{r}_J = J_r \wedge P_0 = \frac{1}{r} D \wedge P_0 - L_0^0 \wedge P_0. \tag{34}$$

Again we refer the reader for the full form of deformed Hopf algebra to [20, 22] as quantum symmetry in this case. It is worth to mention that for the choice of $r = -1$ we can reduce the symmetry group from the whole $\mathcal{U}_{\mathrm{igl}}$ to the the minimal one-generator extension of the Poincare algebra which includes dilatation generator, i.e. Poincare-Weyl algebra [20, 22].

Interestingly, for the minimal extension, i.e. Poincare-Weyl algebra there exists also symmetric version of above Jordanian twist (see e.g. [26, 27, 28]) which can be written in compact way as [28]:

$$\mathcal{F}^{-1} = \sum_{m=0}^{\infty} \frac{1}{m!} \left(-\frac{h}{2}\right)^m \sum_{k=0}^{m} (-1)^k \binom{m}{k} P_0^{m-k} J^{<k>} \otimes P_0^k J^{<m-k>}, \tag{35}$$

where $J^{<0>} = 1, J^{<1>} = J, J^{<2>} = J(J+1), ...,$ and $J = x^\mu P_\mu$ (in realization) and still $[J, P_0] = P_0$. It provides slightly modified realization for noncommutative coordinates with respect to non-covariant family (30):

$$\hat{x}^i = x^i \phi(A); \qquad \hat{x}^0 = x^0 \psi(A) + \imath h x^k \partial_k \gamma(A) + \xi(A) \tag{36}$$

where $\phi = \left(1 - \frac{1}{2}A\right)$, $\psi = \left(1 - \frac{1}{4}A^2\right)$, $\gamma = \frac{1}{2}\left(1 - \frac{1}{2}A\right)$ and $\xi = \frac{ih}{4}A$. In general $\xi(A)$ could be any function satisfying $\xi(0) = 0$ (to provide the classical limit) and still the equation (31) will be satisfied. Which means that this twist via start product (8) also provides κ-Minkowski algebra $\mathcal{A}^\kappa$ with the commutation relations (3).

4. Conclusions

The structure of spacetime, at the scale where quantum gravity effects take place, is one of the most important questions in fundamental physics. Below the quantum gravity scale the symmetry of spacetime should also be deformed. Noncommutative spacetimes as one of the approaches to the description of Planck scale physics allow to describe the geometry from the quantum mechanical point of view. At the Planck scale the idea of size or distance in classical terms is not valid any more, because one has to take into account quantum uncertainty which naturally arise via noncommutative coordinates. Moreover their realizations as differential operators play the role of observables.

The aim of the review was to introduce in a compact way the variety of noncommutative spacetimes and their deformed relativistic symmetries which have been widely used by many authors in the view of applications in physical theories like quantum field theory or gravity. The deformation of spacetime requires a generalisation of its symmetry group and one deals with the θ- or κ-Poincare Quantum Group as well as with twist-deformations of inhomogeneous linear group.

Acknowledgments

The author would like to thank A. Borowiec for pointing out the references [26, 27]. This work was supported by Polish National Science Center grant 2011/01/B/ST2/03354.

References

[1] Gelfand I M, Naimark M A 1943 *Math. Sbornik* **12** 2
[2] Doplicher S, Fredenhagen K, Roberts J E 1994 *Phys. Lett.* B **331** 39
[3] Chaichian M, Kulish P, Nishijima K, Tureanu A 2004 *Phys. Lett.* B **604** 98
[4] Lukierski J, Lyakhovsky V, Mozrzymas M 2002 *Phys. Lett.* B **538** 375
[5] Zakrzewski S 1994 *J. Phys.* A **27** 2075
[6] Klimyk A, Schmüdgen K 1997 *Quantum groups and their representations* (Berlin: Springer-Verlag)

[7] Kassel C 1995 *Quantum groups* (New York: Springer-Verlag)
[8] Drinfeld V 1985 *Sov. Math. Dokl.* **32** 254
[9] Chari V, Pressley A 1994 *A guide to quantum groups* (Cambridge, UK: Cambridge Univ. Press)
[10] Tolstoy V N 2004 *Proc. of International Workshop "Supersymmetries and Quantum Symmetries (SQS'03)"* 242 *Preprint* math/0402433
[11] Lukierski J, Nowicki A, Ruegg H, Tolstoy V N 1991 *Phys. Lett.* B **264** 331
[12] Lukierski J, Ruegg H, Zakrzewski W J 1995 *Ann. Physics* **243** 90
[13] Majid S, Ruegg H 1994 *Phys. Lett.* B **334** 348
[14] Borowiec A, Pachoł A 2010 *J. Phys.* A **43** 045203
[15] Daszkiewicz M, Lukierski J, Woronowicz M 2008 *Mod. Phys. Lett.* A **23** 653
[16] Kowalski-Glikman J, Nowak S 2002 *Phys.Lett.* B **539** 126
[17] Amelino-Camelia G, Arzano M, Kowalski-Glikman J, Rosati G, Trevisan G 2012 *Class. Quantum Grav.* **29** 075007
[18] Bu J G, Kim H C, Lee Y., Vac C H, Yee J H 2008 *Phys. Lett.* B **665**
[19] Meljanac S, Samsarov A, Stojić M, Gupta K S 2008 *Eur. Phys. J.* C **53** 295
[20] Borowiec A, Pachoł A 2009 *Phys. Rev.* D **79** 045012
[21] Borowiec A, Pachoł A 2010 *SIGMA* **6** 086
[22] Pachoł A 2011 *κ -Minkowski spacetime: mathematical formalism and applications in Planck scale physics* PhD Thesis (Wrocław University Archives) *Preprint* hep-th/1112.5366
[23] Meljanac S, Stojić M 2006 *Eur. Phys. J.* C **47** 531
[24] Govindarajan T R, Gupta K S, Harikumar E, Meljanac S, Meljanac D 2008 *Phys.Rev.* D **77** 105010
[25] Dimitrijevic M, Jonke L 2011 *JHEP* **2011** 80
[26] Ohn Ch 1992 *Lett. Math. Phys.* **25** 85
[27] Tolstoy V N 2006 *Proceedings of XXII Max Born Symposium "Quantum, Super and Twistors" Preprint*
[28] Giaquinto A, Zhang J J 1998 *JPAA* **128** 133

Superconductivity, superfluidity and holography

Alberto Salvio

Scuola Normale Superiore and INFN, Piazza dei Cavalieri 7, 56126 Pisa, Italy

Departamento de Física Teórica, Universidad Autónoma de Madrid and
Instituto de Física Teórica IFT-UAM/CSIC, Cantoblanco, 28049 Madrid, Spain

E-mail: alberto.salvio@uam.es

Abstract. This is a concise review of holographic superconductors and superfluids. We highlight some predictions of the holographic models and the emphasis is given to physical aspects rather than to the technical details, although some references to understand the latter are systematically provided. We include gapped systems in the discussion, motivated by the physics of high-temperature superconductivity. In order to do so we consider a compactified extra dimension (with radius R), or, alternatively, a dilatonic field. The first setup can also be used to model cylindrical superconductors; when these are probed by an axial magnetic field a universal property of holography emerges: while for large R (compared to the other scales in the problem) non-local operators are suppressed, leading to the so called Little-Parks periodicity, the opposite limit shows non-local effects, e.g. the uplifting of the Little-Parks periodicity. This difference corresponds in the gravity side to a Hawking-Page phase transition.

1. Introduction: motivations, background material and conclusions

Holographic dualities have become useful tools to explore strongly coupled theories; in particular, in the last few years systems at finite temperature T and density of a U(1) charge have been extensively studied. The main motivation comes from condensed matter physics (for a review see [1]): if the U(1) charge is the electric charge and the U(1) symmetry is spontaneously broken one identifies these systems with superconductors (SCs); when the U(1) symmetry is global, instead, the spontaneously broken phase corresponds to superfluidity.

One of the main ideas behind the application of holography to superconductivity is the hope to shed light on the physics of unconventional SCs (such as high-T SCs), which are not completely described by the weakly coupled theory: the Bardeen-Cooper-Schrieffer theory. Cuprate high-T SCs, the prototype of unconventional materials, are obtained by doping a Mott insulator and as such they typically show in the (T versus "dopant concentration") phase diagram an insulator phase close to the region where the U(1) symmetry is spontaneously broken [2]. This calls for a way to describe both conductor/SC and insulator/SC transitions. Since any insulating material eventually conducts if it is probed by a strong enough external current, this in turn requires a method to study gapped phases.

In this article we will briefly review holographic SCs [3, 4] and superfluids (SFs) [5, 4]. The emphasis will be given to physical aspects rather than to the technical details, for which we will refer to the original papers. Given that all SF systems can be considered as SCs in the limit of non-dynamical electromagnetic (EM) gauge fields, for the sake of definiteness, we will use the terminology adopted in the literature of superconductivity, rather than that of superfluidity (unless otherwise stated). We will focus on the best understood version of holographic duality, the anti de Sitter/conformal field theory (AdS/CFT) correspondence. The discussion of the previous paragraph then motivates us to introduce a conformal symmetry breaking in the infrared (IR), which is needed to have a gap. Moreover, since much of the interesting physics of cuprate SCs is layered, we shall restrict our analysis to 2+1 dimensional systems, dual to gravitational theories on four dimensional spacetimes. Although beyond the scope of this review, it is useful to mention that other regions of cuprate phase diagrams require a departure from the theory of Fermi liquids, a challenging property which can be realized in holography [6].

In the rest of this section we shall introduce the concepts and general properties of superconductivity which are needed to understand the holographic results discussed in the following sections, providing at the same time the reader with the organization of the paper and the conclusions. We assume the minimal field content to describe superconductivity: a charged scalar $\Phi_{\rm cl}$, responsible for the breaking of the U(1) symmetry and a gauge field a_μ. The quantum effective action is a generic functional of gauge invariant operators

$$\Gamma = \int d^3x\, \mathcal{L}_{\rm eff}(\mathcal{F}_{\mu\nu}\,, \partial_\mu \arg(\Phi_{\rm cl}) - a_\mu\,, \psi_{\rm cl}) + \Gamma_{n.l.}\,, \tag{1}$$

where we have introduced the field strength $\mathcal{F}_{\mu\nu} = \partial_\mu a_\nu - \partial_\nu a_\mu$ and $\psi_{\rm cl} = |\Phi_{\rm cl}|$. The first term in the expression above represents the contribution of *local* gauge invariant terms. $\Gamma_{n.l.}$ depends instead on *non-local* gauge invariant objects, which may be required if the material has a non-trivial topology. Although the action in (1) is very general, its form simplifies in some limits. This allows us to extract model independent properties and, at the same time, identify the predictions of specific models, such as the holographic ones that will be discussed here.

One of these limits is the small field limit, in which Γ is approximated by the Ginzburg-Landau (GL) action:

$$\Gamma_{\rm GL} = \int d^3x \left(-\frac{1}{4g_0^2}\mathcal{F}_{\mu\nu}^2 - |D_\mu \Phi_{\rm GL}|^2 + \frac{1}{2\xi_{\rm GL}^2}|\Phi_{\rm GL}|^2 - b_{\rm GL}|\Phi_{\rm GL}|^4 \right)\,, \tag{2}$$

where g_0, $\xi_{\rm GL}$ and $b_{\rm GL}$ are real and positive parameters and we have rescaled the scalar field, $\Phi_{\rm GL} \propto \Phi_{\rm cl}$, to have a canonically normalized kinetic term for $\Phi_{\rm GL}$. Also $D_\mu \Phi_{GL} = (\partial_\mu - i a_\mu)\Phi_{GL}$. This theory holds if the system is close enough to the symmetry breaking transition, where $\Phi_{\rm cl}$ is small. Another case in which the action simplifies is the limit of slowly varying fields, such that Γ can be approximated by a two-derivative functional

$$\Gamma \simeq \int d^3x\, h(\psi_{\rm cl})\Big\{ -\frac{1}{4g^2(\psi_{\rm cl})}\mathcal{F}_{\mu\nu}^2 - |D_\mu \Phi_{\rm cl}|^2 - W(\psi_{\rm cl})\Big\}\,, \tag{3}$$

where h, g and W are unspecified functions of $\psi_{\rm cl}$. When (3) is a good approximation the SF limit can be taken as $g \to 0$: this corresponds to taking the gauge field non-dynamical and thus the corresponding U(1) symmetry global. Deep inside a uniform SC, where ψ is close to the minimum ψ_∞ of the potential $V = hW$ and the massive gauge field $a_\mu - \partial_\mu \arg(\Phi_{\rm cl})$ is suppressed, (3) reduces to a functional that is quadratic in the fields; this allows us to define the inverse masses of $\psi_{\rm cl}$ and a_μ, respectively ξ and λ:

$$\frac{1}{\xi^2} = \frac{1}{2h(\psi_\infty)}\frac{\partial^2 V}{\partial \psi_{\rm cl}^2}(\psi_\infty) > 0\,, \quad \lambda = \frac{1}{\sqrt{2}e(\psi_\infty)\,\psi_\infty}\,. \tag{4}$$

Note that ξ and λ cannot be computed in a model-independent way as their values strongly depend on the form of h, g and W. The distinctive property of the superconductive phase is $\psi_\infty \neq 0$, as opposed to the normal phase in which $\psi_\infty = 0$.

Starting from the general action in (1) it is possible to show in very general terms some of the most famous properties of SCs, such as the Meissner effect, the infinite DC conductivity and the Josephson effect [7], which have been obtained in holography in Refs. [4, 8, 9, 3, 10].

In this article (unless otherwise stated) we will focus on static configurations, in which everything is constant in time and the temporal component of the gauge field is identified with the chemical potential of the U(1) charge: $a_0 = \mu$. Also, in order to study the thermodynamics of these systems we introduce the free energy $F = -T\,\Gamma$.

In section 3 we will holographically study the normal phase and explain how to describe conductors and insulators. The homogeneous phase, in which $\psi_{\rm cl} = \psi_\infty$, will be studied in section 4 through the AdS/CFT correspondence. As we will see, this provides a rationale for the fact that superconductivity is suppressed at large rather than at small T.

However, inhomogeneous configurations are also of great importance in SCs; e.g. they generically occur when the system is probed by an external EM field. Among the most famous properties of SCs we can certainly include the existence of vortex solutions, which are indeed strongly inhomogeneous configurations. The simplest example of such configurations are straight vortex lines, which can be described by an ansatz of the form $\Psi_{\rm cl} = \psi_{\rm cl}(r)e^{in\phi}$ and $a_\phi = a_\phi(r)$ and the other components of the vector potential set to zero. Here r, ϕ are the usual polar coordinates which parametrize the Euclidean two dimensional space. By using the action in (3) one obtains the following large r field behavior [4]

$$\psi_{\rm cl} \simeq \psi_\infty + \frac{\psi_1}{\sqrt{r}} e^{-r/\xi'} \,, \quad a_\phi \simeq n + a_1\sqrt{r} e^{-r/\lambda'} \,, \tag{5}$$

for SCs, while for SFs, where the magnetic field $B = \partial_r a_\phi / r$ is frozen to an external constant value, $a_\phi = Br^2/2$, the condensate approaches the homogeneous value with a power law $\psi_{\rm cl} - \psi_\infty \sim n^2/r^2$ [4]. In Eq. (5) ψ_1 and a_1 are unspecified constants, while the action in (3) implies that $\lambda' = \lambda$ and $\xi' = \xi$. However, the exponential behavior in Eq. (5) tells us that higher derivative terms cannot be neglected as they are generically of the same order of the two-derivative terms. Fortunately, the only effect of the higher derivatives is to modify the values of λ' and ξ', such that we generically have $\lambda' \neq \lambda$ and $\xi' \neq \xi$, as discussed in [4]. λ' and ξ' are respectively called penetration depth and coherence length and are important to characterize SCs. Although the field behavior in (5) is model independent, λ' and ξ', like λ and ξ, can only be fixed once the model is specified. Thus their actual values are predictions of the particular model one considers; in section 5 we will discuss how to extract them from holography.

While vortex solutions exist for any SC, it is not always the case that there is a range of the external magnetic field H such that the vortex phase is energetically favorable; when this is true the SC is of Type II and such range is denoted with $H_{c1} \leq H \leq H_{c2}$ (if such range does not exist we have instead a Type I SC). H_{c2} is the value of the external magnetic field above which the system is always in the normal phase. The vortex configuration for Type II SCs and H slightly smaller than H_{c2} is known to be a triangular lattice of vortices independently on the specific model one considers: for those high magnetic fields the condensate is small and the GL theory can be applied to predict that configuration [11]. H_{c1} is instead the value of H below which the system is in the homogeneous superconducting phase. It can be computed through the model independent formula[1] (see e.g. [4])

$$H_{c1} = \frac{g_0^2}{2\pi}(F_1 - F_0) \,, \tag{6}$$

where F_1 and F_0 are the free energies for the $n = 1$ vortex and $n = 0$ superconducting phase respectively. We note that the actual value of H_{c1} is a prediction of the specific model one

[1] Here we use the normalization of H such that the external current $\vec{J}_{ext}$ that generates H through $\nabla \times \vec{H} = g_0^2 \vec{J}_{ext}$ is coupled to the gauge field by means of the interaction term $a_\mu J^\mu_{ext}$ in the Lagrangian.

considers because g_0, F_1 and F_0 are model dependent. However, in the SF limit, $g_0 \to 0$, we always have $H_{c1} \to 0$; in other words H_{c1} is non-trivial only if the magnetic field is dynamical.

In section 5 we will study the vortex phase in holography and illustrate how to compute the critical magnetic fields and show that the holographic SC is of type II. We will emphasize in particular the differences between the case in which there are no sources of conformal symmetry breaking, other than T and μ, and the case in which conformal symmetry is broken.

Up to now we have not discussed the role of the non-local terms in Eq. (1). For reasons which will be clear soon, we would like to spend the rest of this introductory section to describe the simplest setup in which $\Gamma_{n.l.}$ is relevant: cylindrical SCs threaded by an external magnetic field along the symmetry axis of the cylinder. To render the discussion even simpler we will take the deep Type II limit, $g_0 \to 0$, in which the (total) magnetic field coincides with the external one. Since there is a non-contractible loop on this geometry we can construct non-local gauge invariant objects:

$$W \equiv \exp\left(ei \oint dx^\mu a_\mu\right), \quad m \equiv \frac{1}{2\pi} \oint dx^\mu \partial_\mu \mathrm{arg}(\Phi_{\rm cl}) = \text{integer}, \quad \ldots, \tag{7}$$

where e is the charge of the fundamental charge carriers (say the electrons for real world SCs) and the integrals are performed along the non-contractible loop. We will refer to W and m as the Wilson line and the fluxoid respectively and the dots represent other non-local objects. While the classical action does not depend on $(W, m, \ldots)$, quantum effects, such as the Aharonov-Bohm one (or Sagnac one [12, 13] in the SF literature), could introduce a dependence of Γ on these quantities. A simple (but still general) way to visualize a dependence of this sort is to think that the coefficients of the local part of Γ vary with $(W, m, \ldots)$: for example, in the domain of validity of the GL theory in (2), this corresponds to thinking of ξ_{GL} and b_{GL} as functions of $(W, m, \ldots)$. Such dependence ought to be suppressed in the classical limit and/or when the typical scale of the non-contractible loops, R, is large compared to the other scales in the problem (such as $1/T$ and $1/\mu$) and therefore we should stay away from these limits to see any interest effect. Moreover, *since quantum corrections are small in a weakly coupled theory, the biggest effects of the non-local quantities are expected in strongly coupled theories.* This justifies the use of holography in this setup (see section 6). Before moving to holography, however, let us identify model-independent effects.

We shall consider the simplest case in which the magnetic field is constant so that it can be represented by a constant vector potential along χ, which parametrizes the compact spatial dimension, $\chi \sim \chi + 2\pi R$. Since everything is static and homogeneous an appropriate ansatz is

$$\Phi_{\rm cl} = \psi_{\rm cl} e^{im\chi/R}, \quad a = \mu dt + a_\chi d\chi \quad \text{with } \psi_{\rm cl} \text{ and } a_\chi \text{ constant}. \tag{8}$$

Inserting this ansatz in the quantum effective action, Eq. (1), we obtain

$$\Gamma = \int d^3x\, \mathcal{L}_{\rm eff}(m/R - a_\chi\,, -\mu\,, \psi_{\rm cl}) + \Gamma_{n.l.}\,; \tag{9}$$

in other words, *modulo non-local terms*, the system must be periodic in the magnetic flux[2] $\Phi(B) \equiv \oint dx^\mu a_\mu / g_0$ with period $\Delta\Phi^{LP} \equiv 2\pi/g_0$, because $\Phi(B) \to \Phi(B) + \Delta\Phi^{LP}$ can be compensated by a unit shift of the integer m. This is known as the Little-Parks effect [14]. This phenomenon has been observed in experiments, which give $g_0 = 2e$, and thus it is considered as an evidence for Cooper pairing. If the non-local terms in (9) are non-negligible the Little-Parks does not generically occur; indeed nothing forces $\Gamma_{n.l}$ to be a function of the local combination $m/R - a_\chi$ only, but it may depend on m and W separately. The conclusion is that if the system is

[2] The factor $1/g_0$ has been introduced in order to have a canonically normalized kinetic term for a_μ in Eq. (2).

strongly coupled, far away from the classical limit and R is small enough we could see an uplifting of the Little-Parks periodicity to an enhanced periodicity, set by the fundamental charge: $2\pi/e$. This would corresponds to resolving the internal structure of the *composite* condensing operator. Whether this really occurs is not model-independent.

Remarkably, we will see (in section 6) that holography *predicts* an uplifting of the Little-Parks periodicity for small values of R and that this uplifting corresponds in the gravity side to the celebrated Hawking-Page phase transition [15].

2. The holographic model

The first step to define a holographic model based on the standard AdS/CFT correspondence is to introduce fields living on an asymptotically AdS space. The minimal field content required to describe SCs holographically is a charged scalar field Ψ dual to a condensing operator $\mathcal{O}$ ($\langle\mathcal{O}\rangle = \Phi_{\rm cl}$) and a gauge field A_α dual to the electric current operator J_μ. As we have stated in the introduction, another ingredient we are interested in is a mechanism for conformal symmetry breaking in the IR; this can be achieved by introducing a real scalar ϕ (a *dilaton*) which acquires a non-trivial VEV. The class of actions we consider is [9]

$$S = \int d^4x\,\sqrt{-g}\left\{\frac{1}{16\pi G_N}\left[\mathcal{R} - (\partial_\alpha\phi)^2 - V(\phi)\right] - \frac{Z_A(\phi)}{4g_4^2}\mathcal{F}_{\alpha\beta}^2 - \frac{Z_\psi(\phi)}{L^2 g_4^2}|D_\alpha\Psi|^2\right\}, \tag{10}$$

where $\mathcal{R}$ is the Ricci scalar, G_N the Newton constant, V the dilaton potential, g_4 the U(1) gauge coupling and L is the radius of AdS. Also, Z_A and Z_ψ are generic functions of the dilaton, which do not fulfill particular properties, besides the fact that they never vanish (in order for the semiclassical approximation in the bulk to be justified). Also, $D_\alpha\Psi = (\partial_\alpha - iA_\alpha)\Psi$.

We are interested in describing a static system with two dimensional rotation and translation invariance, for which the metric and the dilaton have the form

$$ds^2 = W(z)\left(-f(z)dt^2 + dy^2 + \frac{dz^2}{f(z)}\right), \quad \phi = \phi(z), \tag{11}$$

where dy^2 is the two dimensional Euclidean metric. The temperature can be introduced by requiring the presence of a black hole; so there is a value of the holographic coordinate, z_0, such that $f(z_0) = 0$. Then $T = |f'(z_0)|/4\pi$, where the prime denotes the derivative with respect to z.

Another physical situation that can be described with this setup is the case of multiply connected SCs, more precisely of cylindrical geometry: this can be simply achieved by compactifying one of the two y-coordinates: $\chi \sim \chi + 2\pi R$. Then there is necessarily another set of solutions [16]: those obtained from the black holes by exchanging the Euclidean time $t_E = it$ with χ; these configurations have no event horizons and are commonly called AdS solitons. One reason the compactification of χ is interesting is because this procedure allows us to introduce a scale R (other than the T and μ) and thus breaks conformal invariance even in the absence of the dilaton. For this reason from now on the case in which χ is compact will be discussed with a simplified field content, where ϕ is removed and $Z_A = Z_\psi = 1$. The AdS soliton (black hole) geometry is energetically favorable with respect to the black hole (AdS soliton) at sufficiently small (large) temperatures, $T \le 1/(2\pi R)$ ($T \ge 1/(2\pi R)$). Another reason why we are interested in compactifying a spatial coordinate is to study what happens when cylindrical SCs are threaded by magnetic fields: in these physical setups indeed a universal prediction of holography emerges, as it will be discussed in section 6 and anticipated in the introduction.

Although we include the possibility of conformal symmetry breaking in the IR, in this article we will always assume that the ultraviolet (UV) is conformally invariant, or in other words that the fields are approximated by an AdS configuration close to a value of z, say $z = 0$: for z close to zero, $W(z) \simeq L^2/z^2$, $f(z) \simeq 1$ and $\phi(z) \simeq 0$. Thus we can use the standard AdS/CFT dictionary, which relates the properties of a gravitational theory with those of a CFT. In particular the $z = 0$ values of Ψ and A_μ are the sources of $\mathcal{O}$ and J_μ in the CFT. Therefore, if one solves the bulk field equations with boundary conditions

$$\Psi|_{z=0} = \Psi_0 \,, \quad A_\mu|_{z=0} = a_\mu \,, \tag{12}$$

the Green's function for $\mathcal{O}$ and J_μ are given by differentiating the on-shell action with respect to Ψ_0 and a_μ; e.g. the vacuum expectation values are given by

$$\langle J_\mu \rangle = \frac{1}{g_4^2} \mathcal{F}_{z\mu}|_{z=0} \,, \quad \langle \mathcal{O} \rangle = \frac{1}{g_4^2 z^2} D_z \Psi|_{z=0} \,. \tag{13}$$

In order to keep the discussion as simple as possible we assume that Ψ and A_α do not backreact on the geometry and the dilaton; this can be consistently achieved by taking the limit $G_N \to 0$. A generalization to the case $G_N \neq 0$ can be found in [17, 18]. A general form of asymptotically AdS dilatonic black hole solutions with the assumed symmetries was derived in [19]. One important aspect of these black holes is that the dilaton is not generically constant but runs with the energy:

$$\phi(z) = \sqrt{\frac{\nu^2 - 1}{2}} \ln(1 + z/L) \,, \tag{14}$$

where ν is a real parameter of the potential such that $\nu \geq 1$ and $\nu = 1$ recovers the Schwarzschild black hole (S-BH) of Einstein's gravity. Eq. (14) tells us that although the theory has a UV fixed point, $\phi(0) = 0$, conformal invariance is broken and maximally in the IR, $z = z_0$. In order for the configurations of Ref. [19] to be a solution, V has to be appropriately chosen. In particular the requirement of an asymptotically AdS configuration space with cosmological constant Λ implies that at the conformal point (conventionally $\phi = 0$) where $V(0) = \Lambda$ we have $\partial V / \partial \phi = 0$. As a result, for each value of ν, even $\nu \neq 1$, the S-BH with cosmological constant Λ is always a solution. One finds [9] that the S-BH is favorable at high temperatures, while the dilaton-BH, Eq. (14), dominates in the low temperature region.

3. Conductivity

Having described the solutions of the dilatonic gravity systems we now want to understand the type of materials they correspond to. Some information can be gained by studying the conductivity, σ. This will also shed light on the nature of the SF phase transitions, which will be discussed in the next section.

To compute σ let us consider, on top of these geometries, a small plane wave along a spatial coordinate x,

$$A_x(t, z) = \mathcal{A}(z) e^{i\omega(p(z) - t)} \,, \tag{15}$$

which is induced by a small electromagnetic field, $A_x(t, z)|_{z=0} = a_x(t)$. Here $\mathcal{A}$ and p are real functions of z. The system responds creating a current which is linear in the electric field E_x: that is $\langle J_x \rangle = \sigma E_x$. Using the AdS/CFT dictionary, the first equation in (13), we have

$$\sigma = \frac{p'(0)}{g_4^2} - i \frac{\mathcal{A}'(0)}{g_4^2 \omega \mathcal{A}(0)} \,. \tag{16}$$

When an event horizon is present regularity of the solution implies that the plane wave should be ingoing rather than outgoing from the horizon. It can be shown that this results in a non-vanishing Re$[\sigma]$ and DC conductivity [9]

$$\lim_{\omega \to 0} \mathrm{Re}[\sigma] = \frac{1}{g_4^2} Z_A|_{z=z_0} \neq 0 \,. \tag{17}$$

For an AdS soliton, which has no event horizon, regularity allows for a vanishing DC conductivity, which has been identified with an insulating behavior [20, 18]. Therefore, compactifying a spatial dimension χ allows us to realize a conductor/insulator transition as the temperature is lowered. Eq. (17), however, tells us that there is another way to suppress the DC conductivity and obtain such transition if a dilaton is present in the spectrum: if one chooses Z_A sufficiently small for large values of ϕ, the running in Eq. (14) implies that $\lim_{\omega\to 0}\mathrm{Re}[\sigma]$ is small, especially in the low temperature limit where z_0 is large. In this setup the transition (discussed in the previous section) between the S-BH and dilaton-BH, which is obtained by lowering the temperature, represents a conductor/insulator transition. It is worth noting that insulating systems corresponds to solids in the fluid mechanical interpretation [8].

4. Superfluid phase transition

We now move on and study the simplest example of superconducting state: a static, homogeneous and isotropic material in which the U(1) symmetry is spontaneously broken. The ansatz is

$$\Psi = \psi(z)\,, \quad A = A_0(z)dt\,. \tag{18}$$

The reader may wonder why we have introduced the temporal component of the gauge field, because having a non-vanishing $\psi(z)$ seems already enough to describe the state we are interested in. The reason is that no regular solution is found when $A_0 = 0$ in (18), as shown in [9], and we want to exclude singular solutions. Physically this corresponds to the fact that there is only one parameter breaking conformal symmetry (T for the black hole and R for the AdS soliton) in the absence of the dilaton[3], thus no phase transition can occur in this case. Therefore we set the UV boundary conditions

$$\Psi_0 = 0\,, \quad a_0 = \mu\,. \tag{19}$$

The first condition guarantees that the U(1) symmetry is spontaneously broken, while the second one, with a non-vanishing chemical potential μ, keeps the profile of A_0 different from zero and allows us to find a regular solution.

An immediate consequence of this argument is that when the temperature is large compared to the chemical potential a black hole suppresses the superconductivity. At small temperatures instead the system turns into a SC [3]. In AdS/CFT there is therefore a reason for the fact that a metal becomes a SC at small rather than at large temperatures. For the AdS soliton the same conclusion can be reached, but with T substituted by $1/R$.

According to the results of the previous section these SF phase transitions can be either of the conductor/SC or insulator/SC type, depending on the behavior of the conductivity in the normal phase: the two cases correspond to a non-vanishing or vanishing DC conductivity respectively. In the fluid mechanical interpretation, the SF phase of a holographic system with a solid normal phase has been identified with a supersolid [21].

5. Dynamical gauge fields in AdS/CFT and superconductivity

In the holographic results we have discussed so far the dynamics of the EM field is not important. Therefore, according to the discussion presented in the introduction, they can be applied equally well to SFs and SCs. In this section instead we discuss some important effects of superconductivity, which crucially rely on the dynamics of the EM field.

However, we note that imposing the first boundary condition in (12) treats a_μ as an external source, which does not participate in the dynamics of the system. In order to have a dynamical

[3] The dilaton cannot change this conclusion as it is regular for any T and the function Z_ψ, Eq. (10), is assumed to be regular and non-zero.

a_μ we should integrate over all possible field configurations: (working in the Euclidean space)

$$Z[J_{ext}] = \int Da\, e^{-S_E[a_\mu]+\int d^3x\left(-\frac{1}{4g_b^2}\mathcal{F}_{\mu\nu}^2+a_\mu J^\mu_{ext}\right)}, \tag{20}$$

where $S_E[a_\mu]$ is the Euclidean bulk action computed on a solution of the field equations with boundary condition in (12). Also, we have introduced for generality a kinetic term for a_μ and an external current J_{ext}. Then $Z[J_{ext}]$ defines as usual the generating functional for the Green functions of a_μ. Of course, if there are other operators in the theory, besides the gauge field, $S_E[a_\mu]$ in Eq. (20) will depend on the corresponding sources as well; for example in the case of superconductivity discussed in the previous sections $S_E[a_\mu]$ also depends on Ψ_0, which we introduced in (12).

In the semiclassical limit this procedure reduces to solving the Maxwell equations,

$$\langle J^\mu\rangle + \frac{1}{g_b^2}\partial_\nu \mathcal{F}^{\nu\mu} + J^\mu_{ext} = 0\,, \tag{21}$$

where we have used $\langle \hat{J}^\mu\rangle = -\delta S_E/\delta a_\mu$. The semiclassical limit corresponds to taking a large external current, $g_b \to 0$ and a limit on the parameters of the bulk theory such that $S_E[a_\mu]$ becomes large; for example, for the theory defined in (10) this limit is $G_N \to 0$ and $g_4 \to 0$.

Now, using Eq. (13), we can see that the Maxwell equations in (21) can viewed as a Neumann type boundary condition in the gravity side:

$$\frac{1}{g_4^2}\mathcal{F}_z{}^\mu\Big|_{z=0} + \frac{1}{g_b^2}\partial_\nu\mathcal{F}^{\nu\mu}\Big|_{z=0} + J^\mu_{ext} = 0\,. \tag{22}$$

The lesson is therefore that switching from the Dirichlet boundary condition (12) to the condition above promotes a_μ to a dynamical field.

As far as the holographic SC model of section 2 is concerned, this procedure has been applied to find the Meissner effect and genuine SC vortices [4, 9], at least in the simplest case of a straight vortex line:

$$\Psi = \psi(z,r)e^{in\phi}\,, \quad A_0 = A_0(z,r)\,, \quad A_\phi = A_\phi(z,r)\,. \tag{23}$$

r, ϕ are the usual polar coordinate parametrizing the two dimensional Euclidean space and so here we assume that there are at least two non-compact dimensions in the CFT. This is not the case for the four dimensional AdS soliton; however, vortex solutions have been found on top of the five dimensional AdS soliton in [8]. Using the ansatz in (23) the Neumann-like boundary condition in (22) becomes

$$\frac{1}{g_4^2}\partial_z A_\phi\Big|_{z=0} + \frac{1}{g_b^2}r\partial_r\left(\frac{1}{r}\partial_r A_\phi\right)\Big|_{z=0} + J_{ext\,\phi} = 0\,, \tag{24}$$

while the requirement of spontaneous symmetry breaking and the presence of a finite charge density again fixes the other $z = 0$ boundary conditions, Eq. (19). Regularity of the solutions instead fixes the conditions at $z = z_0$ and at the center of the vortex $r = 0$ and physical conditions on the behavior at infinity, $r \to \infty$, imposes constraints on the remaining boundary.

This results in genuine SC profiles for the total magnetic field $B = \partial_r a_\phi/r$: for example for $n = 0$ one recovers the Meissner effect, while for $n \neq 0$ one observes the exponential damping of B far away from the center of the vortex [4, 8, 9], Eq. (5), allowing for a holographic prediction for λ' and ξ'. Such properties should be contrasted with the Dirichlet boundary condition for the magnetic field, which in polar coordinates is $A_\phi|_{z=0} = Br^2/2$ (with B constant), corresponding to SF systems (see e.g. [21, 4, 8, 9, 22]).

Once the vortex solutions are obtained we can compute H_{c1} through Eq. (6). Indeed selecting the $n = 1$ vortex and $n = 0$ homogeneous superconducting phase we can compute $F_1 - F_0$ while, integrating over the bulk *à la* Kaluza-Klein gives g_0:

$$\frac{1}{g_0^2} = \frac{1}{g_4^2}\int_0^{z_0} dz\, Z_A(\phi)\,. \tag{25}$$

The fact that the dilaton appears in this formula has an important consequence on the behavior of H_{c1} at low temperatures [9], as we now explain. In the absence of ϕ Eq. (25) implies that $g_0 \to 0$ as $T \to 0$ (in this limit the horizon is removed, $z_0 \to \infty$) and Eq. (6) forces H_{c1} to vanish. The physical reason why this emerges is because the theory without the dilaton is scale invariant and so g_0 should go to zero as $T \to 0$. The presence of the dilaton breaks scale invariance and can avoid this conclusion, providing a non-vanishing H_{c1}: g_0 remains non-zero all the way down to $T = 0$ if $Z_A(\phi)$ goes to zero fast enough as $\phi \to \infty$, a limit that always occurs close to the horizon as $T \to 0$, Eq. (14). Interestingly, when H_{c1} remains non-zero at zero temperature, Eq. (16) tells us that the corresponding normal phase must have a suppressed DC conductivity in the low temperature region, resembling an insulator.

The second critical field H_{c2} can be instead computed as the value of H at which the condensate goes to zero. Up to now all the holographic models of superconductivity have turned out to be of Type II [4, 8, 9], like all known high temperature SCs.

6. Multiply connected holographic superconductors and superfluids

As explained in section 1 an interesting setup to apply the holographic techniques are multiply connected SCs probed by magnetic fields. We shall consider the simplest case described at the end of section 1. In the presence of a compactified dimension scale invariance is broken and we will assume for simplicity that the bulk field content does not include the dilaton. Also, the ansatz in (8) corresponds holographically to

$$\Psi = \psi(z)e^{im\chi/R}\,, \quad A_0 = A_0(z)\,, \quad A_\chi = A_\chi(z)\,, \tag{26}$$

with UV boundary conditions $\Psi|_{z=0} = 0\,, A|_{z=0} = \mu\,, A_\chi|_{z=0} = a_\chi$. As already discussed, the system is in the S-BH phase at large radii and in the AdS soliton otherwise and, as shown in [23], such transition corresponds to an uplifting of the Little-Parks periodicity. This is a universal prediction of holography as it uniquely relies on the geometrical properties of the S-BH and the AdS soliton. The simplest way to understand this point is to look at the IR boundary condition (at $z = z_0$) for A_χ and ψ, which regularity of the bulk solutions requires:

$$A_\chi' + \frac{2\,\psi^2}{3\,z_h}\left(A_\chi - \frac{m}{R}\right)\bigg|_{z=z_h} = \frac{3\psi'}{z_0} + \left(A_\chi - \frac{m}{R}\right)^2\psi\bigg|_{z=z_0} = 0 \tag{27}$$

for the S-BH and

$$A_\chi|_{z=z_0} = 0\,, \quad \begin{cases} \frac{3\psi'}{z_0} - A_0^2\psi\Big|_{z=z_0} = 0 & \text{for } m = 0\,, \\ \psi|_{z=z_0} = 0 & \text{for } m \neq 0 \end{cases} \tag{28}$$

for the AdS soliton. While the conditions in (27) only depend on the local combination of m and W, that is $m/R - a_\chi$, the conditions in (28) depend separately on the fluxoid and the Wilson line. This means that the non-local terms in the quantum effective action are suppressed (unsuppressed) in the S-BH (AdS soliton) phase. Correspondingly, according to the model-independent discussion of section 1, the system should be characterized by the Little-Parks periodicity for the S BH and an uplifting of such periodicity for the AdS soliton as shown in [23, 8].

Acknowledgements

We would like to thank Oriol Domènech, Marc Montull, Alex Pomarol, Oriol Pujolàs and Pedro J. Silva for collaborations and Daniele Dorigoni and Massimo Mannarelli for useful discussions. This work was partly supported by the EU ITN "Unification in the LHC Era", contract PITN-GA-2009-237920 (UNILHC) and by MIUR under contract 2006022501.

References

[1] Hartnoll S A 2009 *Class. Quant. Grav.* **26** 224002 (*Preprint* arXiv:0903.3246 [hep-th]). Herzog C P 2009 *J. Phys.* A **42** 343001 (*Preprint* arXiv:0904.1975 [hep-th]). McGreevy J 2010 *Adv. High Energy Phys.* 723105 (*Preprint* arXiv:0909.0518 [hep-th]). Hartnoll S A 2009 *Preprint* arXiv:0909.3553 [cond-mat.str-el]. Sachdev S 2010 *Preprint* arXiv:1002.2947 [hep-th]. Pires A S T 2010 *Preprint* arXiv:1006.5838 [cond-mat.str-el]. Horowitz G T 2011 *Class. Quant. Grav.* **28** 114008 (*Preprint* arXiv:1010.2784 [gr-qc]). Hartnoll S A 2011 *Preprint* arXiv:1106.4324 [hep-th]

[2] Lee P A, Nagaosa N and Wen X 2006 *Rev. Mod. Phys.* 78 17–85

[3] Hartnoll S A, Herzog C P and Horowitz G T 2008 *Phys. Rev. Lett.* **101** 031601 (*Preprint* arXiv:0803.3295 [hep-th])

[4] Domènech O, Montull M, Pomarol A, Salvio A and Silva P J 2010 *JHEP* **1008** 033 (*Preprint* arXiv:1005.1776 [hep-th])

[5] Herzog C P, Kovtun P K and Son D T 2009 *Phys. Rev.* D **79** 066002 (*Preprint* arXiv:0809.4870 [hep-th])

[6] Liu H, McGreevy J and Vegh D 2011 *Phys. Rev.* D **83** 065029 (*Preprint* arXiv:0903.2477 [hep-th]). Hartnoll S A, Polchinski J, Silverstein E and Tong D 2010 *JHEP* **1004** 120 (*Preprint* [arXiv:0912.1061 [hep-th]). Pal S S 2012 *Preprint* arXiv:1209.3559 [hep-th]

[7] Weinberg S 1986 *Prog. Theor. Phys. Suppl.* **86** 43

[8] Montull M, Pujolàs O, Salvio A and Silva P J 2012 *JHEP* **1204** 135 (*Preprint* arXiv:1202.0006 [hep-th])

[9] Salvio A 2012 *JHEP* **1209** 134 (*Preprint* arXiv:1207.3800 [hep-th])

[10] Horowitz G T, Santos J E and Way B 2011 *Phys. Rev. Lett.* **106** 221601 (*Preprint* arXiv:1101.3326 [hep-th])

[11] Kleiner W H, Roth L M and Autler S H 1964 *Phys. Rev.* A **133** 1226

[12] Anandan J 1981 *Phys. Rev. Lett.* **47** 7

[13] Sato Y and Packard R E 2012 *Rep. Prog. Phys.* 75 016401

[14] Little W A, Parks R D 1962 *Phys. Rev. Lett.* **9** 9

[15] Hawking S W, Page D N 1983 *Commun. Math. Phys.* **87** 577

[16] Witten E 1998 *Adv. Theor. Math. Phys.* **2** 505-532 (*Preprint* hep-th/9803131). Horowitz G T and Myers R C 1998 *Phys. Rev.* D **59** 026005 (*Preprint* hep-th/9808079)

[17] Hartnoll S A, Herzog C P and Horowitz G T 2008 *JHEP* **0812** 015 (*Preprint* arXiv:0810.1563 [hep-th])

[18] Horowitz G T and Way B 2010 *JHEP* **1011** 011 (*Preprint* arXiv:1007.3714 [hep-th])

[19] Anabalon A 2012 *JHEP* **1206** 127 (*Preprint* arXiv:1204.2720 [hep-th])

[20] Nishioka T, Ryu S and Takayanagi T 2010 *JHEP* **1003** 131 (*Preprint* arXiv:0911.0962 [hep-th])

[21] Montull M, Pomarol A and Silva P J 2009 *Phys. Rev. Lett.* **103** 091601 (*Preprint* arXiv:0906.2396 [hep-th]). Keranen V, Keski-Vakkuri E, Nowling S and Yogendran K P 2010 *Phys. Rev.* D **81** 126012 (*Preprint* arXiv:0912.4280 [hep-th])

[22] Roychowdhury D 2012 *Phys. Rev.* D **86** 106009 (*Preprint* arXiv:1211.0904 [hep-th]) Roychowdhury D 2013 *Phys. Lett.* B **718** 1089 (*Preprint* arXiv:1211.1612 [hep-th])

[23] Montull M, Pujolàs O, Salvio A and Silva P J 2011 *Phys. Rev. Lett.* **107** 181601 (*Preprint* arXiv:1105.5392 [hep-th])

De Donder-Weyl Hamiltonian formulation and precanonical quantization of vielbein gravity

I V Kanatchikov

National Center of Quantum Information in Gdansk, 81-824 Sopot, Poland

E-mail: kanattsi@gmail.com

Abstract. The De Donder-Weyl (DW) covariant Hamiltonian formulation of the Palatini first-order Lagrangian of vielbein (tetrad) gravity and its precanonical quantization are presented. No splitting into space and time is required in this formulation. Our recent generalization of Dirac brackets is used to treat the second class primary constraints appearing in the DW Hamiltonian formulation and to find the fundamental brackets. Quantization of the latter yields the representation of vielbeins as differential operators with respect to the spin connection coefficients and the Dirac-like precanonical Schrödinger equation on the space of spin connection coefficients and space time variables. The transition amplitudes on this space describe the quantum geometry of space-time. We also discuss the Hilbert space of the theory, the invariant measure on the spin connection coefficients, and point to the possible quantum singularity avoidance built in in the natural choice of the boundary conditions of the wave functions on the space of spin connection coefficients.

1. Introduction

There are several dominating approaches in the literature which aim at quantization of gravity or, more generally, a synthesis of general relativity and quantum theory. They can be conditionally classified according to their main strategies:

1) application of standard QFT techniques to the Lagrangians of general relativity theory or its alternatives (canonical QG, path integral, asymptotic safety),
2) adaptation of the classical GR to the technical requirements or limitations of QFT (LQG, shape dynamics),
3) postulating the fundamental microscopic dynamics so that classical GR would appear as an effective or emergent low energy theory (string theory, GFT, induced gravity, quantum/non-commutative space-times, causal networks).

However, considerably less efforts have been devoted to the fourth logical possibility:

4) a modification or improvement of quantum theoretic formalism and its adaptation to the geometric context of general relativity.

The distinguished role of time in the formalism and interpretation of quantum theory is one of the aspects to be overcome in the quantum formalism adapted to the goal of quantization of general relativity. As this feature of quantum description can be seen as inherited from the canonical Hamiltonian formalism whose structures underlie canonical quantization, one could try to find a generalization of canonical Hamiltonian formalism and its quantization in which all space-time variables would be treated on an equal footing. Fortunately, such generalizations are known in the mathematical theory of the calculus of variations of multiple integrals and there is an infinite variety of covariant finite-dimensional Hamiltonian-like formulations given by different Lepage equivalents of the Poincaré-Cartan form [1–5]) which, from the point of view of physics, implement exactly this idea. The simplest of these formulations is known as the De Donder-Weyl (DW) theory (see e.g. [1,6,7]).

The DW Hamiltonian formulation of a field theory given by the first order Lagrangian $L = L(y^a, y^a_\mu, x^\nu)$ uses the covariant Legendre transformation to the new set of variables: *polymomenta*

$$p^\mu_a := \frac{\partial L}{\partial y^a_\mu}$$

and the *DW Hamiltonian function*

$$H(y^a, p^\mu_a, x^\mu) := y^a_\mu(y,p)p^\mu_a - L,$$

which, for regular theories with

$$\det \left|\left|\partial^2 L/\partial y^\mu_a \partial y^\nu_b\right|\right| \neq 0,$$

enable us to write the field equations in the DW covariant Hamiltonian form:

$$\partial_\mu y^a(x) = \frac{\partial H}{\partial p^\mu_a}, \quad \partial_\mu p^\mu_a(x) = -\frac{\partial H}{\partial y^a}. \tag{1}$$

The latter look like a multidimensional field theoretic analogue of the Hamilton equations with all space-time variables treated on an equal footing.

A generalization of Poisson brackets to the DW Hamiltonian formulation [8–10], which is suitable for quantization, is defined on semi-basic forms on the polymomentum phase space (with the space-time being the base manifold and the space of field and polymomentum variables being the fiber) and leads to the Poisson-Gerstenhaber algebra structure with respect to the graded Lie bracket and a special $\bullet$-product of forms:

$$A \bullet B := *^{-1}(*A \wedge *B), \tag{2}$$

called co-exterior.

Just as the canonical quantization proceeds from the mathematical structures of canonical Hamiltonian formalism, *precanonical quantization* starts from the mathematical structures underlying the DW Hamiltonian formulation: the polysymplectic form, Poisson-Gerstenhaber brackets, DW Hamilton-Jacobi theory, etc.

It was found in our earlier work [11–13] that the quantization of the subalgebra of precanonically conjugate variables (similar to the Heisenberg algebra) leads to the following representation of the operators of polymomenta:

$$\hat{p}^\nu_a = -i\hbar\varkappa\gamma^\nu \frac{\partial}{\partial y^a}, \tag{3}$$

which act on the Clifford-valued wave functions $\Psi(y,x)$ on the finite dimensional covariant configuration space of field and space-time variables y and x.

The constant $\varkappa$ of the dimension ℓ^{1-n} in n space-time dimensions appears in precanonical quantization on the dimensional grounds. Its meaning as the inverse of a very small "elementary volume" is obvious e.g. in the representation of the basic $(n-1)$-forms

$$\varpi_\nu := \partial_\nu \lrcorner (dx^0 \wedge ... \wedge dx^{n-1})$$

in terms of the space-time Clifford algebra elements:

$$\widehat{\varpi}_\nu = \frac{1}{\varkappa}\gamma_\nu.$$

Note that the approach of precanonical quantization does not modify the microscopic structure of space-time by any *ad hoc* assumptions.

The covariant analogue of the Schrödinger equation in precanonical quantization reads

$$i\hbar\varkappa\gamma^\mu\partial_\mu\Psi = \hat{H}\Psi, \tag{4}$$

where $\hat{H}$ is the DW Hamiltonian operator composed from the partial differential operators with respect to the field variables, c.f. (3). For the free scalar field y

$$H = \frac{1}{2}p^\mu p_\mu + \frac{1}{2}\frac{m^2}{\hbar^2}y^2$$

and the operator $\hat{H}$ corresponds to the harmonic oscillator along the field dimension y [11,12,14]:

$$\hat{H} = -\frac{1}{2}\,\hbar^2\varkappa^2\frac{\partial^2}{\partial y^2} + \frac{1}{2}\frac{m^2}{\hbar^2}y^2. \tag{5}$$

The self-adjointness of $\hat{H}$ with respect to the inner product

$$\langle\Phi|\Psi\rangle := \int dy\ \overline{\Psi}\Psi, \tag{6}$$

where $\overline{\Psi} := \Psi^\dagger\gamma^0$, and Eq. (4) lead to the conservation law

$$\partial_\mu \int dy \overline{\Psi}\gamma^\mu\Psi = 0 \tag{7}$$

which makes the probabilistic interpretation of $\Psi(y,x)$ possible. Note however, that in pseudoeuclidean space-times the inner product in (6) is indefinite, while the conserved quantity

$$\int d\mathbf{x} \int dy\ \overline{\Psi}\gamma^0\Psi \tag{8}$$

is positive definite (here the notation $x^\mu = (\mathbf{x}, t)$ is used). Hence, the approach of precanonical quantization implies a generalization of mathematical formalism of quantum theory with an indefinite metric Hilbert space, where γ^0 plays the role of J-operator (see e.g. [15]).

The particle interpretation of the free scalar field is suggested by the spectrum of DW Hamiltonian operator in (5): $\varkappa m(N+\frac{1}{2})$ with $N \in \mathbb{N}$, which implies that free particles of mass m correspond to the transitions between the neighbouring eigenstates of DW Hamiltonian operator.

The relation between the precanonical Schrödinger equation (4) and the functional differential Schrödinger equation following from canonical quantization:

$$i\hbar\partial_t\mathbf{\Psi} = \widehat{\mathbf{H}}\mathbf{\Psi},$$

where $\mathbf{\Psi} = \mathbf{\Psi}([y(\mathbf{x})], t)$ is the Schrödinger wave functional and $\widehat{\mathbf{H}}$ is the functional derivative operator of the canonical Hamiltonian functional, is established by assuming that there is a relation between $\mathbf{\Psi}$ and the precanonical wave function $\Psi(y,x)$ restricted to the subspace $\Sigma : (y = y(\mathbf{x}), t = \text{const})$:

$$\mathbf{\Psi}([y(\mathbf{x})],t) = \mathbf{\Psi}([\Psi_\Sigma(y(\mathbf{x}),\mathbf{x},t)],[y^a(\mathbf{x})]),$$

and substituting the equation for $\partial_t \Psi_\Sigma$ following from the restriction of (4) to Σ into the chain rule differentiation

$$i\partial_t \mathbf{\Psi} = \int d\mathbf{x}\ \mathsf{Tr}\left\{ \frac{\delta \mathbf{\Psi}}{\delta \Psi^T_\Sigma(y^a(\mathbf{x}),\mathbf{x},t)} i\partial_t \Psi_\Sigma(y^a(\mathbf{x}),\mathbf{x},t)\right\}. \tag{9}$$

Then, in the limiting case $\gamma^0\varkappa \rightarrow \delta(\mathbf{0})$, we are able to obtain the functional differential Schrödinger equation as the consequence of (4) and the expression of the Schrödinger wave functional in terms of the continuum product of precanonical wave functions [16,17] (c.f. [14,18]):

$$\mathbf{\Psi} = \mathsf{Tr}\left\{\prod_{\mathbf{x}} e^{-iy(\mathbf{x})\gamma^i\partial_i y(\mathbf{x})/\varkappa}\Psi_\Sigma(y(\mathbf{x}),\mathbf{x},t)\right\}\Bigg|_{\gamma^0\varkappa\rightarrow\delta(\mathbf{0})}. \tag{10}$$

The existence of such a relation between precanonical quantization and functional Schrödinger representation suggests that the standard QFT based on canonical quantization is a singular limit of QFT based on precanonical quantization when the "elementary volume" $1/\varkappa$ is vanishing. Note also that the map $\gamma^0\varkappa \rightarrow \delta(\mathbf{0})$ is actually the inverse of the "quantization map" from the exterior forms to Clifford numbers: $\widehat{\varpi}_0 = \frac{1}{\varkappa}\gamma_0$, which underlie precanonical quantization, in the limit of infinite $\varkappa$.

2. DW Hamiltonian formulation of vielbein/tetrad gravity

Because the Dirac operator enters in the precanonical analogue of the covariant Schrödinger equation (4), the vielbein formulation of gravity is a more natural framework for precanonical quantization than the metric formulation used in our earlier work [19]. The latter essentially led to a hybrid quantum-classical theory (c.f. [20,21]) because a part of the spin connection term in the curved space-time Dirac operator in (4) can not be expressed and quantized in terms of the variables of the metric formulation.

Let us consider the first order Palatini type Lagrangian density of Einstein's gravity with the cosmological term:

$$\mathfrak{L} = \frac{1}{\kappa_E}\mathfrak{e}e^{[\alpha}_I e^{\beta]}_J(\partial_\alpha\omega_\beta{}^{IJ} + \omega_\alpha{}^{IK}\omega_{\beta K}{}^J) + \frac{1}{\kappa_E}\Lambda\mathfrak{e}, \tag{11}$$

where e^μ_I are the vielbein components, ω^{IJ}_α are torsion-free spin connection coefficients, $\kappa_E := 8\pi G$, and $\mathfrak{e} := \det||e^a_\mu||$.

The polymomenta associated with the vielbein and spin connection field variables treated as independent dynamical variables:

$$\mathfrak{p}^\alpha_{e^I_\beta} = \frac{\partial\mathfrak{L}}{\partial_\alpha e^I_\beta} \text{ and } \mathfrak{p}^\alpha_{\omega^{IJ}_\beta} = \frac{\partial\mathfrak{L}}{\partial_\alpha\omega^{IJ}_\beta},$$

yield the primary constraints of the DW Hamiltonian formalism, viz.

$$\mathfrak{p}^\alpha_{e^I_\beta} \approx 0, \quad \mathfrak{p}^\alpha_{\omega^{IJ}_\beta} \approx \frac{1}{\kappa_E}\mathfrak{e}e^{[\alpha}_I e^{\beta]}_J. \tag{12}$$

Consequently, not all space-time gradients of vielbein and spin connection fields can be expressed as functions of polymomenta and fields and we need to develop an analogue of the constraints analysis within the DW formalism.

Notwithstanding the fact that a mathematical literature related to the DW Hamiltonian theory with constraints exists [22–28], the analysis suitable for the purposes of quantization, though incomplete, seems to be found only in our paper [29]. The idea of that paper is to use the $(n-1)$-forms constructed from the constraints and their Poisson-Gerstenhaber brackets found within the DW formalism in our earlier papers [8–10], and to try to find a generalization of the Dirac's treatment of constrains in the Hamiltonian formalism of mechanics to the DW Hamiltonian formalism in field theory.

Following this line and using the primary constraints (12), let us write down an extended DW Hamiltonian density

$$\mathfrak{H} = -\frac{1}{\kappa_E}\mathfrak{e} e_I^{[\alpha} e_J^{\beta]} \omega_\alpha{}^{IK} \omega_{\beta K}{}^J - \frac{1}{\kappa_E}\Lambda\mathfrak{e} + \mu^I_{\alpha\beta}\mathfrak{p}^\alpha_{e^I_\beta} + \lambda^{IJ}_{\alpha\beta}\Big(\mathfrak{p}^\alpha_{\omega^{IJ}_\beta} - \frac{1}{\kappa_E}\mathfrak{e} e_I^{[\alpha} e_J^{\beta]}\Big), \tag{13}$$

where μ and λ are the Lagrange multipliers. The DW Hamiltonian equations given by $\mathfrak{H}$ yield:

$$\partial_\alpha e^I_\beta = \mu^I_{\alpha\beta}, \quad \partial_{[\alpha}\omega^{IJ}_{\beta]} = \lambda^{IJ}_{\alpha\beta}, \tag{14}$$

$$\partial_\alpha \mathfrak{p}^\alpha_{e^I_\beta} = -\frac{\partial\mathfrak{H}}{\partial e^I_\beta}, \quad \partial_\alpha \mathfrak{p}^\alpha_{\omega^{IJ}_\beta} = -\frac{\partial\mathfrak{H}}{\partial \omega^{IJ}_\beta}. \tag{15}$$

The first equation in (15) and the second one in (14) reproduce, on the constraints subspace, the Einstein equations. The second equation in (15) and the first one in (14) lead to the covariant constancy condition:

$$\nabla_\beta(\mathfrak{e} e_I^{[\alpha} e_J^{\beta]}) = 0, \tag{16}$$

which can be transformed into the expression of the spin connection in terms of vielbeins and their derivatives.

Equations (14), (15) are equivalent to the preservation of semi-basic $(n-1)-$forms constructed from the constraints (12):

$$\mathfrak{C}_{e^I_\beta} := \mathfrak{p}^\alpha_{e^I_\beta}\varpi_\alpha, \quad \mathfrak{C}_{\omega^{IJ}_\beta} := \mathfrak{p}^\alpha_{\omega^{IJ}_\beta}\varpi_\alpha - \frac{1}{\kappa_E}\mathfrak{e} e_I^{[\alpha} e_J^{\beta]}\varpi_\alpha. \tag{17}$$

By calculating the brackets of those forms using the local coordinate expression of the *polysymplectic form* introduced in our papers [8–10]:

$$\boldsymbol{\Omega} = d\mathfrak{p}^\alpha_{e^I_\beta}\wedge de^I_\beta\wedge\varpi_\alpha + d\mathfrak{p}^\alpha_{\omega^{IJ}_\beta}\wedge d\omega^{IJ}_\beta\wedge\varpi_\alpha, \tag{18}$$

we obtain:

$$\begin{aligned} \{\![\mathfrak{C}_e, \mathfrak{C}_{e'}]\!\} &= 0, \\ \{\![\mathfrak{C}_\omega, \mathfrak{C}_{\omega'}]\!\} &= 0, \\ \{\![\mathfrak{C}_{e^K_\gamma}, \mathfrak{C}_{\omega^{IJ}_\beta}]\!\} &= -\frac{1}{\kappa_E}\frac{\partial}{\partial e^K_\gamma}\left(\mathfrak{e} e_I^{[\alpha} e_J^{\beta]}\right)\varpi_\alpha, \end{aligned} \tag{19}$$

were the bracket of two semi-basic Hamiltonian $(n-1)$-forms F and G is defined as follows:

$$\{\![F, G]\!\} := -X_F \rfloor dG, \tag{20}$$

where

$$X_F \rfloor \boldsymbol{\Omega} := dF.$$

From (19) we conclude that the primary constraints in (12) are second class.

We use our generalization of Dirac bracket to the singular DW Hamiltonian formalism [29]:

$$\{\![F,G]\!\}^D := \{\![F,G]\!\} - \{\![F,\mathfrak{C}_U]\!\} \bullet \left(\mathfrak{C}^{-1}_{UV} \wedge \{\![\mathfrak{C}_V, G]\!\}\right), \tag{21}$$

where the indices U, V run over all components of e and ω and the inverse of the form-valued matrix $\mathfrak{C}_{UV} := \{\![\mathfrak{C}_U, \mathfrak{C}_V]\!\}$ is defined by

$$\mathfrak{C}^{-1}_{UV} \wedge \mathfrak{C}_{VU'} := \sigma\varpi\delta_{UU'}, \tag{22}$$

where $\sigma = (-)1$ is the (pseudo)euclidean signature of the metric, $dx^\mu \wedge \varpi_\nu =: \delta^\mu_\nu \varpi$, and $\sigma\varpi\bullet$ is the unit operator when acting on any semi-basic form, Then we can obtain the following brackets of precanonically conjugate variables on the subalgebra of $(n-1)$-forms:

$$\{\![\mathfrak{p}^\alpha_e \varpi_\alpha, e'\varpi_{\alpha'}]\!\}^D = 0, \tag{23}$$
$$\{\![\mathfrak{p}^\alpha_\omega \varpi_\alpha, \omega'\varpi_{\alpha'}]\!\}^D = \{\![\mathfrak{p}^\alpha_\omega \varpi_\alpha, \omega'\varpi_{\alpha'}]\!\} = \delta^{\omega'}_\omega \varpi_{\alpha'}, \tag{24}$$
$$\{\![\mathfrak{p}^\alpha_e \varpi_\alpha, \mathfrak{p}_\omega]\!\}^D = \{\![\mathfrak{p}^\alpha_e \varpi_\alpha, \omega'\varpi_{\alpha'}]\!\}^D = \{\![\mathfrak{p}^\alpha_\omega \varpi_\alpha, e'\varpi_{\alpha'}]\!\}^D = 0, \tag{25}$$

and similar precanonical brackets of $(n-1)$- and 0-forms, which also constitute a subalgebra with respect to the Poisson-Gerstenhaber bracket operation:

$$\{\![\mathfrak{p}^\alpha_e \varpi_\alpha, e']\!\}^D = 0, \tag{26}$$
$$\{\![\mathfrak{p}^\alpha_\omega \varpi_\alpha, \omega']\!\}^D = \{\![\mathfrak{p}^\alpha_\omega \varpi_\alpha, \omega']\!\} = \delta^{\omega'}_\omega, \tag{27}$$
$$\{\![\mathfrak{p}^\alpha_e \varpi_\alpha, \mathfrak{p}_\omega]\!\}^D = \{\![\mathfrak{p}^\alpha_e \varpi_\alpha, \omega]\!\}^D = \{\![\mathfrak{p}^\alpha_\omega \varpi_\alpha, e]\!\}^D = 0, \tag{28}$$

and

$$\{\![\mathfrak{p}^\alpha_e, e'\varpi_{\alpha'}]\!\}^D = 0, \tag{29}$$
$$\{\![p^\alpha_\omega, \omega'\varpi_\beta]\!\}^D = \{\![p^\alpha_\omega, \omega'\varpi_\beta]\!\} = \delta^\alpha_\beta \delta^\omega_{\omega'}, \tag{30}$$
$$\{\![\mathfrak{p}^\alpha_e, \mathfrak{p}_\omega \varpi_{\alpha'}]\!\}^D = \{\![\mathfrak{p}^\alpha_e, \omega\varpi_{\alpha'}]\!\}^D = \{\![\mathfrak{p}^\alpha_\omega, e'\varpi_{\alpha'}]\!\}^D = 0. \tag{31}$$

The following remarks regarding the above calculation are in order. Note that the formula in (21) assumes that $\mathfrak{C}^{-1}_{UV}$ exists in the sense of (22) [29]. However, it is not the case for the matrix defined by (19):

$$\mathfrak{C}_{UV} := \begin{Vmatrix} 0 & \mathfrak{C}_{e\omega} \\ \mathfrak{C}_{\omega e} & 0 \end{Vmatrix}, \tag{32}$$

where $\mathfrak{C}_{e\omega} := \{\![\mathfrak{C}_e, \mathfrak{C}_\omega]\!\}$ is a rectangular matrix (16×24 in $n = 4$ dimensions). In the usual Dirac's Hamiltonian formalism it would signal that not all of the second class constraints are found. However, it is not necessarily the case here, because the number of polymomenta is different from the number of field variables and the analogues of the symplectic matrix in the polysymplectic formalism are singular matrices similar to the higher dimensional Duffin-Kemmer-Petiau matrices (c.f. [30]) whose algebraic definition is actually tantamount to the statement that, up to a sign factor, they are generalized Moore-Penrose inverse to themselves. More generally than in (22) we can understand $\mathfrak{C}^{-1}_{UV}$ as a generalized inverse such that

$$\mathfrak{C}_{UU'} \bullet (\mathfrak{C}^{-1}_{U'V'} \wedge \mathfrak{C}_{V'V}) = \mathfrak{C}_{UV}. \tag{33}$$

Then the specific structure of (32) ensures that the Moore-Penrose-type generalized inverse of $\mathfrak{C}_{UV}$ has the same matrix block structure as (32) with $\mathfrak{C}_{e\omega}$ replaced by $\mathfrak{C}^{-1}_{e\omega}$, so that (22) is fulfilled on the e-subspace, viz.

$$\mathfrak{C}^{-1}_{eU'} \wedge \mathfrak{C}_{U'e'} = \delta_{ee'}\sigma\varpi. \tag{34}$$

This is the e-subspace inverse which is needed in order to calculate the brackets in (23), (26), and (29). For example, by denoting $\mathfrak{p}_e := \mathfrak{p}_e^\alpha \varpi_\alpha$, we obtain:

$$\{\![e', \mathfrak{p}_e]\!\}^D = \{\![e', \mathfrak{p}_e]\!\} - \{\![e', \mathfrak{C}_{e''}]\!\} \bullet (\mathfrak{C}^{-1}_{e''\omega} \wedge \{\![\mathfrak{C}_\omega, \mathfrak{p}_e]\!\}) = -\delta^e_{e'} + \delta^{e''}_{e'} \bullet (\mathfrak{C}^{-1}_{e''w} \wedge \mathfrak{C}_{we}) = 0, \quad (35)$$

which is the result in (26). All other brackets in (23)-(31) are vanishing as a consequence of the specific matrix block structure of $\mathfrak{C}_{UV}$ and its generalized inverse.

The brackets in (23)-(31) are assumed to be the analogue of the fundamental Dirac brackets of canonical variables in constrained mechanics and they underlie the quantization procedure below.

3. Quantization

Usually, quantization of systems with second class constraints is performed by transforming the Dirac brackets into commutators according to the Dirac's quantization rule. However, in the present approach the latter has to be modified in order to make sure that densities are quantized as density valued operators, viz.

$$[\hat{A}, \hat{B}] = -i\hbar\mathfrak{e}\widehat{\{\![A,B]\!\}^D}, \quad (36)$$

where (the operator of) $\mathfrak{e}$ appears due to the fact that the polysymplectic form and polymonenta are densities.

Now, quantization of brackets in (23), (25), (26), (28), (29), (31) and the constraint $p_e \approx 0$ lead us to the conclusion that the operators of the conjugate polymomenta of vielbeins are zero: $\hat{p}_e = 0$. We can, therefore, set our precanonical wave function to depend only on the spin connection and space-time variables, i.e. $\Psi = \Psi(\omega^{IJ}_\alpha, x^\mu)$.

Further, quantization of the Dirac bracket in (27) yields

$$\widehat{p^\alpha_{\omega^{IJ}_\beta}\varpi_\alpha} = -i\hbar\hat{\mathfrak{e}}\frac{\partial}{\partial\omega^{IJ}_\beta}. \quad (37)$$

Moreover, by quantizing (30), which coincides with the familiar bracket of polymomenta and field variables that underlies precanonical quantization in flat space-time [11, 12], we obtain the formal representation of polymomenta:

$$\hat{\mathfrak{p}}^\alpha_{\omega^{IJ}_\beta} = -i\hbar\varkappa \,\vdots\, \hat{\mathfrak{e}}\,\hat{\gamma}^{[\alpha}\frac{\partial}{\partial\omega^{IJ}_{\beta]}} \,\vdots\,, \quad (38)$$

where the density $\hat{\mathfrak{e}}$ and the curved space-time Dirac matrices $\hat{\gamma}^\alpha$ are yet unknown operators, and $\vdots$ stands for a potential operator ordering ambiguity. Note that when obtaining (38) we still assumed that

$$\widehat{\varpi}_\nu = \frac{1}{\varkappa}\widehat{\gamma}_\nu, \quad (39)$$

which is just a formal generalization of the relation known from precanonical quantization in flat space-time [11, 12], as far as the explicit operator representation of $\widehat{\gamma}_\nu$ is not known explicitly.

Next, let us insert the precanonical operator representation of $\hat{\mathfrak{p}}^\alpha_\omega$, Eq. (38), into the strong operator version of the second constraint in (12) and contract it with flat $\bar{\gamma}^{IJ}$-s:

$$(\mathfrak{e}e^{[\alpha}_I e^{\beta]}_J \bar{\gamma}^{IJ})^{op} = \widehat{\mathfrak{e}\gamma}^{\alpha\beta} = \kappa_E(\mathfrak{p}^\alpha_{\omega^{IJ}_\beta}\bar{\gamma}^{IJ})^{op}, \quad (40)$$

where $()^{op}$ replaces the hat over the longer expressions and

$$\hat{\gamma}^\nu := \hat{e}^\nu_I\bar{\gamma}^I,$$

where $\bar{\gamma}^I\bar{\gamma}^J + \bar{\gamma}^J\bar{\gamma}^I = 2\eta^{IJ}$, η^{IJ} is a fiducial flat Minkowski metric with the signature $+--...-$, and $\bar{\gamma}^{IJ} := \bar{\gamma}^{[I}\bar{\gamma}^{J]}$. A comparison with (38) yields the operator representation of the curved space-time Dirac matrices:

$$\widehat{\gamma}^\beta = -i\hbar\varkappa\kappa_E\bar{\gamma}^{IJ}\frac{\partial}{\partial\omega_\beta^{IJ}}\,, \tag{41}$$

vielbeins:

$$\hat{e}_I^\beta = -i\hbar\varkappa\kappa_E\bar{\gamma}^J\frac{\partial}{\partial\omega_\beta^{IJ}}\,, \tag{42}$$

and the polymomenta conjugate to spin connection:

$$\hat{\mathfrak{p}}^\alpha_{\omega_\beta^{IJ}} = -\hbar^2\varkappa^2\kappa_E\,\hat{\mathfrak{e}}\,\bar{\gamma}^{KL}\frac{\partial}{\partial\omega_{[\alpha}^{KL}}\frac{\partial}{\partial\omega_{\beta]}^{IJ}}\,, \tag{43}$$

where the operator of $\mathfrak{e}$ can now be constructed from (42):

$$\hat{\mathfrak{e}} = \left(\frac{1}{n!}\epsilon^{I_1\ldots I_n}\epsilon_{\mu_1\ldots\mu_n}\hat{e}_{I_1}^{\mu_1}\ldots\hat{e}_{I_n}^{\mu_n}\right)^{-1}. \tag{44}$$

We can also obtain the operators of $(n-1)$-volume elements (39), that leads to a rather complicated non-local expression:

$$\widehat{\omega}_\nu = \frac{1}{\varkappa(n-1)!}\,\hat{\mathfrak{e}}\;\epsilon_{\nu\mu_1\ldots\mu_{n-1}}\widehat{\gamma}^{\mu_1}\ldots\widehat{\gamma}^{\mu_{n-1}}, \tag{45}$$

and the operator of the metric tensor $g^{\mu\nu}$:

$$\widehat{g^{\mu\nu}} = -\hbar^2\varkappa^2\kappa_E^2\eta^{IJ}\eta^{KL}\frac{\partial^2}{\partial\omega_\mu^{IK}\partial\omega_\nu^{JL}}\,. \tag{46}$$

Finally, using (43) we construct the DW Hamiltonian operator $\widehat{H}$ which corresponds to the DW Hamiltonian density restricted to the subspace of constraints (12), $\mathfrak{e}H := \mathfrak{H}|_C$:

$$\widehat{H} = \hbar^2\varkappa^2\kappa_E\bar{\gamma}^{IJ}\!:\omega_{[\alpha}{}^{KM}\omega_{\beta]M}{}^L\frac{\partial}{\partial\omega_\alpha^{IJ}}\frac{\partial}{\partial\omega_\beta^{KL}}\!: - \frac{1}{\kappa_E}\Lambda\,. \tag{47}$$

4. Covariant Schrödinger equation for quantum gravity

The precanonical covariant Schrödinger equation which generalizes Eq. (4) to the context of quantum gravity will have the form

$$i\hbar\varkappa\widehat{\nabla\!\!\!/}\Psi = \widehat{H}\Psi, \tag{48}$$

were $\widehat{\nabla\!\!\!/} := (\gamma^\mu(\partial_\mu + \omega_\mu))^{op}$ with the spin connection term $\omega_\mu := \frac{1}{4}\omega_{\mu IJ}\bar{\gamma}^{IJ}$ is what we called the "quantized Dirac operator", because the Dirac matrices and the spin connection term in it are now operators themselves. Using the operator representation of the curved space γ-matrices in (41) we obtain:

$$\widehat{\nabla\!\!\!/} = -i\hbar\varkappa\kappa_E\bar{\gamma}^{IJ}\!:\frac{\partial}{\partial\omega_\mu^{IJ}}\left(\partial_\mu + \frac{1}{4}\omega_{\mu KL}\bar{\gamma}^{KL}\right)\!:. \tag{49}$$

Therefore, the precanonical counterpart of the Schrödinger equation for quantum gravity takes the form

$$\bar{\gamma}^{IJ}\!:\left(\partial_\mu + \frac{1}{4}\omega_{\mu KL}\bar{\gamma}^{KL} - \omega_{\mu M}^K\omega_\beta^{ML}\frac{\partial}{\partial\omega_\beta^{KL}}\right)\frac{\partial}{\partial\omega_\mu^{IJ}}\!:\Psi + \frac{\Lambda}{\hbar^2\varkappa^2\kappa_E^2}\Psi = 0 \tag{50}$$

and determines the wave function $\Psi(\omega, x)$ or, more generally, the transition amplitudes $\langle\omega, x|\omega', x'\rangle$. The latter provide an inherently quantum description of the geometry of space-time which generalizes classical differential geometry with its smooth connection fields $\omega(x)$.

Let us note that the combination of the constants including $\varkappa$ in the last term of (50) is dimensionless. By fixing the operator ordering in (50) we would generate a dimensionless constant of the order $\sim n^6$ (the number of components of ω-s is $\sim n^3$) which can be interpreted as the cosmological constant Λ devided by $\hbar^2\varkappa^2\kappa_E^2$. In this case, however, the observable value of Λ is obtained (at $n = 4$) only if $\varkappa$ is roughly at the nuclear energy scale, which is far away from our original expectation that $\varkappa$ is at about the Planck scale and contradicts the experimental evidence that the usual relativistic space-time holds even at TeV scale. On the other hand, if we take $\varkappa$ at the Planck scale then we arrive at the familiar 120 orders of magnitude error in the estimation of the cosmological constant, which is usually obtained by using the Planck scale cutoff in the momentum space integration of the zero point energies. This rather confirms that the constant $\varkappa$ of precanonical quantization is related to the ultra-violet cutoff scale and that the cosmological constant problem is not related to the ground state of pure quantum gravity but rather to the particle composition of the universe.

5. Hilbert space

It is natural to assume that the wave functions $\Psi(\omega, x)$ vanish at large values of ω-s. Then the probability amplitude of observing the regions of space-time with very large curvature is very small, so that the quantum gravitational singularity avoidance is essentially built in in the choice of the boundary condition in ω-space.

The scalar product is expected to have the form:

$$\langle\Phi|\Psi\rangle := \int [d\omega]\overline{\Phi}\Psi,$$

where $[d\omega]$ is an invariant measure on the space of spin connection coefficients. Using the arguments similar to those used by Misner to obtain the invariant measure on the space of metrics [31], we found:

$$[d\omega] = \mathfrak{e}^{-n(n-1)} \prod_{\mu, I<J} d\omega_\mu^{IJ}. \tag{51}$$

Because in the present picture $\mathfrak{e}$ is an operator given by (44), the measure $[d\omega]$ is operator valued and the scalar product of the theory has the form

$$\langle\Phi|\Psi\rangle := \int \overline{\Phi}\, \widehat{[d\omega]}\, \Psi. \tag{52}$$

Then the most natural definition of the expectation values of operators using the scalar product with the operator valued measure implies the Weyl ordering, viz.

$$\langle\hat{O}\rangle := \int \overline{\Psi} \left(\widehat{[d\omega]}\hat{O}\right)_W \Psi. \tag{53}$$

When discussing the specific physical problems using the formalism of this paper we will have to distinguish between the physical aspects and those attributed to the choice of coordinates. The latter are a macroscopic notion due to the observer's choice and, therefore, can be implemented on the average. For example, the choice of the harmonic coordinates on the average leads to the following condition on the wave function $\Psi(\omega, x)$:

$$\partial_\mu \left\langle \Psi(\omega, x) \left| \widehat{\mathfrak{e}g^{\mu\nu}} \right| \Psi(\omega, x) \right\rangle = 0, \tag{54}$$

which should be solved together with the covariant Schrödinger equation, Eq. (50), and that makes the problem more complicated.

6. Conclusion

Quantization of vielbein gravity using the approach of precanonical quantization, which is based on the De Donder-Weyl covariant Hamiltonian formulation, is discussed. All space-time variables are treated on an equal footing as generalizations of the time parameter in non-relativistic mechanics. No global splitting to space and time is required.

The DW Hamiltonian formulation of the first order Palatini action of vielbein gravity leads to the second class primary constraints which are treated using our recent generalization of Dirac bracket to the DW formalism [29]. The consideration of the fundamental generalized Dirac brackets of precanonically conjugate variables, Eqs. (23)-(31), leads to the conclusion that the quantum dynamics of gravity can be formulated using the wave functions on the space of spin connection coefficients and space-time variables. The operators of vielbeins, metric tensor, DW Hamiltonian operator and the quantized Dirac operator which enters the covariant precanonical Schrödinger equation of quantum gravity are explicitly constructed. We also discuss the Hilbert space of the theory and the invariant operator-valued integration measure on the space of spin connection coefficients. Let us note that the resulting (still tentative) formulation of quantum gravity is non-perturbative, covariant and background independent.

However, it is not clear at this stage if the consideration of the fundamental Dirac brackets in (23)-(31) is sufficient, because on the subalgebra of 0- and $(n-1)$-forms we can also calculate brackets between the forms composed from vielbeins and spin connection coefficients, such as $\{\![e, \omega\, \varpi_\mu]\!\}^D \sim \partial_\mu \lrcorner\, \mathfrak{C}^{-1}_{ew}$, which explicitly depend on the complicated nonlinear expression of the generalized inverse of the rectangular matrix $\mathfrak{C}_{ew}$ in (19) and cannot be quantized directly using the Dirac's quantization rule (36).

Among the issues left beyond the scope of the paper there are the details of the choice of the coordinate (gauge) conditions on the average (c.f. [32]), which are not sufficiently clear to us, and the issues related to the indefinite inner product Hilbert space appearing in the formalism of the theory. We were also unable so far to demonstrate that the present formulation reproduces the Einstein equations on the average or in the classical limit. We hope to elaborate on those issues in the forthcoming publications.

Acknowledgments

I thank R. Sverdlov for his critical remarks and M. Reginatto and A. Khugaev for encouraging discussions.

References

[1] Kastrup H 1983 Canonical theories of dynamical systems in physics *Phys. Rep.* **101** 1-156
[2] Krupka D 1973 Some geometric aspects of variational problems in fibred manifors *Folia Fac. Sci. Nat. Univ. Purk. Brunensis, Physica* **XIV** (*Preprint* math-ph/0110005)
[3] Olver P J 1993 Equivalence and the Cartan form *Acta Applicandae Math.* **31** 99-136
[4] Krupková O 2002 Hamiltonian field theory *J. Geom. Phys.* **43** 93-132
[5] Hélein F and Kouneiher J 2004 Covariant Hamiltonian formalism for the calculus of variations with several variables: Lepage-Dedecker versus de Donder–Weyl *Adv. Theor. Math. Phys.* **8** 565601 (*Preprint* math-ph/0401046)
[6] De Donder Th 1935 *Théorie invariantive du calcul des variations* (Paris: Gauthier-Villars)
[7] Rund H 1973 *The Hamilton-Jacobi theory in the calculus of variations* (Nuntington. N.Y.: Robert E. Krieger Publ. Co.)
[8] Kanatchikov I V 1993 On the canonical structure of De Donder-Weyl covariant Hamiltonian formulation of field theory. 1. Graded Poisson brackets and equations of motion *Preprint* hep-th/9312162
[9] Kanatchikov I V 1997 On field theoretic generalizations of a Poisson algebra *Rep. Math. Phys.* **40** 225-234 (*Preprint* hep-th/9710069)
[10] Kanatchikov I V 1998 Canonical structure of classical field theory in the polymomentum phase space *Rep. Math. Phys.* **41** 49-90 (*Preprint* hep-th/9709229)
[11] Kanatchikov I V 1998 On quantization of field theories in polymomentum variables *AIP Conf. Proc.* **453** 356-367 (*Preprint* hep-th/9811016)

[12] Kanatchikov I V 1999 Donder-Weyl theory and a hypercomplex extension of quantum mechanics to field theory *Rept. Math. Phys.* **43** 157-170 (*Preprint* hep-th/9810165)
[13] Kanatchikov I V 2001 Geometric (pre)quantization in the polysymplectic approach to field theory *Preprint* hep-th/0112263
[14] Kanatchikov I V 1998 Towards the Born-Weyl quantization of fields *Int. J. Theor. Phys.* **37** 333-342 (*Preprint* quant-ph/9712058)
[15] Azizov T Ya and Iokhvidov I S 1989 *Linear operators in spaces with an indefinite metric* (Chichester: John Wiley & Sons)
[16] Kanatchikov I V 2011 Precanonical quantization and the Schrödinger wave functional revisited *Preprint* arXiv:1112.5801
[17] Kanatchikov I V 2013 Precanonical structure of the Schrödinger wave functional (*in preparation*)
[18] Kanatchikov I V 2001 Precanonical quantization and the Schrödinger wave functional *Phys. Lett.* **A283** 25-36 (*Preprint* hep-th/0012084)
[19] Kanatchikov I V 2001 Precanonical quantum gravity: quantization without the space-time decomposition *Int. J. Theor. Phys.* **40** 1121-1149 (*Preprint* gr-qc/0012074); See also: Kanatchikov I V, From the De Donder-Weyl Hamiltonian formalism to quantization of gravity *Preprint* gr-qc/9810076; Quantization of gravity: yet another way *Preprint* gr-qc/9912094; Precanonical perspective in quantum gravity *Nucl. Phys. Proc. Suppl.* **88** 326-330 (2000) (*Preprint* gr-qc/0004066).
[20] Elze H T 2012 Linear dynamics of quantum-classical hybrids *Phys. Rev.* A **85** 052109 (*Preprint* arXiv:1111.2276)
[21] Elze H T 2012 Proliferation of observables and measurement in quantum-classical hybrids *Int. J. Quantum Inform.* **10** 1241012 (*Preprint* arXiv:1212.2380)
[22] Gràcia X, Martín R and Román-Roy N 2009 Constraint algorithm for k-presymplectic Hamiltonian systems, Application to singular field theories *Int. J. Geom. Methods Mod. Phys.* **06** 851 [*Preprint* arXiv:0903.1791]
[23] Castrillón López M and Marsden J E 2003 Some remarks on Lagrangian and Poisson reduction for field theories *J. Geom. Phys.* **48** 52-83
[24] Sardanashvily G 2000 Constraints in polysymplectic (covariant) Hamiltonian formalism *Preprint* math-ph/0008024; See also: Sardanashvily G 1995 *Generalized Hamiltonian Formalism for Field Theory: Constraint Systems* (Singapore, World Scientific)
[25] Gràcia X, Marín-Solano J and Muñoz-Lecanda M-C 2003 Some geometric aspects of variational calculus in constrained mechanics *Rep. Math. Phys.* **51** 127-48 (*Preprint* math-ph/0004019)
[26] Campos C M, de Leon M and de Diego D M 2010 Constrained variational calculus for higher order classical field theories *Preprint* arXiv:1005.2152
[27] García P L, García A and Rodrigo C 2006 Cartan forms for first order constrained variational problem *J. Geom. Phys.* **56** 571-610
[28] Castrillón López M 2012 Constraints in Euler-Poincaré reduction of field theories *Acta Applicandae Math.* **120** 87-99
[29] Kanatchikov I V 2007 On a generalization of the Dirac bracket in the De Donder-Weyl Hamiltonian formalism, In: *Differential Geometry and its Applications*, Proc. 10th Int. Conf. on Diff. Geom. & Appl., Olomouc, August 2007, O. Kowalski, D. Krupka, O. Krupková and J. Slovák (Eds.) (Singapore: World Scientific) pp 615-625 (*Preprint* arXiv:0807.3127)
[30] Kanatchikov I V 2000 On the Duffin-Kemmer-Petiau formulation of the covariant Hamiltonian dynamics in field theory *Rep. Math. Phys.* **46** 107-112 (*Preprint* hep-th/9911175)
[31] Misner C 1957 Feynman quantization of General Relativity *Rev. Mod. Phys.* **29** 497-509
[32] Kanatchikov I V 2012 On precanonical quantization of gravity in spin connection variables *AIP Conf. Proc.* **1514** 73-76 (*Preprint* arXiv:1212.6963)

An example of numerical simulation in causal set dynamics

Alexey L Krugly

Scientific Research Institute for System Analysis of the Russian Academy of Science, 117218, Nahimovskiy pr., 36, k. 1, Moscow, Russia

E-mail: akrugly@mail.ru

Ivan A Tserkovnikov

The Russian Presidential Academy of National Economy and Public Administration, Faculty of Economics, 119571, Vernadskogo pr., 82, Moscow, Russia

E-mail: tserkovnikov@hotmail.com

Abstract. The model of a discrete pregeometry on a microscopic scale is an x-graph. This is a directed acyclic graph. An outdegree and an indegree of each vertex are not more than 2. The sets of vertices and edges of x-graph are particular cases of causal sets. The sequential growth of a graph is an addition of new vertices one by one. A simple stochastic algorithm of sequential growth of x-graph are considered. It is based on a random walk at the x-graph. The particles in this model must be self-organized repetitive structures. We introduce the method of search of such repetitive structures. It is based on a discrete Fourier transformation. An example of numerical simulation is introduced.

1. Introduction

A causal set is a directed model of spacetime at a microscopic level. A causal set approach to quantum gravity was introduced by G. 't Hooft [1] and J. Myrheim [2] in 1978 (see e.g. [3]). Usually we try to find an approximation of a causal set as a continuous spacetime. But if we consider a causal set as a model of reality, a causal set must describe matter. Particles must be some topological fiches of a causal set.

Our particular model is a directed dyadic acyclic graph. The dyadic graph means that each vertex has two incident incoming edges and two incident outgoing edges. This model was introduced by Devid Finkelstein [4] in 1988. The acyclic graph means that there is not a directed loop. Hereinafter only such graph is considered, and it is called an x-graph. The set of vertices and the set of edges of x-graph are causal sets.

The model of the universe is an infinite x-graph. But any observer can only actually know a finite number of facts. Then we consider only finite x-graphs. In a graph theory, by definition, an edge is a relation of two vertices. Consequently some vertexes of finite x-graph have less than four incident edges. These free valences are called external edges as external lines in Feynman diagrams. They a figured as edges that are incident only to one vertex. External edges are not real edges. They are a property of vertices. But often it is useful to consider external edges as edges. There are incoming and outgoing external edges. We can prove that in the x-graphs

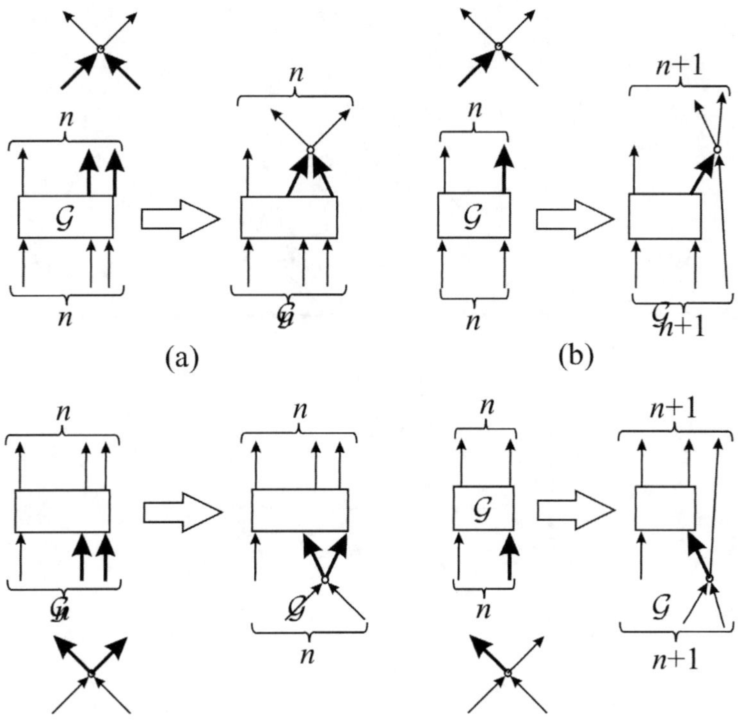

Figure 1. The types of elementary extensions: (a) the first type, (b) the second type, (c) the third type, and (d) the fourth type.

the number of incoming external edges is equal to the number of outgoing external edges [5, Lemma 5].

Each such graph is a model of a part of some process. In this model, the particles must be repetitive processes. We can make a lot of repetitive graphs by hand. But such repetitive structures must be a consequence of a dynamics. These structures must be self-organized structures.

What is a dynamics of this model? The task of any dynamics is to predict the future stages of the process or to reconstruct the past stages. We can reconstruct the x-graph step by step. The minimal part is a vertex. We can add new vertices only to external edges. This procedure is called an elementary extension. There are four types of elementary extensions [6]. There are two types of elementary extensions to outgoing external edges (Fig. 1 (a) and 1 (b)). This is a reconstruction of the future of the process. In this and following figures the x-graph $\mathcal{G}$ is represented by a rectangle because it can have an arbitrary structure. The edges that take part in the elementary extension are figured by bold arrows. First type is an elementary extension to two outgoing external edges (Fig. 1 (a)). Second type is an elementary extension to one outgoing external edge (Fig. 1 (b)). Similarly, there are two types of elementary extensions to incoming external edges (Fig. 1 (c) and 1 (d)). These elementary extensions reconstruct the past evolution of the process. Third type is an elementary extension to two incoming external edges (Fig. 1 (c)). Fourth type is an elementary extension to one incoming external edge (Fig. 1 (d)). We can prove that we can get every connected x-graph by a sequence of elementary extensions of these four types [5, Teorem 2].

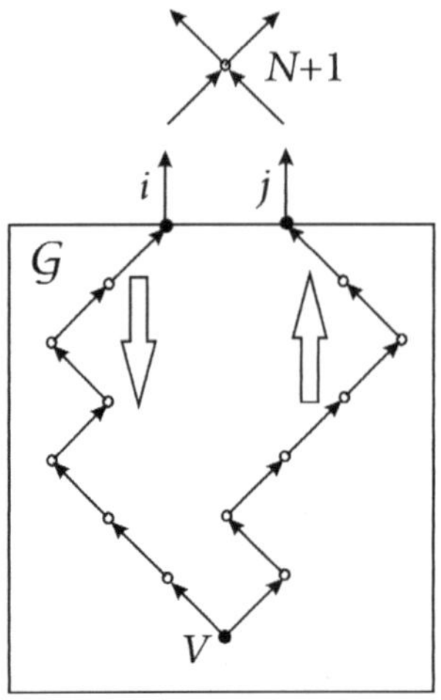

Figure 2. The random walk at the x-graph $\mathcal{G}$ from the outgoing external edge number i to the outgoing external edge number j with the U-turn in the vertex number V, and the addition of a new vertex number $N+1$.

By assumption, the dynamics of this model is a stochastic dynamics. We can only calculate probabilities of different variants of elementary extensions. In general case, these probabilities depend on the structure of the existed x-graph. In this model, causality is defined as the order of vertices and edges. But the causality has a real physical meaning only if the dynamics agrees with causality. The probability to add a new vertex to the future can only depend on the x-subgraph that precedes this vertex. Similarly, the probability to add a new vertex to the past can only depend on the x-subgraph that follows this vertex.

5 simple algorithms are introduced in [7]. The first algorithm is considered in [8, 9]. The second algorithm is considered in [10]. In this paper the third example is considered.

2. The algorithm

2.1. The random walk at the x-graph

We can describe the algorithm to calculate probabilities for this model as 3 steps. The first step is the choice of the elementary extension to the future or to the past. We assume the probability of this choice is 1/2. The second step is the choice of the first external edge that takes part in the elementary extension. We assume the equiprobable choice. The probability of this choice is $1/n$, where n is the number of outgoing (and incoming) external edges in the x-graph. The third step is the choice of second external edge that takes part in the elementary extension.

By assumption, the third step is based on a random walk at the x-graph (Fig. 2). Consider the elementary extension to the future. We have chosen the first outgoing external edge at the second step. Start the walk from this edge in the opposite direction of the edges. In each vertex we choose the U-turn or the continuation of the walk. We assume the probability of this choice is 1/2. We make the first choice in the vertex that is incident to the first external edge. If we choose the continuation of the walk, we must choose one next edge. We assume the probability of this choice is 1/2. If the next edge is an internal edge, we go to the next vertex and repeat this procedure. If the next edge is an incoming external edge, we must do the U-turn. After the U-turn we continue the walk in the direction of the edges without the second U-turn. In each vertex, we choose the next edge with the probability 1/2. The walk is finished in some outgoing external edge. By definition, this is the second outgoing external edge that take part in the elementary extension. If the first and the second outgoing external edges coincide, this is the elementary extension of the second type. The case of the elementary extension to the past is the same if we turn the direction of all edges in the x-graph.

In this algorithm, the area of the x-graph that is near to the first external edge has the main influence. The influence of the distant area exponentially decreases.

2.2. *The probabilities of elementary extensions*

Consider an analytic formulation of probabilities. Number outgoing external edges by lowercase Latin indices. Number incoming external edges by lowercase Greek indices. These Latin and Greek indices range from 1 to n, where n is the number of outgoing (and incoming) external edges in the x-graph. Number vertices by capital Latin indices. These indices range from 1 to N, where N is the number of vertices in the x-graph.

Introduce the probabilities $a_{i\alpha}$ and a_{iV}.

$$a_{i\alpha} = \sum_{m=1}^{M_{i\alpha}} 2^{-k(m)}, \quad a_{iV} = \sum_{m=1}^{M_{iV}} 2^{-r(m)}. \tag{1}$$

$a_{i\alpha}$ is a probability to reach the edge number i by the random walk in the direction of the edges if we start from the edge number α. $M_{i\alpha}$ is the number of directed paths from the incoming external edge number α to the outgoing external edge number i, and $k(m)$ is the number of vertices in the path number m. a_{iV} is a probability to reach the edge number i by the random walk in the direction of the edges if we start from the vertex number V. M_{iV} is the number of directed paths from the vertex number V to the outgoing external edge number i, and $r(m)$ is the number of vertices in the path number m including the vertex number V.

Similarly, introduce the probabilities $b_{i\alpha}$ and b_{iV}.

$$b_{i\alpha} = \sum_{m=1}^{M_{i\alpha}} 2^{-2k(m)}, \quad b_{iV} = \sum_{m=1}^{M_{iV}} 2^{-2r(m)+1}. \tag{2}$$

$b_{i\alpha}$ is a probability to reach the edge number α by the random walk in the opposite direction of the edges if we start from the edge number i, and we can interrupt the walk in each vertex with probability 1/2. b_{iV} is a probability to reach the vertex number V by the random walk in the opposite direction of the edges if we start from the edge number i, and we can interrupt the walk in each vertex with probability 1/2.

Using (1) - (2) we get the probability p_{ij} to choose the outgoing external edge number j at the third step if we have chosen the outgoing external edge number i at the second step.

$$p_{ij} = \sum_{V=1}^{N} b_{iV} a_{jV} + \sum_{\alpha=1}^{n} b_{i\alpha} a_{j\alpha}. \tag{3}$$

By definition, if the outgoing external edges number i and j coincides, this is the probability p_{ii} to add a new vertex to one outgoing external edge number i. Multiplying the probabilities of all three steps we get the probability P_{ii} of the elementary extension of the second type.

$$P_{ii} = \frac{1}{2n} \left(\sum_{V=1}^{N} b_{iV} a_{iV} + \sum_{\alpha=1}^{n} b_{i\alpha} a_{i\alpha} \right). \tag{4}$$

We get the same elementary extension of the first type if we choose the outgoing external edge number i at the second step and the outgoing external edge number j at the third step or if we choose the outgoing external edge number j at the second step and the outgoing external edge number i at the third step. Summing these two possibilities and multiplying the probabilities of all three steps we get the probability P_{ij} of the elementary extension of the first type.

$$P_{ij} = \frac{1}{2n}(\sum_{V=1}^{N} b_{iV}a_{jV} + \sum_{\alpha=1}^{n} b_{i\alpha}a_{j\alpha} + \sum_{V=1}^{N} b_{jV}a_{iV} + \sum_{\alpha=1}^{n} b_{j\alpha}a_{i\alpha}). \quad (5)$$

Similarly, we get the probabilities of the third and fourth types.

$$P_{\alpha\beta} = \frac{1}{2n}(\sum_{V=1}^{N} b_{V\alpha}a_{V\beta} + \sum_{i=1}^{n} b_{i\alpha}a_{i\beta} + \sum_{V=1}^{N} b_{V\beta}a_{V\alpha} + \sum_{i=1}^{n} b_{i\beta}a_{i\alpha}), \quad (6)$$

$$P_{\alpha\alpha} = \frac{1}{2n}(\sum_{V=1}^{N} b_{V\alpha}a_{V\alpha} + \sum_{i=1}^{n} b_{i\alpha}a_{i\alpha}), \quad (7)$$

The iterative procedure to calculate probabilities is considered in Appendix A.

3. The numerical simulation

3.1. The generation of x-graphs

We consider the numerical simulation of the sequential growth from 1 vertex during 2000 steps. In Fig. 3, the result of only the first 200 steps is figured because the whole x-graph is very tangled.

In the book [11], deterministic algorithms are considered for different discrete models. The deterministic algorithms for the causal sets is considered in [12]. Some algorithms generates repetitive structures. On the contrary we consider stochastic algorithms. Such algorithms cannot generate exact repetitive structures. The x-graph in Fig. 3 looks like random structure. But it can has some approximately repetitive properties. We need the methods to search such properties.

3.2. The analysis of structures

Suppose the considered dynamics generates repetitive structures. We can analyze such structures by a Fourier transform [13]. Consider some directed path $\mathcal{P}$ from some incoming external edge number α to some outgoing external edge number i. We consider $\mathcal{P}$ as a set of edges. Number the edges of $\mathcal{P}$ by the index t. If $\mathcal{P}$ consists of $T+1$ edges, $t=0$ for the incoming external edge number α, and $t=T$ for the outgoing external edge number i. Consider the probability $a_{t\alpha}$ to reach the edge number t by the random walk in the direction of the edges if we start from the edge number α. In this case, $i=t$ in left-hand definition (1). Number the vertexes of $\mathcal{P}$ by the index t too. By definition, the vertex has the number t if it has the incident incoming edge number t. Consider $a_{t\alpha}$ as a function $a(t)$ of the number t of vertex in the path.

In general case, $a(t)$ exponentially decreases. Consider $\log_2 a(t)$. This function describes properties of an Alexandrov set $\mathcal{A}(0,\ t)$ of the vertices number 0 and t. By definition, $\mathcal{A}(a,\ b) = \{c \mid a \prec c \prec b\}$ for the elements of any partially ordered set $\{a,\ b,\ldots\}$.

We found longest paths in our example of the x-graph with 2001 vertices. These paths consist of 65 vertices ($T = 65$). We chose one longest path and calculate $\log_2 a(t)$ at this path (Fig. 4). If there is only one directed path from the vertex number 0 to the vertex number t, $a(t) = a(t-1)/2$. In the considered path, such dependence is typical near both ends from the vertex number 0 to the vertex number 22 and from the vertex number 48 to the vertex number 65. If there are several directed paths from the vertex number 0 to the vertex number t, $a(t)$ is a sum over these paths. In this case, it is possible that $a(t) > a(t-1)$. We can see such dependence in the central part of the path. If the vertices number $t-1$ and t are connected by

Journal of Physics: Conference Series **442** (2013) 012042 doi:10.1088/1742-6596/442/1/012042

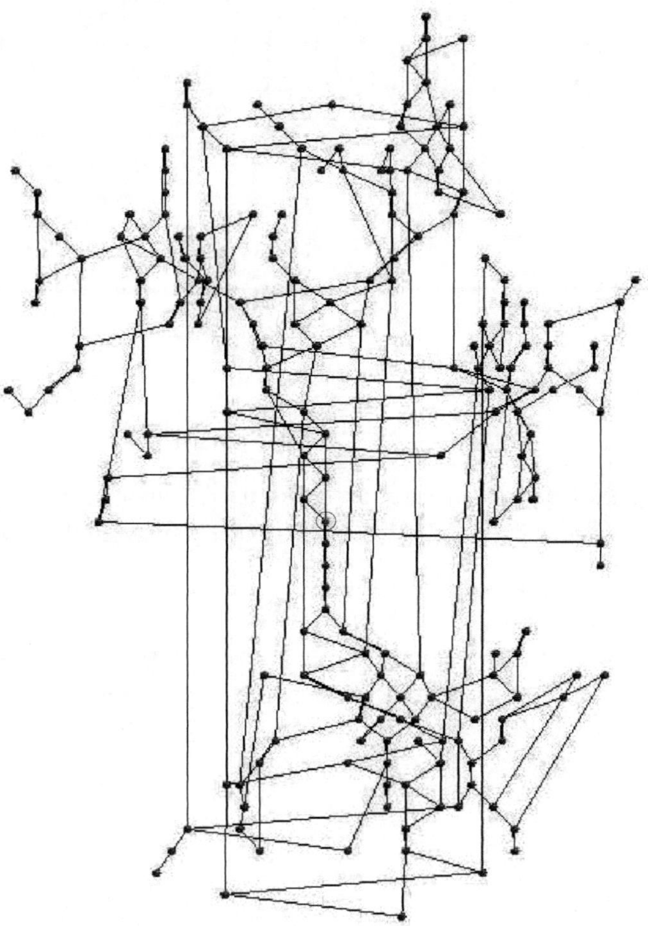

Figure 3. The example of the x-graph that consists of 201 vertices. The initial vertex is figured as big vertex. For simplicity, the internal edges are shown as lines without arrows. The external edges are not shown. The up direction is a direction from the past to the future. Other spatial coordinates of the vertexes have not physical meaning.

a double edge, $a(t) = a(t-1)$. For example, we have a sequence of double edges between the vertices number 2 and 6.

Consider a discrete Fourier transform of $\log_2 a(t)$.

$$g(f) = \frac{1}{T}\sum_{t=0}^{T-1} \log_2 a(t) \exp(-i\frac{2\pi ft}{T}), \tag{8}$$

If the directed path belongs to the repetitive structure the dominant frequency must exist. We get the power spectrum for the considered path (Fig. 5). The zero frequency is not figured because it has very high value: $|g(0)|^2 = 5424.145$. We get the high value for the zero frequency because $\log_2 a(t)$ cannot be positive. Denote by $\tau(f)$ the period of the mode with the frequency f. By definition, $\tau(f) = T/f$. The mode with the lowest frequency $f = 1$ has $\tau(f) = 65$ edges. This is the length of the path. The lowest possible period is equal to 2 edges. There are not

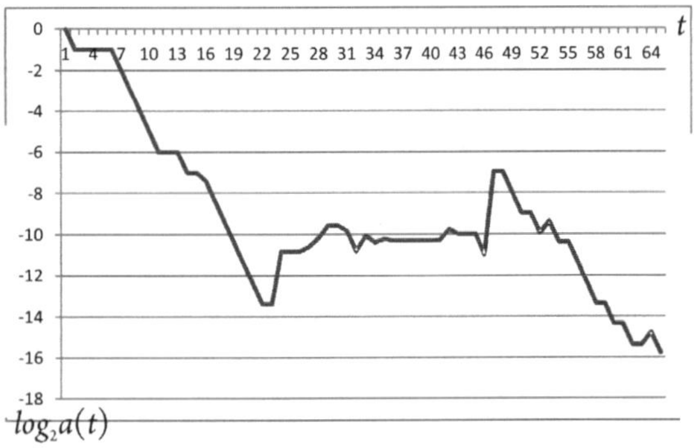

Figure 4. $\log_2 a(t)$ at the longest path.

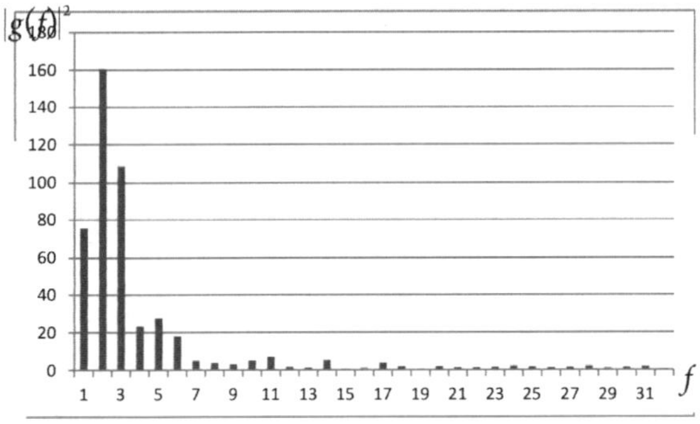

Figure 5. $|g(f)|^2$ for the longest path from $f = 1$ to the $f = 32$.

repetitive structures with short periods at the considered path. There is only one high maximum with $\tau(2) = 32.5$ edges. This is not a repetitive structure. This low frequency describes global properties of the x-graph. The second maximum with $\tau(5) = 13$ edges can be a consequence of some repetitive property of the x-graph.

4. Conclusion

In the book [11], a very big set of simple deterministic algorithms are investigated. Only few algorithms generate interesting self-organized structures. Probably only few stochastic algorithms generate interesting self-organized structures too. We must investigate a big set of simple stochastic algorithms to find such interesting cases. This is a task for further investigations.

Acknowledgments

I am grateful to Alexander V. Kaganov and Tommaso Bolognesi for extensive discussions on this subject.

Appendix A. The iterative procedure for the probabilities

The number of direct calculations of all probabilities at each step of sequential growth is proportional to $n \times N$. But if we use the probabilities at the previous step, the number of calculations is proportional to $n \times n$.

Consider the probabilities p_{ij}, $p_{\alpha\beta}$, $a_{i\alpha}$, and $b_{i\alpha}$ as elements of square matrices $\mathbf{p}_f$, $\mathbf{p}_p$, $\mathbf{a}$, and $\mathbf{b}$ respectively. The size of these matrices is n. Denote by $\mathcal{G}_{N,n}$ the x-graph that includes N vertex and n outgoing (and incoming) external edges. The x-graph $\mathcal{G}_{1,2}$ includes one vertex ($N = 1$, $n = 2$). We have

$$\mathbf{p}_f(\mathcal{G}_{1,2}) = \mathbf{p}_p(\mathcal{G}_{1,2}) = \mathbf{a}(\mathcal{G}_{1,2}) = \begin{pmatrix} 1/2 & 1/2 \\ 1/2 & 1/2 \end{pmatrix}, \tag{A.1}$$

$$\mathbf{b}(\mathcal{G}_{1,2}) = \begin{pmatrix} 1/4 & 1/4 \\ 1/4 & 1/4 \end{pmatrix}. \tag{A.2}$$

If we know the probabilities for $\mathcal{G}_{N,n}$, we can calculate probabilities after any addition of a new vertex number $N + 1$.

First type is an elementary extension to the future (Fig. 1 (a)). Two outgoing external edges numbers i and j become internal edges. We get two free numbers of outgoing external edges: i

and j. Two new outgoing external edges appear. Number these new outgoing external edges by i and j. We have

$$p_{ij}(\mathcal{G}_{N+1,n}) = p_{ji}(\mathcal{G}_{N+1,n}) = p_{ii}(\mathcal{G}_{N+1,n}) = p_{jj}(\mathcal{G}_{N+1,n}) = \\ = \tfrac{1}{4} + \tfrac{1}{8}(p_{ij}(\mathcal{G}_{N,n}) + p_{ji}(\mathcal{G}_{N,n}) + p_{ii}(\mathcal{G}_{N,n}) + p_{jj}(\mathcal{G}_{N,n})), \tag{A.3}$$

$$p_{si}(\mathcal{G}_{N+1,n}) = p_{sj}(\mathcal{G}_{N+1,n}) = \frac{1}{2}(p_{si}(\mathcal{G}_{N,n}) + p_{sj}(\mathcal{G}_{N,n})), \tag{A.4}$$

$$p_{is}(\mathcal{G}_{N+1,n}) = p_{js}(\mathcal{G}_{N+1,n}) = \frac{1}{4}(p_{is}(\mathcal{G}_{N,n}) + p_{js}(\mathcal{G}_{N,n})), \text{ where } s \neq i \text{ and } s \neq j. \tag{A.5}$$

$$p_{rs}(\mathcal{G}_{N+1,n}) = p_{rs}(\mathcal{G}_{N,n}), \text{ where } r \neq i,\ r \neq j,\ s \neq i, \text{ and } s \neq j. \tag{A.6}$$

$$p_{\alpha\beta}(\mathcal{G}_{N+1,n}) = p_{\alpha\beta}(\mathcal{G}_{N,n}) + \tfrac{1}{2}(b_{i\alpha}(\mathcal{G}_{N,n})a_{j\beta}(\mathcal{G}_{N,n}) + \\ + b_{j\alpha}(\mathcal{G}_{N,n})a_{i\beta}(\mathcal{G}_{N,n}) - b_{i\alpha}(\mathcal{G}_{N,n})a_{i\beta}(\mathcal{G}_{N,n}) - b_{j\alpha}(\mathcal{G}_{N,n})a_{j\beta}(\mathcal{G}_{N,n})). \tag{A.7}$$

$$a_{i\alpha}(\mathcal{G}_{N+1,n}) = a_{j\alpha}(\mathcal{G}_{N+1,n}) = \frac{1}{2}(a_{i\alpha}(\mathcal{G}_{N,n}) + a_{j\alpha}(\mathcal{G}_{N,n})), \tag{A.8}$$

$$b_{i\alpha}(\mathcal{G}_{N+1,n}) = b_{j\alpha}(\mathcal{G}_{N+1,n}) = \frac{1}{4}(b_{i\alpha}(\mathcal{G}_{N,n}) + b_{j\alpha}(\mathcal{G}_{N,n})). \tag{A.9}$$

Second type is an elementary extension to the future too (Fig. 1 (b)). One outgoing external edge number i becomes an internal edge. We get i as free number of an outgoing external edge. Two new outgoing external edges and one new incoming external edge appear. Number these new outgoing external edges by i and $n+1$, and new incoming external edge by $n+1$. We have new column number $n+1$ and new row number $n+1$ in $\mathbf{p}_f$, $\mathbf{p}_p$, $\mathbf{a}$, and $\mathbf{b}$. We have

$$p_{ii}(\mathcal{G}_{N+1,n+1}) = p_{i(n+1)}(\mathcal{G}_{N+1,n+1}) = \\ = p_{(n+1)i}(\mathcal{G}_{N+1,n+1}) = p_{(n+1)(n+1)}(\mathcal{G}_{N+1,n+1}) = \tfrac{1}{4} + \tfrac{1}{8} + \tfrac{1}{8}p_{ii}(\mathcal{G}_{N,n}), \tag{A.10}$$

$$p_{si}(\mathcal{G}_{N+1,n+1}) = p_{s(n+1)}(\mathcal{G}_{N+1,n+1}) = \frac{1}{2}p_{si}(\mathcal{G}_{N,n}), \tag{A.11}$$

$$p_{is}(\mathcal{G}_{N+1,n+1}) = p_{(n+1)s}(\mathcal{G}_{N+1,n+1}) = \frac{1}{4}p_{is}(\mathcal{G}_{N,n}), \text{ where } s \neq i \text{ and } s \neq n+1. \tag{A.12}$$

$$p_{rs}(\mathcal{G}_{N+1,n+1}) = p_{rs}(\mathcal{G}_{N,n}), \text{ where } r \neq i,\ r \neq n+1,\ s \neq i, \text{ and } s \neq n+1. \tag{A.13}$$

$$p_{\alpha\beta}(\mathcal{G}_{N+1,n+1}) = p_{\alpha\beta}(\mathcal{G}_{N,n}) - \tfrac{1}{2}b_{i\alpha}(\mathcal{G}_{N,n})a_{i\beta}(\mathcal{G}_{N,n}) = \\ = p_{\alpha\beta}(\mathcal{G}_{N,n}) - 4b_{i\alpha}(\mathcal{G}_{N+1,n+1})a_{i\beta}(\mathcal{G}_{N+1,n+1}), \tag{A.14}$$

$$p_{\alpha(n+1)}(\mathcal{G}_{N+1,n+1}) = \frac{1}{2}b_{i\alpha}(\mathcal{G}_{N,n}) = 2b_{i\alpha}(\mathcal{G}_{N+1,n+1}), \tag{A.15}$$

$$p_{(n+1)\alpha}(\mathcal{G}_{N+1,n+1}) = \frac{1}{2}a_{i\alpha}(\mathcal{G}_{N,n}) = a_{i\alpha}(\mathcal{G}_{N+1,n+1}). \tag{A.16}$$

If $\alpha = n+1$,

$$p_{(n+1)(n+1)}(\mathcal{G}_{N+1,n+1}) = \frac{1}{2}. \tag{A.17}$$

$$a_{i\alpha}(\mathcal{G}_{N+1,n+1}) = a_{(n+1)\alpha}(\mathcal{G}_{N+1,n+1}) = \frac{1}{2}a_{i\alpha}(\mathcal{G}_{N,n}), \tag{A.18}$$

$$b_{i\alpha}(\mathcal{G}_{N+1,n+1}) = b_{(n+1)\alpha}(\mathcal{G}_{N+1,n+1}) = \frac{1}{4}b_{i\alpha}(\mathcal{G}_{N,n}), \text{ where } \alpha \neq n+1. \tag{A.19}$$

$$a_{i(n+1)}(\mathcal{G}_{N+1,n+1}) = a_{(n+1)(n+1)}(\mathcal{G}_{N+1,n+1}) = \frac{1}{2}, \tag{A.20}$$

$$b_{i(n+1)}(\mathcal{G}_{N+1,n+1}) = b_{(n+1)(n+1)}(\mathcal{G}_{N+1,n+1}) = \frac{1}{4}, \tag{A.21}$$

$$a_{s(n+1)}(\mathcal{G}_{N+1,n+1}) = 0, \ b_{s(n+1)}(\mathcal{G}_{N+1,n+1}) = 0, \text{ where } s \neq i \text{ and } s \neq n+1. \tag{A.22}$$

If we interchange the Latin and Greek indices in (A.1) - (A.22), we get the equations for the elementary extensions of the third and fourth types.

Third type is an elementary extension to the past (Fig. 1 (c)). Two incoming external edges numbers α and β become internal edges. We get two free numbers of incoming external edges: α and β. Two new incoming external edges appear. Number these new incoming external edges by α and β. We have

$$\begin{aligned} p_{\alpha\beta}(\mathcal{G}_{N+1,n}) &= p_{\beta\alpha}(\mathcal{G}_{N+1,n}) = p_{\alpha\alpha}(\mathcal{G}_{N+1,n}) = p_{\beta\beta}(\mathcal{G}_{N+1,n}) = \\ &= \tfrac{1}{4} + \tfrac{1}{8}(p_{\alpha\beta}(\mathcal{G}_{N,n}) + p_{\beta\alpha}(\mathcal{G}_{N,n}) + p_{\alpha\alpha}(\mathcal{G}_{N,n}) + p_{\beta\beta}(\mathcal{G}_{N,n})), \end{aligned} \tag{A.23}$$

$$p_{\gamma\alpha}(\mathcal{G}_{N+1,n}) = p_{\gamma\beta}(\mathcal{G}_{N+1,n}) = \frac{1}{2}(p_{\gamma\alpha}(\mathcal{G}_{N,n}) + p_{\gamma\beta}(\mathcal{G}_{N,n})), \tag{A.24}$$

$$p_{\alpha\gamma}(\mathcal{G}_{N+1,n}) = p_{\beta\gamma}(\mathcal{G}_{N+1,n}) = \frac{1}{4}(p_{\alpha\gamma}(\mathcal{G}_{N,n}) + p_{\beta\gamma}(\mathcal{G}_{N,n})), \text{ where } \gamma \neq \alpha \text{ and } \gamma \neq \beta. \tag{A.25}$$

$$p_{\delta\gamma}(\mathcal{G}_{N+1,n}) = p_{\delta\gamma}(\mathcal{G}_{N,n}), \text{ where } \delta \neq \alpha, \ \delta \neq \beta, \ \gamma \neq \alpha, \text{ and } \gamma \neq \beta. \tag{A.26}$$

$$\begin{aligned} p_{ij}(\mathcal{G}_{N+1,n}) &= p_{ij}(\mathcal{G}_{N,n}) - \tfrac{1}{2}(b_{i\alpha}(\mathcal{G}_{N,n})a_{j\beta}(\mathcal{G}_{N,n}) + \\ &+ b_{i\beta}(\mathcal{G}_{N,n})a_{j\alpha}(\mathcal{G}_{N,n}) - b_{i\alpha}(\mathcal{G}_{N,n})a_{j\alpha}(\mathcal{G}_{N,n}) - b_{i\beta}(\mathcal{G}_{N,n})a_{j\beta}(\mathcal{G}_{N,n})). \end{aligned} \tag{A.27}$$

$$a_{i\alpha}(\mathcal{G}_{N+1,n}) = a_{i\beta}(\mathcal{G}_{N+1,n}) = \frac{1}{2}(a_{i\alpha}(\mathcal{G}_{N,n}) + a_{i\beta}(\mathcal{G}_{N,n})), \tag{A.28}$$

$$b_{i\alpha}(\mathcal{G}_{N+1,n}) = b_{i\beta}(\mathcal{G}_{N+1,n}) = \frac{1}{4}(b_{i\alpha}(\mathcal{G}_{N,n}) + b_{i\beta}(\mathcal{G}_{N,n})). \tag{A.29}$$

Fourth type is an elementary extension to the past too (Fig. 1 (b)). One incoming external edge number α becomes an internal edge. We get α as free number of an incoming external edge. Two new incoming external edges and one new outgoing external edge appear. Number these new incoming external edges by α and $n+1$, and new outgoing external edge by $n+1$. We have new column number $n+1$ and new row number $n+1$ in $\mathbf{p}_p$, $\mathbf{p}_f$, $\mathbf{a}$, and $\mathbf{b}$. We have

$$\begin{aligned} p_{\alpha\alpha}(\mathcal{G}_{N+1,n+1}) &= p_{\alpha(n+1)}(\mathcal{G}_{N+1,n+1}) = \\ &= p_{(n+1)\alpha}(\mathcal{G}_{N+1,n+1}) = p_{(n+1)(n+1)}(\mathcal{G}_{N+1,n+1}) = \tfrac{1}{4} + \tfrac{1}{8} + \tfrac{1}{8}p_{\alpha\alpha}(\mathcal{G}_{N,n}), \end{aligned} \tag{A.30}$$

$$p_{\gamma\alpha}(\mathcal{G}_{N+1,n+1}) = p_{\gamma(n+1)}(\mathcal{G}_{N+1,n+1}) = \frac{1}{2}p_{\gamma\alpha}(\mathcal{G}_{N,n}), \tag{A.31}$$

$$p_{\alpha\gamma}(\mathcal{G}_{N+1,n+1}) = p_{(n+1)\gamma}(\mathcal{G}_{N+1,n+1}) = \frac{1}{4}p_{\alpha\gamma}(\mathcal{G}_{N,n}), \text{ where } \gamma \neq \alpha \text{ and } \gamma \neq n+1. \tag{A.32}$$

$$p_{\delta\gamma}(\mathcal{G}_{N+1,n+1}) = p_{\delta\gamma}(\mathcal{G}_{N,n}), \text{ where } \delta \neq \alpha, \ \delta \neq n+1, \ \gamma \neq \alpha, \text{ and } \gamma \neq n+1. \tag{A.33}$$

$$p_{ij}(\mathcal{G}_{N+1,n+1}) = p_{ij}(\mathcal{G}_{N,n}) - \tfrac{1}{2} b_{i\alpha}(\mathcal{G}_{N,n}) a_{j\alpha}(\mathcal{G}_{N,n}) = \\ = p_{ij}(\mathcal{G}_{N,n}) - 4 b_{i\alpha}(\mathcal{G}_{N+1,n+1}) a_{j\alpha}(\mathcal{G}_{N+1,n+1}). \tag{A.34}$$

$$p_{i(n+1)}(\mathcal{G}_{N+1,n+1}) = \frac{1}{2} b_{i\alpha}(\mathcal{G}_{N,n}) = 2 b_{i\alpha}(\mathcal{G}_{N+1,n+1}), \tag{A.35}$$

$$p_{(n+1)i}(\mathcal{G}_{N+1,n+1}) = \frac{1}{2} a_{i\alpha}(\mathcal{G}_{N,n}) = a_{i\alpha}(\mathcal{G}_{N+1,n+1}). \tag{A.36}$$

If $i = n+1$,

$$p_{(n+1)(n+1)}(\mathcal{G}_{N+1,n+1}) = \frac{1}{2}. \tag{A.37}$$

$$a_{i\alpha}(\mathcal{G}_{N+1,n+1}) = a_{i(n+1)}(\mathcal{G}_{N+1,n+1}) = \frac{1}{2} a_{i\alpha}(\mathcal{G}_{N,n}), \tag{A.38}$$

$$b_{i\alpha}(\mathcal{G}_{N+1,n+1}) = b_{i(n+1)}(\mathcal{G}_{N+1,n+1}) = \frac{1}{4} b_{i\alpha}(\mathcal{G}_{N,n}), \text{ where } i \neq n+1. \tag{A.39}$$

$$a_{(n+1)\alpha}(\mathcal{G}_{N+1,n+1}) = a_{(n+1)(n+1)}(\mathcal{G}_{N+1,n+1}) = \frac{1}{2}, \tag{A.40}$$

$$b_{(n+1)\alpha}(\mathcal{G}_{N+1,n+1}) = b_{(n+1)(n+1)}(\mathcal{G}_{N+1,n+1}) = \frac{1}{4}, \tag{A.41}$$

$$a_{(n+1)\gamma}(\mathcal{G}_{N+1,n+1}) = 0,\ b_{(n+1)\gamma}(\mathcal{G}_{N+1,n+1}) = 0, \text{ where } \gamma \neq \alpha \text{ and } \gamma \neq n+1. \tag{A.42}$$

References

[1] 't Hooft G 1979 Quantum gravity: a fundamental problem and some radical ideas, in *Recent Development in Gravitation, Proceedings of the 1978 Cargese Summer Institute* ed M Levy and S Deser (New York, London: Plenum) 323-345
[2] Myrheim J 1978 Statistical Geometry *Preprint* CERN TH-2538
[3] Sorkin R D 2005 Causal sets: Discrete gravity (notes for the Valdivia summer school), in *Lectures on Quantum Gravity, Proceedings of the Valdivia Summer School, Valdivia, Chile, January 2002* ed A Gomberoff and D Marolf. (Berlin, New York: Springer) 305-327 (*Preprint* arXiv:gr-qc/0309009)
[4] Finkelstein D 1988 "Superconducting" causal net *Int. J. Theor. Phys.* **27** 473-519
[5] Krugly A L 2010 Discrete mechanics: a kinematics for a particular case of causal sets *Preprint* arXiv:1008.5169 [gr-qc]
[6] Krugly A L 2010 Discrete mechanics: a sequential growth dynamics for causal sets, and a self-organization of particles *Preprint* arXiv:1004.5077 [gr-qc]
[7] Krugly A L 2012 Simple algorithms of the sequential growth of the x-graph, talk at *2-nd International Conference on Theoretical Physics, July 2-6, 2012, Moscow, Russia*
[8] Krugly A L 2011 A sequential growth dynamics for a directed acyclic dyadic graph *Preprint* arXiv: 1112.1064 [gr-qc]
[9] Krugly A L and Stepanian I V 2012 An example of the stochastic dynamics of a causal set, in *"Foundations of Probability and Physics-6" (FPP6), 12-15 June 2011, the Linnaeus University, Växjö, Sweden, AIP Conference Proceedings* **1424** ed M D'Ariano, S-M Fei, E Haven, B Hiesmayr, G Jaeger, A Khrennikov and J-Å Larsson, pp. 206-210 (*Preprint* arXiv:1111.5474 [gr-qc])
[10] Koganov A V and Krugly A L 2012 Algoritm rosta x-grafa i printsypy fisiki *Programnye producty i sistemy* **3** 96-102 (Algorithm of X-graph growth and principles of physics. *Software products and systems* 3, 96-102, in Russian).
[11] Wolfram S 2002 *A New Kind of Science* (Wolfram Media, Inc.)
[12] Bolognesi T 2010 Causal Sets from simple models of computation *Int. J. Unconventional Computing* **6**(6) (OCP Science) 489 - 524 (*Preprint* arXiv:1004.3128 [physics.comp-ph])
[13] Krugly A L 2012 Complex amplitudes as a consequence of classical sequential growth dynamics for a particular case of a causal set, in *XLVIII all-russia conference on problems of physics of particles, physics of plasma and the condensed matter, optoelectronics. Dedicated to the 100th anniversary of Professor Ya P Terletsky, Moscow, Russia, 15-18 May 2012* (Moscow: PFUR) 3-6

Realist model approach to quantum mechanics

P Hájíček

Institute for Theoretical Physics, University of Bern, Sidlerstrasse 5, CH-3012 Bern, Switzerland

E-mail: hajicek@itp.unibe.ch

Abstract. The paper proves that quantum mechanics is compatible with the constructive realism of modern philosophy of science. The proof is based on the observation that properties of quantum systems that are uniquely determined by their preparations can be assumed objective without the difficulties that are encountered by the same assumption about values of observables. The resulting realist interpretation of quantum mechanics is made rigorous by studying the space of quantum states—the convex set of state operators. Prepared states are classified according to their statistical structure into indecomposable and decomposable instead of pure and mixed. Simple objective properties are defined and showed to form a Boolean lattice.

1. Introduction

If our aim is to propose a realist interpretation of quantum mechanics, we ought to ask first, what is realism. The abstract realist doctrine, "The world is not just a construction of our mind, but does really exist", cannot be falsified because it is vague, but then, of course, not of much use. The so-called naive realism, "The real world is as we perceive it", is clearly false. The metaphysical realism [1], "There is exactly one true and complete description of 'the way the world is' ", is surely near to a physicist's heart, but it seems to be rather different from what the history of physics teaches us.

In the contemporary philosophy of science, there is a stream that can be characterised by the doctrine "Our knowledge is dominated by a family of models". It was founded and is represented eg by Patrick Suppes, Bas van Fraassen, Ronald Giere, Wolfgang Stegmüller and Nancy Cartwright. Very schematically, it can be described as follows. Each grand theory of physics (such as Newtonian mechanics or quantum mechanics) can be divided into two parts:

- Treasure of successful models. The set is open in the sense that it grows with time.
- General language part, which contains mathematical structure of state space, of dynamical equations, of symmetries and the form of observables. It is considered as secondary, because it is only obtained through study of models, or serves as an instrument of model construction as well as of unification and classification of different models.

A particular philosophy within the stream is the so-called Constructive Realism by R. N. Giere [2], which can be roughly characterised as follows:

- Each model is constructed with the aim to give an approximative representation of some aspects of a real object.

- Concepts of the general language part are connected to the reality only via models.

It is a relatively strong kind of realism, and it seems to correspond very well to how physics works in practice. The present paper follows [3], where this philosophy is accepted and its application to quantum mechanics is described in more details.

In particular, each model has two components: an empirical one being an identifiable real object, and a language one, being a theoretical construct that models the object.

Our work on quantum mechanics [4, 5, 6, 7, 8, 9, 10, 3] is aimed at constructing quantum models of classical world. Within the constructive realism, the relation between quantum and Newtonian mechanics, say, ought to be studied by comparing quantum and Newtonian models of one and the same real object rather than looking for a relation between their general language parts.

To prevent misunderstandings, let us stress that we are not trying to construct any universal map between states of quantum mechanics and those of Newtonian mechanics, such as Wigner-Weyl-Moyal map (see eg [11], p 85). Our aim is to construct quantum models of classical world, not vice versa. We start by distinguishing three things: a real object, its Newtonian and its quantum model. Then, only such real objects are considered that are macroscopic and possess successful classical Newtonian models. Thus, a hydrogen atom does not qualify. (It can be a real object within our interpretation.) Moreover, the quantum model will, in most cases, be much richer, than the classical one. For instance, properties of a Newtonian model of a free real macroscopic solid body are eg isotropic continuum distribution of matter with a given mass density, elastic coefficient, total momentum and angular momentum. Properties of a quantum model can be eg all particle numbers that form the body, their masses and spins, their Hamiltonian with an interaction potential and a suitable quantum state of the whole system.

2. An heuristic principle of quantum realism

The project runs into three well-known problems:

(i) Classical properties are objective and robust, see [12]. How can they emerge from quantum mechanics which does not seem to allow such properties?

(ii) Realism leads quite generally to contradiction with the linear quantum evolution in the measurement theory (Schrödinger cat, see [12]).

(iii) What are real quantum objects to construct models of? It seems correct to require that language parts of models ascribe enough objective properties to their empirical parts. However, there is an apparent lack of objective properties in the language part of quantum models. This makes implausible to assume the reality of quantum systems, see [13].

The problems ought to be dealt with simultaneously (see [3]), but we shall focus just on the third problem here.

We define objective properties as follows. Let $\mathcal{S}$ be a real object and $\mathcal{S}_m$ its theoretical model. $\mathcal{S}_m$ contains two kinds of information. First, it describes $\mathcal{S}_m$ in sufficient detail by values of some quantities that can be ascribed to $\mathcal{S}$ alone independently of any observations. Second, it enables to calculate what will be observed on $\mathcal{S}$ if some measurement is carried out. An objective property of $\mathcal{S}$ is the value of a quantity of the first kind. Thus, the values of observables of quantum system $\mathcal{S}$ are *not* objective properties of $\mathcal{S}$. To ascribe values of observables to $\mathcal{S}$ leads to contradictions such as contextuality (for more detail, see [3]). Here, we accept the old tenet: each such value is only created during registration of $\mathcal{S}$ by some apparatus $\mathcal{A}$ and it is an objective property of the composite $\mathcal{S} + \mathcal{A}$ rather than of $\mathcal{S}$ alone. It gives only an indirect information about $\mathcal{S}$.

Fortunately, there are other observable properties of quantum systems. They can be described by the following heuristic principle [4].

Basic Ontological Hypothesis of QM *A property is objective if its value is uniquely determined by a preparation according to the rules of standard QM. The "value" is the value of the mathematical expression that describes the property and it may be more sophisticated than just a real number. To observe an objective property, many registrations of one or more observables are necessary.*

In fact, the Hypothesis just states explicitly the meaning that is tacitly given to preparation by standard quantum mechanics. More discussion on the meaning of preparation is in [6, 9]. In any case, prepared properties can be ascribed to the prepared system without any problems or paradoxes. Often, we shall use the Hypothesis as an heuristic principle: it will just help to find some specific properties and then it will be forgotten, that is, an independent assumption will be made that these properties can be objective and each of them will be further studied.

It is easy to give examples of such objective properties. We can classify them into *structural* and *dynamical*. The structural ones are eg mass, charge, spin, structure of Hamiltonian of an isolated system, etc. The dynamical ones are eg state operator, average value and variance of an observable, etc.

Often, the Hypothesis meets one of the following two questions. First, how can it be applied to cosmology, when there was nobody there at the Big Bang to perform any state preparation? Second, a state preparation is an action of some human subject; how can it result in an objective property? Both objections result from the view that preparations are just some manipulations in human laboratory. This is too narrow: even textbooks (see eg [13]) notice that preparations can also run spontaneously as natural processes, such as the preparations of neutrinos that arrive at the Earth from the core of the Sun. Moreover, the second objection is not much more than a pun. It is not logically impossible that a human manipulation of a system results in an objective property of the system. For example, pushing a snooker ball imparts it a certain momentum and angular momentum that can be assumed to be objective properties of the pushed ball.

3. The space of quantum states

Quantum states of objects that we meet in our everyday life are very different from wave functions. Thus, the usual focus on wave functions is misleading if we want to understand classical properties [5]. We have to work with general states, which are described by what is usually called "density matrices" [13] or "state operators" [14].

3.1. Mathematical preliminaries

Let us collect some mathematics on quantum states that can be found in literature but is not generally known.

A bounded self-adjoint operator a is called *positive* if

$$\langle\psi|\mathsf{a}|\psi\rangle \geq 0 \quad \forall|\psi\rangle \in \mathbf{H}$$

and *trace-class* if

$$tr[\sqrt{\mathsf{a}^2}] < \infty \ .$$

The left-hand side of the inequality can be shown to be a norm and the space of trace-class operators completed with respect to this norm is a Banach space. With the partial ordering $\mathsf{a} > \mathsf{b}$ defined by operator $\mathsf{a} - \mathsf{b}$ being positive, it is an ordered Banach space. (For more details, see [15], Appendix III and IV.)

What is the true state space of quantum mechanics?

Definition 1 *Let $\mathcal{S}$ be a quantum system with Hilbert space $\mathbf{H}$. Let $\mathbf{T}(\mathbf{H})$ denote the ordered Banach space of trace-class operators on $\mathbf{H}$. State space $\mathbf{T}(\mathbf{H})_1^+$ of $\mathcal{S}$ is the set of all positive elements of $\mathbf{T}(\mathbf{H})$ with trace 1.*

Journal of Physics: Conference Series **442** (2013) 012043 doi:10.1088/1742-6596/442/1/012043

Hence, the space of states is the intersection of the unit sphere and the positive cone of the Banach space $\mathbf{T}(\mathbf{H})$. As the positive cone has a non-trivial boundary in $\mathbf{T}(\mathbf{H})$, $\mathbf{T}(\mathbf{H})_1^+$ has a rich face structure.

The space $\mathbf{T}(\mathbf{H})_1^+$ is not a linear space but it is *convex*, that is, invariant with respect to *convex combination*,

$$w\mathsf{T}_1 + (1-w)\mathsf{T}_2 = \mathsf{T} ,$$

where $\mathsf{T}_{1,2} \in \mathbf{T}(\mathbf{H})_1^+$ and $w \in [0,1]$. The states T_1 and T_2 are called *convex components* of T.

Definition 2 *Face* $\mathbf{F}$ *is a norm-closed subset of* $\mathbf{T}(\mathbf{H})_1^+$ *that is invariant with respect to convex combinations and contains all convex components of any* $\mathbf{T} \in \mathbf{F}$.

"Face" is an important notion of the mathematical theory of convex sets.

Theorem 1 *Every face* $\mathbf{F} \subset \mathbf{T}(\mathbf{H})_1^+$ *can be written as* $\mathbf{F}(\mathsf{T})$ *for a suitably chosen* $\mathsf{T} \in \mathbf{F}(\mathsf{T})$ *where* $\mathbf{F}(\mathsf{T})$ *is the smallest face which contains* T.

For proof, see [15], p 76. Thus, each state operator lies in some face. There is a useful relation between faces and projections:

Theorem 2 *To each face* $\mathbf{F}$ *of* $\mathbf{T}(\mathbf{H})_1^+$ *there is a unique projection* $\mathsf{P} : \mathbf{H} \mapsto \mathbf{H}'$, *where* $\mathbf{H}'$ *is a closed subspace of* $\mathbf{H}$, *for which* $\mathsf{T} \subset \mathbf{F}$ *is equivalent to*

$$\mathsf{T} = \mathsf{PTP} .$$

The map so defined between the set of faces and the set of projections is an order isomorphism ie it is invertible and $\mathsf{P}_1 < \mathsf{P}_2$ *is equivalent to* $\mathbf{F}_1 \subset \mathbf{F}_2$.

For proof, see [15], p 77. We shall denote the face that corresponds to a projection P by $\mathbf{F}_\mathsf{P}$.

Clearly, intersection of two faces, if non-empty, is a face and a unitary map of a face is a face. The next theorem shows that $\mathbf{F}(\mathsf{T})$ is not necessarily the set of all convex components of T.

Theorem 3 *Let* $\mathsf{P}(\mathbf{H})$ *be infinite-dimensional. Let* $\mathsf{T}_1, \mathsf{T}_2 \in \mathbf{F}_\mathsf{P}$ *and be positive definite on* $\mathsf{P}(\mathbf{H})$. *Then*

$$\mathbf{F}(\mathsf{T}_1) = \mathbf{F}(\mathsf{T}_2) = \mathbf{F}_\mathsf{P} .$$

Let $\{|k\rangle\}$ *be an orthonormal basis of* $\mathsf{P}(\mathbf{H})$ *and let*

$$\sup_k \frac{\langle k|\mathsf{T}_1|k\rangle}{\langle k|\mathsf{T}_2|k\rangle} = \infty . \qquad (1)$$

Then T_1 *is not a convex component of* T_2.

For proof, see [3].

Definition 3 $\mathsf{T} \in \mathbf{T}(\mathbf{H})_1^+$ *is called* *extremal* *if it lies in a zero-dimensional face.*

For extremal states, we have:

Theorem 4 T *is extremal iff* $\mathsf{T} = |\psi\rangle\langle\psi|$, *where* ψ *is a unit vector of* $\mathbf{H}$.

For proof, see [15], p 78. Thus, extremal state can be described by elements of $\mathbf{H}$, and only the q-representation of such an element is a "wave function".

3.2. Physical interpretation of the mathematics

Here, the mathematical consequences of our heuristic principle will be worked out. The physical meaning of elements of $\mathbf{T}(\mathbf{H})_1^+$ is given by

Assumption 1 *Let $\mathcal{S}$ be a quantum system with Hilbert space* $\mathbf{H}$. *Every* $\mathsf{T} \in \mathbf{T}(\mathbf{H})_1^+$ *can be prepared as a state of $\mathcal{S}$ and is then an objective property of $\mathcal{S}$.*

Next, we start to give a possible physical meaning to convex combinations.

Definition 4 *Let $\mathcal{P}_1$ and $\mathcal{P}_2$ be two preparations of $\mathcal{S}$ and $w \in [0,1]$. Statistical mixture,*

$$\mathcal{P} = \{(w, \mathcal{P}_1), ((1-w), \mathcal{P}_2)\} \ , \tag{2}$$

of $\mathcal{P}_1$ and $\mathcal{P}_2$ is the following preparation: Let $\mathcal{S}$ be prepared either by $\mathcal{P}_1$ or by $\mathcal{P}_2$ in a random way so that $\mathcal{P}_1$ is used with probability w and $\mathcal{P}_2$ with probability $1-w$.

Assumption 2 *Let $\mathcal{P}_1$ and $\mathcal{P}_2$ be two preparations of $\mathcal{S}$ and let they prepare states T_1 and T_2. Then the statistical mixture (2) prepares state*

$$\mathsf{T} = w\mathsf{T}_1 \ (+)_p \ (1-w)\mathsf{T}_2 \ .$$

We call the right-hand side statistical decomposition of T.

The purpose of sign "$(+)_p$" on the right-hand side is to stress that this convex combination is a statistical decomposition. To be aware of the distinction is very important for the understanding of the theory of quantum measurement, see [3]. For example, the theory of quantum decoherence can achieve that the final state of the apparatus is a convex combination of distinct pointer states but cannot conclude that it is a statistical decomposition and must, therefore, resort to further assumptions such as Everett interpretation [16]. Let us stress that a statistical decomposition of state T is not determined by the mathematical structure of state operator T but by a preparation of T.

Sometimes, one meets the objection that states $w\mathsf{T}_1 \ (+)_p \ (1-w)\mathsf{T}_2$ and $w\mathsf{T}_1 + (1-w)\mathsf{T}_2$ of system $\mathcal{S}$ cannot be distinguished by any measurement. But this is only true if the measurements are limited to registrations of observables of $\mathcal{S}$. If observables of arbitrary composite systems containing $\mathcal{S}$ are also admitted, then the difference between a statistical decomposition and a convex combination can be found by measurements [7]. This is exactly the argument against the decoherence theory described in [17], p 171. Let us also emphasise that quantum state statistics has nothing to do with the statistics of values of observables.

As a mathematical operation, $(+)_p$ is commutative and associative. Thus, the definitions and assumptions can be generalised to more than two preparations and states. Moreover, state decomposition is invariant with respect to system composition and unitary evolution [6].

Definition 5 *Prepared state T of the form $w\mathsf{T}_1 \ (+)_p \ (1-w)\mathsf{T}_2$ with $w \in (0,1)$ is called decomposable.*

The meaning of a prepared state T being indecomposable is that the ensemble $\mathbf{E}$ defined by repeating the preparation $\mathcal{P}$ of T has no sub-ensemble $\mathbf{E}'$ that can be obtained by repeating preparation $\mathcal{P}'$ of state $\mathsf{T}' \neq \mathsf{T}$.

Thus, to be decomposable or indecomposable are *physical* properties of prepared states that are determined by the preparations of the states, not just by *mathematical* existence of convex combinations. The faces just restrict possible statistical decomposition of states. For example, any extremal state is *indecomposable*. Decomposable states has been also called "proper mixtures" [17], "direct mixtures" [15] or "gemenges" [14, 7].

There are prepared non-extremal states that are indecomposable. Consider system $\mathcal{S}$ prepared in the EPR experiment (see eg [13], p 150). $\mathcal{S}$ is a spin 0 system composite of two spin 1/2 systems $\mathcal{S}_1$ and $\mathcal{S}_2$. $\mathcal{S}$ is prepared in extremal state $|\psi\rangle\langle\psi|$,

$$|\psi\rangle = \frac{1}{\sqrt{2}}(|1+\rangle \otimes |2-\rangle - |1-\rangle \otimes |2+\rangle) ,$$

where $|1+\rangle$ is the state of the first particle with spin up etc, so that the z components of the spins of the two subsystems are anti-correlated. Then the state of $\mathcal{S}_1$ is

$$tr_{\mathcal{S}_2}[|\psi\rangle\langle\psi|] = \frac{1}{2}(|1+\rangle\langle 1+| + |1-\rangle\langle 1-|) , \tag{3}$$

which is not extremal but the right-hand side on Eq. (3) cannot be a statistical decomposition because this would imply that Bell inequality were satisfied. Here, $tr_{\mathcal{S}_2}[|\psi\rangle\langle\psi|]$ is a partial trace of $|\psi\rangle\langle\psi|$ over the degrees of freedom of $\mathcal{S}_2$, see eg [13].

By the way, the experiment can also be considered as a preparation of $\mathcal{S}_1$ in state (3). This state is then an objective property of $\mathcal{S}_1$ irrespectively of the entanglement of $\mathcal{S}_1$ with $\mathcal{S}_2$. One can also say that $|\psi\rangle\langle\psi|$ is an entangled state,

$$|\psi\rangle\langle\psi| \neq tr_{\mathcal{S}_2}[|\psi\rangle\langle\psi|] \otimes tr_{\mathcal{S}_1}[|\psi\rangle\langle\psi|] ,$$

but (3) is not.

We observe that a quantum state is conceptually very different from a state in Newtonian mechanics. It may be helpful to look at some important differences. Let us define a state of a Newtonian system as point p of the phase space of the system, $\mathbf{\Gamma}$. Newtonian state defined in this way is generally assumed to satisfy:

(i) *Objectivity*: a state is an objective property.

(ii) *Universality*: any system is always in some state.

(iii) *Exclusivity*: a system cannot be in two different states simultaneously.

(iv) *Completeness*: any state of an isolated system represents the maximum information that can exist about this system.

(v) *Locality*: the state of a systems determines the position of the system.

An incomplete information about the state of a system can be described by a probability distributions on $\mathbf{\Gamma}$. Such a distribution is sometimes called *statistical state*. In any case, we distinguish a state from a statistical state.

A quantum state is an element of $\mathbf{T}(\mathbf{H})_1^+$ and the comparison to Newtonian states is described by:

(i) There is objectivity: a prepared state is an objective property.

(ii) There is no universality: a system need not be in any state (an example is a particle $\mathcal{S}$ in a system $\mathcal{S}'$ of identical particles and we assume that a state of $\mathcal{S}'$ has been prepared but that of $\mathcal{S}$ has not [9]).

(iii) There is no exclusivity: a system can be in several states simultaneously (such as T_1 and $w\mathsf{T}_1\ (+)_p\ (1-w)\mathsf{T}_2$ above).

(iv) There is no completeness: a state operator alone does not contain any information on the statistical decomposition of the prepared state. However, if an indecomposable state of an isolated system is given, no more knowledge on the system can objectively exist than that given by the state.

(v) Quantum states are non-local: most states of a single particle do not determine its position, but two detectors at different positions will give anti-correlated results ([9]).

In particular, we have the following correspondences:

$$\begin{array}{lcl} \text{quantum indecomposable state} & \longleftrightarrow & \text{Newtonian state} \\ \text{quantum decomposable state} & \longleftrightarrow & \text{Newtonian statistical state} \end{array}$$

There is, of course, the difference that a decomposable state is described by an element of $\mathbf{T}(\mathbf{H})_1^+$, whereas a Newtonian statistical state is not an element of $\mathbf{\Gamma}$. From the point of view of both statistics of states and statistics of values of observables, textbooks focus on the difference between *pure* (extremal) and *mixed* (non-extremal) states is rather misleading.

An important consequence of our interpretation is the following. We consider any indecomposable state as a complete description of the reality of the system. Thus, the collapse of wave function and analogous processes must be viewed as physical processes, not just as changes of our information about the system [6, 9, 3].

3.3. Simple objective properties

In quantum mechanics, simple objective properties can be defined as follows.

Definition 6 *Let $f : \mathbf{T}(\mathbf{H})_1^+ \mapsto \mathbb{R}$. A simple objective property is defined by proposition "f has value a".*

As states are uniquely determined by preparations, and values of f by the states, the simple properties are objective. Each simple property is equivalent to a subset of $\mathbf{T}(\mathbf{H})_1^+$ described by

$$\{\mathsf{T} \in \mathbf{T}(\mathbf{H})_1^+ \mid f(\mathsf{T}) = a\} \ .$$

Quantum simple properties form a Boolean lattice with respect to logical operations of union and intersection that is isomorph to the lattice of subsets of $\mathbf{T}(\mathbf{H})_1^+$. This is clearly different from the so-called "quantum logic", which holds for properties defined by propositions "Projection P has value η" with $\eta = 0,\ 1$ (see eg [18]). The reason is that our simple properties are defined by preparations while those of quantum logic by registrations because P is an observable and η is its value.

Quantum simple properties are analogous to Newtonian properties that can be defined as real functions on $\mathbf{\Gamma}$. For example, a system having energy E is a property equivalent to the subset of points of $\mathbf{\Gamma}$ at which the energy has value E. Newtonian properties also form a Boolean lattice.

A very important example of quantum simple property is average $tr[\mathsf{AT}]$ of observable A in state T. One could ask how comes that average of A is an objective property while its values are not? But this is simple: the average can be considered as a predetermined condition on the values that are to be registered: in large numbers these values must add to the average. Thus, the average is an example of a property that can be measured by a large number of individual registrations.

Clearly there are enough simple objective properties to describe the dynamical situation of any quantum object completely. Thus, our theoretical models of real objects can ascribe them enough objective properties, as required. More discussion is in [4, 7, 3].

4. Conclusion and outlook

- Constructive Realism by Giere—a kind of relatively strong realism—is shown to be compatible with quantum mechanics.
- The resulting realist interpretation is indispensable for our method to construct quantum models of classical properties [5, 3] and of measurement processes [6, 8, 3].

Acknowledgments
The author is indebted to Petr Jizba, Karel Kuchař and Jiří Tolar for discussion.

References
[1] Putnam H 1983 *Realism and Reason*, Philosophical Papers, Vol. 3 (Cambridge, UK: Cambridge Univ. Press)
[2] Giere R N 1988 *Explaining Science: A Cognitive Approach* (Chicago : The University of Chicago Press)
[3] Hájíček P 2012 Quantum models of classical world *Preprint* arXiv:1212.2400
[4] Hájíček P and Tolar J 2009 *Found. Phys.* **39** 411
[5] Hájíček P 2009 *Found. Phys.* **39** 1072
[6] Hájíček P 2011 *Found. Phys.* **41** 640
[7] Hájíček P and Tolar J 2010 *Acta Phys. Slovaca* **60** 613
[8] Hájíček P 2011 *J. Phys.: Conf. Ser.* **306** 012035
[9] Hájíček P 2012 *Found. Phys.* **42** 555
[10] Hájíček P 2012 *J. Phys.: Conf. Ser.* **301** 012035
[11] Schroeck F E Jr. 1996 *Quantum Mechanics on Phase Space* (Dordrecht: Kluwer)
[12] Leggett A J (2002) *J. Phys.: Condens. Matter* **14** R415
[13] Peres A 1995 *Quantum Theory: Concepts and Methods* (Dordrecht: Kluwer)
[14] Bush P, Lahti P J and Mittelstaedt P 1996 *The Quantum Theory of Measurement* (Berlin: Springer)
[15] Ludwig G 1983 *Foundations of Quantum Mechanics* Vol I (New York: Springer)
[16] Schlosshauer M 2004 *Rev. Mod. Phys.* **76** 1267
[17] d'Espagnat B 1995 *Veiled Reality* (Reading: Addison-Wesley)
[18] Bub J 1999 *Interpreting the Quantum World* (Cambridge, UK: Cambridge Univ. Press)

The coevent formulation of quantum theory

Petros Wallden

1. University of Athens, Physics Department, Nuclear & Particle Physics Section, Panepistimiopolis 157-71, Ilissia Athens, Greece
2. Technological Educational Institute of Chalkida, General Science Department, Psahna-34400, Greece

E-mail: petros.wallden@gmail.com, pwallden@phys.uoa.gr

Abstract. Understanding quantum theory has been a subject of debate from its birth. Many different formulations and interpretations have been proposed. Here we examine a recent novel formulation, namely the coevents formulation. It is a histories formulation and has as starting point the Feynman path integral and the decoherence functional. The new ontology turns out to be that of a coarse-grained history. We start with a quantum measure defined on the space of histories, and the existence of zero covers rules out single-history as potential reality (the Kochen Specker theorem casted in histories form is a special case of a zero cover). We see that allowing coarse-grained histories as potential realities avoids the previous paradoxes, maintains deductive non-contextual logic (alas non-Boolean) and gives rise to a unique classical domain. Moreover, we can recover the probabilistic predictions of quantum theory with the use of the Cournot's principle. This formulation, being both a realist formulation and based on histories, is well suited conceptually for the purposes of quantum gravity and cosmology.

1. Motivation

Quantum theory is undoubtedly one of the most successful theories. The understanding and interpretation of quantum theory, has been a subject of debate from the birth of the theory until now. There is no unique, widely accepted interpretation of quantum theory. In the recent years, due to the developments in the field of quantum information and the search of a quantum theory of gravity and cosmology, there is a new wave of interest for the foundations of quantum theory. This contribution, is a review of a novel formulation of quantum theory namely the "coevents formulation". The motivation for such a formulation, is twofold.

First is the need for a realist interpretation of quantum theory. The standard view of quantum theory, is that of an instrumentalist. Avoids to refer to the actual ontology, and makes assertions only related with some idealised measurements carried out by an external observer. However, particularly for the field of quantum cosmology where one treats the full universe as a quantum system, the need for an interpretation that does not require an external observer becomes vital. In this sense the need for a realist interpretation, becomes more than a philosophical investigation.

Second is the need for a formulation that treats space and time in equal footing. In constructing a quantum theory of gravity, one is led to the observation that while space and time are totally different objects in quantum theory, they are more-or-less the same in general relativity. This leads to a tension that has both technical and philosophical difficulties (e.g. see the problem of time [1]). We can claim that better suited for the purpose of quantum gravity should be a formulation of quantum theory that is based on full histories of the system (such as in the Feynman path integral) and *not* a canonical formulation (that relies on the Hamiltonian and uses one-moment-of-time propositions). The latter may not even be applicable for certain approaches to quantum gravity such as the causal sets.

The coevents formulation was introduced by Rafael Sorkin in [2] following his earlier work in [3] (see section 3). It is a realist interpretation, that is based on histories. In particular, the starting point is the Feynman amplitudes for histories, and interpreting those amplitudes in an observer-free, context-independent way is the aim of the formulation. In section 2 we will give the standard histories view for classical physics, state which is the importance of precluded sets and stress which is the problem in adopting a "naive" realist view in quantum physics. In section 3 we will introduce the coevents formulation, giving two different (but equivalent) views. In section 4 we will explore (briefly) certain developments of the formulation. In particular we will define what a classical domain is, show that there exist a unique, context-independent such domain and point out that this is not the case for the consistent histories approach[1] (section 4.1). We will give an account of how probabilistic predictions arise (section 4.2), and point out the role and the status of the initial state in the formulation (section 4.3). In section 5 we will summarise and conclude.

2. Histories, classical physics and quantum measure

We are taking a histories view. Here we will review classical physics casted in histories perspective and stressing certain important features that become important in quantum physics.

2.1. General Structure

In describing physics with histories, there are three important mathematical structures.
(1) The first, is the space of all finest grained descriptions Ω. In probability theory, it is called sample space, while in histories formulation, is called *history space*. The histories space, in classical physics is the space of potential realities. However this picture does not carry over in quantum theory, as we will see, where potential realities are no longer fine-grained descriptions.

Each element of this space $h \in \Omega$, corresponds to a full description of the system, specifying every detail and property. For example, a fine grained history gives the exact position of the system along with the specification of any internal degree of freedom, for every moment of time. For a single non-relativistic particle, Ω would be the space of all trajectories in the physical space.

Along with Ω, we define the space $\mathcal{U}$, which consists of all the subsets $A \subseteq \Omega$ and the Boolean algebra associated with them (called events algebra), where the addition is defined as the symmetric difference between subsets $A + B := A \triangle B$ and the multiplication given by the intersection of subsets $A \cdot B := A \cap B$. Each subset $A \subseteq \Omega$ is called *event*, as in probability theory. Note, that all physical questions that one can ask, correspond to one of those subsets. If, for example, one wishes to ask "was the system at the region Δ at time t?", it corresponds

[1] The approach is also known as *decoherent histories*.

to the subset A defined as $\{A : h_i \in A \text{ iff } h_i(t) \in \Delta\}$, i.e. all histories that the system at time t is in the region Δ.
(2) The second structure, is the space of truth values (for which we will use the notation $\mathcal{T}$), and the algebra associated with them. In the case of classical physics, the truth values are simply the two elements set $\mathcal{T} := \{\text{True, False}\}$ (or simply $\{1, 0\}$), and the Boolean algebra, called the truth values algebra, associated with them. We have $1 + 1 := 0, 0 + 0 := 0, 1 + 0 := 1, 0 \cdot 1 := 0, 1 \cdot 1 := 1, 0 \cdot 0 := 0$. It is possible to use as truth values a more general algebra.
(3) The third structure, is the space of possible *truth valuations maps* ($\phi : \mathcal{U} \to \mathcal{T}$). These maps assign a truth value, to each of the possible questions (events). We shall use the notation $\mathcal{M}$ for this space. However, to be able to reason, we need to be able to make deductions, in other words to be able to obtain the truth value of some event from the truth values of some other events. The most strict condition, that holds in classical physics, is that the allowed truth valuation maps ϕ are homomorphisms between the events algebra $\mathcal{U}$ and the truth values algebra $\mathcal{T}$. In other words we require the map ϕ to obey:

(a) Multiplicativity

$$\phi(A \cdot B) = \phi(A \cap B) = \phi(A)\phi(B) \tag{1}$$

(b) Additivity (Linearity)

$$\phi(A + B) = \phi(A \triangle B) = \phi(A) + \phi(B) \tag{2}$$

An important observation, is that maps ϕ that are homomorphisms[2], are in a one-to-one correspondence with single elements h of the space of histories Ω. In particular, maps that are the characteristic function for a particular history h, are homomorphisms between the Boolean event algebra and the Boolean truth values algebra.

$$\begin{aligned} \phi_h(A) &= 1 \text{ if } h \in A \\ \phi_h(A) &= 0 \text{ otherwise} \end{aligned} \tag{3}$$

Moreover, *all* homomorphisms are of this type. There exists one homomorphic map for each one of the (single) elements h of Ω. Due to the one-to-one correspondence of homomorphic maps and single elements h, we could adopt a dual view, and state that classical reality is a homomorphic map ϕ_h between the event algebra $\mathcal{U}$ and the truth values algebra $\mathcal{T}$. To sum up, for classical physics, potential realties can be viewed either as single histories h elements of Ω or as homomorphic maps ϕ_h. The potential realities are further restricted by the dynamics. As we will see later in section 3, the ontology in quantum physics changes, but the existence of a dual picture of reality as an event or as a map is maintained.

We have explored, so far, the "kinematic" part of the theory. We have not mentioned the dynamics (Hamiltonian), or initial condition. In classical deterministic physics, given the initial state and the dynamics, we know all the evolution of the system, and thus the full history. In other words, we know which history h from the space of histories Ω is actually realised. Given the initial state and dynamics, there is only one possible potential reality h, the one that is actually realised. This is the meaning of determinism. Obtaining predictions, becomes trivial, since an event is true if and only if it contains the one realised history. We can define a classical measure μ_c on $\mathcal{U}$, such that $\mu_c(A) = 1$ if $h \in A$ and $\mu_c(A) = 0$ otherwise. This gives

[2] In the rest paper, when we refer to homomorphisms or homomorphic maps, it should be understood as homomorphisms between the event algebra $\mathcal{U}$ and the truth values algebra $\mathcal{T}$.

the probability that an event A occurs, which is always either one or zero. This simple measure, coincides with the truth valuation ϕ_h, of the single history that is realised, but should *not* be confused, since the analogy holds only for deterministic classical physics.

In classical stochastic physics, the picture is different. We are not given which history is actually realised, but given the initial state and dynamics, we can obtain a classical measure μ_c (non-trivial this time) on the space $\mathcal{U}$ of subsets of Ω. The measure of each event, corresponds to the probability that this event occurs. The measure is no longer related with the valuations $\phi \in \mathcal{M}$. It is important to note here, that the actual realised reality in stochastic physics, will still be one fine grained history h element of Ω. Since our theory is no longer deterministic, the space of potential such histories has many elements and in particular (for finite histories space) all fine grained histories h_i that have non-zero measure, $\mu_c(h_i) \neq 0$. The role of the measurement in stochastic classical physics, is restricting further the potential realities. For example, before tossing a coin, two outcome were possible, but after performing the measurement and looking the outcome, it is determined wether the outcome was heads or tails. In histories formulation, the latter corresponds to the experimenter restricting the set of possible histories in the universe to those that are compatible with his new observation. To sum up, in stochastic classical physics, the ontology remains the same but the mechanism to obtain predictions changes. We will see more on probabilistic predictions for closed classical or quantum systems, in section 4.2 after having introduced the coevents formulation.

2.2. Precluded sets

As we have seen, the set of potential realities, both in deterministic and stochastic classical physics, is determined by the classical measure that is defined on Ω, which in its term is fixed by the initial state and dynamics. More specifically, for finite Ω, it is the measure zero sets that determine which are the allowed potential realities/histories [3]. We define *precluded event* to be an event $P \subset \Omega$ such that its measure is zero $\mu_c(P) = 0$. The precluded events do not occur. If an event P is precluded, any subset P_1 of P, should also be ruled out. The latter is a vital condition that needs to be satisfied, if one wishes to make any deductive argument[4]. In classical physics, this condition is guaranteed by the fact that we require the maps/elements of $\mathcal{M}$, to be homomorphisms of the Boolean algebras. Simply requiring the condition $\mu_c(P) = 0 \Rightarrow \phi(P) = 0$ leads directly to $\phi(P_1) = 0$ if $P_1 \subseteq P$ and $\phi(P) = 0$. While this discussion seems trivial for classical physics, it will become apparent that it is important for quantum physics.

A final interesting remark, concerning precluded sets, is that one can fully reconstruct all probabilities using only the set of precluded events provided that he has $n \to \infty$ identical copies of the system. Heuristically, any distribution of outcomes of the identical copies, that differs from the one given by probability, has small chance of occurring, and as the number of copies tends to infinity, this chance of occurring also tends to zero. Technically, let A an event and A_n independent copies. There exists a unique number p such that the following event to be precluded (I_A is indicator function)

$$\{s \in \Omega^* : (I_A(s_1) + I_A(s_2) + \cdots + I_A(s_n))/n \text{ does not tend to } p \text{ when } n \to \infty\} \tag{4}$$

See for example [4]. In this sense, all the content of relative frequency interpretation of

[3] For infinite Ω more care is needed, and definition of measurable sets is required. However, most of the arguments carry over.

[4] This is related with the modus ponens rule of inference. E.g. "if all humans have two hands" and "Plato is human", we wish to be able to deduce that "Plato has two hands".

probability, is included in the precluded sets. Moreover, the precluded sets can be used to recover some predictions via the Cournot principle (see later), even in the absence of many identically prepared copies. This is one more reason why, we choose to give specific importance to precluded events.

2.3. Quantum measure

The picture described above, cannot be (fully) carried over to quantum theory. The histories space Ω and its subsets/events $\mathcal{U}$ remains the same. The main difference arises, mathematically, from the fact that we no longer have a classical measure on $\mathcal{U}$ but rather a *quantum measure*, which we will shortly define. Given an initial condition and dynamics the quantum measure is totally fixed. The initial condition and the dynamics can be given either in form of some initial condition on a path integral along with an action S, or as an initial wavefunction in a Hilbert space along with a Hamiltonian operator.

To define the quantum measure (which was first done by Sorkin in [5]) we need to introduce amplitudes to histories. Starting from Feynman's path integral approach, one can assign an amplitude (complex number) to each history.

$$\alpha(h_i) = \exp iS(h_i) \tag{5}$$

which depends on the initial state and on the dynamics of the system encoded in the action S. Obtaining the transition amplitudes from (x_1, t_1) to (x_2, t_2), is done by summing through all the paths $\mathcal{P}$ obeying the initial and final condition, i.e.

$$\alpha(x_1, t_1; x_2, t_2) = \int_{\mathcal{P}} \exp(iS[x(t)])\mathcal{D}x(t) \tag{6}$$

The mod square of this amplitude is the transition probability. One can extend this to any event $A \subseteq \Omega$ and proceed to define a quantum measure μ:

$$\mu(A) = |\int_{A\subseteq\Omega} \exp(iS[x(t)])\mathcal{D}x(t)|^2 \tag{7}$$

Alternatively, one can take a Hamiltonian view, and obtain the quantum measure in the following way. Define the class operator C_A

$$C_A = P_{A_n}U(t_n - t_{n-1})\cdots U(t_3 - t_2)P_{A_2}U(t_2 - t_1)P_{A_1} \tag{8}$$

Where $U(t)$ is the unitary evolution operator that relates to the Hamiltonian via $U(t) = \exp(-iHt)$, and P_{A_i} is the subspace that history A lied at time t_i. This class operator corresponds to the history that the system is at P_{A_1} at t_1 *and* at P_{A_2} at t_2 etc. The projection operators can be at any subspace of the Hilbert space. The quantum measure is then defined to be

$$\mu(A) = \mathrm{Tr}(C_A^\dagger \rho C_A) \tag{9}$$

For cases where there is a well defined time, such as in non-relativistic quantum mechanics, the two definitions are equivalent and one uses the more convenient one. For finite moments of time histories, the operator expression is usually easier to deal with[5].

Note, that the quantum measure is closely related with the decoherence functional [6], since it arises from the diagonal parts of the latter.

[5] Introducing more moments of time is simply a fine graining of the previous histories.

Since the quantum measure is a non-negative function that is also normalised ($\mu(\Omega) = 1$), one could be tempted to interpret the quantum measure as probability. However this is not possible, due to interference. The additivity condition of probabilities[6], is not satisfied:

$$\mu(A \sqcup B) \neq \mu(A) + \mu(B) \tag{10}$$

However a weaker condition holds that shows that there is no three-paths interference, that cannot be deduced from pairwise interference:

$$\mu(A \sqcup B \sqcup C) = \mu(A \sqcup B) + \mu(A \sqcup C) + \mu(B \sqcup C) - \mu(A) - \mu(B) - \mu(C) \tag{11}$$

In other worlds, the quantum measure is fully determined once we know the quantum measure of single histories and of pairs of histories. The specific expression is given for example in [7]:

$$\mu(A = \{h_1, h_2, \cdots, h_n\}) = (2-n)\sum_{i=1}^{n} \mu(h_i) + \frac{1}{2}\sum_{i,j=1}^{n} \mu(\{h_i, h_j\}) \tag{12}$$

Recently, experiments [8] have confirmed that indeed the condition of eq. (11) holds in nature.

To fully characterise a histories theory, we need the following triplet: The histories space, the events algebra and the quantum measure $(\Omega, \mathcal{U}, \mu)$.

A *partition* $\mathcal{P}$ of the histories space Ω, is a collection of events $\{P_1, P_2, \cdots\}$ such that $P_i \cap P_j = \emptyset$ for all $i \neq j$ and $\bigcup_i P_i = \Omega$. It is an exhaustive and exclusive collection. Each event of the partition is called *cell.* A *coarse-graining* of a histories theory $(\mathcal{P}, \mathcal{U}_\mathcal{P}, \mu_\mathcal{P})$ is defined in the following way:

(i) The cells of a partition $\mathcal{P}$ are the elements of the coarse-grained histories space.

(ii) The Boolean algebra generated by addition and multiplication of these cells $\mathcal{U}_\mathcal{P}$, is the coarse grained event algebra.

(iii) The quantum measure defined on them should be compatible with the fine grained quantum measure, i.e. $\mu_\mathcal{P}(A) = \mu(A)$ for all $A \in \mathcal{U}_\mathcal{P}$

2.4. The trouble with quantum theory

It is already clear, that since the quantum measure cannot be understood as probability, the picture of classical stochastic physics presented in the previous section might need modification. However, the main reason why one cannot interpret the quantum measure in the same way as the classical measure, appears from consideration of the precluded events. Note that from now on, when we mention precluded events we mean that they have *quantum* measure zero.

In many cases (see below for examples), one can find a collection of precluded events $\mathcal{Z} = \{P_1, P_2, \cdots\}$, where $\mu(P_i) = 0$, that also cover the full histories space Ω, i.e. that $\bigcup_i P_i = \Omega$. We will call such collection of precluded events as a *zero cover.* Note, that since in general $P_i \cap P_j \neq \emptyset$, this collection does *not* constitute a partition.

However, as we mentioned earlier, in order to be able to construct deductive arguments, the truth value of *any* subset of a precluded set has to be "False". Having the full Ω covered with precluded sets, imply that no single history $h_i \in \Omega$ can have a truth value "True". This also leads to no homomorphic map $\phi \in \mathcal{M}$ existing.

We will first consider a simplified example, with a three-slits interference experiment. Consider a point at the screen, where crossing slit A destructively interferes with crossing through slit B and slit B destructively interferes with slit C, i.e. $\mu(\{h_1, h_2\}) = 0, \mu(\{h_2, h_3\}) = 0$, however $\mu(\{h_1, h_2, h_3\}) \neq 0$ (see Figure 1).

[6] Additivity of disjoint regions of the sample space.

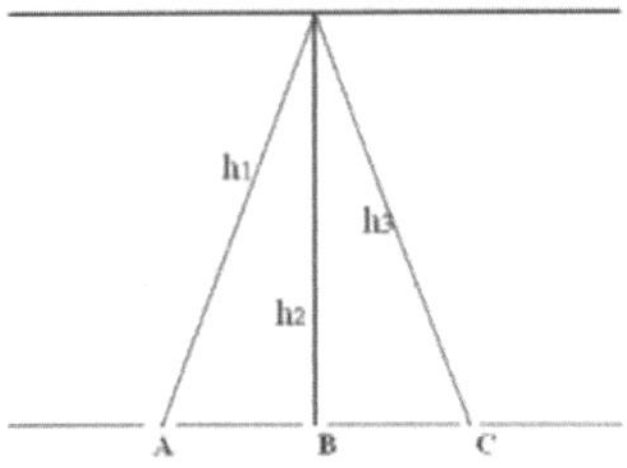

Figure 1. Three slits, h_1 is the history that passes from slit A and hits the central point on the screen, etc

Since the quantum measure for hitting that point on the screen is non-zero, we know that particles do hit the screen at this point some times. In these cases we run into paradox, by assuming that the particle followed a single trajectory. Crossing from any slit, is a subset of an event of quantum measure zero and thus cannot occur no matter what possible hidden variables determine the realised trajectory. However, in this example, one can say that we did not cover all Ω with precluded events, because the particle could have been detected at other points of the screen where the above analysis does not hold. Even though the nature of the problem that arose in this example was essentially the same as that for zero covers, this example is not a proper zero cover.

We now give, an example of a proper zero cover for a single qubit with initial state $|0\rangle$. We measure it in three-moments-of-time[7]. We assume that the qubit evolves in the following (unitary) way:

$$U = \frac{1}{\sqrt{2}} \begin{pmatrix} 1 & i \\ i & 1 \end{pmatrix} \tag{13}$$

which can arise for some particular Hamiltonian H and time interval t between measurements.

We evolve and measure in the $\{|0\rangle, |1\rangle\}$ basis, three times. The amplitudes for different histories are given by the expression using the class operators $|\psi_{final}\rangle = C_\alpha|0\rangle = P_{t_3}UP_{t_2}UP_{t_1}U|0\rangle$, with P_{t_i} being the projection either $|0\rangle\langle 0|$ or $|1\rangle\langle 1|$ depending on what was measured at the moment t_i:

$$\begin{aligned}
\alpha(h_1 = 000) &= \frac{1}{2\sqrt{2}} \quad , \quad \alpha(h_2 = 001) = -\frac{1}{2\sqrt{2}} \\
\alpha(h_3 = 010) &= -\frac{1}{2\sqrt{2}} \quad , \quad \alpha(h_4 = 011) = -\frac{1}{2\sqrt{2}} \\
\alpha(h_5 = 100) &= \frac{1}{2\sqrt{2}}i \quad , \quad \alpha(h_6 = 101) = -\frac{1}{2\sqrt{2}}i \\
\alpha(h_7 = 110) &= \frac{1}{2\sqrt{2}}i \quad , \quad \alpha(h_8 = 111) = \frac{1}{2\sqrt{2}}i
\end{aligned} \tag{14}$$

The labels indicate outcomes of the $\{|0\rangle, |1\rangle\}$ measurements from right to left. For example h_4 is the history that we get the outcome $|1\rangle$ in the two first measurements and $|0\rangle$ in the third. This leads to the following zero cover (collection of precluded events that their union is Ω):

[7] Having more moments of time, is simply a fine-graining and does not alter the coarse-grained cover we give.

$$\mathcal{Z} = \{\{h_1, h_2\}, \{h_1, h_3\}, \{h_1, h_4\}, \{h_5, h_6\}, \{h_6, h_7\}, \{h_6, h_8\}\} \tag{15}$$

This example is simple and very general, since with little modifications, it can hold for any Hamiltonian for the qubit (choosing different moments of time).

More complicated examples can also be constructed for all systems (initial states, Hamiltonian). The Kochen-Specker theorem [9], when casted in the histories language, is precisely a special case of a zero cover (see [7, 10]). It is the existence of these zero covers, that rules out the possibility of the reality being a single-history and thus rules out hidden variables theories. We have to stress however, that the conclusion holds, provided one accepts as starting point the Feynman path integral as the amplitude for all histories, being fine or coarse grained[8].

One attempt to overcome the zero covers paradox, would be to avoid taking *all* events with quantum measure zero as precluded, i.e. allow for some events attaining truth value "True", even if they have quantum measure zero. This runs to, at least two problems. One is that we need a physical/natural criterion for when an event of quantum measure zero is actually precluded and when it is not. The second comes from consistent histories approach, and the observation, that quantum measure zero events, decohere (and thus behave classically) with their complement. Having, a zero cover, leads necessarily either to a context-dependent view or to the need of selecting a preferred classical domain for the consistent histories (see section 4.1).

Other attempts involve modifying one of the structures we had in classical physics $(\mathcal{U}, \mathcal{T}, \mathcal{M})$.

(i) Instead of considering the full set of possible questions the event algebra $\mathcal{U}$ (subsets of Ω), to somehow select a preferred set of classical questions, in other words a subalgebra of the event algebra. This is essentially done in the consistent histories approach [6]. However, this approach allows an arbitrary choice of which "Boolean subalgebra" to consider (see the critic in [11]). In this sense, fails to give a satisfactory account of what actually occurs, unless some physical principle is discovered that selects a preferred classical domain (see section 4.1).

(ii) Alter the space of truth values (and the algebra associated) $\mathcal{T}$. Instead of using a two values Boolean logic ({True, False}) one could use truth valuation maps that take truth values on a subobject classifier of a category, and the associated algebra is a Heyting algebra (see for example [12]). The resulting logic is intuitionistic logic which is deductive logic[9] but most importantly, contextual (the truth value, depends on the context/question asked). One can also attempt to interpret consistent histories in this contextual view [13].

(iii) Finally, one could alter the set of allowed maps ϕ from $\mathcal{U}$ to $\mathcal{T}$, while keeping both the event algebra and the truth values algebra, same as in classical physics. This is achieved by weakening the requirement that ϕ is a homomorphism. This is essentially done in the coevent formulation which is the topic of this paper.

3. The coevent formulation

We are now in position to give the technical definition of the coevents formulation. The approach was pioneered by Sorkin [2, 3] and one can find older reviews in [14, 15]. There are two, essentially equivalent, ways of viewing the coevents formulation. Here we will introduce both,

[8] In De Broglie-Bohm theory, the problem of zero covers is avoided, by deviating from the Feynman amplitudes in general and only agreeing when necessary to recover standard non-relativistic QM results.

[9] Proofs by contradiction are not allowed, and only constructive/deductive proofs are acceptable.

starting with what appears as more intuitive, leaving as second the one that was historically developed first. In particular, the first view, is that the reality (ontology of the theory) is a coarse-grained history, in direct analogy with the picture in classical physics, that reality was a fine-grained history h_i. The second view, is the dual one, that considers the reality as being the (generalised) valuation map ϕ, which is in general anhomomorphic, and it is analogous with the view in classical physics, that reality is a homomorphic map ϕ_{h_i} that is the characteristic function of the history h_i.

3.1. The coarse-grained history view

Instead of viewing as possible realities the fine-grained histories h, elements of Ω, we could allow as potential realities, coarse-grained histories, i.e. (as starting point) any event R element of $\mathcal{U}$. However, it is clear from classical physics, that this would lead to allowing potential realities that are unphysical. We need to impose conditions, that among other things, will reproduce the classical picture for the case of classical measure. There are two such requirements.

(i) The allowed potential realities (events), are not precluded, and moreover, they are not included in any precluded event. We define an event A to be *non-preclusive* if

$$\nexists \quad P \text{ such that } A \subseteq P \text{ and } \mu(P) = 0 \tag{16}$$

The set of all non-preclusive events we call it $\mathcal{NP}$. We require that an allowed reality R is a non-preclusive event ($R \in \mathcal{NP}$).
This condition is satisfied by potential realties in classical physics. However, for a classical measure, all subsets of a measure zero set, are also measure zero sets. It follows, that in classical physics is sufficient to require that potential realities h_i (single-histories events) are not precluded ($\mu_c(h_i) \neq 0$).
For a quantum measure, it is possible to have an event A that is not precluded ($\mu(A) \neq 0$) but *is* subset of a precluded event P. As we have seen earlier, we wish to rule out such events from potential realties, and this is done by requiring the allowed coevents to be non-preclusive (rather than being simply not precluded). To this point, we stress again the reason to rule out such events. We wish to recover deductive logic, and we cannot have an event given truth value "True", while one of its supersets is precluded and thus given the truth value "False".

(ii) The second condition is that the potential realities R are the finest grained events that are non-preclusive, i.e. obey eq. (16). Mathematically:

$$R \in \mathcal{NP} \text{ and } \nexists \quad A \in \mathcal{NP} \text{ such that } A \subset R \tag{17}$$

This requirement states, that from all the non-preclusive events, we consider as potential realities *only* those that give the most detailed description. This is called the "maximum detail" condition. It is precisely this condition, that in the case of a classical measure, restricts the possible realties to be fine-grained histories that are not precluded, and thus agree with the picture developed in the previous section. For a quantum measure, the condition of eq. (16) forces us to consider potential realities that are not single-histories, as we will see in examples below.

The potential realities are events that are non-preclusive and of maximum detail and we will call them *covents*[10]. The set of all coevents, given a quantum measure, is $\mathcal{R}$.

[10] Note that the term "coevent", is used usually for the dual picture and refers to the corresponding maps, while the term "support of the coevent" is used for the events described here. We will not adopt this convention in

In order to find the set of possible coevents, we only need to know which events have zero quantum measure. Let us see, which are the allowed coevents for the examples of zero covers we gave in the previous section.

In the three-slit example, the precluded events, were $\{h_1, h_2\}$ and $\{h_2, h_3\}$. The non-preclusive events (those not included in precluded events) are $\{h_1, h_3\}$ and $\{h_1, h_2, h_3\}$ which are the elements of $\mathcal{NP}$. From those, only the event $\{h_1, h_3\}$ satisfies eq.(17) since the other event $\{h_1, h_2, h_3\}$ has as subset the non-preclusive event $\{h_1, h_3\}$. We therefore have a single coevent as potential reality and $\mathcal{R} = \{\{h_1, h_3\}\}$.

In the qubit example with the three-moments-of-time, we have many precluded events, namely:

$$\begin{aligned}&\{h_1,h_2\},\{h_1,h_3\},\{h_1,h_4\},\{h_5,h_6\},\{h_6,h_7\},\{h_6,h_8\},\{h_1,h_2,h_5,h_6\},\\&\{h_1,h_2,h_6,h_7\},\{h_1,h_2,h_6,h_8\},\{h_1,h_3,h_5,h_6\},\{h_1,h_3,h_6,h_7\},\\&\{h_1,h_3,h_6,h_8\},\{h_1,h_4,h_5,h_6\},\{h_1,h_4,h_6,h_7\},\{h_1,h_4,h_6,h_8\}\end{aligned} \tag{18}$$

One can find that there are the following six coevents:

$$\mathcal{R} = \{\{h_2,h_3\},\{h_2,h_4\},\{h_3,h_4\},\{h_5,h_7\},\{h_5,h_8\},\{h_7,h_8\}\} \tag{19}$$

This is an example where the potential realities, are not single-histories and moreover there are many (six) such potential realties.

After having defined the new ontology of the theory, as being that of a coevent, we wish to be able to make assertions. Given one realised coevent, which physical questions/events occur, i.e. take truth value "True"? The truth value "True" is given to all events that have the realised coevent R as a subset and the rest take truth value "False". However, this might lead us to a truth valuation map ϕ that is *not* homomorphic. We will analyse this in the second view of the coevent formulation that follows.

Note that the above, analysis holds for finite histories space Ω. Maintaining this view for infinite histories space is possible but needs more further analysis in the definitions in order to be rigorously formulated.

3.2. The anhomomorphic valuation view

In classical physics, we mentioned a dual view of reality. Instead of considering the single-history h as the ontology of the theory, one could consider the homomorphic truth valuation map ϕ_h as the reality. These valuation maps are characteristic function of single histories h, i.e. gives truth value "True" to any event A that includes the history h, and "False" to all the other events.

$$\phi_h(A) = 1 \text{ iff } h \in A \tag{20}$$

The second view of the coevents formulation, is to try and generalise the truth valuations ϕ's. It should be done in such a way, that is compatible with the paradoxes arising from using a quantum rather than a classical measure. In particular it should be compatible with the presence of zero covers. The modification that we adopt, is weakening the requirement that the truth valuation maps are homomorphisms between the Boolean algebras of $\mathcal{U}$ and $\mathcal{T}$. In

this paper, but will comment on this later.

particular we still require the maps ϕ to be multiplicative i.e. obey eq.(1), but no longer require them to be additive.

$$\begin{aligned}\phi(A\cdot B) &= \phi(A\cap B)=\phi(A)\phi(B)\\ \phi(A+B) &\neq \phi(A)+\phi(B)\end{aligned} \tag{21}$$

We should note, that the first attempt of weakening the homomorphism to allow for the zero covers, was keeping the additive property and dropping the multiplication [3] but it was ruled out by some gedanken experiments that provided counter examples where no such map existed at all. There are also other attempts to weaken the homomorphism. All different attempts were coined "schemes" and the one we have adopted and present in this paper, is the multiplicative scheme. Other than concrete examples that alternative schemes fail, there are several advantages in adopting the multiplicative scheme: (i) being able to have the coarse-grained view, (ii) being guaranteed that there always exists allowed maps no matter which is the quantum measure, (iii) recovering deductive logic and (iv) having a unique classical domain (we will analyse (ii),(iii) and (iv) later).

It can be shown, that all multiplicative maps, are in fact characteristic maps of some (non-trivial) event S

$$\phi_S(A)=1 \text{ iff} S\subseteq A \tag{22}$$

We call this event S *support* of the multiplicative map. Two things to note. First, is that the choice of allowed maps as characteristic maps is very similar with what was done in classical physics with the only difference that we now allow for characteristic maps of events that may consist of many fine grained histories $A=\{h_1,h_2,\cdots\}$. Second thing to note, is that precisely this property of multiplicative maps allows us to have both views of the coevent formulation. There is a one-to-one correspondence of anhomomorphic (but multiplicative) maps and coarse-grained histories (their supports).

Other than requiring the allowed coevents to be multiplicative, we need to impose two further conditions to potential realties, as it was done in the coarse-grained history view.

(i) The maps are preclusive, in other words they give truth value "False", for any event that has zero quantum measure

$$\mu(P)=0\Rightarrow\phi(P)=0 \tag{23}$$

This condition together with the multiplicativity condition, guarantees that an event A being subset of a precluded event P, gets truth value "False". This plays the same role as the first requirement in the previous section (non-preclusive).

(ii) The condition that corresponds to the maximum detail condition of section 3.1, is that the map needs to be primitive. We say that a multiplicative map ψ dominates a multiplicative map ϕ if

$$\phi(A)=1\Rightarrow\psi(A)=1 \quad \forall \quad A\in\mathcal{U} \tag{24}$$

A preclusive multiplicative map is *primitive* if it is not dominated by any other preclusive multiplicative map. For a classical measure, the primitivity condition, leads to maps that are homomorphisms and thus correspond to characteristic maps of single histories, as expected.

The potential realities are (also) called coevents, and are valuation maps that are (i) multiplicative, (ii) preclusive and (iii) primitive.

We should make here a comment on the terminology used to avoid confusion. As coevent, we defined both the multiplicative, preclusive and primitive map ϕ_R and the support of this map R which is a non-preclusive, maximum detail event. In other words we called coevents, the potential realities of the coevents formulation, no matter if we adopted the coarse grained view or the anhomomorphic valuation view. In literature, the term coevent was used only for the (in general anhomomorphic) maps. However, when restricting attention in the multiplicative scheme as we are in this paper, the coarse grained history view is equivalent. In this case, even referring to the anhomomorphic map is easier done using its support.

Since, we no longer have homomorphism between the events Boolean algebra $\mathcal{U}$ and the truth values Boolean algebra $\mathcal{T}$ the laws of inference do not follow directly. The main deductive inference rule, is the modus ponens that essentially means if A is "True" and A implies B, $(A \to B)$, then B also is "True". In [16] it was shown, that modus ponens, is applicable for multiplicative maps. In particular, it was shown that if we require that our valuation maps obey modus ponens, we are necessarily led to multiplicative maps. This is one further motivation for the choice of the multiplicative scheme. It is clear however, that due to the fact that we no longer require the maps to be homomorphisms, some of the structure of the Boolean algebra is lost. In particular, we cannot use proofs by contradiction. There exist events A such that both $\phi(A) = 0$ and $\phi(\neg A) = 0$.

In reality, one of the possible coevents is realised. Given that coevent, we should be able to answer whether any given question/event is realised ("True") or not realised ("False"). As long as the event in question A, either includes the realised coevent R_1, i.e. $R_1 \subseteq A$ or does not intersect it at all $R_1 \cap A = \emptyset$, it is easy to see that either $\phi_{R_1}(A) = 1$ or $\phi_{R_1}(\neg A) = 1$. However, if the event A intersects non-trivially R_1, then both A and $\neg A$ take truth value "False".

For example, in the three-slit case (see Figure 1), we have a single allowed coevent $\{h_1, h_3\}$ which is realised. Asking the question "did the system cross from any of slit A or B" $(A = \{h_1, h_2\})$ gets answer "No". However, answer "No" gets the question "did it cross slit C" $(\neg A = \{h_3\})$. This type of paradox is natural for the quantum world, but should not persist in classical world (see later). In other words, our anhomomorphic logic, allows for deductive proofs but not for proofs by contradiction which is precisely what happens in intuitionistic logic[11]. The latter is partly the motivation for the field of "constructive mathematics" where mathematics are reformulated using solely deductive (constructive) proofs.

Note here that while the rule

$$\phi(A) = 0 \Rightarrow \phi(\neg A) = 1 \tag{25}$$

does *not* hold in the coevents formulation, and we cannot use proofs by contradiction, the opposite rule holds. Namely, if A is "True" its complement is indeed "False".

$$\phi(A) = 1 \Rightarrow \phi(\neg A) = 0 \tag{26}$$

3.3. Remarks

The problem in applying the classical picture of histories in the quantum world arises because we have a quantum rather than a classical measure. The quantum measure allows the existence of zero covers. We therefore can no longer maintain the picture of reality as a single fine-grained history or equivalently as a homomorphic map.

[11] Note however, that our logic is *not* intuitionistic.

We are led to extend the potential realities from single fine-grained histories to coevents (either viewed as a coarse-grained histories or as multiplicative maps). After this generalisation, we are guaranteed to have some potential realities that are compatible with *any* quantum measure (see [7]).

The initial conditions and dynamics fix the quantum measure. From that we obtain the precluded events which determine the allowed coevents. Given the quantum measure, we have, in general, many allowed coevents. In this sense, quantum theory appears as a generalisation of classical stochastic physics rather than generalisation of classical deterministic physics. Measurements are treated in a very similar way as in stochastic classical physics. They are used to rule out some of the potential realities/coevents, alas without altering the system (unlike standard Copenhagen quantum mechanics). The new ontology of quantum theory, is that of a coevent, is not affected by measurements (in the way it does in Copenhagen quantum mechanics) and is not contextual. We can therefore conclude that the coevents formulation is realist.

A final point to raise here, concerns the role of logic. As we have seen, we retain deductive proofs. However, the precise form of logic, i.e. when other rules of inference can be applied, depends crucially on the set of possible coevents, which in their turn depend on the particular quantum measure. It therefore appears that the logic of quantum world is no longer fixed, but becomes dynamical. To quote Sorkin "Logic is to quantum, as geometry is to gravity" [14].

4. Developments

Having given the definition of the coevent formulation, we will now proceed analysing certain important aspects.

4.1. Classical domain and consistent histories

A very important question for quantum theory of a closed system is how one recovers, at some level/coarse-graining, classical physics. The finest grained description is not expected to behave classically, and thus some counter-intuitive properties such as the anhomomorphisms described earlier (e.g. end of section 3.2) are natural. We expect, though, that if we coarse-grain sufficiently, the arising structure behaves classically. In our case, the non-classical behaviour is the anhomomorphisms of the valuation maps. We call classical coarse graining one that all the allowed coevents, elements of $\mathcal{R}$ give rise to homomorphism between the coarse-grained event algebra and the truth values Boolean algebra.

Given a quantum measure μ we have many "allowed" coevents $R_i \in \mathcal{R}$. A classical coarse graining $\mathcal{C}$ is a coarse-graining (see end of section 2.3) such that *all* the coevents R_i give rise to homomorphism. To do so, all coevents R_i should be subsets of some cell of the partition:

$$\begin{aligned} \mathcal{C} &= \{C_1, C_2, \cdots\} \text{ partition} \\ \text{such that} \quad \forall &\quad R_i \in \mathcal{R} \quad \exists \quad j \text{ such that } R_i \subseteq C_j \end{aligned} \tag{27}$$

One of the most important results, is that there exists a *unique* finest grained classical partition, which we will call *principle classical partition.* This means that *all* classical partitions arise as coarse grainings of this unique finest grained partition (see [17] the Appendix). The construction of the principle classical partition is done as follows[12]:

[12] A more rigorous proof can be found in the original reference.

(i) If no pair of coevents intersects $R_i \cap R_j = \emptyset$, the finest partition, is one that has each of R_i as a cell C_i of the partition, and the rest of the history space $\Omega \setminus \bigcup_i R_i$ is covered by single-history cells $C_k = \{h_k\}$ for all $h_k \in \Omega \setminus \bigcup_i R_i$.

(ii) If the coevents overlap ($R_i \bigcap R_j \neq \emptyset$), we take the union of the intersecting coevents and consider this as a cell C_i of the classical partition[13]. The rest space $\Omega \setminus \bigcup_i R_i$ is again covered by single-history cells $C_k = \{h_k\}$ for all $h_k \in \Omega \setminus \bigcup_i R_i$.

Any measurement we can carry out with "classical" apparatuses, will correspond to some coarse-graining of the principle classical partition. Therefore, it is not possible to distinguish between two intersecting coevents with any classical measurement. One could view these "fat" coevents (union of the intersecting ones), as more physical, or treat this property as a short of "coevents uncertainty".

The importance of this result, for histories formulations, becomes more apparent if we compare the situation with the consistent histories approach [6]. The consistent histories approach, with the use of the decoherence functional, finds partitions $\mathcal{P}_i = \{P_1^i, P_2^i, \cdots\}$ of the histories space Ω, such that there is no interference between pairs of cells of the partition. The induced quantum measure becomes classical[14]. Each of these partitions, is called a *consistent set*, and one could attempt to view each consistent set as a classical domain. However, there does *not* exist a single finest grained consistent set that all the others arise from. This leads to properties in one consistent set being incompatible with properties on another, and we are forced to one of the following two solutions. Either with some physical mechanism (not known yet) to select a preferred consistent set, or to adopt a contextual view, that the answers to propositions depend on which consistent set one wishes to refer to (for further details see [11]).

We have, so far, examined the meaning, existence and uniqueness of a classical domain. What remains, is to examine particular physical examples and how this classical domain emerges from the dynamics and whether it agrees with our classical intuition (pointers in a lab being classical, etc). This is essentially matter of calculation (see section 7 of [14]) and is subject of ongoing research.

4.2. Probabilities

The analysis we did so far, concerned the potential realities. However, there is much more to a physical theory, than solely stating which are the possible realities. One needs to be able to make predictions, and in quantum theory, these predictions are in general probabilistic. We should be able, for example, to recover the double slit pattern, if we consider as system a large number of particles crossing the double slit apparatus.

However, the concept of probabilistic prediction for a single non-repeatable closed system is problematic already in the classical level. For example a statement such as "the universe has property A with probability 1/3" cannot be falsified by any experiment. Both results, possessing or not possessing the property, are compatible with the "prediction". This fails the criterion for a scientific proposition, which is that there should exist possible outcomes that falsies the assumption (see Popper). In our case, the system is closed and ideally non-repeatable (we construct such a theory to be able to treat the whole universe as a quantum system). We therefore need to understand the meaning of probability and its connection to the physical world, and to do so we have to resort to the founders of probability theory.

[13] We need to take the transitive closure under intersection. If for example R_1 intersects non-trivially with R_2 and R_2 with R_3, we need to consider the union of all three coevents, even if $R_1 \cap R_3 = \emptyset$.

[14] The quantum measure defined on the coarse graining event algebra, is additive and thus classical.

We can see, that in a non-repeatable system, the only testable predictions, are those that concern things occurring with probability one or zero. We have also seen in section 2.2 that in the limit of infinite identical copies, one can recover probabilities from precluded events[15]. Considering the above, one is lead to the Cournot principle [19] (see also Kolmogorov's view [20] and the references in [17]):

> In a repeated trial, an event A singled out in advance, of small measure ($\leq \epsilon$), rarely occurs.

There exist a stronger formulation of the principle where the "rarely occurs" is replaced with "does not occur", but we adopt the one above. First thing to note in this definition, is the term "small measure". In the coevents formulation we will call such events *approximately precluded.* It is clear, that this is a subjective choice, and depending on how small ϵ is, different prediction may arise. Once the theoretician makes the choice of ϵ, the rest follows. Second thing to stress, is the importance of selecting the event A in advance. Before exploring the latter, we will give an example of how Cournot's principle does work.

Consider tossing a coin 1000 times, and start with the assumption that the coin is fair. Any sequence of 1000 outcomes heads/tails is a fine grained history h_i. Suppose we select the event A, which is the history that has all outcomes to be heads[16]. The probability (with the assumption of a fair coin) is clearly negligible ($1/2^{1000}$). We can thus predict that the event A will not occur, for most reasonable choices of ϵ. However, it is conceivable that such an event does occur. In that very rare occasion, that such event occurs, we would be led to believe that the coin was not fair and thus falsify our initial assummption[17]. This is precisely what is done in actual experimental practise, with the only exception, that if our belief for the assumption is strong, we may convince ourselves to further repeat the experiment if that is possible. In a fundamentally closed system however, such option is not available.

We return now to the importance of pre-selecting the event to be tested. All the time in nature, things of arbitrary small probability do occur. In the earlier example, of tossing a coin 1000 times, any sequence of outcomes heads/tails, has extremely small measure and it is identically the same with the outcome "1000 times heads" that we dismissed earlier. However, the difference between the two is the following. In the one case (1000 times heads) we could ask whether this will occur before carrying the experiment. For some other random sequence of outcomes (the one that actually happened), we did not ask in advance. If we had given a sequence of 1000 outcomes in advance, and then this precise sequence did occur when we carried out the experiment, we would be very sceptical and troubled, and likely assume that somehow there was a correlation of our mind and the tossing of the coin. If the number of tosses were greater than 1000, then we would be even more surprised. In general, we would dismiss the assumption of a fully fair coin and attempt to find which was the pattern that gave this series of outcomes (in that example, it would be easy to perform few more measurements to test the hypothesis, but this is not always an option for all physical systems).

We therefore see that in the use of the weak Cournot principle, there is a dichotomy between the ontology and the prediction of the theory. There exist things of small measure, that despite

[15] Note that the expression at that section, concerns a classical probability measure, and not a quantum measure. However, recovering probabilistic predictions from approximately precluded events can be done for a quantum measure as well, as was done for example, for the double slit example in [18].

[16] The argument holds for other non-trivial events, such as the event that has all series of outcomes, with more than 990 heads outcomes.

[17] Actually, we might even check whether the coin had at both sides head!

being conceivable realities, one predicts that they will not occur. And if they do occur, the theoretician is led to the wrong conclusion that his assumptions were wrong[18].

Note here, that the Cournot principle was supported by many of the founders of probability theory, such as Bernoulli, Cournot, Markov, Borel, Levy, Kolmogorov and many others. Borel stated that Cournot's principle "is the *only* law of chance", while Levy stated that "Cournot's principle is the *only* connection between probability and the empirical world". The use of the principle in recent times has fallen out of fashion, mainly due to the dominance of the subjective view of probabilities such as the approach of De Finetti where probabilities are treated as betting strategies. However, both the objective view of probabilities and the use of precluded sets as the essence of prediction, suits very well the coevents formulation. This is because the coevents formulation deals with a single closed system, adopts a realist view and uses the preclusion as the central feature in selecting the potential realities/coevents.

In coevents formulation the dichotomy between ontology and predictions persists. In particular, the ontology of the theory is the coevents themselves, while to recover probabilistic predictions one takes the quantum measure and applies the weak Cournot principle as described above. In [18] for example, one can see how a discrete version of the double slit pattern is recovered using the quantum measure and applying the Cournot principle[19].

4.3. The role and status of the initial state

In the coevent formulation, the role of the initial state, is that it affects the quantum measure and thus determines both the allowed coevents, and the probabilistic predictions (see previous section). An interesting question to ask, would be whether the initial state characterises the system (universe) in the following way. Given the coevent realised[20], can we uniquely determine which was the initial state? Or in other words, is a universe with initial state Ψ_1 different from one with Ψ_2? The answer to those questions is closely related with the interpretation of the (initial) state, as was analysed by Pusey Barrett and Rudolph in [21]. If one views different states as ontological different entities, the answer is "yes", while if one views them as statistical distributions, the answer is "no".

Given an initial state Ψ_1, we can find the set of allowed coevents $\mathcal{R}_1 = \{R^1_1, R^1_2, \cdots\}$ and similarly for initial state Ψ_2, we have $\mathcal{R}_2 = \{R^2_1, R^2_2, \cdots\}$. If there are no common covents $\mathcal{R}_1 \cap \mathcal{R}_2 = \emptyset$ for any pair of initial states Ψ_1, Ψ_2, then we can say the universe starting with Ψ_1 is different from that with Ψ_2. Otherwise, it is possible that the actual realised coevent is compatible with two (or more) initial states.

For the coevents formulation, in [22] was given evidence[21] that if one looks at sufficiently fine-grained description, the sets of coevents starting from different pure states is disjoint, i.e. the universe of Ψ_1 *is* different from that of Ψ_2. It follows that one is at least in principle, able to retrodict uniquely the initial state. While the initial state per se, has no ontological status in histories formulations, it obtains such status indirectly. Any possible reality/coevent, can arise from a unique initial state, and in this sense the initial state becomes a property possessed by the system. This observation has also interesting consequences for the field of (quantum)

[18] For example, it is conceivable that the fact that the universe in large scale looks homogeneous is *not* due to a common cause, but it happened randomly. In our construction of a theory of cosmology, we dismiss such coincidence and try to justify it with the existence of a common cause.

[19] Note, that using any other classical measure, would *not* recover the double slit pattern.

[20] Assuming for the moment that such thing is possible.

[21] See the appendix of [22]. Note however, that the particular gendanken experiment proposed in [21] did not fully carry over to the coevent formulation, due the failure to satisfy in histories approaches one of their assumption.

cosmology.

4.4. Other developments and outlook

There are several other questions that one can ask in relation with the quantum measure and its interpretation in the coevents formulation. Here we will briefly mention some works that have been done giving references.

One interesting thing to analyse, that is closely related with the question asked in the previous section, is which properties of our system can we deduce from the coevents themselves, or else which are the properties possessed by the coevents. One could wonder for example, what is the meaning of energy, for a given coevent, i.e. if our ontology is just a subset of histories. In a toy model of a three-sites hopper Sorkin showed that the system "does know its own energy" from simply looking the coevents [23], and we can treat the energy of the site-hopper as a property the coevents have.

Other works that have been done, is the attempt to formally extend the quantum measure for infinite histories space. The problems that arise can be found in [24] and directions on how to overcome them in [25]. The derivation of a Hilbert space structure starting purely from histories is given in [26] and the concept of quantum integration and its use is developed in (for example) [27].

There are several direction to further explore. One direction is exploring the conceptual issues that could arise from (i) considering composite systems, (ii) time extending the coevents (in other words, the effects that fine-graining the quantum measure has, to the set of allowed coevents) and (iii) considering large systems that classicality should emerge. Technical issues to be explored, are: (1) the extension of quantum measure and rigorously founding the formulation for infinite histories space and (2) finding ways to compute the allowed coevents for complicated systems (for the moment, it is very hard to do so, even for histories space with only few histories).

5. Summary and conclusions

We presented the coevents formulation of quantum theory. It is a realistic, histories formulation that applies to closed quantum systems and is well suited for quantum gravity. The three-fold structure of applying histories was presented, where we have (i) the space of histories Ω along with the event algebra $\mathcal{U}$, (ii) the space of truth values $\mathcal{T}$ and (iii) the space of allowed truth valuation maps $\mathcal{M}$. The latter in classical physics is the homomorphic maps ϕ from $\mathcal{U} \to \mathcal{T}$. The role of the (classical) measure, the difference of classical deterministic and stochastic physics and the importance of precluded events were analysed. The replacement of the classical measure with quantum measure led us to the need to alter the picture. This was due to the existence of zero covers which ruled out the single-histories as potential realities.

As a solution we introduced the coevents formulation, which can be viewed in two different ways. The new ontology, is either coarse-grained histories (non-trivial events) R_i that are non-preclusive and are of maximum detail or multiplicative maps ϕ_{R_i} (in general anhomomorphic), that are preclusive and primitive. The two pictures are equivalent, since there is a one-to-one correspondence between events and multiplicative maps, and the further conditions are also equivalent. With the coevents as ontology, we avoid the paradoxes described, that arose from the existence of zero covers. This is the case, because there is an existence theorem showing that given a quantum measure, there always exist non-trivial coevents (see [7]). The resulting logic, while it is not Boolean, is deductive (the modus ponens inference rule holds [16]). Proofs by contradiction are not, in general, possible due to the failure of eq. (25). However, the opposite rule given by eq. (26) holds.

The coevents formulation, appears as a generalisation of classical stochastic physics, where there are many potential realities. The ontology (coevents) is not affected by measurement and is not contextual. It is in this sense that we can call the formulation *realist*. Furthermore, it was shown that there exists a unique classical domain (unlike consistent histories), and that probabilistic predictions can be recovered by the use of Cournot's principle. Other recent developments and future directions were stated in section 4.4.

Acknowledgments
The author is grateful to Rafael Sorkin for introducing and developing the coevents formulation. He also wants to thank him and Fay Dowker, Sumati Surya and Yousef Ghazi-Tabatabai for many discussions. This work is partly supported by COST Action MP1006 "Fundamental Problems in Quantum Physics".

References

[1] Isham C J 1993 in *Integrable Systems, Quantum Groups, and Quantum Field Theories* ed L A Ibort and M A Rodrguez (London: Kluwer Academic Publishers); Anderson E 2012 Chapter 4 in *Classical and Quantum Gravity: Theory, Analysis and Applications* ed V R Frignanni (New York: Nova)
[2] Sorkin R D 2007 *J. Phys. Conf. Ser.* **67** 012018
[3] Sorkin R D 2007 *J. Phys. A* **40** 3207
[4] Cameron P 2012 http://cameroncounts.wordpress.com/2012/10/19/creating-modern-probability/
[5] Sorkin R D 1994 Mod. Phys. Lett. A **9** 3119 ; Sorkin R D 1997 pp 229-251 in *Quantum Classical Correspondence: Proceedings of the 4th Drexel Symposium on Quantum Nonintegrability* ed D H Feng and B-L Hu (Cambridge: International Press)
[6] Griffiths R B 1984 *J. Stat. Phys.* **36** 219; Omnès R 1988 *J. Stat. Phys.* **53** 893; Gell-Mann M and Hartle J 1990 in *Complexity, Entropy and the Physics of Information, SFI Studies in the Science of Complexity, Vol. VIII* ed W Zurek (Reading: Addison-Wesley); Gell-Mann M and Hartle J 1993 *Phys. Rev. D* **47** 3345
[7] Surya S and Wallden P 2010 *Found. Phys.* **40** 585
[8] Sinha U, Couteau C, Jennewein T,Laflamme R and Weihs G 2010 *Science* **329** 418
[9] Kochen S and Specker E 1967 *J. Math. Mech.* **17** 59
[10] Dowker F and Ghazi-Tabatabai Y 2008 *J. Phys. A* **41** 105301
[11] Dowker F and Kent A 1995 *Phys. Rev. Lett.* **75** 3038; Dowker F and Kent A 1996 *J. Statist. Phys.* **82** 1575; Kent A 1997 *Phys. Rev. Lett.* **78** 2874
[12] Isham C J and Butterfield J 2002 *Int. J. Theor. Phys.* **41** 613 and previous papers; Doering A and Isham C J 2008 *J. Math. Phys.* **49** 053518 and previous papers
[13] Isham C J 1997 *Int. J. Theor. Phys.* **36** 785
[14] Sorkin R 2010 Logic is to the quantum as geometry is to gravity in *Foundations of Space and Time* ed G F R Ellis, J Murugan and A Weltman (Cambridge: Cambridge University Press) (*Preprint* arXiv:1004.1226)
[15] Ghazi-Tabatabai Y 2009 PhD thesis (*Preprint* arXiv:0906.0294)
[16] Clements K, Dowker F and Wallden P 2012 Modus Ponens in Physics *Preprint* arXiv:1201.6266; Wallden P 2011 *J. Phys. Conf. Ser.* **306** 012044
[17] Ghazi-Tabatabai Y and Wallden P 2009 *J. Phys. A: Math. Theor.* **42** 235303
[18] Ghazi-Tabatabai Y and Wallden P 2009 *J. Phys.: Conf. Ser.* **174** 012054
[19] Cournot A 1843 *Exposition de la theorie des chances et des probabilites* (Paris: Hachet)
[20] Kolmogorov A N 1956 *Foundations of the Theory of Probability* (New York: Chelsea Pub. Co.)
[21] Pusey M, Barrett J and Rudolph T 2012 *Nature Phys.* **8** 476
[22] Wallden P 2012 Distinguishing initial state-vectors from each other in histories formulations and the PBR argument *Preprint* arXiv:1211.2084
[23] Sorkin R D 2012 Does a quantum particle know its own energy? talk in conference "Foundations of QM and Relativistic Spacetime" Athens http://www.pwallden.gr/mp1006/Sorkin.pdf
[24] Dowker F, Johnston S and Surya S 2010 *J. Phys. A* **43** 505305
[25] Sorkin R D 2011 Towards a fundamentak theorem of quantum measure theory *Preprint* arXiv:1104.0997; Gudder S and Sorkin R D 2011 Two-site quantum random walk *Preprint* arXiv:1105.0705

[26] Dowker F, Johnston S and Sorkin R D 2010 *J. Phys.* A **43** 275302; Gudder S 2010 Hilbert space representations of decoherence functionals and quantum measures *Preprint* arXiv:1011.1694
[27] Gudder S 2009 *J. Math. Phys.* **50** 123509; Gudder S 2010 *J. Phys. A* **43** 485301

Journal of Physics: Conference Series **442** (2013) 012045 doi:10.1088/1742-6596/442/1/012045

Is the consistent histories approach to quantum mechanics consistent?[1]

Elias Okon

Instituto de Investigaciones Filosóficas, Universidad Nacional Autónoma de México.
Circuito Maestro Mario de la Cueva s/n, C. U., México D.F. 04510, México.

E-mail: eokon@filosoficas.unam.mx

Abstract. The Consistent Histories (CH) formalism attempts to construct a quantum framework which can be used without the need to introduce observers external to the studied system. The prime motivation in mind is the application of the formalism to the universe as a whole. In order to achieve this, CH maintains that a formulation of quantum mechanics should allow for the assignment of probabilities to alternative histories of a system. Therefore, it provides an observer-independent criterion to decide which sets of histories can be given probabilities and states rules to determine them. The framework establishes that each realm, that is, each set of histories to which probabilities can be assigned, provides a valid quantum-mechanical account of a system. Furthermore, the version of CH first presented in [1, 2] proposes an "evolutionary" explanation of our existence in the universe and of our preference for quasiclassical descriptions of nature. The present work critically evaluates claims to the effect that the formalism offered in [1, 2] solves many interpretational problems in quantum mechanics. In particular, it is pointed out that the interpretation of the proposed framework leaves vague two crucial points, namely, whether realms are fixed or chosen and the link between measurements and histories. The claim of this work is that by doing so, CH overlooks the main interpretational problems of quantum mechanics. Furthermore, we challenge the evolutionary explanation offered and we critically examine the proposed notion of a realm-dependent reality.

1. Introduction

The Consistent Histories (CH) program tries to develop a version of quantum mechanics, which can be used in the study of closed systems. That is, one that, in contrast to the standard interpretation, can be applied without the need to divide the world into a system and an observer. In particular, CH tries to develop a formulation of quantum mechanics appropriate for the universe as a whole. CH stresses the importance of histories for quantum mechanics, as opposed to measurements, and maintains that a satisfactory formulation of quantum mechanics allows one to assign probabilities to alternative histories of the universe. It claims however that not all histories can be assigned probabilities, therefore, it provides an observer-independent criterion for deciding which sets of histories can be so endowed and gives rules to tell what these probabilities are.

The CH approach was originally developed by Griffiths[2] but this paper will be concerned with a particular version of the formalism first developed in [1, 2]. Such interpretation proposes that

[1] A longer exposition of these ideas appears in [3].

[2] The interpretation of he formalism offered by Griffiths can be consulted in [4, 5].

each *realm*, that is, each set of histories to which probabilities can be assigned, provides a valid quantum-mechanical account of a given system. It insists, however, that different realms can be mutually incompatible, in the sense that quantum-mechanical statements are meaningful only relative to a particular realm. Moreover, it maintains that the standard measurement situation is no more than a special case of setting for which the CH formalism allows for probabilities to be assigned. Consequently, it claims that the standard interpretation of quantum mechanics must be seen as a limiting case of the CH framework. References [1, 2] also use the CH formalism to offer an "evolutionary" explanation for the existence in the universe of complex systems, like humans, and for their almost exclusive preference for quasiclassical descriptions of nature.

The purpose of this work is to critically evaluate arguments offered by CH proponents asserting that the CH formalism solves many of the interpretational problems of quantum mechanics. In particular, it is pointed out that the interpretation proposed in [1, 2] leaves vague two crucial points, namely, whether complex systems can choose or not a realm, and the link between measurements and histories, overlooking on the way the challenging interpretational problems involved. Furthermore, the evolutionary explanation offered is disputed and the proposed notion of a realm-dependent reality is critically examined. The plan for doing the above is as follows: section 2 presents the CH formalism and section 3 describes its interpretation developed in [1, 2]. Section 4 then presents four objections against the CH formulation and evaluates its overall viability. Finally, section 5 presents some conclusions.

2. Consistent Histories

As was mentioned in the introduction, the CH formulation of quantum mechanics aims at a quantum-mechanical framework which, in contrast with the standard interpretation, does not critically depend on the notion of measurement. This feature would allow such interpretation to be applied to closed systems and, in particular, to the entire universe. In order to achieve this, proponents of CH propose as the most general objective of quantum mechanics the prediction of probabilities for time histories of a system. The CH formalism then provides an observer-independent criterion to tell what sets of alternative histories of a given system can be assigned probabilities and allows for these probabilities to be computed.

In order to achieve the above, the CH formalism takes as inputs an initial state $|\psi\rangle$ and a Hamiltonian $\hat{H}$, which are supposed to be given by an external theory. It then introduces the notion of an exhaustive and exclusive set of alternatives at a time. Such sets are represented (in the Heisenberg picture) by a set of projection operators:

$$\{P_\alpha(t)\}, \qquad \alpha = 1, 2, 3, \ldots \tag{1}$$

In order to implement the exhaustiveness and the exclusiveness, respectively, the projections of each set must satisfy the following equations:

$$\sum\nolimits_\alpha P_\alpha(t) = 1, \qquad \text{and} \qquad P_\alpha(t)P_\beta(t) = \delta_{\alpha\beta}P_\alpha(t). \tag{2}$$

Employing these sets, it is possible to construct *sets of histories*. These are sequences in time of the above constructed sets of alternatives, and are represented by

$$\{P^1_{\alpha_1}(t_1)\}, \{P^2_{\alpha_2}(t_2)\}, \ldots, \{P^n_{\alpha_n}(t_n)\}, \quad \text{at times} \quad t_1 < t_2 < \ldots < t_n. \tag{3}$$

Individual histories are then collected by selecting one projection form each set: $(\bar{\alpha}_1, \bar{\alpha}_2, \ldots, \bar{\alpha}_n)$. Such histories are represented by the corresponding chain of projections $C_{\bar{\alpha}} \equiv P^n_{\bar{\alpha}_n}(t_n) \ldots P^1_{\bar{\alpha}_1}(t_1)$ and each history gets assigned a branch state vector: $|\psi_{\bar{\alpha}}\rangle = C_{\bar{\alpha}}|\psi\rangle$. Sets of histories are generally coarse-grained in the sense that sets of projections are only given at particular times and individual projections are not required to be one-dimensional. Operations of fine- and coarse-graining can be defined by, for example, removing or adding sets of projections at additional times, or by compounding or splitting individual projections within a set.

The next step is to assign probabilities to individual histories within a set, but, according to the formalism, this cannot be done in general. Probabilities can be assigned to histories within a set only when there is *negligible* interference between its branches. That is, when the following relation holds:

$$\langle \psi_\alpha | \psi_\beta \rangle \approx 0. \tag{4}$$

The above condition guarantees that the assigned probabilities (approximately) satisfy the axioms of probability (sets satisfying the condition above are said to (medium) decohere). Then, according to CH, only for these sets it is possible to make predictions, and the corresponding probabilities are given by $p(\bar{\alpha}) = \|C_{\bar{\alpha}}|\psi\rangle\|^2$. A set of alternative (coarse-grained) histories which decoheres is called a *realm*.

Finally, in order to apply the formalism, the following prescriptions are postulated: given data d at time t_0, represented by a projection operator $P_d(t_0)$, predictions for the probability of the future history α_f are given by

$$p(\alpha_f|d) = \frac{\|C_{\alpha_f} P_d(t_0)|\psi\rangle\|^2}{\|P_d(t_0)|\psi\rangle\|^2}, \tag{5}$$

with C_{α_f} an exhaustive set of alternative histories to the future of t_0. Similarly, retrodictions for the past history β_p are given by

$$p(\alpha_f|d) = \frac{\|P_d(t_0) C_{\beta_p}|\psi\rangle\|^2}{\|P_d(t_0)|\psi\rangle\|^2}, \tag{6}$$

with C_{β_p} an exhaustive set of alternative histories to the past of t_0.

3. An interpretation of the formalism

In this section we will explore four core aspects of the interpretation of the CH formalism presented in [1, 2]: i) the concept of incompatible realms, ii) the connection between quasiclassical realms and complex adaptive systems, iii) the way in which standard quantum mechanics is supposed to be contained in CH and iv) the notion of a realm-dependent reality.

3.1. Inconsistent realms

As was mentioned already, a realm is a set of histories which allows a consistent assignment of probabilities. It turns out that, given a generic system, many different realms can be constructed and that the theory does not distinguish between all these different realms. The problem is that not all realms are compatible, in the sense that two different realms of the same system may lead to *contrary inferences*. That is, it can be shown (see [6]) that using two realms, both compatible with the same given data, it is possible to arrive at inconsistent stories of what actually happened; i.e., it is possible to retrodict, with *certainty* in each realm, two inconsistent facts about the past. References [1, 2] are well aware of this complication and in order to avoid inconsistencies impose the following rule: *inferences may not be drawn by combining probabilities from incompatible realms.* Doing so is just something you are not allowed to do while using the formalism.

3.2. Quasiclassical domains and IGUSes

A *quasiclassical domain* is defined in [1] as a realm that is maximally refined and that contains individual histories exhibiting as much patterns of classical correlation in time as possible. The world we perceive is supposed to be a great example of such a domain. In addition, humans are taken to be complex adaptive systems, and, in particular, special types of IGUSes (i.e., information gathering and utilizing systems). The foremost characteristic of an IGUS is taken to be the fact that it uses a (possibly primitive) physical theory in order to navigate, and make predictions, about its environment.

Given these two notions, references [1, 2] claim that the existence of IGUSes is to be explained in *evolutionary* terms. Concretely, the proposal is that:

(i) IGUSes evolved to make predictions because it is adaptive to do so;

(ii) they focus on quasiclassical domains because these present enough regularity to permit predictions by primitive methods.

As a result, the formalism explains the existence of IGUSes, and their preference for quasiclassical realms.

3.3. Recovering the standard interpretation

In order to display the fact that the CH formalism contains, as a special case, standard quantum mechanics, reference [1] first defines a *measurement* as a correlation between values of operators of a quasiclassical domain. Then, it states that this definition implies that the standard measurement situation (i.e., one with a system, a measuring apparatus and an observer) is only a special case of situation in which CH allows for an assignment of probabilities. Furthermore, it is claimed that the standard quantum-mechanical probabilities coincide with the ones supplied through the CH formulation. The conclusion is that the standard interpretation of quantum mechanics is no more than the application of the CH formalism to the particular setting of a standard measurement situation.

3.4. Reality relativism

Another element of the interpretation in [1, 2] that deserves our attention is the proposal of what could be called a *reality relativism*, i.e., the ontological claim that the notion of reality, or what is real, is meaningful only relative to a realm. Furthermore, [7] holds that our problem for accepting such an idea arises from shortcomings in our everyday language, i.e., from tension between colloquial language and the language of physics.

The idea of a reality relativism is of course a strong and controversial assertion, one that needs to be evaluated carefully. The next section scrutinizes this and other components of the interpretation of the formalism proposed in [1, 2].

4. Objections

This section presents four objections[3] against the formulation of CH developed in [1, 2, 7]. The first one asserts that the fact that the formalism does not provide a mechanism for realm selection represents a serious omission that greatly hinders its viability. The second objection arises from the fact that the formalism seems to deny the empirical possibility of determining the Hamiltonian and initial state of a system, both of them necessary for making predictions. Next, the claim that standard quantum mechanics is contained in CH is challenged and, finally, the evolutionary explanation for the existence of IGUSes, and their relation to quasiclassical realms, is disputed.

4.1. Realm-dependent reality

As was seen in the previous section, references [1, 2, 7] quite explicitly propose a reality relativism. They are asking us to consider the idea that reality, or what is real, is relative to a

[3] An early thorough evaluation of conceptual issues in CH is contained in [8].

realm. There is no doubt that the mere idea that reality is relative is controversial. However, the qualm considered here is not about the idea itself but about the self-consistency of the realm-dependent reality proposed in [1, 2, 7]. However, before developing the objection, it will be useful to examine the assertion in [7] regarding the connection between problems with colloquial language and problems for accepting a reality relativism.

The history of science is full of concepts that at some time are considered absolute but that latter turn out to be relative. A standard example of this, specifically mentioned in [7], is the time order of events, which was considered absolute in Newtonian mechanics but turns out to be relative (on the frame of reference) according to Special Relativity (SR). Therefore, if true, the reality-relativism proposal would surely not be the first time science discovers something to be relative. Furthermore, human language seems perfectly capable of dealing with relative concepts; we do it all the time with notions like big, far, cold, etc. Therefore, it is not clear how the uneasiness with the idea of a realm-dependent reality could have anything to do with shortcomings or excess baggage in human language; it seems more like an ontological problem than a language-related one.

What about the self-consistency of the proposal for a realm-dependent reality developed in [1, 2, 7]? The proposal is that in order to make sense of the notion of reality, and in order to use the formalism to make predictions, etc., it is necessary to fix a realm. The problem is that it isn't clear how IGUSes are supposed to select realms (among all the possible options offered by the formalism) because the formalism does not offer any mechanism for doing so. Actually, the sources are ambiguous about it:

> "...we could adopt a subjective point of view... and say that the IGUS "chooses" its coarse graining of histories and, therefore, "chooses" a particular quasiclassical domain... It would be better, however, to say that the IGUS evolves to exploit a particular quasiclassical domain or set of such domains." [1].

It is not clear, then, if an IGUS chooses a realm or if it is the realm that limits or constrains the possibility of existence and characteristics of IGUSes living in it. In any case, there are only two options: either IGUSes can or cannot "choose" realms. The problem is that neither option is satisfactory. On the one hand, if selecting a realm is beyond our capacities as IGUSes, talk of multiple realms seems extravagant and serves no real purpose in the theory (other realms being empirically inaccessible). On the other hand, if an IGUS can choose a realm, proponents of the formalism owe us an explanation of how this could be so, especially after noticing that it involves fixing projections everywhere in the universe, and at all times, and considering that this projections might radically change our current experience, or even our present existence! The problem then is not only that a mechanism for selecting a realm is missing; the problem is that the formalism seems to lack the resources for providing it.

The standard reply of CH proponents with respect to a realm-choosing mechanism within CH is that *realms must be chosen according to the questions one is trying to answer and the predictions one is interested in obtaining.* Nevertheless, that reply misses the point considered here because the issue is not about a recipe for applying the formalism in standard measurement situations. That is, a recipe for coming up with predictions for experiments in which we can take for granted a myriad of things, like a distinct system, a measuring apparatus, well defined observables, observers, etc. The issue is rather about evaluating the formalism as a theory of all this things together, which of course is the central motivation for taking CH seriously in the first place. The point is that there are two different levels of discourse that get entangled, one about how IGUSes use the theory to make predictions and the other about what the theory tells us with respect to reality. Maintain these two levels of discourse separate is essential for an adequate assessment of the CH approach.

4.2. Initial conditions

The second objection we will consider has to do with the (seemingly innocuous) fact that the formalism takes as inputs an initial state $|\psi\rangle$ and a Hamiltonian $\hat{H}$. The problem is that these objects, which are of course essential for applying the formalism (see for example equations 5 and 6), are supposed to be fixed and absolute, i.e., *realm-independent.* However, if the idea of a realm-dependent reality is taken seriously, it is far from clear how one could have access to this absolute elements. In other words, how are we supposed to choose initial conditions for a theory that holds that the past is relative? As a way out of this state of affairs, [1, 2] insist that we need to use a separate and external theory for choosing the initial conditions. However, if the past really is relative, nothing at all that we observe can ever count as evidence for such an external theory. That is, we cannot, even in principle, test these theories.

4.3. Measurements

We turn next to two related issues, both having to do with the treatment of the concept of measurements in CH. The first is the debatable way in which the formalism is supposed to contain standard quantum theory and the second is the fact that [1, 2] fail to determine the connection between actual measurements and the projection operators of the formalism.

As was remarked in section 3, references [1, 2] sustain that the CH formalism incorporates Copenhagen quantum theory as a limiting scenario. If this is true, it of course implies that CH is consistent with experiments (to the extent that standard quantum mechanics is). However, the situation is a bit more complicated than what [1, 2] suggest. The first problem is that the proposal remains silent about what is the ontological status of the different histories within a given realm. Once again, two options are available.

(i) The first one is that *only one of the histories within the realm is actual*, in which case we would have to conclude that the formalism is descriptively incomplete since, as it lacks a projection postulate, does not offer a mechanism to explain the preference for the chosen history from among all of the available choices.

(ii) The remaining option is that *all the histories within the realm are actual*, in which case there are two problems. First, it is not possible to interpret as probabilities the numbers generated by the framework since all options are realized. Second, it sharply conflicts with our everyday experience of obtaining determined outcomes when we perform measurements, (these of course are the standard objections against many-world scenarios).

As was mentioned before, the other important omission in [1, 2] with respect to measurements is that the proposal does not make explicit what is the relation between actual measurements performed by IGUSes and the projection operators of the formalism. That is, it does not specify how are we to connect the mathematical formalism provided, with experimental practices. The only rule in this respect that the formalism provides is the following: "realms are to be chosen according to the questions one is trying to answer." But, is this rule useful in practice? That is, given a standard measurement situation, would it help us select a realm? Which one would it be? An initial (and partial) response could be that the chosen realm would contain, as a minimum, a projection corresponding to the measurement to be realized. This, however, is deeply problematic because *the CH formalism does not specify under what conditions one is allowed to conclude that a measurement is taking place.* In other words, it does not solve the measurement problem.

The situation is even worse for the rest of the projections that comprise the realm (remember that realms contain sets of projections at various times). Clearly, these cannot be associated with measurements performed by IGUSes, even if that link could be established, regardless of what was said above. That is because for a measurement situation to arise, with an IGUS, and experimental apparatus, etc., specific projections must had occurred early on in the history of the universe (those that permitted the formation of galaxies, for example), before any IGUS was around to perform measurements. The remaining option is to disassociate the projections of the formalism with measurement but this solution is also unacceptable. The reason is that, without a connection between our measurements and the formalism, we would loose all predictive power. A possible last resource would be to hold that free will is an illusion: that whenever we (falsely) believe that we decide to perform a measurement, a projection occurs. In this case the issue of the two different levels of discourse mentioned above resurfaces because it would turn out that IGUSes are in fact *incapable* of using the formalism to make predictions. The discussion would then have to be entirely in terms of CH as seen as a the theory of everything, but as we stressed already, that issue is far from being satisfactorily addressed by its proponents.

4.4. IGUSes and quasiclassicality

The last critique we will consider questions the viability of the explanation for the existence of IGUSes, and there preference for quasiclassical realms, offered in [1, 2]. As was mentioned in section 3, such explanation is an evolutionary one: IGUSes exist because i) they are good at making predictions and ii) it is evolutionarily advantageous to be able to do so. Furthermore, they prefer quasiclassical realms because such environments present enough regularities for predictions to be generated.

The explanation displayed above might seem reasonable at first. Nevertheless, when examined with some care, starts looking less promising. To see why, it is useful to start by recalling the basic mechanism behind evolutionary theory, which, in a nutshell, can be stated as follows: *impact of the environment on reproductive success*. Therefore, the essential elements in an evolutionary scenario include: a varied initial population, an external environment, heredity and selection. However, none of these elements seem to be present in the CH context. In particular, there is no environment for IGUSes to evolve since, according to the proposed explanation, such environment (i.e., the realm in which the IGUSes dwell) is an essential part of what is supposed to be adaptively selected (the result purportedly being a quasiclassical realm). In other words, one cannot argue that evolution takes place unless the scenario in which it is supposed to happen permits the operation of mechanisms such as the fact that failure to obtain resources results in death, or that systems that are unfit do not reproduce. Those rules presuppose a quasiclassical realm, and it is thus clear that they cannot be used to argue that they play a role in selecting one such realm over something else.

Getting back to the explanation offered in [1, 2], we notice that, in effect, what it says is the following: we experience a quasiclassical realm because it is the only one that allows for IGUSes like ourselves to exist. Written this way, such explanation can easily be recognized as an instance of the, so called, *anthropic principle*. That is, of the principle that holds that *features of the world are what they are because, otherwise, we wouldn't be here to remark on it*. Therefore, it seems clear that what [1, 2] consider to be an evolutionary argument is, essentially, no more than an invocation of the anthropic principle.

5. Conclusions

The CH program aims at a generalization of the standard quantum formalism that does not employ the notions of measurements or observers in any fundamental way. It also attempts to construct a purely unitary version of quantum mechanics, where temporal evolution is implemented solely in terms of Schrödinger's equation, with no mention of a projection postulate. Certainly, these two goals are worthy and interesting. Unfortunately, the formulations of CH presented in [1, 2] suffer from too many conceptual problems in order to be considered satisfactory.

On the other hand, it is very likely that the problems encountered while trying to apply quantum theory to the universe as a whole, arise not from our inability to interpret quantum theory, as advanced by CH proponents, but from the (bold) assumption that quantum theory is universally valid.

Acknowledgements
This work has been partially supported by project PAPIIT IA400312-2.

References
[1] Gell-Mann M and Hartle J 1990 *Complexity, Entropy, and the Physics of Information* ed Zurek W H (New York: Addison Wesley)
[2] Hartle J 1991 *Quantum Cosmology and Baby Universes* ed Coleman S (Singapore: World Scientific)
[3] Okon E and Sudarsky D 2013 On the Consistency of the Consistent Histories Approach to Quantum Mechanics *Preprint* arXiv:1301.2586
[4] Griffiths RB 2000 *Consistent Quantum Theory* (Cambridge, UK: Cambridge Univ. Press)
[5] Griffiths RB 2011 A Consistent Quantum Ontology *Preprint* arXiv:1105.3932
[6] Kent A 1997 *Phys. Rev. Lett.* **87** 15
[7] Hartle J 2007 *J. Phys.* A **40** 3101
[8] Dowker F and Kent A 1996 *J. Stat. Phys.* **82** 5

Journal of Physics: Conference Series **442** (2013) 012046 doi:10.1088/1742-6596/442/1/012046

Beyond Bell's theorem: Admissible hidden-variable models for the spin-singlet

Antonio Di Lorenzo

Instituto de Física, Universidade Federal de Uberlândia, 38400-902 Uberlândia, Minas Gerais, Brazil

E-mail: dilorenzo@infis.ufu.br

Abstract. Assuming that quantum mechanics is obeyed exactly after averaging over hidden variables, and considering models that obey both the hypotheses of free will and locality, we establish the form of all possible hidden-variable models that reproduce the spin-singlet.

1. Introduction

Bell [1] established an inequality, which was later extended and reformulated [2, 3], obeyed by a large family of theories, deterministic or statistical, based on the assumption that the quantum state is not the ultimate limit, but that there may be additional parameters, the hidden variables, providing a finer description. Experiments do not only violate the inequality, but they also reproduce the quantum mechanical predictions with high precision and accuracy [4–17]. Furthermore, the empirical data show an increasing agreement with quantum mechanics as experimental techniques become more refined.

Here, we shall extrapolate this trend, and assume that quantum mechanics is obeyed exactly, in absence of any knowledge about the hidden variable. This puts further restrictions on the admissible hidden-variable theories, as they must not only violate Bell inequality, but also reproduce quantum mechanics after the hidden variables are averaged out. Under the hypotheses that the hidden-variable models satisfy the locality condition and the free will assumption, it is possible to determine their shape, thus going beyond Bell's theorem.

The main result of this work was published in Ref. [18]. Here, we just give a more compact account of the result, while dedicating some more space to epistemological issues.

2. Hypotheses at the basis of Bell inequality

Bell inequality follows easily under the assumption that the joint probability $P_{\sigma,\tau}(\boldsymbol{a},\boldsymbol{b},\psi)$ of observing an outcome $\sigma \in \{-1,1\}$ at a detector characterized by $\boldsymbol{a}$ and $\tau \in \{-1,+1\}$ at a detector characterized by $\boldsymbol{b}$, for a fixed preparation ψ, can be decomposed as

$$P_{\sigma,\tau}(\boldsymbol{a},\boldsymbol{b},\psi) = \int d\mu(\lambda|\psi) M_\sigma(\boldsymbol{a},\lambda,\psi) M_\tau(\boldsymbol{b},\lambda,\psi), \tag{1}$$

with μ a positive distribution normalized to one, and M marginal probability conditioned on λ. In the following, we shall consider implicit for brevity the dependence on ψ, as the preparation does not change. We remark that applying Bayes' rule, the probability generally factors as

$$P_{\sigma,\tau}(\boldsymbol{a},\boldsymbol{b}) = \int \mathrm{d}\mu(\lambda|\boldsymbol{a},\boldsymbol{b}) M_\sigma(\boldsymbol{a},\boldsymbol{b},\lambda) Q_\tau(\sigma,\boldsymbol{a},\boldsymbol{b},\lambda), \tag{2}$$

with Q conditional probability for given σ. In order to comply with Eq. (1), the models considered by Bell must satisfy three hypotheses: $\mu(\lambda|\boldsymbol{a},\boldsymbol{b}) = \mu(\lambda)$, often justified on the basis of "free will", but which we shall address with the technical term Uncorrelated Choice; $M_\sigma(\boldsymbol{a},\boldsymbol{b},\lambda) = M_\sigma(\boldsymbol{a},\lambda)$ and $M_\tau(\boldsymbol{a},\boldsymbol{b},\lambda) = M_\tau(\boldsymbol{b},\lambda)$, called "Setting-Independence", sometimes identified with the no-signaling principle, mistakenly; $Q_\tau(\sigma,\boldsymbol{a},\boldsymbol{b},\lambda) = M_\tau(\boldsymbol{a},\boldsymbol{b},\lambda)$, called in the literature Outcome-Independence, which we shall call "Reducibility of Correlations".

Let us discuss these assumptions.

2.1. Uncorrelated Choice (UC)

This assumption means that the choice of measurement and the distribution of the hidden variables λ are uncorrelated, so that there is no causal connection between the two. Let us assume momentarily that UC does not hold. If one thinks of λ as a set of parameters attached to the physical system, and assumes that the choice of measurement $\Sigma = \{\boldsymbol{a},\boldsymbol{b}\}$ is the cause and the change in the distribution $\mu(\lambda|\Sigma)$ is the effect, then a violation of Uncorrelated Choice would result in non-locality. However, if one inverts the cause-effect relationship, then a violation of Uncorrelated Choice corresponds to a limitation of free will [19–22]. As an example, think of an astronaut on a distant planet, who is video-communicating with an engineer at NASA. You notice that, after subtracting the time delay due to the finite speed of communication, when the astronaut raises her left arm, the engineer does as well, and so when she scratches her head, etc. What could you conclude by observing such correlations? Perhaps the astronaut has access to an image of the engineer through some superluminal channel, and is just mimicking the gestures she can see in advance, or vice versa; perhaps the astronaut and the engineer have agreed to make those gestures at those times, in the first interplanetary flashmob ever; or perhaps they have no free will, and the same forces that determine his movements are determining hers. Thus, as pointed out e.g. in Refs. [21, 22], there is no way one can decide whether the correlations are due to conspiracy, non-locality, or slave will, unless a physical theory is formulated that describes how the decisions to do a movement (i.e., out of metaphor, to choose a measurement) are taken.

Marketing experts and statisticians can confirm that people who are exposed to the same advertisements or who share the same social conditions tend to buy the same products or to vote for the same candidate. Thus, we could accept a limitation of free will. However, with respect to the hypothesis of Uncorrelated Choice, we make the following consideration: the experimenters deciding the settings Σ can delegate the choice to third parties, as for instance the fluctuations in the voltage of an electric socket, the variations in the albedo of a distant star, etc. As there is empirical evidence [23] that Bell inequality is violated in this case as well, the only possible conclusion, if one insists that UC is violated, is that there is a leibnitzian pre-established harmony, where the hidden variables determine not only the choice of Σ, but also the choice of an apparent random process among a multitude of possible ones. Furthermore, the outcome of this apparently random process is determined as well by λ. Thus, while we cannot rule out a violation of free will as an explanation, we can say, based on partial experimental evidence, that if this is the case then we should admit a cosmic choreography. As far as a theory of the intentions and preferences of the choreographer is not put forward, the assumption is untestable, as it can explain anything and predict nothing, and thus we should store it away, together with solipsism and intelligent design. In conclusion, we do assume that Uncorrelated Choice holds.

2.2. Setting-Independence (SI)

This hypothesis requires that the marginal probability of observing an event σ in a space-time region A, for a given λ, does not depend on the settings Σ' at space-time regions outside of the lightcone (past and future) of A. Often, SI is considered as synonymous of no-signaling, a form of locality,[1] in the sense that if SI holds, then it is impossible to send superluminal signals from a region B to a region A. Actually SI is weaker than no-signaling, as SI may hold but signaling still be possible. Let us see how: suppose that the hidden variables λ can be divided in λ_A, a set of parameters localized in region A, λ_B, localized in region B, and possibly λ_G, a set of global parameters that cannot be associated to any particular region (the wave function of entangled particles is an example of a global parameter). The assumption SI allows the marginal probability at A to depend on λ_B, $M_\sigma(\boldsymbol{a},\boldsymbol{b},\lambda_A,\lambda_B,\lambda_G) = M_\sigma(\boldsymbol{a},\lambda_A,\lambda_B,\lambda_G)$. Thus, if λ_B changes, so does the marginal probability at A. By receiving a large number of particles in parallel and estimating the marginal probability by the frequency of observing σ, an observer at A can thus receive an instantaneous message from an observer at B, provided the latter has a way to influence λ_B. To me, the existence of this non-locality is dubious, and therefore I shall assume that SI holds, and furthermore that the marginal probability does not depend on λ_B.

A special case of SI is the compliance with Malus's law, postulated by Leggett in the derivation of a new inequality [24]. It requires that the hidden-variables consist in two unit vectors $\boldsymbol{u}$ and $\boldsymbol{v}$, such that the marginal probabilities are $M_\sigma(\boldsymbol{u},\boldsymbol{a},\boldsymbol{b}) = (1+\sigma\boldsymbol{a}\cdot\boldsymbol{u})/2$ and $M_\tau(\boldsymbol{v},\boldsymbol{a},\boldsymbol{b}) = (1+\sigma\boldsymbol{b}\cdot\boldsymbol{v})/2$. There is a result [18,25,26], which we may call the *trivial-marginals theorem*, establishing that all theories that obey UC and SI, and that also reproduce the quantum mechanical predictions for a spin singlet must have a marginal probability $M_\sigma(\lambda,\boldsymbol{a},\boldsymbol{b}) = 1/2$. Thus, the compliance with Malus's law is excluded, and in the following we shall use this theorem, which also guarantees the locality of the models.

2.3. Reducibility of Correlations (RC)

This assumption, called outcome-independence by Shimony [27], states that $Q_\tau(\sigma,\boldsymbol{a},\boldsymbol{b},\lambda) = M_\tau(\boldsymbol{a},\boldsymbol{b},\lambda)$, i.e. the conditional probability of obtaining an outcome τ, for fixed λ and given that the outcome of the remote measurement along $\boldsymbol{a}$ is σ, does not depend on the latter outcome. Hence, after applying Bayes' rule, the joint probability factorizes symmetrically $P_{\sigma,\tau}(\boldsymbol{a},\boldsymbol{b},\lambda) = M_\sigma(\boldsymbol{a},\boldsymbol{b},\lambda)M_\tau(\boldsymbol{a},\boldsymbol{b},\lambda)$. Hence, for fixed parameters λ, there would be no correlations, and these would appear only after averaging over λ. Thus the assumption of Reducibility of Correlations amounts to state that the quantum correlations are a result of the ignorance of more fundamental parameters.

The hypothesis RC is implied by determinism, in the sense that, if by knowing λ one may predict with certainty whether the outcome τ will occur or not, then the knowledge of the outcome σ is superfluous. While the vice versa is not true, however, a theorem established by M. Hall [28] shows that any model obeying RC admits a natural deterministic completion, i.e. a description in terms of additional variables μ such that $M_\sigma(\boldsymbol{a},\boldsymbol{b},\lambda,\mu) \in \{0,1\}$. Thus, RC generalizes the concept of determinism.

In order to verify whether Reducibility of Correlations holds, an observer in the region A should ask the observer at B about the τ he measured and the $\boldsymbol{b}$ he used, then evaluate the conditional probability $Q_\sigma(\tau,\boldsymbol{a},\boldsymbol{b},\lambda)$, and finally check whether this quantity changes with τ, with the remaining parameters fixed. In order to do so, A must have access to the remote information $\boldsymbol{b},\tau$, which B can send only at a speed not exceeding the speed of light. Hence, violating Reducibility of Correlations does not imply action-at-a-distance, nor the possibility of instantaneous communication. Yet, the locution "quantum nonlocality" is widely used. The

[1] Here, we consider locality and no-signaling as synonymous, all other definitions of locality being unsatisfactory, as discussed below in the main text.

reason is that originally Bell [1] did assume determinism, thus justifying RC, and locality. Later generalizations of Bell inequality [3, 29] dropped the hypothesis of determinism, but kept assuming RC, in the mistaken belief that this was a consequence of locality, not of determinism. Jarrett [30], and subsequently Shimony [27], pointed out that RC does not follow from the notion of locality, as commonly meant. This resulted in some new terminology being introduced to refer to both the hypothesis SI and RC: strong locality [30], Bell locality, local causality [31]. After a while, this distinction fell, and unfortunately many studies of Bell inequality refer to the two hypotheses SI and RC (and sometimes UC) as "locality", while the reader of the present manuscript should be aware that this is just a misnomer: there is no factual evidence of an intrinsic non-locality in Nature. Furthermore, the presumed non-locality of Nature (or of quantum mechanics) would be of a very peculiar kind, as it would manifest in the kinematics. By this, we mean that, since ordinary quantum mechanics is, as classical mechanics, a non-relativistic theory, it can accept non-local forces (think, e.g., to Newtonian gravitation or Coulomb's law). However, the "non-locality" of entangled states manifests when there is no known interaction between the two subsystems, and thus it would be a feature of the very kinematics of the theory, i.e. we would have "non-locality" without interaction, a patent absurdity.

In the following, we shall assume that the hypothesis RC does not hold, as this is the only assumption not fully justified by physical considerations.

3. Models for the spin singlet

As one or more of the three hypotheses discussed in the previous section are dropped, it is possible to violate Bell inequality. However, this is a necessary but not sufficient condition in order to reproduce experimental data, as these agree well with the predictions of quantum mechanics. In the Appendix, we review some significant models that reproduce the probability for the spin singlet after averaging out the hidden variables. None of these models relies solely on the violation of Reducibility of Correlations.

4. Main theorem

As our goal is to find models that reproduce the quantum mechanical predictions, $P_{\sigma,\tau}(\boldsymbol{a},\boldsymbol{b}) = (1-\sigma\tau\boldsymbol{a}\cdot\boldsymbol{b})/4$, and that obey both UC and SI, we can apply the trivial-marginals theorem, so that the joint probability can be written as

$$P_{\sigma,\tau}(\boldsymbol{a},\boldsymbol{b}) = \int \mathrm{d}\mu(\lambda)\frac{1}{4}\left[1+\sigma\tau D(\lambda,\boldsymbol{a},\boldsymbol{b})\right]. \tag{3}$$

The question that we ask is: what functions D are admissible (i.e. give a probability between 0 and 1) so that the quantum mechanical predictions are reproduced after averaging out λ? The main idea is of using the tools of Calculus to approach this problem, while former approaches rely on algebra. Without loss of generality we can write

$$D(\lambda,\boldsymbol{a},\boldsymbol{b}) = -\boldsymbol{a}\cdot\boldsymbol{b} + C(\lambda,\boldsymbol{a},\boldsymbol{b}) \tag{4}$$

Clearly, $\int \mathrm{d}\mu(\lambda)C(\lambda,\boldsymbol{a},\boldsymbol{b}) = 0$ identically in $\boldsymbol{a},\boldsymbol{b}$, so that C must take both positive and negative values.

Furthermore, let us take $\sigma=\tau$ and $\boldsymbol{a}=\boldsymbol{a}$. Then the probability, conditioned on λ, reduces to $P_{\sigma,\sigma}(\lambda,\boldsymbol{a},\boldsymbol{a}) = C(\lambda,\boldsymbol{a},\boldsymbol{a})$, and this implies that $C(\lambda,\boldsymbol{a},\boldsymbol{a}) = 0, \forall\boldsymbol{a},\lambda$. Analogously, you can prove that $C(\lambda,\boldsymbol{a},-\boldsymbol{a}) = 0, \forall\boldsymbol{a},\lambda$. Thus, perfect correlations and anti-correlations limit the admissible C to the form

$$C(\lambda,\boldsymbol{a},\boldsymbol{b}) = \left[1+\boldsymbol{a}\cdot\boldsymbol{b}\right]^{s_+}\left[1-\boldsymbol{a}\cdot\boldsymbol{b}\right]^{s_-}G(\lambda,\boldsymbol{a},\boldsymbol{b}), \tag{5}$$

with $s_+ > 0, s_- > 0$ and $0 < |G(\lambda, \boldsymbol{a}, \pm\boldsymbol{a})| < \infty$. This is but a Frobenius expansion, with the indices $s_\pm$ determining how fast the function vanishes close to the perfect (anti-)correlations configuration $\boldsymbol{a} \cdot \boldsymbol{b} = \mp 1$. Compatibility with quantum mechanics requires

$$\int \mathrm{d}\mu(\lambda) G(\lambda, \boldsymbol{a}, \boldsymbol{b}) = 0, \forall \boldsymbol{a}, \boldsymbol{b}. \tag{6}$$

Now, let us consider a configuration close to the perfect anticorrelation $\boldsymbol{a} \cdot \boldsymbol{b} = 1 - \varepsilon$, with $\varepsilon \ll 1$. Then

$$P_{\sigma,\sigma}(\boldsymbol{a}, \boldsymbol{b}, \lambda) \simeq \frac{\varepsilon}{4} \left[1 - 2^{s_+} \varepsilon^{s_- - 1} G(\lambda, \boldsymbol{a}, \boldsymbol{a})\right], \tag{7}$$

The positivity of the probability implies that $s_- \geq 1$. The inequality $s_+ \geq 1$ is proved analogously.

5. Conclusions

In conclusion, we have established the form of all the hidden variable models able to reproduce the quantum mechanics of a spin-singlet by satisfying both the assumptions of "free will" and no-signaling, which correspond, respectively, to Uncorrelated Choice (Measurement-Independence) and Setting-Independence. By contrast, we have assumed the violation of Reducibility of Correlations, as this can never result in superluminal signaling, since it consists in the dependence of a conditional probability on a remote outcome, and as such it requires the communication of said outcome through means that are necessarily subluminal. Rather, the violation of Reducibility of Correlations implies that the quantum correlations cannot be attributed to the ignorance of the hidden parameters.

Acknowledgments

This work was performed as part of the Brazilian Instituto Nacional de Ciência e Tecnologia para a Informação Quântica (INCT–IQ) and it was supported by Fundação de Amparo à Pesquisa do Estado de Minas Gerais through Processes No. APQ-02804-10 and PEE-00421-12.

Appendix: Review of models reproducing the quantum-mechanical correlations

In the literature there is a handful of hidden-variable models able to reproduce quantum mechanics. These models must violate at least one of Bell's hypotheses and one of Leggett's. However, as it is not always clear which hypotheses are dropped, we shall briefly review the existing (to the best of our knowledge) models. We recall that a hidden-variable theory predicts the decomposition of the quantum probability as

$$\int \mathrm{d}\mu(\lambda|\boldsymbol{a}, \boldsymbol{b}) P_{\sigma,\tau}(\lambda, \boldsymbol{a}, \boldsymbol{b}) = \frac{1}{4} \left[1 - \sigma\tau \boldsymbol{a} \cdot \boldsymbol{b}\right]. \tag{A.1}$$

Inspection of $\mu(\lambda|\boldsymbol{a}, \boldsymbol{b})$, depending whether it depends on $\boldsymbol{a}, \boldsymbol{b}$ or not, allows to establish if Uncorrelated Choice holds. After extracting the marginal probability $P_\sigma(\lambda, \boldsymbol{a}, \boldsymbol{b})$ from $P_{\sigma,\tau}(\lambda, \boldsymbol{a}, \boldsymbol{b})$, one can check whether Setting-Independence (and possibly Malus's law) is satisfied. Finally, obtaining the conditional probability $P_\sigma(\lambda, \boldsymbol{a}, \boldsymbol{b}, \tau)$, the hypothesis of Reducibility of Correlations can be checked.

Appendix A.1. Brans model.

Brans [19] proposed the following model using four discrete binary hidden-variables and two unit-vectors:

$$\mu(j, k, \alpha, \beta, \boldsymbol{u}, \boldsymbol{v}|\boldsymbol{a}, \boldsymbol{b}) = \frac{jk}{2} \delta(\boldsymbol{u} - \boldsymbol{a}) \delta(\boldsymbol{v} - \boldsymbol{b}) \langle j_{\hat{\boldsymbol{z}}}|\alpha_{\boldsymbol{u}}\rangle \langle \alpha_{\boldsymbol{u}}|k_{\hat{\boldsymbol{z}}}\rangle \langle -j_{\hat{\boldsymbol{z}}}|\beta_{\boldsymbol{v}}\rangle \langle \beta_{\boldsymbol{v}}| - k_{\hat{\boldsymbol{z}}}\rangle, \tag{A.2}$$

$$P_{\sigma,\tau}(j, k, \alpha, \beta, \boldsymbol{u}, \boldsymbol{v}, \boldsymbol{a}, \boldsymbol{b}) = \frac{1}{4}(1 + \sigma\alpha)(1 + \tau\beta). \tag{A.3}$$

The symbol $|\alpha_{\boldsymbol{u}}\rangle$ denotes the eigenstate $\alpha = \pm 1$ of the spin component $\boldsymbol{u}\cdot\boldsymbol{\sigma}$. Thus, Brans's model violates only the Uncorrelated Choice hypothesis. Using the standard definition of spin eigenstates that can be found in Sakurai's textbook [32], after some algebra one gets

$$\mu(j,k,\alpha,\beta,\boldsymbol{u},\boldsymbol{v}|\boldsymbol{a},\boldsymbol{b}) = \frac{1}{2}\delta(\boldsymbol{u}-\boldsymbol{a})\delta(\boldsymbol{v}-\boldsymbol{b}) \times$$
$$\times\Bigg\{\delta_{j,k}\left[\delta_{j,\alpha}\cos^2(\theta_{\boldsymbol{u}}/2)+\delta_{j,-\alpha}\sin^2(\theta_{\boldsymbol{u}}/2)\right]\left[\delta_{j,-\beta}\cos^2(\theta_{\boldsymbol{v}}/2)+\delta_{j,\beta}\sin^2(\theta_{\boldsymbol{v}}/2)\right]$$
$$-\delta_{j,-k}\frac{\alpha\beta}{4}\sin\theta_{\boldsymbol{u}}\sin\theta_{\boldsymbol{v}}e^{-ij(\phi_{\boldsymbol{u}}-\phi_{\boldsymbol{v}})}\Bigg\}, \tag{A.4}$$

with $0 \le \theta_{\boldsymbol{u}}, \theta_{\boldsymbol{v}} \le \pi$ and $\phi_{\boldsymbol{u}}, \phi_{\boldsymbol{v}}$ the azimuthal and polar angles of the unit vectors in the arbitrary coordinate system fixing the phase of the eigenstates. Clearly, Eq. (A.4) is non-real, and gives a negative probability for $\alpha = \beta$, $j = -k$, $\boldsymbol{u} = \boldsymbol{v}$. We notice however that j, k do not appear in the conditional probability Eq. (A.3), thus, after summing over them, we have

$$\begin{aligned}\mu(\alpha,\beta,\boldsymbol{u},\boldsymbol{v}|\boldsymbol{a},\boldsymbol{b}) &= \delta(\boldsymbol{u}-\boldsymbol{a})\delta(\boldsymbol{v}-\boldsymbol{b})\frac{1-\alpha\beta\left[\cos\theta_{\boldsymbol{u}}\cos\theta_{\boldsymbol{v}}-\sin\theta_{\boldsymbol{u}}\sin\theta_{\boldsymbol{v}}\cos(\phi_{\boldsymbol{u}}-\phi_{\boldsymbol{v}})\right]}{4} \\ &= \frac{1}{4}\delta(\boldsymbol{u}-\boldsymbol{a})\delta(\boldsymbol{v}-\boldsymbol{b})\left[1-\alpha\beta\boldsymbol{u}\cdot\boldsymbol{v}\right]. \end{aligned} \tag{A.5}$$

This is the correct, positive-definite distribution to replace Eq. (A.2). In conclusion, Brans's model, properly formulated in terms of a real and positive distribution, turns out to be trivial: there are four hidden variables $\alpha, \beta, \boldsymbol{u}, \boldsymbol{v}$, distributed according to Eq. (A.5), so that the conditional distribution of α, β for given $\boldsymbol{u}, \boldsymbol{v}$ mimics by construction the quantum mechanical distribution, $\mu(\alpha,\beta,\boldsymbol{u},\boldsymbol{v}) = (1-\alpha\beta\boldsymbol{u}\cdot\boldsymbol{v})/4$. The variables α, β are fixed by Eq. (A.3) to coincide with the observed values of σ, τ, while $\boldsymbol{u}, \boldsymbol{v}$, in virtue of Eq. (A.5) coincide with $\boldsymbol{a}, \boldsymbol{b}$.

Appendix A.2. Toner and Bacon model.
In Ref. [33], the hidden variables are a pair of unit vectors $(\boldsymbol{u}, \boldsymbol{v})$, each uniformly distributed over the unit-sphere. The probabilities characterizing the model are

$$\mu(\boldsymbol{u},\boldsymbol{v}|\boldsymbol{a},\boldsymbol{b}) = \frac{1}{(4\pi)^2}, \tag{A.6}$$

$$P_{\sigma,\tau}(\boldsymbol{u},\boldsymbol{v},\boldsymbol{a},\boldsymbol{b}) = \frac{1}{4}\left[1+\sigma\mathrm{sgn}(\boldsymbol{u}\cdot\boldsymbol{a})\right]\left\{1+\tau\mathrm{sgn}\left[\boldsymbol{u}\cdot\boldsymbol{b}+\mathrm{sgn}(\boldsymbol{u}\cdot\boldsymbol{a})\mathrm{sgn}(\boldsymbol{v}\cdot\boldsymbol{a})\boldsymbol{v}\cdot\boldsymbol{b}\right]\right\}. \tag{A.7}$$

The model is clearly deterministic (we wrote $\delta_{a,b} = (1+ab)/2$ for $a, b = \pm 1$ to avoid clumsy indices) and asymmetric, since it can be realized by A making a measurement which decides the output σ depending on the sign of the projection of $\boldsymbol{u}$ over $\boldsymbol{a}$, and by subsequently A sending a bit

$$c = \mathrm{sgn}(\boldsymbol{u}\cdot\boldsymbol{a})\mathrm{sgn}(\boldsymbol{v}\cdot\boldsymbol{a}) \tag{A.8}$$

to B, where the output τ is determined by a suitable combination of $\boldsymbol{b}, \boldsymbol{u}, \boldsymbol{v}$ and c. The asymmetry yields one marginal probability obeying Setting-Independence, and the other not obeying:

$$M_\sigma^{(A)}(\boldsymbol{u},\boldsymbol{v},\boldsymbol{a},\boldsymbol{b}) = \frac{1}{2}\left[1+\sigma\mathrm{sgn}(\boldsymbol{u}\cdot\boldsymbol{a})\right], \tag{A.9}$$

$$M_\tau^{(B)}(\boldsymbol{u},\boldsymbol{v},\boldsymbol{a},\boldsymbol{b}) = \frac{1}{2}\{1+\tau\mathrm{sgn}\left[\boldsymbol{u}\cdot\boldsymbol{b}+\mathrm{sgn}(\boldsymbol{u}\cdot\boldsymbol{a})\mathrm{sgn}(\boldsymbol{v}\cdot\boldsymbol{a})\boldsymbol{v}\cdot\boldsymbol{b}\right]\}. \tag{A.10}$$

Equation (A.6) guarantees that Uncorrelated Choice holds, while the factorizability of Eq. (A.7) implies the validity of Reducibility of Correlations.

Appendix A.3. Cerf et al. model.
Cerf and collaborators [34] proposed a model relying on the Popescu-Rohrlich box [35]. The latter is a hypothetical machine taking two binary inputs $x = \pm 1$ and $y = \pm 1$ from location A and B, resp., and giving correlated outputs at locations A and B, according to

$$4 - 2|o_A + o_B| = (1 - x)(1 - y). \tag{A.11}$$

The model of Cerf *et al.* assumes as hidden variables x, y and two unit vectors $\boldsymbol{u}, \boldsymbol{v}$, both vectors being accessible through shared randomness[2] by A and B. The model relies on an anthropomorphic description, but can be stated in a neutral form: the observer at A checks $\boldsymbol{u}, \boldsymbol{v}$, arbitrarily chooses $\boldsymbol{a}$, and then inputs

$$x = \text{sgn}(\boldsymbol{u} \cdot \boldsymbol{a})\ \text{sgn}(\boldsymbol{v} \cdot \boldsymbol{a}), \tag{A.12}$$

where $\text{sgn}(t) = 1$ for non-negative t, $\text{sgn}(t) = -1$ for negative t; analogously the observer at B, after arbitrarily choosing $\boldsymbol{b}$, will input

$$y = \text{sgn}(\boldsymbol{n}_+ \cdot \boldsymbol{b})\ \text{sgn}(\boldsymbol{n}_- \cdot \boldsymbol{b}), \tag{A.13}$$

with $\boldsymbol{n}_\pm = \boldsymbol{u} \pm \boldsymbol{v}$. The Popescu-Rohrlich box will give then outputs according to Eq. (A.11), which are transformed to the simulated outcomes $\sigma = -o_A\ \text{sgn}(\boldsymbol{u} \cdot \boldsymbol{a})$, $\tau = o_B\ \text{sgn}(\boldsymbol{n}_+ \cdot \boldsymbol{b})$. Assuming a uniform distribution of $\boldsymbol{u}, \boldsymbol{v}$, the quantum mechanical correlator is recovered.

Restating the formulation in a neutral form, we have that Cerf *et al.* model is

$$\begin{aligned} \mu(x, y, \boldsymbol{u}, \boldsymbol{v}|\boldsymbol{a}, \boldsymbol{b}) &= \frac{1}{(4\pi)^2}\, \delta[x - \text{sgn}(\boldsymbol{u}\cdot\boldsymbol{a})\ \text{sgn}(\boldsymbol{v}\cdot\boldsymbol{a})]\, \delta\left[y - \text{sgn}(\boldsymbol{n}_+\cdot\boldsymbol{b})\ \text{sgn}(\boldsymbol{n}_-\cdot\boldsymbol{b})\right], && \text{(A.14)} \\ P_{\sigma,\tau}(x, y, \boldsymbol{u}, \boldsymbol{v}, \boldsymbol{a}, \boldsymbol{b}) &= \frac{1}{4}\left\{1 - \sigma\tau \frac{1 + x + y - xy}{2} \text{sgn}(\boldsymbol{u} \cdot \boldsymbol{a})\text{sgn}(\boldsymbol{n}_+ \cdot \boldsymbol{b})\right\}. && \text{(A.15)} \end{aligned}$$

Thus the model considered violates both Uncorrelated Choice and Reducibility of Correlations. The presentation of the model, however, might obscure its characteristics, since it is stated that the choices $\boldsymbol{a}, \boldsymbol{b}$ are made freely. We remark that Eqs. (A.12) and (A.13) constrain the values of x, y. Mathematics and probability theory are time-symmetric and acausal, the first because of the symmetry of the equality sign, the second because it provides correlations. Thus, the interpretation of what is the cause and what the effect is purely metaphysical. One may interpret Eqs. (A.12) and (A.13), or Eq. (A.14), as limiting the possible choices of $\boldsymbol{a}, \boldsymbol{b}$ for given x, y. In this case the model can be realized through classical resources, since the inputs x, y are given, and the observers are instructed, or programmed, to choose only the $\boldsymbol{a}, \boldsymbol{b}$ satisfying Eqs. (A.12) and (A.13).

The model can be reduced to one violating only the Reducibility of Correlations hypothesis after integrating out x and y (thus the PR box is just a mathematical artifact to justify the procedure):

$$\begin{aligned} \mu(\boldsymbol{u}, \boldsymbol{v}|\boldsymbol{a}, \boldsymbol{b}) &= \frac{1}{(4\pi)^2}, && \text{(A.16)} \\ P_{\sigma,\tau}(\boldsymbol{u}, \boldsymbol{v}, \boldsymbol{a}, \boldsymbol{b}) &= \frac{1}{4}\left\{1 - \sigma\tau\text{sgn}(\boldsymbol{u} \cdot \boldsymbol{a})\text{sgn}(\boldsymbol{n}_+ \cdot \boldsymbol{b}) \frac{1 + x_{\boldsymbol{a},\boldsymbol{u},\boldsymbol{v}} + y_{\boldsymbol{b},\boldsymbol{u},\boldsymbol{v}} - x_{\boldsymbol{a},\boldsymbol{u},\boldsymbol{v}} y_{\boldsymbol{a},\boldsymbol{u},\boldsymbol{v}}}{2}\right\}, && \text{(A.17)} \end{aligned}$$

[2] We notice, however, that classical resources cannot provide true shared randomness, but only a simulation of it by means of two identical random generators synchronized and shared between A and B. The Kolmogorov complexity of the sequence of outputs would result less than the one of true random events to an observer intelligent enough to guess the algorithm.

where $x_{\boldsymbol{a},\boldsymbol{u},\boldsymbol{v}}$ and $y_{\boldsymbol{b},\boldsymbol{u},\boldsymbol{v}}$ are the right-hand-sides of Eqs. (A.12) and (A.13). This corresponds to a C function (see main text):

$$C(\boldsymbol{u},\boldsymbol{v},\boldsymbol{a},\boldsymbol{b}) = \boldsymbol{a}\cdot\boldsymbol{b} - \mathrm{sgn}(\boldsymbol{u}\cdot\boldsymbol{a})\mathrm{sgn}(\boldsymbol{n}_+\cdot\boldsymbol{b})\frac{x_{\boldsymbol{a},\boldsymbol{u},\boldsymbol{v}} + y_{\boldsymbol{b},\boldsymbol{u},\boldsymbol{v}} - x_{\boldsymbol{a},\boldsymbol{u},\boldsymbol{v}}y_{\boldsymbol{b},\boldsymbol{u},\boldsymbol{v}} - 1}{2}. \tag{A.18}$$

We leave to the reader the tedious verification that $C(\boldsymbol{u},\boldsymbol{v},\boldsymbol{a},\pm\boldsymbol{a}) = 0$. Furthermore, because of the piecewise constancy of the sgn function, $C(\boldsymbol{u},\boldsymbol{v},\boldsymbol{a},\boldsymbol{b}) \underset{\boldsymbol{b}\to\pm\boldsymbol{a}}{=} \boldsymbol{a}\cdot\boldsymbol{b}\mp 1$, identically for almost all values of $\boldsymbol{a}$ (the exception are the values $\boldsymbol{a} = \boldsymbol{u},\boldsymbol{v},\boldsymbol{u}+\boldsymbol{v},\boldsymbol{u}-\boldsymbol{v}$) so that we have, with reference to our main theorem, $s_+ = s_- = 1$.

Appendix A.4. Gröblacher et al. model.
In Ref. [36] the following model, relying on two unit vectors as hidden variables,[3] was proposed

$$\mu(\boldsymbol{u},\boldsymbol{v}|\boldsymbol{a},\boldsymbol{b}) = 0, \text{ for } |\boldsymbol{u}\cdot\boldsymbol{a}\pm\boldsymbol{v}\cdot\boldsymbol{b}| > 1\mp\boldsymbol{a}\cdot\boldsymbol{b}, \tag{A.19}$$

$$P_{\sigma,\tau}(\boldsymbol{u},\boldsymbol{v},\boldsymbol{a},\boldsymbol{b}) = \frac{1}{4}\left[1+\sigma\boldsymbol{u}\cdot\boldsymbol{a}+\tau\boldsymbol{v}\cdot\boldsymbol{b} - \sigma\tau\boldsymbol{a}\cdot\boldsymbol{b}\right]. \tag{A.20}$$

The distribution μ, besides satisfying (A.19), should also comply with $\int d\boldsymbol{u}d\boldsymbol{v}\mu(\boldsymbol{u},\boldsymbol{v}|\boldsymbol{a},\boldsymbol{b})\boldsymbol{u} = 0$ and $\int d\boldsymbol{u}d\boldsymbol{v}\mu(\boldsymbol{u},\boldsymbol{v}|\boldsymbol{a},\boldsymbol{b})\boldsymbol{v} = 0$ in order to reproduce the correct quantum mechanical probability, and is otherwise arbitrary. The model violates both Uncorrelated Choice and Reducibility of Correlations, and satisfies Setting-Independence and Compliance with Malus's Law.

Appendix A.5. Hall model.
Hall [20] proposed the following deterministic model

$$\mu(\boldsymbol{u},\boldsymbol{v}|\boldsymbol{a},\boldsymbol{b}) = \delta(\boldsymbol{u}+\boldsymbol{v})\frac{1-f(\boldsymbol{u},\boldsymbol{v},\boldsymbol{a},\boldsymbol{b})}{8\arccos f(\boldsymbol{u},\boldsymbol{v},\boldsymbol{a},\boldsymbol{b})}, \tag{A.21}$$

$$P_{\sigma,\tau}(\boldsymbol{u},\boldsymbol{v},\boldsymbol{a},\boldsymbol{b}) = \frac{1}{4}\left[1+\sigma\mathrm{sgn}(\boldsymbol{u}\cdot\boldsymbol{a})\right]\left[1+\tau\mathrm{sgn}(\boldsymbol{v}\cdot\boldsymbol{b})\right], \tag{A.22}$$

where

$$f(\boldsymbol{u},\boldsymbol{v},\boldsymbol{a},\boldsymbol{b}) = \mathrm{sgn}(\boldsymbol{u}\cdot\boldsymbol{a})\mathrm{sgn}(\boldsymbol{v}\cdot\boldsymbol{b})\ \boldsymbol{a}\cdot\boldsymbol{b}. \tag{A.23}$$

Hall's model clearly violates only the Uncorrelated Choice hypothesis, and it does so using the minimum amount of correlation [20]. The model does not comply with Malus's law since $P_\sigma(\boldsymbol{u},\boldsymbol{v},\boldsymbol{a},\boldsymbol{b}) = [1+\sigma\mathrm{sgn}(\boldsymbol{u}\cdot\boldsymbol{a})]/2$.

Appendix A.6. Di Lorenzo model.
We proposed [21] the following model:

$$\mu(\boldsymbol{u},\boldsymbol{v}|\boldsymbol{a},\boldsymbol{b}) = \frac{1}{4}\sum_{\boldsymbol{n}=\pm\boldsymbol{a},\pm\boldsymbol{b}}\delta(\boldsymbol{u}-\boldsymbol{n})\delta(\boldsymbol{v}+\boldsymbol{n}) \tag{A.24}$$

$$P_{\sigma,\tau}(\boldsymbol{u},\boldsymbol{v},\boldsymbol{a},\boldsymbol{b}) = \frac{1}{4}\left(1+\sigma\boldsymbol{u}\cdot\boldsymbol{a}\right)\left(1+\tau\boldsymbol{v}\cdot\boldsymbol{b}\right). \tag{A.25}$$

The model can be generalized by breaking the symmetry in Eq. (A.24),

$$\mu(\boldsymbol{u},\boldsymbol{v}|\boldsymbol{a},\boldsymbol{b}) = \sum_{\boldsymbol{n}=\pm\boldsymbol{a},\pm\boldsymbol{b}} w_{\boldsymbol{n}}\delta(\boldsymbol{u}-\boldsymbol{n})\delta(\boldsymbol{v}+\boldsymbol{n}), \tag{A.26}$$

[3] The original formulation is more contrived, since it introduces additional variables which, in the end, are integrated over.

with $\sum_n w_n = 1$ and w non-negative weights. This model is the only one[4] that satisfies at the same time Setting-Independence, Malus's law, and Reducibility of Correlations, while it violates only Uncorrelated Choice.

References

[1] Bell J S 1964 *Physics* **1** 195–200
[2] Clauser J F, Horne M A, Shimony A and Holt R A 1969 *Phys. Rev. Lett.* **23**(15) 880–884 URL http://link.aps.org/doi/10.1103/PhysRevLett.23.880
[3] Clauser J F and Horne M A 1974 *Phys. Rev.* D**10**(2) 526–535 URL http://link.aps.org/doi/10.1103/PhysRevD.10.526
[4] Freedman S J and Clauser J F 1972 *Phys. Rev. Lett.* **28** 938–941 URL http://link.aps.org/doi/10.1103/PhysRevLett.28.938
[5] Aspect A, Grangier P and Roger G 1982 *Phys. Rev. Lett.* **49** 91–94 URL http://link.aps.org/doi/10.1103/PhysRevLett.49.91
[6] Aspect A, Dalibard J and Roger G 1982 *Phys. Rev. Lett.* **49** 1804–1807 URL http://link.aps.org/doi/10.1103/PhysRevLett.49.1804
[7] Tapster P R, Rarity J G and Owens P C M 1994 *Phys. Rev. Lett.* **73** 1923–1926 URL http://link.aps.org/doi/10.1103/PhysRevLett.73.1923
[8] Weihs G, Jennewein T, Simon C, Weinfurter H and Zeilinger A 1998 *Phys. Rev. Lett.* **81** 5039–5043 URL http://link.aps.org/doi/10.1103/PhysRevLett.81.5039
[9] Tittel W, Brendel J, Gisin B, Herzog T, Zbinden H and Gisin N 1998 *Phys. Rev.* A**57** 3229–3232 URL http://link.aps.org/doi/10.1103/PhysRevA.57.3229
[10] Rowe M A, Kielpinski D, Meyer V, Sackett C A, Itano W M, Monroe C and Wineland D J 2001 *Nature* **409** 791–794 URL http://dx.doi.org/10.1038/35057215
[11] Gröblacher S, Paterek T, Kaltenbaek R, Brukner Č, Żukowski M, Aspelmeyer M and Zeilinger A 2007 *Nature* **446** 871–875 URL http://dx.doi.org/10.1038/nature05677
[12] Paterek T, Fedrizzi A, Gröblacher S, Jennewein T, Żukowski M, Aspelmeyer M and Zeilinger A 2007 *Phys. Rev. Lett.* **99** 210406 URL http://link.aps.org/doi/10.1103/PhysRevLett.99.210406
[13] Branciard C, Ling A, Gisin N, Kurtsiefer C, Lamas-Linares A and Scarani V 2007 *Phys. Rev. Lett.* **99** 210407 URL http://link.aps.org/doi/10.1103/PhysRevLett.99.210407
[14] Eisaman M D, Goldschmidt E A, Chen J, Fan J and Migdall A 2008 *Phys. Rev.* A**77** 032339 URL http://link.aps.org/doi/10.1103/PhysRevA.77.032339
[15] Romero J, Leach J, Jack B, Barnett S M, Padgett M J and Franke-Arnold S 2010 *New J. Phys.* **12** 123007 URL http://dx.doi.org/10.1088/1367-2630/12/12/123007
[16] Paternostro M and Jeong H 2010 *Phys. Rev.* A**81** 032115 URL http://dx.doi.org/10.1103/PhysRevA.81.032115
[17] Lee C W, Paternostro M and Jeong H 2011 *Phys. Rev.* A**83**(2) 022102 URL http://link.aps.org/doi/10.1103/PhysRevA.83.022102
[18] Di Lorenzo A 2012 *Phys. Rev.* A**86**(4) 042119 URL http://link.aps.org/doi/10.1103/PhysRevA.86.042119
[19] Brans C H 1988 *Int. J. Theor. Phys.* **27**(2) 219–226 ISSN 0020-7748 URL http://dx.doi.org/10.1007/BF00670750
[20] Hall M J W 2010 *Phys. Rev. Lett.* **105**(25) 250404 URL http://link.aps.org/doi/10.1103/PhysRevLett.105.250404
[21] Di Lorenzo A 2012 *J. Phys. A: Math. Theor.* **45** 265302 URL http://stacks.iop.org/1751-8121/45/i=26/a=265302
[22] Di Lorenzo A 2013 *Int. J. Mod. Phys.* B**27** 1345016 URL http://www.worldscientific.com/doi/abs/10.1142/S0217979213450161
[23] Scheidl T, Ursin R, Kofler J, Ramelow S, Ma X S, Herbst T, Ratschbacher L, Fedrizzi A, Langford N K, Jennewein T and Zeilinger A 2010 *Proc. Natl. Acad. Sci. USA* **107** 19708–19713 ISSN 0027-8424
[24] Leggett A 2003 *Found. Phys.* **33**(10) 1469–1493 URL http://dx.doi.org/10.1023/A:1026096313729
[25] Colbeck R and Renner R 2008 *Phys. Rev. Lett.* **101** 050403 URL http://link.aps.org/doi/10.1103/PhysRevLett.101.050403
[26] Branciard C, Brunner N, Gisin N, Kurtsiefer C, Lamas-Linares A, Ling A and Scarani V 2008 *Nat. Phys.* **4** 681–685 URL http://dx.doi.org/10.1038/nphys1020

[4] A former attempt was made in Ref. [37], where however the distribution μ was not normalized [38].

[27] Shimony A 1990 *Sixty-Two Years of Uncertainty: Historical, Philosophical, and Physical Inquiries into the Foundations of Quantum Mechanics.* ed Miller A I (Plenum, New York) pp 33–43
[28] Hall M J W 2011 *Phys. Rev.* **84**(2) 022102 URL http://link.aps.org/doi/10.1103/PhysRevA.84.022102
[29] Bell J S 1971 *49th International School of Physics "Enrico Fermi": Foundations of quantum mechanics, Varenna, Italy* ed D'Espagnat B (Academic Press, New York) pp 171–181
[30] Jarrett J P 1984 *Noûs* **18** pp. 569–589 ISSN 00294624 URL http://www.jstor.org/stable/2214878
[31] Bell J S 1990 *Between Science and Technology* ed Sarlemijn A and Kroes P (Elsevier/North-Holland) pp 97–115
[32] Sakurai J J and Napolitano J 2010 *Modern Quantum Mechanics* 2nd ed (Addison-Wesley) ISBN 9780805382914
[33] Toner B F and Bacon D 2003 *Phys. Rev. Lett.* **91**(18) 187904 URL http://link.aps.org/doi/10.1103/PhysRevLett.91.187904
[34] Cerf N J, Gisin N, Massar S and Popescu S 2005 *Phys. Rev. Lett.* **94**(22) 220403 URL http://link.aps.org/doi/10.1103/PhysRevLett.94.220403
[35] Popescu S and Rohrlich D 1994 *Found. Phys.* **24**(3) 379–385 URL http://dx.doi.org/10.1007/BF02058098
[36] Gröblacher S, Paterek T, Kaltenbaek R, Brukner Č, Żukowski M, Aspelmeyer M and Zeilinger A 2007 *Nature* **446** 871–875 URL http://www.nature.com/nature/journal/v446/n7138/suppinfo/nature05677.html
[37] De Zela F 2008 *J. Phys. A: Math. Theor.* **41** 505301 URL http://stacks.iop.org/1751-8121/41/i=50/a=505301
[38] Di Lorenzo A 2012 *J. Phys. A: Math. Theor.* **45** 268001 URL http://stacks.iop.org/1751-8121/45/i=26/a=268001

Journal of Physics: Conference Series **442** (2013) 012047 doi:10.1088/1742-6596/442/1/012047

Modeling "quantum" field theory through "classical" cellular automaton

Roman Sverdlov

IISER Mohali, Knowledge City, Sector 81, SAS Nagar, Manauli

E-mail: `romansverdlov @iisermohali.ac.in`

Abstract. The purpose of this paper is to illustrate how cellular automaton can be used in order to arrive at discrete, but "classical", model leading to the emergent "quantum" field theory.

1. Introduction

It is commonly assumed that one of the main problems in interpretation of quantum mechanics is incompatibility of wavefunction collapse with relativity. This, however, is not entirely true. It is possible to claim that our spacetime is fundamentally non-relativistic but there is a "rule" that the "allowed" Lagrangians are the ones that can be written in relativistic form. Such Lagrangians would still produce relativistic phenomenology; at the same time the superluminal signals that are responsible for wavefunction collapse will not be subject to said Lagrangians which will allow them to propagate faster than light. Thus, there will be a preferred frame which will allow us to avoid grandfather paradox. It is true that non-locality is counter-intuitive regardless of presence or absence of relativity. But such non-locality can be avoided by realizing that measurement takes finite amount of time, δt; thus, if the size of the universe (L) is also finite, then superluminal signals that are moving faster than $L/\delta t$ would account for entanglement, while their speed is still finite. Finally, the collapse itself can be described by means of either Bohmian or GRW models, modified to account for "finite" speed of above signals.

There is, however, another issue that *is* problematic: what is the physical meaning of ψ, regardless of presence or absence of its collapse? If there were no interference, we could have said that ψ refers to probability. Since there is interference, we would like to instead say that it is analogous to *classical* electromagnetic field (which is also subject to interference). What stops us from sticking with the second option is the fact that ψ is evaluated over Fock space while classical electromagnetic field lives in $\mathbb{R}^3$. Apart from that, ψ taken over Fock space also forces us to accept strictly-infinite speed of superluminal signals, while in single particle case it could have been "very large but finite" per earlier argument. While some might assume that in QFT case locality is restored, this is not true either: despite the fact that propagators between spacelike separated points are zero, the "definitions" of in- and out- states that we are ultimately computing are still non-local.

The ultimate solution to these issues is to "encode" functions over Fock space in terms of functions over $\mathbb{R}^3$. This in principle is possible: one can envision a computer, built in $\mathbb{R}^3$-based universe, which is programmed to model quantum field theory. We can go around the seeming obstacle that Fock space is "larger" than $\mathbb{R}^3$ through discretization. In the context of QFT each point in space has its own set of degrees of freedom. Thus, the finer the discretization scale, the more degrees of freedom we have. Therefore, we can trade these extra degrees of freedom with the ones needed for creation/annihilation of particles. In other words, Fock space with "rougher" space discretization can be modeled through wave functions living in "finer-discretized" $\mathbb{R}^3$. Such "wave functions", however, are *not* well behaved since we would like every single new point in space to add non-trivial piece of information that is to be utilized. In what follows we will describe how the encoding is done.

2. Quantum states as $\psi(\phi)$-functionals

We recall that spin-0 QFT is simply a "quantum mechanics" of infinitely dimensional harmonic oscillator. Furthermore, n-dimensional harmonic oscillator establishes one to one correspondence between the "Fock space" over n points with a wave function over $\mathbb{R}^n$. Therefore, Fock space in QFT should correspond to wave function over $\mathbb{R}^\infty$, where ∞ is the number of points in $\mathbb{R}^3$. The role of $\vec{x}_k$ in Harmonic oscillator is taken up by $\phi(\vec{x})$ in spin-0 QFT (integer index k is replaced by "label" $\vec{x}$, while "variable" $\vec{x}$ is replaced by ϕ). Thus, $\psi(\vec{x},t)$ is replaced with $\psi(\phi,t)$ and $\mathbb{R}^\infty$ describes the set of all possible $\phi\colon \mathbb{R}^3 \to \mathbb{R}$. A "point" in $\mathbb{R}^\infty$ is one specific $\phi\colon \mathbb{R}^3 \to \mathbb{R}$, while its x_0-th coordinate is the value $\phi(\vec{x}_0)$. One should note that domain of ϕ is $\mathbb{R}^3$ rather than $\mathbb{R}^4$; in other words we assume an existence of "preferred frame".

The "space derivative" terms imply the coupling between $\vec{x}$-th coordinate and $(\vec{x}+\delta\vec{x})$-th. It is easy to see that the coupled harmonic oscillators can be represented as decoupled ones in "rotated" reference frame; here, we roate $\{\phi(\vec{x}_1),\phi(\vec{x}_2),\cdots\}$-frame into $\{\tilde{\phi}(\vec{p}_1),\tilde{\phi}(\vec{p}_2),\cdots\}$. Of course, this rotation corresponds to Fourier transform. We will discretize our theory and assume that the only "allowed" momenta are $(\vec{p}_1,-\vec{p}_1,\cdots,\vec{p}_n,-\vec{p}_n)$. Thus, for each given k we have two-dimensional harmonic oscillator with coordinates $(\vec{p}_k,-\vec{p}_k)$. The resulting $\psi(\phi)$ will simply be a product of n such 2D harmonic oscillators. General state becomes

$$|\psi(t)\rangle = \sum_{\{c\},\{d\}} K_{c_1\cdots c_n d_1\cdots d_n}(t)(a^\dagger_{p_1})^{c_1}(a^\dagger_{-p_1})^{d_1}\cdots(a^\dagger_{p_n})^{c_n}(a^\dagger_{-p_n})^{d_n}|0\rangle\,. \tag{1}$$

which, through Fourier series, corresponds to

$$\psi(\phi,t) = \sum_{\{c\},\{d\}}\left[K_{c_1\cdots c_n d_1\cdots d_n}(t)\prod_a \lambda_{c_a d_a}\left(\int \mathrm{d}^3x\,\phi(\vec{x})\cos(\vec{p}\cdot\vec{x}),\int \mathrm{d}^3x\,\phi(\vec{x})\sin(\vec{p}\cdot\vec{x})\right)\right], \tag{2}$$

where

$$\lambda_{cd}(x,y) = \left[\prod_{j=1}^{n}\left(\left(\frac{m^2+p_j^2}{4}\right)^{1/4}(x+\mathrm{i}y)+\frac{1}{\sqrt{2}(m^2+p^2)^{1/4}}\left(\frac{\mathrm{d}}{\mathrm{d}x}+\mathrm{i}\frac{\mathrm{d}}{\mathrm{d}y}\right)\right)^{c_j}\right]\times \tag{3}$$
$$\times\left[\prod_{j=1}^{n}\left(\left(\frac{m^2+p_j^2}{4}\right)^{1/4}(x-\mathrm{i}y)+\frac{1}{\sqrt{2}(m^2+p^2)^{1/4}}\left(\frac{\mathrm{d}}{\mathrm{d}x}-\mathrm{i}\frac{\mathrm{d}}{\mathrm{d}y}\right)\right)^{d_j}\right]\mathrm{e}^{-\frac{\sqrt{m^2+p^2}}{2}(x^2+y^2)}\,.$$

3. Encoding quantum states in terms of functions of $\vec{x}$

The definition of quantum states above is "non-classical"; after all, "classical" fields (such as electromagnetic field) are functions over set of points rather than set of functions. Therefore, our next challenge is to replace $\psi(\phi,t)$ with $\psi(\vec{x},t)$. In order to illustrate how it is done, let us consider a simple example where $\psi[\phi(x)=\sin x]=i\sqrt{3}/2$ and $\psi[\phi(x)=\cos x]=1/2$. We can do the encoding by postulating

$$(q(x), \phi(x), \psi(x)) = \begin{cases} (1, \cos x, 1/2) & \text{if} x = (2k+1)\delta x \\ (2, \sin x, i\sqrt{3}/2) & \text{if} x = 2k\delta x. \end{cases} \tag{4}$$

Thus, we have replaced a functional $\psi(\phi)$ with functions $q(x)$, $\phi(x)$ and $\psi(x)$. This restores classical realism since we now have functions of points (similar to classical electromagnetic field) as opposed to functions of functions (corresponding to quantum probability amplitudes). None of these three functions are "well behaved" (although they "jump" between two different "well-behaved" trajectories). In some sense, "not well behaved" function is "more complex" than "well behaved" one (in case of "well behaved" function most of the information about its value at a point can be inferred from its values at neighboring points which means that the former carries less nontrivial information). This "extra complexity" is what allows us to use a single "not well behaved" function in order to "encode" probability amplitudes of two different "well behaved" ones at the same time; namely, $\psi[\sin x]$ and $\psi[\cos x]$. This, however, comes with a price: each of these two functions is represented with the discretization interval $2\delta x$ instead of δx. This price, however, is not a big loss if the functions we are trying to "encode" (which in this case are $\sin x$ and $\cos x$) are well behaved. If, on the other hand, we try to encode functions that are *not* well behaved, the information loss due to $\delta x \to 2\delta x$ becomes more important. However, we can still "not care" about it if we assume δx is smaller than Planck scale.

We can now relax conditions on q and rewrite the above expression as

$$(\phi(x), \psi(x)) = \begin{cases} (\cos x, 1/2) & \text{if} q = 1 \\ (\sin x, i\sqrt{3}/2) & \text{if} q = 2. \end{cases} \tag{5}$$

where $q\colon S \to \{1, 2\}$ is some unknown function over discrete set $S \subset \mathbb{R}$. In order to get "sample" of both $\sin x$ and $\cos x$ over any given small region, we need to assume that $q(x)$ takes values of both 1 and 2 within any such region:

$$\forall x_0 \exists \{x_1, x_2\} \subset (x_0 - \Delta x, x_0 + \Delta x)(q(x_1) = 1, q(x_2) = 2) \ , \ \delta x \ll \Delta x \ll 1 \tag{6}$$

in other words, $q(x)$ is not well behaved. Let us now generalize the above to the situation where we want to define $\psi(\phi_{(1)}), \cdots, \psi(\phi_{(M)})$ while replacing $M = 2$ with $M \gg 2$, $x \in \mathbb{R}$ with $\vec{x} \in \mathbb{R}^3$ and stationary $\psi(\phi)$ with evolving $\psi(\phi, t)$. The condition on $q(x)$ generalizes to

$$\forall \vec{x}_0 \exists \{\vec{x}_1, \cdots, \vec{x}_M\} \subset n_{\Delta x}(\vec{x}_0)(q(\vec{x}_k) = k) \ , \ n_{\Delta x}(\vec{x}_0) = \{\vec{x} | |\vec{x} - \vec{x}_0| < \Delta x\} \ , \ \delta x \ll \Delta x \ll 1 \tag{7}$$

We will then define ϕ and ψ according to

$$\phi(\vec{x}_k) = \phi_{(q(\vec{x}_k))}(\vec{x}_k) \ , \ \psi(\vec{x}_k, t) = \psi(\phi_{(q)}, t) \tag{8}$$

We can now use the Eq 8 in order to "convert" Eq 2 into a "local" framework:

$$\psi(\vec{x}_k) = \sum_{\{c\},\{d\}} \left[K_{c_1 \cdots c_n d_1 \cdots d_n} \prod_a \lambda_{c_a d_a} \left(\sum_{l=1}^N \delta^{q(\vec{x}_k)}_{q(\vec{x}_l)} \phi(\vec{x}_l) \cos(\vec{p} \cdot \vec{x}_l), \sum_{l=1}^N \delta^{q(\vec{x}_k)}_{q(\vec{x}_l)} \phi(\vec{x}_l) \sin(\vec{p} \cdot \vec{x}_l) \right) \right], \tag{9}$$

The above is "realistic" definition of general Fock space state described in Eq 1. The domain of ψ is now $\{\vec{x}_1, \cdots, \vec{x}_N\}$, which is a subset of $\mathbb{R}^3$, as desired!

Now, in order for the above to approximate functional $\psi(\phi)$ we need to make sure that $\{\phi_{(1)}, \cdots, \phi_{(M)}\}$ sprinkles the range of ϕ densely enough. We accomplish this in the following way. First, we impose a constraint $\phi_{min} \leq \phi(\vec{x}) \leq \phi_{max}$, where ϕ_{min} and ϕ_{max} are chosen in such a way that the "tails" of $\psi(\phi)$ that are being "cut off" are "very small". After that, we will discretize the set of trajectories by imposing a restriction $\phi \in \{\phi_{(1)}, \cdots, \phi_{(M)}\}$. We also assume that the *spacelike* volume of the universe, V, is finite, and the following set of conditions holds:

$$1 \ll \left(\frac{\phi_{max} - \phi_{min}}{\delta\phi}\right)^{V/\delta v} \approx M \approx \frac{MV}{\delta v} \ll N, \tag{10}$$

where $\delta v \ll V$ is much smaller than Plank's volume, $\delta\phi \ll \phi_{max} - \phi_{min}$ is the smallest detectable variation in ϕ, and $N \gg MV/\delta v$ is the total number of lattice points. This inequality statistically guarantees that any conceivable ϕ that obeys specified upper and lower bounds can, in fact, be approximated by some sample of M "choices" of ϕ.

By referring back to the analogy of harmonic oscillator, we see that expectation value of ϕ grows when we have larger number of particles. Thus, imposing boundaries on ϕ amounts to making the probability amplitude of "very large" particle numbers "even smaller" than it would have been otherwise. Such probability amplitude will still be non-zero: after all, even if "most" of the support of $\psi(\phi)$ is outside (ϕ_{min}, ϕ_{max}) range, the latter is still a subset of the former. On the other hand, if we take a few particle state, $\psi(\phi)$ would still have very small "tails" outside (ϕ_{min}, ϕ_{max}) so it would still be affected. Nevertheless, it is still correct to say that probability amplitudes of "very large" particle numbers get lowered a lot more than the ones of few particle states. We can adjust ϕ_{min} and ϕ_{max} to make sure that the only states whose probability amplitudes are significantly affected are the ones that are never reached, such as us having enough particles to fill the entire universe with solid matter.

Let us now go back and reflect on what our encoding has accomplished. It is easy to see that Eq 8 can be satisfied if and only if

$$\forall \vec{x}_i, \vec{x}_j [q(\vec{x}_i) = q(\vec{x}_j) \Longrightarrow \psi(\vec{x}_i, t) = \psi(\vec{x}_j, t)] \tag{11}$$

Since both $\psi(\vec{x}_i, t)$ and $\psi(\vec{x}_j, t)$ evolve in time, the above correlation can only hold in one preferred frame. Such preferred frame is enforced by superluminal signals. We assume that superluminal signals move with finite speed but fast enough to pass our universe within very small time (the finite size of the universe follows from the fact that number of points, N, is finite). Thus, the above correlation is approximate, but its error is too small to be detected.

Intuitively, we have a hologram. A lattice, consisting of N points, is divided into M different sublattices: (q)-th sublattice is $\{k | q(\vec{x}_k) = q\}$. If we were to make an assumption $\psi(\vec{x}_k, t) = \delta^{q(\vec{x}_k)}_{q(t)}$ (to be abandoned shortly), we would be "looking" at sublattice number $q(t) \in \mathbb{N}$ while "ignoring" all other sublattices. Changing the sublattice we are "looking" at will result in perceived change of picture while, in reality, the pictures drawn on each sublattice are the same. This is in direct analogy to hologram.

Now, if we were to abandone $\psi(\vec{x}_k, t) = \delta^{q(\vec{x}_k)}_{q(t)}$ and replace it with "less restrictive" condition of Eq 11, we will "quantize" the hologram in a sense that we can be "looking" at several different "sub-pictures" *at the same time*, similarly to particle going through both slits "at the same time". For any given k, the value of $\psi(\vec{x}_k)$ is a "probability amplitude" that we are looking at the *entire* sublattice in which the point number k happens to reside. Eq 11 has to hold simply because the questions "are we looking at sublattice in which point i resides?" and "are we looking at the sublattice in which point j resides?" are equivalent if i and j reside in the same sublattice; thus, the "probability amplitudes" assigned to corresponding answers "yes" should be the same. Now, from Section 2, we know that any "fixed" particle state in Fock space corresponds to "several sublattices looked at the same time", as well. But if the "linear combination" of "sublattices" we are considering is different from the ones prescribed by any given state in Fock space, we will end up with linear combination of different Fock space states, which is what we ultimately want.

Finally, similar to what we explained about δx and $2\delta x$ in example of $\sin x$ and $\cos x$ will now be replaced with $\delta v/M$ and δv. However, this information loss wont be a problem if we assume that δv is below Planck scale.

4. Desired "nonlocal" dynamics of $\psi(\vec{x}_k, t)$

We recall that Fock space and path integral formulations of QFT are equivalent. Therefore, after having convinced ourselves that $\{q(\vec{x}_k), \psi(\vec{x}_k), \phi(\vec{x}_k)\}$ describes particle states, we can make them evolve per Hamiltonian evolution if and only if we can design $\psi(\vec{x}_k, t)$ in such a way that it reproduces path integral. In light of the fact that we only have M allowed $\phi\colon \mathbb{R} \to \mathbb{R}^3$, any trajectory is bound to look like step function in time. This implies discretization of time. For simplicity, let us for now assume continuous space and discrete time. If we postulate the process

$$\psi(\phi_{(q)}, a\delta t) = \sum_{q'} \psi(\phi_{(q')}, (a-1)\delta t) \exp\left(i\delta t \int d^3x d^3y d^3z \mathcal{K}(\vec{x}, \vec{y}, \vec{z}, \phi_{(q')}(\vec{x}), \phi_{(q')}(\vec{y}), \phi_{(q)}(\vec{z}))\right) \tag{12}$$

this will reproduce an action

$$S = \sum_a \delta t \int d^3z \mathcal{L}(\vec{z}, a\delta t; \phi) \tag{13}$$

where the Lagrangian density $\mathcal{L}$ is given by

$$\mathcal{L}(\vec{z}, a\delta t; \phi) = \int d^3x d^3y \mathcal{K}(\vec{x}, \vec{y}, \vec{z}, \phi(\vec{x}, (a-1)\delta t), \phi(y, (a-1)\delta t), \phi(z, a\delta t)) \tag{14}$$

For this reason we will refer to $\mathcal{K}$ as *Lagrangian generator.* It turns out that there is a direct way of "reading off" the Lagrangian generator from the Lagrangian. In particular, a Lagrangian

$$\mathcal{L} = A^{\mu\nu}\partial_\mu\phi\partial_\nu\phi + B^\mu\partial_\mu\phi + C \tag{15}$$

can be produced (up to $0(\delta t\alpha^{-1/2})$) from the Lagrangian generator

$$\begin{aligned}\mathcal{K}_\alpha(\vec{x}, \vec{y}, \vec{z}, \phi_1, \phi_2, \phi_3) &= A^{\mu\nu}\Gamma_{\mu\alpha}(\vec{x}, \vec{z}, \phi_1, \phi_3)\Gamma_{\nu\alpha}(\vec{y}, \vec{z}, \phi_2, \phi_3) + \\ &+ B^\mu\Gamma_{\mu\alpha}(\vec{x}, \vec{z}, \phi_1, \phi_3) I_\alpha(\vec{y}, \vec{z}) + C I_\alpha(\vec{x}, \vec{z}) I_\alpha(\vec{y}, \vec{z})\end{aligned} \tag{16}$$

where α is very large and

$$\Gamma_{0\alpha}(\vec{x}, \vec{z}, \phi_1, \phi_3) = \frac{\alpha^{d/2}}{(2\pi)^{d/2}\delta\tau} e^{-\alpha|\vec{z}-\vec{x}|^2/2} (\phi_3 - \phi_1) \tag{17}$$

$$\Gamma_{p\alpha}(\vec{x}, \vec{z}, \phi_1, \phi_3) = \frac{2\alpha^{(d+2)/2}}{(2\pi)^{(d+2)/2}} e^{-\alpha|\vec{z}-\vec{x}|^2/2} (z^p - x^p)(\phi_3 - \phi_1) \tag{18}$$

$$I_\alpha(\vec{x}, \vec{z}) = \left(\frac{\alpha}{2\pi}\right)^{d/2} e^{-\alpha|\vec{z}-\vec{x}|^2/2}\ , \tag{19}$$

We have thus established that Eq 12 is our version of path integral. But, of course, that equation involves $\psi(\phi)$ which we "don't like". Therefore, we will use the previous section in "replacing" $\psi(\phi)$ with $\psi(\vec{x}_k)$ which we "like". Thus, we have

$$\psi(\vec{x}_p, a\delta t) = \sum_r \Bigg[\psi(\vec{x}_r, (k-1)\delta t) \times \tag{20}$$

$$\times \exp\left(i\delta t \sum_{\{(i,j,k|q(\vec{x}_i)=q_j=q(\vec{x}_r);q(\vec{x}_k)=q(\vec{x}_p)\}} (\delta v)^3 \mathcal{K}(\vec{x}_i,\vec{x}_j,\vec{x}_k,\phi(\vec{x}_i),\phi(\vec{x}_j),\phi(\vec{x}_k)))\right)\Bigg]$$

The last equation refers only to quantities living in $\mathbb{R}^3$ and, at the same time, it reproduces mathematical information of QFT up to desired approximation.

5. "Local" model of emergence of "nonlocal" properties of $\psi(\vec{x}_k, t)$

So far we have solved part of the problem: Eq 20 has "classical ontology" in a sense that it lives in $\mathbb{R}^3$. The above equation, however, is nonlocal. So the next part of the problem is to come up with "local" mechanism of its emergence. In order to do that, we will picture cellular automaton. Each lattice point can perform only very simple calculation and pass the result of that simple step to other lattice points which perform other steps (the simplicity of work of each given lattice point is due to very small number of "parameters" it can operate with; after all, we would like to view the latter as discretized, non-well behaved, "classical fields"; and the number of fields in "classical physics" is small). The eventual outcome will be the emergence of Eq 20. The process is "local" in a sense that a point i can only "pass" the information to point j if $i \leftrightarrow j$ (in other words, i and j are connected by the same edge of the lattice) and it can do so within time period δt or greater. However, it is superluminal; in other words, the "finite" amount of time for the above calculation over entire lattice (which far exceeds $N\delta t$) is still very small.

Let us now go ahead and describe the algorithm of "calculation". First, lattice points "evaluate" the sum inside the exponent, "store" it in their "memory", and after that they "exponentiate" it. Thus, we will introduce additional parameter, q'_p, in addition to $q(\vec{x}_p)$. We then introduce "dynamical" parameters $\mathcal{L}(\vec{x}_p, t)$ and $S(\vec{x}_p, t)$. *Our goal* is to define their dynamics in such a way that, at the equilibrium,

$$\lim_{t\to\infty} S(\vec{x}_p,t) = \sum_{\{(i,j,k)|q(\vec{x}_i)=q(\vec{x}_j)=q'(\vec{x}_p);q(\vec{x}_k)=q(\vec{x}_p)\}} (\delta v)^3 \mathcal{K}(\vec{x}_i,\vec{x}_j,\vec{x}_k,\phi(\vec{x}_i),\phi(\vec{x}_j),\phi(\vec{x}_k)) \quad (21)$$

$$\lim_{t\to\infty} \mathcal{L}(\vec{x}_p,t) = \sum_{\{(i,j,k)|q(\vec{x}_i)=q_j=q'_p;q(\vec{x}_k)=q(\vec{x}_p);|\vec{x}_k-\vec{x}_p|<\epsilon\}} (\delta v)^3 \mathcal{K}(\vec{x}_i,\vec{x}_j,\vec{x}_k,\phi(\vec{x}_i),\phi(\vec{x}_j),\phi(\vec{x}_k)) \quad (22)$$

where it is understood that $\mathcal{K}$ is "very small" *unless* $\vec{x}_i$, $\vec{x}_j$ and $\vec{x}_k$ are "very close" (due to the $e^{-\alpha x^2}$ coefficients in 17, 18 and 19) ; thus, $|\vec{x}_k - \vec{x}_p| < \epsilon$ implies that non-negligible contributions occur when x_i, x_j and x_k are *all* close to p. The q'_o tells us that $S(\vec{x}_p)$ and $\mathcal{L}(\vec{x}_p)$ specifically refer to "transitions" *from* $\phi_{(q'(\vec{x}_p))}$ *to* $\phi_{(q(\vec{x}_p))}$. The counterpart of Eq 11 is

$$\forall \vec{x}_i, \vec{x}_j[(q(\vec{x}_i), q'(\vec{x}_i)) = (q(\vec{x}_j), q'(\vec{x}_j)) \Longrightarrow (\mathcal{L}(\vec{x}_i), S(\vec{x}_i)) = (\mathcal{L}(\vec{x}_j), S(\vec{x}_j))] \quad (23)$$

In order to expect large number of lattice points with any given (q, q') within volume δv, we assume

$$\Delta v = M\delta v \Longrightarrow^{Eq\ 10} \frac{N}{M^2} \approx \frac{V}{\Delta v} \quad (24)$$

Now, we know from "conventional" physics that ψ evolves while $\mathcal{L}$ does not. Therefore, we would like $\psi(\vec{x}_p, t)$ to change while $(\mathcal{L}(\vec{x}_p), S(\vec{x}_p))$ not to. *But* our goal is to make everything "mechanical". Thus, instead of simply postulating desired values of $(\mathcal{L}(\vec{x}_p), S(\vec{x}_p))$, we will introduce a mechanism through which said values are reached. In other words, $(\mathcal{L}(\vec{x}_p), S(\vec{x}_p))$ *does* evolve *at first*, but *eventually* the evolution becomes negligibly small *once equilibrium is reached.* Lets start from a toy example of such evolution,

$$A_{k+1} = (1 - \epsilon_1)A_k + \epsilon_2 a_k \ , \ a_k \in \{b_1, \cdots, b_p\} \quad (25)$$

for some aforegiven $B = \{b_1, \cdots, b_p\}$. It can be shown by induction that

$$A_n = (1-\epsilon_1)^{n-m} A_m + \epsilon_2 \sum_{k=m}^{n-1} (1-\epsilon_1)^{n-1-k} b_k \tag{26}$$

If we assume $n - m \gg p$, the first term on right hand side will go to 0; the element b_l would appear approximately $(n-m)/p$ times in the sum, leading to

$$A_n \approx \epsilon_2 \sum_{l=1}^{p} \left(b_l \sum_{k=1}^{[(n-m)/p]} (1-\epsilon_1)^{pk} \right) \approx \frac{\epsilon_2}{\epsilon_1} (1 - (1-\epsilon_1)^{(n-m)/p}) \sum_{l=1}^{p} b_l \tag{27}$$

If we "wait long enough" we would reach the situation

$$n - m \gg \frac{p}{\epsilon} \Longrightarrow (1-\epsilon)^{(n-m)/p} \approx 0 \Longrightarrow A_n \approx \frac{\epsilon_2}{\epsilon_1} \sum_{l=1}^{p} b_l \tag{28}$$

which produces desired result if we set $\epsilon_1 = \epsilon_2$.

Let us now return to $\mathcal{L}(\vec{x}_p)$. Our first goal is to "generate" Eq 22. This immediately tells us that $\{b_1, \cdots, b_p\}$ should be replaced with $\{\mathcal{K}(\vec{x}_i, \vec{x}_j, \vec{x}_k, \phi(\vec{x}_i), \phi(\vec{x}_j), \phi(\vec{x}_k)) || \vec{x}_k - \vec{x}_p| < \epsilon\}$. Thus, at any given time, point p adds *one* of these $\mathcal{K}$-s per prescription in Eq 25. Which of the $\mathcal{K}$-s point p adds is determined by the three memory boxes point p has in which it stores information about three other points. While, at any given time, the point p can remember at most three points, it systematically deletes some points from its memory and replaces them with others. Thus, over time, it ends up adding all possible $\mathcal{K}$-s just like Eq 25 prescribes.

Let us now be more specific. The three memory boxes that point p posesses are defined in terms of internal parameters attached to point p. Thus, memory boxes 1, 2 and 3 are represented by the left hand side of the equations below while the information stored in these boxes is represented by right hand side:

$$\text{Box 1 of Point p} = (q_1(\vec{x}_p, t), q_1'(\vec{x}_p, t), \mathcal{L}_1(\vec{x}_p, t), S_1(\vec{x}_p, t), \psi_1(\vec{x}_p, t), \phi_1(\vec{x}_p, t), \vec{x}_1(\vec{x}_p, t)) =^{\text{want}}$$
$$=^{\text{want}} (q(\vec{x}_i), q'(\vec{x}_i), \mathcal{L}(\vec{x}_i, t'), S(\vec{x}_i, t'), \psi(\vec{x}_i, t'), \phi(\vec{x}_i, t'), \vec{x}_i - \vec{x}_p) = \text{Point i}$$
$$\text{Box 2 of Point p} = (q_2(\vec{x}_p, t), q_2'(\vec{x}_p, t), \mathcal{L}_2(\vec{x}_p, t), S_2(\vec{x}_p, t), \psi_2(\vec{x}_p, t), \phi_2(\vec{x}_p, t), \vec{x}_2(\vec{x}_p, t)) =^{\text{want}}$$
$$=^{\text{want}} (q(\vec{x}_j), q'(\vec{x}_j), \mathcal{L}(\vec{x}_j, t'), S(\vec{x}_j, t'), \psi(\vec{x}_j, t'), \phi(\vec{x}_j, t'), \vec{x}_j - \vec{x}_p) = \text{Point j}$$
$$\text{Box 3 of Point p} = (q_3(\vec{x}_p, t), q_3'(\vec{x}_p, t), \mathcal{L}_3(\vec{x}, t), S_3(\vec{x}_p, t), \psi_3(\vec{x}_p, t), \phi_3(\vec{x}_p, t), \vec{x}_3(\vec{x}_p, t)) =^{\text{want}}$$
$$=^{\text{want}} (q(\vec{x}_k), q'(\vec{x}_k), \mathcal{L}(\vec{x}_k, t'), S(\vec{x}_k, t'), \psi(\vec{x}_k, t'), \phi(\vec{x}_k, t'), \vec{x}_k - \vec{x}_p) = \text{Point k}$$

where t' refers to "slightly earlier" time than t due to delays of superluminal signals, and

$$q(\vec{x}_i) = q(\vec{x}_j) = q'(\vec{x}_k) \ , \ q(\vec{x}_p) = q(\vec{x}_k) \tag{29}$$

but the specific choices of i, j and k that satisfy the above vary over time. Thus, the 25 is produced per the following algorithm:

Rule 1: If $\vec{x}_3(\vec{x}_p, t) < \Delta x$ then
$\mathcal{L}(\vec{x}_p, t+\delta t) = (1-\epsilon)\mathcal{L}(\vec{x}_p, t) + \epsilon(\delta v)\mathcal{K}(\phi_1(\vec{x}_p, t), \phi_2(\vec{x}_p, t), \phi_3(\vec{x}_p, t), \vec{x}_1(\vec{x}_p, t), \vec{x}_2(\vec{x}_p, t), \vec{x}_3(\vec{x}_p, t))$.

Now, the information stored in the memory boxes changes whenever point $\vec{x}_p$ receives a signal from some other point $\vec{x}_l$ such that $q(\vec{x}_l) \in \{q(\vec{x}_p), q'(\vec{x}_p)\}$. If $q(\vec{x}_l) = q'(\vec{x}_p)$, then, in $\vec{x}_p$-s mind, j will replace i and l will replace j. If $q(\vec{x}_l) = q(\vec{x}_p)$ then in $\vec{x}_p$-s mind l will replace k. All these operations, of course, are performed by point $\vec{x}_p$. Therefore, point $\vec{x}_p$

needs to know the information about l *before* it performs the above operations. In fact, even if $q(\vec{x}_l) \notin \{q(\vec{x}_p), q'(\vec{x}_p)\}$, point p needs to *first* know the parameters of l in order to "realize" that it has to "ignore" l. Such "knowledge" that point $\vec{x}_p$ possesses is represented by left hand side of the following equation:

$$(q_0(\vec{x}_p), q'_0(\vec{x}_p), \mathcal{L}_0(\vec{x}_p), S_0(\vec{x}_p), \psi_0(\vec{x}_p), \phi_0(\vec{x}_p), \vec{x}_0(\vec{x}_p)) =^{\text{want}}$$

$$=^{\text{want}} (q(\vec{x}_l), q'(\vec{x}_l), \mathcal{L}(\vec{x}_l), S(\vec{x}_l), \psi(\vec{x}_l), \phi(\vec{x}_l), \vec{x}_l - \vec{x}_p)$$

and the mechanism of "storing" it into one of the memory boxes is described by the following algorithm:

Rule 2: If $q_0(\vec{x}_p, t) = q'(\vec{x}_p, t)$ then
a) $(q_1(\vec{x}_p), q'_1(\vec{x}_p), \mathcal{L}_1(\vec{x}_p), S_1(\vec{x}_p), \psi_1(\vec{x}_p), \phi_1(\vec{x}_p), \vec{x}_1(\vec{x}_p))(t+\delta t) =$
$= (q_2(\vec{x}_p), q'_2(\vec{x}_p), \mathcal{L}_2(\vec{x}_p), S_2(\vec{x}_p), \psi_2(\vec{x}_p), \phi_2(\vec{x}_p), \vec{x}_2(\vec{x}_p))(t)$
b) $(q_2(\vec{x}_p), q'_2(\vec{x}_p), \mathcal{L}_2(\vec{x}_p), S_2(\vec{x}_p), \psi_2(\vec{x}_p), \phi_2(\vec{x}_p), \vec{x}_2(\vec{x}_p))(t+\delta t) =$
$= (q_0(\vec{x}_p), q'_0(\vec{x}_p), \mathcal{L}_0(\vec{x}_p), S_0(\vec{x}_p), \psi_0(\vec{x}_p), \phi_0(\vec{x}_p), \vec{x}_0(\vec{x}_p))(t)$
c) $(q_3(\vec{x}_p), q'_3(\vec{x}_p), \mathcal{L}_3(\vec{x}_p), S_3(\vec{x}_p), \psi_3(\vec{x}_p), \phi_3(\vec{x}_p), \vec{x}_3(\vec{x}_p))(t+\delta t) =$
$= (q_3(\vec{x}_p), q'_3(\vec{x}_p), \mathcal{L}_3(\vec{x}_p), S_3(\vec{x}_p), \psi_3(\vec{x}_p), \phi_3(\vec{x}_p), \vec{x}_3(\vec{x}_p))(t)$

Rule 3: If $q_0(\vec{x}_p, t) = q(\vec{x}_p, t)$ then
a) $(q_1(\vec{x}_p), q'_1(\vec{x}_p), \mathcal{L}_1(\vec{x}_p), S_1(\vec{x}_p), \psi_1(\vec{x}_p), \phi_1(\vec{x}_p), \vec{x}_1(\vec{x}_p))(t+\delta t) =$
$= (q_1(\vec{x}_p), q'_1(\vec{x}_p), \mathcal{L}_1(\vec{x}_p), S_1(\vec{x}_p), \psi_1(\vec{x}_p), \phi_1(\vec{x}_p), \vec{x}_1(\vec{x}_p))(t)$
b) $(q_2(\vec{x}_p), q'_2(\vec{x}_p), \mathcal{L}_2(\vec{x}_p), S_2(\vec{x}_p), \psi_2(\vec{x}_p), \phi_2(\vec{x}_p), \vec{x}_2(\vec{x}_p))(t+\delta t) =$
$= (q_2(\vec{x}_p), q'_2(\vec{x}_p), \mathcal{L}_2(\vec{x}_p), S_2(\vec{x}_p), \psi_2(\vec{x}_p), \phi_2(\vec{x}_p), \vec{x}_2(\vec{x}_p))(t)$
c) $(q_3(\vec{x}_p), q'_3(\vec{x}_p), \mathcal{L}_3(\vec{x}_p), S_3(\vec{x}_p), \psi_3(\vec{x}_p), \phi_3(\vec{x}_p), \vec{x}_3(\vec{x}_p))(t+\delta t) =$
$= (q_0(\vec{x}_p), q'_0(\vec{x}_p), \mathcal{L}_0(\vec{x}_p), S_0(\vec{x}_p), \psi_0(\vec{x}_p), \phi_0(\vec{x}_p), \vec{x}_0(\vec{x}_p))(t)$

Rule 4: If $q_0(\vec{x}_p, t) \notin \{q(\vec{x}_p, t), q'(\vec{x}_p, t)\}$ then
a) $(q_1(\vec{x}_p), q'_1(\vec{x}_p), \mathcal{L}_1(\vec{x}_p), S_1(\vec{x}_p), \psi_1(\vec{x}_p), \phi_1(\vec{x}_p), \vec{x}_1(\vec{x}_p))(t+\delta t) =$
$= (q_1(\vec{x}_p), q'_1(\vec{x}_p), \mathcal{L}_1(\vec{x}_p), S_1(\vec{x}_p), \psi_1(\vec{x}_p), \phi_1(\vec{x}_p), \vec{x}_1(\vec{x}_p))(t)$
b) $(q_2(\vec{x}_p), q'_2(\vec{x}_p), \mathcal{L}_2(\vec{x}_p), S_2(\vec{x}_p), \psi_2(\vec{x}_p), \phi_2(\vec{x}_p), \vec{x}_2(\vec{x}_p))(t+\delta t) =$
$= (q_2(\vec{x}_p), q'_2(\vec{x}_p), \mathcal{L}_2(\vec{x}_p), S_2(\vec{x}_p), \psi_2(\vec{x}_p), \phi_2(\vec{x}_p), \vec{x}_2(\vec{x}_p))(t)$
c) $(q_3(\vec{x}_p), q'_3(\vec{x}_p), \mathcal{L}_3(\vec{x}_p), S_3(\vec{x}_p), \psi_3(\vec{x}_p), \phi_3(\vec{x}_p), \vec{x}_3(\vec{x}_p))(t+\delta t) =$
$= (q_3(\vec{x}_p), q'_3(\vec{x}_p), \mathcal{L}_3(\vec{x}_p), S_3(\vec{x}_p), \psi_3(\vec{x}_p), \phi_3(\vec{x}_p), \vec{x}_3(\vec{x}_p))(t)$

Now we have to fill in a gap and explain how do the said signals (described by parameters with index 0) come about. In order for Rule 1 to agree with process described in Eq 25, the information in the three memory boxes needs to appear at random. This can be achieved by letting each point emit a signal with its own period, which would imply that the sequence of emitted signals changes over time. This can be enforced by assigning a "timer" $\theta(\vec{x}_k, t)$ to a lattice point $\vec{x}_k$, and it emits a signal whenever $\theta(\vec{x}_k, t)$ crosses $2\pi n$. The difference in periodicity can be produced by having a "step size" $(\delta\theta)(\vec{x}_k)$ differ for each point k. Thus, we propose the following algorithm:

Rule 5: $\theta(\vec{x}_k, t+\delta t) = \theta(\vec{x}_k, t) + (\delta\theta)(\vec{x}_k)$, where $(\delta\theta)(\vec{x}_k)$ is different for each k.

The act of emitting a signal will be described by the parameter $\sigma(\vec{x}_k, t)$. Normally, $\sigma(\vec{x}_k, t)$ is 0, but during the time of signal emission it becomes $-\sigma_{max}$ (Rule 6) at the point of emission. On the other hand, $\sigma(\vec{x}_l, t)$ becomes $+\sigma_{max}$ (Rule 7) at the points past which the signal passes. In both cases, $\sigma(\vec{x}, t)$ will go on to return to 0 afterwards; but it can only do so in small steps (Rule 8). Now, a point l is not receptive to any signals *unless* $\sigma(\vec{x}_l, t) = 0$ (Rules 9, 10, 11). The fact that it "takes time" for $\sigma(\vec{x}, t)$ to "become" zero implies that once a signal passes by a certain point it would only be able to return to that point after said period of time is over. Now, if the total number of points (N) is smaller than the number of steps needed for recovery ($\sigma_{max}/\delta\sigma$), the signal will cover all the points and "have nowhere to go". Thus, after passing through each point once, it will disappear. On the other hand, if $\sigma/\delta\sigma$ is much smaller than the average time between signal emissions $((1/N) \times \min(2\pi/(\delta\theta)_1, \cdots, 2\pi/(\delta\theta)_N))$, then by the time some other point emits the signal, all the σ-s will have already recovered back to 0, which is why the new signal will pass without any trouble. Thus, we assume that

$$N \ll \frac{\sigma_{\max}}{\delta\sigma} \ll \frac{1}{N} \min\left(\frac{2\pi}{(\delta\theta)_1}, \cdots, \frac{2\pi}{(\delta\theta)_N}\right) \tag{30}$$

Now, as will be seen in Rules 13-15, ψ can only change in response to the above signal. Thus, the fact that we have one signal at a time implies that ψ does not change during the time delay of said signal. This is the reason why we have false impression that the signal moves with infinite speed while, in fact, its speed is finite. It should be admitted, however, that in light of differences in $(\delta\theta)(\vec{x}_k)$, the phase differences keep changing and from time to time different signals can overlap in time "by accident". We assume, however, that such events are rare enough in order for their implications to be negligible.

Let us now write out explicitly the algorithm that will produce the above. What we will write will be a bit more complicated than it could have been because we would like to preserve discretized version of "locality"; that is, k and l are not allowed to communicate unless they are connected by an edge of discrete structure; the latter will be denoted by $k \leftrightarrow l$. However, as we said previously, the speed of signal propagation is still superluminal, despite being finite. Apart from the above, we also need to account for the possibility of spacetime curvature. Thus, we will instruct point k to only measure the position of l if l is close enough for curvature to be negligible; otherwise, k will record l as infinitely far away (Rule 10b). It should be understood that whenever one of the distances is ∞, we have $\mathcal{K} = 0$. With this being said, here is the list of rules:

Rule 6: If $\theta(\vec{x}_k, t) \in (2\pi n, 2\pi n + (\delta\theta)_k)$, then $\sigma(\vec{x}_k, t + \delta t) = -\sigma_{max}$

Rule 7 If there exists $\vec{x}_l \leftrightarrow \vec{x}_k$ s.t. $\sigma(\vec{x}_l, t) \in \{-\sigma_{max}, \sigma_{max}\}$, then $\sigma(\vec{x}_k, t + \delta t) = \sigma_{max}$

Rule 8: If $\theta(\vec{x}_k, t) \in (2\pi n + (\delta\theta)_k, 2\pi(n + 1))$ and $\sigma(\vec{x}_l, t) \notin \{-\sigma_{max}, \sigma_{max}\}$ for all $\vec{x}_l \leftrightarrow \vec{x}_k$, then

a) If $\sigma(\vec{x}_k, t) =\in (-\delta\sigma, \delta\sigma)$ then $\sigma(\vec{x}_k, t + \delta t) = 0$

b) If $\sigma(\vec{x}_k, t) < -\delta\sigma$ then $\sigma(\vec{x}_k, t + \delta t) = \sigma(\vec{x}_k, t) + \delta\sigma$

c) If $\sigma(\vec{x}_k, t) > \delta\sigma$ then $\sigma(\vec{x}_k, t + \delta t) = \sigma(\vec{x}_k, t) - \delta\sigma$

Rule 9: If $\vec{x}_l \leftrightarrow \vec{x}_k$, $\sigma(\vec{x}_k, t) = -\sigma_{max}$, and $\sigma_l = 0$, then $(\mathcal{L}_0(\vec{x}_l), S_0(\vec{x}_l), \psi_0(\vec{x}_l), \phi_0(\vec{x}_l), \vec{x}_0(\vec{x}_l))(t + \delta t) = (\mathcal{L}(\vec{x}_k), S(\vec{x}_k), \psi(\vec{x}_k), \phi(\vec{x}_k), x_k - x_l)(t)$

Rule 10: If $\vec{x}_l \leftrightarrow \vec{x}_k$, $\sigma(\vec{x}_k, t) > \delta\sigma$, and $\sigma(\vec{x}_l, t) = 0$, then

a) $(\mathcal{L}_0(\vec{x}_l), S_0(\vec{x}_l), \psi_0(\vec{x}_l), \phi_0(\vec{x}_l))(t + \delta t) = (\mathcal{L}_0(\vec{x}_k), S_0(\vec{x}_k), \psi_0(\vec{x}_k), \phi_0(\vec{x}_k))(t)$

b) If $|x_k - x_l + \mu_{xk}(t)| < \delta x$ then $\vec{x}_0(\vec{x}_l, t + \delta t) = \vec{x}_k - \vec{x}_l + \vec{x}_0(\vec{x}_k, t)$; if the former is not the case then $x_0(\vec{x}_l, t + \delta t) = \infty$.

Rule 11: If $\sigma_l(t) \notin (-\delta\sigma, \delta\sigma)$, then $(\mathcal{L}_0(\vec{x}_l), S_0(\vec{x}_l), \psi_0(\vec{x}_l), \phi_0(\vec{x}_l), x_0(\vec{x}_l))(t + \delta t) = 0$

To recap, we have written Rule 1 in order to allow point p to compute $\mathcal{L}(\vec{x}_p)$ per mechanism of Eq 25, and then we have written Rules 2-11 in order to come up with a mechanism of point p obtaining all the necessary information in order to perform Step 1. Now, going back to what we were doing, after obtaining $\mathcal{L}(\vec{x}_p)$ we have to "integrate" it to obtain $S(\vec{x}_p)$. This can be done in a way analogous to Rule 1:

Rule 12: If $q_0(\vec{x}_p, t) = q(\vec{x}_p)$ and $q_0'(\vec{x}_p, t) = q'(\vec{x}_p, t)$ then $S(\vec{x}_p, t + \delta t) = (1 - \epsilon)S(\vec{x}_p, t) + \epsilon(\Delta v)\mathcal{L}_0(\vec{x}_p, t)$. If, on the other hand, one or both of the conditions are not true, then $S(\vec{x}_p, t + \delta t) = S(\vec{x}_p, t)$.

At this point, we have successfully evaluated the sum in the exponential of Eq 20 *provided that* $q(\vec{x}_r) = q'(\vec{x}_p)$. This means that point p can utilize an algorithm analogous Rules 1 and 12 in order to evaluate *part* of the overall sum in Eq 20 (Rule 13) and after that it has to pass the result to other lattice points (Rule 14) in order for them to add other parts of Eq 20:

Rule 13: If $q_0(\vec{x}_p, t) = q'(\vec{x}_p)$ then $\psi(\vec{x}_p, t + \delta t) = \psi(\vec{x}_p, t)e^{i\delta t S(\vec{x}_p, t)}$

Rule 14: If $q_0(\vec{x}_p, t) = q(\vec{x}_p)$ then $\psi(\vec{x}_p, t + \delta t) = \psi_0(\vec{x}_p, t)$
Rule 15: If $q_0(\vec{x}_p, t) \notin \{q(\vec{x}_p), q'(\vec{x}_p)\}$ then $\psi(\vec{x}_p, t + \delta t) = \psi(\vec{x}_p, t)$

Apart from the above, the Rule 14 is also needed in order to enforce Eq 11. However, one of the negative "side effects" of Rule 14 is that a point p might "update" its value of ψ per rule 13 and then "undo" whatever has happened after it receives a signal from particle l which has not "updated" ψ yet. In other words, the only situation in which the effect of Rule 13 survives is when point p emits a signal *before* any other point with the same q does (although it is fine if a point with q other than $q(\vec{x}_p)$ emits a signal before p since in this case its signal is ignored per rule 15). Now the number of points with $q = q(\vec{x}_p)$ is N/M; thus the probability of the above happening is M/N which is very small. Nevertheless, on the scales

$$\Delta t \gg \frac{N\delta t}{M} \tag{31}$$

we would still have the expected time evolution of states; we have simply "rescaled" time by M/N.

Another issue that needs to be addressed is that $(\delta\theta)(\vec{x}_k)$ is different for each point in Rule 5 (which was needed in order to perform "random sum") yet we need δt to be the same for the purposes of Eq 20. The way we get around this is by saying that $(\delta\theta)(\vec{x}_k)$ are only *slightly* different from each other; in other words,

$$\min((\delta\theta)(\vec{x}_1), \cdots, (\delta\theta)(\vec{x}_N)) > (1 - \epsilon)\max((\delta\theta)(\vec{x}_1), \cdots, (\delta\theta)(\vec{x}_N)) \tag{32}$$

Thus, for the purposes of integration δt-s will be close enough, while for the purposes of randomness ϵ would be "large enough" for us to cover all possible patterns within the scale $N\delta t/M$.

6. Conclusion

In this paper we have shown a way to "encode" the mathematical information of second quantization within the discretized "classical" $\mathbb{R}^3$. This prescription is by no means unique. For example, it is possible to embed the above discrete structure into continuum and then write down differential equations that produce some of the above rules (see [1]). This might allow some shortcuts: for example, we might allow direct communication between any $\vec{x}_k$ and $\vec{x}_l$ without assuming that they are direct neighbors since such communication will still propagate "locally" through continuum. It is also possible to explore various modifications of above process. For example in current paper we were assuming that any given signal passes the universe before another signal is sent and, as we explained, this is the ultimate reason for the false appearance that the signal's speed is infinite when in reality it is finite (see discussion surrounding Eq 30). If we drop that assumption by relaxing the left hand side of Eq 30, then we would relax the effects of entanglement as well. Thus, instead of a single Hilbert space we might end up producing several different Hilbert spaces over different overlapping domains, each of which is small enough not to feel the effects of delay of superluminal signal while, on a larger scale, its delays are felt and therefore the very notion of ψ breaks down. Such might be interesting questions to ask for future research.

References

[1] Sverdlov R 2011 Can quantum lattice be generated through several classical ones superimposed in spacetime continuum?'*Preprint* arXiv:1108.5729

Journal of Physics: Conference Series **442** (2013) 012048 doi:10.1088/1742-6596/442/1/012048

Intrinsic periodicity: the forgotten lesson of quantum mechanics[1]

Donatello Dolce
University of Melbourne, Parkville 3010 VIC, Australia.
University of Camerino, Piazza Cavour 19F, 62032 Camerino, Italy.
E-mail: `donatello.dolce@coepp.org.au`

Abstract. Wave-particle duality, together with the concept of elementary particles, was introduced by de Broglie in terms of intrinsically periodic phenomena. However, after nearly 90 years, the physical origin of such undulatory mechanics remains unrevealed. We propose a natural realization of the de Broglie periodic phenomenon in terms of harmonic vibrational modes associated to space-time periodicities. In this way we find that, similarly to a vibrating string or a particle in a box, the intrinsic recurrence imposed as a constraint to elementary particles represents a fully consistent quantization condition. The resulting cyclic dynamics formally match ordinary relativistic Quantum Mechanics in both the canonical and Feynman formulations. Interactions are introduced in a geometrodynamical way, similarly to general relativity, by simply considering that variations of kinematical state can be equivalently described in terms of modulations of space-time recurrences, as known from undulatory mechanics. We present this novel quantization prescription from an historical prospective.

1. Introduction

We give some historical motivation for a formulation of Quantum Mechanics (QM) in terms of elementary space-time cycles [1, 2, 3]. Since its earliest days, QM was characterized by an assumption intrinsic periodicity. We may think to the wave-particle duality, introduced by de Broglie in this terms: "we proceed with the assumption of the existence of a certain periodic phenomenon of a yet to be determined character, which is to be attributed to each and every isolated energy parcel" [4]. That is to say, every elementary particle is characterized by an intrinsic recurrence in time and space. As known from undulatory mechanics, every local retarded variation of the kinematical state of a particles (i.e. the local four-momentum) can be equivalently expressed, by means of the Planck constant, in terms of the corresponding local retarded modulation of space-time recurrence. According to the Standard Model, every system in physics is described by a set of elementary particles and their relativistic interactions, thus the wave-particle duality implicitly says that every system in physics can be described in terms of elementary modulated cycles.

The intrinsic periodicity of elementary particles can be used as a quantization condition, similarly to the walls of a "particle in a box". This is intuitive if we think to Bohr's hydrogen atom in which, similarly to the harmonic spectrum of a vibrating 'string', the energy levels are determined by requiring that the only possible orbits are those with an integer number of

[1] Dedicated to Ignaz Philipp Semmelweis, the "savior of mothers" (Buda 1818 — Döbling 1865).

wavelengths. Indeed this is the idea at the base of the Bohr-Sommerfeld quantization: in a semi-classical description of QM, the energy spectrum can be determined by requiring closed orbits associated to a given potential. It has been shown recently that such a semi-classical description can be used to solve complex quantum systems such as atoms with more electrons or the Zeeman effect, considered as some of the main limitations of this approach.

More generally, as mathematically proven in recent papers [1, 2, 3], we will show how the semi-classical description based on the assumption of intrinsic periodicity of elementary systems, when correctly implemented in a relativistic theory, can be used to formally derive the modern descriptions of relativistic QM.

2. The model

We introduce the formalism by following the same arguments used by de Broglie to derive the wave-particle duality [4]. In his paper he pointed out that every elementary particle is characterized by recurrences in time and space. These are fully determined, though Lorentz transformations, by the recurrence along the proper time, fixed by the mass of the particle $\bar{M}$. The periodicity of this rest recurrence is the Compton time of the particle

$$T_\tau = \frac{h}{\bar{M}c^2}.$$

Equivalently, every particle has a quantum recurrence λ_s of along its worldline $s = c\tau$, where λ_s is the Compton wavelength $\lambda_s = cT_\tau$. When observed in a generic reference frame this corresponds to instantaneous temporal and spatial periodicities, T_t and $\vec{\lambda}_x$ respectively, as can be seen by using the Lorentz transformation

$$cT_\tau = c\gamma T_t - \gamma\vec{\beta}\cdot\vec{\lambda}_x.$$

They can be written in a contravariant tangent four-vector

$$T^\mu = \{T_t, \vec{\lambda}_x\}.$$

In fact, from the relativistic definition of four-momentum $\bar{p}_\mu = \{\bar{E}, -\bar{\mathbf{p}}\} = \{\gamma\bar{M}c^2, -\gamma\vec{\beta}\bar{M}c\}$, it is easy to derive the following relativistic invariant (de Broglie phase harmony)

$$T_\tau\bar{M}c^2 \equiv h \quad \longleftrightarrow \quad T^\mu\bar{p}_\mu \equiv h.$$

In other words, the four-momentum of a particle can be equivalently encoded in corresponding modulations of the instantaneous space-time periodicity T_μ. According to undulatory mechanics, particles can be described in terms of periodic phenomena (phasors or waves) in which the space-time coordinates enter as angular variables; their periodicities define their kinematical state. Thus, every system in nature can be consistently described in terms of modulations of elementary space-time cycles. Since they are completely characterized by proper time periodicity, we can consider a minimal topologies $\mathbb{S}^1$ for elementary bosonic cycles[2].

It must be noted that these recurrences are typically extremely fast with respect to the Cs_{133} atomic clock, whose time scale is (by definition of "second") about $10^{-10}s$, or to the present resolution in time, about $10^{-17}s$ (using laser pulses). For instance, simple electrodynamics system are characterized by the the proper time periodicity of the electron $T_{\tau(e)} = h/m_{(e)}c^2 = 8.09329972 \times 10^{-21}s$ (Compton time), corresponding to the *zitterbewegung*. The heavier the mass or the energy of the particles, the faster the corresponding temporal periodicities. For

[2] We neglect the spin which however can be regarded as another manifestation of the intrinsic periodicity of elementary particles. See Schrödinger's *zitterbewegung* and related semiclassical derivation of the Dirac equation.

instance, LHC is exploring "indirectly"[3] time cycles of the order of $T_t \sim 10^{-27}s$ (corresponding to energies scales of the TeV).

We want to impose the intrinsic periodicity T^μ of elementary particles as a constraint. This represents a semiclassical quantization condition. A particle with intrinsic periodicity is similar to a "particle in a box". Through discrete Fourier transform the periodicity T^μ directly implies a quantization of the conjugate spectrum $p_n^\mu = n\bar{p}^\mu$; n is the single quantum number associated to the topology $\mathbb{S}^1$. For instance the quantization of the energy spectrum associated to the time periodicity T_t is the harmonic spectrum

$$E_n = n\bar{E} = nh/T_t.$$

A bosonic particle can be therefore represented as a one dimensional bosonic 'string' $\Phi(x)$ vibrating in compact space-time dimensions of length T^μ and Periodicity Boundary Conditions (PBCs — denoted by the circle in $\oint$):

$$\mathcal{S}^{\lambda_s} = \oint^{T^\mu} d^4x \mathcal{L}(\partial_\mu \Phi(x), \Phi(x)) \,. \tag{1}$$

As known from string theory or extra-dimensional theories, PBCs (or combinations of Dirichlet and Neumann BCs) minimizes the action at the boundary so that all the relativistic properties of (1) are preserved. This is a consequence of the fact that relativity fixes the differential structure of space-time whereas the only requirement for the BCs is to fulfill the variational principle. The expansion in harmonics (discrete Fourier expansion with coefficients a_n and normalization A_n) of a field/string vibrating with persistent periodicity is

$$\Phi(x) = \sum_n \phi_n(x) = \sum_n A_n a_n(\bar{p}_\mu) e^{-\frac{i}{\hbar} p_{n\mu} x^\mu}.$$

Such a description in terms of elementary space-time cycles is covariant as prescribed by relativity. This can be easily seen by using Lorentz transformations. The action (1) describing a particle in a given reference frame, after the generic transformation of variables $x_\mu \to x'_\mu = \Lambda_\mu^\nu x_\nu$, transforms to the action

$$\mathcal{S}^{\lambda_s} = \oint^{T'^\mu = \Lambda^\mu_\nu T^\nu} d^4x' \mathcal{L}(\partial'_\mu \Phi'(x'), \Phi'(x')) \,. \tag{2}$$

This turns out to have a transformed boundary, such that the resulting solution has transformed periodicity

$$T^\mu \to T'^\mu = \Lambda^\mu_\nu T^\nu.$$

According to $\bar{p}'_\mu c T'^\mu = h$, this is the periodicity associated to the four-momentum of the particle in the new frame

$$\bar{p}_\mu \to \bar{p}'_\mu = \Lambda_\mu^\nu \bar{p}_\nu \,.$$

Indeed T^μ is a contravariant tangent four-vector satisfying the relativistic constraint

$$\frac{1}{T_\tau^2} = \frac{1}{T_\mu} \frac{1}{T^\mu},$$

[3] With "direct" observation we mean to compare these fast dynamics with a clock of resolution higher than this time scale. For instance, with the modern clocks, a direct observation of these quantum recurrences are possible for neutrinos and other coherent quantum phenomena in condensed matter such as lasers, superconductivity or carbon nanotubes.

which is the geometric counterpart of the relativistic dispersion relation of the particle $\bar{M}c^2 = \bar{p}_\mu \bar{p}^\mu$. By considering the relativistic modulations of temporal periodicity (relativistic Doppler effect) associated to variations of reference frames, the resulting energy spectrum is

$$E_n(\bar{\mathbf{p}}) = n\frac{h}{T_t(\bar{\mathbf{p}})} = n\sqrt{\bar{\mathbf{p}}^2 c^2 + \bar{M}^2 c^4}.$$

In other words we have obtained semi-classically the energy spectrum prescribed by ordinary second quantization (after normal ordering) for bosonic particles[4].

3. Interactions

Interactions can be introduced in this formalism by considering that, as known in undulatory mechanics, relativistic variations of four-momentum associated to a given interaction scheme can be equivalently described in terms of relativistic modulations of periodicities. That is, in every point $x = X$, a relativistic interaction can be characterized by the relativistic variations of four-momenta of a particle with respect to the free case

$$\bar{p}_\mu \to \bar{p}'_\mu(X) = e^a_\mu(x)|_{x=X}\bar{p}_a.$$

Through $\hbar$, this generic interaction can be equivalently encoded by corresponding relativistic modulations of quantum recurrence

$$T^\mu \to T'^\mu(X) \sim e^\mu_a(x)|_{x=X}T^a.$$

According to our formalism, this corresponds to local and retarded "stretching" the compactified space-time dimensions of (1), and thus by corresponding local and retarded deformations of the metric

$$\eta_{\mu\nu} \to g_{\mu\nu}(X) = [e^a_\mu(x)e^b_\nu(x)]|_{x=X}\eta_{ab}.$$

This geometrodynamical description of interaction can be easily checked by imposing the local transformation of reference frame

$$dx_\mu \to dx'_\mu(X) = e^a_\mu(x)|_{x=X}dx_a$$

as substitution of variables in the free action (1), see [2, 3]. Indeed, the resulting action turns out to have locally deformed metric $g_{\mu\nu}(X)$ and local transformations of boundary. In this way, the solution of this transformed action correctly describes the local modulations of periodicity $T'^\mu(X)$ associated to the interaction scheme. That is, we pass from a free solution of persistent type

$$\phi(x) \propto e^{-\frac{i}{\hbar}p_\mu x^\mu}$$

to the interacting solution with modulated periodicity of type

$$\phi'(x) \propto e^{-\frac{i}{\hbar}\int^{x_\mu} dx'^\mu p_\mu}. \quad (3)$$

Note also that in our formalism the (integral of the) kinematics associated to the given interaction scheme turns out to be encoded on the boundary of the theory. The holographic principle has thus a natural realization in this theory.

[4] In fact, second quantization prescribes that every mode with angular frequency $\bar{\omega}(\bar{\mathbf{p}}) = \sqrt{\bar{\mathbf{p}}^2c^2 + \bar{M}^2c^4}/\hbar$ of a Klein Gordon fields has a quantized energy spectrum $E_n(\bar{\mathbf{p}}) = n\hbar\bar{\omega}(\bar{\mathbf{p}})$, after normal ordering. By equivalently assuming in (1) anti-PBCs we get $E_n(\bar{\mathbf{p}}) = (n + \frac{1}{2})\hbar\bar{\omega}(\bar{\mathbf{p}})$, avoiding normal ordering.

Such a geometrodynamical description of generic interactions is of the same type of the one prescribed by General Relativity (GR). In a weak Newtonian interaction, the corresponding variation energy $\bar{E} \to \bar{E}' \sim \left(1 + GM_{\odot}/|\mathbf{x}|c^2\right) \bar{E}$ implies, through $\hbar$, a modulation of time periodicity $T_t \to T_t' \sim \left(1 - GM_{\odot}/|\mathbf{x}|c^2\right) T_t$, i.e. we have redshift and time dilatation. If we also consider the variation of momentum and the corresponding modulation of spatial periodicity, the resulting metric encoding the Newtonian interaction is actually the linearized Schwarzschild metric. Thus we have obtained linearized gravity, from which GR follows by assuming self-interaction.

We also mention another remarkable consequence of our description of elementary cycles, which can be regarded as a realization of Weyl's, Kaluza's, Wheeler's original proposals. In [2, 3] we have found that gauge interaction has a geometrodynamical description similar to that of gravitational interaction in GR. This possibility is provided by the fact that transformations of flat reference frames, which in ordinary QFT do not vary the solution of the theory[5], in our formulation induce local transformations of the boundary, and thus local transformations of the solution of the theory. The resulting transformation of the field solution is formally equivalent to the internal transformation of ordinary gauge theories. For instance, it is possible to show that the local transformation of flat reference frame[6]

$$dx^{\mu}(x) \to dx'^{\mu} \sim dx^{\mu} - e dx^{a} \omega_{a}^{\ \mu}(x)$$

can be parametrized parametrizing by introducing a vectorial field

$$\bar{A}_{\mu}(x) \equiv \omega^{a}_{\ \mu}(x) \bar{p}_{a},$$

so that the interaction scheme is actually given by the minimal substitution

$$\bar{p}'_{\mu}(x) \sim \bar{p}_{\mu} - e\bar{A}_{\mu}(x).$$

Thus the transformed solution associated to this interaction is of the type

$$\phi'(x) \propto e^{-\frac{i}{\hbar}\int^{x_\mu} dx'^{\mu} A_{\mu}} e^{-\frac{i}{\hbar} p_{\mu} x^{\mu}}.$$

The gauge connection, which must be postulated in ordinary gauge theory, turns out to describe the modulations of periodicity associated to the local transformations of flat reference frame with respect to the free solution (we say that the gauge field "tunes" the periodicity). From this, ordinary Maxwell dynamics (and more in general Yang-Mills theories) can be derived, see [2] for more details.

4. Quantization

From an historical point of view, the modern mathematical formulation of QM is inspired to the theory of sound and Reyleigh studies. Indeed the theory described so far can be regarded as a fully relativistic generalization of the theory of sound (we assume vibrations in compact time, and not only in compact space as in ordinary classical sound sources). Therefore the theory inherits fundamental aspects of the ordinary quantum formalism. The assumption of periodicity at the base of our formulation is a quantization condition (similarly to the quantization of a particle in a box). In particular, when the recurrences of elementary particles are imposed as a constraint, the resulting (classical) cyclic dynamics reproduce formally ordinary relativistic QM.

[5] In ordinary QFT the BCs have a marginal role: the KG field used in computation is the most general solution of the KG equation.

[6] Intuitively we want to describe the trembling motion of a particle interacting electromagnetically as local transformation of reference frame, similarly to the equivalence principle (applied to the boundary) of GR.

In this section we summarize basic aspects of this remarkable correspondence. It is known that a vibrating 'string' is the typical classical system that can be described locally in a Hilbert space. Even if we consider modulations of periodicities, the harmonics of such an interacting 'string' form locally a complete set with respect to the corresponding local inner product

$$\langle\phi|\chi\rangle \equiv \int_0^{\lambda_x(X)} \frac{dx}{\lambda_x(X)}\phi^*(x)\chi(x)\,. \tag{4}$$

These harmonics define locally a Hilbert base

$$\langle x|\phi'_n\rangle = \phi'_n(x).$$

Thus a modulated vibrating 'string', generic superposition of harmonics, is represented by the generic Hilbert state

$$|\phi'\rangle = \sum a_n\,|\phi_n\rangle\,.$$

The non-homogeneous Hamiltonian $\mathcal{H}'$ and momentum $\mathcal{P}'_i$ operator are introduced as the operators associated to the four-momentum spectrum of the locally modulate 'string':

$$\mathcal{P}'_\mu\,|\phi'_n\rangle = p'_{n\mu}\,|\phi'_n\rangle\,, \quad \text{where } \mathcal{P}'_\mu = \{\mathcal{H}', -\mathcal{P}'_i\}\,.$$

From the modulated wave equation, the temporal and spatial evolution of every modulated harmonics satisfies

$$i\hbar\partial_\mu\phi'_n(x) = p'_n\phi'_n(x),$$

thus the time evolution of our modulated 'string' $|\phi'\rangle$ is given by the ordinary Schrödinger equation

$$i\hbar\partial_t\,|\phi'\rangle = \mathcal{H}'\,|\phi\rangle\,. \tag{5}$$

Moreover, since we are assuming intrinsic periodicity, this classical-relativistic theory implicitly contains the ordinary commutation relations of QM. This can be seen by evaluating the expectation value of a total derivative $\partial_x F(x)$, and considering that the boundary terms of the integration by parts cancel each other owing the assumption of intrinsic periodicity. For generic Hilbert states we *obtain* $[F(x), \mathcal{P}'_i] = i\hbar\partial_x F(x)$ and thus, from $F(x) = x_j$,

$$[x_j, \mathcal{P}'_i] = i\hbar\delta_{j,i}\,.$$

The formal correspondence to ordinary relativistic QM is confirmed by the fact that, remarkably, the classical evolution of such a classical vibrating 'string' with all its modulated harmonics is described by the ordinary Feynman Path Integral (we are integrating over a sufficiently large number N of spatial periods so that the $V_{\rm x} = N\lambda_x$ is bigger than the interaction region)

$$\mathcal{Z} = \int_{V_{\rm x}} \mathcal{D}{\rm x} e^{\frac{i}{\hbar}\mathcal{S}'(t_f,t_i)}\,. \tag{6}$$

The action $\mathcal{S}'$ is, by construction, the classical action of the corresponding interaction scheme, with lagrangian $\mathcal{L}' = \mathcal{P}'x - \mathcal{H}'$. This result has a very intuitive justification in the fact that in a cyclic geometry such as that associated to the topology $\mathbb{S}^1$, the classical evolution of $\phi(x)$ from an initial configuration to a final configuration is given by the interference of all the possible classical paths with different windings numbers; without relaxing the classical variational principle. Thus the harmonics of the vibrating string/field $\phi(x)$ are interpreted as quantum excitations. We have also proven that this classical to quantum correspondence pinpoints a fundamental aspect of Maldacena's duality [2]. Therefore, the assumption of intrinsic periodicity which has characterized quantum theory in its early days has a renewed interest in modern physics. This correspondence to relativistic QM can be interpreted in analogy with 't Hooft determinism: "there is a deep relationship between a particle moving very fast in a circle [of periodicity T_t] and a quantum harmonic oscillator [with the same periodicity]" [5]; or the stroboscopic quantization [6]). QM emerges as a statistical description associated to the extremely fast recurrences of the elementary systems in nature: this is the meaning of the Hilbert notation above. Our actual experimental time resolution, about $\Delta T_{exp} \sim 10^{-17}$, is too low to resolve the small time scales of ordinary quantum systems, though the internal clock of an electron has been observed in a recent interference experiment [7]. As for a dice rolling too fast with respect to our resolution in time, the outcomes can be described only in a statistical way. Loosely speaking, an observer with infinite resolution in time can in principle resolve exactly the underling deterministic cyclic dynamics: it would have no fun playing dice ("God doesn't play dice", Einstein).

It is instructive to interpret the black-body radiation in terms of "periodic phenomena". In this case it is natural to assume that the population a_n of the nth energy levels is fixed by the Boltzmann distribution. For the IR components of the radiation, i.e. massless periodic phenomena with long periodicity, the PBCs can be neglected and the energy spectrum can be approximated to a continuum (classical limit) since many energy levels are populated (the thermal noise, i.e. the euclidean time periodicity, destroys the intrinsic periodicity in a sort of decoherence). For the UV components, however, the fundamental energy $\bar{E}$ is big with respect to the thermal energy so that only a few energy levels can be populated. That is to say, these modes have very short periodicity (not destroyed by the thermal noise), so that the PBCs are important and there is a manifest quantization of the energy spectrum (quantum limit). The UV catastrophe is avoided according to Planck.

We have a consistent interpretation of quantum to classical transition as $\hbar \to 0$. For massive particles, in the non-relativistic limit $\bar{p} \ll \bar{M}c$, only the fundamental energy level is largely populated as the gap between the energy levels goes to infinity, $\bar{M} \to \infty$. The cyclic field can therefore be approximated as $\Phi(x) \sim \exp[-i\frac{\bar{M}c^2}{\hbar}t + i\frac{\bar{M}}{\hbar}\frac{\mathbf{x}^2}{2t}]$, see [1]. Neglecting the proper tim recurrence, it is possible to see by plotting the $||^2$ that the wavefunction of a massive periodic phenomenon is centered along the path of the corresponding classical particle and its width is of the order of the Compton wavelength λ_s. Thus, in the non-relativistic limit, the Dirac delta distribution describing a classical particle is reproduced as in the usual Feynman description. Similarly, the spatial compactification length tends to infinity whereas the time compactification tends to zero, so that in the non-relativistic limit we have a point-like distribution in $\mathbb{R}^3$, i.e. the ordinary three-dimensional description of a classical particle. Indeed, the corpuscular description therefore arises at high frequencies. On the other hand, in the relativistic limit the non-local nature of a massive periodic phenomena can not be neglected (the distribution width is of the order of the Compton wavelength). This gives an intuitive interpretation of the wave-particle duality and of the double slit experiment. In particular, if probed with high energy or observed with good resolution, more and more energy levels turn out to be excited, i.e. more and more harmonics can be resolved. In this way we can figure out that the energy excitations play the role of the quantum excitations of the same fundamental elementary system, so that we have an analogy with the *virtual* particles of ordinary relativistic QM. The modes can have in general with either positive or negative frequencies: this means that the relativistic theory has Hamiltonia operator positively defined (contrarily to 't Hooft model).

We conclude this section by noting that the assumption of intrinsic periodicity implicitly contains the Heisenberg uncertain relation of ordinary QM. Briefly, in the Hilbert notation (statistical description of a "periodic phenomenon") the phase of the periodicity can not be determined — in (4) only the square of the field has physical meaning . To determine the frequency and thus the energy $\bar{E}(\bar{\mathbf{p}}) = \hbar\bar{\omega}(\bar{\mathbf{p}})$ with good accuracy $\Delta\bar{E}(\bar{\mathbf{p}})$ we must count a large number of cycles. That is to say we must observe the system for a long time $\Delta t(\bar{\mathbf{p}})$. The phase is defined modulo factors π which can be written as a energy or a temporal uncertainty $\exp[-i\frac{\bar{E}t}{\hbar}+\pi] = \exp[-i\frac{(\bar{E}+\Delta\bar{E})t}{\hbar}] = \exp[-i\frac{\bar{E}(t+\Delta t)}{\hbar}]$. Considering that every cycles is such that $t < T_t$ (for time longer than a cycle we must consider the general phase invariance $n\pi$) we find that this uncertain is described by the famous Heisenberg relation $\Delta\bar{E}(\bar{\mathbf{p}})\Delta t(\bar{\mathbf{p}}) \gtrsim \hbar/2$, [1].

5. Schrödinger problems

We continue with the historical overview of our assumption of intrinsic periodicity — aspects concerning modern physics are discussed in [1, 2, 3]. In particular we what to show how to solve simple non-relativistic quantum problems in terns of the vibrational modes associated to the quantum recurrences[7].

First of all we must consider that the assumption of intrinsic periodicity is equivalent to the Bohr-Sommerfeld condition: it in fact prescribes that the only possible modes are those with a finite number of cycles as in a vibrating 'string'. For instance, the assumption of intrinsic periodicity in the free case (homogeneous periodicity) leads to the harmonic spectrum $p_{n\mu}T^\mu = n\bar{p}_\mu T^\mu = nh$. More in general, as shown in [2], in case of interaction the quantized spectrum associated with the corresponding modulation of periodicity is given by the condition $\oint p_{n\mu}(x)dx^\mu = h(n+v)$ (the so-called Morse index v interpreted as a twist factor in the PBCs, see also [1, 2, 9]). This can be seen by imposing periodicity to the modulated solution of the type (3); note also the correspondences with the WKB method. To obtain the the Bohr-Sommerfeld we quantized only the spatial momentum $\mathbf{p}_n(\mathbf{x})$; the corresponding energy spectrum is obtained through the equations of motion of the interaction. The quantization condition is therefore

$$\oint \mathbf{p}_n(\mathbf{x})d\mathbf{x} = h(n+v).$$

To this spatial momentum we associate a potential, which in the non-relativistic limit can be defined such that $E_n = \frac{\mathbf{p}_n^2(\mathbf{x})}{2m} + V(\mathbf{x})$. The Hamiltonian operator is so that the Schrödinger equation (5), mode by mode, turns out to be given by

$$-\frac{\hbar^2}{2m}\partial^2_{\mathbf{x}}\phi_n(\mathbf{x}) + V(\mathbf{x})\phi_n(\mathbf{x}) = E_n\phi_n(\mathbf{x}) \;. \tag{7}$$

Taking into account all previous arguments, the quantized harmonic oscillator turns out to have an immediate solution[8] with respect to the usual formulation. It is sufficient to consider the (Galileo) isochronism of the pendulum: every orbit of an harmonic oscillator has the same time (homogeneous) periodicity. We have just to impose that the solution is a wave with the characteristic periodicity $T_t = 2\pi/\bar{\omega}$ for every energy level. That is the quantization condition of the energy spectrum is simply $\oint E_{n\mu}dt = E_n T_t = h(n+v)$. Therefore the solution has an harmonic energy spectrum $E_n = \frac{h}{T_t}(n+\frac{1}{2})$ and can be decomposed as

$$\Phi(\mathbf{x},t) = \sum_n \phi_n(\mathbf{x},t) = \sum_n e^{i(n+1/2)\bar{\omega}t}\phi_n(\mathbf{x}) \;. \tag{8}$$

where we have assumed a twist of half a period (anti-periodicity) in the BCs in order to reproduce the Morse factor $v = \frac{1}{2}$.[9] The Schrödinger differential equations of the problem can be written

[7] More in general, by quantum recurrence we mean that quantum system can be described semi-classically by imposing appropriate BCs to relativistic wave equations. Trivial examples are the quantization of a particle in a box, a particle with Dirac delta potentials, and similar problems. However this approach can be extended to non-obvious problems such as the Casimir effect [8], the tunnel effect or atomic physics [9]

[8] This avoides to the complication of the Fock space. In this approach the creation and annihilation operators can be regarded as describing the excitation or de-excitation of the different levels of the harmonic system.

[9] Intuitively the assumption of intrinsic antiperiodicity is natural for fermion in order to reproduce the correct spin statistics

by defining $\mathbf{x} = \rho\mathbf{y}$ with $\rho = \sqrt{\frac{\hbar}{m\bar{\omega}}}$ and $h_n(\mathbf{y}) = \phi_n(\mathbf{y})e^{\frac{\mathbf{y}^2}{2}}$. Thus the solution of the n-th harmonic mode of the vibration associated to the harmonic potential is

$$\partial^2_{\mathbf{y}} h_n(\mathbf{y}) - 2\mathbf{y}\partial_{\mathbf{y}} h_n(\mathbf{y}) = \left[2\left(n + \frac{1}{2}\right) - 1\right] h_n(\mathbf{y}) \; . \tag{9}$$

We have solved the quantum harmonic oscillator by simply assuming intrinsic periodicity. This example is extremely important since the harmonic oscillator is the basic ingredient of ordinary (bosonic) quantum field theory. That is, a second quantized Klein Gordon field is the integral over the quantum harmonic oscillators with all the possible periodicities, from zero to infinity. The creation and annihilation operators in this approach describe the excitation of the different n-th mode of this harmonic 'string'. This suggests that our correspondence to QM can be fully extended to ordinary quantum field theory.

In a similar way it is easy to solve other Schrödinger problems. For instance here we report the energy spectrum of simple one-dimensional potentials. This method applied to the linear potential $V(x) = mgx$ actually gives the correct quantized energy spectrum

$$E_n = \frac{1}{2}[3\pi(n + 1/4)]^{2/3}(\hbar^2 m g^2)^{1/3}$$

where we have assumed $v = 1/4$. Another example is the solution of the quantum anharmonic oscillator, i.e. we assume a perturbation $\epsilon x^4/l^4$ of the harmonics potential, with $l = \sqrt{\hbar/m\omega}$. In this way we obtain (modulo Morse factors) the deviation for the harmonic energy spectrum $E_n = \hbar\bar{\omega}n$ of the same quantity prescribed by the ordinary methods:

$$\Delta E_n = \epsilon\frac{3}{2}(n^2 + n).$$

In this way it is easy to see that the vibrational modes associated to these one-dimensional potentials reproduce the correct solutions of the Schrödinger problems.

From a historical point of view this approach to quantum problems is the same adopted by Bohr to solve the energy spectrum of the hydrogen atom. Indeed, according to the above prescription, the harmonic modes of our 'string' with the modulated periodicities allowed by the Coulomb potential yields the resulting energy spectrum (modulo Morse factors)

$$E_n = -\frac{13.6\text{eV}}{n^2}.$$

This famous results is obtained without necessarily requiring circular orbits. The single space-time periodicity of topology $\mathbb{S}^1$ associated to the orbits describes correctly the principal quantum number n. Furthermore, in spherical problems we must also consider the quantization associated to the spherical periodicity, i.e. the quantized vibrational modes associated to the additional periodicities in the spherical angles $\theta \in [0, 2\pi)$ and $\psi \in [0, \pi)$. As known from ordinary QM these lead to the ordinary quantization of angular momentum. In fact, the spherical vibrational modes of solution are described in terms of the two additional angular and magnetic quantum numbers $\{m, l\}$. In other words, the atomic orbitals can be represented as the vibrational modes of an harmonic system of topology $\mathbb{S}^1 \otimes \mathbb{S}^2$ in a Coulomb potential. This harmonic description of atoms has been recently re-proposed by eminent scientists such as the Nobel laureate Prof. F. Wilczek. It has been successfully generalized to achieve classical descriptions of quantum properties of atoms with more electrons (e.g. Helium atom), the Zeeman effect [9]; these were considered the major limitation of the approach based on the assumption of intrinsic periodicity.

The example of the Hydrogen atom provides us the last element of the formal correspondence to QM: different harmonic sets characterized by different fundamental periodicity are described by the tensor products of their different Hilbert spaces. For instance, in the Hilbert notation, the orbitals $\phi_{n,l,m}$, composition of the harmonics associated to the space-time vibration $|\phi_n\rangle$, or more simply $|n\rangle$, and the spherical harmonics $|l, m\rangle$, can be written in a Hilbert space with base $|n\rangle \otimes |l, m\rangle = |n, l, m\rangle$ associated to the topology $\mathbb{S}^1 \otimes \mathbb{S}^2$. This property plays an important role in the demonstration of the Bells' theorem. In particular, as it will be shown explicitly in a dedicated paper, the formal correspondence of our approach with canonical QM strongly suggests the statistical description of elementary space-time cycles satisfies the same inequalities of ordinary QM. In particular it must be noted that the theory has not local-hidden-variables; the assumption of intrinsic periodicity is a element of non-locality (though this type non-locality is consistent with relativity, as the periodicity varies according to the retarded potentials).

6. Conclusions

We have shown how ordinary relativistic QM can be formally derived from an effective statistical description of the classical mechanics associated to the space-time recurrences of elementary systems [1, 2, 3]. In this paper we have given some historical motivations of this approach. It has its origin in Bohr's description of hydrogen atom, Bohr-Sommerfeld quantization, de Broglie's wave particle-duality, Schrödinger's *zitterbewegung*, and in other foundational contributions to modern QM (not mentioned here for brevity). These ideas are the base of modern quantum field theory, but the attempt of a semi-classical description of QM has been abandoned after Heisenberg and Bell. Nevertheless, we have seen that these limitations can be avoided by thinking in terms of elementary cycles. The formal correspondence to ordinary QM obtained by imposing intrinsic periodicity to elementary particles strongly suggests that relativistic QM can be interpreted as an effective (statistical) description of semi-classical mechanics associated to the fast cyclic dynamics of elementary particles, as also suggested for instance by 't Hooft — and not the contrary. We have proven that fundamental aspects of QM can be consistently described in terms of the harmonic modes of four-dimensional vibrating 'strings'. The resulting theory is a relativistic generalization of the theory of sound (where sources can vibrate along the time dimension). This aspect anticipates the study of QM to Pythagoras, who first investigated the "quantized" spectrum of harmonic system, as also recently suggested by Wilczek. The resulting theory conciliates together important aspects of modern physics: it can be regarded as a (purely 4D) string theory (defined on the single compact word-parameter s) similarly to original Veneziano's proposal; it provides a geometrodynamical description of gauge interaction, similarly to original Weyl's proposal; it pinpoints fundamental aspects (e.g. classical to quantum correspondence) of Maldacena's duality, originally pointed out by Witten; it justifies the mathematical beauty of extra-dimensional theories as original proposed by Nördstrom, Kaluza and Klein; and so on. We conclude that physical origin (the "missing link" [10]) of QM must be found in the assumption of intrinsic periodicity, as originally proposed by the father's of QM and then forgotten. The resulting formulation of physics in terms of elementary space-time cycles brings important new elements to face the open problems of modern physics [1, 2, 3].

References

[1] Dolce D 2011 *Found. Phys.* **41** 178 (*Preprint* 0903.3680v5)
[2] Dolce D 2012 *Ann. Phys.* **327** 1562 (*Preprint* 1110.0315)
[3] Dolce D 2012 *Ann. Phys.* **327** 2354 (*Preprint* 1110.0316)
[4] Broglie L d 1924 *Phil. Mag.* **47** 446
[5] 't Hooft G 2003 *Int. J. Theor. Phys.* **42** 355 (*Preprint* hep-th/0104080)
[6] Elze H T 2004 *Lect. Notes Phys.* **633** 196 (*Preprint* gr-qc/0307014)
[7] Catillon P *et al.* 2008 *Found. Phys.* **38** 659
[8] Jaffe R L 2005 *Phys. Rev.* D **72** 021301 (*Preprint* hep-th/0503158)
[9] Solov'ev E 2011 *Eur. Phys. J.* D **65** 331–351
[10] Ferber R. 1996 *Found. Phys. Lett.* **9** 6, 575

Low temperature motion of hydrogen on metal surfaces signals breakdown of quantum mechanics in 3+1 dimensions

D Drakova and G Doyen

Faculty of Chemistry, University of Sofia, Bulgaria
Ludwig-Maximilians Universität, München, Germany

E-mail: nhdd@chem.uni-sofia.bg

Abstract. The low temperature motion of hydrogen on solid metal surfaces displays some unexplained experimental features: in the quantum diffusion regime more than nine orders of magnitude difference between the diffusion rates on different metal surfaces have been measured, the lowest diffusion rates being established in the low temperature scanning tunnelling microscope. Furthermore telegraph-signal-like adsorption site change, rather than Rabi oscillations predicted by Schrödinger equation in 3+1 dimensions, is observed, signaling the breakdown of quantum mechanics in 3+1 dimensions. A theory is presented to resolve these problems, involving the entanglement of the adsorbate motion to gravitons in high-dimensional spacetime. Soft local massive gravonons, induced in the presence of the adsorbate, determine the time scale for surface diffusion. The $\gamma\eta$-model is used for the evaluation of the soft gravonon modes. Weak and local entanglement of the adsorbate motion with a nearly degenerate graviton continuum of high density of states are the conditions for the telegraph-signal-like time development of adsorption site change. In contrast to the Copenhagen interpretation of quantum mechanics, this apparent "classical" behaviour of the adsorbate in 3+1 dimensional spacetime is the result of the solution of Schrödinger's time dependent equation in high-dimensional spacetime.

1. Introduction

The low-temperature regime of hydrogen and deuterium diffusion on metal surfaces is often discussed as being due to adsorbate tunnelling, both temperature independent or thermally-assisted. In a logarithmic plot the diffusion rates show regions of thermally activated, overbarrier diffusion where the slope of the curve allows to evaluate the barrier height for diffusion using Arrhenius relation. In the low temperature region, below the temperature for transition from overbarrier to tunnelling diffusion T_{trans}, the diffusion rate does not depend on temperature and quantum diffusion of the adsorbate is the assumed mechanism. The quantum diffusion rates vary over many orders of magnitude depending on the adsorption system and the experimental technique used. As it is illustrated in fig. 1 (left panel) with the examples of ^{1}H, ^{2}H and ^{3}H on W(110) [1] and ^{1}H and ^{2}H on Cu(001) [2], the hopping rate ν_o in the quantum regime varies over 6 orders of magnitude, the lowest hopping rate being measured in the low temperature scanning tunnelling microscope (STM) [2], higher values being due to the field-emission microscope (FEM) experiments of Gomer et al. [1], [3]-[6]. Even higher values result from the optical diffraction experiments performed by Zhu et al. on adsorbate gratings [7]-[10],

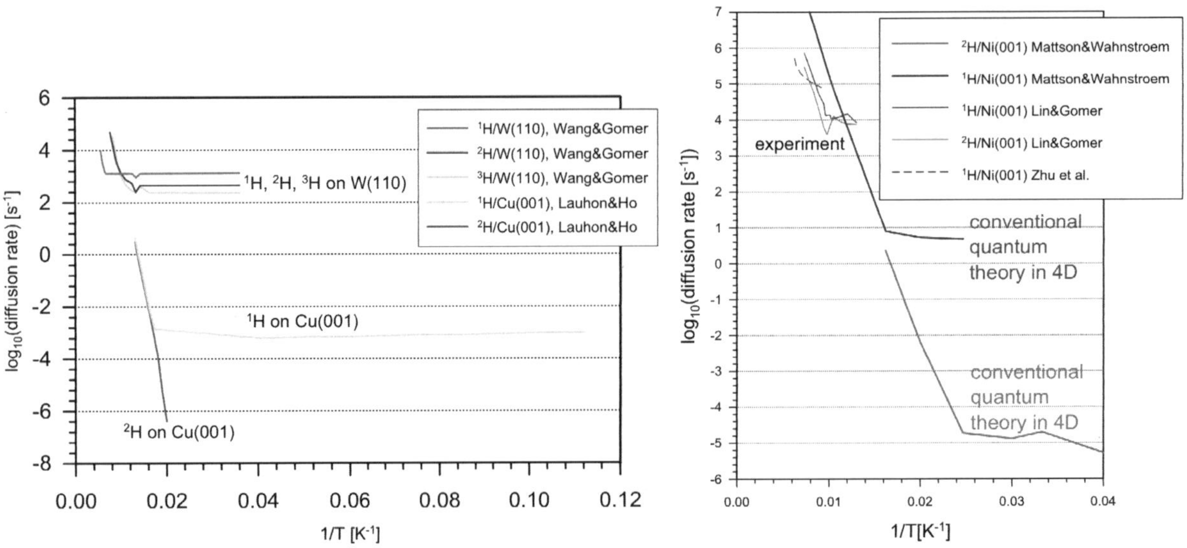

Figure 1. Temperature dependence of the diffusion rate ν_o of ^{1}H, ^{2}H and ^{3}H on W(110) and Cu(001) (Wang, Gomer [1] and Lauhon, Ho [2]) (left panel). Temperature dependence of the diffusion rate ν_o of hydrogen and deuterium on Ni(001) in a field emission microscope (FEM) measurement by Lin and Gomer [3] and for hydrogen on Ni(001) in an optical diffraction measurement on adsorbate gratings by Zhu et al. [9]. The numerical results for ^{1}H and ^{2}H diffusion on Ni(001) are shown in the right panel [11].

where high electric field and laser radiation are used in the measurements (cf. fig. 1, right panel).

Using the idea of tunnelling through a static barrier we can estimate the diffusion rate in the quantum regime. Diffusion barrier height for ^{3}H/W(110) of 208 meV is reported in ref. [1], a lower value of 197 meV for the diffusion barrier height of ^{1}H on Cu(100) is estimated in the low temperature STM experiment [2]. The mass of tritium is larger than that of hydrogen, therefore if tunnelling through a static barrier were the diffusion mechanism we would expect tritium to diffuse slower than hydrogen. However, in experiment the diffusion rate of ^{3}H on W(110) is nearly six orders of magnitude faster than that of ^{1}H on Cu(100). In a more sophisticated treatment of hydrogen and deuterium quantum diffusion on Ni(001), based on conventional quantum mechanics in 3+1 dimensional spacetime, the reproduction of the experimentally established trends fails as well (cf. the plots $\log_{10}\nu_o$ in fig. 1, right panel, due to Mattson and Wahnström [11]). Roughly nine orders of magnitude is the discrepancy between the experimental and the theoretical values of the diffusion rates in the quantum tunnelling regime. We have to conclude that tunnelling through static barriers cannot be the diffusion mechanism of hydrogen in the quantum regime.

An even more intricate feature of hydrogen motion on solid surfaces at low temperature is that it occurs as telegraph-signal-like jumps between different sites. In a low temperature STM experiment (T=5 K) on the tautomerization reaction of two imino hydrogen atoms in the central cavity of the molecule naphthalocyanine, adsorbed on a thin insulating bilayer of RbI on a Cu single crystal surface, the site change of the hydrogen atoms appears as telegraph-signal-like changes of the tunnelling current over a time interval of 20 seconds [12]. The telegraph-like site change appears very different from the Rabi oscillations between two binding sites, as predicted by Schrödinger's time dependent equation of an isolated system in 4 dimensional spacetime.

Why is hydrogen diffusion in the low temperature STM on Cu(100) so slow? Changes in the dynamics of a local system may be due to entanglement to its environment. Permanent "measurement" of a quantum particle by the environment is the decoherence mechanism of localization proposed within the decoherence theory by Joos and Zeh [13]. It is tempting to associate the difference between the hopping rates in quantum diffusion of six orders of magnitude with decoherence and localization of the adsorbate due to entanglement with the phonons, tomonagons, photons, the rest gas in the experimental equipment, etc.. However, the evaluated time for the decoherence of the correlated motion of small molecules is too long because the effective cross section for scattering is proportional to the squared extension of the molecules [13]. Furthermore this idea does not work at low experimental temperatures because the environmental excitations, needed to decohere an atom or a small molecule, should have very small wavelength, hence very high energy, to be able to localize a hydrogen atom on a site with atomic dimensions. At low temperature $T \to 0$ K both phonons and tomonagons have long wavelengths of the order of 100-200 Å. Therefore phonons and tomonagons at low T cannot couple to the motion of a hydrogen atom on a lengthscale of 2-3 Å.

We suggest **entanglement to gravitons** in high spacetime dimensions as a mechanism for telegraph-signal-like changes of the state of a quantum system, as it appears in 3+1 dimensional spacetime structure. In the next section we present the idea and a summary of a formalism for the treatment of the gravitational field within quantum field theory (QFT) and general relativity (GR) in the limit of weak gravity and linearization of Einstein's field equations [14]. In the presence of the adsorbate soft local graviton modes are created, termed gravonons. The formalism leads to a nonrelativistic Schrödinger equation for the gravonons. A model with two parameters γ and η, used to evaluate the soft modes, is briefly outlined, followed by the presentation of the results and their interpretation.

2. Gravitons in high dimensional spacetime

In 10+1 dimensional spacetime (10 spatial dimensions) the gravitational potential has the form:

$$V^{(11)}_{grav}(r) = -\frac{G^{(11)}M}{\pi^7 r^8}. \tag{1}$$

Up to distances of 0.05 mm = 5×10^{-5}m $\approx 10^6$ bohr the classical gravitational law of Newton has been proved to be valid, therefore at this distance both laws have to yield the same value:

$$-\frac{G}{10^6} = -\frac{1}{\pi^7}\frac{G^{(11)}}{10^{48}} \to G^{(11)} = 10^{39}G \tag{2}$$

This means that at distance of 1 bohr the 10+1 dimensional gravitational interaction is 10^{39} times as strong as the classical interaction in 3+1 dimensional spacetime and it is very localized. The additional seven spatial dimensions are compactified to avoid discrepancy with the classical law, an estimate of the dimension of the compactification we need to explain many experimental observations being of the order of a = 10000 bohr [15, 16]. Arkani-Hamed et al. [17]-[21] were the first to claim that the hidden dimensions need not be compactified on the Planck scale. The hidden dimensions can be as large as a fraction of a millimeter or even infinite [21]. No experimental observation comes into conflict with this claim.

We use linearized weak field gravity as described within QFT (see e.g. [22]) and apply the usual procedure for deriving nonrelatistic quantum mechanics from QFT. In QFT the interaction Lagrangian includes terms due to the interaction between the gravitons and all matter fields (electromagnetic field, massive scalar boson field - the adsorbate, massless boson fields - tomonagons etc.).

2.1. Emerging nonrelativistic matter particles

The adsorbate is treated as a massive boson field. Since the kinetic energy ϵ of the moving adsorbate is much smaller than its mass m the relativistic energy E is written as ($c = 1$):

$$E = m + \epsilon. \tag{3}$$

A matter boson field ψ varying slowly with time can then be defined as:

$$\phi(x) = e^{-imt}\psi(x). \tag{4}$$

Starting from the Klein-Gordon equation, Schrödinger's equation for $\psi(x)$ for a free adsorbate is obtained:

$$i\frac{\partial}{\partial t}\psi = -\frac{\nabla^2}{2m}\psi \tag{5}$$

with $\nabla = (\frac{\partial}{\partial x^1}, \frac{\partial}{\partial x^2}, ..., \frac{\partial}{\partial x^N})$. More details on the derivations will be presented in a forthcoming paper [14].

2.2. Emerging massive gravitons (gravonons)

The field of the gravitons is treated in a similar way. As diffusing particles of molecular size couple only to short wavelength gravitons, we write the relativistic energy E of a graviton as:

$$E = kc + \varepsilon \tag{6}$$

where ε is a tiny modification, induced by the adsorbate, $k =\mid \vec{k} \mid$ and $\vec{k}$ labeling the Fourier component of the graviton field $h_{\mu\nu}$. The value of k is fixed by the condition that the wavelength should be of the order of the diffusion distance between two adsorption sites. For adsorbate entanglement to gravitons we need wavelength of the gravitons $\lambda \approx 1$ bohr, i.e. $k \approx 10$ bohr^{-1}, energy $E \approx 10^3$ Hartree [15, 16]. ε characterizes the energy interval of the gravitons which couple to the scalar boson field ϕ. It is of the order of 10^{-12} Hartree. This motivates us to write for the gravitational field:

$$h_{\mu\nu} = e^{-ikct}\zeta_{\mu\nu}(x) \tag{7}$$

leading to a Schrödinger equation for the field quanta $\zeta_{\mu\nu}$, termed gravonons:

$$i\frac{\partial}{\partial t}\zeta_{\mu\nu} = (-\frac{\nabla^2}{2m_g} + V_o)\zeta_{\mu\nu} \tag{8}$$

with $m_g = \frac{k}{c}$ the mass of the gravonon and $V_o = -\frac{kc}{2}$ a constant potential cancelling the zero point energy $\omega_o = \frac{1}{2}kc$ of the light cone graviton $h_{\mu\nu}(\vec{k})$. Thus massive free gravonons have emerged resembling a quantum particle. With $k = 10$ bohr^{-1} the mass m_g of the gravonon is roughly a tenth of the mass of an electron. The derivation of eq. (8) will be presented in a future publication [14].

The massive local gravonons can be mathematically treated as in standard quantum mechanics. An interaction potential can be added to eq. 8, for instance a harmonic potential due to springs. The upper panel in fig. 2 illustrates this situation for the local gravonons induced by the metal atoms in the absence of the adsorbate. In the presence of the adsorbate, soft gravonon modes

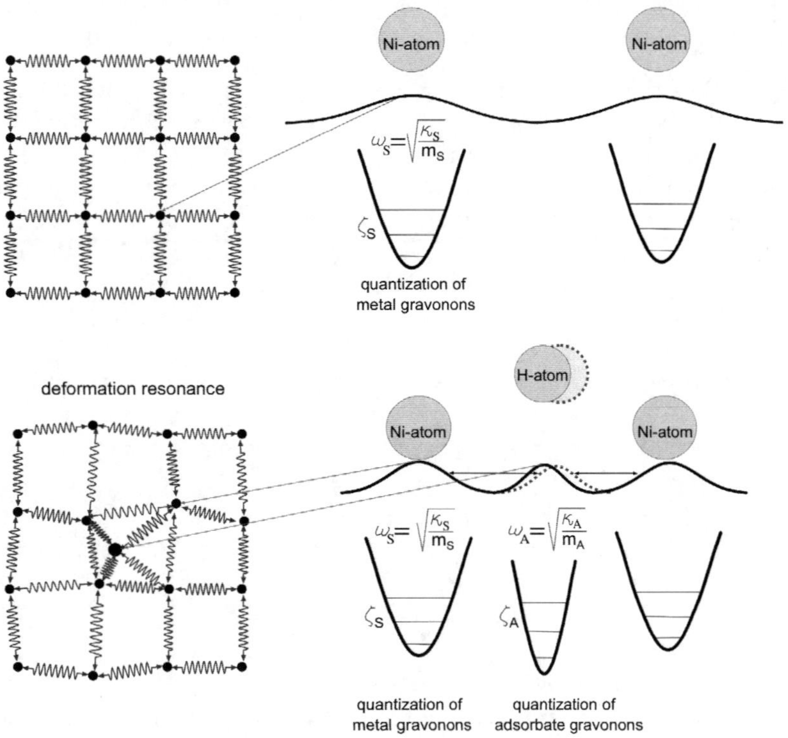

Figure 2. The upper panel illustrates the local gravonon structure induced by metal atoms in a clean metal surface approximated by harmonic potentials. In the presence of the adsorbate the local gravonon structure changes, giving rise to soft gravonon modes (lower panel). The coherence of the adsorbate - gravonon motion is retained within a small region, the so called deformation resonance (drawing on the l.h.s. in the lower panel). The wavy curves in both panels symbolize the local spacetime deformations due to the presence of the atoms, described by excitation of local gravonons.

are induced, which are entangled with the adsorbate motion, determining the time scale of this motion. Adsorbate-induced soft gravonons of lower frequency determine slower adsorbate movement on the metal surface since within the gravonon lifetime the adsorbate cannot change the adsorption site. The adsorbate-induced effects are confined to a small region, called a deformation resonance, and affect some 5-15 metal atoms. The deformation resonance is defined as consisting of the matter particles and the deformation of spacetime due to the adsorbate and the metal atoms and it is schematically represented by the drawing on the l.h.s. in the lower panel of fig. 2. Within the deformation resonance the adsorbate motion is coherently entangled with the gravonons.

The effective Hamiltonian of the total system in D spatial dimensions reads [14]:

$$H_{eff} = \int d^D x \left[-\frac{\nabla^2}{2m}\psi^+\psi - \frac{\nabla^2}{2m_g}\zeta_{00}^+\zeta_{00} + V_o\zeta_{00}^+\zeta_{00} \right] - L_{int} \tag{9}$$

with the following Lagrangian L_{int} of the matter - gravonon interaction:

$$L_{int} = \int \mathrm{d}^D x \left[\frac{G^{(D)} mM}{r^{D-2}} \psi^+ \psi + 2m \left(\frac{G^{(D)} M}{r^{D-2}} \right)^2 \zeta_{00}^+ \zeta_{00} \psi^+ \psi \right] \tag{10}$$

The first term in eq. (10) is the Schrödinger-Newton term and it corresponds to the $\frac{1}{r^8}$ power law in 10+1 dimensional spacetime. The second term describes the interaction between the adsorbate field and the gravonons (m: adsorbate mass, M: metal atom mass). Being a second order term it satisfies a $\frac{1}{r^{16}}$ power law in 10 spatial dimensions [14]. This is the quantum field theoretical Hamiltonian which is solved in a many-body basis [16]. The many-body basis configurations involved have components, describing the states of the gravonon continuum and the adatom diffusion states or states within the deformation resonance, where the local coupling to gravonons is effective.

The time dependent Schrödinger equation for the world wave function $\Psi(t)$ is solved in the many boby basis basis:

$$\mathrm{i}\hbar \frac{\partial \Psi(t)}{\partial t} = H_{eff} \Psi(t). \tag{11}$$

$\Psi(t)$ describes a wave packet consisting of the adsorbate particle and gravonons excited out of the vacuum state. The time development of the squared projection of $\Psi(t)$ on the many-particle basis states, confined to a particular adsorption site, is evaluated to provide data on the diffusive movement of the adsorbate on the solid surface.

2.3. Model for the non-interacting gravonon structure

The gravonon field in the second and third term of the effective Hamiltonian eq. (9) is a free field in a constant potential. The adsorbate motion, which we want to investigate, occurs, however, in the presence of other masses, which have already induced a structure in the gravonon field. This paragraph describes, how the pre-existing gravonon structure is modelled.

The field $\zeta_{00}(\vec{x})$ is expanded in functions centered and localized near the solid atoms:

$$\zeta_{00}(\vec{x}) = \sum_i q_i g(\vec{x} - \vec{x}_i) \tag{12}$$

$g(\vec{x} - \vec{x}_i)$ is a gaussian centered at $\vec{x}_i$. $\vec{x}_i$ is the D-dimensional position of the solid atom with $x_4 = \ldots = x_D = 0$. q_i are expansion coefficients, which become operators upon quantization. $g(\vec{x} - \vec{x}_i)$ is normalized such that:

$$\int \mathrm{d}^D x \, g^2(\vec{x} - \vec{x}_i) = 1, \tag{13}$$

i.e. $g^2(\vec{x} - \vec{x}_i)$ tends to $\delta(\vec{x} - \vec{x}_i)$ for strong localization.

The product of the gravonon field becomes then

$$\zeta_{00}^+(\vec{x})\zeta_{00}(\vec{x}) = \sum_{i,j} q_i q_j g(\vec{x} - \vec{x}_i) g(\vec{x} - \vec{x}_j). \tag{14}$$

The gaussians are assumed to be δ-function-like so that

$$\zeta_{00}^+(\vec{x})\zeta_{00}(\vec{x}) \rightarrow \sum_i q_i^2 \delta(\vec{x} - \vec{x}_i). \tag{15}$$

The gravonon integrand in the Hamiltonian without any adsorbate induced gravonon-gravonon interaction is then transformed as follows:

$$-\frac{\nabla^2}{2m_g}\zeta_{00}^+\zeta_{00} + V_o\zeta_{00}^+\zeta_{00} \rightarrow \frac{1}{2}\omega_A^2 q_A^2 \delta(\vec{x}-\vec{x}_A) + \sum_i \frac{1}{2}\omega_S^2 q_i^2 \delta(\vec{x}-\vec{x}_i) \tag{16}$$

where ω_A and ω_S are the frequencies of the non-interacting adsorbate induced and metal atom induced gravonons. Performing the integration $\int \mathrm{d}^D x$ yields the Hamiltonian of the non-interacting gravonons:

$$\int \mathrm{d}^D x(-\frac{\nabla^2}{2m_g}\zeta_{00}^+\zeta_{00} + V_o\zeta_{00}^+\zeta_{00}) \rightarrow H_A + H_S = \frac{1}{2}\omega_A^2 q_A^2 + \sum_i \frac{1}{2}\omega_S^2 q_i^2. \tag{17}$$

2.4. Adsorbate-induced gravonon structure
Now we add a component to $\zeta_{00}(\vec{x})$ to describe the influence of the adsorbate on the gravonon structure:

$$\zeta_{00}(\vec{x}) = \sum_i \left[q_i g(\vec{x}-\vec{x}_i) + q_i f(\vec{x},\vec{x}_i)\right]. \tag{18}$$

$f(\vec{x},\vec{x}_i)$ is delocalized so that it has amplitude on all defined sites $\vec{x}_i$. It is normalized such that

$$\int \mathrm{d}^D x \; f^2(\vec{x},\vec{x}_i) = 1. \tag{19}$$

In this way:

$$\zeta_{00}^+(\vec{x})\zeta_{00}(\vec{x}) = \sum_i \left[q_i^2\delta(\vec{x}-\vec{x}_i) + q_i^2 f^2(\vec{x},\vec{x}_i)\right] + \sum_{i,j} q_i q_j g(\vec{x}-\vec{x}_i) f(\vec{x},\vec{x}_j). \tag{20}$$

This has to be inserted in eq. (10) and the integration has to be performed.

In order to evaluate the gravonon structure we replace the adsorbate field $\psi^+\psi$ by its expectation value $\langle\psi^+\psi\rangle$. Among the sites we distinguish a particular one q_S as the one which is nearest to the adsorbate site q_A. The interaction terms involving the function $f(\vec{x},\vec{x}_i)$ are then:

$$\begin{aligned} V_{A-S} &= \int \mathrm{d}^D x \; J(\vec{x}) q_S^2 f^2(\vec{x},\vec{x}_S) + \sum_{m \neq S} \int \mathrm{d}^D x \; J(\vec{x}) q_m^2 f^2(\vec{x},\vec{x}_m) \\ &+ \int \mathrm{d}^D x \; J(\vec{x}) q_S g(\vec{x}-\vec{x}_S) \sum_{m \neq S} q_m f(\vec{x},\vec{x}_m) \\ &+ \int \mathrm{d}^D x \; J(\vec{x}) \sum_{m,n \neq S} q_m q_n g(\vec{x}-\vec{x}_m) f(\vec{x},\vec{x}_n) \\ &+ \int \mathrm{d}^D x \; J(\vec{x}) q_S q_A g(\vec{x}-\vec{x}_S) f(\vec{x},\vec{x}_A) \end{aligned} \tag{21}$$

with

$$J(\vec{x}) = 2m \sum_i \frac{G^{(D)} M}{\mid \vec{x}-\vec{x}_i \mid^{D-2}} \langle\psi^+\psi\rangle. \tag{22}$$

Introducing two parameters γ and η, the coupling terms between the gravonons are parameterized as follows:

$$\int \mathrm{d}^D x \; J(\vec{x}) q_S^2 f^2(\vec{x},\vec{x}_S) = \frac{1}{2}\eta\gamma\omega_S^2 q_S^2 \tag{23}$$

$$\int \mathrm{d}^D x \; J(\vec{x}) \sum_{m \neq S} q_m^2 f^2(\vec{x}, \vec{x}_m) = -\frac{1-\eta}{N} \gamma \omega_S^2 \sum_{m \neq S} q_m^2 \quad \text{for } m = 1, \dots, N \tag{24}$$

$$\int \mathrm{d}^D x \; J(\vec{x}) q_S g(\vec{x} - \vec{x}_S) \sum_{m \neq S} q_m f(\vec{x}, \vec{x}_m) = \frac{1-\eta}{N} \gamma \omega_S^2 q_S \sum_{m \neq S} q_m \quad \text{for } m = 1, \dots, N \tag{25}$$

$$\int \mathrm{d}^D x \; J(\vec{x}) q_m q_n g(\vec{x} - \vec{x}_m) f(\vec{x}, \vec{x}_n) = 0 \quad \text{for } m \neq n \tag{26}$$

$$\int \mathrm{d}^D x \; J(\vec{x}) q_S q_A g(\vec{x} - \vec{x}_S) f(\vec{x}, \vec{x}_A) = -\gamma \omega_S^2 q_S q_A \tag{27}$$

With this the operator of the adsorbate induced gravonon interactions V_{A-S} assumes the form:

$$V_{A-S} = \frac{1}{2} \eta \gamma \omega_S^2 q_S^2 - \frac{1-\eta}{N} \gamma \omega_S^2 \sum_{m=1}^{N} q_m^2 + \frac{1-\eta}{N} \gamma \omega_S^2 q_S \sum_{m=1}^{N} q_m - \gamma \omega_S^2 q_S q_A \tag{28}$$

The adsorbate induced gravonon interactions V_{A-S} depends on the following assumptions:

- Mean field approximation with respect to $\psi^+ \psi$.
- $\psi^+ \psi$ describes only the adsorbate. All other matter fields are static.
- Gravonon structure is warped only at certain sites near the matter particles.

After the gravonon structure has been determined, the mean field approximation $\psi^+ \psi \to \langle \psi^+ \psi \rangle$ is no longer used. The determined gravonon structure is plugged into eqs. (9,10) and the time-dependent many-body problem is solved with this gravonon structure.

The gravonon coupling eq. (28) includes two parameters γ and η. γ is the coupling strength defined as:

$$\gamma = \frac{\omega_A^2}{\omega_S^2} \tag{29}$$

the ratio of the frequencies of gravonons induced by the adsorbate and metal atoms, respectively. Since the masses of the adsorbate and metal atom induced gravonons are the same, eq. (29) implies that γ is defined as the ratio of the spring constants. η is responsible for the formation of the soft gravonon mode. Smaller values of η correspond to enhanced weight of the many-body forces, signaling deviation from the spring model and coherence of the adsorabate-gravonon system within the deformation resonance, leading to the generation of soft local gravonons of lower frequency ω_{soft}. Hence, $1 - \eta$ correlates with the strength of many-body forces between substrate gravonons.

The soft mode in the $\gamma\eta$ - gravonon model is an adsorbate induced coherent motion within the deformation resonance involving typically 5 to 15 spacetime deformations which are assumed to be perfectly localized before adsorbate interaction sets in (cf. fig. 2, upper panel). As it is obvious from eq. (28) the smaller the value of η is, the more important the many-particle forces are. Hence, $1 - \eta$ is a measure whether the coherence of particle movement is of long range nature. For larger values of $1 - \eta$, i.e. smaller values of η, the coherent dynamics of the adsorbate-gravonon system is extended over a broader deformation resonance and results in soft gravonons of lower frequency ω_{soft}.

The local intrinsic gravonon coherence is retained unless it is distroyed by external effects. The deformation resonance is embedded in the rest of the surface and its induced gravonon structure and in external fields of the rest universum, which can be the origin of external disturbances, leading to destruction of the gravonon coherence within the deformation resonance. Since softer gravonons correspond to lower values of η, and then to lower values of the coupling strength γ, this implies that the soft gravonon mode is easier to destroy by external influences. Therefore the transition temperature T_{trans} from quantum to overbarrier diffusion must be lower if η is smaller.

This is similar to the effect of temperature increase in experiments with, e.g. scattering of noble gas beams from periodic solid surfaces, which is described by the Debye-Waller factor. In order to make contact with experiment we phenomenologically introduce a similar Debye-Waller-like factor $exp(-A\mathrm{T}_{trans})$ for $1-\eta$:

$$1-\eta = exp(-A\mathrm{T}_{trans}) \approx 1 - A\mathrm{T}_{trans} \tag{30}$$

which makes η proportional to T_{trans}, the temperature for transition from the coherent quantum tunnelling diffusion to statistical non-coherent overbarrier diffusion of adsorbates.

3. Results and discussion

3.1. Telegraph-signal-like change of adsorption site with time

The model of hydrogen diffusion on a solid surface has two adsorption sites and two warp resonances where the interaction with the gravonons is effective. The changes with time of the squared projection of the total world wave function eq. (11) on the initial state with the adsorbate on site 1 of the two-site model are plotted in fig. 3. The plot shows indeed that the adsorbate resides long time on the initial adsorption site before it jumps to a neighbouring one. *Telegraph-signal-like site changes* occur at time intervals, determined by the lifetime, i.e. the energy, of the *entangled gravonons of lowest frequency.* In the lifetime of the soft gravonon modes, which entangle to the adsorbate on a definite adsorption site, the particle cannot change site. This lifetime depends on the energy of the soft gravonon modes ω_{soft}. Therefore the modes of the gravonon subspace determine the diffusion time - low energy modes determine slow diffusion, high energy modes detemine fast diffusion. *The jump time is the inverse of the frequency of the softest gravonon mode.* Of course, the modifications of the gravonon spectrum, induced by the diffusing particle, depend upon the matter fields involved (adsorbate and metal atoms fields), their physical nature and their interaction.

The calculation whose results are shown in fig. 3 is aimed at reproducing the slow diffusion of hydrogen on Cu(100) from the low temperature STM experiments of Lauhon and Ho [2], i.e. a very slow quantum diffusion, slower than the diffusion in the experiments with FEM and optical diffraction (cf. the left panel in fig. 1).

Though the site change appears as a sudden collapse, this is a solution of the coherent time dependent Schrödinger equation, without collapse and state reduction. The physics of the telegraph signal and how the telegraph signal is generated is discussed in the papers by Doyen and Drakova refs. [23, 24].

3.2. Landscape of the soft gravonons

The landscape of the $\gamma\eta$ model is displayed in fig. 4 as a contour plot of the lines of *constant logarithm of the frequency of the softest gravonon mode,* $\log_{10}\omega_{soft}$, versus the coupling strength γ and the temperature for transition from quantum diffusion to statistical overbarrier diffusion T_{trans}, i.e η ($\eta = 0.95$ corresponds to 200 K and varies linearly with temperature from $\eta = 0$ at $T = 0$). It is a contour plot of constant diffusion rates ν_o in the quantum regime. Close to the hyporbolic boundary we have the softest modes. The hyperbolic boundary between the coloured part and the white part of the plot corresponds to the transition from very soft gravonon modes (at the hyperbolic boundary) to zero-frequency modes (within the white part). This parameter

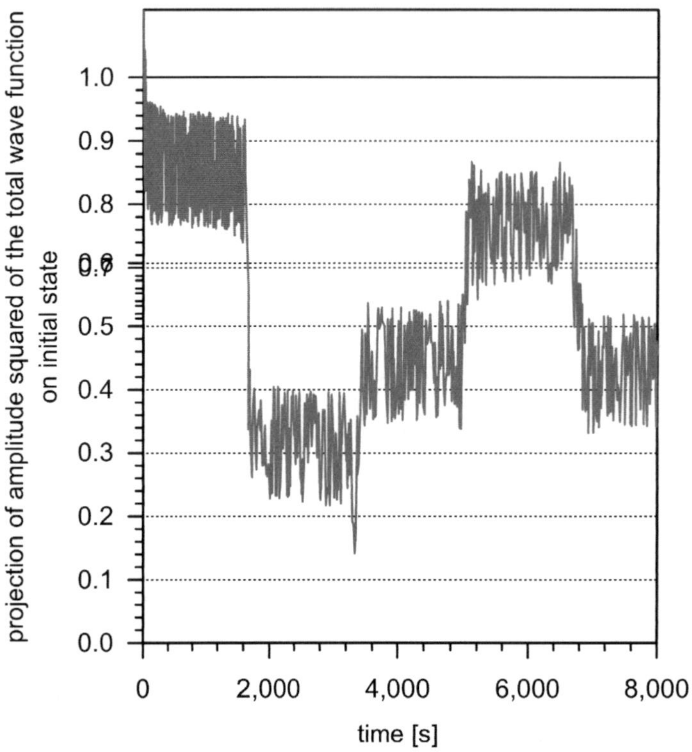

Figure 3. Changes of the squared projection of the world wave function $\Psi(t)$ on the initial state with the adsorbate on site 1 of a two-site model with time. Telegraph-signal-like site changes occur at time intervals, determined by the lifetime, i.e. the energy, of the entangled soft gravonon modes ω_{soft}.

space yields zero-frequency gravonon modes, the so called Nambu-Goldstone bosons.

The experimental diffusion rate of ^{1}H, ^{2}H and ^{3}H on different metal surfaces is plotted on the landscape of the adsorbate-induced soft gravonon frequency, since we know both the diffusion rate and the transition temperature from experiment. We can read off the coupling strength to the gravonons γ which reproduces the experimental diffusion rate at the temperature of transition from quantum to overbarrier diffusion. Hence, in this plot we see the "experimental" values of the coupling strength γ_{exp} to the gravonons, and the "experimental" value of the transition temperature to quantum diffusion, both needed to reproduce the experimenatal hopping rate on a map of hopping rates, evaluated theoretically. The experimentally measured adsorption systems lie in a narrow region on the boundary between the parameter spaces, typical for zero quantum diffusion rate and non-zero quantum diffusion rate (red points in fig. 4).

Since in 11 dimensional spacetime the gravitational interaction is very sensitive to the distance, varying with the inverse 16-th power of the distance between the hydrogen atom and the metal atom in the deformation resonance, the theoretical coupling strength γ_{th} is very sensitive to this distance as well. Having in mind the strong distance dependence of the interaction between the adsorbate field and gravonons in D spatial dimensions with $r^{-2(D-2)}$ (cf. eq. 10) and of the coupling strength as $\gamma_{th} \propto \frac{1}{r^{32}}$ in 11D spacetime, we make the best fit γ_{th} to the experimental

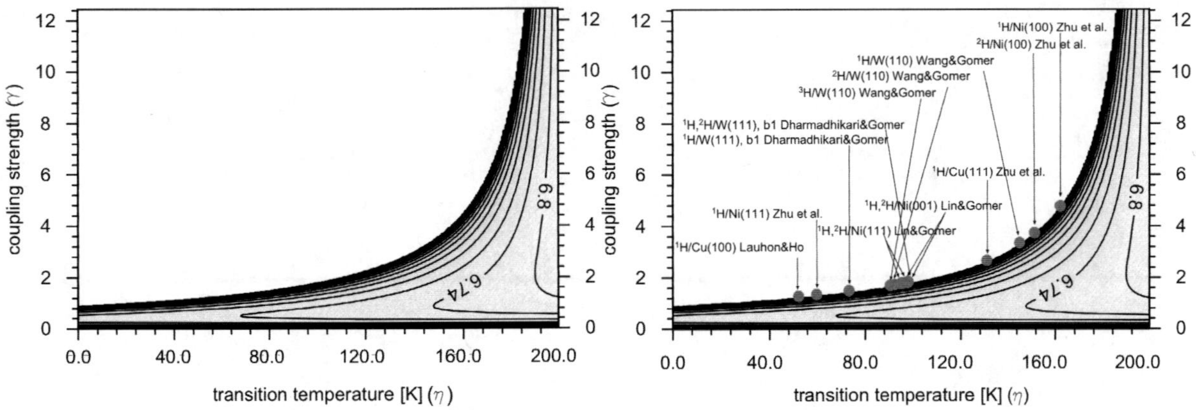

Figure 4. Left panel: Contour plot with lines of constant frequency of the softest gravonon mode $\log_{10}\omega_{soft}$=const, which means contour lines of constant diffusion rate, versus the coupling strength γ and the transition temperature from quantum to overbarrier diffusion T_{trans}. (η = 0.95 corresponds to 200 K and varies linearly with temperature). Right panel: In the $\gamma\eta$ landscape on the l.h.s. the experimental data on the logarithm of the diffusion rate $\log_{10}\nu_o$ in the quantum regime at the experimental transition temperature T_{trans} are plotted with red points for hydrogen, deuterium and tritium on different metal surfaces. The data are gained from experimental plots of the kind displayed in fig. 1.

values of the coupling strength γ_{exp} by varying the distance between the hydrogen atom and the metal atom in the deformation resonance in 11D spacetime (fig. 5). The labels below each point in the curve refer to the deformation resonance configuration of the adsorbate and indicate which values of R_{ads-M} have been chosen for the fit. The theoretical values of γ_{th} evaluated in 11D spacetime are fitting best to the experimental values of the diffusion rates, estimates within 10D and 12D are also not so bad, however, results for γ_{th} in 8D and 26D spacetime diverge.

The best fit to the experimental coupling strength γ_{exp} provides the distances R_{ads-M} at which entanglement to gravonons in the deformation resonance yields the experimental diffusion rates. In fig. 6 we plot R_{ads-M} versus the distance between a hydrogen atom and a metal atom due to ab initio electronic structure calculations for the same adsorption geometries. The agreement is exceptionally good, having in mind the strong distance dependence of the results in 11D-theory and the fact that the two theories have no contact points at all.

4. Conclusion

The set of experimental data on diffusion of hydrogen on transition metal surfaces cannot be rationalized by conventional quantum mechanics in 3+1 dimensions. Entanglement between the adsorbate motion with environmental fields at very low temperatures cannot be achieved in 3+1 dimensions by the common excitations (phonons, tomonagons, etc.). If entanglement with the environment plays a role in these experiments, the environmental excitations have to be gravitons in $D+1$ dimensional spacetime (D=8,9,10,11).

In this theory gravonons emerge as nonrelativistic massive bosons describing spacetime deformations very near to the adparticle. The frequency of adsorbate induced soft local gravonon

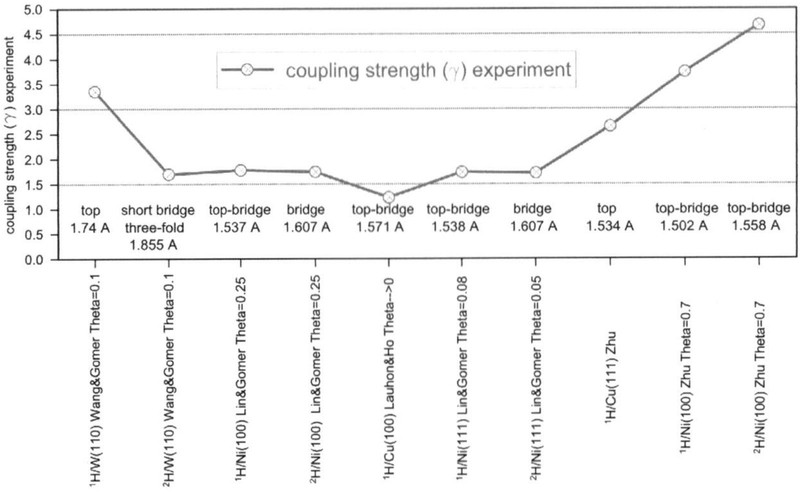

Figure 5. Theoretical values of the coupling strenght evaluated by varying the distance between the adsorbate and a metal atom in the deformation resonance to yield the best fit to γ_{exp} and the experimental diffusion rate. The labels of each point in the curve refer to the configuration in the deformation resonance and indicate which value of R_{ads-M} has been chosen for the fit.

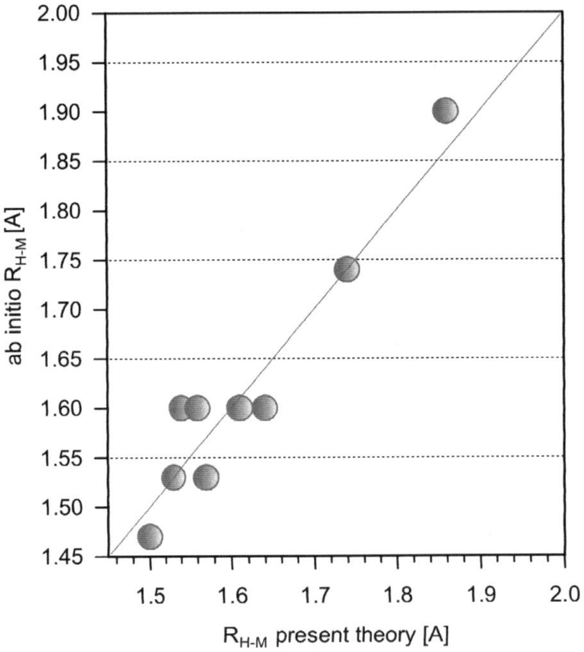

Figure 6. Comparison of the distance [Å] between a hydrogen atom and a metal atom in the deformation resonance, derived from the present theory, and ab initio (DFT) distances [Å] between H and metal atoms for the same adsorption geometries.

Journal of Physics: Conference Series **442** (2013) 012049 doi:10.1088/1742-6596/442/1/012049

modes determines the lifetime of the entangled state on a given site in the local region and the diffusion rate. While the gravonons propagate in the hidden dimensions of 10+1 dimensional spacetime the adsorbate is localized on one adsorption site entangled to the gravonons and cannot change site. Adsorption site changes and surface quantum diffusion occur in a telegraph-signal-like manner with time and not as Rabi oscillations, as a result of the solution of Schrödinger's time dependent equation. The conditions for time variation of the adsorption site similar to telegraph signals are weak and local coupling of the adsorbate motion to a continuum of soft local graviton modes (gravonons) of high density of states in high spacetime dimensions. The frequency of the soft gravonons determines the lifetime of the adsorbate-gravonon entangled state and the diffusion rate of the adsorbate to a neighbouring adsorption site.

References

[1] Wang S C and Gomer R 1985 *J. Chem. Phys.* **83** 4193
[2] Lauhon L J and Ho W 2000 *Phys. Rev. Lett.* **85** 4566; Erratum: Lauhon L J and Ho W 2002 *Phys. Rev. Lett.* **89** 079901
[3] Lin T S and Gomer R 1991 *Surface Sci.* **255** 41
[4] Auerbach A, Freed K F and Gomer R 1987 *J. Chem. Phys.* **86** 2356
[5] DiFoggio R and Gomer R 1982 *Phys. Rev.* B **25** 3490
[6] Dharmadhikari C and Gomer R 1984 *Surface Sci.* **143** 223
[7] Cao G X, Nabighian E and Zhu X D 1997 *Phys. Rev. Lett.* **79** 3696
[8] Wang X, Fei Y Y and Zhu X D 2009 *Chem. Phys. Lett.* **481** 58
[9] Zhu X D, Lee A, Wong A and Linke U 1992 *Phys. Rev. Lett.* **68** 1862
[10] Zhu X D 1992 *Mod. Phys. Lett.* B **6** 1217
[11] Mattsson Th and Wahnström G 1997 *Phys. Rev.* B **56** 14944
[12] Liljeroth P, Repp J and Meyer G 2007 *Science* **317** 1203
[13] Joos E and Zeh H D 1985 *J. Phys. Cond. Matter* **59** 223; Joos E and Zeh H D 1985 *Z. Phys. B: Cond. Matter* **59** 223-243
[14] Doyen G and Drakova D, in preparation
[15] Doyen G and Drakova D 2009 *Surface Sci.* **603** 1579
[16] Doyen G and Drakova D 2011 *J. Phys.: Conf. Series* **306** 012033
[17] Arkani-Hamed N, Dimopoulos S and Dvali G 1998 *Phys. Lett.* B **429** 263 (*Preprint* arXiv: hep-ph/9803315v1 11 Mar 1998.
[18] Antoniadis I, Arkani-Hamed N, Dimopoulos S and Dvali G 1998 *Phys. Lett.* B **436** 257
[19] Arkani-Hamed N, Dimopoulos S and Dvali G 1999 *Phys. Rev.* D **59** 086004
[20] Arkani-Hamed N, Dimopoulos S and Dvali S 2000 *Scientific American* **283** 62
[21] Arkani-Hamed N, Dimopoulos S, Dvali G and Kaloper N 2000 *Phys. Rev. Lett.* **84** 586
[22] Zee A 2003 *Quantum field theory in a nutshell* (Princeton and Oxford: Princeton Univ. Press)
[23] Doyen G and Drakova D, present volume
[24] Drakova D and Doyen G *Preprint* arXiv:1204.5606v1 [quant-ph] 25 Apr 2012

Quantum mechanics and permutation invariants of finite groups

V V Kornyak

Laboratory of Information Technologies, Joint Institute for Nuclear Research, 141980 Dubna, Russia

E-mail: `kornyak@jinr.ru`

Abstract. We study quantum behavior from a constructive "finite" point of view, since the introduction of continuum or other actual infinities into physics poses serious conceptual and technical difficulties without any need for these concepts in physics as an empirical science. Taking this approach, we can show that the quantum-mechanical problems can be formulated in the invariant subspaces of permutation representations of finite groups, while the quantum interferences occur as phenomena that are observable in these subspaces. The scalar products in the invariant subspaces (which are needed for formulating the Born rule — the main postulate of quantum mechanics that links mathematical description with experiment) are linear combinations of independent bilinear invariant forms of the permutation representation. A complete set of such forms for any permutation group can be easily calculated by a simple algorithm. Slightly more sophisticated algorithms are required for expressing quantum observables in terms of these forms.

1. Introduction

Quantum behavior is a natural consequence of symmetries of physical systems. In fact, it is a manifestation of the fundamental impossibility to trace the identity of indistinguishable objects in the process of their evolution. The only that can be extracted from any physical observation is information about invariant relations and quantities which describe ensembles of such objects. Thus, the fundamental problem is to study invariants of groups of symmetries of physical systems. Invariant bilinear forms, i.e., scalar products are particularly important for quantum physics. These forms in accordance with the Born rule provide the link between mathematical description and experimental observations.

Physics, as an empirical science, is insensitive to whether finite or infinite concepts are used in formulating its laws. Moreover, effective modeling is possible only when the problem is formulated in finite terms. From the theory of quantum computing it is known that any unitary operator can be presented with arbitrary precision by a finite combination of matrix operators from a finite set of universal quantum gates. Such universal sets of gates generate an infinite group G_∞ which is everywhere dense in the full unitary group. By the theorem of Mal'cev [1], the group G_∞ — as being a finitely generated matrix group — is *residually finite* [2], i.e., there is a rich set of nontrivial homomorphisms from G_∞ to finite groups. Combining these facts, we can replace the full unitary group by some finite group G in any particular problem without distorting its physical content.

It is well known that any linear representation of a finite group is unitary.[1] Any unitary representation of a finite group is a subrepresentation of some permutation representation. Let U be a representation of the group G in a K-dimensional Hilbert space $\mathcal{H}_{\mathsf{K}}$. Then U can be embedded into a permutation representation P of G in an N-dimensional Hilbert space $\mathcal{H}_{\mathsf{N}}$, where $\mathsf{N} \geq \mathsf{K}$. The representation P is equivalent to an action of G on a set of things $\Omega = \{\omega_1, \ldots, \omega_{\mathsf{N}}\}$ by permutations. In the proper case, i.e., when $\mathsf{N} > \mathsf{K}$, the embedding has the structure

$$\mathrm{T}^{-1}\mathrm{P}\mathrm{T} = \begin{pmatrix} \mathbf{1} & & \\ & \mathrm{V} & \\ & & \mathrm{U} \end{pmatrix} \begin{matrix} \left.\right\} \mathcal{H}_{\mathsf{N}-\mathsf{K}} \\ \left.\right\} \mathcal{H}_{\mathsf{K}} \end{matrix}, \qquad \mathcal{H}_{\mathsf{N}} = \mathcal{H}_{\mathsf{N}-\mathsf{K}} \oplus \mathcal{H}_{\mathsf{K}},$$

where $\mathbf{1}$ is the trivial one-dimensional representation, mandatory for any permutation representation; V is a complementary subrepresentation which may or may not be. T is a matrix of transition from the basis of the permutation representation P to the basis in which the permutation space $\mathcal{H}_{\mathsf{N}}$ is split into the invariant subspaces $\mathcal{H}_{\mathsf{N}-\mathsf{K}}$ and $\mathcal{H}_{\mathsf{K}}$. For brevity, we will refer to this basis as "quantum basis". The data in the spaces $\mathcal{H}_{\mathsf{K}}$ and $\mathcal{H}_{\mathsf{N}-\mathsf{K}}$ are *independent* since both spaces are invariant subspaces of the space $\mathcal{H}_{\mathsf{N}}$. So we can consider the data in $\mathcal{H}_{\mathsf{N}-\mathsf{K}}$ as "hidden parameters" with respect to the data in $\mathcal{H}_{\mathsf{K}}$.

A trivial approach would be to set arbitrary (e.g., zero) data in the complementary subspace $\mathcal{H}_{\mathsf{N}-\mathsf{K}}$. This approach is not interesting since it is not falsifiable by means of the standard quantum mechanics. In fact, it leads to the standard quantum mechanics *modulo* the empirically unobservable distinction between the "finite" and the "infinite". The only difference is technical: we can replace the linear algebra in the K-dimensional space $\mathcal{H}_{\mathsf{K}}$ by permutations of N things.

A more promising approach requires some changes in the concept of quantum amplitudes. We assume [3, 4] that they are projections onto invariant subspaces of the vectors of multiplicities ("occupation numbers") of elements of the set Ω on which the group G acts by permutations. The vectors of multiplicities

$$|n\rangle = \begin{pmatrix} n_1 \\ \vdots \\ n_{\mathsf{N}} \end{pmatrix}$$

are elements of the *module* $\mathsf{H}_{\mathsf{N}} = \mathbb{N}^{\mathsf{N}}$, where $\mathbb{N} = \{0, 1, 2, \ldots\}$ is the *semiring of natural numbers*. We start with the natural permutation representation of G in the module H_{N}. In order to turn the module H_{N} into the Hilbert space $\mathcal{H}_{\mathsf{N}}$, it is sufficient just to add roots of unity — an algebraic incarnation of periodicity — to the natural numbers. Linear combinations of $\mathcal{P}$th roots of unity with natural coefficients form a *semiring* $\mathbb{N}_{\mathcal{P}}$. The period $\mathcal{P}$ (it is called also *conductor*) depends on the structure of the group G. Generally, it is a multiple of some divisor of the *exponent* of G, which is defined as the least common multiple of the orders of elements of G. We will always assume that $\mathcal{P} \geq 2$. In this case, using the standard identities for the roots of unity, we can introduce the negative numbers, and thus, $\mathbb{N}_{\mathcal{P}}$ becomes a *ring of cyclotomic integers*. To complete the conversion of the module H_{N} into the Hilbert space $\mathcal{H}_{\mathsf{N}}$, we introduce the *cyclotomic field* $\mathbb{Q}_{\mathcal{P}}$ as a field of fractions of the ring $\mathbb{N}_{\mathcal{P}}$. If $\mathcal{P} \geq 3$, then $\mathbb{Q}_{\mathcal{P}}$ is a dense subfield of the field of complex numbers $\mathbb{C}$. In particular, this gives a trivial explanation of why the complex numbers are so important in quantum mechanics. In fact, the algebraic properties of elements of $\mathbb{Q}_{\mathcal{P}}$ are quite sufficient for all our purposes, so we can forget the possibility to embed $\mathbb{Q}_{\mathcal{P}}$ into $\mathbb{C}$.

The connection between mathematical description and observation is provided by the Born rule: the probability to register a particle described by the amplitude $|\psi\rangle$, by an apparatus tuned

[1] An additional advantage of the transition to finite groups is a quite natural explanation of the "unitarity" of quantum mechanics: it follows from this trivial property of finite groups.

to the amplitude $|\phi\rangle$, is

$$\mathbf{P}(\phi,\psi)=\frac{|\langle\phi\mid\psi\rangle|^2}{\langle\phi\mid\phi\rangle\,\langle\psi\mid\psi\rangle}\ .$$

In the "finite" background the only reasonable interpretation of probability is the *frequency interpretation*: the probability is the ratio of the number of "favorable" combinations to the total number of combinations. So we expect that $\mathbf{P}(\phi,\psi)$ must be a *rational number* if everything is arranged correctly.

To calculate via the Born rule quantum interferences of projections of natural amplitudes, it is sufficient to know the expressions for scalar products in invariant subspaces of the permutation representation. These expressions can be obtained by straightforward computation using the matrix of transition from the "permutation" to "quantum" basis. Unfortunately, there is no satisfactory algorithm for splitting a module over an associative algebra into irreducible submodules — and, hence, for computing the corresponding transition matrix — over important to us number systems of zero characteristic.[2] However, we could try to cope with these difficulties if we could get the expressions for the scalar products directly, i.e., without the use of transition matrices. We describe here one of the possible approaches. This approach can be used for splitting invariant forms of representations for which one can easily calculate the centralizer ring. Permutation representations belong just to this type of representations.

2. Basis of permutation invariant forms

For a start we construct the set of all invariant bilinear forms in the permutation basis. Suppose that the group $\mathsf{G}\equiv\mathsf{G}(\Omega)$, that acts on the set $\Omega=\{\omega_1,\dots,\omega_{\mathsf{N}}\}$, is generated by K elements $g_1,\dots,g_k,\dots,g_K$. Clearly, it suffices to verify invariance of the form with respect to these elements. For convenience, the set Ω is identified canonically with the set of indices of its elements: $\Omega=\vec{\mathsf{N}}=\{1,\dots,\mathsf{N}\}$. Denote matrices of permutation representation of the generators by symbols $P_k=\mathrm{P}(g_k)$. Obviously, $P_k^{\mathrm{T}}=\mathrm{P}\left(g_k^{-1}\right)$. The condition of invariance of the bilinear form $A=\left(a_{ij}\right)$ under the group G can be written as a system of matrix equations

$$A=P_kAP_k^{\mathrm{T}},\quad 1\le k\le K.$$

It is easy to check that in terms of components these equations have the form

$$a_{ij}=a_{ig_kjg_k}. \tag{1}$$

Thus, the basis of all invariant bilinear forms is in one-to-one correspondence with the set of orbits $\Delta_1,\dots,\Delta_{\mathrm{R}}$ of the action of G on the Cartesian product $\Omega\times\Omega$. Orbits of the group on the product $\Omega\times\Omega$, called *orbitals*, play an important role in the theory of permutation groups and their representations [6, 7]. If the group is transitive, then there is a single orbital, which consists of all pairs of the form (i,i). Such an orbital is called *trivial* or *diagonal*. The number of orbitals R is called the *rank* of the group G. Each orbital Δ_r can be associated with

(i) a directed graph, whose vertices are the elements of Ω and edges are pairs $(i,j)\in\Delta_r$;

(ii) an $\mathsf{N}\times\mathsf{N}$ matrix $A(\Delta_r)$, which is constructed according to the rule

$$A\left(\Delta_r\right)_{ij}=\begin{cases}1, & \text{if } (i,j)\in\Delta_r,\\ 0, & \text{if } (i,j)\notin\Delta_r.\end{cases}$$

Properties of the graphs of orbitals reflect important properties of the groups. For example, a transitive permutation group is primitive if and only if the graphs of all its non-trivial orbitals are connected.

[2] The most practical and popular algorithm called ***MeatAxe*** [5] is designed only for algebras over finite fields.

Matrices of orbitals form a basis of the *centralizer ring* of permutation representation of G. This ring plays an important role in the theory of group representations. We will denote it by $\mathrm{ZR} \equiv \mathrm{ZR}\left(\mathsf{G}\left(\Omega\right)\right)$. The multiplication table of basis elements of the ring ZR has the form

$$A\left(\Delta_p\right) A\left(\Delta_q\right) = \sum_{r=1}^{\mathrm{R}} a_{pq}^{r} A\left(\Delta_r\right),$$

where all the structure constants a_{pq}^{r} are natural numbers.

Note that the centralizer ring can be obtained via slightly different approach [8] which is originated from the works of Schur. The approach is based on the study of orbits of stabilizers of points in the set Ω. Recall that the stabilizer of the element $i \in \Omega$ is the maximal subgroup G_i of the group G that fixes this element, i.e., $g \in \mathsf{G}_i \Leftrightarrow ig = i$. In the general case, the stabilizer G_i of a transitive group G acts on the set Ω intransitively, generating a set of orbits $\Lambda_1 = \{i\}, \Lambda_2, \ldots, \Lambda_{\mathrm{R}}$. There is a trivial one-to-one correspondence between the orbits of the stabilizer (sometimes they are called suborbits) and the orbitals (see, e.g., [6, 7]). Therefore the number R of orbits of the stabilizer G_i coincides with the rank of the permutation group $\mathsf{G}\left(\Omega\right)$. Via the orbits of stabilizers the basis matrices of the centralizer ring are constructed as follows. We consider the orbits of the stabilizer G_1. Each orbit Λ of the stabilizer G_1 is associated with the matrix $A\left(\Lambda\right)$ by the rule

$$A\left(\Lambda\right)_{ij} = \begin{cases} 1, & \text{if there exist } g \in \mathsf{G} \text{ and } l \in \Lambda \text{ such that } 1g = i \text{ and } lg = j, \\ 0, & \text{otherwise.} \end{cases}$$

Technical advantage of this approach is the use of the isomorphism between the orbits of stabilizers of different points: initially, we construct the first row of the matrix using the orbits of the stabilizer of the point 1, and then, the remaining rows are created via the group translations.

We will call the matrices of orbitals $\mathcal{A}_r \equiv A\left(\Delta_r\right)$ *basic forms*, since any permutation invariant bilinear form can be presented by their linear combination.

Algorithm for computing the basic forms is reduced to construction of orbitals in accordance with (1). In a few words, it scans the elements of the set $\Omega \times \Omega$ in some, say, lexicographic, order and distributes these elements over the equivalence classes. The output of the algorithm is a complete basis $\mathcal{A}_1, \ldots, \mathcal{A}_{\mathrm{R}}$ of permutation invariant bilinear forms. The algorithm is quite simple. Our implementation is just a few lines in C.

The following identity follows directly from the construction

$$\mathcal{A}_1 + \mathcal{A}_2 + \cdots + \mathcal{A}_{\mathrm{R}} = L^{\mathrm{T}} L = \mathrm{J}_{\mathsf{N}} \equiv \begin{pmatrix} 1 & 1 & \cdots & 1 \\ 1 & 1 & \cdots & 1 \\ \vdots & \vdots & \vdots & \vdots \\ 1 & 1 & \cdots & 1 \end{pmatrix},$$

where L is a covector of the form $\overbrace{(1, 1, \ldots, 1)}^{\mathsf{N}}$, and J_{N} is an $\mathsf{N} \times \mathsf{N}$ *"matrix of ones"*.

Let us illustrate the algorithm with the example of the group $\mathbb{Z}_3$ acting on the set $\Omega = \{1, 2, 3\}$. The group is generated by a single element, for example, $g_1 = (1, 2, 3)$. We need to distribute the set of pairs of indices

$$\begin{matrix} (1,1) & (1,2) & (1,3) \\ (2,1) & (2,2) & (2,3) \\ (3,1) & (3,2) & (3,3) \end{matrix} \tag{2}$$

over the equivalence classes in accordance with (1). If we start from the top left corner of table (2) and will look for untreated pairs in the lexicographic order, the orbitals will be constructed in the following order:

$$\Delta_1 = \{(1,1),(2,2),(3,3)\}, \Delta_2 = \{(1,2),(2,3),(3,1)\}, \Delta_3 = \{(1,3),(2,1),(3,2)\}.$$

The corresponding complete basis of invariant forms reads

$$\mathcal{A}_1 = \begin{pmatrix} 1 & \cdot & \cdot \\ \cdot & 1 & \cdot \\ \cdot & \cdot & 1 \end{pmatrix}, \quad \mathcal{A}_2 = \begin{pmatrix} \cdot & 1 & \cdot \\ \cdot & \cdot & 1 \\ 1 & \cdot & \cdot \end{pmatrix}, \quad \mathcal{A}_3 = \begin{pmatrix} \cdot & \cdot & 1 \\ 1 & \cdot & \cdot \\ \cdot & 1 & \cdot \end{pmatrix}.$$

Note that in general, if we start the algorithm with the pair $(1,1)$ and G is transitive, then the first basis form will always be the identity matrix $\mathcal{A}_1 = \mathrm{I}_{\mathsf{N}}$, corresponding to the trivial orbital.

3. Permutation invariant forms and decomposition into irreducible components

Despite the ease of obtaining the permutation invariant forms, they are sufficiently informative. Consider the decomposition of permutation representation into irreducible components using the transformation matrix T. This decomposition for transitive groups has the form

$$\mathrm{T}^{-1}\mathrm{P}(g)\mathrm{T} = \begin{pmatrix} \mathbf{1} & & & & & \\ & \mathrm{I}_{m_2} \otimes \mathrm{U}_2(g) & & & & \\ & & \ddots & & & \\ & & & \mathrm{I}_{m_k} \otimes \mathrm{U}_k(g) & & \\ & & & & \ddots & \\ & & & & & \mathrm{I}_{m_{N_{\mathrm{Irr}}}} \otimes \mathrm{U}_{N_{\mathrm{Irr}}}(g) \end{pmatrix}, \qquad g \in \mathsf{G}\,.$$

Here N_{Irr} is the total number of different irreducible representations U_k $(\mathrm{U}_1 \equiv \mathbf{1})$ of the group G, which are contained in the permutation representation P; m_k is the multiplicity of the subrepresentations U_k in the representation P; $\otimes$ denotes the *Kronecker product* of matrices; I_m is an $m \times m$ identity matrix.

The most general permutation invariant form is a linear combination of the basic forms

$$A = a_1\mathcal{A}_1 + a_2\mathcal{A}_2 + \cdots + a_{\mathrm{R}}\mathcal{A}_{\mathrm{R}}\,, \tag{3}$$

where the coefficients a_i are elements of some abelian number field[3] $\mathcal{F}$ which is defined concretely in the computations to be described below.

It is easy to show (see [6, 8]) that in a basis, which splits the permutation representation into irreducible components, the form in (3) takes the form

$$\mathrm{T}^{-1}A\mathrm{T} = \begin{pmatrix} \mathcal{B}_1 & & & & & \\ & \mathcal{B}_2 \otimes \mathrm{I}_{\mathrm{d}_2} & & & & \\ & & \ddots & & & \\ & & & \mathcal{B}_k \otimes \mathrm{I}_{\mathrm{d}_k} & & \\ & & & & \ddots & \\ & & & & & \mathcal{B}_{N_{\mathrm{Irr}}} \otimes \mathrm{I}_{\mathrm{d}_{N_{\mathrm{Irr}}}} \end{pmatrix}. \tag{4}$$

[3] An *abelian number field* is an extension of the field $\mathbb{Q}$ with abelian Galois group. Due to the Kronecker–Weber theorem any such field is a subfield of some cyclotomic field $\mathbb{Q}_{\mathcal{P}}$.

Here $\mathcal{B}_k$ is an $m_k \times m_k$ matrix, whose elements are linear combinations of the coefficients a_i from (3), while m_k is a multiplicity of the irreducible component U_k; d_k is the dimension of U_k. The structure of matrix (4) implies that the rank R of the group (i.e., the dimension of the centralizer ring) is equal to the sum of the squares of the multiplicities: $\mathrm{R} = 1+m_2^2+\cdots+m_{N_{\mathrm{Irr}}}^2$.

Let us now consider the determinant $\det(A)$. In terms of the variables $a_1, \ldots, a_{\mathrm{R}}$ this determinant is a homogeneous polynomial of degree N. Since the determinant of a form does not depend on the choice of basis, and the determinant of a block diagonal matrix is the product of determinants of its blocks, from decomposition (4) it follows

$$\det(A) = \det(\mathcal{B}_1)\det(\mathcal{B}_2)^{\mathrm{d}_2}\cdots\det(\mathcal{B}_k)^{\mathrm{d}_k}\cdots\det(\mathcal{B}_{N_{\mathrm{Irr}}})^{\mathrm{d}_{N_{\mathrm{Irr}}}}.$$

Here we have used the identity $\det(X \otimes Y) = \det(X)^m \det(Y)^n$ for the Kronecker product of an $n \times n$ matrix X and an $m \times m$ matrix Y.

It is clear that $\det(\mathcal{B}_k)$ is a homogeneous polynomial of degree m_k in the variables $a_1, \ldots, a_{\mathrm{R}}$: $\det(\mathcal{B}_k) = E_k(a_1, \ldots, a_{\mathrm{R}})$. Thus, we have the following

Proposition. *The determinant of linear combination* (3) *has the following decomposition into factors over a certain ring of cyclotomic integers*:

$$\det\sum_{i=1}^{\mathrm{R}} a_i\mathcal{A}_i = \prod_{k=1}^{N_{\mathrm{Irr}}} E_k(a_1, \ldots, a_{\mathrm{R}})^{\mathrm{d}_k}, \quad \deg E_k(a_1, \ldots, a_{\mathrm{R}}) = m_k\,, \tag{5}$$

where E_k is an irreducible polynomial corresponding to the irreducible component U_k in the representation P. Recall that N_{Irr} is the number of different irreducible components in P; d_k is dimension of U_k; m_k is the multiplicity of U_k in P.

From this proposition the idea of an algorithm to compute the invariant forms in irreducible subspaces of the permutation representation follows.

We must first calculate the polynomial $\det(A)$. This is a relatively simple task. In particular, algorithms based on the Gaussian elimination have a cubic complexity in the size of the matrix.

Then, the polynomial $\det(A)$ must be decomposed into a maximum number of irreducible factors. Note that advanced algorithms of polynomial factorization automatically determine an algebraic extension of rationals that guarantees maximal factorization. Such algorithms are called algorithms of "*absolute factorization*". There are many practical implementations of polynomial factorization algorithms with different estimates of the complexity.[4]

Thus, constructing decomposition (5) is an algorithmically realizable task. Solving it, we obtain the complete information about the dimensions and multiplicities of all irreducible subrepresentations.

The next natural step is to try to compute explicitly invariant forms $\mathcal{B}_k$ in the irreducible subspaces of the permutation representation. To do this, we exclude from consideration the factor $E_k(a_1, \ldots, a_{\mathrm{R}})$, related to the component $\mathcal{B}_k$, and equate to zero the other factors. That is, we write the system of equations

$$E_1 = \cdots = \widehat{E_k} = \cdots = E_{N_{\mathrm{Irr}}} = 0. \tag{6}$$

3.1. Example with irreducible components of multiplicity one.

If all the multiplicities $m_i = 1$ (under this condition $\mathrm{R} = N_{\mathrm{Irr}}$ and the centralizer ring ZR is commutative), then all the polynomials E_i are linear. In this case, the computation of the scalar products in the invariant subspaces can easily be completed.

[4] To improve efficiency we can try to write a specialized algorithm that takes into account the fact that in our case the polynomials are factored over the rings of cyclotomic integers, whose conductors are divisors of group exponents. However, the benefits of such work require a separate study.

As an example, consider the group $\mathsf{SL}(2,3)$ defined as a group of special linear transformations of two-dimensional space over the field of three elements $\mathbb{F}_3$. This group is used in particle physics where it is often denoted by T' or 2T since it is a double cover of the symmetry group $\mathrm{T} \cong \mathsf{A}_4$ of a tetrahedron. We will consider its faithful permutation action of degree 8, which can be generated, for example, by the following two permutations

$$g_1 = (1,5,3,2,6,4)\,(7,8) \quad \text{and} \quad g_2 = (1,3,7,2,4,8)\,(5,6)\,.$$

The 8-dimensional permutation representation will be denoted by $\underline{\mathbf{8}}$.

The following four matrices — obtained via constructing orbitals — form a basis of the ring $\mathrm{ZR} \equiv \mathrm{ZR}\left(\mathsf{SL}(2,3)\left(\vec{8}\right)\right)$ of permutation invariant forms:

$$\mathcal{A}_1 = \begin{pmatrix} 1 & \cdot & \cdot & \cdot & \cdot & \cdot & \cdot & \cdot \\ \cdot & 1 & \cdot & \cdot & \cdot & \cdot & \cdot & \cdot \\ \cdot & \cdot & 1 & \cdot & \cdot & \cdot & \cdot & \cdot \\ \cdot & \cdot & \cdot & 1 & \cdot & \cdot & \cdot & \cdot \\ \cdot & \cdot & \cdot & \cdot & 1 & \cdot & \cdot & \cdot \\ \cdot & \cdot & \cdot & \cdot & \cdot & 1 & \cdot & \cdot \\ \cdot & \cdot & \cdot & \cdot & \cdot & \cdot & 1 & \cdot \\ \cdot & \cdot & \cdot & \cdot & \cdot & \cdot & \cdot & 1 \end{pmatrix}, \quad \mathcal{A}_2 = \begin{pmatrix} \cdot & 1 & \cdot & \cdot & \cdot & \cdot & \cdot & \cdot \\ 1 & \cdot & \cdot & \cdot & \cdot & \cdot & \cdot & \cdot \\ \cdot & \cdot & \cdot & 1 & \cdot & \cdot & \cdot & \cdot \\ \cdot & \cdot & 1 & \cdot & \cdot & \cdot & \cdot & \cdot \\ \cdot & \cdot & \cdot & \cdot & \cdot & 1 & \cdot & \cdot \\ \cdot & \cdot & \cdot & \cdot & 1 & \cdot & \cdot & \cdot \\ \cdot & \cdot & \cdot & \cdot & \cdot & \cdot & \cdot & 1 \\ \cdot & \cdot & \cdot & \cdot & \cdot & \cdot & 1 & \cdot \end{pmatrix},$$

$$\mathcal{A}_3 = \begin{pmatrix} \cdot & \cdot & 1 & \cdot & 1 & \cdot & 1 & \cdot \\ \cdot & \cdot & \cdot & 1 & \cdot & 1 & \cdot & 1 \\ \cdot & 1 & \cdot & \cdot & \cdot & 1 & 1 & \cdot \\ 1 & \cdot & \cdot & \cdot & 1 & \cdot & \cdot & 1 \\ \cdot & 1 & 1 & \cdot & \cdot & \cdot & \cdot & 1 \\ 1 & \cdot & \cdot & 1 & \cdot & \cdot & 1 & \cdot \\ \cdot & 1 & \cdot & 1 & 1 & \cdot & \cdot & \cdot \\ 1 & \cdot & 1 & \cdot & \cdot & 1 & \cdot & \cdot \end{pmatrix}, \quad \mathcal{A}_4 = \begin{pmatrix} \cdot & \cdot & \cdot & 1 & \cdot & 1 & \cdot & 1 \\ \cdot & \cdot & 1 & \cdot & 1 & \cdot & 1 & \cdot \\ 1 & \cdot & \cdot & \cdot & 1 & \cdot & \cdot & 1 \\ \cdot & 1 & \cdot & \cdot & \cdot & 1 & 1 & \cdot \\ 1 & \cdot & \cdot & 1 & \cdot & \cdot & 1 & \cdot \\ \cdot & 1 & 1 & \cdot & \cdot & \cdot & \cdot & 1 \\ 1 & \cdot & 1 & \cdot & \cdot & 1 & \cdot & \cdot \\ \cdot & 1 & \cdot & 1 & 1 & \cdot & \cdot & \cdot \end{pmatrix}.$$

The determinant of their linear combination $A = a_1\mathcal{A}_1 + a_2\mathcal{A}_2 + a_3\mathcal{A}_3 + a_4\mathcal{A}_4$ is decomposed into linear factors over the ring of cyclotomic integers $\mathbb{N}_3$:

$$\begin{aligned} \det A = {} & (a_1 + a_2 + 3a_3 + 3a_4) \\ & \{a_1 - a_2 + (1+2\mathsf{r})\,a_3 - (1+2\mathsf{r})\,a_4\}^2 \\ & \{a_1 - a_2 - (1+2\mathsf{r})\,a_3 + (1+2\mathsf{r})\,a_4\}^2 \\ & (a_1 + a_2 - a_3 - a_4)^3\,, \end{aligned}$$

where r is a third primitive root of unity. From this formula the structure of decomposition of the representation $\underline{\mathbf{8}}$ into irreducible components is seen immediately:

$$\underline{\mathbf{8}} = \mathbf{1} \oplus \mathbf{2} \oplus \mathbf{2}' \oplus \mathbf{3}.$$

Excluding sequentially the linear factors corresponding to the representations $k = \mathbf{1}, \mathbf{2}, \mathbf{2}', \mathbf{3}$, and equating to zero the other factors, we obtain four systems of three linear equations.

Consider for example the subrepresentation $\mathbf{2}$. For this component, the system of equations (6) takes the form

$$a_1 + a_2 + 3a_3 + 3a_4 = 0, \tag{7}$$

$$a_1 - a_2 - (1+2\mathsf{r})\,a_3 + (1+2\mathsf{r})\,a_4 = 0, \tag{8}$$

$$a_1 + a_2 - a_3 - a_4 = 0. \tag{9}$$

The linear systems of this type consist of $\mathsf{R}-1$ equations, but contain R variables. Since the bilinear form (3) describes a non-degenerate scalar product, the coefficient a_1 at the diagonal basis form can not vanish. Thus, one can treat a_1 as a parameter and solve the system of equations for the remaining variables. In principle, the coefficient a_1 can be taken to be an arbitrary (nonzero) parameter, since Born's probability is independent of its value. However, it is reasonable to choose the value

$$a_1 = \mathrm{d}_k/\mathsf{N} \tag{10}$$

for each irreducible component. Under such a normalization the sum of scalar products in invariant subspaces will be equal to the standard scalar product in the space of permutation representation.

Solving linear system (7—9) and setting $a_1 = 2/8 = 1/4$, we obtain

$$a_2 = -\frac{1}{4},\ a_3 = -\frac{1+2\mathsf{r}}{12},\ a_4 = \frac{1+2\mathsf{r}}{12}.$$

Applying the same procedure to all other irreducible components we come to the following set of forms defining scalar products in the invariant subspaces

$$\begin{aligned}
\mathcal{B}_{\mathbf{1}} &= \frac{1}{8}\left(\mathcal{A}_1 + \mathcal{A}_2 + \mathcal{A}_3 + \mathcal{A}_4\right) \equiv \frac{1}{8}\mathrm{J}_8,\\
\mathcal{B}_{\mathbf{2}} &= \frac{1}{4}\left(\mathcal{A}_1 - \mathcal{A}_2 - \frac{1+2\mathsf{r}}{3}\mathcal{A}_3 + \frac{1+2\mathsf{r}}{3}\mathcal{A}_4\right),\\
\mathcal{B}_{\mathbf{2}'} &= \frac{1}{4}\left(\mathcal{A}_1 - \mathcal{A}_2 + \frac{1+2\mathsf{r}}{3}\mathcal{A}_3 - \frac{1+2\mathsf{r}}{3}\mathcal{A}_4\right),\\
\mathcal{B}_{\mathbf{3}} &= \frac{3}{8}\left(\mathcal{A}_1 + \mathcal{A}_2 - \frac{1}{3}\mathcal{A}_3 - \frac{1}{3}\mathcal{A}_4\right).
\end{aligned}$$

It is easy to check that normalization (10) ensures the identity $\mathcal{B}_{\mathbf{1}} + \mathcal{B}_{\mathbf{2}} + \mathcal{B}_{\mathbf{2}'} + \mathcal{B}_{\mathbf{3}} = \mathcal{A}_1 \equiv \mathrm{I}_8$. Recall that in all such tasks $\mathcal{B}_{\mathbf{1}}$ does not require calculation, since the inner product in the subspace of the trivial representation always has the form $\frac{1}{\mathsf{N}}\mathrm{J}_{\mathsf{N}}$.

Let $\left|n\right\rangle = \left(n_1, \cdots, n_8\right)^{\mathrm{T}}$ and $\left|m\right\rangle = \left(m_1, \cdots, m_8\right)^{\mathrm{T}}$ be natural vectors in the permutation basis, and $\left|\Psi_k\right\rangle$ and $\left|\Phi_k\right\rangle$ their projections onto the invariant subspaces, where $k = \mathbf{1}, \mathbf{2}, \mathbf{2}', \mathbf{3}$. The scalar products of these projections in the invariant subspaces can be expressed in terms of the natural vectors as

$$\left\langle \Phi_k \,\middle|\, \Psi_k \right\rangle = \left\langle m \left| \mathcal{B}_k \right| n \right\rangle.$$

For example, for the subrepresentation **1** we have

$$\left\langle \Phi_{\mathbf{1}} \,\middle|\, \Psi_{\mathbf{1}} \right\rangle = \frac{1}{8}\left(m_1 + m_2 + \cdots + m_8\right)\left(n_1 + n_2 + \cdots + n_8\right).$$

It is clear that this expression never vanishes for natural vectors $\left|n\right\rangle$ and $\left|m\right\rangle$. In fact, in the general case the trivial one-dimensional subrepresentation contained in any permutation representation can be interpreted as the *"counter of particles"*, since the permutation invariant $\sum_{i=1}^{\mathsf{N}} n_i$ corresponding to this subrepresentation is the total number of elements of the set Ω in the ensemble.

As for the other subrepresentations, we can observe non-trivial "quantum behavior" in their invariant subspaces. In particular, the equations $\left\langle m \left| \mathcal{B}_k \right| n \right\rangle = 0$ — the conditions for destructive interference — have infinitely many solutions in natural vectors for $k = \mathbf{2}, \mathbf{2}', \mathbf{3}$.

Note also, that Born's probabilities computed via the forms $\mathcal{B}_{\mathbf{2}}$ and $\mathcal{B}_{\mathbf{2}'}$ take irrational values for natural vectors. This contradicts the idea that any probability in the finite background must be rational. The contradiction is resolved by the fact that the permutation action, that we consider, is *imprimitive*, i.e., it moves some nontrivial subsets of Ω as single units. Such subsets are called *blocks*. In our example, the block system (called also the *system of imprimitivity*) is the following $\{[1,2],[3,4],[5,6],[7,8]\}$. Thus, we can not take the subrepresentations $\mathbf{2}$ and $\mathbf{2}'$ separately and must consider their sum $\mathbf{2}\oplus\mathbf{2}'$ instead. The invariant form for this sum $\mathcal{B}_{\mathbf{2}\oplus\mathbf{2}'}=\frac{1}{2}\left(\mathcal{A}_1-\mathcal{A}_2\right)$ does not contain irrationalities.

3.2. The case of multiple subrepresentations

In the case of multiple subrepresentations the situation becomes more complicated because of the nonlinearity of relations. The question of whether it is possible to develop a general algorithm for the case of multiple subrepresentations requires a deeper additional study. In many concrete examples, using analogy to the case of single subrepresentations and some *ad hoc* tricks, it is possible to carry out the complete decomposition of scalar products into irreducible components.

The source of nonlinearity is an excess in the number of parameters, the definite values of which are inessential for computing the Born probabilities. As one can see from the structure of decomposition (4), any block of multiple components $\mathcal{B}_k\otimes\mathrm{I}_{d_k}$ contains $m_k\times m_k$ such parameters. Our convention (10) allows to fix m_k diagonal elements of the matrix $\mathcal{B}_k$. So we need to fix somehow the remaining $m_k^2-m_k$ parameters. In some cases, the examination of the structure of the centralizer ring helps to do this.

As an illustrative example, let us consider the Coxeter group A_2, which is also the Weyl group of the Lie group, say, $\mathsf{SU}(3)$. We will consider the natural action of A_2 on its root system $\Omega=\{1,2,\cdots,6\}$. The vectors of this root system form a two-dimensional regular hexagon. The generators of the action are, for example, $g_1=(1,4)(2,3)(5,6)$ and $g_2=(1,3)(2,5)(4,6)$.

Computation of orbitals gives the following basis of the centralizer ring

$$\mathcal{A}_1=\begin{pmatrix}1&\cdot&\cdot&\cdot&\cdot&\cdot\\\cdot&1&\cdot&\cdot&\cdot&\cdot\\\cdot&\cdot&1&\cdot&\cdot&\cdot\\\cdot&\cdot&\cdot&1&\cdot&\cdot\\\cdot&\cdot&\cdot&\cdot&1&\cdot\\\cdot&\cdot&\cdot&\cdot&\cdot&1\end{pmatrix},\quad \mathcal{A}_2=\begin{pmatrix}\cdot&1&\cdot&\cdot&\cdot&\cdot\\\cdot&\cdot&\cdot&\cdot&\cdot&1\\\cdot&\cdot&\cdot&\cdot&1&\cdot\\\cdot&\cdot&1&\cdot&\cdot&\cdot\\\cdot&\cdot&\cdot&1&\cdot&\cdot\\1&\cdot&\cdot&\cdot&\cdot&\cdot\end{pmatrix},\quad \mathcal{A}_3=\begin{pmatrix}\cdot&\cdot&1&\cdot&\cdot&\cdot\\\cdot&\cdot&\cdot&1&\cdot&\cdot\\1&\cdot&\cdot&\cdot&\cdot&\cdot\\\cdot&1&\cdot&\cdot&\cdot&\cdot\\\cdot&\cdot&\cdot&\cdot&\cdot&1\\\cdot&\cdot&\cdot&\cdot&1&\cdot\end{pmatrix},$$

$$\mathcal{A}_4=\begin{pmatrix}\cdot&\cdot&\cdot&1&\cdot&\cdot\\\cdot&\cdot&\cdot&\cdot&1&\cdot\\\cdot&\cdot&\cdot&\cdot&\cdot&1\\1&\cdot&\cdot&\cdot&\cdot&\cdot\\\cdot&1&\cdot&\cdot&\cdot&\cdot\\\cdot&\cdot&1&\cdot&\cdot&\cdot\end{pmatrix},\quad \mathcal{A}_5=\begin{pmatrix}\cdot&\cdot&\cdot&\cdot&1&\cdot\\\cdot&\cdot&1&\cdot&\cdot&\cdot\\\cdot&1&\cdot&\cdot&\cdot&\cdot\\\cdot&\cdot&\cdot&\cdot&\cdot&1\\1&\cdot&\cdot&\cdot&\cdot&\cdot\\\cdot&\cdot&\cdot&1&\cdot&\cdot\end{pmatrix},\quad \mathcal{A}_6=\begin{pmatrix}\cdot&\cdot&\cdot&\cdot&\cdot&1\\1&\cdot&\cdot&\cdot&\cdot&\cdot\\\cdot&\cdot&\cdot&1&\cdot&\cdot\\\cdot&\cdot&\cdot&\cdot&1&\cdot\\\cdot&\cdot&1&\cdot&\cdot&\cdot\\\cdot&1&\cdot&\cdot&\cdot&\cdot\end{pmatrix}.$$

Applying the algorithm of absolute factorization to the generic linear combination

$$A=a_1\mathcal{A}_1+a_2\mathcal{A}_2+a_3\mathcal{A}_3+a_4\mathcal{A}_4+a_5\mathcal{A}_5+a_6\mathcal{A}_6\ ,$$

we obtain the following decomposition:

$$\begin{aligned}\det A=&\left(a_1+a_2-a_3-a_4-a_5+a_6\right)\\&\left(a_1+a_2+a_3+a_4+a_5+a_6\right)\\&\left\{a_1^2+a_2^2-a_3^2-a_4^2-a_5^2+a_6^2-a_1a_2-a_1a_6-a_2a_6+a_3a_4+a_3a_5+a_4a_5\right\}^2.\end{aligned}\tag{11}$$

The structure of the action of the Weyl group A_2 on its roots in terms of the permutation representation follows immediately from (11): $\underline{\mathbf{6}}=\mathbf{1}\oplus\mathbf{1}'\oplus\mathbf{2}\oplus\mathbf{2}$. Here we have four irreducible components and a six-dimensional centralizer ring. Thus, we need to eliminate the two extra degrees of freedom. Basis matrices derived from the orbitals are (0,1)-matrices with disjoint sets of unit entries. A result of addition of such matrices belongs to the same type of matrices. So, the natural idea is to sum up some of the matrices $\mathcal{A}_i$ in order to reduce their total number. Recall that if the centralizer ring is commutative, then all the multiplicities are equal to one. So we retain the commutative subset of matrices unchanged. In our case, such a subset consists of the matrices $\mathcal{A}_1$, $\mathcal{A}_2$ and $\mathcal{A}_6$. "Killing noncommutativity" by replacing the remaining non-commuting matrices by their sum $\mathcal{A}_\Sigma = \mathcal{A}_3 + \mathcal{A}_4 + \mathcal{A}_5$, we write the linear combination of four matrices

$$A' = a_1\mathcal{A}_1 + a_2\mathcal{A}_2 + a_6\mathcal{A}_6 + a_\Sigma\mathcal{A}_\Sigma \ .$$

Now the determinant is factorizable to the linear factors over the cyclotomic integer ring $\mathbb{N}_3$:

$$\begin{aligned}\det A' = &(a_1 + a_2 + a_6 - 3a_\Sigma)\\ &(a_1 + a_2 + a_6 + 3a_\Sigma)\\ &\{a_1 + \mathsf{r}a_2 - (1+\mathsf{r})\,a_6\}^2\\ &\{a_1 - (1+\mathsf{r})\,a_2 + \mathsf{r}a_6\}^2 ,\end{aligned}$$

where r is a third root of unity. After the same manipulations as above, we come to the following set of scalar product forms in the invariant subspaces

$$\begin{aligned}\mathcal{B}_\mathbf{1} &= \frac{1}{6}\left(\mathcal{A}_1 + \mathcal{A}_2 + \mathcal{A}_6 + \mathcal{A}_\Sigma\right) \equiv \frac{1}{6}\mathrm{J}_6,\\ \mathcal{B}_{\mathbf{1}'} &= \frac{1}{6}\left(\mathcal{A}_1 + \mathcal{A}_2 + \mathcal{A}_6 - \mathcal{A}_\Sigma\right),\\ \mathcal{B}_\mathbf{2} &= \frac{1}{3}\left\{\mathcal{A}_1 - (1+\mathsf{r})\,\mathcal{A}_2 + \mathsf{r}\mathcal{A}_6\right\},\\ \widetilde{\mathcal{B}}_\mathbf{2} &= \frac{1}{3}\left\{\mathcal{A}_1 + \mathsf{r}\mathcal{A}_2 - (1+\mathsf{r})\,\mathcal{A}_6\right\}.\end{aligned}$$

Here $\mathcal{B}_\mathbf{2}$ and $\widetilde{\mathcal{B}}_\mathbf{2}$ are different coordinate presentations of the same form, associated with the irreducible representation **2**.

4. Conclusions

Elimination of actual infinities from the description of physical reality removes many technical difficulties. This allows to focus on the content aspects of physical problems. We have considered some implications of the idea that quantum mechanics, as any reasonable physical theory, must allow effective modeling by finite means.

Using mathematical arguments of general nature we can show that any quantum problem can be reduced to permutations. If we assume also that the entities, which are subject to the permutations, have a physical meaning, we come to a very simple and self-consistent picture of the quantum behavior. To study the consequences of this assumption, we need to know the inner products in the invariant subspaces of permutation representations.

In this paper, we have considered a possible algorithmic approach to calculating these inner products. With this approach, we can for any permutation representation obtain the full information about the structure of its decomposition into irreducible components. If all irreducible components have unit multiplicities then the invariant inner products can be easily computed. There are many observations which indicate that in the case of multiple subrepresentations a reasonable algorithm is also possible.

Acknowledgments
The work was partially supported by the grants 10-01-00200 from the Russian Foundation for Basic Research and 3802.2012.2 from the Ministry of Education and Science of the Russian Federation.

References

[1] Mal'cev A 1940 On isomorphic matrix representations of infinite groups *Mat. Sbor.* **8(50)**(3) 405–422 (Russian)
[2] Magnus W 1969 Residually finite groups *Bull. Amer. Math. Soc.* **75**(2) 305–316
[3] Kornyak V V 2012 Permutation interpretation of quantum mechanics *J. Phys.: Conf. Ser.* **343** 012059 doi:10.1088/1742-6596/343/1/012059
[4] Kornyak V V 2013 Classical and quantum discrete dynamical systems *Phys. Part. Nucl.* **44**(1) 47-91 (*Preprint* arXiv:1208.5734 [quant-ph])
[5] Holt D F, Eick B and O'Brien E A 2005 *Handbook of Computational Group Theory* (Boca-Raton—London—New-York—Washington-D.C.: Chapman & Hall/CRC)
[6] Cameron P J 1999 *Permutation Groups* (Cambridge: Cambridge University Press)
[7] Dixon J D and Mortimer B 1996 *Permutation Groups* (Berlin: Springer)
[8] Wielandt H 1964 *Finite Permutation Groups* (New-York—London: Academic Press)

Classical and quantum-mechanical axioms with the higher time derivative formalism

Timur Kamalov

Physics Department, Moscow State Open University, P. Korchagina street, 22, Moscow, 107996, Russia
Moscow Polytechnic Institute in Moscow State University of Mechanical Engineering, P. Korchagina street, 22, Moscow, 107996, Russia

E-mail: TimKamalov@gmail.com

Abstract. A Newtonian mechanics model is essentially the model of a point body in an inertial reference frame. How to describe extended bodies in non-inertial (vibration) reference frames with the random initial conditions? One of the most generalized ways of descriptions (known as the higher derivatives formalism) consists in taking into account the infinite number of the higher temporal derivatives of the coordinates in the Lagrange function. Such formalism describing physical objects in the infinite dimensions space does not contradict to the quantum mechanics and infinite dimensions Hilbert space.

1. Introduction

Classical Newtonian mechanics is essentially the simplest way of mechanical system description with second-order differential equations, when higher order time derivatives of coordinates can be neglected. The extended model of Newtonian mechanics with higher time derivatives of coordinates is based on generalization of Newton's Laws onto arbitrary reference systems (both inertial and non-inertial ones) with body dynamics being described with higher order differential equations. Newton's Laws, constituting, from the mathematical viewpoint, the axiomatic of classical physics, actually postulate the assertion that the equations describing the dynamics of bodies in inertial reference frames are second-order differential equations. However, the actual time-space is almost without exception non-inertial, as it is almost without exception that there exist (at least weak) fields, waves, or forces perturbing an ideal inertial reference frames. It corresponds to Mach's principle [1] with a general statement "Local physical laws are determined by the large-scale structure of the Universe." Non-inertial nature of the actual time-space is also supported by observations of the practical astronomy that expansion of the Universe occurs with an acceleration. In other words, actually any real reference frame is a non-inertial one, and such physical reality can be described by a differential equation with time derivatives of coordinates of an order exceeding two, which play the role of additional variables. Aristotle's physics considered velocity to be proportional to the applied force, hence the body dynamics is described by a first-derivative differential equation. Classical Physics in the inertial reference frames describe a free body maintains the constant velocity of the translational motion. In this case, the body dynamics is described by a second-order differential equation, with acceleration being proportional to the force [2]. This corresponds to the Lagrange function depending on coordinates and their first derivatives (velocities) of the body, and the Euler–Lagrange equation resulting from the principle of the least action. This model of the physical reality describes macrocosm fairly well, but it fails to describe microparticles. Both Newtonian axiomatic and the second Newton law are not valid in the microcosm. Only averaged values of observable physical quantities yield in the microcosm an approximate analogue of the second Newton law; this is the so-called Ehrenfest's theorem. The Ehrenfest's equation yields an averaged, rather than precise, ratio between the second time derivative of coordinate and the force, while to describe the scatter of quantum observables, the probability theory apparatus is required. Since Newtonian dynamics is restricted to the second-order derivatives, while microobjects must be described by equations with additional variables, making the Planck's constant tend to zero corresponds to neglecting these variables. Hence, offering the model of extended Newtonian dynamics, we consider the classical and quantum theories with additional variables, describing the body dynamics with higher-order differential equations. In our model, the Lagrangian is considered to be dependent not only on coordinates and their first time derivatives, but also on higher-order time derivatives of coordinates. Classical dynamics of a test particle's motion with higher-order time derivatives of the coordinates was first described in 1850 by Ostrogradski [3] and is known as Ostrogradski's Formalism. Being a mathematician, Ostrogradski considered the coordinate systems rather than the reference frames. This is just the case corresponding to a real reference frame comprising both inertial and non-inertial reference frames. In the general case, the Lagrangian takes on the form

$$L = L(q, \dot{q}, \ddot{q}, ..., \dot{q}^n, ...). \tag{1}$$

Attempts to build a unified theory for both quantum and classical mechanics, exemplified by the Hybrid theory of classical and quantum mechanics [4], are natural and make sense. However, constructing a theory without an axiomatic consistency with both theories resembles a construction without foundation. The point is that the systems of axioms of classical and quantum theories are mutually incompatible and even contradictory. Obviously, the mathematical constructions of united theory may prove questionable without presenting a framework which is conceptually consistent both with classical and quantum mechanics. For example, a natural question arises: "Can the phase space be used, and can the momentum and the coordinate exist simultaneously in quantum mechanics?" The Heisenberg uncertainty principle tells us that it is not possible.

At the same time, quantum theory describes objects in Hilbert space, i.e. in terms of an infinite number of variables, and thus it gives a more detailed description as compared to classical theory. Thus, the question of quantum-mechanical description being incomplete should be answered so that any theory as a model of physical reality is incomplete following Godel's theorem.

The classical description of physical reality contains an incomparably fewer number of variables. This raises the question: "How the classical description can be completed?" While a possibility of supplementing the quantum mechanical description with additional ("hidden") variables has been debated for long, the question as to how to complete the classical description to make it compatible with the quantum mechanical one has not received a due attention.

The classical and quantum theory could be based on the following common axioms:

1. Any reference frame is subject to random external influences. Hence, every reference frame is individual and a transition from one to another reference frame may lead to jump like changes. The notion of the inertial frame in classical mechanics is valid only on the average and, correspondingly, the Galilean relativity is an average notion as well. Then there are many trajectories of a particle corresponding to different reference frames; the Heisenberg uncertainty can be understood as a consequence of the nonexistence of ideal inertial frames; and the Ehrenfest theorem can be seen as a consequence of the inertial frame being an average notion.

Correspondingly, *the free body preserves the same order of its time derivative as the constant kinematic characteristics of the class of reference frames*, e.g. in the uniformly accelerating reference frame the free body preserves its acceleration.

2. In Hybrid theory [4] the generalized Ehrenfest relations for the QM observables are defined by coordinates average. Compare the averaging procedure in this paper with the time averaging. Within the above framework were the ideal inertial frames are non-existent, we can consider the averaging of the classical equations of motion over the time interval Δt:

$$-\frac{\partial U}{\partial q} = \frac{d}{dt}\frac{[p(t+\Delta t)+p(t-\Delta t)]}{2}.$$

Using the Taylor expansion

$$p(t \pm \Delta t) = p(t) \pm \dot{p}\Delta t + \frac{1}{2!}\ddot{p}(t)\Delta t^2 + ... + \frac{(-1)^n}{n!}\overset{.(n)}{p}(t)\Delta t^n + ..$$

the function $F = -\frac{\partial U}{\partial q}$ can be expanded as follows:

$$F(q.\dot{q}, \ddot{q}, \dddot{q}..., \overset{.(k)}{q}) = \dot{p}(t) + \frac{1}{2!}\overset{.(3)}{p}(t)\Delta t^2 + \frac{1}{4!}\overset{.(5)}{p}(t)\Delta t^4 + ...,$$

where $\overset{.(n)}{p}$ denotes n-th time derivative of momentum p. *It is the Extended Law of Dynamics in arbitrary reference frames* including the case of the vibration non-inertial reference frames. Correspondingly, the free body preserves the same order of its time derivative like the constant kinematic characteristics of the reference frames. For example, in the uniformly accelerating reference frame the free body preserves its acceleration.

3. The de-Broglie waves $\psi = \psi_0 \exp(-iS/\hbar)$ with the actions functions $S = S(q, \dot{q}, \ddot{q}, ..., \dot{q}^n, ...)$ can be considered as having the gravity-inertial nature following from the fact that every *reference frame is vibrational due to the influence of random gravitational fields and waves* so that every free particle appears to be oscillating.

4. As the action function $S = S(q, \dot{q}, \ddot{q}, ..., \dot{q}^n, ...)$ is a convergent series in high derivatives of q the difference $S(q, \dot{q}, \ddot{q}, ..., \dot{q}^n, ...) - S(q, \dot{q}) = h$ is finite and can be identified with the constant h. Within the presented framework the variables of *the (high order extension of the) phase space do describe the completed dynamics of a particle*, but they cannot be measured because the ideal inertial reference frames do not exist in really. The infinite dimensionality of Hilbert space can also be understood as a consequence of all high order time derivatives being taken into account in the description of the dynamics.

2. How to complete the quantum-mechanical description?

If the statement by Einstein, Podolsky and Rosen on incompleteness of Quantum-Mechanical description of nature is correct, then we can regard Quantum Mechanics as a Method of Indirect Computation. The problem is, whether the theory is incomplete or the nature itself does not allow complete description? And if the first option is correct, how is it possible to complete the Quantum-Mechanical description? Here we try to complement de-Broglie's idea on wave-pilot the stochastic inertial-gravitation gives origin to. We assume that de-Broglie's wave-pilots are inertial-gravitational stochastic ones with the high derivatives, and we shall regard micro-objects as test classical particles being subject to the influence of de-Broglie's waves stochastic inertial-gravitation. The Quantum Theory exists for many decades. But is everything OK with it completeness[5]? To our opinion, it is not just so. The incompleteness of Quantum-Mechanical description gives rise to various paradoxes, such as Einstein-Podolsky-Rosen (EPR) one, the paradox of the Schrodinger's cat, the Paradox of Quantum Non-locality and Paradox of the Quantum Teleportation. In this study we shall call the phenomena of quantum nonlocal behavior and teleportation of the quantum states as paradoxes because they follow from Stochastic

Gravitation Model of Quantum Mechanics. It can be easily seen that these are paradoxes, and indeed they are brought about by the drawbacks in the Quantum Theory rather than being actual properties of nature. This is due to the fact that time in Quantum Theory plays the role not conforming to the physical reality. In particular, the Quantum Theory employs the concept of Hilbert Space, in which time acts as a parameter. Henceforth, this parameter (i.e. the time) may be the same in different points of the Hilbert Space. This property of time in the Hilbert Space brings about the effects of simultaneous quantum states of microobjects at different space points (or transfer of the state from one Hilbert Space point to another with velocities exceeding the velocity of light). These effects of the Quantum Theory that are apparently real we call here the Quantum Non-locality Paradox. The Paradox of Quantum Teleportation is a sort of Quantum Non-locality Paradox. These paradoxes do not exist in the Classical Physics and in the Stochastic Gravitation Model of the Quantum Mechanic, and General Relativity Theory (employing the 4-dimensional space), in which different points of time-space correspond to different values of the time. And another question is whether the quantum-mechanical wave-function interpretation of micro-objects is complete? Let us select harmonic coordinates (the condition of harmonicas of coordinates mean selection of concomitant frame $\frac{\partial h^m_n}{\partial x^m} = \frac{1}{2}\frac{\partial h^m_m}{\partial x^n}$) and let us take into consideration that $h_{\mu\nu}$ satisfies the gravitational field equations

$$\Box h_{mn}(j) = -16\pi G S_{mn}(j), \tag{2}$$

which follow from the General Theory of Relativity; here S_{mn} is energy-momentum tensor of gravitational field sources with d'Alemberian $\Box$ and gravity constant G. Then, the solution shall acquire the form

$$h_{\mu\nu}(j) = e_{\mu\nu}(j)\exp(ik_\gamma(j)x^\gamma) + e^*_{\mu\nu}(j)\exp(-ik_\gamma(j)x^\gamma), \tag{3}$$

where the value $h_{\mu\nu}(j)$ is called metric perturbation, $e_{\mu\nu}(j)$ polarization, and $k_\gamma(j)$ is 4-dimensional wave vector.

We shall assume that this metric perturbation $h_{\mu\nu}(j)$ is distributed in space with an unknown distribution function $\rho = \rho(h_{\mu\nu})$. Relative oscillations ℓ of two particles in classic gravitational fields are described in the General Theory of Relativity by deviation equations, which we can write for the stochastic case as

$$\frac{D^2}{D\tau^2}\ell^\mu(j) + R^\mu_{\nu mn}(j)\ell^m \frac{dx^\nu}{d\tau}\frac{dx^n}{d\tau} = F(j), \tag{4}$$

being $R^\mu_{\nu mn}(j)$ the gravitational field Riemann's tensor with gravitational field number j of the stochastic gravitational fields and $F(j)$ is the stochastic constant (for the non-stochastic case this constant is zero $F(j) = 0$).

Specifically, the deviation equations give the equations for two particles oscillations

$$\ddot{\ell}^1 + c^2 R^1_{010}\ell^1 = 0, \quad \omega = c\sqrt{R^1_{010}}. \tag{5}$$

The solution of this equation has the form

$$\ell^1(j) = \ell_0 \exp(k_a x^a + i\omega(j)t), \tag{6}$$

being $a = 1, 2, 3$. Each gravitational field or wave with index j and Riemann's tensor $R^\mu_{\nu mn}(j)$ shall be corresponding to the value $\ell^\mu(j)$ with stochastically modulated phase $\Phi(j) = \omega(j)t$. If we to sum the all fields, we can write the stochastic phase $\Phi(t) = \omega(t)t$, where t is the time coordinate.

3. Corrected Bell inequalities in random gravity-inertial fields

We shall consider the physical model with the Stochastic Inertial-Gravitational Background [i.e. with the background of inertial-gravitational fields and waves]. This means that we assume the non-inertial vibration reference frames due to existence of fluctuations in inertial-gravitational waves and fields expressed mathematically by metric fluctuations.

Describing entangled photons in the stochastic curved space, we shall take into consideration the fact that the scalar product of two 4-vectors A^{μ} and B^{ν} equals $g_{\mu\nu}A^{\mu}B^{\nu}$, where for weak inertial-gravitational fields one can use the value $h_{\mu\nu}$, which is the solution of Einstein's equations for the case of weak inertial-gravitational field in harmonic coordinates.

Correlation factor M of random variables λ^{i} [6] which correspondent to higher-order time derivatives of coordinates are projections onto directions A^{ν} and B^{n} defined by polarizers (all these vectors being unit) is

$$|M_{AB}| = |\langle AB\rangle| = \left|\langle \lambda^{i}A^{k}g_{ik}\lambda^{m}B^{n}g_{mn}\rangle\right|. \tag{7}$$

The differential geometry gives

$$\cos\phi = \frac{g_{ik}\lambda^{i}A^{k}}{\sqrt{\lambda^{i}\lambda_{i}}\sqrt{A^{k}A_{k}}},$$
$$\cos(\phi+\theta) = \frac{g_{mn}\lambda^{m}B^{n}}{\sqrt{\lambda^{m}\lambda_{m}}\sqrt{B^{n}B_{n}}}.$$

Here i, k, m, n possess 0,1,2,3; θ is angle between polarizers, then

$$|M_{AB}| = \left|\frac{1}{\pi}\int_{0}^{\pi}\rho(\phi)\cos\phi\cos(\phi+\theta)\,d\phi\right| = |\cos\theta|, \tag{8}$$

here the integral of the distribution function of the metric ρ is $\left|\frac{1}{\pi}\int_{0}^{\pi}\rho(\phi)d\phi\right| = 1$.
Finally, the real part of the correlation factor is

$$|M_{AB}| = |\cos\theta|.$$

Then, we obtain the maximum value the Bell's observable S in Rieman's space for $\theta = \frac{\pi}{4}$

$$|< S >| = |[\langle M_{AB}\rangle + \langle M_{A'B}\rangle + \langle M_{AB'}\rangle - \langle M_{A'B'}\rangle]| =$$
$$= \left|[\cos(-\tfrac{\pi}{4}) + \cos(\tfrac{\pi}{4}) + \cos(\tfrac{\pi}{4}) - \cos(\tfrac{3\pi}{4})]\right| = \left|\sqrt{2}\right|,$$

which agrees fairly with the experimental data. The Bell in equality in Rieman's space shall take on the form $|\langle S\rangle| \leq \sqrt{2}$.

Therefore, we have shown that the Classical Physics with the Stochastic Inertial-Gravitational Background gives the value of the Bell's observable matching both the experimental data and the quantum mechanical value of the Bell's observable. To sum it up, the description of microobjects by the classical physics accounting for the effects brought about by the Inertial-Gravitational Background is equivalent to the Quantum-Mechanical descriptions, both agreeing with the experimental data.

4. Conclusions

We are regarding the inertial-gravitational background of isotopic fields and waves as hidden variables. The inertial-gravitational background could be considered negligible and not affecting the behaviour of quantum microobjects. We are verified whether this is correct. The quantitative assessments of the inertial-gravitational background influence on the quantum microobjects' behaviour have not been performed due to the former having never been examined. The quantum effects are small as well, but their quantitative limits are known and are determined by the Heisenberg inequality. We have demonstrated the inertial-gravitational background being random and isotropic to affect the phases of microobjects. Our case corresponds to the Lagrange function $L(t, q, \dot{q}, \ddot{q}, ..., \dot{q}^{(n)}, ...)$, depending on coordinates, velocities and higher time derivatives, which we call additional variables, extra addends, or hidden variables. In arbitrary reference systems (including non-inertial ones) additional variables (addends) appear in the form of higher time derivatives of coordinates, which complement both classical and quantum physics. We call these additional addends, or variables, constituting the higher time derivatives of coordinates, hidden variables or hidden parameters, complementing the description of particles. The contemporary physics presupposes employment of predominantly inertial reference systems; however, such a system is very hard to obtain, as there always exist external perturbative effects, for example, gravitational forces, fields, or waves. In this case, the relativity principle enables transfer from the gravitational forces or waves to inertial forces. If the fact that the real reference frames is non-inertial and hence there exist additional variables in the form of inertial-gravitation effects is ignored, then non-local correlation of quantum states and quantum non-locality would seem surprising. The inertial-gravitational origin of quantum-mechanical wave functions in the form of non-local hidden variables described [7-9].

References

[1] Mach E 1897 *Die Mechanik in ihrer Entwicklung: Historisch-Kritisch Dargestellt* (Leipzig: F A Brockhaus)
[2] Newton I S 1687 *Philosophiae naturalis principia mathematica* (London: S Pepys)
[3] Ostrogradski M V 1850 *Memoires de l'Academie Imperiale des Sciences de Saint-Peterbourg* **6** 385
[4] Elze H T 2012 *Phys. Rev.* A **85** 052109;
Hall M J W and Reginatto M 2005 *Phys. Rev.* A **72** 062109;
Radonjic M, Prvanovic S and Buric N 2012 *Phys. Rev.* A **85** 064101;
Heslot A 1985 *Phys. Rev.* D **31** 1341
[5] Einstein A, Podolsky B and Rosen N 1935 *Phys. Rev.* **47** 777
[6] Bell J S 1964 *Physics* **1** (3) 195
[7] Kamalov T F 2001 *Journal of Russian Laser Research* **22** (5) 475
[8] Kamalov T F 2009 *Journal of Russian Laser Research* **30** (5) 466
[9] Kamalov T F 2003 in *Quantum Theory: Reconsideration of Foundations-2* ed A Khrennikov (Sweden: Vaxjo Univ. Press) pp 315-322

Noncommutative $\mathbb{C}P^N$ and $\mathbb{C}H^N$ and their physics

Akifumi Sako, Toshiya Suzuki and Hiroshi Umetsu
Kushiro National College of Technology
Otanoshike-Nishi 2-32-1, Kushiro 084-0916, Japan

Abstract. We study noncommutative deformation of manifolds by constructing star products. We start from a noncommutative $\mathbb{R}^d$ and discuss more genaral noncommutative manifolds. In general, star products can not be described in concrete expressions without some exceptions. In this article we introduce new examples of noncommutative manifolds with explicit star products. Karabegov's deformation quantization of $\mathbb{C}P^N$ and $\mathbb{C}H^N$ with separation of variables gives explicit calulable star products represented by gamma functions. Using the results of star products between inhomogeneous coordinates, we find creation and anihilation operators and obtain the Fock representation of the noncommutative $\mathbb{C}P^N$ and $\mathbb{C}H^N$.

1. Introduction

Phyisical theories in noncommutative spaces are realized in several situations. For example, in string theories, low energy effective theories on D-branes in a constant background NS-NS B fields are given as noncommutative gauge theories. In the theories, B field plays a role of an inverse of a noncommutative parameter matrix that represents noncommutativity of space coordinates. Another example in string theories is a series of matrix models which are constructed so that all dynamical variables are given by matricies. Spacetime coordinates themselves are also described as matrices, and then the spacetime becomes a noncommutative space.

By the way, how can we obtain the noncommutative space? Let us consider a simple example, the noncommutative $\mathbb{R}^d$ using the Moyal product. The Moyal product in $\mathbb{R}^d$ is defined as

$$(f * g)(x) = e^{\frac{i}{2}\theta^{ij}\partial_i^x \partial_j^y} f(x)g(y)\Big|_{y=x}.$$

Here θ^{ij} is an element of a skew symmetric constant matrix θ and is called a noncommutative parameter. The right hand side is denoted by $f(x)e^{\frac{i}{2}\theta^{ij}\overleftarrow{\partial}_i \overrightarrow{\partial}_j}g(x)$, for simplicity. Note that the commutation relation of the coordinates are given by

$$[x^i,\ x^j]_* = x^i * x^j - x^j * x^i = i\theta^{ij}.$$

In the usual commutative case, the ordinary commutative product is employed to define a ring of functions. If we chose the Moyal product to define a ring of functions, then the considering space become noncommutative because all products between arbitrary functions are replaced by the Moyal product. This is known as noncommutative $\mathbb{R}^d$.

In the following, we move to more general spaces. To make a notion of noncommutative deformation of space be mathematically rigor, we introduce a definition of deformation quantization.

Definition 1. Deformation Quantization (strong sense)
Deformation Quantization is defined as follows. $\mathcal{F}$ is defined as a set of formal power series of $\hbar$:

$$\mathcal{F} := \Big\{ f \;\Big|\; f = \sum_k f_k \hbar^k \Big\}.$$

A star product of functions f and g is defined as

$$f * g = \sum C_k(f,g)\hbar^k$$

such that the product satisfies the following conditions.

(i) $*$ is associative product.
(ii) C_k is a bidifferential operator.
(iii) C_0 and C_1 are defined as

$$C_0(f,g) = fg, \qquad C_1(f,g) = \frac{i}{2}\{f,g\},$$

where $\{f,g\}$ is the Poisson bracket.
(iv) $f * 1 = 1 * f$.

Deformation quantization of manifolds are given by this star product in a form of a formal power series in a deformation parameter $\hbar$. The power series is obtained as solutions of an infinite system of differential equations in general. The existence of the solution is proved for a wide class of manifolds, however explicit expressions of deformation quantizations are constructed for few kinds of manifolds. For example, Euclidean spaces are deformed by using the Moyal product as we saw above, and on manifolds with spherically symmetric metrics explicit star products are given as concrete calcurable forms in the context of the Fedosov's deformation quantization [1]. The Fedosov's Deformation Quantization is constructed as the Moyal product on the Weyl algebra bundle in essentials.

As mentioned above, there are few noncommutative manifolds that have explicit expression of star products, for example 2 dimensional manifolds with circular symmetric metrics. Following three examples are given by the methods of Fedosov's Deformation Quantization.

(1). $\mathbb{R}^2$

$$f(r,\varphi) * g(r,\varphi) = f(\sqrt{r^2 + 2y^1 r}, \varphi + \frac{y^2}{r}) \exp\Big(-\frac{i\hbar}{2}\frac{\overleftarrow{\partial}}{\partial y^i}\omega^{ij}\frac{\overrightarrow{\partial}}{\partial y^j}\Big)$$
$$g(\sqrt{r^2 + 2y^1 r}, \varphi + \frac{y^2}{r})\Big|_{y=0},$$

where (r,φ) is a polar coordinate and $\omega^{12} = -\omega^{21} = 1, \omega^{11} = \omega^{22} = 0$.
(2). $\mathbb{C}P^1$

$$f(r,\varphi) * g(r,\varphi) = f(\sqrt{\frac{2y^1 r(1+r^2) + r^2}{-2y^1 r(1+r^2) + 1}}, \varphi + \frac{y^2}{r})$$
$$\exp\Big(-\frac{i\hbar}{2}\frac{\overleftarrow{\partial}}{\partial y^i}\omega^{ij}\frac{\overrightarrow{\partial}}{\partial y^j}\Big) g(\sqrt{\frac{2y^1 r(1+r^2) + r^2}{-2y^1 r(1+r^2) + 1}}, \varphi + \frac{y^2}{r})\Big|_{y=0}.$$

(3). $\mathbb{C}H^1$

$$
\begin{aligned}
&f(r,\varphi)*g(r,\varphi)\\
=&f\Big(\sqrt{\frac{2y^1r(1-r^2)+r^2}{2y^1r(1-r^2)+1}},\varphi+\frac{y^2}{r}\Big)\exp\Big(-\frac{i\hbar}{2}\overleftarrow{\frac{\partial}{\partial y^i}}\omega^{ij}\overrightarrow{\frac{\partial}{\partial y^j}}\Big)\\
&g\Big(\sqrt{\frac{2y^1r(1-r^2)+r^2}{2y^1r(1-r^2)+1}},\varphi+\frac{y^2}{r}\Big)\Big|_{y=0}.
\end{aligned}
$$

The purpose of this article is to introduce another way to have explicit expressions of star products, in particular for Kähler manifolds, and to construct star products for $\mathbb{C}P^N$ and $\mathbb{C}H^N$ by using special functions.

2. Deformation quantization of Kähler manifolds

In this article, noncommutative space is constructed by using deformation quantization. In the following, we employ slight differenct deformation quantization from the one in the previous section. The deformation quantization is defined as follows.

Definition 2. Let $\mathcal{F}$ be a set of formal power series of $\hbar$:

$$\mathcal{F} := \Big\{ f \;\Big|\; f = \sum_k f_k \hbar^k \Big\}.$$

A star product of $f, g \in \mathcal{F}$ is defined as

$$f * g = \sum C_k(f,g)\hbar^k,$$

such that the product satisfies the following conditions.

(i) $*$ is associative product.

(ii) C_k is a bidifferential operator.

(iii) C_0 and C_1 is defined as

$$C_0(f,g) = fg, \qquad C_1(f,g) - C_1(g,f) = i\{f,g\}.$$

where $\{f,g\}$ is the Poisson bracket.

(iv) $f * 1 = 1 * f$.

The difference from the definition of the deformation quantization in the previous section is in the condition for C_1.

There is a star product called star product with separation of variables, which is defined on Kähler manifolds. A Kähler manifold have a Kähler potential Φ and a Kähler 2-form ω

$$\omega = i g_{i\bar{j}} dz^i \wedge d\bar{z}^j, \qquad g_{i\bar{j}} = \frac{\partial^2 \Phi}{\partial z^i \partial \bar{z}^j}.$$

$*$ is called a star product with separation of variables when

$$a * f = af$$

for holomorphic function a and

$$f * b = fb$$

for anti-holomorphic function b. Karabegov showed that for arbitrary ω, there exists a star product with separation variables $*$ [2] (see also [3, 4]). In this method of deformation qunatization, a star product is constructed as a formal power series of differential operators. Let L_f be a differential operator corresponding to a left $*$ multiplication by f:

$$L_f g := f * g.$$

Then L_f has the following form:

$$L_f = \sum_{n=0}^{\infty} \hbar^n A_n,$$

, where formal power series of differential operators

$$A_n = a_{n,\alpha}(f) \prod_i \left(D^{\bar{i}}\right)^{\alpha_i}, \qquad (D^{\bar{i}} = g^{\bar{i}j}\partial_j).$$

It is required that L_f satisfies

$$L_f 1 = f * 1 = f,$$
$$L_f (L_g h) = f * (g * h) = (f * g) * h = L_{L_f g} h.$$

L_f which has the properties described above is determined by the following condition.

$$[L_f, \partial_{\bar{i}}\Phi + \hbar\partial_{\bar{i}}] = 0,$$

and $A_0 = f$. This condition is equivalent to the recursion relations

$$[A_n,\ \partial_{\bar{i}}\Phi] = [\partial_{\bar{i}},\ A_{n-1}].$$

If one obtains the operator $L_{\bar{z}^i}$ $(L_{\bar{z}^l} f = \bar{z}^l * f)$, L_f is given by

$$L_f = \sum_{\alpha} \frac{1}{\alpha} \left(\frac{\partial}{\partial \bar{z}}\right)^{\alpha} f \left(L_{\bar{z}} - \bar{z}\right)^{\alpha}.$$

Here, α is a multi-index, $\alpha = (\alpha_1, \cdots, \alpha_m)$. It is not easy to derive explicit expressions of star products in all order of $\hbar$ by solving the recursion relation. In the next section, this recursion relation is solved by using gamma functions for the case of $\mathbb{C}P^N$.

3. Noncommutative deformation of $\mathbb{C}P^N$

From this section to the end of this article, we studied noncommutative $\mathbb{C}P^N$ and $\mathbb{C}H^N$. The detail derivations of the following results are found in [5].

Let z^i $(i = 1, 2, \cdots, N)$ be inhomogeneous coordinates of $\mathbb{C}P^N$. Then, the Kähler potential of $\mathbb{C}P^N$ is given by

$$\Phi = \ln\left(1 + |z|^2\right), \qquad (|z|^2 = \sum_i z^i \bar{z}^i).$$

The complex metric $(g_{i\bar{j}})$ is derived from the Kähler potential as

$$ds^2 = 2g_{i\bar{j}}dz^i d\bar{z}^j,$$
$$g_{i\bar{j}} = \partial_i \partial_{\bar{j}} \Phi = \frac{(1+|z|^2)\delta_{ij} - z^j \bar{z}^i}{(1+|z|^2)^2}$$

Its inverse metric $(g^{\bar{i}j})$ is

$$g^{\bar{i}j} = (1+|z|^2)\left(\delta_{ij} + z^j \bar{z}^i\right).$$

The following relations simplify our calculations of L_f in the case of $\mathbb{C}P^N$,

$$\partial_{\bar{i}_1}\partial_{\bar{i}_2}\cdots\partial_{\bar{i}_n}\Phi = (-1)^{n-1}(n-1)!\; \partial_{\bar{i}_1}\Phi\partial_{\bar{i}_2}\Phi\cdots\partial_{\bar{i}_n}\Phi,$$
$$\text{Riemann tensor:}\ \ R_{i\bar{j}k\bar{l}} = -g_{i\bar{j}}g_{k\bar{l}} - g_{i\bar{l}}g_{k\bar{j}}.$$

Let us construct $L_{\bar{z}^l}$ $(L_{\bar{z}^l} f = \bar{z}^l * f)$. Recall that $L_{\bar{z}^l}$ is described as

$$L_{\bar{z}^l} = \bar{z}^l + \hbar D^{\bar{l}} + \sum_{n=2}^{\infty} \hbar^n A_n,$$

where A_n $(n \geq 2)$ is a formal series of $D^{\bar{k}}$. Then we would like to have the concrete form of A_n. We assume that A_n has the following form,

$$A_n = \sum_{m=2}^{n} a_m^{(n)} \partial_{\bar{j}_1}\Phi \cdots \partial_{\bar{j}_{m-1}}\Phi D^{\bar{j}_1}\cdots D^{\bar{j}_{m-1}} D^{\bar{l}},$$

where the coefficients $a_m^{(n)}$ do not depend on z^i and $\bar{z}^i$. From $[L_{\bar{z}^l}, \partial_{\bar{i}}\Phi + \hbar\partial_{\bar{i}}] = 0$, A_n are recursively determined by

$$[A_n, \partial_{\bar{i}}\Phi] = [\partial_{\bar{i}}, A_{n-1}], \qquad (n \geq 2)$$

where $A_1 = D^{\bar{l}}$. After some calculations, we found the following recursion relation

$$a_m^{(n)} = a_{m-1}^{(n-1)} + (m-1)a_m^{(n-1)}.$$

and $a_2^{(n)} = a_2^{(n-1)} = \cdots = a_2^{(2)} = 1$. To solve these equations, we introduce a generating function

$$\alpha_m(t) \equiv \sum_{n=m}^{\infty} t^n a_m^{(n)}, \qquad (m \geq 2).$$

From the recursion relation, $\alpha_m(t)$ is determined as

$$\alpha_2(t) = \sum_{n=2}^{\infty} t^n a_2^{(n)} = \sum_{n=2}^{\infty} t^n = \frac{t^2}{1-t},$$
$$\alpha_m(t) = t^m \prod_{n=1}^{m-1} \frac{1}{1-nt} = \frac{\Gamma(1-m+\frac{1}{t})}{\Gamma(1+\frac{1}{t})}, \qquad (m \geq 2).$$

The coefficient $a_m^{(n)}$ is related to the Stirling number of the second kind $S(n,k)$,

$$a_m^{(n)} = S(n-1, m-1).$$

Summarizing the above calculations, $L_{\bar{z}^l}$ becomes

$$\begin{aligned} L_{\bar{z}^l} &= \bar{z}^l + \hbar D^{\bar{l}} + \sum_{n=2}^{\infty} \hbar^n \sum_{m=2}^{n} a_m^{(n)} \partial_{\bar{j}_1}\Phi \cdots \partial_{\bar{j}_{m-1}}\Phi D^{\bar{j}_1} \cdots D^{\bar{j}_{m-1}} D^{\bar{l}} \\ &= \bar{z}^l + \sum_{m=1}^{\infty} \alpha_m(\hbar) \partial_{\bar{j}_1}\Phi \cdots \partial_{\bar{j}_{m-1}}\Phi D^{\bar{j}_1} \cdots D^{\bar{j}_{m-1}} D^{\bar{l}}. \end{aligned}$$

Using these results, star products among z^i and $\bar{z}^i$ are obtained as

$$z^i * z^j = z^i z^j, \qquad z^i * \bar{z}^j = z^i \bar{z}^j, \qquad \bar{z}^i * \bar{z}^j = \bar{z}^i \bar{z}^j, \tag{1}$$

$$\begin{aligned} \bar{z}^i * z^j &= \bar{z}^i z^j + \hbar \delta_{ij} (1+|z|^2) {}_2F_1\left(1,1;1-1/\hbar;-|z|^2\right) \\ &\quad + \frac{\hbar}{1-\hbar} \bar{z}^i z^j (1+|z|^2) {}_2F_1\left(1,2;2-1/\hbar;-|z|^2\right). \end{aligned} \tag{2}$$

Furthermore, we can derive an explicite formula for L_f for an arbitrary function f

$$L_f = \sum_{n=0}^{\infty} \frac{\alpha_n(\hbar)}{n!} g_{j_1\bar{k}_1} \cdots g_{j_n\bar{k}_n} \left(D^{j_1} \cdots D^{j_n} f\right) D^{\bar{k}_1} \cdots D^{\bar{k}_n}.$$

This satisfies $[L_f, \partial_{\bar{i}}\Phi + \hbar\partial_{\bar{i}}] = 0$. This star product on $\mathbb{C}P^N$ is characterized by a function of $\hbar$, $\alpha_n(\hbar)$.

4. Fock representation of $\mathbb{C}P^N$

We have considered noncommutative $\mathbb{C}P^N$ in the formalism of deformation quantization until the previous section. In the context of deformation quantization, all objects in noncommutative $\mathbb{C}P^N$ are described as formal power series. In physical theories, phyisical quantities schould be expressed as convergent power seiries. Since much of finiteness or convergency in deformation quantization are not known, we change our strategy here. To introduce a calcurable framework, we consider a Fock representaion of $\mathbb{C}P^N$ in this section.

$\{z^i, \partial_j\Phi \mid i,j = 1,2,\cdots,N\}$ and $\{\bar{z}^i, \partial_{\bar{j}}\Phi \mid i,j = 1,2,\cdots,N\}$ constitute $2N$ sets of the creation-annihilation operators under the star product,

$$\begin{aligned} &\left[\partial_i\Phi,\ z^j\right]_* = \hbar\delta_{ij}, \qquad \left[z^i,\ z^j\right]_* = 0, \qquad \left[\partial_i\Phi,\ \partial_j\Phi\right]_* = 0, \\ &\left[\bar{z}^i,\ \partial_{\bar{j}}\Phi\right]_* = \hbar\delta_{ij}, \qquad \left[\bar{z}^i,\ \bar{z}^j\right]_* = 0, \qquad \left[\partial_{\bar{i}}\Phi,\ \partial_{\bar{j}}\Phi\right]_* = 0. \end{aligned}$$

So we can ascribe $\partial_i\Phi$, $\bar{z}^j$ as annihilation operators, and z^i, $\partial_{\bar{j}}\Phi$ as creation operators. Then, $e^{-\Phi/\hbar} = (1+|z|^2)^{-1/\hbar}$ plays the role of the vacuume projection :

$$\begin{aligned} &\partial_i\Phi * e^{-\Phi/\hbar} = \bar{z}^j * e^{-\Phi/\hbar} = 0, \quad e^{-\Phi/\hbar} * z^i = e^{-\Phi/\hbar} * \partial_{\bar{j}}\Phi = 0, \\ &e^{-\Phi/\hbar} * e^{-\Phi/\hbar} = e^{-\Phi/\hbar}. \end{aligned}$$

Let us introduce a class of functions which is constructed by acting the creation-operators on the vacuume projection:

$$\begin{aligned} M_{i_1\cdots i_m; j_1\cdots j_n} &:= c_{mn} z^{i_1} * \cdots * z^{i_m} * e^{-\Phi/\hbar} * \bar{z}^{j_1} * \cdots * \bar{z}^{j_n} \\ &= c_{mn} z^{i_1} \cdots z^{i_m} \bar{z}^{j_1} \cdots \bar{z}^{j_n} e^{-\Phi/\hbar}, \end{aligned}$$

where we choose $c_{mn} = 1/\sqrt{m!n!\alpha_m(\hbar)\alpha_n(\hbar)}$. These functions form a closed algebra:

$$M_{i_1\cdots i_m;j_1\cdots j_n} * M_{k_1\cdots k_r;l_1\cdots l_s} = \delta_{nr}\delta^{k_1\cdots k_n}_{j_1\cdots j_n} M_{i_1\cdots i_m;l_1\cdots l_s},$$

where $\delta^{k_1\cdots k_n}_{j_1\cdots j_n}$ is defined as

$$\delta^{k_1\cdots k_n}_{j_1\cdots j_n} = \frac{1}{n!}\left[\delta^{k_1}_{j_1}\cdots\delta^{k_n}_{j_n} + \text{permutations of } (j_1,\cdots,j_n)\right].$$

We list the results of the operation of creation and anihilation operators to $M_{i_1\cdots i_m;j_1\cdots j_n}$ here.

$$\begin{aligned}
z^k * M_{i_1\cdots i_m;j_1\cdots j_n} &= \sqrt{\frac{m+1}{-m+1/\hbar}} M_{ki_1\cdots i_m;j_1\cdots j_n},\\
\partial_{\bar{k}}\Phi * M_{i_1\cdots i_m;j_1\cdots j_n} &= \hbar\sqrt{(m+1)(-m+1/\hbar)} M_{ki_1\cdots i_m;j_1\cdots j_n},\\
\partial_k\Phi * M_{i_1\cdots i_m;j_1\cdots j_n} &= \hbar\sqrt{\frac{-m+1+1/\hbar}{m}} \sum_{l=1}^{m} \delta_{ki_l} M_{i_1\cdots\hat{i}_l\cdots i_m;j_1\cdots j_n},\\
\bar{z}^k * M_{i_1\cdots i_m;j_1\cdots j_n} &= \frac{1}{\sqrt{m(-m+1+1/\hbar)}} \sum_{l=1}^{m} \delta_{ki_l} M_{i_1\cdots\hat{i}_l\cdots i_m;j_1\cdots j_n},\\
M_{i_1\cdots i_m;j_1\cdots j_n} * z^k &= \frac{1}{\sqrt{n(-n+1+1/\hbar)}} \sum_{l=1}^{n} \delta_{kj_l} M_{i_1\cdots i_m;j_1\cdots\hat{j}_l\cdots j_n},\\
M_{i_1\cdots i_m;j_1\cdots j_n} * \partial_{\bar{k}}\Phi &= \hbar\sqrt{\frac{-n+1+1/\hbar}{n}} \sum_{l=1}^{n} \delta_{kj_l} M_{i_1\cdots i_m;j_1\cdots\hat{j}_l\cdots j_n}\\
M_{i_1\cdots i_m;j_1\cdots j_n} * \partial_k\Phi &= \hbar\sqrt{(n+1)(-n+1/\hbar)} M_{i_1\cdots i_m;j_1\cdots j_n k},\\
M_{i_1\cdots i_m;j_1\cdots j_n} * \bar{z}^k &= \sqrt{\frac{n+1}{-n+1/\hbar}} M_{i_1\cdots i_m;j_1\cdots j_n k}.
\end{aligned}$$

Using this, we can construct field theories on noncommutative $\mathbb{C}P^N$ by replacing all field elements of usual commutative field theories in $\mathbb{C}P^N$ by matrices valued on these $M_{i_1\cdots i_m;j_1\cdots j_n}$. For example, we can define scalar field theories in this noncommutative $\mathbb{C}P^N$, and it is possible to construct scalar soliton solutions like GMS soliton in this formulation.

5. Noncommutative deformation of $\mathbb{C}H^N$

In this section, we consider the noncommutative defromation of $\mathbb{C}H^N$ by using Karabegov's deformation quantization with separation of varaibles. Any explicit expression of noncommutative $\mathbb{C}H^N$ ($N \geq 2$) is not given until now.

The Kähler potential of $\mathbb{C}H^N$ is given by

$$\Phi = -\ln\left(1 - |z|^2\right).$$

The metric $g_{i\bar{j}}$ and the inverse metric $g^{\bar{i}j}$ are defined by

$$\begin{aligned}
g_{i\bar{j}} &= \partial_i\partial_{\bar{j}}\Phi = \frac{(1-|z|^2)\delta_{ij} + \bar{z}^i z^j}{(1-|z|^2)^2},\\
g^{\bar{i}j} &= (1-|z|^2)\left(\delta_{ij} - \bar{z}^i z^j\right).
\end{aligned}$$

These are similar to the ones of $\mathbb{C}P^N$. The way to construct the star product is also as same as the one in section 3, so we skip the detail of the derivation. As similar to (1)-(2), star products between inhomogeneous coordinates are given as

$$\begin{aligned} z^i * z^j &= z^i z^j, \\ z^i * \bar{z}^j &= z^i \bar{z}^j, \\ \bar{z}^i * \bar{z}^j &= \bar{z}^i \bar{z}^j, \\ \bar{z}^i * z^j &= \bar{z}^i z^j + \hbar\delta_{ij}(1-|z|^2)_2F_1\left(1,1;1+1/\hbar;|z|^2\right) \\ &\quad - \frac{\hbar}{1+\hbar}\bar{z}^i z^j(1-|z|^2)_2F_1\left(1,2;2+1/\hbar;|z|^2\right). \end{aligned}$$

The explicit representation of the star product with separation of variables on $\mathbb{C}H^N$ is given by

$$L_{\bar{z}^l} = \bar{z}^l + \sum_{m=1}^{\infty}(-1)^{m-1}\beta_m(\hbar)\partial_{\bar{j}_1}\Phi\cdots\partial_{\bar{j}_{m-1}}\Phi D^{\bar{j}_1}\cdots D^{\bar{j}_{m-1}}D^{\bar{l}},$$

with

$$\beta_n(t) = (-1)^n\alpha_n(-t) = \frac{\Gamma(1/t)}{\Gamma(n+1/t)}.$$

A Fock representation of $\mathbb{C}H^N$ is also given, similary. As in the case of $\mathbb{C}P^N$, sets $\{z^i, \partial_j\Phi\}$ and $\{\bar{z}^i, \partial_{\bar{j}}\Phi\}$ satisfy the commutation relations for the creation-annihilation operators. Also $e^{-\Phi/\hbar}$ is the vacuum projection operator,

$$\partial_i\Phi * e^{-\Phi/\hbar} = 0, \tag{3}$$

$$\bar{z}^i * e^{-\Phi/\hbar} = 0, \tag{4}$$

$$e^{-\Phi/\hbar} * \partial_{\bar{i}}\Phi = 0, \tag{5}$$

$$e^{-\Phi/\hbar} * z^i = 0, \tag{6}$$

and

$$e^{-\Phi/\hbar} * e^{-\Phi/\hbar} = e^{-\Phi/\hbar}. \tag{7}$$

As in the case of $\mathbb{C}P^N$, we consider a class of functions

$$N_{i_1\cdots i_m;j_1\cdots j_n} = \frac{z^{i_1}\ldots z^{i_m}\bar{z}^{j_1}\ldots\bar{z}^{j_n}}{\sqrt{m!n!\beta_m(\hbar)\beta_n(\hbar)}}e^{-\Phi/\hbar}$$

$N_{i_1\cdots i_m;j_1\cdots j_n}$ is totally symmetric under permutations of i's and j's, respectively. Then we can show that these functions form a closed algebra

$$N_{i_1\cdots i_m;j_1\cdots j_n} * N_{k_1\cdots k_r;l_1\cdots l_s} = \delta_{nr}\delta^{k_1\cdots k_n}_{j_1\cdots j_n}N_{i_1\cdots i_m;l_1\cdots l_s}. \tag{8}$$

Moreover, the star products between $N_{i_1\cdots i_m;j_1\cdots j_n}$ and one of $z^k, \partial_k\Phi, \bar{z}^k$ and $\partial_{\bar{k}}\Phi$ are calculated as follows,

$$z^k * N_{i_1\cdots i_m;j_1\cdots j_n} = \sqrt{\frac{m+1}{m+1/\hbar}} N_{ki_1\cdots i_m;j_1\cdots j_n}, \tag{9}$$

$$\partial_k\Phi * N_{i_1\cdots i_m;j_1\cdots j_n} = \hbar\sqrt{\frac{m-1+1/\hbar}{m}} \sum_{l=1}^{m} \delta_{ki_l} N_{i_1\cdots \hat{i_l}\cdots i_m;j_1\cdots j_n}, \tag{10}$$

$$\bar{z}^k * N_{i_1\cdots i_m;j_1\cdots j_n} = \frac{1}{\sqrt{m(m-1+1/\hbar)}} \sum_{l=1}^{m} \delta_{ki_l} N_{i_1\cdots \hat{i_l}\cdots i_m;j_1\cdots j_n}, \tag{11}$$

$$\partial_{\bar{k}}\Phi * N_{i_1\cdots i_m;j_1\cdots j_n} = \hbar\sqrt{(m+1)(m+1/\hbar)} N_{ki_1\cdots i_m;j_1\cdots j_n}, \tag{12}$$

$$N_{i_1\cdots i_m;j_1\cdots j_n} * z^k = \frac{1}{\sqrt{n(n-1+1/\hbar)}} \sum_{l=1}^{n} \delta_{kj_l} N_{i_1\cdots i_m;j_1\cdots \hat{j_l}\cdots j_n}, \tag{13}$$

$$N_{i_1\cdots i_m;j_1\cdots j_n} * \partial_k\Phi = \hbar\sqrt{(n+1)(n+1/\hbar)} N_{i_1\cdots i_m;j_1\cdots j_n k}, \tag{14}$$

$$N_{i_1\cdots i_m;j_1\cdots j_n} * \bar{z}^k = \sqrt{\frac{n+1}{n+1/\hbar}} N_{i_1\cdots i_m;j_1\cdots j_n k}, \tag{15}$$

$$N_{i_1\cdots i_m;j_1\cdots j_n} * \partial_{\bar{k}}\Phi = \hbar\sqrt{\frac{n-1+1/\hbar}{n}} \sum_{l=1}^{n} \delta_{kj_l} N_{i_1\cdots i_m;j_1\cdots \hat{j_l}\cdots j_n}. \tag{16}$$

6. Summary

In this article, we studied noncommutative deformation of several manifolds by using deformation quantization. We started from a noncommutative $\mathbb{R}^d$ by using the Moyal product and we moved into more genaral manifolds deformed with star products. For most manifolds, star products can not be described in explicit expression, but they are determined recursively. As exceptions, $\mathbb{R}^2$, $\mathbb{C}P^1$ and $\mathbb{C}H^1$ cases were reviewd, and their star products are given by the method of Fedosov's deformation quantization. In this article we introduced new examples of noncommutative manifolds with explicit star products, that is Karabegov's deformation quantization of $\mathbb{C}P^N$ and $\mathbb{C}H^N$ with separation of variables. Explicit calulable star products were given by using gamma functions. Using the results of star products between the inhomogeneous coordinates, we found the creation and anihilation operators and obtained the Fock representations of the noncommutative $\mathbb{C}P^N$ and $\mathbb{C}H^N$. Noncommutative $\mathbb{C}P^N$ is also studied in [6, 7, 8, 9, 10], and noncommutative $\mathbb{C}H^1$ is constructed in [11]. This is the first time to construct an explicit noncommutative $\mathbb{C}H^N$ for $N \geq 2$.

How can we construct physical theories on the noncommutative $\mathbb{C}P^N$ and $\mathbb{C}H^N$? There are a lot of problems about finiteness, convergency and so on, in deformation quantization. Fock representations discussed in this article provide one of good ways to make physical field theories on the noncommutative manifolds. Indeed, few soliton solutions are given on such noncommutative manifolds. It is important to construct frameworks of physics in the noncommutative manifolds defined by deformation quantization.

Acknowledgments

A.S. is supported by KAKENHI No.23540117 (Grant-in-Aid for Scientific Research (C)). H.U. is supported by KAKENHI No.21740197 (Grant-in-Aid for Young Scientists (B)).

References

[1] Fedosov B 1994 *J. Differential Geom.* **40** 213

Journal of Physics: Conference Series **442** (2013) 012052 doi:10.1088/1742-6596/442/1/012052

[2] Karabegov A V 1996 *Commun. Math. Phys.* **180** 745 [arXiv:hep-th/9508013].
[3] Karabegov A V 1996 *Funct. Anal. Appl.* **30** 142
[4] Karabegov A V 2011 "An explicit formula for a star product with separation of variables," [arXiv:1106.4112 [math.QA]].
[5] Sako A, Suzuki T and Umetsu H 2012 *J. Math. Phys.* **53** 073502 [arXiv:1204.4030 [math-ph]].
[6] Bordemann M, Brischle M, Emmrich C, Waldmann S 1996 *Lett. Math. Phys.* **36** 357.
[7] Balachandran A P, Dolan B P, Lee J H, Martin X, and O'Connor D 2002 *J. Geom. Phys.* **43**, 184 [hep-th/0107099].
[8] Kitazawa Y 2002 *Nucl. Phys.* B **642**, 210 [hep-th/0207115].
[9] Karabali D, Nair V P and Randjbar-Daemi S In Shifman M (ed.) et al.: *From fields to strings*, vol. 1 831-875 [hep-th/0407007].
[10] Karabegov A 2008 *Contemporary Math.* **450** 163.
[11] Bieliavsky P, Detournay S and Spindel P 2009 *Commun. Math. Phys.* **289**, 529 [arXiv:0806.4741 [math-ph]].

R^2 theory of gravity

Farrin Payandeh
Department of Physics, Payame Noor University, PO BOX 19395-3697, Tehran, Iran
E-mail: f_payandeh@pnu.ac.ir

Mohsen Fathi
Department of Physics, Islamic Azad University, Central Tehran Branch, Tehran, Iran
E-mail: mohsen.fathi@gmail.com

Abstract. In this article, we review our recent works on fourth order Weyl gravity, and try to obtain same results for cosmological coefficients, using a different approach. Indeed we shall impose a de Sitter like condition on the original components of field equations. Afterwards using the reduced equations on a definite geometric background, we derive the same results as in [1], for same cosmological constraints.

1. Introduction

Einstein's general theory of relativity seems to be a perfect theory, at least at the classical level where it is confirmed by classical tests. This theory has been obtained from the Hilbert-Einstein action

$$I_{HE} = \frac{1}{16\pi G} \int dx^4 \sqrt{-g} R.$$

However, the theory faces with difficulties like cosmological constant, dark matter and dark energy problems. In a series of papers, Mannheim and Kazanas have suggested conformal Weyl gravity instead of Einstein's gravity. With this theory, they could explain two outstanding astrophysical issues namely the cosmological constant and galactic rotation curve problems [2, 3]. Therefore a large amount of works can be found in higher order theories including the Weyl gravity. In the Weyl theory of gravity the Hilbert-Einstein action is replaced by the square of the conformal Weyl tensor

$$I_W = -\alpha \int dx^4 \sqrt{-g} C_{\mu\nu\rho\lambda} C^{\mu\nu\rho\lambda}, \tag{1}$$

where $C_{\mu\nu\rho\lambda}$ is the Weyl tensor

$$C_{\mu\nu\lambda\rho} = R_{\mu\nu\lambda\rho} - \frac{1}{2}\left(g_{\mu\lambda}R_{\nu\rho} - g_{\mu\rho}R_{\nu\lambda} - g_{\nu\lambda}R_{\mu\rho} + g_{\nu\rho}R_{\mu\lambda}\right) + \frac{R}{6}\left(g_{\mu\lambda}g_{\nu\rho} - g_{\mu\rho}g_{\nu\lambda}\right).$$

Since there is no accepted system of sign convention in general relativity, different articles use different sign conventions. In this work we first review Weyl gravity with the usual sign convention of Ref. [4], and then we translate it to a de Sitter like background. Finally, taking the cosmological "constant" and the Hubble parameter as functions of r and t respectively, we derive some estimations for their values, which are in very good agreement with the known values of the parameters for the current state of the universe.

2. Weyl gravity

2.1. The solution

The action (1) can be written as follows

$$I_W = -\alpha \int d^4x\sqrt{-g}(R^{\mu\nu\rho\lambda}R_{\mu\nu\rho\lambda} - 2R^{\mu\nu}R_{\mu\nu} + \frac{1}{3}R^2),$$

since $\sqrt{-g}(R^{\mu\nu\rho\lambda}R_{\mu\nu\rho\lambda} - 4R^{\mu\nu}R_{\mu\nu} + R^2)$ is a total divergence, so it does not contribute to the equation of motion and one can simplify the action as follows [5]

$$I_W = -2\alpha \int d^4x\sqrt{-g}\Big(R_{\alpha\beta}R^{\alpha\beta} - \frac{1}{3}R^2\Big) \equiv -2\alpha \int d^4x\Big(\mathcal{S}_1 - \frac{1}{3}\mathcal{S}_2\Big). \qquad (2)$$

The variation of the action with respect to $g_{\alpha\beta}$ results in

$$\delta(\mathcal{S}_1) = 2\sqrt{-g}R\delta R + R^2\delta\sqrt{-g} =$$

$$\sqrt{-g}\nabla_\gamma A^\gamma + \sqrt{-g}\Big(2\nabla^\alpha\nabla^\beta R - 2g^{\alpha\beta}\nabla_\lambda\nabla^\lambda R - 2RR^{\alpha\beta} + \frac{1}{3}g^{\alpha\beta}R^2\Big)\delta g_{\alpha\beta}, \qquad (3)$$

where

$$A^\gamma \equiv 2g^{\alpha\beta}g^{\gamma\lambda}R(\nabla_\alpha\delta g_{\beta\lambda} - \nabla_\lambda\delta g_{\alpha\beta}) - 2g^{\gamma\beta}g^{\alpha\beta}\nabla_\alpha R\delta g_{\beta\lambda} + 2g^{\alpha\beta}g^{\gamma\lambda}\nabla_\lambda R\delta g_{\alpha\beta}.$$

Similarly, we can write

$$\delta(\mathcal{S}_2) = \sqrt{-g}\nabla_\gamma B^\gamma +$$

$$\sqrt{-g}\Big(\nabla_\rho\nabla^\alpha R^{\beta\rho} + \nabla_\rho\nabla^\beta R^{\alpha\rho} - \nabla_\lambda\nabla^\lambda R^{\alpha\beta} - g^{\alpha\beta}\nabla_\rho\nabla_\lambda R^{\rho\lambda} - 2R^\beta_\rho R^{\rho\alpha} + \frac{1}{2}g^{\alpha\beta}R_{\rho\lambda}R^{\rho\lambda}\Big)\delta g_{\alpha\beta}, \quad (4)$$

where

$$B^\gamma = g^{\gamma d}R^{\alpha\beta}\Big(\nabla_\alpha\delta g_{\beta\rho} + \nabla_\beta\delta g_{\alpha\rho} - \nabla_\rho\delta g_{\alpha\beta}\Big) - g^{\alpha l\rho}\Big(\nabla_\alpha R^{\gamma\beta}\delta g_{\beta\rho} - \nabla_\beta R^{\gamma\rho}\delta g_{\alpha\rho}\Big)$$

$$-g^{\beta\rho}\Big(R^{\alpha\gamma}\nabla_\alpha\delta g_{\beta\rho} + \nabla_\beta R^{\alpha\gamma}\delta g_{\alpha\rho}\Big) + g^{\rho\gamma}\nabla_\rho R^{\alpha\beta}\delta g_{\alpha\beta}.$$

Note that when the equation of motion is considered, the first terms in the right hand side of Eq.s (3-4) have no contribution. Therefore we obtain

$$S_{\alpha\beta} \equiv S^{(2)}_{\alpha\beta} - \frac{1}{3}S^{(1)}_{\alpha\beta} = \nabla^\rho\nabla_\alpha R_{\beta\rho} + \nabla^\rho\nabla_\beta R_{\alpha\rho} - \nabla_\lambda\nabla^\lambda R_{\alpha\beta} - g_{\alpha\beta}\nabla_\rho\nabla_\lambda R^{\rho\lambda}$$

$$-2R_{\rho\beta}R^\rho_\alpha + \frac{1}{2}g_{\alpha\beta}R_{\rho\lambda}R^{\rho\lambda} - \frac{1}{3}\Big(2\nabla_\alpha\nabla_\beta R - 2g_{\alpha\beta}\nabla_\lambda\nabla^\lambda R - 2RR_{\alpha\beta} + \frac{1}{2}g_{\alpha\beta}R^2\Big). \qquad (5)$$

$S_{\alpha\beta} = 0$ is the vacuum Weyl field equation. It is worth to mention that relation (5) has been written in the following sign convention

$$R^\alpha_{\beta\gamma\rho} = \partial_\gamma\Gamma^\alpha_{\beta\rho} + \Gamma^\alpha_{\lambda\gamma}\Gamma^\lambda_{\beta\rho} - \gamma \leftrightarrow \rho,$$

$$R_{\beta\rho} = +R^\alpha_{\beta\alpha\rho}, \text{ and } R = +R^\alpha_\alpha.$$

Note that in Mannheim's works the opposite sign for the Ricci tensor and Ricci scalar has been chosen.

Now let us choose the line element which is static and spherically symmetric

$$ds^2 = -A(r)dt^2 + B(r)dr^2 + r^2d\Omega^2. \qquad (6)$$

Having this metric, one can reach exactly the same relation for S^{rr} as it was reported in [6].

$$S^{rr} = -\frac{1}{48B^5A^4r^4}(-8B^2A^2A'A'''r^4 + 4BB''A^2A'^2r^4 - 7B^2A'^4r^4$$
$$+4B^2A^2A''^2r^4 + 12B^2AA'^2A''r^4 + 8BB'A^2A''A'r^4 - 6BB'AA'^3r^4$$
$$-7B'^2A^2A'^2r^4 - 48B^2A^2A'A''r^3 + 16B^2A^3A'''r^3 - 16BB'A^3A''r^3$$
$$+20B^2AA'^3r^3 + 16BB'A^2A'^2r^3 - 16BB''A^3A'r^3 + 28B'^2A^3A'r^3$$
$$+32B^2A^3A''r^2 + 4B^2A^2A'^2r^2 - 8BB'A^3A'r^2 + 16BB''A^4r^2$$
$$-28B'^2A^4r^2 - 32B^2A^3A'r - 16B^4A^4 + 16B^2A^4), \quad (7)$$

in which the "Primes" denote differentiation with respect to r; as we expected the results are similar.

2.2. The effective potential

Weyl theory of gravity is a theory of fourth order and this makes it problematic. In order to find the vacuum solution for this theory, we must attain a solution for $S^{rr} = 0$. The differential equation has been solved ([6]) and the metric parameter $A(r)$ and $B(r)$ have been derived as:

$$A(r) = -\frac{\beta(2-3\beta\gamma)}{r} + (1-3\beta\gamma) + \gamma r - kr^2 \quad (8)$$

where it has been adopted that

$$B(r) = A^{-1}(r) \qquad \gamma, \beta, k = \text{constant}.$$

Since this theory is of fourth order, its solution possesses three constants of integration (the familiar Schwarzschild solution contains only one constant, namely β in (8)). These constants become very important in cosmology.

As a first step let us consider the geodesics in Weyl gravity. The geodesic equations ($\frac{d^2x^\mu}{d\tau^2} + \Gamma^\mu_{\nu\rho}\frac{dx^\nu}{d\tau}\frac{dx^\rho}{d\tau} = 0$ and τ is the affine parameter) are:

$$\frac{d^2t}{d\tau^2} + \frac{(2\beta - 3\beta^2\gamma + \gamma r^2 - 2kr^3)(\frac{dt}{d\tau})(\frac{dr}{d\tau})}{r(r - 2\beta + 3\beta^2\gamma - 3\beta\gamma r + \gamma r^2 - kr^3)} = 0;$$

$$\frac{d^2r}{d\tau^2} + \frac{1}{2}\frac{(r - 2\beta + 3\beta^2\gamma - 3\beta\gamma r + \gamma r^2 - kr^3)(2\beta - 3\beta^2\gamma + \gamma r^2 - 2kr^3)(\frac{dt}{d\tau})^2}{r^3}$$
$$-\frac{1}{2}\frac{(2\beta - 3\beta^2\gamma + \gamma r^2 - 2kr^3)(\frac{dr}{d\tau})^2}{r(r - 2\beta + 3\beta^2\gamma - 3\beta\gamma r + \gamma r^2 - kr^3)} + (2\beta - r - 3\beta^2\gamma + 3\beta\gamma r - \gamma r^2 + kr^3)(\frac{d\theta}{d\tau})^2$$
$$-(r - 2\beta + 3\beta^2\gamma - 3\beta\gamma r + \gamma r^2 - kr^3)\sin^2(\theta)(\frac{d\phi}{d\tau})^2 = 0;$$

$$\frac{d^2\theta}{d\tau^2} + \frac{2(\frac{dr}{d\tau})(\frac{d\theta}{d\tau})}{r} - \sin(\theta)\cos(\theta)(\frac{d\phi}{d\tau})^2 = 0;$$

$$\frac{d^2\phi}{d\tau^2} + \frac{2(\frac{dr}{d\tau})(\frac{d\phi}{d\tau})}{r} + 2\cot(\theta)(\frac{d\theta}{d\tau})(\frac{d\phi}{d\tau}) = 0. \quad (9)$$

Let us take

$$\epsilon \equiv -g_{\mu\nu}\frac{dx^\mu}{d\tau}\frac{dx^\nu}{d\tau}. \tag{10}$$

For $\theta = \pi/2$ along the geodesics, the two conserved quantities, namely the total energy and the total angular momentum, are:

$$A(r)\frac{dt}{d\tau} \equiv E, \quad \text{and} \quad r^2\frac{d\phi}{d\tau} \equiv L.$$

Expanding (10) for ϵ and maintaining the expression for $A(r)$, we get

$$-\Big(1 - \frac{\beta(2-3\beta\gamma)}{r} - 3\beta\gamma + \gamma r - kr^2\Big)\Big(\frac{dt}{d\tau}\Big)^2 + \Big(1 - \frac{\beta(2-3\beta\gamma)}{r}$$

$$-3\beta\gamma + \gamma r - kr^2\Big)^{-1}\Big(\frac{dr}{d\tau}\Big)^2 + r^2\Big(\frac{d\phi}{d\tau}\Big)^2 = -\epsilon. \tag{11}$$

Multiplying (11) by $A(r)$ yields

$$-E^2 + \Big(\frac{dr}{d\tau}\Big)^2 + \Big(1 - \frac{\beta(2-3\beta\gamma)}{r} - 3\beta\gamma + \gamma r - kr^2\Big)\Big\{\frac{L^2}{r^2} + \epsilon\Big\} = 0. \tag{12}$$

Rewrite (12) as

$$\frac{1}{2}E^2 = \frac{1}{2}\Big(\frac{dr}{d\tau}\Big)^2 + V(r), \tag{13}$$

where

$$V(r) = \frac{1}{2}\Big(1 - \frac{\beta(2-3\beta\gamma)}{r} - 3\beta\gamma + \gamma r - kr^2\Big)\Big(\frac{L^2}{r^2} + \epsilon\Big). \tag{14}$$

Equation (14) gives us the opportunity to define the effective potential as

$$V_{eff} = \frac{1}{2}\{\epsilon - 3\beta\gamma\epsilon - kL^2\} - \frac{1}{2}\{\epsilon\beta(2-3\beta\gamma) - \gamma L^2\}\frac{1}{r} - \frac{1}{2}\{L^2(3\beta\gamma - 1)\}\frac{1}{r^2}$$

$$+\frac{1}{2}\{L^2\beta(3\beta\gamma - 2)\}\frac{1}{r^3} + \frac{1}{2}\gamma\epsilon r - \frac{1}{2}k\epsilon r^2, \tag{15}$$

which is the Weyl effective potential for a particle, supposed to move in Weyl geometry. This potential has some extra terms which may have significant physical meaning, for example in large scale the last two terms play important roles. Now taking $\beta = GM$, and $\epsilon = 1, \gamma = k = 0$, one obtains the famous effective potential of general relativity, namely:

$$V_{eff} = \frac{1}{2} - \frac{GM}{r} + \frac{1}{2}\frac{L^2}{r^2} - \frac{GML^2}{r^3}. \tag{16}$$

Other values of these constants give us various kinds of potentials. For example, taking $\epsilon, \gamma, k = 0$ and $\beta = GM$ we obtain:

$$V_{eff} = \frac{1}{2}\frac{L^2}{r^2} - \frac{GM}{r^3}. \tag{17}$$

This potential has two terms. The term $\frac{1}{2}\frac{L^2}{r^2}$ denotes the Newtonian effective potential for which $L = mr^2\dot{\phi}^2$ and $m = 1$. The potential covers the central force problem. The second, namely $\frac{GM}{r^3}$, denotes the Schwarzchild effective potential. Thus equation (17) with the above conditions indicates the effective potential due to general relativity. For $\epsilon = 0$ but $\gamma, k \neq 0$ and $\beta = GM$, equation (15) yields

$$V_{eff} = -\frac{1}{2}kL^2 + \frac{\gamma L^2}{2r} - (\frac{3}{2}\gamma GML^2 - \frac{1}{2}L^2)\frac{1}{r^2} + (\frac{3}{2}\gamma G^2M^2L^2 - L^2)\frac{1}{r^3}. \quad (18)$$

This effective potential has two important additional terms. The term $\frac{\gamma L^2}{2r}$ in (18) denotes the potential due to the term γr in the metric. This is a repulsive potential which is proportional to the inverse of distance. The term $\frac{1}{2}kL^2$ denotes the potential due to the term kr^2 in the metric. This is a constant potential with respect to r. In analogy to general relativity, the terms contain $\frac{1}{r^2}$ and $\frac{1}{r^3}$ have also two additional terms say $-\frac{3\gamma GML^2}{2r^2}$ and $\frac{3\gamma G^2M^2L^2}{2r^3}$ respectively.

Note that the coefficient γ, is one of the most important constants, introduced in [6], in order to explain the strange flat galactic rotation curves without maintaining dark matter. Its value has been calculated to be $2 \times 10^{-28}\ cm^{-1}$. Also it has been semi-analytically obtained in [7], resulting in same value.

3. Geometric background and the de Sitter like condition

We start with the line element (6) by imposing the following de Sitter like condition on it:

$$R_{\mu\nu} = \frac{1}{4}g_{\mu\nu}R, \quad (19)$$

in which R may have a r or t dependence. Note that this essentially requires a constant total curvature, i.e. $R = 4\Lambda$, which is the de Sitter condition (this is also mentioned in the appendix). However, in this article the question we want to examine is: if we assume this condition is retained for variable total curvature in a definite geometric background, would it be possible to regain our previous results for cosmological parameters or not? Therefore from now on, we shall consider a de Sitter like background geometry, and equation (19) will help us to reduce the original field equations.

We begin with (19). Using the 00 component of metric (6) taking $B = A^{-1}$. We have:

$$\left(r^2A'' + 4\,A'r - 2 + 2\,A\right) = 2r\left(rA'' + 2\,A'\right), \quad (20)$$

The solution can be written as follows

$$A(r) = 1 + \frac{b_1}{r} + b_2r^2. \quad (21)$$

Substituting $b_1 = -2GM$ and $b_2 = -\frac{1}{3}\Lambda$, we obtain the Schwarzschild-de Sitter metric. Now let us work with the following component of metric, forming a de Sitter like geometric background, in which Λ has been chosen to be a function of r:

$$g_{00} = -c^2\left(1 - 2\frac{GM}{rc^2} - \frac{1}{3}\frac{\Lambda(r)\,r^2}{c^2}\right),$$

$$g_{11} = \left(1 - 2\frac{GM}{rc^2} - \frac{1}{3}\frac{\Lambda(r)\,r^2}{c^2}\right)^{-1},$$

$$g_{22} = r^2,$$

$$g_{33} = r^2\sin^2(\theta). \quad (22)$$

The components of the Ricci tensor, due to metric (22) become as follows:

$$R_{00} = \frac{1 - 2\frac{GM}{rc^2} - \frac{1}{3}\frac{\Lambda r^2}{c^2}}{4rc^2}\Big[2\,rc^4\left(-4\frac{GM}{r^3c^2} - \frac{1}{3}\frac{\Lambda'' r^2}{c^2} - \frac{4}{3}\frac{\Lambda' r}{c^2} - \frac{2}{3}\frac{\Lambda}{c^2}\right)$$
$$+4\,c^4\left(2\frac{GM}{r^2c^2} - \frac{1}{3}\frac{\Lambda' r^2}{c^2} - \frac{2}{3}\frac{\Lambda\, r}{c^2}\right)\Big],$$

$$R_{11} = \frac{1}{4c^4\left(1 - 2\frac{GM}{rc^2} - \frac{1}{3}\frac{\Lambda r^2}{c^2}\right)r}\Big[-2\,rc^4\left(-4\frac{GM}{r^3c^2} - \frac{1}{3}\frac{\Lambda'' r^2}{c^2} - \frac{4}{3}\frac{\Lambda' r}{c^2} - \frac{2}{3}\frac{\Lambda}{c^2}\right)$$
$$-4\,c^4\left(2\frac{GM}{r^2c^2} - \frac{1}{3}\frac{\Lambda' r^2}{c^2} - \frac{2}{3}\frac{\Lambda\, r}{c^2}\right)\Big],$$

$$R_{22} = \frac{1 - 2\frac{GM}{rc^2} - \frac{1}{3}\frac{\Lambda r^2}{c^2}}{2c^2}\Big[-2\left(2\frac{GM}{r^2c^2} - \frac{1}{3}\frac{\Lambda' r^2}{c^2} - \frac{2}{3}\frac{\Lambda\, r}{c^2}\right)rc^2\left(1 - 2\frac{GM}{rc^2} - \frac{1}{3}\frac{\Lambda r^2}{c^2}\right)^{-1}$$
$$+2\,c^2\left(1 - 2\frac{GM}{rc^2} - \frac{1}{3}\frac{\Lambda r^2}{c^2}\right)^{-1} - 2\,c^2\Big] = \frac{1}{\cos^2(\theta)}R_{33}. \tag{23}$$

And the Ricci scalar ($R = R^{\mu}_{\mu}$) would be:

$$R = \frac{1}{3}\frac{\Lambda'' r^2 + 8\,\Lambda' r + 12\,\Lambda}{c^2}. \tag{24}$$

Note that, a constant Λ results in $R = \frac{1}{4c^2}\Lambda$, which is the de Sitter condition.

3.1. *Estimating the cosmological constant*

The 00 component of the Weyl field equations (5), using metric (22) and the values in (23) and (24), gives

$$S_{00} = \frac{1}{324r^3c^4}\Big[396\,r^4c^4\Lambda''' + 36\,r^5c^4\Lambda'''' + 720\,GM\Lambda\,r^3\Lambda' + 1440\,GM\Lambda\,r^4\Lambda'' + 48\,\Lambda' r^5GM\Lambda''$$
$$-1548\,GMr^3c^2\Lambda''' + 540\,GM\Lambda\,r^5\Lambda''' + 12\,\Lambda' r^6GM\Lambda''' - 144\,r^4c^2\Lambda''''GM + 48\,GM\Lambda'''' r^6\Lambda$$
$$-3816\,GMr^2c^2\Lambda'' - 1584\,GMrc^2\Lambda' + 864\,G^2M^2\Lambda' - 24\,(\Lambda')^2r^5c^2 + 8\,(\Lambda')^2r^7\Lambda$$
$$+96\,\Lambda^2r^6\Lambda' + 576\,r^2c^4\Lambda' + 3\,(\Lambda'')^2r^7c^2 - (\Lambda'')^2r^9\Lambda + 144\,\Lambda^2r^7\Lambda'' + 48\,\Lambda^2r^8\Lambda'''$$
$$+1044\,r^3c^4\Lambda'' + 4\,\Lambda^2r^9\Lambda'''' + 48\,(\Lambda')^2r^4GM - 480\,\Lambda\,r^4c^2\Lambda' + 3456\,G^2M^2\Lambda'' r$$
$$-780\,\Lambda\,r^5c^2\Lambda'' - 6\,(\Lambda'')^2r^6GM - 276\,\Lambda\,r^6c^2\Lambda''' + 144\,G^2M^2\Lambda'''' r^3 + 8\,\Lambda' r^8\Lambda\,\Lambda''$$
$$+1512\,G^2M^2\Lambda''' r^2 - 6\,\Lambda' r^7c^2\Lambda''' + 2\,\Lambda' r^9\Lambda\,\Lambda''' - 24\,r^7c^2\Lambda''''\Lambda - 24\,\Lambda' r^6c^2\Lambda''\Big]. \tag{25}$$

In order to find an analytical solution for $\Lambda(r)$, we should set $S_{00} = 0$. This evidently generates a rather complicated equation. Hence, we shall reduce the vacuum Weyl equations of (5), in a de Sitter like background, by imposing the condition (19). This results in the following field equation:

$$S^{(dS)}_{\alpha\beta} = -\frac{1}{12} g_{\alpha\beta}\nabla_\lambda\nabla^\lambda R - \frac{1}{6}\nabla_\alpha\nabla_\beta R. \tag{26}$$

Using (22) and (24) in (26) yields:

$$\begin{aligned}
S^{(dS)}_{00} &= \frac{1}{27r^2c^4}(-3rc^2 + 6GM + \Lambda r^3)(2r^3\Lambda'^2 + 6r^2\Lambda\Lambda' - 3rc^2\Lambda'' + 6GM\Lambda'' + r^3\Lambda\Lambda'' - 6c^2\Lambda'),\\
S^{(dS)}_{11} &= -\frac{1}{3(-3rc^2 + 6GM - r^3\Lambda)}(2r^3\Lambda'^2 + 6r^2\Lambda\Lambda' - 9rc^2\Lambda'' + 18GM\Lambda'' + 3r^3\Lambda\Lambda'' - 6c^2\Lambda'),\\
S^{(dS)}_{22} &= \frac{1}{9c^4}(18GM\Lambda' + r^4\Lambda'^2 + 6r^3\Lambda\Lambda' - 3r^2c^2\Lambda'' + 6GMr\Lambda'' + r^4\Lambda\Lambda'' - 12rc^2\Lambda') = \frac{1}{\sin^2(\theta)} S^{(dS)}_{33}.
\end{aligned} \tag{27}$$

And the trace due to these components will be:

$$S^{\alpha(dS)}_{\alpha} = \frac{1}{9r^2c^4}(4r^4\Lambda'^2 + 18r^3\Lambda\Lambda' - 9r^2c^2\Lambda'' + 18GMr\Lambda'' + 3r^4\Lambda\Lambda'' - 30rc^2\Lambda'$$

$$+2r^4c^2\Lambda'^2 + 6r^3c^2\Lambda\Lambda' - 9r^2c^4\Lambda' + 18GMrc^2\Lambda'' + 3r^4c^2\Lambda\Lambda'' - 6rc^4\Lambda' + 36GM\Lambda'). \tag{28}$$

Since the conformal Weyl gravity is a traceless theory, therefore we can obtain a differential equation with respect to $\Lambda(r)$:

$$64r^4\Lambda'^2 + 18r^3\Lambda\Lambda' - 9r^2c^2\Lambda'' + 18GMr\Lambda'' + 3r^4\Lambda\Lambda'' - 30rc^2\Lambda' + 2r^4c^2\Lambda'^2 + 6r^3c^2\Lambda\Lambda'$$

$$-9r^2c^4\Lambda' + 18GMrc^2\Lambda'' + 3r^4c^2\Lambda\Lambda'' - 6rc^4\Lambda' + 36GM\Lambda' = 0, \tag{29}$$

for which a proper solution is

$$\Lambda(r) = \frac{3c^2}{r^2} - \frac{6GM}{r^3}. \tag{30}$$

This can be applied to obtain an estimation for the current value of the cosmological constant Λ_0. To do this, we consider

$$\Lambda_0 = \Lambda(r_{ob}),$$

in which $r_{ob} = 4.39937 \times 10^{28}\,cm$ is the distance to the particle horizon [8]. Other important measurable value in (30) is the mass M, which for the current state of the universe, is the mass of the observable universe, calculated by Sir Fred Hoyle to be $M_{ob} = 8 \times 10^{52}\,kg$ [9]. Also the speed of light is $c = 299792458\ \frac{m}{s}$ and the gravitational constant $G = 6.67 \times 10^{-11}\ m^3kg^{-1}s^{-2}$. These values, all together in (30) result in

$$\Lambda_0 = 1.020631735 \times 10^{-36}\quad s^{-2}. \tag{31}$$

Note that the cosmological constant has been measured to be of order $10^{-35}\ s^{-2}$ or $10^{-29}\ \frac{gr}{cm^3}$ [10]. This latter one, has been also mentioned in [1], which means that we have achieved the same results, using a different method.

3.2. Estimating the Hubble constant
Another important cosmological parameter is the Hubble parameter, and here we pursue the previous method, to evaluate it. Now Let us assume a flat and time dependent metric as:

$$ds^2 = -c^2dt^2 + f(t)dx_i^2, \tag{32}$$

where $i = 1, 2, 3$. We try to find a definite expression for $f(t)$ under the condition (19). The Ricci tensor components due to the metric (32) are:

$$R_{00} = -\frac{3}{4}\frac{2\ddot{f}(t)f(t) - \dot{f}(t)^2}{f(t)^2},$$

$$R_{11} = R_{22} = R_{33} = \frac{1}{4}\frac{2\ddot{f}(t)f(t) + \dot{f}(t)^2}{c^2 f(t)}, \tag{33}$$

where dot stands for the time derivative, and the Ricci scalar will be

$$R = \frac{3\ddot{f}(t)}{c^2 f(t)}. \tag{34}$$

By imposing (19) we obtain

$$\ddot{f}(t)f(t) - \dot{f}(t)^2 = 0. \tag{35}$$

A proper solution to (35) may be found as

$$f(t) = f_1 e^{f_2 t}, \tag{36}$$

in which f_1 and f_2 are integration constants. Considering $f_1 = 1$ and $f_2 = 2H$, turns the metric in (32) to

$$ds^2 = -c^2dt^2 + e^{2Ht}dx_i^2, \tag{37}$$

which is exactly the de Sitter's flat slicing metric. To investigate the metric in (37) we firstly suppose that H is itself a function of t. Therefore the Ricci tensor components and the Ricci scalar change to

$$R_{00} = -3\Big(\ddot{H}t + 2\dot{H} + (\dot{H}t + H)^2\Big),$$

$$R_{ij} = \frac{e^{2Ht}}{c^2}\Big(\ddot{H}t + 2\dot{H} + 3(\dot{H}t + H)^2\Big),$$

$$R = \frac{6}{c^2}\Big(\ddot{H}t + 2\dot{H} + 2\dot{H}^2t^2 + 4\dot{H}Ht + 2H^2\Big). \tag{38}$$

Note that if H be a constant then we will also get the de Sitter condition for the Ricci scalar. Now using the de Sitter like condition in (19) for the Ricci tensor components in (38) we have

$$\frac{3}{2}\ddot{H}t + 3\dot{H} = 0, \tag{39}$$

which has a solution of the form:

$$H(t) = c_1 + \frac{c_2}{t}. \tag{40}$$

Substituting the Ricci tensor components of (38) in (5) vanishes all of the components, i.e.

$$S_{\alpha\beta} = 0. \tag{41}$$

This means that (37) even when $H = H(t)$, is a vacuum solution to Weyl gravity.

Now let us compute these relations in de Sitter like background. The components of $S_{\mu\nu}^{(dS)}$:

$$S_{00}^{(dS)} = -\frac{3}{2}(H^{(4)}t + 4H^{(3)} + 4\ddot{H}^2t^2 + 31\dot{H}\ddot{H}t + 5\dot{H}H^{(3)}t^2 + 16\dot{H}^2 + 5H^{(3)}Ht$$
$$+15\ddot{H}H + 4\dot{H}^2\ddot{H}t^3 + 8\dot{H}^3t^2 + 8\dot{H}\ddot{H}Ht^2 + 16H\dot{H}^2t + 4\ddot{H}H^2t + 8H^2\dot{H}), \quad (42)$$

$$S_{11}^{(dS)} = S_{22}^{(dS)} = S_{33}^{(dS)} = \frac{e^{2Ht}}{2c^4}(H^{(4)}t + 4H^{(3)} + 4\ddot{H}^2t^2 + 43\dot{H}\ddot{H}t + 9\dot{H}H^{(3)}t^2$$
$$+16\dot{H}^2+9H^{(3)}Ht+27\ddot{H}H+20\dot{H}^2\ddot{H}t^3+40\dot{H}^3t^2+40\dot{H}\ddot{H}Ht^2+80H\dot{H}^2t+20\ddot{H}H^2t+40H^2\dot{H}). \quad (43)$$

Requiring the traceless condition results in the following equation:

$$\frac{c^4}{3}S^{\mu(dS)}_{\mu} = H^{(4)}t + 4H^{(3)} + 4\ddot{H}^2t^2 + 37\dot{H}\ddot{H}t + 7\dot{H}H^{(3)}t^2 + 16\dot{H}^2 + 7H^{(3)}Ht$$
$$+21\ddot{H}H + 12\dot{H}^2\ddot{H}t^3 + 24\dot{H}^3t^2 + 24\dot{H}\ddot{H}Ht^2 + 48H\dot{H}^2t + 12\ddot{H}H^2t + 24H^2\dot{H} = 0. \quad (44)$$

This fourth order differential equation has a special solution as follows

$$H(t) = \frac{c_1}{t} + c_2. \quad (45)$$

A glance on equation (45) leads us to conclude that the Hubble parameter has a hyperbolic behavior with respect to time. As an example we try to find the recently derived Hubble parameter by taking the simplest guess namely $c_1 = 1$, $c_2 = 0$ and using the age of the universe. The age of the universe is now calculated to be $13.75 \times 10^9 \pm 0.17 \quad yrs$ [11]. Substituting this value in (45) yields the Hubble constant for this time.

$$H_0 = 2.31 \times 10^{-18}s^{-1}. \quad (46)$$

The value for the Hubble constant, derived in [1], was the same. The measured Hubble constant in 2006 was $2.5 \times 10^{-18} \pm 0.15s^{-1}$ [12].

4. Conclusion

There have been intense research activities in higher order theories of gravity since a couple of decades ago. In this context, much attention has been given to Weyl gravity, where the usual Einstein-Hilbert action is replaced by the square of the conformal Weyl tensor. This leads to a gravitational theory of fourth order. Within a fourth order theory of gravity, we first reviewed this theory in the sign convention of Ref. [4]. Moreover, the added terms in the effective potential have been reviewed. We studied the time and r dependence of the Hubble and cosmological parameters in de Sitter background and finally we presented an estimation for them. In this article we used a different method, consisting of imposing a de Sitter like condition on the original Weyl field equations, to reach the same values which have been derived formerly in our previous articles. It turned out that they are in very good agreement with other approaches and observational results.

Appendix A. de Sitter spacetime

The de Sitter space-time is a vacuum solution for both Einstein and Weyl gravity. Recent Astrophysical data indicate that our universe has a positive and non-vanishing acceleration, so it might be currently in de Sitter phase. de Sitter space-time is a maximally symmetric space so we can write

$$R_{abcd} = H^2(g_{ac}g_{bd} - g_{ad}g_{bc}),\ R_{ab} = 3H^2 g_{ab},\ R = 12H^2 = 4\Lambda.$$

In de Sitter background the following variations become important

$$\delta g_{ab} = -\delta g^{ab},$$

$$\delta\Gamma^a_{bd} = \frac{1}{2}g^{ae}\Big(\nabla_b\delta g_{ed} + \nabla_d\delta g_{be} - \nabla_e\delta g_{bd}\Big),$$

$$\delta R^a_{bcd} = \frac{1}{2}g^{ae}\Big(\nabla_c\nabla_b\delta g_{de} + \nabla_c\nabla_d\delta g_{be} - \nabla_c\nabla_e\delta g_{bd} - \nabla_d\nabla_b\delta g_{ce} - \nabla_d\nabla_c\delta g_{be} + \nabla_d\nabla_e\delta g_{bc}\Big),$$

$$\delta R_{ab} = \frac{1}{2}g^{cd}\Big(\nabla_c\nabla_a\delta g_{bd} + \nabla_c\nabla_b\delta g_{ad} - \nabla_c\nabla_d\delta g_{ab} - \nabla_b\nabla_a\delta g_{cd}\Big),$$

$$\delta R = \Big(\nabla^a\nabla^b\delta g_{ab} - g^{ab}\nabla_\lambda\nabla^\lambda\delta g_{ab}\Big) - 3H^2 g^{ab}\delta g_{ab}.$$

References

[1] Tanhayi M R, Fathi M, Takook M V 2011 *Mod. Phys. Lett.* **A 26** 32
[2] Mannheim P D 1997 *Are Galactic Rotation Curves Really Flat?, Astrophys. J.* **479** 659
[3] Mannheim P D 1990 *Conformal cosmology with no cosmological constant, General Relativity and Gravitation* **22** 289
[4] Misner C W, Thorne K S, Wheeler J A 1973 *Gravitation*, (New York, Freeman)
[5] Kazanas D, Mannheim P D 1991 *Astrophysical Journal. Supplement Series* **76** 431 ,76, 431(1991).
[6] Mannheim P D, Kazanas D 1989 *Astrophysical Journal* **342** 635
[7] Payandeh F, Fathi M 2012 *Int. J. Theor. Phys.* **51** 2227
[8] Lineweaver C, Tamara M D 2005 *Scientific American* http://www.sciam.com/article.cfm?id=misconceptions-about-the-2005-03,P.5, Retrieved 2008-11-06.
[9] Kragh H 1999 *Cosmology and Controversy: The Historical Development of Two Theories of the Universe*, (Princeton, University Press) ISBN 0-691-00546-X.
[10] Tegmark M et al. 2004 *Cosmological parameters from SDSS and WMAP. Phys. Rev. D* **69** 103501
[11] Suyu S H et al. *Astrophysical Journal* **711** 201
[12] Riess Adam G and et., al. 2011 *Astrophysical Journal* **730** 119

Granular universe: statistical origin of special and doubly special relativity

Petr Jizba[†] **and Fabio Scardigli**[‡]

[†] FNSPE, Czech Technical University in Prague, Břehová 7, 115 19 Praha 1, Czech Republic
[‡] Dipartimento di Matematica, Politecnico di Milano, Piazza Leonardo da Vinci 32, 20133 Milano, Italy

E-mail: p.jizba@fjfi.cvut.cz, fabio@phys.ntu.edu.tw

Abstract. We present a new theoretical evidence that a relativistically invariant quantum dynamics at large enough space-time scales can be derived from two inter-correlated genuinely non-relativistic stochastic processes that operate at different energy scales. This leads to Feynman amplitudes that are, in the Euclidean regime, identical to transition probability of a Brownian particle propagating through a granular space. Our observation implies a preferred frame and can have distinct experimental signatures. Ensuing implications for special and doubly-special relativity, quantum field theory, quantum gravity and cosmology are discussed.

1. Introduction

In recent years, the notion of *emergence* have been of vital importance to quantum field theory community for many reasons. Especially, in particle and condensed matter physics it embodies the essential feature of systems with several inter-correlated time or energy scales. In these systems, the observed macroscopic-scale dynamics and related degrees of freedom differ drastically from the actual underlying microscopic-scale physics [1]. *Superstatistics* provides an explicit realization of this paradigm: It predicts that the emergent behavior can be often regarded as a *superposition* of several statistical systems that operate at different spatio-temporal scales [2, 3]. In particular, many applications have recently been reported, in hydrodynamic turbulence [4], turbulence in quantum liquids [5], pattern forming systems [6] or scattering processes in high-energy physics [7].

The key conceptual premiss of the superstatistics is the existence of sufficient spatio-temporal scale separations between relevant dynamics, so that the system has enough time to relax to a local equilibrium state and to stay within it for some time. In practical applications we are typically concerned with two scales. Following Ref. [2], we consider an intensive parameter v that fluctuates on a much larger time scale than the typical relaxation time of the local dynamics. The random variable v can be in practice identified, e.g., with the inverse temperature [2, 3], coupling constant [2], friction constant [8], volatility [9] or einbein [10]. On an intuitive ground, one may understand the superstatistics by using the *adiabatic Ansatz.* Namely, the system under consideration, during its evolution, travels within its state space $\mathfrak{X}$ (described by state variable $\mathfrak{a} \in \mathfrak{X}$) which is partitioned into small cells characterized by a sharp value of v. Within each cell, the system is described by the conditional distribution $p(\mathfrak{a}|v)$. As v varies adiabatically from cell to cell, the joint distribution of finding the system with a sharp value of v in the state A is $p(\mathfrak{a}, v) = p(\mathfrak{a}|v)p(v)$ (Bayes theorem). The resulting macro-scale (or emergent) statistical distribution $p(\mathfrak{a})$ for finding a system in the state $\mathfrak{a}$ is obtained by marginalizing over the distribution of the variable being discarded, i.e.,

$$p(\mathfrak{a}) = \int \mathrm{d}v\, p(\mathfrak{a}|v)p(v)\,. \tag{1}$$

In mathematical language is $p(\mathfrak{a})$ known as the marginal distribution. From technical point of view, the sufficient time scale separation between two relevant dynamics in a studied system allows to qualify superstatistics as a form of slow modulation [11].

In this paper, we recast the Feynman transition amplitude of a relativistic scalar particle into a form, which (after being analytically continued to imaginary times) coincides with a superstatistics marginal probability (1). The derivation is based on the Lévy–Khinchine theorem for infinitely divisible distributions [13, 12], and for illustration we consider the dynamics and the propagator of a Klein–Gordon (i.e., neutral spin-0) particle. Our reasonings can be also extended to charged spin-0, spin$-\frac{1}{2}$, Proca's spin-1 particles and to higher-spin particles phrased via the Bargmann–Wigner wave equation [10]. Further generalization to external electromagnetic potential has been reported in Refs. [10, 14]. We further demonstrate that the above formulation can be looked at as if the particle would randomly propagate (in the sense of Brownian motion) through inhomogeneous or granular medium ("vacuum") [14].

Our argument is based upon a recent observation [9, 10, 14] that the Euclidean path integral (PI) for relativistic particles may be interpreted as describing a doubly-stochastic process that operates at two separate spatio-temporal scales. The short spatial scale, which is much smaller than particle's Compton length $\lambda_C = 1/mc$ ($\hbar = 1$ is assumed throughout this paper), describes a Wiener (i.e., Galilean relativity) process with a sharp Newtonian mass. The large spatial scale, which is of order λ_C, corresponds to distances over which the fluctuating Newtonian mass changes appreciably. At scales much larger than λ_C the particle evolves according to a genuine relativistic dynamics, with a sharp value of the mass coinciding with the Einstein rest mass. Particularly striking is the fact that when we average the particle's velocity over the structural correlation distance (i.e., over particle's λ_C) we obtain the velocity of light c. So the picture that emerges from this analysis is that the particle (with a non-zero mass!) propagates over the correlation distance λ_C with an average velocity c, while at larger distance scales (i.e., when a more coarse grained view is taken) the particle propagates as a relativistic particle with a sharp mass and an average velocity that is subluminal. Quite remarkably, one can observe an identical behavior in the well-known Feynman's checkerboard PI [16, 17] to which the transition amplitude (1) reduces in the case of a relativistic Dirac fermion in $1+1$ dimensions [10, 14].

A considerably expanded presentation including the issue of reparametrization invariance, bibliography, and proofs of the main statements and formulas is presented in Ref. [14].

2. Superstatistics path integrals

When a conditional probability density function (PDF) is formulated through a PI, then it satisfies the Chapman–Kolmogorov equation (CKE) for continuous Markovian processes, namely [15]

$$P(x_b, t_b|x_a, t_a) = \int_{-\infty}^{\infty} \mathrm{d}x\, P(x_b, t_b|x, t)P(x, t|x_a, t_a)\,, \tag{2}$$

with t being any time between t_a and t_b. Conversely, any transition probability satisfying CKE possesses a PI representation [16]. In physics one often encounters probabilities formulated as a superposition of PI's,

$$\bar{P}(x_b, t_b|x_a, t_a) = \int_0^{\infty} \mathrm{d}v\; \omega(v, t_{ba}) \int_{x(t_a)=x_a}^{x(t_b)=x_b} \mathcal{D}x\mathcal{D}p\; e^{\int_{t_a}^{t_b} \mathrm{d}\tau (\mathrm{i}p\dot{x} - vH(p,x))}\,. \tag{3}$$

Here $\omega(v, t_{ba})$ with $t_{ba} = t_b - t_a$ is a normalized PDF defined on $\mathbb{R}^+ \times \mathbb{R}^+$. The form (3) typically appears in non-perturbative approximations to statistical partition functions, in polymer physics, in financial markets, in systems with reparametrization invariance, etc. The random variable v is then related to the inverse temperature, coupling constant, volatility, vielbein, etc.

The existence of different time scales and the flow of the information from slow to fast degrees of freedom typically creates the irreversibility on the macroscopical level of the description. The corresponding information thus is not lost, but passes in a form incompatible with the Markovian description. The most general class of distributions $\omega(v, t_{ba})$ on $\mathbb{R}^+ \times \mathbb{R}^+$ for which the superposition of Markovian processes remain Markovian, i.e., when also $\bar{P}(\boldsymbol{x}_b, t_b|\boldsymbol{x}_a, t_a)$ satisfies the CKE (2), was found in Ref. [9]. The key is to note that in order to have (2) satisfied by $\bar{P}$, the rescaled PDF $w(v, t) \equiv \omega(v/t, t)/t$ should satisfy the CKE for homogeneous Markov process

$$w(v, t_1 + t_2) = \int_0^v \mathrm{d}v' \, w(v', t_1) w(v - v', t_2) \, . \tag{4}$$

Consequently the Laplace image fulfills the functional equation $\tilde{w}(p_v, t_1 + t_2) = \tilde{w}(p_v, t_2)\tilde{w}(p_v, t_1)$, with $t_1, t_2 \in \mathbb{R}^+$. By assuming continuity in t, it follows that the multiplicative semigroup $\tilde{w}(p_v, t)_{t \geq 0}$ satisfies $\tilde{w}(p_v, t) = \{\tilde{w}(p_v, 1)\}^t$. From this we see that the distribution of v at t is completely determined by the distribution of v at $t = 1$. In addition, because $\tilde{w}(p_v, 1) = \{\tilde{w}(p_v, 1/n)\}^n$ for any $n \in \mathbb{N}^+$, $w(v, 1)$ is infinitely divisible. The Lévy–Khinchine theorem [13, 12] then ensures that $\log \tilde{w}(p_v, t) \equiv -tF(p_v)$ must have the generic form

$$\log \tilde{w}(p_v, t) = -t \left(a p_v + \int_0^\infty (1 - e^{-p_v u}) \nu(\mathrm{d}u) \right), \tag{5}$$

where $a \geq 0$ is a drift constant and ν is some non-negative measure on $(0, \infty)$ satisfying $\int_{\mathbb{R}^+} \min(1, u)\nu(\mathrm{d}u) < \infty$. Finally the Laplace inverse of $\tilde{w}(p_v, t)$ yields $\omega(v, t)$. Once $\omega(v, t)$ is found, then $\bar{P}(\boldsymbol{x}_b, t_b|\boldsymbol{x}_a, t_a)$ possesses a PI representation on its own. What is the form of the new Hamiltonian? To this end we rewrite (3) in Dirac operator form as [9]

$$\begin{aligned} \bar{P}(x_b, t_b|x_a, t_a) &= \langle x_b| \int_0^\infty \mathrm{d}v \, w(v, t_{ab}) e^{-v\hat{H}} |x_a\rangle = \langle x_b|\{\tilde{w}(\hat{H}, 1)\}^{t_{ab}}|x_a\rangle \\ &= \langle x_b|e^{-t_{ab}F(\hat{H})}|x_a\rangle \, . \end{aligned} \tag{6}$$

From this we see that the identification $\bar{H}(\boldsymbol{p}, \boldsymbol{x}) = F(H(\boldsymbol{p}, \boldsymbol{x}))$ can be made. Here one might worry about the operator-ordering problem, not knowing in which temporal order $\boldsymbol{p}$ and $\boldsymbol{x}$ must be taken in F. For our purpose it suffices to note that when H is $\boldsymbol{x}$-independent, the former relation is exact. In more general instances the Weyl ordering is a natural choice since because in this case the required mid-point rule follows automatically and one does not need to invoke the gauge invariance [9, 29]. In situations when other non-trivial configuration space symmetries (such as non-holonomic symmetry) are required, other orderings might be more physical [9].

3. Statistical origin of relativistic quantum mechanics — granular universe

The Feynman transition amplitudes (or better its Euclidean version — transition probabilities) naturally fits into the structure of superstatistics PI's discussed above.

Note first that the choice $a = 0$ and $\nu(\mathrm{d}u) = 1/(2\sqrt{\pi}u^{3/2})\mathrm{d}u$ leads to $F(p_v) = \sqrt{p_v}$. This identifies $w(v, t_{ab})$ with the (unshifted) Lévy distribution with the scale parameter $t_{ab}^2/2$.

Moreover, when $H(\boldsymbol{p}, \boldsymbol{x}) = \boldsymbol{p}^2c^2 + m^2c^4$ then (3) can be cast into the form (cf. also Refs. [9, 10, 14])

$$\int_{\boldsymbol{x}(0)=\boldsymbol{x}'}^{\boldsymbol{x}(t)=\boldsymbol{x}} \mathcal{D}\boldsymbol{x}\frac{\mathcal{D}\boldsymbol{p}}{(2\pi)^D} \exp\left\{\int_0^t d\tau \left[i\boldsymbol{p}\cdot\dot{\boldsymbol{x}} - c\sqrt{\boldsymbol{p}^2 + m^2c^2}\right]\right\}$$
$$= \int_0^\infty d\mathfrak{m}\, f_{\frac{1}{2}}\left(\mathfrak{m}, tc^2, tc^2m^2\right) \int_{\boldsymbol{x}(0)=\boldsymbol{x}'}^{\boldsymbol{x}(t)=\boldsymbol{x}} \mathcal{D}\boldsymbol{x}\frac{\mathcal{D}\boldsymbol{p}}{(2\pi)^D} \exp\left\{\int_0^t d\tau \left[i\boldsymbol{p}\cdot\dot{\boldsymbol{x}} - \frac{\boldsymbol{p}^2}{2\mathfrak{m}} - mc^2\right]\right\}, \quad (7)$$

where $t_b - t_a \equiv t - 0 = t$, and

$$f_p(z, a, b) = \frac{(a/b)^{p/2}}{2K_p(\sqrt{ab})}\, z^{p-1}\, e^{-(az+b/z)/2}, \quad (8)$$

is the generalized inverse Gaussian distribution [13] (K_p is the modified Bessel function of the second kind with index p). The left-hand side of (7) represents the PI for the free relativistic particle in the Newton–Wigner representation [18]. The genuine Klein–Gordon (KG) kernel which also contains negative-energy spectrum (reflecting the existence of the charge-conjugated solution, i.e., the antiparticle) can be received from Eq. (7) by considering the Feshbach–Villars representation of the KG equation and making the substitution [10]

$$f_{\frac{1}{2}}\left(\mathfrak{m}, tc^2, tc^2m^2\right) \mapsto \frac{1 + \text{sgn}(t)\sigma_3}{2}\, f_{\frac{1}{2}}\left(\mathfrak{m}, |t|c^2, |t|c^2m^2\right). \quad (9)$$

The 2×2 matrix nature of the smearing distribution (σ_3 is the usual 3rd Pauli matrix) naturally includes the Feynman–Stuckelberg causal boundary condition and accommodate thus both particles and antiparticles in a symmetric way [10, 19]. Note, that when partition function is to be calculated the trace will get rid of sgn(t) term and 1/2 is turned to 1.

The structure of (7) implies that $\mathfrak{m}$ can be interpreted as a Newtonian mass which takes on continuous values distributed according to $f_{\frac{1}{2}}\left(\mathfrak{m}, tc^2, tc^2m^2\right)$ with $\langle\mathfrak{m}\rangle = m + 1/tc^2$ and $\text{var}(\mathfrak{m}) = m/tc^2 + 2/t^2c^4$. Fluctuations of the Newtonian mass can be then depicted as originating from particle's evolution in an inhomogeneous or granular (i.e., "polycrystalline") medium. Granularity, as known, for example, from solid-state systems, typically leads to corrections in the local dispersion relation [20] and hence to alterations in the local *effective mass*. The following picture thus emerges: on the short-distance scale, a non-relativistic particle can be envisaged as propagating via classical Brownian motion through a single grain with a local mass $\mathfrak{m}$. This fast-time process has a time scale $\sim 1/\mathfrak{m}c^2$. An averaged value of the local time scale represents a transient temporal scale $\langle 1/\mathfrak{m}c^2\rangle = 1/mc^2$ which coincides with particle's Compton time t_C. At time scales much longer than t_C (large-distance scale), the probability that the particle encounters a grain which endows it with a mass $\mathfrak{m}$ is $f_{\frac{1}{2}}\left(\mathfrak{m}, tc^2, tc^2m^2\right)$. As a result one may view a single-particle relativistic theory as a single-particle non-relativistic theory where the particle's Newtonian mass $\mathfrak{m}$ represents a fluctuating parameter which approaches on average the Einstein rest mass m in the large t limit. We stress that t should be understood as the *observation time*, a time after which the observation (position measurement) is made. In particular, during the period t the system remains unperturbed. One can thus justly expect that in the long run all mass fluctuations will be washed out and only a sharp time-independent effective mass will be perceived. The form of $\langle\mathfrak{m}\rangle$ identifies the time scale at which this happens with $t \sim 1/mc^2$, i.e. with the Compton time t_C. It should be stressed that above mass fluctuations have nothing to do with the *Zitterbewegung* which is caused by interference between positive- and negative-energy wave components. In our formulation both regimes are, however, decoupled.

We may also observe that by coarse-graining the velocity over the correlation time t_C we have

$$\langle|\boldsymbol{v}|\rangle_{t_C} \;=\; \left.\frac{\langle|\boldsymbol{p}|\rangle}{\langle\mathfrak{m}\rangle}\right|_{t_C} \;=\; c\,. \tag{10}$$

So on a short-time scale of order λ_C the Klein–Gordon particle propagates with an averaged velocity which is the speed of light c. But if one checks the particle's position at widely separated intervals (much larger than λ_C), then many directional reversals along a typical PI trajectory will take place, and the particle's net velocity will be then less than c — as it should be for a massive particle. The particle then acquires a sharp mass equal to Einstein's mass, and the process (not being hindered by fluctuating masses) is purely Brownian. This conclusion is in line with the well-known Feynman checkerboard picture [14, 17] to which it reduces in the case of $(1+1)$ dimensional relativistic Dirac particle.

4. Robustness of the emergent special relativity

Understanding the robustness of the emergent Special Relativity under small variations in the mass-smearing distribution function $f_{1/2}$ can guide the study of the relation between Einsteinian SR and other deformed variants of SR, such as Magueijo–Smolin and Amelino-Camelia's doubly special relativity [21, 22], or (quantum) κ-Poincaré deformation of relativistic kinematics [23]. In DRS models a further invariant scale ℓ is introduced, besides the usual speed of light c, and ℓ is typically considered to be of order of the Planck length. A small variation $\delta f_{1/2}$ of the smearing function originates the new Hamiltonian

$$\bar{H} \;=\; \frac{\epsilon_1}{4} + \left(1+\frac{\epsilon_0}{2}\right)\sqrt{\boldsymbol{p}^2c^2 + m^2c^4 + \frac{\epsilon_2}{4}}\,, \tag{11}$$

with $\epsilon_1 = -2\,(1+\epsilon_0/2)\,\sqrt{\epsilon_2}$ (see Ref. [14] for details). By setting

$$\epsilon_1 = 2\left(\sqrt{\frac{1}{1-c^2m^2\ell^2}} - 1\right), \quad \epsilon_2 = \frac{4c^6m^4\ell^2}{1-c^2m^2\ell^2}\,,$$

the new Hamiltonian $\bar{H}$ can be easily identified with

$$\bar{H} \;=\; c\,\frac{-m^2c^2\ell \mp \sqrt{\boldsymbol{p}^2(1-m^2c^2\ell^2)+m^2c^2}}{1-m^2c^2\ell^2}\,, \tag{12}$$

which coincides with the Magueijo–Smolin's doubly special relativistic Hamiltonian, in, say, its version [24]. For the last Hamiltonian a relation analog to (7) holds, where now the smearing function has the form $f_{1/2}\big(\mathfrak{m}, tc^2\lambda, tc^2m^2\lambda\big)$ with $\lambda = 1/(1-m^2c^2\ell^2)$. The correlation distance $1/mc\lambda$ can be naturally assumed as the minimal size $L_{\rm GRAIN}$ of the "grain of space" of the polycrystalline medium, which is linked to the new invariant scale ℓ by

$$L_{\rm GRAIN} := \frac{1}{mc\lambda} = \lambda_C(1-m^2c^2\ell^2)\,. \tag{13}$$

By tuning the size $L_{\rm GRAIN}$ of these "grains of space" it is possible to pass continuously from Lorenz symmetry to other different symmetries, as those enjoyed by DSR models. We can in principle speculate that each large scale symmetry could originate from a specific kind of space(time) granularity.

5. Generalization to quantum fields

The transition probability presented in (6) was constructed under assumption that the Hamiltonian H represents a single particle. Of course, the single-particle relativistic quantum theory is logically untenable, since a multi-particle production is allowed whenever the particle reaches the threshold energy for pair production. In addition, Leutwyler's no-interaction theorem [25] prohibits interaction for any finite number of particles in the context of relativistic mechanics. To evade the no-interaction theorem it is necessary to have an infinite number of degrees of freedom to describe interaction. The latter is typically achieved via local quantum field theories (QFT).

It should be underlined in this context that the PI for a single relativistic particle is still a perfectly legitimate building block even in QFT. Recall that in the standard perturbative treatment of, say, *generating functional* for a *scalar field* each Feynman diagram is composed of integrals over product of free correlation functions (Feynman's correlators):

$$\Delta_F(\boldsymbol{y}, ct_y; \boldsymbol{z}, ct_z) \;=\; \frac{1}{4}\int_{-\infty}^{\infty} \mathrm{d}\tau \, \mathrm{sgn}(\tau - t_y) \bar{P}(\boldsymbol{y}, \tau | \boldsymbol{z}, t_z), \tag{14}$$

and may thus be considered as a functional of the PI $\bar{P}(\mathbf{x}', t'|\mathbf{x}, t)$. In fact, QFT in general, can be viewed as a grand-canonical ensemble of fluctuating particle histories (worldlines) where Feynman diagrammatic representation of quantum fields depicts directly the pictures of the world-lines in a grand-canonical ensemble. This the so-called "worldline quantization" of particle physics is epitomized, e.g., in Feynman's worldline representation of the one-loop affective action in quantum electrodynamics [26], in Kleinert's disorder field theory [27] or in the Bern–Kosower and Strassler "string-inspired" approaches to QFT [28].

Because of (14), the relationship between bosonic Bern–Kosower Green's function $G_B(\tau_1, \tau_2)$ and the PI $\bar{P}(\mathbf{x}', t'|\mathbf{x}, t)$ can be found easily through the known functional relation between G_B and Δ_F, cf. Refs. [28].

6. Implications in cosmology and gravity

When space-time is curved, a metric tensor enters in both PI's in (7) in a different way, yielding different "counterterms" [15, 29]. For instance, in Bastianelli–van Nieuwenhuizen's time slicing regularization scheme [29] one has (when $\hbar$ is reintroduced)

$$\begin{aligned}
&\frac{\boldsymbol{p}^2}{2\mathfrak{m}} \;\mapsto\; \frac{g^{ij}p_ip_j}{2\mathfrak{m}} + \frac{\hbar^2}{8\mathfrak{m}}(R + g^{ij}\Gamma^m_{il}\Gamma^l_{jm}) \, , \\
&\sqrt{\boldsymbol{p}^2 + m^2c^2} \mapsto \sqrt{g^{ij}p_ip_j + \frac{\hbar^2}{4}(R + g^{ij}\Gamma^m_{il}\Gamma^l_{jm}) + m^2c^2} \; + \; \hbar^4\Phi(R, \partial R, \partial^2 R) \; + \; \mathcal{O}(\hbar^6) \, .
\end{aligned} \tag{15}$$

Here g^{ij}, R and Γ^j_{kl} are the (space-like) pull-back metric tensor, the scalar curvature and Christoffel symbol respectively. The non-vanishing counterterm $\Phi(\ldots)$ has the explicit form

$$\begin{aligned}
\Phi(R, \partial R, \partial^2 R) \;\; &= \;\; \frac{1}{16}\left(\partial_i\partial_j g^{ij}\right)^2 \; - \; \frac{1}{16}\partial_i\partial_j\partial_k\partial_l\left(g^{ij}g^{kl}\right) \; - \; \frac{1}{8}\left(\partial_i\partial_j g^{ij}\right)\left(R + g^{mn}\Gamma^l_{mk}\Gamma^k_{nl}\right) \\
&+ \;\; \frac{1}{8}\partial_i\partial_j\left[g^{ij}\left(R + g^{mn}\Gamma^l_{mk}\Gamma^k_{nl}\right)\right] .
\end{aligned} \tag{16}$$

When (15) is inserted into (7), the superstatistics identity breaks down as can be explicitly checked to the lowest order in $\hbar$. The respective sides of the equation will thus have differnt physical implications. Because the Einstein *equivalence principle* requires that the *local* spacetime structure can be identified with the Minkowski spacetime possessing Lorentz symmetry, one might assume the validity of (7) at least locally. However, in different space-time points one has, in general, a different typical length scale of the local inertial frames, depending on the gravitational field. The characteristic size of the local inertial (i.e. Minkowski) frame is of order $1/|K|^{1/4}$ where $K = R_{\alpha\beta\gamma\delta}R^{\alpha\beta\gamma\delta}$ is the Kretschmann invariant and $R_{\alpha\beta\gamma\delta}$ is the Riemann curvature. Relation (7) tells us that the special relativistic description breaks down in regions of size smaller than λ_C. For curvatures large enough, namely for strong gravitational fields, the size of the local inertial frame can become smaller than λ_C, that is $1/|K|^{1/4} \lesssim \lambda_C$. In such regions the special relativistic description is no more valid, and according to (7) must be replaced by a Newtonian description of the events. For instance, in Schwarzschild geometry we have $K = 12\, r_s^2/r^6$, and the breakdown should be expected at radial distances $r \lesssim (\lambda_C^2 r_s)^{1/3}$ (r_s is the Schwarzschild radius) which are — apart from the hypothetical case of micro-black holes (where $\lambda_C \simeq r_s$) — always deeply buried below the Schwarzschild event horizon. In the cosmologically relevant Friedmann-Lemaître-Robertson-Walker (FLRW) geometry, we have $K = 12\,(\dot{a}^4 + a^2\ddot{a}^2)/(ac)^4$, and the breakdown happens when $(\dot{a}^4 + a^2\ddot{a}^2) \gtrsim (ac/\lambda_C)^4$, where $a(t)$ is the FLRW scale factor of the Universe and $\dot{a} = da/dt$. Applying the well-known Vilenkin–Ford model [30] for inflationary cosmology, where $a(t)$ is given by: $a(t) = A\sqrt{\sinh(Bt)}$ with $B = 2c\sqrt{\Lambda/3}$ (Λ is the cosmological constant), we obtain a temporal bound on the validity of local Lorentz invariance, which, expressed in FLRW time, is

$$t \lesssim \frac{1}{B}\,\text{arcsinh}\left[\frac{B\lambda_C}{(8c^4 - (B\lambda_C)^4)^{1/4}}\right] \equiv \bar{t}\,. \tag{17}$$

By using the presently known [31] value $\Lambda \simeq 10^{-52}\text{m}^{-2}$ and the τ-lepton Compton's wavelength $\lambda_C^\tau \simeq 6.7 \times 10^{-16}$m (yielding the tightest upper bound on t), we obtain $\bar{t} \simeq 4 \times 10^{-24}$s. Note that, since $B\lambda_C \ll c$, then $\bar{t} \simeq \lambda_C/c = t_C$. Such a violation of the local Lorentz invariance naturally breaks the particle-antiparticle symmetry since there is no unified theory of particles and antiparticles in the non-relativistic physics — formally one has two separate theories. If the resulting matter-antimatter asymmetry provides a large enough CP asymmetry then this might have essential consequences in the early Universe, e.g., for leptogenesis. In this respect, $\bar{t}$ is compatible with the *nonthermal* leptogenesis period that typically dates between 10^{-26}–10^{-12}s after the Big Bang.

7. Conclusions and perspectives

The presented picture of emergent special relativity, where the observed space-time symmetries have a superstatistics basis, suggests an alternative reading of relativistic signals phrased in terms of quantum-mechanical probabilities. As such, it also makes explicit how the SR invariance is encoded in the grain smearing distribution. Notably, the exact LS of a space-time has no fundamental significance in our analysis, as it is only an accidental symmetry of the coarse-grained configuration space in which a particle executes a standard Wiener process. In passage from grain to grain particles's Newtonian mass fluctuates according to an inverse Gaussian distribution. The observed inertial mass of the particle is thus not a fundamental constant, but it reflects the particle's interaction with the granular vacuum (cosmic field). This, in a sense, supports Mach's view of the phenomenon of inertia.

Interactions can be included in our framework in two different ways. The interaction with a background field (such as electromagnetic field) can be directly treated with the superstatistics prescription (7), see [10]. On the other hand, the multi-particle interactions can be consistently formulated by "embedding" the relativistic PI in QFT via the worldline quantization. Such an embedding may help to study several cosmological implications of systems with granular space. If any of such systems quickly flows to the infrared fixed point, any direct effect due to the space discreteness, and related SR violation, might be insignificant on cosmological scales (where Lorentz and diffeomorphism invariance are restored), while it might be crucial in the early Universe, e.g., for leptogenesis and the ensuing baryogenesis. Consequences on the detailed structure of the Cosmic Microwave Background spectrum will be explored in future work.

The presented approach implies a preferred frame. In this connection it is worth of noting that, despite the fact that (7) is not manifestly LS invariant, one may use the Stückelberg trick and introduce a new fictitious variable into the PI (7), in such a way that the new action will have the reparametrization symmetry, but will still be dynamically equivalent to the original action. For relevant details see Ref. [14]. By not knowing the source, one may then view this artificial gauge invariance as being a fundamental or even a defining property of the relativistic theory. One might, however, equally well, proclaim the "polycrystalline" picture as being a basic (or primitive) edifice of SR and view the reparametrization symmetry as a mere artefact of an artificial redundancy that is allowed in our description. It is this second view that we favored here.

The observant reader could notice that the presented scenario cannot directly accommodate the massless particles such as photons (identity (7) holds true only for $m \neq 0$). One possibility would be to use the PI representation of Polyakov–Wheeler for massless particles and try to construct a similar superstatistics duality as in the case of massive particles. This procedure is, however, not without technical difficulties and currently is under investigation. Conceptually is far more simpler to assume that the photon has a small mass. At present, there are a number of experimental limits to the mass of the photons [32]. For instance, tests based on Coulomb's law and the galactic vector potential set the upper limit of $m_\gamma \lesssim 10^{-18}$ eV/c$^2 \simeq 10^{-57}$g. This gives the domain correlation distance for the photon $\simeq 1/m_\gamma c^2 \simeq 10^{43}$m which is bigger than the radius of observable Universe ($\simeq 10^{26}$ m) and so in this picture the photon mass does not fluctuate — it is a quasi-invariant.

Finally, it is hoped that our approach should reinforce the links between the superstatistics paradigm and the approach to quantum gravity based on stochastic quantization [33]. In particular, the outlined granular space could be a natural model for the noise terms in a Parisi–Wu stochastic-like quantization approach to gravity.

Acknowledgments

The writers are grateful to H. Kleinert, Z. Haba, M. Sakellariadou and L.S. Schulman for discussions and useful feedbacks. This work was supported by GAČR Grant No. P402/12/J077 and by RVO68407700 Grant.

References

[1] see, e.g., Anderson P 1972 *Science* **177** 393
Laughlin R B 2005 *A Different Universe: Reinventing Physics from the Bottom Down* (Basic Books: London)
Licata I and Sakaji A 2008 *Physics of Emergence and Organization* (WS: London)
[2] Beck C 2001 *Phys. Rev. Lett.* **87** 180601
Beck C and Cohen E G D 2003 *Physica* A **322** 267
Wilk G and Wlodarczyk Z 2000 *Phys. Rev. Lett.* **85** 2770
[3] Hanel R, Thurner S and Gell-Mann M 2011 *Proc. Natl. Acad. Sci. (USA)* **108** 6390
[4] Reynolds A 2003 *Phys. Rev. Lett.* **91**, 084503
[5] Beck C and Miah S 2012 (*Preprint* arXiv:1207.4062)
[6] Daniels K E, Beck C and Bodenschatz E 2004 *Physica* D **193** 208
[7] Wilk G and Wlodarczyk Z 2009 *Eur. Phys. J.* A **40** 299
Beck C 2009 *Eur. Phys. J.* A **40** 267
[8] Beck C 2007 in Radons G *et al* (eds): *Anomalous Transport: Foundations and Applications* (John Wiley: London)

[9] Jizba P and Kleinert H 2008 *Phys. Rev.* E **78** 031122
[10] Jizba P and Kleinert H 2010 *Phys. Rev.* D **82** 085016
[11] Allegrini P, Barbi F, Grigolini P and Paradisi P 2006 *Phys. Rev.* E **73** 046136
[12] Bertoin J 1996 *Lévy Processes* (CUP: Cambridge)
[13] Feller W 1966 *An Introduction to Probability Theory and its Applications, Vol. II* (John Wiley: London)
[14] Jizba P and Scardigli F 2012 *Phys. Rev.* D **86** 025029
Jizba P and Scardigli F 2012 *Int. J. Mod. Phys.* B **26** 1241003
Jizba P and Scardigli F 2012 *J. Phys.: Conf. Ser.* **361** 012026
Jizba P and Scardigli F 2012 *J. Phys.: Conf. Ser.* **343** 012048
Jizba P and Scardigli F 2013 (*Preprint* arXiv:1301.4091)
[15] Kleinert H 2009 *Path Integrals in Quantum Mechanics, Statistics, Polymer Physics and Financial Markets* (WS: Singapore 2009).
[16] Feynman R P and Hibbs A R 1965 *Quantum Mechanics and Path Integrals* (McGraw-Hill: New York)
[17] Gaveau B, Jacobson T, Kac M and Schulman L S 1984 *Phys. Rev. Lett.* **53** 419
Jacobson T and Schulman L S 1984 *J. Phys.* A **17** 375
[18] Newton T and Wigner E 1949 *Rev. Mod. Phys.* **21** 400
[19] Feshbach H and Villars F 1958 *Rev. Mod. Phys* **30** 24
[20] see, e.g., Johnson E A and MacKinnon A 1993 *J. Phys.: Condens. Matter* **5** 5859
[21] Amelino-Camelia G 2002 *Int. J. Mod. Phys.* D **11** 35
Magueijo J and Smolin L 2002 *Phys. Rev. Lett.* **88** 190403
Amelino-Camelia G 2002 *Nature* **418** 34
[22] Kowalski-Glikman J 2005 *Introduction to Doubly Special Relativity* (Springer: Berlin)
[23] Daszkiewicz M, Lukierski J and Woronowicz M 2008 *Phys. Rev.* D **77** 105007
[24] Magueijo J and Smolin L 2003 *Phys. Rev.* D **67** 044017
[25] Leutwyler H 1965 *Nuovo Cimento* **37** 556
Marmo G, Mukunda N and Sudarshan E C G 1984 *Phys. Rev.* D **30** 2110
[26] Feynman R P 1950 *Phys. Rev.* **80** 440
Feynman R P 1951 *Phys. Rev.* **84** 108
[27] Kleinert H 1989 *Gauge Fields in Condensed Matter, Vol. I Superflow and Vortex Lines* (WS: Singapore)
[28] Bern Z and Kosower D A 1991 *Phys. Rev. Lett.* **66** 1669
Bern Z and Kosower D A 1992 *Nucl. Phys.* B **379** 451
Strassler M J 1992 *Nucl. Phys.* B **385** 145 Schubert C 2001 *Phys. Rep.* **335** 73
[29] Bastianelli F and van Nieuwenhuizen P 2005 *Path Integrals and Anomalies in Curved Space* (CUP: Cambridge)
[30] Vilenkin A and Ford L H 1982 *Phys. Rev.* D **26** 1231
Scardigli F, Gruber C and Chen P 2011 *Phys. Rev.* D **83** 063507
[31] see, e.g., Tegmark M et al. 2006 *Phys. Rev.* D **74** 123507
[32] Chibisov G V 1976 *Soviet Physics Uspekhi* **19** 624
Lakes R 1998 *Phys. Rev. Lett.* **80** 1826
[33] Rumpf H 1986 *Phys. Rev.* D **33** 185
Rumpf H 1993 *Prog. Theor. Phys. Suppl.* **111** 63
de Aguiar T C, Menezes G and Svaiter N F 2009 *Class. Quantum Grav.* **26** 075003
Ambjorn J, Loll R, Westra W and Zohren S 2009 *Phys. Lett.* B **680** 359

An introduction to the quantum backflow effect

J M Yearsley[1] **and J J Halliwell**[2]

[1] Centre for Quantum Information and Foundations, DAMTP, Centre for Mathematical Sciences, University of Cambridge, Wilberforce Road, Cambridge CB3 0WA, UK
[2] Theoretical Physics Group, Blackett Laboratory, Imperial College, London SW7 2BZ, UK

E-mail: jmy27@cam.ac.uk

Abstract. We present an introduction to the backflow effect in quantum mechanics – the phenomenon in which a state consisting entirely of positive momenta may have negative current and the probability flows in the opposite direction to the momentum. We show that the effect is present even for simple states consisting of superpositions of gaussian wave packets, although the size of the effect is small. Inspired by the numerical results of Penz et al, we present a wave function whose current at any time may be computed analytically and which has periods of significant backflow, with a backwards flux equal to about 70 percent of the maximum possible backflow, a dimensionless number $c_{bm} \approx 0.04$, discovered by Bracken and Melloy. This number has the unusual property of being independent of $\hbar$ (and also of all other parameters of the model), despite corresponding to a quantum-mechanical effect, and we shed some light on this surprising property by considering the classical limit of backflow. We conclude by discussing a specific measurement model in which backflow may be identified in certain measurable probabilities.

1. Introduction

A striking but little-appreciated phenomenon in quantum mechanics is the backflow effect. This is the fact that, for a free particle described by a wave function centred in $x < 0$ consisting entirely of positive momenta, the probability of remaining in $x < 0$ may nevertheless *increase* with time. That is, the quantum-mechanical current at the origin can be negative and the probability flows "backwards".

This effect was first noted by Allcock in his early work on the arrival time problem in quantum theory [1], but was not studied in detail until 1994 when Bracken and Melloy carried out the first systematic study [2]. Their most important discovery was that there exists a limit on the total amount of backflow. Backflow means that the probability of remaining in $x < 0$ may increase with time, but this increase is bounded by an amount c_{bm}, a dimensionless number computed numerically to be approximately 0.04. Furthermore although backflow is clearly a nonclassical effect, the quantity c_{bm} is independent of $\hbar$ (and also of the mass of the particle and the time duration of the effect.) For this reason c_{bm} has been described as a new quantum number.

The effect has been further investigated by a number of authors. Better estimates for c_{bm} were given in Refs.[3, 4], with the latter also giving a numerical estimate of the maximum backflow state. Some analytic examples of backflow states and the measurability of the effect were explored in Ref.[5]. Connections between backflow and the phenomenon of superoscillations were noted in Ref.[6]. The effect has also been explored in the Dirac equation [7], for a particle in a linear potential [8] and for angular momentum [9]. Most recently the present authors have investigated a number of aspects of the backflow effect, including the classical limit [10]. Finally, the effect is often noted in connection with the arrival time problem in quantum theory [11]

Apart from the studies cited above, there has been little discussion of the effect in the literature to date. The purpose of this contribution is to provide an introduction to the backfow effect both as a reference for researchers encountering it in their studies of quantum theory and also hopefully as an invitation to encourage others to take an interest in the effect.

We begin in Section 2 with a detailed formulation of the problem. We define the current and the flux and consider the properties of the spectrum of the flux operator, which allows us to give the clearest statement of the backflow effect.

In Section 3 we give a simple example of a state with backflow consisting of a superposition of gaussian states. The backflow for such a state is, however, rather small.

In Section 4, we review the numerical computation of the maximal backflow state and eigenvalue. We also present an analytic wave function with backflow which appears to match closely the numerical solutions for the maximal backflow state by Penz et al [4]. The current at arbitrary times for this wave function may be computed analytically and we find that it has a backflow of approximately 70 percent of the maximal value. This is a much larger backflow than any analytically tractable states previously discovered.

In Section 5, we consider the naive classical limit $\hbar \to 0$ of backflow, and in particular, we address the fact that the bound on backflow c_{bm} discovered by Bracken and Melloy appears to be independent of $\hbar$. We show that the expected dependency on $\hbar$ reappears in realistic measurement models, where measurements are described not by exact projectors but by quasiprojectors involving parameters characterizing the imprecision of real measurements. Under these conditions the naive classical limit is restored.

In Section 6, we consider a simple measurement model and discuss the ways in which backflow may be seen in the probabilities for measurements.

We summarize and conclude in Section 7.

2. Detailed formulation of the problem

2.1. The flux

We consider a free particle with initial wave function $\psi(x)$ concentrated in $x < 0$ and consisting entirely of positive momenta. We consider the probability flux $F(t_1, t_2)$ crossing the origin during the time interval $[t_1, t_2]$, defined by

$$\begin{aligned} F(t_1, t_2) &= \int_{-\infty}^{0} dx\, |\psi(x, t_1)|^2 - \int_{-\infty}^{0} dx\, |\psi(x, t_2)|^2 \\ &= \int_{t_1}^{t_2} dt\ J(t) \end{aligned} \tag{1}$$

where $J(t)$ is the usual quantum-mechanical current at the origin

$$J(t) = -\frac{i\hbar}{2m} \left(\psi^*(0,t) \frac{\partial \psi(0,t)}{\partial x} - \frac{\partial \psi^*(0,t)}{\partial x} \psi(0,t) \right) \tag{2}$$

The flux may also be written in terms of the Wigner function [12] at time t, $W_t(p,q)$,

$$F(t_1, t_2) = \int_{t_1}^{t_2} dt \int dpdq\ \frac{p}{m} \delta(q) W_t(p,q) \tag{3}$$

(For a useful review of the properties of the current and related phase space distribution functions, see Ref.[13]). It is useful to write these expressions in an operator form. We introduce projection operators onto the positive and negative x-axis, $P = \theta(\hat{x})$, and, $\bar{P} = 1 - P = \theta(-\hat{x})$ respectively. The flux may then be written in terms of the operator $\hat{F}(t_1, t_2)$ defined by

$$\begin{aligned} \hat{F}(t_1,t_2) &= P(t_2) - P(t_1) = \int_{t_1}^{t_2} dt\ \dot{P}(t) \\ &= \int_{t_1}^{t_2} dt\ \frac{i}{\hbar}[H, \theta(\hat{x})] = \int_{t_1}^{t_2} dt \hat{J}(t) \end{aligned} \tag{4}$$

where the current operator is given by

$$\hat{J} = \frac{1}{2m}\left(\hat{p}\delta(\hat{x}) + \delta(\hat{x})\hat{p}\right) \tag{5}$$

So Eq.(1) may also be written

$$F(t_1, t_2) = \langle \hat{F}(t_1, t_2) \rangle = \int_{t_1}^{t_2} dt\ \langle \psi | \hat{J}(t) | \psi \rangle. \tag{6}$$

The flux Eq.(1) is a difference between two probabilities and is positive when those probabilities behave according to classical intuition, i.e., when the probability of remaining in $x < 0$ decreases monotonically.

As indicated, in the full quantum-mechanical case, the flux can be negative. One way to see this is to note that the Wigner function need not be positive for general states [12], and thus Eq.(3) need not be either. Another way is from the current operator Eq.(5): both $\hat{p}$ and $\delta(\hat{x})$ are non-negative operators on states with positive momentum, but since they do not commute, the current operator $\hat{J}$ is not a positive operator.

2.2. *Most negative flux as an eigenvalue problem*

Following Bracken and Melloy [2], a useful way to investigate the backflow effect is to look at the spectrum of the flux operator Eq.(4) (restricted to positive momenta). We thus look for solutions to the eigenvalue problem

$$\theta(\hat{p})\hat{F}(t_1, t_2)|\Phi\rangle = \lambda|\Phi\rangle \tag{7}$$

where the states $|\Phi\rangle$ consist only of positive momenta. (We choose an opposite sign convention to [2] so that the backflow states have $\lambda < 0$). The most negative value of the flux $F(t_1, t_2)$ is then given by the most negative eigenvalue of Eq.(7).

By time evolving the state we may choose the time interval $[t_1, t_2]$ to be $[-T/2, T/2]$ and the eigenvalue equation in momentum space then reads

$$\frac{1}{\pi}\int_0^\infty dk\ \frac{\sin[(p^2 - k^2)T/4m\hbar]}{(p-k)}\ \Phi(k) = \lambda\Phi(p) \tag{8}$$

Defining rescaled variables u and v by $p = 2\sqrt{m\hbar/T}u$ and $k = 2\sqrt{m\hbar/T}v$ Eq.(8) becomes

$$\frac{1}{\pi}\int_0^\infty dv\ \frac{\sin(u^2 - v^2)}{(u-v)}\phi(v) = \lambda\phi(u) \tag{9}$$

where $\phi(u) = (m\hbar/4T)^{1/4}\Phi(p)$ and is dimensionless. Crucially all physical constants have dropped out of this equation so that the λ are dimensionless and independent of $\hbar, m$ and T.

Journal of Physics: Conference Series **442** (2013) 012055 doi:10.1088/1742-6596/442/1/012055

This eigenvalue equation has been studied by a number of authors, both numerically and analytically [2, 3, 4]. The eigenvalues lie in the range

$$-c_{bm} \leq \lambda \leq 1 \tag{10}$$

where c_{bm} was computed numerically and found to be

$$c_{bm} \approx 0.038452. \tag{11}$$

It was conjectured in Ref.[2] that the spectrum is discrete in the interval $[-c_{bm}, 0]$ but continuous in the interval $[0, 1]$. The extremizing state was given numerically by Penz et al [4]. We will display this maximising state and also a good analytic expression approximating the numerical results in what follows. At present there is no analytic account of the properties Eqs.(10), (11). Note that the eigenvalues are independent of T which means that the duration of a period of backflow can be arbitrarily long.

3. Backflow for superpositions of gaussians

In this section we show that the backflow effect arises in the familiar, and also potentially experimentally realisable, setting of a superposition of two gaussian wavepackets. However gaussian wavepackets have support on both positive and negative momentum, so we will also have to show that this is not the result of any initially negative momentum. We will see that whilst superpositions of gaussian states do indeed give rise to backflow, the size of the effect is considerably smaller than the theoretical maximum. In this Section we work in units in which $\hbar = 1$ and we set the particle mass $m = 1$.

Bracken and Melloy demonstrated that the backflow effect may be observed in the simple case of a superposition of two plane waves [2]. This state is unnormalisatble, but it can be turned into a more physical state by replacing the plane waves with gaussians tightly peaked in momentum, without affecting the basic conclusion that the state displays backflow for well chosen values of the various parameters.

Consider the normalised state

$$\psi(x,t) = \sum_{k=1,2} A_k \frac{1}{\sqrt{4\sigma^2 + 2it}} \exp\left(ip_k(x - p_k t) - \frac{(x - p_k t)^2}{4\sigma^2 + 2it} \right). \tag{12}$$

This is a sum of two initial gaussian wavepackets with equal spatial width σ, evolved for a time t. If we let $\sigma \to \infty$ we essentially recover a sum of two plane waves. Instead of giving the rather complex expression for the current, we will simply plot the current at the origin and the probability of remaining in $x < 0$ as functions of time for the state in Eq.(12) and for the following set of parameters;

$$p_1 = 0.3, \quad p_2 = 1.4, \quad \sigma = 10, \quad A_1 = 1.8, \quad A_2 = 1. \tag{13}$$

We clearly see from these plots that there are several intervals during which the current is negative. A magnification of one of these backflow region is shown in Fig.(3). The effect is robust with respect to small changes to these parameters.

The parameters in Eq.(13) give rise to the greatest amount of backflow we have been able to find, although we have not searched the entire parameter space. The value of the flux during the largest period of backflow is

$$F = \int_{t_1}^{t_2} dt J(t), \tag{14}$$

Journal of Physics: Conference Series **442** (2013) 012055 doi:10.1088/1742-6596/442/1/012055

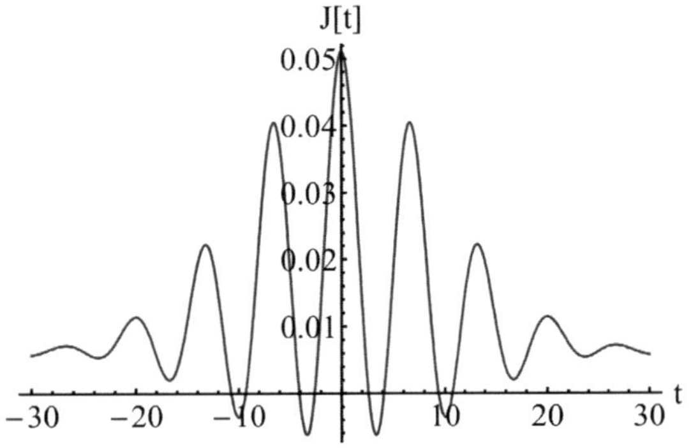

Figure 1. Plot of the current for a wavefunction consisting of a superposition of two gaussians, with the parameters given in Eq.(13).

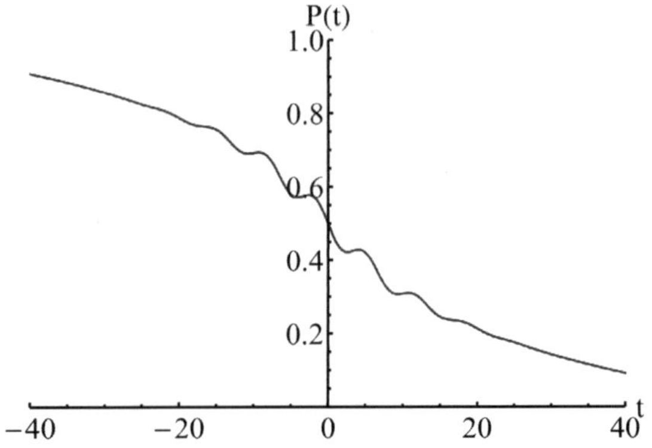

Figure 2. Plot of the probability for remaining in $x < 0$ for a wavefunction consisting of a superposition of two gaussians, with the parameters given in Eq.(13).

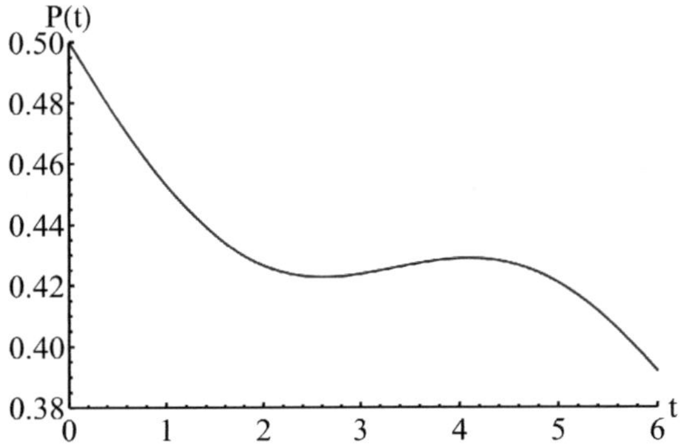

Figure 3. Close up of Fig.(2). $P(t)$ is clearly seen to increase between $t \approx 2$ and $t \approx 4$.

where $J(t)$ is the current, Eq.(2), and the interval $[t_1, t_2]$ is chosen such that the current is negative for the whole of this time. Computing the flux during this interval gives,

$$F \approx -0.0061, \tag{15}$$

or about 16 percent of c_{bm}.

It is important to check that this probability backflow cannot be explained by the tiny probability of having negative momentum in this gaussian state. It can be shown that the probability that a measurement of the momentum of this state would yield a negative answer is of order 10^{-10}, so the negative flow of probability is entirely due to the backflow effect [10].

4. An approximation to the backflow maximising state

Backflow states may be found by solving the eigenvalue equation Eq.(9). The numerical work of Penz et al [4] yielded an approximate eigenstate satisfying Eq.(9) giving the most negative eigenvalue $-c_{bm}$, i.e., the largest amount of backflow. It is of interest to find analytic expressions for wave functions matching these numerical results closely and for which the current may be computed analytically. We give such a state in this section. To be clear, we are not giving an approximate analytic solution to the eigenvalue problem, Eq.(9). Rather, inspired by the numerical solution, we will exhibit an analytic expression for a wave function which closely matches the backflow maximising state, and show that is has significant negative flux.

4.1. Numerical results

We first review the numerical results of the computation of the backflow state. We have repeated the numerical analysis of Penz et al.[4] of the optimizing state, ϕ_{max}, and its current in order to compare with our analytic approximation. In Fig.(4) we plot the numerically computed maximum backflow state, ϕ_{max}, together with the function

$$\phi_{as}(u) = -\frac{1}{10}\frac{\cos(u^2)}{u}, \tag{16}$$

which seems to match well the asymptotic form of ϕ_{max}.

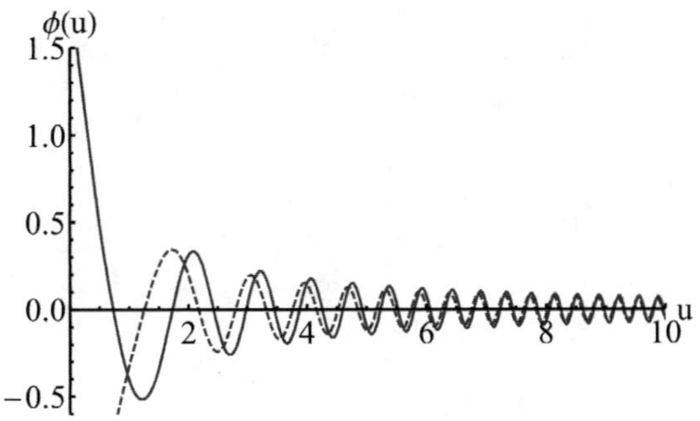

Figure 4. Plot of ϕ_{max} (solid line) together with ϕ_{as} (dashed line).

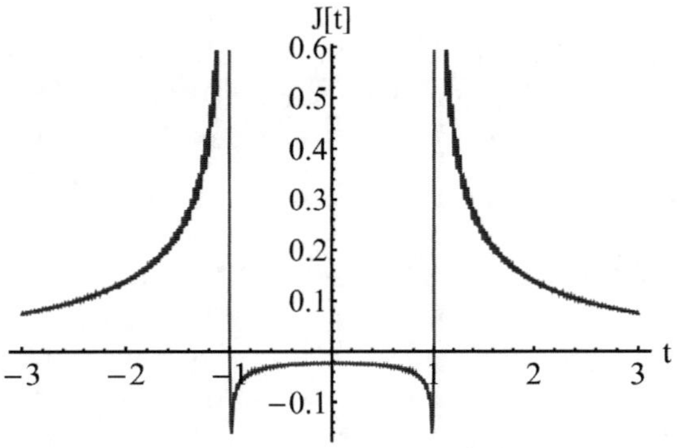

Figure 5. The current, $J(t)$, as computed from ϕ_{max}.

We plot in Fig.(5) the current computed from ϕ_{max}. Note that the current appears to have a specific singularity structure at $t = \pm 1$, where it jumps between $\pm\infty$. This is presumably related to some properties of the flux operator, but it also seems to depend on the asymptotic behavior of ϕ_{max} [14]. These two plots are in agreement, in general shape, with the numerical results of Penz et al [4] and we will compare our analytic results with these plots in what follows.

4.2. Analytic approximation

Consider the momentum space wavefunction

$$\phi_A(u) = N\left[ae^{-bu} + (\frac{1}{2} - C(u))\right], \quad a, b \in R \tag{17}$$

where N is a normalisation factor. Here

$$C(u) = \text{FresnelC}\left(\sqrt{\frac{2}{\pi}}u\right) = \sqrt{\frac{2}{\pi}}\int_0^u dx \cos(x^2). \tag{18}$$

This has the asymptotic form,

$$\phi_A(u) \sim N\frac{\sin(u^2)}{u}. \tag{19}$$

Note that this does not in fact match the asymptotic form of ϕ_{max}.

Journal of Physics: Conference Series **442** (2013) 012055 doi:10.1088/1742-6596/442/1/012055

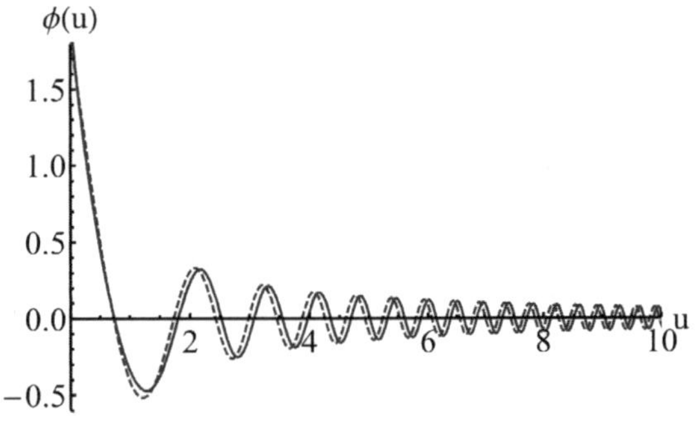

Figure 6. $\phi_A(u)$, for $a = 0.6$, $b = 2.8$ (solid line), with $\phi_{max}(u)$ for comparison (dashed line).

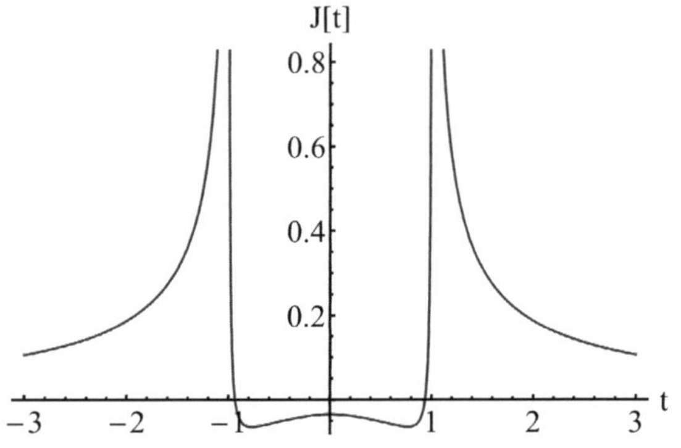

Figure 7. Current, $J(t)$ for $a = 0.6$ and $b = 2.8$.

We plot ϕ_A in Fig.(6) for the values of a and b which maximize backflow. We see good agreement with the numerical result. The maximum amount of negative flux we can generate is

$$F = -0.02757, \tag{20}$$

which occurs for the parameters $a = 0.6$, $b = 2.8$. This corresponds to about 70 percent of c_{bm}. We plot the current $J(t)$ for these parameters in Fig.(7). The current is close to the numerical result, Fig.(5) away from $t = \pm 1$ but lacks the correct behavior as $t \to \pm 1$. This is presumably related to the fact that the asymptotic behaviour of ϕ_A does not match that of ϕ_{max}. It would be interesting to know if there exists a better analytic approximation to the maximum backflow state that does have the correct asymptotic behavior and still gives rise to an analytically computable current. This will be explored elsewhere [14].

5. The classical limit of backflow

Some insights into the properties of backflow may be found by looking at its classical limit. Strictly one should do this via the usual framework of open system dynamics, and a detailed discussion of the current for an open system was given in Ref.[15]. In this work the resulting positivity of the current, after finite time, is clearly seen.

However it is nevertheless interesting to look at the naive classical limit of backflow characterized by the limit $\hbar \to 0$. As seen from Eq.(9) the eigenvalues of the flux operator are independent of $\hbar$. Thus although the existence of negative eigenvalues (negative flux) is clearly a quantum phenomenon, backflow does not go away in the naive classical limit, $\hbar \to 0$, as one might have expected. The origin of the independence of the eigenvalues of the flux operator on $\hbar$ is the fact that there is no way to construct a dimensionless number from the parameters in the problem, $\hbar$, m and T. What is missing is a length scale.

In any realistic experimental set up, measurements have finite resolution. It is therefore often more appropriate to model the measurement process using POVMs rather than exact projection operators. With this in mind, instead of defining the flux operator in terms of exact projection operators $P = \theta(\hat{x})$, instead let us define it in terms of a quasiprojector Q. This seems reasonable since, as discussed earlier, backflow can be measured by measuring whether the particle is in $x > 0$ at two different times and, due to the inevitable imprecision of real measurements, such measurements are best modeled by quasiprojectors. A convenient choice of quasiprojector is

$$Q = \int_0^\infty dy\ \delta_\sigma(\hat{x} - y) \tag{21}$$

where $\delta_\sigma(\hat{x}-y)$ is a smoothed out δ-function,

$$\delta_\sigma(\hat{x}-y) = \frac{1}{(2\pi\sigma^2)^{1/2}} \exp\left(-\frac{(\hat{x}-y)^2}{2\sigma^2}\right) \tag{22}$$

This goes to the usual δ-function as $\sigma \to 0$ and then $Q \to \theta(\hat{x})$. By replacing the exact projector in the definition of the flux with this new quasi-projector we may compute modified expressions for the current and the flux [10]. In particular the eigenvalue equation Eq.(9) becomes,

$$\frac{1}{\pi}\int_0^\infty dv \, \frac{\sin(u^2-v^2)}{(u-v)} \, e^{-a^2(u-v)^2} \, \phi(v) = \lambda\phi(u) \tag{23}$$

where the dimensionless number a is given by $a^2 = 2m\sigma^2/\hbar T$. This means that the eigenvalues λ will now depend on a, so we write $\lambda = \lambda(a)$. Through a they will therefore depend on $\hbar$ and the "limit" $\hbar \to 0$ now clearly means the regime $a \gg 1$, that is, $\hbar \ll 2m\sigma^2/T$. Hence, in a more realistic measurement situation, the bound on the total backflow – the most negative eigenvalue of Eq.(23) – *will* depend on $\hbar$ and the limit $\hbar \to 0$ may now be more meaningful.

A reasonable conjecture is that the negative eigenvalues will increase with a and also that

$$\lambda(a) \geq -c_{bm} \tag{24}$$

for all a, so that $-c_{bm}$ emerges as a lower bound on the eigenvalues, achievable only in the limit $a \to 0$. It seems unlikely, however, that all the negative eigenvalues will all become positive or zero, except perhaps in the limit $a \to \infty$. We have not been able to solve Eq.(23) analytically, so instead we have obtained numerical estimates for $\lambda(a)$ for various values of a, and we plot the result in Fig.(8). The value of $\lambda(a)$ does indeed increase with a, tending to zero asymptotically. In fact, numerical solutions are consistent with the asymptotic form,

$$\lambda(a) \sim -\frac{1}{a^2} \tag{25}$$

for large a, which can be understood by examining the asymptotic form of Eq.(23) for large a [10].

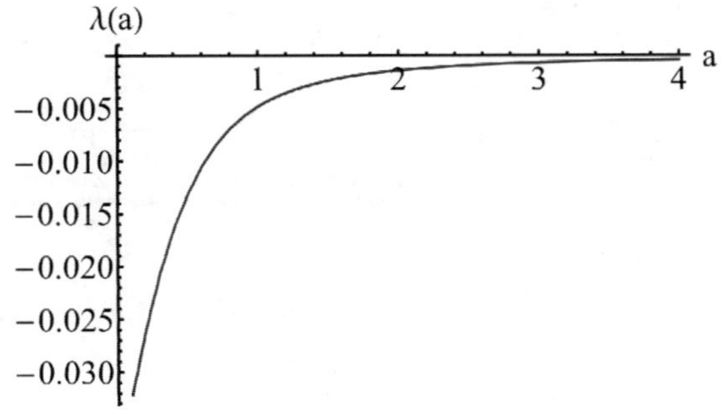

Figure 8. The most negative eigenvalue of Eq.(23), $\lambda(a)$ as a function of a.

In summary, for quasi-projectors, Eq.(21), which are more realistic models for measurements than exact projectors, the eigenvalues, in particular the lowest eigenvalue representing the most negative flux, do depend on $\hbar$ (and m and T). The lowest eigenvalue appears to go to zero for $a \to \infty$. This indicates that all negative eigenvalues go to zero (or become positive) in the naive classical limit $\hbar \to 0$. By contrast the positive eigenvalues are not significantly affected [10].

6. Backflow and measurement models

In this section we relate the above results on backflow to measurements. The aim is to begin to address the practical question of how backflow may be measured.

Many more elaborate and realistic models for the measurement of the arrival time (involving model detectors, for example) naturally lead to an arrival time probability defined with a complex potential. This is described in detail in many places [16]. These models typically yield an arrival time probability distribution which is closely related to the current and from which the current may be extracted, even when negative, thereby leading to a possible measurement of backflow.

Consider an initial wave packet starting in $x < 0$ with positive momentum. We seek the arrival time probability distribution $\Pi(\tau)d\tau$ for crossing the origin between τ and $\tau + d\tau$. We consider a complex absorbing potential of step function form in $x > 0$ so the Hamiltonian is $H_0 - iV_0\theta(\hat{x})$, where H_0 is the free Hamiltonian. We define the survival probability $N(\tau)$ to be the norm of the state at time τ after evolution with this complex Hamiltonian. Then

$$\begin{aligned} \Pi(\tau) &= -\frac{dN}{d\tau} \\ &= 2V_0 \langle \psi | e^{(iH_0 - V_0\theta(\hat{x}))\tau} \theta(\hat{x}) e^{(-iH_0 - V_0\theta(\hat{x}))\tau} | \psi \rangle \end{aligned} \tag{26}$$

This expression may be simplified considerably in the usual weak measurement approximation (V_0 small compared with the energy of the state), yielding [10],

$$\Pi(\tau) \approx 2V_0 \int_{-\infty}^{\tau} dt \ e^{-2V_0(\tau - t)} \ \langle \psi_t | \hat{J} | \psi_t \rangle \tag{27}$$

where $|\psi_t\rangle = e^{-iH_0 t}|\psi\rangle$. This is the expected semiclassical result [1, 17, 16]. Note that Eq.(27) is not necessarily positive, due to the negativity of the current in certain states. However this is an artifact of the approximation and should not matter for sufficiently small V_0, and we will assume that Eq.(27) is positive.

The quantity $\Pi(\tau)$ corresponds to the arrival time distribution measured by a realistic measurement so can in principle be determined experimentally. The current can then be extracted from Eq.(27) by deconvolution [17]. This therefore gives a method of measuring the current and the flux, and checking for backflow. In Fig.(9) we plot the measurement probability Eq.(27) for two values of V_0 and also the original numerically computed current, to see how the time-smearing affects the backflow. We see that positive regions of the current are not qualitatively changed very much, in keeping with semiclassical expectations, but negative regions of the current become positive as a result of the smearing, as they must, since the measured probability is positive. Details of other possible measurements of backflow are given in Ref.[10].

7. Summary and conclusions

The purpose of this paper was to introduce various aspects of the backflow effect. After setting up the problem in Section 2, in Section 3, we showed that backflow may occur for a superposition of gaussian wavepackets. These states are important since they are experimentally realizable, but the maximum amount of flux is very small, only about 16 percent of the maximum possible.

In Section 4, we gave an analytic expression for a state matching rather closely the numerically computed states giving maximal backflow, computed by Penz et al [4]. The plot of the current shows reasonably good agreement with the numerical solution, except at the end points $t = \pm 1$ of the backflow region. We computed the most negative flux for this state and found it to be about 70 percent of the numerically computed maximum backflow, significantly better than any previous analytic expression for a backflow state.

In Section 5, we discussed the classical limit of backflow. The most interesting aspect of this is the issue is that the eigenvalues of the flux operator are independent of $\hbar$. This appears to

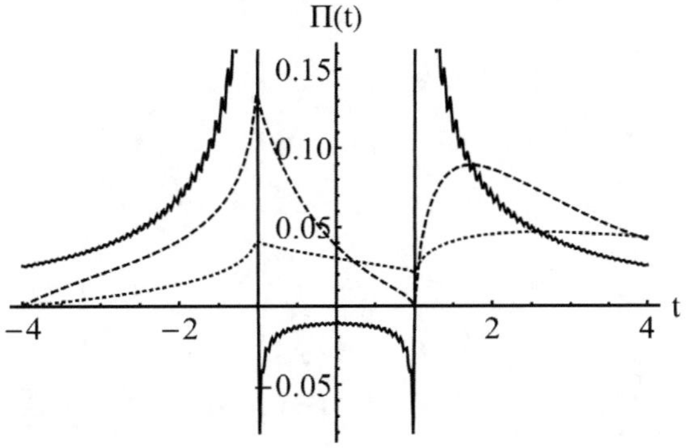

Figure 9. A plot of the current (solid line) and time-smeared current Eq.(27) for $V_0 = 0.5$ (dashed line) and $V_0 = 0.1$ (dotted line).

mean that backflow does not go away in the naive classical limit $\hbar \to 0$. We showed that this situation starts to appear more reasonable when the projectors used in the definition of the flux operator are replaced by quasiprojectors, which include a physical parameter characterizing the imprecision of real measurements. The eigenvalues then do depend on $\hbar$ and there is evidence that all the negative eigenvalues become zero or positive as $\hbar \to 0$, restoring the naive classical limit. However, there are clearly more issues to explore around this question.

In Section 6 we discussed a simple measurement model in which the current can be obtained from the measured probability by deconvolution, and from this result the negative current could in principle be obtained. The features of backflow elucidated here may be of value in designing experiments to test backflow. These and related ideas with be explored elsewhere.

Acknowledgements
We are very grateful to Gonzalo Muga for many useful conversations about the topic of this paper. JMY was supported by the John Templeton Foundation.

[1] Allcock G R 1969 *Ann. Phys.* **53** 253; *do.* 1969 *Ann. Phys.* **53** 286; *do.* 1969 *Ann. Phys.* **53** 311
[2] Bracken A J and Melloy G F 1994 *J. Phys.* A **27** 2197
[3] Eveson S P, Fewster C J and Verch R 2005 *Ann. Inst. H. Poincaré* **6** 1
[4] Penz M, Grübl G, Kreidl S and Wagner P 2006 *J. Phys.* A **39** 423
[5] Muga J G, Palao J P and Leavens C R 1999 *Phys. Lett.* A **253** 21
[6] Berry M V 2010 *J. Phys.* A **43** 415302
[7] Bracken A J and Melloy G F 1998 *Found. Phys.* **28** 505
[8] Bracken A J and Melloy G F 1998 *Ann. Phys. (Leipzig)* **7** 726
[9] Strange P 2012 *Eur. J. Phys.* **33** 1147
[10] Yearsley J M, Halliwell J J, Hartshorn R and Whitby A 2012 *Phys. Rev.* A **86** 042116
[11] For a good introduction see: Muga J G, Sala Mayato R and Egusquiza I L (eds) 2002 *Time in Quantum Mechanics* (Berlin: Springer); Muga J G and Leavens C R 2000 *Phys. Rep.* **338** 353
[12] Balazs N and Jennings B K 1984 *Phys. Rep.* **104** 347; Hillery M, O'Connell R F, Scully M O and Wigner E P 1984 *Phys. Rep.* **106** 121
[13] Loughlin P and Cohen L 2003 *J. Mod. Optics* **50** 2305
[14] Halliwell J J and Yearsley J M, in preparation
[15] Yearsley J M 2010 *Phys. Rev.* A **82** 012116
[16] Halliwell J J and Yearsley J M 2009 *Phys. Rev.* A**79** 062101; Muga J G, Brouard S and Macias D 1995 *Ann. Phys. (NY)* **240** 355; Blanchard P and Jadczyk A 1996 *Helv. Phys. Acta* **69** 613; Palao J P, Muga J G, Brouard S and Jadczyk A 1997 *Phys. Lett.* A**233** 227; Halliwell J J 1999 *Prog. Th. Phys.* **102** 707
[17] Damborenea J A, Egusquiza I L, Hegerfeld G C and Muga J G 2002 *Phys. Rev.* A**66** 052104; Hegerfeldt G C, Seidel D and Muga J G 2003 *Phys. Rev.* A **68** 022111

Graphene: QFT in curved spacetimes close to experiments

Alfredo Iorio

Faculty of Mathematics and Physics, Charles University in Prague, V Holešovičkách 2, 18000 Prague 8, Czech Republic

E-mail: iorio@mff.cuni.cz

Abstract. A recently proposed step-by-step procedure, to merge the low-energy physics of the π-bonds electrons of graphene, and quantum field theory on curved spacetimes, is recalled. The last step there is the proposal of an experiment to test a Hawking-Unruh effect, emerging from the model, that manifests itself as an exact (within the model) prediction for the electronic local density of states, in the ideal case of the graphene membrane shaped as a Beltrami pseudosphere. A discussion about one particular attempt to experimentally test the model on molecular graphene is presented, and it is taken as an excuse to solve some basic issues that will help future experiments. In particular, it is stated that the effect should be visible on generic surfaces of constant negative Gaussian curvature, that are infinite in number.

1. Introduction

Recently (see [1], [2] and, for a review, [3]), in order to realize in practise a quantum Dirac field theory on curved spacetimes, there has been interest in exploiting certain special features of the low-energy physics of the π-bonds electrons of graphene, Weyl symmetry [4] being among them [1]. The clearest signature of quantum field theory (QFT) on curved spacetime is the Hawking effect [5]. In [2] the best conditions for the simplest realization of the latter on graphene were proposed.

On one side, we have graphene, a one-atom-thick membrane of carbon atoms arranged in a hexagonal lattice, see Fig. 1, first discussed theoretically in [6, 7], then found experimentally in [8]. Its very peculiar transport properties are summarized by the following low-energy tight-binding Hamiltonian, see, e.g., [9], valid near the Fermi *points* $\vec{k}^D_\pm = (\pm\frac{4\pi}{3\sqrt{3}}, 0)$ in the Brillouin zone, see Figs. 2 and 3,

$$H = \frac{3t\ell}{2}\sum_{\vec{p}}\left(\psi^\dagger_+\vec{\sigma}\cdot\vec{p}\,\psi_+ + \psi^\dagger_-\vec{\sigma}^*\cdot\vec{p}\,\psi_-\right) = -i\hbar v_F\int d^2x\left(\psi^\dagger_+\vec{\sigma}\cdot\vec{\partial}\,\psi_+ + \psi^\dagger_-\vec{\sigma}^*\cdot\vec{\partial}\,\psi_-\right), \quad (1)$$

where $t \simeq 2.8$eV, is the electron energy for hopping to the nearest-neighbor, $\ell \simeq 1.4$Å is the carbon to carbon distance in the hexagonal honeycomb lattice, $v_F \equiv 3t\ell/(2\hbar) \simeq c/300 \simeq 10^6$m/s, and with $\vec{\sigma} \equiv (\sigma_1, \sigma_2)$, $\vec{\sigma}^* \equiv (-\sigma_1, \sigma_2)$, σ_i the Pauli matrices and $\psi_\pm \equiv (a_\pm, b_\pm)^{\rm T}$ Dirac spinors, where $a_\pm \equiv a(\vec{k}^D_\pm)$ ($b_\pm \equiv b(\vec{k}^D_\pm)$) is the annihilation operator for the electron in the L_A (L_B) sublattice, evaluated at the Fermi points.

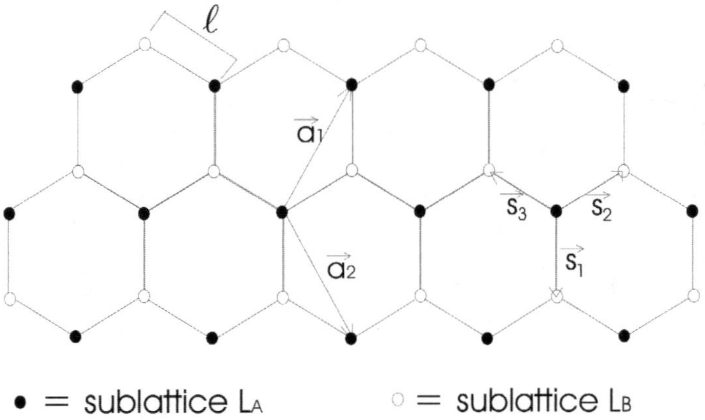

Figure 1. The honeycomb graphene lattice. The basis vectors, $\vec{a}_1 = \frac{\ell}{2}(\sqrt{3}, 3)$, $\vec{a}_2 = \frac{\ell}{2}(\sqrt{3}, -3)$, link sites of the sublattice L_A, but cannot reach the sublattice L_B. For the latter sites the extra vectors, $\vec{s}_1 = \ell(0, -1)$, $\vec{s}_2 = \frac{\ell}{2}(\sqrt{3}, 1)$, $\vec{s}_3 = \frac{\ell}{2}(-\sqrt{3}, 1)$, are necessary.

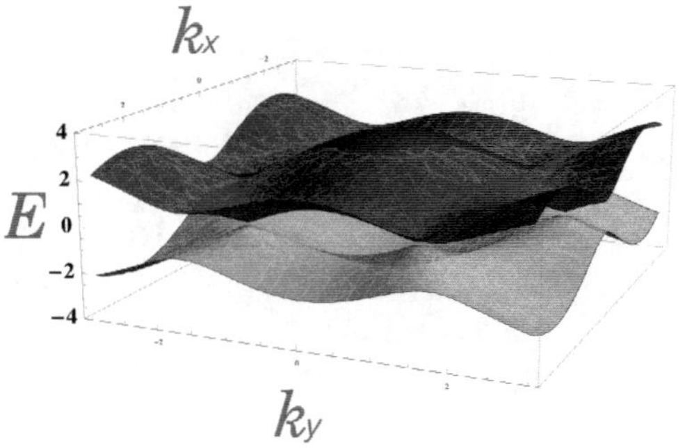

Figure 2. The dispersion relations of graphene, $E(k_x, k_y)$. The energy E is in units of the hopping, $t \simeq 2.8$ eV. The conductivity band (red) and the valence band (green) touch in six *points* (where $E = 0$), reflecting, in the Brillouin zone, the hexagonal symmetry of the lattice. Only the two points $\vec{k}^D_\pm$ are inequivalent.

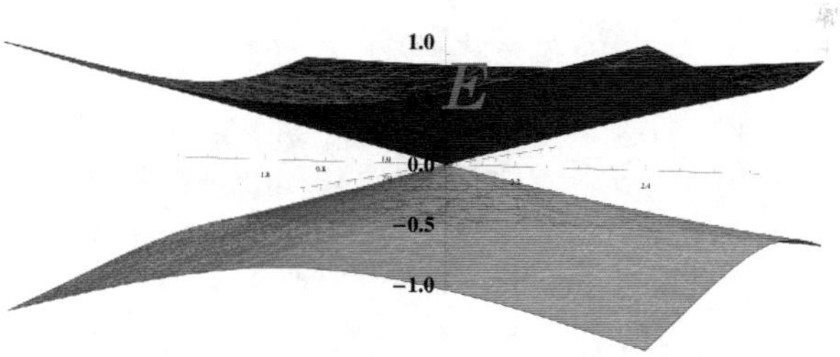

Figure 3. The linear dispersion relations near one of the Dirac points, showing the typical behavior of a relativistic-like system (the "v_F-light-cone"). Note that the energy range is of the order of one unit of the hopping energy t, i.e., here $E \in [-3\text{eV}, +3\text{eV}]$. Within this range, the continuum field approximation is tenable.

On the other side we have QFT in curved spacetime. This is by now an old theory [5], with no ultimate experimental evidence, hence with many open problems, and interesting issues to investigate in relation to the practical implementation of a fundamental theory. The open questions range from the structure of the quantum vacuum, and the related role of the inequivalent quantizations (think, for instance, to the *querelle* between the young and the old Hawking, about the information loss in the presence of a black hole), to the very meaning of particles in a curved spacetime. There are then open questions relative to the Hawking phenomenon itself, predicted but to-date impossible to detect from astrophysical sources, and the associated black-hole thermodynamics. These questions, although old, are all still on the table and permeate the contemporary research on high energy theory, as, for instance, one direction pursued these days is to see gravity, and the spacetime itself, as an emergent phenomenon.

Graphene appears to be a system where many of the issues above could be experimentally addressed. In the following, I will propose arguments and experiments in favor of this.

2. The step-by-step merging

The preliminary step to start the merging is to consider the dimensionality of the problem, that is (2+1) dimensions. I will take this duly into account. For instance, it is by now well known that many features of (3+1) dimensions need to be changed on the gravity side [10], [11], [12], [13], that a black-hole is possible [14], and that certain features, on the matter side, such as the "statistical swapping" [15], can take place.

• The 1$^{\rm st}$ step is to take time on board, hence we move from the Hamiltonian (1) to the corresponding action

$$A = i\hbar v_F \int d^3x \bar{\psi}\, \gamma^a \partial_a\, \psi \,, \tag{2}$$

where $a = 0, 1, 2$, $\gamma^0 = \sigma_3$, $\gamma^1 = i\sigma_2$, $\gamma^2 = -i\sigma_1$, and, since we do not consider phenomena mixing the two Fermi points, we focus on one point only, e.g., $\psi \equiv \psi_+$. When the graphene sheet is curved, and within certain limits [3], the action to consider is

$$\mathcal{A} = i\hbar v_F \int d^3x \sqrt{g}\; \bar{\psi}\gamma^\mu \nabla_\mu \psi \,, \tag{3}$$

where $\nabla_\mu = \partial_\mu + \Omega_\mu$, and $\Omega_\mu \equiv \frac{1}{2}\omega_\mu^{\ ab} J_{ab}$ is the gauge field able to take into account intrinsic curvature. Here $\omega_\mu^{\ ab}$ is the spin connection, and J_{ab} are the generators of local Lorentz transformations, with $\mu, ...$ responding to diffeomorphisms, and $a, b, ...$ flat indices. Within the limits indicated in [3] and [16] (essentially, in the absence of torsion and of strong scatters [17], and for small curvatures, i.e. large radius of curvature w.r.t. the lattice, $r > \ell$), this action well describes the physics of the very long wavelength π electrons on curved graphene sheets. The only effects it needs to take care of are the ones induced by intrinsic curvature, i.e., *inelastic* effects that we ascribe to disclination defects [18].

• The 2$^{\rm nd}$ step is to exploit the absence of mass in the action (3), that, in the case in point of the Dirac field, guarantees full local Weyl symmetry [4]. In the case of conformally flat metrics, $g_{\mu\nu} = \phi^2 \eta_{\mu\nu}$, Weyl symmetry gives

$$i \int d^3x \sqrt{g} \bar{\psi} \gamma^\mu \nabla_\mu \psi = i \int d^3x \bar{\psi}' \gamma^a \partial_a \psi' \tag{4}$$

where, under Weyl's transformations, $\psi = \phi^{-1}\psi'$. Namely, the flat space theory is classically equal to a curved space theory. We have to consider quantum effects, though. As we are in a odd-dimensional spacetime, this does not refer to quantum trace anomalies, absent here [19]. What we refer to here (see [1, 20]) is that one can, essentially, do two type of things: (a) perform measurements in a frame where the quantum effects of curvature disappear, e.g.,

$$S(x_1, x_2) \equiv \langle 0|\psi(x_1)\bar{\psi}(x_2)|0\rangle = {}'\langle 0|\psi'(x_1)\bar{\psi}'(x_2)|0\rangle' = S'(x_1, x_2) \,, \tag{5}$$

where $S(x_1, x_2)$ is the two-point function in the curved case, and $S'(x_1, x_2)$ is the two-point function in the flat case; or (b) perform the measurements in a frame where, due to Weyl symmetry, the effects are seen as a simple modification of the flat case (conformal triviality [5])

$$S(x_1, x_2) \equiv {}'\langle 0|\psi(x_1)\bar{\psi}(x_2)|0\rangle' = \phi^{-1}(x_1)\phi^{-1}(x_2)\; S'(x_1, x_2) \,. \tag{6}$$

In our case, due to the obvious difficulty to realize a frame where the measuring apparatus is in a gravitating field whose spacetime coincides with the graphene curved spacetime, the second choice will be the right one.

• The 3rd step is to stick to a metric that is the most easy to realize in a lab, namely

$$g_{\mu\nu}^{(2+1)}(x,y) = \begin{pmatrix} 1 & \\ & g_{\alpha\beta}^{(2)}(x,y) \end{pmatrix} , \tag{7}$$

i.e., with all the curvature in the spatial part $g_{\alpha\beta}^{(2)}$, and no time-dependence. Can this be conformally flat, rich and still interesting? To answer all these questions we need to solve [1, 12]

$$C_{\mu\nu} = \epsilon_{\mu\lambda\kappa}\nabla^\lambda R^{(2+1)}{}^\kappa_\nu + \epsilon_{\nu\lambda\kappa}\nabla^\lambda R^{(2+1)}{}^\kappa_\mu = 0 \, . \tag{8}$$

Since all surfaces of constant Gaussian curvature $\mathcal{K}$ solve this equation [1], they give rise to conformally flat (2+1)-dimensional spacetimes $g_{\mu\nu}^{(2+1)}$. Hence the answer to the above questions is yes: by simply curving the graphene sheet one can have rich enough situations (surfaces of constant $\mathcal{K}$) implying a conformally flat spacetime.

• The 4th step. The latter result is intrinsic, but to implement Weyl symmetry we need the explicit conformal factor, and this might mean to change the frame

$$q^\mu \equiv (t,x,y) \to \mathcal{Q}^\mu \equiv (T,X,Y) \tag{9}$$

so that

$$g_{\mu\nu}^{(2+1)}(\mathcal{Q}) = \phi^2(\mathcal{Q})\, g_{\mu\nu}^{\rm flat}(\mathcal{Q}) \, . \tag{10}$$

This can be technically difficult (a system of 6 partial differential equations), but the most important here are the physical meaning, global predictability, and practical feasibility of this coordinate frame.

The surfaces of $\mathcal{K} = +1/r^2 =$ const. can be described by the spatial line element [21]

$$d\ell^2 = du^2 + c^2\cos^2(u/r)dv^2 \tag{11}$$

where u is meridian coordinate, and v the parallel coordinate, and with the real constant c such that $c = r$ (sphere), $c < r$, $c > r$. These $\mathcal{Q}^\mu$ are difficult to find.

Interesting objects, like, e.g., the black-hole of [14], are related to spacetimes of negative constant curvature. For the surfaces of $\mathcal{K} = -1/r^2 =$ const. the following theorem by Hilbert holds: *No non-singular surface of constant* $\mathcal{K} = -1/r^2$ *can be embedded in* $\mathbf{R}^3$. If we can live (and we can) with singularities, such surfaces are many more than the positive curvature counterparts (there are, actually, an infinite number of them). Among them there are the surfaces of revolution, called pseudospheres, and they can be described by the spatial line element [21]

$$d\ell^2 = du^2 + \left(c_1^2\cosh^2(u/r) + c_2^2\sinh^2(u/r)\right) dv^2 \tag{12}$$

with $c_1, c_2 \in \mathbf{R}$. For all these surfaces (and also for those not included in the above line element)

$$ds^2_{\rm graphene} = dt^2 - \frac{r^2}{\tilde{y}^2}(d\tilde{x}^2 + d\tilde{y}^2) = \frac{r^2}{\tilde{y}^2}\left[\frac{\tilde{y}^2}{r^2}dt^2 - d\tilde{x}^2 - d\tilde{y}^2\right] \tag{13}$$

with $\tilde{x}, \tilde{y}$ the abstract coordinates of upper-half plane model of the Lobachevsky geometry (see [16] for the case in point of graphene, and [22] for a general introduction to the subject). To make sense in a laboratory, these coordinates need to be expressed in terms of measurable coordinates, say the (u,v) coordinates above, and this amounts to choose the surface. The line element in square bracket is flat, but this is not yet the solution of our problem, because we need to see, case by case, to which physical $\mathcal{Q}^\mu$ the coordinates $(t, \tilde{x}, \tilde{y})$ correspond to.

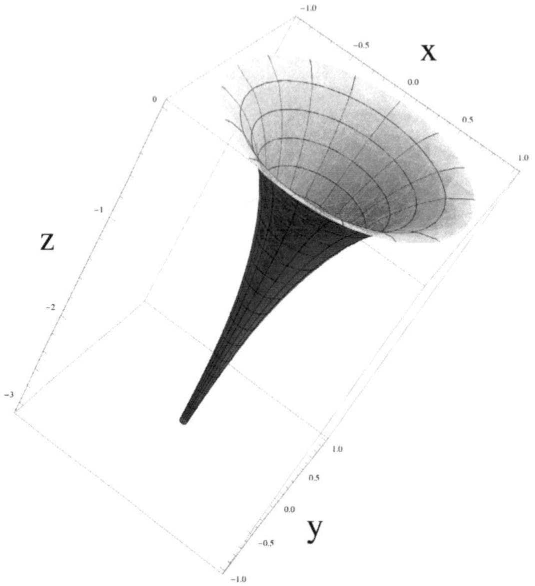

Figure 4. The $\mathbf{R}^3$ coordinates of the Beltrami pseudosphere, in the canonical form, are [23] $x(u,v) = R(u)\cos v$, $y(u,v) = R(u)\sin v$, $z(u) = r(\sqrt{1 - R^2(u)/r^2} - \text{arctanh}\sqrt{1 - R^2(u)/r^2})$, with $R(u) = c\, e^{u/r}$, $c > 0$ and $r = \sqrt{-\mathcal{K}^{-1}} > 0$ where $\mathcal{K}$ is the constant negative Gaussian curvature. We can choose $c = r$, thus $R(u) \in [0, r]$ as $u \in [-\infty, 0]$. The surface is not defined for $R > r$ (z becomes imaginary), an instance of the Hilbert theorem. In the plot $r = 1$ and $u \in [-3.37, 0]$, $v \in [0, 2\pi]$.

As a counter-example, consider the case of the hyperbolic pseudosphere ($c_1 \neq 0$ and $c_2 = 0$ in (12))

$$\tilde{x} = e^{cv/r} \tanh(u/r) \, , \quad \tilde{y} = e^{cv/r} / \cosh(u/r) \, . \tag{14}$$

Since the conformal factor $r^2/\tilde{y}^2$ in (13) is multivalued (v is an angle), this solution as no global predictability in $q^\mu = (t, u, v)$, hence one must look for a different frame where this does not happen, but the frame in point might be difficult to realize in practise (and surely it is not the lab frame q^μ, where all that needs to be done is to curve the surface).

It is the Beltrami pseudosphere (see Fig. 4) that solves the problem

$$\tilde{x} = v/r \, , \quad \tilde{y} = e^{-u/r}/r \tag{15}$$

i.e. $r^2/\tilde{y}^2$ makes sense over the whole surface/spacetime. Hence, the physical $\mathcal{Q}^\mu$ are the lab's q^μ. A physically doable $g_{\mu\nu}^{\text{graphene}} = \phi^2 \, g_{\mu\nu}^{\text{flat}}$ is already there in the frame q^μ:

$$ds^2_{\text{graphene}} \equiv ds^2_{(B)} = e^{2u/r} \left[e^{-2u/r}(dt^2 - du^2) - r^2 dv^2 \right] \, . \tag{16}$$

Now we can say that the line element in square brackets is the Rindler line element, $ds^2_{(R)}$, and exploit this fact in our quest for measurable effects of the QFT in curved space interpretation of the physics of graphene π electrons.

• The 5th step consists in implementing the Unruh requests on this setting. In particular it is important the choice of the quantum vacuum of reference, taken here to be the Minkowski vacuum $|0\rangle' \equiv |0\rangle_M$, thus the Green's function of interest are of the form (see (6) and [1, 2, 16])

$$S^B(x_1, x_2) \equiv \langle 0_M | \psi^B(x_1) \bar{\psi}^B(x_2) | 0_M \rangle = \phi^{-1}(x_1) \phi^{-1}(x_2) \, \langle 0_M | \psi^R(x_1) \bar{\psi}^R(x_2) | 0_M \rangle \, , \tag{17}$$

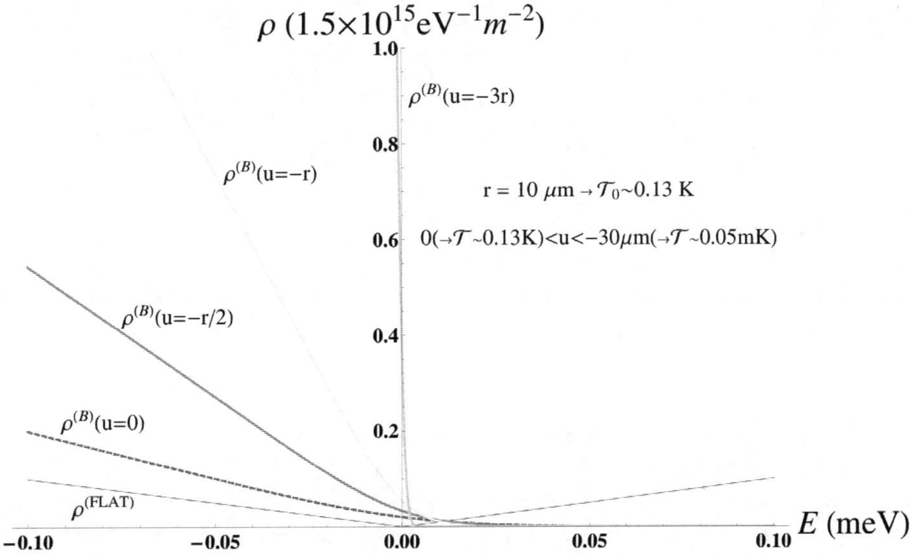

Figure 5. Series of dI/dV spectra ("line-cut") from an STM. The model applies to energies below the natural scale, $E^* = \hbar v_F/r \simeq 6.6 \times 10^{-7}\text{eV} \times [r \text{ mm}]^{-1}$, where $[r \text{ mm}]$ is the numerical value of r measured in mm. In this plot $E \in [-0.1, 0.1]$meV, corresponding to $r = 10\mu$m. The flat LDOS is shown in blue. $u \in]0, -3r]$, $u = 0$ (red dashed curve) corresponds to the singular boundary of the pseudosphere, beyond the upper limit $u = -3r$ (green curve) the pseudosphere is too sharp. To each spectrum corresponds a u-dependent temperature $\mathcal{T}$ while the constant temperature, same for the whole line-cut, is $\mathcal{T}_0 = e^{-u/r}\mathcal{T} \sim 0.13$K.

where B stands for "Beltrami", R for "Rindler", and the Green's function on the right side is S^R, the customary two-point function used in the computation of the Unruh effect. The power spectrum is customarily computed by taking S^R at the same point in space, $\vec{q}$, and at two different times, with $t_2 - t_1 = e^{u/r}\tau$, and τ the proper time (the factor $e^{u/r}$ is related here to the proper acceleration, and needs be constant, to fit the requests for the standard Unruh effect). From this, by Fourier transforming on the time variable, we get [15]

$$F^R(\omega, \vec{q}) \equiv \frac{1}{2}\text{Tr}\left[\gamma^0 \int_{-\infty}^{+\infty} d\tau e^{-i\omega\tau} S^R(\tau, \vec{q}, \vec{q})\right] , \tag{18}$$

and it coincides here with the (not yet physical) electronic local density of states (LDOS) [9]

$$\rho^R(\omega, \vec{q}) \equiv \frac{g}{\pi} F^R(\omega, \vec{q}) . \tag{19}$$

As this is a massless case, the result of the computation is exact

$$F^{(R)}(\omega) = \frac{1}{2} \frac{\omega}{e^{\omega/\mathcal{T}} - 1} \tag{20}$$

where $\mathcal{T} \equiv \alpha(u)/(2\pi) = e^{u/r}/(2\pi r)$. We are now in the position to make a verifiable prediction, by going back to the physical spacetime. Due to the Weyl symmetry, this is easy (see (17))

$$\rho^B(\omega, \vec{q}) = \phi^{-2}(\vec{q})\rho^R(\omega, \vec{q}) . \tag{21}$$

Thus, the LDOS predicted for a graphene sheet shaped as a Beltrami pseudosphere is [2]

$$\rho^B(E,u,r) = \frac{4}{\pi}\frac{1}{(\hbar v_F)^2}\frac{E\,e^{-2u/r}}{\exp\left[E/(k_B \mathcal{T}_0 e^{u/r})\right]-1}\,, \tag{22}$$

i.e., a Hawking phenomenon with Hawking temperature

$$\mathcal{T}_0 \equiv \frac{1}{2\pi}\frac{\hbar v_F}{k_B\,r}\,. \tag{23}$$

In Fig. 5 a plot of what we expect from a dedicated experiment with a Scanning Tunneling Microscope [2].

3. Paving the way to future experimental set-ups

Lately, I have been interacting with experimentalists from the Manoharan Lab of Stanford University[24], who asked whether it is possible to see effects, of the kind described above, on their "molecular graphene" [25], that is a planar hexagonal lattice of carbon monoxide, CO, molecules that mimics graphene. Unfortunately, there were severe limitations in their experimental settings. The first is the flat setting, the second is the small number of points, the third is that molecular graphene is a simulator of the real thing that, although a very good one to test certain properties, it has basic parameters (e.g., the neighbor-to-neighbor distance, ℓ) that make our task more difficult.

To write things in a list, what I had to address are the following issues:

(i) How to tile a Beltrami pseudosphere?

(ii) Would a "planar development" (the result of mapping the surface on a plane, hence with curvature traded for strain), still captures the features necessary for the Hawking-Unruh phenomenon to happen? Which one, among the infinitely many, is the best?

(iii) Is the effect still there for a *generic surface* of negative constant $\mathcal{K}$? If yes ...

(iv) ... can we solve issues (i) and (ii) for the given surface?

(v) Can we spot any Hawking-Unruh in a planar configuration of molecular graphene, that mimics the planar development of a *portion* of a generic graphene surface of (not even necessarily constant) negative $\mathcal{K}$?

My conclusions for the last point (the best meeting point between Manoharan Lab's and this model's needs), are negative, as I will explain below. This analysis gave extra arguments in favor of a dedicated experiment, with a Beltrami pseudosphere, as the best way to test this model. Nonetheless, the attempt was worth the effort. First, I learned that point (iii) has a positive solution. Thus, the realization of a generic graphene surface of constant negative Gaussian curvature, although not being the best possible setting, should be enough to test a modified formula (22), at least in an approximate way. On this I will briefly comment below, and will write in detail elsewhere. Second, for whatever dedicated experiment one would like to carry on, matters like the issue (i) in the list are crucial. In the following, I will address in detail this latter issue, and will also briefly comment about the other issues.

Therefore, let me focus on the geometric and topological constraints on building a Beltrami pseudosphere with the hexagonal graphene membrane. There are two issues here. The first is the infinite (non-compact) surface ($u \in [-\infty, 0]$). The second is how many heptagonal defects are necessary. Let me introduce, briefly, the matter by recalling how to tile a compact surface Σ (see, e.g., [26]). Suppose we want to tile Σ with regular s-sided polygons assembled in such a way that each vertex is shared by 3 polygons, and each edge is shared by 2 polygons. If n_s is the number of polygons used (there are, say, n_5 pentagons, n_6 hexagons, n_7 heptagons, etc.) to tile Σ, the resulting *polyhedron* P has $V_P = 1/3 \sum_n n_s s$ vertices, $E_P = 1/2 \sum_n n_s s$ edges, and $F_P = \sum_n n_s$ faces, giving for the Euler characteristic

$$\chi(\Sigma) = V_P - E_P + F_P \ , \tag{24}$$

the following expression

$$\sum_n n_s(6-s) = 6\chi(\Sigma) \ . \tag{25}$$

For the sphere, the Gauss-Bonnet theorem[23], that links the topological number, χ, to the geometric object that is total curvature, gives

$$K_{tot} \equiv \int \mathcal{K} \, d^2\mu = 2\pi\chi \ , \tag{26}$$

and, being $\mathcal{K} = 1/r^2 =$ constant

$$\chi = \frac{\text{Area}}{2\pi r^2} \ , \tag{27}$$

or, being Area $= 4\pi r^2$, we have $\chi = 2$. Hence, if we try to wrap the hexagonal lattice of graphene, to form a tiling of the sphere, we cannot match $\sum_n n_s(6-s) = 12$, only with hexagons (whose number is not constrained by this argument), but need something else. Considering that the creation of each defect costs energy, the minimum number of defects we need are 12 pentagons, $\sum_n n_s(6-s) = n_5 \ (6-5) + n_6 \ (6-6) = 12$, with which we could make the smallest (and "minimum energy") tiling of the sphere, the dodecahedron ($F = 12$, $V = 20$, $E = 30$). Adding 20 hexagons to it, one makes a buckyball (fullerene). One can increase n_6 further (according to the rule $n_6 = 10(T-1)$, with $T = m^2 + l^2 + ml = 0, 1, 3, 4, 7, 9, \ldots$, and excluding $T = 0$), making "inflated" buckyballs. Once we trade the continuous sphere for its tiling polyhedron, the curvature resides in the vertices. For the sake of simplicity, let us focus on the icosahedron, dual ($F \leftrightarrow V$, E the same) to the dodecahedron, so that $K_{tot} = 12 \ K_V^{ico} = 4\pi$, and $K_V^{ico} \equiv K_5$. The curvature associated to each pentagon is then

$$K_5 = +\pi/3 \ , \tag{28}$$

while no curvature is associated to the hexagons, $K_6 = 0$. Of course, the same results can be obtained using the dodecahedron, although the argument becomes more involved. A direct way to obtain this result is with the *defect/excess angle* argument. The curvature at one vertex of the polyhedron is

$$K_V \equiv 2\pi - \sum_i \alpha_i \ , \tag{29}$$

where α_i is the angle between edges of each face at the vertex. For a pentagon (formed with equilateral triangles in the icosahedron), $\alpha_i = \pi/3$, and $i = 1, \ldots, 5$, hence, $K_V = 2\pi - 5\pi/3 = +\pi/3$ (defect angle). The same argument, for an hexagon gives $K_V = 2\pi - 6\pi/3 = 0$.

Notice that the radius of curvature, r, does not enter these formulae, hence the curvature associated to each defect is fixed. If we want to increase the curvature of the sphere (make r smaller), we do not have to increase the *number* of defects, but their *density* (i.e., we have to decrease the number of hexagons).

From the elastic point of view, it has been shown that, for the sphere, the creation of *heptagonal* defects is energetically favored in certain conditions [27], but this does not violate the Euler-Poincaré result (24), provided they come in pairs with pentagonal defects, $n_5 - n_7 = 12$, i.e. $n_5 = 12 + x$ and $n_7 = x$, and that is indeed the case (scars [27]). The same argument descending from the Gauss-Bonnet theorem (26) (and the excess angle argument of (29)) gives

$$K_7 = -\pi/3 \ , \tag{30}$$

as the curvature associated to each one of these defects. Thus, when we want to build (tiling of) surfaces of negative $\mathcal{K}$ we need heptagonal defects.

If we want to wrap the graphene membrane into a Beltrami pseudosphere, we first need to consider that the issue of an infinite (non-compact) surface is actually minor, from a practical (lab) point of view. One can truncate the Beltrami, without changing the topology, and still has a surface of constant negative curvature. (Notice, also, that the elliptic and hyperbolic pseudospheres are finite surfaces, so, for them, this problem is absent). So one can think of tiling the surface in the standard way illustrated above. The issue here is the number and types of defects necessary, and, in turn, how the Euler-Poincaré and Gauss-Bonnet theorems apply here, as they have been conceived for compact surfaces.

Having the topology of an annulus, the Euler number for Beltrami is [28]

$$\chi_{\rm Beltrami} = 0 \; . \tag{31}$$

(The same is true for the other two pseudospheres, not clear is the matter for more exotic pseudospheres [28]). If one blindly applies (24), $V - E + F = 0 \to \sum(6-s)n_s = 0$, the implications do not make sense. Would that mean that $s = 6$? So I can tile with hexagons all over? But that is true for the plane, the cylinder, or the cone, not for the pseudosphere. Would that mean that $n_7 = n_5$? But then I can satisfy the constraint with even 1 heptagon and 1 pentagon, and this makes no sense either. The unsolved (and intriguing) problem here is that, as far as I know, it is not known the negative curvature counterpart of the dodecahedron (or, of its dual, the icosahedron), the "elementary" polyhedron one obtains from the Euler-Poincaré for the sphere. There must be a fixed number of heptagons, that take into account the basic geometric properties, just like the 12 pentagons do for the sphere. The beauty and intricacies of the Lobachevsky geometry come in here, and make changes to the sphere story, and then introduce the singular boundaries, but the basic object should be there to give a unifying "Platonic-Lobachevsky" solid for all the infinite pseudospheres (not just the three that clearly have $\chi = 0$, but all the others). A partial solution to this problem, and in turn to our main problem (wrapping graphene to make a Beltrami), is to consider that the Gauss-Bonnet for surfaces with boundaries is [23]

$$\int_\Sigma \mathcal{K} d^2\mu + \int_{\partial\Sigma} \mathcal{K}_g dl = 2\pi\chi \; , \tag{32}$$

with $\mathcal{K}_g$ the geodesic curvature that measures how far is a curve from a geodesic (e.g., $\mathcal{K}_g = 0$ for the equator of a sphere). In (32) there are two parts, one for the bulk that is the total curvature, K_{tot}, and one for the boundaries. It is K_{tot} that tells about the number of defects, the boundary term does not. Indeed, from Area $= 2\pi r^2$, we have $K_{tot} = -\text{Area}/r^2 = -2\pi$, (the same is true for the other two pseudospheres), hence, using (30), $K_{tot} = -2\pi = n_7 \, K_7 = n_7 \, (-\pi/3)$, that gives

$$n_7 = 6 \; . \tag{33}$$

In other words, the Euler-Poincaré and the Gauss-Bonnet theorems "decouple" for pseudospheres, and we cannot apply (24) to compute the number of defects, but need the curvature arguments just given. That said, though, everything else is like for the sphere: the negative curvature counterpart of the icosahedron/dodecahedron, at least for the three classic pseudospheres, has 6 heptagons, arranged to form a non-compact "polyhedron"(actually, I did not succeeded to use only heptagons, and a certain number of hexagons are necessary even for the basic object), and, elastic energy allowing, one can add pentagon-heptagon pairs. This is my solution of the issue (i) in the list above. Let me now face the other issues.

The procedure to map a surface onto a plane is called "planar development" [23]. For the intrinsically flat surfaces, this can be done isometrically (that is why those surfaces are called "developable"). For intrinsically curved surfaces, planar development comes at the price of deforming the distance among points (think of any map of the Earth), hence of trading curvature for strain, and there are infinitely many different "planar versions" of the curved surface, see, e.g., [29]. At first sight, the effects we are describing with (22), being due to intrinsic curvature, could never be seen on a planar development. In the case in point of the graphene membrane, though, there is some hope to still have some sort of signal of the Hawking-Unruh effect, although we can expect it to be substantially deformed by the planar development. It is so, because, in our model, the intrinsic curvature of the graphene membrane is encoded in the heptagonal defects, hence, retaining these defects in the developed Beltrami surface, in principle, could be the way out to test the effect on a flat sample. Unfortunately, the geometric constraints on the development, and the limits of the experimental set-up of the device of Manoharan Lab, conspired to make this test impossible. Let me explain in a bit more detail.

The actual values of various important quantities differ, for graphene and for molecular graphene (for the latter I rely on [31], see also [24] and [25]). In the table 1, I summarize the ones relevant for this discussion. There, I give two energy scales, $E_\ell \simeq \hbar v_F/\ell$ (below which the continuum Dirac field approximation holds), and $E_r \simeq \hbar v_F/r$ (below which the field theory in curved space time holds). The formula (22) has a chance to be tested only if $r >> \ell$, and it holds for energies below E_r. On this see [3] and [2, 16]. Essentially, what goes wrong with the Manoharan Lab set-up is that, even to just include all the necessary 6 heptagons, the limited number of points makes the density of defects (hence, the corresponding curvature) too close to the limit for the model to hold. Given the values of v_F and ℓ, and the maximum sensitivity of the STM machine of 1-10 meV, to a small enough curvature (big r) corresponds an energy too small to detect the effect in that range (see the entry for $r \sim 100\ell$ in table 1, and consider that the electrons need to have an energy below the corresponding E_r).

Table 1. Values of various important quantities, for graphene and for molecular graphene. Here $v_F = \frac{3}{2\hbar} t\ell$, with t hopping parameter, ℓ nearest-neighbor distance, and $E_\ell \simeq \hbar v_F/\ell$, $E_r \simeq \hbar v_F/r$.

	Graphene	Molecular Graphene
ℓ (m)	1.4×10^{-10}	20×10^{-10}
v_F (m/s)	9×10^5	6.5×10^5
E_ℓ (eV)	4.2	0.2
$E_r(r \sim 100\ell)$ (meV)	42	2.1
$E_r(r \sim 50\ell)$ (meV)	84	4
$E_r(r \sim 20\ell)$ (meV)	200	10

Indeed, the number of defects per unit area is $d^{(7)} \equiv n_7/A = 3/\pi\ r^{-2}$, as $n_7 = 6$ and $A = 2\pi r^2$ for Beltrami. So the number of defects in the area a of the surface is $n_7^a = d^{(7)}\ a = 3/\pi\ a/r^2$, hence

$$r(n_7^a) = \sqrt{\frac{3a}{\pi n_7^a}}\ . \tag{34}$$

The number of CO molecules is limited to 500, hence an estimate of the area is: 170 hexagons, each with area $3\sqrt{3}/2\ell^2$, giving $a \sim 450\ell^2 \sim 9000\mathring{A}^2$. With these, $E_r \simeq \hbar v_F/r = \alpha^{-1} E_\ell = \alpha^{-1} \times 0.2$eV, where $\alpha = r/\ell$, all the quantities are for molecular graphene. In table 2, I report the estimate of the values of r as a function of the defects present in the lattice, and the corresponding E_rs. It is clear from there that we are right at the limit of validity of the model, (see the entries for $n_7^a = 6$).

Furthermore, a planar development of the Beltrami, without any meridian cut, is a highly complicated problem, as is well known to shoes, ships and aircrafts makers [32]. Our goal was to "squash" it into a deformed annulus [30], without cutting the surface along a meridian (the truncation along some parallel on the "tail", instead, was required, as explained earlier). This would help retaining the necessary global properties of the Beltrami spacetime. Unfortunately, we realized that the strain would be too big [30], and would break the σ-bonds [31].

Table 2. Estimates of radii of curvature as function of the number of defects, and corresponding threshold energies, for a 500 points lattice of molecular graphene. The last values correspond to include all the six heptagonal defects, required to mimic a Beltrami (and other) pseudosphere. The corresponding curvature is too close to the limit to have reliable results.

n_7^a	r/ℓ	E_r (meV)
1	21	10
2	15	14
3	12	17
4	10.5	20
5	9.5	22
6	8.5	25

For all these reasons, the last proposal that was left to test the model, within these limits, was to see whether some relics of the effect are still there, for a configuration of planar molecular graphene that includes few heptagonal defects (with this, at least the condition of small curvature of the original surface is better satisfied/less unsatisfied, see table 2). In this fashion, the sample could mimic the planar development of a *portion* of a generic (not necessarily constant) negatively curved surface. The answer, as said, is negative. Let me explain why.

One piece of good news here arrived from the discovery that the issue (iii) of the list has positive answer. Namely, it is indeed possible to have a modified and approximate version of formula (22) at work, for small neighbors of a point of a graphene membrane shaped into a generic surface of constant negative curvature. Indeed, it is true that, as discussed in the 5$^{\rm th}$ step of the previous Section (see [2] and [16]), that the global predictability for the given surface in the lab frame is lost, but, as long as local considerations are concerned, some predictive power of (22) is still at work. Full details of this result will be given elsewhere. For the moment let me say that the main reason for this is that the horizon, in this model, is reached "time-wise", like in the standard Unruh effect, and not "space-wise", i.e. the horizon is not a specific curve on the manifold. Being the time t the same for all surfaces in the lab frame q^μ, the effect is still there for all, although modified and only valid portion by portion of the surface.

Nonetheless (leaving aside the issue of constant vs nonconstant $\mathcal{K}$), one thing is to *focus* on one portion (neighbor of a point) of a whole surface, for which the global conditions for the existence of an horizon (hence the Hawking) are realized, and another thing is to have a *detached* (cut-away) portion of that surface. For instance, in order to feel the curvature, the electrons need ho have enough time to travel back and forth on the *whole* surface, an instance that also means that the electrons that feel the curvature are those of small energy. When this takes place on a surface, the electrons come back to the initial point (for a non-compact surface, at least some of them), while for a piece cut-away from the surface, obviously this cannot happen. Thus, there is no hope to see any effect that originates from global properties, as the Hawking-Unruh of (22).

To conclude on a positive note: although it was impossible to test these ideas in the experimental settings of Manoharan's lab (that were not meant to test these kind of effects, in the first place), the fact that we can test them on a generic surface of constant negative $\mathcal{K}$ is very promising for future, dedicated experiments.

Acknowledgments
I thank P. Azariadis, G. Gibbons, and S. Sen, for interesting correspondence, and G. Lambiase and D. Rastawicki, for collaboration on some of the matters discussed in Section 2 and in Section 3, respectively.

References

[1] Iorio A *Ann. Phys.* 2011 **326** 1334
[2] Iorio A and Lambiase G 2012 *Phys. Lett. B* **716** 334
[3] Iorio A 2012 *Eur. Phys. J. Plus* **127** 156
[4] Iorio A, O'Raifeartaigh L, Sachs I and Wiesendanger C 1997 *Nucl. Phys. B* **495** 433
[5] Birrell N D and Davies P C W 1984 *Quantum Fields in Curved Space* (Cambridge: Cambridge Univ Press)
[6] Wallace P R 1947 *Phys. Rev.* **71** 622
[7] Semenoff G W 1984 *Phys. Rev. Lett.* **53** 2449
[8] Novoselov K S *et al.* 2004 *Science* **306** 666
[9] Castro Neto A H *et al.* 2009 *Rev. Mod. Phys.* **81** 109
[10] Deser S, Jackiw R and 't Hooft G 1984 *Ann. Phys.* **152** 220
[11] Deser S, Jackiw R and Templeton S 1982 *Ann. Phys.* **140** 372
[12] Guralnik G, Iorio A, Jackiw R and Pi S-Y 2003 *Ann. Phys.* **308** 222
[13] Carlip S 1998 *Quantum gravity in 2+1 dimensions* (Cambridge: Cambridge Univ Press)
[14] Bañados M, Teitelboim C and Zanelli J 1992 *Phys. Rev. Lett.* **69** 1849; Bañados M, Henneaux, Teitelboim C and Zanelli J 1993 *Phys. Rev. D* **48** 1506
[15] Takagi S 1986 *Prog. Theor. Phys. Suppl.* **88** 1
[16] Iorio A and Lambiase G 2013 in preparation
[17] Peres N M R 2010 *Rev. Mod. Phys.* **82** 2673
[18] Katanaev M O and Volovich I V 1992 *Ann. Phys.* **216** 1; Kleinert H 1989 *Gauge fields in condensed matter, Vol II* (Singapore: World Scientific)
[19] Christensen S M 1978 *Phys. Rev. D* **17** 946
[20] Iorio A, Lambiase G and Vitiello G 2004 *Ann. Phys.* **309** 151
[21] Eisenhart L P 1909 *A treatise on the differential geometry of curves and surfaces* (Princeton: Princeton Univ Press)
[22] Penrose R 2005 *The road to reality* (New York: Knopf)
[23] Špivak M 1999 *A comprehensive introduction to differential geometry. Part III* (Houston: Publish or Perish Inc)
[24] http://mota.stanford.edu/
[25] Gomes K K, Mar W, Ko W, Guinea F and Manoharan H C 2012 *Nature* **483** 307
[26] Iorio A and Sen S 2006 *Phys. Rev. B* **74** 052102
[27] Bowick M, Nelson D and Travesset A 2000 *Phys. Rev. B* **62** 8738; Bausch A *et al.* 2003 *Science* **299** 1716
[28] Gibbons G W 2012, private communications; Gibbons G W and Kallosh R E 1995 *Phys. Rev. D* **51** 2839
[29] Azariadis P and Sapidis N 2003 *Geometric Modeling* ed M Sarfraz (Chichester: John Wiley) p 19
[30] Azariadis P 2012, private communications
[31] Rastawicki D 2012, private communications
[32] Cai Z-Y, Li M-Z and Zhang H-M 2007 *Finite Elements in Analysis and Design* **43** 301

The PVLAS experiment: detecting vacuum magnetic birefringence

G Zavattini[3], F Della Valle[1], U Gastaldi[2], G Messineo[3], E Milotti[1], R Pengo[2], L Piemontese[4] and G Ruoso[2]

[1] INFN, Sez. di Trieste and Dip. di Fisica, Università di Trieste, via A. Valerio 2, I-34127 Trieste, Italy
[2] INFN, Lab. Naz. di Legnaro, viale dell'Università 2, I-35020 Legnaro, Italy
[3] INFN, Sez. di Ferrara and Dip. di Fisica e di Scienze della Terra, Università di Ferrara, via Saragat 1, Edificio C, I-44122 Ferrara, Italy
[4] INFN, Sez. di Ferrara, via Saragat 1, Edificio C, I-44122 Ferrara, Italy

E-mail: guido.zavattini@unife.it

Abstract. The PVLAS collaboration is presently assembling a new apparatus to detect vacuum magnetic birefringence. This property is related to the structure of the QED vacuum and is predicted by the Euler-Heisenberg-Weisskopf effective Lagrangian. It can be detected by measuring the ellipticity acquired by a linearly polarised light beam propagating through a strong magnetic field. Here we report results of a scaled-down test setup and briefly describe the new PVLAS apparatus. This latter one is in construction and is based on a high-sensitivity ellipsometer with a high-finesse Fabry-Perot cavity ($> 4\times10^5$) and two 0.8 m long 2.5 T rotating permanent dipole magnets. Measurements with the test setup have improved by a factor 2 the previous upper bound on the parameter A_e, which determines the strength of the nonlinear terms in the QED Lagrangian: $A_e^{(\rm PVLAS)} < 3.3\times10^{-21}$ T^{-2} 95% c.l.

1. Introduction

In the absence of matter, Maxwell's equations can be obtained from the classical electromagnetic Lagrangian density $\mathcal{L}_{\rm Cl}$ (in S.I. units)

$$\mathcal{L}_{\rm Cl} = \frac{1}{2\mu_0}\left(\frac{E^2}{c^2} - B^2\right) \tag{1}$$

where μ_0 is the magnetic permeability of vacuum and c is the speed of light in vacuum. A quadratic Lagrangian leads to linear partial differential equations for the fields, and the superposition principle holds, thereby excluding light-light scattering and other non linear electromagnetic effects in vacuum.

With the introduction of Dirac's equation (1928) for electrons and Heisenberg's Uncertainty Principle (1927), Euler, Heisenberg and Weisskopf in 1936 [1] derived an effective Lagrangian density which leads to electromagnetic non linear effects even in vacuum. For photon energies well below the electron mass and fields much smaller than their critical values, $B \ll B_{\rm crit} = m_e^2c^2/e\hbar = 4.4\times10^9$ T, $E \ll E_{\rm crit} = m_e^2c^3/e\hbar = 1.3\times10^{18}$ V/m, the Euler-Heisenberg-Weisskopf

Journal of Physics: Conference Series **442** (2013) 012057 doi:10.1088/1742-6596/442/1/012057

effective Lagrangian can be written as

$$\mathcal{L}_{\rm EHW} = \mathcal{L}_{\rm Cl} + \frac{A_e}{\mu_0}\left[\left(\frac{E^2}{c^2} - B^2\right)^2 + 7\left(\frac{\vec{E}}{c}\cdot\vec{B}\right)^2\right] \tag{2}$$

where

$$A_e = \frac{2}{45\mu_0}\frac{\alpha^2 \lambda\!\!\!^{-}_e{}^3}{m_e c^2} = 1.32 \times 10^{-24}\ \mathrm{T}^{-2} \tag{3}$$

with $\lambda\!\!\!^{-}_e = \hbar/mc$ being the Compton wavelength of the electron, $\alpha = e^2/(\hbar c 4\pi\epsilon_0)$ the fine structure constant, m_e the electron mass.

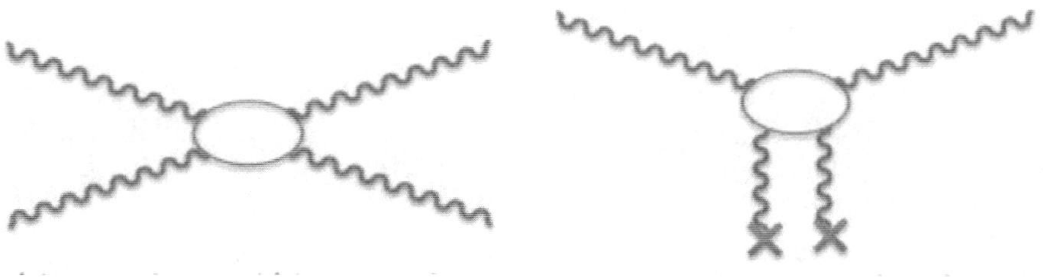

a) b)

Figure 1. Feynman diagrams for four electromagnetic field interactions. Left: elastic photon-photon scattering. Right: Vacuum magnetic birefringence

This Euler-Heisenberg-Weisskopf Lagrangian allows four field interactions that can be represented, to first order, by the Feynman diagrams shown in Figure 1 a) and b). Figure 1 a) represents light by light scattering whereas Figure 1 b) represents the interaction of real photons with a classical field leading to vacuum magnetic birefringence. Quantum Electrodynamics (QED) also predicts non-linear effects in vacuum leading to the same birefringence and light-by-light (LbL) scattering as the Euler-Heisenberg Lagrangian through the four-photons box diagram [2, 3, 4, 5, 6, 7, 8, 9].

Vacuum magnetic linear birefringence and light-light interaction in vacuum at very low energies have yet to be observed. Several experimental efforts are underway to detect such effects [10, 11, 12, 13, 14, 15, 16, 17, 18, 19]. Before the measurements presented here (and in [20]), the previous bound on four photon interaction was set by the PVLAS collaboration [12] with an upper bound on vacuum magnetic birefringence $\Delta n^{(\rm PVLAS)}$ measured with a magnetic field $B = 2.3$ T

$$\Delta n^{(\rm PVLAS)} < 1.0 \times 10^{-19} \qquad @\ 2.3\ \mathrm{T\ and\ 95\%\ c.l.} \tag{4}$$

The measurement was performed at $\lambda = 1064$ nm, corresponding to a photon energy $\hbar\omega =$ 1.17 eV. This limit translates into an upper bound for the unpolarised photon-photon elastic cross section [5, 6, 7, 8, 9] as

$$\sigma_{\gamma\gamma}^{(\rm PVLAS)}(1.17\ \mathrm{eV}) < 4.6 \times 10^{-62}\ \mathrm{m}^2. \tag{5}$$

The predicted QED value of the vacuum magnetic birefringence (see below) $\Delta n^{(\rm EHW)}$ and photon-photon elastic scattering cross section $\sigma_{\gamma\gamma}^{(\rm EHW)}$ are

$$\Delta n^{(\rm EHW)} = 2.1 \times 10^{-23} \qquad @\ 2.3\ \mathrm{T} \tag{6}$$

$$\sigma_{\gamma\gamma}^{(\rm EHW)}(1.17\ \mathrm{eV}) = 1.8 \times 10^{-69}\ \mathrm{m}^2 \tag{7}$$

1.1. Electrodynamics

To calculate the magnetic birefringence of vacuum one can proceed by determining the electric displacement vector $\vec{D}$ and the magnetic intensity vector $\vec{H}$ from the Lagrangian density $\mathcal{L}_{\rm EHW}$ of Eq. (2) by using the constitutive relations [3]

$$\vec{D} = \frac{\partial \mathcal{L}}{\partial \vec{E}} \quad \text{and} \quad \vec{H} = -\frac{\partial \mathcal{L}}{\partial \vec{B}}. \tag{8}$$

From these one obtains

$$\vec{D} = \epsilon_0 \vec{E} + \epsilon_0 A_e \Big[4\Big(\frac{E^2}{c^2} - B^2\Big)\vec{E} + 14\big(\vec{E}\cdot\vec{B}\big)\vec{B}\Big] \tag{9}$$

$$\vec{H} = \frac{\vec{B}}{\mu_0} + \frac{A_e}{\mu_0}\Big[4\Big(\frac{E^2}{c^2} - B^2\Big)\vec{B} - 14\Big(\frac{\vec{E}}{c}\cdot\vec{B}\Big)\frac{\vec{E}}{c}\Big]. \tag{10}$$

Let's consider a linearly polarised beam of light propagating perpendicularly to an external magnetic field $\vec{B}_{\rm ext}$. With $\vec{D}$ and $\vec{H}$, and using Maxwell's equations in media, one can describe light propagation in an external field. It is evident that the equations for light propagation will no longer be linear due to the non linear dependence of $\vec{D}$ and $\vec{H}$ on $\vec{E}$ and $\vec{B}$, respectively. We therefore substitute $\vec{E} = \vec{E}_\gamma$ and $\vec{B} = \vec{B}_\gamma + \vec{B}_{\rm ext}$ in (9) and (10), where the index γ refers to the light, assuming $|\vec{B}_{\rm ext}| \gg |\vec{B}_\gamma|$. One finds the following relations for the relative dielectric constant and magnetic permeability of vacuum:

$$\begin{cases} \epsilon_\parallel^{\rm (EHW)} &= 1 + 10A_e B_{\rm ext}^2 \\ \mu_\parallel^{\rm (EHW)} &= 1 + 4A_e B_{\rm ext}^2 \\ n_\parallel^{\rm (EHW)} &= 1 + 7A_e B_{\rm ext}^2 \end{cases} \qquad \begin{cases} \epsilon_\perp^{\rm (EHW)} &= 1 - 4A_e B_{\rm ext}^2 \\ \mu_\perp^{\rm (EHW)} &= 1 + 12A_e B_{\rm ext}^2 \\ n_\perp^{\rm (EHW)} &= 1 + 4A_e B_{\rm ext}^2 \end{cases} \tag{11}$$

where the indices $\parallel$ and $\perp$ refer to light polarisation parallel and perpendicular to $\vec{B}_{\rm ext}$, respectively.

From these sets of equations two important consequences are apparent: the velocity of light in the presence of an external magnetic field is no longer c and vacuum is birefringent with

$$\Delta n^{\rm (EHW)} = n_\parallel^{\rm (EHW)} - n_\perp^{\rm (EHW)} = 3A_e B_{\rm ext}^2. \tag{12}$$

Magnetised vacuum behaves like a uniaxial crystal. Numerically this leads to the value given in equation (6).

2. Experimental method

A birefringence Δn induces an ellipticity ψ on a linearly polarised beam of light given by

$$\psi(\vartheta) = \frac{\Delta\varphi}{2}\sin 2\vartheta = \pi\frac{L\Delta n}{\lambda}\sin 2\vartheta \tag{13}$$

where L is the optical path length within the birefringent region, λ is the wavelength of the light traversing it and ϑ is the angle between the light polarisation and the slow axis, defined in our case by the magnetic field direction. The optical path length through a region of length L can be increased by using a Fabry-Perot cavity; in this case we think of an effective path length $L_{\rm eff}$ and use a capital letter Ψ to indicate the total induced ellipticity. In fact, given a birefringent region of length L within a Fabry-Perot cavity of finesse $\mathcal{F}$, the effective path length is $L_{\rm eff} = (2\mathcal{F}/\pi)L$. Today finesses $\mathcal{F} > 400000$ can be obtained.

In the Euler-Heisenberg-Weisskopf case which we are discussing here, Δn depends quadratically on the magnetic field $\vec{B}$. High magnetic fields can be obtained with superconducting magnets but, as we will see below, it is desirable to have a time dependent field either by modulating its intensity, thereby changing Δn, or by rotating the field direction, thereby changing ϑ. This makes superconducting magnets far less appealing than permanent magnets which, today, can reach fields above 2.5 T. In the BFRT (Brookhaven, Fermilab, Rochester, Trieste) experiment the first scheme was adopted [21] by ramping the field intensity, whereas PVLAS chose to rotate an entire superconducting magnet [10, 11, 12] thereby rotating the magnetic field direction. At present the PVLAS collaboration has abandoned the superconducting magnet for 2.5 T permanent dipole magnets which are relatively inexpensive, have no running costs and have in principle 100% duty cycle, allowing very long integration times.

We are working with a Nd:YAG laser emitting radiation at 1064 nm. Frequency doubled versions exist and could double the induced ellipticity, but at the moment the highest finesses have been obtained for infrared light.

The expected ellipticity must be extracted from the noise within the maximum available integration time. For this reason, the magnetic field is made time dependent, to move away from DC and limit $1/f$ noise. Homodyne and heterodyne detections are two possible schemes. This second technique has been adopted in the PVLAS apparatus.

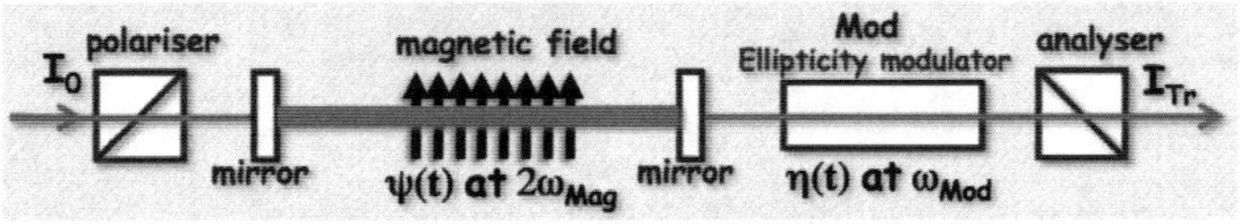

Figure 2. Scheme of the PVLAS ellipsometer.

A scheme of the PVLAS ellipsometer is shown in Figure 2. The input polariser linearly polarises the laser beam of intensity I_0 which then enters the sensitive region delimited by the Fabry-Perot cavity mirrors where the magnetic field is present. The laser is phase locked to this cavity. After the cavity the laser beam passes through an ellipticity modulator which adds a known time dependent ellipticity $\eta(t)$ to the beam. This modulator ellipticity adds to the ellipticity $\Psi(t)$ acquired within the magnetic field region. After the modulator, the beam passes through the analyser which selects the polarisation perpendicular to the input polarisation and finally a photodiode detects $I_{\rm Tr}$. The sought for information can be extracted from the Fourier spectrum of $I_{\rm Tr}$ and from the value of the intensity $I_{\rm out}$ before the analyser.

2.1. Estimate of the effect

To better understand what follows, it is useful to present some numerical values of the different quantities involved in the PVLAS experiment. Considering the vacuum magnetic birefringence due to the Euler-Heisenberg-Weisskopf Lagrangian, let us determine the ellipticity we expect in the apparatus under construction. The magnets have a total magnetic field length $L = 1.6$ m with a measured field intensity $|\vec{B}_{\rm ext}| = 2.5$ T resulting in $\int B_{\rm ext}^2\, dL = 10$ T^{2}m. In the estimate we use the the maximum finesse value we have reached: $\mathcal{F} = 414000$ [22]. Putting these numbers together leads to

$$\Psi_{\rm PVLAS} = 2\mathcal{F}\,\frac{3A_e \int B_{\rm ext}^2\, dL}{\lambda} = 3\times 10^{-11}. \tag{14}$$

Assuming a maximum integration time $T_{\rm max} = 10^6$ s and a signal to noise ratio SNR $= 1$ implies that the sensitivity must be

$$s_{\rm PVLAS} < \Psi_{\rm PVLAS}\sqrt{T_{\rm max}} = 3 \times 10^{-8}\ {\rm Hz}^{-1/2}. \tag{15}$$

In principle this is well above the shot noise limit (see formula (27) below). Unfortunately, as will be briefly discussed below, other noise sources are present which limit the ellipsometer sensitivity. We now discuss in detail various issues of the measurement technique and of the setup.

2.2. Heterodyne technique and optical path multiplier

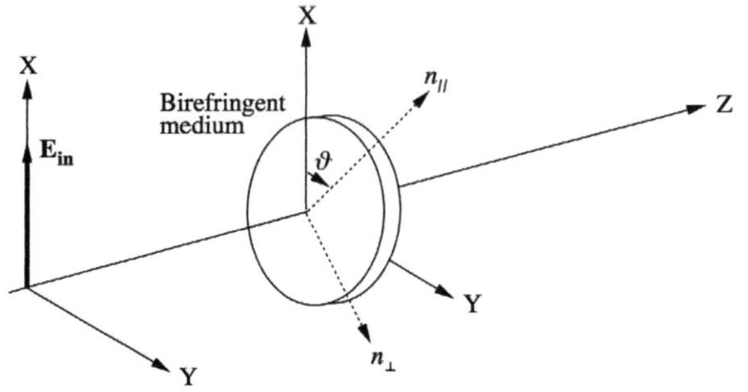

Figure 3. Reference frame for the calculations using the Jones matrix formalism. The birefringent medium has a thickness L.

Considering the coherence of the light source, a full treatment of the system can be done with the Jones matrix formalism [23]. For the purpose of our discussion let the laser beam propagate along the Z axis and let the incoming (linear) polarisation define the X axis (Figure 3). The Jones matrix for a uniaxial birefringent element of thickness L is given by

$$\mathbf{BRF}(\vartheta) = \begin{pmatrix} 1 + \imath\psi\cos 2\vartheta & \imath\psi\sin 2\vartheta \\ \imath\psi\sin 2\vartheta & 1 - \imath\psi\cos 2\vartheta \end{pmatrix} \tag{16}$$

where $\psi = \Delta\varphi/2$ ($\psi \ll 1$) is now the maximum ellipticity acquired by the light and ϑ represents the angle between the slow axis ($n_{\parallel} > n_{\perp}$) of the medium and the X axis.

In the heterodyne technique an ellipticity modulator is introduced after the birefringent region (in our case a Photoelastic Modulator, PEM), which adds a known sinusoidal ellipticity $\eta(t)$. This results in the linearisation of ψ in the intensity signal. In fact, the Jones matrix for the modulator is the same as **BRF** with ϑ set at an angle of $\pi/4$ ($\psi \ll \eta \ll 1$):

$$\mathbf{MOD} = \begin{pmatrix} 1 & \imath\eta(t) \\ \imath\eta(t) & 1 \end{pmatrix}. \tag{17}$$

Given an input electric field vector $\vec{E}_{\rm in} = E_0\binom{1}{0}$ the resulting vector describing the electric field after the modulator will be

$$\vec{E}_{\rm Tr} = E_0 \cdot \mathbf{MOD} \cdot \mathbf{BRF} \cdot \begin{pmatrix} 1 \\ 0 \end{pmatrix} = E_0 \begin{pmatrix} 1 + \imath\psi\cos 2\vartheta - \eta(t)\psi\sin 2\vartheta \\ \imath\eta(t) + \imath\psi\sin 2\vartheta - \eta(t)\psi\cos 2\vartheta \end{pmatrix}. \tag{18}$$

Neglecting second order terms, the intensity $I_{\rm Tr}$ after the analyser will be

$$I_{\rm Tr}(t) = I_0 \left| \imath\eta(t) + \imath\psi \sin 2\vartheta \right|^2 \simeq I_0 \left[\eta(t)^2 + 2\eta(t)\psi \sin 2\vartheta \right] \tag{19}$$

which now depends linearly on the ellipticity ψ.

To further improve the sensitivity of the ellipsometer one can increase the number of passes through the birefringent region. Either a multi-pass cavity or a Fabry-Perot cavity can be used for this purpose. In the PVLAS experiment a Fabry-Perot has been chosen. Let $\mathcal{T}$ and $\mathcal{R}$ be the transmittivity and reflectivity of the Fabry-Perot mirrors, and p their losses such that $\mathcal{T} + \mathcal{R} + p = 1$. In this case it can be shown that the electric field vector after the analyser is

$$\vec{E}_{\rm Tr} = E_0 \frac{\mathcal{T}}{\mathcal{T}+p} \begin{pmatrix} 0 \\ \imath\eta(t) + \imath \left(\frac{1+\mathcal{R}}{1-\mathcal{R}} \right) \psi \sin 2\vartheta \end{pmatrix} \tag{20}$$

and, indicating the output intensity, including mirror losses, as $I_{\rm out} = \left(E_0 \frac{\mathcal{T}}{\mathcal{T}+p} \right)^2$, the detected intensity will be

$$I_{\rm Tr}(t) = I_{\rm out} \left| \imath\eta(t) + \imath \left(\frac{1+\mathcal{R}}{1-\mathcal{R}} \right) \psi \sin 2\vartheta \right|^2. \tag{21}$$

The ellipticity ψ is multiplied by a factor $\frac{1+\mathcal{R}}{1-\mathcal{R}} \approx 2\mathcal{F}/\pi$ (typically $\mathcal{R} \approx 99.999\%$) where $\mathcal{F}$ is defined as the finesse of the Fabry-Perot. To complete the discussion, one finds experimentally that static and slowly varying ellipticities, indicated as $\alpha(t)$, are always present in a real apparatus and that two crossed polarisers have an intrinsic extinction ratio σ^2, mainly due to imperfections in the calcite crystals. Therefore, taking into account an additional spurious ellipticity term $\alpha(t)$ (since $\alpha, \psi, \eta \ll 1$ these terms commute and therefore add up algebraically) and a term proportional to σ^2, the total intensity at the output of the analyser will be

$$\begin{aligned} I_{\rm Tr}(t) &= I_{\rm out} \left[\sigma^2 + \left| \imath\eta(t) + \imath \left(\frac{2\mathcal{F}}{\pi} \right) \psi \sin 2\vartheta + \imath\alpha(t) \right|^2 \right] \simeq \\ &\simeq I_{\rm out} \left[\sigma^2 + \eta(t)^2 + \alpha(t)^2 + 2\eta(t) \left(\frac{2\mathcal{F}}{\pi} \right) \psi \sin 2\vartheta + 2\eta(t)\alpha(t) \right]. \end{aligned} \tag{22}$$

In PVLAS, to be able to distinguish the large term $\eta(t)\alpha(t)$ from the term $\eta(t) \left(\frac{2\mathcal{F}}{\pi} \right) \psi \sin 2\vartheta$, $\psi \sin 2\vartheta$ is also modulated in time by rotating the magnetic field direction (varying ϑ). The final expression, explicitly indicating the time dependence of ϑ, for the intensity at the output of the analyser is therefore

$$I_{\rm Tr}(t) = I_{\rm out} \left[\sigma^2 + \eta(t)^2 + \alpha(t)^2 + 2\eta(t) \left(\frac{2\mathcal{F}}{\pi} \right) \psi \sin 2\vartheta(t) + 2\eta(t)\alpha(t) \right]. \tag{23}$$

This expression is at the basis of the ellipsometer in the PVLAS apparatus [24]. Small ellipticities add up algebraically and the Fabry-Perot multiplies the single pass ellipticity $\psi \sin 2\vartheta$, generated within the cavity, by a factor $(1+\mathcal{R}) / (1-\mathcal{R}) \approx 2\mathcal{F}/\pi$, where $\mathcal{F}$ is the finesse of the cavity. The ellipticity signal to be detected is therefore $\Psi(\vartheta) = (2\mathcal{F}/\pi)\psi \sin 2\vartheta$.

The Fabry-Perot as an optical path amplifier works well until the accumulated phase difference due to birefringences within the cavity is less than $\pi/2$. Beyond this value the interferometer has two separate resonances for two orthogonal polarisations [24] and therefore elliptically polarised light cannot be resonant. As total birefringences also include the intrinsic birefringence of the cavity mirrors, two mirrors with similar characteristics must be used and their birefringence axes aligned.

2.3. Fourier components

In the PVLAS experiment, $\eta(t) = \eta_0 \cos(2\pi\nu_{\rm Mod}t + \theta_{\rm Mod})$ and the magnetic field direction is rotated at a frequency $\nu_{\rm Mag}$: $\vartheta(t) = 2\pi\nu_{\rm Mag}t + \vartheta_{\rm Mag}$. A Fourier analysis of the intensity $I_{\rm Tr}(t)$ of equation (21) results in four main frequency components each with a definite amplitude and phase. These are reported in table 1.

Table 1. Intensity of the frequency components of the signal after the analyser.

Frequency	Fourier component	$I_{\rm Tr}/I_{\rm out}$	Phase
dc	$I_{\rm dc}$	$\sigma^2 + \alpha_{\rm dc}^2 + \eta_0^2/2$	–
$\nu_{\rm Mod}$	$I_{\nu_{\rm Mod}}$	$2\alpha_{\rm dc}\eta_0$	$\theta_{\rm Mod}$
$\nu_{\rm Mod} \pm 2\nu_{\rm Mag}$	$I_{\nu_{\rm Mod}\pm 2\nu_{\rm Mag}}$	$\eta_0 \dfrac{2\mathcal{F}}{\pi}\psi$	$\theta_{\rm Mod} \pm 2\vartheta_{\rm Mag}$
$2\nu_{\rm Mod}$	$I_{2\nu_{\rm Mod}}$	$\eta_0^2/2$	$2\theta_{\rm Mod}$

The presence of a component at $\nu_{\rm Mod} \pm 2\nu_{\rm Mag}$ in the signal identifies an ellipticity induced by the rotating magnetic field. Furthermore the phase of this component must satisfy the value in table 1. In the presence of a signal above background with the correct Fourier phase, the ellipticity $\Psi = (2\mathcal{F}/\pi)\psi$ can be calculated from $I_{\rm out}$, from the Fourier components $I_{\nu_{\rm Mod}\pm 2\nu_{\rm Mag}}$, and from $I_{2\nu_{\rm Mod}}$ as the average of the two sideband signals:

$$\Psi = \frac{1}{2}\left(\frac{I_{\nu_{\rm Mod}+2\nu_{\rm Mag}}}{\sqrt{2I_{\rm out}I_{2\nu_{\rm Mod}}}} + \frac{I_{\nu_{\rm Mod}-2\nu_{\rm Mag}}}{\sqrt{2I_{\rm out}I_{2\nu_{\rm Mod}}}}\right). \tag{24}$$

2.4. Noise considerations

Indicating with $R_{\nu_{\rm Mod}\pm 2\nu_{\rm Mag}}$ the peak (as opposed to rms) noise spectral density at the signal frequencies, and assuming $R_{\nu_{\rm Mod}+2\nu_{\rm Mag}} = R_{\nu_{\rm Mod}-2\nu_{\rm Mag}}$ but uncorrelated, the sensitivity spectral density s of the ellipsometer for a unity signal to noise ratio is

$$s = \frac{1}{2}\frac{\sqrt{R^2_{\nu_{\rm Mod}+2\nu_{\rm Mag}} + R^2_{\nu_{\rm Mod}-2\nu_{\rm Mag}}}}{\sqrt{2I_{\rm out}I_{2\nu_{\rm Mod}}}} = \frac{1}{\sqrt{2}}\frac{R_{\nu_{\rm Mod}+2\nu_{\rm Mag}}}{\sqrt{2I_{\rm out}I_{2\nu_{\rm Mod}}}}. \tag{25}$$

In principle, the noise limit for such a system is determined by the rms shot noise $i_{\rm shot}$ of the current $i_{\rm dc}$:

$$i_{\rm shot} = \sqrt{2ei_{\rm dc}\Delta\nu} = \sqrt{2eI_{\rm out}q\left(\sigma^2 + \frac{\eta_0^2}{2} + \alpha_{\rm dc}^2\right)\Delta\nu} \tag{26}$$

where q is the efficiency of the photodetector, and $\Delta\nu$ the bandwidth. In the case $\eta_0^2 \gg \sigma^2, \alpha_{\rm dc}^2$, the dc current will depend only on η_0. By substituting $R_{\nu_{\rm Mod}+2\nu_{\rm Mag}} = \sqrt{2}\, i_{\rm rms}/(q\sqrt{\Delta\nu})$ with $i_{\rm rms} = i_{\rm shot}$ into Eq. (25), the peak shot-noise sensitivity spectral density $s_{\rm shot}$ becomes

$$s_{\rm shot} = \sqrt{\frac{e}{I_{\rm out}q}}. \tag{27}$$

Assuming an intensity $I_{\rm out} = 5$ mW and the efficiency of the diode $q = 0.7$ A/W, equation above gives $s_{\rm shot} \sim 7 \times 10^{-9}$ Hz$^{-1/2}$.

Other intrinsic noise sources are rms photodiode dark current noise $i_{\rm dark} = V_{\rm dark}\sqrt{\Delta\nu}/G$ where $V_{\rm dark}$ is the measured rms voltage noise spectral density and G is the transimpedance gain of the photodetection system, Johnson rms current noise in the transimpedance resistor in the amplifier of the photodiode $i_J = \sqrt{4k_BT\Delta\nu/G}$, and relative laser intensity current noise $i_{\rm RIN} = I_{\rm Tr}\, q\, {\rm RIN}(\nu)\sqrt{\Delta\nu}$, where RIN($\nu$) is the rms relative amplitude noise spectral density of the laser light. These noises must be kept below $i_{\rm shot}$ at a frequency near $\nu_{\rm Mod}$ in order to reach the theoretical sensitivity. Complete expressions for these noise contributions to the ellipticity peak spectral noise density can be obtained from Eq. (25) as functions of the modulation amplitude η_0:

$$s_{\rm shot} = \sqrt{\frac{2e}{I_{\rm out}q}\left(\frac{\sigma^2+\eta_0^2/2}{\eta_0^2}\right)} \tag{28}$$

$$s_{\rm dark} = \frac{V_{\rm dark}}{G}\,\frac{1}{I_{\rm out}\,q\,\eta_0} \tag{29}$$

$$s_{\rm J} = \sqrt{\frac{4k_BT}{G}}\,\frac{1}{I_{\rm out}\,q\,\eta_0} \tag{30}$$

$$s_{\rm RIN} = {\rm RIN}(\nu_{\rm Mod})\,\frac{\sqrt{(\sigma^2+\eta_0^2/2)^2+(\eta_0^2/2)^2}}{\eta_0}. \tag{31}$$

3. Experimental studies

3.1. PVLAS - LNL: main features and limitations

The previous PVLAS apparatus was set up at LNL in Legnaro, Italy and took data from 2002 to 2007. It featured a vertical assembly of the ellipsometer with a 6 m long Fabry-Perot cavity, the injection and detection optics installed on different optical benches resting on the ground and on the top of a granite tower nearly 8 m high, respectively. A superconductive dipole magnet about 1 m long, installed in a cryostat resting on a turntable, provided a rotating magnetic field orthogonal to the light path in the Fabry-Perot cavity. The magnet was operated with magnetic fields up to 5.5 T. The magnet support and the optics tower had different foundations, and were hence mechanically decoupled. Although this apparatus has set best limits on magnetic vacuum birefringence and photon-photon elastic scattering at low energies [12], it suffered from several limitations:

- high stray field when operating the superconducting magnet above 2.3 T, due to saturation of the iron return yoke;
- limited running time due to liquid Helium consumption;
- high running costs for liquid Helium;
- the granite tower could not be seismically insulated, due to its size and configuration;
- with a single magnet, a zero ellipticity measurement was only possible with field off, hence not in the same experimental condition as in the magnetic birefringence measurements.

To solve the first three problems, we have now chosen to work with permanent magnets instead of superconducting ones. In permanent magnets the stray field can be made much smaller and the duty cycle can be as high as 100%. Seismic noise also limited the sensitivity of the LNL setup; instead, a large improvement of this parameter can be obtained by installing the whole ellipsometer on a single seismically isolated optical bench, as shown in ref. [22]. The last point in the list above is extensively addressed in a separate section.

A completely new setup has been designed according to these lines and is being installed in a clean room of the Ferrara section of INFN. The new set up is built using a single 4.8 m long granite optical bench actively isolated, two identical rotating permanent dipole magnets in Halbach configuration, each 0.8 m in length with magnetic field strength $B = 2.5$ T and 2 cm bore. The design of the magnets features an integrated magnetic field shield strongly reducing the stray field. We plan to rotate them at $\nu_{\rm Mag} = 10$ Hz, thanks to their reduced mass and good balancing. The optical setup features a 2 W tuneable Nd:YAG laser frequency-locked to a 3.2 m long Fabry-Perot cavity using a modified Pound-Drever-Hall scheme [25] and an entirely non magnetic optomechanical setup. The motion of the in-vacuum optical elements is based on stepper piezo motors which maintain their position when powered off and are non magnetic. The whole ellipsometer is kept in a vacuum chamber pumped with non evaporable getters, which are intrinsically non magnetic. Two 1 m long glass tubes traverse the bore of the magnets; each tube is pumped at both ends.

3.2. Two magnet configuration

As reported in ref. [11], the rotating stray magnetic field and the vibrations associated to the masses in rotation can act on the optics and fake ellipticity signals. It is therefore desirable to perform measurements of the zero of the ellipticity scale in conditions as close as possible to those in which an authentic signal is expected. This was not possible in the PVLAS - LNL setup with one single superconducting magnet because when there is no field in the magnet bore also the stray field is absent. A similar problem exists when using a single rotating permanent magnet: the field cannot be turned off. By using two identical magnets and orienting their fields at 90° the net ellipticity generated by the magnetic birefringence is zero. Running the system with the magnets parallel and perpendicular will allow the identification of a real physical signal with respect to some spurious signal due to stray field.

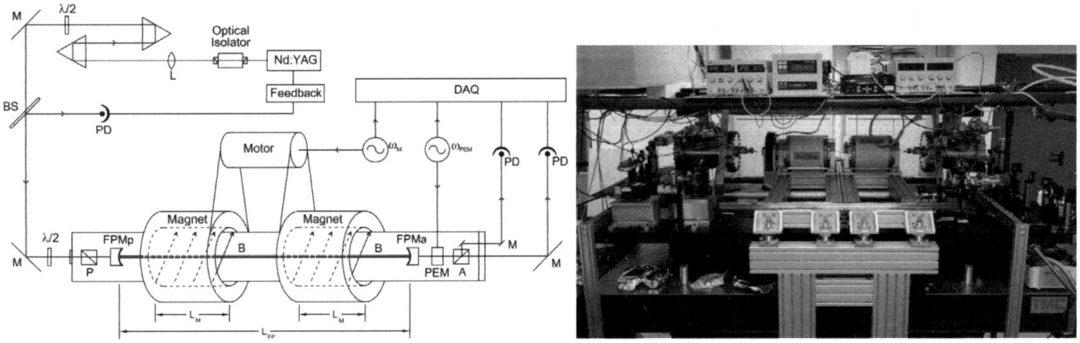

Figure 4. Optical scheme (left panel) and photograph (right panel) of the test apparatus in Ferrara. At the centre one can see the two dipole permanent magnets. The optics is supported by antivibration stages whereas the magnet supports are on the floor.

3.3. Vacuum ellipticity measurements with the test setup

With the test setup (Figure 4), featuring two small magnets, measurements have been performed to understand its limits and optimise the new apparatus in construction. Two different noise sources exist and are under study: wide-band noise and spurious signals at the magnet rotation frequency and its harmonics. Below we briefly report some results in the two cases.

3.3.1. Sensitivity - wide-band noise Noise measurements were first performed without the Fabry-Perot cavity. In this configuration we successfully excluded noise sources from readout electronics and optical elements other than the cavity mirrors, practically reaching the expected sensitivity of $s_{\rm no\ cavity} = 6 \times 10^{-9}$ Hz$^{-1/2}$, dominated by $s_{\rm RIN}$, and a noise floor of $\Psi_{\rm floor} = 1.5 \times 10^{-10}$ with 1600 s integration time.

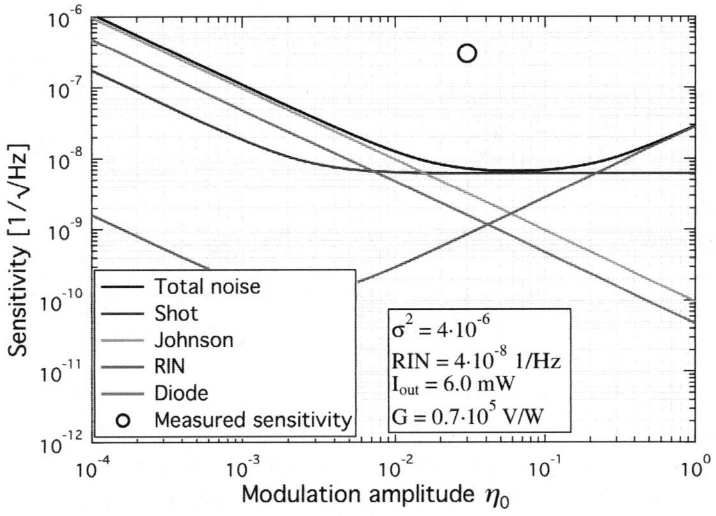

Figure 5. Calculated noise contributions for the PVLAS test apparatus (solid curves). The green curve is the Johnson noise contribution from the transimpedance resistor in the current to voltage converter, the red curve is the dark current noise contribution from the photodiode, the blue curve is the shot noise contribution and the brown curve is the relative intensity noise contribution measured at the output of the cavity. The circle represents the measured ellipticity sensitivity of the apparatus.

With the introduction of a cavity with finesse $\mathcal{F} = 240000$ the noise increased to $s_{\rm cavity} = 3 \times 10^{-7}$ Hz$^{-1/2}$ at about 6 Hz. This was significantly more than what was expected from the reduction of $I_{\rm out}$ due to cavity losses (see Eq. (27)). The actual ellipticity noise budget situation is depicted in Figure 5 as a function of the modulation amplitude η_0 [12]; in the figure, Relative Intensity Noise (RIN), Johnson noise in the amplifier transimpedance, intrinsic current noise in the photodiode and shot noise are summed quadratically. The circle marks the observed noise level. This unexplained noise is under study and we suspect variations of the intrinsic birefringence of the mirrors.

We also observed that the magnet rotation did not contribute to the wide-band noise, indicating a good isolation between the magnet support and the optical setup.

3.3.2. Spurious peaks With the magnets in rotation we often observe ellipticity peaks varying from a few 10^{-8} to a few 10^{-7} whereas sometimes such peaks are not present. The frequencies of these peaks are at harmonics of the magnet rotation frequency. The variability of these peaks from one run to another seems to depend (in a non reproducible way) on the adjustment of the input and output polarisers which is done with motorised stages. To study the dependence of such peaks on the magnet orientation, a field probe is present near the output side magnet. Changing the relative orientation of the two magnets does not change the amplitude of these peaks but does change their phase; the result of this study indicates that the more sensitive part of the apparatus is the entrance optics. All the motorised stages have small electric motors which might couple to the rotating stray magnetic field and may introduce beam jitter and therefore ellipticity. The substitution of all these stages is under way.

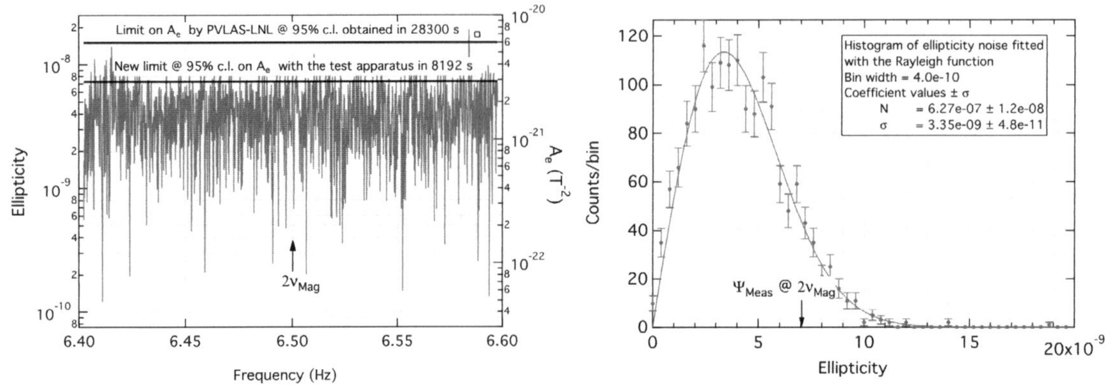

Figure 6. Left: ellipticity noise spectrum in a 0.20 Hz frequency band around $2\nu_{\rm Mag}$. The spectrum is demodulated with a lock-in amplifier with respect to the $\nu_{\rm Mod}$ carrier. The integration time was $T = 8192$ s. Right: histogram of the ellipticity noise values plotted above. A fit with the Rayleigh distribution is superimposed with an ellipticity standard deviation $\sigma = 3.35 \times 10^{-9}$. The arrow indicates the value of the ellipticity Fourier spectrum at exactly $2\nu_{\rm Mag} = 6.5$ Hz.

3.3.3. Noise floor measurements With the test apparatus in a condition in which the peak at $2\nu_{\rm Mag}$ was not present, measurements of a few hours have been done. The magnet rotation frequency was $\nu_{\rm Mag} = 3.25$ Hz. In the left panel of Figure 6 we report the 1600 bin amplitude Fourier noise spectrum of the ellipticity in a frequency band of 0.20 Hz (precisely 1600 bins/8912 s) centred at $2\nu_{\rm Mag}$ for the best measurement; phase information is not reported. The total integration time was 8192 s acquired in a single time record.

In the right panel of the figure, a 50 bin histogram of the same data is given. A vertical arrow in the same figure indicates the ellipticity value of the Fourier spectrum bin at exactly $2\nu_{\rm Mag} = 6.5$ Hz. The probability density function for a noise signal with equal standard deviations σ for the 'in phase' and quadrature components is the Rayleigh function: $P(r) = N\frac{r}{\sigma^2}e^{-\frac{r^2}{2\sigma^2}}$. A fit with the Rayleigh function is superimposed to the histogram. The resulting ellipticity standard deviation is $\sigma = 3.35 \times 10^{-9}$ which translates, at 95% c.l., in a birefringence limit induced by the magnetic field of ($\mathcal{F} = 240000, \lambda = 1064$ nm, $L = 0.4$ m)

$$\Delta n < \sqrt{-2\ln(1-0.95)}\,\frac{\sigma\lambda}{2\mathcal{F}L} = 4.5 \times 10^{-20}. \tag{32}$$

The limits obtained for the parameter A_e and the photon-photon total elastic scattering cross section (at 95 % c.l.) are:

$$A_e < \frac{\Delta n}{3\langle B_{\rm ext}^2\rangle} = 2.45\,\frac{\sigma\lambda}{6\mathcal{F}\int B_{\rm ext}^2\,dL} = 3.3 \times 10^{-21}\ {\rm T}^{-2} \tag{33}$$

$$\sigma_{\gamma\gamma} < 1.2 \times 10^{-62}\ {\rm m}^2\ @\ 1064\ {\rm nm} \tag{34}$$

4. Other experimental efforts

Three other experiments aimed at studying vacuum magnetic birefringence exist: the Q & A experiment, the BMV experiment and the OSQAR experiment. All of them study the propagation of a laser beam through a magnetic field region and use a Fabry-Perot interferometer to increase the optical path length. A number of other proposals have been put forward to measure vacuum magnetic birefringence [27, 15, 26] or other related quantities [28] with different techniques: they will not be described here. As far as the OSQAR experiment [18] is concerned, the information available in the literature at the time this article was written was rather incomplete.

Vacuum magnetic birefringence appears not only in QED but in a number of different theories. In the case of the Euler-Heisenberg-Weisskopf Lagrangian discussed here, Δn is proportional to $B^2_{\rm ext}$ through the parameter $A_e = \Delta n/(3B^2_{\rm ext})$. On the other hand, the optical quantity the experiments are measuring is the ellipticity generated by the induced birefringence. The relation between birefringence and ellipticity is given in formula (13). The three experiments have therefore two distinct parts: the magnetic field source to generate the birefringence and the optical ellipsometer to measure the induced ellipticity. Figures of merit for the optical setup are the ellipticity sensitivity s at the frequency of interest, the signal amplification factor defined by the Fabry-Perot cavity finesse and the wavelength used.

Table 2. Main parameters of the three ongoing experiments to measure vacuum magnetic birefringence. Effect modulation frequency $f_{\rm mod}$ is $2\nu_{\rm Mag}$ for PVLAS and Q & A, and $1/T_{\rm pulse}$ for BMV.

Experiment	PVLAS	Q & A	BMV
Status	Achieved/Planned	Achieved/Planned	Achieved/Planned
Wavelength (nm)	1064	1064/532	1064
Magnetic dipole	permanent	permanent	pulsed
$\int B^2\,dL$ (T^{2}m)	1.85/10	3.2/19	25/600
Average $B_{\rm ext}$ (T)	2.15/2.5	2.3/2.3	14/30
Finesse	$2.4\times10^5/>4\times10^5$	$3\times10^4/1\times10^5$	$5\times10^5/1\times10^6$
QED ellipticity (Eq. 13)	$3\times10^{-12}/3\times10^{-11}$	$7\times10^{-13}/3\times10^{-11}$	$9\times10^{-11}/5\times10^{-9}$
Detection scheme	heterodyne	heterodyne	homodyne
Effect mod. freq. $f_{\rm mod}$	6 Hz/20 Hz	26 Hz	500 Hz
Duty cicle D_t	~ 1	~ 1	3×10^{-6}
s @ $f_{\rm mod}$ (Hz$^{-1/2}$)	$3\times10^{-7}/3\times10^{-8}$	$1\times10^{-6}/1\times10^{-8}$	$5\times10^{-8}/7\times10^{-9}$
$s_{\rm eff}$ @ $f_{\rm mod}$ (Hz$^{-1/2}$)	$3\times10^{-7}/3\times10^{-8}$	$1\times10^{-6}/1\times10^{-8}$	$3\times10^{-5}/4\times10^{-6}$
$\Delta n_{\rm eff}$ sensitivity (Hz$^{-1/2}$)	$1.7\times10^{-18}/2.5\times10^{-20}$	$3.0\times10^{-17}/7.4\times10^{-21}$	$2.6\times10^{-16}/3.0\times10^{-18}$
$A_{\rm e}$ sensitivity (T^{-2}Hz$^{-1/2}$)	$1.2\times10^{-19}/1.3\times10^{-21}$	$1.8\times10^{-18}/4.7\times10^{-22}$	$4.4\times10^{-19}/1.1\times10^{-21}$
Time for SNR = 1	260 yr/12 d	63×10^3 yr/1.4 d	3.6×10^3 yr/8.3 d

In Table 4 the main features of the ongoing experiments are summarised with an estimated running time necessary to reach the goal of measuring $A_{\rm e}$. As the scheme of the Q & A experiment [17] closely resembles PVLAS's one, the interpretation of its parameters is straightforward. The BMV experiment deserves a few more words: it employs an intense pulsed magnetic field lasting $T_{\rm pulse} = 2$ ms FWHM (total duration about 4 ms), with a repetition rate of five shots per hour to allow the magnets to cool down. Therefore on average the effective duty cycle with the magnetic field ON is $D_t = 2\ {\rm ms}\times(5/3600\ {\rm s}) \approx 3\times10^{-6}$. The analysis correlates the measured ellipticity during each pulse, assumed to be proportional to $B^2_{\rm ext}$, with the magnetic pulse shape, also taking into account the dynamical response of the Fabry-Perot whose decay time is of the order of the magnetic pulse rise time [29]. The analysis gives $\Delta n_{\rm B} = \Delta n/B^2_{\rm filtered}$ for each pulse, where $B_{\rm filtered}$ is the effective field filtered by the first order response of the Fabry-Perot cavity. The BMV collaboration therefore presents results in terms of the birefringence per tesla squared per pulse. At present BMV has obtained a best $\Delta n_B = 5\times10^{-20}$ T^{-2} per pulse and claims that a future upgrade will allow them to reach SNR = 1 in the QED measurement with 1000 pulses [29]. Given a number of pulses $N_{\rm pulses}$ with equivalent noise, the limit on $\Delta n_{\rm B}$ will scale with $1/\sqrt{N_{\rm pulses}}$. In the detection of a birefringence much smaller than $\Delta n_{\rm B}$ many pulses will be needed and therefore, to compare their method with the Q & A or PVLAS capability, it is reasonable to rescale $\Delta n_{\rm B}$ per pulse to the average sensitivity in Δn achievable in 1 s: $\Delta n_{\rm eff} = \Delta n_{\rm B} B_{\rm ext}^2 \sqrt{T_{\rm pulse}/D_t}$. Since BMV has a small duty cycle D_t their effective optical sensitivity reduces to $s_{\rm eff} = s/\sqrt{D_t}$.

It is apparent that all the experiments are still far from their goal, the chance of their eventual success relying on improvements of the optical sensitivity that, unlike hardware upgrades, cannot be given for granted, as they would require complete understanding of the noise sources limiting the sensitivity. Indeed the shot noise limited sensitivity of all the experiments, given by formula (27), is about a factor between 10 to 100 below the current achieved sensitivities. No clear explanation to this discrepancy has yet been given by any of the collaborations. Note that the best value of sensitivity in ellipticity in ellipsometers using high finesse Fabry-Perot interferometers has been reported by PVLAS in a small test ellipsometer without the implementation of a magnetic field [22] of 2×10^{-8} Hz$^{-1/2}$ above 5 Hz. The best reported sensitivity of a complete apparatus is by BMV with a sensitivity of 5×10^{-8} Hz$^{-1/2}$ over a time span of 2 ms. If in Table 4 the QED ellipticity planned by each experiment is combined with the presently *achieved* sensitivities, the last line in the table would become (3.2, 39, 1.5) years for the three experiments, respectively, instead of (12, 1.4, 8.3) days.

5. Conclusions

We have presented the main physics goal of the PVLAS experiment and have briefly discussed the experimental method. Noise sources are being studied on a bench-top small test apparatus in Ferrara, Italy, in view of the construction of the final apparatus with which we hope to measure for the first time the magnetic birefringence of vacuum due to vacuum fluctuations.

We have discussed the importance, when using rotating magnets, of using not one but two dipole magnets whose relative directions can be made perpendicular to each other in order to have a zero effect condition with the magnetic field present. This is vital to study and eliminate spurious signals generated by the rotating field.

Noise floor measurement results have been presented. A new limit on the parameter A_e describing non linear electrodynamic effects in vacuum has been obtained: $A_e < 3.3 \times 10^{-21}$ T^{-2}. This value, obtained by using two small compact permanent magnets, improves the previous limit by a factor 2.

Finally, the perspectives of the three current existing experimental efforts based on similar techniques active on this subject have been discussed and the experiments compared. Ellipticity noise sources are under study by all groups so as to improve their optical sensitivity. Without such improvements the measurement of vacuum magnetic birefringence is still out of reach.

References

[1] Euler H and Kochel B 1935 *Naturwiss.* **23** 246; Heisenberg W and Euler H 1936 *Z. Phys.* **98** 718; Weisskopf V S 1936 *Kgl. Danske Vid. Sels., Math.-fys. Medd.* **14** 6
[2] Schwinger J 1951 *Phys. Rev.* **82** 664
[3] Baier R and Breitenlohner P 1967 *Acta Phys. Austriaca* **25** 212; Baier R and Breitenlohner P 1967 *Nuovo Cimento* **47**, 261; Adler S L 1971 *Ann. Phys.* **67** 559; Bialynicka-Birula Z and Bialynicki-Birula I 1970 *Phys. Rev. D* **2** 2341
[4] Iacopini E and Zavattini E 1979 *Phys. Lett. B* **85** 151
[5] De Tollis B 1965 *Nuovo Cimento* **35** 1182; De Tollis B 1964 *Nuovo Cimento* **32** 757
[6] Karplus R *et al* 1951 *Phys. Rev.* **83** 776
[7] Dicus D A *et al* 1998 *Phys. Rev. D* **57** 2443
[8] Bernard D *et al* 2000 *Eur. Phys. J. D* **10** 141
[9] Moulin F *et al* 1996 *Z. Phys. C* **72** 607
[10] Bakalov D *et al* 1998 *Hyperfine Interactions* **114** 103; Bakalov D *et al* 1998 *Quantum Semicl. Opt.* **10** 239
[11] Zavattini E *et al* 2008 *Phys. Rev. D* **77** 032006
[12] Bregant M *et al* 2008 *Phys Rev. D* **78** 032006
[13] Lundström *et al* 2006 *Phys. Rev. Lett.* **96** 083602
[14] Tommasini D *et al* 2008 *Phys. Rev. A* **77** 042101

[15] Luiten A N and Petersen J C 2004 *Phys. Lett. A* **330** 429; Luiten A N and Petersen J C 2004 *Phys. Rev. A* **70** 033801
[16] Battesti R *et al* 2008 *Eur. Phys. J. D* **46** 323
[17] Ni W-T 1996 *Chinese J. Phys.* **34** 962
[18] Pugnat P *et al* 2005 Czech. J. Phys. **55** A389
[19] Heinzl T *et al* 2006 *Optics Comm.* **267** 318
[20] Zavattini G *et al* 2012 *Int. J. Mod. Phys. A* **27** 1260017-1
[21] Cameron R *et al* 1993 *Phys. Rev. D* **47** 3707
[22] Della Valle F *et al* 2010 *Optics Comm.* **293** 4194
[23] Jones R C 1948 *J. Opt. Soc. Am.* **38** 671; Hecht E 1987 *Optics* 2nd ed. (San Francisco: Addison-Wesley)
[24] Zavattini G *et al* 2006 *Appl. Phys. B* **83** 571
[25] Ruoso G and Zavattini G 1996 *Proceedings of the Seventh Marcel Grossman Meeting* Jantzen R T and Mac Kaiser G eds. (Singapore: World Scientific) p 1379; Cantatore G *et al* 1995 *Rev. Sci. Instrum.* **66** 2785
[26] Zavattini G and Calloni E 2009 *Eur. Phys. J. C* **62** 459
[27] Hall J L *et al* 2000 *Phys Rev. A* **62** 013815
[28] Partovi M H 1993 *Preprint* hep-ph/9308293v1
[29] Berceau P *et al* 2012 *Phys. Rev. A* **85** 013837

Generalized creation and annihilation operators via complex nonlinear Riccati equations

Dieter Schuch

Institut für Theoretische Physik, J.W. Goethe-Universität Frankfurt am Main, Max-von-Laue-Str. 1, D-60438 Frankfurt am Main, Germany

E-mail: `Schuch@em.uni-frankfurt.de`

Octavio Castaños

Instituto de Ciencias Nucleares, Universidad National Autónoma de México, Apdo. Postal 70-543, 04510 México D.F.

E-mail: `ocasta@nucleares.unam.mx`

Oscar Rosas-Ortiz

Departamento de Física, Cinvestav, Apdo. Postal 14-740, 07000 México, D.F.

E-mail: `orosas@fis.cinvestav.mx`

Abstract.

Based on Gaussian wave packet solutions of the time-dependent Schrödinger equation, a generalization of the conventional creation and annihilation operators and the corresponding coherent states can be obtained. This generalization includes systems where also the width of the coherent states is time-dependent as they occur for harmonic oscillators with time-dependent frequency or systems in contact with a dissipative environment. The key point is the replacement of the frequency ω_0 that occurs in the usual definition of the creation/annihilation operator by a complex time-dependent function that fulfils a nonlinear Riccati equation. This equation and its solutions depend on the system under consideration and on the (complex) initial conditions. Formal similarities also exist with supersymmetric quantum mechanics. The generalized creation and annihilation operators also allow to construct exact analytic solutions of the free motion Schrödinger equation in terms of Hermite polynomials with time-dependent variable.

1. Introduction

In classical as well as quantum mechanics, emphasis is placed on Lagrangians, Hamiltonians and potentials, that is, on quantities with the dimension of energy. However, already for the harmonic oscillator (HO) with time-dependent (TD) frequency $\omega = \omega(t)$, the Hamiltonian is no longer a constant of motion. Not to mention dissipative systems with friction forces where from the onset definition of an appropriate Hamiltonian is problematic. However, the quantity that is actually quantized in quantum mechanics is not energy but action since Planck's constant h (or $\hbar = h/2\pi$) is a quantum of action. For example, the fact that the energy of the HO can be quantized is only a consequence of the constant frequency of oscillation, $\omega = \omega_0$.

In the following, it will be shown that a *dynamical* (so-called Ermakov) *invariant* [1] with the dimension of *action* can still exist for systems like the HO with TD frequency or for open systems with a friction force linear proportional to velocity (or momentum), in other words, where the energy or the Hamiltonian is no longer a constant of motion. Factorization of the corresponding operator (in terms of a complex variable fulfilling a nonlinear (NL) Riccati equation) yields generalized creation and annihilation operators for these systems that also allow for construction of the associated coherent states (CS).

In section 2, the standard factorization method, Gaussian wave packets (WP) and CS definitions will be recollected in order to introduce our notation. For treatment of the above-mentioned systems one must go beyond the minimum uncertainty CS with constant width and consider WPs with time-dependent width. The corresponding equations of motion for the width, as well as for the maximum, of the WPs and the resulting dynamical invariant will be discussed in detail in section 3. Here formal similarities with supersymmetric quantum mechanics will also be mentioned. In section 4 generalized creation/annihilation operators are obtained by factorizing the operator corresponding to the Ermakov invariant and generalized CS can be constructed in the usual way.

For some effective models an exact Ermakov invariant also exists for the description of dissipative systems. As an example, an approach based on a modified Lagrange–Hamilton formalism and another using a nonlinear modification of the time-dependent Schrödinger equation (TDSE) will be presented in section 5 along with their interrelation. Section 6 illustrates how generalized creation/annihilation operators (here in the conservative case) can be applied to obtain exact solutions of the TDSE for the free motion in terms of Hermite polynomials with a TD argument. A short summary concludes the paper.

2. Factorization method, wave packets and coherent states

An algebraic method to solve the SE of the HO was proposed by Schrödinger himself [2] although, in principle, this can already be found in Dirac's seminal book [3]. For this purpose, the Hamiltonian operator of the (one-dimensional) HO is expressed in terms of the operators a and a^+ (which is the adjoint of a) as

$$H_{op} = \frac{1}{2m}\, p_{op}^2 + \frac{m}{2}\, \omega_0^2\, x^2 = \hbar\omega_0 \left(a^+ a + \frac{1}{2} \right) \tag{1}$$

where these operators fulfil the commutator relation $[a, a^+] = 1$ and $a^+ a$ is called number operator. This name becomes comprehensible when rewriting Eq. (1) in the conventional way by dividing it by $\hbar\omega_0$ to yield the dimensionless operator

$$\hat{H} = \frac{H_{op}}{\hbar\omega_0} = \left(a^+ a + \frac{1}{2} \right) . \tag{2}$$

Since $\frac{H}{\omega_0}$ has the dimension of *action* (!), the number operator just counts the quanta of $\hbar$.

The annihilation operator a and the creation operator a^+ (in position representation with $p_{op} = \frac{\hbar}{i}\frac{\partial}{\partial x}$) can then easily be specified as

$$a = i \sqrt{\frac{m}{2\hbar\omega_0}} \left(\frac{p_{op}}{m} - i\,\omega_0\, x \right) \tag{3}$$

$$a^+ = -\, i \sqrt{\frac{m}{2\hbar\omega_0}} \left(\frac{p_{op}}{m} + i\,\omega_0\, x \right) . \tag{4}$$

The ground state wave function $\psi_0(x)$ can be obtained from the relation $a\psi_0(x) = 0$, the excited states from the property of the creation operator $a^+\psi_n(x) = \sqrt{n+1}\psi_{n+1}(x)$, yielding $\psi_n(x) = \frac{1}{\sqrt{n!}}(a^+)^n\psi_0(x)$.

Soon after he proposed his equation, Schrödinger himself [4] already found that by superimposing the solutions $\psi_n(x)$ with appropriate weight factors, a stable Gaussian WP solution $\Psi_{WP}(x,t)$ can be obtained. Generalizations of Schrödinger's approach were achieved in the description of coherent light beams emitted by lasers [5, 6, 7] leading to what is now known as CS. There are at least three different formal definitions of these CS in the literature [8] where the minimum uncertainty CS means Gaussian WPs that minimize the Heisenberg uncertainty relation. However, this can only be fulfilled by WPs with constant width. Since we wish to go further and must also consider TD widths, the definition of annihilation operator CS, i.e., eigenstate of a with eigenvalue z, will be applied in our context. In position space, $< x|z >= \Psi_{WP}(x,t)$ is valid. Therefore, in the next section, the dynamics of the two parameters specifying Gaussian WPs, maximum and width, will be examined.

3. Wave packet dynamics, Ermakov invariant and supersymmetric quantum mechanics

In general, Gaussian WP solutions of the TDSE exist for any Hamiltonian that is at most quadratic (or bilinear) in position and momentum operators, e.g.,

$$i\hbar\frac{\partial}{\partial t}\Psi_{WP}(x,t) = \left\{-\frac{\hbar^2}{2m}\frac{\partial^2}{\partial x^2} + V(x,t) + W(x,p_{op};t,\Psi_{WP})\right\}\Psi_{WP}(x,t) \tag{5}$$

where $W(x,p_{op};t,\Psi_{WP})$ occurs, for instance, in effective models for dissipative systems (see section 5). First we want to consider (one-dimensional) systems without dissipation but with TD WP width or position uncertainty $< \tilde{x}^2 >=< x^2 > - < x >^2$ where $< x >= \int_{-\infty}^{+\infty} dx\Psi^* x\Psi = \eta(t)$ is the mean value of position. Well-known examples of this kind are

a) $V = 0$: free motion with

$$< \tilde{x}^2 > (t) \;=\; < \tilde{x}^2 >_0 \left[1 + (\beta_0 t)^2\right] \tag{6}$$

where the frequency-type quantity β_0 is used to abbreviate $\beta_0 = \frac{\hbar}{2m<\tilde{x}^2>_0}$;

b) $V = \frac{m}{2}\omega_0^2 x^2$: HO, but for the case $\beta_0 \neq \omega_0$,

$$< \tilde{x}^2 > (t) =< \tilde{x}^2 >_0 \left\{\cos^2\omega_0 t + \left(\frac{\beta_0}{\omega_0}\sin\omega_0 t\right)^2\right\} =< \tilde{x}^2 >_0 \left\{1 + \left[\left(\frac{\beta_0^2-\omega_0^2}{\omega_0^2}\right)\sin^2\omega_0 t\right]\right\}, \tag{7}$$

i.e., WPs with oscillating width;

c) $V = \frac{m}{2}\omega^2(t)x^2$: HO with TD frequency; the explicit form of $< \tilde{x}^2 > (t)$ depends on the specific time-dependence of $\omega(t)$.

The question we wish to pose is: in these cases, do (generalized) creation and annihilation operators also exist that provide the corresponding CSs with TD width?

For this purpose we study the time-evolution of the maximum and width of a Gaussian WP with the general form

$$\Psi_{WP}(x,t) = N(t)\,\exp\left\{i\left[y(t)\tilde{x}^2 + \frac{1}{\hbar} < p > \tilde{x} + K(t)\right]\right\} \tag{8}$$

with $y(t) =$ complex, $\tilde{x} = x - \eta(t)$, i.e., WP maximum at $x = \eta(t)$. The purely TD normalization factor $N(t)$ and phase factor $K(t)$ are not relevant at the moment and will be specified later.

Inserting ansatz (8) into the TDSE for the HO (with arbitrary frequency ω) yields, as equation of motion for the WP maximum, the classical Newton-type equation

$$\ddot{\eta} + \omega^2(t)\eta = 0 \tag{9}$$

and, for the WP width, one obtains the *complex* NL Riccati equation

$$\left(\frac{2\hbar}{m}\dot{y}\right) + \left(\frac{2\hbar}{m}y\right)^2 + \omega^2 = 0\ . \tag{10}$$

There are different ways of treating this equation. In order to obtain the afore-mentioned action-type invariant, we transform Eq. (10) into a *real* NL differential equation. For this purpose we introduce a new variable $\alpha(t)$ that is related to the imaginary part of the complex Riccati variable via $\frac{2\hbar}{m}y_I = \frac{1}{\alpha^2(t)}$ where $\alpha(t)$ is directly proportional to the position uncertainty or WP width since $\alpha = \sqrt{\frac{2m}{\hbar} < \tilde{x}^2 >}$. The real part of this variable is related to the (relative) change in time of the width what can be expressed as $\frac{2\hbar}{m}y_R = \frac{\frac{d}{dt}<\tilde{x}^2>}{2<\tilde{x}^2>} = \frac{\dot{\alpha}}{\alpha}$ and vanishes for WPs with constant width. Inserting this into the Riccati equation (10), finally turns it into the so-called Ermakov equation for $\alpha(t)$,

$$\ddot{\alpha} + \omega^2(t)\ \alpha = \frac{1}{\alpha^3}\ . \tag{11}$$

This equation was rediscovered (and renamed) several times and studied also in a quantum mechanical context [9].

Eliminating ω^2 from Eq. (11) by using Eq.(9) for the WP maximum, leads to the dynamical invariant

$$I_L = \frac{1}{2}\left[(\dot{\eta}\alpha - \dot{\alpha}\eta)^2 + \left(\frac{\eta}{\alpha}\right)^2\right] = \text{const.} \tag{12}$$

This form, expressed in terms of η, α and their time-derivatives, was used by Hartley and Ray [10] to obtain generalized creation/annihilation operators for the HO with TD frequency but is no longer applicable for open dissipative systems. A further generalization can be achieved to also include these systems if the Ermakov invariant is expressed in terms of η, $\dot{\eta}$ and $\left(\frac{2\hbar}{m}y\right)$ as

$$\begin{aligned} I_L &= \frac{1}{2}\alpha^2\left[\left(\dot{\eta} - \frac{\dot{\alpha}}{\alpha}\eta\right)^2 + \left(\frac{\eta}{\alpha^2}\right)^2\right] = \frac{1}{2}\alpha^2\left[\left(\dot{\eta} - \left(\frac{2\hbar}{m}y_R\right)\eta\right)^2 + \left(\left(\frac{2\hbar}{m}y_I\right)\eta\right)^2\right] \\ &= \frac{1}{2}\alpha^2\left[\left(\dot{\eta} - \left(\frac{2\hbar}{m}y\right)\eta\right)\left(\dot{\eta} - \left(\frac{2\hbar}{m}y^*\right)\eta\right)\right] \end{aligned} \tag{13}$$

where mI_L has the dimension of *action*.

After we have obtained the invariant, now in terms of $\left(\frac{2\hbar}{m}y\right)$ and in a form that can easily be factorized, let us return to the Riccati equation (10) and give some more reasons why a formulation in these terms is advantageous over the one using the real variable $\alpha(t)$. For this, we make use of the fact that the inhomogeneous Riccati equation can be transformed into a homogeneous Bernoulli equation, providing a particular solution $\left(\frac{2\hbar}{m}\tilde{y}\right)$ is known. The general solution of (10) is then given by $\frac{2\hbar}{m}y = \frac{2\hbar}{m}\tilde{y} + \frac{2\hbar}{m}v(t)$, where $\frac{2\hbar}{m}v(t)$ fulfils the Bernoulli equation

$$\left(\frac{2\hbar}{m}\dot{v}\right) + 2\left(\frac{2\hbar}{m}\tilde{y}\right)\left(\frac{2\hbar}{m}v\right) + \left(\frac{2\hbar}{m}v\right)^2 = 0 \tag{14}$$

and the coefficient of the linear term, $A = 2\left(\frac{2\hbar}{m}\tilde{y}\right)$, depends on the particular solution. This equation can be linearized using $\left(\frac{2\hbar}{m}v\right) = \frac{1}{\kappa(t)}$ to yield $\dot{\kappa} - A\ \kappa = 1$.

Finally, the general solution of the Riccati equation can then be given in the form

$$\frac{2\hbar}{m}y(t) = \frac{2\hbar}{m}\tilde{y} + \frac{d}{dt}\ \ln\left[\kappa_0 + \mathcal{I}(t)\right]\ , \tag{15}$$

with $\kappa_0 = \left(\frac{2\hbar}{m}v_0\right)^{-1}$ and using the abbreviated form $\mathcal{I}(t) = \int^t dt'\ \mathrm{e}^{-\int^{t'} dt'' A(t'')}$. It is worth mentioning that the solution (15) depends on the (complex) initial condition κ_0 and since $\frac{2\hbar}{m}y(t)$ fulfills a NL differential equation, the dependence on the initial condition can have quite drastic qualitative consequences.

This can be seen by applying a formal comparison with supersymmetric quantum mechanics [11] [12], showing that (15) is formally identical to the most general superpotential

$$W(x) \ = \ \tilde{W}(x) \ + \ \frac{d}{dx} \ \ln\left[\lambda_1 + \mathcal{I}_1(x)\right] \ , \tag{16}$$

leading to the one-parameter family of complex isospectral potentials

$$V_1(x) \ = \ W^2 - \frac{d}{dx}W \ = \ \tilde{V}_1 - 2\frac{d^2}{dx^2} \ \ln\left[\lambda_1 + \mathcal{I}_1(x)\right] \tag{17}$$

that have the same supersymmetric partner potential $\tilde{V}_2(x)$ (see, e.g., [12][13]). The shape of the isospectral potentials strongly depends on the parameter λ_1 and is far different from the original one.

In this case, the integral $\mathcal{I}_1(x)$ is defined as $\mathcal{I}_1(x) = \int_{-\infty}^{x} dx'\ \Psi_1^2(x')$ where $\Psi_1(x)$ is the normalized ground state wave function of the SE with potential $\tilde{V}_1(x) = \tilde{W}^2(x) - \frac{d}{dx}\tilde{W}(x)$ and λ_1 is a (usually real) parameter.

A major difference between this supersymmetric situation and the one in our case (apart from replacing the spatial variable by a temporal one) is the fact that the variables of the nonlinear Eqs.(10) and (14) are *complex*, whereas $\tilde{W}(x), W(x)$ and $\mathcal{I}_1(x)$ are real quantities; also the parameter κ_0 in our case is generally complex.

This provides a larger variety but, nevertheless, certain methods and results can be transferred from one system to the other (which will be discussed elsewhere).

4. Generalized creation and annihilation operators and corresponding coherent states

As mentioned in section 2, the usual creation and annihilation operators (3) and (4) provide minimum uncertainty CSs with constant width, corresponding to the particular solution of the Riccati equation with $\frac{2\hbar}{m}\tilde{y} = i\frac{2\hbar}{m}\tilde{y}_I = i\frac{2\hbar}{m}y_I = i\omega_0$. Replacing ω_0 by $\frac{1}{\alpha_0^2}$ or $\frac{2\hbar}{m}y_I$, respectively, the operators a and a^+ can be rewritten as

$$a \ = \ i\ \sqrt{\frac{m}{2\hbar}}\ \alpha_0\ \left(\frac{p_{op}}{m}\ -\ i\ \left(\frac{2\hbar}{m}y_I\right)\ x\right) \tag{18}$$

$$a^+ \ = \ -\,i\ \sqrt{\frac{m}{2\hbar}}\ \alpha_0\ \left(\frac{p_{op}}{m}\ +\ i\ \left(\frac{2\hbar}{m}y_I\right)\ x\right)\ . \tag{19}$$

As we have shown, already for $\omega = \omega_0 =$ const., solutions with TD WP width exist, i.e., α_0 turns into $\alpha(t)$ and $\dot{\alpha} \neq 0$, hence $\frac{2\hbar}{m}y_R = \frac{\dot{\alpha}}{\alpha}$, must be taken into account. Obviously, the same also applies for the HO with TD frequency $\omega = \omega(t)$. Therefore, in Eqs.(18) and (19) α_0 must be replaced by $\alpha(t)$ and $i\ \left(\frac{2\hbar}{m}y_I\right)$ by $\left(\frac{2\hbar}{m}y\right)$, thus leading to

$$\tilde{a}(t) \ = \ i\ \sqrt{\frac{m}{2\hbar}}\ \alpha\ \left(\frac{p_{op}}{m}\ -\ \left(\frac{2\hbar}{m}y\right)\ x\right) \tag{20}$$

$$\tilde{a}^+(t) \ = \ -\,i\ \sqrt{\frac{m}{2\hbar}}\ \alpha\ \left(\frac{p_{op}}{m}\ -\ \left(\frac{2\hbar}{m}y^*\right)\ x\right)\ . \tag{21}$$

It is easy to check that they satisfy the standard commutation relations.

Since, at least for TD frequency ω, the corresponding Hamiltonian is no longer a constant of motion, one might ask if $\tilde{a}(t)$ and $\tilde{a}^+(t)$, as defined above, are constants of motion.

It turns out that

$$\frac{\partial}{\partial t}\tilde{a} + \frac{1}{i\hbar}\left[\tilde{a}, H\right]_- = -i\,\frac{1}{\alpha^2}\,\tilde{a} \neq 0\,. \tag{22}$$

So, $\tilde{a}(t)$ and $\tilde{a}^+(t)$ are no constants of motion but can be turned into such by simply introducing a phase factor according to

$$a(t) = \tilde{a}(t)\,\mathrm{e}^{i\int^t dt'\frac{1}{\alpha^2}} \tag{23}$$

$$a^+(t) = \tilde{a}^+(t)\,\mathrm{e}^{-i\int^t dt'\frac{1}{\alpha^2}}\,. \tag{24}$$

Next, the CSs that can be obtained using these generalized operators. For this purpose, the phase factor shall be omitted since it can be absorbed in the purely TD function $K(t)$ in the exponent of the WP/CS or into $N(t)$.

To create a generalized CS $|z>$ we use the property that it is an eigenstate of $a(t)$ with complex eigenvalue z, i.e., $a(t)\,|z> \;=\; z\,|z>$.

The meaning of the complex eigenvalue can be determined using this property and expressing $<x>_z=\eta$ and $<p>_z=m\dot{\eta}$ in terms of z and z^* as

$$<x>_z=\sqrt{\frac{\hbar}{2m}}\,\alpha\,(z^*+z) = \sqrt{\frac{2\hbar}{m}}\,\alpha\,z_R \quad\text{and}\quad <p>_z=\sqrt{\frac{\hbar}{2m}}\,\alpha\,m\left[\left(\frac{2\hbar}{m}y\right)z^* + \left(\frac{2\hbar}{m}y^*\right)z\right] \tag{25}$$

where from it follows that

$$z_R = \frac{1}{\sqrt{2}}\sqrt{\frac{m}{\hbar}}\,\alpha\,\left(\frac{2\hbar}{m}y_I\right)\eta \quad\text{and}\quad z_I = \frac{1}{\sqrt{2}}\sqrt{\frac{m}{\hbar}}\,\alpha\,\left[\dot{\eta} - \left(\frac{2\hbar}{m}y_R\right)\eta\right] \tag{26}$$

or

$$z = \sqrt{\frac{m}{2\hbar}}\left[\left(\frac{\eta}{\alpha}\right) + i\,(\dot{\eta}\alpha - \eta\dot{\alpha})\right] = \frac{1}{\sqrt{2}}\sqrt{\frac{m}{\hbar}}\,\alpha\left[\left(\frac{2\hbar}{m}y_I\right)\eta + i\left(\dot{\eta} - \left(\frac{2\hbar}{m}y_R\right)\eta\right)\right]\,. \tag{27}$$

Therefore, the complex eigenvalue z is directly related to the Ermakov invariant via

$$I_L = \frac{\hbar}{m}\left(z_I^2 + z_R^2\right) = \frac{\hbar}{m}zz^* = \frac{\hbar}{m}|z|^2\,. \tag{28}$$

The operator corresponding to this invariant can be written in terms of the generalized creation/annihilation operators similar to Eq. (2) as $I_{L,op} = \frac{\hbar}{m}\left[a^+(t)\,a(t) + \frac{1}{2}\right]$ (taking into account the commutation relation between a and a^+), what is in agreement with the result of Hartley and Ray [10].

In position space representation, the generalized CS can be obtained from $<x|a(t)|z> \;=\; z<x|z> \;=\; z\,\Psi_z(x,t)$, and finally be written as

$$\Psi_z(x,t) = \left(\frac{m}{\pi\hbar}\right)^{1/4}\left(\frac{1}{\lambda}\right)^{1/2}\exp\left\{i\left[y\tilde{x}^2 + \frac{1}{\hbar}<p>\tilde{x} + \frac{1}{2\hbar}<p><x>\right]\right\}\,, \tag{29}$$

corresponding to our above Gaussian WP $\Psi_{WP}(x,t)$ with $N(t) = \left(\frac{m}{\pi\hbar}\right)^{1/4}\left(\frac{1}{\lambda}\right)^{1/2}$ and $K(t) = \frac{1}{2\hbar}<p><x>$. For $\alpha^2 = \alpha_0^2 = \omega_0^{-1}$, $N(t)$ turns into $N(t) = \left(\frac{m\omega_0}{\pi\hbar}\right)^{1/4}\mathrm{e}^{-i\omega_0 t/2}$, where the exponential contributes the ground state energy to the energy of the WP.

5. Dissipative systems with effective Hamiltonians

There exist several approaches for describing dissipative quantum systems using modified effective one-body SEs where the effect of the environment on the observable system is taken into account by dissipative friction terms without considering the individual degrees of freedom of the environment. Several of these approaches are discussed in the literature [14][15][16]. In this context it was also investigated [17] if Ermakov invariants also exist for these approaches. An explicitly TD Hamiltonian [14] and a NLSE [16] for which this is the case will be discussed subsequently.

The explicitly TD approach of Caldirola and Kanai [14] starts on the classical level with a *non-canonical* transformation between the *physical* variables x and $p = m\dot{x}$ and the *canonical* variables $\hat{x} = x$, $\hat{p} = p\ \mathrm{e}^{\gamma}t$.

Expressed in the canonical variables, the Hamiltonian can be formulated as

$$\hat{H}_{CK} = \frac{1}{2m}\mathrm{e}^{-\gamma t}\ \hat{p}^2 + \mathrm{e}^{\gamma t} V(x) \tag{30}$$

which yields the proper equation of motion including a linear momentum (or velocity) dependent friction force.

The transition to quantum mechanics is achieved by replacing, as usual, the *canonical* momentum with a differential operator according to $\hat{p} \longrightarrow \hat{p}_{op} = \frac{\hbar}{i}\frac{\partial}{\partial x}$, leading to the corresponding explicitly TD Hamiltonian operator and, thus, to a modified SE.

For the systems considered in this work, this equation also possesses exact Gaussian WP solutions like (8). The equation of motion for the WP maximum is just the one for the classical trajectory $x(t)$ including the friction force (in our notation, $x(t)$ corresponds to $\eta(t)$),

$$\ddot{\eta} + \gamma\dot{\eta} + \omega^2\eta = 0\ . \tag{31}$$

The modified Riccati equation for the complex variable $\left(\frac{2\hbar}{m}\hat{y}\right)_{CK}$ reads

$$\left(\frac{2\hbar}{m}\dot{\hat{y}}\right)_{CK} + \mathrm{e}^{-\gamma t}\left(\frac{2\hbar}{m}\hat{y}\right)^2_{CK} + \omega^2\ \mathrm{e}^{\gamma t} = 0\ . \tag{32}$$

The imaginary part of this variable is again connected with the WP width and the real variable α_{CK} in the same way as in the conservative case. This again allows for the transformation of the Riccati equation into a (real) Ermakov-type equation,

$$\ddot{\alpha}_{CK} + \gamma\dot{\alpha}_{CK} + \omega^2\alpha_{CK} = \frac{\mathrm{e}^{-2\gamma t}}{\alpha^3_{CK}}\ . \tag{33}$$

This equation, together with Eq.(31) for η, forms the required system of equations that possesses an exact Ermakov-type invariant, here given in the form

$$\hat{I}_{CK} = \frac{1}{2}\left[e^{2\gamma t}\left(\dot{\eta}\ \alpha_{CK} - \eta\ \dot{\alpha}_{CK}\right)^2 + \left(\frac{\eta}{\alpha_{CK}}\right)^2\right] = \mathrm{const.} \tag{34}$$

The NL approach [16] was motivated by the attempt to break the time-reversal symmetry on all levels to include irreversibility, a phenomenon usually associated also with dissipative systems (but not necessarily occurring only in those systems).

This can be achieved by introducing a diffusion term into the continuity equation thus turning it into a Fokker–Planck-type equation (in particular a Smoluchowski equation). Following a method described in [16], the (real) Smoluchowski equation can be separated into two complex equations, namely a (modified) SE for the wave function $\Psi(x,t)$ and its complex conjugate, $\Psi^*(x,t)$, if a separation condition concerning the diffusion term is fulfilled.

This leads to the NLSE

$$i\hbar\frac{\partial}{\partial t}\Psi_{NL}(x,t) \;=\; \left\{-\frac{\hbar^2}{2m}\frac{\partial^2}{\partial x^2} + V(x) + \gamma\frac{\hbar}{i}\left(\ln\Psi_{NL} - <\ln\Psi_{NL}>\right)\right\}\Psi_{NL}(x,t) \qquad (35)$$

with a complex logarithmic nonlinearity (for details, see e.g., [16]).

The equation of motion for the WP maximum is again identical to Eq.(31). The equation for the time-dependence of the width can now be obtained from the modified complex Riccati equation

$$\left(\frac{2\hbar}{m}\dot{y}\right)_{NL} + \gamma\left(\frac{2\hbar}{m}y\right)_{NL} + \left(\frac{2\hbar}{m}y\right)_{NL}^2 + \omega^2(t) \;=\; 0\;. \qquad (36)$$

Also in this case, the imaginary part has the same relation to the WP width and the real variable α_{NL} as in the conservative case. The real part of the complex Riccati-variable now takes the modified form $\frac{2\hbar}{m}y_{R,NL} = \frac{\dot{\alpha}_{NL}}{\alpha_{NL}} - \frac{\gamma}{2}$.

With the help of these relations, the Riccati equation now turns into the Ermakov-type equation

$$\ddot{\alpha}_{NL} + \left(\omega^2 - \frac{\gamma^2}{4}\right)\alpha_{NL} = \frac{1}{\alpha_{NL}^3}\quad . \qquad (37)$$

Together with the dissipative equation for the WP maximum, Eq.(31), this leads to the exact Ermakov invariant

$$I_{NL} == \frac{1}{2}e^{\gamma t}\,\alpha_{NL}^2\left[\left(\dot{\eta} - \left(\frac{\dot{\alpha}_{NL}}{\alpha_{NL}} - \frac{\gamma}{2}\right)\eta\right)^2 + \left(\frac{1}{\alpha_{NL}^2}\,\eta\right)^2\right] \;=\; \text{const.} \qquad (38)$$

To establish the connection between the explicitly TD Caldirola–Kanai approach and the logarithmic NLSE (35) we refer to Schrödinger's first communication on wave mechanics [18] where he starts from the Hamilton–Jacobi equation with the action function S. He introduced the wave function $\Psi(x,t)$ via $S_c = \frac{\hbar}{i}\ln\Psi$, where the subscript c (added by us) indicates that this action is a complex quantity, since Ψ is, in general, a complex function. Via a variational ansatz, Schrödinger arrived at the Hamiltonian operator $H_L = -\frac{\hbar^2}{2m}\frac{\partial^2}{\partial x^2} + V(x)$.

We now reverse Schrödinger's procedure, starting with Eq. (35) (divided by Ψ, which causes no problems for Gaussian WPs) and, using the definition of S_c, arrive at $\left(\frac{\partial}{\partial t} + \gamma\right)S_c + H \;=\; -\gamma < S_c >$. The purely TD term $-\gamma < S_c >$ is necessary mainly for normalization purposes (can therefore be absorbed by the normalization coefficient) and shall be neglected in the following.

Multiplying the remaining equation by $e^{\gamma t}$ and using the definitions $\hat{S}_c \;=\; e^{\gamma t}S_c$ and $\hat{H} \;=\; e^{\gamma t}H$, it can be rewritten as Hamilton–Jacobi equation in terms of $\hat{S}_c$ and $\hat{H}$.

From the definition of the action function, it follows that the wave function $\hat{\Psi}(x,t)$ in the transformed (canonical) system is connected with the wave function $\Psi(x,t)$ in the physical system via the *non-unitary* relation

$$\ln\hat{\Psi} \;=\; e^{\gamma t}\ln\Psi\;\;. \qquad (39)$$

Consequently, the (complex) momenta in the two systems are connected via

$$\hat{p}_c \;=\; \frac{\hbar}{i}\frac{\partial}{\partial x}\ln\hat{\Psi} \;=\; e^{\gamma t}\,\frac{\hbar}{i}\frac{\partial}{\partial x}\ln\Psi_{NL} \;=\; e^{\gamma t}p_c\;\;, \qquad (40)$$

which is equivalent to the connection between the canonical and the kinetic momentum in the Caldirola–Kanai approach. The $non-canonical$ connection between the classical variables (x,p) and $(\hat{x},\hat{p})$ corresponds to the $non-unitary$ transformation between Ψ and $\hat{\Psi}$ (see also [19]).

Expressing $\hat{H}$ in terms of the canonical momentum $\hat{p}_c$ and following Schrödinger's quantization procedure, finally yields the modified SE of the Caldirola–Kanai approach. Because the WP solution of this equation, $\hat{\Psi}_{CK}(x,t)$, is related to the WP solution $\Psi_{NL}(x,t)$ of the NLSE (35) via (39), this means particularly for the corresponding Riccati and Ermakov equations that

$$\left(\frac{2\hbar}{m}\hat{y}\right)_{CK} = e^{\gamma t}\left(\frac{2\hbar}{m}y\right)_{NL} \quad \text{and} \quad \alpha_{CK} = e^{-\gamma t/2}\alpha_{NL} \, . \tag{41}$$

Taking this into account, Eqs.(32) and (33) turn into Eqs.(36) and (37) and the Ermakov invariant $\hat{I}_{CK}$ turns exactly into I_{NL}. Factorization of the operator corresponding to (34) would provide the creation/annihilation operators on the canonical level, therefore, factorization of the operator corresponding to (38) without the exponential factor (because of relation (39)) but now in terms of $\left(\frac{2\hbar}{m}y\right)$, provides the generalized creation/annihilation operators for the dissipative case on the physical level in the same form as in the conservative case.

6. Application of $a(t)$ and $a^+(t)$ to obtain new exact solutions of the TDSE

The generalized creation and annihilation operators will now be used to obtain exact solutions of the TDSE that are different from the standard textbook solutions. As an example the free motion, i.e. $V = 0$, without dissipation shall be considered, generalizations to other quadratic Hamiltonians and the inclusion of dissipation are, according to the results obtained above, straightfotward.

The exact solutions of the free motion TDSE are usually given by plane waves. A different complete set of solutions in terms of Hermite polynomials with TD variable can be obtained via the generalized creation and annihilation operators (what is in agreement with results of [20] obtained in a different context). The ground state is obtained via $a(t)|0>= 0$ and (in position representation) given by

$$< x|0 > = \Psi_0(x,t) = \left(\frac{m}{\pi\hbar}\right)^{1/4}\left(\frac{1}{\alpha}\right)^{1/2} e^{iy(t)x^2} \exp\left\{-\frac{i}{2}\int^t dt'\frac{1}{\alpha^2}\right\} . \tag{42}$$

The excited states are obtained by applying $a^+(t)$ as

$$\Psi_n(x,t) = \left[\left(\frac{m}{\pi\hbar}\right)^{1/2}\frac{1}{\alpha}\,\frac{1}{n!\;2^n}\right]^{1/2} \mathrm{H}_n\left(\frac{x}{x_0}\right)\, e^{-\frac{1}{2}\left(\frac{x}{x_0}\right)^2} \exp\left\{-i\left(n+\frac{1}{2}\right)\int^t dt'\frac{1}{\alpha^2}\right\} \tag{43}$$

with the TD variable $\xi = \frac{x}{x_0} = \xi(t)$ where $x_0 = \sqrt{\frac{\hbar}{m}}\alpha(t)$ and $\mathrm{H}_n(\xi)$ are Hermite polynomials. The functions $\Psi_n(x,t)$ are *exact solutions* of the TDSE for $V = 0$, but *no eigenfunctions* of $H_{L,0}$, the free motion Hamiltonian (but of $I_{L,0}$), since

$$\begin{aligned} H_{L,0}\,\Psi_n &= -\frac{\hbar^2}{2m}\frac{\partial^2}{\partial x^2}\Psi_n = \left\{\hbar\,\frac{1}{\alpha^2}\left(n+\frac{1}{2}\right) + \frac{m}{2}\left(\frac{2\hbar}{m}y\right)^2 x^2 + \frac{\hbar}{2i}\left(\frac{2\hbar}{m}y_R\right)\right\}\Psi_n \\ &+ \left[\left(\frac{m}{\pi\hbar}\right)^{1/2}\frac{1}{\alpha}\,\frac{1}{n!\;2^n}\right]^{1/2} e^{-\frac{1}{2}\left(\frac{x}{x_0}\right)^2}\left\{-i\hbar\left(\frac{2\hbar}{m}y_R\right)\left(\frac{x}{x_0}\right)\mathrm{H}_n'\left(\frac{x}{x_0}\right)\right\} \end{aligned} \tag{44}$$

with $\mathrm{H}_n'\left(\frac{x}{x_0}\right) = \frac{d}{d\xi}\mathrm{H}_n(\xi)$.

However, using the relations $\xi\mathrm{H}_n' = 2\xi^2\mathrm{H}_n - \mathrm{H}_{n+1} = [2\xi^2-(n+1)]\,\mathrm{H}_n - \frac{1}{2}\mathrm{H}_{n+2}$ and $< n|n+2 >= 0$, it can be shown that for $\left(\frac{2\hbar}{m}y\right)$ obeying Eq. (13) with $\omega = 0$ the mean values fulfill

$$\left\langle -\frac{\hbar^2}{2m}\frac{\partial^2}{\partial x^2}\right\rangle = \frac{\hbar}{2}\left(n+\frac{1}{2}\right)\left[\dot{\alpha}^2+\frac{1}{\alpha^2}\right] = \left\langle i\hbar\frac{\partial}{\partial t}\right\rangle . \tag{45}$$

7. Conclusions

In this paper a generalization of the creation/annihilation operators is investigated for cases, where the corresponding CSs or WPs have TD width and the energy (generally) is no longer a constant of motion. Nevertheless, for these systems still a dynamical invariant can exist. This can be obtained from the Newtonian equation of motion for the WP maximum and a complex NL Riccati equation for the WP width. Factorization of the corresponding operator supplies the generalized creation/annihilation operators in terms of position and momentum operators and the complex variable $\left(\frac{2\hbar}{m}y\right)$ obeing the Riccati equation. These results are also in agreement with the work of Malkin et al. [21] where these operators were expressed in terms of a complex variable that can be obtained by linearizing the Riccati equation. In our case, however, new insight is gained from the information contained in the nonlinearity of the Riccati equation and the relation to supersymmetric quantum mechanics.

Using the definition of the CS as eigenstate of the annihilation operator, it is possible to determine the complex eigenvalue $z(t)$ in terms of the classical quantities η and $\dot{\eta}$ and of real and imaginary parts of $\left(\frac{2\hbar}{m}y\right)$. The CS in position space obtained in this way is identical to the Gaussian WP that is a solution of the corresponding TDSE.

In the second part, the formalism is extended to include dissipative environmental effects. For this purpose effective descriptions in terms of an explicitly TD Hamiltonian based on a modified canonical approach and a logarithmic NLSE are compared. In both cases, the equations of motion for the corresponding WP maximum and width again supply an Ermakov invariant. Furthermore, both approaches are related via the non-unitary transformation (39). Factorization of the corresponding operator again provides the modified creation/annihilation operators that have, expressed ind terms of $\left(\frac{2\hbar}{m}y\right)$, the same form as in the conservative case.

Finally, these generalized operators are used to construct exact analytic solutions of the free motion Schrödinger equation in terms of Hermite polynomials with time-dependent variable.

References

[1] Ermakov V P 1880 *Univ. Izv. Kiev* **20** 1
[2] Schrödinger E 1940 *Proc. Roy. Irish Acad.* **46** A 9
[3] Dirac P A M 1935 *The Principles of Quantum Mechanics* 34 (Oxford: Oxford Univ. Press)
[4] Schrödinger E 1926 *Naturwissenschaften* **14** 664
[5] Glauber R J 1963 *Phys. Rev.* **130** 2529
[6] Sudarshan E C G 1963 *Phys. Rev. Lett.* **10** 227
[7] Klauder J R 1963 *J. Math. Phys.* **4** 1055
[8] Nieto M M and Simmons Jr L M 1979 *Phys. Rev.* D **20** 1321
[9] Lewis H R and Riesenfeld W B 1969 *J. Math. Phys.* **10** 1458
[10] Hartley J G and Ray J R 1982 *Phys. Rev.* D **25** 382
[11] Mielnik B and Rosas-Ortiz O 2004 *J. Phys. A: Math. Gen.* **37** 10007
[12] Cooper F, Khare A and Sukhatme U 2001 *Supersymmetry in Quantum Mechanics* (Singapore: World Scientific) Chap. 6.1
[13] Rosas-Ortiz O and Muñoz R 2003 *J. Phys. A: Math. Gen.* **36** 8497; Fernández D J, Muñoz R and Ramos A 2003 *Phys. Lett.* A **308** 11
[14] Caldirola P 1941 *Nuovo Cimento* **18** 393; Kanai E 1948 *Progr. Theor. Phys.* **3** 440
[15] Hasse R W 1975 *J. Math. Phys.* **16** 2005
[16] Schuch D, Chung K-M and Hartmann H 1983 *J. Math. Phys.* **24** 1652; ibid 1984 *J. Math. Phys.* **25** 3086; Schuch D and Chung K-M 1986 *Int. J. Quantum Chem.* **29** 1561
[17] Nassar A B 1986 *J. Math. Phys.* **27** 755; ibid 1986 *J. Math. Phys.* **27** 2949
[18] Schrödinger E 1926 *Ann. d. Phys.* **79** 361
[19] Schuch D 1997 *Phys. Rev.* A **55** 935
[20] Guerrero J, López-Ruiz F F, Aldaya V and Cossío F 2011 *J. Phys. A: Math. Theor.* **44** 445307
[21] Malkin I A, Man'ko V I and Trifonov D A 1970 *Phys. Rev.* D **2** 1371

Solving the nonlocality riddle by conformal quantum geometrodynamics

Enrico Santamato

Dipartimento di Scienze Fisiche, Università di Napoli "Federico II", Complesso Universitario di Monte S. Angelo, 80126 Napoli, Italy

E-mail: enrico.santamato@na.infn.it

Francesco De Martini

Accademia Nazionale dei Lincei, via della Lungara 10, 00165 Roma, Italy

E-mail: rancesco.demartini@uniroma1.it

Abstract. A rigorous *ab initio* derivation of the quantum mechanics of a single particle with spin is presented starting by conformally Weyl-gauge-invariant principle. The particle is described as a relativistic top with six Euler's angles and quantum effects are introduced by assuming that the Weyl's curvature of the particle configuration space acts on it as an external scalar potential. Weyl's conformal covariance is made explicit in all steps of the theory. It is shown that metric of the configuration space accounts for non-quantum relativistic effects, while the affine connections account for quantum relativistic effects. In this way, classical and quantum features acquire well distinguished geometrical origin. A scalar wave function is also introduced to recover the connection with the standard quantum description based on Dirac's four-component spinors. Finally, the case of two entangled spins is considered in the nonrelativistic limit and it is found that the nonlocality rests on the entanglement of the spin internal orientational variables, playing the role of "hidden variables". The theory was carried out in the Minkowski space-time, but it can be easily extended to a space with nonzero Riemann curvature.

1. Introduction

Spin is a peculiar yet fundamental feature of the quantum world. Consequently, it is usually argued that no approach based on classical concepts is able to account for quantum spin. This seems particularly true for the spin 1/2, because a classical system, even when quantized, can carry only integer values of angular momentum (in units of $\hbar$). In this work, we show that the mechanics of quantum spin may be derived from the mechanics of the relativistic spherical top in a conformally curved configuration space without having recourse to explicit quantization rules or operator algebra. Many attempts were made in the past to interpret quantum physics under the language of classical physics, dating from the hidden variables approach [1] to the Madelung-Bohm hydrodynamic approach [2, 3, 4], to the Fényes-Nelson stochastic approach [5, 6] and others. The present work is based on the approach proposed some time ago to derive the Schrödinger [7, 8] and the Klein-Gordon [9] wave equations from space conformal geometry. Here beside space-time translational degrees of freedom, also orientational degrees of freedom

have been added to the particle so to extend the theory to include spin [10, 11]. The orientational coordinates are internal to the particle and play the role of hidden variables. In the case of two identical spatially distant spin $1/2$ particles, the results of the present theory are in full agreement with standard quantum mechanics and violation of Bell's inequalities [12, 13] is found, when the two particles are prepared in the EPR state [14]. Quantum correlations are shown to derive from nonseparable coupling among the internal angular variables of the two particles in the Weyl curvature and appear to be manifestly nonlocal. However, we studied the EPR entanglement in the nonrelativistic framework only, so we cannot say, at this stage of the theory, if this residual orientational coupling is really superluminal as some experiments seems to confirm [13].
Quantum effects are ascribed to the presence of a not trivial parallel transport law in the configuration space of the relativistic top. The metric plays no role and can be arbitrarily prescribed. However, we assume the Minkowsi metric with signature $(-,+,+,+)$. The particle motion and the affine connections for parallel transport are derived by a single action principle using Palatini's variation with fixed metric tensor. As a result, an integrable Weyl's connection is found and the theory acquires full invariance in a conformal change of the metric. The particle motion and the space affine connections must be consistent each other and they are part of a unique problem. The overall physical picture is analogous to the situation prevailing in the general relativity: geometry is not prescribed; rather it is determined by the physical reality. In turn, geometry acts as a "guidance field" for matter. However, although gravitational and quantum phenomena share a common geometrical origin, they rely on *independent* geometrical objects: the metric tensor for the former and the affine connections for the latter. As in the general relativity, the geometric approach forces to describe matter as a fluid or as a bundle of elementary trajectories in the configuration space rather than as a single point particle moving along one trajectory. In this respect, the Hamilton-Jacobi approach to mechanics plays a much more important role more than Newton's law in describing the particle motion.

2. The conformal quantum geometrodynamics

We consider a generic quantum system with n-dimensional configuration space V_n, spanned by generalized coordinates q^i $(i = 1, ..., n)$. The space V_n is endowed with a metric tensor $g_{ij}(q)$ that we assumed to be arbitrarily prescribed and with some affine connection fields $\Gamma^i_{jk}(q)$. The mechanics is governed by the multiple-integral variational principle

$$\delta\left[\int d^n q\sqrt{g}\rho(g^{ij}\partial_i\sigma\partial_j\sigma + R_W)\right] \tag{1}$$

where $g = |\det(g_{ij})|$, $\rho(q)$ and $\sigma(q)$ are scalar fields and R_W is the scalar curvature built up from the connections $\Gamma^i_{jk}(q)$. The variation with respect to $\rho(q)$, $\sigma(q)$ yields, respectively [9]

$$g^{ij}\partial_i\sigma\partial_j\sigma + R_W = 0, \tag{2}$$

$$\frac{1}{\sqrt{g}}\partial_i(\sqrt{g}\rho g^{ij}\partial_j\sigma) = 0. \tag{3}$$

The variation of (1) with respect to the connections $\Gamma^i_{jk}(q)$ yields Weyl's conformal connection [9]

$$\Gamma^i_{jk} = -\left\{ {i \atop jk} \right\} + \delta^i_j\phi_k + \delta^i_k\phi_j + g_{jk}\phi^i, \tag{4}$$

where $\left\{ {i \atop jk} \right\}$ are the Cristoffel symbols out of the metric g_{ij}, $\phi^i = g^{il}\phi_l$, and ϕ_i is Weyl's vector given by

$$\phi_i = -\frac{1}{n-2}\partial_i(\ln\rho). \tag{5}$$

Equation (5) shows that the Weyl connection is *integrable* with Weyl's scalar potential ϕ given by $\phi = -\frac{1}{n-2}\ln\rho$. Equations (2–5) are the main equations of the theory. From Eq. (5) we see that ρ can be used in place of ϕ as Weyl's potential to calculate the affine connections. Thus, the scalar curvature derived from connections (4) can be written as

$$R_W = R + \left(\frac{n-1}{n-2}\right)\left[\frac{g^{ij}\partial_i\rho\partial_j\rho}{\rho^2} - \frac{2\partial_i(\sqrt{g}g^{ij}\partial_j\rho)}{\rho\sqrt{g}}\right]. \tag{6}$$

The affine Weyl connections (4) are invariant under Weyl's conformal changes

$$g_{ij} \rightarrow \lambda g_{ij}, \tag{7}$$

$$\phi_i \rightarrow \phi_i - \frac{\partial_i\lambda}{2\lambda}. \tag{8}$$

Tensors T which under Weyl's conformal gauge transformations (7) change as $T \rightarrow \lambda^{w(T)}T$ are said to transform simply. The exponent $w(T)$ is the Weyl weight of tensor T. Examples are the metric tensor with $w(g_{ij}) = 1$ and Weyl's curvature R_W with $w(R_W) = -1$. Weyl-gauge-invariant quantities as the affine connections (4) have $w = 0$. The action principle (1) is Weyl-gauge invariant if $w(\sigma) = 0$ and $w(\rho) = -(n-2)/2$, which ensures Weyl-gauge invariance of all equations derived from this principle. Weyl's geometry is connected with the so-called co-covariant derivative D_i, defined as $D_iT = \nabla_iT - 2w(T)\phi_iT$, where T is a tensor, ∇_i is the covariant derivative associated to the metric tensor g_{ij}, and ϕ_i is Weyl's vector. The metric tensor obeys $D_ig_{jk} = 0$ and hence indices can be lowered and raised in the co-covariant derivative in Weyl's geometry as they are in the covariant derivative in Riemann's geometry. The Weyl-gauge covariance of Eqs. (2), (3), (5) is then made manifest by writing them as $D_i\sigma D^i\sigma + R_W = 0$, $D_iD^i\sigma = 0$, $D_i\rho = 0$, respectively.

Equation (2) has the form of the classical Hamilton-Jacobi (HJ) equation with scalar potential R_W. We may then associate to the function $\sigma(q)$ the bundle of canonical trajectories defined by $dq^i/d\tau = \partial_i\sigma$, where τ is an arbitrary parameter. Each trajectory obeys the Euler-Lagrange equations derived by the Weyl-gauge invariant and parameter invariant single-integral variational principle

$$\delta\int L\,d\tau = \delta\left[\int\sqrt{-R_W g_{ij}\dot{q}^i\dot{q}^j}\,d\tau\right] = 0 \tag{9}$$

where the dot means derivative with respect to the parameter τ. The HJ Eq. (2) and the variational principle (9) provide the mechanical features of the theory. These equations have the same form of the relativistic equations for a free particle, but the particle mass is replaced here by Weyl's curvature R_W as required for Weyl-gauge invariance. The Lagrangian L in Eq. (9) fully characterizes the system under study and usually provides the easiest way to find out the metric tensor g_{ij}.

Also Eq. (3) has a dynamical interpretation as continuity equation for the Weyl-gauge invariant current density j^i given by

$$j^i = \rho\partial_i\sigma\sqrt{g}. \tag{10}$$

The continuity equation and the dependence of R_W on ρ shown in Eq. (6) are in agreement with the hydrodynamical picture of quantum mechanics first proposed by Madelung [2] and then developed by Bohm [3, 4]. In the Madelung-Bohm approach the particle trajectories are deterministically governed by classical mechanics and quantum effects are due to a "quantum potential" of quite mysterious origin, whose gradient acts as a newtonian force on the particle. According to the present approach the active potential originates from geometry, as does gravitation, and arises from the space curvature (6) due to the presence of the nontrivial affine connections (4) with Weyl's vector field given by Eq. (5).

The particle dynamics (governed by $\sigma(q)$) and the affine connections (governed by $\rho(q)$) must be compatible, however. We must solve Eqs. (2) and (3) together. The nonlinear problem implied by Eqs. (2) and (3) looks very hard at first glance. However, by introducing the complex scalar function ψ given by

$$\psi(q) = \sqrt{\rho} e^{i\frac{S(q)}{\hbar}} \tag{11}$$

with $S(q) = \xi\hbar\sigma(q)$, and

$$\xi = \sqrt{\frac{n-2}{4(n-1)}}, \tag{12}$$

one converts Eqs. (2) and (3) into the *linear* differential equation

$$\nabla_k \nabla^k \psi - \xi^2 \hbar^2 R \psi = 0 \tag{13}$$

where $\nabla_k \nabla^k$ is the Laplace-Beltrami operator built from the metric tensor g_{ij} and R is the Riemann curvature appearing in Eq. (6). This is a striking result as it demonstrates that the Hamilton-Jacobi equation, applied to a general dynamical problem can be transformed into a linear wave equation, the foremost ingredient of the formal structure of quantum mechanics and of the Hilbert space theory. The value of ξ given by Eq. (12) ensures that Eq. (13) is conformally invariant. The wave function is not gauge invariant, however, because its modulus has Weyl weight $w(|\psi|) = -(n-2)/4$. The phase of ψ is conformally invariant, instead. Equation (13) resembles the covariant quantum Klein-Gordon wave-equation with the mass term replaced by the Riemann curvature term $\hbar^2\xi^2 R(q)$. What is more surprising is that any explicit reference to the Weyl vector field $\phi_i(q)$ and to the Weyl curvature R_W has been cancelled out from Eq. (13). In fact, as said above, the curvature $R(q)$ and the covariant derivatives ∇_i in Eq. (13) are calculated using the Cristoffel symbols derived from the metric g_{ij}.
Finally, we observe that external electromagnetic vector fields $A_i(q)$ can be added to the theory by simply replacing $\partial_i\sigma$ with $\partial_i\sigma - A_i$ in all equations and adding a term $A_i\dot{q}^i$ to the Lagrangian in Eq. (9) so to include also invariance under electromagnetic gauge transformations $A_i \to A_i + \partial_i\chi$, $\sigma \to \sigma + \chi$. In this case, as expected, the wave equation (13) is modified by replacing the covariant derivative ∇_i with the operator $\nabla_i - A_i$ and the phase of the wave function (11) is affected by the electromagnetic gauge in the usual way. Therefore, Weyl's and electromagnetic gauges act separately on the modulus and phase of the wave function, respectively.

3. The single spin 1/2 particle

In this section, we specialize the theory to the case of a single spin 1/2 particle – one electron, say. We start considering the simplest model for the relativistic spinning particle, namely the spherical top described by six Euler angles, as made, for example, by Frenkel [15], Thomas [16] and in the classic work by Hanson and Regge [17]. Subsequent important works on relativistic spinning particles can be found in many textbooks [18, 19, 20]. Here, we simply imagine that the particle follows a path $x^\mu = x^\mu(\tau)$ in space-time, where τ is an arbitrary parameter along the path, and that it carries along with itself a moving fourleg $e_a^\mu = e_a^\mu(\tau)$ $(\mu, a = 0, \ldots, 3)$ with $e_0^0 > 0$. The fourleg vectors e_a^μ are normalized according to $g_{\mu\nu} e_a^\mu e_b^\nu = g_{ab}$ where $g_{\mu\nu} = g_{ab} = \mathrm{diag}(-1, 1, 1, 1)$ is the Minkowski metric tensor. Setting $\Lambda = \{e_a^\mu\}$ and $G = \{g_{\mu\nu}\}$, the normalization relations can be cast in the matrix form as $\Lambda^T G \Lambda = G$, showing that the 4×4 matrix $\Lambda \in SO(3,1)$ is a proper orthochronous Lorentz matrix. The fourleg e_a^μ is parametrized by six Euler's angles $\theta^A = \{\alpha, \beta, \gamma; \chi, \theta, \phi\}$ $(A = 1, ..., 6)$: first three angles for space rotations and last three angles for boosts. Then, the configuration space of our relativistic top has $n = 10$ coordinates: the four coordinates x^μ of the "center of mass" and the six Euler's angles θ^A for "orientation". The derivative of $e_a^\mu(\tau)$ with respect to τ can be written as $de_a^\mu/d\tau = \omega_\nu^\mu e_a^\nu$. The contravariant tensor $\omega^{\mu\nu} = \omega_\sigma^\mu g^{\sigma\nu}$ is skewsymmetric and can be considered as the "angular

velocity" of the top in space-time. The three space components of $\omega^{\mu\nu}$ are associated to rotations and the three time components to boosts.
The metric of the relativistic top configuration space is fixed by the Weyl-gauge and parameter invariant Lagrangian

$$\begin{aligned} L &= \sqrt{-R_W(g_{\mu\nu}\dot{x}^\mu\dot{x}^\nu + a^2 g_{\mu\nu}g^{ab}\dot{e}^\mu_a\dot{e}^\nu_b)} = \\ &= \sqrt{-R_W(g_{\mu\nu}\dot{x}^\mu\dot{x}^\nu - a^2\omega_{\mu\nu}\omega^{\mu\nu})}. \end{aligned} \tag{14}$$

When written in full as a function of the ten generalized coordinates $q^i = \{x^\mu, \theta^A\}$ and their derivatives, the Lagrangian (14) reduces to the canonical form of Eq. (9). In the presence of external electromagnetic fields a term $A_\mu\dot{x}^\mu + \kappa a^2 F_{\mu\nu}\omega^{\mu\nu}$ with constant κ must be added to the Lagrangian [10, 11]. Once the metric is fixed by the dynamical system, we must find its dynamical state and compatible Weyl's connections. As said above, this is conveniently made by solving the wave equation (13). A long but straightforward calculation shows that the wave function $\psi(q)$ which solves Eq. (13) can be written as the Dirac product $\psi(q) = \overline{\Phi}(\theta)\Psi(x) = \Phi^\dagger(\theta)\gamma^0\Psi(x)$, where $\gamma^0 = \begin{pmatrix} 0 & 1 \\ 1 & 0 \end{pmatrix}$ is Dirac's matrix in the spinor representation and $\Phi(\theta)$ and $\Psi(x)$ are the four-component Dirac's spinors

$$\Phi(\theta) = \begin{pmatrix} D^\sigma(\theta) \\ D^{\dot\sigma}(\theta) \end{pmatrix}, \tag{15}$$

$$\Psi(x) = \begin{pmatrix} \psi^\sigma(x) \\ \psi^{\dot\sigma}(x) \end{pmatrix}. \tag{16}$$

Here $\psi^\sigma(x)$ and $\psi^{\dot\sigma}(x)$ are undotted and dotted two-component Lorentz spinors depending on the spacetime coordinates x^μ only, and $D^\sigma(\theta)$ and $D^{\dot\sigma}(\theta)$ are the first columns of the D-matrices belonging to the (0,1/2) and (1/2,0) irreducible representation of Lorentz group, respectively. These matrices are related by $D^{(0,1/2)} = [(D^{(1/2,0)})^\dagger]^{-1}$ and depend on the Euler's angles θ^A only. To satisfy the wave equation, the four-component spinor $\Psi(x)$ must be a solution of the standard Klein-Gordon equation in space-time[1] with mass m given by

$$m^2 = \frac{3\hbar^2}{2a^2c^2} + \frac{6\hbar^2\xi^2}{a^2c^2}. \tag{17}$$

The first term in Eq. (17) comes from the spin (it vanishes for zero spin) and the second term comes from the constant Riemann curvature $R = 6/a^2$ of the configuration space. If we take $m = m_e$, the electron mass, the gyration radius a of the associated relativistic top turns out to be of the order of the electron Compton length: $a \simeq \hbar/(m_e c)$. Notice that the spinor field $\Psi(x)$ has a constant mass in spite of the fact that in the Weyl-gauge invariant variational principle (1) and in the Lagrangian (14) Weyl's curvature field $R_W(q)$ replaces the mass. As it is well-known, the Klein-Gordon equation (and the second-order Dirac's equation) have unphysical four-component solutions, corresponding to a negative mass in the first-order Dirac's equation [21]. Any four-component solution of the Klein-Gordon (or second-order Dirac) equation can be written as a superposition of solutions of the first-order Dirac's equation with positive and negative mass [11]. The latter must be discarded, which can be made also systematically [21]. Four-component field $\Psi(x)$ has been used here to preserve invariance under parity. However, the second-order Klein-Gordon (or Dirac) equation has also (left and right) chiral solutions corresponding to the two-component Lorentz spinor fields $\psi^\sigma(x)$ or $\psi^{\dot\sigma}$ in Eq. (16). As shown by Brown [22], the two-component solutions, beside reproducing the same physical result's of Dirac's equation when parity is restored, are also computationally easier to work with.

[1] When electromagnetic fields are present, the Klein-Gordon equation is replaced by the second-order Dirac equation [10, 11].

4. The nonrelativistic limit

A deeper insight into the scalar wave function $\phi(q) == \overline{\Phi}(\theta)\Psi(x)$ is obtained when the matrix $D(\theta)$ is factorized into the boost $B(e_0)$ associated to the time-like vector $e_0 = \{e_0^\mu(\chi, \theta, \phi)\}$ of the particle fourleg and a rotation $R(\alpha, \beta, \gamma) \in \mathrm{SU}(2)$, viz $D(\theta) = B(e_0)R(\alpha, \beta, \gamma)$. The rotation R belongs to the little Poincaré group of the time-like vector e_0 and, under generic Lorentz transformation Λ, transforms with the Wigner rotation $B(\bar{e}_0)^{-1}D(\Lambda)B(e_0)$, where $\bar{e}_0^\mu = \Lambda^\mu_\nu e_0^\nu$. According to our relativistic top picture, the rotation R accounts for particle spin and the boost $B(e_0)$ accounts for the motion of the particle "center of energy" around the world line $x^\mu(\tau)$ of its "center-of-mass" [20, Chap. 20].

Having introduced the rotational degrees of freedom explicitly makes easy to go into the low velocity nonrelativistic limit of the theory. In this limit, the dotted and undotted Lorentz spinors $\psi^\sigma(x)$ and $\psi^{\dot{\sigma}}(x)$ coalesce into one two-component Pauli spinor $w^\sigma(x)$ and the fourleg angle $\chi \to 0$ [2]. Then, in the low velocity limit we obtain

$$\begin{aligned}\psi(q) &= R_\sigma^{-1}(\alpha,\beta,\gamma)w^\sigma(x) = \\ &= e^{\frac{i\gamma}{2}}\left(e^{\frac{i\alpha}{2}}\cos\frac{\beta}{2}w_\uparrow(\boldsymbol{r},t) + e^{-\frac{i\alpha}{2}}\sin\frac{\beta}{2}w_\downarrow(\boldsymbol{r},t)\right) \qquad (18)\end{aligned}$$

where $R_\sigma^{-1}(\alpha,\beta,\gamma)$ is the first row of the inverse of the SU(2) rotation matrix $R(\alpha,\beta,\gamma)$. As expected, the two-component Pauli spinor $w^\sigma(x) = \{w_\uparrow, w_\downarrow\}$ obeys the non relativistic Schrödinger equation and its components correspond to spin up or down along the fixed z-axis, respectively [23]. To better see the role played by the wave function in the present approach, we consider the simple case of spin up state. From Eq. (18) with $w_\downarrow = 0$ we calculate the mechanical action S and the Weyl curvature R_W as

$$S = \frac{\hbar}{2}(\gamma+\alpha) + \arg(w_\uparrow(\boldsymbol{r},t)), \qquad (19)$$

$$R_W = -\frac{5}{2a^2(1+\cos\beta)} + R_W^\uparrow(\boldsymbol{r},t) + \text{const.} \qquad (20)$$

where $R_W^\uparrow(\boldsymbol{r},t)$ is the contribution of $w_\uparrow(\boldsymbol{r},t)$ to Weyl's curvature. From Eq. (19) we see that the β coordinate is cyclic, and, hence, β is a constant of motion. From Eq. (20) we see that the particle is not free, but is subjected to a self-force proportional to the gradient of Weyl's curvature. This self-force has a geometric origin and cannot be eliminated: it is needed to have Weyl-gauge invariance. However, its existence is hidden in the standard quantum mechanics based on the space-time spinor $w_\uparrow(\boldsymbol{r},t)$, which obeys the Schrödinger equation for the free particle. In the next section, we will demonstrate that the geometric force due to Weyl's curvature is the source of quantum entanglement and of the EPR paradox.

5. The EPR paradox

The calculation to obtain Eqs. (19) and (20) from the wave function (18) can be repeated when two identical spin 1/2 particle are considered. The configuration space is now the product space spanned by the 20 coordinates given by the 8 space-time coordinates and the 12 angular coordinates of the two particles. To clarify the source of EPR quantum correlations, we consider here two cases: a) the two particles have opposite spin along the z-axis; b) the two particles are in the EPR state. In the quantum notation, case (a) corresponds to the spin product state $|\uparrow\rangle|\downarrow\rangle$ and case (b) to the entangled state $(1/\sqrt{2})(|\uparrow\rangle|\downarrow\rangle - |\downarrow\rangle|\uparrow\rangle)$.

[2] Beside the center of mass velocity, also the velocity v_e of the center of energy given by $\tanh\chi = v_e/c$ is small in the nonrelativistic limit.

5.1. (a) The two opposite spins

The wave function of the state $|\uparrow\rangle|\downarrow\rangle$ is easily written by taking the product of the two terms on the right of Eq. (18) and S and R_W are then calculated from modulus and phase of this wave function (for details see Ref. ([23]). The result is

$$S = S^{(A)}(\boldsymbol{r}_A, \theta_A) + S^{(B)}(\boldsymbol{r}_B, \theta_B) \tag{21}$$

$$R_W = R_W^{(A)}(\boldsymbol{r}_A, \theta_A) + R_W^{(B)}(\boldsymbol{r}_B, \theta_B) \tag{22}$$

where $S^{(A,B)}(\boldsymbol{r}_{A,B}, \theta_{A,B})$ and $R_W^{(A,B)}(\boldsymbol{r}_{A,B}, \theta_{A,B})$ are given by Eqs. (19) calculated for particle A and B separately[3]. From Eqs. (21,22) we see that in this case the particles have independent motions. In particular, the Weyl curvature reduces to the sum of the two Weyl curvatures so that each particle is affected only by its own self-force.

5.2. (b) The two spin EPR state

The same procedure can be applied to the EPR state of the two spins. The result is [23]

$$S = \hbar\left[\frac{\gamma_A+\gamma_B}{2} + \arctan\left(\csc\frac{\beta_A-\beta_B}{2}\sin\frac{\beta_A+\beta_B}{2}\tan\frac{\alpha_B-\alpha_A}{2}\right) + \right.$$
$$\left. + \arg(w_\uparrow^{(A)}(\boldsymbol{r}_A, t)) + \arg(w_\downarrow^{(B)}(\boldsymbol{r}_B, t))\right] \tag{23}$$

and

$$R_W = \frac{22}{5a^2(1-\cos\beta_A\cos\beta_B - \cos\Delta\alpha\sin\beta_A\sin\beta_B)} + R_W^{(A)}(\boldsymbol{r}_A, t) + R_W^{(B)}(\boldsymbol{r}_B, t). \tag{24}$$

In this case, although the particle motions are independent in space-time, they are still coupled by the Weyl curvature in the angular variables and, beside the self-force, the particles exert a force on each other. As we have shown elsewhere by considering in detail the behavior of the two particles on a Stern-Gerlach apparatus [23], the features of this geometrical interaction reproduce all results of the standard quantum mechanics and, in particular, violate Bell's inequalities.

6. Conclusions

We proposed an alternative approach to quantum mechanics of spin 1/2 particles based on conformal Weyl-gauge invariance. No quantization rules appear explicitly and quantum effects are derived from the existence of a force from Weyl's scalar curvature of the configuration space of the system. A scalar wave function is introduced as a useful *ansatz* to solve the nonlinear equations of the theory. The particle mass is replaced by Weyl's scalar curvature field and a constant mass appears only at the level of wave equation. In the case of spin 1/2 a one-to-one correspondence is established between the results of the theory and the results of the standard quantum mechanics based on Dirac's spinors and on the second-order Dirac equation [11]. However, the underlying Weyl-gauge symmetry is lost when standard quantum mechanics is exploited. Beside space-time coordinates, internal variables are introduced corresponding to orientational degrees of freedom. These extra coordinates are hidden to experiment and are absent in the standard quantum mechanics, which is based on space-time degrees of freedom only. However, the internal hidden variables and Weyl's geometry of the configuration space are the source of quantum effects, including entanglement. The occurrence of EPR paradox, in particular, can be traced back to a geometrical nonlocal interaction among the particles due to Weyl's curvature. At the present stage we cannot say if this interaction is superluminal, because we studied the EPR state in the nonrelativistic approximation. Future work will be addressed in this direction.

[3] Except for a common factor due to the changed dimensionality of the configuration space.

References

[1] Belinfante F 1951 *A Survay of Hidden Variable Theories* (Englewood Cliffs, NJ: Prentice Hall)
[2] Madelung E 1926 Quantentheorie in hydrodynamischer Form *Z. Phys.* **40** 332
[3] Bohm D 1952 A suggested interpretation of the quantum theory in terms of "hidden" variables I *Phys. Rev.* **85** 166
[4] Bohm D 1952 A suggested interpretation of the quantum theory in terms of "hidden" variables II *Phys. Rev.* **85** 180
[5] Fényes I 1952 Eine wahrscheinlichkeitstheoretische Begrundung und Interpretation der Quantenmechanik *Z. Phys.* **132** 81
[6] Nelson E 1966 Derivation of the Schrödinger equation from Newtonian mechanics *Phys. Rev.* **150** 1079
[7] Santamato E 1984 Geometric derivation of the Schrödinger equation from classical mechanics in curved Weyl spaces *Phys. Rev.* D **29** 216
[8] Santamato E 1984 Statistical interpretation of the Klein-Gordon equation in terms of the space-time Weyl curvature *J. Math. Phys.* **25** 2477
[9] Santamato E 1985 Gauge-invariant statistical mechanics and average action principle for the Klein-Gordon particle in geometric quantum mechanics *Phys. Rev.* D **32** 2615
[10] De Martini F and Santamato E 2011 Derivation of Dirac's equation from conformal differential geometry *Quantum Foundations* ed A Khrennikov (Melville: AIP) (*Preprint* quant-ph/1107.3168v1)
[11] Santamato E and De Martini F 2013 Derivation of the Dirac equation by conformal differential geometry *Found. Phys.* to appear
[12] Bell J S 1987 *Speakable and Unspeakable in Quantum Mechanics* (Cambridge, UK: Cambridge Univ. Press)
[13] Redhead M 1987 *Incompleteness, Nonlocality and Realism: a Prolegomenon to the Philosophy of Quantum Mechanics* (Oxford: Clarendon Press)
[14] Einstein A, Podolsky B and Rosen N 1936 Can quantum-mechanical description of physical reality be considered complete? *Phys. Rev.* **47** 777
[15] Frenkel J 1926 Die Elektrodynamik des rotierenden Elektrons *Z. Phys.* **37** 243–262
[16] Thomas L H 1926 The motion of the spinning electron *Nature* **117** 514
[17] Hanson A J and Regge T 1974 The relativistic spherical top *Ann. Phys. (NY)* **87** 498–566
[18] Souriau J M 1969 *Structure des systèmes dynamiques* (Paris: Dunod Université)
[19] Corben H C 1968 *Classical and quantum theory of spinning particles* (San Francisco: Holden-Day)
[20] Sudarshan E C G and Mukunda N 1974 *Classical dynamics: a modern perspective* (New York: John Wiley & Sons)
[21] Landau L D and Lifshitz E M 1960 *Relativistic Quantum Theory* (New York: Pergamon Press)
[22] Brown L M 1958 Two-component fermion theory *Phys. Rev.* **111**(3) 957–964
[23] Santamato E and De Martini F 2012 Solving the quantum nonlocality enigma by Weyl's conformal geometrodynamics *Preprint* arXiv:1203.0033v1 [quant-ph], submitted to Found. Phys.

Is the de Broglie-Bohm interpretation of quantum mechanics really plausible?

Kurt Jung
Fachbereich Physik, Technische Universität Kaiserslautern,
Erwin-Schrödinger-Str. 56, D-67663 Kaiserslautern, Germany,
E-mail: jung@physik.uni-kl.de

Abstract. Bohmian mechanics also known as de Broglie-Bohm theory is the most popular alternative approach to quantum mechanics. Whereas the standard interpretation of quantum mechanics is based on the complementarity principle Bohmian mechanics assumes that both particle and wave are concrete physical objects. In 1993 Peter Holland has written an ardent account on the plausibility of the de Broglie-Bohm theory. He proved that it fully reproduces quantum mechanics if the initial particle distribution is consistent with a solution of the Schrödinger equation. Which may be the reasons that Bohmian mechanics has not yet found global acceptance? In this article it will be shown that predicted properties of atoms and molecules are in conflict with experimental findings. Moreover it will be demonstrated that repeatedly published ensembles of trajectories illustrating double slit diffraction processes do not agree with quantum mechanics. The credibility of a theory is undermined when recognizably wrong data presented frequently over years are finally not declared obsolete.

1. Introduction

Ever since Einstein, Podolski and Rosen [1] asked whether quantum mechanics is complete, physicists have tried to find hidden parameters which are revealed by measurement processes. In 1964 John Bell [2] made it feasible to discriminate local hidden-parameter theories from quantum mechanics. Sophisticated experiments on spin correlations [3] proved that local hidden variables cannot exist. John Bell himself emphasized that Bohmian mechanics [4] is exempted from this verdict because this theory uses non-local hidden parameters. In fact, for a given quantum state Bohmian trajectories are derived from the associated Schrödinger wave function, which depends not only on the local potential but on its overall shape. Bohmian mechanics belongs to the coexistence models. Particles are localized and incorporate nearly all energy of the particle-wave system. If there is any charge it is almost completely concentrated in the particle. The Schrödinger wave functions are the primary physical entities. They guide the associated particles. More precisely, the momentum $\boldsymbol{p}$ of a particle is given by $\boldsymbol{p}(\boldsymbol{x}) = \nabla S(\boldsymbol{x})$ when the corresponding Schrödinger wave function $\psi(\boldsymbol{x})$ is written in the form

$$\psi(\boldsymbol{x}) = R(\boldsymbol{x})e^{iS(\boldsymbol{x})/\hbar}. \tag{1}$$

with the real amplitude function $R(\boldsymbol{x}) \geq 0$ and the real phase function $S(\boldsymbol{x})$.

As shown by Holland [5] the local flux distribution of the Schrödinger wave function agrees with the flux of the associated ensemble of Bohmian trajectories. The consistency of the fluxes has been taken as proof that the results of quantum mechanics are fully reproduced by Bohmian mechanics. Therefore Bohmian mechanics has often been viewed as a causal interpretation of quantum mechanics providing detailed information about particle trajectories.

Despite these considerable successes there are conspicuous drawbacks. In fact, in the literature one finds extensive discussions arguing against the uniqueness and validity of Bohmian mechanics [6–10]. Besides these intricate considerations there are less elusive objections against the de Broglie-Bohm theory. Especially if the Schrödinger wave function contains standing waves Bohmian mechanics predicts properties of atoms or molecules which are not observed in nature. These results suggest that Bohmian trajectories describe streamlines of the local flow but not the movement of real particles. In their book on trajectory based approaches to quantum mechanics Sanz and Miret-Artés [11] also emphasize the hydrodynamical aspects of Bohmian mechanics. They consider Bohmian trajectories as the paths of tracer particles indicating the flux of the Schrödinger wave function.

2. Failure of Bohmian mechanics in case of purely real wave functions

In case of scattering problems continuous density distributions of the Schrödinger wave function can be reproduced by an ensemble of trajectories representing subsequent scattering events. However, Bohmian mechanics has a conceptual problem if only single particles are involved as in the case of a hydrogen atom. If the atomic electron is associated with a purely real wave function Bohmian mechanics postulates that the electron is at rest. This is the case for all s-states and for all magnetic sublevels with zero magnetic quantum number.

Holland [5] shows that a resting electron is prevented from falling down by a quantum potential which balances the Coulomb potential. It is also clear that such a resting electron will not radiate and does not cause a magnetic moment. But Holland does not consider the fact that an electron resting outside the nucleus would cause a strong electric dipole moment. This contradicts to parity considerations. Because the absolute square of an atomic wave function is even the odd dipole operator will lead to a zero expectation value of the electric dipole moment.

Strong dipole moments would have been found in many experiments when atoms traverse inhomogenous electric fields. Atom-atom scattering experiments would exhibit a strong dipole-dipole interaction which decreases much slower with increasing distance between the atoms as the usual van-der-Waals interaction. The long range forces between atoms also strongly influence the recombination rates used to simulate the ratio of atoms and molecules in the high atmosphere. It is impossible that such electric dipole moments would not have been observed. The postulate of resting electrons must be wrong.

Actually Bohmian mechanics predicts electric dipole moments not only for all s-states and magnetic sublevels with vanishing magnetic quantum number but for all other states of atomic hydrogen because if the magnetic quantum number is not zero Bohmian mechanics postulates that the associated electrons move on circles with constant latitude (constant polar angle θ). In general the centers of these orbits will not coincide with the location of the nucleus.

Discrepancies between Bohmian predictions and experimental results are also found for vibrating diatomic molecules. The wave functions describing molecular vibrations are purely real. That means the distance between the two nuclei is assumed to be constant. Individual molecules would generally not be in the equilibrium distance. On the other hand the energies of the vibrational levels depend strongly on the distance of the two nuclei. Vibrational spectra prove that all molecules have the same moment of inertia. Only for large vibrational states the anharmonicity of the innermolecular potential leads to a small but clearly observable increase of the internuclear distance. That means the nuclei cannot be at rest for diatomic molecules in vibrationally excited states.

3. Quantum and Bohmian results of double slit diffraction

Bohmian mechanics got a new impetus in 1979 when Philippidis, Dewdney and Hiley [12] published an ensemble of trajectories simulating the diffraction on double slits. Inspired by Jönsson's electron diffraction experiment [13] Philippidis, Dewdney and Hiley have chosen the distance of the slits d to be ten times larger than the slit width b. They claimed that the predicted intensity distribution of the diffraction pattern agrees with the experimental findings. Strange enough that this statement has never been rigorously checked especially because the intensity ratio of the three inner peaks seems to contradict experimental results.

Actually in the scientific literature one can find several ensembles of Bohmian trajectories which all can be traced back to the same original data set. In Holland's book [5] two ensembles of trajectories with different intensity distributions in the slits are shown. In one figure the density of trajectories in the slits is uniform, in the other figure it is Gaussian. Holland explicitely refers to the original publication of Philippidis et al [12] and emphasizes that in both cases the wave function is taken from the Gaussian intensity distribution case. In the following the Bohmian results will be compared with quantum mechanics for both alternatives.

If a particle passes through a slit its position is restricted to the slit width. The Fourier transformation of the associated wave packet reveals the momentum distribution of the particle, which is finally transformed into the angular width of the diffraction pattern. For uniform particle flux in front of a double slit aperture and small diffraction angles ϑ ($\ll 1$) quantum mechanics asymptotically predicts the angular dependence

$$I(\vartheta) = I_0 \cos^2(zd/b)\sin^2(z)/z^2 = I_0\cos^2(zd/b)\mathrm{sinc}^2(z) \tag{2}$$

with $z = \pi\vartheta b/\lambda$. Thereby I_0 and λ denote the maximum intensity reached at $\vartheta = 0$ and the wave length of the particle, respectively. For light this intensity distribution is already known from classical wave optics.

The periodic $\cos^2(zd/b)$ term describes the rapidly oscillating double slit interference. The envelope function $\mathrm{sinc}^2(z) = \sin^2(z)/z^2$, which is the Fourier transform of a rectangular intensity profile, denotes the diffraction pattern of a single slit. The zeroings limiting the central peak in the single slit diffraction pattern are located at $\vartheta_{zero} = \pm\lambda/b$. The period of the double slit oscillations is $\Delta\vartheta = \lambda/d$. That means the central peak of the single slit distribution incorporates $2d/b$ peaks of the double slit oscillations.

Formula (2) has been confirmed for photons, electrons, atoms, neutrons, and even for large molecules like fullerenes [14–17]. All experiments with large d/b-ratio exhibit several central peaks with almost equal intensity.

Figure 1 shows the quantum mechanically result (full line) and the intensity of the Bohmian calculation (black bars) determined from the trajectory ensemble of Philippidis, Dewdney and Hiley [12] with uniform density of the trajectories in the slits. Because the trajectories have apparently not fully reached the Fraunhofer limit it makes no sense to plot a continuous probability distribution. The Bohmian intensities are determined by counting the trajectories ending in the corresponding peaks. Therefore they are shown as block diagram in the figure.

Quantum mechanics predicts a central peak surmounting the neighboured peaks by only 3%. On the contrary the central peak in Bohmian mechanics is 3.25 times larger than the neighboured peaks. The intensity ratios of quantum and Bohmian distribution for second and third peak apparently also disagree by a factor 1.44. Whereas Bohmian mechanics predicts that 53 % of the total intensity is concentrated in the three central peaks the corresponding percentage of quantum mechanics is less than 35 %. Obviously the ensemble of trajectories of Philippidis, Dewdney and Hiley [12] does not provide the theoretically expected diffraction pattern.

If the intensity distribution in the slits is assumed to be proportional to the Gaussian function $f(x) = \exp(-x^2/2\sigma^2)/\sqrt{2\pi}\sigma$ with the standard variation σ, the $\mathrm{sinc}^2(z)$-function has to be replaced by the Fourier transform of the Gaussian function $f(x)$. The full width $a = 2\sigma$ of

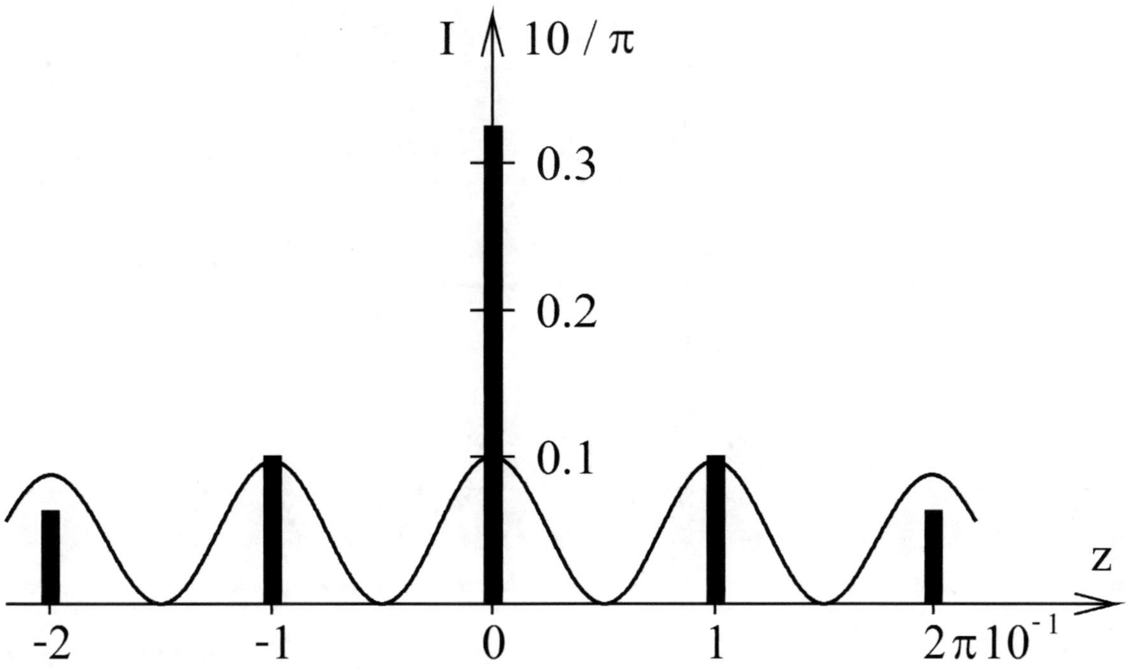

Figure 1. Quantum mechanically calculated intensity profile of a double slit diffraction for uniform intensity distribution inside the slits. The corresponding Bohmian intensities, which are indicated by vertical bars, are derived from the trajectories of Philippidis, Dewdney and Hiley [12]. Both data sets are independently normalized. Details on the experimental parameters and on the data analysis are given in the text.

the Gaussian distribution can be associated with the slit width b because the majority namely 68.3 % of the total intensity is concentrated in the interval $-\sigma = -a/2 < x < a/2 = \sigma$. On this condition the double slit intensity pattern has the form

$$I(\vartheta) = I_0 \cos^2(zd/a)e^{-z^2/2} \tag{3}$$

with $z = \pi\vartheta a/\lambda$.

It is rather difficult to reliably extract the interference pattern from the figures showing ensembles of Bohmian trajectories with Gaussian density distribution in the slits. The normal deviation of the Gaussian distribution is not explicitly given in the publications and can only be roughly determined from the figures. In addition the intensity contained in the outer wings of the normal distribution is not represented by trajectories.

Therefore the diffraction pattern for Gaussian density distribution in the slits is derived from the ensemble of trajectories with uniform intensity in the slits. For this purpose all trajectories in the two slit regions $-d/2 - a/2 < x < -d/2 + a/2$ and $d/2 - a/2 < x < d/2 + a/2$ have to be weighted by the local probability in the entrance slits. The intensities contained in the right wing of the left slit $(-d/2 + a/2 \leq x < 0)$ and the left wing of the right slit $(0 < x \leq d/2 - a/2)$ contribute to the central peak. The intensities contained in the opposite wings $(x \leq -d/2 - a/2$ and $d/2 + a/2 \leq x)$ merely have to be taken into account in the normalization procedure.

Figure 2 shows the probability densities of Bohmian and quantum mechanics for Gaussian intensity distributions in the entrance slits. Similar as in figure 1 the Bohmian intensity of the central peak is much too large in comparison with the quantum results. The intensity ratios of the second and third peak also disagree by a factor 1.38. Moreover, 51.8 % of the total Bohmian

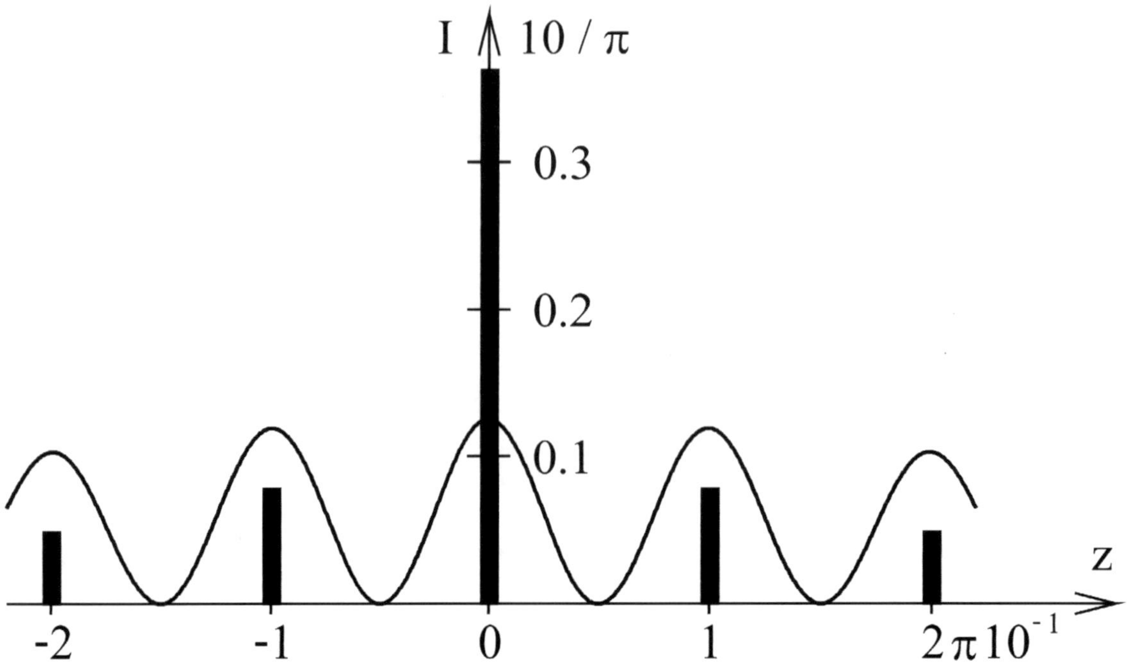

Figure 2. Quantum mechanically calculated intensity profile of a double slit diffraction for Gaussian intensity distribution inside the slits. The corresponding Bohmian intensities, which are indicated by vertical bars, are derived from the trajectories of Philippidis, Dewdney and Hiley [12]. Both data sets are independently normalized. Details on the experimental parameters and on the data analysis are given in the text.

intensity is concentrated in the three inner peaks, whereas for quantum mechanics only 36.2% of the total intensity is found in this region. All tests indicate that the often reproduced diagrams with Bohmian trajectories significantly deviate from quantum mechanics. Thus the statement of Philippidis, Dewdney and Hiley [12] 'If we assume a Gaussian distribution at the slits, then the intensity distribution in the Fraunhofer limit agrees with that expected from the usual considerations' must be wrong.

The peak intensities of the double slit diffraction pattern for quantum and Bohmian mechanics are summarized in table I. For both intensity distributions inside the slits the central peak is much too large whereas the flux contributing to the side peaks is too small.

One could argue that in the figures showing ensembles of trajectories the Fraunhofer limit is not yet reached because still some trajectories change from one bundle to neighboured ones. But actually the trajectories only change from outer bundles to inner bundles. That means the intensity in the central region of the diffraction pattern is increased on the cost of the intensity in the outer regions. Thus the discrepancy stated above will even become more pronounced if the trajectories would be traced to the Fraunhofer limit.

Another often expressed objection is that in the diagrams the intensity inside the slits might not be uniform. But in this case one has to admit even higher Fourier components in the momentum distribution which would lead to a broader envelope of the diffraction pattern thus again increasing the discrepancy.

Indeed the Gaussian distribution inside the slits provides the sharpest peak of the envelope function. If the distance of the slits d is ten times larger than the full width $a = 2\sigma$ of the Gaussian distribution the intensity ratio of the central peak to the sides peaks is 1.05. That

Theoretical approach	Quantum mechanics		Bohmian mechanics	
Intensity distribution in the entrance slits	Uniform	Gaussian	Uniform	Gaussian
Central peak	0.100	0.124	0.324	0.362
First side peak	0.097	0.119	0.100	0.078
Second side peak	0.088	0.103	0.063	0.049
Further side peaks	0.265	0.216	0.175	0.192

Table 1. Quantum and Bohmian intensities of the double slit diffraction pattern for uniform and Gaussian intensity distributions in the entrance slits. The distance of the slits is assessed to be ten times larger than the slit width. In case of Gaussian distribution the slit width is associated with twice the standard deviation of the normal distribution function. Bohmian intensities are taken from the trajectories of Philippidis, Dewdney and Hiley [12]. All four data sets are independently normalized. That means, the intensity of the central peak plus twice the intensities of the side peaks add up to unity.

means even for the flattest possible distribution inside the slits the intensities of the three inner peaks of the diffraction pattern are practically equal. One can never reach an intensity ratio of three by modifying the intensity profile in the slits. Therefore the extreme accumulation of Bohmian trajectories in the central peak is far from being realistic.

Actually in case of Gaussian distribution the product of the uncertainties of position x and momentum p_x in ths slits just reaches the minimum possible value compatible with Heisenberg's uncertainty relation. Thus the much narrower envelope function predicted by Philippidis, Dewdney and Hiley clearly contradicts to the uncertainty relation.

There is another question closely related with the possible violation of the uncertainty relation. In quantum mechanics it is uncontroversial that the question which slit has been traversed by a single particle cannot be answered without destroying the diffraction pattern. In contrast Bohmian mechanics allows to answer this question. Because of symmetry reasons particles cannot cross the middle plane. All particles found on the right side of the diffraction pattern must have gone through the right slit and vice versa.

4. The escort wave concept – a more realistic coexistence model

From the failure of Bohmian mechanics one cannot conclude that all theories assuming trajectories must be wrong. The recently presented escort wave concept [18, 19] removes the deficiencies of Bohmian mechanics discussed above. Whereas Bohmian mechanics is based on the pilot wave theory [20] proposed by de Broglie in 1927, the escort wave concept goes back to the phase wave model [21] introduced by de Broglie in 1924. Both theories assume that particle and wave are concrete physical objects coupled to each other. However, cause and effect are interchanged. Bohmian mechanics derives deterministic trajectories of particles from the presumed Schrödinger wave function. That means the Schrödinger equation is assumed to be the basic equation of physics. Classical physics is only a special case of quantum mechanics. The escort wave concept derives the Schrödinger wave function from classical trajectories. Quantum phenomena arise from the interplay of particles with the waves accompanying them.

Bohmian mechanics especially the constituting guiding equation (1) has already been presented in the introduction. The escort wave concept adopts the main aspects of the phase wave theory, namely that each particle is accompanied by a wave and that angular frequency ω and wave vector $\boldsymbol{k}$ of the wave are linked to energy E and momentum $\boldsymbol{p}$ of the particle via the well known relations $E = \hbar\omega$ and $\boldsymbol{p} = \hbar\boldsymbol{k}$.

In addition the escort wave concept assumes that particles are not only subjected to external

	Bohmian mechanics	Escort wave concept
Trajectories	deterministic far from classical	indeterministic almost classical
Derivation of trajectories from wave function	yes	no
Derivation of wave function from trajectories	no	yes
Meaning of standing wave	particle at rest	sum of two counterpropagating waves periodic movement of the particle
Born's rule	generally valid	only asymptotically valid

Table 2. Comparison of properties of Bohmian mechanics and escort wave concept

potentials but also to vacuum fluctuations. A particle thus can never be regarded as being isolated from the rest of the universe. That means a stochastic motion is superimposed to the classical kinematics. Even a free particle is not moving with constant velocity. Therefore the coherence length of the accompanying wave is finite. The interaction of particles with vacuum fluctuations is the deeper reason for the emergence of Heisenberg's uncertainty relation.

Bound particles in periodic motion are associated with counterpropagating waves. The two travelling waves interfere with each other thus forming standing waves. That means the occurence of knots in the wave function is a pure wave effect. There is no counterpart in the probability density of the associated particles. A particle is only coupled to the travelling wave moving in the same direction as itself. The knots of the standing wave have no influence on the motion of the particles. Thus in case of standing waves Born's rule is not valid. Only the probability of finding a particle at a given scattering angle is asymptotically given by the appropriately normalized intensity of the outgoing spherical wave. In fact, originally Born [22] has only postulated his rule in this restricted sense in order to allow the interpretation of scattering and diffraction experiments in terms of the Schrödinger wave function.

Actually, Bohmian mechanics and escort wave concept are quite different. In order to allow a sound comparison the main differences of the two coexistence models are compiled in table 2.

In section 2 of this publication several problems have been discussed, where the strict application of Bohmian mechanics leads to conflicts with experimental observations. For example Bohmian mechanics erraneously predicts that all states of atomic hydrogen generally have strong permanent dipole moments. The escort wave concept assumes that an atomic electron is continuously moving on quasi-periodic orbits. In combination with the deviation from periodicity the fast movement of the electron leads on average to the formation of an extended charge cloud.

For the ground state of atomic hydrogen the electron motion is purely stochastic thus leading to a spherically symmetric charge cloud. With increasing principle quantum number the regular motion becomes more and more important. For highly excited s-states the elliptical trajectories almost degenerate to straight lines in radial direction. The stochastic motion prevents that the electrons hit the nucleus and causes a random orientation of the orbits. Because the centroid of the resulting spherically symmetric charge cloud just coincides with the nucleus the electric dipole moment is zero.

If one applies a homogenous external magnetic field magnetic sublevels are no longer degenerate. The wave functions of excited states with zero magnetic quantum number are purely real. The momentary magnetic moment produced by the movement of the electron on its elliptical orbit is always perpendicular to the external field. On the influence of the fluctuating waves the orientation of the momentary magnetic moment with respect to the azimuthal angle is at random. Thus the resulting magnetic moment is zero. On average the charge cloud is rotationally symmetric with respect to the direction of the magnetic field. As for s-states the centroid of the charge cloud coincides with the nucleus.

Obviously the Bohmian postulate that particles associated with purely real wave function are permanently at rest leads to consequences which are in contradiction to experimental observations. The escort wave concept avoids this problem because it superimposes stochastic motions to the regular movement and associates standing waves with moving particles.

The same arguments remove the problems with vibrating molecules. In the escort wave concept the particles perform harmonic oscillations. Therefore in time average the molecule is in the equilibrium distance. The problem with the nuclei at rest disappears.

If one assumes that Bohmian mechanics fully agrees with quantum mechanics the wrong prediction of Philippidis, Dewdney and Hiley [12] about the interference pattern for the double slit diffraction must be caused by some error in the calculation. It seems to be quite difficult to reproduce the correct envelope function. Because the trajectories cannot cross the symmetry plane the ensemble of trajectories passing through one of the two slits must be totally different from the ensemble of trajectories describing the diffraction on a single slit.

The escort wave concept does not have to face such difficulties because particle trajectories can easily cross the symmetry axis. Particles going through one of the two slits will reach both sides of the diffraction pattern. Thus the envelope of the diffraction pattern for two slits is given by the interference structure of a single slit. The double slit interference only causes a redistribution of the local intensity.

There are good reasons to assume that quantum phenomena are based on the interplay of particles and waves accompanying them. The escort wave concept is a realistic alternative to Bohmian mechanics. Both models agree with respect to the local flux distribution. In view of numerous inconsistencies of Bohmian mechanics the escort wave concept looks more promising to find a logical explanation for the emergence of quantum phenomena.

References

[1] Einstein A, Podolsky B and Rosen N 1935 *Phys. Rev.* **47** 777-780
[2] Bell J S 1964 *Physics* **1** 195-200
[3] Aspect A, Grangier P and Roger G 1982 *Phys. Rev. Lett.* **49** 91-94
[4] D. Bohm, Phys. Rev. **85**, 166-179 (1952), Phys. Rev. **85**, 180-193 (1952)
[5] Holland P R 1993 *The Quantum Theory of Motion* (Cambridge, UK: Cambridge Univ. Press)
[6] Englert B G, Scully M O, Süssmann G and Walther H 1992 *Z. Naturforschung* **47**a 1175-1186
[7] Deotto E and Ghirardi GC 1998 *Found. Phys.* **28** 1-30
[8] Ghose P 2003 *Preprint* arxiv:quant-ph/0103126
[9] Golshani M and Akhavan O 2002 *Preprint* arxiv:quant-ph. 0103101
[10] Matzkin A 2007 *Preprint* arxiv:quant-ph/0609172
[11] Sanz A S and Miret-Artés S 2012 *A Trajectory Description of Quantum Processes I* (Heidelberg: Springer)
[12] Philippidis C, Dewdney C and Hiley B J 1979 *Nuovo Cimento* B **52** 15-28
[13] Jönsson C 1961 *Z. Phys.* **161** 454-474; English translation: *do.* 1974 *Am. J. Phys.* **42** 4-11
[14] Tonomura A, Endo J, Matsuda T, Kawasaki T and Ezawa H 1961 *Am. J. Phys.* **57** 117-120
[15] Shimizu F, Shimizu K and Takuma H 1992 *Phys. Rev.* A **46** R17-R20
[16] Zeilinger A, Göhler R, Shull C G, Treimer W and Mampe W 1988 *Rev. Mod. Phys.* **60** 1067-1073
[17] Arndt M, Nairz O, Voss-Andreae J, Keller C, van der Zouw G and Zeilinger A 1999 *Nature* **401** 680-682
[18] Jung K 2009 *Ann. Fond. Louis de Broglie* **34** 143-164
[19] Jung K 2011 *J. Phys.: Conf. Ser.* **306** 012071
[20] de Broglie L 1927 *J. Physique Radium* **8** 225-241
[21] de Broglie L 1925 *Annales de Physique* **3** 22-128
[22] Born M 1926 *Z. Phys.* A **37** 863-867

Entanglement and Complexity of Measurement

Constantin V Usenko

Theoretical physics department, Physics faculty, Kyiv Shevchenko university, Kyiv 01680, Ukraine

E-mail: usenko@univ.kiev.ua

Abstract. Notion about complexity of composite systems is applied to the problem of measurement of quantum states. An expression for the determination of the value of complexity of measurement is given. Its practical meaning is illustrated in the example how this value differs when measuring complexity of authorized and non-authorized receiving of information transferred by quantum channels in protocol of direct coding.

Unconditional security of information in direct coding protocol with information masking by uniformly distributed state of carrier is shown.

1. Introduction

In the algorithm theory, the concept of computational complexity [1] that characterizes the number of typical actions that form the computational algorithm is actively used.

This value makes it possible to analyze cryptographic stability of coding algorithms with asymmetric key that are actively used today and to evaluate the key length that provides unreasonably long time for unauthorized decoding with existing computational resources.

Adaptation of the notion of complexity to physical problems, proposed recently [2], is restricted by classical mechanics. Better correspondence with the notion of complexity one finds in the case of quantum systems, with the simplest ones, like spin degree of freedom of electron, or polarization degree of freedom of photon, among those: both are characterized by two-dimensional Hilbert space and are combined by common name "qubit". Complexity of qubit leads to impossibility of state cloning [3]: one measurement of an observable is not enough for reconstruction of state. Values needed for reconstruction can be measured as a result of long enough series of measurements only. Length of series can be used as a measure of complexity of a quantum system.

Hereinafter expression for determination of the value of complexity of measurement is proposed, and its values that characterize different processes of measurement of states of quantum systems are considered.

Measurement of state of quantum system is a component part of the problem on reconstruction of quantum state [4] that is actively studied because of practical realization of information transfer by quantum carriers.

The present article has the following structure:

In the section "Measurement of state: observables" information about observables measurement of which makes it possible to reconstruct a quantum state, pure or mixed, completely, is given. In particular, entangled states are considered.

In the section"Process of measurement of state" measurement of a state is represented as a sequence of simple events of measurement that are characterized by preparation and detection of pure states that in aggregate represent the state under consideration.

In the section "Complexity of measurement" expression for determination of this value is given, with evaluation of complexity for measurement of probability of some event. Main result that is in evaluation of complexity of measurement for the problem of reconstruction of a quantum state is formulated as well. Explicit expression for dependence of complexity of measurement on dimension of state space is obtained, and it is shown that at non-demolition measurements complexity of measurement tends to zero.

In the section "Direct coding", as an illustration of effectiveness of evaluation of the complexity of measurement, difference between the complexities of the Eve's and Bob's measuring tasks in the protocol of direct coding with feedback is considered. It is shown that masking of the message transferred by a sequence with an equilibrium density matrix makes the problem of intercept of information unsolvable.

2. Measurement of state: observables

Quantum systems with N-dimensional Hilbert state are considered. The state is described by density matrix $\rho = \sum \rho_{n,n'} |n\rangle \langle n'|$, components of which are determined by $N^2 - 1$ real numbers. Reconstruction of state can be realized only in the case if the results of measurements make it possible to calculate the values of all those numbers.

2.1. Measuring basis of algebra of observables

Measurement of an observable is mathematically modeled by a system of projectors D_n, giving resolution of identity $\sum D_n = \hat{I}$. Operator of observable has representation by a weighted sum of projectors $\hat{O} = \sum O_n D_n$. Each of projectors projects a state into eigenstate $|n\rangle$ of the observable. As the result of measurements, one obtains $N-1$ independent real numbers – probabilities $p_n = \langle n |\rho| n\rangle$ of registration of eigenstates of observable. To calculate the values of all the components of density matrix, one has to measure $N+1$ different incompatible observables. Set of observables enough for reconstruction of state can be constructed [5] by the method of ladder operators.

Let $\{|n\rangle, n = 1 \dots N\}$ is an orthonormal basis. Let us define the ladder operators

$$J_- = \sum_{n=1}^{N} \sqrt{n(N-n+1)}\, |n-1\rangle \langle n|, \quad J_+ = J_-^+, \quad J_3 = \frac{1}{2}[J_+ J_-] \tag{1}$$

and the measuring basis of the algebra of observables

$$B_0 = J_3, \quad B_m = \frac{e^{i\pi(m-1)/2N}}{2} J_- + \frac{e^{-i\pi(m-1)/2N}}{2} J_+ : \quad m = 1 \dots N. \tag{2}$$

The results of measurement of averages and moments up to $N-1$ make it possible to calculate all the components of density matrix as a solution of a system of linear equations:

$$R_{m,p} \equiv \langle \rho B_m^p \rangle = \sum_{n,n'=1}^{N} \rho_{n,n'} \langle n' |B_m^p| n\rangle, \quad m = 0 \dots N, \quad p = 1 \dots N-1. \tag{3}$$

If calculated density matrix is not degenerate, i.e all its eigenvalues p_n are different, one can calculate as well the basis (unique) in which the density matrix is diagonal. In the case of degenerate density matrix the basis in which the density matrix is diagonal is determined up to the group of orthogonal transformations of degeneration subspace. In the specific case of equilibrium state $\rho = \frac{1}{N}\hat{I}$ the density matrix is diagonal in any basis.

2.2. Measuring basis of a composite system

Each subsystem of a composite system is characterized by algebra of local observables $\mathcal{A}_k$ operating in the space of local states $\mathcal{H}_k$.

Sets of local operators $\left\{L_x^{\{k\}},\ x = 3, +, -\right\}$, similar to (1), generate measuring bases of local observables $\left\{B_{n_k}^{\{k\}},\ n_k = 0, \ldots, N_k\right\}$, the sum of operators

$$J_x = \sum_{\forall k} L_x^{\{k\}} \tag{4}$$

generates the measuring basis $\{B_n,\ n = 0, \ldots, N_1 \cdot \ldots \cdot N_k\}$ of the algebra of observables of composite system.

Arbitrary state of a composite system belongs to one of three types of states:

- States of independent subsystems: density matrix of a composite state is equal to direct product of density matrices of subsystems $\rho = \rho_A \otimes \rho_B$;
- States of a mix of independent subsystems: density matrix of a composite state is equal to weighted mix of direct products of density matrices of subsystems $\rho = \sum p_k \rho_A^{\{k\}} \otimes \rho_B^{\{k\}}$;
- All the other states are the states of entangled subsystems (entangled states).

Measurement of the average values and moments of the measuring basis of local observables is enough for calculation of the density matrix of the simplest type of composite system states – states of independent subsystems only. More complicated state of a mix of independent subsystems can be calculated if measurement of local observables is completed by measurement of correlation of non-demolition local observables.

Reconstruction of entangled states is possible as a result of measurement of correlation of over-classical observables, in addition to solving the problem on reconstruction of the subsystem states. This correlation exceeds the correlation possible by probability theory (Bell's inequalities).

3. Process of measurement of a state

Statistical observable is measured in sequence of events of measurement (trials). For each event a source constructed in respective way produces the measured system in a given state (prepares the state). The state prepared in each event is a pure state. Prepared sequence of states

$$\mathcal{S}_n = \{|n_1)(n_1|, \ldots, |n_k)(n_k|, \ldots, |n_K)(n_K|\} \tag{5}$$

is characterized by length K, by the set of states (alphabet) $\mathcal{S} = \{S_n = |n)(n|, n = 1, \ldots, N\}$ and the set of initial repetition rates $\nu_n = \frac{K_n}{K}$ of each state.

Density matrix of prepared sequence is a weighted sum of density matrices of prepared states:

$$\rho = \frac{1}{K} \sum_{k=1}^{K} |n_k)(n_k| = \sum_{n=1}^{N} \nu_n |n)(n|. \tag{6}$$

Sequences that have same repetition rates and differ by the order of states only, are characterized by same density matrices, and thus are physically indistinguishable. For each given set of repetition rates there exist

$$K_I = \frac{K!}{\prod_{n=1}^{N} K_n!} \tag{7}$$

of different sequences that are characterized by the same density matrix, and thus the same average values of all the observables.

Sequences differing only by the order of states can be used for transfer of K_I different messages, and include amount of information equal to entropy

$$I = \log_2 K_I = -\sum_{n=1}^{N} \nu_n \log_2 \nu_n = -\sum_{n=1}^{N} \rho \log_2 \rho = S. \tag{8}$$

Sequences of states can differ only by information, if all the occupation numbers K_n are the same; only by physical values of observables, if sequences of the numbers of states are the same and alphabets are different; or by both of those. Physical and information components of the process of measurement are complementary.

3.1. Simple measurement, reduction of state

In each trial the state is registered by detector that consists of a set of counters that are mathematically represented by a set of projectors $\mathcal{D} = \{D_m = |m\rangle \langle m|, m = 1, \ldots, N\}$. Probability of registration of a state $|n)$ by a counter $|m\rangle$ is given by expression:

$$P(m|n) = \langle m |\rho_n| m\rangle = |\langle m \mid n)|^2 . \tag{9}$$

In each event of measurement density matrix is reduced to a pure state density matrix of which is equal to the projector of respective counter:

$$\rho_k^{\{reduced\}} = |m\rangle \langle m| . \tag{10}$$

As the result of measurement one gets a sequence of the counter numbers

$$\mathcal{R} = \{m_1, \ldots, m_k, \ldots, m_K\}, \tag{11}$$

that can be described by repetition rates $\mu_m = \frac{K_m}{K}$. The most probable values of repetition rates of counter numbers are the values of conditional probabilities weighted by repetition rates of the numbers of prepared states:

$$\mu_m \approx \langle p_m\rangle = \sum_{n=1}^{N} P(m|n)\, \nu_n. \tag{12}$$

Reduced density matrix is a weighted sum of density matrices of reduced pure states:

$$\rho^{\{reduced\}} = \frac{1}{K} \sum_{k=1}^{K} |m_k\rangle \langle m_k| = \sum_{m=1}^{N} \mu_m |m\rangle \langle m| . \tag{13}$$

Reduced density matrix differs from the initial one, sequences of the numbers of states differ as well. This difference can be revealed by comparison of the initial and resultant sequences.

3.2. Non-demolition measurements

A measurement is a non-demolition one, if the set $\mathcal{D}$ of projectors that represent the detector is equal to the set $\mathcal{S}$ of density matrices of states prepared by the detector. In this case probability $P(m|n)$ is equal to unity, if the numbers of states are equal, and to zero, if not. So, the resulting sequence of numbers of counters is equal to the initial sequence of the numbers of prepared states

$$\mathcal{R}_{nd} = \{m_1 = n_1, \ldots, m_k = n_k, \ldots, m_K = n_K\}, \tag{14}$$

the message prepared is received by the detector without distortions, and reduced density matrix is equal to the initial one.

In a more general case the measurement is a non-demolition one, if the reduced density matrix of the system does not differ from the initial matrix. This is possible, if the projectors of counters D_n are along with that the projectors to the eigenvectors of the density matrix $\rho\,|n\rangle = p_n\,|n\rangle$.

Special attention is to be paid to non-demolition measurements for the system with a degenerate density matrix (6). In this case there exists a set of not commuting non-demolition observables, the series of measurements for each of those gives the same reduced density matrix. Along with that, in each specific event of measurement reduced state can differ from the initial one. Such weakly non-demolition measurements make it possible to reconstruct the density matrix on average, though lead to a sequence of results (11) differing from the initial sequence (5).

Example of a weakly non-demolition measurement. The source prepares the qubit states in the basis $\mathcal{S} = \{|0\rangle\langle 0|, |1\rangle\langle 1|\}$.

The detector consists of two counters represented in the basis of the source by matrices:

$$\mathcal{D}_{nd} = \left\{ |0)\,(0| = \begin{pmatrix} \frac{1}{2} & -\frac{1}{2} \\ -\frac{1}{2} & \frac{1}{2} \end{pmatrix}, |1)\,(1| = \begin{pmatrix} \frac{1}{2} & \frac{1}{2} \\ \frac{1}{2} & \frac{1}{2} \end{pmatrix} \right\}.$$

Probability $P_{m|n}$ of operation of the detector m under condition that the state n is prepared is equal to $\frac{1}{2}$.

Table 1. Density matrices for weakly non-demolition measurement.

S_n	0	0	1	1	
ρ	$\begin{pmatrix} 1 & 0 \\ 0 & 0 \end{pmatrix}$	$\begin{pmatrix} 1 & 0 \\ 0 & 0 \end{pmatrix}$	$\begin{pmatrix} \frac{2}{3} & 0 \\ 0 & \frac{1}{3} \end{pmatrix}$	$\begin{pmatrix} \frac{1}{2} & 0 \\ 0 & \frac{1}{2} \end{pmatrix}$	
$\mathcal{R}_n$	0	1	0	1	
$\rho^{\{res\}}$	$\begin{pmatrix} 1 & 0 \\ 0 & 0 \end{pmatrix}$	$\begin{pmatrix} \frac{1}{2} & 0 \\ 0 & \frac{1}{2} \end{pmatrix}$	$\begin{pmatrix} \frac{2}{3} & 0 \\ 0 & \frac{1}{3} \end{pmatrix}$	$\begin{pmatrix} \frac{1}{2} & 0 \\ 0 & \frac{1}{2} \end{pmatrix}$	D-basis
$\rho^{\{res\}}$	$\begin{pmatrix} \frac{1}{2} & -\frac{1}{2} \\ -\frac{1}{2} & \frac{1}{2} \end{pmatrix}$	$\begin{pmatrix} \frac{1}{2} & 0 \\ 0 & \frac{1}{2} \end{pmatrix}$	$\begin{pmatrix} \frac{1}{2} & -\frac{1}{6} \\ -\frac{1}{6} & \frac{1}{2} \end{pmatrix}$	$\begin{pmatrix} \frac{1}{2} & 0 \\ 0 & \frac{1}{2} \end{pmatrix}$	S-basis

An example of a series of measurements is given in table (1). The upper row includes the numbers of the prepared states, in the second row current values of the initial density matrix are given. The third row includes the numbers of the states registered by the detector, in the fourth one the current values of the reduced density matrix in the detector basis are given, in the fifth row – the values for the matrix in the source basis. From the table it follows that the reduced density matrix of intermediate states in the source basis can have non-diagonal components (the first and the third columns). This results from weak reduction of state. Degeneration of the initial density matrix (the fourth column) is such that any measurement is a weakly non-demolition one, and the reduced density matrix is equal to the initial matrix up to random deviation.

3.3. Complex measurement

Complex measurement is carried out at measurement of incompatible observables. In the process of complex measurement a set of simple detectors $\{\mathcal{D}_r\}$ is applied, each of detectors represents an observable

$$\left\{\mathcal{D}_r = \{D_{m:r} = |m:r\rangle\langle m:r|, m=1,\ldots,N\}, \sum_{m=1}^{N} D_{m:r} = \hat{I}\right\}.$$

The observables represented by different detectors are incompatible, if at least some of projectors of one detector to not commute with the projectors of another one. Application of a measuring basis (2) as a set of detectors makes it possible to reconstruct all the components of density matrix by the results of the measurement.

In each measurement event (event number k) the experimenter performing the measurement chooses one of simple detectors (one of incompatible observables); let the number of this detector is r_k. The measured system is registered by a counter with number m_k, so the result of the sequence of measurements is a sequence of pairs that consists of the number of the counter m and the number of the detector r:

$$\mathcal{R} = \{[m_1 : r_1], \ldots, [m_k : r_k], \ldots, [m_K : r_K]\}. \tag{15}$$

Number of the counter is determined with probability

$$p_{m|r,k} = \langle m : r_k || n_k \rangle \langle n_k || m : r_k \rangle$$

that depends on the number of simple detector r_k chosen by experimenter and the number of state n_k prepared by the source.

3.4. Measurement of the density matrix

In the process of measurement of the density matrix the average values and moments (3) of all the observables of the measuring basis are to be obtained with enough accuracy. Number of the events of measurement is to be enough for determination of distribution of probabilities of the values of each observable of the measuring basis (2). At that the measurements of N observables from $N+1$ basic observables are not non-demolition ones, so the sequence of the results (15) differs from the sequence of the source (5). Even at occasional coincidence of eigenvectors of one of the observables with the source basis, only $\frac{1}{N+1}$ part of the resultant sequence can coincide with the sequence of the source.

4. Complexity of measurement

Complexity of measurement is characterized by the number of measurements needed for reconstruction of statistical characteristics of the measured value with given accuracy. Calculation of the components of density matrix by the results of measurement of probability distributions is accompanied by accumulation of ratio error that characterizes the number of significant bits in binary notation of the results.

4.1. Definition of the value

We define the value of complexity of measurement as a logarithm of the number of events of measurement $M(\epsilon)$ needed to obtain the result with given ratio error $s = \frac{\sigma_X}{X}$.

$$C_M(s) = \log_2 M(s). \tag{16}$$

Logarithmic scale is convenient for evaluation of complexity of measurement of composite systems: in general case the complexities of measurement are added, in specific cases (for instant, at cloning, when one of sub-systems is a true copy of another one) complexity of measurement can be substantially smaller.

Complexity of measurement is easily calculated in the case of non-demolition measurement for a pure state. In such a case the density matrix of the measured state $|\psi\rangle\langle\psi|$ coincides with the projector into one of eigenvectors $D_N = |m\rangle\langle m|$ of the operator of observable $O = \sum O_n |n\rangle\langle n|$. The result of each event of measurement $\langle O\rangle = \mathrm{Tr}\, O |\psi\rangle\langle\psi|$ is the same value O_m, this result is found in the first event of measurement, and all the following events can not decrease the error. So, at non-demolition measurement one event of measurement is enough, and complexity of measurement is equal to zero.

4.2. Complexity of measurement of probability

Unbiased evaluation of probability of registration of a state $|n)(n|$ is the repetition rate $p_n \approx \nu_n = \frac{K_n}{K}$. Probability K_n of repetitions in a series of K events of measurement is described by Bernoulli law

$$P(K_n) = C_K^{K_n} p_n^{K_n} (1-p_n)^{(K-K_n)}.$$

Error of evaluation of probability by repetition rate is calculated by means of expression for dispersion $V_n = \frac{p_n(1-p_n)}{K}$, so the ratio error is

$$\epsilon^2 = \frac{V_n}{p_n^2} \approx \frac{(1-p_n)}{p_n K}, \tag{17}$$

and the number of events of measurement needed to provide the error s ($\epsilon \leq s$) is given by the expression $K \geq \frac{(1-p_n)}{p_n s^2}$ that leads to the expression for complexity of measurement of probability of a given state:

$$C_M = \log_2\left(\frac{1}{p_n} - 1\right) - 2\log_2 s. \tag{18}$$

In Fig. 1 the graphs of dependences of the complexity of measurement of probability on the expected value of probability for several values of ratio error are given. The value of complexity of measurement 10 corresponds to $2^{10} = 1024$ needed events of measurement.

4.3. Complexity of measurement of probability distribution

The events of registration of a state from a set of mutually orthogonal states are complementary, thus the complexity of measurement of probability distribution is less than the sum of complexities of measurement of each probability. All the events that do not contribute to registration of one state, contribute to registration of some other state, and provide a positive result for measurement of probability of that other state. Uncertainties of probability of registration of each state are given by the same expression (17).

Thus, the expression (18) gives the value of complexity of measurement of probability distribution, if as p_n the minimal non-zero value of repetition rate is used:

$$C_M = \log_2\left(\frac{1}{p_{min}} - 1\right) - 2\log_2 s. \tag{19}$$

Numerical evaluation of the value of complexity of measurement is achieved for uniform distribution of probabilities by N possible outcomes, and gives

$$C_{M,u} = \log_2(N-1) - 2\log_2 s. \tag{20}$$

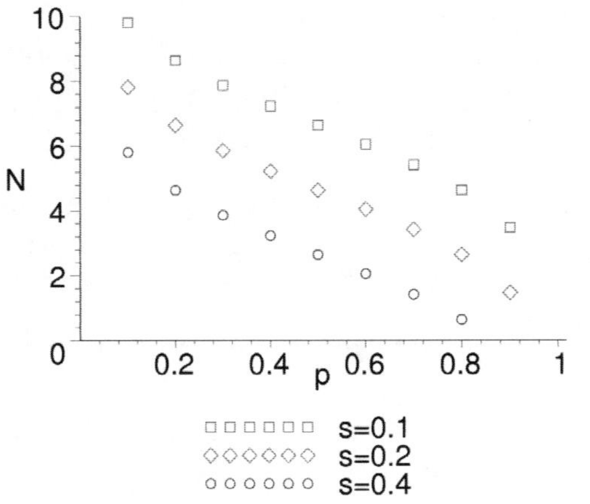

Figure 1. Dependence of complexity of measurement on expected probability p for several values of ratio error s.

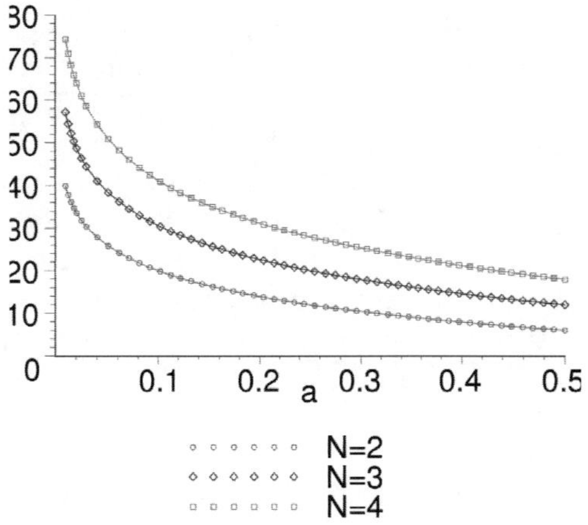

Figure 2. Dependence of complexity of measurement on ratio error for three values of dimension of state space.

4.4. Complexity of measurement of density matrix

The measurement of density matrix is realized (see 2.1,3.4) through measurement of distributions of repetition rates $p_{m|n}$ for not less than $N+1$ incompatible detectors. Each of incompatible detectors is used in a separate series of measurements, thus the number of measurements needed for determination of density matrix is given by the sum of the numbers of measurements required for each detector, and complexity of measurement is:

$$C_M\left(N,s\right) = \log_2\left(\sum_{n=0}^{N}\left(\frac{1}{p_{\min|n}} - 1\right)\right) - 2\log_2 s. \qquad (21)$$

Numerical evaluation of the lower boundary of complexity of measurement can be found under supposition on quasi-uniform distribution for which all the probabilities being measured are of one order $p_{m|n} \approx 1/N$:

$$C_M\left(N,s\right) \approx \log_2\left(N^2 - 1\right) - 2\log_2 s. \qquad (22)$$

The graph of dependence of complexity of measurement on dimension of state space for three values of ratio error is given in Fig. 2. One shall note that complexity of measurement is above 17 for common values of state space dimension ($N = 8$) and the ratio error ($\epsilon = 0.02$). This value corresponds to needed number of measurements 160000, thus practical realization of experimental determination of density matrix of an arbitrary state is not an easy problem.

Complexity of non-demolition measurements is quite different from the complexity of measurement considered above.

At exactly non-demolition measurement of density matrix $N-1$ probabilities p_m are to be measured, though in each separate event conditional probability equal to zero or unity is measured, and the resultant sequence $\mathcal{R}$ is equal to the initial one $\mathcal{S}$. Thus complexity of an exactly non-demolition measurement of a pure state is zero, and complexity of non-demolition measurement of a mix is given by the same expression (19), this determines the complexity of measurement of classical probability distribution.

Measurement of density matrix of an entangled state is realized through measurement of local observables. For entangled states there is no non-demolition combination of local observables, thus the measurement is complete, and its complexity of measurement is given by (21).

5. Direct coding protocol

In this section evaluation of the value of complexity of measurement is used for a practical problem – analysis of differences in capabilities to obtain information from measurements carried out by an authorized (Bob) and unauthorized (Eve) participants of the process of information transfer by means of the direct coding protocol.

The direct coding protocol [6] is in use of the same sets of states $\mathcal{A} = \{|1)(1|, \ldots, |n)(n|, \ldots, |N)(N|\}$ in the Alice's source and the Bob's detector. Alice produces entangled states, and Bob registers covariation of special pair of local observables. Thus, complexity of measurement at the Bob's side is zero.

The backward channel from Bob to Alice is used for compensation of errors.

5.1. Data transfer

Transferred information is masked by the state of carrier with the same repetition rates of all the symbols. For masking one uses the following procedure:

- Alice sends the next symbol of message;
- Bob sends each obtained symbol by backward channel with the same security;
- Alice resends erroneous symbol until returned symbol is correct, or sends the next symbol of the message;
- Having come by the repetition number of a symbol to a given number K, Alice accomplishes the sentence with all the other symbols up to the same number of repetitions K, and so the resultant density matrix of the sequence with length of KN symbols is equivalent to density matrix of white noise $\rho = \frac{1}{N}\hat{I}$.

Measurement with the set of detectors $\mathcal{D} = \{|1)(1|, \ldots, |n)(n|, \ldots, |N)(N|\}$ reproduces in each event of data transfer the symbol, generated by Alice.

5.2. Eavesdropping

It is supposed that the person who makes attempts of eavesdropping is restricted in capabilities by physical laws only, though is not restricted by lack of possibilities of technical realization. There exist several common strategies of eavesdropping, here we list the most typical ones:

- Non-demolition measurements. If Eve knows the set of states used for coding, she can measure a non-demolition observable. Absence of reduction of states prevents reveal of intervention to the process of data transfer.
- Intercept-resent attack. Eve completely intercepts the carrier, measures its state and resends a copy of the carrier in same state to Bob.
- Partitioning. Eve segregates a part of the beam (for instance with a semi-transparent mirror) and measures the state of segregated part. By the results of measurements Eve makes efforts for calculation of the set of carrier states used by Alice. In case of success Eve can adapt her detector like Bob's detector and copy information.

Use of the backward channel by Alice and Bob means that Eve can attack the backward channel as well.

5.3. Making eavesdropping senseless

- Eavesdropping with non-demolition measurements is made senseless by use of entangled states of the carrier. Measurement of local observables carried out at eavesdropping is accompanied by reduction of states of the carrier, and can be revealed by Bob.

- Intercept-resent attack. Control of time of carrier propagation from Alice to Bob makes this strategy inconsistent.
- Partitioning. Eve carries out a series of measurements of incompatible observables and calculates the density matrix. Complexity of measurement for this problem is given by expression (22) and is large enough. If the density matrix is degenerate (its eigenvalues are equal within inaccuracy of measurements), it is not possible to calculate the states used by Alice and Bob as alphabet. Till Eve does not know the alphabet, the results of her measurements do not represent transferred information.

6. Conclusions

The new physical value – complexity of measurement, proposed here, makes it possible to prove unconditional security of direct coding protocol with masking information in uniform mixed state of carrier. Complexity of measurement of a quantum state depends on dimension of Hilbert space, and on entanglement of states of the sub-systems of composite quantum system.

References

[1] Dijkstra E W 1976 *A Discipline of Programming* Prentice-Hall Series in Automatic Computation (USA: Prentice-Hall)
[2] Beggs E, Costa J F, Loff B and Tucker J 2008 *Theory and Applications of Models of Computation, Lecture Notes in Computer Science* **4978** 20
[3] Wootters W K and Zurek W H 1982 *Nature* **299** 802
[4] Buzek V, Drobny G, Derka R, Adam G and Wiedemann H 1999 *Chaos, Solitons & Fractals* **10** 981
[5] Usenko C V2011 *Optics and Spectroscopy* **111** 678
[6] Usenko C V 2012 *Journal of Russian Laser Research* **33** 283

Quantum driving of a two level system: quantum speed limit and superadiabatic protocols – an experimental investigation

N Malossi[a], M G Bason[b,1], M Viteau[b,2], E Arimondo[a,b,c], D Ciampini[a,b,c], R Mannella[a] and O Morsch[b]

[a] Università di Pisa Dipartimento di Fisica 'E. Fermi', Lgo Pontecorvo 3, I-56127 Pisa, Italy
[b] INO-CNR, Dipartimento di Fisica 'E. Fermi', Lgo Pontecorvo 3, I-56127 Pisa, Italy
[c] CNISM UdR Pisa, Dipartimento di Fisica 'E. Fermi', Lgo Pontecorvo 3, I-56127 Pisa, Italy

E-mail: malossi@df.unipi.it

Abstract. A fundamental requirement in quantum information processing and in many other areas of science is the capability of precisely controlling a quantum system by preparing a quantum state with the highest fidelity and/or in the fastest possible way. Here we present an experimental investigation of a two level system, characterized by a time-dependent Landau-Zener Hamiltonian, aiming to test general and optimal high-fidelity control protocols. The experiment is based on a Bose-Einstein condensate (BEC) loaded into an optical lattice, then accelerated, which provides a high degree of control over the experimental parameters. We implement generalized Landau-Zener sweeps, comparing them with the well-known linear Landau-Zener sweep. We drive the system from an initial state to a final state with fidelity close to unity in the shortest possible time (quantum brachistochrone), thus reaching the ultimate speed limit imposed by quantum mechanics. On the opposite extreme of the quantum control spectrum, the aim is not to minimize the total transition time but to maximize the adiabaticity during the time-evolution, the system being constrained to the adiabatic ground state at any time. We implement such transitionless superadiabatic protocols by an appropriate transformation of the Hamiltonian parameters. This transformation is general and independent of the physical system.

1. Introduction

Controlling the dynamics of quantum systems is a key issue in many fields of science [1] (for instance quantum computation [2], metrology [3], atomic and molecular physics [4]) and could be a valuable instrument for testing fundamental issues in quantum mechanics.

A perfect tool for studying quantum control is the "simplest not-simple quantum system" [5], i.e. the evolution of a two level system governed by a time dependent Landau-Zener Hamiltonian [6, 7, 8]

$$\mathcal{H} = \Gamma(t)\sigma_z + \omega(t)\sigma_x, \tag{1}$$

written in the basis of the two diabatic states $|0\rangle$, $|1\rangle$ with the Pauli matrices $\sigma_{x,z}$ acting on the two-dimensional basis. Here $\Gamma(t)$ and $\omega(t)$ are respectively the energy and the coupling term

[1] present address: Department of Physics and Astronomy, Aarhus University Aarhus, Denmark
[2] present address: Orsay Physics, 95 av. des Monts Aurliens, ZAC Saint Charles, 13710 Fuveau - France

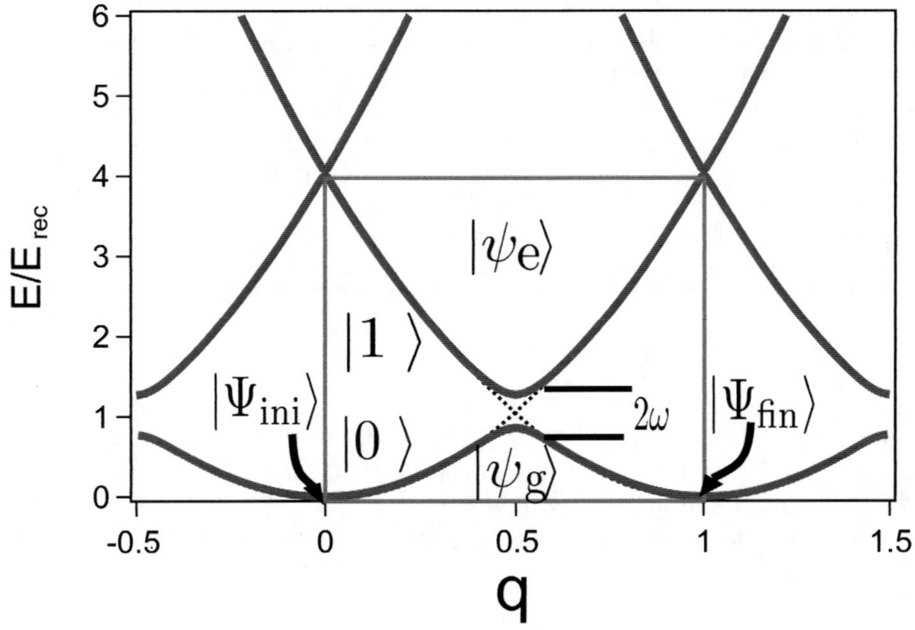

Figure 1. Band structure of an effective two level system. The states $|0\rangle$ and $|1\rangle$ are the diabatic basis (dashed line), $|\psi_{g,e}\rangle$ are the adiabatic basis states (solid line) and $\omega = V_0/4$ is the energy gap between the bands where V_0 is the lattice depth. $|\Psi_{ini}\rangle$ and $|\Psi_{fin}\rangle$ represent the initial state and the final state of a protocol, respectively.

between these states. It is possible to identify an adiabatic basis $|\psi_{g,e}(t)\rangle$ which instantaneously diagonalizes the Hamiltonian at any time t.

The present paper reports the basis of the experimental realization of this quantum system together with its dynamics and control protocols in specific regimes, for a detailed analysis see [9, 10].

The central idea is to drive an initial state $|\psi_{in}\rangle$ for a sweep time T through an avoided crossing shown in Fig. 1, to a chosen final state $|\Psi_{fin}\rangle$ as close as possible to the adiabatic ground state $|\psi_g(T)\rangle$, realizing a fidelity $\mathcal{F} = |\langle\Psi_{fin}|\psi_g(T)\rangle|^2$ close to unity. An infinite number of protocols, i.e., paths in the Hilbert space, connects the two states. We are interested in two specific ones. The first protocol minimizes the transition time T (keeping the coupling term ω constant) between the two states, reaching the quantum speed limit imposed by quantum mechanics and related to the Heisenberg uncertainty principle on energy and time. On the other hand, by lifting the constraint on the coupling term of the first protocol we realise protocols that allow a perfect following of the adiabatic ground state, $|\psi(t)\rangle = |\psi_g(t)\rangle$ for any $0 \le t \le T$. In the following, we will call such protocols 'superadiabatic', inspired by Berry's work on the superadiabatic basis [11]. For the case of superadiabatic protocols, also called counter-adiabatic or transitionless, the construction of an auxiliary Hamiltonian that cancels the non adiabiatic terms has been widely studied theoretically [11, 12, 13, 14].

As a paradigmatic and benchmark protocol for other protocols, we choose the well-known Landau-Zener problem, where the energy term $\Gamma(t)$ in the Hamiltonian is linear in time. Before addressing the quantum speed limit or superadiabticity, we considered a generalized version of the Landau Zener problem (general LZ) in order to test the experimental technique and theoretical models. We experimentally investigated protocols where the diagonal term is governed by a power law in time [10], which have been theoretically studied in the past [15, 16].

Both the quantum speed limit problem and the superadiabatic problem are connected to the optimal control of a quantum system which has a long history in different fields [17] such as nuclear magnetic resonance (NMR) [18], adiabatic quantum computation [19] and reverse engineering using the Lewis-Riesenfeld invariants [20, 21]. Experimentally a number of quantum protocols have been tested in NMR [22] for quantum computation algorithms. For atomic systems an experiment on transforming a non-adiabatic transition into an adiabatic one has been performed by [23], while in the field of Bose-Einstein Condensation in optical lattices the nearly perfect preparation of a quantum state by non-adiabatic transformation was achieved [24].

The case of the quantum speed limit has been extensively studied theoretically [25, 26, 27, 28, 29] but has not been explicitly experimentally tested before our work.

2. Experimental realization and results

In our experiment a two-level system is realized using a Bose-Einstein condensate of rubidium atoms loaded into an optical lattice [30, 31]. The optical lattice is created by two-counterpropagating linearly polarized laser beams controlled by two independent acousto-optic modulators. Bose-Einstein condensates of around 5×10^4 rubidium-87 atoms are loaded into a periodic potential by ramping up the power of the lattice beams in 100 ms. The lattice constant is given by $d_{\rm L} = \lambda/2 = 421\,{\rm nm}$ with $\lambda = 842\,{\rm nm}$ the laser wavelength.

By the Bloch theorem a periodic potential generates an energy level structure which can be described by energy bands. Under specific conditions on the optical lattice potential depth, the energy structure can be restricted to the ground state band and the first excited band.

Using acousto-optic modulators we introduced time-dependent frequency differences between the two counterpropagating beams, leading to a moving interference pattern and hence to a moving periodic potential which, in the rest-frame of the lattice, can be interpreted as a time dependent force on the atoms $F_{\rm LZ} = Ma_{\rm LZ}$ with $a_{\rm LZ} = d_{\rm L}\frac{d\Delta\nu(t)}{dt}$. Close to the edge of the Brillouin zone, where an energy gap between the bands is present, the system is well approximated by the Landau-Zener Hamiltonian, and the time dependent force can be used to explore the band structure. Varying the laser power changes the lattice depth V_0 and the time dependent force changes the quasi-momentum q. The Hamiltonian elements $\Gamma(t)$ and $\omega(t)$ are related to the lattice parameters as

$$\Gamma(t) = 2\Gamma_0\left(\frac{q(t)}{\hbar k} - \frac{1}{2}\right) \quad (2)$$

$$\omega = \frac{V_0}{4} \quad (3)$$

$$(4)$$

In the following all the physical quantities are written in the natural energy ($E_{\rm rec} = \hbar\omega_{\rm rec} = \pi^2\hbar^2/2Md_{\rm L}^2$) and time units $\tau_{rec} = 1/\omega_{\rm rec}$ of the system.

After loading the BEC into the lattice, we manipulated the system by applying time dependent sweeps on the control parameters and we performed a projection measurement on the diabatic or adiabatic basis. Measurements in the free-particle diabatic basis were performed by instantaneously switching off the lattice beams within less than 1 μs. The number of atoms in the $q = 0$ and $q = 2p_{rec}$ ($p_{rec} = \hbar\pi/d_{\rm L}$) momentum classes were measured after a time-of-flight by absorption imaging and the transition probability $\mathcal{P}^{\rm diab}(\tau)$ was given by the ratio between the number of atoms in the $q = 2p_{rec}$ momentum class and the total number of atoms. In order to project on the *adiabatic* basis the lattice depth was adiabatically decreased to zero by reducing the laser power within 400 μs. The ratio between the fraction of atoms in $q = 0$, (the atoms in the lower band) and the total number of atoms gave an experimental measurement in the adiabatic basis and hence of the protocol fidelity. The noise in our imaging system does not

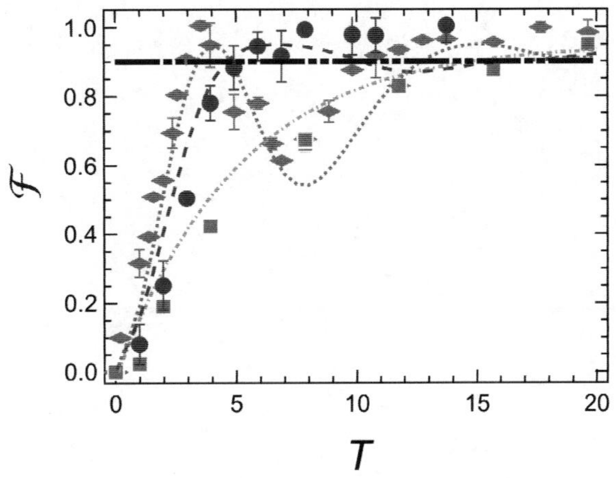

Figure 2. Fidelity as a function of the duration T of the power law sweep for different α; red diamonds $\alpha = 4$, blue circles $\alpha = 2$ and green squares for the linear LZ with $\alpha = 1$. The experimental results are compared with numerical simulations, given by the dotted lines (red, blue and green). The dotted black line is the threshold fidelity determining the transition time T_{thr}.

allow us to reliably measure fidelities $\mathcal{F}$ larger than 0.98. In addition, the non-adiabaticity of the preparation and measurement protocols contributes with an infidelity on the order of few percent.

2.1. Generalized Landau-Zener: power law

As a first step of our investigation we considered a generalized LZ sweep with the following form:

$$\Gamma(\tau) = \begin{cases} -2^{\alpha+1}(1/2-\tau)^{\alpha} & \text{for } 0 \leq \tau \leq 1/2 \\ 2^{\alpha+1}(\tau-1/2)^{\alpha} & \text{for } 1/2 \leq \tau \leq 1 \, , \end{cases} \tag{5}$$

where $\tau = t/T$ is the rescaled time and T the sweep time. The boundary conditions $\Gamma(0) = -2$ and $\Gamma(1) = 2$ are satisfied, α is the power law parameter and $\omega(\tau) = \omega$ is constant during the sweep. For $\alpha = 1$ the sweep is linear, corresponding to the standard Landau-Zener sweep.

In Fig. 2, together with a numerical simulation, we show the measured fidelity as function of total time T for different power law parameters $\alpha = 1$, $\alpha = 2$ and $\alpha = 4$. The numerical simulation and the experimental data are in good agreement. Since for Linear Landau Zener the time to reach fidelity $\mathcal{F} = 1$ diverges, in order to compare the speed of different sweeps, we fix a threshold fidelity at 0.9 and from experimental results as those in Fig. 2 we derive the time T_{thr} at which that fidelity is reached (see Fig. 3). The dependency of T_{thr} on the parameter α is well demonstrated. For increasing α the limiting time decreases, approaching the minimum time of the quantum speed limit (the dashed red line, $T_{QSL} = 2.75$ for the experimental parameters). An intuitive explanation for that behavior is that for large α the power laws sweep approaches the composite pulse protocol which, as shown below, reaches the quantum speed limit.

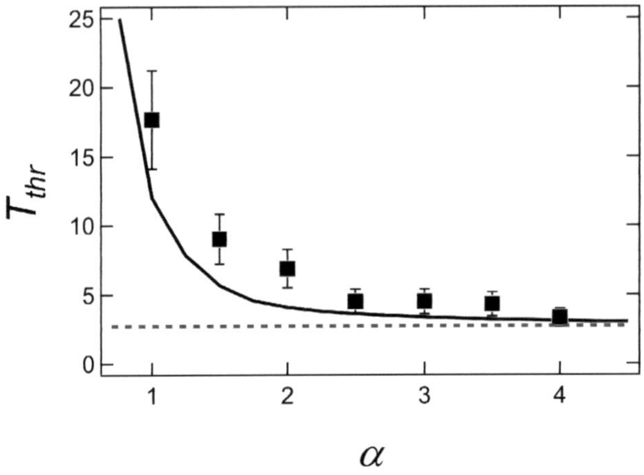

Figure 3. Minimum transition time T_{thr} to reach $\mathcal{F}_{\rm thr} = 0.9$. Comparison between the experimental points (black solid squares) and a numerical simulation (black solid line) as a function of the power law parameter α. The dashed red line presents the quantum speed limit for the experimental parameters.

2.2. Quantum speed limit: composite pulse protocol

As shown experimentally and theoretically in [9], we have found that the minimum time to perform the transition with fidelity 1 is reached by a composite pulse protocol of the form:

$$\Gamma(\tau) = \begin{cases} -\Gamma_0 & \text{for } t = 0 \\ \Gamma_M & \text{for } t \in [0, t_0] \\ 0 & \text{for } t \in [t_0, T - t_0] \\ -\Gamma_M & \text{for } t \in [T - t_0, 1] \\ +\Gamma_0 & \text{for } t = T\,, \end{cases} \tag{6}$$

where ω remains constant during the sweep; Γ_M and t_0 satisfy the relation $\Gamma_M t_0 = \pi/4$ while Γ_M is asymptotically large and t_0 small. This protocol is analogous to the NMR composite protocol consisting of half a Rabi Oscillation with frequency ω and $\Gamma = 0$ which is preceded and followed by two short pulses of area $\pi/4$.

In Fig. 4 we show experimental results and numerical calculations of the fidelity as a function of the total sweep time T for the composite pulse protocol (black triangles) and for a linear Landau-Zener protocol (red squares). For the composite pulse protocol the minimum time T_{thr} for reaching optimal fidelidy is extracted from the position of the maxiumum of the fidelity $\mathcal{F}$ (Fig. 4), tipically 0.95 ± 0.03. For the linear Landau Zener protocol we measure the threshold time T_{thr} for reaching 0.9 fidelity since the needed time for reaching $\mathcal{F} = 1$ diverges. In figure 5 we show the minimum time for the composite pulse (black triangles) and the threshold time for the linear Landau Zener protocol (red squares) as a function of the coupling parameter ω

While the threshold time for the linear Landau-Zener sweep is always larger than the optimal minimum time for the composite pulse protocol (by an order of magnitude for small ω), the minimum time for the composite pulse protocol approaches the quantum speed limit (Fleming's or Bhattacharyya's bound [26]) T_{qsl} given by:

$$T_{qsl} = \frac{\arccos |\langle \Psi_{\rm fin} | \Psi_{\rm ini} \rangle|}{\omega}\,. \tag{7}$$

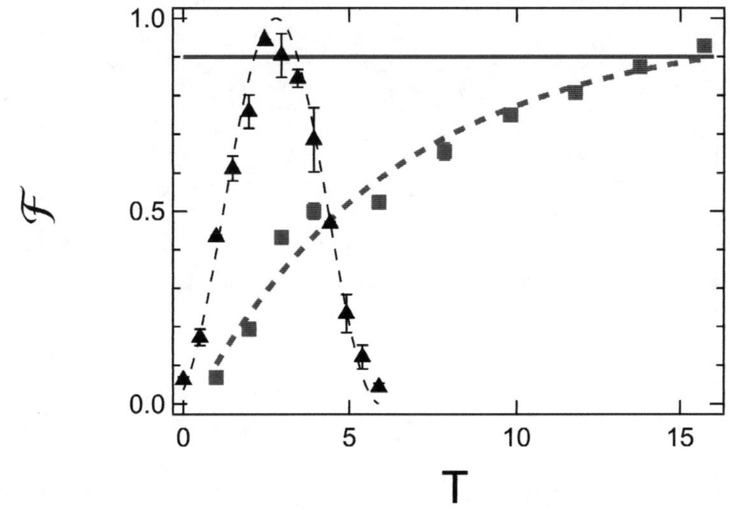

Figure 4. Fidelity as a function of the total sweep time T for a composite pulse (black triangle) and linear Landau-Zener sweep (red squared). The dashed lines are numerical results for the composite pulse (black line) and the linear Landau-Zener sweep (red line). The horizontal black line denotes the 0.9 threshold value.

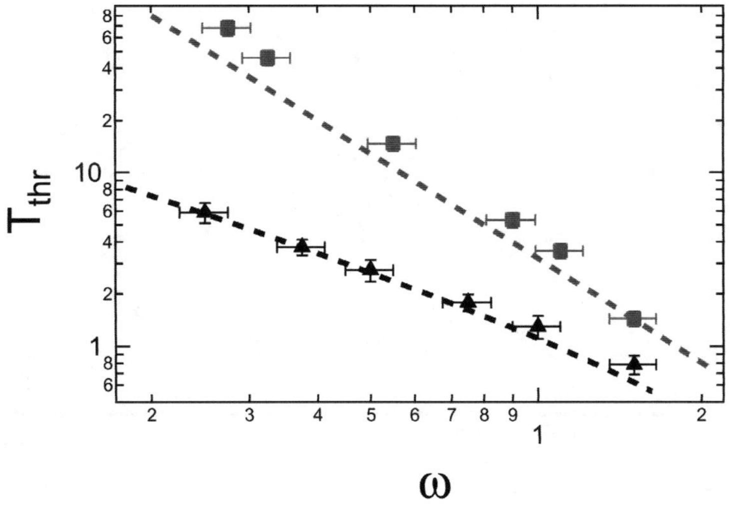

Figure 5. Time T_{thr} needed to reach optimal fidelity for the composite pulse protocol (black triangles) and to reach 0.9 fidelity for a linear Landau-Zener sweep (red squares)as a function of the coupling term ω. The dashed lines are numerical results as for Fig. 4.

2.3. Superadiabatic protocol

We now turn to the task of engineering a protocol that drives the system through the anticrossing point in such a way that it remains in the adiabatic ground state at any time t. The existence of transitionless or superadiabatic protocols follows directly from the properties of quantum mechanics. For any given time dependent Hamiltonian $H(t)$ it is always possible to write a new

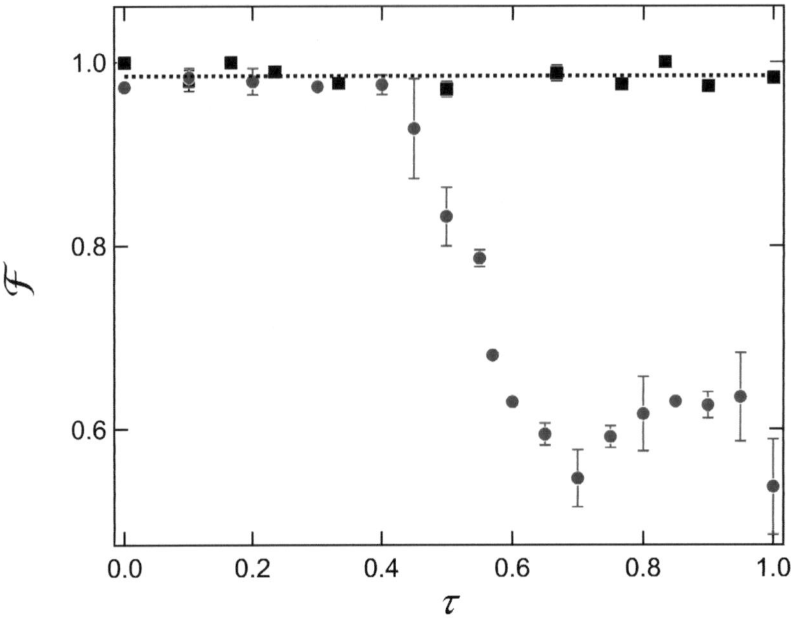

Figure 6. Fidelity as a function of the normalized time τ, for a superadiabatic linear sweep (black squared) and a linear Landau-Zener sweep (red circles). The horizontal line at $\mathcal{F} = 0.98$ denotes a fit through the superadiabatic protocol data.

Hamiltonian

$$H'(t) = H(t) + H_{aux}(t) \tag{8}$$

by constructing an (ad-hoc) additional control H_{aux} that cancels the non-adiabatic terms of H

$$H_{aux} = i\hbar \sum_n (|\partial n(t)\rangle\langle n(t)| - |n(t)\rangle\langle \partial n(t)|)/2 \tag{9}$$

where $|n(t)\rangle$ are the eigenstates of the original Hamiltonian H [11]. In the case of the two level system Hamiltonian of Eq. 1, H_{aux} can be cast into the form:

$$H_{aux}(t) = \frac{\hbar}{2}\frac{\partial \psi}{\partial t}\sigma_y \tag{10}$$

where $\psi = \arctan(\omega(t)/\Gamma(t))$,

As we have shown in [9], for any given time dependent Hamiltonian H satisfying the above prescription, we can find a transformation $\{\Gamma(t), \omega(t)\} \rightarrow \{\Gamma'(t), \omega\prime(t)\}$ so that the new Hamiltonian $H'(\Gamma'(t), \omega'(t))$ realizes a superadiabatic protocol. Applying that transformation to the linear Landau-Zener protocol $\Gamma(t)$, we obtain:

$$\begin{aligned} \Gamma'(\tau) &= \Gamma(\tau) - \frac{4(\tau-\frac{1}{2})}{T^2\left[(\tau-\frac{1}{2})^2+\frac{1}{2}\omega^2\right]^2+1}, \\ \omega'(\tau) &= \omega\sqrt{1+\frac{1}{T^2(8(\tau-\frac{1}{2})^2+\frac{1}{2}\omega^2)^2}}. \end{aligned} \tag{11}$$

with $\tau = t/T$ and where the discontinuities in the derivative of $\Gamma(t)$ at the beginning and at the end of the sweep are realised experimentally by introducing a large finite correction $\Delta\Gamma_M$ for a short time $\Delta\tau$, satisfying:

$$\Delta\tau\Delta\Gamma_M = \mp\frac{1}{2}\arctan\left(\frac{\Gamma\dot{\omega} - \omega\dot{\Gamma}}{2\omega(\Gamma^2+\omega^2)}\right) \tag{12}$$

where the sign $\mp$ is for the correction at the beginning or at the end of the protocol. From Eq. 11 it follows taht the coupling term ω' is now time dependent. This time dependence is experimentally achieved by varying the laser power.

Fig. 6 reports a time resolved measurement of the fidelity $\mathcal{F}(\tau)$ during the sweep for the two protocols. For the superadiabatic linear protocol the fidelity is overall constant with $\mathcal{F} = 0.98 \pm 0.02$ indicating that the system remians in the ground adiabatic state during the entire protocol. In the case of the usual linear Landau-Zener protocol, at the band gap (around $\tau = 0.5$) we observe tunneling between the bands.

We have numerically tested the sensitivity of the superadiabatic tangent protocol to the variations in the control parameters by varying both ω and T around the optimum value and measuring the fidelity $\mathcal{F}$ in each case [9]. We clearly showed that the superadiabatic tangent protocol is extremely robust with respect to an increase in T or ω, while for large (more than twenty percent) reductions in T or ω the fidelity naturally drops sharply as otherwise the quantum speed limit would be violated.

3. Conclusions

We have reported the development of experimental techniques, based on Bose-Einstein condensates in optical lattices, for the investigation of high fidelity quantum control protocols. For the case of a two level system, we have shown the possibility of implementing protocols for reaching the quantum speed limit on the one hand or superadiabatic following on the other hand. For the latter case the generality of the transformation on the Hamiltonian suggests the possibility of applying this technique to other fields of research.

Acknowledgments

This work was supported by the E.U. grant No. 265031-ITN-COHERENCE and by the MIUR PRIN-2009 QUANTUM GASES BEYOND EQUILIBRIUM. The authors thank Paul Huillery for assistance in the experimental realization.

References

[1] Walmsley I and Rabitz H 2003 *Phys. Today* **56**, 43
[2] Nielsen M and Chuang I 2000 *Quantum Comunication and Quantum Communication* (Cambridge, UK: Cambridge Univ. Press)
[3] Hänsch T W 2006 *Rev. Mod. Phys.* **78** 1297
[4] Rice S A and Zhao M 2000 *Optical Control of Molecular Dynamics* (New York: Wiley)
[5] Berry M V 1995 *Ann N.Y. Acad. Sci.* **755** 303
[6] Landau L 1932 *Phys. Z. Sowjetunion* **2** 46
[7] Zener C 1932 *Proc. R. Soc.* A **137** 696
[8] Majorana E 1932 *Nuovo Cimento* **9** 4350
[9] Bason M G, Viteau M, Malossi N, Huillery P, Arimondo E, Ciampini D, Fazio R, Giovannetti V, Mannella R and Morsch O 2012 *Nat. Phys.* **8** 145
[10] Malossi N, Bason M G, Viteau M, Arimondo E, Mannella R, Morsch O and Ciampini D 2013 *Phys. Rev.* A **87**, 012116
[11] Lim R and Berry M V 1991 *J. Phys. A: Math. Gen.* **24** 3255
[12] Berry M V 2009 *J. Phys. A: Math. Theor* **42** 365303
[13] Demirplak M and Rice S A 2003 *J. Phys. Chem.* A **107** 9937
[14] Demirplak M and Rice S A 2008 *J. Chem. Phys.* **129** 154111
[15] Garanin D A and Schilling R 2002 *Europhys. Lett.* **59** 7
[16] Garanin D A and Schilling R 2002 *Phys. Rev.* B **66** 174438

[17] Garraway B M and Suominen K A 1995 *Rep. Prog. Phys.* **58** 365
[18] Levitt M H 1996 *Composite Pulses* (Encyclopedia of Nuclear Magnetic Resonance, Vol. **2**) eds Grant D M and Harris K R (New York: Wiley)
[19] Roland J and Cerf N J 2002 *Phys. Rev.* A **65** 042308
[20] Muga J G , Chen X, Ruschhaupt A and Guéry-Odelin D 2009 *J. Phys. B: At. Mol. Opt. Phys.* **42** 241001
[21] Ruschhaupt A, Chen X, Alonso D and Muga J G 2012 *New J. Phys.* **14** 093040
[22] Mitra A, Ghosh A, Das R, Patel A and Kumar A 2005 *J. Mag. Res.* **177** 285
[23] Lu T, Miao X and Metcalf H 2007 *Phys. Rev.* A **75** 063422
[24] Mellish A S, Duffy G, McKenzie C, Geursen R and Wilson A C 2003 *Phys, Rev.* A **68** 051601(R)
[25] Levitin L B 1982 *Int. J Theor. Phys.* **21** 299
[26] Bhattacharyya K J 1985 *Phys. A: Math. Gen.* **16** 2993
[27] Peres A 1993 *Quantum Theory: Concepts and Methods* (Kluwer)
[28] Giovannetti V, Lloyd S and Maccone L 2003 *Phys. Rev.* A **67** 052109
[29] Caneva T, Murphy M, Calarco T, Fazio R, Montangero S, Giovannetti V and Santoro G E 2009 *Phys. Rev. Lett.* **103**, 240501
[30] Zenesini A, Lignier H, Tayebirad G, Radogostowicz J, Ciampini D, Mannella R, Wimberger S, Morsch O and Arimondo E 2009 *Phys. Rev. Lett.* **103** 090403
[31] Jona-Lasinio M, Morsch O, Cristiani M, Malossi N, Müller J H, Courtade E, Anderlini M and Arimondo E 2003 *Phys. Rev. Lett.* **91** 230406

Journal of Physics: Conference Series **442** (2013) 012063 doi:10.1088/1742-6596/442/1/012063

Resonant driving of a nonlinear Hamiltonian system

Carlo Palmisano, Gianpiero Gervino
Dipartimento di Fisica, Università di Torino, Via P. Giuria, I–10125 Torino

Massimo Balma, Dorina Devona
SELEX Galileo, San Maurizio Canavese (TO), Italy

Sandro Wimberger
Institut für Theoretische Physik and Center for Quantum Dynamics, Universität Heidelberg, D–69120 Heidelberg

E-mail: s.wimberger@thphys.uni-heidelberg.de

Abstract. As a proof of principle, we show how a classical nonlinear Hamiltonian system can be driven resonantly over reasonably long times by appropriately shaped pulses. To keep the parameter space reasonably small, we limit ourselves to a driving force which consists of periodic pulses additionally modulated by a sinusoidal function. The main observables are the average increase of kinetic energy and of the action variable (of the non-driven system) with time. Applications of our scheme aim for driving high frequencies of a nonlinear system with a fixed modulation signal.

1. Introduction

Driving a nonlinear Hamiltonian system with a fixed form of drive is never effective over long times. The local intrinsic frequency of the system changes with increasing energy, or more generally, with the increasing action of the system. This is the very definition of nonlinearity [1]. Consequently, to keep the driving in resonance with the system the external frequencies must be gradually adapted.

In this paper we show that a nonlinear system can nevertheless be pumped with energy from an external drive even if the latter is fixed. The condition for our approach to work is that the driven system allows for global transport in phase space. A very rough, yet effective criterion for the onset of chaos in that sense has been introduced by Chirikov and is by now known as Chirikov (or also resonance overlap) criterion [2]. The overlapping of nonlinear resonances in phase space is the origin of dynamical stochasticity in classical Hamiltonian systems [1]. In section 2 we introduce a model system which was studied before in reference [3], where the effectiveness of the overlap criterion was theoretically predicted. The following section 3 reports major results from ample numerical investigations of the model, which verify the predictions of [3]. Section 4 concludes the paper.

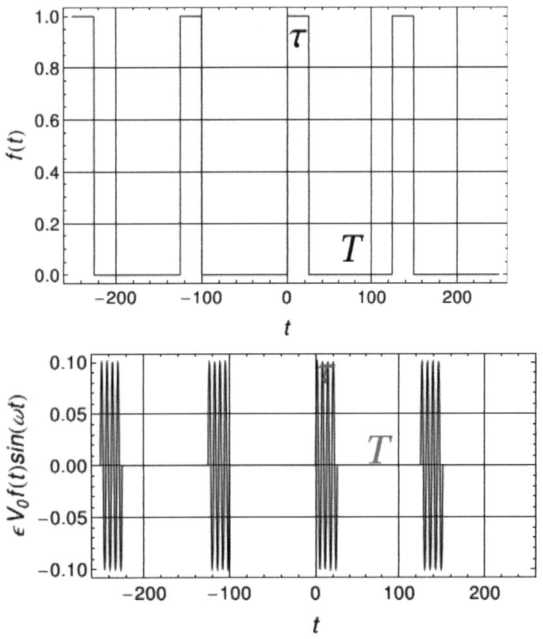

Figure 1. Upper panel: Form of the pulse train $f(t)$ which is periodic with period $T+\tau$ and the single pulses have a width τ. Lower panel: Modulation signal $\epsilon V_0 f(t)\sin(\omega t)$. The scales are enhanced for better visibility compared to the parameters chosen latter.

2. Nonlinear model system

Throughout we will stick to the following model for a classical Hamiltonian system, given by an harmonic oscillator H_0, a quartic nonlinear potential $H_{\rm nl}$, and a periodic driving $H_{\rm f}$:

$$H(x,p,t) = H_0(x,p) + H_{\rm nl}(x) + H_{\rm f}(x,t)\,, \tag{1}$$

with the following parts

$$\begin{aligned} H_0(x,p) &= \frac{p^2}{2m} + \frac{1}{2}m\omega_0^2 x^2\,, &(2)\\ H_{\rm nl}(x) &= \frac{1}{4}mbx^4\,, &(3)\\ H_{\rm f}(x,t) &= -\epsilon V_0 x f(t)\sin(\omega t)\,. &(4)\end{aligned}$$

Similar models (also known as Lorentz models) have been studied over many decades, in particular in the context of nonlinear optics [4]. Instead of integrating directly Newton's equation corresponding to the Hamiltonian (1) in the variables (x,p), we follow reference [3] and rewrite the problem in action-angle variables of the harmonic oscillator (θ, J) which are connected to the previous ones by the canonical transformation

$$\begin{aligned} x(\theta,J) &= \sqrt{\frac{2J}{m\omega_0}}\sin\theta\,, &(5)\\ p(\theta,J) &= \sqrt{2Jm\omega_0}\cos\theta\,. &(6)\end{aligned}$$

The advantage of using the action-angle variables lies not only in their use in perturbation theory as exercised in [3], but for us, first and foremost, in the fact that we are interested in transport in phase space. The latter is characterized best by the temporal evolution of the action J of the "unperturbed system". In the new variables we obtain the new Hamiltonian

$$H(x,p,t) \longrightarrow H'(\theta,J,t) = J\omega_0 + \frac{J^2 b\sin^4\theta}{m\omega_0^2} - \epsilon V_0\sqrt{\frac{2J}{m\omega_0}}\sin\theta f(t)\sin(\omega t)\,. \tag{7}$$

The corresponding equations of motion for (θ, J) are then simply given by Hamilton's equations:

$$\begin{aligned}\frac{\mathrm{d}\theta}{\mathrm{d}t}(t) &= \frac{\partial H'}{\partial J}(\theta, J, t) \\ &= \omega_0 + 2\frac{Jb\sin^4\theta}{m\omega_0^2} - \epsilon V_0 \frac{1}{2\sqrt{J}}\sqrt{\frac{2}{m\omega_0}}\sin\theta f(t)\sin(\omega t) \qquad (8)\\ \frac{\mathrm{d}J}{\mathrm{d}t}(t) &= -\frac{\partial H'}{\partial\theta}(\theta, J, t) \\ &= -4\frac{J^2 b\sin^3\theta\cos\theta}{m\omega_0^2} + \epsilon V_0\sqrt{\frac{2J}{m\omega_0}}\cos\theta f(t)\sin(\omega t)\,. \qquad (9)\end{aligned}$$

In the next section we present results obtained by integrating these equations of motion over time. Before we motivate our choice of the driving signal given by H_{f}. It essentially contains two frequency scales, one given by the sinusoidal part oscillating with frequency ω and a second one determined by a period $T + \tau$. The detailed form of the drive is best seen pictorially in Fig. 1. Now we fix this form of the external driving and try to investigate the behavior of the time-averaged action $\langle J(t)\rangle_t = 1/t\int_0^t J(t')dt'$ for different parameters of the model.

3. Numerical results

For the following computations, we fix the initial conditions $x(0) = 0, mx'(0) = p(0) = -2$ and the parameters

$$m = 1, \omega_0 = 20, \tau = 1, \epsilon = 0.5, V_0 = 100, T = 10\,\tau, \omega = \frac{16\pi}{\tau}. \qquad (10)$$

Units have been suppressed throughout (and maybe easily reintroduced in case having in mind the meaning of all parameters). This leaves us only one control parameter, the effective nonlinearity b, which characterizes the nonlinear response of the system with respect to the external drive.

We do not repeat the analytical estimates for the widths of corresponding nonlinear resonances in the phase space spanned by the variables (θ, J). Reference [3] derives from them a critical parameter K which separates motion without overlapping resonances, for $K < 1$, from the one with overlapping resonances, implying chaotic transport for $K > 1$. Written in our parameters, we obtain from [3]

$$K(J) = \frac{\tau + T}{2\pi}\sqrt{\frac{\partial H'_{\mathrm{nl}}(J)}{\partial J}\frac{\epsilon}{4}\frac{2\pi}{\tau + T}V_0\sqrt{\frac{2J}{m\omega_0}}} > 1\,. \qquad (11)$$

K is time-dependent through the dependence on the action variable J. Its initial value at time $t = 0$ indeed determines the increase of the mean action as the following figures show.

Figs. 2 and 3 compare the evolution of the actions for various nonlinearities $b = b_1 = 3200$ (left panel in Fig. 2), $b = b_1/2$ (left panel of Fig. 3), and $b = b_1/4$ (right panel of Fig. 3). These parameters imply for Eq. (11) $K(J(t = 0)) = 1.44$, 1.02, and 0.72, respectively. The shown results confirm in practice that the initial value of the parameter K indeed determines whether transport in action space occurs (see the left panels of Figs. 2 and 3) or is suppressed by transport barriers in phase space (see the right panel of Fig. 3).

4. Conclusions

We studied a nonlinear model system for controlling the temporal evolution by an appropriate external driving. Exploiting the nonlinear response of the system to the drive, we confirmed

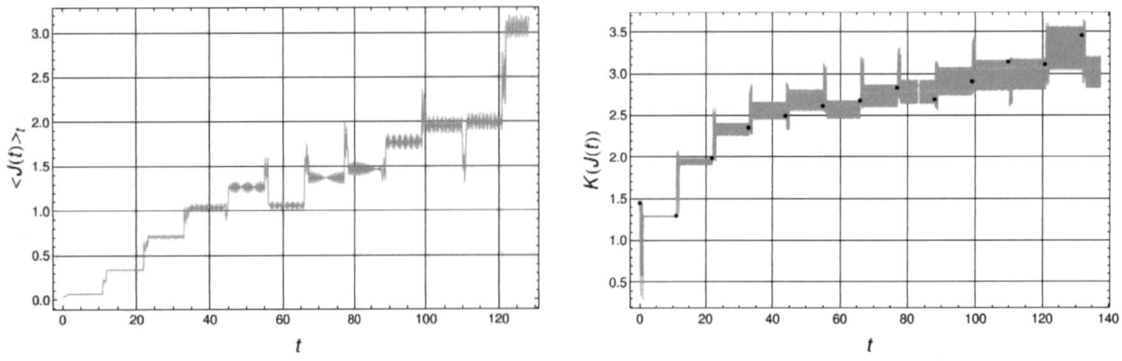

Figure 2. Time averaged action $\langle J(t)\rangle_t$ (left panel) and critical parameter K (right panel) vs. time t for $b = 3200$, implying $K(J(t=0)) = 1.44$. The overall increase of both quantities with time is clearly visible confirming the theoretical predictions.

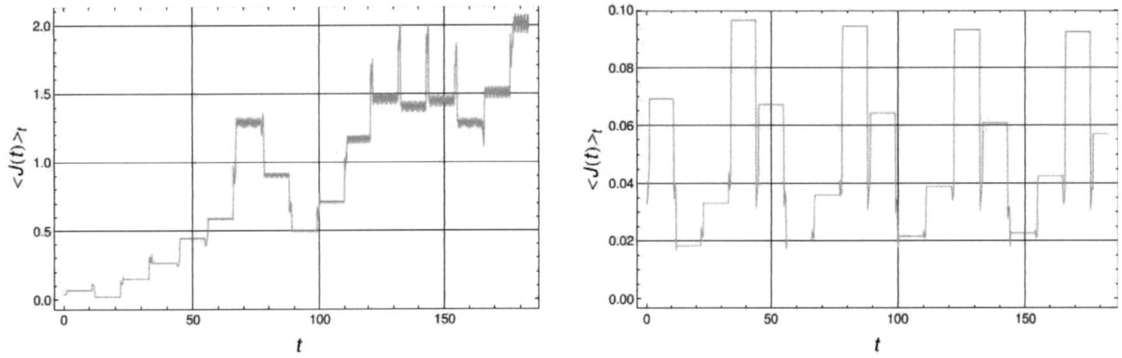

Figure 3. Time averaged actions $\langle J(t)\rangle_t$ vs. time t for two sets of parameters: $b = 1600$ and $K(J(t=0)) = 1.02$ (left panel) and $b = 800$ and $K(J(t=0)) = 0.72$ (right panel). While the left plot is similar to the previous Fig. 2, the right panel shows a case below the criterion for transport, and hence the action just oscillates around small values.

that resonant driving is possible as long as there are no transport barriers hindering diffusion in phase space. The transporting and the non-transporting regimes are separated by a critical effective nonlinearity which determine the value of a control parameter introduced in [3].

Our results give a proof of principle of the applied method. In the future we want to extend the idea to nonlinear systems in the presence of additional noise, as suggested in the recent reference [5].

Acknowledgments

S W and G G thank with pleasure the organiser Thomas Elze for his kind invitation to DICE2012. We acknowledge also the help of the Heidelberg Graduate School of Fundamental Physics (DFG grant: GSC 129/1).

References

[1] Lichtenberg A J and Lieberman M A 1992 *Regular and chaotic dynamics* (Berlin-Heidelberg: Springer)
[2] Chirikov B V 1972 *Phys. Rep.* **52** 265
[3] Ugulava A, Chotorlishvili L and Nickoladze K 2003 *Phys. Rev.* E **68** 026216

[4] Boyd R B 1992 *Nonlinear Optics* (San Diego: Academic Press)
[5] Chotorlishvili L, Sukhov A, Wimberger S and Berakdar J 2013 *Fluctuation and Noise Letters* **12** 1350003; published online DOI: 10.1142/S021947751350003X

Microscopic theory of warm ionized gases: equation of state and kinetic Schottky anomaly

A Capolupo, S M Giampaolo, F Illuminati

Dipartimento di Ingegneria Industriale, Universitá di Salerno, Via Ponte don Melillo, I-84084 Fisciano (SA), Italy

Abstract. Based on accurate Lennard-Jones type interaction potentials, we derive a closed set of state equations for the description of warm atomic gases in the presence of ionization processes. The specific heat is predicted to exhibit peaks in correspondence to single and multiple ionizations. Such kinetic analogue in atomic gases of the Schottky anomaly in solids is enhanced at intermediate and low atomic densities. The case of adiabatic compression of noble gases is analyzed in detail and the implications on sonoluminescence are discussed.

1. Introduction

The recent progresses in stellar astrophysics [1]–[3], in experiments on inertial confinement in plasmas [4]–[6], and in plasma diagnostics have triggered renewed interest in the determination of accurate state equations for mixed systems of atoms and ions in ionized gases [7, 8, 9]. At the same time, much recent effort has been devoted to investigating the role of plasmas of ionized atomic gases in sonoluminescence (SL) [10]–[19]. Sonoluminescence consists in the emission of short flashes of light from collapsing bubbles in a liquid excited by ultrasonic waves. The light-emitting bubbles contain mainly noble gas atoms plus small quantities of water vapor and in the last part of the collapse they can reach densities comparable to those of solids. According to recent results [19] the conditions inside the bubbles may be very extreme, the temperature may far exceed those estimated from the emitted light, and the plasma electron density can reach the same order of magnitude as that of densities created in laser-driven fusion experiments. A detailed understanding of the noble-gas thermodynamics in the high energy regime is then crucial in order to explain the experimental results and uncover the basic mechanisms of sonoluminescence.

In the present work we analyze the thermodynamics of atomic gases for different regimes of atomic densities and temperatures characteristic of warm plasmas, and consider applications to the case of noble gases by studying regimes that can be typical of the last stage of collapsing bubbles in sonoluminescence, i.e. the stage of light emission. We derive a set of state equations based on accurate microscopic interaction potentials of the Lennard-Jones type and we predict the presence of anomalous peaks in the specific heats. These peaks are explained in terms of a kinetic extension of the Schottky anomaly observed in solids with a limited number of energy levels at very low temperature [20]. The peaks appear in correspondence to ionization processes as the temperature is increased. This implies that the different degrees of ionization play in plasma physics a role similar to the one played by the finite number of energy levels in the solid state case, the fundamental difference lying in the fact that the predicted nontrivial contribution to the specific heat in ionized gases is provided by the unbounded nature of the atomic kinetic energy. The existence of kinetic Schottky anomalies bears important consequences on the collective heating of atomic gases. In particular, it is strongly suppressed when the specific heats undergo a sharp increase in correspondence of atomic ionizations. Coming to concrete settings, we will discuss in detail the case of noble gases, which is especially relevant for sonoluminescence. Considering the late stage of adiabatic compression, we determine self-consistently the number density of plasma electrons and find good agreement with recently reported experimental results [19]. Finally, we remark that our approach can be readily generalized beyond the case of atomic gases by including in the description the presence of chemical bonds.

The paper is organized as follows. In the Sec. 2 we define a set of state equations useful for warm plasmas, and we analyze the ultra-low density and the high density plasma regimes in Secs. 3 and 4, respectively. In Sec. 5 we consider an adiabatic transformation and compare the results obtained by using the state equations introduced in Sec. 2 with new experimental data on the sonoluminescence. Sec. 6 is devoted to the conclusions.

2. Equations of state

We begin by introducing the constitutive equations for the energy of a generic particle system in the presence of ionization. By taking into account the kinetic, interaction, and ionization contributions, the energy density of the atomic ensemble can be expressed as

$$E = \frac{3}{2} K_b T \sum_{i=-1}^{N_A} n_i + \sum_{i=-1}^{N_A} n_i (E_i + h_i) , \tag{1}$$

where the index i classifies the particles present in the systems (electrons, atoms, and ions of different charge multiplicity) by their charge expressed in units of $|e|$. Explicitly, $i = -1$ for free electrons, $i = 0$ for neutral atom and $i \in [1, N_A]$ for ions of different charge, ranging up to the maximum allowed value corresponding to the atomic number N_A. The number density of the $i-$th particle species is denoted by n_i, while E_i is the energy needed to ionize an atom i times, with $E_{-1} = E_0 = 0$. Finally, h_i is the potential energy density of the i-th particle species due to the interaction with the other particles of the system. Taking the virial expansion up to second order (higher orders may be considered when necessary), h_i takes the form

$$h_i = \sum_{j=-1}^{N_A} 2\pi n_j \int_0^\infty r^2 U_{i,j}(r) \exp\left(-\frac{U_{i,j}(r)}{K_b T}\right) dr , \tag{2}$$

where $U_{i,j}(r)$ is the interaction energy between the i-th and the j-th particles at distance r. In Eq. (1) we have neglected the presence of chemical bonds between the atoms. Such assumption holds, for instance, in the case of noble gas atoms. The approach can be readily extended to other systems by taking into account the energy associated to the chemical bond.

The atom and ion densities n_i are not fixed *a priori*, but are determined by the underlying quantum dynamics through the self-consistent coupling of Eq. (2) and the Saha equations [21]:

$$\frac{n_{i+1}\, n_{-1}}{n_i} = \left(\frac{m_e K_b T}{2\pi\hbar^2}\right)^{\frac{3}{2}} \frac{2 g_{i+1}}{g_i} \frac{e^{-\frac{E_{i+1}+h_{i+1}+h_{-1}}{K_b T}}}{e^{-\frac{E_i+h_i}{K_b T}}} , \tag{3}$$

where m_e is the electron mass and g_i is the ground-state degeneracy of an atom ionized i-th times. Eq. (3) is a set of N_A equations in $N_A + 2$ variables represented by the densities n_i. This set of equations can be closed by taking into account that the electronic density is related to the ion density by $n_{-1} = \sum_{i=1}^{N_A} i n_i$ and that n_0 is related to the total atomic density n by $n_0 = n - \sum_{i=1}^{N_A} n_i$.

We now provide an expression for the pressure P. In agreement with Eq. (1), P can be written as the sum of a kinetic term and a term associated to the interaction of any particle with the surrounding ones,

$$P = K_b T \sum_{i=-1}^{N_A} n_i - 2\pi \sum_{i=-1}^{N_A} n_i h'_i \,, \tag{4}$$

where the interaction term h'_i, considering the virial expansion up to second order, reads

$$h'_i = \sum_{j=-1}^{N_A} n_j \int_0^\infty r^3 \frac{dU_{i,j}(r)}{dr} \exp\left(-\frac{U_{i,j}(r)}{K_b T}\right) dr \,. \tag{5}$$

By means of Eq. (1) we can compute the specific heat at constant volume, $c_v = \left(\frac{\partial E}{\partial T}\right)_V$. Eq. (4) can be used to obtain the specific heat at constant pressure by using the relation $c_p - c_v = -T \left(\frac{\partial P}{\partial V}\right)_T^{-1} \left(\frac{\partial P}{\partial T}\right)_V^2$.

Eqs.(1)–(5) identify a complete set of state equations for a gas in which the degree of ionization is determined self-consistently from thermal equilibrium. In the following we will apply such equations to the description of warm plasmas.

3. Non interacting models: warm plasma at low density

We consider first ultra-low density regimes such that the interaction effects can be neglected, i.e. we can put ($U_{i,j} = 0 \,, \forall i, j$).

The main outcomes are presented in Figs. 1 and 2 in which the dependence of P, c_v and c_p, as well as the dependence of free electron density n_{-1} on the temperature are shown for different densities of a system of Helium atoms.

Let us start from the top panel of both the figures, which refers to an extremely low total density $n = 10^{10} cm^{-3}$. In Fig. 1 we note the presence of two peaks in both the specific heats in correspondence of the two different ionization processes, which are signaled by steps in the free-electron density n_{-1}. In Fig. 2 the pressure shows a step-wise behavior associated to the increment of free-electrons released in the ionization process.

Comparing now the top panel with the central one, in which the density is higher, we may observe that the peaks of the specific heats becomes lower, and the step-wise shapes of n_{-1} and P are wider and less pronounced. This is due to the fact that the ranges of temperature in which the processes of first and second ionization occur start to merge together. The smoothing of the steps continues up to the step-wise shape disappears almost completely when the density approaches to values comparable with those typical of metals (bottom panel $n = 10^{22} cm^{-3}$).

The peculiar behavior of the specific heats can be explained, as in the Schottky anomaly case, by taking into account the ratio of the change of the system entropy. For temperatures relatively low with respect to the one at which ionizations become relevant, the entropy of the system is small because all the atoms are in the same state. As the temperature approaches the difference between the energy levels, the different ionization states become significantly populated and hence there is an enormous increment of the system entropy due to the uncertainty on the degree of ionization. This implies the presence of peaks in the profiles of the specific heats. At the difference of the Schottky anomaly in the solid state physics, in the plasma case the entropies are related also to the kinetic energy of the atoms.

4. Interacting models: warm dense plasma

We now take into account the interactions between different particles by introducing potentials $U_{i,j}$ which can describe the physics of warm dense plasmas.

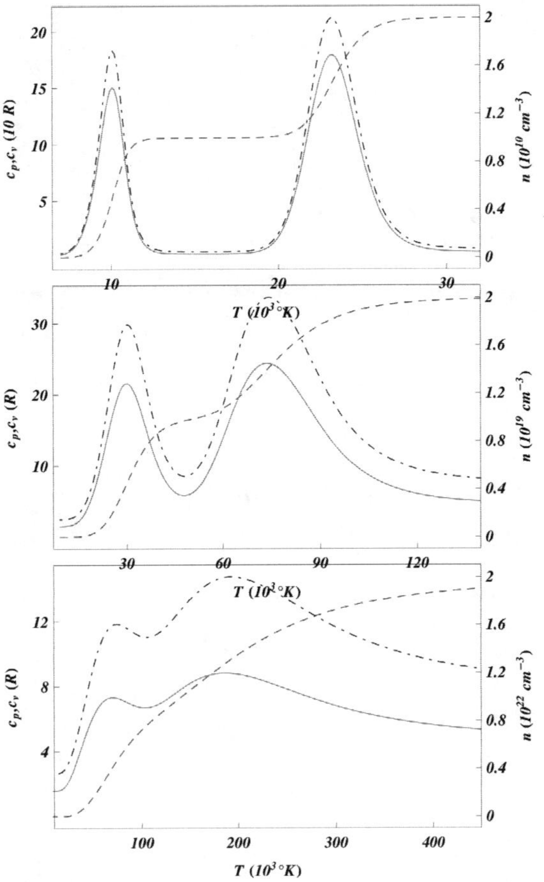

Figure 1. Dependence of the molar specific heat at constant volume c_v (Red solid line), and at constant pressure c_p (black dot-dashed line) on the temperature compared with the electron density n_{-1} (blue dashed line) for an ensemble of Helium atoms at three different densities n. In top panel $n = 10^{10} cm^{-3}$, in central panel $n = 10^{19} cm^{-3}$, and in bottom panel $n = 10^{22} cm^{-3}$. Both the specific heats are expressed as function of the universal constant of gas R in order to simplify their comparison with the result obtained with different assumptions.

Let us start by assuming that the interaction between two neutral atoms of noble gas can be represented by Lennard-Jones potential $U_{0,0}(r) = 4\varepsilon \left[\left(\frac{\sigma}{r}\right)^{12} - \left(\frac{\sigma}{r}\right)^{6} \right]$ in which ε stands for the depth of the potential well and σ is the finite distance at which the inter-particle potential vanishes. Lennard-Jones potential can be seen as the sum of two terms of which the first one is repulsive and represents the short range Pauli repulsion due to the partial overlap of the electron clouds of the two atoms, while the second one represents the long range dipole-dipole attraction. The interaction energy between a neutral atom and an ion can be represented by adding to $U_{0,0}$ the charge-dipole attractions $U_{0,i}(r) = U_{0,0}(r) - i^2 \alpha E_l^2(r)$, where α is the polarizability of the neutral atom and $E_l(r)$ is the electric field associated to the electrostatic potential $W_l(r)$ that solve the Debye-Hückel equation [21]. In a similar way, we describe ion-ion interaction by adding to the standard Lennard-Jones term both the charge-dipole interactions and the electrostatic repulsion between charges, $U_{i,j}(r) = U_{0,0}(r) - (i^2 + j^2)\alpha E_l^2(r) + ijW_l(r)$.

The interaction involving one or two free-electrons can be described by considering any free

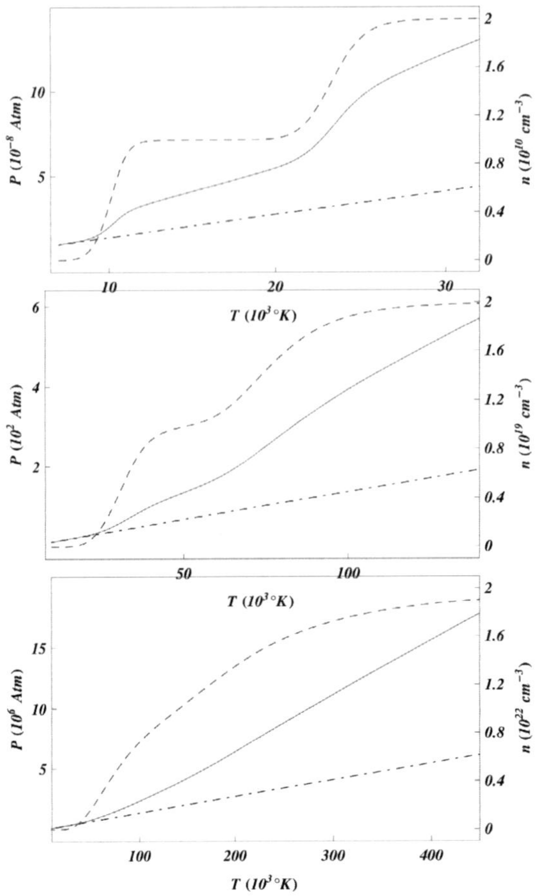

Figure 2. Dependence of the pressure (red line) on the temperature compared with the electron density n_{-1} (blue dashed line) and with the pressure obtained from the ideal gas law neglecting the ionization (black dot-dashed line) for an ensemble of atoms of Helium at the same three different densities showed in Fig. 1.

electron as a cloud similar to the one surrounding an Hydrogen atom, but with a radius equal to the thermal De Broglie length λ and a polarizability equal to 0. With these assumptions free-electrons have interactions similar to the ionic ones. Indeed, the interaction terms between a free-electron and an ion can be described by replacing in the Lennard Jones term, σ with $\frac{\sigma+\lambda}{2}$ and ε with $\sqrt{\varepsilon\,\varepsilon_H}$ (being ε_H the depth of the potential well of the Lennard-Jones potentials between two atoms of Hydrogen), and the $U_{-1,-1}(r)$ term can be represented by substituting σ with λ and ε with ε_H.

In order to show the difference of behavior between the dense interacting and the non-interacting plasmas we consider a system of Xenon atoms at high density and neglect the dependence of σ and ε on the degree of ionization. This approximation does not affects the results in the case in which only of the first ionizations are considered. The plots in Fig. 3 are obtained by considering the first two degrees of ionization of the Xenon and $n = 10^{22}cm^{-3}$. Such plots show that the presence of particle interactions in the state equations does not change significantly the value of c_v, even at high densities. Indeed, the fields h_i in Eq. (2) are about an order of magnitude lower than the ionization energy and therefore their effects on the solution of Saha Equation Eq. (3) are almost negligible. On the contrary, the behaviors of P and of c_p

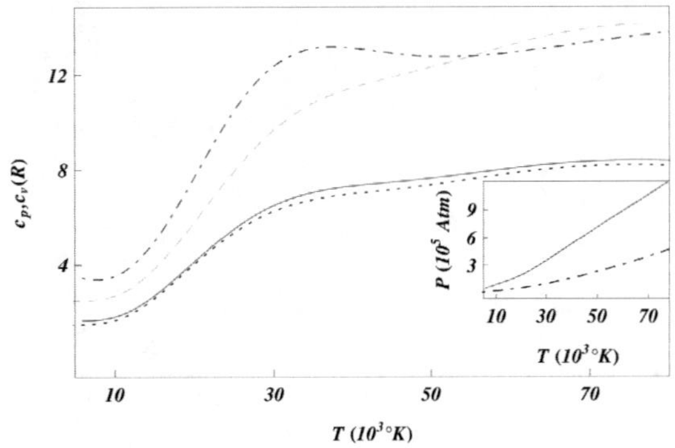

Figure 3. Comparison between interacting and non interacting cases for a mole of xenon. In the main plot it is shown the comparison between the specific heats. Dotted black line stands for c_v in the non interacting case, and solid red line shows c_v in the interacting framework. Green dashed line denotes c_p in the non interacting case, and Blue dot-dashed line represents c_p in the interacting case. In the inset we present the comparison between the pressures. Black dot dashed line refers to the pressure in the non interacting case while the red solid line shows the same quantity in the interacting framework. We considered $n = 10^{22}cm^{-3}$ and only the first two degrees of ionization.

are strongly influenced by the interactions.

5. Adiabatic compression and Schottky like anomaly

In this section we show the influence of the Schottky like anomaly on the temperature in an adiabatic compression. We consider such particular transformation because it plays a fundamental role in different phenomena as for example in the sonoluminescence. Indeed, in such phenomenon the gas inside the bubble follows an isothermal transformation in the bubble expansion phase and an adiabatic compression during the bubble collapse, which represents the more characterizing phase for the emission of light [13].

We study in particular a quasi-static adiabatic compression of a sphere containing 1 mole of atoms which is initially in thermal equilibrium under the standard conditions of pressure and temperature ($P = 1atm$, $T = 300^oK$), and assume that the compression stops when the density in the sphere becomes comparable with the one present in sonoluminescing bubble during the last part of the collapse.

The results of our evaluation are displayed in Fig. 4. Notice that when the radius of the sphere R is large, the solutions obtained for an ideal gas are indistinguishable from our ones. When R becomes smaller than 0.2 R_a ($R_a \simeq 18.04$ cm stands for the radius of sphere at standard condition), gas temperatures obtained by our state equations are lower than the ones obtained for ideal gas. The lowering of the temperature, that depends on the kind of atoms in the sphere, is due to the fact that a part of the energy provided to the system by the compression is used by some atoms to ionize themselves. Atoms of Xenon are the ones in which this effect is more relevant because their ionization energies are lower than the ones of the Argon and of the Helium. Hence the ionization process becomes relevant at higher radii (lower temperatures). Even if the number of free-electrons is very low compared to the number of the atoms, and hence we are only at the onset of the Schottky anomaly, the effect is so strong to halve the temperature in the Helium case at $R = 0.05R_a$. The correction is even stronger if we compare our results with the temperature obtained for van der Waals gases which diverges when the volume of the bubble approaches the covolume of the species.

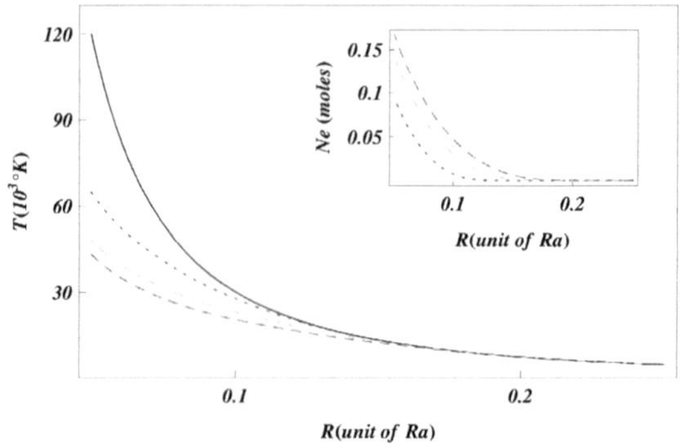

Figure 4. Main plot: Dependence of the temperature on the bubble radius (expressed in unit of the ambient radius R_a), in the case of an adiabatic compression, for the ideal gas (black line) compared with the one that we have obtained using our set of state equations for Helium (blue dotted line), Argon (green dot-dashed line) and Xenon (red dashed-line). Inset: total number of free-electrons as function of the the bubble radius, in the case of an adiabatic compression for the same atomic species of the main plot.

Finally, we note that for densities comparable with the one present in sonoluminescing bubble, the electronic density is in good agreement with the recent results on sonoluminescence [19] according to which the plasma electron density in the bubble exceeds $10^{21} cm^{-3}$. In a forthcoming paper we will analyze in detail the dynamic of sonoluminescing bubbles.

6. Conclusions

We have presented a set of state equations useful for plasma physics, in which the ionization degrees are not fixed *a priori* but are obtained self-consistently using the Saha equation. We derive state equations that can describe thermodynamical properties of warm plasmas both with low and high density. We have focused our analysis on the noble gas atoms, because of their importance in many physical phenomena such as sonoluminescence, but our work can be easily extended also to all kinds of atoms, molecules and mixtures of them by considering also chemical bounds. Using this new approach, we revealed a Schottky like anomaly in the specific heats of noble gas atoms at low densities, and a step-wise behavior of pressure and density of the free-electrons associated to ionization processes.

The anomalies in specific heats become less evident at high densities and disappear when all the atoms are fully ionized. Such anomalies can play a very relevant role in many phenomena of actual interest as for example in sonoluminescence in which collapsing bubbles close to their minimum radius reach temperatures of few tens of thousands of K, densities of order of the solid one and are almost completely filled with noble gas atoms [13, 14]. In such phenomenon, during the last part of the collapse phase, the temperature and the radial speed of the bubble depend on c_p and c_v, then an increment of them can influence strongly the bubble dynamic. In the present paper we have not directly considered the problem of the dynamic of sonoluminescence, which will be analyzed in detail elsewhere, but we have only treated the simple problem of an adiabatic compression of a sphere that reaches a final density comparable with the one of sonoluminescence during the emission. However, even with this simple dynamic, we have obtained electronic densities for the gas in the sphere which is in good agreement with some recent experimental results on sonoluminescence.

Acknowledgement
We acknowledge financial support from the European Commission of the European Union under the FP7 STREP Project HIP (Hybrid Information Processing), Grant Agreement n. 221889.

References
[1] Guillot T 1999 *Science* **286** 72
[2] Remington B A, Drake R P and Ryutov D D 2006 *Rev. Mod. Phys.* **78** 755
[3] Van Horn H M 1991 *Science* **252** 384
[4] Koenig M et al. 2005 *Plasma Phys. Contr. Fusion* **47** B441
[5] Atzeni S and Meyer-ter-Vehn J 2004 *The Physics of Inertial Fusion*, Clarendon Press, Oxford
[6] Key M H 2007 *Phys. Plasmas* **14** 055502
[7] Hummer D G and Mihalas D 1988 *Astroph. J.* **331** 794
[8] Rogers F J, Swenson F J, and Iglesias C A 1996 *Astroph. J.* **456** 902
[9] Cardona O, Martinez-Arroyo M, and Lopez-Castillo M A 2010 *Astroph. J.* **711** 239
[10] Barber B P and Putterman S J 1991 *Nature* **352** 318
[11] Barber B P and Putterman S J 1992 *Phys. Rev. Lett.* **69** 3839
[12] Hiller R, Weninger K, Putterman S J and Barber B P 1994 *Science* **266** 248
[13] Brenner M P, Hilgenfeldt S and Lohse D 2002 *Rev. Mod. Phys.* **74** 425
[14] Lohse D, *Nature* 2002 **418** 381
[15] Flannigan D J and Suslick K S 2005 *Nature* **434** 52
[16] Suslick K S and Flannigan D J 2008 *Annu. Rev. Phys. Chem.* **59** 659
[17] Kurcz A, Capolupo A and Beige A 2009 *New J. Phys.* **11** 053001
[18] Xu H X and Suslick K S 2010 *Phys. Rev. Lett.* **104** 244301-1-4
[19] Flannigan D J and Suslick K S 2010 *Nature Physics* **6** 598-601
[20] Tari A 2003 *The Specific Heat of Matter at Low Temperatures*, (London: Imperial College Press)
[21] Zeldovich Y B and Raizer Y P 1966 *Physics of shock waves and high-temperature hydrodynamic phenomena, Vols. I and II*, (New York: Academic Press)

Power-law quantum distributions in protoneutron stars

G Gervino[1,3], A Lavagno[2,3] and D Pigato[2,3]

[1]Department of Physics, Università di Torino, I-10126 Torino, Italy
[2]Department of Applied Science and Technology, Politecnico di Torino, I-10129 Torino, Italy
[3]Istituto Nazionale di Fisica Nucleare (INFN), Sezione di Torino, I-10126 Torino, Italy

E-mail: andrea.lavagno@polito.it

Abstract. We investigate the bulk properties of protoneutron stars in the framework of a relativistic mean field theory based on nonextensive statistical mechanics, originally proposed by C. Tsallis and characterized by power-law quantum distributions. We study the relevance of nonextensive statistical effects on the β-stable equation of state at fixed entropy per baryon, for nucleonic and hyperonic matter. We concentrate our analysis in the maximum heating and entropy per baryon $s = 2$ stage and $T \approx 40 \div 80$ MeV. This is the phase, at high temperature and high baryon density, in which the presence of nonextensive effects may alter more sensibly the thermodynamical and mechanical properties of the protoneutron star. We show that nonextensive power-law effects could play a crucial role in the structure and in the evolution of the protoneutron stars also for small deviations from the standard Boltzmann-Gibbs statistics.

1. Introduction

A protoneutron star (PNS) is born immediately after the gravitational collapse of a massive star ($M \approx 10 \div 20\ M_\odot$) and in the first seconds of its evolution it is a very hot, lepton rich and β-stable object, with a temperature of a few tens of MeV and a lepton concentration typical of the pre-supernova matter [1].

It is well known that the knowledge of the nuclear equation of state (EOS) plays a crucial role in the determination of the structure and in the evolution of the PNS [2]. The processes related to strong interaction should in principle be described by quantum chromodynamics. However, in the energy density range reached in the compact stars, strongly non-perturbative effects in the complex theory of QCD are not negligible. In the absence of a converging method to approach QCD at finite density one often turns to effective and phenomenological model investigations.

In the last years there is an increasing evidence that the nonextensive statistical mechanics, originally proposed by C. Tsallis and characterized by power-law quantum equilibrium distributions, can be considered as an appropriate physical and mathematical basis to deal with physical phenomena where strong dynamical correlations, long-range interactions, anomalous diffusion and microscopic memory effects take place [3, 4, 5, 6, 7].

In this framework, several authors have outlined the relevance of nonextensive statistical mechanics effects in high energy physics and astrophysical problems [8, 10, 9, 11, 12, 13, 14, 15, 16, 17]. This feature results to be particular valid in the central core of such compact object, where hadrons may be strongly correlated and long range interaction may take place.

The existence of nonextensive statistical effects, should sensibly affect the finite temperature and density nuclear EOS [18, 19, 20, 21]. In fact, by varying temperature and density, the EOS reflects in terms of the macroscopic thermodynamical variables the microscopic interactions in different nuclear matter regimes.

In this work we limit ourselves to consider only small variation from the standard Boltzmann-Gibbs (BG) statistics (from $q = 0.97$ to $q = 1.03$). In particular, we concentrate our study in the stage immediately after the supernova explosion (called deleptonizazion era), when the PSN assumes the maximum heating and entropy per baryon ($s = 2$) and the presence of nonextensive effects may alter more sensibly the thermodynamical and mechanical proprieties of the PNS.

2. Nonextensive hadronic equation of state

The hadronic EOS is calculated in the framework of the nonextensive statistical mechanics introduced by Tsallis [3, 4, 5]. The nonextensive statistics represents a physical mathematical tool in several physical fields and it is based on the following definition of q-deformed entropy functional

$$S_q[f] = \frac{1}{q-1}\left(1 - \int [f(\mathbf{x})]^q \, d\Omega\right), \left(\int f(\mathbf{x})\, d\Omega = 1\right), \tag{1}$$

where $f(\mathbf{x})$ stands for a normalized probability distribution, $\mathbf{x}$ and $d\Omega$ denoting, respectively, a generic point and the volume element in the corresponding phase space. The nonextensive statistics is, therefore, a generalization of the common BG statistical mechanics and for $q \to 1$ it reduces to the standard BG entropy. Furthermore, nonextensive statistical effects vanishes approaching to zero temperature.

In this context, the nonextensive statistics entails a sensible difference on the power-law particle distribution shape in the high energy region with respect to the standard statistic. In particular, for a dilute gas of particles and for small deviations from the standard statistical mechanics ($q \approx 1$), it can be written as [22]

$$n_i = \frac{1}{\tilde{e}_q(\beta(E_i - \mu_i)) \pm 1}, \tag{2}$$

where $\beta = 1/T$ and the sign (± 1) is for fermions and bosons respectively. Following Ref. [22], in Eq.(2), for $q > 1$, we have

$$\tilde{e}_q(x) = \begin{cases} [1 + (q-1)x]^{1/(q-1)} & \text{if } x > 0\,; \\ [1 + (1-q)x]^{1/(1-q)} & \text{if } x \le 0\,, \end{cases} \tag{3}$$

whereas, for $q < 1$,

$$\tilde{e}_q(x) = \begin{cases} [1 + (q-1)x]^{1/(q-1)} & \text{if } x \le 0\,; \\ [1 + (1-q)x]^{1/(1-q)} & \text{if } x > 0\,. \end{cases} \tag{4}$$

In this work, we study the proprieties of hot and dense nuclear medium, using the nonlinear Walecka model (GM3 parameter set), based on a relativistic mean-field model (RMF) approach, where the nuclear force is mediated by the exchange of virtual isoscalar-scalar (σ), isoscalar-vector (ω) and isovector-vector (ρ) mesons fields [23, 24, 25, 26, 27].

The vector (ρ_B) and scalar (ρ_S) baryon density are given respectively by [28]

$$\rho_B = 2\sum_{i=B} \int \frac{\mathrm{d}^3k}{(2\pi)^3} [n_i(k) - \overline{n}_i(k)]\,, \tag{5}$$

$$\rho_S = 2\sum_{i=B} \int \frac{\mathrm{d}^3k}{(2\pi)^3} \frac{M_i^*}{E_i^*} [n_i^q(k) + \overline{n}_i^q(k)]\,, \tag{6}$$

where $n_i(k)$ and $\overline{n}_i(k)$ are the q-deformed fermion particle and antiparticle distributions function given in Eq.(2); For $q > 1$ and $\beta(E_i^* - |\mu_i^*|) > 0$, we have, for example,

$$n_i(k) = \frac{1}{[1 + (q-1)\,\beta(E_i^*(k) - \mu_i^*)]^{1/(q-1)} + 1}\,. \tag{7}$$

The nucleon effective energy is defined, as usual, by $E_i{}^*(k) = \sqrt{k^2 + M_i{}^{*2}}$, where $M_i{}^* = M_i - g_{\sigma B}\sigma$ and the effective chemical potentials is: $\mu_i^* = \mu_i - g_{\omega B}\omega - \tau_{3iB}g_{\rho B}\rho$.

The β-stability condition and the charge neutrality are given by

$$\mu_\Lambda = \mu_{\Sigma^0} = \mu_{\Xi^0} = \mu_n\,, \tag{8}$$
$$\mu_{\Sigma^-} = \mu_{\Xi^-} = \mu_n + \mu_e\,, \tag{9}$$
$$\mu_p = \mu_{\Sigma^+} = \mu_n - \mu_e\,, \tag{10}$$
$$\rho_p + \rho_{\Sigma^+} - \rho_{\Sigma^-} - \rho_{\Xi^-} - \rho_e = 0\,. \tag{11}$$

The total entropy per baryon is calculated as

$$s = \frac{1}{T\rho_B}(S_B + S_l)\,, \tag{12}$$

with

$$S_B = P_B + \epsilon_B - \sum_{i=B}\mu_i\rho_i\,, \tag{13}$$
$$S_l = P_l + \epsilon_l - \sum_{i=l}\mu_i\rho_i\,, \tag{14}$$

where the sums runs over all the baryons and leptons species. In the considered stage ($s = 2, Y_\nu = 0$), neutrinos have already escaped from the PNS and the lepton fraction ($Y_L = Y_e + Y_{\nu_e} = (\rho_e + \rho_{\nu_e})/\rho_B$) reduces to $Y_L = Y_e$, where ρ_e, ρ_{ν_e} and ρ_B are the electron, neutrino and baryon number densities, respectively.

Finally, the thermodynamical quantities can be obtained from the thermodynamic potential in the standard way. More explicitly, the baryon pressure P_B and the energy density ϵ_B can be written as

$$P_B = \frac{2}{3}\sum_i\int\frac{\mathrm{d}^3k}{(2\pi)^3}\frac{k^2}{E_i^*(k)}[n_i^q(k) + \overline{n}_i^q(k)] - \frac{1}{2}m_\sigma^2\sigma^2 - U(\sigma) + \frac{1}{2}m_\omega^2\omega^2 + \frac{1}{2}m_\rho^2\rho^2\,, \tag{15}$$

$$\epsilon_B = 2\sum_i\int\frac{\mathrm{d}^3k}{(2\pi)^3}E_i^*(k)[n_i^q(k) + \overline{n}_i^q(k)] + \frac{1}{2}m_\sigma^2\sigma^2 + U(\sigma) + \frac{1}{2}m_\omega^2\omega^2 + \frac{1}{2}m_\rho^2\rho^2\,. \tag{16}$$

The implementation of the hyperon degrees of freedom comes from determination of the corresponding meson-hyperon coupling constants that have been fitted to hypernuclear properties and their specific values are taken from Ref. [26].

In Fig. 1 and in Fig. 2, we report some of the most important effects in the structure and in the thermodynamical properties of the PNS in the framework of nonextensive statistical mechanics, during the maximum heating phase ($s = 2$ and $Y_\nu = 0$). In the left panel of Fig. 1, we show the variation of the maximum baryonic mass as a function of the central baryon density ρ_c, for different values of q, both for nucleonic (np) and hyperonic (npH) stars. We observe a remarkable increase (reduction) of the maximum baryonic mass when $q > 1$ ($q < 1$). The corresponding mass-radius relation (right panel) is also sensibly modified in presence of a nonextensive statistics.

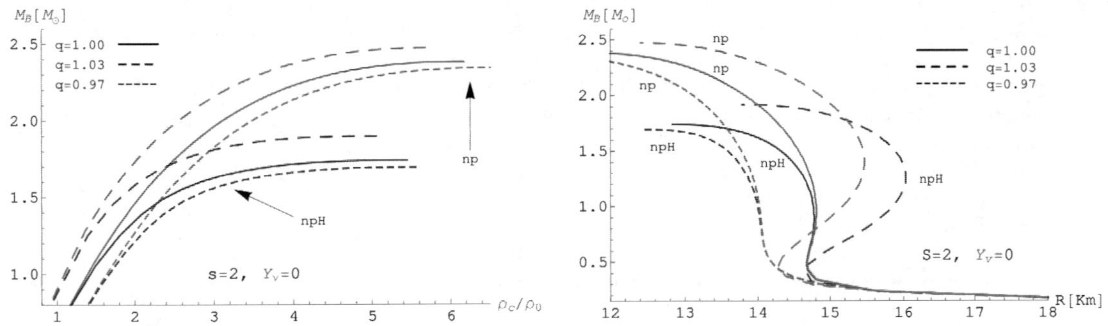

Figure 1. Maximum baryonic mass M_B in units of solar mass $M_\odot$ (left panel) and the corresponding stellar radius profile (right panel), as a function of the central baryon density ρ_c (in units of the nuclear saturation density ρ_0) for different values of q and for nucleons and hyperons stars.

The above behaviors are substantially due to the softening of the EOS in presence sub-extensive statistics ($q < 1$), together with a remarkable reduction of the maximum temperature both for nucleonic and hyperonic PNS, as reported in Fig. 2. Contrariwise, when $q > 1$, we observe an increase of the maximum temperature. These remarkable differences in the stellar temperature have important consequences in the PNS evolution and, consequently, in the cooling of the PNS, making it longer when $q > 1$, and shorter when $q < 1$, with important astrophysical implications. All of these effects are also present when neutrinos are trapped in the PNS ($s = 1$ and $Y_L = 0.4$), however, due to the lower temperature achieved in this phase, they are less pronounced.

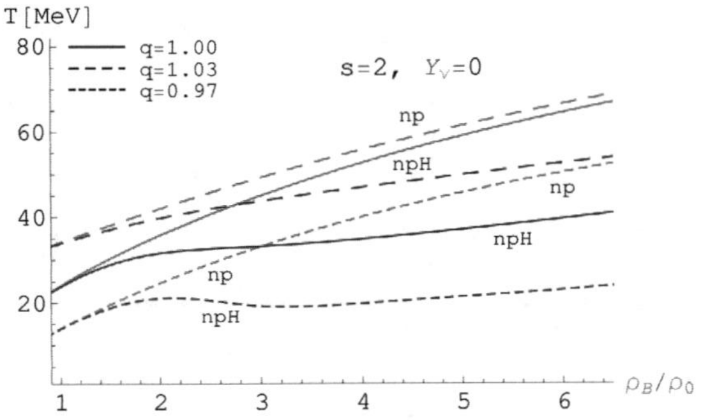

Figure 2. Temperature as a function of the baryon density (in units of the nuclear saturation density ρ_0) for nucleonic and hyperonic PNS, for different values of q.

In conclusion, we have shown that the nonextensive statistical mechanics, characterized by power-law quantum distributions, can play a crucial role on the physical properties of the PNS and can be considered as an effective mathematical basis to investigate the complex structure and evolution of the PNS. From a phenomenological point of view, we have considered the nonextensive index q as a free parameter, even if, in principle, it should depend on the physical conditions inside the PNS, on the fluctuation of the temperature and be related to microscopic quantities (such as, for example, the mean interparticle interaction length). The variation of the maximum temperature, together with a significant variation of the maximum mass and stellar radius, could be considered as a relevant phenomenological behavior in order to observe some astrophysical evidences of nonextensive statistical effects in PNS.

References

[1] Prakash M, Lattimer J M, Pons J A, Steiner A W and Reddy S 2001 *lect. Notes Phys.* **578** 364
[2] Drago A, Lavagno A and Parenti I 2007 *Astrophys. J.* **659** 1519
[3] Tsallis C 1988 *J. Stat. Phys.* **52** 479; Tsallis C, Mendes R S and Plastino A R 1998 *Physica* A **261** 534
[4] Gell-Mann M, Tsallis C 2004 *Nonextensive Entropy: Interdisciplinary Applications* (New York, Oxford University Press)
[5] Tsallis C 2009 *Introduction to Nonextensive Statistical Mechanics* (New York, Springer-Verlag)
[6] Kodama T, Elze H -T, Aiguiar C E and Koide T 2005 *Europhys. Lett.* **70** 439
Elze H -T and Kodama T 2005 *Phys. Lett.* A **335** 363
[7] Lavagno A and Swamy P N 2002 *Phys. Rev.* E **65** 036101
Lavagno A and Swamy P N 2000 *Phys. Rev.* E **61** 1218
[8] Bediaga I, Curado E M F and de Miranda J M 2000 *Physica* A **286** 156
[9] Wilk G and Włodarczyk Z 2009 *Eur. Phys. J.* A **40** 299
[10] Biró T S, Purcsel G and Ürmössy K 2009 *Eur. Phys. J.* A **40** 325
[11] Alberico W M and Lavagno A 2009 *Eur. Phys. J.* A **40** 313
[12] Ferro F., Lavagno A and Quarati P 2004 *Eur. Phys. J.* A **21**, 529 (2004)
[13] Lavagno A, Scarfone A M and Swamy P N 2007 *J. Phys. A: Math. Theor.* **40** 8635
Lavagno A, Scarfone A M and Swamy P N 2006 *Eur. Phys. J.* C **47** 253
[14] Lavagno A and Quarati P 2001 *Phys. Lett.* B **498** 47
Lavagno A 2008 *J. Phys. A: Math. Theor.* **41** 244014
[15] Cleymans J 2010 *J. Phys. G: Nucl. Part. Phys.* **37** 094015
Cleymans J and Worku D 2012 *J. Phys. G: Nucl. Part. Phys.* **39** 025006
[16] Chinellato D D, Takahashi T and Bediaga I 2010 *J. Phys. G: Nucl. Part. Phys.* **37** 094042
[17] Alberico W M et al. 2008 *Physica* A **387** 467
Alberico W M, Lavagno A and Quarati P 2000 *Eur. Phys. J.* C **12** 499
[18] Pereira F I M, Silva R and Alcaniz J S 2009 *Phys. Lett.* A **373** 4214
[19] Lavagno A, Pigato D and Quarati P 2010 *J. Phys. G: Nucl. Part. Phys.* **37** 115102
[20] Lavagno A and Pigato D 2011 *Eur. Phys. J.* A **47** 101140
[21] Lavagno A and Pigato D 2012 *J. Phys. G: Nucl. Part. Phys.* **39** 125106
[22] Rozynek J and Wilck G 2009 *J. Phys. G: Nucl. Part. Phys.* **36** 125108
[23] Glendenning N K and Moszkowski S A 1991 *Phys. Rev. Lett.* **67** 2414
[24] Di Toro M, Drago A, Gaitanos T, Greco V and Lavagno A 2006 *Nucl. Phys.* A **775** 102
[25] Bonanno L, Drago A and Lavagno A 2007 *Phys. Rev. Lett.* **99** 242301
[26] Lavagno A 2010 *Phys. Rev.* C **81** 044909
[27] Lavagno A and Pigato D 2012 *Phys. Rev.* C **86** 024917
[28] Lavagno A 2002 *Phys. Lett.* A **301** 13
Lavagno A 2002 *Physica* A **305** 238

DICE2012 IOP Publishing
Journal of Physics: Conference Series **442** (2013) 012066 doi:10.1088/1742-6596/442/1/012066

Primordial Planck mass black holes (PPMBHs) as candidates for dark matter?

Gustav M Obermair

Department of Physics, University of Regensburg, Germany
presently: Swakopmund, Namibia

E-mail: obermair.namibia@gmail.com

Abstract. It is proposed here that due to rapid cooling at the end of the Planck-era a second order phase transition with Planck-size space-time-energy fluctuations may have led to a precipitation of Planck-dimension real objects with masses around m_P and hence a Schwarzschild radius $\Re = 2Gm_P/c^2$. This equals twice their spatial dimension l_P; hence they are black holes. They could have survived through inflation and expansion because of their extremely small cross section for the fluctuations that are supposed to cause the hypothetical Hawking decay. In the case of our galaxy they could thus be the missing WIMPS of the dark mass halo, consisting of $\sim 10^{50}$ PPMBHs – in contrast to the $\sim 10^{68}$ baryons in the stars – both bound together by gravitation with negligible other interaction between the two. They would constitute a storage of gravitational energy three orders of magnitude larger than the energy stock of $H \rightarrow He$ fusion, an energy to be released somehow when they very slowly coalesce to form big black holes.

1. The birth of quantum theory and of Planck units

Six years after his epoch-making discovery of the quantization of radiation energy and a first remarkably accurate estimate of his constant h, Max Planck devotes a chapter of his 1906 book Theory of Thermal Radiation to the construction of a system of basic natural units of length, time and mass, today called Planck units. They are constructed from the three fundamental constants of nature: c, G and h (or $\hbar = h/2\pi$); much of today's cosmology uses these three alleged invariables to extrapolate the evolution of our universe back to a plausible model for the famous first three minutes, thus tacitly assuming these constants had their presently measured values throughout the evolution of our universe.

Planck units can be constructed because speed of light c, gravitational constant G and Planck constant $\hbar$ all contain in their definition different arithmetic combinations of length L, time T and mass M: $c\,[L/T]$, $G\,[L^3/MT^2]$, $\hbar\,[ML^2/T]$.

Thus a seemingly trivial reshuffling leads to the desired new units: Planck length l_P, Planck time t_P, and Planck mass m_P, given here with their numerical values [1] in our conventional SI units:

$$l_P = \sqrt{(\hbar G/c^3)} = 1.62 \times 10^{-35}\ \text{m}\ , \tag{1}$$

$$t_P = l_p/c = 5.40 \times 10^{-44}\ \text{s}\ , \tag{2}$$

$$m_P = \hbar/l_p c = 2.17 \times 10^{-8}\ \text{kg}\ . \tag{3}$$

The simple construction, though, implies a much deeper insight: if c, G and $\hbar$ express universal relations within the dynamics of the universe, the Planck units should give us a universal natural metric for this universe, spanning the cosmological gap between microphysics and macrophysics, in contrast to our familiar SI units that – like foot, hour, and pound – are strictly incidental and geocentristic, based on an arbitrary choice of equally arbitrary fractions of spatio-temporal dimensions of our little home-planet.

In spite of this plausible argument for Planck units, the idea was not followed up and nearly forgotten in the next half-century: not only are the Planck dimensions rather impractical, more than 25 orders of magnitude smaller than any quantities considered in that period, there were (and are) also no physical effects for which a theory predicts results containing both $\hbar$ and G. In other words: quantum gravity as a branch of the discipline did not exist. Only 50 years later, when astrophysics and cosmology had become respectable and promising fields, one came back to the attempt to bridge the gulf separating the two extremely successful theories involved - general relativity as the theory of the very large, ruled by G - and quantum theory as the theory of the very small, ruled by $\hbar$; for a thorough account of these efforts of the last 50 years see [2].

It may have been noticed before, but not much discussed, that if objects of Planck dimensions did exist, they would be black holes: not really surprisingly, since we only have c, G and $\hbar$ to play around with: the Schwarzschild radius $\Re = 2Gm_P/c^2$ of a Planck mass object equals twice the Planck length:

$$2Gm_P/c^2 = 2\sqrt{G\hbar/c^3} = 2l_P \ . \quad (4)$$

Such objects of semi-classical Planck dimension, from here on called Planck mass black holes – PMBHs – would then obviously be the materialization of the missing link between QT and GR: both $\hbar$ and G determine their properties.

2. The Planck era and the putative birth of Planck mass primordial black holes

Within the widely accepted scenario of the very early universe it is assumed that in the first extremely hot stage, a few Planck times after the bang, the state is homogeneous up to space - time - energy fluctuations of Planck dimension $(l_P, t_P, m_P c^2)$; this stage hence has been called the Planck era.

The energy involved in the fluctuations – $m_P c^2$ – is $\sim 10^{19}$ GeV, the corresponding Planck temperature is:

$$T_P = m_P c^2/k_B \sim 10^{32}\ \mathrm{K} \ . \quad (5)$$

Within quantum field theory this stage is understood as the high temperature phase of QCD; forty years ago George Gamow created the wonderful word YLEM for the substance of this primordial state [3].

Now one may imagine the following scenario of a second order phase transition brought about by the dominance of gravitation between masses of Planck order at distances of Planck length order: Above a critical temperature T_c of the order T_P the primordial soup Ylem is nearly homogeneous; as it cools down to approach T_c one expects critical fluctuations on the Planck scale. At further cooling Ylem divides up into two phases: a precipitation of grains of the black hole phase immersed in what is to become the quark-gluon-plasma: this is where "God divided the light from the darkness" (Gen.1 verse 4) - Planck mass black holes floating around in the future ordinary matter ("light") with none but gravitational interaction between the two, except for the elusive quantum relativity effects that may or may not occur near the horizon of the PMBHs to couple them to all other fields.

The PMBH-grains need of course not be uniform in size; their size and mass distribution around l_P and m_P should be determined by factors like the compressibility of the high temperature phase and the speed of cooling. In the extremely rapid expansion and cooling phase only some part of the grains may have had a chance to coagulate to become the seeds of macroscopic black holes, the others will have remained as an extremely rarified gas; see sec. 4.

3. Could Planck mass black holes be (meta)stable?

The answer is obviously connected to the general question of the coupling of black holes to all other, non-gravitational fields, more specifically to the possible effect of the Hawking process on Planck-size objects. To quote Kip Thorne, an important contributor to the Black Hole Physics development in the 1970ies, who expresses his doubts 20 years later: "Maybe we physicists understand the quantum fields in a curved space-time much less than we believe. Our assumption that black holes vaporize might turn out to be a complete error." [4]

Both empirical and theoretical arguments nourish these doubts still now:

- The intensive gamma-radiation from Hawking decay could up to now not be extracted and identified within the total gamma spectrum obtained with ever more sensitive gamma telescopes.
- A fundamental theory to fully and correctly describe the process by which one photon of a pair is falling into the hole to deliver the escape energy for the other one: that is precisely the theory of Quantum Gravity which – in spite of half a century's work of the physics elite – we still do not have.

There are, e.g., recent string theory considerations indicating the possibility of a very slow decay [5].

But even assuming the Hawking process does exist, one may argue that it is nearly forbidden for Planck size objects that can only interact with quantum fluctuations of wavelengths around l_P; the phase space segment of such fluctuations is vanishingly small and hence also the probability of PMBH vaporization. Accepting these arguments for the near stability of PMBHs and the evidence that galaxies are immersed in a halo of dark matter that may consist of WIMPs – massive particles slow enough to be bound gravitationally in this halo and with (nearly) vanishing other interactions – we arrive at the hypothesis that these mysterious WIMPs could be the primordial PMBHs.

4. Our galaxy's dark matter halo

The proposition is that the dark matter halo in which our galaxy appears to be embedded is made up of black holes with undiminished masses of the order of Planck masses, i.e. some 10^{-8} kg each. With an estimated 2×10^{41} kg of baryonic mass in the galaxy, roughly 10^{68} baryons, and apparently at least 5 times more dark mass, this amounts to an estimate of 10^{50} PMBHs within the halo of about 10^5 light years diameter, in other words roughly one PMBH in 10^{13} m^3.

Since they appear to be gravitationally bound to the galaxy, their velocities must be in the range up to the escape velocity from the galaxy's outskirts, that is up to 500 km/s. In contrast to the nondescript WIMPs - after all nothing more than a nice word for something unknown - the proposed PMBHs would at least have a cosmological history, a plausible origin at the end of the Planck era. But how could any empirical evidence of their survival and their individual existence be obtained and a search for them organized, beyond the collective gravitational effect ascribed to them?

This question can convincingly only be answered once we have a valid concept of quantum gravitation, that is better understanding of the interaction of quantum fields with curved space-time.

In detail we would have to be able to answer questions like this:

What happens if the classical trajectory of a PMBH traverses a proton which - classically speaking - is twenty orders of magnitude larger than the hole's Schwarzschild radius? (The purely gravitational classical interaction of the two would lead to a transfer of a tiny fraction of kinetic energy of the order $m_p/m_P \sim 10^{-19}$, not much of a moderation.)

What happens in a rare close encounter of two PMBHs? In a fusion at least part of the gravitational potential energy of the Newtonian order $E_{pot} = Gm_P^2/l_P = m_P c^2$, i.e. of the order of their rest energy, might be emitted as some kind of radiation.

What composition would this emission have – gravitons, ultrahard gamma rays, particle showers? How efficient would the different decay channels be?

One might announce a competition for a study: Potential graviton emission from the fusion of primordial Planck mass black holes.

To sum it up: if the speculations developed here find any attention, they might motivate some specialists for such investigations. In any case the gravitational fusion energy E_{pot} stored in the PMBHs that make up all the dark matter in the galaxies would be three orders of magnitude larger than the $H \rightarrow He$-fusion energy of all protons. This gives rise to a further daring speculation: Might such energy, extremely slowly released as radiation over many billions of years since the formation of dense regions in galaxies, play a role as a source of dark energy?

References

[1] Kiefer C 2004 in *Decoherence and Entropy in Complex Systems* Springer Lecture Notes in Physics, LNP **633** ed Elze H-T (Berlin: Springer)
[2] Rovelli C 2004 *Quantum Gravity* (Cambridge, UK: Cambridge Monographs)
[3] Gamow G 1970 *My World Line: An informal autobiography* (New York: Viking)
[4] Thorne K S 1993 *Black Holes and Time Warps* (New York: W. W. Norton & Co)
[5] Barrau A *et al.* 2005 Experimental searches for primordial black holes *Astrophys. J.* **630** 1015

Acceleration factor for propagation of a stationary wave in its wave medium: movement of energy in the 3-D space[1]

Shigeto Nagao[2]
Fujizuka, Kohoku-ku, Yokohama, Kanagawa, 222-0012 Japan
E-mail: snagao@lilac.plala.or.jp

Abstract. According to the formerly reported 4-D spherical model of the time and universe, any energy in the 3-dimensional space is a vibration of the intrinsic space energy. There is a special frame stationary to the space energy and the principle of relativity is no longer valid. Accordingly, abandonment of the Special Relativity and then introduction of a factor of acceleration for energy in the 3-D space are proposed.

1. Introduction

The author of this article has been proposing a model of the time and universe, in which the space energy is spread with expansion in a 3-dimensional surface of a 4-dimensional sphere (referred to as the "4-D Spherical Model" or "4DS Model") [1, 2]. What we detect as energy or mass in the 3-D space is a vibration of the intrinsic space energy in the 3-D space. What we observe as the time passing at a constant speed (referred to as the "Time") is the radius dimension of the 4-D sphere for the distribution of the space energy. The most notable consequence expected from the 4DS Model is the perfect linearity of the space expansion by our observed Time. The model also insists on the changing speed of light propagation in accordance with the expansion of the universe. By adjusting the light speed, the Hubble plots from the Supernova Cosmology Project has successfully turned to show a perfect linearity, which is evident for the fact that the universe has been expanding at a constant speed by our observed Time [2].

The 4DS Model for the universe insists the wave medium for any energy in the 3-D space. Therefore there is a special frame, which is stationary to the medium, differentiated from other frames in an inertial motion to the medium. Furthermore, for energy in the 3-D space, there should be special requirements in its mechanics for motion, which are not applied for the expansion of the universe, because it is a movement of a vibration in its wave medium.

In this article, firstly abandonment of the Special Relativity and then respective mechanics for energy in the 3-D space and for the space energy in the 4-D sphere are discussed.

[1] This paper is a private work independent from the company.
[2] Business Development and Licensing, Nippon Boehringer Ingelheim Co., Ltd.

2. Special Relativity

2.1. Propositions of Special Relativity

The Special Relativity (referred to as the "SR") is based on the two propositions; one is the Principle of Relativity and the other is the Principle of Invariant Light Speed [3]. If the light is a vibration of the intrinsic space energy (referred to as the "Spacia") as I strongly propose, there is a special frame stationary to the medium Spacia, differentiated from other frames in an inertial motion to the Spacia. The Principle of Relativity is no longer valid. The Principle of Invariant Light Speed implies that the light propagating speed is independent from the sate of motion of the emitting body. This is rather rationale as a feature of wave that the propagation speed to the medium is constant independent from the emitter's speed. However, combined with the Principle of Relativity, it becomes to signify that the light speed is constant even when the observer is moving. The light speed invariance by selection of frame to measure was believed to be proven by the Michelson-Morley experiment [4]-[6]. However, no change in the location of interference fringe can be expected theoretically from the Michelson-Morley experiment. In my previous article [2], there were wrong explanations for the invalidity of the Michelson-Morley experiment. New interpretation for it is as follows.

In the experiment, a coherent light beam is split by a half-mirror into a straight beam and a reflected right angle beam. Both beams are reflected at the respective ends of arms, return to the half-mirror and are combined for detection. The apparatus is moving at speed v to the light medium. Take the reference frame stationary to the medium. Propagation image of the two beams are shown in the Figure 1.

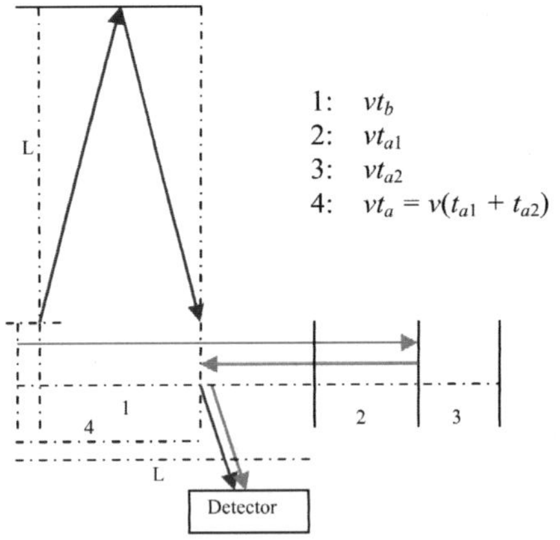

Figure 1. Michelson-Morley experiment by the frame stationary to the medium

The split beam A parallel to the apparatus movement takes t_a to return to the half-mirror since released from it. The beam B perpendicular to the movement takes t_b to return. The traveled distance of the beam B is given as

$$D_b = \sqrt{4L^2 + v^2 t_b^2}\ . \tag{1}$$

That of the beam A is as follows:

$$D_a = L + vt_{a1} + L - vt_{a2} = 2L + 2vt_{a1} - vt_a \tag{2}$$

From $t_a/t_{a1} = D_a/(L + vt_{a1})$, t_{a1} is eliminated and D_a is written as

$$D_a = L + \sqrt{L^2 + v^2 t_a^2} \tag{3}$$

Light speed is equal for both beams by the frame of medium, $c = D_b/t_b = D_a/t_a$. Therefore we get the following formula.

$$t_b^2 = \frac{2Lt_a^2}{L + \sqrt{L^2 + v^2 t_a^2}} \tag{4}$$

Unless $v = 0$ or $t_a = 0$, $t_b < t_a$. Put $\Delta t \equiv t_a - t_b$.

The displacement of the beam A to be combined with the beam B is released Δt earlier at the splitter than the beam B is released. Respective amplitudes are given as follows.

$$U_a = A\sin(kx - \omega t) \tag{5}$$

$$U_b = A\sin(kx - \omega(t + \Delta t)) \tag{6}$$

At the point where the two beams are combined to interfere, the sum of amplitudes is given as follows.

$$U_a + U_b = -A(\sin\omega t + \sin(\omega t + \omega\Delta t)) \tag{7}$$

At the interfering point, the frequencies are the same for both the beam A and beam B. Therefore, no change in the location of fringe is theoretically expected by variation of Δt or v, whereas the maximum amplitude and the phase of the combined wave are varied by Δt or v.

To make sure, let's check measurement by a different frame fixed to the apparatus. The traveled distance is $2L$ for the both beams. The average light speed of the beam A is $2L/t_a$ and that of the beam B is $2L/t_b$. They are different unless $v = 0$. However, the respective directions of propagation of the two beams are same at the detector (interferometer). Therefore, the light speed, frequency and wave length are common for the two beams at the detector even by the apparatus frame, whereas those values depend on the apparatus speed v to the medium.

2.2. Invariance of the rest mass proposed by Special Relativity

One of the most important outcomes from the SR is the energy-mass equivalence shown by $E = mc^2$ [3][7]-[10]. This equation can be induced even without using the combination of the two propositions of the SR or other consequences from the SR, for instance from the conservation of momentum in a process that light is released from one end and captured at the other end [11].

On the other hand, the SR insisted the invariance of 4-dimensional momentum in the Minkowski space-time, and proposed the following relativistic energy-momentum equation [9][10].

$$E^2 - (pc)^2 = (mc^2)^2 \tag{8}$$

It insisted that the quantity $E^2 - (pc)^2$ is invariant from selection of the frame of reference in which it is measured, while the energy E and the momentum p depend on the frame. The mass in the equation is called the rest mass or the invariant mass, which is conserved and invariant for all observers [9]. However, is there any basis which can theoretically support the 4-D momentum or the invariance of the quantity $E^2 - (pc)^2$? I strongly insist that the total energy should be invariant for selection of a reference frame, but its breakdown to the rest mass energy and the kinetic energy should depend on a reference coordinate to express the subject.

2.3. Electromagnetics measured by a moving coordinate

A trigger of the Lorentz transformation and the Special Relativity was the covariance of Maxwell equations toward the selection of coordinates to express. However, Maxwell and other equations in the electromagnetics should be applied as they are even without Lorentz transformation if the light speed c is replaced by a relative light speed c_r, which is variable by selection of a coordinate to express. Let's check up the light speed as an example.

The light speed is given as follows from Maxwell equations for vacuum where the electric charge density $\rho = 0$ and the electric current density $j = 0$.

$$\nabla^2 \boldsymbol{E} - \mu_0 \epsilon_0 \frac{\partial^2 \boldsymbol{E}}{\partial t^2} = 0 \ , \quad \nabla^2 \boldsymbol{B} - \mu_0 \epsilon_0 \frac{\partial^2 \boldsymbol{B}}{\partial t^2} = 0 \tag{9}$$

$\boldsymbol{E}$, $\boldsymbol{B}$, μ_0 and ϵ_0 denote electric field, magnetic field (or magnetic flux density), vacuum permeability and vacuum permittivity, respectively. A solution of the wave equation is

$$\Psi\left(x, t\right) = A \sin\left(x \pm \frac{1}{\sqrt{\mu_0 \epsilon_0}} t\right) = A \sin\left(x \pm ct\right) \ . \tag{10}$$

$$\mu_0 \epsilon_0 = \frac{1}{c^2} \tag{11}$$

By a frame in an inertial movement at a speed $\boldsymbol{v}$ toward the Spacia, the relative light speed c_r should be given as

$$c_r = c - v \cos\theta \tag{12}$$

, where θ is the angle between the light propagation and $\boldsymbol{v}$. By replacing c , μ_0 and ϵ_0 respectively by relative ones c_r, μ_{0r} and ϵ_{0r} measured in a given coordinate system, the equations (9), (10) and (11) are applicable in same form without using Lorentz transformation, because c_r becomes a constant once a coordinate system and a direction of light propagation are given.

3. The 4-D spherical model of the universe

The 4DS Model of the universe previously reported is summarized in the Figure 2 as follows [1][2]. Any energy is a vibration in multiple dimensions. The gravitational force works between

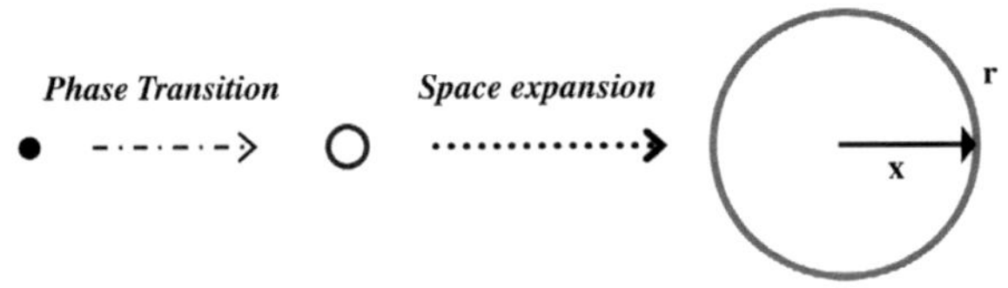

Figure 2. Expansion of the energy distribution of the space

any pieces of energy in accordance with the Newtonian equation of gravity. At the Big Bang, the energy distribution transformed to be spread with expansion in a 3-dimensional surface of a 4-dimensional sphere in tracing by the original tracing time dimension (referred to as the "time" or "t"). The time we observe passing constantly (referred to as the "Time" or "T") is the movement of the space energy in the radius dimension (x in the Figure 2) of the 4-D sphere. The "Tracing" dimension (or coordinate) is defined as a dimension, by variance of which a value-change (movement) of the resting other dimensions is expressed. To work as a Tracing dimension, any single value of it should correspond to a single value of any resting dimension for a movement [1].

The Tracing dimension can be expressed as an imaginary number component of a spatial movement. A movement in $\boldsymbol{e}_k$ space traced by $\boldsymbol{e}_0$ can be expressed as

$$d\boldsymbol{A}^k = idx^0 + dx^k \boldsymbol{e}_k \ . \tag{13}$$

Its corresponding movement in $\boldsymbol{e}_{k-1}$ space traced by $\boldsymbol{e}_1$ can be expressed as follows if $\boldsymbol{e}_1$ shows the Imaginary order of freedom when it is traced by $\boldsymbol{e}_0$. Here the "Imaginary order of freedom" is defined that there is no freedom of selecting a value and furthermore the value is not a constant but is moving in a single direction [1].

$$d\boldsymbol{A}^{k-1} = idx^1 + dx^{k-1}\boldsymbol{e}_{k-1} \tag{14}$$

For the universe, a movement in the 4-D space traced by the original tracing dimension t can be expressed as

$$d\boldsymbol{A}(4)_t = idt + dx\,\boldsymbol{e}_x + dr\,\boldsymbol{e}_r \ , \tag{15}$$

where $\boldsymbol{e}_x$ and $\boldsymbol{e}_r$ denote a base vector for the radius vector $\boldsymbol{x}$ and that for a position vector $\boldsymbol{r}$ in the 3-D surface space, respectively. Multiplying the imaginary component by $v_x \equiv dx/dt$, we get the following vector

$$d\boldsymbol{A}(4)_t = iv_x dt + dx\,\boldsymbol{e}_x + dr\,\boldsymbol{e}_r \tag{16}$$

so that its metric for the space expansion in the radius x becomes zero as shown by

$$ds^2 = -\left(\frac{dx}{dt}\right)^2 dt^2 + dx^2 = 0 \ . \tag{17}$$

A movement in the 3-D space $\boldsymbol{r}$ can be expressed as follows with tracing respectively by t and T.

$$d\boldsymbol{A}(3)_t = iv_x dt + dr\,\boldsymbol{e}_r \ , \quad d\boldsymbol{A}(3)_T = idx + dr\,\boldsymbol{e}_r = idT + dr\,\boldsymbol{e}_r \tag{18}$$

Respective speeds of the movement traced by t and T are

$$v_r \equiv \frac{dr}{dt} \ , \quad V_r \equiv \frac{dr}{dT} = \frac{dr}{dx} \ . \tag{19}$$

The speed traced by t can be expressed by T as follows.

$$v_r = \frac{dr}{dt} = \frac{dx}{dt}\frac{dr}{dx} = v_x V_r \tag{20}$$

In our usual observation in the 3-D apace, the expansion speed of universe v_x can be regarded as a constant because the observation period is extremely short compared with the time scale of the expansion of universe. Making the current v_x be v_{xc}, we can regard $v_x \approx v_{xc}$ for our usual observation in the 3-D space. We obtain the following correspondence between tracing by t and T.

$$d\boldsymbol{A}(3)_t \approx iv_{xc} dt + dr\,\boldsymbol{e}_r \tag{21}$$

$$d\boldsymbol{A}(3)_T = idT + dr\,\boldsymbol{e}_r \tag{22}$$

$$v_r \approx v_{xc} V_r \tag{23}$$

4. Mechanics of energy in the 3-D space in a stationary frame to the Spacia

From now, let us discuss mechanics in a fixed frame stationary to the space energy Spacia. In this section 4, use just "time" or "t" for our observed Time or T, and call the original tracing dimension as the "original tracing time".

Even if gravitational interaction is same for both the expansion of the universe in the 4-D sphere and the energy in the 3-D space, the motion equations for them may differ because there should be limitations in kinetics for energy in 3-D space as it is a propagation of a wave in its medium. There are important features as follows for our consideration:

- Any non-zero energy can not be accelerated to the maximum speed of light speed c.

- In addition that the light speed is not accelerated, it is not decelerated by gravitational force.
- Light receives acceleration by a massive star to a direction perpendicular from its propagation, that is, gravitational lens effect.

In order to comply with them, it is reasonable to expect the followings:

- Any inertial movement of a stationary wave in the medium receives no resistance from the medium.
- However, acceleration of a stationary wave by force would be affected by a factor depending on its speed to the medium. The maximum speed is the propagation speed of a phase in the medium.

I propose that the Newtonian equation of motion shown by (24) is no longer valid for energy in 3-D space, but acceleration is multiplied by a certain factor (referred to as the "acceleration factor" or "f_a") as shown by (25).

$$F = m\alpha \tag{24}$$

$$\alpha = \frac{F}{m} f_a \ , \quad F = \frac{m}{f_a}\alpha \tag{25}$$

From the fact that light is not decelerated and the maximum speed is c, the following formula would be expected as the acceleration factor.

$$f_a = \left(1 - \frac{v^2}{c^2}\right)^n \quad \texttt{(}n\texttt{ is a positive real number)} \tag{26}$$

I previously proposed that the working mass for gravitational interaction would be corresponding to the total energy because light receives the gravitational lens effect. However, there is a possibility that the rest mass may be for gravity if f_a exists. Before searching what n is, let's clarify mass for which, the rest energy or the total energy, is applied for the Newtonian gravitational equation and the equation of motion (25). Take a case of a subject with rest mass of m_0 being accelerated by a huge mass M. At $r = \infty$, $v = v_0$. m_t is the mass for total energy of the subject. In the movement from $r = \infty$ to $r = r$ the total energy H is preserved. The rest mass energy is also conserved as the observation frame is fixed to be stationary to the Spacia.

$$H = E_k + E_p + E_r = m_t c^2 \tag{27}$$

$$E_r = m_0 c^2 \tag{28}$$

$$|H|_{\infty}^{r} = |H|_{v0}^{v} = |E_k|_{v0}^{v} + |E_p|_{\infty}^{r} + |E_r|_{\infty}^{r} \tag{29}$$

$$|E_k|_{v0}^{v} + |E_p|_{\infty}^{r} + 0 = 0 \tag{30}$$

$$|E_k|_{v0}^{v} = -\,|E_p|_{\infty}^{r} \tag{31}$$

In case of light, $v_0 = c$ and $v = c$. Therefore, from the equation (31),

$$0 = -\left| -\frac{GM}{r} m \right|_{\infty}^{r} = \frac{GM}{r} m \ . \tag{32}$$

The potential energy at any r is zero. Therefore, $m = 0$. Because $m_t \neq 0$ for light, the working mass m should be the rest mass $m = m_0$. This is not only for light but for any subject. We can conclude as follows:

- Working mass for force is the rest mass.

- In case of light, the rest mass is zero and the potential energy is zero. Light does not receive gravitational force in the direction of propagation. The total energy is equal to the kinetic energy.

$$m_0 = 0\ ,\quad E_p = 0\ ,\quad E_r = 0 \tag{33}$$

$$H = E_k = m_t c^2 \tag{34}$$

Next let's check a possibility of n for f_a. For any positive real number n except $n = 1$, the kinetic energy is given as

$$E_k = \frac{m_0 c^2}{2(1-n)} \left(1 - \left(1 - \frac{v^2}{c^2}\right)^{1-n}\right) . \tag{35}$$

The integration constant is given from $E_k = 0$ at $v = 0$. If $n < 1$, the kinetic energy for light ($v = c$) becomes zero as $m_0 = 0$. Therefore $n < 1$ for f_a is denied, and at least $n \geq 1$ is required.

In case $n = 1$ for f_a ,

$$f_a = 1 - \frac{v^2}{c^2} \tag{36}$$

$$F = \frac{m_0}{f_a}\alpha = m_0 \left(1 - \frac{v^2}{c^2}\right)^{-1} \alpha . \tag{37}$$

The kinetic energy is given as follows.

$$\begin{aligned} E_k &= \int F dr = \int m_0 \left(1 - \frac{v^2}{c^2}\right)^{-1} \frac{dv}{dt} dr = \int m_0 \left(1 - \frac{v^2}{c^2}\right)^{-1} v dv \\ &= -\frac{m_0 c^2}{2} \log \left(1 - \frac{v^2}{c^2}\right) + k \end{aligned} \tag{38}$$

At $v = 0$, $E_k = 0$. Therefore, $k = 0$.

$$E_k = -\frac{m_0 c^2}{2} \log \left(1 - \frac{v^2}{c^2}\right) \tag{39}$$

By Taylor's expansion, the equation (39) is rewritten as

$$E_k = \frac{1}{2} m_0 v^2 \left(1 + \frac{1}{2}\frac{v^2}{c^2} + \frac{1}{3}\frac{v^4}{c^4} + \frac{1}{4}\frac{v^6}{c^6} + \cdots\cdots\right) . \tag{40}$$

The equation for the total energy is given as

$$\begin{aligned} H &= E_k + E_p + E_r \\ &= -\frac{m_0 c^2}{2} \log \left(1 - \frac{v^2}{c^2}\right) - \frac{GMm_0}{r} + m_0 c^2 = m_t c^2 . \end{aligned} \tag{41}$$

If $m_0 \neq 0$, $E_{k(v \to c)} = \infty$ in the equation (39). It signifies that non-zero energy can not be accelerated to the light speed c. If $m_0 = 0$, that is light, $E_{k(v \to c)} = -0 \cdot (-\infty)$ in (39). It does not conflict with $E_{k(v=c)} = m_t c^2$. $n = 1$ for f_a is not yet proven but I speculate a possibility of $n = 1$.

What we discussed above is the case that force direction is parallel to the wave propagation. Force in a perpendicular direction to the light propagation, for instance, works to light in a different manner as its speed to the force direction is not the light speed c. Comparison of the two directions of force to light is given as follows.

(i) In the parallel direction to light propagation

$$v_{(//)} = \pm c\ , \quad m_0 = 0\ , \quad E_r = 0\ , \quad E_p = 0\ , \quad E_k = m_t c^2$$

(ii) In a perpendicular direction to light propagation
The light speed in the perpendicular direction to the light propagation $v_{(\perp)}$ should always be zero from the definition of the perpendicular direction even if light propagation is bent. It should be reminded that the perpendicular direction is not constant to the medium but changes by time.

$$v_{(\perp)} = 0 \tag{42}$$

The momentum of light to the perpendicular direction is also zero. Therefore, the kinetic energy of light can be regarded as rest mass energy of light for force in the perpendicular direction. We can confirm it as follows.
The kinetic energy of light, which is equal to the total energy, can be divided into the kinetic energy and the rest mass of light for the perpendicular direction to the propagation.

$$E_k = E_{r(\perp)} + E_{k(\perp)} \tag{43}$$

Because $v_{(\perp)} = 0$,

$$E_{k(\perp)} = 0 \tag{44}$$

Therefore,

$$E_k = m_t c^2 = E_{r(\perp)} + E_{k(\perp)} = E_{r(\perp)} + 0 \tag{45}$$

$$E_{r(\perp)} = m_t c^2 \tag{46}$$

We can conclude that the rest mass of light for the perpendicular direction is equal to the total energy mass.

$$m_{0(\perp)} = m_t \tag{47}$$

In the perpendicular direction to propagation, light has the rest mass corresponding to the total energy and receives gravitational attraction, which changes the direction of propagation. This is the process of gravitational lens effect. Light cannot be accelerated in propagation speed but can be bent by a massive substance.

Not only for light, but the relation shown by

$$E_{r(\perp)} = E_{r(//)} + E_{k(//)} \tag{48}$$

should be effective also for any substance or energy in 3-D space since $E_{k(\perp)} = 0$. In a case of circular motion like a satellite orbiting around the earth, the satellite has the rest mass for the centripetal force equal to the sum of the rest mass and kinetic energy mass for the direction of the orbit.

5. Mechanics of the space energy in the 4-D sphere

Subsequently think of mechanics of the space energy traced by the original tracing time. Let's come back to use time or t for the original tracing time and Time or T for our observed Time in the 3-D space.

The movement of the space energy in the radius dimension of the 4-D sphere is not a propagation of a wave in a medium but a movement of the space energy itself without a medium. Therefore some limitations posed to wave propagation in the 3-D space should not be applied for it. We can expect key features for mechanics of the space energy in the 4-D sphere as follows:

- There is not a maximum limit of speed like light speed c.

- The acceleration factor f_a for wave propagation is not applied.

Accordingly, the Newtonian equation of motion (24) should be applied. The formula of Einstein's energy mass equation $E = mc^2$ is no longer valid, while energy mass equivalence remains valid. For the process of space expansion-shrink, the light speed is not constant but varies by time [2]. Put k for the proportionality factor of energy mass equivalence.

$$E = mk \tag{49}$$

Take an area of universe defined by a 4-D spherical coordinate system $(x, \theta, \varphi_1, \varphi_2)$. A vector $\boldsymbol{r}$ in the 3-D space is expressed by $(x\theta, \varphi_1, \varphi_2)$. The total energy of a unit angular area defined by constant angles of coordinates remains constant in the expansion of universe. The rest mass, which is defined from a frame stationary to the center of the universe, is also conserved in the process. Let the rest mass of the area be m_0 and the mass for the whole universe be M. The kinetic, potential and rest mass energies are expressed as follows.

$$\begin{aligned} H &= E_k + E_p + E_r \\ &= \frac{1}{2} m_0 v_x^2 - \frac{GMm_0}{x} + m_0 k = m_t k \end{aligned} \tag{50}$$

The total energy mass m_t should be very close to the rest mass m_0 because the maximum x, at which $v_x = 0$, is very huge.

The expansion speed of the universe formerly reported [1] is as follows. Let the initial values of x and v_x at the Big Bang be x_0 and v_{x0} . The expansion speed v_x of the radius of universe is given as follows.

$$\frac{1}{2} m_0 v_{x0}^2 - \frac{1}{2} m_0 v_x^2 = GMm_0 \left(\frac{1}{x_0} - \frac{1}{x} \right) \tag{51}$$

$$\begin{aligned} v_x &= \pm \sqrt{v_{x0}^2 + 2GM \left(\frac{1}{x} - \frac{1}{x_0} \right)} = \pm \sqrt{\frac{2GM}{x} + v_{x0}^2 - \frac{2GM}{x_0}} \\ &\equiv \pm \sqrt{\frac{2GM}{x} + k} \end{aligned} \tag{52}$$

Use the unit for x as 1 at its maximum.

$$v_x = \pm \sqrt{2GM + k} = 0 \tag{53}$$

Therefore,

$$v_x = \pm \sqrt{2GM \left(\frac{1}{x} - 1 \right)} \quad (x_0 \leq x \leq 1) \ . \tag{54}$$

From $x = x_0$ to $x = 1$, the universe expands and then shrinks in accordance with the formula (54) in the radius.

3-D space expansion is given as

$$r = x\theta$$

$$v_r = \frac{dr}{dt} = v_x \theta = \pm \theta \sqrt{2GM \left(\frac{1}{x} - 1 \right)} \ . \tag{55}$$

Lastly let's confirm the expansion of the universe by our observed Time. Because $dT = dx$, the expansion speed of the 3-D space is constant as shown by

$$V_r \equiv \frac{dr}{dT} = \frac{dr}{dx} = \theta \ . \tag{56}$$

An evidence of the constant speed of the 3-D space expansion was shown in my previous article from the Hubble plots of the Supernova Cosmology Project [12] with adjustment for changing light speed by Time [2] .

Journal of Physics: Conference Series **442** (2013) 012067 doi:10.1088/1742-6596/442/1/012067

6. Conclusion

The 4-D Spherical Model of the universe insists the presence of wave medium for any energy in the 3-D space and the changing speed of light along with space expansion. The Special Relativity is denied from the disaffirmance of its propositions and that of the Michelson-Morley experiment. There would be arguments that the time prolongation according to the Special Relativity is experimentally proven. However, once the invariance of light speed by selection of frame becomes invalid, the results can be interpreted as combination of constant time and different light speed in a moving frame to the space energy.

As a feature of wave, the acceleration factor for movement of a wave in its medium shown by the formula (26) is newly introduced and the new equation of motion expressed by the formula (25) is proposed. The number n for the acceleration factor is not yet determined, while I speculate $n = 1$ without a clear ground. The kinetic energy is accordingly expressed by the formula (39) if $n = 1$ and by the formula (35) if $n > 1$. The new motion equation accords nicely with various features of light and gravitational lens effect even based on the Newtonian equation of gravity.

The process of expansion of the universe based on the original tracing time complies with the Newtonian equation of motion without the acceleration factor because it is not a wave-propagation but a direct movement of the space energy itself without a medium. The universe is proposed to repeat the process of expansion and shrink.

By our observed Time, which is the radius of the 4-D sphere of the universe, the 3-D space expands and then shrinks at a constant speed. Its evidence from the Supernova Cosmology Project data [12] was reported in the previous article [2]. According to this 4-D Spherical Model of the universe, we can exclude the Dark Energy.

References

[1] Nagao S 2009 Nature of tracing dimension, imaginary order of freedom and our observed time passing at constant speed *J. Phys.: Conf. Ser.* **174** 012069 doi: 10.1088/1742-6596/174/1/012069

[2] Nagao S 2011 Light speed and the expansion of the universe *J. Phys.: Conf. Ser.* **306** 012073 doi: 10.1088/1742-6596/306/1/012073

[3] Einstein A 1905 *Annalen der Phys.* **17** 891
(English translation: http://www.fourmilab.ch/etexts/einstein/specrel/www/)

[4] Michelson A A 1881 *American Journal of Science* **22** 120

[5] Michelson A A and Morley E W 1887 *American Journal of Science* **34** 333

[6] Michelson-Morley experiment *Wikipedia*
http://en.wikipedia.org/wiki/Michelson-Morley_experiment

[7] Einstein A 1905 *Annalen der Phys.* **18** 639
(English translation: http://www.fourmilab.ch/etexts/einstein/E_mc2/www/)

[8] Einstein A 1935 *Am. Math. Soc. Bull.* **41** 223
(http://www.ams.org/journals/bull/1935-41-04/S0002-9904-1935-06046-X/S0002-9904-1935-06046-X.pdf)

[9] Special Relativity *Wikipedia*
http://en.wikipedia.org/wiki/Special_relativity

[10] Mass-energy equivalence *Wikipedia*
http://en.wikipedia.org/wiki/Mass%E2%80%93energy_equivalence

[11] Sunakawa S 1993 *Sotaisei Riron no Kangaekata (Vision of the Theory of Relativity)* (Tokyo: Iwanami Shoten) pp 56-58

[12] Cosmology from Type Ia Supernovae *Supernova Cosmology Project*
http://www-supernova.lbl.gov/public/papers/aasposter198dir/wwwposter2b.jpg

Exciting peculiarities of the Planck scale physics

Paolo Di Sia

Faculty of Education, Free University of Bolzano-Bozen, Viale Ratisbona 16, 39042 Bressanone-Brixen, Italy

E-mail: paolo.disia@gmail.com

Abstract. By the study of physics at the Planck scale interesting and unusual peculiarities emerge, which make this sector an extremely fascinating field of theoretical investigation; in this paper it is given an overview concerning quantum foam, space rips, branes, duality, mirror simmetry, appeared with the study of the Planck scale physics.

1. Introduction

Physics attempts to describe in the most general and mathematical way the laws of nature; during the history one of the objectives of this discipline has been the will to build unifying theories, able to provide an overall description of the universe.

In 1974 scientists discovered a new symmetry, called "supersymmetry" (often abbreviated SUSY). It appeared immediately mathematically very elegant and able to resolve several problems. Supersymmetry has quite different properties with respect to the ordinary known ones and can be used as the basis of a geometric theory of gravity, called "supergravity" (often abbreviated SUGRA); it incorporates and extends the Einstein general relativity. In this theory the number of space-time dimensions is fixed; supergravity theories (there are on fact more versions of it) are not valid for a number of space-time dimensions higher than 11. If that number is exceeded, it does not seem possible to find a mathematically consistent way for correlating the bosonic with the fermionic fields. The extra dimensions would form the analogous of a sphere in 7 dimensions, which is one of the most symmetrical and simple structures. The equations derived from supergravity offer however quantities with infinite values, not having therefore physical interpretation.

Among the attempts to overcome this problem, great importance had in the present extreme physics the so-called "string theories", or "superstrings", in their supersimmetric version. In such theories the coordinates of a point (understood as zero-dimensional) are replaced with the coordinates of a 1-D structure, called string [1].

The most promising string theories are formulated in 10 dimensions. The dimensions exceeding the 3+1 spacetime ones are compactified, closed on themselves, giving rise in each point to an internal space. These theories appear to be finite, solving the problem of infinities, which afflicts the supergravity theories; theorists have great expectations on them as latest theories describing the reality in the natural language of quantum field theories. The other followed way was the quantum gravity. At this level very small lengths and times are considered, of order of the Planck length unit L_{Planck} ($L_{Planck} = 1.61 \cdot 10^{-33}\, cm$) and Planck time t_{Planck} ($t_{Planck} = 5.36 \cdot 10^{-44}\, s$). At such scale the distinction between past and future becomes uncertain and spacetime is chaotic and fluctuating [2]; special and unusual features appear, which are the subject of current research in the field of speculative physics, also called "extreme physics", with interesting and deep relations with philosophy [3].

2. The quantum foam

The two pillars of modern physics are Einstein General Relativity (GR), which concerns the gravitation, and Quantum Mechanics (QM), which describes the world at micro-level. These theories had a lot of experimental confirmations and all their predictions have been experimentally confirmed with high accuracy. But there is a big problem afflicting them, their incompatibility.

One of the basis of QM is the uncertainty principle; according to it, everything is subject to quantum fluctuations, the gravitational field too. According to classical physics, the value of the gravitational field in empty space is zero; with QM it becomes oscillating, due to the quantum fluctuations. The uncertainty principle leads to an increase of such oscillations, reducing the size of the regions of considered space. Since the gravitational field is reflected in the space curvature, quantum fluctuations lead to distortions of the space structure, increasing with respect to the decrease of distances. The magnitude order under which the quantum fluctuations are very important is the Planck length L_{Planck}; at such length the structure of spacetime is not smooth and regular, the random fluctuations due to quantum effects are so pronounced to give a geometry with irregular curvature. The structure of spacetime results twisted and the conventional notions of displacement in the three ordinary dimensions lose their classical meaning. Theorists call this environment "quantum foam" [4].

Therefore, at ultramicroscopic level, two of the most significant elements of GR and QM, the geometric model of spacetime and the uncertainty principle respectively, come into direct conflict.

String theories, in which the concept of point-like particle is replaced with that of 1-D extended vibrating object, allow an overcoming of the problem of these strong quantum fluctuations.

The string extension implies the impossibility to probe the structure of objects shorter than its length, i.e. to go under L_{Planck}. Strings, as elementary constituents of the universe, cannot probe lengths smaller than L_{Planck}, therefore are not influenced by strong quantum fluctuations; so the infinities of quantum theories of gravitation based on the concept of point-like particles are not present at string level. String theories establish that the spatial dimensions cannot become less than L_{Planck}; beyond this length the common notion of distance loses meaning, it is studied with a new geometry, the quantum geometry [5,6].

3. Space rips

According to GR, it does not possible that the geometric texture of the space tears; the metric relations of this theory require a smooth space, i.e. without sharp tips and folds, framed distinct pieces, rips. At the Planck scale, holes and rips might on the contrary represent a rather common characteristic of the space structure. String theories show that particular physical circumstances can tear the space. In 1987 it was discovered that it is possible to transform some Calabi-Yau (C-Y) spaces (spaces in which are compactified the extra dimensions) in others by making a puncture on their surface and then mending the hole according to a precise mathematical process.

The procedure can be simply described as follows: it shrinks a sphere inside a C-Y space to the size of a point, operating a sort of bottleneck in space; space tears and produces a sphere which, by swelling, makes again smooth its surface. The original (initial) sphere has in this way undergone a procedure called "flop". In some cases the new C-Y space produced by a flop is topologically different by the initial one, which means that it is impossible to obtain a final

C-Y space deforming the initial one, without producing rips in any intermediate step of the previously indicated operation. There occur therefore transitions with topology change via a flop; flop transitions produce rips in the space texture. The theorist Edward Witten, among the first scientists to study this peculiarity, has proved that the universe sheet described by a string provides a shield which erases the potentially damage effects associated with a rip of the space texture. A string can perform two different types of motion near a rip, whereas a particle can move in one way only; the string can on fact move also forming a circumference around the rip, surrounding it. So the world sheet of the string constitutes a protective barrier able to neutralize exactly the consequences of the geometric space degeneration. According to the Feynman's quantum mechanics formulation, the string moves "sniffing" all possible trajectories; the observed motion results by the combination of all possible paths. Among all the possible string trajectories, it must be considered also the ones wrapping around the rip. QM takes into account the physical effects arising from all possible trajectories, resulting so infinite "protective" ways surrounding the rip. These pathways contribute neutralizing the possible "cosmic catastrophe" produced by the rip.

The transformations which produce the above described rips are technically called "topology changing transitions". As a further consequence, the geometric degeneration predicts no particular effect from the physical point of view. Rips can occur also in the 3 extended dimensions, not only in the extra dimensions of the C-Y compactification. It happens that the mass values of the individual particles change, i.e. the energies of the possible vibration modes, but without discontinuity. Experimental measures of the masses of elementary particles in the most important experimental research centers show actually that these masses are significantly stable in time. In the first instants after the Big Bang the masses of particles changed in time; so there is the probability that during this period occurred rips, with consequent change in topology. If actually the space texture is undergoing a rip, it should happen very slowly, so that the effects on the masses are less than the resolving power of the current experimental devices [2,4].

4. The branes

String theories involve a different spacetime with respect to the common ideas, with very particular characteristics, such as a greater number of dimensions, some of which "rolled" on themselves forming particular subspaces, which can have incredible transformations, such as to shoot up, tear, selfsaw.

Advanced research on the M-theory, result of the second superstring revolution, theory that unifies the previous indicated string theories into a comprehensive conceptual framework, provides an 11-D spacetime. The branes are extended objects appearing in string theories, p-branes in general (p-dimensional objects), basic constituents behaving for some aspects as point particles at large scales, but with different properties at microscopic level. Increasing the string coupling α', fundamental parameter of string theories, related to the string tension by the relation $T_{string} = 1/2\pi\alpha'$, a new dimension becomes visible. The string expands into a membrane, whose width is controlled by the value of the coupling constant. The perturbative scheme (with $\alpha' < 1$) leads to a 10-D universe with 1-D strings; this picture seems to be the approximation of a 11-D universe with 2-D membranes [7].

The configuration studies of such theories (the BPS states) (with properties determined by means of arguments based on symmetry) led to the evaluation of their masses, charges and geometrical shape. Some BPS states are 1-D strings, other states are 2-D membranes and there are also higher dimensional objects. The possibilities match the number of spatial dimensions.

Despite the presence of these extended objects with a different number of spatial dimensions, strings (or 1-branes) maintain a particular role; it has been shown that the mass of the extended objects of any size, with the only exception of strings, is inversely proportional to the value of the coupling constant of the respective theory.

This implies that in the case of "weakly coupled" theories all the extended objects (except strings) have "enormous" masses (of order of $10^{-4} \div 10^{-5}\,g$, i.e. heavier than the Planck mass $m_{Planck} = 2.18 \cdot 10^{-5}\,g$). It follows that very high energies are required for producing them, thus they have a negligible effect on a great part of physics. Increasing the constant, on the contrary, the multidimensional branes become lighter and therefore more important.

The appearance of the new dimension changes the structure of the string, transforming it in a 2-D "belt". At dimensionality level, an interesting analogy can be done:

(i) the membrane has a so small "width" to seem a string;

(ii) the 11th dimension is so small to be undetectable using the perturbative equations.

P-branes play an important role in string theory, related to the fact that the approximate equations (result of perturbative methods) indicate as possible the collapse of a 4-sphere in a C-Y space. It would give rise to serious negative consequences, i.e. the creation of a bottleneck in the space structure with consequent production of unwanted infinities. P-branes can wrap and completely cover a part of a (p-1)-D space, avoiding therefore this problem [8,9].

5. Mirror symmetry and duality

String theories increase the connections between physics and geometry; the properties of vibrating strings are largely determined by the properties of the compactified space components. In such theories it is possible that a circle with radius R and another with radius $1/R$ are undistinguished from a physical point of view; the particles masses and charges in a universe of radius R are the same as those of a universe of radius $1/R$. This particular aspect derives by considerations regarding the string states spectrum. The two universes (with supposed circular dimension) are geometrically different, but physically undistinguishable.

If R is of order of $15 \cdot 10^9\,l\,y$, such scenario results to be physically equivalent to a universe with the 3 spatial dimensions as ordinary circumferences of radius R of order of $10^{-61}\,L_{Planck}$. The pairs of C-Y spaces, physically equivalent but mathematically different, are indicated as "mirror manifolds". Two spaces with this feature are not the mirror images of each other in the ordinary sense; they have different geometrical properties, but give rise to the same universe when used for the compactification of the extra dimensions of string theories; in this case there is a "mirror symmetry".

The concept of duality is also referred to theoretical models which, although quite distinct, describe exactly the same physics, for example two universes with a circular dimension of radius R and $1/R$ respectively: the properties of strings imply the same physical situation.

This duality symmetry, which prevent us from distinguishing between large and small distance scales, is called "T-duality"; it comes from the compactification of extra space dimensions in a 10-D superstring theory. In general, we can exchange the compactification radius R with a $\alpha'\,/\,R$ radius, exchanging the winding modes of a string with the quantized momentum modes. T-duality is referred only to string physics; for point-like particles it is not possible, because in this case there aren't winding modes. String theory implies that the separation between large vs. small distance scales in physics is not a fixed separation, but dependent upon the type of probe used for measuring the distances.

In relation to the mirror symmetry we can think to two different compactification ways of the extra dimensions in a C-Y space; the two universes will have the same physical properties.

Theorists showed that the five developed string theories are different ways to describe the same fundamental physical theory. The duality implies that, considering two of them, one changes into another if we vary their coupling constants. Considering also the dualities concerning the spacetime geometry, the five theories and M-theory are connected by a duality network.

Journal of Physics: Conference Series **442** (2013) 012068 doi:10.1088/1742-6596/442/1/012068

The six theories can be graphically visualized with a pentagonal structure, with M-theory at the center.

Among the SUGRA theories, as possible reconciliation elements between GR and QM, the closest to this big objective are formulated in 10 or 11 dimensions. A greater number of dimensions was excluded for both theoretical and experimental reasons. In the "low energy" processes, where the "extended" character of the string is not detectable, it is possible to approximate a string with a point-like particle without internal structure, using the usual scheme of quantum field theories.

Supergravity theory is thus the quantum field theory which best approximates the string theory. Considering this theory too, the unified scheme becomes an hexagonal-type structure, always with the M-theory at the center [6,8,10].

6. Closing remarks

String theories have allowed extraordinary advances in the description of an ultimate theory of the universe. They have motivated an understanding of black holes in higher dimensions and of extended objects such as strings and branes.

The study of these theories has opened the way to exciting scenarios, in which unusual situations, as quantum foam, space rips, branes, duality, mirror symmetry appear, characterizing the Planck scale physics.

Perhaps spacetime geometry is not something of a fundamental nature in string theory, but it emerges in the theory at large distance scales or weak coupling. This and all the previously mentioned Planck scale phenomena are ideas with fascinating philosophical implications.

7. References

[1] Hawking S 2010 *The Grand Design* (USA: Bantam Books)

[2] Greene B 2003 *The Elegant Universe: Superstrings, Hidden Dimensions, and the Quest for the Ultimate Theory* (New York: W.W. Norton & Company)

[3] Mohaupt T 2003 *Lect. Notes Phys.* **631** 173

[4] Greene B 2004 *The Fabric of the Cosmos: Space, Time, and the Texture of Reality* (New York: Alfred A. Knopf)

[5] Halpern P 2004 *The Great Beyond: Higher Dimensions, Parallel Universes, and the Extraordinary Search for a Theory of Everything* (Hoboken, New Jersey: John Wiley & Sons, Inc.)

[6] Szabo R J 2002 *BUSSTEPP Lectures on String Theory* HWM-02-18, EMPG-02-11 *Preprint* arXiv:hep-th/0207142v1

[7] Clifford J 2003 *D-branes* (Cambridge, UK: Cambridge Univ. Press)

[8] Becker K, Becker M and Schwarz J H 2007 *String Theory and M-Theory: A Modern Introduction* (Cambridge, UK: Cambridge Univ. Press)

[9] Dine M 2007 *Supersymmetry and String Theory: Beyond the Standard Model* (Cambridge, UK: Cambridge Univ. Press)

[10] Hull C M 1999 *Duality and Strings, Space and Time* QMW-PH-99-19 *Preprint* arXiv:hep-th/9911080v1

Geometric invariants as detector of Hawking and Unruh effects and quantum field theory in curved space

Antonio Capolupo[1] **and Giuseppe Vitiello**[2]
[1]Dipartimento di Ingegneria Industriale, Universitá di Salerno, Fisciano (SA) - 84084, Italy
[2]Dipartimento di Fisica E.R.Caianiello Universitá di Salerno, and INFN Gruppo collegato di Salerno, Fisciano (SA) - 84084, Italy

Abstract. We report on the recent results revealing the presence of geometric invariants in all the phenomena in which vacuum condensates appear and we show that Aharonov–Anandan phase can be used to provide the evidence of phenomena like Hawking and Unruh effects and to test some behavior of quantum field theory in curved space. A very precise quantum thermometer can be also built by using geometric invariants.

1. Introduction

Disparate physical systems show the presence of vacuum condensate and are effectively described in a similar way by Bogoliubov transformations. Examples are represented by gravitational effects like Unruh [1], Hawking [2] and Parker [3] ones, phenomena like Schwinger effect [4], superconductivity [5], Thermo Field Dynamics [6], particle mixing [7, 8, 9], quantum dissipative systems [10] and graphene physics [11]. The formal differences among these systems are contained in the Bogoliubov coefficients.

Some of these phenomena, like Hawking, Unruh and Parker effects, which consist in the nonperturbetive production of particle-antiparticle pairs from vacuum, are very hard to be detected and intensive study has been devoted to the tentative to observe such phenomena in analogous atomic systems and to test quantum field theory in curved space-times in laboratory.

An apparently separated research line is represented by the study of geometric phases [12]–[16] appearing in the evolution of many physical systems. Such phases have been observed in a variety of systems [17, 18, 19, 20] and their potential applications in particle physics has been also studied [16].

In the present paper we report the results of Ref.[21] where the presence of geometric phase in all the systems effectively described by a Bogoliubov transformation has been unveiled. We show that Aharonov–Anandan (A-A) invariant can be used to prove experimentally, in atomic systems, the existence of Hawking radiation and Unruh effect, and to reveal some aspects of quantum field theory in curved background in particular graphene morphologies. A-A invariant can be also used to build a very precise thermometer.

The paper is structured as follows. In Sec.II we recall some basic facts about Bogoliubov transformations and A-A invariant and we derive the expression for A-A phase for systems underlying to a Bogoliubov transformation. In Sec.III we present the geometric invariant arising in thermal states and study its application to the Hawking effect in acoustic black hole. The application of A-A invariant to graphene physics and to the study of Unruh effect are presented in Secs.IV and V, respectively. The possibility to build a quantum thermometer by using A-A phase is shown in Sec.VI and Sec.VII is devoted to conclusions.

2. Aharonov–Anandan invariant and vacuum condensate

The Bogoliubov transformation [6] has the form: $\tilde{\alpha}^r_{\mathbf{k}}(\xi,t) = U^{\psi}_{\mathbf{k}}\, \alpha^r_{\mathbf{k}}(t) + V^{\psi}_{-\mathbf{k}}\, \alpha^{r\dagger}_{-\mathbf{k}}(t)$. It is required to be a canonical transformation, i.e. it must leave the canonical commutation relations (CCRs) invariant. This request translates in the condition $|U_{\mathbf{k}}|^2 \pm |V_{\mathbf{k}}|^2 = 1$, with $+$ for fermions and $-$ for bosons. The Bogoliubov transformation can be rewritten in terms of the generator $J(\xi,t)$ as: $\tilde{\alpha}^r_{\mathbf{k}}(\xi,t) = J^{-1}(\xi,t)\, \alpha^r_{\mathbf{k}}(t) J(\xi,t)$. In the case of the phenomena described by Bogoliubv transformations, the states physically relevant are the states $|\widetilde{\psi}(\xi,t)\rangle$ related to the original states $|\psi(t)\rangle$ by the transformation $|\widetilde{\psi}(\xi,t)\rangle = J^{-1}(\xi,t)|\psi(t)\rangle$. For these states the variance of the energy is always different from zero. This fact implies that A-A invariant appears in a similar way in all the phenomena representable by a Bogoliubov transformation. Indeed, in order to generate the A-A phase [13] in the time evolution of an isolated system it is necessary and sufficient that the state of the system $|\phi(t)\rangle$ is not a stationary state, i.e. it has a nonzero value of the uncertainty $\Delta E(t)$ in energy, $\Delta E^2(t) = \langle\phi(t)|H^2|\phi(t)\rangle - (\langle\phi(t)|H|\phi(t)\rangle)^2$. The A-A invariant is then defined as $s = \frac{2}{\hbar}\int_0^t \Delta E(t')\, dt'$.

In the case of phenomena in Refs.[1]–[10], the geometric phase, expressed without making explicit the expressions of the Bogoliubov coefficients, is formally the same for all the systems. In particular, for neutral scalar field, it is given by $S_B(t) = 2\sqrt{2}\int_0^t \omega_{\mathbf{k}} U^B_{\mathbf{k}}(t') V^B_{\mathbf{k}}(t') dt'$. Similar relation holds for fermions. We now study different specific cases.

3. Aharonov–Anandan invariant and Hawking effect

Thermal state – Let us start by considering the thermal state defined in Thermo Field Dynamics [6]. The parameter ξ characteristic of the Bogoliubov transformation describing the relation between the thermal and the non-thermal state depends on temperature. The relevant vacuum is the finite temperature one, and the coefficients U and V for bosons are $U^T_{\mathbf{k}} = \sqrt{e^{\beta\hbar\omega_{\mathbf{k}}}/(e^{\beta\hbar\omega_{\mathbf{k}}} - 1)}$ and $V^T_{\mathbf{k}} = \sqrt{1/(e^{\beta\hbar\omega_{\mathbf{k}}} - 1)}$, respectively, with $\beta = 1/k_B T$. Similar coefficients are found for fermions. The Aharonov–Anandan invariant is $S_T(t) = 2\sqrt{2}\omega_{\mathbf{k}} t\, e^{\beta\hbar\omega_{\mathbf{k}}/2}/(e^{\beta\hbar\omega_{\mathbf{k}}} - 1)$, then, the phase difference existing between two thermal states at different temperature is

$$\Delta S_T(t) = 2\sqrt{2}\omega_{\mathbf{k}} t \left[\frac{e^{\hbar\omega_{\mathbf{k}}/2k_B T_1}}{e^{\hbar\omega_{\mathbf{k}}/k_B T_1} - 1} - \frac{e^{\hbar\omega_{\mathbf{k}}/2k_B T_2}}{e^{\hbar\omega_{\mathbf{k}}/k_B T_2} - 1}\right]. \tag{1}$$

Hawking effect – An example in which the state $|\widetilde{0}(\xi,t)\rangle$ represents a thermal vacuum is given by the Hawking effect. Then the invariant (1) can help to detect the Hawking radiation.

Such an effect consists in black body radiation that is predicted to be emitted by black holes, due to quantum effects near the event horizon. The thermal bath observed outside the event horizon of a black hole has a temperature depending on the black hole mass M, $T_H = \hbar c^3/8\pi G M k_B$, where G is the gravitational constant. This effect can be described by a thermal state with temperature T_H.

Recently, an acoustic black hole has been generated by accelerating a Bose-Einstein condensate of 10^5 atoms of ^{87}Rb to velocities which exceed the sound velocity [22]. The sonic event horizon is represented by the point where the flow velocity equals the speed of sound. Thus sound waves, rather than light waves, cannot escape the event horizon. In this experiment the flow velocity is in only one direction. The effective temperature of Hawking radiation for such a system is given by $T_H = \hbar g_H/2\pi k_B c_H$, where c_H is the sound speed at the black hole horizon,

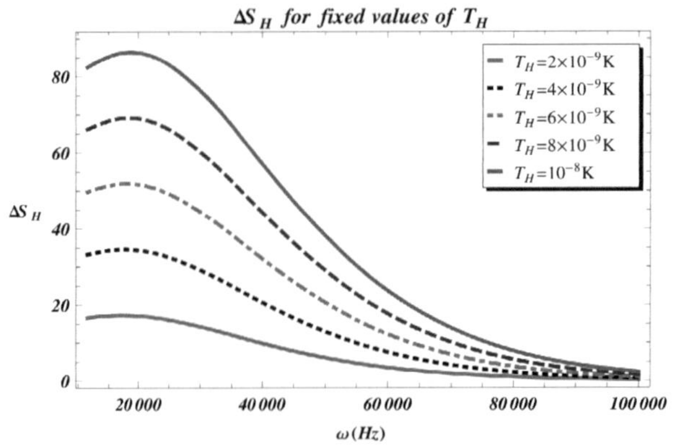

Figure 1. *Plots of ΔS_H as a function of excitation energy ω, for a time interval $t = 20ms$ and for sample values of $T_H \in [2, 10]nK$, as indicated in the inset and $T_{cond} = 40nK$.*

and g_H represents the effective surface gravity [22]. Such a temperature can be also written as $T_H = \mu/k_B\pi\lambda$, where $\mu = mc^2$ is the chemical potential of the condensate with m atomic mass, c speed of sound in the condensate, and λ represents a parameter corresponding to the number of correlation lengths need to have a thermal spectrum for the Hawking radiation. $\lambda = 7$ in [22], moreover, the temperature of the acoustic black hole is $T_H \in (2-10)nK$. Since the temperature of the condensate T_{cond} is $(20-170)nK$, the Hawking radiation is indistinguishable from the thermal noise and very difficult to identify also in such a system. An interferometric study could then be crucial in the detection of Hawking radiation in acoustic black holes. Indeed, a geometric phase similar to the thermal one is associated to Hawking radiation with $T \equiv T_H$.

Such phase could be detected via interferometry, measuring the difference of geometric phase between two flows of one-dimensional Bose-Einstein condensate, one in which is realized an acoustic horizon and the other in which the stream is subsonic. Such a configuration can be obtained by using two devices like the one presented in [22]. In this way the Hawking radiation can be distinguished from thermal noise, by means of the presence of a difference of geometric phases ΔS_H between the supersonic and the subsonic condensate. ΔS_H is given by Eq.(1) where the relevant temperatures are $T_1 \equiv T_H + T_{cond}$ and $T_2 \equiv T_{cond}$.

By considering the limit imposed by the dimension of the acoustic black hole [22, 23] and the fact that the black hole horizon is maintained for about $20ms$ in the experiments [22], we plot in Fig.1 ΔS_H as function of excitation energy for different black hole temperatures. Such pictures show that A-A phases are in principle detectable, thus they can be used to reveal Hawking effect in Bose-Einstein condensate.

4. Aharonov–Anandan invariant and graphene

A system in which the presence of A-A invariant can help to verify quantum field theory in curved space in table top experiment is represented by graphene. Recent studies on graphene have shown that morphology of the sample modifies its electronic properties. In Ref.[24] it has be shown that a sheet of graphene shaped as Beltrami pseudosphere, $dl^2 = du^2 + r^2 e^{2u/r} dv^2$, with $v \in [0, 2\pi]$, $u \in [-\infty, 0]$ and r radius of curvature, displays a finite temperature electronic local density of states, thus revealing characteristic properties of quantum field theory in curved spaces. The temperature found in [24] is $T = T_0 e^{u/r}$ with $T_0 = \hbar v_F/(2\pi k_B r)$ where v_F is the Fermi velocity. In [24] an experiment with a Scanning Tunneling Microscope (STM) has been also proposed to detect the effect there provided. Here we show that such effect could be revealed easily by using the A-A invariant.

Journal of Physics: Conference Series **442** (2013) 012069 doi:10.1088/1742-6596/442/1/012069

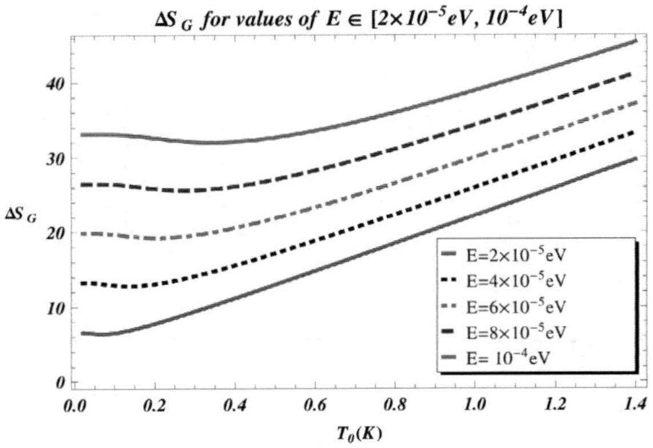

Figure 2. *Plots of ΔS_G vs T_0, for a time interval $t = 10^{-10}s$ and for sample values of $E \in [2\times 10^{-5}eV, 10^{-4}eV]$, as indicated in the inset.*

By using the local Weyl symmetry and the formalism presented in [24], we find that the geometric phase in graphene shaped as Beltrami pseudosphere is given by

$$S_g(t) = 2\sqrt{2}\omega_{\mathbf{k}} t \frac{e^{\hbar\omega_{\mathbf{k}}/2k_BT} e^{-2u/r}}{e^{\hbar\omega_{\mathbf{k}}/k_BT} - 1}, \tag{2}$$

Since the zero-curvature limit of finite temperature electronic local density of states does not match the flat electronic local density of states [24], for $r \to \infty$, Eq. (2) does not match the geometric phase in the case of flat space.

A phase difference of the tunneling current along two different meridians u and u' can be measured by fixing the energy $E = \hbar\omega$ and the curvature r of the surface. Then, from the value of ΔS_g one derives the value of the temperature T_0. In Fig. 2 we plot ΔS_g as function of T_0 for values of the electron energy shown in the inset and for $u = -r/2$ and $u' = -r/3$, with variable r. We see that $\Delta S_g \neq 0$ in graphene for a wide range of values of T_0.

5. Aharonov–Anandan and Unruh effect

Another example in which the state $|\widetilde{0}(\xi, t)\rangle$ represents a thermal vacuum is given by the Unruh effect. In this phenomenon, the ground state for an inertial observer is seen as in thermodynamic equilibrium with a temperature different from zero by the uniformly accelerated observer. The Bogoliubov coefficients allow to express the Minkowski vacuum in terms of Rindler states. The temperature of thermal bath depends on the acceleration a of the observer, $T_U = \hbar a/2\pi c k_B$. The detection of such phenomenon is very hard, thus, also in this case, the geometric phase (1) could represent an useful instrument.

Since, in free field case, the inertial systems has no A-A phase, the geometric phase difference between the accelerated and inertial systems coincides with the phase of the accelerated system and Eq.(1) reduces to $\Delta S_U(t) = 2\sqrt{2}\omega_{\mathbf{k}} t \left[e^{\pi\omega_{\mathbf{k}}c/a}/(e^{2\pi\omega_{\mathbf{k}}c/a} - 1)\right]$. Such a equation describes completely the geometric phase produced in the gravitational Unruh phenomenon. In this regard, let us note that many proposal have been suggested to detect Unruh effect in atomic device by means of Berry phase [25, 26]. It has be shown that such a phenomenon could be demonstrate by analyzing the phase variation due to the acceleration of a two level atom, which can be observed through interference with an inertial atom.

We note that the Berry phase is defined only for systems which have an adiabatic and cyclic evolution. Here, we compute the A-A phase that is independent by the particular time evolution of the system and generalize the Berry phase to the noncyclic case. Thus A-A invariant can

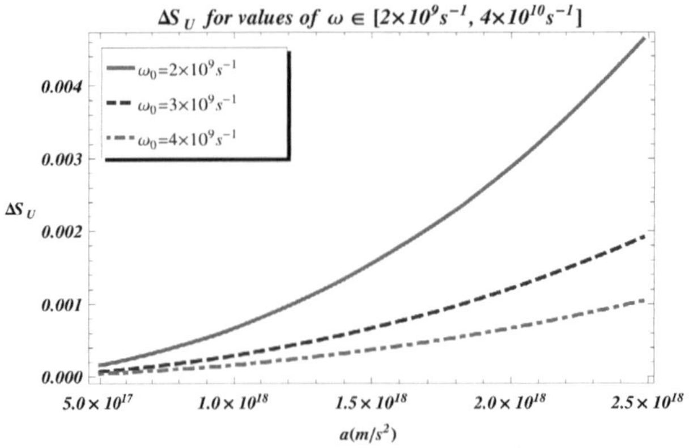

Figure 3. *Plots of ΔS_U as a function of acceleration a, for a time interval $t = 4 \times 2\pi/\omega_0$ and for sample values of $\omega_0 \in [2 \times 10^9 s^{-1}, 4 \times 10^9 s^{-1}]$, as indicated in the inset.*

be more useful than Berry phase in the study of Unruh radiation. We consider the interacting Hamiltonian describing the interaction between the atom and the electromagnetic field in the multipolar scheme, $H = (1/2)\,\hbar\,\omega_0\,\sigma_3 + H_\phi - e\sum_{mn}\mathbf{r}_{mn}\cdot\mathbf{E}(x(t))\sigma_{mn}$, where ω_0 is the energy level spacing of the atom, H_ϕ is the Hamiltonian of the electromagnetic field, and $\mathbf{E}$ is the electric field strength. Assuming a weak interaction between atom and field, one has the effective hamiltonian $H_{eff} = (1/2)\,\hbar\,\Omega\,\sigma_3$, with Ω renormalized energy level spacing which contains the Lamb shift term. Considering a two level atom state, $|\phi(t)\rangle = \sin(\theta(t)/2)|+\rangle + \cos(\theta(t)/2)|-\rangle$ (with $\theta(t)$ defined in [21, 25]), the A-A invariant generated in the evolution of this state is $S = \Omega\int_0^T \sin\theta(t)d\tau$. For a two-level atom uniformly accelerated in the x direction with acceleration a, the phase S_a is

$$S_a = \pm\int_0^T \frac{e^{2A_a t}\sin\theta}{\sqrt{e^{4A_a t}\sin^2\theta + (R - R\,e^{4A_a t} + \cos^2\theta)^2}}\,\Omega_a\,dt\,, \tag{3}$$

where $R_a = B_a/A_a$, with $A_a = (1/4)\,\gamma_0\,(1 + a^2/c^2\omega_0^2)\,(e^{2\pi c\omega_0/a} + 1)/(e^{2\pi c\omega_0/a} - 1)$, and $B_a = (1/4)\,\gamma_0\,(1 + a^2/c^2\omega_0^2)$, with $\gamma_0 = e^2\,|\langle -|\mathbf{r}|+\rangle|^2\omega_0^3/3\pi\varepsilon_0\hbar c^3$ spontaneous emission rate and Ω_a effective level spacing of the atoms [21, 25].

In the case of an inertial atom, $a = 0$, the phase $S_{a=0}$ assume the same form of Eq.(3), with A_a, B_a, R_a replaced by $A_0 = B_0 = (1/4)\,\gamma_0$ and $R_0 = 1$, respectively. The phase difference between the phases of accelerated and inertial atoms, $\Delta S_U = S_a - S_{a=0}$, gives the geometric phase due purely to the atom acceleration. We neglect the very small contribution given by the Lamb shift term, $\Omega_a \sim \omega_0$, and assume $|\langle -|\mathbf{r}|+\rangle| \sim a_0$, where a_0 is the Bohr radius, and $\omega_0 \approx -e^2/8\pi\hbar\varepsilon_0 a_0$, so that $\gamma_0/\omega_0 \approx 10^{-6}$ [25]. By considering an initial state with angle $\theta = \pi/5$ and $\omega_0 \sim 10^9 s^{-1}$, we obtain detectable phase differences for $a \in (10^{17} - 10^{18})m/s^2$, as shown in Fig.3. These values of the accelerations are much less than the one required to detect Unruh radiation ($\sim 10^{26}m/s^2$) and are physically accessible with current technology.

6. Quantum thermometer

A further application of Aharonov–Anandan phase is represented by the possibility to obtain very precise measurement of temperature by means of phase difference of two atomic systems at different temperatures. Indeed, a phase similar to the one in Eq.(3) is acquired by the atom also if it interacts with a thermal state. In this case the coefficients A_a and B_a are replaced by $A_T = (1/4)\,\gamma_0(1+4\pi^2k_B^2T^2/\hbar^2\omega_0^2)\,(e^{E_0/k_BT}+1)/(e^{E_0/k_BT}-1)$, and $B_T = (1/4)\,\gamma_0(1+4\pi^2k_B^2T^2/\hbar^2\omega_0^2)$, respectively, with $E_0 = \hbar\omega_0$.

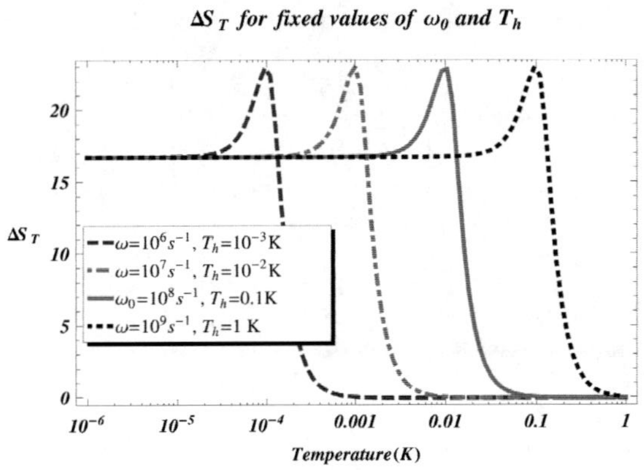

Figure 4. *Plots of ΔS_T as a function of temperatures of cold sources T_c, for sample values of $\omega \in [10^6 s^{-1} - 10^9 s^{-1}]$, time intervals $t = 4 \times 2\pi/\omega\, s$, and temperatures of the hot source of $T_h \in [10^{-3}K - 1K]$, as shown in the inset.*

Thus, an atomic interferometer in which a single atom follows two different paths and interacts with two thermal states at different temperatures can represent a new type of very precise thermometer. The difference of geometric invariants between the two paths permits to determine the temperature of one sample once known the temperature of the other sample. If it is known the temperature T_h of the hotter source, for fixed values of ω_0 and of time, one can measure the temperature T_c of the colder cavity. Considering the atomic transition frequencies ω_0, values of T_h reported in Fig.4, and time intervals of order of $t = 4 \times 2\pi/\omega\ s$, we obtain measurement of temperatures of cold sources of about 2 orders of magnitude below the reference temperature of the hot source, see Fig.4.

7. Conclusions

We have revealed the presence of non-cyclic geometric A-A invariants in all the phenomena where vacuum condensates appear and shown their possible use in different physical phenomena. We have indeed shown that A-A invariant can be utilized as novel tool in laboratory detection of Hawking and Unruh effects, as instrument to test QFT in curved space in graphene. Also, we have suggested that a very precise quantum thermometer can be built by using geometric invariants properties.

Acknowledgement

Partial financial support from MIUR is acknowledged.

References

[1] Unruh W G 1976 *Phys. Rev. D* **14** 870
[2] Hawking S W 1975 *Commun. Math. Phys.* **43** 199 [1976 Erratum-ibid. **46** 206].
[3] Parker L 1968 *Phys. Rev. Lett.* **21** 562
[4] Schwinger J S 1951 *Phys. Rev.* **82** 664
[5] Bardeen J, Cooper L N, Schrieffer J R 1957 *Phys. Rev.* **108** 1175
[6] Takahasi Y, Umezawa H 1975 *Collect. Phenom.* **2** 55; Umezawa H 1993 *Advanced field theory: Micro, macro, and thermal physics* (New York: AIP)
[7] Blasone M, Henning P A and Vitiello G 1999 *Phys. Lett. B* **451** 140; Blasone M, Capolupo A, Vitiello G 2002 *Phys. Rev. D* **66** 025033; Blasone M, Capolupo A, Romei O, Vitiello G 2001 *Phys. Rev. D* **63** 125015; Capolupo A, Ji C R, Mishchenko Y, Vitiello G 2004 *Phys. Lett. B* **594** 135; Blasone M, Capolupo A, Ji C R and Vitiello G 2010 *Int. J. Mod. Phys. A* **25** 4179; Blasone M, Capolupo A, Terranova F and Vitiello G 2005 *Phys. Rev. D* **72** 013003

[8] Capolupo A, Capozziello S, Vitiello G 2007 *Phys. Lett. A* **363**, 53; 2009 *Phys. Lett. A* **373** 601; 2008 *Int. J. Mod. Phys. A* **23**, 4979; Blasone M, Capolupo A, Capozziello S, Vitiello G 2008 *Nucl. Instrum. Meth. A* **588** 272; Blasone M, Capolupo A, Vitiello G 2010 *Prog. Part. Nucl. Phys.* **64** 451; Blasone M, Capolupo A, Capozziello S, Carloni S, Vitiello G 2004 *Phys. Lett. A* **323**, 182
[9] Capolupo A, Di Mauro M, Iorio A 2011 *Phys. Lett. A* **375** 3415; Capolupo A and Di Mauro M 2012 *Phys. Lett. A* **376**, 2830; Capolupo A and Di Mauro M 2013 *Acta Phys. Pol. B* **44**, 81
[10] Celeghini E, Rasetti M, Vitiello G 1992 *Annals Phys.* **215** 156
[11] Iorio A 2011 *Annals Phys.* **326** 1334
[12] Berry M V 1984 *Proc. Roy. Soc. Lond. A* **392** 45
[13] Anandan J and Aharonov Y 1990 *Phys. Rev. Lett.* **65**, 1697
[14] Bruno A, Capolupo A, Kak S, Raimondo G, Vitiello G 2011 *Mod. Phys. Lett. B* **25** 1661
[15] Blasone M, Capolupo A, Celeghini E, Vitiello G 2009 *Phys. Lett. B* **674** 73
[16] Capolupo A 2011 *Phys. Rev. D* **84** 116002
[17] Tomita A, Chiao R Y 1986 *Phys. Rev. Lett.* **57** 937
[18] Jones J A, Vedral V, Ekert A, and Castagnoli G 2000 *Nature* **403** 869
[19] Leek P J, et al. 2007 *Science* **318** 1889
Neeley M, et al. 2009 *Science* **325** 722
[20] Pechal M, et al. 2012 *Phys. Rev. Lett.* **108** 170401
[21] Capolupo A, Vitiello G 2013 *Probing Hawking and Unruh effects and quantum field theory in curved space by geometric phases,* Submitted
[22] Lahav O, Itah A, Blumkin A, Gordon C, Rinott S, Zayats A, and Steinhauer J 2010 *Phys. Rev. Lett.* **105** 240401
[23] Garay L J, Anglin J R, Cirac J I, and Zoller P 2000 *Phys. Rev. Lett.* **85** 4643
[24] Iorio A, Lambiase G 2012 *Phys. Lett. B* **716** 334
[25] Hu J. and Yu H 2012 *Phys. Rev. A* **85** 032105
[26] Martin-Martnez E, Fuentes I, Mann R B 2011 *Phys. Rev. Lett.* **107** 131301

Journal of Physics: Conference Series **442** (2013) 012070 doi:10.1088/1742-6596/442/1/012070

Entanglement in Quantum Field Theory: particle mixing and oscillations

M Blasone[1,3], **F Dell'Anno**[2], **S De Siena**[2,3,4] **and F Illuminati**[2,3,4]

[1] Dipartimento di Fisica, Università degli Studi di Salerno, Via Ponte don Melillo, I-84084 Fisciano (SA), Italy
[2] Dipartimento di Ingegneria Industriale, Università degli Studi di Salerno, Via Ponte don Melillo, I-84084 Fisciano (SA), Italy
[3] INFN Sezione di Napoli, Gruppo collegato di Salerno, Italy
[4] CNISM Unità di Salerno, Italy

E-mail: blasone@sa.infn.it, dellanno.fabio@gmail.com, desiena@sa.infn.it, illuminati@sa.infn.it

Abstract. The phenomena of particle mixing and flavor oscillations in elementary particle physics are associated with multi-mode entanglement of single-particle states. We show that, in the framework of quantum field theory, these phenomena exhibit a fine structure of quantum correlations, as multi-mode multi-particle entanglement appears. Indeed, the presence of anti-particles adds further degrees of freedom, thus providing nontrivial contributions both to flavor entanglement and, more generally, to multi-partite entanglement. By using the global entanglement measure, based on the linear entropies associated with all the possible bipartitions, we analyze the entanglement in the multiparticle states of two-flavor neutrinos and anti-neutrinos. A direct comparison with the instance of the quantum mechanical Pontecorvo single-particle states is also performed.

1. Introduction

Quantum entanglement is a fundamental resource in quantum information and computation science. Topics concerning the study of quantum correlations in paradigmatic quantum systems have been addressed in several branches of condensed matter, atomic physics, and quantum optics [1]. Recently, some attention has been devoted to the investigation of entanglement in the context of elementary particles physics [2, 3, 4, 5, 6, 7, 8, 9]. In particular, in Refs. [5, 6], it has been studied the behavior of single-particle, multi-mode entanglement associated to particle mixing and oscillations. More specifically, in Ref. [6], by considering the physically relevant cases of two- and three-flavor neutrino oscillations, it has been shown that multi-mode single-particle entanglement can be expressed in terms of flavor transition probabilities. In the same paper, it is proposed a scheme for the transfer of the quantum information encoded in neutrino states to spatially delocalized two-flavor charged lepton states. It is worth to remark that, very recently, beams of neutrinos have been used to experimentally investigate the performance of a low-rate communications link, thus providing the feasibility of digital communication using neutrinos [10]. Therefore, at least in principle, the entangled states of oscillating neutrinos are legitimate physical resources for quantum information tasks. In the present work we extend the results of Ref. [6], obtained in the context of quantum mechanics (QM), to the framework of quantum field theory (QFT). By exploiting tools of quantum information theory, we quantify the content of multi-particle flavor entanglement in the QFT system of oscillating neutrinos. Flavor states of mixed particles have been extensively discussed and a proper QFT framework for their description has been worked out in Refs. [11, 12, 13]. In this framework we show that the phenomena of flavor oscillations exhibit a richer structure of quantum correlations with respect to the corresponding QM (Pontecorvo) analog. Indeed, while the QM system of neutrino states are single-particle states possessing flavor entanglement, the QFT system of neutrino states are flavor-entangled multi-particle states, as the presence of anti-neutrino particle species provides further degrees of freedom to the oscillating neutrino system. It is important to notice that entanglement is a relative physical quantity, possessing a specific operational meaning according to the reference quantum observables, and to the quantum subsystems selected as suitable parties of the physical system. By assuming the particle-antiparticle species as further quantum modes, in section 2 we investigate the multiparticle flavor-species entanglement associated with flavor oscillations of the QFT neutrino system. Moreover, in the appendix, we also study the bipartite flavor entanglement associated with flavor oscillations, assuming, for instance, species-non-resolving detectors.

In the following we briefly review the background of the present analysis, i.e. the formalism and the achieved results associated with quantum mechanical instance. Flavor mixing of neutrinos for two generations is described by the 2×2 rotation matrix $\mathbf{U}(\theta)$ [14, 15]

$$\mathbf{U}(\theta) = \begin{pmatrix} \cos\theta & \sin\theta \\ -\sin\theta & \cos\theta \end{pmatrix}, \qquad (1)$$

where θ is the mixing angle. The two-flavor neutrino states are defined as

$$|\underline{\nu}^{(f)}\rangle = \mathbf{U}(\theta)\,|\underline{\nu}^{(m)}\rangle \qquad (2)$$

where $|\underline{\nu}^{(f)}\rangle = (|\nu_e\rangle, |\nu_\mu\rangle)^T$ are the states with definite flavors e, μ and $|\underline{\nu}^{(m)}\rangle = (|\nu_1\rangle, |\nu_2\rangle)^T$ those with definite masses m_1, m_2. Both $|\nu_\alpha\rangle$ ($\alpha = e, \mu$) and $|\nu_j\rangle$ ($j = 1, 2$) are orthonormal. By describing the free propagation of the neutrino mass-eigenstates with plane-waves of the form $|\nu_j(t)\rangle = e^{-i\omega_j t}|\nu_j\rangle$, ω_j denoting the frequency associated with the mass-eigenstate $|\nu_j\rangle$, the time evolution of the flavor states is given by

$$|\underline{\nu}^{(f)}(t)\rangle = \mathbf{U}(t)|\underline{\nu}^{(f)}\rangle \equiv \mathbf{U}(\theta)\mathbf{U}_0(t)\mathbf{U}^{-1}(\theta)|\underline{\nu}^{(f)}\rangle \qquad (3)$$

where $|\underline{\nu}^{(f)}\rangle$ are the flavor states at $t = 0$, and $\mathbf{U}_0(t) = diag(e^{-i\omega_1 t}, e^{-i\omega_2 t})$. At time t the average neutrino number of the state $|\nu_\alpha(t)\rangle$ in the mode β is

$$\langle N_\beta \rangle_\alpha \equiv \langle \nu_\alpha(t)|N_\beta|\nu_\alpha(t)\rangle = |\mathbf{U}_{\alpha\beta}(t)|^2, \ (\alpha, \beta = e, \mu). \qquad (4)$$

By assuming the neutrino occupation number associated with a given flavor (mode) as reference quantum number, one can establish the following correspondence with two-qubit states: $|\nu_e\rangle \equiv |1\rangle_{\nu_e}|0\rangle_{\nu_\mu}$, $|\nu_\mu\rangle \equiv |0\rangle_{\nu_e}|1\rangle_{\nu_\mu}$, where $|j\rangle_{\nu_\alpha}$ stands for a j-occupation number state of a neutrino in mode α. Entanglement is thus established among flavor modes, in a single-particle setting. For instance, the free evolution of the electron-neutrino state $|\nu_e(t)\rangle$ can then be written in the form

$$|\nu_e(t)\rangle = \mathbf{U}_{ee}(t)|10\rangle + \mathbf{U}_{e\mu}(t)|01\rangle, \qquad (5)$$

where $|ij\rangle$ denotes the two-qubit vector $|i\rangle_{\nu_e}|j\rangle_{\nu_\mu}$, and $|\mathbf{U}_{ee}(t)|^2 + |\mathbf{U}_{e\mu}(t)|^2 = 1$ due to normalization. Thus the time-evolved states $|\underline{\nu}^{(f)}(t)\rangle$ are entangled Bell-like superpositions of the two flavor eigenstates with time-dependent coefficients. Such an entanglement is in principle

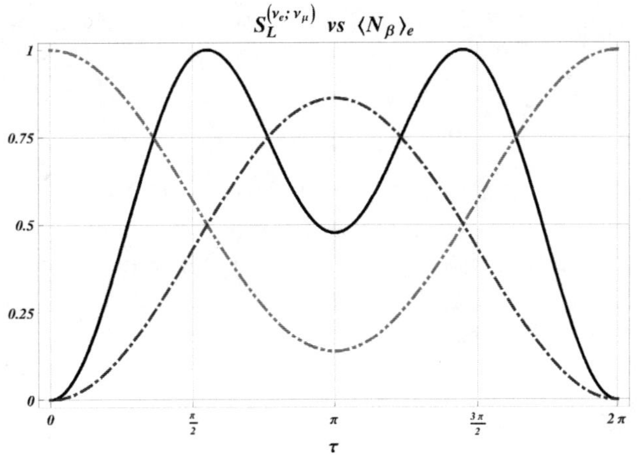

Figure 1. (Color online) QM instance: The linear entropy $S_L^{(\nu_e;\nu_\mu)}$ (full line) as a function of the scaled time $\tau = (\omega_2 - \omega_1)t$. The mixing angle θ is fixed at the experimental value $\sin^2\theta = 0.314$. The average neutrino numbers $\langle N_e \rangle_e$ (double-dot-dashed line) and $\langle N_\mu \rangle_e$ (dot-dashed line) are also reported.

experimentally accessible, throughout a scheme for its transfer from single-neutrino states to two-flavor charged lepton states [6].

The von Neumann entropy, or any other monotonic function of it, quantifies the bipartite entanglement of pure states [16, 17]. In this paper we use a set of linear entropies to describe the quantum correlations of the neutrino oscillating system. Indeed, the linear entropy can be exploited to construct entanglement measure for multipartite systems [18, 19, 20]. For example, the global entanglement is associated to the set of linear entropies associated to all possible bi-partitions of the whole system. An alternative (geometric) characterization of multipartite entanglement is given in Refs. [21, 22] For completeness, we recall the definition of such an entanglement measure. Let $\rho = |\psi\rangle\langle\psi|$ be the density operator corresponding to a pure state $|\psi\rangle$, describing the system S partitioned into N parties. Consider the bipartition of the N-partite system S in two subsystems S_{A_n}, constituted by n parties $(1 \leq n < N)$, and $S_{B_{N-n}}$, constituted by the remaining $N-n$ parties. Let $\rho_{A_n} \equiv Tr_{B_{N-n}}[\rho]$ denote the reduced density matrix of subsystem S_{A_n} after tracing over subsystem $S_{B_{N-n}}$. The linear entropy associated to such a bipartition is defined as

$$S_L^{(A_n;B_{N-n})}(\rho) = \frac{d}{d-1}(1 - Tr_{A_n}[\rho_{A_n}^2]), \tag{6}$$

where the d is the Hilbert-space dimension given by $d = \min\{\dim S_{A_n}, \dim S_{B_{N-n}}\} = \min\{2^n, 2^{N-n}\}$. Finally, we introduce the average linear entropy

$$\langle S_L^{(n:N-n)}(\rho)\rangle = \begin{pmatrix} N \\ n \end{pmatrix}^{-1} \sum_{A_n} S_L^{(A_n;B_{N-n})}(\rho), \tag{7}$$

where the sum is intended over all the possible bi-partitions of the system in two subsystems, respectively with n and $N-n$ elements $(1 \leq n < N)$ [20]. The linear entropies (6), (7) can be easily computed for the two-qubit Bell state $|\nu_e(t)\rangle$, i.e. Eq. (5), with density matrix $\rho_e = |\nu_e(t)\rangle\langle\nu_e(t)|$.

The linear entropy $S_L^{(\nu_e;\nu_\mu)} \equiv S_L^{(\nu_e;\nu_\mu)}(\rho_e)$ associated to the reduced state after tracing over one mode (flavor) writes:

$$S_L^{(\nu_e;\nu_\mu)} = S_L^{(\nu_\mu;\nu_e)} = 4|\mathbf{U}_{ee}(t)|^2\,|\mathbf{U}_{e\mu}(t)|^2 = 4|\mathbf{U}_{e\mu}(t)|^2(1-|\mathbf{U}_{e\mu}(t)|^2)\,. \tag{8}$$

Therefore, the linear entropy (8) is strictly related to the average neutrino numbers, i.e. $S_L^{(\nu_e;\nu_\mu)} = \langle N_e\rangle_e\langle N_\mu\rangle_e$. Specifically the linear entropies $S_L^{(\nu_e;\nu_\mu)}$ coincide, apart from a constant factor, to the variance associated with the average neutrino number, i.e.

$$\langle(\Delta N_{\nu_e})^2\rangle \equiv \langle N_e^2\rangle_e - \langle N_e\rangle_e^2 = \langle N_e\rangle_e\,(1-\langle N_e\rangle_e) = 4^{-1}\,S_L^{(\nu_e;\nu_\mu)}\,. \tag{9}$$

In Fig. 1 we show the behavior of $S_L^{(\nu_e;\nu_\mu)}$ and of $\langle N_\beta\rangle_e$ as functions of the scaled, dimensionless time $\tau = (\omega_2-\omega_1)t$. At time $\tau = 0$, the two flavors are not mixed, the entanglement is zero, and the global state of the system is factorized. For $\tau > 0$, flavor oscillations occur; the entanglement is maximal at largest mixing: $\langle N_e\rangle_e = \langle N_\mu\rangle_e = 1/2$.

2. QFT flavor entanglement

In order to present a generalization of the above analysis to the QFT framework, we recall the essential features of a specific QFT model of particle mixing describing the phenomena of neutrino oscillations [11, 12]. The neutrino fields $\nu_e(x)$ and $\nu_\mu(x)$ are defined through the mixing relations:

$$\nu_e(x) = \cos\theta\;\nu_1(x) + \sin\theta\;\nu_2(x)\,, \tag{10}$$
$$\nu_\mu(x) = -\sin\theta\;\nu_1(x) + \cos\theta\;\nu_2(x)\,, \tag{11}$$

where, in standard notation, x stands for the four-vector $x \equiv (t, x_1, x_2, x_3) \equiv (t, \mathbf{x})$, and the free fields $\nu_1(x)$ and $\nu_2(x)$ denote the neutrino mass eigenstates. The generator of the mixing transformations is given by:

$$G_\theta(t) = \exp\left[\theta\int d^3\mathbf{x}\left(\nu_1^\dagger(x)\nu_2(x) - \nu_2^\dagger(x)\nu_1(x)\right)\right] \tag{12}$$

so that

$$\nu_\sigma^\alpha(x) = G_\theta^{-1}(t)\;\nu_i^\alpha(x)\;G_\theta(t), \tag{13}$$

where $(\sigma, i) = (e, 1), (\mu, 2)$, and the superscript $\alpha = 1, \ldots, 4$ denotes the spinorial component. At finite volume, $G_\theta(t)$ is a unitary operator, i.e. $G_\theta^{-1}(t) = G_{-\theta}(t) = G_\theta^\dagger(t)$, preserving the canonical anticommutation relations. The generator $G_\theta^{-1}(t)$ maps the Hilbert space for free fields $\mathcal{H}_{1,2}$ to the Hilbert space for mixed fields $\mathcal{H}_{e,\mu}$: $G_\theta^{-1}(t) : \mathcal{H}_{1,2} \mapsto \mathcal{H}_{e,\mu}$. In particular, the flavor vacuum is given by $|0(t)\rangle_{e,\mu} = G_\theta^{-1}(t)\;|0\rangle_{1,2}$ at finite volume V. We denote by $|0\rangle_{e,\mu}$ the flavor vacuum at $t = 0$. It is worth noticing that, in the infinite volume limit, the flavor and the mass vacua are unitarily inequivalent. The free fields $\nu_i(x)$ $(i = 1, 2)$ are given by the following expansions

$$\nu_i(x) = \frac{1}{\sqrt{V}}\sum_{\mathbf{k},r}\left[u_{\mathbf{k},i}^r\,\alpha_{\mathbf{k},i}^r(t) + v_{-\mathbf{k},i}^r\,\beta_{-\mathbf{k},i}^{r\dagger}(t)\right]e^{i\mathbf{k}\cdot\mathbf{x}}, \tag{14}$$

where $\mathbf{k} \equiv (k_1, k_2, k_3)$ is the momentum vector, $r = 1, 2$ denotes the helicity, $\alpha_{\mathbf{k},i}^r(t) = \alpha_{\mathbf{k},i}^r\,e^{-i\omega_{k,i}t}$, $\beta_{\mathbf{k},i}^{r\dagger}(t) = \beta_{\mathbf{k},i}^{r\dagger}\,e^{i\omega_{k,i}t}$, and $\omega_{k,i} = \sqrt{\mathbf{k}^2 + m_i^2}$. The operators $\alpha_{\mathbf{k},i}^r$ and $\beta_{\mathbf{k},i}^r$ are the annihilation operators for the vacuum state $|0\rangle_m \equiv |0\rangle_1 \otimes |0\rangle_2$, i.e. $\alpha^r_{\mathbf{k},i}|0\rangle_m = \beta^r_{\mathbf{k},i}|0\rangle_m = 0$. The anticommutation relations are the usual ones; for further details, e.g. orthonormality and completeness relations, see Refs. [11, 12]. By use of $G_\theta(t)$, the flavor fields can be expanded as:

$$\nu_\sigma(\mathbf{x}) = \frac{1}{\sqrt{V}}\sum_{\mathbf{k},r}\left[u^r_{\mathbf{k},i}\,\alpha^r_{\mathbf{k},\sigma}(t) + v^r_{-\mathbf{k},i}\,\beta^{r\dagger}_{-\mathbf{k},\sigma}(t)\right]e^{i\mathbf{k}.\mathbf{x}}. \tag{15}$$

The flavor annihilation operators are defined as $\alpha^r_{\mathbf{k},\sigma}(t) \equiv G^{-1}_\theta(t)\alpha^r_{\mathbf{k},i}G_\theta(t)$ and $\beta^{r\dagger}_{\mathbf{k},\sigma}(t) \equiv G^{-1}_\theta(t)\beta^{r\dagger}_{\mathbf{k},i}G_\theta(t)$. Without any loss of generality, let us choose the reference frame such that $\mathbf{k} = (0, 0, |\mathbf{k}|)$, we have

$$\alpha^r_{\mathbf{k},e}(t) = \cos\theta\;\alpha^r_{\mathbf{k},1}(t) \;+\; \sin\theta\;\left(|U_\mathbf{k}|\;\alpha^r_{\mathbf{k},2}(t) \;+\; \epsilon^r\;|V_\mathbf{k}|\;\beta^{r\dagger}_{-\mathbf{k},2}(t)\right), \tag{16}$$

where $\epsilon^r = (-1)^r$ and

$$|U_\mathbf{k}| \equiv u^{r\dagger}_{\mathbf{k},i}u^r_{\mathbf{k},j} = v^{r\dagger}_{-\mathbf{k},i}v^r_{-\mathbf{k},j} = \frac{|\mathbf{k}|^2 + (\omega_{k,1}+m_1)(\omega_{k,2}+m_2)}{2\sqrt{\omega_{k,1}\omega_{k,2}(\omega_{k,1}+m_1)(\omega_{k,2}+m_2)}}, \tag{17}$$

$$|V_\mathbf{k}| \equiv \epsilon^r\;u^{r\dagger}_{\mathbf{k},1}v^r_{-\mathbf{k},2} = -\epsilon^r\;u^{r\dagger}_{\mathbf{k},2}v^r_{-\mathbf{k},1} = \frac{(\omega_{k,1}+m_1)-(\omega_{k,2}+m_2)}{2\sqrt{\omega_{k,1}\omega_{k,2}(\omega_{k,1}+m_1)(\omega_{k,2}+m_2)}}\,|\mathbf{k}|, \tag{18}$$

with $i,j = 1,2,\quad i \neq j$, $|U_\mathbf{k}|^2+|V_\mathbf{k}|^2 = 1$. The coefficient $|V_\mathbf{k}|$ is associated with the condensation density of the flavor vacuum and is responsible for several phenomenological consequences [13, 23, 24, 25]. The explicit expression for the flavor states $|\nu^r_{\mathbf{k},e}\rangle$ at time $t = 0$ is

$$\begin{aligned}|\nu^r_{\mathbf{k},e}\rangle \equiv\;& \alpha^{r\dagger}_{\mathbf{k},e}(0)|0\rangle_{e,\mu}\\ =\;& [\cos\theta\,\alpha^{r\dagger}_{\mathbf{k},1} + |U_\mathbf{k}|\;\sin\theta\;\alpha^{r\dagger}_{\mathbf{k},2} - \epsilon^r\;|V_\mathbf{k}|\;\sin\theta\;\alpha^{r\dagger}_{\mathbf{k},1}\alpha^{r\dagger}_{\mathbf{k},2}\beta^{r\dagger}_{-\mathbf{k},1}]G^{-1}_{\theta,\mathbf{k},r}\prod_{\mathbf{p}\neq\mathbf{k}}G^{-1}_{\theta,\mathbf{p},s}|0\rangle_{1,2},\end{aligned} \tag{19}$$

where $G_\theta = \prod_\mathbf{p}\prod_{s=1}^2 G_{\theta,\mathbf{p},s}$. In the state (19), a multiparticle component is present, disappearing in the relativistic limit $|\mathbf{k}| \gg \sqrt{m_1m_2}$: indeed, for large $|\mathbf{k}|$, since one gets $|U_\mathbf{k}|^2 \longrightarrow 1$ and $|V_\mathbf{k}|^2 \longrightarrow 0$, the (quantum-mechanical) Pontecorvo states are recovered. In order to simplify the notation, we omit the superscript r (by fixing $r = 2$) and the subscript $\mathbf{k}$, thus restricting the analysis to the flavor neutrino state $|\nu_e\rangle$ of fixed momentum and helicity. Let us consider again the free evolution of the electron-neutrino state (19):

$$|\nu_e(t)\rangle = e^{-iH_{free}t}|\nu_e\rangle, \tag{20}$$

where H_{free} is the standard QFT free Hamiltonian. In the Hilbert space $\mathcal{H}_{e,\mu}$, Eq. (20) can be rewritten in the form:

$$|\nu_e(t)\rangle = [\mathbf{U}_{ee}(t)\,\alpha^\dagger_e + \mathbf{U}_{e\mu}(t)\,\alpha^\dagger_\mu + \mathbf{U}^{e\bar{e}}_{e\mu}(t)\,\alpha^\dagger_e\alpha^\dagger_\mu\beta^\dagger_e + \mathbf{U}^{\mu\bar{\mu}}_{ee}(t)\,\alpha^\dagger_e\alpha^\dagger_\mu\beta^\dagger_\mu]\,|0\rangle_{e,\mu} \tag{21}$$

where the time-dependent coefficients are given by:

$$\begin{aligned}&\mathbf{U}_{ee}(t) = e^{-i\omega_1t}[\cos^2\theta + \sin^2\theta(e^{-i(\omega_2-\omega_1)t}|U|^2 + e^{-i(\omega_2+\omega_1)t}|V|^2)]\\ &\mathbf{U}_{e\mu}(t) = e^{-i\omega_1t}U\cos\theta\sin\theta(e^{-i(\omega_2-\omega_1)t} - 1)\\ &\mathbf{U}^{e\bar{e}}_{e\mu}(t) = e^{-i\omega_1t}V\cos\theta\sin\theta(1 - e^{-i(\omega_2+\omega_1)t})\\ &\mathbf{U}^{\mu\bar{\mu}}_{ee}(t) = e^{-i\omega_1t}UV\sin^2\theta(e^{-i(\omega_2+\omega_1)t} - e^{-i(\omega_2-\omega_1)t}),\\ &|\mathbf{U}_{ee}(t)|^2 + |\mathbf{U}_{e\mu}(t)|^2 + |\mathbf{U}^{e\bar{e}}_{e\mu}(t)|^2 + |\mathbf{U}^{\mu\bar{\mu}}_{ee}(t)|^2 = 1.\end{aligned} \tag{22}$$

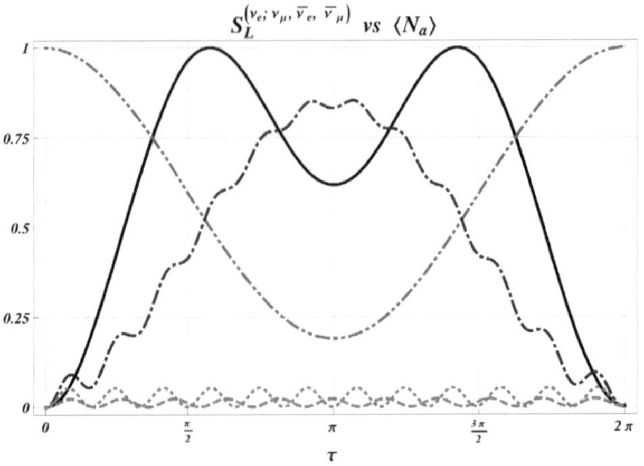

Figure 2. (Color online) QFT instance: The linear entropy $S_L^{(\nu_e;\nu_\mu,\bar{\nu}_e,\bar{\nu}_\mu)}$ (full line) as a function of the scaled time $\tau = (\omega_2 - \omega_1)t$. The mixing angle θ is fixed at the experimental value $\sin^2\theta = 0.314$; the parameters x and p are fixed as $x = 10$ and $p = 5$. The average particle numbers $\langle N_{\nu_e} \rangle$ (double-dot-dashed line), $\langle N_{\nu_\mu} \rangle$ (dot-dashed line), $\langle N_{\bar{\nu}_e} \rangle$ (dotted line), and $\langle N_{\bar{\nu}_\mu} \rangle$ (dashed line) are also reported.

In the following, in order to conveniently parameterize the neutrino masses m_1 and m_2, momentum $|\mathbf{k}|$, and the evolution time t, we use the real parameters $x = m_2/m_1$, $p = |\mathbf{k}|/\sqrt{m_1 m_2}$, and $\tau = (\omega_2 - \omega_1)t$. Therefore, x represents the ratio between the two masses eigenvalues; p expresses the ratio between the momentum and the masses geometrical mean and corresponds to the relativistic limit for $p \gg 1$. Evidently, the time-evolved flavor state (21) is a multi-particle entangled state. Analogously with the Pontecorvo states (5), we assume the neutrino occupation number as reference quantum number. However, with respect to Eq. (5), we have still two flavors, but we have a further degree of freedom that is the neutrino species, i.e. particles and anti-particles. In particular, the present analysis focuses on the multiparticle entanglement both in flavors and in species (detectors resolving both the flavors and the species). Thus, we have truly multipartite entanglement in a four-qubit state that can be written in the form:

$$|\nu_e(t)\rangle = \mathbf{U}_{ee}(t)|1000\rangle + \mathbf{U}_{e\mu}(t)|0100\rangle + \mathbf{U}_{e\mu}^{e\bar{e}}(t)|1110\rangle + \mathbf{U}_{ee}^{\mu\bar{\mu}}(t)|1101\rangle, \tag{23}$$

where $|ijkh\rangle$ denotes the four-qubit vector $|i\rangle_{\nu_e}|j\rangle_{\nu_\mu}|k\rangle_{\bar{\nu}_e}|h\rangle_{\bar{\nu}_\mu}$ with $i, j, k, h = 0, 1$. For simplicity of notation, in Eq. (23) the time dependence has been omitted. In the appendix, we also consider the following instance: multiparticle flavor entanglement (detectors resolving only the flavors independently of the species). Let us analyze the multipartite entanglement possessed by $|\nu_e\rangle$, i.e. Eq. (23), by using the global entanglement defined by Eqs. (6) and (7). Specifically, we compute the linear entropies $S_L^{(a;b,c,d)}$ associated with the bipartition of a single-particle subsystem and a three-particles subsystem, and the corresponding average linear entropy $\langle S_L^{(1:3)} \rangle$.

The linear entropies $S_L^{(a;b,c,d)}$ associated with $|\nu_e(t)\rangle$ are given by simple generalizations of Eq. (8); we have:

$$S_L^{(\nu_e;\nu_\mu,\bar{\nu}_e,\bar{\nu}_\mu)} = 4|\mathbf{U}_{e\mu}(t)|^2(1 - |\mathbf{U}_{e\mu}(t)|^2)\,, \tag{24}$$

$$S_L^{(\nu_\mu;\nu_e,\bar{\nu}_e,\bar{\nu}_\mu)} = 4|\mathbf{U}_{ee}(t)|^2(1 - |\mathbf{U}_{ee}(t)|^2)\,, \tag{25}$$

$$S_L^{(\bar{\nu}_e;\nu_e,\nu_\mu,\bar{\nu}_\mu)} = 4|\mathbf{U}_{e\mu}^{e\bar{e}}(t)|^2(1 - |\mathbf{U}_{e\mu}^{e\bar{e}}(t)|^2)\,, \tag{26}$$

$$S_L^{(\bar{\nu}_\mu;\nu_e,\nu_\mu,\bar{\nu}_e)} = 4|\mathbf{U}_{ee}^{\mu\bar{\mu}}(t)|^2(1 - |\mathbf{U}_{ee}^{\mu\bar{\mu}}(t)|^2)\,. \tag{27}$$

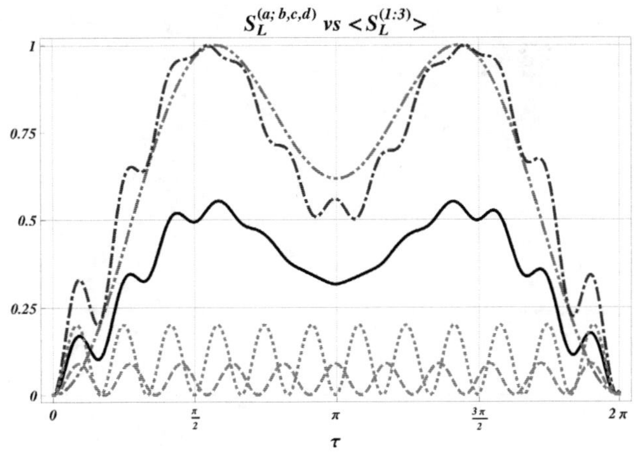

Figure 3. (Color online) QFT instance: The linear entropies $S_L^{(\nu_e;\nu_\mu,\bar{\nu}_e,\bar{\nu}_\mu)}$ (double-dot-dashed line), $S_L^{(\nu_\mu;\nu_e,\bar{\nu}_e,\bar{\nu}_\mu)}$ (dot-dashed line), $S_L^{(\bar{\nu}_e;\nu_e,\nu_\mu,\bar{\nu}_\mu)}$ (dotted line), $S_L^{(\bar{\nu}_\mu;\nu_e,\nu_\mu,\bar{\nu}_e)}$ (dashed line), and the average linear entropy $\langle S_L^{(1:3)}\rangle$ (full line) as functions of the scaled time $\tau = (\omega_2 - \omega_1)t$. The mixing angle θ is fixed at the experimental value $\sin^2\theta = 0.314$; the parameters x and p are fixed as $x = 10$ and $p = 5$.

Of course, in the quantum mechanical limit, Eqs. (24) and (25) reduce to the Pontecorvo analogs, while Eqs. (26) and (27) go to zero. It is important to remark that, analogously to the quantum mechanics instance, the linear entropies $S_L^{(a;b,c,d)}$ are proportional to the variances associated with the particle number:

$$\langle(\Delta N_a)^2\rangle \equiv \langle N_a^2\rangle - \langle N_a\rangle^2 = 4^{-1}\, S_L^{(a;b,c,d)}\,. \tag{28}$$

Eq. (28) furnishes a clear operational meaning for the entanglement measures (24) to (27), throughout a direct connection with the average particle number $\langle N_a\rangle$, $\quad a = \nu_e, \nu_\mu, \bar{\nu}_e, \bar{\nu}_\mu$.

In Fig. 2 we plot the quantity (24) as a function of the scaled time τ for $x = 10$ and $p = 5$; it is worth noticing that, such a choice of the parameters corresponds to the following assumptions: mass m_2 greater than mass m_1 of one order of magnitude, and momentum of the same order of magnitude as the masses geometrical mean. In the same plot, the average particle numbers $\langle N_a\rangle$ are also reported. Comparing the QM and QFT instances, i.e. Figs. 1 and 2 respectively, the linear entropies $S_L^{(\nu_e;\nu_\mu)}$ and $S_L^{(\nu_e;\nu_\mu,\bar{\nu}_e,\bar{\nu}_\mu)}$ show a very similar shape and behavior; however, the common minimum at $\tau = \pi$ is greater in the QFT instance. Moreover, the presence of antineutrinos gives rise to rapidly oscillating components characterizing the average particle numbers $\langle N_{\nu_\mu}\rangle$, $\langle N_{\bar{\nu}_e}\rangle$, and $\langle N_{\bar{\nu}_\mu}\rangle$. In Fig. 3 we plot the linear entropies of the form $S_{L,e}^{(a;b,c,d)}$, i.e. associated with the bipartition formed by a single party versus the other three parties; the average linear entropy (7) is also plotted. All the linear entropies, except $S_L^{(\nu_e;\nu_\mu,\bar{\nu}_e,\bar{\nu}_\mu)}$, exhibit rapidly oscillating components, due to the phenomenon of oscillations involving antineutrinos.

In conclusion we have investigated the multiparticle entanglement in the phenomenon of the neutrino flavor oscillations in the arena of QFT. The analysis represents an extension of the results reported in Refs. [6, 26]. An interesting outlook for the future developments concerns the study of quantum information protocols exploiting the entanglement associated with neutrino oscillations.

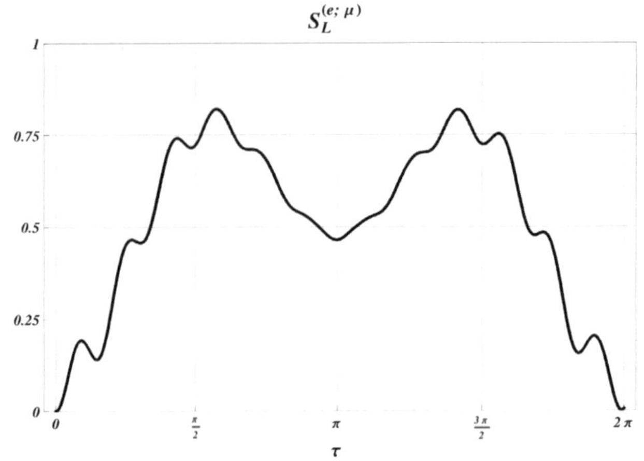

Figure A1. (Color online) QFT instance: The linear entropy $S_L^{(e;\mu)}$ (full line) as a function of the scaled time $\tau = (\omega_2 - \omega_1)t$. The mixing angle θ is fixed at the experimental value $\sin^2\theta = 0.314$; the parameters x and p are fixed as $x = 10$ and $p = 5$.

Appendix

We present a further operational configuration: only flavor entanglement is considered and the species represents an internal degree of freedom; thus, one has bipartite entanglement in a two-qutrit state that can be written in the form:

$$|\nu_e\rangle = \mathbf{U}_{ee}(t)|10\rangle + \mathbf{U}_{e\mu}(t)|01\rangle + \mathbf{U}_{e\mu}^{e\bar{e}}(t)|21\rangle + \mathbf{U}_{ee}^{\mu\bar{\mu}}(t)|12\rangle\,, \tag{A.1}$$

where $|ij\rangle$ denotes the two-qutrit vector $|i\rangle_{\nu_e}|j\rangle_{\nu_\mu}$ with $i, j = 0, 1, 2$, the level 2 representing the anti-neutrino species. In this case, it is assumed that the detection does not resolve the species (particle or antiparticle); then the entanglement is established between the flavors e and μ. It is easy to show that the linear entropy associated with the bipartition $(e;\mu)$ is given by:

$$S_L^{(e;\mu)} = (|\mathbf{U}_{ee}(t)|^2 + |\mathbf{U}_{e\mu}(t)|^2)\left[1 - (|\mathbf{U}_{ee}(t)|^2 + |\mathbf{U}_{e\mu}(t)|^2)\right]. \tag{A.2}$$

$S_L^{(e;\mu)}$, i.e. Eq. (A.2), is plotted in Fig. A1. In this case, the neutrino oscillations associated with the antineutrino channels strongly affects the behavior of the linear entropy which exhibits oscillating components.

References

[1] Nielsen M A and Chuang I L 2001 *Quantum Computation and Quantum Information* (Cambridge: CUP)
[2] Bertlmann R A and Grimus W 1997 *Phys. Lett.* B **392** 426
[3] Bertlmann R A and Grimus W 2001 *Phys. Rev.* D **64** 056004
[4] Bertlmann R A and Hiesmayr B C 2001 *Phys. Rev* A **63** 062112
[5] Blasone M, Dell'Anno F, De Siena S, Di Mauro M and Illuminati F 2008 *Phys. Rev.* D **77** 096002
[6] Blasone M, Dell'Anno F, De Siena S and Illuminati F 2009 *EPL* **85** 50002
[7] Kayser B, Kopp J, Hamish Robertson R G and Vogel P 2010 *Phys. Rev.* D **82** 093003
[8] Friis N, Bertlmann R A, Huber M and Hiesmayr B C 2011 *EPL* **95** 20002
[9] Akhmedov E K and Smirnov A Y 2011 *Found. Phys.* **41** 1279
[10] Stancil D D et al. 2012 *Mod. Phys. Lett.* A **27** 1250077
[11] Blasone M and Vitiello G 1995 *Ann. Phys. (N.Y.)* **244** 283
[12] Blasone M, Henning P A and Vitiello G 1999 *Phys. Lett.* B **451** 140
[13] Blasone M, Jizba P and Vitiello G 2001 *Phys. Lett.* B **517** 471
[14] Cheng T and Li L 1989 *Gauge Theory of Elementary Particle Physics* (Clarendon Press)

[15] Giunti C 2007 *J. Phys. G: Nucl. Part. Phys.* **34** R93
[16] Amico L, Fazio R, Osterloh A, and Vedral V 2008 *Rev. Mod. Phys.* **80** 517
[17] Horodecki R, Horodecki P, Horodecki M and Horodecki K 2009 *Rev. Mod. Phys.* **81** 865
[18] Meyer D A and Wallach N R 2002 *J. Math. Phys.* **43** 4273
[19] Brennen G K 2003 *Quantum Inf. Comp.* **3** 619
[20] de Oliveira T R, Rigolin G and de Oliveira M C 2006 *Phys. Rev.* A **73** 010305(R)
[21] Wei T C and Goldbart P M 2003 *Phys. Rev.* A **68** 042307
[22] Blasone M, Dell'Anno F, De Siena S and Illuminati F 2008 *Phys. Rev.* A **77** 062304
[23] Blasone M, Capolupo A, Capozziello S, Carloni S and Vitiello G 2004 *Phys. Lett.* A **323** 182
[24] Blasone M, Di Mauro M and Vitiello G 2011 *Phys. Lett.* B **697** 238
[25] Blasone M 2011 *J. Phys. Conf. Ser.* **306** 012037
[26] Blasone M, Dell'Anno F, De Siena S and Illuminati F 2010 *J. Phys. Conf. Ser.* **237** 012007

Journal of Physics: Conference Series **442** (2013) 012071 doi:10.1088/1742-6596/442/1/012071

The SSC formalism and the collapse hypothesis for inflationary origin of the seeds of cosmic structure

Daniel Sudarsky

Institute for Nuclear Sciences, UNAM, México D.F. 04510, México

E-mail: sudarsky@nucleares.unam.mx

Abstract.
This article resumes recent advances in the development of a formalism that allows the incorporation of a hypothetical spontaneous reduction or collapse of wave function of matter fields within the context of semiclassical gravity. The proposal is applied to the inflationary scenario for the emergence of the primordial inhomogeneities and anisotropies out of the initially homogeneous and isotropic quantum state, to which, the early stages of inflation are supposed to drive the universe. In previous works we have argued that a scheme of this kind is required if we want to satisfactorily account for the emergence of the seeds of cosmic structure.

1. Introduction

Although the inflationary account for the primordial power spectrum is phenomenologically very successful, there is an obscure part in our understanding of the emergence of the seeds of cosmic structure: How does a universe[1], which at one point in time, is described by a state that is fully homogeneous and isotropic (H&I), evolve into a state that is not, given that the dynamics does not contain any source for the undoing of such symmetry? The issue has been treated in [1], and further elaborated in [2]. We have been pursuing the proposal that the resolution of such problem is tied to the generic resolution of the measurement problem in quantum theory, based on the ideas about spontaneous (i.e. measurement unrelated) collapse of the wave function. The fact that the situation at hand is one where gravity plays an important role, induces us to be particularly sympathetic to the ideas of L. Diósi and R. Penrose that such collapse is somehow tied to fundamental quantum aspects of gravitation, or more generically, that the issue must have its roots in the quantum gravity interface. Previous studies have concentrated on the phenomenological implementation of such ideas and to the uncovering of the essential features that would be required for the scheme to match the now detailed observations of the cosmic microwave background (CMB) and matter distribution spectra.

This work is devoted, instead, to the formal characterization of the process of collapse within the semiclassical treatment of the problem, an issue that can have applications in dealing with

[1] The argument can be applied also to any relevant and causally connected part of such universe. In fact most inflationary cosmologists would say that the Bunch Davies (BD) vacuum, or a state closely related to it, characterizes only a region that, as a result of the relatively propitious initial conditions, undergoes a successful period of inflation. The other regions, being causally disconnected with the former, play no role in the discussion, and, according to the inflationary paradigm, do not, in any way, prevent the characterization of the relevant region in terms of the pure BD state.

other problems related to the quantum gravity interface.

Before entering the main discussion it seems appropriate to point out the connection of the work with several other various lines of work motivated by closely related issues. Those issues are discussed in Sec. 2 to 5 while Sec. 6 is devoted to the conceptual cosmological problem. Section 7 describes our proposed formalism and Sec. 8 illustrates its application to a baby version of the cosmological problem. We end in Sec. 8 with a brief discussion of the work, its problematic aspects and the prospects for its future development.

2. The measurement problem in quantum theory

The extent of the existing work on the subject [11] clearly puts it beyond what we can even begin to address here. However, we should just point out that the great majority of said work can be divided into two great lines: I) works on interpretation of the theory, and II) works based on modifications of the theory.

The basic issue is the fact that the standard presentation of quantum mechanics involves two distinct evolution processes for the quantum state of a system. In Penrose's words, these are the U-Process (characterized by the unitary and deterministic Schrödinger equation), and the R–Process (characterized by the non-deterministic change of the state into one of the eigenstates of a "measured" observable with probabilities dictated by Born's rule).

With regards to I), we will just characterize the situation concerning the Interpretation-Measurement Problem with the following two quotes:

"Either the wave function as given by Schrödinger equation is not everything, or it is not right" [6] in J. Bell's Are There Quantum Jumps? and *" There is now in my opinion no entirely satisfactory interpretation of quantum mechanics"* by S. Weinberg [7].

Regarding II) the main proposals trace back to the works [4], [5] and related developments [31, 32, 33], which we will generically call *Collapse Theories.* These are based on the unification of R and U evolution processes into a single evolution equation, which is supposed to hold universally. Moreover, the proposals remove the special role for the concept of measurement which becomes some particular case of generic interactions between subsystems of larger systems. In this work, we will not adhere to the early suggestions of the collapse occurring at discrete jumps, or the more modern continuous versions where the collapse is controlled by a stochastic function, but will often present the discussion in terms of the former simply because it is easier to handle mathematically. The basic aspect we will take from these proposals is that the evolution of a quantum state is controlled by a modified Schrödinger's equation that includes some spontaneous and unpredictable jumps

$$|\Psi\rangle \rightarrow |\Psi'\rangle \tag{1}$$

which are, moreover, independent of any external notion of measurement. This removes the anthropomorfization usually associated with the measurement concept, and makes the theory applicable to closed systems. The ultimate closed system being, of course, the subject of study of cosmology. The study of any regime where quantum and gravity come together becomes even more alluring given the intriguing suggestion by Diósi [8] and R. Penrose [9] that the fundamental modification of quantum theory envisioned by the collapse theories might trace its origin to the gravity/quantum interface.

The argument: The stability of a quantum superposition of two different stationary mass distributions is examined, where the perturbing effect of each distribution on the space-time structure is taken into account, in accordance with the principles of general relativity. It is argued that the definition of the time-translation operator for the superposed space-times involves an inherent ill-definedness, leading to an essential uncertainty in the energy of the superposed state which, in the Newtonian limit, is proportional to the gravitational self-energy E_Δ of the difference between the two mass distributions. This is consistent with a suggested finite lifetime of the order of $t \approx \hbar/E_\Delta$ for the superposed state, in agreement with a certain proposal made by the author for a gravitationally induced spontaneous quantum state reduction, and with closely related earlier suggestions by Diósi and by Ghirardi et al.

This is one of the general subjects where our work could be relevant: If gravity is tied to the origin of the modifications in quantum theory envisioned by the collapse theories, it becomes of interest how to incorporate gravitation in any such scheme. One possibility, that can not be discarded, is that the whole enterprise would have to wait for a fully satisfactory and workable theory of quantum gravity. The other possibility, based on the work [3] and which we want to describe here, is to do so in terms of an effective semiclassical scheme where matter is treated quantum mechanically and gravitation is treated in terms of the space-time metric and the semiclassical Einstein's equations.

3. Semi-classical gravity

Semiclassical gravity is based on the notion that the energy-momentum tensor expectation value is what should play the source term in Einstein's equations when the quantum nature of matter is taken into account. Thus, one writes:

$$G_{\mu\nu} = 8\pi G \langle|\hat{T}_{\mu\nu}|\rangle \ . \tag{2}$$

The resulting theory, and its potential problems, have been considered for instance in [29], where it is argued that it is, in fact, far from clear that the theory should be regarded as being unviable. One of the works that is taken as most damaging is that of [30], describing an actual experiment involving mass distributions controlled by a quantum mechanical device capable, in principle, of leading to a macroscopic superposition of the mass distribution, a Schrödinger's cat state. The results are in contradiction with the above formula, if one assumes there is nothing like an effective collapse of the wave function. The connection with our problem lies in the question: how would we treat the situation if there was one such process?

3.1. Relation to the stochastic gravity formalism

One of the criticisms often presented against the idea of semiclassical gravity is that Eq. (2) fails to take into account the so called "quantum fluctuations". Dealing with such criticisms is the objective of the stochastic gravity formalism [35].

The basic idea behind such approach is to add to the Eq. (2) a term characterizing those "quantum fluctuations":

$$G_{\mu\nu}(x) = 8\pi G(\langle|\hat{T}_{\mu\nu}(x)|\rangle + \xi_{\mu\nu}(x)) \tag{3}$$

where $\xi_{\mu\nu}$ is a stochastic classical field, corresponding to a distribution with ensemble averages (which we denote by an overline, to distinguish them from the standard quantum expectation values) such that,

$$\overline{\xi_{\mu\nu}(x)} = 0, \qquad \overline{\xi_{\mu\nu}(x)\xi_{\rho\sigma}(y)} = \langle|\hat{T}_{\mu\nu}(x)\hat{T}_{\rho\sigma}(y)|\rangle. \tag{4}$$

Those properties characterize the ensemble from where the individual representatives of the field $\xi_{\mu\nu}(x)$ are drawn. That is, they characterize the random field representing the "quantum fluctuations". What is implicitly assumed in this approach is that the quantum uncertainties might be well characterized by these stochastic fields. The world out there would correspond to one such realization of the stochastic variable and, as the selection of the particular one that becomes realized, is not determined by the setting, we have a certain degree of unpredictability. That unpredictability would somehow correspond to the one which is often associated with the R process of standard quantum theory. Therefore, it should not be surprising that one can relate this approach to the collapse theories, which, as we explained before, do incorporate that R-process within the modified evolution equations.

In order to see this, let us consider one of such jumps or collapses in the state of the matter fields. To simplify the discussion we will be assuming that the space-time is foliated by hyper surfaces labeled by the time function "t" and that the quantum fields are described in the Heisenberg picture. The states would therefore, not evolve with time except in those special instances associated with a collapse[2]. Then collapse is described by an instantaneous jump: $|\psi(t)\rangle = \theta(t_0 - t)|0\rangle + \theta(t - t_0)|\xi\rangle$. Now, let us consider its gravitational effects. In this case Einstein's semiclassical equations read:

$$G_{\mu\nu} = 8\pi G\langle\psi(t)|\hat{T}_{\mu\nu}|\psi(t)\rangle \tag{5}$$

which we can write as

$$G_{\mu\nu} = 8\pi G\langle 0|\hat{T}_{\mu\nu}|0\rangle + 8\pi G\xi_{\mu\nu} \tag{6}$$

where

$$\xi_{\mu\nu} \equiv \theta(t - t_0)(\langle\xi|\hat{T}_{\mu\nu}|\xi\rangle - \langle 0|\hat{T}_{\mu\nu}|0\rangle) \tag{7}$$

might be seen as corresponding to an individual stochastic step. Stochastic gravity might correspond, according to this, to a continuous version of dynamical collapses (like CSL [5]).

4. Emergent gravity

There are many arguments suggesting that gravitation might be an emergent phenomenon. The arguments range from ideas based on string theory developments [36, 37] to those motivated directly by the thermodynamical characteristics of the laws of black hole mechanics [38]; see also the discussion in [39].

It has been argued that in that case the quantization of variables, directly connected with the space-time metric, would be as inappropriate as the quantization of the Navier-Stokes equations. Moreover, if one must take into account the interaction of gravitation with some matter fields described in a quantum mechanical language (as in inflationary cosmology for instance), the semiclassical treatment would be the most appropriate. However if one wants to further consider the modifications of the quantum theory associated with any sort of collapse theory, one finds that one lacks a formalism to do so.

The fact is that, while the semiclassical gravity equation (2) might be valid in the absence of quantum jumps, it is clear that it can not be valid on the jump itself. We take the view, motivated in part by the ideas about the resolution of black hole singularities in Loop Quantum Gravity [40], that during the jump the degrees of freedom of the quantum space-time are excited. In the fluid analogy, this might be thought as corresponding to some chemical reaction or phase transition occurring in the fluid. It is clear that during such processes, which generally involve energy flux between the atomic or molecular degrees of freedom to the macroscopic degrees of freedom characterized in terms of the fluid variables, the Navier-Stokes' equations can not be valid. If, however, the phase transition takes place in a time period that is very short compared with the other temporal scales characteristic of the situation, one can assume that such effective equation is valid before and after the chemical reaction or phase transition.

[2] The idea is similar to the juncture conditions characterizing thin shells in general relativity [23]. If one knows the energy-momentum tensor outside and inside the shell, then Einstein's equations can be taken to hold on both sides. At the juncture, corresponding to the infinitely thin shell's location, one has a singular energy-momentum tensor, and the usual, smooth version of Einstein's equations must be replaced by something else which takes into account the singular nature of such term.

5. Recovering space-time notions from a quantum gravity theory

The recovering of the classical notions of space-time, and Einstein's general relativity in terms of the corresponding variables from a fundamental theory of quantum gravity is far from trivial [41]. One of the most notorious difficulties is the dealing with the so called "Problem of Time" in theories of quantum gravity. The fact is that, generically, in canonical theories of quantum gravity time disappears from the framework and the theory becomes, essentially, a timeless theory [26]. The problem is then how to recover an appropriate characterization of the state of gravitational degrees of freedom, and to do so in a way that respects the standard covariant space-time notions associated with general relativity. The most promising paths in this respect involve the notion of relational time [27], i.e., the selection of a certain variable constructed out of the matter and gravity degrees of freedom, which plays the role of a time variable. Although very interesting proposals of this sort have been considered, those are tied to specific examples, and the general resolution of the problem through this path is still lacking. Moreover, it is not clear if the recovery of time, and more importantly of a unified space-time can be achieved while retaining a fully quantized description.

We, thus, take the conservative view that the characterization of the gravitational degrees of freedom in terms of space-time notions should be done in the context of approximated, and classical level, even if the correct underlying theory is a theory of quantum gravity.

Moreover, it appears very likely that in attempts to incorporate a notion of time by selecting a variable to act as a clock, one ends with a theory where the remaining degrees of freedom are characterized by an evolution equation that involves subtle deviations from the expected Schrödinger's equation. In fact, as discussed in [28], the attempts to recover time in this fashion seem to lead to equations involving loss of unitarity.

6. Cosmology

Now, let us consider the problem we have mentioned concerning the emergence of the seeds of structure during inflation.

The problem is that, according to the inflationary ideas, the very early stages of inflation drive our universe into a situation that is described in terms of a background metric and scalar field configuration that is perfectly H&I, and by a perturbation field constructed out of metric and scalar field perturbations in a state closely connected to the Bunch Davies vacuum, and which is also exactly H&I. The issue is the inability of our physical theories to account for the emergence, out of this situation, of the primordial inhomogeneities and anisotropies. We note that several arguments have been put forward in this direction, but as discussed in [2], they all fail to be fully satisfactory. The difficulties are also alluded to in page 476 of S. Weinberg's book on cosmology [14] where we find *"... the field configurations must become locked into one of an ensemble of classical configurations with ensemble averages given by quantum expectation values... It is not apparent just how this happens....".*

The most popular of those arguments rely on the ideas of "decoherence", but that alone is insufficient as reflected in the following quote: *" .. However decoherence is not enough to explain the breakdown of translational invariance.."* found in page 348 of the well known book [15]. Moreover, even W. Zurek says [16]: *"The interpretation based on the ideas of decoherence and ein-selection has not really been spelled out to date in any detail. I have made a few half-hearted attempts in this direction, but, frankly, I was hoping to postpone this task, since the ultimate questions tend to involve such anthropic attributes of the observership as perception, awareness, or consciousness, which, at present, cannot be modeled with a desirable degree of rigor."*

In the cosmological situation at hand, we can not appeal to "observers" or "measuring apparatuses", as the emergence of the seeds of structure is not only prior to observers or apparatuses, but is actually a prerequisite for the conditions where such things are possible.

This might lead us to the general concern: Is quantum theory applicable to cosmology? If not, where is the boundary of applicability of Quantum theory? Is this an issue that is resolved with some suitable reformulation or interpretation of quantum theory?

Such approach was considered in [42], and led to the development of the Consistent Decohering Histories approach as the only one applicable to cosmology. As we have pointed out in [2] and [17], that approach does not offer a fully satisfactory resolution of the issues. As we said, we need to account for the breaking of the H&I character of the initial state (i.e. the situation corresponding to the early states of inflation), in a context where the dynamics does not contain anything capable of generating such breaking[3]. As already pointed out, our approach is based on the idea that such breaking is to find an explanation within the so called **collapse theories**.

7. The proposal

Based on the discussions above, we will look at Einstein's general relativity as an emergent phenomenon, or as "hydrodynamical level of description" of some underlying theory of quantum gravity. We will assume that such description validity requires $R << 1/l^2_{Plank}$. However, we will not assume that this condition is sufficient. For instance, the situations considered in Penrose's experimental proposals, which clearly satisfy this latter condition, might be ones where important modifications associated with the quantum gravity interface become relevant. We will assume, however, that such situation includes, with the appropriate caveats to be discussed below, the inflationary regime. Of course, that situation is one where the matter fields (particularly the inflation) still require a full quantum treatment.

The setting will thus naturally be semiclassical Einstein's gravity with the **collapse** reflecting some remanent signature from quantum aspects of the gravitation (here, we are following general ideas by Penrose and Diósi): i.e., besides U, we have, sometimes, spontaneous jumps of the quantum state:

$$....|0\rangle_{k_1} \otimes |0\rangle_{k_2} \otimes |0\rangle_{k_3} \otimes .... \to ....|\Xi\rangle_{k_1} \otimes |0\rangle_{k_2} \otimes |0\rangle_{k_3} \otimes ....$$

More precisely, we will rely on the notion of *Semiclassical Self-consistent Configuration* (SSC).

DEFINITION: The set $g_{\mu\nu}(x), \hat{\varphi}(x), \hat{\pi}(x), \hat{H}, |\xi\rangle \in \hat{H}$ represents a SSC if and only if $\hat{\varphi}(x)$, $\hat{\pi}(x)$ and $\hat{H}$ correspond to a quantum field theory constructed over a space-time with metric $g_{\mu\nu}(x)$ (as described in, say, [10]), and the state $|\xi\rangle$ in $\mathcal{H}$ is such that

$$G_{\mu\nu}[g(x)] = 8\pi G \langle\xi|\hat{T}_{\mu\nu}[g(x), \hat{\varphi}(x), \hat{\pi}(x)]|\xi\rangle.$$

It is, in a sense, the general relativistic version of Schrödinger-Newton equation [34].

This, however, can not describe the transition from a H&I SSC to one that is not. For that we need to add a collapse. We must describe that as a transition from one SSC to another, not simply from one state to another. So, we need the collapse to be represented by a full transition of the type SSC-I→SSC-II. In particular, they will describe a transition form an H&I SSC to one that is not H&I SSC. That involves changing the state, and thus the space-time, and thus the Hilbert space where the state "lives".

We take the view that, at the jump, the degrees of freedom of the underlying theory of quantum gravity, which are not susceptible of a metric description, are excited. In the fluid analogy, this might be thought as corresponding to a chemical reaction or phase transition occurring in underlying matter constituents of the fluid.

[3] People sometimes find attractive to relate this issue with that of spontaneous symmetry breaking. In this regard, we turn the reader to [44] where some of the myths connected with those ideas are clarified.

8. Example: a baby version of inflation

As an example of the application of the formalism, we consider the collapse associated with the generation of a primordial perturbation characterizing the breakdown of homogeneity and isotropy (H&I) at a single wave number $\vec{k_0}$. The discussion here is a shorthand version of that presented in [3] in which some inessential complications have been omitted[4].

As we indicated in the SSC formalism, space-time is thus treated as classical, and, in our case (working in a specific gauge and ignoring the tensor perturbations), the relevant metric is

$$ds^2 = a^2(\eta)\left[-(1+2\psi)d\eta^2 + (1-2\psi)\delta_{ij}dx^i dx^j\right], \qquad \psi(\eta,\vec{x}) \ll 1 \tag{8}$$

with $a(\eta)$ the scale factor and $\psi(\eta, x)$ the Newtonian potential. Later we will limit ourselves to the case where this metric is perfectly H&I, and to the transition to the case where it develops inhomogeneity and anisotropy characterized by $\vec{k_0}$. The scalar field with action $S = 1/2\int d^4x(\nabla_\mu\phi\nabla^\mu\phi - m^2\phi^2)$ is treated at the quantum field theory on curved space-times level, so we write:

$$\hat{\phi}(x) = \sum_\alpha \left(\hat{a}_\alpha u_\alpha(x) + \hat{a}^\dagger_\alpha u^*_\alpha(x)\right), \tag{9}$$

with the functions $u_\alpha(x)$ a complete set of normal modes, satisfying the field equation on the corresponding space–time:

$$(g^{\mu\nu}\nabla_\mu\nabla_\nu + m^2)u_\alpha(x) = 0, \tag{10}$$

and orthonormal w.r.t. the symplectic product:

$$((\phi_1,\pi_1),(\phi_2,\pi_2))_{\text{Sympl}} \equiv -i\int_\Sigma [\phi_1\pi_2^* - \pi_1\phi_2^*]\, d^3x. \tag{11}$$

For simplicity, we'll set everything in a co-moving coordinate box of size L.

Finally, one constructs the state such that Einstein's semiclassical equations hold. This is nontrivial, but is a well defined problem. We will, in fact, consider two such constructions and their matching at the collapse time.

8.1. The homogeneous and isotropic case: SSC-I

This situation corresponds to choosing $\psi(\eta,\vec{x}) \equiv 0$. We assume an almost de Sitter expansion (characterized by the parameters $H_0^{(I)}$ and $\epsilon^{(I)}$). That is, $a^{(\mathrm{I})}(\eta) = -1/(H_0\eta)^{1+\epsilon^{(\mathrm{I})}}$. More precisely, the conformal expansion factor is defined as $\mathcal{H}^{(\mathrm{I})} \equiv \frac{\dot{a}^{(\mathrm{I})}(\eta)}{a^{(\mathrm{I})}(\eta)}$ and the slow-roll parameter $\epsilon^{(\mathrm{I})} \equiv 1 - \dot{\mathcal{H}}^{(\mathrm{I})}/\mathcal{H}^{2(\mathrm{I})}$.

The quantum field construction will be based on a complete set of modes of the form $u^{(\mathrm{I})}_{\vec{k}}(x) = v^{(\mathrm{I})}_{\vec{k}}(\eta)e^{i\vec{k}\cdot\vec{x}}/L^{3/2}$. These must then satisfy the evolution equation:

$$\ddot{v}^{(\mathrm{I})}_{\vec{k}} + 2\mathcal{H}^{(\mathrm{I})}\dot{v}^{(\mathrm{I})}_{\vec{k}} + \left(k^2 + a^{2(\mathrm{I})}m^2\right)v^{(\mathrm{I})}_{\vec{k}} = 0, \tag{12}$$

and be normalized:

$$v^{(\mathrm{I})}_{\vec{k}}\dot{v}^{(\mathrm{I})*}_{\vec{k}} - \dot{v}^{(\mathrm{I})}_{\vec{k}}v^{(\mathrm{I})*}_{\vec{k}} = i\hbar a^{-2(\mathrm{I})}. \tag{13}$$

For the modes with $k \neq 0$, the most general solution to the evolution equation is a linear combination of the Hankel functions of first and second kind, $\eta^{3/2}H^{(1)}_\nu(-k\eta)$ and $\eta^{3/2}H^{(2)}_\nu(-k\eta)$

[4] One such simplification is the avoidance of the separation of the modes of the quantum field into symmetric and antisymmetric, or real and imaginary, and denoted by *R&I* in [3].

with $(\nu^{(\mathrm{I})})^2 = (9/4) - (m/H_0^{(\mathrm{I})})^2$. We are only limited by the requirement that the norm be positive, which is easily achieved.

However, the Hankel functions are not well behaved at the origin, and thus the zero mode is not included. For $k = 0$, the general solution to the equation is a linear combination of the functions $\eta^{(3-2\nu)/2}$ and $\eta^{(3+2\nu)/2}$. The choice is again arbitrary, provided it has positive symplectic norm. In this case, we take

$$v_0^{(\mathrm{I})}(\eta) = \sqrt{\frac{\hbar}{H_0^{(\mathrm{I})}}}\left[1 - \frac{i}{6}\left(-H_0^{(\mathrm{I})}\eta\right)^3\right]\left(-H_0^{(\mathrm{I})}\eta\right)^{m^2/3H_0^{2(\mathrm{I})}}. \tag{14}$$

For the $k \neq 0$, we make the Bunch-Davis choice, i.e., we use modes which, in the asymptotic past, behave as purely "positive frequency solutions". This fixes $\hat{H}^{(\mathrm{I})}$ as the Fock space of that construction.

To complete the SSC, we still need a state $|\xi^{(\mathrm{I})}\rangle$ in $\hat{H}^{(\mathrm{I})}$ such that its expectation value for the energy-momentum tensor leads to the desired nearly de Sitter expansion. Consider a state in which all the modes with $k \neq 0$ are in their vacuum state, while the zero mode is excited in a coherent state:

$$|\xi^{(\mathrm{I})}\rangle = c e^{\xi_0^{(\mathrm{I})}\hat{a}_0^{(\mathrm{I})\dagger}}|0^{(\mathrm{I})}\rangle. \tag{15}$$

The semiclassical Einstein's equation now reads

$$3\mathcal{H}^{2(\mathrm{I})} = 4\pi G\left((\dot{\phi}^{2(\mathrm{I})})_{\xi,0} + a^{2(\mathrm{I})}m^2(\phi^{2(\mathrm{I})})_{\xi,0}\right), \tag{16}$$

$$\mathcal{H}^{2(\mathrm{I})} + 2\dot{\mathcal{H}}^{(\mathrm{I})} = -4\pi G\left((\dot{\phi}^{2(\mathrm{I})})_{\xi,0} - a^{2(\mathrm{I})}m^2(\phi^{2(\mathrm{I})})_{\xi,0}\right) \tag{17}$$

where we have used a normal ordering recipe for the energy-momentum tensor, and defined:

$$(\dot{\phi}^{2(\mathrm{I})})_{\xi,0} \equiv \langle\xi^{(\mathrm{I})}| : (\partial_\eta\hat{\phi}^{(\mathrm{I})})^2 : |\xi^{(\mathrm{I})}\rangle, \qquad (\hat{\phi}^{2(\mathrm{I})})_{\xi,0} \equiv \langle\xi^{(\mathrm{I})}| : (\phi^{(\mathrm{I})})^2 : |\xi^{(\mathrm{I})}\rangle. \tag{18}$$

In general, the "classical relations" $(\dot{\phi}^{2(\mathrm{I})})_{\xi,0} = (\dot{\phi}^{(\mathrm{I})}_{\xi,0})^2$ and $(\phi^{2(\mathrm{I})})_{\xi,0} = (\phi^{(\mathrm{I})}_{\xi,0})^2$ will not hold; they, however, do for our simple coherent a state $|\xi^{(\mathrm{I})}\rangle$.

Writing scale factor as $a(\eta) = (-1/H_0\eta)^{1+\epsilon_1}$ and substituting in Einstein's equation up to $\epsilon^{(\mathrm{I})}$, we obtain $3(\dot{\phi}^{(\mathrm{I})}_{\xi,0})^2 \approx \epsilon^{(\mathrm{I})}a^{2(\mathrm{I})}m^2(\phi^{(\mathrm{I})}_{\xi,0})^2$. This leads to

$$\phi^{(\mathrm{I})}_{\xi,0}(\eta) \propto \eta^{\sqrt{\epsilon^{(\mathrm{I})}m^2/3(H_0^{(\mathrm{I})})^2}}. \tag{19}$$

On the other hand, taking the parameter $\xi_0^{(\mathrm{I})}$ as real, we find

$$\langle\xi^{(\mathrm{I})}|\hat{\phi}^{(\mathrm{I})}(x)|\xi^{(\mathrm{I})}\rangle = \frac{2\xi_0^{(\mathrm{I})}}{L^{3/2}}\sqrt{\frac{\hbar}{H_0^{(\mathrm{I})}}}\left(-H_0^{(\mathrm{I})}\eta\right)^{m^2/3H_0^{2(\mathrm{I})}}, \tag{20}$$

That is, we have compatibility if we set

$$\epsilon^{(\mathrm{I})} = \frac{m^2}{3H_0^{2(\mathrm{I})}}, \quad H_0^{(\mathrm{I})} = 16\pi G\hbar\epsilon^{(\mathrm{I})}\frac{(\xi_0^{(\mathrm{I})})^2}{L^3}. \tag{21}$$

This was the explicit SSC-I construction representing a H&I state, corresponding to the early stages of inflation.

8.2. An inhomogeneous and anisotropic case: SSC-II

Next, we want to consider a situation where the universe is no longer H&I , but has been excited in the $\vec{k}_0$ mode: we will denote this new SSC by SSC-II. It will be characterized by the parameters $H_0^{(\mathrm{II})}$ and $\epsilon^{(\mathrm{II})}$ (which might, in principle, differ slightly from those corresponding to the SSC-I discussed in the previous section), and a Newtonian potential described by $\psi(\eta,\vec{x}) = \varepsilon P(\eta)Cos(\vec{k}_0.\vec{x})$, where $P(\eta)$ is an (in principle) arbitrary function of η, and ε is a small (expansion) parameter (not to be confused with the slow roll parameter ϵ) .

The strategy: We first construct the "generic" Hilbert space assuming that $P(\eta)$ is given. Then make an "educated" guess for the form of the quantum state, and then, by requiring that our construction be an SSC (i.e., that Einstein's semiclassical equations hold) we will find what the function $P(\eta)$ ought to be.

The first step is to find the complete set of modes, which we write as

$$u_{\vec{k}}^{(\mathrm{II})}(x) = \frac{1}{L^{3/2}}[v_{\vec{k}}^{(\mathrm{II})0}(\eta)\, e^{i\vec{k}\cdot\vec{x}} + \varepsilon(\delta v_{\vec{k}}^{(\mathrm{II})-}(\eta)\, e^{i(\vec{k}-\vec{k}_0)\cdot\vec{x}} + \delta v_{\vec{k}}^{(\mathrm{II})+}(\eta)\, e^{i(\vec{k}+\vec{k}_0)\cdot\vec{x}})] \tag{22}$$

to the zeroth order in ε. The evolution equation is given, at the lowest order, by

$$\ddot{v}_{\vec{k}}^{(\mathrm{II})0} + 2\mathcal{H}^{(\mathrm{II})}\dot{v}_{\vec{k}}^{(\mathrm{II})0} + \left(k^2 + a^{2(\mathrm{II})}m^2\right) v_{\vec{k}}^{(\mathrm{II})0} = 0, \tag{23}$$

with normalization condition

$$v_{\vec{k}}^{(\mathrm{II})0}\dot{v}_{\vec{k}}^{(\mathrm{II})0*} - \dot{v}_{\vec{k}}^{(\mathrm{II})0}v_{\vec{k}}^{(\mathrm{II})0*} = i\hbar a^{-2(\mathrm{I})}, \tag{24}$$

which is identical to the construction we have already done. Thus, we take $v_{\vec{k}}^{(\mathrm{II})0}(\eta)$ as before. At first order in ε the corresponding evolution equation takes the form

$$\delta\ddot{v}_{\vec{k}}^{(\mathrm{II})\pm} + 2\mathcal{H}^{(\mathrm{II})}\delta\dot{v}_{\vec{k}}^{(\mathrm{II})\pm} + \left[(\vec{k}\pm\vec{k}_0)^2 + a^{2(\mathrm{II})}m^2\right]\delta v_{\vec{k}}^{(\mathrm{II})\pm} = F_{\vec{k}}(\eta) \tag{25}$$

where

$$F_{\vec{k}}(\eta) \equiv 4\dot{v}_{\vec{k}}^{(\mathrm{II})0}\dot{P} - 2\left(2k^2 + a^{2(\mathrm{II})}m^2\right)v_{\vec{k}}^{(\mathrm{II})0}P. \tag{26}$$

The normalization condition (needed only at one time) is

$$\begin{aligned}\dot{v}_{\vec{k}+\vec{k}_0}^{(\mathrm{II})0*}\delta v_{\vec{k}}^{(\mathrm{II})+} &- v_{\vec{k}+\vec{k}_0}^{(\mathrm{II})0*}\delta\dot{v}_{\vec{k}}^{(\mathrm{II})+} - \dot{v}_{\vec{k}}^{(\mathrm{II})0}\delta v_{\vec{k}+\vec{k}_0}^{(\mathrm{II})-*} + v_{\vec{k}}^{(\mathrm{II})0}\delta\dot{v}_{\vec{k}+\vec{k}_0}^{(\mathrm{II})-*} \\ &= 4\left(v_{\vec{k}}^{(\mathrm{II})0}\dot{v}_{\vec{k}+\vec{k}_0}^{(\mathrm{II})0*} - \dot{v}_{\vec{k}}^{(\mathrm{II})0}v_{\vec{k}+\vec{k}_0}^{(\mathrm{II})0*}\right)P.\end{aligned} \tag{27}$$

If we had $P(\eta)$ and the initial conditions satisfying (27), the Eq. (25) would define a unique solution for the functions $\delta v_{\vec{k}}^{(\mathrm{II})\pm}$. As we said, we will assume that $P(\eta)$ is given and take the initial conditions to be

$$\delta\dot{v}_{\vec{k}}^{(\mathrm{II})\pm}(\eta_c) = 0, \quad \delta v_{\vec{k}}^{(\mathrm{II})\pm}(\eta_c) = 4v_{\vec{k}}^{(\mathrm{II})0}(\eta_c)P(\eta_c). \tag{28}$$

This completes the construction of the generic Hilbert space representation $\hat{H}^{(\mathrm{II})}$ of the quantum field. Next, we need to find the state $|\zeta^{(\mathrm{II})}\rangle \in \mathcal{H}^{(\mathrm{II})}$ that completes the SSC construction.

The symmetries of the space-time lead us to consider the multi coherent state "ansatz" :

$$|\zeta^{(II)}\rangle = \ldots|\zeta^{(II)}_{-2\vec{k}_0}\rangle \otimes |\zeta^{(II)}_{-\vec{k}_0}\rangle \otimes |\zeta^{(II)}_{0}\rangle \otimes |\zeta^{(II)}_{\vec{k}_0}\rangle \otimes |\zeta^{(II)}_{2\vec{k}_0}\rangle \ldots. \tag{29}$$

The vector in Fock space is characterized by the specific modes that are excited (all other modes are assumed to be in the vacuum of the corresponding oscillator) and the parameters $\zeta^{(II)}_{\vec{k}}$ indicate the coherent state for the mode $\vec{k}$. The expectation value of the field operator in such a state is given by

$$\phi^{(II)}_{\zeta}(x) = \phi^{(II)}_{\zeta,0}(\eta) + \left(\delta\phi^{(II)}_{\zeta,\vec{k}_0}(\eta)e^{i\vec{k}_0\cdot\vec{x}}\right) + \left(\delta\phi^{(II)}_{\zeta,2\vec{k}_0}(\eta)e^{i2\vec{k}_0\cdot\vec{x}}\right) + \ldots. \tag{30}$$

In fact, for a general state of the form considered above, we have

$$\begin{aligned}
\phi^{(II)}_{\zeta,0}(\eta) &= \zeta^{(II)}_0 v^{(II)0}_0(\eta) + \varepsilon[\zeta_{-\vec{k}_0}\delta v^{(II)+}_{-\vec{k}_0}(\eta) + \zeta_{\vec{k}_0}\delta v^{(II)-}_{\vec{k}_0}(\eta)] + c.c.,\\
\delta\phi^{(II)}_{\zeta,\vec{k}_0}(\eta) &= \zeta^{(II)}_{\vec{k}_0} v^{(II)0}_{\vec{k}_0}(\eta) + \varepsilon[\zeta^{(II)}_0 \delta v^{(II)+}_0(\eta) + \zeta^{(II)}_{2\vec{k}_0}\delta v^{(II)-}_{2\vec{k}_0}(\eta)] + \ldots..,\\
\delta\phi^{(II)}_{\zeta,2\vec{k}_0}(\eta) &= \zeta^{(II)}_{2\vec{k}_0} v^{(II)0}_{2\vec{k}_0}(\eta) + \varepsilon[\zeta^{(II)}_{\vec{k}_0} \delta v^{(II)+}_{\vec{k}_0}(\eta) + \zeta^{(II)}_{3\vec{k}_0}\delta v^{(II)-}_{3\vec{k}_0}(\eta)] + \ldots.,
\end{aligned}$$

We set $\delta\phi^{(II)}_{\zeta,n\vec{k}_0}(\eta) = 0$ for all $n \geq 2$, simply by imposing the required relations between the parameters $\zeta^{(II)}_{\pm\vec{k}_0}$, $\zeta^{(II)}_{\pm 2\vec{k}_0}$, $\zeta^{(II)}_{\pm 3\vec{k}_0}$, etc. It is easy to see that $|\zeta^{(II)}_{\pm n\vec{k}_0}| \sim \varepsilon^n |\zeta^{(II)}_0|$.

The conditions above ensure that there are no terms in $e^{\pm in\vec{k}_0\cdot\vec{x}}$ (with $n \geq 2$) appearing in the expectation value of the energy-momentum tensor. That is necessary for compatibility of our state ansatz with the semiclassical Einstein's equations. Let us consider these in detail up to the first order in ε. The zero order equations are just the same as the equations we found in SSC-I , and are used to fix the construction of SSC up to that order, i.e., to determine the relation between $a^{(II)}$ and $\zeta^{(II)}_0$.

The key result, and the aspect that enables us to carry out the construction in a complete manner, is the following fact (see [3] for the details): the first order semiclassical Einstein's equations for the metric and the above state can be combined into a simple dynamical equation for the Newtonian potential, which is independent of the first order quantities $\delta\tilde{\phi}^{(II)}_{\zeta,\vec{k}_0}$ and $\delta\dot{\tilde{\phi}}^{(II)}_{\zeta,\vec{k}_0}$. To the level of precision we are working, the equation above becomes simply

$$\ddot{P} + \epsilon^{(II)}\mathcal{H}^{(II)}\dot{P} + \left[k_0^2 - \epsilon^{(II)}\mathcal{H}^{2(II)}\right]P = 0. \tag{31}$$

The general solution reads:

$$P(\eta) = C_1\,\eta^{\frac{1}{2}[1+\epsilon^{(II)}]}J_\alpha(-k\eta) + C_2\,\eta^{\frac{1}{2}[1+\epsilon^{(II)}]}Y_\alpha(-k\eta), \tag{32}$$

where $J_\alpha(-k\eta)$ and $Y_\alpha(-k\eta)$ are the Bessel functions of first and second kind, $\alpha = [1+3\epsilon^{(II)}]/2$.

The remaining Einstein's equations to this order take the form of two constraints (again, see [3] for the details). Using these, we express the initial values that would determine the specific solution $P(\eta)$:

$$\begin{pmatrix} P \\ \dot{P} \end{pmatrix} = \tag{33}$$

$$\frac{\sqrt{4\pi G\epsilon^{(II)}}\mathcal{H}^{(II)}}{k_0^2 - \mathcal{H}^{2(II)}\epsilon^{(II)}}\begin{pmatrix} (3\mathcal{H}^{(II)} - am\sqrt{3/\epsilon^{(II)}}) & 1 \\ (am\sqrt{3/\epsilon^{(II)}}\mathcal{H}^{(II)} - k_0^2 + (\epsilon^{(II)} - 3)\mathcal{H}^{2(II)}) & -\mathcal{H}^{(II)} \end{pmatrix}\begin{pmatrix} \delta\tilde{\phi}^{(II)}_{\zeta,\vec{k}_0} \\ \delta\dot{\tilde{\phi}}^{(II)}_{\zeta,\vec{k}_0} \end{pmatrix}.$$

Therefore, given the two numbers $\delta\tilde{\phi}^{(\mathrm{II})}_{\zeta,\vec{k}_0}(\eta_c)$ and $\delta\dot{\tilde{\phi}}^{(\mathrm{II})}_{\zeta,\vec{k}_0}(\eta_c)$, we would have a completely determined space-time metric. That is the condition above would determine P and $\dot{P}$ at η_c, and these would determine the constants C_1 and C_2 in (32). And then, the Eq. (25), together with the initial conditions (27) (which depend on $P(\eta_c)$ which, as we saw was fixed), would fix the mode functions for all η. Furthermore, those two numbers would determine the state parameters $\zeta_{\vec{k}_0}$ (and thus the rest as well). Thus, we have a complete SSC-II (to this order in ε).

8.3. The actual collapse process or the matching of SSC-I and SSC-II

Next, we want to consider the process of collapse. That is represented by the transition from the SSC-I to the SSC-II, and, in particular, the **matching** of the H&I region of space-time with the region where such symmetry has disappeared. We will be considering here that the transition corresponds to the hypersurfaces $\eta = \eta_c$ of SSC-I and SSC-II. Note that this gives such hypersurface Σ_c a preferred status in the resulting space-time, and is not something to be thought as related to a gauge freedom: to the past of that hypesrurface Σ_c, the space-time is H&I, and to the future it is not. We will assume here the induced metric is continuous on Σ_c. As shown in [3], the continuity can be obtained as a conclusion of a natural assumption regarding the energy-momentum tensor. In any event, continuity requires $P(\eta_c) = 0$ and thus

$$(3\mathcal{H}^{(\mathrm{II})} - am\sqrt{3/\epsilon^{(\mathrm{II})}})\delta\tilde{\phi}^{(\mathrm{II})}_{\zeta,\vec{k}_0}(\eta_c) + \delta\dot{\tilde{\phi}}^{(\mathrm{II})}_{\zeta,\vec{k}_0}(\eta_c) = 0, \tag{34}$$

and therefore

$$\dot{P} = -\sqrt{4\pi G\epsilon^{(\mathrm{II})}}\mathcal{H}^{(\mathrm{II})}\delta\tilde{\phi}^{(\mathrm{II})}_{\zeta,\vec{k}_0}(\eta_c). \tag{35}$$

Finally, we want to use the conditions prior to the collapse to help determine the situation just after it. Following the early work in [1], we will assume that the collapse is characterized by a loose analogy with something like an imprecise measurement (of the operators $\hat{\phi}^{(\mathrm{I})}_{\vec{k}_0}(\eta)$) in standard quantum theory: Before the collapse the operator had an uncertainty $\Delta\hat{\phi}^{(\mathrm{I})}_{\vec{k}_0}(\eta_c)$, and after the collapse, the new state will have an expectation value for that operator given by a random number within the range of uncertainty prior to collapse. The final result is then:

$$\varepsilon\delta\tilde{\phi}^{(\mathrm{II})}_{\zeta_t,\vec{k}_0}(\eta_c) = x_{\vec{k}_0}\sqrt{\langle 0^{(\mathrm{I})}_{\vec{k}_0}|\left[\Delta\hat{\phi}^{(\mathrm{I})}_{\vec{k}_0}(\eta_c)\right]^2|0^{(\mathrm{I})}_{\vec{k}_0}\rangle} \approx x_{\vec{k}_0}a(\eta_c)^{-1}\sqrt{\frac{\hbar}{2k}} \tag{36}$$

with $x_{\vec{k}_0}$ taken to be a random variable distributed according to a Gaussian function centered at zero with unit-spread. Thus, a specific choice of the random number $x_{\vec{k}_0}$ fully determines the SSC-II.

This finishes the analysis of the emergence of an actual anisotropic and inhomogeneous perturbation associated with a single mode, trough a single discrete collapse in that mode. We note that the scheme leads to a description of space-time in terms of two regions, one corresponding to the homogeneous and isotropic metric of SSC-I, and one of the slightly inhomogeneous and anisotropic metric of SSC-II. The matching conditions on the hypersurface Σ_c are such that the metric is continuous, but the extrinsic curvature is not. This issue is closely tied to the jump of the energy-momentum tensor across such surface. This in turn is closely connected to the issue of conservation laws in any collapse theory (see however [43]).

Journal of Physics: Conference Series **442** (2013) 012071 doi:10.1088/1742-6596/442/1/012071

9. Discussion

We have presented a well defined framework where one could, in principle, carry out all the analysis of the collapse approach to the inflationary origin of the seeds of cosmic structure. However, the fact that it is so complicated has led us, in the past to seek and use some shortcuts in the actually phenomenological analysis such as those in [20, 21]. But having a clear framework opens the path to the investigations of topics of more fundamental nature.

Of course, the fact that the scheme is well defined does not mean that it is free of problems, or that all questions have been successfully addressed. As far as we see, the most serious problems, afflicting this proposal are: a) the fact that during the collapse or jump from one SSC to another the energy-momentum tensor fails to be conserved (i.e. its four-divergence jumps); b) the fact that the collapse, as described here, takes place on a certain space-like hypersurface, and is thus a highly nonlocal process.

Regarding issue a), as we already indicated, the source is the generic violation of energy conservation that afflicts all collapse theories. The first thing to note in this regard is the fact that new developments in such collapse proposals provide hope for the possibility of modifications curing that problem, as discussed for instance in [33]. On the other hand, if we do take the view that space-time is emergent, and thus not a fundamental concept, it seems natural to question whether energy-momentum and its conservation should still be regarded as fundamental. If we take the view that Einstein's equations are a sort of Navier-Stokes effective equations, then their absolute validity would be limited, and of course so should be the conservation of energy-momentum. After all, as we already noted, nobody would be surprised if a violation of Navier-Stokes equations was observed in a liquid undergoing some process where the fundamental microscopic degrees of freedom of the fluid in question were being excited. In such situation, exemplified by one where the fluid was undergoing a phase transition, we would expect the fluid dynamics description to fail or at least to require some correction.

Regarding the issue b) one might be concerned with the nonlocality being tied to violations of causality. The first thing we should put out is that is easy to imagine processes that look highly nonlocal, when viewed in the context of an effective theory, and which are seen to be essentially local when viewed within the context of a fundamental theory. Consider, for instance, a very large lake that freezes essentially instantaneously as a result of a rapid drop in temperature. It is clear that such instantaneous freezing might, at first sight, look problematic in the context of special relativity, but a moment's thought clarifies that there is nothing odd going on. The particular preferential frame in which the freezing is instantaneous is determined globally (on the scale of the lake's size), as the frame where the temperature drops is uniform and homogeneous. The microscopic process is clearly controlled by local physics, and the instantaneous freeze-out is the result of the change in external conditions taking place in a correlated fashion over large distances. Second, when one considers collapse taking place in a stochastic fashion, and for all modes, it is quite possible that the effective collapse can be described in terms of something more local, as it is explicitly shown to occur in the inflationary problem when addressed in terms of the theory known as continuous spontaneous localization as discussed in [25], a work related to that of [24], but taking a slightly different point of view.

Furthermore, non locality and violation of causality, although superficially connected, are two distinct concepts. In fact, any analysis of a B-EPR system (and its experimental confirmation as the breakdown of Bell's inequalities) indicates that the system possesses nonlocal aspects. However, it is also clear that it does not violate causality (one can not use the EPR setups to send information faster than light, see for instance [45]).

In fact, the standard Copenhagen interpretation, where the collapse is the result of an "observer" performing a measurement, is, if anything, more susceptible to causality violations, as discussed for instance in [46], simply because there, the will of the experimenter plays a central role in determining the time of collapse. The current proposal falls within the class of ideas known as "spontaneous collapse", where the actions of experimental physicists do not determine the time of collapse. This fact seriously limits the ability of anyone to use the underlying physics to produce causality violations. Of course, as the final theory about the dynamical collapse is not at hand, we can not say that there is a proof that such a theory exists where all problems related to causality will be avoided. In any case, it is our view that that this line of studies which is motivated by the conceptual difficulties facing the inflationary account of the emergence of the seeds of cosmic structure, can offer one of the few sources of guidance in the search for such theory. It is worth pointing out that the approach does have phenomenological implications, as shown by the studies in [20, 21, 22], and the particular development presented here has lead to suggestions to look for novel kinds of statistical anomalies in then large scale structure and in the cosmic microwave background, which go beyond the usual kind of non-gaussianities [47]. Such analysis might be used to constrain the kind of viable collapse theories, if one takes the view that the phenomenological success of inflation must be backed by the corresponding conceptual clarity.

Finally, it is worth noting that, when considering and assuming the inflationary problem (ignoring the difficulties discussed in [2]) one adopts one of the more popular postures regarding the emergence of classicality, or more precisely, the generation of primordial inhomogeneities and anisotropies. The effective characterization of the actual emergence, if one truly attempts it, is bound to rely on a closely related formalism. These popular postures include, for instance, i) the notion that after a given mode exits the horizon (its physical wavelength, as seen in a co-moving frame, becomes larger than the Hubble radius), the fluctuation corresponding to that mode becomes classical, or ii) that, due to some decoherence effect, we can, at a certain point, adopt the "many worlds interpretation" of quantum theory, and consider the state of the quantum field as characterizing not our universe, but an ensemble of universes of which ours is just a typical element. Therefore, let us say that in case i) we want to produce a description (even an approximate one) of our universe concentrating, for simplicity, on a single mode, and want the characterization to be valid both before and after the mode exits the horizon. In that case, the approach presented here seems to be the best one can do, as long as we do not have a workable theory of quantum gravity which allows us to characterize space-time in a full quantum language. In case ii), we might also be interested in putting together the characterization of our space-time before the decoherence is taken to be effective, and the description of the particular branch of the many worlds (the particular element of the ensemble of universes) in which we happen to find ourselves. Again, in that situation the present analysis offers perhaps the furthest one can go in achieving the said goal, given the present stage of the development of candidate theories of quantum gravity.

The inflationary account for the emergence of the seeds of structure, represents the only situation where general relativity and quantum theory come together in the explanation of observable phenomena. We can let the phenomenological success blind us to its shortcomings [2] but that would mean discarding what might be perhaps the only source of empirical clues available to the enterprise of clarifying the way these two pillars of modes physics are to be reconciled. If instead we try to address these shortcomings we will have to face very difficult questions, and in doing so we would be clearly at risk of making tremendous mistakes. However in this case we can take consolation from Sir Francis Bacon's profound observation[5] about scientific methodology: *"truth emerges more readily from error that from confusion"*.

Acknowledgments
This work is supported in part by the CONACYT grant No 101712. and by UNAM-PAPIIT grant IN107412.

[5] As cited in Thomas Kuhn's "The structure of Scientific Revolutions"

References

[1] Perez A, Sahlmann H and Sudarsky D 2006 On the Quantum Mechanical Origin of the Seeds of Cosmic Structure *Class. Quant. Grav.* **23** 2317 (*Preprint* arXiv:gr-qc/0508100)
[2] Sudarsky D 2011 Shortcomings in the Understanding of Why Cosmological Perturbations Look Classical *Int. J. Mod. Phys.* D **20** 509 (*Preprint* arXiv:gr-qc/0906.0315v1)
[3] Diez-Tejedor A and Sudarsky D 2012 Towards a formal description of the collapse approach to the inflationary origin of the seeds of cosmic structure *JCAP* **1207** 045 (*Preprint* arXiv:1108.4928 [gr-qc])
[4] Ghirardi G C, Rimini A and Weber T 1986 A Unified Dynamics for Micro and MACRO Systems *Phys. Rev.* D **34** 470; 1987 On Disentanglement Of Quantum Wave Functions: Answer To A Comment On 'unified Dynamics For Microscopic And Macroscopic Systems *Phys. Rev.* D **36** 3287
[5] Pearle P 1989 Combining Stochastic Dynamical State Vector Reduction With Spontaneous Localization *Phys. Rev.* A **39** 2277; Ghirardi G C, Pearle P and Rimini A 1990 Markov Processes In Hilbert Space And Continuous Spontaneous Localization Of Systems Of Identical Particles *Phys. Rev.* A **42** 78
[6] Bell J S 1987 Are there quantum jumps? in *Speakable and unspeakable in quantum mechanics* (Cambridge: Cambridge University Press)
[7] Weinberg S 2012 Collapse of the State Vector *Phys. Rev.* A **85** 062116 (*Preprint* arXiv:1109.6462 [quant-ph])
[8] Diósi L 1987 A universal master equation for the gravitational violation of quantum mechanics *Phys. Lett.* A **120** 377; *do.* 1989 Models for universal reduction of macroscopic quantum fluctuations *Phys. Lett.* A **40** 1165
[9] Penrose R 1989 *The Emperor's New Mind* (Oxford: Oxford University Press); 1996 On Gravity's Role in Quantum State Reduction *Gen. Rel. Grav.* **28** 581
[10] Wald R M 1994 *Quantum Field Theory in Curved Spacetime and Black Hole Thermodynamics* (Chicago: University Of Chicago Press)
[11] For reviews abut the various approaches to the measurement problem in quantum mechanics see for instance the classic references, Jammer M 1974 *Philosophy of quantum mechanics. The interpretations of quantum mechanics in historical perspective* (New York: John Wiley and Sons); D'Espagnat B 1970 *Conceptual Foundations Of Quantum Mechanics* (Menlo Park Ca.: W.A. Benjamin); Peres A 1993 *Quantum Theory: Concepts and Methods* (Dordrecht: Kluwer Academic Publishers); Omnes R 1994 *The Interpretation of Quantum Mechanics* (Princeton: Princeton University Press); and the more specific critiques, Adler S L 2003 Why Decoherence has not Solved the Measurement Problem: A Response to P. W. Anderson *Stud. Hist. Philos. Mod. Phys.* **34** 135-142 (*Preprint* arXiv:quant-ph/0112095); Elby A 1993 Why modal interpretations of quantum mechanics don't solve the measurement problem *Found. of Phys. Lett.* **6** 5-19; and the review, Bassi A and Ghirardi G C 2003 *Phys. Rep.* **379** 257 (*Preprint* arXiv:quant-ph/0302164)
[12] Bell J 1990 in *Sixty-Two years of Uncertainty* ed A Miller (New York: Plenum) pp 17-31
[13] Bohm D and Bub J 1966 *Rev. Mod. Phys.* **38** 453; Károlyházy F 1966 *Nouvo Cim.* **XLII** 1506; Pearle P 1976 *Phys. Rev.* D **13** 857
[14] Weinberg S 2008 *Cosmology* (Oxford: Oxford University Press)
[15] Mukhanov V 2005 *Physical Foundations of Cosmology* (Cambridge: Cambridge University Press)
[16] Zurek W H 1998 Decoherence Einselection and the existencial interpretation *Phil. Trans. R. Soc. Lond.* A **356** 1793-1821 (*Preprint* arXiv:quant-ph/9805065)
[17] Okon E and Sudarsky D 2013 On the Consistency of the Consistent Histories Approach to Quantum Mechanics *Preprint* arXiv:1301.2586
[18] Penrose R 1989 *The Emperor's New Mind* (Oxford: Oxford University Press)
[19] Callender C 2001 *Physics meets philosophy at the Planck Scale* (Cambridge, UK: Cambridge University Press) pp 290-304
[20] De Unanue A and Sudarsky D 2008 Phenomenological Analysis of Quantum Collapse as Source of the Seeds of Cosmic Structure *Phys. Rev.* D **78** 043510 (*Preprint* arXiv:gr-qc/0801.4702v2)
[21] Landau S J, Scoccola C G and Sudarsky D 2012 Cosmological constraints on nonstandard inflationary quantum collapse models *Phys. Rev.* D **85** 123001 (*Preprint* arXiv:1112.1830 [astro-ph.CO])
[22] León G and Sudarsky D 2010 The Slow roll condition and the amplitude of the primordial spectrum of cosmic fluctuations: Contrasts and similarities of standard account and the 'collapse scheme' *Class. Quant. Grav.* **27** 225017 (*Preprint* arXiv:1003.5950 [gr-qc])
[23] Israel W 1966 Singular Hypersurfaces and Thin Shells in General Relativity *Nuovo Cim.* **B44** 1 (1966); *do.* 1967 Erratum *ibid.* **B48** 463
[24] Martin J, Vennin V and Peter P 2012 Cosmological Inflation and the Quantum Measurement Problem *Preprint* arXiv:1207.2086 [hep-th]
[25] Cañate P, Pearle P and Sudarsky D 2012 CSL Wave Function Collapse Model as a Mechanism for the Emergence of Cosmological Asymmetries in Inflation *Preprint* arXiv:1211.3463 [gr-qc]
[26] De Witt B S 1967 *Phys. Rev.* **160** 1113; Wheeler J A 1987 in *Battelle Reencontres 1987* eds C De Witt and J

A Wheeler (New York: Benjamin); Isham C J 1992 Canonical Quantum Gravity and the problem of Time *GIFT Semminar-* 0157228 (*Preprint* arXiv:qr-qc/9210011)

[27] See for instance, Page D N and Wootters W K 1983 *Phys. Rev.* D **27** 2885; Wootters W K 1984 *Int. J. Theor Phys.* D **23** 701; See also Rovelli C 1991 Time in Quantum Gravity: An Hypothesis *Phys. Rev.* D **43** 442 and references therein

[28] Gambini R, Porto R A and Pullin J 2004 Realistic Clocks, Universal Decoherence and the Black Hole information paradox *Phys. Rev. Lett.* **93** 240401 (*Preprint* hep-th/0406260); *do.* 2004 Fundamental Decoherence from relational Time in Discrete Quantum Gravity: Galilean Covariance *Phys. Rev.* D **70** 124001 (*Preprint* arXiv:gr-qc/0408050)

[29] Carlip S 2008 Is Quantum Gravity Necessary? *Class. Quant. Grav.* **25** 154010 (*Preprint* arXiv:0803.3456 [gr-qc])

[30] Kibble T W B 1981 in *Quantum Gravity 2: A Second Oxford Symposium* eds C J Isham, R Penrose and D W Sciama (Oxford: Clarendon Press) p 63; Page D N and Geilker C D 1981 *Phys. Rev. Lett.* **47** 979

[31] Tumulka R 2006 On spontaneous wave function collapse and quantum field theory *Proc. Roy. Soc. Lond.* A **462** 1897-1908

[32] Pearle P 2008 Stress tensor for quantized random field and wave-function collapse *Phys. Rev.* A **78** 022107

[33] Bedingham D J 2011 Relativistic state reduction dynamics *Found. Phys.* **41** 686-704

[34] Diósi L 1984 Gravitation and quantum mechanical localization of macro-objects *Phys. Lett.* A **105** 199-202; Penrose R 1998 Quantum computation, entanglement and state reduction *Phil. Trans. R. Soc. Lond.* A **356** 1927

[35] Hu B L and Verdaguer E 2008 Stochastic Gravity: Theory and Applications *Living Rev. Rel.* **11** 3 (*Preprint* arXiv:0802.0658 [gr-qc])

[36] Hu B L 2005 Can spacetime be a condensate? *Int. J. Theor. Phys.* **44** 1785-1806 (*Preprint* arXiv:gr-qc/0503067)

[37] Seiberg N 2007 Emergent Spacetime *Proceedings of the 23rd Solvay Conference in Physics: The Quantum Structure of Space and Time* eds D Gross, M Henneaux, A Sevrin Hackensack (Singapore: World Scientific) p 272 (*Preprint* arXiv:hep-th/0601234v1)

[38] Jacobson T 1995 Thermodynamics of space-time: The Einstein equation of state *Phys. Rev. Lett.* **75** 1260-1263 (*Preprint* arXiv:gr-qc/9504004)

[39] Wüthrich C 2005 To quantize or not to quantize: fact and folklore in quantum gravity *Phil. Sci.* **72** 777-788

[40] Ashtekar A and Bojowald M 2006 Quantum geometry and the Schwarzschild singularity *Class. Quant. Grav.* **23** 391-411

[41] Sudarsky D 2008 Unspeakables and the Epistemological path towards Quantum Gravity *Int. J. Mod. Phys.* D **17** 425-443

[42] Hartle J B 1997 Quantum cosmology: Problems for the 21st century *Preprint* arXiv:gr-qc/9701022; *do.*2006 Generalizing quantum mechanics for quantum gravity *Int. J. Theor. Phys.* **45** 1390-1396 (*Preprint* arXiv:gr-qc/0510126)

[43] Pearle P M 2000 Wave function collapse and conservation laws *Found. Phys.* **30** 1145-1160

[44] Perez A and Sudarsky D 2011 On the symmetry of the vacuum in theories with spontaneous symmetry breaking *Int. J. Mod. Phys.* A **26** 1493 (*Preprint* arXiv:0811.3181 [hep-th])

[45] Peres A and Terno D 2004 Quantum Information and Relativity Theory *Rev. Mod. Phys.* **76** 93-123 (*Preprint* arXiv:quant-ph/0212023); Hall M J W 1987 Imprecise measurements and non-locality in quantum mechanics *Phys. Lett.* A **125** 89-91; Ghirardi G C et al. 1988 Experiments of the EPR Type Involving CP-Violation Do not Allow Faster-than-Light Communication between Distant Observers *Europhys. Lett.* **6** 95-100; Florig M and Summers S J 1997 On the statistical independence of algebras of observables *J. Math. Phys.* **38** 1318-1328

[46] Sorkin R D 1993 Impossible measurements on quantum fields *Directions in general relativity* vol. 2, 293-305 SU-GP-92-12-4A (*Preprint* gr-qc/9302018)

[47] León G and Sudarsky D 2012 Novel possibility of nonstandard statistics in the inflationary spectrum of primordial inhomogeneities *Sigma* **8** 024 (*Preprint* arXiv:1109.0052)

Environment induced time arrow and the Closed Time Path method

Janos Polonyi
Strasbourg University, High Energy Theory Group, CNRS-IPHC, 23 rue du Loess, BP28 67037 Strasbourg Cedex 2 France
E-mail: polonyi@iphc.cnrs.fr

Abstract. It is shown in the framework of a harmonic system that the thermodynamical time arrow is induced by the environmental initial conditions in a manner similar to spontaneous symmetry breaking. The Closed Time Path formalism is introduced in classical mechanics to handle Green functions for initial condition problems by the action principle, in a systematic manner. The application of this scheme for quantum systems shows the common dynamical origin of the thermodynamical and the quantum time arrows. It is furthermore conjectured that the quantum-classical transition is strongly coupled.

1. Introduction

The fundamental equations of motions are time reversal invariant when weak interaction is ignored and the emergence of past and future, an orientation of time is an interesting, open question [1, 2, 3]. One usually distinguishes radiational (use of retarded Green functions in classical electrodynamics), thermodynamical (second law of thermodynamics), quantum (information loss at the quantum-classical crossover) and gravitational (singularity generated time reversal asymmetry) time arrows. It is argued below that the thermodynamical time arrow, irreversibility, is generated by a spontaneous breakdown of time reversal invariance and has the same dynamical origin as decoherence, thereby the thermodynamical and quantum time arrows are identical. These points will be demonstrated in the framework of harmonic systems. To render this argument valid for weakly interactive systems one needs the Closed Time Path (CTP) formalism which will briefly be introduced in classical mechanics [4].

The construction presented here is part of a long time project, the derivation of classical electrodynamics from QED [5]. Due to the large number of particles involved, the description of the quantum-classical crossover must be based on quantum field theory. The best way to cross the quantum-classical transition is to follow the renormalization group trajectory which has to be set up for the density matrix, rather than transition amplitudes. This is the point where the CTP formalism appears in the discussion. It is not difficult to find the CTP renormalization group equation [5], however its solution, more precisely the construction of a reliable, treatable approximate solution requires the clarification of several details. One needs a handle on the dynamical build up of decoherence [6], the equation of motion for the expectation value and the decoherence of composite operators for free particles [7], the quantum corrections to the equation of motion of elementary operators [8], the characterization of the classicality of a scattering process [10] to mention some of them. A further issue, the relation between irreversibility and decoherence [9, 4] is discussed briefly here.

The point where the quantum field theoretical description becomes important is the generation of new scales by the help of the number of particles involved, $N \sim 10^{23}$. The probability distributions become very narrow, $\Delta p/p \sim 10^{-12}$ and the central limit theorem extended by perturbation expansion may produce deterministic laws for its peak around an IR fix point. It is no longer necessary to assume the collapse of the wave function, N may generate short time scales and the state reduction can take place in extremely short time, requiring the use of relativistic quantum field theory. The way a charged particle is detected in the Wilson cloud chamber suggests that the number of particles, participating in a measurement may depend strongly on time, $N_{eff}(t)$ and can produce highly non-linear effects.

The presentation starts by pointing to the initial conditions for the environment, as the origin of the thermodynamical time in Section 2 and continues with a simple, classical harmonic system displaying irreversibility in Section 3. The view of irreversibility as a dynamical symmetry breakdown is motivated in Section 4. In Section 5 the CTP formalism is introduced in classical mechanics, quantum systems are addressed in Section 6. Finally, Section 7 contains a brief summary.

2. Thermodynamical time arrows

The thermodynamical time arrow, irreversibility, is easiest to understand in terms of coarse graining, or blocking in the language of renormalization group. Let us consider a classical dynamical system, described by the coordinates x and y, called system and its environment. The former component comprises the degrees of freedom observed. The complete trajectory is supposed to obey a set of reversible equations of motion,

$$\ddot{x} = F(x, y), \qquad \ddot{y} = G(x, y). \tag{1}$$

The time arrow is inferred for each degree of freedom by observation: One introduces an external perturbation of the degree of freedom in question which acts at a given time t_p and if the response is seen for $t > t_p$ or $t < t_p$ then the experimental time arrow is forward or backward, respectively.

Such a a time arrow can be set either in a trivial, kinematical or a highly non-trivial, dynamical manner. The kinematical way is the imposition of some boundary conditions in time which renders the solution unique. In fact, by imposing initial or final conditions, independently on the external perturbations, on a certain degree of freedom we construct its motion forward or backward in time. It may happen that the kinematical time arrows, introduced in this manner for several degrees of freedom are in conflict and the complete system has no causal behavior and well defined time arrow.

A time arrow is set dynamically and the corresponding dynamics is irreversible if the time arrow is formed even before imposing the boundary conditions on time on the system coordinate. This may happen if the environment time arrows, as external agents manage to break the time reversal invariance of the effective system dynamics. The setting of a kinematical system time arrow, identical to the environment leads to the usual irreversible dynamics but the opposite kinematical time arrow generates acausal effective system dynamics, without well defined system time arrow. It is assumed in what follows that the kinematical time arrow is identical, forward pointing, for all degree of freedom.

The system dynamics is constructed by eliminating the environment by means of its equation of motion, $y \to y[x; y(0), \dot{y}(0)]$, $y(0)$ and $\dot{y}(0)$ denoting the environment initial conditions, imposed at the initial time $t_i = 0$. The insertion of this solution into the system equation of motion yields the effective system dynamics,

$$\ddot{x} = F_{eff}[x; y(0), \dot{y}(0)]. \tag{2}$$

Time reversal of the trajectories for the time interval $0 < t < t_f$ acts as $Tx(t) = x(t_f - t)$, it exchanges initial conditions into final conditions, $Ty(0) = y(t_f)$ and the time reversed effective equation of motion (2) can be written as

$$T\ddot{x} = F_{eff}[Tx; y(t_f), \dot{y}(t_f)]. \tag{3}$$

Since $y(t_f) \neq y(0)$ and $\dot{y}(t_f) \neq \dot{y}(0)$ the thermodynamical system time arrow is set by the environment initial conditions.

It will be found below that a necessary condition for irreversibility is a condensation point in the normal mode spectrum and the emergence of the thermodynamical time arrow may lead to acausality, the response to an external effect may show up before the effect. This latter phenomenon is rather puzzling because causality, the precedence of an external perturbation of its response, seems to be granted automatically when Newton equations are integrated in time.

3. Classical harmonic systems

Let us consider a harmonic dynamics, governed by the Lagrangian

$$L = \frac{m}{2}\dot{x}^2 - \frac{m\omega_0^2}{2}x^2 + jx + \sum_n \left(\frac{m}{2}\dot{y}_n^2 - \frac{m\omega_n^2}{2}y_n^2 + g_n x y_n\right), \tag{4}$$

where the condition $\sum_n g_n^2/m\omega_n^2 < m\omega_0^2$ guarantees the stability of the motion. We impose the environment initial conditions $y_n(0) = \dot{y}_n(0) = 0$ and write the system trajectory in the form

$$x(t) = -\int_0^\infty dt' D^r(t - t')j(t') \tag{5}$$

where the retarded Green function,

$$D^r(\omega) = \frac{1}{m[(\omega + i\epsilon)^2 - \omega_0^2] - \Sigma^r(\omega)}, \tag{6}$$

is given in terms of the retarded self energy

$$\Sigma^r(\omega) = \sum_n \frac{g_n^2}{m(\omega + i\epsilon)^2}. \tag{7}$$

The sign of ϵ reflects the environment time arrow, one needs $\epsilon = 0^+$ or $\epsilon = 0^-$ for initial or final environment conditions. We use forward environment time arrow below, $\epsilon = 0^+$. The spectrum, defined by the equation $0 = D^{r-1}(\omega_j + i\epsilon)$ consists of frequencies with negative, $\mathcal{O}(\epsilon)$ imaginary part, hence the effective system dynamics is reversible (infinite life-time) and causal (poles below the real axis).

It is advantageous to introduce the spectral function

$$\rho(\Omega) = \sum_n \frac{g_n^2}{2m\omega_n}\delta(\omega_n - \Omega) \tag{8}$$

whose phenomenological, Ohmic form,

$$\rho(\Omega) = \frac{\Theta(\Omega)g^2\Omega}{m\Omega_D(\Omega_D^2 + \Omega^2)} \tag{9}$$

leads to

$$\begin{aligned}\Sigma^r(\omega) &= \frac{g^2}{m\Omega_D}\int_0^\infty d\Omega\frac{\Omega}{\Omega_D^2+\Omega^2}\frac{2\Omega}{(\omega+i\epsilon)^2-\Omega^2} \\ &= -\frac{i\pi g^2}{\Omega_D(\omega+i\Omega_D)}\end{aligned} \tag{10}$$

and

$$D^r(\omega) = \frac{1}{m[(\omega+i\epsilon)^2-\omega_0^2]+\frac{i\pi g^2}{\Omega_D(\omega+i\Omega_D)}}. \tag{11}$$

The lesson of this Green function is that in case of continuous spectrum, in particular for the Ohmic spectral function the effective dynamics is irreversible (finite life-time) and acausal (poles above the real axis).

4. Irreversibility

The goal of this section is to underline the similarities between spontaneous symmetry breaking and irreversibility. There are two ways to detect spontaneous symmetry breaking, a dynamical and a static procedure.

4.1. Slow order parameter

We see a finite part of our Universe which requires a more refined understanding of spontaneous symmetry breaking in large but finite systems. Spontaneous symmetry breaking can be understood dynamically as the slowing down of the order parameter. In fact, consider the rotation of a macroscopic, rigid body consisting of a large number, $N \sim 10^{23}$, of particles and ignore first its environment. The rigid rotation is driven by the Hamiltonian

$$H_{rot} = \frac{1}{2}L^j(\Theta^{-1})^{jk}L^k + \mathcal{O}(L^4) \tag{12}$$

where the inertia tensor is $\Theta^{jk} = m\sum_n x_n^j x_n^k = \mathcal{O}(N)$. The ground state is non-degenerate but a macroscopic systems cannot be isolated from their environment hence we have to assume $L \sim n\hbar$ where n is a not a too large number. The angular velocity, $\omega^j = (\Theta^{-1})^{jk}L^k = \mathcal{O}(N^{-1})$, is extremely slow, a full rotation takes approximately $10^{10}/n$ times the age of our Universe. Therefore, the spontaneous breakdown of rotational symmetry is the applicability of the adiabatic approximation for the slow order parameters, the Euler angles in this case.

Such a slowing down appears as a violation of Fubini theorem in perturbation expansion, the non-commutativity of the thermodynamical limit, $N \to \infty$ and the limit of long observation time, $T \to \infty$,

$$\lim_{N\to\infty}\lim_{T\to\infty} A(N,T) \neq \lim_{T\to\infty}\lim_{N\to\infty} A(N,T). \tag{13}$$

The discrete spectral lines of a large but finite system are resolved in the first case but the infinite system, observed in finite time allows us the use of continuous spectrum in the second order of limits and the scenario described in Section 3 is recovered.

A mathematical ambiguities, such as the non-commutativity of limits always correspond to a physical alternative, the possibility of different preparation or observation of the system. Irreversibility is detected by observations which are limited in time and which, as a result, can not resolve the discrete spectrum. This happens if the spectral lines have an unexpectedly high density. This is reminiscent of quantum anomalies, the finite, observable UV cutoff effects of renormalizable quantum field theories due to the non-uniform convergence of the bare, regulated Feynman loop-integrals, except that irreversibility arises due to the smearing out an IR cutoff effect, the discreteness of spectral lines. In this sense spontaneous symmetry breaking, in particular irreversibility is an IR anomaly.

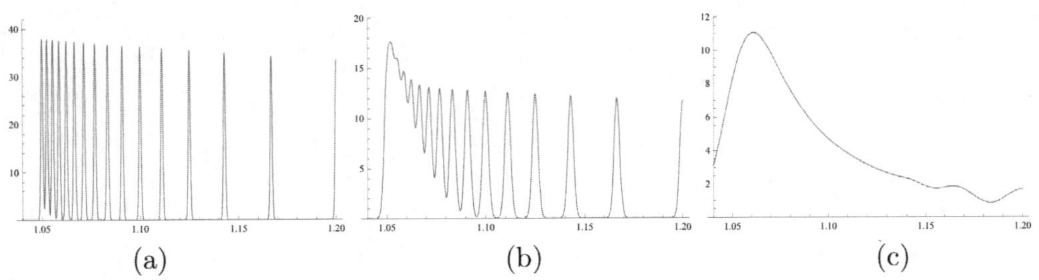

Figure 1. The apparent spectral density, (16) for observation times (a): $T = 2000$, (b) $T = 700$ and (c) $T = 100$. The spectrum $\omega_j = 1 + 1/j$ is constructed by $N = 20$, $a_{kj} = 1/\sqrt{N}$, $m = 1$ and the IR cutoff is $c(t) = \exp(-t^2/2T^2)$.

4.2. Observations with IR cutoff

It is instructive to follow what happens in a simple harmonic system given by the coordinates x_n if the limitation on observation time is taken into account by writing an observed normal mode $\tilde{x}_j = \sum_n a^{-1}_{jn} x_n$ as $x_{obs}(t) = c(t)x(t)$ where the function $c(t)$, satisfying the conditions $c(0) = 1$, $\lim_{t\to\infty} c(t) = 0$ acts as an IR cutoff. It is well known that the finite observation time spreads the discrete spectral lines of the harmonic system. In fact, an external source $j(t) = j_k\delta(t)$, coupled linearly to normal mode $\tilde{x}_k$ yields the response

$$\tilde{x}_{k\ obs}(\omega) = -j_k \int \frac{d\Omega \rho_m(\Omega)}{\omega + i\epsilon - \Omega}, \tag{14}$$

where the spectrum weight,

$$\rho_m(\Omega) = \sum_{j=1}^{N} \delta(\Omega - \tilde{\omega}_j) \frac{a^2_{kj}}{2m\tilde{\omega}_j}, \tag{15}$$

inferred by infinitely long observations for a discrete system is replaced by the apparent spectral function

$$\rho_m(\Omega) = \frac{1}{2\pi} \sum_{j=1}^{N} c(\Omega - \tilde{\omega}_j) \frac{a^2_{kj}}{2m\tilde{\omega}_j} \tag{16}$$

in the integral of Eq. (14). The apparent spectral function, (16), depicted in Fig. 1 shows that if a discrete spectrum has a condensation point then arbitrarily long, but finite time observations leave infinitely many spectrum lines unresolved. The dense spectral lines of the unresolved normal modes make the continuous spectrum approximation applicable at the condensation point and these normal modes act as a sink for energy, generating dissipation and irreversibility if they have sufficient spectral weight. If the spectrum is uniformly dense in the thermodynamical limit, as in field theories then dissipation to the unresolved normal modes and irreversibility takes place in the form of Bremsstrahlung at any energy.

Acausality can be understood in a similar manner, as a result of uncertainties in determining the time when the external source $j(t)$ acts on the system. The relation

$$x^{obs}(\omega) = -\int \frac{d\omega'}{2\pi} c(\omega - \omega') D^r(\omega') j(\omega') \tag{17}$$

between the external source and the limited observation allows us to infer an apparent source

$$\begin{aligned} j^{obs}(\omega) &= -D^{r-1}(\omega) x^{obs}(\omega) \\ &= \int \frac{d\omega'}{2\pi} c(\omega - \omega') \frac{D^r(\omega')}{D^r(\omega)} j(\omega'), \end{aligned} \tag{18}$$

The phenomenological form: $D^r(\omega) = \sum_p Z_p/(\omega - \omega_p)$ gives the pole contribution

$$j_p^{obs}(\omega) \approx \int \frac{d\omega'}{2\pi} c(\omega - \omega') \frac{\omega - \omega_p}{\omega' - \omega_p} j(\omega') \tag{19}$$

for $\omega \sim \omega_p$. The IR cutoff $c(\omega) = 2\eta/(\omega^2 + \eta^2)$, with $\eta = 2\pi/T$ and the source: $j(t) = j_0\delta(t)$ yields the contribution

$$j_p^{obs}(t) = j_0 \left[\delta(t) - \frac{2\pi}{T} e^{-i\omega_p t - 2\pi\frac{|t|}{T}}\right] \tag{20}$$

to the apparent source. Hence the apparent acausality results from difficulties in reconstructing a sharply localized external source in time due to the abundant soft modes. When a finite time step Δt is used to integrate Newton equations then acausality appears as an ambiguity in the order of the steps $N \to \infty$ and $t \to t + \Delta t$.

4.3. Weak explicit symmetry breaking

Spontaneous symmetry breaking can be detected in a static, equilibrium state, too: One introduces an external symmetry breaking and lets its strength tend to zero. Spontaneous symmetry breaking takes place if the system does not "forget" the external symmetry breaking in this limit. The analogue of external symmetry breaking in case of irreversibility is the environment boundary condition. Thus it is illuminating to follow the impact of environment time arrow in the effective system dynamics.

Let us consider the response of the system coordinate to a source $j(t)$, either an environment coordinate or an external diagnostic tool, given in the form of Eq. (5). The symmetric or antisymmetric part of the retarded Green function with respect to time reversal is the near or far Green function, respectively, hence we have $D^r = D^n + D^f$, with $D^n(t) = D^n(-t)$ and $D^f(t) = -D^f(-t)$. In case of a harmonic oscillator of mass m and frequency ω_0 we have

$$D^n(\omega) = P\frac{1}{m(\omega^2 - \omega_0^2)}, \qquad D^f(\omega) = -\frac{\pi i}{m}\text{sign}(\omega)\delta(\omega^2 - \omega_0^2), \tag{21}$$

where P denotes the principal part. The support of D^f is the mass-shell, the null-space $D^{-1}x = 0$ of the linear equation of motion, $D^{-1}x = j$, where D^n is vanishing. Therefore the time reversal symmetric response to $j(t)$ is generated by the off-mass shell modes of D^n and the time arrow is set by the mass shell modes of D^f. Such a transmutation of the negative time parity couplings can be seen in the self energy (10). In fact, the $i\epsilon$ term in the environment Green function can be viewed as a weak, external time reversal invariance breaking because it changes sign when the environment initial conditions are replaced by final conditions. The response of the infinitesimal $i\epsilon$ term is the finite imaginary part of the system self energy, $\Im\Sigma^r(\omega)$, which changes sign under time reversal, too.

5. Classical Closed Time Path formalism

The CTP formalism has quantum origin, it has been introduced first by J. Schwinger to establish perturbation expansion in the Heisenberg representation [11]. It has been reinvented in several other contexts such as to derive relaxation in many-body systems [12], to find the mixed state contributions to the density matrix by path integral [13, 14], to develop perturbation expansion for retarded Green-functions [15], to find manifestly time reversal invariant description of quantum mechanics [16], to describe finite temperature effects in quantum field theory [17, 18, 19], to formulate the consistency of histories [20, 21, 22], to derive equations of motion for the expectation value of local operators [24, 5], to follow non-equilibrium processes [23] and to describe scattering processes with non-equilibrium final states

[10]. This scheme will be introduced below in classical mechanics to allow us to recast initial problems as variational equations and to facilitate the introduction of retarded Green functions. The distinguishing feature of CTP is the redoubling of the degrees of freedom, coming from the double occurrence of the time evolution operator $U(t) = \exp(-iHt/\hbar)$ in the Heisenberg representation, $A(t) = U^\dagger(t)A(0)U(t)$. Apart of this physical origin, the doubling facilitates the handling of dissipative forces by variation principle, as well [25, 26, 4].

5.1. Action

The variational principle can not be used for initial condition problems because the variation at the final time cancels the momentum, $\partial S/\partial x_f = p(t_f) = 0$. The remedy of this problem is the flipping the time arrow at t_f and the sending of the whole system which obeys reversible dynamics back to its initial conditions. We impose the initial condition $x(0) = \dot{x}(0) = 0$ and use the corresponding trajectory $\tilde{x}(t)$, $0 < t < 2t_f$ to construct a CTP doublet, $\hat{x}(t) = (x^+(t), x^-(t)$ with $x^+(t) = \tilde{x}(t)$ and $x^-(t) = \tilde{x}(2t_f - t)$. The action for the CTP doublet trajectory is

$$S_{CTP}[\hat{x}] = \int_0^{t_f} dt[L_{CTP}(x^+(t), \dot{x}^+(t)) - L^*_{CTP}(x^-(t), \dot{x}^-(t))] \tag{22}$$

where the minus sign and the complex conjugation are due to the flipping of the time arrow.The Lagrangian

$$L_{CTP}(x, \dot{x}) = L(x, \dot{x}) + i\frac{m\epsilon}{2}x^2 \tag{23}$$

contains an infinitesimal imaginary part to remove the degeneracy of the action for $x^+ = x^-$. Note that the condition $x^+(t_f) = x^-(t_f)$ closes the trajectories and renders the variational equation trivial, $p(t_f) - p(t_f) = 0$, at the final time. The possibility of irreversible effective system dynamics, signaled by $x^-(t) \neq x^+(t)$ arises because the environment time arrow is not flipped at the final time and the system and environment time arrows end up in conflict.

There is an important symmetry in the CTP formalism, the exchange of the trajectories, $\tau x^\pm = x^\mp$, in particular $\tau S_{CTP}[\hat{x}] = -S^*_{CTP}[\hat{x}]$.

5.2. Green function

The action of a harmonic system,

$$S_{CTP}[\hat{\phi}] = \int dx \left[\frac{1}{2}\hat{\phi}(x)\hat{K}\hat{\phi}(x) + \hat{j}(x)\hat{\phi}(x)\right], \tag{24}$$

identifies the CTP Green function $\hat{D} = \hat{K}^{-1}$. The CTP symmetry τ requires the form

$$\hat{K} = \begin{pmatrix} K^n + i\bar{K}_1 & K^f - i\bar{K}_2 \\ -K^f - i\bar{K}_2 & -K^n + i\bar{K}_1 \end{pmatrix}, \tag{25}$$

which results in the parametrization

$$\hat{D} = \begin{pmatrix} D^n + i\bar{D}_1 & -D^f + i\bar{D}_2 \\ D^f + i\bar{D}_2 & -D^n + i\bar{D}_1 \end{pmatrix}, \tag{26}$$

of the Green function in terms of four real functions, $\hat{K}^{tr} = \hat{K}$, $\hat{D}^{tr} = \hat{D}$, $\bar{D}_1^{tr} = \bar{D}_1$, $\bar{D}_2^{tr} = \bar{D}_2$, $(D^n)^{tr} = D^n$, $(D^f)^{tr} = -D^f$. A physical source, $j^\pm(t) = \pm j(t)$ generates the trajectory

$$x^\sigma(t) = -\sum_{\sigma'=\pm 1}\int_0^{t_f} dt' D^{\sigma\sigma'}(t-t')\sigma' j(t') \tag{27}$$

and the condition $x^+ = x^-$, expressed as

$$D^{++} - D^{+-} = D^{-+} - D^{--}, \tag{28}$$

reduces the number of functions in the Green function to three, $\bar{D}_1 = \bar{D}_1 = \bar{D}$. It is easy to see that the retarded and advance Green functions are defined by $D^{\overset{r}{a}} = D^n \pm D^f$. Such an origin of the Green function explains that $\bar{D}$ drops out from Eq. (27) and remains arbitrary in classical dynamics. Nevertheless $\bar{D}$ plays an important role in quantum dynamics as we shall see below.

6. Quantum systems

The CTP formalism is needed in quantum physics when the density matrix $\rho(t) = U(t,0)\rho(0)U^\dagger(t,0)$ or expectation values, $\langle A\rangle(t) = \mathrm{Tr}[A\rho(t)]$ are sought. The reduplication of the degrees of freedom is due to the need of the independent construction of the time evolution operator U and $U^\dagger$, achieved by means of the generator functional

$$e^{iW[\hat{j}]} = \mathrm{Tr}T[e^{-\frac{i}{\hbar}\int dt'[H(t')-j^+(t')x(t')]}]\rho_i T^* e^{\frac{i}{\hbar}\int dt'[H(t')+j^-(t')x(t')]}], \tag{29}$$

where T^* is anti-time ordering. The path integral representation,

$$e^{iW[\hat{j}]} = \int D[\hat{x}]e^{\frac{i}{\hbar}S_{CTP}[\hat{x}]+\frac{i}{\hbar}\int dt\hat{j}(t)\hat{x}(t)}, \tag{30}$$

contains the action (22). The propagator $\hat{D}$ is of the form

$$\hat{D}(t-t') = -i\begin{pmatrix}\langle T[x(t)x(t')]\rangle & \langle x(t')x(t)\rangle\\ \langle x(t)x(t')\rangle & -\langle T[x(t)x(t')]\rangle^*\end{pmatrix} \tag{31}$$

and one encounters the same equations of motion and Green functions in harmonic classical and quantum systems.

When the density matrix is sought then the trace is omitted in the generator functional (29) and the corresponding path integral expression contains open time paths (OTP) [13, 7].

6.1. Decoherence

Decoherence, the suppression of the off-diagonal elements of the density matrix in a given base, refers to the state of the system at a given time and the OTP path integral offers a dynamical view of the establishment of this decohered state. In fact, parametrize the OTP trajectories by means of the physical and the decoherence trajectories, $x(t)$ and $y(t)$, respectively, as $x^\pm(t) = x(t) \pm y(t)/2$ and write the open time path path integral of a harmonic system as

$$\int D[\hat{x}]e^{\frac{i}{2\hbar}\hat{x}\hat{K}\hat{x}} = \int D[x]D[y]e^{\frac{i}{2\hbar}(xK^a y + yK^r x + \frac{1}{2}yK^n y) - \frac{1}{2\hbar}y\bar{K}y} \tag{32}$$

where $\hat{K}$ is given by Eq. (25) and $K^{\overset{r}{a}} = K^n \pm K^f$. The block matrix $\bar{K}$ plays a remarkable role in this path integral:

(i) *Decoherence:* The imaginary part of the CTP action governs the suppression of the off diagonal elements of the density matrix. Therefore $\bar{K}$ is responsible of the gradual building up the decoherence in the final state, namely the consistency of trajectories [20, 21, 22].

(ii) *Irreversibility:* $\bar{K}(\omega)$, being the imaginary part of the inverse Feynman propagator, is actually the inverse life-time of the mode of frequency ω. Due to the CTP consistency equation (28) decoherence (suppression of paths with $y \neq 0$ in Eq. (32)) and irreversibility (normal modes of finite life-time) have the same dynamical origin and the thermodynamical and quantum time arrows are identical.

(iii) *Quantum and classical fluctuations:* Classical fluctuations are encoded in the common initial conditions of the CTP doublet trajectory $\hat{x}$, by means of $x(0)$ and $\dot{x}(0)$ and are independent of $\bar{K}$. Quantum fluctuations emerge during the time evolution as the difference between the two members of the CTP doublet, $x^+(t) - x^-(t)$, and are governed by $\bar{K}$.

6.2. Soft quantum-classical transition

The CTP formalism suggests an unusual property of the quantum-classical transition. The classical limit of the path integral representation of the transition amplitude,

$$\langle x_f | e^{-\frac{i}{\hbar} H t_f} | x_i \rangle = \int_{x(0)=x_i}^{x(t_f)=x_f} D[x] e^{\frac{i}{\hbar} S[x]}, \tag{33}$$

used to be identified by the limit $\hbar \to 0$. The transition amplitude becomes dominated in this limit by the trajectories within a small neighborhood of the classical trajectory, $x(t) \to x_{cl}(t)$, a feature which can be summarized by saying that the trajectory is rigid, it supports only small fluctuations in this limit.

The picture of the transition, displayed by the OTP path integral for the density matrix,

$$\rho(x^+, x^-) = \int_{x^\pm(t_f)=x^\pm} D[\hat{x}] e^{\frac{i}{\hbar} S[x^+] - \frac{i}{\hbar} S^*[x^-] + \frac{i}{\hbar} S_i[\hat{x}]} \tag{34}$$

where the influence functional $S_i[\hat{x}]$ represents the environment is just the opposite. In fact, by assuming that the influence functional induces decoherence in the coordinate representation, $x^+(t) \sim x^-(t)$ we can approximate the diagonal elements of the density matrix as a path integral over the common trajectory $x(t) = x^\pm(t)$,

$$\rho(x_f, x_f) \sim \int_{x(t_f)=x_f} D[x] e^{\frac{i}{\hbar} S_i[x,x]}. \tag{35}$$

We have weak dependence on the trajectory in the exponent of integrand, $S_i[\hat{x}] = \mathcal{O}(g)$ where g is the system-environment coupling strength. A wide range of trajectories contribute to the transition probabilities, in other words the classical limit is soft. One encounters here a delicate situation, known from Statistical Physics, where the system-environment interactions must be strong enough to drive the system to equilibrium but weak enough to ignore the environment in equilibrium. In our case the dependence of the influence functional $S_i[\hat{x}]$ on $\hat{x}$ is supposed to be strong enough to decohere the system but the numerical value, $S_i[\hat{x}] = \mathcal{O}(g)$ is small enough for weakly coupled system-environment to observe large fluctuations in the path integral (35).

I believe that the resolution of this apparent conflict is to recall that the classical limit can not be reached by means of transition amplitudes between pure states, (33), rather we need the reduced density matrix, (34). The formal limit $\hbar \to 0$ of (33) is not the classical one. Another indication of the softness of the classical limit comes from operator formalism: The classical limit is supposed to be reached when the excitation level density of the whole system is high. We need then small energy to orthogonalize the whole system by a jump from one stationary state to another. This is the hallmark of softness.

The softness of the trajectories in the path integral (35) indicates that the quantum-classical transition is in the strong coupling regime. But one must be able to recover the traditional weakly coupled classical systems from quantum mechanics. For this to happen the trajectories should become stiff as we enter into the classical domain. Such a strongly coupled crossover beside a weakly coupled regime may hide a change of degrees of freedom. In fact, this is just what happens when the asymptotically free quark-gluon description of the strong interaction changes into the hadronic scheme around a characteristic length scale, the nucleon size. This possibility raises the question whether the degrees of freedom of the classical and quantum domains are the same.

7. Summary

The emergence of the thermodynamical and quantum time arrow was demonstrated in the framework of a simple harmonic system. The time arrow is induced by the environment in the same manner as the symmetry is broken spontaneously by infinitesimal explicit symmetry breaking. To generalize this mechanism for realistic many-body systems one needs the CTP formalism which yields a systematic treatment of initial condition problems both on classical and on quantum level. The harmonic oscillators, used in this work become quasi particles when interactive quantum field theoretical models are considered, an unavoidable step in view of the large number of particles, involved in macroscopic bodies.

The structure of the CTP propagator assures the identity of the thermodynamical and quantum time arrow, an issue similar to the conjecture that statistical physics exists in the quantum domain only [27].

It was argued that the quantum-classical crossover is strongly coupled. This possibility represents a wonderful challenge to revisit the correspondence principle and to derive classical physics in a quantum field theoretical framework.

References

[1] H. D. Zeh, *The Direction of Time*, Springer-Verlag, Berlin, 1989.
[2] S. F. Savit, ed. *Time's Arrows Today*, Cambridge university Press, Cambridge, 1995.
[3] M. C. Mackey, *Time's Arrow: The origins of Thermodynamic Behavior*, Springer, New York, 1993.
[4] J. Polonyi, in *The Time-arrow and Anthropic Principle*, 2012 Open Academic Press, Berlin.
[5] J. Polonyi, 2006 *Phys. Rev.* **D74** 065014.
[6] J. Polonyi, 2008 *Phys. Rev.* **D77** 125018.
[7] J. Polonyi, 2010 *Europhysics Letters* **91** 67003.
[8] M. Planat, J. Polonyi, 2010 *Phys. Rev.* **D82** 045021.
[9] J. Polonyi, 2011 *Phys. Rev.* **D84** 105021.
[10] J. Polonyi, K. Zazoua, 2012 *Phys. Rev.* **D85** 065013.
[11] J. Schwinger, 1961 *J. Math. Phys.* **2** 407.
[12] L. P. Kadanoff, G. Baym, *Quantum Statistical Mechanics*, Benjamin, New-York (1962).
[13] R. P. Feynman, F. L. Vernon, 1963 *Ann. Phys.* **24** 118.
[14] L. Diosi, 1990 *Found. Phys.* **20** 63.
[15] L. V. Keldysh, 1964 *Zh. Eksp. Teor. Fiz.* **47** 1515 (1965 *Sov. Phys. JETP* **20** 1018).
[16] Y. Aharonov, P. G. Bergmann, J. L. Lebowitz, . *Phys. Rev.* **134** B1410
[17] H. Umezawa, *Advanced Field Theory: Micro, Macro and Thermal Physics* Am. Inst. of Phys. (1993).
[18] A. J. Niemi, G. W. Semenoff, 1984 *Ann. Phys.* **152** 105.
[19] A. J. Niemi, G. W. Semenoff, 1984 *Nucl. Phys.* **B230** 181.
[20] R. B. Griffith, 1984 *J. Stat. Phys.* **36** 219.
[21] R. Omnès, 1988 *J. Stat. Phys.* **53** 893.
[22] M. Gell-Mann, J. B. Hartle, 1993 *Phys. Rev.* **D47** 3345.
[23] E. A. Calzetta, B. L. A. Hu, *Nonequilibrium Quantum Field Theory*, Cambridge University press, 2008.
[24] R. D. Jordan, 1986 *Phys. Rev.* **D33** 444.
[25] H. Bateman, 1931 *Phys. Rev.* **38** 815.
[26] S. De Martino, S. De Siena, G. Vitiello, 1996 *Int. J. Mod. Phys.* **B10** 1615.
[27] J. Gemmer, M. Michel, G. Mahler, *Quantum Thermodynamics* Springer, Berlin 2004.

AUTHOR INDEX

Arimondo, E 594
Balma, M 603
Bason, M G 594
Berchera, I Ruo 230
Blasone, M 647
Brida, G 230
Bruschi, D E 242
Buric, N 264
Capolupo, A 608, 640
Castaños, O 558
Castellano, A C 290
Chakraborty, B 167
Champion, T 28
Ciampini, D 594
Cocciaro, B 34
De Córdoba, P F 311
De Martini, F 568
Degiovanni, I P 230
Dell'Anno, F 647
Devona, D 603
Dharma-Wardana, M W C 282
Di Lorenzo, A 433
Di Sia, P 635
Diósi, L 1
Dolce, D 453
Doyen, G 299, 463
Drakova, D 299, 463
Dunningham, J A 257
Elze, H-T 58
Emelyanov, S A 323
Faccio, D 242
Faetti, S 34
Fathi, M 503
Fronzoni, L 34
Fussy, S 114
Gastaldi, U 544
Genovese, M 230
Gervino, G 603, 616
Ghirardi, G C 8
Giampaolo, S M 608
Giudice, E D 272, 290
Grössing, G 114
Guerra, F 129
Gupta, K S 167
Hájícek, P 398
Halliwell, J J 181, 522
Hu, B L 90
Illuminati, F 608, 647
Iorio, A 532
Isidro, J M 311
Jaksch, D 28
Jin, X-M 28
Jizba, P 513
Jung, K 576
Kamalov, T 487
Kanatchikov, I V 377
Khrennikov, A 101
Kiefer, C 250
Kleinert, H 197
Knott, P A 257
Kordas, G 277
Kornyak, V V 476
Krugly, A L 388
Langford, N 28
Lavagno, A 616
Lee, K C 28
Lopaeva, E D 230
Louko, J 242
Malossi, N 594
Man'Ko, M A 65
Man'Ko, V I 65
Mannella, R 594
Mavromatos, N E 206
Meda, A 230
Medinaceli, E 223
Melas, E 344
Mendaš, I 264
Messineo, G 544
Michelberger, P 28
Milotti, E 544
Minguzzi, E 319
Morsch, O 594
Nagao, S 625
Ninno, A D 290
Nunn, J 28
Obermair, G M 621
Okon, E 425
Pachol, A 359
Palmisano, C 603
Parra-Murillo, C A 277
Pascasio, J M 114
Payandeh, F 503
Pengo, R 544
Perea, M H 311
Petruccione, F 18
Piemontese, L 544
Pigato, D 616
Polonyi, J 671
Popovic, D B 264
Prati, E 237
Prvanovic, S 264
Radonjic, M 264
Reginatto, M 75
Reim, K 28
Reverberi, L 334
Romano, R 8
Rosas-Ortiz, O 558
Ruoso, G 544
Sakellariadou, M 145, 156
Sako, A 493

AUTHOR INDEX

Salvio, A 367
Santamato, E 568
Sarkar, S 206
Scardigli, F 513
Schuch, D 558
Schwabl, H 114
Sen, S 167
Siena, S D 647
Sinayskiy, I 18
Sindt, J 257
Sorkin, R D 136
Sprague, M 28
Stabile, A 156
Subasi, Y 90
Sudarsky, D 656
Sussman, B 28
Suzuki, T 493
Sverdlov, R 443
Tedeschi, A 272
Tserkovnikov, I A 388
Uk, D 28
Umetsu, H 493
Usenko, C V 584
Valle, F D 544
Viteau, M 594
Vitiello, G 156, 272, 640
Voeikov, V 272
Wallden, P 406
Walmsley, I A 28
Wimberger, S 277, 603
Wu, B 50
Yearsley, J M 181, 522
Zavattini, G 544
Zhan, F 50
Zonetti, S 352